your **learning** today!

D1030534

Get more out of biopsychology with MyPsychKit, the exciting online companion to this textbook that combines multimedia and practice tools to take learning to the next level! Packaged with this book and *so easy to use*, MyPsychKit gives students access to **weblinks, book-specific chapter summaries, flashcards, practice tests,** and **animations.** MyPsychKit also includes **Research Navigator™, a suite of powerful and reliable research tools** that help students through every step of the research paper writing process. MyPsychKit is ideal for instructors too. . . the **media assets** enhance any classroom presentation, and the **Grade Tracker** feature helps track student progress.

Enter your access code at www.mypsychkit.com and start exploring this rich and dynamic resource today!

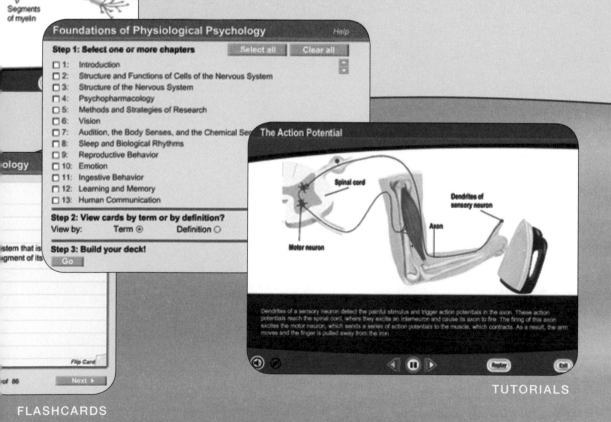

FLASHCARDS

TUTORIALS

mypsychkit™

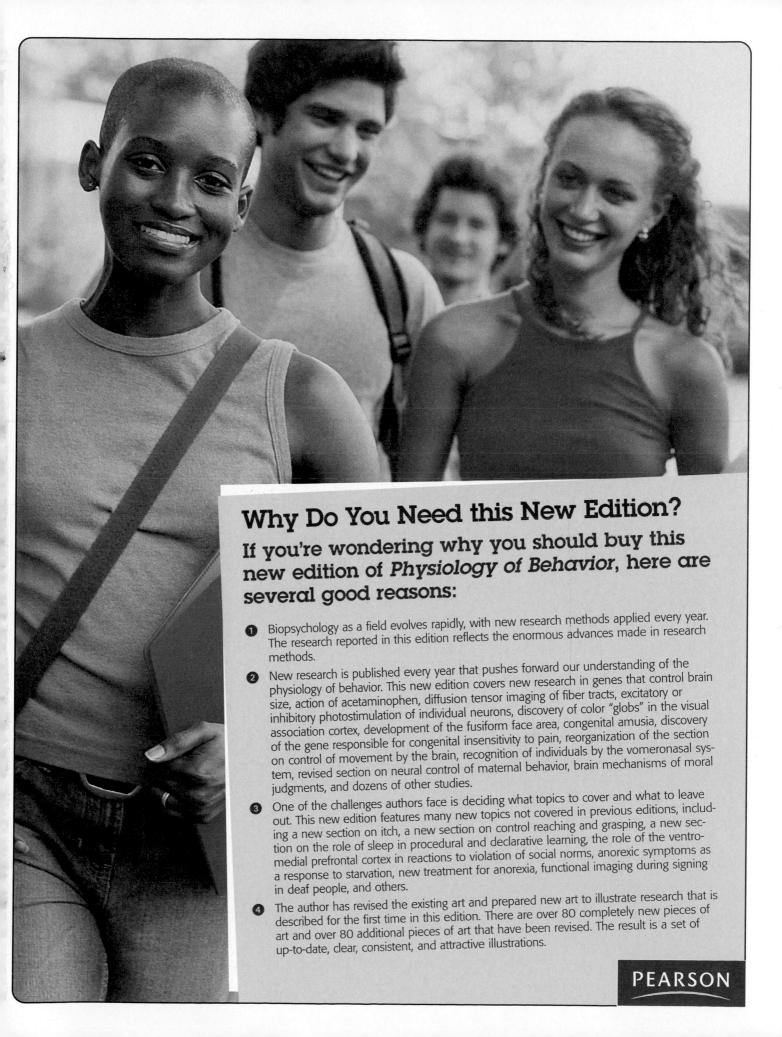

Why Do You Need this New Edition?

If you're wondering why you should buy this new edition of *Physiology of Behavior*, here are several good reasons:

❶ Biopsychology as a field evolves rapidly, with new research methods applied every year. The research reported in this edition reflects the enormous advances made in research methods.

❷ New research is published every year that pushes forward our understanding of the physiology of behavior. This new edition covers new research in genes that control brain size, action of acetaminophen, diffusion tensor imaging of fiber tracts, excitatory or inhibitory photostimulation of individual neurons, discovery of color "globs" in the visual association cortex, development of the fusiform face area, congenital amusia, discovery of the gene responsible for congenital insensitivity to pain, reorganization of the section on control of movement by the brain, recognition of individuals by the vomeronasal system, revised section on neural control of maternal behavior, brain mechanisms of moral judgments, and dozens of other studies.

❸ One of the challenges authors face is deciding what topics to cover and what to leave out. This new edition features many new topics not covered in previous editions, including a new section on itch, a new section on control reaching and grasping, a new section on the role of sleep in procedural and declarative learning, the role of the ventromedial prefrontal cortex in reactions to violation of social norms, anorexic symptoms as a response to starvation, new treatment for anorexia, functional imaging during signing in deaf people, and others.

❹ The author has revised the existing art and prepared new art to illustrate research that is described for the first time in this edition. There are over 80 completely new pieces of art and over 80 additional pieces of art that have been revised. The result is a set of up-to-date, clear, consistent, and attractive illustrations.

PEARSON

tenth edition

Physiology
of Behavior

NEIL R. CARLSON

University of Massachusetts, Amherst

Allyn & Bacon

Boston   New York   San Francisco
Mexico City   Montreal   Toronto   London   Madrid   Munich   Paris
Hong Kong   Singapore   Tokyo   Cape Town   Sydney

Senior Acquisition Editor: Stephen Frail
Associate Editor: Angela Pickard
Series Editorial Assistant: Kate Motter
Marketing Manager: Nicole Kunzman
Editorial Production Service: Nesbitt Graphics, Inc./Barbara Gracia
Copy Editor: Barbara Willette
Manufacturing Buyer: JoAnne Sweeney
Electronic Composition: Nesbitt Graphics, Inc.
Illustrator: Jay ALexander, I-Hua Graphics
Interior Design: Nesbitt Graphics, Inc./Gina Hagen
Photo Researcher: Nesbitt Graphics, Inc./Gina Hagen (chapter openers)
Cover Administrator: Kristina Mose-Libon

Photo Credits: p. 1: © 2010 iStockphoto.com/Tatraholiday; p. 28: © 2010
Image99/Image100/JupiterImages; p. 68: © 2010 Creatas Images/Ime Icons/
JupiterImages; p. 102: © 2010 iStockphoto.com/imagemonkey; p. 134: © 2010
Corbis Image/JupiterImages; p. 169: © 2010 JupiterImages/Thinkstock Images/
JupiterImages; p. 212: © 2010 Image Source Black/JupiterImages; p. 262: © 2010
Steve Allen/Brank X Pictures/JupiterImages; p. 295: © Purestock/Sacred Symbols/
JupiterImages; p. 329: © 2010 Hemera Technologies/AbleStock.com/JupiterImages;
p. 366: © 2010 PhotoAlto/James Hardy/JupiterImaages; p. 400: © 2010 Hermera
Technologies/JupiterImages; p. 439: © 2010 Corbis/JupiterImages; p. 485: © 2010
Steve Allen/Brand X Pictures/World Destinations 2/JupiterImages; p. 520: © 2010
Image99/Image100/JupiterImages; p. 554: © 2010 Clark Dunbar/Rubberball
Images/3-D Stock Volume 1/JupiterImages; p. 585: © 2010 Goodshoot/Attractive
Landscapes/JupiterImages; p. 618: © Photo 24/Brand X Pictures/JupiterImages

Library of Congress Cataloging-in-Publication Data
Carlson, Neil R., 1942-
  Physiology of behavior / Neil R. Carlson. – 10th ed.
    p. cm.
    Includes bibliographical references and index.
    ISBN 0-205-66627-2 (hardcover : alk. paper)
    1. Psychophysiology. I. Title.
      QP360.C35 2010
      612.8–dc22
                            2008044812

10 9 8 7 6 5 4 3 2 1   Q-WC-V   13 12 11 10 09

Allyn & Bacon
is an imprint of

ISBN-13: 978-0-205-66627-0
ISBN-10: 0-205-66627-2

www.pearsonhighered.com

FOR
GARTH JOHNSON THOMAS, 1916–2008
TEACHER, MENTOR, FRIEND

■■■

Brief Contents

Contents

CHAPTER 3

Structure of the Nervous System 68

CHAPTER 4

Psychopharmacology 102

CHAPTER 5
Methods and Strategies of Research 134

CHAPTER 6
Vision 169

CHAPTER 7
Audition, the Body Senses, and the Chemical Sense 212

CHAPTER 8
Control of Movement 262

CHAPTER 9
Sleep and Biological Rhythms 295

CHAPTER 10
Reproductive Behavior 329

CHAPTER 11

Emotion 366

CHAPTER 12

Ingestive Behavior 400

CHAPTER 13
Learning and Memory 439

CHAPTER 14
Human Communication 485

CHAPTER 15

Neurological Disorders 520

CHAPTER 16

Schizophrenia and the Affective Disorders 554

CHAPTER 17

Anxiety Disorders, Autistic Disorder, Attention-Deficit/ Hyperactivity Disorder, and Stress Disorders 585

CHAPTER 18

Drug Abuse 613

Preface

I wrote the first edition of *Physiology of Behavior* over thirty years ago. When I did so, I had no idea I would someday be writing the tenth edition. I'm still having fun, so I hope to do a few more. The interesting work coming out of my colleagues' laboratories—a result of their creativity and hard work—has given me something new to say with each edition. Because there was so much for me to learn (there are over 400 new references in this edition), I enjoyed writing this edition just as much as the first one. That is what makes writing new editions interesting: learning something new and then conveying that information to the reader.

NEW TO THIS EDITION

The research reported in this edition reflects the enormous advances made in research methods. Nowadays, as soon as a new method is developed in one laboratory, it is adopted by other laboratories and applied to a wide range of problems. And more and more, researchers are combining techniques that converge upon the solution to a problem. In the past, individuals tended to apply their particular research method to a problem; now they are more likely to use many methods, often in collaboration with other laboratories.

The art in this book continues to evolve. With the collaboration of Jay Alexander of I-Hua Graphics, I have revised the existing art and prepared new art to illustrate research that is described for the first time in this edition. The result is a set of up-to-date, clear, consistent, and attractive illustrations.

The animations that previously were provided on a CD are now included in MyPsychKit, www.mypsychkit .com, which also includes practice tests, customized study plans, tutorials, and Web links.

The first part of the book is concerned with foundations: the history of the field, the structure and functions of neurons, neuroanatomy, psychopharmacology, and research methods. The second part is concerned with inputs and outputs: the sensory systems and the motor system. The third part deals with classes of species–typical behavior: sleep, reproduction, emotional behavior, and ingestion. The chapter on reproductive behavior includes parental behavior as well as mating. The chapter on emotion includes a discussion of fear, anger and aggression, communication of emotions, and feelings of emotions. The chapter on ingestive behavior covers the neural and metabolic bases of drinking and eating.

The fourth part of the book deals with learning, including research on synaptic plasticity, the neural mechanisms that are responsible for perceptual learning and stimulus-response learning (including classical and operant conditioning), human amnesia, and the role of the hippocampal formation in relational learning. The final part of the book deals with verbal communication and mental and behavioral disorders. The latter topic is covered in three chapters; the first discusses schizophrenia and the affective disorders; the second discusses the anxiety disorders, autism, attention deficit disorder, and stress disorders; and the third discusses drug abuse.

Each chapter begins with a *Case History*, which describes an episode involving a neurological disorder or an issue in neuroscience. Other case histories are included in the text of the chapters. *Interim Summaries* follow each major section of the book. They not only provide useful reviews, but also break each chapter into manageable chunks. *Definitions of Key Terms* are printed in the margin near the places where the terms are first discussed. *Pronunciation Guides* for terms that might be difficult to pronounce are also found there. Each chapter ends with a list of *Suggested Readings* that provide more information about the topics discussed in the chapter.

The following list includes some of the information that is new to this edition.

New Research

- Genes that control brain size
- Action of acetaminophen
- Diffusion tensor imaging of fiber tracts
- Excitatory or inhibitory photostimulation of individual neurons
- Discovery of color "globs" in the visual association cortex
- Development of the fusiform face area
- Congenital amusia
- Discovery of the gene responsible for congenital insensitivity to pain

Reorganization of the section on control of movement by the brain

Recognition of individuals by the vomeronasal system

Revised section on neural control of maternal behavior

Brain mechanisms of moral judgments

Emotional judgments of body posture

Bariatric surgery and secretion of anorexic peptide

Role of basal ganglia in nondeclarative learning

Role of the right superior temporal cortex in comprehension of metaphors

Causes of AIDS dementia complex

Possible role of vitamin D deficiency in the development of schizophrenia

Cleansing rituals as atonement for unethical behavior

Brain pathology in autistic disorder

New Topics

New section on itch

Identification of taste receptors for sourness

Mirror neuron system activation and expertise in dancing

New section on control of reaching and grasping

New section on the role of sleep in procedural and declarative learning

Brain mechanisms of REM sleep: the REM flip-flop

Role of the ventromedial prefrontal cortex in reactions to violation of social norms

Role of the mirror neuron system in emotional imitation and empathy

Anorexic symptoms as a response to starvation

New treatment for anorexia

Transfer of memories from the hippocampus to the neocortex

Functional imaging during signing in deaf people

Identification of the visual word-form area

Ketogenic diet and development of 2-DG trials for the treatment of seizure disorders

Mirror neurons and therapy after stroke

Brain-computer interfaces to operate computer-controlled devices

Trial of gene delivery to the basal ganglia to treat Parkinson's disease

Animal research on the delivery of siRNA to treat Huntington's disease

Trial of ketamine for treatment-resistant depression

Reorganization of the section on anxiety disorders

Mirror neuron system and autistic disorder

Role of 5-HT transporters in posttraumatic stress disorder

Role of the dorsal striatum in addictive behavior

Role of the insular cortex in nicotine addiction

New experimental drugs for the treatment of addiction

Role of airway sensations in cigarette addiction

Besides updating my discussion of research, I keep updating my writing. Writing is a difficult, time-consuming endeavor, and I find that I am still learning how to do it well. I have said this in the preface of every edition of this book, and it is still true. I have worked with copy editors who have ruthlessly marked up my manuscript, showing me how to do it better the next time. I keep thinking, "This time there will be nothing for the copy editor to do," but I am always proved wrong: Many pages contain notes showing me how to improve my prose. But I do think that each time the writing is better organized, smoother, and more coherent.

Good writing means including all steps of a logical discourse. My teaching experience has taught me that an entire lecture can be wasted if the students do not understand all of the "obvious" conclusions of a particular experiment before the next one is described. Unfortunately, puzzled students sometimes write notes feverishly, in an attempt to get the facts down so they can study them—and understand them—later. A roomful of busy, attentive students tends to reinforce the lecturer's behavior. I am sure all my colleagues have been dismayed by a question from a student that reveals a lack of understanding of details long since passed, accompanied by quizzical looks from other students that confirm that they have the same question. Painful experiences such as these have taught me to examine the logical steps between the discussion of one experiment and the next and to make sure they are explicitly stated. A textbook writer must address the students who will read the book, not simply colleagues who are already acquainted with much of what the writer will say.

Because research on the physiology of behavior is an interdisciplinary effort, a textbook must provide the student with the background necessary for understanding a variety of approaches. I have been careful to provide enough biological background early in the book that students without a background in physiology can understand what is said later, while students with such a background can benefit from reviewing details that are familiar to them.

I designed this text for serious students who are willing to work. In return for their efforts, I have endeavored to provide a solid foundation for further study. Those students who will not take subsequent courses in this or related fields should receive the satisfaction of a much better understanding of their own behavior. Also, they will have a greater appreciation for the forthcoming advances in medical practices related to disorders that affect a person's perception, mood, or behavior. I hope that students who read this book carefully will henceforth perceive human behavior in a new light.

SUPPLEMENTS FOR STUDENTS

For the Tenth Edition of *Physiology of Behavior*, the supplements author team has collaborated on a new online resource called MyPsychKit (found at www. mypsychkit.com; an access code is required), which replaces the CD-ROM published with previous editions, and which contains all of the content, including the animated *Figures* and *Diagrams* and the *Simulations* previously found on that CD-ROM. The animations demonstrate some of the most important principles of neuroscience through movement and interaction. They include modules on neurophysiology (*Neural Communication, The Action Potential, Synapses,* and *Postsynaptic Potentials*), neuroanatomy, psychopharmacology, audition, sleep, emotion, ingestive behavior, memory, and verbal communication. The interactive *Computerized Study Guide*, accessible through the same menu, contains a set of *Self Tests* that include multiple-choice questions and an on-line review of *Terms and Definitions*. The questions and list of terms and definitions present questions and allow students to keep track of their progress, presenting missed items until they have answered all of them correctly. The computerized study guide also includes interactive *Figures* and *Diagrams* from the book that will help students learn terms and concepts.

The *Study Guide,* revised by Eric P. Wiertelak Ph. D., Macalaster College, is also available. This workbook provides a framework for guiding study behavior. It promotes a thorough understanding of the principles of physiological psychology through active participation in the learning process. The study guide contains a set of *Concept Cards*. An important part of learning about physiological psychology is acquiring a new vocabulary, and the concept cards will help with this task. Terms are printed on one side of these cards, and definitions are printed on the other. A crossword puzzle, based on the vocabulary introduced in the text, finishes each chapter.

SUPPLEMENTS FOR INSTRUCTORS

Several supplements are available for instructors who adopt *Physiology of Behavior*. The *Instructor's Manual* was written by Dr. Scott Wersinger, University of Buffalo, SUNY. Each chapter includes an Integrated Teaching Outline with information about other supplements, teaching objectives, lecture material, demonstrations and activities, videos, suggested readings, and web resources. An appendix contains a set of student handouts.

Dr. Grant McLaren, Edinboro University of Pennsylvania, has prepared a set of *PowerPoint Presentations* specifically for the tenth edition of the book. These are available on the Instructor Resource Center (for instructors only, please contact your Pearson representative for assistance with downloading these files) at www.pearsonhighered.com. They contain images from the textbook and provide a framework for lecture outlines.

Paul Wellman, Texas A&M University, has prepared a *Test Bank*. Available in print and also in an electronic test-generating software format, the test bank includes over 2500 thoroughly reviewed multiple-choice, completion, short answer, and essay questions, each with answer skill justification, page references, difficulty rating, and skill type designation.

The *Digital Media Archive* CD-ROM provides a comprehensive source for images that are useful in classroom presentations. A set of 145 full-color acetate transparencies is also available.

ACKNOWLEDGMENTS

Although I must accept the blame for any shortcomings of the book, I want to thank the many colleagues who helped me by responding to my requests for reprints of their work, suggesting topics that I should cover, permitting me to reproduce their diagrams and photographs in this book, and pointing out deficiencies in the previous edition.

I also want to thank the people at my publisher, Allyn and Bacon. Stephen Frail, Senior Acquisitions Editor, provided assistance, support, advice, and encouragement. Katharine Motter, editorial assistant, helped to gather comments and suggestions from my colleagues. Judith Fiske, production editor, assembled the team that designed and produced the book. Barbara Gracia demonstrated her masterful organization skills in managing the details of the book's production. She got everything done on time, despite an extremely tight schedule.

Few people realize what a difficult, demanding, and time-consuming job it is to produce a project such as this, with hundreds of illustrations and an author who tends to procrastinate, but I do, and I thank her for all she has done. Barbara Willette served as copy editor. Her attention to detail surprised me again and again; she found inconsistencies in my terminology and awkwardness in my prose and gave me a chance to fix them before anyone else saw them in print.

I must also thank my wife Mary for her support. Writing is a lonely pursuit, because one must be alone with one's thoughts for many hours of the day. I thank her for giving me the time to read, reflect, and write without feeling that I was neglecting her too much. I also thank her for the superb job she did preparing the study guide.

I was delighted to hear from many students and colleagues who read previous editions of my book, and I hope that the dialogue will continue. Please write to me and tell me what you like and dislike about the book. My address is Department of Psychology, Tobin Hall, University of Massachusetts, Amherst, Massachusetts 01003. My e-mail is nrc@psych.umass.edu. When I write, I like to imagine that I am talking with you, the reader. If you write to me, we can make the conversation a two-way exchange.

Introduction

outline

Miss S. was a sixty-year-old woman with a history of high blood pressure, which was not responding well to the medication she was taking. One evening she was sitting in her reclining chair reading the newspaper when the phone rang. She got out of her chair and walked to the phone. As she did, she began feeling giddy and stopped to hold onto the kitchen table. She has no memory of what happened after that.

The next morning, a neighbor, who usually stopped by to have coffee with Miss S., found her lying on the floor, mumbling incoherently. The neighbor called an ambulance, which took Miss S. to a hospital.

Two days after her admission, I visited her in her room, along with a group of neuropsychologists and neurological residents being led by the chief of neurology. We had already been told by the neurological resident in charge of her case that Miss S. had had a stroke in the back part of the right side of the brain. He had attached a CT scan to an illuminated viewer mounted on the wall and had showed us a white spot caused by the accumulation of blood in a particular region of her brain. (You can look at the scan yourself if you like; it is shown in Figure 5.19.)

About a dozen of us entered Miss S.'s room. She was awake but seemed a little confused. The resident greeted her and asked how she was feeling. "Fine, I guess," she said. "I still don't know why I'm here."

"Can you see the other people in the room?"

"Why, sure."

"How many are there?"

She turned her head to the right and began counting. She stopped when she had counted the people at the foot of her bed. "Seven," she reported. "What about us?" asked a voice from the left of her bed. "What?" she said, looking at the people she had already counted. "Here, to your left. No, toward your left!" the voice repeated. Slowly, rather reluctantly, she began turning her head to the left. The voice kept insisting, and finally, she saw who was talking. "Oh," she said, "I guess there are more of you."

The resident approached the left side of her bed and touched her left arm. "What is this?" he asked. "Where?" she said. "Here," he answered, holding up her arm and moving it gently in front of her face.

"Oh, that's an arm."

"An arm? Whose arm?"

"I don't know. . . . I guess it must be yours."

"No, it's yours. Look, it's a part of you." He traced with his fingers from her arm to her shoulder.

"Well, if you say so," she said, still sounding unconvinced.

When we returned to the residents' lounge, the chief of neurology said that we had seen a classic example of unilateral neglect, caused by damage to a particular part of the right side of the brain. "I've seen many cases like this," he explained. "People can still perceive sensations from the left side of their body, but they just don't pay attention to them. A woman will put makeup on only the right side of her face, and a man will shave only half of his beard. When they put on a shirt or a coat, they will use their left hand to slip it over their right arm and shoulder, but then they'll just forget about their left arm and let the garment hang from one shoulder. They also don't look at things located toward the left or even the left halves of things. Once I visited a man in his hospital room who had just finished eating breakfast. He was sitting in his bed, with a tray in front of him. There was half of a pancake on his plate. 'Are you all done?' I asked. 'Sure,' he said. When he wasn't looking, I turned the plate around so that the uneaten part was on his right. He saw it, looked startled, and said, 'Where the hell did that come from?'"

The last frontier in this world—and perhaps the greatest one—lies within us. The human nervous system makes possible all that we can do, all that we can know, and all that we can experience. Its complexity is immense, and the task of studying it and understanding it dwarfs all previous explorations our species has undertaken.

One of the most universal of all human characteristics is curiosity. We want to explain what makes things happen. In ancient times, people believed that natural phenomena were caused by animating spirits. All moving objects—animals, the wind and tides, the sun, moon, and stars—were assumed to have spirits that caused them to move. For example, stones fell when they were dropped because their animating spirits wanted to be reunited with Mother Earth. As our ancestors became more sophisticated and learned more about nature, they abandoned this approach (which we call *animism*) in favor of physical explanations for inanimate moving objects. But they still used spirits to explain human behavior.

From the earliest historical times, people have believed that they possess something intangible that animates them: a mind, or a soul, or a spirit. This belief stems from the fact that each of us is aware of his or her own existence. When we think or act, we feel as though something inside us is thinking or deciding to act. But what is the nature of the human mind? We have physical bodies, with muscles that move it and sensory organs such as eyes and ears that perceive information about the world around us. Within our bodies the nervous system plays a central role, receiving information from the sensory organs and controlling the movements of the muscles. But what role does the mind play? Does it *control* the nervous system? Is it a *part of* the nervous system? Is it physical and tangible, like the rest of the body, or is it a spirit that will always remain hidden?

This puzzle has historically been called the *mind–body question.* Philosophers have been trying to answer it for many centuries, and more recently, scientists have taken up the task. Basically, people have followed two different approaches: dualism and monism. **Dualism** is a belief in the dual nature of reality. Mind and body are separate; the body is made of ordinary matter, but the mind is not. **Monism** is a belief that everything in the universe consists of matter and energy and that the mind is a phenomenon produced by the workings of the nervous system.

Mere speculation about the nature of the mind can get us only so far. If we could answer the mind–body question simply by thinking about it, philosophers would have done so long ago. Physiological psychologists take an empirical, practical, and monistic approach to the study of human nature. Most of us believe that once we understand the workings of the human body—and, in particular, the workings of the nervous system—the mind–body problem will have been solved. We will be able to explain how we perceive, how we think, how we remember, and how we act. We will even be able to explain the nature of our own self-awareness. Of course, we are far from understanding the workings of the nervous system, so only time will tell whether this belief is justified. In any event there is no way to study nonphysical phenomena in the laboratory. All that we can detect with our sense organs and our laboratory instruments are manifestations of the physical world: matter and energy.

UNDERSTANDING HUMAN CONSCIOUSNESS: A PHYSIOLOGICAL APPROACH

As you will learn from subsequent chapters, scientists have discovered much about the physiology of behavior: of perception, motivation, emotion, memory, and control of specific movements. But before addressing these problems, I want to show you that a scientific approach to perhaps the most complex phenomenon of all—human consciousness—is at least possible.

The term *consciousness* can be used to refer to a variety of concepts, including simple wakefulness. Thus, a researcher may write about an experiment using "conscious rats," referring to the fact that the rats were awake and not anesthetized. However, in this context I am using the word *consciousness* to refer to the fact that we humans are aware of—and can tell others about—our thoughts, perceptions, memories, and feelings.

FIGURE 1.1 ■ Studying the Brain

Will the human brain ever completely understand its own workings? A sixteenth-century woodcut from the first edition of *De humani corporis fabrica (On the Workings of the Human Body)* by Andreas Vesalius.

(Courtesy of National Library of Medicine.)

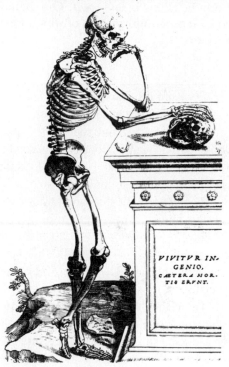

We know that consciousness can be altered by changes in the structure or chemistry of the brain; therefore, we may hypothesize that consciousness is a physiological function, just like behavior. We can even speculate about the origins of this self-awareness. Consciousness and the ability to communicate seem to go hand in hand. Our species, with its complex social structure and enormous capacity for learning, is well served by our ability to communicate: to express intentions to one another and to make requests of one another. Verbal communication makes cooperation possible and permits us to establish customs and laws of behavior. Perhaps the evolution of this ability is what has given rise to the phenomenon of consciousness. That is, our ability to send and receive messages with other people enables us to send and receive our own messages inside our own heads—in other words, to think and to be aware of our own existence. (See *Figure 1.1.*)

dualism The belief that the body is physical but the mind (or soul) is not.

monism (*mahn ism*) The belief that the world consists only of matter and energy and that the mind is a phenomenon produced by the workings of the nervous system.

Blindsight

A particularly interesting phenomenon known as **blindsight** has some implications for our understanding of consciousness (Weiskrantz et al., 1974; Stoerig and Cowey, 2007). This phenomenon suggests that the common belief that perceptions must enter consciousness to affect our behavior is incorrect. Our behavior can be guided by sensory information of which we are completely unaware.

Natalie J. had brought her grandfather to see Dr. M., a neuropsychologist. Mr. J.'s stroke had left him almost completely blind; all he could see was a tiny spot in the middle of his visual field. Dr. M. had learned about Mr. J.'s condition from his neurologist and had asked Mr. J. to come to his laboratory so that he could do some tests for his research project.

Dr. M. helped Mr. J. find a chair and sit down. Mr. J., who walked with the aid of a cane, gave it to his granddaughter to hold for him. "May I borrow that?" asked Dr. M. Natalie nodded and handed the cane to Dr. M. "The phenomenon I'm studying is called blindsight," he said. "Let me see if I can show you what it is.

"Mr. J., please look straight ahead. Keep looking that way, and don't move your eyes or turn your head. I know that you can see a little bit straight ahead of you, and I don't want you to use that piece of vision for what I'm going to ask you to do. Fine. Now, I'd like you to reach out with your right hand and point to what I'm holding."

"But I don't see anything—I'm blind!" said Mr. J., obviously exasperated.

"I know, but please try, anyway."

Mr. J. shrugged his shoulders and pointed. He looked startled when his finger encountered the end of the cane, which Dr. M. was pointing toward him.

"Gramps, how did you do that?" asked Natalie, amazed. "I thought you were blind."

"I am!" he said, emphatically. "It was just luck."

"Let's try it just a couple more times, Mr. J.," said Dr. M. "Keep looking straight ahead. Fine." He reversed the cane, so that the handle was pointing toward Mr. J. "Now I'd like you to grab hold of the cane."

Mr. J. reached out with an open hand and grabbed hold of the cane.

"Good. Now put your hand down, please." He rotated the cane 90 degrees, so that the handle was oriented vertically. "Now reach for it again."

Mr. J. did so. As his arm came up, he turned his wrist so that his hand matched the orientation of the handle, which he grabbed hold of again.

"Good. Thank you, you can put your hand down." Dr. M. turned to Natalie. "I'd like to test your grandfather now, but I'll be glad to talk with you later."

As Dr. M. explained to Natalie afterward, the brain contains not one but several mechanisms involved in vision. To simplify matters somewhat, let's consider two systems, which evolved at different times. The more primitive one, which resembles the visual system of animals such as fish and frogs, evolved first. The more complex one, which is possessed by mammals, evolved later. This second, "mammalian" system seems to be the one that is responsible for our ability to perceive the world around us. The first, "primitive" visual system, is devoted mainly to controlling eye movements and bringing our attention to sudden movements that occur off to the side of our field of vision.

Mr. J.'s stroke had damaged the mammalian visual system: the visual cortex of the brain and some of the nerve fibers that bring information to it from the eyes. Cases like his show that that after the mammalian visual system is damaged, people can use the primitive visual system of their brains to guide hand movements toward an object even though they cannot see what they are reaching for. In other words, visual information can control behavior without producing a conscious sensation. The phenomenon of blindsight suggests that *consciousness is not a general property of all parts of the brain;* some parts of the brain, but not others, play a special role in consciousness. Although we are not sure just where these parts are or exactly how they work, they seem to be related to our ability to communicate—with others and with ourselves. The primitive system, which evolved before the development of consciousness, does not have these connections, so we are not conscious of the visual information it detects. It *does* have connections with the parts of the brain responsible for controlling hand movements. Only the mammalian visual system has direct connections with the parts of the brain responsible for consciousness. (See *Figure 1.2.*)

Split Brains

Studies of humans who have undergone a particular surgical procedure demonstrate dramatically how disconnecting parts of the brain involved with perceptions from parts that are involved with verbal behavior also

---

blindsight The ability of a person who cannot see objects in his or her blind field to accurately reach for them while remaining unconscious of perceiving them, caused by damage to the "mammalian" visual system of the brain.

FIGURE 1.2 ■ An Explanation of the Blindsight Phenomenon

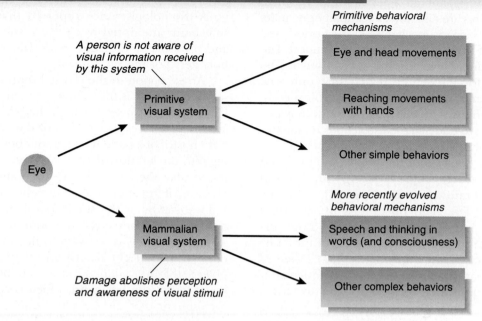

A person is not aware of visual information received by this system

Primitive visual system

Primitive behavioral mechanisms

Eye and head movements

Reaching movements with hands

Other simple behaviors

Eye

Mammalian visual system

Damage abolishes perception and awareness of visual stimuli

More recently evolved behavioral mechanisms

Speech and thinking in words (and consciousness)

Other complex behaviors

disconnects them from consciousness. These results suggest that the parts of the brain involved in verbal behavior may be the ones responsible for consciousness.

The surgical procedure is one that has been used for people with very severe epilepsy that cannot be controlled by drugs. In these people, nerve cells in one side of the brain become overactive, and the overactivity is transmitted to the other side of the brain by the corpus

FIGURE 1.3 ■ The Split-Brain Operation

A "window" has been opened in the side of the brain so that we can see the corpus callosum being cut at the midline of the brain.

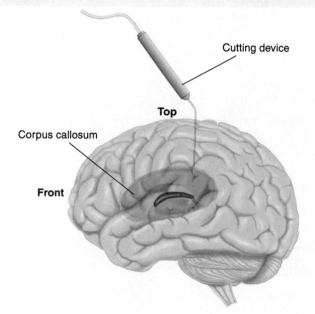

Cutting device

Top

Corpus callosum

Front

callosum. The **corpus callosum** is a large bundle of nerve fibers that connect corresponding parts of one side of the brain with those of the other. Both sides of the brain then engage in wild activity and stimulate each other, causing a generalized epileptic seizure. These seizures can occur many times each day, preventing the patient from leading a normal life. Neurosurgeons discovered that cutting the corpus callosum (the **split-brain operation**) greatly reduced the frequency of the epileptic seizures.

Figure 1.3 shows a drawing of the split-brain operation. We see the brain being sliced down the middle, from front to back, dividing it into its two symmetrical halves. A "window" has been opened in the left side of the brain so that we can see the corpus callosum being cut by the neurosurgeon's special knife. (See *Figure 1.3*.)

Sperry (1966) and Gazzaniga and his associates (Gazzaniga and LeDoux, 1978; Gazzaniga, 2005) have studied these patients extensively. The largest part of the brain consists of two symmetrical parts, called the **cerebral hemispheres,** which receive sensory information from the opposite sides of the body. They also control movements of the opposite sides. The corpus callosum permits the two hemispheres to share information so

---

corpus callosum (*core pus ka **low** sum*) The largest commissure of the brain, interconnecting the areas of neocortex on each side of the brain.

split-brain operation Brain surgery that is occasionally performed to treat a form of epilepsy; the surgeon cuts the corpus callosum, which connects the two hemispheres of the brain.

cerebral hemispheres The two symmetrical halves of the brain; constitute the major part of the brain.

that each side knows what the other side is perceiving and doing. After the split-brain operation is performed, the two hemispheres are disconnected and operate independently; their sensory mechanisms, memories, and motor systems can no longer exchange information. The effects of these disconnections are not obvious to the casual observer, for the simple reason that only one hemisphere—in most people, the left—controls speech. The right hemisphere of an epileptic person with a split brain appears able to understand instructions reasonably well, but it is totally incapable of producing speech.

Because only one side of the brain can talk about what it is experiencing, people speaking with a person who has a split brain are conversing with only one hemisphere: the left. The operations of the right hemisphere are more difficult to detect. Even the patient's left hemisphere has to learn about the independent existence of the right hemisphere. One of the first things that these patients say they notice after the operation is that their left hand seems to have a "mind of its own." For example, patients may find themselves putting down a book held in the left hand, even if they have been reading it with great interest. This conflict occurs because the right hemisphere, which controls the left hand, cannot read and therefore finds holding the book boring. At other times these

patients surprise themselves by making obscene gestures (with the left hand) when they had not intended to. A psychologist once reported that a man with a split brain attempted to beat his wife with one hand and protect her with the other. Did he *really* want to hurt her? Yes and no, I guess.

An exception to the crossed representation of sensory information is the olfactory system. That is, when a person sniffs a flower through the left nostril, only the left brain receives a sensation of the odor. Thus, if the right nostril of a patient with a split brain is closed, leaving only the left nostril open, the patient will be able to tell us what the odors are (Gordon and Sperry, 1969). However, if the odor enters the right nostril, the patient will say that he or she smells nothing. But, in fact, the right brain *has* perceived the odor and *can* identify it. To show that this is so, we ask the patient to smell an odor with the right nostril and then reach for some objects that are hidden from view by a partition. If asked to use the left hand, which is controlled by the hemisphere that detected the smell, the patient will select the object that corresponds to the odor—a plastic flower for a floral odor, a toy fish for a fishy odor, a model tree for the odor of pine, and so forth. But if asked to use the right hand, the patient fails the test because the right hand is connected to the left hemisphere,

FIGURE 1.4 ■ Smelling with a Split Brain

An object is identified in response to an olfactory stimulus by a person with a split brain

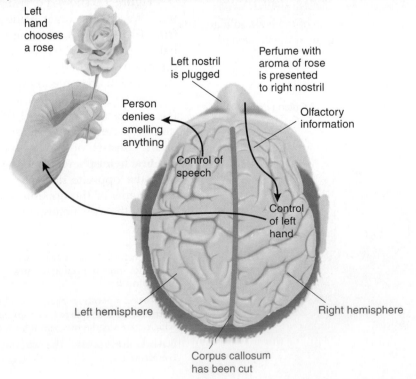

Left hand chooses a rose

Left nostril is plugged

Perfume with aroma of rose is presented to right nostril

Person denies smelling anything

Olfactory information

Control of speech

Control of left hand

Left hemisphere

Right hemisphere

Corpus callosum has been cut

which did not smell the odor presented to the right nostril. (See *Figure 1.4.*)

The effects of cutting the corpus callosum reinforce the conclusion that we become conscious of something only if information about it is able to reach the parts of the brain responsible for verbal communication, which are located in the left hemisphere. If the information does not reach these parts of the brain, then that information does not reach consciousness. We still know very little about the physiology of consciousness, but studies of people with brain damage are beginning to provide us with some useful insights. This issue is discussed in later chapters.

Unilateral Neglect

The phenomenon described in the case history at the beginning of this chapter—failure to notice things located to a person's left—is known as **unilateral neglect** (Husain and Rorden, 2003). Unilateral ("one-sided") neglect is produced by damage to a particular part of the right side of the brain: the cortex of the parietal lobe. (Chapter 3 will describe the location of this region.) The parietal lobe receives information directly from the skin, the muscles, the joints, the internal organs, and the part of the inner ear that is concerned with balance. Thus, it is concerned with the body and its position. But that is not all; the parietal cortex receives auditory and visual information as well. Its most important function seems to be to put together information about the movements and location of the parts of the body with the locations of objects in space around us.

If unilateral neglect simply consisted of blindness in the left side of the visual field and anesthesia of the left side of the body, it would not be nearly as interesting. But individuals with unilateral neglect are neither half blind nor half numb. Under the proper circumstances, they *can* see things located to their left, and they *can* tell when someone touches the left side of their bodies. But normally, they ignore such stimuli and act as if the left side of the world and of their bodies did not exist. In other words, their inattention to things to the left means that they normally do not become conscious of them.

Volpe, LeDoux, and Gazzaniga (1979) presented pairs of visual stimuli to people with unilateral neglect—one stimulus in the left visual field and one stimulus in the right. Invariably, the people reported seeing only the right-hand stimulus. But when the investigators asked the people to say whether or not the two stimuli were identical, they answered correctly, *even though they said that they were unaware of the left-hand stimulus.*

If you think about the story that the chief of neurology told about the man who ate only the right half of a

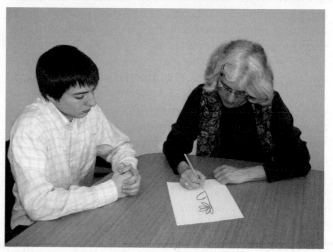

FIGURE 1.5 ■ Unilateral Neglect

When people with unilateral neglect attempt to draw simple objects, they demonstrate their unawareness of the left half of things by drawing only the features that appear on the right.

pancake, you will realize that people with unilateral neglect *must* be able to perceive more than the right visual field. Remember that people with unilateral neglect fail to notice not only things to their left but also the *left halves* of things. But to distinguish between the left and right halves of an object, you first have to perceive the entire object—otherwise, how would you know where the middle was?

People with unilateral neglect also demonstrate their unawareness of the left half of things when they draw pictures. For example, when asked to draw a clock, they almost always successfully draw a circle; but then when they fill in the numbers, they scrunch them all in on the right side. Sometimes they simply stop after reaching 6 or 7, and sometimes they write the rest of the numbers underneath the circle. When asked to draw a daisy, they begin with a stem and a leaf or two and then draw all the petals to the right. (See *Figure 1.5.*)

Bisiach and Luzzatti (1978) demonstrated a similar phenomenon, which suggests that unilateral neglect extends even to a person's own visual imagery. The investigators asked two patients with unilateral neglect to describe the Piazza del Duomo, a well-known landmark in Milan, the city in which they and the patients lived. They asked the patients to imagine that they were standing at the north end of the piazza and to

unilateral neglect A syndrome in which people ignore objects located toward their left and the left sides of objects located anywhere; most often caused by damage to the right parietal lobe.

describe what they saw. The patients duly named the buildings, but only those on the west, to their right. Then the investigators asked them to imagine themselves at the south end of the piazza. This time, they named the buildings on the east—again, to their right. Obviously, they knew about *all* of the buildings and their locations, but they visualized them only when the buildings were located in the right side of their (imaginary) visual field.

As you can see, there are two major symptoms of unilateral neglect: neglect of the left halves of things in the environment and neglect of the left half of one's own body. In fact, although most people with unilateral neglect show both types of symptoms, research indicates that they are produced by damage to slightly different regions of the brain (Hillis et al., 2005).

You might wonder whether damage to the *left* parietal lobe causes unilateral *right* neglect. The answer is yes, but it is very slight, is difficult to detect, and seems to be temporary. For all practical purposes, then, there is no right neglect. But why not? The answer is still a mystery. To be sure, people have suggested some possible explanations, but they are still speculative. Not until we know a lot more about the brain mechanisms of attention will we be able to understand this discrepancy.

Although neglect of the left side of one's own body can be studied only in people with brain abnormalities, an interesting phenomenon seen in people with undamaged brains confirms the importance of the parietal lobe (and another region of the brain) in feelings of body ownership. Ehrsson, Spence, and Passingham (2004) studied the *rubber hand illusion*. Normal subjects were positioned with their left hand hidden out of sight. They saw a lifelike rubber left hand in front of them. The experimenters stroked both the subject's hidden left hand and the visible rubber hand with a small paintbrush. If the two hands were stroked synchronously and in the same direction, the subjects began to experience the rubber hand as their own. In fact, if they were then asked to use their right hand to point to their left hand, they tended to point toward the rubber hand. However, if the real and artificial hands were stroked in different directions or at different times, the subjects did *not* experience the rubber hand as their own. (See *Figure 1.6.*)

While the subjects were participating in the experiment, the experimenters recorded the activity of their brains with a functional MRI scanner. (Brain scanning is described in Chapter 5.) The scans showed increased activity in the parietal lobe and then, as the subjects began to experience the rubber hand as belonging to their body, in the *premotor cortex,* a region of the brain involved in planning movements. When the stroking of the real and artificial hands was uncoordinated and the

FIGURE 1.6 ■ The Rubber Hand Illusion

If the subject's hidden left hand and the visible rubber hand are stroked synchronously in the same direction, the subject will come to experience the artificial hand as his or her own. If the hands are stroked asynchronously or in different directions, this illusion will not occur.

(Adapted from Botwinick, M. *Science*, 2004, *305*, 782–783.)

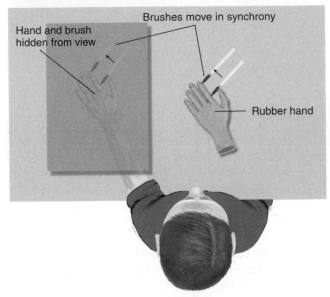

subjects did not experience the rubber hand as their own, the premotor cortex did not become activated. The experimenters concluded that the parietal cortex analyzed the sight and the feeling of brush strokes. When the parietal cortex detected that they were congruent, this information was transmitted to the premotor cortex, which gave rise to the feeling of ownership of the rubber hand.

A second study from the same laboratory provided a particularly convincing demonstration that people experience a genuine feeling of ownership of the rubber hand (Ehrsson et al., 2007). The investigators used the procedure described above to establish a feeling of ownership and then threatened the rubber hand by making a stabbing movement toward it with a needle. (They did not actually touch the hand with the needle.) Brain scans showed increased activity in a region of the brain (the anterior cingulate cortex) that is normally activated when a person anticipates pain and also in a region (the supplementary motor area) that is normally activated when a person feels the urge to move his or her arm (Fried et al., 1991; Peyron and Garcia-Larrea, 2000). So the impression that the rubber hand was about to receive a painful stab from a needle made people react as they would if their own hand were the target of the threat.

Interim Summary

Understanding Human Consciousness

The mind–body body question has puzzled philosophers for many centuries. Modern science has adopted a monistic position—the belief that the world consists of matter and energy and that nonmaterial entities such as minds are not a part of the universe. Studies of the functions of the human nervous system tend to support this position, as two specific examples show. Both phenomena show that brain damage, by damaging conscious brain functions or disconnecting them from the speech mechanisms in the left hemisphere, can reveal the presence of other functions, of which the person is *not* conscious.

Blindsight is a phenomenon that is seen after partial damage to the "mammalian" visual system on one side of the brain. Although the person is, in the normal meaning of the word, blind to anything presented to part of the visual field, the person can nevertheless reach out and point to objects whose presence he or she is not conscious of. Similarly, when sensory information about a particular object is presented to the right hemisphere of a person who has had a split-brain operation, the person is not aware of the object but can nevertheless indicate by movements of the left hand that the object has been perceived. Unilateral neglect—failure to become aware of the left half of one's body, the left half of objects, or items located to a person's left—reveals the existence of brain mechanisms that control our attention to things and hence our ability to become aware of them. These phenomena suggest that consciousness involves operations of the verbal mechanisms of the left hemisphere. Indeed, consciousness may be, in large part, a matter of our "talking to ourselves." Thus, once we understand the language functions of the brain, we may have gone a long way toward understanding how the brain can be conscious of its own existence.

Thought Questions

1. Could a sufficiently large and complex computer ever be programmed to be aware of itself? Suppose that someone someday claims to have done just that. What kind of evidence would you need to prove or disprove this claim?
2. Clearly, the left hemisphere of a person with a split brain is conscious of the information it receives and of its own thoughts. It is not conscious of the mental processes of the right hemisphere. But is it possible that the right hemisphere is conscious too but is just unable to talk to us? How could we possibly find out whether it is? Do you see some similarities between this issue and the one raised in the first question?

THE NATURE OF PHYSIOLOGICAL PSYCHOLOGY

The field of physiological psychology grew out of psychology. Indeed, the first textbook of psychology, written by Wilhelm Wundt in the late nineteenth century, was titled *Principles of Physiological Psychology*. In recent years, with the explosion of information in experimental biology, scientists from other disciplines have become prominent contributors to the investigation of the physiology of behavior. The united effort of physiological psychologists, physiologists, and other neuroscientists is due to the realization that the ultimate function of the nervous system is behavior.

When I ask my students what they think the ultimate function of the brain is, they often say "thinking," or "logical reasoning," or "perceiving," or "remembering things." Certainly, the nervous system performs these functions, but they support the primary one: control of movement. The basic function of perception is to inform us of what is happening in our environment so that our behaviors will be adaptive and useful: Perception without the ability to act would be useless. Of course, once perceptual abilities have evolved, they can be used for purposes other than guiding behavior. For example, we can enjoy a beautiful sunset or a great work of art without the perception causing us to do anything in particular. And thinking can often take place without causing any overt behavior. However, the *ability to think* evolved because it permits us to perform complex behaviors that accomplish useful goals. And whereas reminiscing about things that happened in our past can be an enjoyable pastime, the ability to learn and remember evolved—again—because it permitted our ancestors to profit from experience and perform behaviors that were useful to them.

The modern history of investigating the physiology of behavior has been written by psychologists who have combined the experimental methods of psychology with those of physiology and have applied them to the issues that concern all psychologists. Thus, we have studied perceptual processes, control of movement, sleep and waking, reproductive behaviors, ingestive behaviors, emotional behaviors, learning, and language. In recent years we have begun to study the physiology of human pathological conditions, such as addictions and neurological and mental disorders. All of these topics are discussed in subsequent chapters of this book.

The Goals of Research

The goal of all scientists is to explain the phenomena they study. But what do we mean by *explain?* Scientific explanation takes two forms: generalization and reduction. All scientists deal with **generalization.** For example, psychologists explain particular instances of behavior as examples of general laws, which they deduce from their experiments. For instance, most psychologists would explain a pathologically strong fear of dogs as an example of a particular form of learning called *classical conditioning.* Presumably, the person was frightened earlier in life by a dog. An unpleasant stimulus was paired with the sight of the animal (perhaps the person was knocked down by an exuberant dog or was attacked by a vicious one), and the subsequent sight of dogs evokes the earlier response: fear.

Most physiologists use an additional approach to explanation: **reduction.** They explain complex phenomena in terms of simpler ones. For example, they may explain the movement of a muscle in terms of the changes in the membranes of muscle cells, the entry of particular chemicals, and the interactions among protein molecules within these cells. By contrast, a molecular biologist would explain these events in terms of forces that bind various molecules together and cause various parts of the molecules to be attracted to one another. In turn, the job of an atomic physicist is to describe matter and energy themselves and to account for the various forces found in nature. Practitioners of each branch of science use reduction to call on sets of more elementary generalizations to explain the phenomena they study.

The task of the physiological psychologist is to explain behavior by studying the physiological processes that control it. But physiological psychologists cannot simply be reductionists. It is not enough to observe behaviors and correlate them with physiological events that occur at the same time. Identical behaviors may occur for different reasons and thus may be initiated by different physiological mechanisms. Therefore, we must understand "psychologically" why a particular behavior occurs before we can understand what physiological events made it occur.

Let me provide a specific example: Mice, like many other mammals, often build nests. Behavioral observations show that mice will build nests under two conditions: when the air temperature is low and when the animal is pregnant. A nonpregnant mouse will build a nest

only if the weather is cool, whereas a pregnant mouse will build one regardless of the temperature. The same behavior occurs for different reasons. In fact, nest-building behavior is controlled by two different physiological mechanisms. Nest building can be studied as a behavior related to the process of temperature regulation, or it can be studied in the context of parental behavior. Although the same set of brain mechanisms will control the movements that a mouse makes in building a nest in both cases, these mechanisms will be activated by different parts of the brain. One part receives information from the body's temperature detectors, and the other part is influenced by hormones that are present in the body during pregnancy.

Sometimes, physiological mechanisms can tell us something about psychological processes. This relationship is particularly true of complex phenomena such as language, memory, and mood, which are poorly understood psychologically. For example, damage to a specific part of the brain can cause very specific impairments in a person's language abilities. The nature of these impairments suggests how these abilities are organized. When the damage involves a brain region that is important in analyzing speech sounds, it also produces deficits in spelling. This finding suggests that the ability to recognize a spoken word and the ability to spell it call on related brain mechanisms. Damage to another region of the brain can produce extreme difficulty in reading unfamiliar words by sounding them out, but it does not impair the person's ability to read words with which he or she is already familiar. This finding suggests that reading comprehension can take two routes: one that is related to speech sounds and another that is primarily a matter of visual recognition of whole words.

In practice, the research efforts of physiological psychologists involve both forms of explanation: generalization and reduction. Ideas for experiments are stimulated by the investigator's knowledge both of psychological generalizations about behavior and of physiological mechanisms. A good physiological psychologist must therefore be both a good psychologist *and* a good physiologist.

Biological Roots of Physiological Psychology

Study of (or speculations about) the physiology of behavior has its roots in antiquity. Because its movement was necessary for life and because emotions caused it to

---

generalization A type of scientific explanation; a general conclusion based on many observations of similar phenomena.

reduction A type of scientific explanation; a phenomenon is described in terms of the more elementary processes that underlie it.

beat more strongly, many ancient cultures, including the Egyptian, Indian, and Chinese cultures, considered the heart to be the seat of thought and emotions. The ancient Greeks did too, but Hippocrates (460–370 B.C.E.) concluded that this role should be assigned to the brain. Except for the somewhat flowery language, the following extract from *On the Sacred Disease* (epilepsy) could have been written by a modern neurobiologist:

> Men ought to know that from nothing else but the brain come joys, delights, laughter and sports, and sorrows, griefs, despondency, and lamentations. And by this, in an especial manner, we acquire wisdom and knowledge, and see and hear and know what are foul and what are fair, what are bad and what are good, what are sweet, and what are unsavory. . . . And by the same organ we become mad and delirious, and fears and terrors assail us. . . . All these things we endure from the brain when it is not healthy (Hippocrates, 1952 translation, p. 159).

Not all ancient Greek scholars agreed with Hippocrates. Aristotle did not; he thought the brain served to cool the passions of the heart. But Galen (A.D. 130–200), who had the greatest respect for Aristotle, concluded that Aristotle's role for the brain was "utterly absurd, since in that case Nature would not have placed the encephalon so far from the heart, . . . and she would not have attached the sources of all the senses [the sensory nerves] to it (Galen, 1968 translation, p. 387). Galen thought enough of the brain to dissect and study the brains of cattle, sheep, pigs, cats, dogs, weasels, monkeys, and apes (Finger, 1994).

René Descartes, a seventeenth-century French philosopher and mathematician, has been called the father of modern philosophy. Although he was not a biologist, his speculations concerning the roles of the mind and brain in the control of behavior provide a good starting point in the modern history of physiological psychology. Descartes assumed that the world was a purely mechanical entity that, once having been set in motion by God, ran its course without divine interference. Thus, to understand the world, one had only to understand how it was constructed. To Descartes, animals were mechanical devices; their behavior was controlled by environmental stimuli. His view of the human body was much the same: It was a machine. As Descartes observed, some movements of the human body were automatic and involuntary. For example, if a person's finger touched a hot object, the arm would immediately withdraw from the source of stimulation. Reactions like this did not require participation of the mind; they occurred automatically. Descartes called these actions **reflexes** (from the Latin *reflectere*, "to bend back upon itself"). Energy coming from the outside source would

FIGURE 1.7 ■ Descartes's Explanation of a Reflex Action to a Painful Stimulus

be reflected back through the nervous system to the muscles, which would contract. The term is still in use today, but, of course, we explain the operation of a reflex differently. (See *Figure 1.7*.)

Like most philosophers of his time, Descartes was a dualist; he believed that each person possessed a mind—a uniquely human attribute that was not subject to the laws of the universe. But his thinking differed from that of his predecessors in one important way: He was the first to suggest that a link exists between the human mind and its purely physical housing, the brain. He believed that the mind controlled the movements of the body, while the body, through its sense organs, supplied the mind with information about what was happening in the environment. In particular, he hypothesized that this interaction took place in the pineal body, a small organ situated on top of the brain stem, buried beneath the cerebral hemispheres. He noted that the brain contained hollow chambers (the *ventricles*) that were filled with fluid, and he hypothesized that this fluid was under pressure. When the mind decided to perform an action, it tilted the pineal body in a particular direction like a little joystick, causing fluid to flow from the brain into the appropriate set of nerves. This flow of fluid caused the same muscles to inflate and move. (See *Figure 1.8*.)

---

reflex An automatic, stereotyped movement that is produced as the direct result of a stimulus.

FIGURE 1.8 ■ Descartes's Theory

Descartes believed that the "soul" (what we would today call the *mind*) controlled the movements of the muscles through its influence on the pineal body. His explanation is modeled on the mechanism that animated statues in the Royal Gardens near Paris.

(Courtesy of Historical Pictures Service, Chicago.)

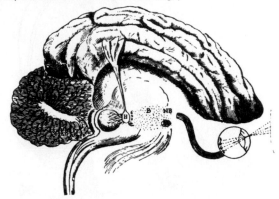

FIGURE 1.9 ■ Johannes Müller (1801–1858)

(Courtesy of National Library of Medicine.)

As a young man, René Descartes was greatly impressed by the moving statues in the grottoes of the Royal Gardens, just west of Paris (Jaynes, 1970). He was fascinated by the hidden mechanisms that caused the statues to move when visitors stepped on hidden plates. For example, as a visitor approached a bronze statue of Diana, bathing in a pool of water, she would flee and hide behind a bronze rose bush. If the visitor pursued her, an imposing statue of Neptune would rise up and bar the way with his trident.

These devices served as models for Descartes in theorizing about how the body worked. The pressurized water of the moving statues was replaced by pressurized fluid in the ventricles; the pipes by nerves; the cylinders by muscles; and, finally, the hidden valves by the pineal body. This story illustrates one of the first times that a technological device was used as a model for explaining how the nervous system works. In science a **model** is a relatively simple system that works on known principles and is able to do at least some of the things that a more complex system can do. For example, when scientists discovered that elements of the nervous system communicate by means of electrical impulses, researchers developed models of the brain based on telephone switchboards and, more recently, computers. Abstract models, which are completely mathematical in their properties, have also been developed.

Descartes's model was useful because, unlike purely philosophical speculations, it could be tested experimentally. In fact, it did not take long for biologists to prove that Descartes was wrong. For example, Luigi

Galvani, a seventeenth-century Italian physiologist, found that electrical stimulation of a frog's nerve caused contraction of the muscle to which it was attached. Contraction occurred even when the nerve and muscle were detached from the rest of the body, so the ability of the muscle to contract and the ability of the nerve to send a message to the muscle were characteristics of these tissues themselves. Thus, the brain did not inflate muscles by directing pressurized fluid through the nerve. Galvani's experiment prompted others to study the nature of the message transmitted by the nerve and the means by which muscles contracted. The results of these efforts gave rise to an accumulation of knowledge about the physiology of behavior.

One of the most important figures in the development of experimental physiology was Johannes Müller, a nineteenth-century German physiologist. Müller was a forceful advocate of the application of experimental techniques to physiology. Previously, the activities of most natural scientists had been limited to observation and classification. Although these activities are essential, Müller insisted that major advances in our understanding of the workings of the body would be achieved only by experimentally removing or isolating animals' organs, testing their responses to various chemicals, and otherwise altering the environment to see how the organs responded. (See *Figure 1.9.*)

model A mathematical or physical analogy for a physiological process; for example, computers have been used as models for various functions of the brain.

His most important contribution to the study of the physiology of behavior was his **doctrine of specific nerve energies.** Müller observed that although all nerves carry the same basic message—an electrical impulse—we perceive the messages of different nerves in different ways. For example, messages carried by the optic nerves produce sensations of visual images, and those carried by the auditory nerves produce sensations of sounds. How can different sensations arise from the same basic message?

The answer is that the messages occur in different channels. The portion of the brain that receives messages from the optic nerves interprets the activity as visual stimulation, even if the nerves are actually stimulated mechanically. (For example, when we rub our eyes, we see flashes of light.) Because different parts of the brain receive messages from different nerves, the brain must be functionally divided: Some parts perform some functions, while other parts perform others.

Müller's advocacy of experimentation and the logical deductions from his doctrine of specific nerve energies set the stage for performing experiments directly on the brain. Indeed, Pierre Flourens, a nineteenth-century French physiologist, did just that. Flourens removed various parts of animals' brains and observed their behavior. By seeing what the animal could no longer do, he could infer the function of the missing portion of the brain. This method is called **experimental ablation** (from the Latin *ablatus,* "carried away"). Flourens claimed to have discovered the regions of the brain that control heart rate and breathing, purposeful movements, and visual and auditory reflexes.

Soon after Flourens performed his experiments, Paul Broca, a French surgeon, applied the principle of experimental ablation to the human brain. Of course, he did not intentionally remove parts of human brains to see how they worked but observed the behavior of people whose brains had been damaged by strokes. In 1861 he performed an autopsy on the brain of a man who had had a stroke that resulted in the loss of the ability to speak. Broca's observations led him to conclude that a portion of the cerebral cortex on the left side of the brain performs functions that are necessary for speech. (See *Figure 1.10.*) Other physicians soon obtained evidence supporting his conclusions. As you will learn in Chapter 14, the control of speech is not localized in a particular region of the brain. Indeed, speech requires many different functions, which are organized throughout the brain. Nonetheless, the method of experimental ablation remains important to our understanding of the brains of both humans and laboratory animals.

As I mentioned earlier, Luigi Galvani used electricity to demonstrate that muscles contain the source of the energy that powers their contractions. In 1870, German physiologists Gustav Fritsch and Eduard Hitzig used electrical stimulation as a tool for understanding the physiology of the brain. They applied weak electrical current to the exposed surface of a dog's brain and observed the effects of the stimulation. They found that stimulation of different portions of a specific region of the brain caused contraction of specific muscles on the opposite side of the body. We now refer to this region as the *primary motor cortex,* and we know that nerve cells there communicate directly with those that cause muscular contractions. We also know that other regions of the brain communicate with the primary motor cortex and thus control behaviors. For example, the region that Broca found necessary for speech communicates with, and controls, the portion of the primary motor cortex that controls the muscles of the lips, tongue, and throat, which we use to speak.

One of the most brilliant contributors to nineteenth-century science was the German physicist and physiologist Hermann von Helmholtz. Helmholtz devised a mathematical formulation of the law of conservation of energy; invented the ophthalmoscope (used to examine the retina of the eye); devised an important and influential theory of color vision and color blindness;

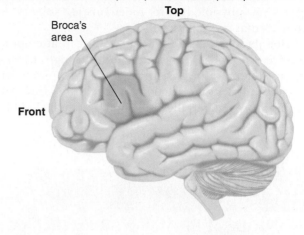

FIGURE 1.10 ■ Broca's Area

This region of the brain is named for French surgeon Paul Broca. Broca discovered that damage to a part of the left side of the brain disrupts a person's ability to speak.

Broca's area

Top

Front

---

doctrine of specific nerve energies Müller's conclusion that because all nerve fibers carry the same type of message, sensory information must be specified by the particular nerve fibers that are active.

experimental ablation The research method in which the function of a part of the brain is inferred by observing the behaviors an animal can no longer perform after that part is damaged.

and studied audition, music, and many physiological processes. Although Helmholtz had studied under Müller, he opposed Müller's belief that human organs are endowed with a vital nonmaterial force that coordinates their operations. Helmholtz believed that all aspects of physiology are mechanistic, subject to experimental investigation.

Helmholtz was also the first scientist to attempt to measure the speed of conduction through nerves. Scientists had previously believed that such conduction was identical to the conduction that occurs in wires, traveling at approximately the speed of light. But Helmholtz found that neural conduction was much slower—only about 90 feet per second. This measurement proved that neural conduction was more than a simple electrical message, as we will see in Chapter 2.

Twentieth-century developments in experimental physiology include many important inventions, such as sensitive amplifiers to detect weak electrical signals, neurochemical techniques to analyze chemical changes within and between cells, and histological techniques to see cells and their constituents. Because these developments belong to the modern era, they are discussed in detail in subsequent chapters.

Interim Summary

The Nature of Physiological Psychology

All scientists hope to explain natural phenomena. In this context the term *explanation* has two basic meanings: generalization and reduction. Generalization refers to the classification of phenomena according to their essential features so that general laws can be formulated. For example, observing that gravitational attraction is related to the mass of two bodies and to the distance between them helps to explain the movement of planets. Reduction refers to the description of phenomena in terms of more basic physical processes. For example, gravitation can be explained in terms of forces and subatomic particles.

Physiological psychologists use both generalization and reduction to explain behavior. In large part, generalizations use the traditional methods of psychology. Reduction explains behaviors in terms of physiological events within the body—primarily within the nervous system. Thus, physiological psychology builds on the tradition of both experimental psychology and experimental physiology.

A dualist, René Descartes, proposed a model of the brain on the basis of hydraulically activated statues. His model stimulated observations that produced important discoveries. The results of Luigi Galvani's experiments eventually led to an understanding of the nature of the message transmitted by nerves between the brain and the sensory organs and the muscles. Johannes Müller's doctrine of specific nerve energies paved the way for study of the functions of specific parts of the brain, through the methods of experimental ablation and electrical stimulation. Hermann von Helmholtz, a former student of Müller's, insisted that all aspects of human physiology were subject to the laws of nature. He also discovered that the conduction through nerves was slower than the conduction of electricity, which meant that it was a physiological phenomenon.

Thought Questions

1. What is the value of studying the history of physiological psychology? Is it a waste of time?
2. Suppose we studied just the latest research and ignored explanations that we now know to be incorrect. Would we be spending our time more profitably, or might we miss something?

NATURAL SELECTION AND EVOLUTION

Müller's insistence that biology must be an experimental science provided the starting point for an important tradition. However, other biologists continued to observe, classify, and think about what they saw, and some of them arrived at valuable conclusions. The most important of these scientists was Charles Darwin. (See *Figure 1.11.*) Darwin formulated the principles of *natural selection* and *evolution,* which revolutionized biology.

Functionalism and the Inheritance of Traits

Darwin's theory emphasized that all of an organism's characteristics—its structure, its coloration, its behavior—have functional significance. For example, the strong talons and sharp beaks that eagles possess permit the birds to catch and eat prey. Caterpillars that eat green leaves are themselves green, and their color makes it difficult for birds to see them against their usual background. Mother mice construct nests, which keep their offspring warm and out of harm's way. Obviously, the

FIGURE 1.11 ■ Charles Darwin (1809–1882)

Darwin's theory of evolution revolutionized biology and strongly influenced early psychologists.

FIGURE 1.12 ■ Bones of the Forelimb

The figure shows the bones of (a) human, (b) bat, (c) whale, and (d) dog. Through the process of natural selection these bones have been adapted to suit many different functions.

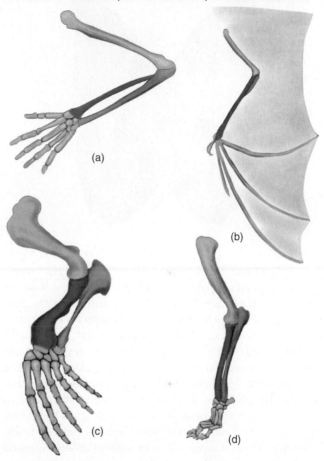

behavior itself is not inherited—how could it be? What *is* inherited is a brain that causes the behavior to occur. Thus, Darwin's theory gave rise to **functionalism,** a belief that characteristics of living organisms perform useful functions. So to understand the physiological basis of various behaviors, we must first understand what these behaviors accomplish. We must therefore understand something about the natural history of the species being studied so that the behaviors can be seen in context.

To understand the workings of a complex piece of machinery, we should know what its functions are. This principle is just as true for a living organism as it is for a mechanical device. However, an important difference exists between machines and organisms: Machines have inventors who had a purpose when they designed the machines, whereas organisms are the result of a long series of accidents. Thus, strictly speaking, we cannot say that any physiological mechanisms of living organisms have a *purpose.* But they do have *functions,* and these we can try to determine. For example, the forelimbs shown in Figure 1.12 are adapted for different uses in different species of mammals. (See *Figure 1.12.*)

A good example of the functional analysis of an adaptive trait was demonstrated in an experiment by Blest (1957). Certain species of moths and butterflies have spots on their wings that resemble eyes—particularly

the eyes of predators such as owls. (See *Figure 1.13.*) These insects normally rely on camouflage for protection; the backs of their wings, when folded, are colored like the bark of a tree. However, when a bird approaches, the insect's wings flip open, and the hidden eyespots are suddenly displayed. The bird then tends to fly away rather than eat the insect. Blest performed an experiment to see whether the eyespots on a moth's or butterfly's wings really disturbed birds that saw them. He placed mealworms on different backgrounds and counted how many worms the birds ate. Indeed, when the worms were placed on a background that contained eyespots, the birds tended to avoid them.

Darwin formulated his theory of evolution to explain the means by which species acquired their adaptive characteristics. The cornerstone of this theory is the

functionalism The principle that the best way to understand a biological phenomenon (a behavior or a physiological structure) is to try to understand its useful functions for the organism.

FIGURE 1.13 ■ The Owl Butterfly

This butterfly displays its eyespots when approached by a bird. The bird usually will fly away.

principle of **natural selection.** Darwin noted that members of a species were not all identical and that some of the differences they exhibited were inherited by their offspring. If an individual's characteristics permit it to reproduce more successfully, some of the individual's offspring will inherit the favorable characteristics and will themselves produce more offspring. As a result, the characteristics will become more prevalent in that species. He observed that animal breeders were able to develop strains that possessed particular traits by mating together only animals that possessed the desired traits. If *artificial selection,* controlled by animal breeders, could produce so many varieties of dogs, cats, and livestock, perhaps *natural selection* could be responsible for the development of species. Of course, it was the natural environment, not the hand of the animal breeder, that shaped the process of evolution.

Darwin and his fellow scientists knew nothing about the mechanism by which the principle of natural selection works. In fact, the principles of molecular genetics were not discovered until the middle of the twentieth century. Briefly, here is how the process works: Every sexually reproducing multicellular organism consists of a large number of cells, each of which contains chromosomes. Chromosomes are large, complex molecules that contain the recipes for producing the proteins that cells need to grow and to perform their functions. In essence, the chromosomes contain the blueprints for the construction (that is, the embryological development) of a particular member of a particular species. If the plans are altered, a different organism is produced.

The plans do get altered; mutations occur from time to time. **Mutations** are accidental changes in the chromosomes of sperms or eggs that join together and develop into new organisms. For example, cosmic radiation might strike a chromosome in a cell of an animal's testis or ovary, thus producing a mutation that affects that animal's offspring. Most mutations are deleterious; the offspring either fails to survive or survives with some sort of deficit. However, a small percentage of mutations are beneficial and confer a **selective advantage** to the organism that possesses them. That is, the animal is more likely than other members of its species to live long enough to reproduce and hence to pass on its chromosomes to its own offspring. Many different kinds of traits can confer a selective advantage: resistance to a particular disease, the ability to digest new kinds of food, more effective weapons for defense or for procurement of prey, and even a more attractive appearance to members of the opposite sex (after all, one must reproduce to pass on one's chromosomes).

Naturally, the traits that can be altered by mutations are physical ones; chromosomes make proteins, which affect the structure and chemistry of cells. But the *effects* of these physical alterations can be seen in an animal's behavior. Thus, the process of natural selection can act on behavior indirectly. For example, if a particular mutation results in changes in the brain that cause a small animal to stop moving and freeze when it perceives a novel stimulus, that animal is more likely to escape undetected when a predator passes nearby. This tendency makes the animal more likely to survive and produce offspring, thus passing on its genes to future generations.

Other mutations are not immediately favorable, but because they do not put their possessors at a disadvantage, they are inherited by at least some members of the species. As a result of thousands of such mutations, the members of a particular species possess a variety of genes and are all at least somewhat different from one another. Variety is a definite advantage for a species. Different environments provide optimal habitats for different kinds of organisms. When the environment changes, species must adapt or run the risk of becoming extinct. If some members of the species possess assortments of genes that provide characteristics permitting them to adapt to the new environment, their offspring will survive, and the species will continue.

An understanding of the principle of natural selection plays some role in the thinking of every scientist

natural selection The process by which inherited traits that confer a selective advantage (increase an animal's likelihood to live and reproduce) become more prevalent in a population.

mutation A change in the genetic information contained in the chromosomes of sperms or eggs, which can be passed on to an organism's offspring; provides genetic variability.

selective advantage A characteristic of an organism that permits it to produce more than the average number of offspring of its species.

who undertakes research in physiological psychology. Some researchers explicitly consider the genetic mechanisms of various behaviors and the physiological processes upon which these behaviors depend. Others are concerned with comparative aspects of behavior and its physiological basis; they compare the nervous systems of animals from a variety of species to make hypotheses about the evolution of brain structure and the behavioral capacities that correspond to this evolutionary development. But even though many researchers are not directly involved with the problem of evolution, the principle of natural selection guides the thinking of physiological psychologists. We ask ourselves what the selective advantage of a particular trait might be. We think about how nature might have used a physiological mechanism that already existed in order to perform more complex functions in more complex organisms. When we entertain hypotheses, we ask ourselves whether a particular explanation makes sense in an evolutionary perspective.

Evolution of the Human Species

To *evolve* means to develop gradually (from the Latin *evolvere*, "to unroll"). The process of **evolution** is a gradual change in the structure and physiology of plant and animal species as a result of natural selection. New species evolve when organisms develop novel characteristics that can take advantage of unexploited opportunities in the environment.

The first vertebrates to emerge from the sea—some 360 million years ago—were amphibians. In fact, amphibians have not entirely left the sea; they still lay their eggs in water, and the larvae that hatch from these eggs have gills and only later transform into adults with air-breathing lungs. Seventy million years later, the first reptiles appeared. Reptiles had a considerable advantage over amphibians: Their eggs, enclosed in a shell just porous enough to permit the developing embryo to breathe, could be laid on land. Thus, reptiles could inhabit regions away from bodies of water, and they could bury their eggs where predators would be less likely to find them. Reptiles soon divided into three lines: the *anapsids,* the ancestors of today's turtles; the *diapsids,* the ancestors of dinosaurs, birds, lizards, crocodiles, and snakes; and the *synapsids,* the ancestors of today's mammals. One group of synapsids, the *therapsids,* became the dominant land animal during the Permian period. Then, about 248 million years ago, the end of the Permian period was marked by a mass extinction. Dust from a catastrophic series of volcanic eruptions in present-day Siberia darkened the sky, cooled the earth, and wiped out approximately 95 percent of all animal species. Among those that survived was a small therapsid known as a *cynodont*—the direct ancestor of the mammal,

which first appeared about 220 million years ago. (See *Figure 1.14.*)

The earliest mammals were small nocturnal predators that fed on insects. Their eyesight was poorer than that of the cynodonts from which they evolved, but their hearing was better. The middle ear of amphibians and reptiles contains a single tiny bone, the stapes ("stirrup"), which transmits sound vibrations to the receptive organ for hearing located in the inner ear. As a result of a series of mutations, the earliest mammals evolved a jaw that did away with two of the bones found in the jaws of reptiles. Rather than becoming altogether lost, these bones became incorporated into the mammalian middle ear. The chain of three tiny bones (the *ossicles*) in the middle ear makes it possible for mammals to hear very high frequencies. Presumably, this ability enabled the earliest mammals to hear sounds made by insects and prey on them at night, when larger predators could not see them. (See *Figure 1.15* on page 19.)

Mammals (and the other warm-blooded animals: birds) were only a modest success for many millions of years. Dinosaurs ruled, and mammals had to remain small and inconspicuous to avoid the large variety of agile and voracious predators. Then, around 65 million years ago, another mass extinction occurred. An enormous meteorite struck the Yucatan peninsula of present-day Mexico, producing a cloud of dust that destroyed many species, including the dinosaurs. Small, nocturnal mammals survived the cold and dark because they were equipped with insulating fur and a mechanism for maintaining their body temperature. The void left by the extinction of so many large herbivores and carnivores provided the opportunity for mammals to expand into new ecological niches, and expand they did.

The climate of the early Cenozoic period, which followed the mass extinction at the end of the Cretaceous period, was much warmer than the climate of today. Tropical forests covered much of the land areas, and in these forests our most direct ancestors, the primates, evolved. The first primates, like the first mammals, were small and preyed upon insects and small cold-blooded vertebrates such as lizards and frogs. They had grasping hands that permitted them to climb about in small branches of the forest. Over time, larger species developed, with larger, forward-facing eyes (and the brains to analyze what the eyes saw), which facilitated arboreal locomotion and the capture of prey.

Plants evolved as well as animals. Dispersal of seeds is a problem inherent in forest life; if a tree's seeds fall at its base, they will be shaded by the parent and will not

---

evolution A gradual change in the structure and physiology of plant and animal species—generally producing more complex organisms—as a result of natural selection.

FIGURE 1.14 ■ Evolution of Vertebrate Species

(Adapted from Carroll, R. *Vertebrate Paleontology and Evolution.* New York: W. H. Freeman, 1988.)

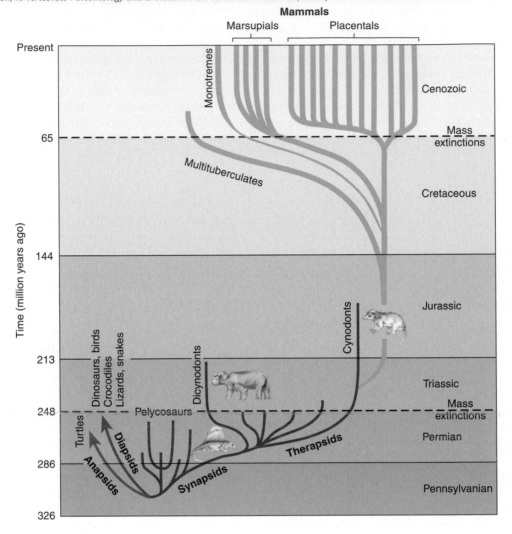

grow. Thus, natural selection favored trees that encased their seeds in sweet, nutritious fruit that would be eaten by animals and dropped on the ground some distance away, undigested, in the animals feces. (The feces even served to fertilize the young plants.) The evolution of fruit-bearing trees provided an opportunity for fruit-eating primates. In fact, the original advantage of color vision was probably the ability to discriminate ripe fruit from green leaves and eat the fruit before it spoiled—or some other animals got to it first. And because fruit is such a nutritious form of food, its availability provided an opportunity that could be exploited by larger primates, which were able to travel farther in quest of food.

The first *hominids* (humanlike apes) appeared in Africa. They appeared not in dense tropical forests but in drier woodlands and in the savanna—vast areas of grasslands studded with clumps of trees and populated by large herbivorous animals and the carnivores that preyed

on them. Our fruit-eating ancestors continued to eat fruit, of course, but they evolved characteristics that enabled them to gather roots and tubers as well, to hunt and kill game, and to defend themselves against other predators. They made tools that could be used to hunt, produce clothing, and construct dwellings; they discovered the many uses of fire; they domesticated dogs, which greatly increased their ability to hunt and helped warn of attacks by predators; and they developed the ability to communicate symbolically, by means of spoken words.

Figure 1.16 shows the primate family tree. Our closest living relatives are the chimpanzees, gorillas, and orangutans. DNA analysis shows very little genetic difference between these four species. (See *Figure 1.16.*) For example, humans and chimpanzees share 98.8 percent of their DNA. (See *Figure 1.17.* on page 20)

The first hominid to leave Africa did so around 1.7 million years ago. This species, *Homo erectus* ("upright

FIGURE 1.15 ■ Evolution of the Mammalian Middle Ear Bones (Ossicles)

The quadrate and articular bones of the reptilian jaw became the incus and the malleus.

(Adapted from Gould, S. J. *The Book of Life.* New York: W. W. Norton, 1993.)

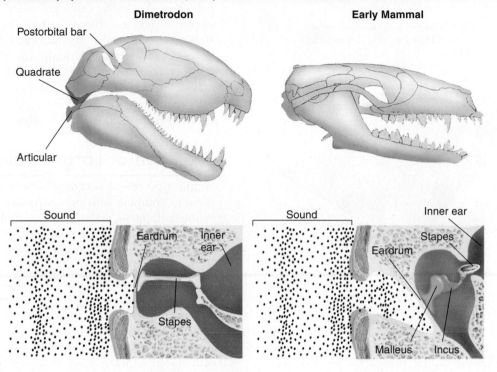

FIGURE 1.16 ■ Evolution of Primate Species

(Redrawn from Lewin, R. *Human Evolution: An Illustrated Introduction*, 3rd ed. Boston: Blackwell Scientific Publications, 1993. Reprinted with permission by Blackwell Science Ltd.)

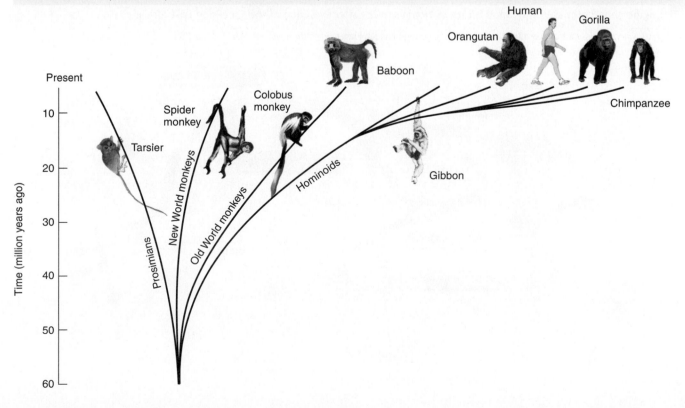

FIGURE 1.17 ■ DNA Among Species of Hominids

The pyramid illustrates the percentage differences in DNA among the four major species of hominids.

(Redrawn from Lewin, R. *Human Evolution: An Illustrated Introduction.* Boston: Blackwell Scientific Publications, 1993. Reprinted with permission by Blackwell Science Ltd.)

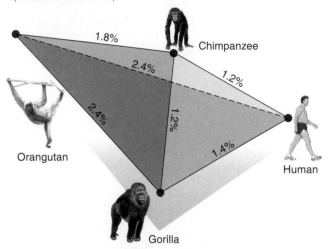

1.8%

2.4%

1.2%

2.4%

1.2%

1.4%

Chimpanzee

Orangutan

Human

Gorilla

man"), scattered across Europe and Asia. One branch of *Homo erectus* appears to have been the ancestor of *Homo neanderthalis*, which inhabited Western Europe between 120,000 and 30,000 years ago. Neanderthals resembled modern humans. They made tools out of

stone and wood and discovered the use of fire. Our own species, *Homo sapiens*, evolved in East Africa around 100,000 years ago. Some of our ancestors migrated to other parts of Africa and out of Africa to Asia, Polynesia, Australia, Europe, and the Americas. They encountered the Neanderthals in Europe around 40,000 years ago and coexisted with them for approximately 10,000 years. Eventually, the Neanderthals disappeared—perhaps through interbreeding with *Homo sapiens*, perhaps through competition for resources. Scientists have not found evidence of warlike conflict between the two species. (See *Figure 1.18.*)

Evolution of Large Brains

Humans possessed several characteristics that enabled them to compete with other species. Their agile hands enabled them to make and use tools. Their excellent color vision helped them to spot ripe fruit, game animals, and dangerous predators. Their mastery of fire enabled them to cook food, provide warmth, and frighten nocturnal predators. Their upright posture and bipedalism made it possible for them to walk long distances efficiently, with their eyes far enough from the ground to see long distances across the plains. Bipedalism also permitted them to carry tools and food with them, which meant that they could bring fruit, roots, and pieces of meat back to their tribe. Their linguistic abilities enabled them to combine the collective

FIGURE 1.18 ■ Migration Routes of *Homo Sapiens*

The figure shows proposed migration routes of *Homo sapiens* after evolution of the species in East Africa.

(Redrawn with permission from Cavalli-Sforza, L. L. Genes, peoples and languages. *Scientific American*, Nov. 1991, p. 75.)

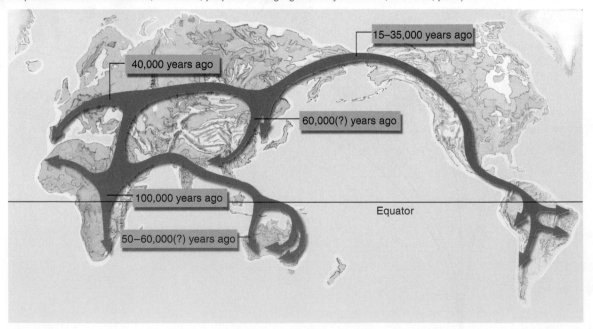

15–35,000 years ago

40,000 years ago

60,000(?) years ago

100,000 years ago

Equator

50–60,000(?) years ago

knowledge of all the members of the tribe, to make plans, to pass information on to subsequent generations, and to establish complex civilizations that established their status as the dominant species. All of these characteristics required a larger brain.

A large brain requires a large skull, and an upright posture limits the size of a woman's birth canal. A newborn baby's head is about as large as it can safely be. As it is, the birth of a baby is much more arduous than the birth of mammals with proportionally smaller heads, including those of our closest primate relatives. Because a baby's brain is not large or complex enough to perform the physical and intellectual abilities of an adult, the brain must continue to grow after the baby is born. In fact, all mammals (and all birds, for that matter) require parental care for a period of time while the nervous system develops. The fact that young mammals (particularly young humans) are guaranteed to be exposed to the adults who care for them means that a period of apprenticeship is possible. Consequently, the evolutionary process did not have to produce a brain that consisted solely of specialized circuits of neurons that performed specialized tasks. Instead, it could simply produce a larger brain with an abundance of neural circuits that could be modified by experience. Adults would nourish and protect their offspring and provide them with the skills they would need as adults. Some specialized circuits were necessary, of course (for example, those involved in analyzing the complex sounds we use for speech), but by and large, the brain is a general-purpose, programmable computer.

How does the human brain compare with the brains of other animals? In absolute size, our brains are dwarfed by those of elephants or whales. However, we might expect such large animals to have large brains to match their large bodies. Indeed, the human brain makes up 2.3 percent of our total body weight, while the elephant brain makes up only 0.2 percent of the animal's total body weight, which makes our brains seem very large in comparison. However, the shrew, which weighs only 7.5 g, has a brain that weighs 0.25 g, or 3.3 percent of its total body weight. Certainly, the shrew brain is much less complex than the human brain, so something is wrong with this comparison.

The answer is that although bigger bodies require bigger brains, the size of the brain does not have to go up proportionally with that of the body. For example, larger muscles do not require more nerve cells to control them. What counts, as far as intellectual ability goes, is having a brain with plenty of nerve cells that are not committed to moving muscles or analyzing sensory information—nerve cells that are available for learning, remembering, reasoning, and making plans. Figure 1.19 shows a graph of the brain sizes and body weights of several hominid species, including the ancestors of our

FIGURE 1.19 ■ Hominid Brain Size

The graph shows average brain size as a function of body weight for several species of hominids.

(Redrawn from Lewin, R. *Human Evolution: An Illustrated Introduction*, 3rd ed. Boston: Blackwell Scientific Publications, 1993. Reprinted with permission by Blackwell Science Ltd.)

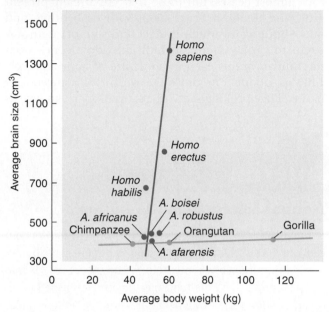

own species. Note that the brain size of nonhuman hominids increases very little with size: A gorilla weighs almost three times as much as a chimpanzee, but their brains weigh almost the same. In contrast, although the body weight of modern humans is only 29 percent more than that of *Australopithecus africanus*, our brains are 242 percent larger. Clearly, some important mutations of the genes that control brain development occurred early in the evolution of the primate line. (See *Figure 1.19.*)

Besides varying is size, brains also vary in the number of neurons found in each gram of tissue. Herculano-Houzel et al. (2007) compared the weight of the brains of several species of rodents and primates with the number of neurons that each brain contained. They found that primate brains—especially large ones—contain many more neurons per gram than rodent brains do. For example, the brain of a capuchin monkey weighs 52 g and contains 3.7 billion neurons, while the brain of a capybara (a very large South American rodent) weighs 76 g but contains only 1.6 billion neurons. The brain of a capuchin monkey (and a human brain, for that matter) contains 70.7 million neurons per gram, while that of a capybara contains only 21 million neurons per gram.

What types of genetic changes were responsible for the evolution of the human brain? This question will be addressed in more detail in Chapter 3, but evidence suggests that the most important principle is a slowing of the process of brain development, allowing more time for growth. As we will see, the prenatal period of cell division in the brain is prolonged in humans, which results in a brain that weighs an average of 350 g and contains approximately 100 billion neurons. After birth the brain continues to grow. Production of new neurons almost ceases, but those that are already present grow and establish connections with each other, and other brain cells, which protect and support neurons, begin to proliferate. Not until late adolescence does the human brain reach its adult size of approximately 1400 g—about four times the weight of a newborn's brain. This prolongation of maturation is known as

neoteny (roughly translated as "extended youth"). The mature human head and brain retain some infantile characteristics, including their disproportionate size relative to the rest of the body. Figure 1.20 shows fetal and adult skulls of chimpanzees and humans. As you can see, the fetal skulls are much more similar than are those of the adults. The grid lines show the pattern of growth, indicating much less change in the human skull from birth to adulthood. (See *Figure 1.20.*)

---

neoteny A slowing of the process of maturation, allowing more time for growth; an important factor in the development of large brains.

Interim Summary

Natural Selection and Evolution

Darwin's theory of evolution, which was based on the concept of natural selection, provided an important contribution to modern physiological psychology. The theory asserts that we must understand the functions that are performed by an organ or body part or by a behavior. Through random mutations, changes in an individual's genetic material cause different proteins to be produced, which results in the alteration of some physical characteristics. If the changes confer a selective advantage on the individual, the new genes will be transmitted to more and more members of the species. Even behaviors can evolve, through the selective advantage of alterations in the structure of the nervous system.

Amphibians emerged from the sea 360 million years ago. One branch, the therapsids, became the dominant land animal until a catastrophic series of volcanic eruptions wiped out most animal species. A small therapsid, the cynodont, survived the disaster and became the ancestor of the mammals. The earliest mammals were small, nocturnal insectivores who lived in trees. They remained small and inconspicuous until the extinction of the dinosaurs, which occurred around 65 million years ago. Mammals quickly filled the vacant ecological niches. Primates also began as small, nocturnal, tree-dwelling insectivores. Larger fruit-eating primates, with forward-facing eyes and larger brains, eventually evolved.

The first hominids appeared in Africa around 25 million years ago, eventually evolving into four major species: orangutans, gorillas, chimpanzees, and humans. Our ancestors acquired bipedalism around 3.7 million years ago and discovered tool making around 2.5 million years ago. The first hominids to leave Africa, *Homo erectus*, did so around

1.7 million years ago and scattered across Europe and Asia. *Homo neanderthalis* evolved in Western Europe, eventually to be replaced by *Homo sapiens*, which evolved in Africa around 100,000 years and spread throughout the world. By 30,000 years ago, *Homo sapiens* had replaced *Homo neanderthalis*.

The evolution of large brains made possible the development of tool making, fire building, and language, which in turn permitted the development of complex social structures. Large brains also provided a large memory capacity and the abilities to recognize patterns of events in the past and to plan for the future. Because an upright posture limits the size of a woman's birth canal and therefore the size of the head that can pass through it, much of the brain's growth must take place after birth, which means that children require an extended period of parental care. This period of apprenticeship enabled the developing brain to be modified by experience.

Although human DNA differs from that of chimpanzees by only 1.2 percent, our brains are more than three times larger, which means that a small number of genes is responsible for the increase in the size of our brains. As we will see in Chapter 3, these genes appear to retard the events that stop brain development, resulting in a phenomenon known as neoteny.

Thought Questions

1. What useful functions are provided by the fact that a human can be self-aware? How was this trait selected for during the evolution of our species?
2. Are you surprised that the difference in the DNA of humans and chimpanzees is only 1.2 percent? How do you feel about this fact?
3. If our species continues to evolve, what kinds of changes do you think might occur?

FIGURE 1.20 ■ Neoteny in Evolution of the Human Skull

The skulls of fetal humans and chimpanzees are much more similar than are those of the adults. The grid lines show the pattern of growth, indicating much less change in the human skull from birth to adulthood.

(Redrawn from Lewin, R. *Human Evolution: An Illustrated Introduction*, 3rd ed. Boston: Blackwell Scientific Publications, 1993. Reprinted with permission by Blackwell Science Ltd.)

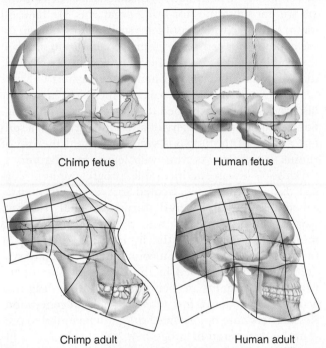

Chimp fetus   Human fetus

Chimp adult   Human adult

ETHICAL ISSUES IN RESEARCH WITH ANIMALS

Most of the research described in this book involves experimentation on living animals. Any time we use another species of animals for our own purposes, we should be sure that what we are doing is both humane and worthwhile. I believe that a good case can be made that research on the physiology of behavior qualifies on both counts. Humane treatment is a matter of procedure. We know how to maintain laboratory animals in good health in comfortable, sanitary conditions. We know how to administer anesthetics and analgesics so that animals do not suffer during or after surgery, and we know how to prevent infections with proper surgical procedures and the use of antibiotics. Most industrially developed societies have very strict regulations about the care of animals and require approval of the experimental procedures that are used on them. There is no excuse for mistreating animals in our care. In fact, the vast majority of laboratory animals *are* treated humanely.

Whether an experiment is *worthwhile* can be difficult to say. We use animals for many purposes. We eat their meat and eggs, and we drink their milk; we turn their hides into leather; we extract insulin and other hormones from their organs to treat people's diseases; we train them to do useful work on farms or to entertain us. Even having a pet is a form of exploitation; it is we—not they—who decide that they will live in our homes. The fact is, we have been using other animals throughout the history of our species.

Pet owning causes much more suffering among animals than scientific research does. As Miller (1983) notes, pet owners are not required to receive permission from a board of experts that includes a veterinarian to house their pets, nor are they subject to periodic inspections to be sure that their home is clean and sanitary, that their pets have enough space to exercise properly, or that their pets' diets are appropriate. Scientific researchers are. Miller also notes that fifty times more dogs and cats are killed by humane societies each year because they have been abandoned by former pet owners than are used in scientific research.

If a person believes that it is wrong to use another animal in any way, regardless of the benefits to humans, there is nothing anyone can say to convince that person of the value of scientific research with animals. For that person the issue is closed from the very beginning. Moral absolutes cannot be settled logically; like religious beliefs they can be accepted or rejected, but they cannot be proved or disproved. My arguments in support of scientific research with animals are based on an evaluation of the benefits the research has to humans. (We should also remember that research with animals often helps other animals; procedures used by veterinarians, as well as those used by physicians, come from such research.)

Before describing the advantages of research with animals, let me point out that the use of animals in research and teaching is a special target of animal rights activists. Nicholl and Russell (1990) examined twenty-one books written by such activists and counted the number of pages devoted to concern for different uses of animals. Next, they compared the relative concern the authors showed for these uses to the numbers of animals actually involved in each of these categories. The results indicate that the authors showed relatively little concern for animals that are used for food, hunting, or furs or for those killed in animal shelters; but although only 0.3 percent of the animals are used for research and education, 63.3 percent of the pages were devoted to this use. In terms of pages per million animals used, the authors devoted 0.08 to food, 0.23 to hunting, 1.27 to furs, 1.44 to killing in pounds—and 53.2 to research and education. The authors showed 665 times more concern for research and education compared with food and 231 times compared with hunting. Even the use of animals for furs (which consumes two-thirds as many animals as research and education) attracted 41.9 times less attention per animal.

The disproportionate amount of concern that animal rights activists show toward the use of animals in research and education is puzzling, particularly because this is the one *indispensable* use of animals. We *can* survive without eating animals, we *can* live without hunting, we *can* do without furs; but without using animals for research and for training future researchers, we *cannot* make progress in understanding and treating diseases. In not too many years our scientists will probably have developed a vaccine that will prevent the further spread of AIDS. Some animal rights activists believe that preventing the deaths of laboratory animals in the pursuit of such a vaccine is a more worthy goal than the prevention of the deaths of millions of humans that will occur as a result of the disease if a vaccine is not found. Even diseases that we have already conquered would take new victims if drug companies could no longer use animals. If they were deprived of animals, these companies could no longer extract hormones used to treat human diseases, and they could not prepare many of the vaccines we now use to prevent disease.

Our species is beset by medical, mental, and behavioral problems, many of which can be solved only through biological research. Let us consider some of the major neurological disorders. Strokes, caused by bleeding or occlusion of a blood vessel within the brain, often leave people partly paralyzed, unable to read, write, or converse with their friends and family. Basic research on the means by which nerve cells communicate with each other has led to important discoveries about the causes of the death of brain cells. This research was not directed toward a specific practical goal; the potential benefits actually came as a surprise to the investigators.

Experiments based on these results have shown that if a blood vessel leading to the brain is blocked for a few minutes, the part of the brain that is nourished by that vessel will die. However, the brain damage can be prevented by first administering a drug that interferes with a particular kind of neural communication. This research is important, because it may lead to medical treatments that can help to reduce the brain damage caused by strokes. But it involves operating on a laboratory animal, such as a rat, and pinching off a blood vessel. (The animals are anesthetized, of course.) Some of the animals will sustain brain damage, and all will be killed so that their brains can be examined. However, you will probably agree that research like this is just as legitimate as using animals for food.

As you will learn later in this book, research with laboratory animals has produced important discoveries about the possible causes or potential treatments of neurological

and mental disorders, including Parkinson's disease, schizophrenia, manic-depressive illness, anxiety disorders, obsessive-compulsive disorders, anorexia nervosa, obesity, and drug addictions. Although much progress has been made, these problems are still with us, and they cause much human suffering. Unless we continue our research with laboratory animals, they will not be solved. Some people have suggested that instead of using laboratory animals in our research, we could use tissue cultures or computers. Unfortunately, tissue cultures or computers are not substitutes for living organisms. We have no way to study behavioral problems such as addictions in tissue cultures, nor can we program a computer to simulate the workings of an animal's nervous system. (If we could, that would mean we already had all the answers.)

This book will discuss some of the many important discoveries that have helped to reduce human suffering. For example, the discovery of a vaccine for polio, a serious disease of the nervous system, involved the use of rhesus monkeys. As you will learn in Chapter 4, Parkinson's disease, an incurable, progressive neurological disorder, has been treated for years with a drug called L-DOPA, discovered through animal research. Now, because of research with rats, mice, rabbits, and monkeys stimulated by the accidental poisoning of several young people with a contaminated batch of synthetic heroin (described in the case study that opens Chapter 5), patients are being treated with a drug that actually slows down the rate of brain degeneration. Researchers have hopes that a drug will be found to prevent the degeneration altogether.

The easiest way to justify research with animals is to point to actual and potential benefits to human health, as I have just done. However, we can also justify this research with a less practical, but perhaps equally important, argument. One of the things that characterizes our species is a quest for an understanding of our world. For example, astronomers study the universe and try to uncover its mysteries. Even if their discoveries never lead to practical benefits such as better drugs or faster methods of transportation, the fact that they enrich our understanding of the beginning and the fate of our universe justifies their efforts. The pursuit of knowledge is itself a worthwhile endeavor. Surely the attempt to understand the universe within us—our nervous system, which is responsible for all that we are or can be—is also valuable.

CAREERS IN NEUROSCIENCE

What is physiological psychology, and what do physiological psychologists do? By the time you finish this book, you will have as complete an answer as I can give to these questions, but perhaps it is worthwhile for me to describe the field—and careers open to those who specialize in it—before we begin our study in earnest.

Physiological psychologists study all behavioral phenomena that can be observed in nonhuman animals. Some study humans as well, using noninvasive physiological research methods. They attempt to understand the physiology of behavior: the role of the nervous system, interacting with the rest of the body (especially the endocrine system, which secretes hormones), in controlling behavior. They study such topics as sensory processes, sleep, emotional behavior, ingestive behavior, aggressive behavior, sexual behavior, parental behavior, and learning and memory. They also study animal models of disorders that afflict humans, such as anxiety, depression, obsessions and compulsions, phobias, psychosomatic illnesses, and schizophrenia.

Although physiological psychology is the original name for this field, several other terms are now in general use, such as *biological psychology, biopsychology, psychobiology*, and *behavioral neuroscience*. Most professional physiological psychologists have received a Ph.D. from a graduate program in psychology or from an interdisciplinary program. (My own university awards a Ph.D. in Neuroscience and Behavior. The program includes faculty members from the departments of psychology, biology, biochemistry, and computer science.)

Physiological psychology belongs to the larger field of *neuroscience*. Neuroscientists concern themselves with all aspects of the nervous system: its anatomy, chemistry, physiology, development, and functioning. The research of neuroscientists ranges from the study of molecular genetics to the study of social behavior. The field has grown enormously in the last few years; the membership of the Society for Neuroscience is currently over thirty-eight thousand.

Most professional physiological psychologists are employed by colleges and universities, where they are engaged in teaching and research. Others are employed by institutions devoted to research—for example, in laboratories owned and operated by national governments or by private philanthropic organizations. A few work in industry, usually for pharmaceutical companies that are interested in assessing the effects of drugs on behavior. To become a professor or independent researcher, one must receive a doctorate—usually a Ph.D., although some people turn to research after receiving an M.D. Nowadays, most physiological psychologists spend two years or more in a temporary postdoctoral position, working in the laboratory of a senior scientist to gain more research experience. During this time they write articles describing their research findings and submit them for publication in scientific journals. These publications are an important factor in obtaining a permanent position.

Two other fields often overlap with that of physiological psychology: *neurology* and *experimental neuropsychology* (often called *cognitive neuroscience*). Neurologists are physicians who are involved in the diagnosis and treatment of diseases of the nervous system. Most neurologists are solely involved in the practice of medicine, but a few engage in research devoted to advancing our understanding of the physiology of behavior. They study the behavior of people whose brains have been damaged by natural causes, using advanced brain-scanning devices to study the activity of various regions of the brain as a subject participates in various behaviors. This research is also carried out by experimental neuropsychologists (or cognitive neuroscientists)—scientists with a Ph.D. (usually in psychology) and specialized training in the principles and procedures of neurology.

Not all people who are engaged in neuroscience research have doctoral degrees. Many research technicians perform essential—and intellectually rewarding—services for the scientists with whom they work. Some of these technicians gain enough experience and education on the job to enable them to collaborate with their employers on their research projects rather than simply working for them.

---

physiological psychologist A scientist who studies the physiology of behavior, primarily by performing physiological and behavioral experiments with laboratory animals.

Interim Summary

Ethical Issues in Research with Animals and Careers in Neuroscience

Research on the physiology of behavior necessarily involves the use of laboratory animals. It is incumbent on all scientists who use these animals to ensure that they are housed comfortably and treated humanely, and laws have been enacted to ensure that they are. Such research has already produced many benefits to humankind and promises to continue to do so.

Physiological psychology (also called biological psychology, biopsychology, psychobiology, and behavioral neuroscience) is a field devoted to our understanding of the physiology of behavior. Physiological psychologists are allied with other scientists in the broader field of neuro-science. To pursue a career in physiological psychology (or in the sister field of experimental neuropsychology), one must obtain a graduate degree and (usually) serve two years or more as a "postdoc"—a junior scientist.

Thought Question
Why do you think some people are apparently more upset about using animals for research and teaching than about using them for other purposes?

STRATEGIES FOR LEARNING

The brain is a complicated organ. After all, it is responsible for all our abilities and all our complexities. Scientists have been studying this organ for a good many years and (especially in recent years) have been learning a lot about how it works. It is impossible to summarize this progress in a few simple sentences; therefore, this book contains a lot of information. I have tried to organize this information logically, telling you what you need to know in the order in which you need to know it. (After all, to understand some things, you sometimes need to understand other things first.) I have also tried to write as clearly as possible, making my examples as simple and as vivid as I can. Still, you cannot expect to master the information in this book by simply giving it a passive read; you will have to do some work.

Learning about the physiology of behavior involves much more than memorizing facts. Of course, there *are* facts to be memorized: names of parts of the nervous system, names of chemicals and drugs, scientific terms for particular phenomena and procedures used to investigate them, and so on. But the quest for information is nowhere near completed; we know only a small fraction of what we have to learn. And almost certainly, many of the "facts" that we now accept will some day be shown to be incorrect. If all you do is learn facts, where will you be when these facts are revised?

The antidote to obsolescence is knowledge of the process by which facts are obtained. In science facts are the conclusions scientists reach about their observations. If you learn only the conclusions, obsolescence is almost guaranteed. You will have to remember which conclusions are overturned and what the new conclusions are, and that kind of rote learning is hard to do. But if you learn about the research strategies the scientists use, the observations they make, and the reasoning that leads to the conclusions, you will develop an understanding that is easily revised when new observations (and new "facts") emerge. If you understand what lies behind the conclusions, then you can incorporate new information into what you already know and revise these conclusions yourself.

In recognition of these realities about learning, knowledge, and the scientific method, this book presents not just a collection of facts, but a description of the procedures, experiments, and logical reasoning that

scientists have used in their attempt to understand the physiology of behavior. If, in the interest of expediency, you focus on the conclusions and ignore the process that leads to them, you run the risk of acquiring information that will quickly become obsolete. On the other hand, if you try to understand the experiments and see how the conclusions follow from the results, you will acquire knowledge that lives and grows.

Enough said. Now let me offer some practical advice about studying. You have been studying throughout your academic career, and you have undoubtedly learned some useful strategies along the way. Even if you have developed efficient and effective study skills, at least consider the possibility that there might be some ways to improve them.

If possible, the first reading of an assignment should be as uninterrupted as you can make it; that is, read the chapter without worrying much about remembering details. Next, after the first class meeting devoted to the topic, read the assignment again in earnest. Use a pen or pencil as you go, making notes. *Don't use a highlighter.* Sweeping the felt tip of a highlighter across some words on a page provides some instant gratification; you can even imagine that the highlighted words are somehow being transferred to your knowledge base. You have selected what is important, and when you review the reading assignment, you have only to read the highlighted words. But this is an illusion.

Be active, not passive. Force yourself to write down whole words and phrases. The act of putting the information into your own words will not only give you something to study shortly before the next exam but also put something into your head (which is helpful at exam time). Using a highlighter puts off the learning until a later date; rephrasing the information in your own words starts the learning process *right then.*

A good way to get yourself to put the information into your own words (and thus into your own brain) is to answer the questions in the study guide. If you cannot answer a question, look up the answer in the book, *close the book,* and write the answer down. The phrase *close the book* is important. If you *copy* the answer, you will get very little out of the exercise. However, if you make yourself remember the information long enough to write it down, you have a good chance of remembering it later. The importance of the study guide is *not* to have a set of short answers in your own handwriting that you can study before the quiz. The behaviors that lead to long-term learning are doing enough thinking about the material to summarize it in your own words, then going through the mechanics of writing those words down.

Before you begin reading the next chapter, let me say a few things about the design of the book that may help you with your studies. The text and illustrations are integrated as closely as possible. In my experience, one of the most annoying aspects of reading some books is not knowing when to look at an illustration. Therefore, in this book you will find figure references in boldfaced italics (like this: *Figure 5.6*), which means "stop reading and look at the figure." These references appear in locations I think will be optimal. If you look away from the text then, you will be assured that you will not be interrupting a line of reasoning in a crucial place and will not have to reread several sentences to get going again. You will find passages like this: "Figure 3.1 shows an alligator and two humans. This alligator is certainly laid out in a linear fashion; we can draw a straight line that starts between its eyes and continues down the center of its spinal cord. (See *Figure 3.1.*)" This particular example is a trivial one and will give you no problems no matter when you look at the figure. But in other cases the material is more complex, and you will have less trouble if you know what to look for before you stop reading and examine the illustration.

You will notice that some words in the text are *italicized* and others are printed in **boldface.** Italics mean one of two things: Either the word is being stressed for emphasis and is not a new term or I am pointing out a new term that is not necessary for you to learn. On the other hand, a word in boldface is a new term that you should try to learn. Most of the boldfaced terms in the text are part of the vocabulary of the physiological psychologist. Often, they will be used again in a later chapter. As an aid to your studying, definitions of these terms are printed at the bottom of the page, along with pronunciation guides for terms whose pronunciation is not obvious. In addition, a comprehensive index at the end of the book provides a list of terms and topics, with page references.

At the end of each major section (there are usually three to five of them in a chapter), you will find an *Interim Summary,* which provides a place for you to stop and think again about what you have just read to make sure that you understand the direction the discussion has gone. Taken together, these sections provide a detailed summary of the information introduced in the chapter. My students tell me that they review the interim summaries just before taking a test.

One more thing. As you have undoubtedly noticed, a CD-ROM is included with your textbook. I prepared this CD-ROM to help you learn some of the material presented in this book. The CD-ROM (which works on both Windows and Apple operating systems) contains exercises and animations. Once you select a chapter from the opening menu, you will see what options are available. The exercises supplement the study guide: They will help you learn and remember definitions of new terms, and they contain multiple-choice questions for you to test yourself. They also contain many of the figures that appear in the book in a format that lets you practice putting the right labels in the right places. The animations contain illustrated presentations of things I talk about in the text, and in some cases they introduce new information. I urge you to put the disk in your computer and see what's there. I always say this to my own students, but some never bother. Others do so late in the semester and then tell me that they wish they had used the CD-ROM earlier, because if they had done so, they probably would have gotten better grades on their exams. I have even received e-mails from students at other schools urging me to add reminders to the text to encourage readers to consult the CD-ROM. As you can see, I have taken their advice.

Okay, the preliminaries are over. The next chapter starts with something you can sink your (metaphorical) teeth into: the structure and functions of neurons, the most important elements of the nervous system.

SUGGESTED READINGS

Allman, J. M. *Evolving Brains.* New York: Scientific American Library, 1999.

Damasio, A. R. *Descartes's Error: Emotion, Reason, and the Human Brain.* New York: G. P. Putnam, 1994.

Finger, S. *Origins of Neuroscience: A History of Explorations into Brain Function.* New York: Oxford University Press, 1994.

Schultz, D., and Schultz, S. E. *A History of Modern Psychology*, 8th ed. Belmont, CA: Wadsworth, 2003.

ADDITIONAL RESOURCES

Visit www.mypsychkit.com for additional review and practice of the material covered in this chapter. Within MyPsychKit, you can take practice tests and receive a customized study plan to help you review. Dozens of animations, tutorials, and Web links are also available. You can even review using the interactive electronic version of this textbook. You will need to register for MyPsychKit. See **www.mypsychkit.com** for complete details.

chapter
2

Structure and Functions of Cells of the Nervous System

outline

Kathryn D. was getting desperate. All her life she had been healthy and active, eating wisely and keeping fit with sports and regular exercise. She went to her health club almost every day for a session of low-impact aerobics followed by a swim. But several months ago, she began having trouble keeping up with her usual schedule. At first, she found herself getting tired toward the end of her aerobics class. Her arms, particularly, seemed to get heavy. Then when she entered the pool and started swimming, she found that it was hard to lift her arms over her head; she abandoned the crawl and the backstroke and did the sidestroke and breaststroke instead. She did not have any flulike symptoms, so she told herself that she needed more sleep and perhaps she should eat a little more.

Over the next few weeks, however, things only got worse. Aerobics classes were becoming an ordeal. Her instructor became concerned and suggested that Kathryn see her doctor. She did so, but he could find nothing wrong with her. She was not anemic, showed no signs of an infection, and seemed to be well nourished. He asked how things were going at work.

"Well, lately I've been under some pressure," she said. "The head of my department quit a few weeks ago, and I've taken over his job temporarily. I think I have a chance of getting the job permanently, but I feel as if my bosses are watching me to see whether I'm good enough for the job." Kathryn and her physician agreed that increased stress could be the cause of her problem. "I'd prefer not to give you any medication at this time," he said, "but if you don't feel better soon we'll have a closer look at you."

She *did* feel better for a while, but then all of a sudden her symptoms got worse. She quit going to the health club and found that she even had difficulty finishing a day's work.

She was certain that people were noticing that she was no longer her lively self, and she was afraid that her chances for the promotion were slipping away. One afternoon she tried to look up at the clock on the wall and realized that she could hardly see—her eyelids were drooping, and her head felt as if it weighed a hundred pounds. Just then, one of her supervisors came over to her desk, sat down, and asked her to fill him in on the progress she had been making on a new project. As she talked, she found herself getting weaker and weaker. Her jaw was getting tired, even her tongue was getting tired, and her voice was getting weaker. With a sudden feeling of fright she realized that the act of breathing seemed to take a lot of effort. She managed to finish the interview, but immediately afterward she packed up her briefcase and left for home, saying that she had a bad headache.

She telephoned her physician, who immediately arranged for her to go to the hospital to be seen by Dr. T., a neurologist. Dr. T. listened to a description of Kathryn's symptoms and examined her briefly. She said to Kathryn, "I think I know what may be causing your symptoms. I'd like to give you an injection and watch your reaction." She gave some orders to the nurse, who left the room and came back with a syringe. Dr. T. took it, swabbed Kathryn's arm, and injected the drug. She started questioning Kathryn about her job. Kathryn answered slowly, her voice almost a whisper. As the questions continued, she realized that it was getting easier and easier to talk. She straightened her back and took a deep breath. Yes, she was sure. Her strength was returning! She stood up and raised her arms above her head. "Look," she said, her excitement growing. "I can do this again. I've got my strength back! What was that you gave me? Am I cured?"

(For an answer to her question, see p. 61.)

The brain is the organ that moves the muscles. That might sound simplistic, but ultimately, movement—or, more accurately, behavior—is the primary function of the nervous system. To make useful movements, the brain must know what is happening outside, in the environment. Thus, the body also contains cells that are specialized for detecting environmental events. Of course, complex animals such as we do not react automatically to events in our environment; our brains are flexible enough that we behave in different ways, according to present circumstances and those we experienced in the past. Besides perceiving and acting, we can remember and decide. All these abilities are made possible by the billions of cells found in the nervous system or controlled by those cells.

This chapter describes the structure and functions of the most important cells of the nervous system. Information, in the form of light, sound waves, odors, tastes, or contact with objects, is gathered from the environment by specialized cells called **sensory neurons.** Movements are accomplished by the contraction of muscles, which are controlled by **motor neurons.** (The term *motor* is used here in its original sense to refer to movement, not to a mechanical engine.) And in between sensory neurons and motor neurons come the **interneurons**—neurons that lie entirely within the central

sensory neuron A neuron that detects changes in the external or internal environment and sends information about these changes to the central nervous system.

motor neuron A neuron located within the central nervous system that controls the contraction of a muscle or the secretion of a gland.

interneuron A neuron located entirely within the central nervous system.

nervous system. *Local interneurons* form circuits with nearby neurons and analyze small pieces of information. *Relay interneurons* connect circuits of local interneurons in one region of the brain with those in other regions. Through these connections, circuits of neurons throughout the brain perform functions essential to tasks such as perceiving, learning, remembering, deciding, and controlling complex behaviors. How many neurons are there in the human nervous system? The most common estimate is around 100 billion, but no one has counted them yet.

To understand how the nervous system controls behavior, we must first understand its parts—the cells that compose it. Because this chapter deals with cells, you need not be familiar with the structure of the nervous system, which is presented in Chapter 3. However, you need to know that the nervous system consists of two basic divisions: the central nervous system and the peripheral nervous system. The **central nervous system (CNS)** consists of the parts that are encased by the bones of the skull and spinal column: the brain and the spinal cord. The **peripheral nervous system (PNS)** is found outside these bones and consists of the nerves and most of the sensory organs.

CELLS OF THE NERVOUS SYSTEM

The first part of this chapter is devoted to a description of the most important cells of the nervous system—neurons and their supporting cells—and to the blood–brain barrier, which provides neurons in the central nervous system with chemical isolation from the rest of the body.

Neurons

Basic Structure

The neuron (nerve cell) is the information-processing and information-transmitting element of the nervous system. Neurons come in many shapes and varieties, according to the specialized jobs they perform. Most neurons have, in one form or another, the following four structures or regions: (1) cell body, or soma; (2) dendrites; (3)

mypsychkit

Animation 2.1
Neurons and Supporting Cells

axon; and (4) terminal buttons. (*MyPsychKit 2.1, Neurons and Supporting Cells,* illustrates the information presented in the following section.)

Soma.   The **soma** (cell body) contains the nucleus and much of the machinery that provides for the life processes of the cell. (See *Figure 2.1*.) Its shape varies considerably in different kinds of neurons.

Dendrites.   *Dendron* is the Greek word for tree, and the **dendrites** of the neuron look very much like trees. (See *Figure 2.1*.) Neurons "converse" with one another, and dendrites serve as important recipients of these messages. The messages that pass from neuron to neuron are transmitted across the **synapse,** a junction between the terminal buttons (described later) of the sending cell and a portion of the somatic or dendritic membrane of the receiving cell. (The word *synapse* derives from the Greek *sunaptein,* "to join together.") Communication at a synapse proceeds in one direction: from the terminal button to the membrane of the other cell. (Like many general rules, this one has some exceptions. As we will see in Chapter 4, some synapses pass information in both directions.)

Axon.   The **axon** is a long, slender tube, often covered by a *myelin sheath.* (The myelin sheath is described later.) The axon carries information from the cell body to the terminal buttons. (See *Figure 2.1*.) The basic message it carries is called an *action potential.* This function is an important one and will be described in more detail later in the chapter. For now, it suffices to say that an action potential is a brief electrical/chemical event that starts at the end of the axon next to the cell body and travels toward the terminal buttons. The action potential is like a brief pulse; in a given axon the action potential is always of the same size and duration. When it reaches a point where the axon branches, it splits but does not diminish in size. Each branch receives a *full-strength* action potential.

Like dendrites, axons and their branches come in different shapes. In fact, the three principal types of neurons are classified according to the way in which their axons and dendrites leave the soma. The neuron depicted in Figure 2.1 is the most common type found in the central nervous system; it is a **multipolar neuron.** In this type of neuron the somatic membrane gives rise to one axon but to the trunks of many dendritic trees. **Bipolar neurons** give rise to one axon and one dendritic

central nervous system (CNS) The brain and spinal cord.

peripheral nervous system (PNS) The part of the nervous system outside the brain and spinal cord, including the nerves attached to the brain and spinal cord.

soma The cell body of a neuron, which contains the nucleus.

dendrite A branched, treelike structure attached to the soma of a neuron; receives information from the terminal buttons of other neurons.

synapse A junction between the terminal button of an axon and the membrane of another neuron.

axon The long, thin, cylindrical structure that conveys information from the soma of a neuron to its terminal buttons.

multipolar neuron A neuron with one axon and many dendrites attached to its soma.

FIGURE 2.1 ■ The Principal Parts of a Multipolar Neuron

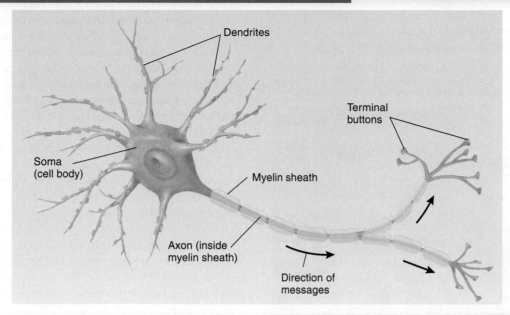

Dendrites

Terminal buttons

Soma (cell body)

Myelin sheath

Axon (inside myelin sheath)

Direction of messages

FIGURE 2.2 ■ Neuron

(a) A bipolar neuron, is found primarily in sensory systems (for example, vision and audition). (b) A unipolar neuron is found in the somatosensory system (touch, pain, and the like).

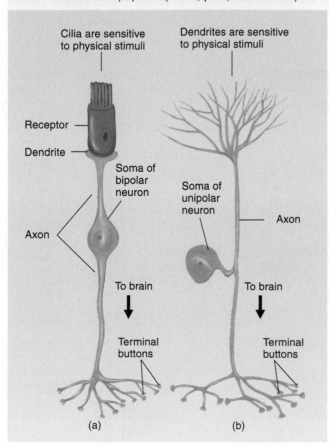

Cilia are sensitive to physical stimuli

Dendrites are sensitive to physical stimuli

Receptor

Dendrite

Soma of bipolar neuron

Soma of unipolar neuron

Axon

Axon

To brain

To brain

Terminal buttons

Terminal buttons

(a)   (b)

tree, at opposite ends of the soma. (See *Figure 2.2a.*) Bipolar neurons are usually sensory; that is, their dendrites detect events occurring in the environment and communicate information about these events to the central nervous system.

The third type of nerve cell is the **unipolar neuron.** It has only one stalk, which leaves the soma and divides into two branches a short distance away. (See *Figure 2.2b.*) Unipolar neurons, like bipolar neurons, transmit sensory information from the environment to the CNS. The arborizations (treelike branches) outside the CNS are dendrites; the arborizations within the CNS end in terminal buttons. The dendrites of most unipolar neurons detect touch, temperature changes, and other sensory events that affect the skin. Other unipolar neurons detect events in our joints, muscles, and internal organs.

The central nervous system communicates with the rest of the body through nerves attached to the brain and to the spinal cord. Nerves are bundles of many thousands of individual fibers, all wrapped in a tough, protective membrane. Under a microscope, nerves look something like telephone cables, with their bundles of wires. (See *Figure 2.3.*) Like the individual wires in a telephone cable, nerve fibers transmit messages through the nerve, from a sense organ to the brain or from the brain to a muscle or gland.

bipolar neuron A neuron with one axon and one dendrite attached to its soma.

unipolar neuron A neuron with one axon attached to its soma; the axon divides, with one branch receiving sensory information and the other sending the information into the central nervous system.

A nerve consists of a sheath of tissue that encases a bundle of individual nerve fibers (also known as axons).
BV = blood vessel; A = individual axons.

(From *Tissues and Organs: A Text-Atlas of Scanning Electron Microscopy*, by Richard G. Kessel and Randy H. Kardon. Copyright © 1979 by W. H. Freeman and Co. Reprinted by permission of Barbara Kessel and Randy Kardon.)

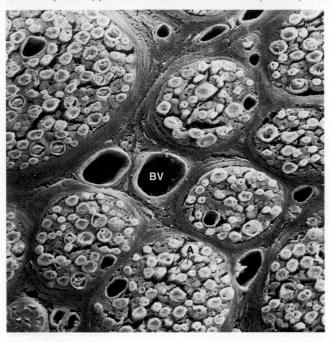

Terminal Buttons. Most axons divide and branch many times. At the ends of the twigs are found little knobs called **terminal buttons.** (Some neuroscientists prefer the original French word *bouton,* and others simply refer to them as *terminals.*) Terminal buttons have a very special function: When an action potential traveling down the axon reaches them, they secrete a chemical called a **neurotransmitter.** This chemical (there are many different ones in the CNS) either excites or inhibits the receiving cell and thus helps to determine whether an action potential occurs in its axon. Details of this process will be described later in this chapter.

An individual neuron receives information from the terminal buttons of axons of other neurons—and the terminal buttons of *its* axons form synapses with other neurons. A neuron may receive information from dozens or even hundreds of other neurons, each of which can form a large number of synaptic connections with it. Figure 2.4 illustrates the nature of these connections. As you can see, terminal buttons can form synapses on the membrane of the dendrites or the soma. (See *Figure 2.4.*)

Internal Structure

Figure 2.5 illustrates the internal structure of a typical multipolar neuron. (See *Figure 2.5.*) The **membrane** defines the boundary of the cell. It consists of a double layer of lipid (fatlike) molecules. Embedded in the

FIGURE 2.4 ■ Synaptic Connections Between Neurons

The arrows in this overview represent the directions of the flow of information.

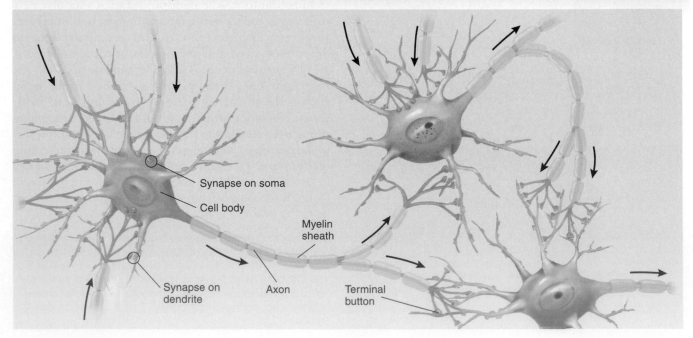

Synapse on soma

Cell body

Myelin sheath

Synapse on dendrite

Axon

Terminal button

FIGURE 2.5 ■ The Principal Internal Structures of a Multipolar Neuron

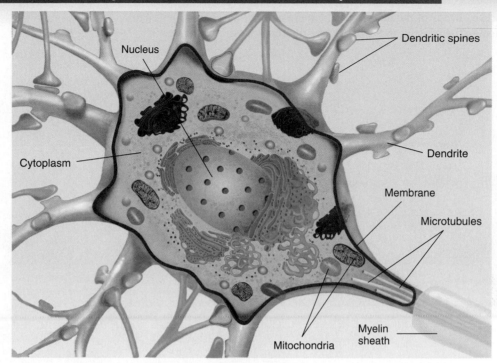

membrane are a variety of protein molecules that have special functions. Some proteins detect substances outside the cell (such as hormones) and pass information about the presence of these substances to the interior of the cell. Other proteins control access to the interior of the cell, permitting some substances to enter but barring others. Still other proteins act as transporters, actively carrying certain molecules into or out of the cell. Because the proteins that are found in the membrane of the neuron are especially important in the transmission of information, their characteristics will be discussed in more detail later in this chapter.

The **nucleus** ("nut") of the cell is round or oval and is enclosed by the nuclear membrane. The nucleolus and

the chromosomes reside here. The **nucleolus** is responsible for the production of **ribosomes,** small structures that are involved in protein synthesis. The **chromosomes,** which consist of long strands of **deoxyribonucleic acid (DNA),** contain the organism's genetic information. When they are active, portions of the chromosomes (**genes**) cause production of another complex molecule, **messenger ribonucleic acid (mRNA),** which receives a copy of the information stored at that location. The mRNA leaves the nuclear membrane and attaches to ribosomes, where it causes the production of a particular protein. (See *Figure 2.6.*)

Proteins are important in cell functions. As well as providing structure, proteins serve as **enzymes,** which

---

terminal button The bud at the end of a branch of an axon; forms synapses with another neuron; sends information to that neuron.

neurotransmitter A chemical that is released by a terminal button; has an excitatory or inhibitory effect on another neuron.

membrane A structure consisting principally of lipid molecules that defines the outer boundaries of a cell and also constitutes many of the cell organelles, such as the Golgi apparatus.

nucleus A structure in the central region of a cell, containing the nucleolus and chromosomes.

nucleolus (*new **clee** o lus*) A structure within the nucleus of a cell that produces the ribosomes.

ribosome (*ry bo soam*) A cytoplasmic structure, made of protein, that serves as the site of production of proteins translated from mRNA.

chromosome A strand of DNA, with associated proteins, found in the nucleus; carries genetic information.

deoxyribonucleic acid (DNA) (*dee ox ee ry bo new **clay** ik*) A long, complex macromolecule consisting of two interconnected helical strands; along with associated proteins, strands of DNA constitute the chromosomes.

gene The functional unit of the chromosome, which directs synthesis of one or more proteins.

messenger ribonucleic acid (mRNA) A macromolecule that delivers genetic information concerning the synthesis of a protein from a portion of a chromosome to a ribosome.

enzyme A molecule that controls a chemical reaction, combining two substances or breaking a substance into two parts.

FIGURE 2.6 ■ Protein Synthesis

When a gene is active, a copy of the information is made onto a molecule of messenger RNA. The mRNA leaves the nucleus and attaches to a ribosome, where the protein is produced.

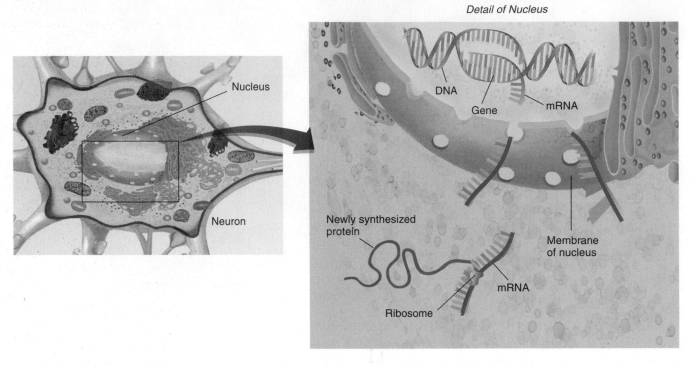

direct the chemical processes of a cell by controlling chemical reactions. Enzymes are special protein molecules that act as catalysts; that is, they cause a chemical reaction to take place without becoming a part of the final product themselves. Because cells contain the ingredients needed to synthesize an enormous variety of compounds, the ones that cells actually do produce depend primarily on the particular enzymes that are present. Furthermore, there are enzymes that break molecules apart as well as enzymes that put them together; the enzymes that are present in a particular region of a cell thus determine which molecules remain intact. For example,

$$A + B \underset{Y}{\overset{X}{\rightleftarrows}} AB$$

In this reversible reaction the relative concentrations of enzymes X and Y determine whether the complex substance AB or its constituents, A and B, will predominate. Enzyme X makes A and B join together; enzyme Y splits AB apart. (Energy may also be required to make the reactions proceed.)

As you undoubtedly know, the sequence of the human genome—along with that of several other plants and animals—has been determined. (The *genome* is the sequence of nucleotide bases on the chromosomes that provide the information needed to synthesize all the proteins that can be produced by a particular organism.) Biologists were surprised to learn that the number of genes was not correlated with the complexity of the organism (Mattick, 2004). For example, *Caenorhabditis elegans,* a simple worm that consists of about 1000 cells, has 19,000 genes, whereas humans have around 25,000 genes. The research also revealed that the genomes of most vertebrates contained much "junk" DNA, which did not contain information needed to produce proteins. For example, only about 1.5 percent of the human genome encodes for proteins. At first, molecular geneticists assumed that "junk" DNA was a leftover from our evolutionary history and that only the sequences of DNA that encoded for proteins were useful. However, further research found that the amount of non-protein-coding DNA did correlate well with the complexity of an organism and that many of these sequences have been conserved for millions of years. In other words, it started looking as though "junk" DNA was not junk after all. (See *Figure 2.7.*)

A study by Woolfe et al. (2005) illustrates the longevity of most non-coding DNA. The researchers compared the genomes of the human and the pufferfish. The common ancestor of these two species lived many millions of years ago, which means that if non-coding DNA is really just

FIGURE 2.7 ■ Non-coding DNA

The figure shows the percentage of DNA that does not code for proteins in various categories of living organisms.

(Adapted from Mattick, J. S. *Scientific American*, 2004, *291*, 60–67.)

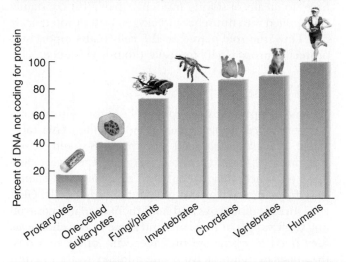

useless, leftover junk, then random mutations should have produced many changes in their sequences. However, Woolfe and his colleagues found 1400 highly conserved non-coding sequences, many of which were over 90 percent identical in humans and pufferfish. In addition, they found that these conserved sequences were located near genes that control development, which is unlikely to be a coincidence. In fact, we will see in Chapter 3 that mutations in non-coding regions of the human genome appear to be responsible for the increased size and complexity of the human brain.

What do these non-coding sequences of DNA do? Although their sequences can be transcribed into RNA, this RNA does not result in the production of protein. Instead, **non-coding RNA (ncRNA)** appears to have functions of it own. For example, when most genes become active, segments of DNA are transcribed into molecules of messenger RNA, and then other molecules cut the mRNA into pieces, discard some parts, and splice the remaining pieces together. The protein is then made from the resulting chunk of mRNA. The cutting and splicing are accomplished by molecular complexes called *spliceosomes,* and one of the constituents of spliceosomes is non-coding RNA. Molecules of ncRNA also attach to—and modify—proteins that regulate gene expression (Szymanski et al., 2003; Storz, Altuvia, and Wassarman, 2005; Satterlee et al., 2007). Thus, the human genome, more broadly defined to include non-coding RNA, is much larger than biologists previously believed.

The bulk of the cell consists of cytoplasm. **Cytoplasm** is complex and varies considerably across types of cells, but it can most easily be characterized as a jellylike,

semiliquid substance that fills the space outlined by the membrane. It contains small, specialized structures, just as the body contains specialized organs. The generic term for these structures is *organelle,* "little organ." The most important organelles are described next.

Mitochondria (singular: mitochondrion) are shaped like oval beads and are formed of a double membrane. The inner membrane is wrinkled, and the wrinkles make up a set of shelves *(cristae)* that fill the inside of the bead. Mitochondria perform a vital role in the economy of the cell; many of the biochemical steps that are involved in the extraction of energy from the breakdown of nutrients take place on the cristae, controlled by enzymes located there. Most cell biologists believe that many eons ago, mitochondria were free-living organisms that came to "infect" larger cells. Because the mitochondria could extract energy more efficiently than the cells they infected could, the mitochondria became useful to the cells and eventually became a permanent part of them. Cells provide mitochondria with nutrients, and mitochondria provide cells with a special molecule—**adenosine triphosphate (ATP)**—that cells use as their immediate source of energy. Mitochondria contain their own DNA and reproduce independently of the cells in which they reside.

Endoplasmic reticulum, which serves as a storage reservoir and as a channel for transporting chemicals through the cytoplasm, appears in two forms: rough and smooth. Both types consist of parallel layers of membrane, arranged in pairs, of the sort that encloses the cell. Rough endoplasmic reticulum contains ribosomes. The protein produced by the ribosomes that are attached to the rough endoplasmic reticulum is destined to be transported out of the cell or used in the membrane. Unattached ribosomes are also distributed around the cytoplasm; the unattached variety appears to produce protein for use within the neuron. Smooth endoplasmic reticulum provides channels for the segregation of molecules involved in various cellular processes. Lipid (fatlike) molecules are produced here.

non-coding RNA (ncRNA) A form of RNA that does not encode for protein but has functions of its own.

cytoplasm The viscous, semiliquid substance contained in the interior of a cell.

mitochondrion An organelle that is responsible for extracting energy from nutrients.

adenosine triphosphate (ATP) (*ah den o seen*) A molecule of prime importance to cellular energy metabolism; its breakdown liberates energy.

endoplasmic reticulum Parallel layers of membrane found within the cytoplasm of a cell. Rough endoplasmic reticulum contains ribosomes and is involved with production of proteins that are secreted by the cell. Smooth endoplasmic reticulum is the site of synthesis of lipids and provides channels for the segregation of molecules involved in various cellular processes.

The **Golgi apparatus** is a special form of smooth endoplasmic reticulum. Some complex molecules, made up of simpler individual molecules, are assembled here. The Golgi apparatus also serves as a wrapping or packaging agent. For example, secretory cells (such as those that release hormones) wrap their product in a membrane produced by the Golgi apparatus. When the cell secretes its products, it uses a process called **exocytosis** (*exo*, "outside"; *cyto*, "cell"; *-osis*, "process"). Briefly stated, the container migrates to the inside of the outer membrane of the cell, fuses with it, and bursts, spilling its contents into the fluid surrounding the cell. As we will see, neurons communicate with one another by secreting chemicals by this means. Therefore, I will describe the process of exocytosis in more detail later in this chapter. The Golgi apparatus also produces **lysosomes,** small sacs that contain enzymes that break down substances no longer needed by the cell. These products are then recycled or excreted from the cell.

If a neuron grown in a tissue culture is exposed to a detergent, the lipid membrane and much of the interior of the cell dissolve away, leaving a matrix of insoluble strands of protein. This matrix, called the **cytoskeleton,** gives the neuron its shape. The cytoskeleton is made of three kinds of protein strands, linked to each other and forming a cohesive mass. The thickest of these strands, **microtubules,** are bundles of thirteen protein filaments arranged around a hollow core.

Axons can be extremely long relative to their diameter and the size of the soma. For example, the longest axon in a human stretches from the foot to a region located in the base of the brain. Because terminal buttons need some items that can be produced only in the soma, there must be a system that can transport these items rapidly and efficiently through the axoplasm (that is, the cytoplasm of the axon). This system is referred to as **axoplasmic transport,** an active process by which substances are propelled along microtubules that run the length of the axon. Movement from the soma to the terminal buttons is called **anterograde** axoplasmic transport. (*Antero-* means "toward the front.") This form of transport is accomplished by molecules of a protein called *kinesin*. In the cell body, kinesin molecules, which resemble a pair of legs and feet, attach to the item being transported down the axon. The kinesin molecule then walks down a microtubule, carrying the cargo to its destination (Yildiz et al., 2004). Energy is supplied by ATP molecules produced by the mitochondria. (See *Figure 2.8.*) Another protein, *dynein*, carries substances from the terminal buttons to the soma, a process known as **retrograde** axoplasmic transport. Anterograde axoplasmic transport is remarkably fast: up to 500 mm per day. Retrograde axoplasmic transport is about half as fast as anterograde transport.

Supporting Cells

Neurons constitute only about half the volume of the CNS. The rest consists of a variety of supporting cells. Because neurons have a very high rate of metabolism but have no means of storing nutrients, they must constantly be supplied with nutrients and oxygen or they will quickly die. Thus, the role played by the cells that support and protect neurons is very important to our existence.

Glia

The most important supporting cells of the central nervous system are the *neuroglia,* or "nerve glue." **Glia** (also called *glial cells*) do indeed glue the CNS together, but they do much more than that. Neurons lead a very sheltered existence; they are buffered physically and chemically from the rest of the body by the glial cells. Glial cells surround neurons and hold them in place, controlling their supply of nutrients and some of the chemicals they need to exchange messages with other neurons; they insulate neurons from one another so that neural messages do not get scrambled; and they even act as housekeepers, destroying and removing the carcasses of neurons that are killed by disease or injury.

There are several types of glial cells, each of which plays a special role in the CNS. The three most important types are *astrocytes, oligodendrocytes,* and *microglia.* **Astrocyte** means "star cell," and this name accurately describes the shape of these cells. Astrocytes (or *astroglia*) provide physical support to neurons and clean up debris within the brain. They produce some chemicals that

---

Golgi apparatus (*goal jee*) A complex of parallel membranes in the cytoplasm that wraps the products of a secretory cell.

exocytosis (*ex o sy toe sis*) The secretion of a substance by a cell through means of vesicles; the process by which neurotransmitters are secreted.

lysosome (*lye so soam*) An organelle surrounded by membrane; contains enzymes that break down waste products.

cytoskeleton Formed of microtubules and other protein fibers, linked to each other and forming a cohesive mass that gives a cell its shape.

microtubule (*my kro too byool*) A long strand of bundles of protein filaments arranged around a hollow core; part of the cytoskeleton and involved in transporting substances from place to place within the cell.

axoplasmic transport An active process by which substances are propelled along microtubules that run the length of the axon.

anterograde In a direction along an axon from the cell body toward the terminal buttons.

retrograde In a direction along an axon from the terminal buttons toward the cell body.

glia (*glee ah*) The supporting cells of the central nervous system.

astrocyte A glial cell that provides support for neurons of the central nervous system, provides nutrients and other substances, and regulates the chemical composition of the extracellular fluid.

FIGURE 2.8 ■ Fast Axoplasmic Transport

(a) Kinesin molecules "walk" down a microtubule, carrying their cargo from the soma to the terminal buttons. Another protein, dynein, carries substances from the terminal buttons to the soma. (b) A photomicrograph of a mouse axon shows an organelle being transported along a microtubule. The arrow points to what appears to be a kinesin molecule.

(From Hirokawa, N. *Science*, 1998, *279*, 519–526. Copyright © 1998 American Association for the Advancement of Science. Reprinted with permission.)

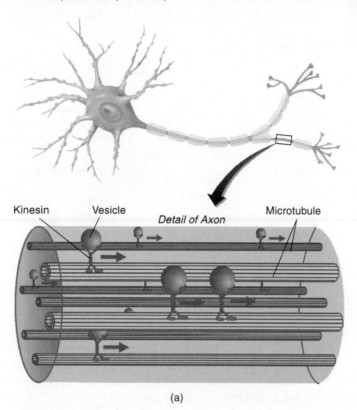

Kinesin   Vesicle   *Detail of Axon*   Microtubule

(a)

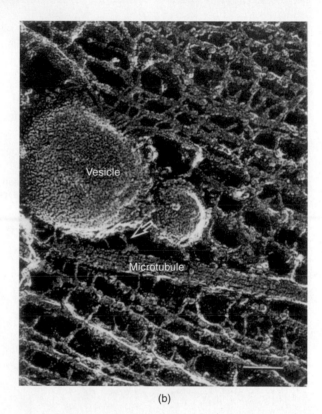

(b)

neurons need to fulfill their functions. They help to control the chemical composition of the fluid surrounding neurons by actively taking up or releasing substances whose concentrations must be kept within critical levels. As we will see later, they are involved in establishing structures responsible for communication between neurons. Finally, astrocytes are involved in providing nourishment to neurons.

Some of the astrocyte's processes (the arms of the star) are wrapped around blood vessels; other processes are wrapped around parts of neurons, so the somatic and dendritic membranes of neurons are largely surrounded by astrocytes. This arrangement suggested to the Italian histologist Camillo Golgi (1844–1926) that astrocytes supplied neurons with nutrients from the capillaries and disposed of their waste products (Golgi, 1903). He thought that nutrients passed from capillaries to the cytoplasm of the astrocytes and then through the cytoplasm to the neurons.

Recent evidence suggests that Golgi was right: Although neurons receive some glucose directly from capillaries, they receive most of their nutrients from astrocytes. Astrocytes receive glucose from capillaries and break it down to *lactate*, the chemical produced during the first step of glucose metabolism. They then release lactate into the extracellular fluid that surrounds neurons, and neurons take up the lactate, transport it to their mitochondria, and use it for energy. (Tsacopoulos and Magistretti, 1996; Brown, Tekkök, and Ransom, 2003; Pellerin et al., 2007). Apparently, this process provides neurons with a fuel that they can metabolize even more rapidly than glucose. In addition, astrocytes store a small amount of a carbohydrate called *glycogen* that can be broken down to glucose and then to lactate when the metabolic rate of neurons in their vicinity is especially high. (See *Figure 2.9.*)

Besides transporting chemicals to neurons, astrocytes serve as the matrix that holds neurons in place—the "nerve glue," so to speak. These cells also surround and isolate synapses, limiting the dispersion of neurotransmitters that are released by the terminal buttons.

When cells in the central nervous system die, certain kinds of astrocytes take up the task of cleaning away

FIGURE 2.9 ■ Structure and Location of Astrocytes

The processes of astrocytes surround capillaries and neurons of the central nervous system.

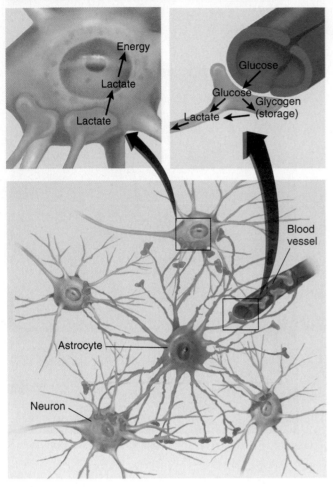

FIGURE 2.10 ■ Oligodendrocyte

An oligondendrocyte forms the myelin that surrounds many axons in the central nervous system. Each cell forms one segment of myelin for several adjacent axons.

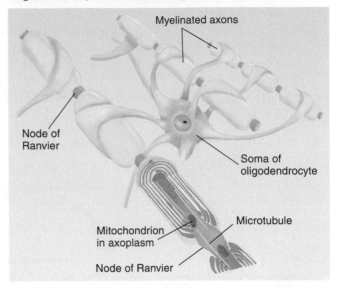

rather, it consists of a series of segments, each approximately 1 mm long, with a small (1–2 μm) portion of uncoated axon between the segments. (A *micrometer,* abbreviated *μm,* is one-millionth of a meter, or one-thousandth of a millimeter.) The bare portion of axon is called a **node of Ranvier,** after its discoverer. The myelinated axon, then, resembles a string of elongated beads. (Actually, the beads are *very much* elongated—their length is approximately 80 times their width.)

A given oligodendrocyte produces up to fifty segments of myelin. During the development of the CNS, oligodendrocytes form processes shaped something like canoe paddles. Each of these paddle-shaped processes then wraps itself many times around a segment of an axon and, while doing so, produces layers of myelin. Each paddle thus becomes a segment of an axon's myelin sheath. (See *Figures 2.10* and *2.11a.*)

As their name indicates, **microglia** are the smallest of the glial cells. Like some types of astrocytes, they act as phagocytes, engulfing and breaking down dead and

the debris. These cells are able to travel around the CNS; they extend and retract their processes (*pseudopodia,* or "false feet") and glide about the way amoebas do. When these astrocytes contact a piece of debris from a dead neuron, they push themselves against it, finally engulfing and digesting it. We call this process **phagocytosis** (*phagein,* "to eat"; *kutos,* "cell"). If there is a considerable amount of injured tissue to be cleaned up, astrocytes will divide and produce enough new cells to do the task. Once the dead tissue has been broken down, a framework of astrocytes will be left to fill in the vacant area, and a specialized kind of astrocyte will form scar tissue, walling off the area.

The principal function of **oligodendrocytes** is to provide support to axons and to produce the **myelin sheath,** which insulates most axons from one another. (Very small axons are not myelinated and lack this sheath.) Myelin, 80 percent lipid and 20 percent protein, is produced by the oligodendrocytes in the form of a tube surrounding the axon. This tube does not form a continuous sheath;

phagocytosis (*fagg o sy toe sis*) The process by which cells engulf and digest other cells or debris caused by cellular degeneration.

oligodendrocyte (*oh li go den droh site*) A type of glial cell in the central nervous system that forms myelin sheaths.

myelin sheath (*my a lin*) A sheath that surrounds axons and insulates them, preventing messages from spreading between adjacent axons.

node of Ranvier (*raw vee ay*) A naked portion of a myelinated axon between adjacent oligodendroglia or Schwann cells.

microglia The smallest of glial cells; act as phagocytes and protect the brain from invading microorganisms.

FIGURE 2.11 ■ Formation of Myelin

During development a process of an oligodendrocyte or an entire Schwann cell tightly wraps itself many times around an individual axon and forms one segment of the myelin sheath. (a) Oligodendrocyte. (b) Schwann cell.

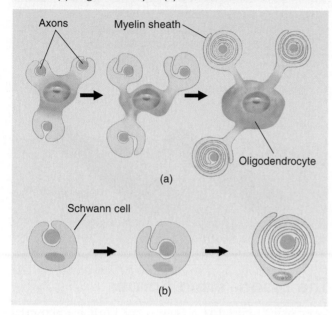

dying neurons. But in addition, they serve as one of the representatives of the immune system in the brain, protecting the brain from invading microorganisms. They are primarily responsible for the inflammatory reaction in response to brain damage.

Dr. C., a retired neurologist, had been afflicted with multiple sclerosis for more than two decades when she died of a heart attack. One evening, twenty-three years previously, she and her husband had had dinner at their favorite restaurant. As they were leaving, she stumbled and almost fell. Her husband joked, "Hey, honey, you shouldn't have had that last glass of wine." She smiled at his attempt at humor, but she knew better—her clumsiness wasn't brought on by the two glasses of wine she had drunk with dinner. She suddenly realized that she had been ignoring some symptoms that she should have recognized.

The next day, she consulted with one of her colleagues, who agreed that her own tentative diagnosis was probably correct: Her symptoms fit those of multiple sclerosis. She had experienced fleeting problems with double vision, she sometimes felt unsteady on her feet, and she occasionally noticed tingling sensations in her right hand. None of these symptoms was serious, and they lasted for only a short while, so she ignored them—or perhaps denied to herself that they were important.

A few weeks after Dr. C.'s death, a group of medical students and neurological residents gathered in an autopsy room at the medical school. Dr. D., the school's neuropathologist, displayed a stainless-steel tray on which were lying a brain and a spinal cord. "These belonged to Dr. C.," he said. "Several years ago she donated her organs to the medical school." Everyone looked at the brain more intently, knowing that it had animated an esteemed clinician and teacher whom they all knew by reputation, if not personally. Dr. D. led his audience to a set of light boxes on the wall, to which several MRI scans had been clipped. He pointed out some white spots that appeared on one scan. "This scan clearly shows some white-matter lesions, but they are gone on the next one, taken six months later. And here is another one, but it's gone on the next scan. The immune system attacked the myelin sheaths in a particular region, and then glial cells cleaned up the debris. MRI doesn't show the lesions then, but the axons can no longer conduct their messages."

He picked up Dr. C.'s brain and cut it in several slices. He picked one up. "Here, see this?" He pointed out a spot of discoloration in a band of white matter. This is a sclerotic plaque—a patch that feels harder than the surrounding tissue. There are many of them, located throughout the brain and spinal cord, which is why the disease is called multiple sclerosis." He picked up the spinal cord, felt along its length with his thumb and forefinger, and then stopped and said, "Yes, I can feel a plaque right here."

Dr. D. put the spinal cord down and said, "Who can tell me the etiology of this disorder?"

One of the students spoke up. "It's an autoimmune disease. The immune system gets sensitized to the body's own myelin protein and periodically attacks it, causing a variety of different neurological symptoms. Some say that a childhood viral illness somehow causes the immune system to start seeing the protein as foreign."

"That's right," said Dr. D. "The primary criterion for the diagnosis of multiple sclerosis is the presence of neurological symptoms disseminated in time and space. The symptoms don't all occur at once, and they can be caused only by damage to several different parts of the nervous system, which means that they can't be the result of a stroke."

Schwann Cells

In the central nervous system the oligodendrocytes support axons and produce myelin. In the peripheral nervous system the **Schwann cells** perform the same functions. Most axons in the PNS are myelinated. The myelin sheath occurs in segments, as it does in the CNS;

Schwann cell A cell in the peripheral nervous system that is wrapped around a myelinated axon, providing one segment of its myelin sheath.

each segment consists of a single Schwann cell, wrapped many times around the axon. In the CNS the oligodendrocytes grow a number of paddle-shaped processes that wrap around a number of axons. In the PNS a Schwann cell provides myelin for only one axon, and the entire Schwann cell—not merely a part of it—surrounds the axon. (See *Figure 2.11b.*)

Schwann cells also differ from their CNS counterparts, the oligodendrocytes, in an important way. As we saw, a nerve consists of a bundle of many myelinated axons, all covered in a sheath of tough, elastic connective tissue. If damage occurs to such a nerve, Schwann cells aid in the digestion of the dead and dying axons. Then the Schwann cells arrange themselves in a series of cylinders that act as guides for regrowth of the axons. The distal portions of the severed axons die, but the stump of each severed axon grows sprouts, which then spread in all directions. If one of these sprouts encounters a cylinder provided by a Schwann cell, the sprout will grow through the tube quickly (at a rate of up to 3–4 mm a day), while the other, nonproductive sprouts wither away. If the cut ends of the nerve are still located close enough to each other, the axons will reestablish connections with the muscles and sense organs they previously served.

Unfortunately, the glial cells of the CNS are not as cooperative as the supporting cells of the PNS. If axons in the brain or spinal cord are damaged, new sprouts will form, as in the PNS. However, the budding axons encounter scar tissue produced by the astrocytes, and they cannot penetrate this barrier. Even if the sprouts could get through, the axons would not reestablish their original connections without guidance similar to that provided by the Schwann cells of the PNS. During development, axons have two modes of growth. The first mode causes them to elongate so that they reach their target, which could be as far away as the other end of the brain or spinal cord. Schwann cells provide this signal to injured axons. The second mode causes axons to stop elongating and begin sprouting terminal buttons because they have reached their target. Liuzzi and Lasek (1987) found that even when astrocytes do not produce scar tissue, they appear to produce a chemical signal that instructs regenerating axons to begin the second mode of growth: to stop elongating and start sprouting terminal buttons. Thus, the difference in the regenerative properties of axons in the CNS and the PNS results from differences in the characteristics of the supporting cells, not from differences in the axons.

There is another difference between oligodendrocytes of the CNS and Schwann cells of the PNS: the chemical composition of the myelin protein they produce. The immune system of someone with multiple sclerosis attacks only the myelin protein produced by oligodendrocytes; thus, the myelin of the peripheral nervous system is spared.

FIGURE 2.12 ■ The Blood–Brain Barrier

(a) The cells that form the walls of the capillaries in the body outside the brain have gaps that permit the free passage of substances into and out of the blood. (b) The cells that form the walls of the capillaries in the brain are tightly joined.

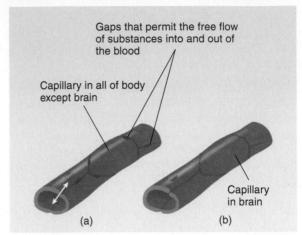

Gaps that permit the free flow of substances into and out of the blood

Capillary in all of body except brain

Capillary in brain

(a)   (b)

The Blood–Brain Barrier

Over one hundred years ago, Paul Ehrlich discovered that if a blue dye is injected into an animal's bloodstream, all tissues except the brain and spinal cord will be tinted blue. However, if the same dye is injected into the fluid-filled ventricles of the brain, the blue color will spread throughout the CNS (Bradbury, 1979). This experiment demonstrates that a barrier exists between the blood and the fluid that surrounds the cells of the brain: the **blood–brain barrier.**

Some substances can cross the blood–brain barrier; others cannot. Thus, it is *selectively permeable* (*per*, "through"; *meare*, "to pass"). In most of the body the cells that line the capillaries do not fit together absolutely tightly. Small gaps are found between them that permit the free exchange of most substances between the blood plasma and the fluid outside the capillaries that surrounds the cells of the body. In the central nervous system the capillaries lack these gaps; therefore, many substances cannot leave the blood. Thus, the walls of the capillaries in the brain constitute the blood–brain barrier. (See *Figure 2.12.*) Other substances must be actively transported through the capillary walls by special proteins. For example, glucose transporters bring the brain its fuel, and other transporters rid the brain of toxic waste products (Rubin and Staddon, 1999).

---

blood–brain barrier A semipermeable barrier between the blood and the brain produced by the cells in the walls of the brain's capillaries.

What is the function of the blood–brain barrier? As we will see, transmission of messages from place to place in the brain depends on a delicate balance between substances within neurons and in the extracellular fluid that surrounds them. If the composition of the extracellular fluid is changed even slightly, the transmission of these messages will be disrupted, which means that brain functions will be disrupted. The presence of the blood–brain barrier makes it easier to regulate the composition of this fluid. In addition, many of the foods that we eat contain chemicals that would interfere with the transmission of information between neurons. The blood–brain barrier prevents these chemicals from reaching the brain.

The blood–brain barrier is not uniform throughout the nervous system. In several places the barrier is relatively permeable, allowing substances that are excluded elsewhere to cross freely. For example, the **area postrema** is a part of the brain that controls vomiting. The blood–brain barrier is much weaker there, permitting neurons in this region to detect the presence of toxic substances in the blood. (A barrier around the area postrema prevents substances from diffusing from this region into the rest of the brain.) A poison that enters the circulatory system from the stomach can thus stimulate area postrema to initiate vomiting. If the organism is lucky, the poison can be expelled from the stomach before causing too much damage.

---

area postrema (*poss **tree** ma*) A region of the medulla where the blood–brain barrier is weak; poisons can be detected there and can initiate vomiting.

InterimSummary

Cells of the Nervous System

Neurons are the most important cells of the nervous system. The central nervous system (CNS) includes the brain and spinal cord; the peripheral nervous system (PNS) includes nerves and some sensory organs.

Neurons have four principal parts: dendrites, soma (cell body), axon, and terminal buttons. They communicate by means of synapses, junctions between the terminal buttons of one neuron and the somatic or dendritic membrane of another. When an action potential travels down an axon, its terminal buttons secrete a chemical that has either an excitatory or an inhibitory effect on the neurons with which they communicate. Ultimately, the effects of these excitatory and inhibitory synapses cause behavior, in the form of muscular contractions.

Neurons contain a quantity of cytoplasm, enclosed in a membrane. Embedded in the membrane are protein molecules that have special functions, such as the detection of hormones or neurotransmitters or transport of particular substances into and out of the cell. The cytoplasm contains the nucleus, which contains the genetic information; the nucleolus (located in the nucleus), which manufactures ribosomes; the ribosomes, which serve as sites of protein synthesis; the endoplasmic reticulum, which serves as a storage reservoir and as a channel for transportation of chemicals through the cytoplasm; the Golgi apparatus, which wraps substances that the cell secretes in a membrane; the lysosomes, which contain enzymes that destroy waste products; microtubules and other protein fibers, which compose the cytoskeleton and help to transport chemicals from place to place; and the mitochondria, which serve as the location for most of the chemical reactions through which the cell extracts energy from nutrients. Recent evidence indicates that only a small proportion of the human genome is devoted to the production of protein; the rest (formerly called "junk" DNA) is involved in the production of non-coding RNA, which has a variety of functions.

Neurons are supported by the glial cells of the central nervous system and the supporting cells of the peripheral nervous system. In the CNS, astrocytes provide support and nourishment, regulate the composition of the fluid that surrounds neurons, and remove debris and form scar tissue in the event of tissue damage. Microglia are phagocytes that serve as the representatives of the immune system. Oligodendrocytes form myelin, the substance that insulates axons, and also support unmyelinated axons. In the PNS, support and myelin are provided by the Schwann cells.

In most organs, molecules freely diffuse between the blood within the capillaries that serve them and the extracellular fluid that bathes their cells. The molecules pass through gaps between the cells that line the capillaries. The walls of the capillaries of the CNS lack these gaps; consequently, fewer substances can enter or leave the brain across the blood–brain barrier.

Thought Question

The fact that the mitochondria in our cells were originally microorganisms that infected our very remote ancestors points out that evolution can involve interactions between two or more species. Many species have other organisms living inside them; in fact, the bacteria in our intestines are necessary for our good health. Some microorganisms can exchange genetic information, so adaptive mutations that develop in one species can be adopted by another. Is it possible that some of the features of the cells of our nervous system were bequeathed to our ancestors by other species?

COMMUNICATION WITHIN A NEURON

This section describes the nature of communication *within* a neuron—the way an action potential is sent from the cell body down the axon to the terminal buttons, informing them to release some neurotransmitter. The details of synaptic transmission—the communication between neurons—will be described in the next section. As we will see in this section, an action potential consists of a series of alterations in the membrane of the axon that permit various substances to move between the interior of the axon and the fluid surrounding it. These exchanges produce electrical currents. (*MyPsychKit 2.2, The Action Potential*, illustrates the information presented in the following section.)

mypsychkit *Where learning comes to life!*

Animation 2.2
The Action Potential

Neural Communication: An Overview

Before I begin my discussion of the action potential, let's step back and see how neurons can interact to produce a useful behavior. We begin by examining a simple assembly of three neurons and a muscle that controls a withdrawal reflex. In the next two figures (and in subsequent figures that illustrate simple neural circuits), multipolar neurons are depicted in shorthand fashion as several-sided stars. The points of these stars represent dendrites, and only one or two terminal buttons are shown at the end of the axon. The sensory neuron in this example detects painful stimuli. When its dendrites are stimulated by a noxious stimulus (such as contact with a hot object), it sends messages down the axon to the terminal buttons, which are located in the spinal cord. (You will recognize this cell as a unipolar neuron; see *Figure 2.13*.) The terminal buttons of the sensory neuron release a neurotransmitter that excites the interneuron, causing it to send messages down its axon. The terminal buttons of the interneuron release a neurotransmitter that excites the motor neuron, which sends messages down its axon. The axon of the motor neuron joins a nerve and travels to a muscle. When the terminal buttons of the motor neuron release their neurotransmitter, the muscle cells contract, causing the hand to move away from the hot object. (See *Figure 2.13*.)

So far, all of the synapses have had excitatory effects. Now let us complicate matters a bit to see the effect of inhibitory synapses. Suppose you have removed a hot casserole from the oven. As you start walking over to the table to put it down, the heat begins to penetrate the rather thin potholders you are using. The pain caused by the heat triggers a withdrawal reflex that tends to make you drop the casserole. Yet you manage to keep hold of it long enough to get to the table and put it down. What prevented your withdrawal reflex from making you drop the casserole on the floor?

The pain from the hot casserole increases the activity of excitatory synapses on the motor neurons, which tends to cause the hand to pull away from the casserole. However, this excitation is counteracted by *inhibition*, supplied by another source: the brain. The brain contains neural circuits that recognize what a disaster it would be if you dropped the casserole on the floor. These neural circuits send information to the spinal

---

FIGURE 2.13 ■ A Withdrawal Reflex

The figure shows a simple example of a useful function of the nervous system. The painful stimulus causes the hand to pull away from the hot iron.

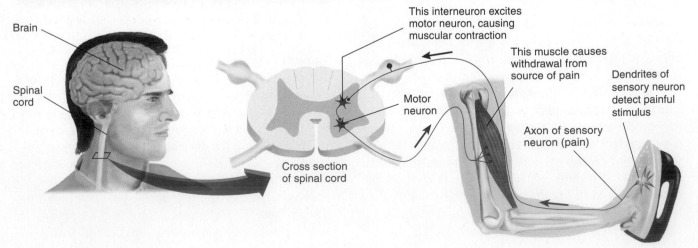

Brain

Spinal cord

Cross section of spinal cord

This interneuron excites motor neuron, causing muscular contraction

Motor neuron

This muscle causes withdrawal from source of pain

Dendrites of sensory neuron detect painful stimulus

Axon of sensory neuron (pain)

FIGURE 2.14 ■ The Role of Inhibition

Inhibitory signals arising from the brain can prevent the withdrawal reflex from causing the person to drop the casserole.

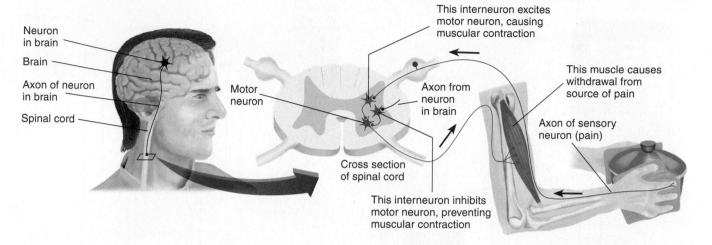

cord that prevents the withdrawal reflex from making you drop the dish.

Figure 2.14 shows how this information reaches the spinal cord. As you can see, an axon from a neuron in the brain reaches the spinal cord, where its terminal buttons form synapses with an inhibitory interneuron. When the neuron in the brain becomes active, its terminal buttons excite this inhibitory interneuron. The interneuron releases an inhibitory neurotransmitter, which *decreases* the activity of the motor neuron, blocking the withdrawal reflex. This circuit provides an example of a contest between two competing tendencies: to drop the casserole and to hold onto it. (See *Figure 2.14.*)

Of course, reflexes are more complicated than this description, and the mechanisms that inhibit them are even more so. And thousands of neurons are involved in this process. The five neurons shown in Figure 2.14 represent many others: Dozens of sensory neurons detect the hot object, hundreds of interneurons are stimulated by their activity, hundreds of motor neurons produce the contraction—and thousands of neurons in the brain must become active if the reflex is to be inhibited. Yet this simple model provides an overview of the process of neural communication, which is described in more detail later in this chapter.

Measuring Electrical Potentials of Axons

Let's examine the nature of the message that is conducted along the axon. To do so, we obtain an axon that is large enough to work with. Fortunately, nature has provided the neuroscientist with the giant squid axon (the giant axon of a squid, not the axon of a giant squid!). This axon is about 0.5 mm in diameter, which is hundreds of

times larger than the largest mammalian axon. (This large axon controls an emergency response: sudden contraction of the mantle, which squirts water through a jet and propels the squid away from a source of danger.) We place an isolated giant squid axon in a dish of seawater, in which it can exist for a day or two.

To measure the electrical charges generated by an axon, we will need to use a pair of electrodes. **Electrodes** are electrical conductors that provide a path for electricity to enter or leave a medium. One of the electrodes is a simple wire that we place in the seawater. The other one, which we use to record the message from the axon, has to be special. Because even a giant squid axon is rather small, we must use a tiny electrode that will record the membrane potential without damaging the axon. To do so, we use a microelectrode.

A **microelectrode** is simply a very small electrode, which can be made of metal or glass. In this case we will use one made of thin glass tubing, which is heated and drawn down to an exceedingly fine point, less than a thousandth of a millimeter in diameter. Because glass will not conduct electricity, the glass microelectrode is filled with a liquid that conducts electricity, such as a solution of potassium chloride.

We place the wire electrode in the seawater and insert the microelectrode into the axon. (See *Figure 2.15a.*) As soon as we do so, we discover that the inside of the axon is negatively charged with respect to the outside; the difference in charge being 70 mV (millivolts, or

---

electrode A conductive medium that can be used to apply electrical stimulation or to record electrical potentials.

microelectrode A very fine electrode, generally used to record activity of individual neurons.

FIGURE 2.15 ■ Measuring Electrical Charge

(a) A voltmeter detects the charge across a membrane of an axon. (b) A light bulb detects the charge across the terminals of a battery.

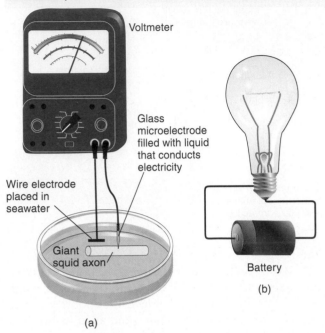

(a)

(b)

FIGURE 2.16 ■ Studying the Axon

The figure illustrates the means by which an axon can be stimulated while its membrane potential is being recorded.

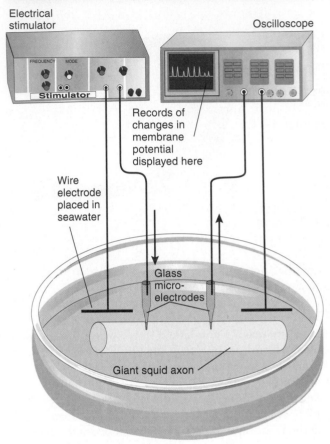

thousandths of a volt). Thus, the inside of the membrane is –70 mV. This electrical charge is called the **membrane potential.** The term *potential* refers to a stored-up source of energy—in this case, electrical energy. For example, a flashlight battery that is not connected to an electrical circuit has a *potential* charge of 1.5 V between its terminals. If we connect a light bulb to the terminals, the potential energy is tapped and converted into radiant energy (light). (See *Figure 2.15b.*) Similarly, if we connect our electrodes—one inside the axon and one outside it—to a very sensitive voltmeter, we will convert the potential energy to movement of the meter's needle. Of course, the potential electrical energy of the axonal membrane is very weak in comparison with that of a flashlight battery.

As we will see, the message that is conducted down the axon consists of a brief change in the membrane potential. However, this change occurs very rapidly—too rapidly for us to see if we were using a voltmeter. Therefore, to study the message, we will use an **oscilloscope.** This device, like a voltmeter, measures voltages, but it also produces a record of these voltages, graphing them as a function of time. These graphs are displayed on a screen, much like the one found in a television. The vertical axis represents voltage, and the horizontal axis represents time, going from left to right.

Once we insert our microelectrode into the axon, the oscilloscope draws a straight horizontal line at –70 mV,

as long as the axon is not disturbed. This electrical charge across the membrane is called, quite appropriately, the **resting potential.** Now let us disturb the resting potential and see what happens. To do so, we will use another device—an electrical stimulator that allows us to alter the membrane potential at a specific location. (See *Figure 2.16.*) The stimulator can pass current through another microelectrode that we have inserted into the axon. Because the inside of the axon is negative, a positive charge applied to the inside of the membrane produces a **depolarization.** That is, it takes away

membrane potential The electrical charge across a cell membrane; the difference in electrical potential inside and outside the cell.

oscilloscope A laboratory instrument that is capable of displaying a graph of voltage as a function of time on the face of a cathode ray tube.

resting potential The membrane potential of a neuron when it is not being altered by excitatory or inhibitory postsynaptic potentials; approximately –70 mV in the giant squid axon.

depolarization Reduction (toward zero) of the membrane potential of a cell from its normal resting potential.

some of the electrical charge across the membrane near the electrode, reducing the membrane potential.

Let us see what happens to an axon when we artificially change the membrane potential at one point. Figure 2.17 shows a graph drawn by an oscilloscope that has been monitoring the effects of brief depolarizing stimuli. The graphs of the effects of these separate stimuli are superimposed on the same drawing so that we can compare them. We deliver a series of depolarizing stimuli, starting with a very weak stimulus (number 1) and gradually increasing their strength. Each stimulus briefly depolarizes the membrane potential a little more. Finally, after we present depolarization number 4, the membrane potential suddenly reverses itself, so the inside becomes *positive* (and the outside becomes negative). The membrane potential quickly returns to normal, but first it overshoots the resting potential, becoming **hyperpolarized**—more polarized than normal—for a short time. The whole process takes about 2 msec (milliseconds). (See *Figure 2.17*.)

This phenomenon, a very rapid reversal of the membrane potential, is called the **action potential**. It constitutes the message carried by the axon from the cell body to the terminal buttons. The voltage level that triggers an action potential—which was achieved only by depolarizing shock number 4—is called the **threshold of excitation**.

FIGURE 2.17 ■ An Action Potential

These results would be seen on an oscilloscope screen if depolarizing stimuli of varying intensities were delivered to the axon shown in Figure 2.16.

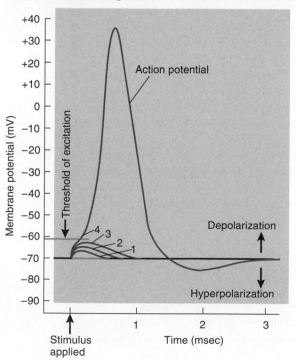

The Membrane Potential: Balance of Two Forces

To understand what causes the action potential to occur, we must first understand the reasons for the existence of the membrane potential. As we will see, this electrical charge is the result of a balance between two opposing forces: diffusion and electrostatic pressure.

The Force of Diffusion

When a spoonful of sugar is carefully poured into a container of water, it settles to the bottom. After a time the sugar dissolves, but it remains close to the bottom of the container. After a much longer time (probably several days) the molecules of sugar distribute themselves evenly throughout the water, even if no one stirs the liquid. The process whereby molecules distribute themselves evenly throughout the medium in which they are dissolved is called **diffusion.**

When there are no forces or barriers to prevent them from doing so, molecules will diffuse from regions of high concentration to regions of low concentration. Molecules are constantly in motion, and their rate of movement is proportional to the temperature. Only at absolute zero [0 K (kelvin) = $-273.15°C = -459.7°F$] do molecules cease their random movement. At all other temperatures they move about, colliding and veering off in different directions, thus pushing one another away. The result of these collisions in the example of sugar and water is to force sugar molecules upward (and to force water molecules downward), away from the regions in which they are most concentrated.

The Force of Electrostatic Pressure

When some substances are dissolved in water, they split into two parts, each with an opposing electrical charge. Substances with this property are called **electrolytes;** the charged particles into which they decompose are called **ions.** Ions are of two basic types: *Cations* have a positive charge, and *anions* have a negative charge. For example, when sodium chloride (NaCl, table salt) is dissolved in

---

hyperpolarization An increase in the membrane potential of a cell, relative to the normal resting potential.

action potential The brief electrical impulse that provides the basis for conduction of information along an axon.

threshold of excitation The value of the membrane potential that must be reached to produce an action potential.

diffusion Movement of molecules from regions of high concentration to regions of low concentration.

electrolyte An aqueous solution of a material that ionizes—namely, a soluble acid, base, or salt.

ion A charged molecule. *Cations* are positively charged, and *anions* are negatively charged.

FIGURE 2.18 ■ Control of the Membrane Potential

The figure shows the relative concentration of some important ions inside and outside the neuron and the forces acting on them.

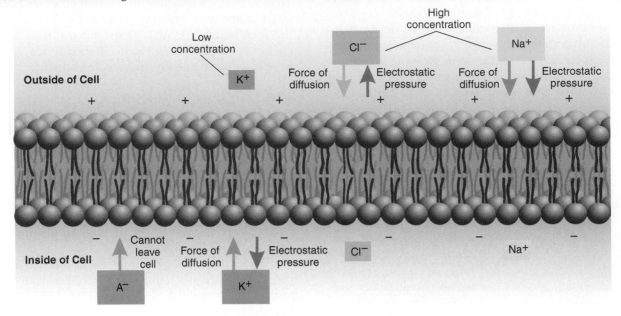

water, many of the molecules split into sodium cations (Na^+) and chloride anions (Cl^-). (I find that the easiest way to keep the terms *cation* and *anion* straight is to think of the cation's plus sign as a cross and remember the superstition of a black *cat* crossing your path.)

As you have undoubtedly learned, particles with the same kind of charge repel each other (+ repels +, and − repels −), but particles with different charges are attracted to each other (+ and − attract). Thus, anions repel anions, cations repel cations, but anions and cations attract each other. The force exerted by this attraction or repulsion is called **electrostatic pressure.** Just as the force of diffusion moves molecules from regions of high concentration to regions of low concentration, electrostatic pressure moves ions from place to place: Cations are pushed away from regions with an excess of cations, and anions are pushed away from regions with an excess of anions.

Ions in the Extracellular and Intracellular Fluid

The fluid within cells (**intracellular fluid**) and the fluid surrounding them (**extracellular fluid**) contain different ions. The forces of diffusion and electrostatic pressure contributed by these ions give rise to the membrane potential. Because the membrane potential is produced by a balance between the forces of diffusion and electrostatic pressures, understanding what produces this potential requires that we know the

concentration of the various ions in the extracellular and intracellular fluids.

There are several important ions in these fluids. I will discuss four of them here: organic anions (symbolized by A^-), chloride ions (Cl^-), sodium ions (Na^+), and potassium ions (K^+). The Latin words for sodium and potassium are *natrium* and *kalium*; hence, they are abbreviated *Na* and *K*, respectively. Organic anions— negatively charged proteins and intermediate products of the cell's metabolic processes—are found only in the intracellular fluid. Although the other three ions are found in both the intracellular and extracellular fluids, K^+ is found in predominantly in the intracellular fluid, whereas Na^+ and Cl^- are found predominantly in the extracellular fluid. The sizes of the boxes in Figure 2.18 indicate the relative concentrations of these four ions. (See *Figure 2.18.*) The easiest way to remember which ion is found where is to recall that the fluid that surrounds our cells is similar to seawater, which is predominantly a solution of salt, NaCl. The primitive ancestors of our cells lived in the ocean; thus, the seawater was their extracellular fluid. Our extracellular fluid thus

---

electrostatic pressure The attractive force between atomic particles charged with opposite signs or the repulsive force between atomic particles charged with the same sign.

intracellular fluid The fluid contained within cells.

extracellular fluid Body fluids located outside of cells.

resembles seawater, produced and maintained by regulatory mechanisms that are described in Chapter 12.

Let us consider the ions in Figure 2.18, examining the forces of diffusion and electrostatic pressure exerted on each and reasoning why each is located where it is. A^-, the organic anion, is unable to pass through the membrane of the axon; therefore, although the presence of this ion within the cell contributes to the membrane potential, it is located where it is because the membrane is impermeable to it.

The potassium ion K^+ is concentrated within the axon; thus, the force of diffusion tends to push it out of the cell. However, the outside of the cell is positively charged with respect to the inside, so electrostatic pressure tends to force the cation inside. Thus, the two opposing forces balance, and potassium ions tend to remain where they are. (See *Figure 2.18*.)

The chloride ion Cl^- is in greatest concentration outside the axon. The force of diffusion pushes this ion inward. However, because the inside of the axon is negatively charged, electrostatic pressure pushes the anion outward. Again, two opposing forces balance each other. (See *Figure 2.18*.)

The sodium ion Na^+ is also in greatest concentration outside the axon, so it, like Cl^-, is pushed into the cell by the force of diffusion. But unlike chloride, the sodium ion is *positively* charged. Therefore, electrostatic pressure does *not* prevent Na^+ from entering the cell; indeed, the negative charge inside the axon *attracts* Na^+. (See *Figure 2.18*.)

How can Na^+ remain in greatest concentration in the extracellular fluid, despite the fact that both forces (diffusion and electrostatic pressure) tend to push it inside? The answer is this: Another force, provided by the *sodium–potassium pump*, continuously pushes Na^+ out of the axon. The sodium–potassium pump consists of a large number of protein molecules embedded in the membrane, driven by energy provided by molecules of ATP produced by the mitochondria. These molecules, known as **sodium–potassium transporters,** exchange Na^+ for K^+, pushing three sodium ions out for every two potassium ions they push in. (See *Figure 2.19*.)

Because the membrane is not very permeable to Na^+, sodium–potassium transporters very effectively keep the intracellular concentration of Na^+ low. By transporting K^+ into the cell, they also increase the intracellular concentration of K^+ a small amount. The membrane is approximately 100 times more permeable to K^+ than to Na^+, so the increase is slight; but as we will see when we study the process of neural inhibition later in this chapter, it is very important. The transporters that make up the sodium–potassium pump use considerable energy: Up to 40 percent of a neuron's metabolic resources are used to operate them. Neurons, muscle cells, glia—in fact, most cells of the body—have sodium–potassium transporters in their membrane.

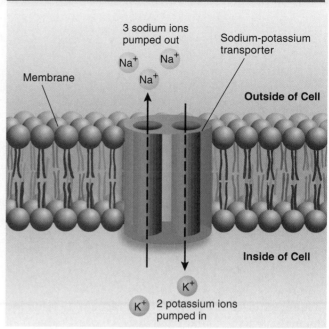

FIGURE 2.19 ■ **A Sodium–Potassium Transporter, Situated in the Cell Membrane**

3 sodium ions pumped out

Na$^+$ Na$^+$ Na$^+$

Sodium-potassium transporter

Membrane

Outside of Cell

Inside of Cell

K$^+$ K$^+$ 2 potassium ions pumped in

The Action Potential

As we saw, the forces of both diffusion and electrostatic pressure tend to push Na^+ into the cell. However, the membrane is not very permeable to this ion, and sodium–potassium transporters continuously pump out Na^+, keeping the intracellular level of Na^+ low. But imagine what would happen if the membrane suddenly became permeable to Na^+. The forces of diffusion and electrostatic pressure would cause Na^+ to rush into the cell. This sudden influx (inflow) of positively charged ions would drastically change the membrane potential. Indeed, experiments have shown that this mechanism is precisely what causes the action potential: A brief increase in the permeability of the membrane to Na^+ (allowing these ions to rush into the cell) is immediately followed by a transient increase in the permeability of the membrane to K^+ (allowing these ions to rush out of the cell). What is responsible for these transient increases in permeability?

We already saw that one type of protein molecule embedded in the membrane—the sodium–potassium transporter—actively pumps sodium ions out of the cell and pumps potassium ions into it. Another type of protein molecule provides an opening that permits ions

sodium–potassium transporter A protein found in the membrane of all cells that extrudes sodium ions from and transports potassium ions into the cell.

FIGURE 2.20 ■ Ion Channels

When ion channels are open, ions can pass through them, entering or leaving the cell.

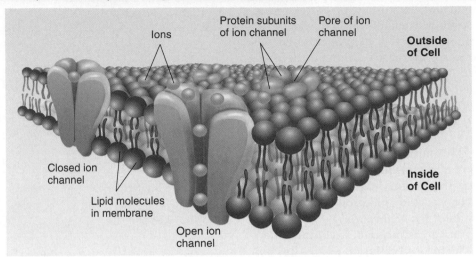

to enter or leave the cells. These molecules provide **ion channels,** which contain passages ("pores") that can open or close. When an ion channel is open, a particular type of ion can flow through the pore and thus can enter or leave the cell. (See *Figure 2.20.*) Neural membranes contain many thousands of ion channels. For example, the giant squid axon contains several hundred sodium channels in each square micrometer of membrane. (There are one million square micrometers in a square millimeter; thus, a patch of axonal membrane the size of a lowercase letter "o" in this book would contain several hundred million sodium channels.) Each sodium channel can admit up to 100 million ions per second when it is open. Thus, the permeability of a membrane to a particular ion at a given moment is determined by the number of ion channels that are open.

The following numbered paragraphs describe the movements of ions through the membrane during the action potential. The numbers in the figure correspond to the numbers of the paragraphs that follow. (See *Figure 2.21.*)

1. As soon as the threshold of excitation is reached, the sodium channels in the membrane open and Na$^+$ rushes in, propelled by the forces of diffusion and electrostatic pressure. The opening of these channels is triggered by reduction of the membrane potential (depolarization); they open at the point at which an action potential begins: the threshold of excitation. Because these channels are opened by changes in the membrane potential, they are called **voltage-dependent ion channels.** The influx of positively charged sodium ions produces a rapid change in the membrane potential, from −70 mV to +40 mV.

2. The membrane of the axon contains voltage-dependent potassium channels, but these channels are less sensitive than voltage-dependent sodium channels. That is, they require a greater level of depolarization before they begin to open. Thus, they begin to open later than the sodium channels.

3. At about the time the action potential reaches its peak (in approximately 1 msec), the sodium channels become *refractory*—the channels become blocked and cannot open again until the membrane once more reaches the resting potential. At this time, no more Na$^+$ can enter the cell.

4. By now, the voltage-dependent potassium channels in the membrane are open, letting K$^+$ ions move freely through the membrane. At this time, the inside of the axon is *positively* charged, so K$^+$ is driven out of the cell by diffusion and by electrostatic pressure. This outflow of cations causes the membrane potential to return toward its normal value. As it does so, the potassium channels begin to close again.

5. Once the membrane potential returns to normal, the sodium channels reset so that another depolarization can cause them to open again.

6. The membrane actually overshoots its resting value (−70 mV) and only gradually returns to normal as the potassium channels finally close. Eventually, sodium–potassium transporters remove the Na$^+$ ions that leaked in and retrieve the K$^+$ ions that leaked out.

ion channel A specialized protein molecule that permits specific ions to enter or leave cells.

voltage-dependent ion channel An ion channel that opens or closes according to the value of the membrane potential.

FIGURE 2.21 ■ Ion Movements During the Action Potential

The shaded box at the top shows the opening of sodium channels at the threshold of excitation, their refractory condition at the peak of the action potential, and their resetting when the membrane potential returns to normal.

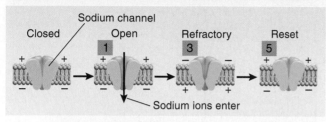

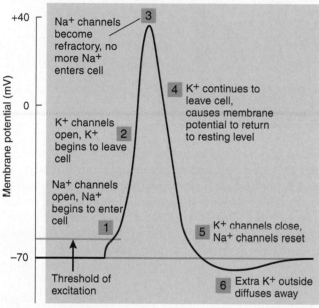

FIGURE 2.22 ■ Permeability

The graph shows changes in the permeability of the membrane to Na⁺ and K⁺ during the action potential.

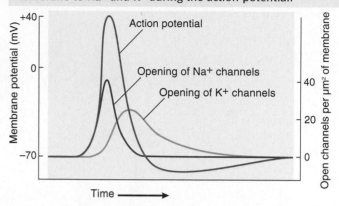

Figure 2.22 illustrates the changes in permeability of the membrane to sodium and potassium ions during the action potential. (See *Figure 2.22.*)

How much ionic flow is there? The increased permeability of the membrane to Na^+ is brief, and diffusion over any appreciable distance takes some time. Thus, when I say, "Na^+ rushes in," I do not mean that the axoplasm becomes flooded with Na^+. At the peak of the action potential a very thin layer of fluid immediately inside the axon becomes full of newly arrived Na^+ ions; this amount is indeed enough to reverse the membrane potential. However, not enough time has elapsed for these ions to fill the entire axon. Before that event can take place, the Na^+ channels close, and K^+ starts flowing out.

Experiments have shown that an action potential temporarily increases the number of Na^+ ions inside the giant squid axon by 0.0003 percent. Although the concentration just inside the membrane is high, the total number of ions entering the cell is very small relative to the number already there. This means that on a

short-term basis, sodium–potassium transporters are not very important. The few Na^+ ions that manage to leak in diffuse into the rest of the axoplasm, and the slight increase in Na^+ concentration is hardly noticeable. However, sodium–potassium transporters are important on a *long-term* basis. Without the activity of sodium–potassium transporters the concentration of sodium ions in the axoplasm would eventually increase enough that the axon would no longer be able to function.

Conduction of the Action Potential

Now that we have a basic understanding of the resting membrane potential and the production of the action potential, we can consider the movement of the message down the axon, or *conduction of the action potential.* To study this phenomenon, we again make use of the giant squid axon. We attach an electrical stimulator to an electrode at one end of the axon and place recording electrodes, attached to oscilloscopes, at different distances from the stimulating electrode. Then we apply a depolarizing stimulus to the end of the axon and trigger an action potential. We record the action potential from each of the electrodes, one after the other. Thus, we see that the action potential is conducted down the axon. As the action potential travels, it remains constant in size. (See *Figure 2.23.*)

This experiment establishes a basic law of axonal conduction: the **all-or-none law.** This law states that an action potential either occurs or does not occur; and once triggered, it is transmitted down the axon to its end. An action potential always remains the same size,

all-or-none law The principle that once an action potential is triggered in an axon, it is propagated, without decrement, to the end of the fiber.

FIGURE 2.23 ■ Conduction of the Action Potential

When an action potential is triggered, its size remains undiminished as it travels down the axon. The speed of conduction can be calculated from the delay between the stimulus and the action potential.

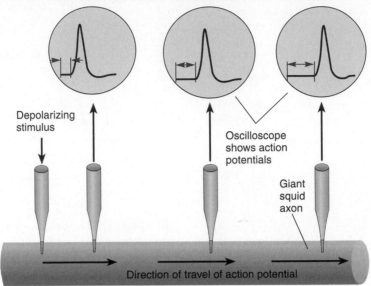

Depolarizing stimulus

Oscilloscope shows action potentials

Giant squid axon

Direction of travel of action potential

without growing or diminishing. And when an action potential reaches a point where the axon branches, it splits but does not diminish in size. An axon will transmit an action potential in either direction, or even in both directions, if it is started in the middle of the axon's length. However, because action potentials in living animals always start at the end attached to the soma, axons normally carry one-way traffic.

As you know, the strength of a muscular contraction can vary from very weak to very forceful, and the strength of a stimulus can vary from barely detectable to very intense. We know that the occurrence of action potentials in axons controls the strength of muscular contractions and represents the intensity of a physical stimulus. But if the action potential is an all-or-none event, how can it represent information that can vary in a continuous fashion? The answer is simple: A single action potential is not the basic element of information; rather, variable information is represented by an axon's *rate of firing*. (In this context, *firing* refers to the production of action potentials.) A high rate of firing causes a strong muscular contraction, and a strong stimulus (such as a bright light) causes a high rate of firing in axons that serve the eyes. Thus, the all-or-none law is supplemented by the **rate law.** (See *Figure 2.24.*)

Action potentials are not the only kind of electrical signals that occur in neurons. As we will see in the last section of this chapter, when a message is sent across a synapse, a small electrical signal is produced in the membrane of the neuron that receives the message. To understand this process and to understand the way in which action potentials are conducted in myelinated

axons (described later in this section), we must see how signals other than action potentials are conducted. To do so, we produce a weak, subthreshold depolarization (too small to produce an action potential) at one end of an axon and record its effects from electrodes placed along the axon. We find that the stimulus produces a disturbance in the membrane potential that becomes smaller as it moves away from the point of stimulation. (See *Figure 2.25.*)

The transmission of the weak, subthreshold depolarization is *passive*. Neither sodium channels nor potassium channels open or close. The axon is acting like an electrical cable, carrying along the current that started at one end. This property of the axon follows laws discovered in the nineteenth century that describe the conduction of electricity through telegraph cables laid along the ocean floor. As a signal passes through an undersea cable, the signal gets smaller because of the electrical characteristics of the cable, including leakage through the insulator and resistance in the wire. Because the signal decreases in size (decrements), it is referred to as *decremental conduction*. We say that the conduction of a weak depolarization by the axon follows the laws that describe the **cable properties** of the axon—the

rate law The principle that variations in the intensity of a stimulus or other information being transmitted in an axon are represented by variations in the rate at which that axon fires.

cable properties The passive conduction of electrical current, in a decremental fashion, down the length of an axon.

FIGURE 2.24 ■ The Rate Law

The strength of a stimulus is represented by the rate of firing of an axon. The size of each action potential is always constant.

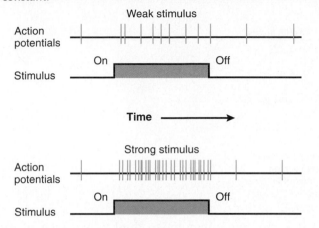

same laws that describe the electrical properties of an undersea cable. And because hyperpolarizations never trigger action potentials, these disturbances, too, are transmitted by means of the passive cable properties of an axon.

Recall that all but the smallest axons in mammalian nervous systems are myelinated; segments of the axons are covered by a myelin sheath produced by the oligodendrocytes of the CNS or the Schwann cells of the PNS. These segments are separated by portions of naked axon, the nodes of Ranvier. Conduction of an

action potential in a myelinated axon is somewhat different from conduction in an unmyelinated axon.

Schwann cells (and the oligodendrocytes of the CNS) wrap tightly around the axon, leaving no measurable extracellular fluid between them and the axon. The only place where a myelinated axon comes into contact with the extracellular fluid is at a node of Ranvier, where the axon is naked. In the myelinated areas there can be no inward flow of Na^+ when the sodium channels open, because there *is* no extracellular sodium. How, then, does the "action potential" travel along the area of axonal membrane covered by myelin sheath? You guessed it: by cable properties. The axon passively conducts the electrical disturbance from the action potential to the next node of Ranvier. The disturbance gets smaller, but it is still large enough to trigger an action potential at the node. The action potential gets retriggered, or repeated, at each node of Ranvier and is passed, by means of cable properties of the axon, along the myelinated area to the next node. Such conduction, appearing to hop from node to node, is called **saltatory conduction,** from the Latin *saltare,* "to leap, to dance." (See *Figure 2.26.*)

Saltatory conduction confers two advantages. The first is economic. Sodium ions enter axons during action potentials, and these ions must eventually be removed. Sodium–potassium transporters must be located along

---

saltatory conduction Conduction of action potentials by myelinated axons. The action potential appears to jump from one node of Ranvier to the next.

FIGURE 2.25 ■ Decremental Conduction

When a subthreshold depolarization is applied to the axon, the disturbance in the membrane potential is largest near the stimulating electrode and gets progressively smaller at distances farther along the axon.

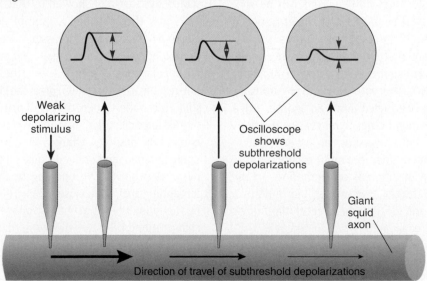

FIGURE 2.26 ■ **Saltatory Conduction**

The figure shows propagation of an action potential down a myelinated axon.

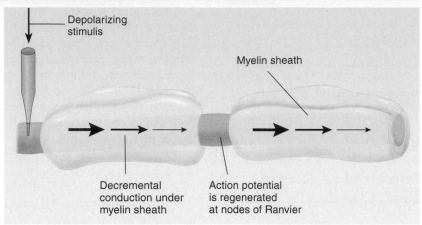

Depolarizing stimulis

Myelin sheath

Decremental conduction under myelin sheath

Action potential is regenerated at nodes of Ranvier

the entire length of unmyelinated axons because Na⁺ enters everywhere. However, because Na⁺ can enter myelinated axons only at the nodes of Ranvier, much less gets in, and consequently, much less has to be pumped out again. Therefore, myelinated axons expend much less energy to maintain their sodium balance.

The second advantage to myelin is speed. Conduction of an action potential is faster in a myelinated axon because the transmission between the nodes, which occurs by means of the axon's cable properties, is very fast. Increased speed enables an animal to react faster and (undoubtedly) to think faster. One of the ways to increase the speed of conduction is to increase size. Because it is so large, the unmyelinated squid axon, with a diameter of 500 μm, achieves a conduction velocity of approximately 35 m/sec (meters per second). However, the same speed is achieved by a myelinated cat axon with a diameter of a mere 6 μm. The fastest myelinated axon, 20 μm in diameter, can conduct action potentials at a speedy 120 m/sec, or 432 km/h (kilometers per hour). At that speed, a signal can get from one end of an axon to the other without much delay.

Interim Summary

Communication Within a Neuron

The withdrawal reflex illustrates how neurons can be connected to accomplish useful behaviors. The circuit responsible for this reflex consists of three sets of neurons: sensory neurons, interneurons, and motor neurons. The reflex can be suppressed when neurons in the brain activate inhibitory interneurons that form synapses with the motor neurons.

The message that is conducted down an axon is called an action potential. The membranes of all cells of the body are electrically charged, but only axons can produce action potentials. The resting membrane potential occurs because various ions are located in different concentrations in the fluid inside and outside the cell. The extracellular fluid (like seawater) is rich in Na⁺ and Cl⁻, and the intracellular fluid is rich in K⁺ and various organic anions, designated as A⁻.

The cell membrane is freely permeable to water, but its permeability to various ions—in particular, Na⁺ and K⁺—is regulated by ion channels. When the membrane potential is at its resting value (−70 mV), the voltage-dependent sodium and potassium channels are closed. The experiment with radioactive seawater showed us that some Na⁺ continuously leaks into the axon but is promptly forced out of the cell again by the sodium–potassium transporters (which also pump potassium *into* the axon). When an electrical stimulator depolarizes the membrane of the axon so that its potential reaches the threshold of excitation, voltage-dependent sodium channels open, and Na⁺ rushes into the cell, driven by the force of diffusion and by electrostatic pressure. The entry of the positively charged ions further reduces the membrane potential and, indeed, causes it to reverse, so the inside becomes positive. The opening of the sodium channels is temporary; they soon close again. The depolarization caused by the influx of Na⁺ activates voltage-dependent potassium channels, and K⁺ leaves the axon, traveling down its concentration gradient. This efflux (outflow) of K⁺ quickly brings the membrane potential back to its resting value.

Because an action potential of a given axon is an all-or-none phenomenon, neurons represent intensity by their rate of firing. The action potential normally begins at one end of the axon, where the axon attaches to the soma. The action potential travels continuously down unmyelinated axons, remaining constant in size, until it reaches the terminal buttons. (If the axon divides, an action potential continues down each branch.) In myelinated axons, ions can flow through the membrane only at the nodes of Ranvier, because the axons are covered everywhere else with myelin, which isolates them from the extracellular fluid. Thus, the action potential is conducted from one node of Ranvier to the next by means of passive cable properties. When the electrical message reaches a node, voltage-dependent sodium channels open, and a new action potential is triggered. This mechanism saves a considerable amount of energy because sodium–potassium transporters are not needed along the myelinated portions of the axon, and saltatory conduction is faster.

Thought Question

The evolution of the human brain, with all its complexity, depended on many apparently trivial mechanisms. For example, what if cells had not developed the ability to manufacture myelin? Unmyelinated axons must be very large if they are to transmit action potentials rapidly. How big would the human brain have to be if oligodendrocytes did not produce myelin? *Could* the human brain as we know it have evolved without myelin?

COMMUNICATION BETWEEN NEURONS

Now that you know about the basic structure of neurons and the nature of the action potential, it is time to describe the ways in which neurons can communicate with each other. These communications make it possible for circuits of neurons to gather sensory information, make plans, and initiate behaviors.

The primary means of communication between neurons is *synaptic transmission*—the transmission of messages from one neuron to another through a synapse. As we saw, these messages are carried by neurotransmitters, released by terminal buttons. These chemicals diffuse across the fluid-filled gap between the terminal buttons and the membranes of the neurons with which they form synapses. As we will see in this section, neurotransmitters produce **postsynaptic potentials**—brief depolarizations or hyperpolarizations—that increase or decrease the rate of firing of the axon of the postsynaptic neuron. (*MyPsychKit 2.3, Synapses,* illustrates the information presented in the following section.)

mypsychkit
Animation 2.3
Synapses

Neurotransmitters exert their effects on cells by attaching to a particular region of a receptor molecule called the **binding site.** A molecule of the chemical fits into the binding site the way a key fits into a lock: The shape of the binding site and the shape of the molecule of the neurotransmitter are complementary. A chemical that attaches to a binding site is called a **ligand,** from *ligare,* "to bind." Neurotransmitters are natural ligands, produced and released by neurons. But other chemicals found in nature (primarily in plants or in the poisonous venoms of animals) can serve as ligands too. In addition, artificial ligands can be produced in the laboratory. These chemicals are discussed in Chapter 4, which deals with drugs and their effects.

Structure of Synapses

As you have already learned, synapses are junctions between the terminal buttons at the ends of the axonal branches of one neuron and the membrane of another. Synapses can occur in three places: on dendrites, on the soma, and on other axons. These synapses are referred to as *axodendritic, axosomatic,* and *axoaxonic.* Axodendritic synapses can occur on the smooth surface of a dendrite or on **dendritic spines**—small protrusions that stud the dendrites of several types of large neurons in the brain. (See *Figure 2.27.*)

Figure 2.28 illustrates a synapse. The **presynaptic membrane,** located at the end of the terminal button, faces the **postsynaptic membrane,** located on the neuron

---

postsynaptic potential Alterations in the membrane potential of a postsynaptic neuron, produced by liberation of neurotransmitter at the synapse.

binding site The location on a receptor protein to which a ligand binds.

ligand (*lye gand* or *ligg and*) A chemical that binds with the binding site of a receptor.

dendritic spine A small bud on the surface of a dendrite, with which a terminal button of another neuron forms a synapse.

presynaptic membrane The membrane of a terminal button that lies adjacent to the postsynaptic membrane and through which the neurotransmitter is released.

postsynaptic membrane The cell membrane opposite the terminal button in a synapse; the membrane of the cell that receives the message.

FIGURE 2.27 ■ Types of Synapses

Axodendritic synapses can occur on the smooth surface of a dendrite (a) or on dendritic spines (b). Axosomatic synapses occur on somatic membrane (c). Axoaxonic synapses consist of synapses between two terminal buttons (d).

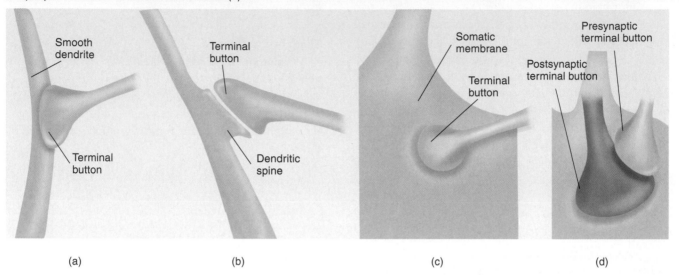

(a)   (b)   (c)   (d)

that receives the message (the *postsynaptic* neuron). These two membranes face each other across the **synaptic cleft,** a gap that varies in size from synapse to synapse but is usually around 20 nm wide. (A nanometer (nm) is one billionth of a meter.) The synaptic cleft contains extracellular fluid, through which the neurotransmitter diffuses. A meshwork of filaments crosses the synaptic cleft and keeps the presynaptic

and postsynaptic membranes in alignment. (See *Figure 2.28.*)

As you may have noticed in Figure 2.28, two prominent structures are located in the cytoplasm of the

synaptic cleft The space between the presynaptic membrane and the postsynaptic membrane.

FIGURE 2.28 ■ Details of a Synapse

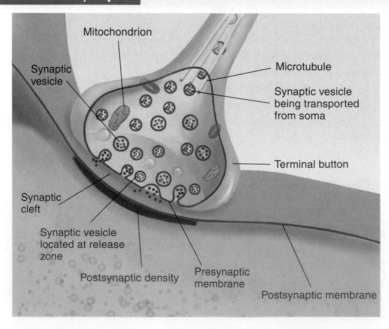

terminal button: mitochondria and synaptic vesicles. We also see microtubules, which are responsible for transporting material between the soma and terminal button. The presence of mitochondria implies that the terminal button needs energy to perform its functions. **Synaptic vesicles** are small, rounded objects in the shape of spheres or ovoids. (The term *vesicle* means "little bladder.") A given terminal button can contain from a few hundred to nearly a million synaptic vesicles. Many terminal buttons contain two types of synaptic vesicles: large and small. Small synaptic vesicles (found in all terminal buttons) contain molecules of the neurotransmitter. They range in number from a few dozen to several hundred. The membrane of small synaptic vesicles consists of approximately 10,000 lipid molecules into which are inserted about 200 protein molecules. *Transport proteins* fill vesicles with the neurotransmitter, and *trafficking proteins* are involved in the release of neurotransmitter and recycling of the vesicles. Synaptic vesicles are found in greatest numbers around the part of the presynaptic membrane that faces the synaptic cleft—near the **release zone,** the region from which the neurotransmitter is released. In many terminal buttons we see a scattering of large, dense-core synaptic vesicles. These vesicles contain one of a number of different peptides, the functions of which are described later in this chapter. (See *Figures 2.28* and *2.29.*)

Small synaptic vesicles are produced in the Golgi apparatus located in the soma and are carried by fast

FIGURE 2.29 ■ Cross Section of a Synapse

This photograph from an electron microscope, shows a cross section of a synapse. The terminal button contains many synaptic vesicles, filled with the neurotransmitter, and a single large dense-core vesicle, filled with a peptide.

(From De Camilli, P., et al., in *Synapses*, edited by W. M. Cowan, T. C. Südhof, and C. F. Stevens. Baltimore, MD: Johns Hopkins University Press, 2001. Reprinted with permission.)

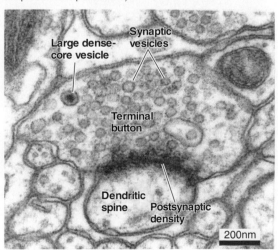

FIGURE 2.30 ■ Cross Section of a Synapse

The omega-shaped figures in this cross section of a synapse are synaptic vesicles fusing with the presynaptic membranes of terminal buttons that form synapses with frog muscle.

(From Heuser, J. E., in *Society for Neuroscience Symposia, Vol. II*, edited by W. M. Cowan and J. A. Ferrendelli. Bethesda, MD: Society for Neuroscience, 1977. Reprinted with permission.)

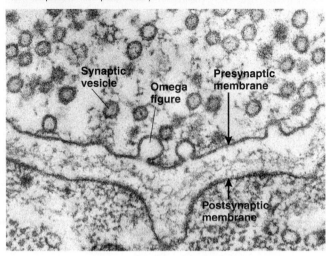

axoplasmic transport to the terminal button. As we will see, some are also produced from recycled material in the terminal button. Large synaptic vesicles are produced only in the soma and are transported through the axoplasm to the terminal buttons.

In an electron micrograph the postsynaptic membrane under the terminal button appears somewhat thicker and more dense than the membrane elsewhere. This postsynaptic density is caused by the presence of receptors—specialized protein molecules that detect the presence of neurotransmitters in the synaptic cleft—and protein filaments that hold the receptors in place. (See *Figures 2.28* and *2.29.*)

Release of Neurotransmitter

When action potentials are conducted down an axon (and down all of its branches), something happens inside all of the terminal buttons: A number of small synaptic vesicles located just inside the presynaptic membrane fuse with the membrane and then break open, spilling their contents into the synaptic cleft. *Figure 2.30* shows a portion of a frog's neuromuscular

synaptic vesicle (*vess i kul*) A small, hollow, beadlike structure found in terminal buttons; contains molecules of a neurotransmitter.

release zone A region of the interior of the presynaptic membrane of a synapse to which synaptic vesicles attach and release their neurotransmitter into the synaptic cleft.

junction—the synapse between a terminal button and a muscle fiber. The axon has just been stimulated, and synaptic vesicles in the terminal button are in the process of releasing the neurotransmitter. Note that some vesicles are fused with the presynaptic membrane, forming the shape of an omega (Ω). (See *Figure 2.30*.)

How does an action potential cause synaptic vesicles to release the neurotransmitter? The process begins when a population of synaptic vesicles become "docked" against the presynaptic membrane, ready to release their neurotransmitter into the synaptic cleft. Docking is accomplished when clusters of protein molecules attach to other protein molecules located in the presynaptic membrane. (See *Figure 2.31*.)

The release zone of the presynaptic membrane contains voltage-dependent calcium channels. When the membrane of the terminal button is depolarized by an arriving action potential, the calcium channels open. Like sodium ions, calcium ions (Ca^{2+}) are located in highest concentration in the extracellular fluid. Thus, when the voltage-dependent calcium channels open, Ca^{2+} flows into the cell, propelled by electrostatic pressure and the force of diffusion. The entry of Ca^{2+} is an

essential step; if neurons are placed in a solution that contains no calcium ions, an action potential no longer causes the release of the neurotransmitter. (Calcium transporters, similar in operation to sodium–potassium transporters, later remove the intracellular Ca^{2+}.)

As we will see later in this chapter and in subsequent chapters of this book, calcium ions play many important roles in biological processes within cells. Calcium ions can bind with various types of proteins, changing their characteristics. Some of the calcium ions that enter the terminal button bind with the clusters of protein molecules that join the membrane of the synaptic vesicles with the presynaptic membrane. This event makes the segments of the clusters of protein molecules move apart, producing a *fusion pore*—a hole through both membranes that enables them to fuse together. The process of fusion takes approximately 0.1 msec. (See *Figure 2.31*.)

Figure 2.32 shows two photomicrographs of the presynaptic membrane, before and after the fusion pores have opened. We see the face of the presynaptic membrane as it would be viewed from the postsynaptic membrane. As you can see, the synaptic vesicles are

FIGURE 2.31 ■ Release of Neurotransmitter

An action potential opens calcium channels. Calcium ions enter and bind with the protein embedded in the membrane of synaptic vesicles docked at the release zone. The fusion pores open, and the neurotransmitter is released into the synaptic cleft.

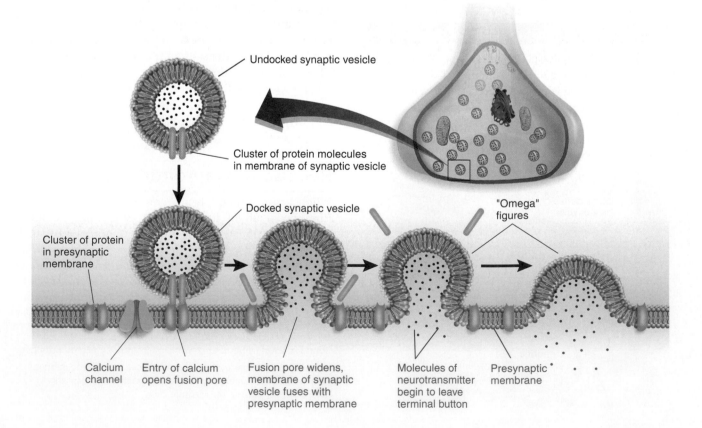

Undocked synaptic vesicle

Cluster of protein molecules in membrane of synaptic vesicle

Docked synaptic vesicle

"Omega" figures

Cluster of protein in presynaptic membrane

Calcium channel

Entry of calcium opens fusion pore

Fusion pore widens, membrane of synaptic vesicle fuses with presynaptic membrane

Molecules of neurotransmitter begin to leave terminal button

Presynaptic membrane

These photomicrographs show the release of neurotransmitter by a terminal button that forms a synapse with a frog muscle. The views are of the surface of the fusion zone of the terminal button. (a) Just before release. The two rows of dots are probably calcium channels. (b) During release. The larger circles are holes in the presynaptic membrane, revealing the contents of the synaptic vesicles that have fused with it.

(From Heuser, J., and Reese, T. *Journal of Cell Biology*, 1981, *88*, 564–580. Reprinted with permission.)

Calcium channels–when open, cause release of neurotransmitter

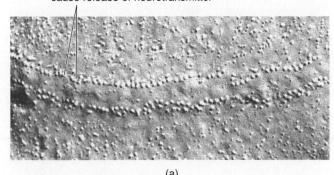

(a)

Synaptic vesicles fused with the presynaptic membrane, releasing the neurotransmitter

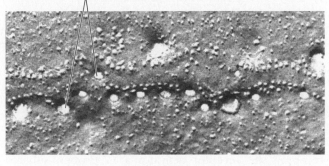

(b)

aligned in a row along the release zone. The small bumps arranged in lines on each side of the synaptic vesicles appear to be voltage-dependent calcium channels. (See *Figure 2.32*.)

Research indicates that there are three distinct pools of synaptic vesicles (Rizzoli and Betz, 2005). *Release-ready vesicles* are docked against the inside of the presynaptic membrane, ready to release their contents when an action potential arrives. These vesicles constitute less than 1 percent of the total number found in the terminal. Vesicles in the *recycling pool* constitute 10–15 percent of the total pool of vesicles, and those in the *reserve pool* make up the remaining 85–90 percent. If the axon fires at a low rate, only vesicles from the release-ready pool will be called on. If the rate of firing increases,

vesicles from the recycling pool and finally from the reserve pool will release their contents.

What happens to the membrane of the synaptic vesicles after they have broken open and released the neurotransmitter they contain? It appears that many vesicles in the ready-release pool use a process known as *kiss and run*. These synaptic vesicles release most or all of their neurotransmitter, the fusion pore closes, and the vesicles break away from the presynaptic membrane and get filled with neurotransmitter again. Other vesicles (primarily those in the recycling pool) merge and recycle and consequently lose their identity. The membranes of these vesicles merge with the presynaptic membrane. Little buds of membrane then pinch off into the cytoplasm and become synaptic vesicles. The appropriate proteins are inserted into the membrane of these vesicles, and the vesicles are filled with molecules of the neurotransmitter. The membranes of vesicles in the reserve pool are recycled through a process of *bulk endocytosis*. (*Endocytosis* means "the process of entering a cell.") Large pieces of the membrane of the terminal button fold inward, break off, and enter the cytoplasm. New vesicles are formed from small buds that break off of these pieces of membrane. The recycling process takes less than a second for the readily releasable pool, a few seconds for the recycling pool, and a few minutes for the reserve pool. (See *Figure 2.33*.)

Activation of Receptors

How do molecules of the neurotransmitter produce a depolarization or hyperpolarization in the postsynaptic membrane? They do so by diffusing across the synaptic cleft and attaching to the binding sites of special protein molecules located in the postsynaptic membrane, called **postsynaptic receptors.** Once binding occurs, the postsynaptic receptors open **neurotransmitter-dependent ion channels,** which permit the passage of specific ions into or out of the cell. Thus, the presence of the neurotransmitter in the synaptic cleft allows particular ions to pass through the membrane, changing the local membrane potential.

Neurotransmitters open ion channels by at least two different methods, direct and indirect. The direct method is simpler, so I will describe it first. Figure 2.34 illustrates a neurotransmitter-dependent ion channel

postsynaptic receptor A receptor molecule in the postsynaptic membrane of a synapse that contains a binding site for a neurotransmitter.

neurotransmitter-dependent ion channel An ion channel that opens when a molecule of a neurotransmitter binds with a postsynaptic receptor.

FIGURE 2.33 ■ Synaptic Vesicles After Release of Neurotransmitter

After synaptic vesicles have released neurotransmitter into the synaptic cleft, the following takes place. In "kiss and run," the vesicle fuses with the presynaptic membrane, releases the neurotransmitter, reseals, leaves the docking site, becomes refilled with the neurotransmitter, and mixes with other vesicles in the terminal button. In "merge and recycle," the vesicle completely fuses with the postsynaptic membrane, losing its identity. Extra membrane from fused vesicles pinches off into the cytoplasm and forms vesicles, which are filled with the neurotransmitter. The membranes of vesicles in the reserve pool are recycled through a process of "bulk endocytosis." Large pieces of the membrane of the terminal button fold inward, break off, and enter the cytoplasm. New vesicles are formed from small buds that break off of these pieces of membrane.

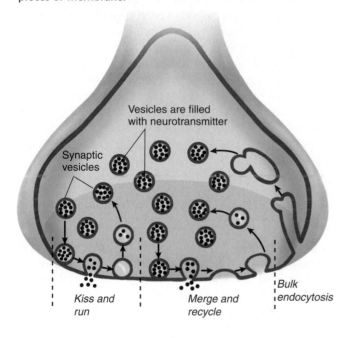

FIGURE 2.34 ■ Ionotropic Receptors

The ion channel opens when a molecule of neurotransmitter attaches to the binding site. For purposes of clarity the drawing is schematic; molecules of neurotransmitter are actually much larger than individual ions.

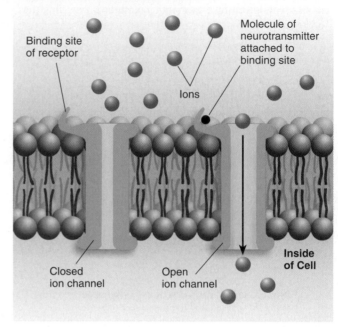

that is equipped with its own binding site. When a molecule of the appropriate neurotransmitter attaches to it, the ion channel opens. The formal name for this combination receptor/ion channel is an **ionotropic receptor.** (See *Figure 2.34.*)

Ionotropic receptors were first discovered in the organ that produces electrical current in *Torpedo,* the electric ray, where they occur in great number. (The electric ray is a fish that generates a powerful electrical current, not some kind of Star Wars weapon.) These receptors, which are sensitive to a neurotransmitter called *acetylcholine,* contain sodium channels. When these channels are open, sodium ions enter the cell and depolarize the membrane.

The indirect method is more complicated. Some receptors do not open ion channels directly but instead

start a chain of chemical events. These receptors are called **metabotropic receptors** because they involve steps that require that the cell expend metabolic energy. Metabotropic receptors are located in close proximity to another protein attached to the membrane—a **G protein.** When a molecule of the neurotransmitter binds with the receptor, the receptor activates a G protein situated inside the membrane next to the receptor. When activated, the G protein activates an enzyme that stimulates the production of a chemical called a **second messenger.**

ionotropic receptor (*eye on oh **trow** pik*) A receptor that contains a binding site for a neurotransmitter and an ion channel that opens when a molecule of the neurotransmitter attaches to the binding site.

metabotropic receptor (*meh tab oh **trow** pik*) A receptor that contains a binding site for a neurotransmitter; activates an enzyme that begins a series of events that opens an ion channel elsewhere in the membrane of the cell when a molecule of the neurotransmitter attaches to the binding site.

G protein A protein coupled to a metabotropic receptor; conveys messages to other molecules when a ligand binds with and activates the receptor.

second messenger A chemical produced when a G protein activates an enzyme; carries a signal that results in the opening of the ion channel or causes other events to occur in the cell.

FIGURE 2.35 ■ Metabotropic Receptors

(a) The ion channel is opened directly by the α subunit of an activated G protein. (b) The α subunit of the G protein activates an enzyme, which produces a second messenger that opens the ion channel.

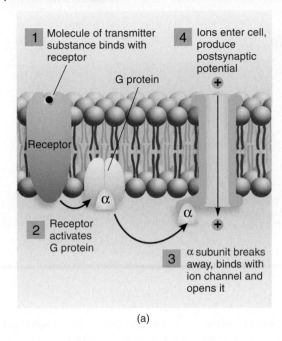

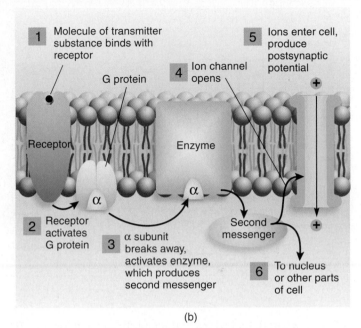

(a)   (b)

(The neurotransmitter is the first messenger.) Molecules of the second messenger travel through the cytoplasm, attach themselves to nearby ion channels, and cause them to open. Compared with postsynaptic potentials produced by ionotropic receptors, those produced by metabotropic receptors take longer to begin and last longer. (See *Figure 2.35.*)

The first second messenger to be discovered was *cyclic AMP*, a chemical that is synthesized from ATP. Since then, several other second messengers have been discovered. As you will see in later chapters, second messengers play an important role in both synaptic and nonsynaptic communication. And they can do more than open ion channels. For example, they can travel to the nucleus or other regions of the neuron and initiate biochemical changes that affect the functions of the cell. They can even turn specific genes on or off, thus initiating or terminating production of particular proteins.

Postsynaptic Potentials

As I mentioned earlier, postsynaptic potentials can be either depolarizing (excitatory) or hyperpolarizing (inhibitory). What determines the nature of the postsynaptic potential at a particular synapse is not the neurotransmitter itself. Instead, it is determined by the characteristics of the postsynaptic receptors—in particular, *by the particular type of ion channel they open.*

As Figure 2.36 shows, four major types of neurotransmitter-dependent ion channels are found in the postsynaptic membrane: sodium (Na^+), potassium (K^+), chloride (Cl^-), and calcium (Ca^{2+}). Although the figure depicts only directly activated (ionotropic) ion channels, you should realize that many ion channels are activated indirectly, by metabotropic receptors coupled to G proteins.

The neurotransmitter-dependent sodium channel is the most important source of excitatory postsynaptic potentials. As we saw, sodium–potassium transporters keep sodium outside the cell, waiting for the forces of diffusion and electrostatic pressure to push it in. Obviously, when sodium channels are opened, the result is a depolarization—an **excitatory postsynaptic potential (EPSP).** (See *Figure 2.36a.*)

We also saw that sodium–potassium transporters maintain a small surplus of potassium ions inside the cell. If potassium channels open, some of these cations

---

excitatory postsynaptic potential (EPSP) An excitatory depolarization of the postsynaptic membrane of a synapse caused by the liberation of a neurotransmitter by the terminal button.

FIGURE 2.36 ■ Ionic Movements During Postsynaptic Potentials

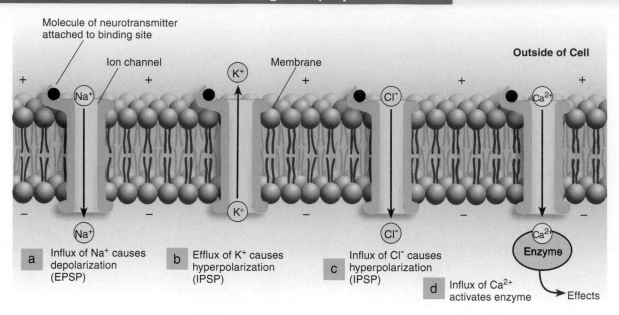

Molecule of neurotransmitter attached to binding site

Ion channel

Membrane

Outside of Cell

a Influx of Na⁺ causes depolarization (EPSP)

b Efflux of K⁺ causes hyperpolarization (IPSP)

c Influx of Cl⁻ causes hyperpolarization (IPSP)

d Influx of Ca²⁺ activates enzyme

Enzyme → Effects

will follow this gradient and leave the cell. Because K^+ is positively charged, its efflux will hyperpolarize the membrane, producing an **inhibitory postsynaptic potential (IPSP).** (See *Figure 2.36b.*)

At many synapses, inhibitory neurotransmitters open the chloride channels, instead of (or in addition to) potassium channels. The effect of opening chloride channels depends on the membrane potential of the neuron. If the membrane is at the resting potential, nothing happens, because (as we saw earlier) the forces of diffusion and electrostatic pressure balance perfectly for the chloride ion. However, if the membrane potential has already been depolarized by the activity of excitatory synapses located nearby, then the opening of chloride channels will permit Cl^- to enter the cell. The influx of anions will bring the membrane potential back to its normal resting condition. Thus, the opening of chloride channels serves to neutralize EPSPs. (See *Figure 2.36c.*)

The fourth type of neurotransmitter-dependent ion channel is the calcium channel. Calcium ions (Ca^{2+}), being positively charged and being located in highest concentration outside the cell, act like sodium ions; that is, the opening of calcium channels depolarizes the membrane, producing EPSPs. But calcium does even more. As we saw earlier in this chapter, the entry of calcium into the terminal button triggers the migration of synaptic vesicles and the release of the neurotransmitter. In the dendrites of the postsynaptic cell, calcium binds with and activates special enzymes. These enzymes have a variety of effects, including the production of

biochemical and structural changes in the postsynaptic neuron. As we will see in Chapter 14, one of the ways in which learning affects the connections between neurons involves changes in dendritic spines initiated by the opening of calcium channels. (See *Figure 2.36d.*)

Termination of Postsynaptic Potentials

Postsynaptic potentials are brief depolarizations or hyperpolarizations caused by the activation of postsynaptic receptors with molecules of a neurotransmitter. They are kept brief by two mechanisms: reuptake and enzymatic deactivation.

The postsynaptic potentials produced by almost all neurotransmitters are terminated by **reuptake.** This process is simply an extremely rapid removal of neurotransmitter from the synaptic cleft by the terminal button. The neurotransmitter does not return in the vesicles that get pinched off the membrane of the terminal button. Instead, the membrane contains special transporter molecules that draw on the cell's energy reserves

inhibitory postsynaptic potential (IPSP) An inhibitory hyperpolarization of the postsynaptic membrane of a synapse caused by the liberation of a neurotransmitter by the terminal button.

reuptake The reentry of a neurotransmitter just liberated by a terminal button back through its membrane, thus terminating the postsynaptic potential.

FIGURE 2.37 ■ Reuptake

Molecules of a neurotransmitter that has been released into the synaptic cleft are transported back into the terminal button.

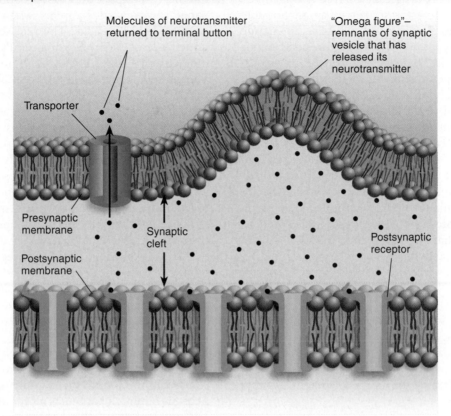

Molecules of neurotransmitter returned to terminal button

"Omega figure"– remnants of synaptic vesicle that has released its neurotransmitter

Transporter

Presynaptic membrane

Synaptic cleft

Postsynaptic membrane

Postsynaptic receptor

to force molecules of the neurotransmitter from the synaptic cleft directly into the cytoplasm—just as sodium–potassium transporters move Na^+ and K^+ across the membrane. When an action potential arrives, the terminal button releases a small amount of neurotransmitter into the synaptic cleft and then takes it back, giving the postsynaptic receptors only a brief exposure to the neurotransmitter. (See *Figure 2.37.*)

Enzymatic deactivation is accomplished by an enzyme that destroys molecules of the neurotransmitter. As far as we know, postsynaptic potentials are terminated in this way for only one neurotransmitter: **acetylcholine (ACh).** Transmission at synapses on muscle fibers and at some synapses between neurons in the central nervous system is mediated by ACh. Postsynaptic potentials produced by ACh are short-lived because the postsynaptic membrane at these synapses contains an enzyme called **acetylcholinesterase (AChE).** AChE destroys ACh by cleaving it into its constituents: choline and acetate. Because neither of these substances is capable of activating postsynaptic receptors, the postsynaptic potential is terminated once the molecules of ACh are broken apart. AChE is an extremely energetic destroyer of ACh; one molecule of AChE will chop apart more that 5000 molecules of ACh each second.

You will recall that Kathryn, the woman featured in the case history that opened this chapter, suffered from progressive muscular weakness. As her neurologist discovered, Kathryn had *myasthenia gravis.* This disease was first described in 1672 by Thomas Willis, an English physician. The term literally means "grave muscle weakness." It is not a very common disorder, but most experts believe that many mild cases go undiagnosed.

enzymatic deactivation The destruction of a neurotransmitter by an enzyme after its release—for example, the destruction of acetylcholine by acetylcholinesterase.

acetylcholine (ACh) (*a see tul koh leen*) A neurotransmitter found in the brain, spinal cord, and parts of the peripheral nervous system; responsible for muscular contraction.

acetylcholinesterase (AChE) (*a see tul koh lin ess ter ace*) The enzyme that destroys acetylcholine soon after it is liberated by the terminal buttons, thus terminating the postsynaptic potential.

In 1934, Dr. Mary Walker remarked that the symptoms of myasthenia gravis resembled the effects of curare, a poison that blocks neural transmission at the synapses on muscles. A drug called *physostigmine,* which deactivates acetylcholinesterase, serves as an antidote for curare poisoning. As we just saw, AChE is an enzyme that destroys the ACh and terminates the postsynaptic potentials it produces. By deactivating AChE, physostigmine greatly increases and prolongs the effects of ACh on the postsynaptic membrane. Thus, it increases the strength of synaptic transmission at the synapses on muscles and reverses the effects of curare. (Chapter 4 will say more about both curare and physostigmine.)

Dr. Walker reasoned that if physostigmine reversed the effects of curare poisoning, perhaps it would also reverse the symptoms of myasthenia gravis. She tried it, and it did within a matter of a few minutes. Later, pharmaceutical companies discovered drugs that could be taken orally and that produced longer-lasting effects. Nowadays, an injectable drug is used to make the diagnosis (as in Kathryn's case), and an oral drug is used to treat it. Unfortunately, no cure has yet been found for myasthenia gravis.

Like multiple sclerosis, myasthenia gravis is an autoimmune disease. For some reason the immune system becomes sensitized against the protein that makes up acetylcholine receptors. Almost as fast as new ACh receptors are produced, the immune system destroys them.

Effects of Postsynaptic Potentials: Neural Integration

We have seen how neurons are interconnected by means of synapses, how action potentials trigger the release of neurotransmitters, and how these chemicals initiate excitatory or inhibitory postsynaptic potentials. Excitatory postsynaptic potentials increase the likelihood that the postsynaptic neuron will fire; inhibitory postsynaptic potentials decrease this likelihood. (Remember, "firing" refers to the occurrence of an action potential.) Thus, the rate at which an axon fires is determined by the relative activity of the excitatory and inhibitory synapses on the soma and dendrites of that cell. If there are no active excitatory synapses or if the activity of inhibitory synapses is particularly high, that rate could be close to zero.

Let us look at the elements of this process. (*MyPsychKit 2.4, Postsynaptic Potentials,* illustrates the material presented in the rest of this chapter.) The interaction of the effects of excitatory and inhibitory synapses on a particular neuron is called

mypsychkit
Where learning comes to life!

Animation 2.4

Postsynaptic Potentials

neural integration. (*Integration* means "to make whole," in the sense of combining two or more functions.) Figure 2.38 illustrates the effects of excitatory and inhibitory synapses on a postsynaptic neuron. The top panel shows what happens when several excitatory synapses become active. The release of the neurotransmitter produces depolarizing EPSPs in the dendrites of the neuron. These EPSPs (represented in red) are then transmitted, by means of passive cable properties, down the dendrites, across the soma, to the *axon hillock* located at the base of the axon. If the depolarization is still strong enough when it reaches this point, the axon will fire. (See *Figure 2.38a.*)

Now let's consider what would happen if, at the same time, inhibitory synapses also become active. Inhibitory postsynaptic potentials are hyperpolarizing—they bring the membrane potential away from the threshold of excitation. Thus, they tend to cancel the effects of excitatory postsynaptic potentials. (See *Figure 2.38b.*)

The rate at which a neuron fires is controlled by the relative activity of the excitatory and inhibitory synapses on its dendrites and soma. If the activity of excitatory synapses goes up, the rate of firing will go up. If the activity of inhibitory synapses goes up, the rate of firing will go down.

Note that *neural* inhibition (that is, an inhibitory postsynaptic potential) does not always produce *behavioral* inhibition. For example, suppose a group of neurons inhibits a particular movement. If these neurons are inhibited, they will no longer suppress the behavior. Thus, inhibition of the inhibitory neurons makes the behavior more likely to occur. Of course, the same is true for neural excitation. *Excitation* of neurons that *inhibit* a behavior suppresses that behavior. For example, when we are dreaming, a particular set of inhibitory neurons in the brain becomes active and prevents us from getting up and acting out our dreams. (As we will see in Chapter 9, if these neurons are damaged, people *will* act out their dreams.) Neurons are elements in complex circuits; without knowing the details of these circuits, one cannot predict the effects of the excitation or inhibition of one set of neurons on an organism's behavior.

Autoreceptors

Postsynaptic receptors detect the presence of a neurotransmitter in the synaptic cleft and initiate excitatory or inhibitory postsynaptic potentials. But the postsynaptic

neural integration   The process by which inhibitory and excitatory postsynaptic potentials summate and control the rate of firing of a neuron.

FIGURE 2.38 ■ Neural Integration

(a) If several excitatory synapses are active at the same time, the EPSPs they produce (shown in red) summate as they travel toward the axon, and the axon fires. (b) If several inhibitory synapses are active at the same time, the IPSPs they produce (shown in blue) diminish the size of the EPSPs and prevent the axon from firing.

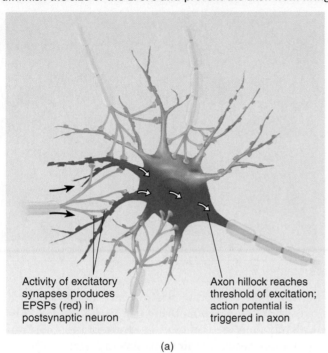

Activity of excitatory synapses produces EPSPs (red) in postsynaptic neuron

Axon hillock reaches threshold of excitation; action potential is triggered in axon

(a)

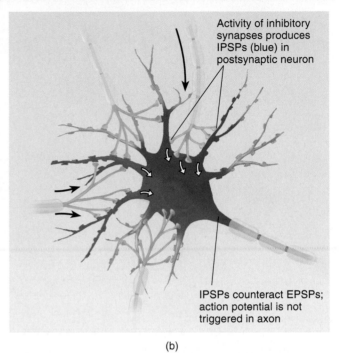

Activity of inhibitory synapses produces IPSPs (blue) in postsynaptic neuron

IPSPs counteract EPSPs; action potential is not triggered in axon

(b)

membrane is not the only location of receptors that respond to neurotransmitters. Many neurons also possess receptors that respond to the neurotransmitter that *they themselves* release, called **autoreceptors.**

Autoreceptors can be located on the membrane of any part of the cell, but in this discussion we will consider those located on the terminal button. In most cases these autoreceptors do not control ion channels. Thus, when stimulated by a molecule of the neurotransmitter, autoreceptors do not produce changes in the membrane potential of the terminal button. Instead, they regulate internal processes, including the synthesis and release of the neurotransmitter. (As you may have guessed, autoreceptors are metabotropic; the control they exert on these processes is accomplished through G proteins and second messengers.) In most cases the effects of autoreceptor activation are inhibitory; that is, the presence of the neurotransmitter in the extracellular fluid in the vicinity of the neuron causes a decrease in the rate of synthesis or release of the neurotransmitter. Most investigators believe that autoreceptors are part of a regulatory system that controls the amount of neurotransmitter that is released. If too much is released, the

autoreceptors inhibit both production and release; if not enough is released, the rates of production and release go up.

Other Types of Synapses

So far, the discussion of synaptic activity has referred only to the effects of postsynaptic excitation or inhibition. These effects occur at axosomatic or axodendritic synapses. Axoaxonic synapses work differently. Axoaxonic synapses do not contribute directly to neural integration. Instead, they alter the amount of neurotransmitter released by the terminal buttons of the postsynaptic axon. They can produce presynaptic modulation: presynaptic inhibition or presynaptic facilitation.

As you know, the release of a neurotransmitter by a terminal button is initiated by an action potential. Normally, a particular terminal button releases a fixed amount of neurotransmitter each time an action potential

autoreceptor A receptor molecule located on a neuron that responds to the neurotransmitter released by that neuron.

FIGURE 2.39 ■ An Axoaxonic Synapse

The activity of terminal button A can increase or decrease the amount of neurotransmitter released by terminal button B.

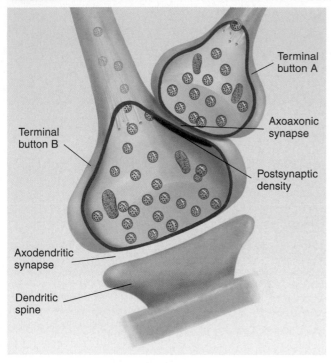

FIGURE 2.40 ■ A Gap Junction

A gap junction permits direct electrical coupling between the membranes of adjacent neurons.

(From Bennett, M. V. L., and Pappas, G. D. *The Journal of Neuroscience*, 1983, *3*, 748–761. Reprinted with permission.)

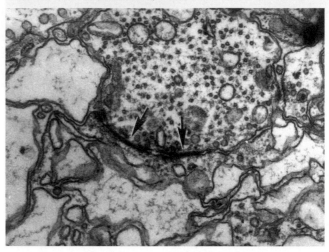

junctions in vertebrate synapses are dendrodendritic, axosomatic and axodendritic gap junctions also occur. Gap junctions are common in invertebrates; their function in the vertebrate nervous system is not known.

Nonsynaptic Chemical Communication

Neurotransmitters are released by terminal buttons of neurons and bind with receptors in the membrane of another cell located a very short distance away. The communication at each synapse is private. **Neuromodulators** are chemicals released by neurons that travel farther and are dispersed more widely than are neurotransmitters. Most neuromodulators are **peptides,** chains of amino

arrives. However, the release of neurotransmitter can be modulated by the activity of axoaxonic synapses. If the activity of the axoaxonic synapse decreases the release of the neurotransmitter, the effect is called **presynaptic inhibition.** If it increases the release, it is called **presynaptic facilitation.** (See *Figure 2.39.*)

Many very small neurons have extremely short processes and apparently lack axons. These neurons form *dendrodendritic synapses,* or synapses between dendrites. Because these neurons lack long axonal processes, they do not transmit information from place to place within the brain. Most investigators believe that they perform regulatory functions, perhaps helping to organize the activity of groups of neurons. Because these neurons are so small, they are difficult to study; therefore, little is known about their function.

Some larger neurons, as well, form dendrodendritic synapses. Some of these synapses are chemical, indicated by the presence of synaptic vesicles in one of the juxtaposed dendrites and a postsynaptic thickening in the membrane of the other. Other synapses are *electrical;* the membranes meet and almost touch, forming a **gap junction.** The membranes on both sides of a gap junction contain channels that permit ions to diffuse from one cell to another. Thus, changes in the membrane potential of one neuron induce changes in the membrane of the other. (See *Figure 2.40.*) Although most gap

presynaptic inhibition The action of a presynaptic terminal button in an axoaxonic synapse; reduces the amount of neurotransmitter released by the postsynaptic terminal button.

presynaptic facilitation The action of a presynaptic terminal button in an axoaxonic synapse; increases the amount of neurotransmitter released by the postsynaptic terminal button.

gap junction A special junction between cells that permits direct communication by means of electrical coupling.

neuromodulator A naturally secreted substance that acts like a neurotransmitter except that it is not restricted to the synaptic cleft but diffuses through the extracellular fluid.

peptide A chain of amino acids joined together by peptide bonds. Most neuromodulators, and some hormones, consist of peptide molecules.

acids that are linked together by chemical attachments called *peptide bonds* (hence their name). Neuromodulators are secreted in larger amounts and diffuse for longer distances, modulating the activity of many neurons in a particular part of the brain. For example, neuromodulators affect general behavioral states such as vigilance, fearfulness, and sensitivity to pain. Chapter 4 discusses the most important neurotransmitters and neuromodulators.

FIGURE 2.41 ■ Action of Steroid Hormones

Steroid hormones affect their target cells by means of specialized receptors in the nucleus. Once a receptor binds with a molecule of a steroid hormone, it causes genetic mechanisms to initiate protein synthesis.

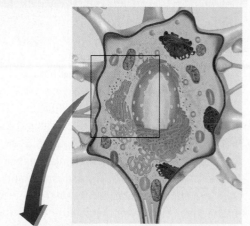

Detail of Cell

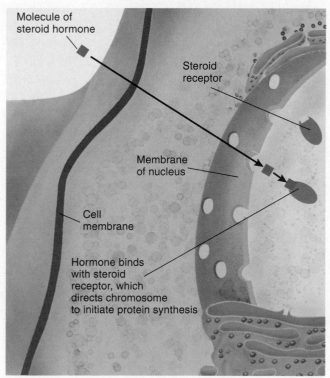

Molecule of steroid hormone

Steroid receptor

Membrane of nucleus

Cell membrane

Hormone binds with steroid receptor, which directs chromosome to initiate protein synthesis

Hormones are secreted by cells of **endocrine glands** (from the Greek *endo-*, "within," and *krinein*, "to secrete") or by cells located in various organs, such as the stomach, the intestines, the kidneys, and the brain. Cells that secrete hormones release these chemicals into the extracellular fluid. The hormones are then distributed to the rest of the body through the bloodstream. Hormones affect the activity of cells (including neurons) that contain specialized receptors located either on the surface of their membrane or deep within their nuclei. Cells that contain receptors for a particular hormone are referred to as **target cells** for that hormone; only these cells respond to its presence. Many neurons contain hormone receptors, and hormones are able to affect behavior by stimulating the receptors and changing the activity of these neurons. For example, a sex hormone, testosterone, increases the aggressiveness of most male mammals.

Peptide hormones exert their effects on target cells by stimulating metabotropic receptors located in the membrane. The second messenger that is generated travels to the nucleus of the cell, where it initiates changes in the cell's physiological processes. **Steroid** hormones consist of very small fat-soluble molecules. (*Steroid* derives from the Greek *stereos*, "solid," and Latin *oleum*, "oil." They are synthesized from chole*sterol*.) Examples of steroid hormones include the sex hormones secreted by the ovaries and testes and the hormones secreted by the adrenal cortex. Because steroid hormones are soluble in lipids, they pass easily through the cell membrane. They travel to the nucleus, where they attach themselves to receptors located there. The receptors, stimulated by the hormone, then direct the machinery of the cell to alter its protein production. (See *Figure 2.41.*)

In the past few years, investigators have discovered the presence of steroid receptors in terminal buttons and around the postsynaptic membrane of some neurons. These steroid receptors influence synaptic transmission, and they do so rapidly. Exactly how they work is still not known.

---

hormone A chemical substance that is released by an endocrine gland that has effects on target cells in other organs.

endocrine gland A gland that liberates its secretions into the extracellular fluid around capillaries and hence into the bloodstream.

target cell The type of cell that is directly affected by a hormone or other chemical signal.

steroid A chemical of low molecular weight, derived from cholesterol. Steroid hormones affect their target cells by attaching to receptors found within the nucleus.

InterimSummary

Communication Between Neurons

Synapses consist of junctions between the terminal buttons of one neuron and the membrane, another neuron, a muscle cell, or a gland cell. When an action potential is transmitted down an axon, the terminal buttons at the end release a neurotransmitter, a chemical that produces either depolarizations (EPSPs) or hyperpolarizations (IPSPs) of the postsynaptic membrane. The rate of firing of the axon of the postsynaptic neuron is determined by the relative activity of the excitatory and inhibitory synapses on the membrane of its dendrites and soma—a phenomenon known as *neural integration*.

Terminal buttons contain synaptic vesicles. Most terminal buttons contain two sizes of vesicles, the smaller of which are found in greatest numbers around the release zone of the presynaptic membrane. When an action potential is transmitted down an axon, the depolarization opens voltage-dependent calcium channels, which permit Ca^{2+} to enter. The calcium ions bind with the clusters of protein molecules in the membranes of synaptic vesicles that are docked at the release zone. The protein clusters spread apart, causing the vesicles to break open and release the neurotransmitter. Vesicles in the ready release pool briefly "kiss" the inside of the presynaptic membrane, release their contents, and then break away to be refilled. Those in the recycling pool and reserve pool completely fuse with the presynaptic membrane and lose their identities. The membrane contributed by these vesicles pinches off into the cytoplasm and is recycled in the production of new vesicles.

The activation of postsynaptic receptors by molecules of a neurotransmitter causes neurotransmitter-dependent ion channels to open, resulting in postsynaptic potentials. Ionotropic receptors contain ion channels, which are directly opened when a ligand attaches to the binding site. Metabotropic receptors are linked to G proteins, which, when activated, open ion channels—usually by producing a chemical called a second messenger.

The nature of the postsynaptic potential depends on the type of ion channel that is opened by the postsynaptic receptors at a particular synapse. Excitatory postsynaptic potentials occur when Na^+ enters the cell. Inhibitory postsynaptic potentials are produced when K^+ leaves the cell or Cl^- enters it. The entry of Ca^{2+} produces EPSPs, but even more important, it activates special enzymes that cause physiological changes in the postsynaptic cell that are involved in learning.

Postsynaptic potentials are normally very brief. They are terminated by two means. Acetylcholine is deactivated by the enzyme acetylcholinesterase. In all other cases (as far as we know), molecules of the neurotransmitter are removed from the synaptic cleft by means of transporters located in the presynaptic membrane. This retrieval process is called reuptake.

The presynaptic membrane, as well as the postsynaptic membrane, contains receptors that detect the presence of a neurotransmitter. Presynaptic receptors, also called autoreceptors, monitor the quantity of neurotransmitter that a neuron releases and, apparently, regulate the amount that is synthesized and released.

Axosomatic and axodendritic synapses are not the only kinds found in the nervous system. Axoaxonic synapses either reduce or enhance the amount of neurotransmitter released by the postsynaptic terminal button, producing presynaptic inhibition or presynaptic facilitation. Dendrodendritic synapses also exist, but their role in neural communication is not yet understood.

Nonsynaptic chemical transmission is similar to synaptic transmission. Peptide neuromodulators and hormones activate metabotropic peptide receptors located in the membrane; their effects are mediated through the production of second messengers. Steroid hormones enter the nucleus, where they bind with receptors that are capable of altering the synthesis of proteins that regulate the cell's physiological processes. These hormones also bind with receptors located elsewhere in the cell, but less is known about their functions.

Thought Questions

1. Why does synaptic transmission involve the release of chemicals? Direct electrical coupling of neurons is far simpler, so why do our neurons not use it more extensively? (A tiny percentage of synaptic connections in the human brain do use electrical coupling.) Normally, nature uses the simplest means possible to a given end, so there must be some advantages to chemical transmission. What do you think they are?
2. Consider the control of the withdrawal reflex illustrated in Figure 2.14. Could you design a circuit using electrical synapses that would accomplish the same tasks?

SUGGESTED READINGS

Aidley, D. J. *The Physiology of Excitable Cells*, 4th ed. Cambridge, England: Cambridge University Press, 1998.

Bean, B. P. The action potential in mammalian central neurons. *Nature Reviews: Neuroscience*, 2007, *8*, 451–465.

Cowan, W. M., Südhof, T. C., and Stevens, C. F. *Synapses*. Baltimore, MD: Johns Hopkins University Press, 2001.

Kandel, E. R., Schwartz, J. H., and Jessell, T. M. *Principles of Neural Science*, 4th ed. New York: McGraw-Hill, 2000.

Nicholls, J. G., Martin, A. R., Fuchs, P. A., and Wallace, B. G. *From Neuron to Brain*, 4th ed. Sunderland, MA: Sinauer, 2001.

ADDITIONAL RESOURCES

Visit www.mypsychkit.com for additional review and practice of the material covered in this chapter. Within MyPsychKit, you can take practice tests and receive a customized study plan to help you review. Dozens of animations, tutorials, and Web links are also available. You can even review using the interactive electronic version of this textbook. You will need to register for MyPsychKit. See www.mypsychkit.com for complete details.

chapter

3

Structure of the Nervous System

Ryan B., a college freshman, had suffered from occasional epileptic seizures since childhood. He had been taking drugs for his seizures for many years, but lately the medication wasn't helping—his seizures were becoming more frequent. His neurologist increased the dose of the medication, but the seizures persisted, and the drug made it difficult for Ryan to concentrate on his studies. He was afraid that he would have to drop out of school.

He made an appointment with his neurologist and asked whether another drug was available that might work better and not affect his ability to concentrate. "No," said the neurologist, "you're taking the best medication we have right now. But I want to send you to Dr. L., a neurosurgeon at the medical school. I think you might be a good candidate for seizure surgery."

Ryan had a focal-seizure disorder. His problems were caused by a localized region of the brain that contained some scar tissue. Periodically, this region would irritate the surrounding areas, triggering epileptic seizures—wild, sustained firing of cerebral neurons that result in cognitive disruption and, sometimes, uncontrolled movements. Ryan's focus was probably a result of brain damage that occurred when he was born. Dr. L. ordered some tests that indicated that the seizure focus was located in the left side of his brain, in a region known as the medial temporal lobe.

Ryan was surprised to learn that he would remain awake during his surgery. In fact, he would be called on to provide information that the surgeon would need to remove a region of his brain that included the seizure focus. As you might expect, he was nervous when he was wheeled into the surgery, but after the anesthesiologist injected something through the tube in one of his veins, Ryan relaxed and thought to himself, "This won't be too bad."

Dr. L. marked something on his scalp, which had previously been shaved, and then made several injections of a local anesthetic. Then he cut the scalp and injected some more anesthetic. Finally, he used a drill and a saw to remove a piece of skull. He then cut and folded back the thick membrane that covers the brain, exposing the surface of the brain.

When removing a seizure focus, the surgeon wants to cut away all the abnormal tissue while sparing brain tissue that performs important functions, such as the comprehension and production of speech. For this reason, Dr. L. began stimulating parts of the brain to determine which regions he could safely remove. To do so, he placed a metal probe against the surface of Ryan's brain and pressed a pedal that delivered a weak electrical current. The stimulation disrupts the firing patterns of the neurons located near the probe, preventing them from carrying out their normal functions. Dr. L. found that stimulation of parts of the temporal lobe disrupted Ryan's ability to understand what he and his associates were saying. When he removed the part of the brain containing the seizure focus, he was careful not to damage these regions.

The operation was successful. Ryan continued to take his medication but at a much lower dose. His seizures disappeared, and he found it easier to concentrate in class. I met Ryan during his junior year, when he took a course I was teaching. I described seizure surgery to the class one day, and after the lecture he approached me and told me about his experience. He received the third highest grade in the class.

The goal of neuroscience research is to understand how the brain works. To understand the results of this research, you must be acquainted with the basic structure of the nervous system. The number of terms introduced in this chapter is kept to a minimum (but as you will see, the minimum is still a rather large number). The MyPsychKit "Figures and Diagrams" exercise for this chapter will help you learn the names and locations of the major structures of the nervous system. (See *MyPsychKit: Figures and Diagrams.*) With the framework you will receive from this chapter and from the MyPsychKit figures and animations, you should have no trouble learning the material presented in subsequent chapters.

Figures and Diagrams

BASIC FEATURES OF THE NERVOUS SYSTEM

Before beginning a description of the nervous system, I want to discuss the terms that are used to describe it. The gross anatomy of the brain was described long ago, and everything that could be seen without the aid of a microscope was given a name. Early anatomists named most brain structures according to their similarity to commonplace objects: amygdala, or "almond-shaped object"; hippocampus, or "sea horse"; genu, or "knee"; cortex, or "bark"; pons, or "bridge"; uncus, or "hook," to give a few examples. Throughout this book I will translate the names of anatomical terms as I introduce them, because the translation makes the terms more memorable. For example, knowing that *cortex* means "bark" (like the bark of a tree) will help you to remember that the cortex is the outer layer of the brain.

FIGURE 3.1 ■ Views of Alligator and Human

These side and frontal views show the terms used to denote anatomical directions.

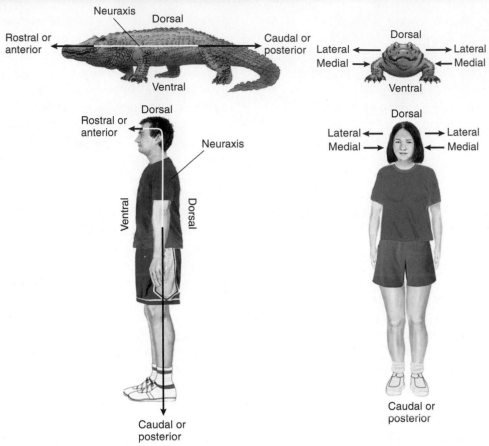

When describing features of a structure as complex as the brain, we need to use terms denoting directions. Directions in the nervous system are normally described relative to the **neuraxis,** an imaginary line drawn through the spinal cord up to the front of the brain. For simplicity's sake, let us consider an animal with a straight neuraxis. Figure 3.1 shows an alligator and two humans. This alligator is certainly laid out in a linear fashion; we can draw a straight line that starts between its eyes and continues down the center of its spinal cord. (See *Figure 3.1.*) The front end is **anterior,** and the tail is **posterior.** The terms **rostral** (toward the beak) and **caudal** (toward the tail) are also employed, especially when referring specifically to the brain. The top of the head and the back are part of the **dorsal** surface, while the **ventral** (front) surface faces the ground. (*Dorsum* means "back," and *ventrum* means "belly.") These directions are somewhat more complicated in the human; because we stand upright, our neuraxis bends, so the top of the head is perpendicular to the back. (You will also encounter the terms *superior* and *inferior.* In referring to the brain, *superior* means "above," and *inferior* means "below." For example, the *superior colliculi* are located above the *inferior colliculi.*) The frontal views of

both the alligator and the human illustrate the terms **lateral** and **medial:** toward the side and toward the midline, respectively. (See *Figure 3.1.*)

---

neuraxis An imaginary line drawn through the center of the length of the central nervous system, from the bottom of the spinal cord to the front of the forebrain.

anterior With respect to the central nervous system, located near or toward the head.

posterior With respect to the central nervous system, located near or toward the tail.

rostral "Toward the beak"; with respect to the central nervous system, in a direction along the neuraxis toward the front of the face.

caudal "Toward the tail"; with respect to the central nervous system, in a direction along the neuraxis away from the front of the face.

dorsal "Toward the back"; with respect to the central nervous system, in a direction perpendicular to the neuraxis toward the top of the head or the back.

ventral "Toward the belly"; with respect to the central nervous system, in a direction perpendicular to the neuraxis toward the bottom of the skull or the front surface of the body.

lateral Toward the side of the body, away from the middle.

medial Toward the middle of the body, away from the side.

FIGURE 3.2 ■ Brain Slices and Planes

Planes of section are shown as they pertain to the human central nervous system.

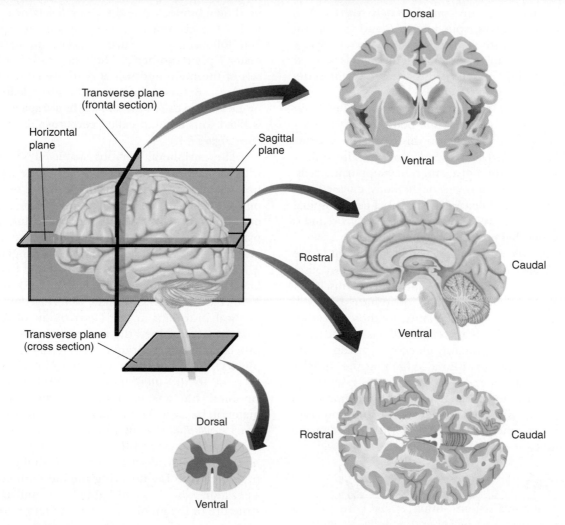

Two other useful terms are *ipsilateral* and *contralateral*. **Ipsilateral** refers to structures on the same side of the body. If we say that the olfactory bulb sends axons to the *ipsilateral* hemisphere, we mean that the left olfactory bulb sends axons to the left hemisphere and the right olfactory bulb sends axons to the right hemisphere. **Contralateral** refers to structures on opposite sides of the body. If we say that a particular region of the left cerebral cortex controls movements of the *contralateral* hand, we mean that the region controls movements of the right hand.

To see what is in the nervous system, we have to cut it open; to be able to convey information about what we find, we slice it in a standard way. Figure 3.2 shows a human nervous system. We can slice the nervous system in three ways:

1. Transversely, like a salami, giving us **cross sections** (also known as **frontal sections** when referring to the brain)
2. Parallel to the ground, giving us **horizontal sections**

3. Perpendicular to the ground and parallel to the neuraxis, giving us **sagittal sections.** The **midsagittal plane** divides the brain into two symmetrical halves. The sagittal section in Figure 3.2 lies in the midsagittal plane.

Note that because of our upright posture, cross sections of the spinal cord are parallel to the ground. (See *Figure 3.2.*)

ipsilateral Located on the same side of the body.

contralateral Located on the opposite side of the body.

cross section With respect to the central nervous system, a slice taken at right angles to the neuraxis.

frontal section A slice through the brain parallel to the forehead.

horizontal section A slice through the brain parallel to the ground.

sagittal section (*sadj i tul*) A slice through the brain parallel to the neuraxis and perpendicular to the ground.

midsagittal plane The plane through the neuraxis perpendicular to the ground; divides the brain into two symmetrical halves.

An Overview

The nervous system consists of the brain and spinal cord, which make up the *central nervous system (CNS)*, and the cranial nerves, spinal nerves, and peripheral ganglia, which constitute the *peripheral nervous system (PNS)*. The CNS is encased in bone: The brain is covered by the skull, and the spinal cord is encased by the vertebral column. (See *Table 3.1*.)

Figure 3.3 illustrates the relationship of the brain and spinal cord to the rest of the body. Do not be concerned with unfamiliar labels on this figure; these structures will be described later. (See *Figure 3.3*.) The brain is a large mass of neurons, glia, and other supporting cells. It is the most protected organ of the body, encased in a tough, bony skull and floating in a pool of cerebrospinal fluid. The brain receives a copious supply of blood and is chemically guarded by the blood–brain barrier.

The brain receives approximately 20 percent of the blood flow from the heart, and it receives it continuously. Other parts of the body, such as the skeletal muscles or digestive system, receive varying quantities of blood, depending on their needs, relative to those of other regions. But the brain always receives its share. The brain cannot store its fuel (primarily glucose), nor can it temporarily extract energy without oxygen, as the muscles can; therefore, a consistent blood supply is essential. A 1-second interruption of the blood flow to the brain uses up much of the dissolved oxygen; a 6-second interruption produces unconsciousness. Permanent damage begins within a few minutes.

Meninges

The entire nervous system—brain, spinal cord, cranial and spinal nerves, and peripheral ganglia—is covered by tough connective tissue. The protective sheaths around the brain and spinal cord are referred to as the **meninges** (singular: *meninx*). The meninges consist of three layers, which are shown in Figure 3.3. The outer layer is thick, tough, and flexible but unstretchable; its name, **dura mater,** means "hard mother." The middle layer of the meninges, the **arachnoid membrane,** gets its name from the weblike appearance of the *arachnoid trabeculae* that protrude from it (from the Greek *arachne*, meaning "spider"; *trabecula* means "track"). The arachnoid membrane, soft and spongy, lies beneath the dura mater. Closely attached to the brain and spinal cord, and following every surface convolution, is the **pia mater** ("pious mother"). The smaller surface blood vessels of the brain and spinal cord are contained within this layer. Between the pia mater and arachnoid membrane is a gap called the **subarachnoid space.** This space is filled with a liquid called **cerebrospinal fluid (CSF).** (See *Figure 3.3*.)

The peripheral nervous system (PNS) is covered with two layers of meninges. The middle layer (arachnoid membrane), with its associated pool of CSF, covers only the brain and spinal cord. Outside the central nervous system, the outer and inner layers (dura mater and pia mater) fuse and form a sheath that covers the spinal and cranial nerves and the peripheral ganglia.

In the first edition of this book I said that I did not know why the outer and inner layers of the meninges were referred to as "mothers." I received a letter from medical historians at the Department of Anatomy at UCLA that explained the name. (Sometimes, it pays to proclaim one's ignorance.) A tenth-century Persian physician, Ali ibn Abbas, used the Arabic term *al umm* to refer to the meninges. The term literally means "mother" but was used to designate any swaddling material, because Arabic lacked a specific term for the word *membrane*. The tough outer membrane was called *al umm al djafiya,* and the soft inner one was called *al umm al rigiga.* When the writings of Ali ibn Abbas were translated into Latin during the eleventh century, the translator, who was probably not familiar with the structure of the meninges, made a literal translation of *al umm.* He referred to the membranes as the "hard mother" and the "pious mother" (*pious* in the sense of "delicate") rather than using a more appropriate Latin word.

TABLE 3.1 ■ The Major Divisions of the Nervous System	
CENTRAL NERVOUS SYSTEM (CNS)	**PERIPHERAL NERVOUS SYSTEM (PNS)**
Brain	Nerves
Spinal cord	Peripheral ganglia

---

meninges (*men in jees*) (**singular: meninx**) The three layers of tissue that encase the central nervous system: the dura mater, arachnoid membrane, and pia mater.

dura mater The outermost of the meninges; tough and flexible.

arachnoid membrane (*a rak noyd*) The middle layer of the meninges, located between the outer dura mater and inner pia mater.

pia mater The layer of the meninges that clings to the surface of the brain; thin and delicate.

subarachnoid space The fluid-filled space that cushions the brain; located between the arachnoid membrane and the pia mater.

cerebrospinal fluid (CSF) A clear fluid, similar to blood plasma, that fills the ventricular system of the brain and the subarachnoid space surrounding the brain and spinal cord.

FIGURE 3.3 ■ The Nervous System

The figure shows (a) the relationship of the nervous system to the rest of the body, (b) detail of the meninges that cover the central nervous system, and (c) a closer view of the lower spinal cord and cauda equina.

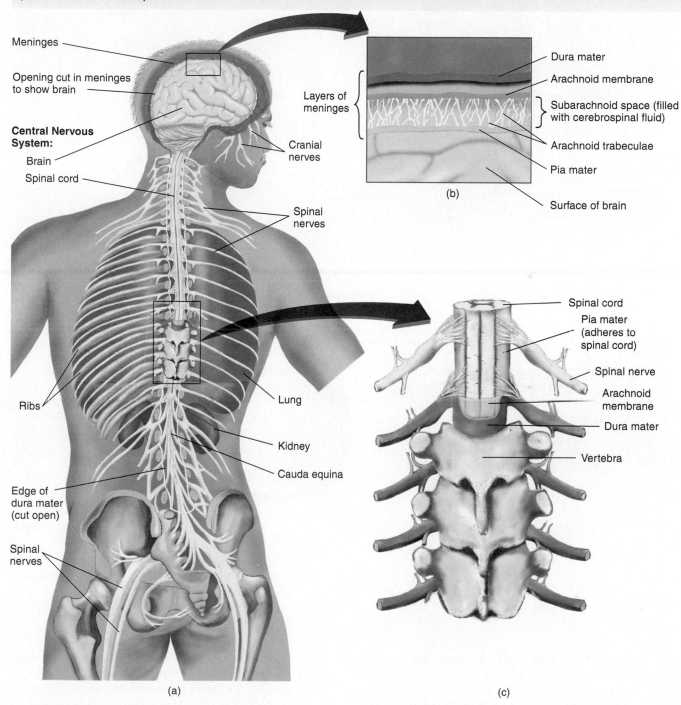

The Ventricular System and Production of CSF

The brain is very soft and jellylike. The considerable weight of a human brain (approximately 1400 g), along with its delicate construction, necessitates that it be

protected from shock. A human brain cannot even support its own weight well; it is difficult to remove and handle a fresh brain from a recently deceased human without damaging it.

Fortunately, the intact brain within a living human is well protected. It floats in a bath of CSF contained

FIGURE 3.4 ■ Ventricular System of the Brain

The figure shows (a) a lateral view of the left side of the brain, (b) a frontal view, (c) a dorsal view, and (d) the production, circulation, and reabsorption of cerebrospinal fluid.

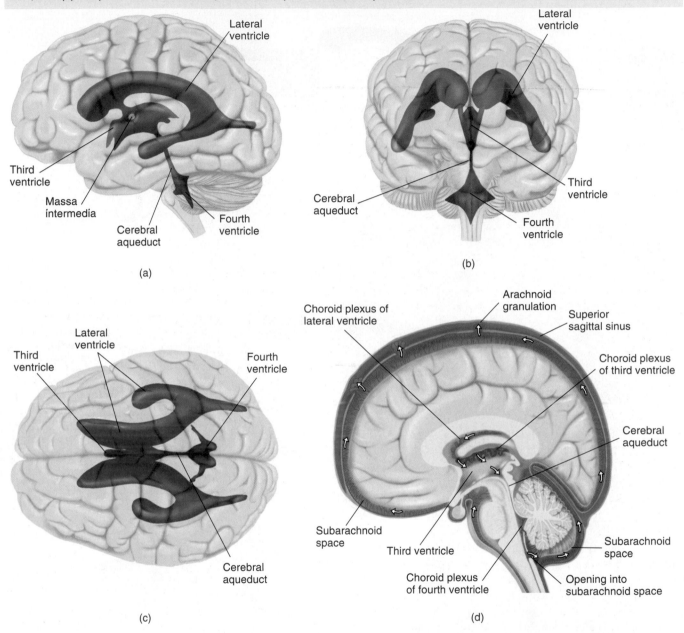

within the subarachnoid space. Because the brain is completely immersed in liquid, its net weight is reduced to approximately 80 g; thus, pressure on the base of the brain is considerably diminished. The CSF surrounding the brain and spinal cord also reduces the shock to the central nervous system that would be caused by sudden head movement.

The brain contains a series of hollow, interconnected chambers called **ventricles** ("little bellies"), which are filled with CSF. (See *Figure 3.4.*) The largest chambers are the **lateral ventricles,** which are connected to the **third ventricle.** The third ventricle is located at the midline of the brain; its walls divide the surrounding

ventricle (*ven trik ul*) One of the hollow spaces within the brain, filled with cerebrospinal fluid.

lateral ventricle One of the two ventricles located in the center of the telencephalon.

third ventricle The ventricle located in the center of the diencephalon.

part of the brain into symmetrical halves. A bridge of neural tissue called the *massa intermedia* crosses through the middle of the third ventricle and serves as a convenient reference point. The **cerebral aqueduct,** a long tube, connects the third ventricle to the **fourth ventricle.** The lateral ventricles constitute the first and second ventricles, but they are never referred to as such. (See *Figure 3.4.*)

Cerebrospinal fluid is extracted from the blood and resembles blood plasma in its composition. CSF is manufactured by special tissue with an especially rich blood supply called the **choroid plexus,** which protrudes into all four of the ventricles. CSF is produced continuously; the total volume of CSF is approximately 125 ml, and the half-life (the time it takes for half of the CSF present in the ventricular system to be replaced by fresh fluid) is about 3 hours. Therefore, several times this amount is produced by the choroid plexus each day. The continuous production of CSF means that there must be a mechanism for its removal. The production, circulation, and reabsorption of CSF are illustrated in *Figure 3.4d.* A scanning electron micrograph of the choroid plexus is shown in *Figure 3.5.*

Figure 3.4(d) shows a slightly rotated midsagittal view of the central nervous system, which shows only the right lateral ventricle (because the left hemisphere has been removed). Cerebrospinal fluid is produced by the choroid plexus of the lateral ventricles, and it flows into the third ventricle. More CSF is produced in this ventricle, which then flows through the cerebral aqueduct to the fourth ventricle, where still more CSF is produced. The CSF leaves the fourth ventricle through small openings that connect with the subarachnoid space surrounding the brain. The CSF then flows through the subarachnoid space around the central nervous system, where it is reabsorbed into the blood supply through the **arachnoid granulations.** These pouch-shaped structures protrude into the **superior sagittal sinus,** a blood vessel that drains into the veins serving the brain. (See *Figure 3.4d* and *MyPsychKit 3.1, Meninges and CSF.*)

mypsychkit

Animation 3.1
Meninges and CSF

Occasionally, the flow of CSF is interrupted at some point in its route of passage. For example, a brain tumor growing in the midbrain may push against the cerebral aqueduct, blocking its flow, or an infant may be born with a cerebral aqueduct that is too small to accommodate a normal flow of CSF. This occlusion results in greatly increased pressure within the ventricles, because the choroid plexus continues to produce CSF. The walls of the ventricles then expand and produce a condition known as **obstructive hydrocephalus** (*hydrocephalus* literally means "water-head"). If the obstruction remains and if nothing is done to reverse the increased

FIGURE 3.5 ■ Micrograph of the Choroid Plexus

On the scanning electron micrograph, BV = blood vessel, CE = choroid plexus, V = ventricle.

(From *Tissues and Organs: A Text-Atlas of Scanning Electron Microscopy,* by Richard G. Kessel and Randy H. Kardon. Copyright © 1979 by W. H. Freeman and Co. Reprinted by permission of Barbara Kessel and Randy Kardon.)

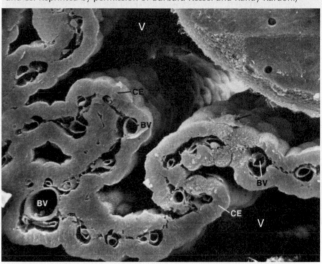

intracerebral pressure, blood vessels will be occluded, and permanent—perhaps fatal—brain damage will occur. Fortunately, a surgeon can usually operate on the person, drilling a hole through the skull and inserting a shunt tube into one of the ventricles. The tube is then placed beneath the skin and connected to a pressure relief valve that is implanted in the abdominal cavity. When the pressure in the ventricles becomes excessive, the valve permits the CSF to escape into the abdomen, where it is eventually reabsorbed into the blood supply. (See *Figure 3.6.*)

cerebral aqueduct A narrow tube interconnecting the third and fourth ventricles of the brain, located in the center of the mesencephalon.

fourth ventricle The ventricle located between the cerebellum and the dorsal pons, in the center of the metencephalon.

choroid plexus The highly vascular tissue that protrudes into the ventricles and produces cerebrospinal fluid.

arachnoid granulation Small projections of the arachnoid membrane through the dura mater into the superior sagittal sinus; CSF flows through them to be reabsorbed into the blood supply.

superior sagittal sinus A venous sinus located in the midline just dorsal to the corpus callosum, between the two cerebral hemispheres.

obstructive hydrocephalus A condition in which all or some of the brain's ventricles are enlarged; caused by an obstruction that impedes the normal flow of CSF.

InterimSummary

Basic Features of the Nervous System

Anatomists have adopted a set of terms to describe the locations of parts of the body. *Anterior* is toward the head, *posterior* is toward the tail, *lateral* is toward the side, *medial* is toward the middle, *dorsal* is toward the back, and *ventral* is toward the front surface of the body. In the special case of the nervous system, *rostral* means toward the beak (or nose), and *caudal* means toward the tail. *Ipsilateral* means "same side," and *contralateral* means "other side." A cross section (or, in the case of the brain, a frontal section) slices the nervous system at right angles to the neuraxis, a horizontal section slices the brain parallel to the ground, and a sagittal section slices it perpendicular to the ground, parallel to the neuraxis.

The central nervous system consists of the brain and spinal cord, and the peripheral nervous system consists of the spinal and cranial nerves and peripheral ganglia. The CNS is covered with the meninges: dura mater, arachnoid membrane, and pia mater. The space under the arachnoid membrane is filled with cerebrospinal fluid, in which the brain floats. The PNS is covered with only the dura mater and pia mater. Cerebrospinal fluid is produced in the choroid plexus of the lateral, third, and fourth ventricles. It flows from the two lateral ventricles into the third ventricle, through the cerebral aqueduct into the fourth ventricle, then into the subarachnoid space, and finally back into the blood supply through the arachnoid granulations. If the flow of CSF is blocked by a tumor or other obstruction, the result is hydrocephalus: enlargement of the ventricles and subsequent brain damage.

FIGURE 3.6 ■ Hydrocephalus in an Infant

A surgeon places a shunt tube in a lateral ventricle, which permits cerebrospinal fluid to escape to the abdominal cavity, where it is absorbed into the blood supply. A pressure valve regulates the flow of CSF through the shunt.

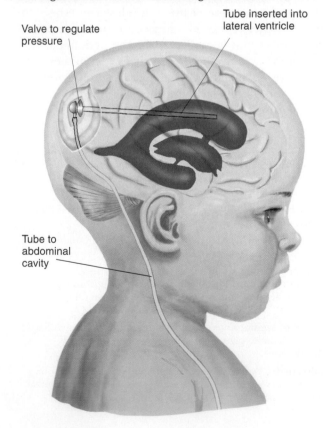

Valve to regulate pressure

Tube inserted into lateral ventricle

Tube to abdominal cavity

THE CENTRAL NERVOUS SYSTEM

Although the brain is exceedingly complicated, an understanding of the basic features of brain development makes it easier to learn and remember the location of the most important structures. With that end in mind, I introduce these features here in the context of development of the central nervous system. Two animations will help you to learn and remember the structure of the brain. *MyPsychKit 3.2, The Rotatable Brain,* is just what the title implies: a drawing of the human brain that you can rotate in three dimensions. You can choose whether to see some internal structures or see specialized regions of the cerebral cortex. *MyPsychKit 3.3, Brain Slices,* is even more comprehensive. It consists of two sets of photographs of human brain slices, taken in the transverse (frontal) and horizontal planes. As you move the cursor across each slice, brain regions are outlined, and their names appear. If you want to know how to pronounce these names, you can click on the region. You can also see magnified views of the slices and move them around by clicking and dragging. Finally, you can test yourself: The computer will present names of the regions shown in each slice, and you try to click on the correct region.

mypsychkit
Animation 3.2
The Rotatable Brain

mypsychkit
Animation 3.3
Brain Slices

Development of the Central Nervous System

The central nervous system begins early in embryonic life as a hollow tube and maintains this basic shape even after it is fully developed. During development, parts of the tube elongate, pockets and folds form, and the tissue around the tube thickens until the brain reaches its final form.

An Overview of Brain Development

Development of the nervous system begins around the eighteenth day after conception. Part of the *ectoderm* (outer layer) of the back of the embryo thickens and forms a plate. The edges of this plate form ridges that curl toward each other along a longitudinal line, running in a rostral–caudal direction. By the twenty-first day these ridges touch each other and fuse together, forming a tube—the **neural tube**—which gives rise to the brain and spinal cord. The top part of the ridges break away from the neural tube and become the ganglia of the autonomic nervous system, described later in this chapter. (See *Figure 3.7.*)

By the twenty-eighth day of development the neural tube is closed, and its rostral end has developed three interconnected chambers. These chambers become ventricles, and the tissue that surrounds them becomes the three major parts of the brain: the forebrain, the midbrain, and the hindbrain. (See *Figures 3.8a* and *3.8c.*) As development progresses, the rostral chamber (the forebrain) divides into three separate parts, which become the two lateral ventricles and the third ventricle. The region around the lateral ventricles becomes the telencephalon ("end brain"), and the region around the third ventricle becomes the diencephalon ("interbrain"). (See *Figures 3.8b* and *3.8d.*) In its final form, the chamber inside the midbrain (mesencephalon) becomes narrow, forming the cerebral aqueduct, and two structures develop in the hindbrain: the metencephalon ("afterbrain") and the myelencephalon ("marrowbrain"). (See *Figure 3.8e.*)

Table 3.2 summarizes the terms I have introduced here and mentions some of the major structures found in each part of the brain. The colors in the table match those in Figure 3.8. These structures will be described in the remainder of the chapter. (See *Table 3.2.*)

Details of Brain Development

Brain development begins with a thin tube and ends with a structure weighing approximately 1400 g (about 3 lb) and consisting of several hundreds of billions of cells. Where do these cells come from, and what controls their growth?

Let's consider the development of the cerebral cortex, about which most is known. The principles

FIGURE 3.7 ■ Neural Plate Development

The development of the neural plate into the neural tube gives rise to the brain and spinal cord. *Left:* Dorsal views. *Right:* Cross section at levels indicated by dashed lines.

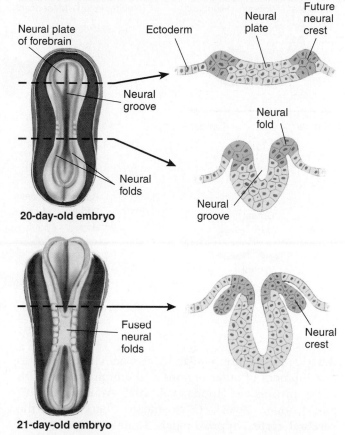

20-day-old embryo

21-day-old embryo

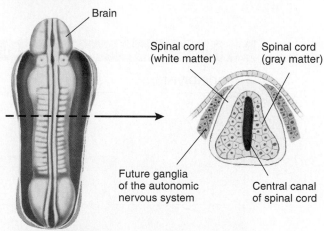

24-day-old embryo

neural tube A hollow tube, closed at the rostral end, that forms from ectodermal tissue early in embryonic development; serves as the origin of the central nervous system.

FIGURE 3.8 ■ Brain Development

This schematic outline of brain development shows its relationship to the ventricles. (a) and (c) show early development. (b) and (d) show later development. (e) A lateral view of the left side of a semitransparent human brain shows the brain stem "ghosted in." The colors of all figures denote corresponding regions.

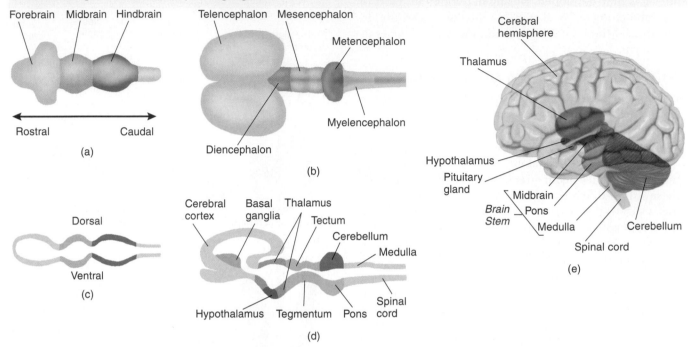

described here are similar to the ones that apply to development of other regions of the brain. (For details of this process, see Honda et al., 2003; Ayala et al., 2007; and Cooper, 2008). *Cortex* means "bark," and the **cerebral cortex**, approximately 3 mm thick, surrounds the cerebral hemispheres like the bark of a tree. Corrected for body size, the cerebral cortex is larger in humans than in any other species. As we will see later in

cerebral cortex The outermost layer of gray matter of the cerebral hemispheres.

TABLE 3.2 ■ Anatomical Subdivisions of the Brain

MAJOR DIVISION	VENTRICLE	SUBDIVISION	PRINCIPAL STRUCTURES
Forebrain	Lateral	Telencephalon	Cerebral cortex
			Basal ganglia
			Limbic system
	Third	Diencephalon	Thalamus
			Hypothalamus
Midbrain	Cerebral aqueduct	Mesencephalon	Tectum Tegmentum
Hindbrain	Fourth	Metencephalon	Cerebellum
			Pons
		Myelencephalon	Medulla oblongata

this book, circuits of neurons in the cerebral cortex play a vital role in cognition and control of movement.

The cells that line the inside of the neural tube—the **ventricular zone**—give rise to the cells of the central nervous system. The cerebral cortex develops from the inside out. That is, the first cells to be produced by the ventricular zone migrate a short distance and establish the first—and deepest—layer. The next wave of newborn cells pass through the first layer and form the second one—and so on, until all six layers of the cerebral cortex are laid down. The last cells to be produced must pass through all the ones born before them.

The cells in the ventricular zone that give rise to the cells of the brain are known as **progenitor cells.** (A *progenitor* is a direct ancestor of a line of descendants.) During the first phase of development, progenitor cells divide, making new progenitor cells and increasing the size of the ventricular zone. This phase is referred to as **symmetrical division,** because the division of each progenitor cell produces two new progenitor cells. This form of division increases the size of the ventricular zone. Then, seven weeks after conception, progenitor cells receive a signal to begin a period of **asymmetrical division.** During this phase, progenitor cells form two different kinds of cells as they divide: another progenitor cell and a brain cell.

The first brain cells produced through asymmetrical division are **radial glia.** The cell bodies of radial glia remain in the ventricular zone, but they extend fibers radially outward from the ventricular zone, like spokes in a wheel. These fibers end in cuplike feet that attach to the pia mater, located at the outer surface of what becomes the cerebral cortex. As the cortex becomes thicker, the fibers of the radial glia grow longer and maintain their connections with the pia mater.

The next set of brain cells produced by asymmetrical division are special neurons known as **Cajal-Retzius (C-R) cells.** C-R cells establish themselves in a layer near the terminals of the radial glia, just inside the pia mater. A second set of neurons is produced that forms a layer just beneath the C-R cells. These neurons constitute the first, innermost of the six layer of the cerebral cortex. As *neurogenesis* (production of new neurons) continues, the cells leave the ventricular zone, pass the first layer of neurons, and establish themselves just inside the layer of C-R cells. Each successive wave of newborn neurons travels past the neurons that were born previously and establishes the next cortical layer. Newborn neurons are guided in their travel by the fibers of radial glial cells. Neurons crawl along radial fibers like amoebas, pushing their way through neurons that were born earlier and finally coming to rest against the layer of C-R cells. Chemicals secreted by these cells causes the neurons to detach from the radial glia fibers and establish themselves in the outmost layer of the cortex. (See *Figures 3.9* and *3.10.*)

FIGURE 3.9 ■ Nervous System Development

This cross section through the nervous system shows it early in its development. Radially oriented glial cells help to guide the migration of newly formed neurons.

(Adapted from Rakic, P. *Trends in Neuroscience*, 1995, *18*, 383–388.)

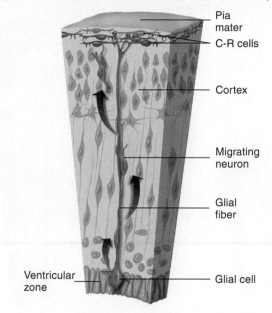

The period of asymmetrical division lasts about three months. Because the human cerebral cortex contains about 100 billion neurons, there are about one billion neurons migrating along radial glial fibers on a given day. The migration path of the earliest neurons is the shortest and takes about one day. The neurons that produce the last, outermost layer have to pass through five layers of neurons, and their migration takes about two weeks. The end of cortical development occurs

ventricular zone A layer of cells that line the inside of the neural tube; contains progenitor cells that divide and give rise to cells of the central nervous system.

progenitor cells Cells of the ventricular zone that divide and give rise to cells of the central nervous system.

symmetrical division Division of a progenitor cell that gives rise to two identical progenitor cells; increases the size of the ventricular zone and hence the brain that develops from it.

asymmetrical division Division of a progenitor cell that gives rise to another progenitor cell and a neuron, which migrates away from the ventricular zone toward its final resting place in the brain.

radial glia Special glia with fibers that grow radially outward from the ventricular zone to the surface of the cortex; provide guidance for neurons migrating outward during brain development.

Cajal-Retzius (C-R) cells Specialized neurons that establish themselves during cortical development in a layer near the terminals of the radial glia, just inside the pia mater; secrete a chemical that controls the establishment of migrating neurons in the layers of the cortex.

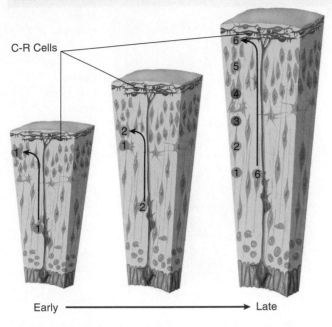

FIGURE 3.10 ■ **Migration of Neurons**

Newly formed neurons migrate from the ventricular zone to their final resting place in the cerebral cortex. Each successive wave of neurons passes neurons that migrated earlier, so the most recently formed neurons occupy layers closer to the cortical surface.

C-R Cells

Early ⟶ Late

when the progenitor cells receive a chemical signal that causes them to die—a phenomenon known as **apoptosis** (literally, a "falling away"). Molecules of the chemical that conveys this signal bind with receptors that activate killer genes within the cells. (All cells have these genes, but only certain cells possess the receptors that respond to the chemical signals that turn them on.) At this time, some radial glia appear to undergo apoptosis, but many are transformed into astrocytes or neurons.

Once neurons have migrated to their final locations, they begin forming connections with other neurons. They grow dendrites, which receive the terminal buttons from the axons of other neurons, and they grow axons of their own. Some neurons extend their dendrites and axons laterally, connecting adjacent columns of neurons or even establishing connections with other neurons in distant regions of the brain. The growth of axons is guided by physical and chemical factors. Once the growing ends of the axons (the *growth cones*) reach their targets, they form numerous branches. Each of these branches finds a vacant place on the membrane of the appropriate type of postsynaptic cell, grows a terminal button, and establishes a synaptic connection. Apparently, different types of cells—or even different parts of a single cell—secrete different chemicals, which attract different types of axons (Benson, Colman, and Huntley, 2001). Of course, the establishment of a synaptic connection also

requires efforts on the part of the postsynaptic cell; this cell must contribute its parts of the synapse, including the postsynaptic receptors. The chemical signals that the cells exchange to tell one another to establish these connections are just now being discovered.

The ventricular zone gives rise to more neurons than are needed. In fact, these neurons must compete to survive. The axons of approximately 50 percent of these neurons do not find vacant postsynaptic cells of the right type with which to form synaptic connections, so they die by apoptosis. This phenomenon, too, involves a chemical signal; when a presynaptic neuron establishes synaptic connections, it receives a signal from the postsynaptic cell that permits it to survive. The neurons that come too late do not find any available space and therefore do not receive this life-sustaining signal. This scheme might seem wasteful, but apparently the evolutionary process found that the safest strategy was to produce too many neurons and let them fight to establish synaptic connections rather than trying to produce exactly the right number of each type of neuron.

As we will see later in this chapter, different regions of the cerebral cortex perform specialized functions. Some receive and analyze visual information, some receive and analyze auditory information, some control movement of the muscles, and so on. Thus, different regions receive different inputs, contain different types of circuits of neurons, and have different outputs. What factors control this pattern of development?

Some of the specialization is undoubtedly programmed genetically. The neurons produced by the asymmetrical division of a particular progenitor cell all follow a particular radial glial fiber, so they end up somewhere in a single column extending outward from the ventricular zone. Thus, if the progenitor cells in different regions of the ventricular zone are themselves different, the neurons they produce will reflect these differences.

Experiments suggest that the specialization of a particular region of the cerebral cortex can also be induced by the axons that provide input to that region. For example, Krubitzer and her colleagues (see Krubitzer, 1998) removed some of the cerebral cortex of an opossum early in development, before the cortex had received its input from the thalamus. (As we will see later in this chapter, the *thalamus* is a structure located in the depths of the brain. Particular groups of neurons in the thalamus send axons to particular regions of the cerebral cortex and provide information from the sense organs.) The investigators used opossums because they are born during an early stage of brain development.

---

apoptosis (*ay po toe sis*) Death of a cell caused by a chemical signal that activates a genetic mechanism inside the cell.

FIGURE 3.11 ■ Cerebral Cortex of the Opossum

The visual, auditory, and somatosensory areas of the opossum (*Monodelphis domestica*) are drawn as if they were flattened out. Removal of the region that normally develops into visual cortex early in cortical development caused the sensory areas to develop in new locations, reduced in size.

(Adapted from Krubitzer, L. in *Brain and Mind: Evolutionary Perspectives*, edited by M. S. Gazzaniga and J. S. Altmann. Strasbourg, France: Human Frontier Science Program, 1998.)

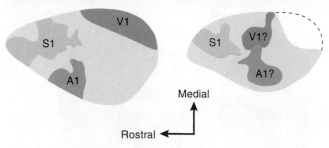

After brain development was complete, the investigators used microelectrodes to record the activity of neurons in various regions of the cortex and examined the neural circuitry in these regions under a microscope. They found that the boundaries of the specialized regions were different from those seen in a normal brain: All regions were present, but they were squeezed into the available space. Thus, the growth of axons from particular regions of the thalamus to particular regions of the cerebral cortex appeared to affect the development of the cortical regions that they served. (See *Figure 3.11*.)

Experience also affects brain development. For example, one cue for depth perception arises from the fact that each eye gets a slightly different view of the world (Poggio and Poggio, 1984). This form of depth perception, *stereopsis* ("solid appearance"), is the kind obtained from a stereoscope or a three-dimensional movie. The particular neural circuits that are necessary for stereopsis, which are located in the cerebral cortex, will not develop unless an infant has experience viewing objects with both eyes during a critical period early in life. If an infant's eyes do not move together properly—if they are not directed toward the same place in the environment (that is, if the eyes are "crossed")—the infant never develops stereoscopic vision, even if the eye movements are later corrected by surgery on the eye muscles. This critical period occurs some time between one and three years of age (Banks, Aslin, and Letson, 1975). Similar phenomena have been studied in laboratory animals and have confirmed that sensory input affects the connections established between cortical neurons.

Evidence indicates that a certain amount of neural rewiring can be accomplished even in the adult brain. For example, after a person's arm has been amputated, the region of the cerebral cortex that previously analyzed

sensory information from the missing limb soon begins analyzing information from adjacent regions of the body, such as the stump of the arm, the trunk, or the face. In fact, the person becomes more sensitive to touch in these regions after the changes in the cortex take place (Elbert et al., 1994; Kew et al., 1994; Yang et al., 1994). In addition, musicians who play stringed instruments have a larger cortical region devoted to analysis of sensory information from the fingers of the left hand (which they use to press the strings), and when a blind person who can read Braille touches objects with his or her fingertips, an enlarged region of the cerebral cortex is activated (Elbert et al., 1995; Sadato et al., 1996).

For many years, researchers have believed that neurogenesis does not take place in the fully developed brain. However, more recent studies have shown this belief to be incorrect—the adult brain contains some stem cells (similar to the progenitor cells that give rise to the cells of the developing brain) that can divide and produce neurons. Detection of newly produced cells is done by administering a small amount of a radioactive form of one of the nucleotide bases that cells use to produce the DNA that is needed for neurogenesis. The next day, the animals' brains are removed and examined with methods described in Chapter 5. Such studies have found evidence for neurogenesis in just two parts of the adult brain: the hippocampus, primarily involved in learning, and the olfactory bulb, involved in the sense of smell (Doetsch and Hen, 2005). Evidence indicates that exposure to new odors can increase the survival rate of new neurons in the olfactory bulbs, and training on a learning task can enhance neurogenesis in the hippocampus. (See *Figure 3.12*.) In addition, as we will see in Chapter 16, depression or exposure to stress can suppress neurogenesis in the hippocampus, and drugs

FIGURE 3.12 ■ Effects of Learning on Neurogenesis

This figure shows sections through a part of the hippocampus of rats that received training on a learning task or were exposed to a control condition that did not lead to learning. Arrows indicate newly formed cells.

(From Leuner, B., Mendolia-Loffredo, S., Kozorovitskiy, Y., Samburg, D., Gould, E., and Shors, T.J. *Journal of Neuroscience*, 2004, *24*, 7477–7481. Reprinted with permission.)

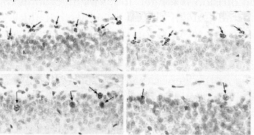

1 day after training

60 days after training

Training task   Control condition

that reduce stress and depression can reinstate neurogenesis. Unfortunately, there is no evidence that growth of new neurons can repair the effects of brain damage, such as that caused by head injury or strokes.

Evolution of the Human Brain

The brains of the earliest vertebrates were smaller than those of later animals and were simpler as well. The evolutionary process brought about genetic changes that were responsible for the development of more complex brains, with more parts and more interconnections. An important factor in the evolution of more complex brains is genetic duplication (Allman, 1999). As Lewis (1992) noted, most of the genes that a species possesses perform important functions. If a mutation causes one of these genes to do something new, the previous function would be lost, and the animal might not survive. However, geneticists have discovered that genes can sometimes duplicate themselves, and if these duplications occur in cells that give rise to ova or sperms, the duplication can be passed on to the organism's offspring. This means that the offspring will have one gene to perform the important functions and another one to "experiment" with. If a mutation of the extra gene occurs, the old gene is still present and its important function is still performed.

As we saw in Chapter 1, the human brain is larger than that of any other large animal when corrected for body size—more than three times larger than that of a chimpanzee, our closest relative. What types of genetic changes are required to produce a large brain?

Rakic (1988, 1995) suggests that the size differences between these two brains could be caused by a very simple process. We just saw that the size of the ventricular zone increases during symmetrical division of the progenitor cells located there. The ultimate size of the brain is determined by the size of the ventricular zone. As Rakic notes, each symmetrical division doubles the number of progenitor cells and thus doubles the size of the brain. The human brain is ten times larger than that of a rhesus macaque monkey. Thus, between three and four additional symmetrical divisions of progenitor cells would account for the difference in the size of these two brains. In fact, the stage of symmetrical division lasts about two days longer in humans, which provides enough time for three more divisions. The period of asymmetrical division is longer, too, which accounts for the fact that the human cortex is 15 percent thicker. Thus, delays in the termination of the symmetrical and asymmetrical periods of development could be responsible for the increased size of the human brain. A few simple mutations of the genes that control the timing of brain development could be responsible for these delays.

We do not yet know what genetic differences between humans and our primate relatives are responsible for our larger brains, but investigators are beginning to discover

FIGURE 3.13 ■ Effects of β-catenin on Development

(a) A normal mouse head. (b) This image shows the head of a genetically engineered mouse that produced excessive amounts of β-catenin. The head and brain are considerably larger, and the cerebral cortex contains convolutions normally seen only in the brains of larger mammals.

(From Chenn, A., and Walsh, C. A. *Science*, 2002, *297*, 365–369. Copyright © 2002 by the American Association for the Advancement of Science. Reprinted with permission.)

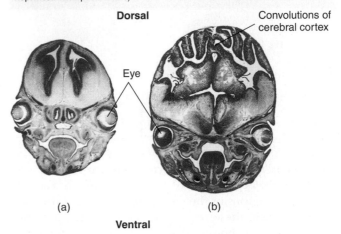

the factors that control brain size. For example, a protein called *β-catenin* is involved in regulation of cell division and tissue growth. Research indicates that this protein also plays a role in regulating the size of the cerebral cortex by controlling symmetrical cell division of progenitor cells. Chenn and Walsh (2002) employed a genetic engineering method that resulted in higher levels of β-catenin in neural progenitor cells in mouse fetuses. As a result, the number of progenitor cells increased dramatically, and the mice developed larger brains—and larger heads to accommodate these brains. The cerebral cortex grew so much that it developed convolutions normally seen only in larger, more complex brains. (See *Figure 3.13*.) A follow-up study (Woodhead et al., 2006) found that interfering with β-catenin signaling in the ventricular zone led to development of a smaller cerebral cortex. It is possible that mutations in the human genome have led to larger brains by affecting the production of β-catenin or some of the other chemicals with which this protein interacts.

The Forebrain

As we saw, the **forebrain** surrounds the rostral end of the neural tube. Its two major components are the telencephalon and the diencephalon.

forebrain The most rostral of the three major divisions of the brain; includes the telencephalon and diencephalon.

FIGURE 3.14 ■ Cross Section Human Brain

This brain slice shows fissures and gyri and the layer of cerebral cortex that follows these convolutions.

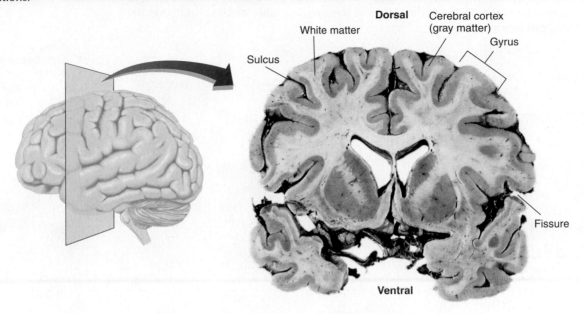

Telencephalon

The telencephalon includes most of the two symmetrical **cerebral hemispheres** that make up the cerebrum. The cerebral hemispheres are covered by the cerebral cortex and contain the limbic system and the basal ganglia. The latter two sets of structures are primarily in the **subcortical regions** of the brain—those located deep within it, beneath the cerebral cortex.

Cerebral Cortex. As we saw in the previous section, the cerebral cortex surrounds the cerebral hemispheres like the bark of a tree. In humans the cerebral cortex is greatly convoluted; these convolutions, consisting of **sulci** (small grooves), **fissures** (large grooves), and **gyri** (bulges between adjacent sulci or fissures), greatly enlarge the surface area of the cortex, compared with a smooth brain of the same size. In fact, two-thirds of the surface of the cortex is hidden in the grooves; thus, the presence of these convolutions triples the area of the cerebral cortex. The total surface area is approximately 2360 cm^2 (2.5 ft^2), and the thickness is approximately 3 mm. The cerebral cortex consists mostly of glia and the cell bodies, dendrites, and interconnecting axons of neurons. Because cell bodies predominate, giving the cerebral cortex a grayish brown appearance, it is referred to as *gray matter*. (See **Figure 3.14.**) Beneath the cerebral cortex run millions of axons that connect the neurons of the cerebral cortex with those located elsewhere in the brain. The large concentration of myelin gives this tissue an opaque white appearance—hence the term *white matter*.

Three areas of the cerebral cortex receive information from the sensory organs. The **primary visual cortex,** which receives visual information, is located at the back of the brain, on the inner surfaces of the cerebral hemispheres—primarily on the upper and lower banks of the **calcarine fissure.** (*Calcarine* means "spur-shaped." See **Figure 3.15.**) The **primary auditory cortex,** which receives auditory information, is located on the lower surface of a deep fissure in the side of the brain—the **lateral fissure.**

cerebral hemisphere (*sa* **ree** *brul*) One of the two major portions of the forebrain, covered by the cerebral cortex.

subcortical region The region located within the brain, beneath the cortical surface.

sulcus (plural: sulci) (*sul kus, sul sigh*) A groove in the surface of the cerebral hemisphere, smaller than a fissure.

fissure A major groove in the surface of the brain, larger than a sulcus.

gyrus (plural: gyri) (*jye russ, jye rye*) A convolution of the cortex of the cerebral hemispheres, separated by sulci or fissures.

primary visual cortex The region of the posterior occipital lobe whose primary input is from the visual system.

calcarine fissure (*kal ka rine*) A fissure located in the occipital lobe on the medial surface of the brain; most of the primary visual cortex is located along its upper and lower banks.

primary auditory cortex The region of the superior temporal lobe whose primary input is from the auditory system.

lateral fissure The fissure that separates the temporal lobe from the overlying frontal and parietal lobes.

FIGURE 3.15 ■ A Lateral View of the Left Side of a Human Brain and Part of the Inner Surface of the Right Side

The inset shows a cutaway of part of the frontal lobe of the left hemisphere, permitting us to see the primary auditory cortex on the dorsal surface of the temporal lobe, which forms the ventral bank of the lateral fissure.

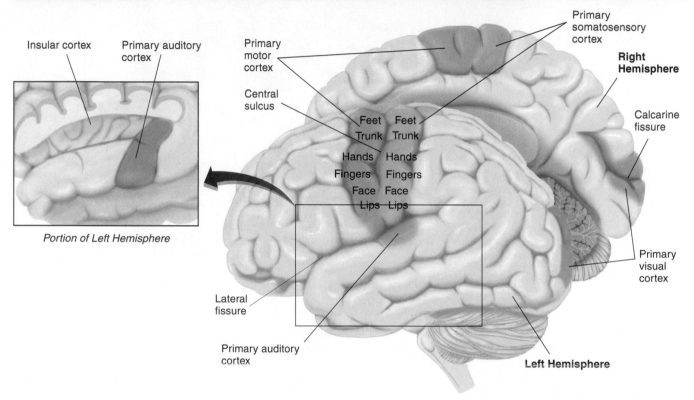

(See the inset in *Figure 3.15.*) The **primary somatosensory cortex,** a vertical strip of cortex just caudal to the **central sulcus,** receives information from the body senses. As Figure 3.15 shows, different regions of the primary somatosensory cortex receive information from different regions of the body. In addition, the base of the somatosensory cortex and a portion of the **insular cortex,** which is normally hidden from view by the frontal and temporal lobes, receives information concerning taste. (See *Figure 3.15.*)

With the exception of olfaction and gustation (taste), sensory information from the body or the environment is sent to primary sensory cortex of the contralateral hemisphere. Thus, the primary somatosensory cortex of the left hemisphere learns what the right hand is holding, the left primary visual cortex learns what is happening toward the person's right, and so on.

The region of the cerebral cortex that is most directly involved in the control of movement is the **primary motor cortex,** located just in front of the primary somatosensory cortex. Neurons in different parts of the primary motor cortex are connected to muscles in different parts of the body. The connections, like

those of the sensory regions of the cerebral cortex, are contralateral; the left primary motor cortex controls the right side of the body and vice versa. Thus, if a surgeon places an electrode on the surface of the primary motor cortex and stimulates the neurons there with a weak electrical current, the result will be movement of a particular part of the body. Moving the electrode to a different spot will cause a different part of the body to move. (See *Figure 3.15.*) I like to think of the strip of primary motor cortex as the keyboard of a piano, with each key controlling a different movement. (We will see shortly who the "player" of this piano is.)

primary somatosensory cortex The region of the anterior parietal lobe whose primary input is from the somatosensory system.

central sulcus (*sul kus*) The sulcus that separates the frontal lobe from the parietal lobe.

insular cortex (*in sue lur*) A sunken region of the cerebral cortex that is normally covered by the rostral superior temporal lobe and caudal inferior frontal lobe.

primary motor cortex The region of the posterior frontal lobe that contains neurons that control movements of skeletal muscles.

FIGURE 3.16 ■ The Four Lobes of the Cerebral Cortex, the Primary Sensory and Motor Cortex, and the Association Cortex

The figure shows (a) a ventral view, from the base of the brain, (b) a midsagittal view with the cerebellum and brain stem removed, and (c) a lateral view.

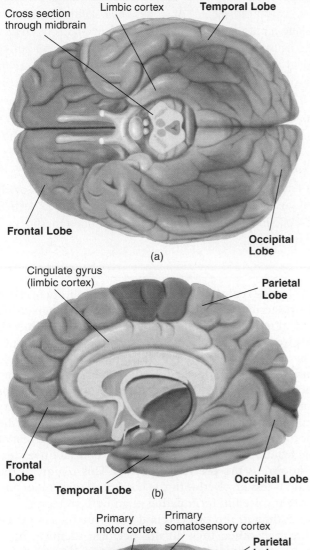

Cross section through midbrain

Limbic cortex

Temporal Lobe

Frontal Lobe

Occipital Lobe

(a)

Cingulate gyrus (limbic cortex)

Parietal Lobe

Frontal Lobe

Temporal Lobe

Occipital Lobe

(b)

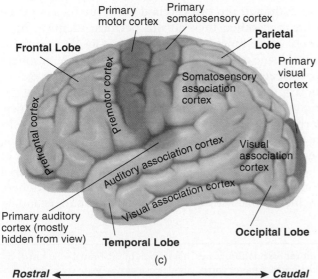

Primary motor cortex

Primary somatosensory cortex

Frontal Lobe

Parietal Lobe

Primary visual cortex

Somatosensory association cortex

prefrontal cortex

Premotor cortex

Visual association cortex

Auditory association cortex

Visual association cortex

Primary auditory cortex (mostly hidden from view)

Temporal Lobe

Occipital Lobe

(c)

Rostral ◀━━━━━▶ *Caudal*

The regions of primary sensory and motor cortex occupy only a small part of the cerebral cortex. The rest of the cerebral cortex accomplishes what is done between sensation and action: perceiving, learning and remembering, planning, and acting. These processes take place in the *association areas* of the cerebral cortex. The central sulcus provides an important dividing line between the rostral and caudal regions of the cerebral cortex. (See *Figure 3.15.*) The rostral region is involved in movement-related activities, such as planning and executing behaviors. The caudal region is involved in perceiving and learning.

Discussing the various regions of the cerebral cortex is easier if we have names for them. In fact, the cerebral cortex is divided into four areas, or *lobes,* named for the bones of the skull that cover them: the frontal lobe, parietal lobe, temporal lobe, and occipital lobe. Of course, the brain contains two of each lobe, one in each hemisphere. The **frontal lobe** (the "front") includes everything in front of the central sulcus. The **parietal lobe** (the "wall") is located on the side of the cerebral hemisphere, just behind the central sulcus, caudal to the frontal lobe. The **temporal lobe** (the "temple") juts forward from the base of the brain, ventral to the frontal and parietal lobes. The **occipital lobe** (from the Latin *ob,* "in back of," and *caput,* "head") lies at the very back of the brain, caudal to the parietal and temporal lobes. Figure 3.16 shows these lobes in three views of the cerebral hemispheres: a ventral view (a view from the bottom), a midsagittal view (a view of the inner surface of the right hemisphere after the left hemisphere has been removed), and a lateral view. (See *Figure 3.16.*)

Each primary sensory area of the cerebral cortex sends information to adjacent regions, called the **sensory association cortex.** Circuits of neurons in the sensory association cortex analyze the information received from the primary sensory cortex; perception takes place there, and memories are stored there. The regions of the sensory association cortex located closest to the primary sensory areas receive information from only one sensory system. For example, the region closest to the primary visual cortex analyzes visual information

frontal lobe The anterior portion of the cerebral cortex, rostral to the parietal lobe and dorsal to the temporal lobe.

parietal lobe (*pa rye i tul*) The region of the cerebral cortex caudal to the frontal lobe and dorsal to the temporal lobe.

temporal lobe (*tem por ul*) The region of the cerebral cortex rostral to the occipital lobe and ventral to the parietal and frontal lobes.

occipital lobe (*ok sip i tul*) The region of the cerebral cortex caudal to the parietal and temporal lobes.

sensory association cortex Those regions of the cerebral cortex that receive information from the regions of primary sensory cortex.

and stores visual memories. Regions of the sensory association cortex located far from the primary sensory areas receive information from more than one sensory system; thus, they are involved in several kinds of perceptions and memories. These regions make it possible to integrate information from more than one sensory system. For example, we can learn the connection between the sight of a particular face and the sound of a particular voice. (See *Figure 3.16.*)

If people sustain damage to the somatosensory association cortex, their deficits are related to somatosensation and to the environment in general; for example, they may have difficulty perceiving the shapes of objects that they can touch but not see, they may be unable to name parts of their bodies (see the case below), or they may have trouble drawing maps or following them. Destruction of the primary visual cortex causes blindness. However, although people who sustain damage to the visual association cortex will not become blind, they may be unable to recognize objects by sight. People who sustain damage to the auditory association cortex may have difficulty perceiving speech or even producing meaningful speech of their own. People who sustain damage to regions of the association cortex at the junction of the three posterior lobes, where the somatosensory, visual, and auditory functions overlap, may have difficulty reading or writing.

Mr. M., a city bus driver, stopped to let a passenger climb board. The passenger asked him a question, and Mr. M. suddenly realized that he didn't understand what she was saying. He could hear her, but her words made no sense. He opened his mouth to reply. He made some sounds, but the look on the woman's face told him that she couldn't understand what he was trying to say. He turned off the engine and looked around at the passengers and tried to tell them to get some help. Although he was unable to say anything, they understood that something was wrong, and one of them called an ambulance.

An MRI scan showed that Mr. M. had sustained an intracerebral hemorrhage—a kind of stroke caused by rupture of blood vessels in the brain. The stroke had damaged his left parietal lobe. Mr. M. gradually regained the ability to talk and understand the speech of others, but some deficits remained. A colleague, Dr. D., and I studied Mr. M. several weeks after his stroke. The dialogue went something like this:

"Show me your hand."

"My hand . . . my hand." Looks at his arms, then touches his left forearm.

"Show me your chin."

"My chin." Looks at his arms, looks down, puts his hand on his abdomen.

"Show me your right elbow."

"My right . . ." (points to the right with his right thumb) "elbow." Looks up and down his right arm, finally touches his right shoulder.

As you can see, Mr. M. could understand that we were asking him to point out parts of his body and could repeat the names of the body parts when we spoke them, but he could not identify which body parts these names referred to. This strange deficit, which sometimes follows damage to the left parietal lobe, is called autotopagnosia, or "poor knowledge of one's own topography." (A better term would be autotopanomia, or "poor knowledge of the names of one's own topography," but, then, no one asked me to choose the term.) The parietal lobes are involved with space: the right primarily with external space and the left with one's body and personal space. I will say more about disorders such as this one in Chapter 14, which deals with brain mechanisms of language.

Just as regions of the sensory association cortex of the posterior part of the brain are involved in perceiving and remembering, the frontal association cortex is involved in the planning and execution of movements. The **motor association cortex** (also known as the *premotor cortex*) is located just rostral to the primary motor cortex. This region controls the primary motor cortex; thus, it directly controls behavior. If the primary motor cortex is the keyboard of the piano, then the motor association cortex is the piano player. The rest of the frontal lobe, rostral to the motor association cortex, is known as the **prefrontal cortex.** This region of the brain is less involved with the control of movement and more involved in formulating plans and strategies.

Although the two cerebral hemispheres cooperate with each other, they do not perform identical functions. Some functions are *lateralized*—located primarily on one side of the brain. In general, the left hemisphere participates in the *analysis* of information—the extraction of the elements that make up the whole of an experience. This ability makes the left hemisphere particularly good at recognizing *serial events*—events whose elements occur one after the other—and controlling sequences of behavior. (In a few people the functions of the left and right hemispheres are reversed.) The serial functions that are performed by the left hemisphere include verbal activities, such as talking, understanding

---

motor association cortex The region of the frontal lobe rostral to the primary motor cortex; also known as the premotor cortex.

prefrontal cortex The region of the frontal lobe rostral to the motor association cortex.

the speech of other people, reading, and writing. These abilities are disrupted by damage to the various regions of the left hemisphere. (I will say more about language and the brain in Chapter 14.)

In contrast, the right hemisphere is specialized for *synthesis;* it is particularly good at putting isolated elements together to perceive things as a whole. For example, our ability to draw sketches (especially of three-dimensional objects), read maps, and construct complex objects out of smaller elements depends heavily on circuits of neurons that are located in the right hemisphere. Damage to the right hemisphere disrupts these abilities.

We are not aware of the fact that each hemisphere perceives the world differently. Although the two cerebral hemispheres perform somewhat different functions, our perceptions and our memories are unified. This unity is accomplished by the **corpus callosum,** a large band of axons that connects corresponding parts of the cerebral cortex of the left and right hemispheres: The left and right temporal lobes are connected, the left and right parietal lobes are connected, and so on. Because of the corpus callosum, each region of the association cortex knows what is happening in the corresponding region of the opposite side of the brain. The corpus callosum also makes a few asymmetrical connections that link different regions of the two hemispheres. *Figure 3.17* shows the bundles of axons that constitute the corpus callosum, obtained by means of *diffusion tensor imaging,* a special scanning method described in Chapter 5.

Figure 3.18 shows a *midsagittal* view of the brain. The brain (and part of the spinal cord) has been sliced down the middle, dividing it into its two symmetrical halves. The left half has been removed, so we see the inner surface of the right half. The cerebral cortex that covers most of the surface of the cerebral hemispheres (including the frontal, parietal, occipital, and temporal lobes) is called the **neocortex** ("new" cortex, because it is of relatively recent evolutionary origin). Another form of cerebral cortex, the **limbic cortex,** is located around the medial edge of the cerebral hemispheres (*limbus* means "border"). The **cingulate gyrus,** an important region of the limbic cortex, can be seen in this figure. (See *Figure 3.18.*) In addition, if you look back at the top two drawings of Figure 3.16, you will see that the limbic cortex occupies the regions that have not been colored in. (Refer to *Figure 3.16.*)

Figure 3.18 also shows the corpus callosum. To slice the brain into its two symmetrical halves, one must slice through the middle of the corpus callosum. (Recall that I described the split-brain operation, in which the corpus callosum is severed, in Chapter 1.) (See *Figure 3.18.*)

As I mentioned earlier, one of the Chapter 3 MyPsychKit animations will permit you to view the brain from various angles and see the locations of the specialized regions of the cerebral cortex. (See *MyPsychKit 3.2, The Rotatable Brain.*)

mypsychkit Where learning comes to life!
Animation 3.2
The Rotatable Brain

Limbic System.   A neuroanatomist, Papez (1937), suggested that a set of interconnected brain structures formed a circuit whose primary function was motivation and emotion. This system included several regions of the limbic cortex (already described) and a set of interconnected structures surrounding the core of the forebrain. A physiologist, MacLean (1949), expanded the system to include other structures and coined the term **limbic system.** Besides the limbic cortex, the most important parts of the limbic system are the **hippocampus**

FIGURE 3.17 ■ Bundles of Axons

Bundles of axons serving different regions of the cerebral cortex that constitute the corpus callosum were obtained by means of diffusion tensor imaging.

(From Hofer, S., and Frahm, J. *NeuroImage,* 2006, *32,* 989–994. Reprinted with permission.)

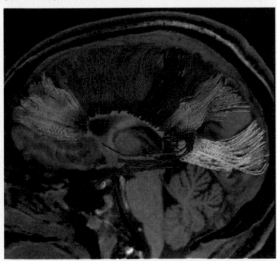

---

corpus callosum (*ka loh sum*) A large bundle of axons that interconnects corresponding regions of the association cortex on each side of the brain.

neocortex The phylogenetically newest cortex, including the primary sensory cortex, primary motor cortex, and association cortex.

limbic cortex Phylogenetically old cortex, located at the medial edge ("limbus") of the cerebral hemispheres; part of the limbic system.

cingulate gyrus (*sing yew lett*) A strip of limbic cortex lying along the lateral walls of the groove separating the cerebral hemispheres, just above the corpus callosum.

limbic system A group of brain regions including the anterior thalamic nuclei, amygdala, hippocampus, limbic cortex, and parts of the hypothalamus, as well as their interconnecting fiber bundles.

hippocampus A forebrain structure of the temporal lobe, constituting an important part of the limbic system; includes the hippocampus proper (Ammon's horn), dentate gyrus, and subiculum.

FIGURE 3.18 ■ A Midsagittal View of the Brain and Part of the Spinal Cord

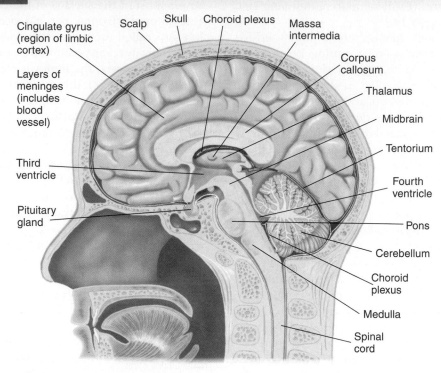

Cingulate gyrus (region of limbic cortex)
Scalp
Skull
Choroid plexus
Massa intermedia
Corpus callosum
Thalamus
Midbrain
Tentorium
Fourth ventricle
Pons
Cerebellum
Choroid plexus
Medulla
Spinal cord
Layers of meninges (includes blood vessel)
Third ventricle
Pituitary gland

("sea horse") and the **amygdala** ("almond"), located next to the lateral ventricle in the temporal lobe. The **fornix** ("arch") is a bundle of axons that connects the hippocampus with other regions of the brain, including the **mammillary** ("breast-shaped") **bodies,** protrusions

FIGURE 3.19 ■ Major Components of the Limbic System

All of the left hemisphere except for the limbic system has been removed.

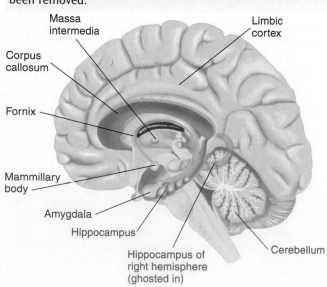

Massa intermedia
Limbic cortex
Corpus callosum
Fornix
Mammillary body
Amygdala
Hippocampus
Hippocampus of right hemisphere (ghosted in)
Cerebellum

on the base of the brain that contain parts of the hypothalamus. (See *Figure 3.19*.)

MacLean noted that the evolution of this system, which includes the first and simplest form of cerebral cortex, appears to have coincided with the development of emotional responses. As you will see in Chapter 13, we now know that parts of the limbic system (notably, the hippocampal formation and the region of limbic cortex that surrounds it) are involved in learning and memory. The amygdala and some regions of limbic cortex are specifically involved in emotions: feelings and expressions of emotions, emotional memories, and recognition of the signs of emotions in other people.

Basal Ganglia.   The **basal ganglia** are a collection of subcortical nuclei in the forebrain, which lie beneath

amygdala (*a mig da la*) A structure in the interior of the rostral temporal lobe, containing a set of nuclei; part of the limbic system.

fornix A fiber bundle that connects the hippocampus with other parts of the brain, including the mammillary bodies of the hypothalamus; part of the limbic system.

mammillary bodies (*mam i lair ee*) A protrusion of the bottom of the brain at the posterior end of the hypothalamus, containing some hypothalamic nuclei; part of the limbic system.

basal ganglia A group of subcortical nuclei in the telencephalon, the caudate nucleus, the globus pallidus, and the putamen; important parts of the motor system.

FIGURE 3.20 ■ Basal Ganglia and Diencephalon

The basal ganglia and diencephalon (thalamus and hypothalamus) are ghosted in to a semitransparent brain.

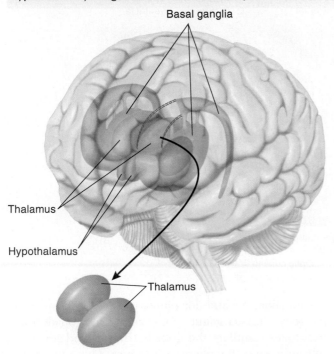

Basal ganglia

Thalamus

Hypothalamus

Thalamus

the anterior portion of the lateral ventricles. **Nuclei** are groups of neurons of similar shape. (The word *nucleus*, from the Greek "nut," can refer to the inner portion of an atom, to the structure of a cell that contains the chromosomes, and—as in this case—to a collection of neurons located within the brain.) The major parts of the basal ganglia are the *caudate nucleus,* the *putamen,* and the *globus pallidus* (the "nucleus with a tail," the "shell," and the "pale globe"). (See *Figure 3.20.*) The basal ganglia are involved in the control of movement. For example, Parkinson's disease is caused by degeneration of certain neurons located in the midbrain that send axons to the caudate nucleus and the putamen. The symptoms of this disease are weakness, tremors, rigidity of the limbs, poor balance, and difficulty in initiating movements.

Diencephalon

The second major division of the forebrain, the **diencephalon,** is situated between the telencephalon and the mesencephalon; it surrounds the third ventricle. Its two most important structures are the thalamus and the hypothalamus. (See *Figure 3.20.*)

Thalamus.   The **thalamus** (from the Greek *thalamos,* "inner chamber") makes up the dorsal part of the diencephalon. It is situated near the middle of the cerebral hemispheres, immediately medial and caudal to the

basal ganglia. The thalamus has two lobes, connected by a bridge of gray matter called the *massa intermedia,* which pierces the middle of the third ventricle. (See *Figure 3.21.*) The massa intermedia is probably not an important structure, because it is absent in the brains of many people. However, it serves as a useful reference point in looking at diagrams of the brain; it appears in Figures 3.4, 3.18, 3.19, 3.20, and 3.21.

Most neural input to the cerebral cortex is received from the thalamus; indeed, much of the cortical surface can be divided into regions that receive projections from specific parts of the thalamus. **Projection fibers** are sets of axons that arise from cell bodies located in one region of the brain and synapse on neurons located within another region (that is, they *project to* these regions).

The thalamus is divided into several nuclei. Some thalamic nuclei receive sensory information from the sensory systems. The neurons in these nuclei then relay the sensory information to specific sensory projection areas of the cerebral cortex. For example, the **lateral geniculate nucleus** receives information from the eye and sends axons to the primary visual cortex, and the **medial geniculate nucleus** receives information from the inner ear and sends axons to the primary auditory cortex. Other thalamic nuclei project to specific regions of the cerebral cortex, but they do not relay sensory information. For example, the **ventrolateral nucleus** receives information from the cerebellum and projects it to the primary motor cortex. Still other nuclei receive information from one region of the cerebral cortex and relay it to another region. And as we will see in Chapter 9, several nuclei are involved in controlling the general excitability of the cerebral cortex. To accomplish this task, these nuclei have widespread projections to all cortical regions.

nucleus (plural: nuclei) An identifiable group of neural cell bodies in the central nervous system.

diencephalon (*dy en seff a lahn*) A region of the forebrain surrounding the third ventricle; includes the thalamus and the hypothalamus.

thalamus The largest portion of the diencephalon, located above the hypothalamus; contains nuclei that project information to specific regions of the cerebral cortex and receive information from it.

projection fiber An axon of a neuron in one region of the brain whose terminals form synapses with neurons in another region.

lateral geniculate nucleus A group of cell bodies within the lateral geniculate body of the thalamus that receives fibers from the retina and projects fibers to the primary visual cortex.

medial geniculate nucleus A group of cell bodies within the medial geniculate body of the thalamus; receives fibers from the auditory system and projects fibers to the primary auditory cortex.

ventrolateral nucleus A nucleus of the thalamus that receives inputs from the cerebellum and sends axons to the primary motor cortex.

FIGURE 3.21 ■ A Midsagittal View of Part of the Brain

This view shows some of the nuclei of the hypothalamus. The nuclei are situated on the far side of the wall of the third ventricle, inside the right hemisphere.

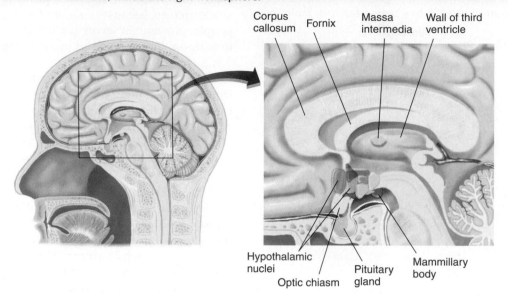

Hypothalamus.   As its name implies, the **hypo-thalamus** lies at the base of the brain, under the thalamus. Although the hypothalamus is a relatively small structure, it is an important one. It controls the autonomic nervous system and the endocrine system and organizes behaviors related to survival of the species—the so-called four F's: fighting, feeding, fleeing, and mating.

The hypothalamus is situated on both sides of the ventral portion of the third ventricle. The hypothalamus is a complex structure, containing many nuclei and fiber tracts. Figure 3.21 indicates its location and size. Note that the pituitary gland is attached to the base of the hypothalamus via the pituitary stalk. Just in front of the pituitary stalk is the **optic chiasm,** where half of the axons in the optic nerves (from the eyes) cross from one side of the brain to the other. (See *Figure 3.21.*) The role of the hypothalamus in the control of the four F's (and other behaviors, such as drinking and sleeping) will be considered in several chapters later in this book.

Much of the endocrine system is controlled by hormones produced by cells in the hypothalamus. A special system of blood vessels directly connects the hypothalamus with the **anterior pituitary gland.** (See *Figure 3.22.*) The hypothalamic hormones are secreted by specialized neurons called **neurosecretory cells,** located near the base of the pituitary stalk. These hormones stimulate the anterior pituitary gland to secrete its hormones. For example, *gonadotropin-releasing hormone* causes the anterior pituitary gland to secrete the *gonadotropic hormones,* which play a role in reproductive physiology and behavior.

Most of the hormones secreted by the anterior pituitary gland control other endocrine glands. Because of

this function, the anterior pituitary gland has been called the body's "master gland." For example, the gonadotropic hormones stimulate the gonads (ovaries and testes) to release male or female sex hormones. These hormones affect cells throughout the body, including some in the brain. Two other anterior pituitary hormones—prolactin and somatotropic hormone (growth hormone)—do not control other glands but act as the final messenger. The behavioral effects of many of the anterior pituitary hormones are discussed in later chapters.

The hypothalamus also produces the hormones of the **posterior pituitary gland** and controls their secretion. These hormones include oxytocin, which stimulates ejection of milk and uterine contractions at the time of childbirth, and vasopressin, which regulates urine output by the kidneys. They are produced by neurons in the

hypothalamus The group of nuclei of the diencephalon situated beneath the thalamus; involved in regulation of the autonomic nervous system, control of the anterior and posterior pituitary glands, and integration of species-typical behaviors.

optic chiasm (*kye az'm*) An X-shaped connection between the optic nerves, located below the base of the brain, just anterior to the pituitary gland.

anterior pituitary gland The anterior part of the pituitary gland; an endocrine gland whose secretions are controlled by the hypothalamic hormones.

neurosecretory cell A neuron that secretes a hormone or hormonelike substance.

posterior pituitary gland The posterior part of the pituitary gland; an endocrine gland that contains hormone-secreting terminal buttons of axons whose cell bodies lie within the hypothalamus.

FIGURE 3.22 ■ The Pituitary Gland

Hormones released by the neurosecretory cells in the hypothalamus enter capillaries and are conveyed to the anterior pituitary gland, where they control its secretion of hormones. The hormones of the posterior pituitary gland are produced in the hypothalamus and carried to the gland in vesicles by means of axoplasmic transport.

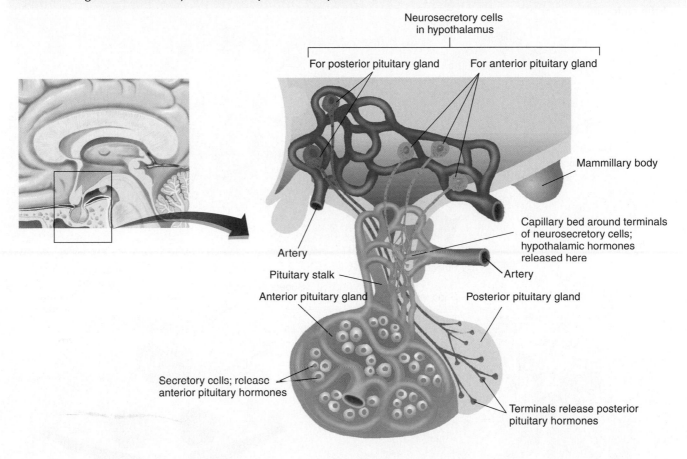

hypothalamus whose axons travel down the pituitary stalk and terminate in the posterior pituitary gland. The hormones are carried in vesicles through the axoplasm of these neurons and collect in the terminal buttons in the posterior pituitary gland. When these axons fire, the hormone contained within their terminal buttons is liberated and enters the circulatory system.

The Midbrain

The **midbrain** (also called the **mesencephalon**) surrounds the cerebral aqueduct and consists of two major parts: the tectum and the tegmentum.

Tectum

The **tectum** ("roof") is located in the dorsal portion of the mesencephalon. Its principal structures are the **superior colliculi** and the **inferior colliculi,** which appear as four bumps on the dorsal surface of the **brain stem.** The brain stem includes the midbrain and the

hindbrain, and it is called the brain stem because it looks just like that: a stem. Figure 3.23 shows several views of the brain stem: lateral and posterior views of the brain stem inside a semitransparent brain, an enlarged view of the brain stem with part of the cerebellum cut away to reveal the inside of the fourth ventricle, and a

midbrain The mesencephalon; the central of the three major divisions of the brain.

mesencephalon (*mezz en **seff** a lahn*) The midbrain; a region of the brain that surrounds the cerebral aqueduct; includes the tectum and the tegmentum.

tectum The dorsal part of the midbrain; includes the superior and inferior colliculi.

superior colliculi (*ka **lik** yew lee*) Protrusions on top of the midbrain; part of the visual system.

inferior colliculi Protrusions on top of the midbrain; part of the auditory system.

brain stem The "stem" of the brain, from the medulla to the midbrain, excluding the cerebellum.

FIGURE 3.23 ■ The Cerebellum and Brain Stem

(a) A lateral view of a semitransparent brain shows the cerebellum and brain stem ghosted in. (b) A view from the back of the brain. (c) A dorsal view of the brain stem. The left hemisphere of the cerebellum and part of the right hemisphere have been removed to show the inside of the fourth ventricle and the cerebellar peduncles. (d) A cross section of the midbrain.

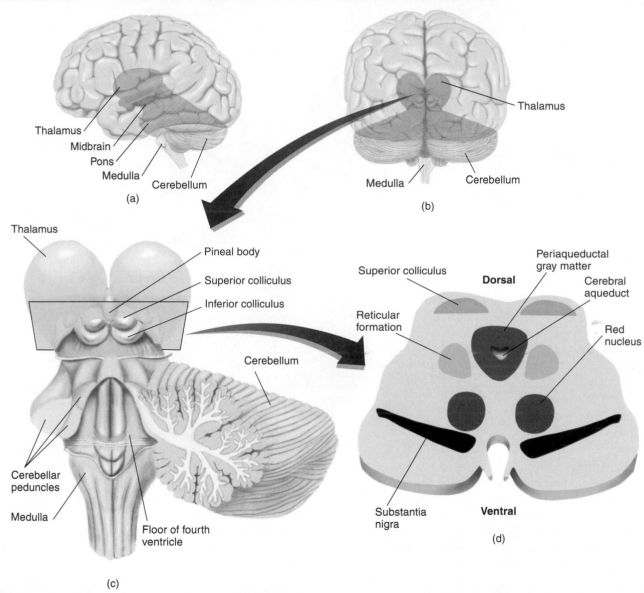

cross section through the midbrain. (See *Figure 3.23.*) The inferior colliculi are a part of the auditory system. The superior colliculi are part of the visual system. In mammals they are primarily involved in visual reflexes and reactions to moving stimuli.

Tegmentum

The **tegmentum** ("covering") consists of the portion of the mesencephalon beneath the tectum. It includes the rostral end of the reticular formation, several nuclei controlling eye movements, the periaqueductal gray

matter, the red nucleus, the substantia nigra, and the ventral tegmental area. (See *Figure 3.23d.*)

The **reticular formation** is a large structure consisting of many nuclei (over ninety in all). It is also characterized

tegmentum The ventral part of the midbrain; includes the periaqueductal gray matter, reticular formation, red nucleus, and substantia nigra.

reticular formation A large network of neural tissue located in the central region of the brain stem, from the medulla to the diencephalon.

by a diffuse, interconnected network of neurons with complex dendritic and axonal processes. (Indeed, *reticulum* means "little net"; early anatomists were struck by the netlike appearance of the reticular formation.) The reticular formation occupies the core of the brain stem, from the lower border of the medulla to the upper border of the midbrain. (See *Figure 3.23d.*) The reticular formation receives sensory information by means of various pathways and projects axons to the cerebral cortex, thalamus, and spinal cord. It plays a role in sleep and arousal, attention, muscle tonus, movement, and various vital reflexes. Its functions will be described more fully in later chapters.

The **periaqueductal gray matter** is so called because it consists mostly of cell bodies of neurons ("gray matter," as contrasted with the "white matter" of axon bundles) that surround the cerebral aqueduct as it travels from the third to the fourth ventricle. The periaqueductal gray matter contains neural circuits that control sequences of movements that constitute species-typical behaviors, such as fighting and mating. As we will see in Chapter 7, opiates such as morphine decrease an organism's sensitivity to pain by stimulating receptors on neurons located in this region.

The **red nucleus** and **substantia nigra** ("black substance") are important components of the motor system. A bundle of axons that arises from the red nucleus constitutes one of the two major fiber systems that bring motor information from the cerebral cortex and cerebellum to the spinal cord. The substantia nigra contains neurons whose axons project to the caudate nucleus and putamen, parts of the basal ganglia. As we will see in Chapter 4, degeneration of these neurons causes Parkinson's disease.

The Hindbrain

The **hindbrain,** which surrounds the fourth ventricle, consists of two major divisions: the metencephalon and the myelencephalon.

Metencephalon

The metencephalon consists of the pons and the cerebellum.

Cerebellum.   The **cerebellum** ("little brain"), with its two hemispheres, resembles a miniature version of the cerebrum. It is covered by the **cerebellar cortex** and has a set of **deep cerebellar nuclei.** These nuclei receive projections from the cerebellar cortex and themselves send projections out of the cerebellum to other parts of the brain. Each hemisphere of the cerebellum is attached to the dorsal surface of the pons by bundles of axons: the superior, middle, and inferior **cerebellar peduncles** ("little feet"). (See *Figure 3.23c.*)

Damage to the cerebellum impairs standing, walking, or performance of coordinated movements. (A virtuoso pianist or other performing musician owes much to his or her cerebellum.) The cerebellum receives visual, auditory, vestibular, and somatosensory information, and it also receives information about individual muscle movements being directed by the brain. The cerebellum integrates this information and modifies the motor outflow, exerting a coordinating and smoothing effect on the movements. Cerebellar damage results in jerky, poorly coordinated, exaggerated movements; extensive cerebellar damage makes it impossible even to stand. Chapter 8 discusses the anatomy and functions of the cerebellum in more detail.

Pons.   The **pons,** a large bulge in the brain stem, lies between the mesencephalon and medulla oblongata, immediately ventral to the cerebellum. *Pons* means "bridge," but it does not really look like one. (Refer to *Figures 3.18* and *3.23a.*) The pons contains, in its core, a portion of the reticular formation, including some nuclei that appear to be important in sleep and arousal. It also contains a large nucleus that relays information from the cerebral cortex to the cerebellum.

Myelencephalon

The myelencephalon contains one major structure, the **medulla oblongata** (literally, "oblong marrow"), usually

periaqueductal gray matter The region of the midbrain surrounding the cerebral aqueduct; contains neural circuits involved in species-typical behaviors.

red nucleus A large nucleus of the midbrain that receives inputs from the cerebellum and motor cortex and sends axons to motor neurons in the spinal cord.

substantia nigra A darkly stained region of the tegmentum that contains neurons that communicate with the caudate nucleus and putamen in the basal ganglia.

hindbrain The most caudal of the three major divisions of the brain; includes the metencephalon and myelencephalon.

cerebellum (*sair a **bell** um*) A major part of the brain located dorsal to the pons, containing the two cerebellar hemispheres, covered with the cerebellar cortex; an important component of the motor system.

cerebellar cortex The cortex that covers the surface of the cerebellum.

deep cerebellar nuclei Nuclei located within the cerebellar hemispheres; receive projections from the cerebellar cortex and send projections out of the cerebellum to other parts of the brain.

cerebellar peduncle (*pee dun kul*) One of three bundles of axons that attach each cerebellar hemisphere to the dorsal pons.

pons The region of the metencephalon rostral to the medulla, caudal to the midbrain, and ventral to the cerebellum.

medulla oblongata (*me doo la*) The most caudal portion of the brain; located in the myelencephalon, immediately rostral to the spinal cord.

just called the *medulla*. This structure is the most caudal portion of the brain stem; its lower border is the rostral end of the spinal cord. (Refer to *Figures 3.16* and *3.23a*.) The medulla contains part of the reticular formation, including nuclei that control vital functions such as regulation of the cardiovascular system, respiration, and skeletal muscle tonus.

The Spinal Cord

The **spinal cord** is a long, conical structure, approximately as thick as our little finger. The principal function of the spinal cord is to distribute motor fibers to the effector organs of the body (glands and muscles) and to collect somatosensory information to be passed on to the brain. The spinal cord also has a certain degree of autonomy from the brain; various reflexive control circuits (some of which are described in Chapter 8) are located there.

The spinal cord is protected by the vertebral column, which is composed of twenty-four individual vertebrae of the *cervical* (neck), *thoracic* (chest), and *lumbar* (lower back) regions and the fused vertebrae that make

FIGURE 3.25 ■ The Spinal Cord

(a) A portion of the spinal cord shows the layers of the meninges and the relationship of the spinal cord to the vertebral column. (b) A cross section through the spinal cord shows ascending tracts in blue and descending tracts in red.

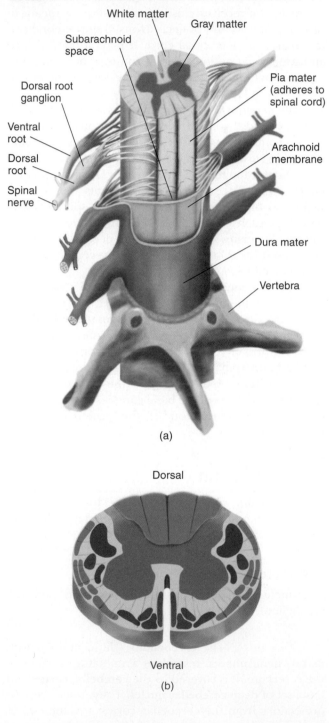

(a)

(b)

spinal cord The cord of nervous tissue that extends caudally from the medulla.

FIGURE 3.24 ■ Spinal Column, Ventral View

This ventral view of the human spinal column, with details shows the anatomy of the vertebrae.

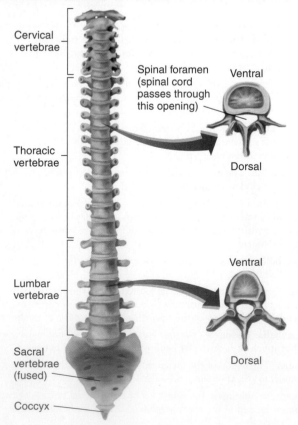

up the *sacral* and *coccygeal* portions of the column (located in the pelvic region). The spinal cord passes through a hole in each of the vertebrae (the *spinal foramens*). Figure 3.24 illustrates the divisions and structures of the spinal cord and vertebral column. (See *Figure 3.24*.) Note that the spinal cord is only about two-thirds as long as the vertebral column; the rest of the space is filled by a mass of **spinal roots** composing the **cauda equina** ("horse's tail"). (Refer to *Figure 3.3c*.)

Early in embryological development the vertebral column and spinal cord are the same length. As development progresses, the vertebral column grows faster than the spinal cord. This differential growth rate causes the spinal roots to be displaced downward; the most caudal roots travel the farthest before they emerge through openings between the vertebrae and thus compose the cauda equina. To produce the **caudal block** that is sometimes used in pelvic surgery or childbirth, a local anesthetic can be injected into the CSF contained within the sac of dura mater surrounding the cauda equina. The drug blocks conduction in the axons of the cauda equina.

Figure 3.25(a) shows a portion of the spinal cord, with the layers of the meninges that wrap it. Small bundles of fibers emerge from each side of the spinal cord in two straight lines along its dorsolateral and ventrolateral surfaces. Groups of these bundles fuse together and become the thirty-one paired sets of **dorsal roots** and **ventral roots**. The dorsal and ventral roots join together as they pass through the intervertebral foramens and become spinal nerves. (See *Figure 3.25a*.)

Figure 3.25(b) shows a cross section of the spinal cord. Like the brain, the spinal cord consists of white matter and gray matter. Unlike the brain's, the spinal cord's white matter (consisting of ascending and descending bundles of myelinated axons) is on the outside; the gray matter (mostly neural cell bodies and short, unmyelinated axons) is on the inside. In Figure 3.25(b), ascending tracts are indicated in blue; descending tracts are indicated in red. (See *Figure 3.25b*.)

---

spinal root   A bundle of axons surrounded by connective tissue that occurs in pairs, which fuse and form a spinal nerve.

cauda equina (*ee kwye na*)   A bundle of spinal roots located caudal to the end of the spinal cord.

caudal block   The anesthesia and paralysis of the lower part of the body produced by injection of a local anesthetic into the cerebrospinal fluid surrounding the cauda equina.

dorsal root   The spinal root that contains incoming (afferent) sensory fibers.

ventral root   The spinal root that contains outgoing (efferent) motor fibers.

Interim Summary

The Central Nervous System

The brain consists of three major divisions, organized around the three chambers of the tube that develops early in embryonic life: the forebrain, the midbrain, and the hindbrain. The development of the neural tube into the mature central nervous system is illustrated in Figure 3.7, and Table 3.2 outlines the major divisions and subdivisions of the brain.

During the first phase of brain development, symmetrical division of the progenitor cells of the ventricular zone, which lines the neural tube, increases its size. During the second phase, asymmetrical division of these cells gives rise to neurons, which migrate up the fibers of radial glial cells to their final resting places. There, neurons develop dendrites and axons and establish synaptic connections with other neurons. Later, neurons that fail to develop a sufficient number of synaptic connections are killed through apoptosis. Although the basic development of the nervous system is genetically controlled, sensory stimulation plays a role in refining the details. In addition, the neural circuitry of even a fully mature brain can be modified through experience.

The duplication of genes—in particular, master genes that control groups of other genes—facilitated the increase in complexity of the brain during the process of evolution. When a gene is duplicated, one of the copies can continue to perform vital functions, leaving the other copy for "experimentation" through mutations. The large size of the human brain, relative to the brains of other primates, appears to be accomplished primarily by lengthening the first and second periods of brain development.

The forebrain, which surrounds the lateral and third ventricles, consists of the telencephalon and diencephalon. The telencephalon contains the cerebral cortex, the limbic system, and the basal ganglia. The cerebral cortex is organized into the frontal, parietal, temporal, and occipital lobes. The central sulcus divides the frontal lobe, which deals specifically with movement and the planning of movement, from the other three lobes, which deal primarily with perceiving and learning. The limbic system, which includes the limbic cortex, the hippocampus, and the amygdala, is involved in emotion, motivation, and learning. The basal ganglia participate in the control of movement. The diencephalon consists of the thalamus, which directs information to and from the cerebral cortex, and the hypothalamus, which controls the endocrine system and modulates species-typical behaviors.

The midbrain, which surrounds the cerebral aqueduct, consists of the tectum and the tegmentum. The tectum is involved in audition and the control of visual reflexes and reactions to moving stimuli. The tegmentum contains the reticular formation, which is important in sleep, arousal, and movement; the periaqueductal gray matter, which controls various species-typical behaviors; and the red nucleus and the substantia nigra, both parts of the motor system. The

hindbrain, which surrounds the fourth ventricle, contains the cerebellum, the pons, and the medulla. The cerebellum plays an important role in integrating and coordinating movements. The pons contains some nuclei that are important in sleep and arousal. The medulla oblongata, too, is involved in sleep and arousal, but it also plays a role in control of movement and in control of vital functions such as heart rate, breathing, and blood pressure.

The outer part of the spinal cord consists of white matter: axons conveying information up or down. The central gray matter contains cell bodies.

THE PERIPHERAL NERVOUS SYSTEM

The brain and spinal cord communicate with the rest of the body via the cranial nerves and spinal nerves. These nerves are part of the peripheral nervous system, which conveys sensory information to the central nervous system and conveys messages from the central nervous system to the body's muscles and glands.

Spinal Nerves

The **spinal nerves** begin at the junction of the dorsal and ventral roots of the spinal cord. The nerves leave the vertebral column and travel to the muscles or sensory receptors they innervate, branching repeatedly as they go. Branches of spinal nerves often follow blood vessels,

spinal nerve A peripheral nerve attached to the spinal cord.

FIGURE 3.26 ■ Spinal Cord Cross Section

This cross section of the spinal cord shows the route taken by afferent and efferent axons through the dorsal and ventral roots.

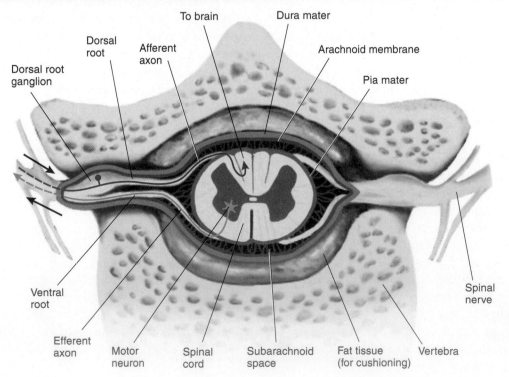

To brain · Dura mater · Dorsal root · Afferent axon · Arachnoid membrane · Dorsal root ganglion · Pia mater · Ventral root · Spinal nerve · Efferent axon · Motor neuron · Spinal cord · Subarachnoid space · Fat tissue (for cushioning) · Vertebra

especially those branches that innervate skeletal muscles. (Refer to *Figure 3.3*.)

Now let us consider the pathways by which sensory information enters the spinal cord and motor information leaves it. The cell bodies of all axons that bring sensory information into the brain and spinal cord are located outside the CNS. (The sole exception is the visual system; the retina of the eye is actually a part of the brain.) These incoming axons are referred to as **afferent axons** because they "bear toward" the CNS. The cell bodies that give rise to the axons that bring somatosensory information to the spinal cord reside in the **dorsal root ganglia**, rounded swellings of the dorsal root. (See *Figure 3.26*.) These neurons are of the unipolar type (described in Chapter 2). The axonal stalk divides close to the cell body, sending one limb into the spinal cord and the other limb out to the sensory organ. Note that all of the axons in the dorsal root convey somatosensory information.

Cell bodies that give rise to the ventral root are located within the gray matter of the spinal cord. The axons of these multipolar neurons leave the spinal cord via a ventral root, which joins a dorsal root to make a spinal nerve. The axons that leave the spinal cord through the ventral roots control muscles and glands. They are referred to as **efferent axons** because they "bear away from" the CNS. (See *Figure 3.26*.)

Cranial Nerves

Twelve pairs of **cranial nerves** are attached to the ventral surface of the brain. Most of these nerves serve sensory and motor functions of the head and neck region. One of them, the *tenth*, or **vagus nerve**, regulates the functions of organs in the thoracic and abdominal cavities. It is called the *vagus* ("wandering") nerve because its branches wander throughout the thoracic and abdominal cavities. (The word *vagabond* has the same root.) Figure 3.27 presents a view of the base of the brain and illustrates the cranial nerves and the structures they serve. Note that efferent (motor) fibers are drawn in red and that afferent (sensory) fibers are drawn in blue. (See *Figure 3.27*.)

As I mentioned in the previous section, cell bodies of sensory nerve fibers that enter the brain and spinal cord (except for the visual system) are located outside the central nervous system. Somatosensory information (and the sense of taste) is received, via the cranial nerves, from unipolar neurons. Auditory, vestibular, and visual information is received via fibers of bipolar neurons (described in Chapter 2). Olfactory information is received via the **olfactory bulbs**, which receive information from the olfactory receptors in the nose. The olfactory bulbs are complex structures that contain a considerable amount of neural circuitry; actually, they are part

of the brain. Sensory mechanisms are described in more detail in Chapters 6 and 7.

The Autonomic Nervous System

The part of the peripheral nervous system that I have discussed so far—which receives sensory information from the sensory organs and that controls movements of the skeletal muscles—is called the **somatic nervous system.** The other branch of the peripheral nervous system—the **autonomic nervous system (ANS)**—is concerned with regulation of smooth muscle, cardiac muscle, and glands. (*Autonomic* means "self-governing.") Smooth muscle is found in the skin (associated with hair follicles), in blood vessels, in the eyes (controlling pupil size and accommodation of the lens), and in the walls and sphincters of the gut, gallbladder, and urinary bladder. Merely describing the organs innervated by the autonomic nervous system suggests the function of this system: regulation of "vegetative processes" in the body.

The ANS consists of two anatomically separate systems: the *sympathetic division* and the *parasympathetic division*. With few exceptions, organs of the body are innervated by both of these subdivisions, and each has a different effect. For example, the sympathetic division speeds the heart rate, whereas the parasympathetic division slows it.

Sympathetic Division of the ANS

The **sympathetic division** is most involved in activities associated with expenditure of energy from reserves that are stored in the body. For example, when an organism is excited, the sympathetic nervous system increases

---

afferent axon An axon directed toward the central nervous system, conveying sensory information.

dorsal root ganglion A nodule on a dorsal root that contains cell bodies of afferent spinal nerve neurons.

efferent axon (*eff ur ent*) An axon directed away from the central nervous system, conveying motor commands to muscles and glands.

cranial nerve A peripheral nerve attached directly to the brain.

vagus nerve The largest of the cranial nerves, conveying efferent fibers of the parasympathetic division of the autonomic nervous system to organs of the thoracic and abdominal cavities.

olfactory bulb The protrusion at the end of the olfactory nerve; receives input from the olfactory receptors.

somatic nervous system The part of the peripheral nervous system that controls the movement of skeletal muscles or transmits somatosensory information to the central nervous system.

autonomic nervous system (ANS) The portion of the peripheral nervous system that controls the body's vegetative functions.

sympathetic division The portion of the autonomic nervous system that controls functions that accompany arousal and expenditure of energy.

FIGURE 3.27 ■ The Twelve Pairs of Cranial Nerves

The figure shows the regions and functions the cranial nerves serve. Red lines denote axons that control muscles or glands; blue lines denote sensory axons.

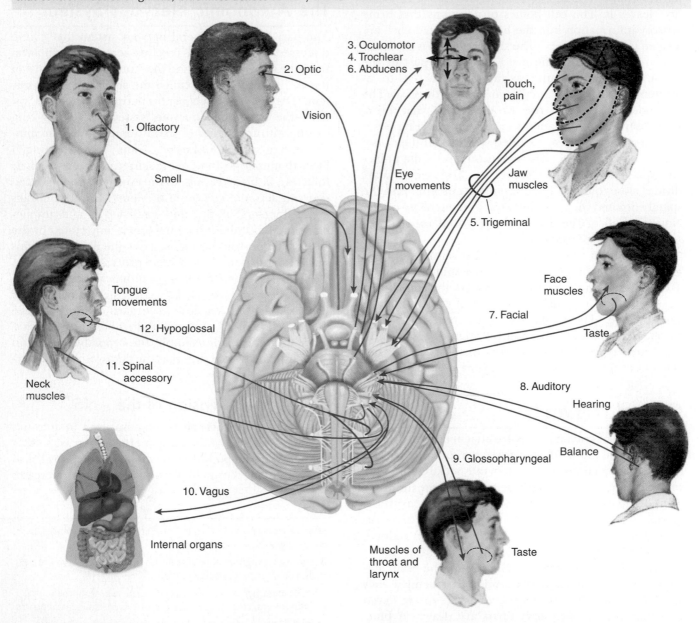

blood flow to skeletal muscles, stimulates the secretion of epinephrine (resulting in increased heart rate and a rise in blood sugar level), and causes piloerection (erection of fur in mammals that have it and production of "goose bumps" in humans).

The cell bodies of sympathetic motor neurons are located in the gray matter of the thoracic and lumbar regions of the spinal cord (hence, the sympathetic nervous system is also known as the *thoracolumbar system*). The fibers of these neurons exit via the ventral roots. After joining the spinal nerves, the fibers branch off and pass into **sympathetic ganglia** (not to be confused with the

dorsal root ganglia). Figure 3.28 shows the relationship of these ganglia to the spinal cord. Note that individual sympathetic ganglia are connected to the neighboring ganglia above and below, thus forming the **sympathetic ganglion chain**. (See *Figure 3.28.*)

sympathetic ganglia Nodules that contain synapses between preganglionic and postganglionic neurons of the sympathetic nervous system.

sympathetic ganglion chain One of a pair of groups of sympathetic ganglia that lie ventrolateral to the vertebral column.

FIGURE 3.28 ■ The Autonomic Nervous System

This schematic figure shows the target organs and functions served by the sympathetic and parasympathetic branches of the autonomic nervous system.

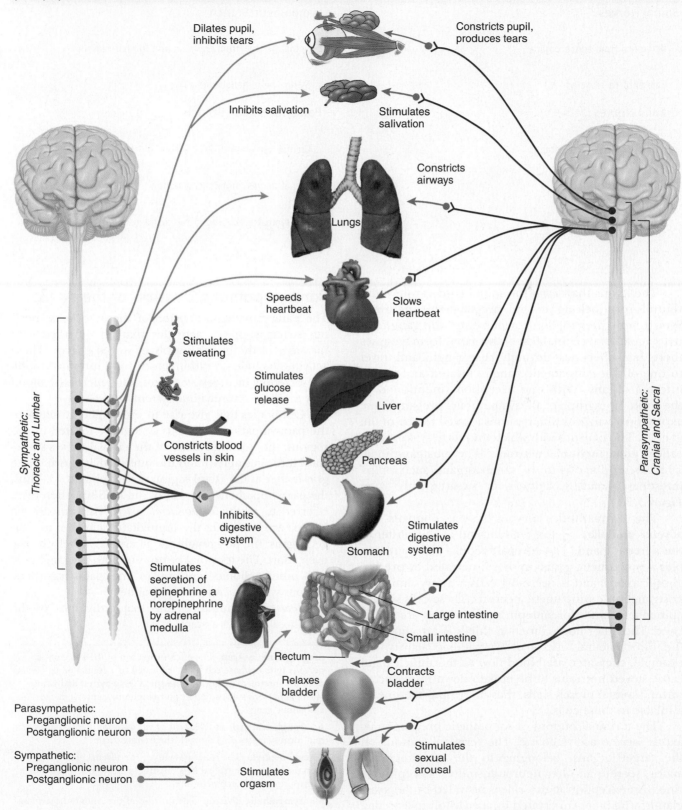

TABLE 3.3 ■ The Major Divisions of the Peripheral Nervous System

SOMATIC NERVOUS SYSTEM	AUTONOMIC NERVOUS SYSTEM (ANS)
Spinal Nerves	Sympathetic Branch
Afferents from sense organs	Spinal nerves (from thoracic and lumbar regions)
Efferents to muscles	Sympathetic ganglia
Cranial Nerves	Parasympathetic Branch
Afferents from sense organs	Cranial nerves (3rd, 7th, 9th, and 10th)
Efferents to muscles	Spinal nerves (from sacral region)
	Parasympathetic ganglia (adjacent to target organs)

The axons that leave the spinal cord through the ventral root belong to the **preganglionic neurons.** Sympathetic preganglionic axons enter the ganglia of the sympathetic chain. Most of the axons form synapses there, but others pass through these ganglia and travel to one of the sympathetic ganglia located among the internal organs. With one exception (mentioned in the next paragraph), all sympathetic preganglionic axons form synapses with neurons located in one of the ganglia. The neurons with which they form synapses are called **postganglionic neurons.** The postganglionic neurons send axons to the target organs, such as the intestines, stomach, kidneys, or sweat glands. (See *Figure 3.28.*)

The sympathetic nervous system controls the **adrenal medulla,** a set of cells located in the center of the adrenal gland. The adrenal medulla closely resembles a sympathetic ganglion. It is innervated by preganglionic axons, and its secretory cells are very similar to postganglionic sympathetic neurons. These cells secrete epinephrine and norepinephrine when they are stimulated. These hormones function chiefly as an adjunct to the direct neural effects of sympathetic activity; for example, they increase blood flow to the muscles and cause stored nutrients to be broken down into glucose within skeletal muscle cells, thus increasing the energy available to these cells.

The terminal buttons of sympathetic preganglionic axons secrete acetylcholine. The terminal buttons on the target organs, belonging to the postganglionic axons, secrete another neurotransmitter: norepinephrine. (An exception to this rule is provided by the sweat glands, which are innervated by acetylcholine-secreting terminal buttons.)

Parasympathetic Division of the ANS

The **parasympathetic division** of the autonomic nervous system supports activities that are involved with increases in the body's supply of stored energy. These activities include salivation, gastric and intestinal motility, secretion of digestive juices, and increased blood flow to the gastrointestinal system.

Cell bodies that give rise to preganglionic axons in the parasympathetic nervous system are located in two regions: the nuclei of some of the cranial nerves (especially the vagus nerve) and the intermediate horn of the gray matter in the sacral region of the spinal cord. Thus, the parasympathetic division of the ANS has often been referred to as the *craniosacral system.* Parasympathetic ganglia are located in the immediate vicinity of the target organs; the postganglionic fibers are therefore relatively short. The terminal buttons of both preganglionic and postganglionic neurons in the parasympathetic nervous system secrete acetylcholine.

Table 3.3 summarizes the major divisions of the peripheral nervous system.

preganglionic neuron The efferent neuron of the autonomic nervous system whose cell body is located in a cranial nerve nucleus or in the intermediate horn of the spinal gray matter and whose terminal buttons synapse upon postganglionic neurons in the autonomic ganglia.

postganglionic neuron Neurons of the autonomic nervous system that form synapses directly with their target organ.

adrenal medulla The inner portion of the adrenal gland, located atop the kidney, controlled by sympathetic nerve fibers; secretes epinephrine and norepinephrine.

parasympathetic division The portion of the autonomic nervous system that controls functions that occur during a relaxed state.

Interim Summary

The Peripheral Nervous System

The spinal nerves and the cranial nerves convey sensory axons into the central nervous system and motor axons out from it. Spinal nerves are formed by the junctions of the dorsal roots, which contain incoming (afferent) axons, and the ventral roots, which contain outgoing (efferent) axons. The autonomic nervous system consists of two divisions: the sympathetic division, which controls activities that occur during excitement or exertion, such as increased heart rate, and the parasympathetic division, which controls activities that occur during relaxation, such as decreased heart rate and increased activity of the digestive system. The pathways of the autonomic nervous system contain preganglionic axons, from the brain or spinal cord to the sympathetic or parasympathetic ganglia, and postganglionic axons, from the ganglia to the target organ. The adrenal medulla, which secretes epinephrine and norepinephrine, is controlled by axons of the sympathetic nervous system.

SUGGESTED READINGS

Diamond, M. C., Scheibel, A. B., and Elson, L. M. *The Human Brain Coloring Book.* New York: Barnes & Noble, 1985.

Gluhbegovic, N., and Williams, T. H. *The Human Brain: A Photographic Guide.* New York: Harper & Row, 1980.

Heimer, L. *The Human Brain and Spinal Cord: Functional Neuroanatomy and Dissection Guide,* 2nd ed. New York: Springer-Verlag, 1995.

Nauta, W. J. H., and Feirtag, M. *Fundamental Neuroanatomy.* New York: W. H. Freeman, 1986.

Netter, F. H. *The CIBA Collection of Medical Illustrations. Vol. 1: Nervous System. Part 1: Anatomy and Physiology.* Summit, NJ: CIBA Pharmaceutical Products Co., 1991.

Woolsey, T. A., Hanaway, J., and Gado, M. H. *The Brain Atlas: A Guide to the Human Central Nervous System,* 2nd ed. Hoboken, NJ: John Wiley & Sons, 2003.

ADDITIONAL RESOURCES

Visit www.mypsychkit.com for additional review and practice of the material covered in this chapter. Within MyPsychKit, you can take practice tests and receive a customized study plan to help you review. Dozens of animations, tutorials, and Web links are also available. You can even review using the interactive electronic version of this textbook. You will need to register for MyPsychKit. See www.mypsychkit.com for complete details.

Several years ago I spent the academic year in a neurological research center affiliated with the teaching hospital at a medical center. One morning as I was having breakfast, I read a brief item in the newspaper about a man who had been hospitalized for botulism. Later that morning, I attended a weekly meeting during which the chief of neurology discussed interesting cases presented by the neurological residents. I was surprised to see that we would visit the man with botulism.

We entered the intensive care unit and saw that the man was clearly on his way to recovery. His face was pale and his voice was weak, but he was no longer on a respirator. There wasn't much to see, so we went back to the lounge and discussed his case.

Just before dinner a few days earlier, Mr. F. had opened a jar of asparagus that his family had canned. He noted right away that it smelled funny. Because his family had grown the asparagus in their own garden, he was reluctant to throw it away. However, he decided that he wouldn't take any chances. He dipped a spoon into the liquid in the jar and touched it to his tongue. It didn't taste right, so he didn't swallow it. Instead, he stuck his tongue out and rinsed it under a stream of water from the faucet at the kitchen sink. He dumped the asparagus into the garbage disposal.

About an hour later, as the family was finishing dinner, Mr. F. discovered that he was seeing double. Alarmed, he asked his wife to drive him to the hospital. When he arrived at the emergency room, he was seen by one of the neurological residents, who asked him, "Mr. F., you haven't eaten some home-canned foods recently, have you?"

Learning that he had indeed let some liquid from a suspect jar of asparagus touch his tongue, the resident ordered a vial of botulinum antitoxin from the pharmacy. Meanwhile, he took a blood sample from Mr. F.'s vein and sent it to the lab for some in vivo testing in mice. He then administered the antitoxin, but already he could see that it was too late: Mr. F. was showing obvious signs of muscular weakness and was having some difficulty breathing. He was immediately sent to the intensive care unit, where he was put on a respirator. Although he became completely paralyzed, the life support system did what its name indicates, and he regained control of his muscles.

What fascinated me the most was the in vivo testing procedure for the presence of botulinum toxin in Mr. F.'s blood. Plasma extracted from the blood was injected into several mice, half of which had been pretreated with botulinum antitoxin. The pretreated mice survived; the others all died. Just think: Mr. F. had touched only a few drops of the contaminated liquid on his tongue and then rinsed it off immediately, but enough of the toxin entered his bloodstream that a small amount of his blood plasma could kill a mouse. By the way, we will examine the pharmacological effect of botulinum toxin later in this chapter.

Chapter 2 introduced you to the cells of the nervous system, and Chapter 3 described its basic structure. Now it is time to build on this information by introducing the field of psychopharmacology. **Psychopharmacology** is the study of the effects of drugs on the nervous system and (of course) on behavior. (*Pharmakon* is the Greek word for "drug.")

But what *is* a drug? Like many words, this one has several different meanings. In one context it refers to a medication that we would obtain from a pharmacist—a chemical that has a therapeutic effect on a disease or its symptoms. In another context the word refers to a chemical that people are likely to abuse, such as heroin or cocaine. The meaning that will be used in this book (and the one generally accepted by pharmacologists) is "an exogenous chemical not necessary for normal cellular functioning that significantly alters the functions of certain cells of the body when taken in relatively low doses." Because the topic of this chapter is *psycho*pharmacology, we will concern ourselves here only with chemicals that alter the functions of cells within the nervous system. The word *exogenous* rules out chemical messengers produced by the body, such as neurotransmitters, neuromodulators, or hormones. (*Exogenous* means "produced from without"—that is, from outside the body.) Chemical messengers produced by the body are not drugs, although synthetic chemicals that mimic their effects are classified as drugs. The definition of a drug also rules out essential nutrients, such as proteins, fats, carbohydrates, minerals, and vitamins that are a necessary constituent of a healthy diet. Finally, it states that drugs are effective in low doses. This qualification is important, because large quantities of almost any substance—even common ones such as table salt—will alter the functions of cells.

As we will see in this chapter, drugs have *effects* and *sites of action*. **Drug effects** are the changes we can observe in an animal's physiological processes and behavior. For example, the effects of morphine, heroin, and other opiates include decreased sensitivity to pain, slowing of the digestive system, sedation, muscular

psychopharmacology The study of the effects of drugs on the nervous system and on behavior.

drug effect The changes a drug produces in an animal's physiological processes and behavior.

relaxation, constriction of the pupils, and euphoria. The **sites of action** of drugs are the points at which molecules of drugs interact with molecules located on or in cells of the body, thus affecting some biochemical processes of these cells. For example, the sites of action of the opiates are specialized receptors situated in the membrane of certain neurons. When molecules of opiates attach to and activate these receptors, the drugs alter the activity of these neurons and produce their effects. This chapter considers both the effects of drugs and their sites of action.

Psychopharmacology is an important field of neuroscience. It has been responsible for the development of psychotherapeutic drugs, which are used to treat psychological and behavioral disorders. It has also provided tools that have enabled other investigators to study the functions of cells of the nervous system and the behaviors controlled by particular neural circuits.

This chapter does not contain all this book has to say about the subject of psychopharmacology. Throughout the book you will learn about the use of drugs to investigate the nature of neural circuits involved in the control of perception, memory, and behavior. In addition, Chapters 16 and 17 discuss the use of drugs to study and treat mental disorders such as schizophrenia, depression, and the anxiety disorders, and Chapter 18 discusses the physiology of drug abuse.

PRINCIPLES OF PSYCHOPHARMACOLOGY

This chapter begins with a description of the basic principles of psychopharmacology: the routes of administration of drugs and their fate in the body. The second section discusses the sites of drug actions. The final section discusses specific neurotransmitters and neuromodulators and the physiological and behavioral effects of specific drugs that interact with them.

Pharmacokinetics

To be effective, a drug must reach its sites of action. To do so, molecules of the drug must enter the body and then enter the bloodstream so that they can be carried to the organ (or organs) on which they act. Once there, they must leave the bloodstream and come into contact with the molecules with which they interact. For almost all of the drugs we are interested in, this means that the molecules of the drug must enter the central nervous system (CNS). Some behaviorally active drugs exert their effects on the peripheral nervous system, but these drugs are less important to us than the drugs that affect cells of the CNS.

Molecules of drugs must cross several barriers to enter the body and find their way to their sites of action. Some molecules pass through these barriers easily and quickly; others do so very slowly. And once molecules of drugs enter the body, they begin to be metabolized—broken down by enzymes—or excreted in the urine (or both). In time, the molecules either disappear or are transformed into inactive fragments. The process by which drugs are absorbed, distributed within the body, metabolized, and excreted is referred to as **pharmacokinetics** ("movements of drugs").

Routes of Administration

First, let's consider the routes by which drugs can be administered. For laboratory animals the most common route is injection. The drug is dissolved in a liquid (or, in some cases, suspended in a liquid in the form of fine particles) and injected through a hypodermic needle. The fastest route is **intravenous (IV) injection**—injection into a vein. The drug enters the bloodstream immediately and reaches the brain within a few seconds. The disadvantages of IV injections are the increased care and skill they require in comparison to most other forms of injection and the fact that the entire dose reaches the bloodstream at once. If an animal is especially sensitive to the drug, there may be little time to administer another drug to counteract its effects.

An **intraperitoneal (IP) injection** is rapid but not as rapid as an IV injection. The drug is injected through the abdominal wall into the *peritoneal cavity*—the space that surrounds the stomach, intestines, liver, and other abdominal organs. IP injections are the most common route for administering drugs to small laboratory animals. An **intramuscular (IM) injection** is made directly into a large muscle, such as those found in the upper arm, thigh, or buttocks. The drug is absorbed into the bloodstream through the capillaries that supply the muscle. If very slow absorption is desirable, the drug can be mixed with another drug (such as ephedrine) that constricts blood vessels and retards the flow of blood

sites of action The locations at which molecules of drugs interact with molecules located on or in cells of the body, thus affecting some biochemical processes of these cells.

pharmacokinetics The process by which drugs are absorbed, distributed within the body, metabolized, and excreted.

intravenous (IV) injection Injection of a substance directly into a vein.

intraperitoneal (IP) injection (*in tra pair i toe nee ul*) Injection of a substance into the *peritoneal cavity*—the space that surrounds the stomach, intestines, liver, and other abdominal organs.

intramuscular (IM) injection Injection of a substance into a muscle.

through the muscle. A drug can also be injected into the space beneath the skin by means of a **subcutaneous (SC) injection.** A subcutaneous injection is useful only if small amounts of drug need to be administered, because injecting large amounts would be painful. Some fat-soluble drugs can be dissolved in vegetable oil and administered subcutaneously. In this case, molecules of the drug will slowly leave the deposit of oil over a period of several days. If *very* slow and prolonged absorption of a drug is desirable, the drug can be formed into a dry pellet or placed in a sealed silicone rubber capsule and implanted beneath the skin.

Oral administration is the most common form of administering medicinal drugs to humans. Because of the difficulty of getting laboratory animals to eat something that does not taste good to them, researchers seldom use this route. Some chemicals cannot be administered orally because they will be destroyed by stomach acid or digestive enzymes or because they are not absorbed from the digestive system into the bloodstream. For example, insulin, a peptide hormone, must be injected. **Sublingual administration** of certain drugs can be accomplished by placing them beneath the tongue. The drug is absorbed into the bloodstream by the capillaries that supply the mucous membrane that lines the mouth. (Obviously, this method works only with humans, who will cooperate and leave the capsule beneath their tongue.) Nitroglycerine, a drug that causes blood vessels to dilate, is taken sublingually by people who suffer the pains of angina pectoris, caused by obstructions in the coronary arteries.

Drugs can also be administered at the opposite end of the digestive tract, in the form of suppositories. **Intrarectal administration** is rarely used to give drugs to experimental animals. For obvious reasons this process would be difficult with a small animal. In addition, when agitated, small animals such as rats tend to defecate, which would mean that the drug would not remain in place long enough to be absorbed. And I'm not sure I would want to try to administer a rectal suppository to a large animal. Rectal suppositories are most commonly used to administer drugs that might upset a person's stomach.

The lungs provide another route for drug administration: **inhalation.** Nicotine, freebase cocaine, and marijuana are usually smoked. In addition, drugs used to treat lung disorders are often inhaled in the form of a vapor or fine mist, and many general anesthetics are gasses that are administered through inhalation. The route from the lungs to the brain is very short, and drugs administered this way have very rapid effects.

Some drugs can be absorbed directly through the skin, so they can be given by means of **topical administration.** Natural or artificial steroid hormones can be administered in this way, as can nicotine (as a treatment to make it easier for a person to stop smoking). The mucous membrane lining the nasal passages also provides a route for topical administration. Commonly abused drugs such as cocaine hydrochloride are often sniffed so that they come into contact with the nasal mucosa. This route delivers the drug to the brain very rapidly. (The technical, rarely used name for this route is *insufflation.* And note that sniffing is not the same as inhalation; when powdered cocaine is sniffed, it ends up in the mucous membrane of the nasal passages, not in the lungs.)

Finally, drugs can be administered directly into the brain. As we saw in Chapter 2, the blood–brain barrier prevents certain chemicals from leaving capillaries and entering the brain. Some drugs cannot cross the blood–brain barrier. If these drugs are to reach the brain, they must be injected directly into the brain or into the cerebrospinal fluid in the brain's ventricular system. To study the effects of a drug in a specific region of the brain (for example, in a particular nucleus of the hypothalamus), a researcher will inject a very small amount of the drug directly into the brain. This procedure, known as **intracerebral administration,** is described in more detail in Chapter 5. To achieve a widespread distribution of a drug in the brain, a researcher will get past the blood–brain barrier by injecting the drug into a cerebral ventricle. The drug is then absorbed into the brain tissue, where it can exert its effects. This route, **intracerebroventricular (ICV) administration,** is used very rarely in humans—primarily to deliver antibiotics directly to the brain to treat certain types of infections.

Figure 4.1 shows the time course of blood levels of a commonly abused drug, cocaine, after intravenous injection, inhalation, oral administration, and sniffing. The amounts received were not identical, but the graph illustrates the relative rapidity with which the drug reaches the blood. (See *Figure 4.1.*)

---

subcutaneous (SC) injection Injection of a substance into the space beneath the skin.

oral administration Administration of a substance into the mouth so that it is swallowed.

sublingual administration (*sub ling wul*) Administration of a substance by placing it beneath the tongue.

intrarectal administration Administration of a substance into the rectum.

inhalation Administration of a vaporous substance into the lungs.

topical administration Administration of a substance directly onto the skin or mucous membrane.

intracerebral administration Administration of a substance directly into the brain.

intracerebroventricular (ICV) administration Administration of a substance into one of the cerebral ventricles.

FIGURE 4.1 ■ Cocaine in Blood Plasma

The graph shows the concentration of cocaine in blood plasma after intravenous injection, inhalation, oral administration, and sniffing.

(Adapted from Feldman, R. S., Meyer, J. S., and Quenzer, L. F. *Principles of Neuropsychopharmacology.* Sunderland, MA: Sinauer Associates,1997; after Jones, R. T. *NIDA Research Monographs,* 1990, *99,* 30–41.)

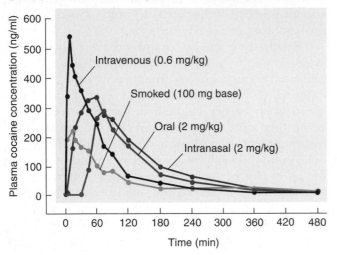

Distribution of Drugs Within the Body

As we saw, drugs exert their effects only when they reach their sites of action. In the case of drugs that affect behavior, most of these sites are located on or in particular cells in the central nervous system. The previous section described the routes by which drugs can be introduced into the body. With the exception of intracerebral or intracerebroventricular administration the routes of drug administration vary only in the rate at which a drug reaches the blood plasma (that is, the liquid part of the blood). But what happens next? All the sites of action of drugs of interest to psychopharmacologists lie outside the blood vessels.

Several factors determine the rate at which a drug in the bloodstream reaches sites of action within the brain. The first is lipid solubility. The blood–brain barrier is a barrier only for water-soluble molecules. Molecules that are soluble in lipids pass through the cells that line the capillaries in the central nervous system, and they rapidly distribute themselves throughout the brain. For example, diacetylmorphine (more commonly known as heroin) is more lipid soluble than morphine is. Thus, an intravenous injection of heroin produces more rapid effects than does one of morphine. Even though the molecules of the two drugs are equally effective when they reach their sites of action in the brain, the fact that heroin molecules get there faster means that they produce a more intense "rush," and this explains why drug addicts prefer heroin to morphine.

Many drugs bind with various tissues of the body or with proteins in the blood—a phenomenon known as **depot binding.** As long as the molecules of the drug are bound to a depot, they cannot reach their sites of action and cannot exert their effects. One source of such binding is **albumin,** a protein found in the blood. Albumin serves to transport free fatty acids, a source of nutrients for most cells of the body, but this protein can also bind with some lipid-soluble drugs. Depot binding can both delay and prolong the effects of a drug. Consider a lipid-soluble drug taken orally. As molecules of the drug are absorbed from the stomach, they begin to bind with albumin in the blood. For a while, very little of the drug reaches the brain. Finally, the albumin molecules can hold no more of the drug, so it begins to enter the brain. Eventually, all the drug is absorbed from the stomach. Then, perhaps over a period of several hours, the albumin molecules gradually release the molecules of the drug as the plasma concentration of the drug falls. (See *Figure 4.2.*)

Other sources of depot binding include fat tissue, bones, muscles, and the liver. Of course, drugs bind with these depots more slowly than they do with albumin, because they must leave the blood vessels to do so. Thus, these sources of binding are less likely to interfere with the initial effects of a drug. For example, thiopental, a barbiturate that is sometimes used to anesthetize the brain, has high lipid solubility. An intravenous injection of this drug reaches the brain within a few seconds after being injected intravenously. The drug also binds very well with muscles and fat tissue, so it soon is taken out of circulation, and the levels of the drug in the brain fall rapidly. Within 30 minutes or so, the drug's anesthetic effect is gone. Eventually, the drug is destroyed by enzymes and excreted by the kidneys.

Inactivation and Excretion

Drugs do not remain in the body indefinitely. Many are deactivated by enzymes, and all are eventually excreted, primarily by the kidneys. The liver plays an especially active role in enzymatic deactivation of drugs, but some deactivating enzymes are also found in the blood. The brain also contains enzymes that destroy some drugs. In some cases, enzymes transform molecules of a drug into other forms that themselves are biologically active. Occasionally, the transformed molecule is *even more* active than the one that is administered. In such cases the effects of a drug can have a very long duration.

depot binding Binding of a drug with various tissues of the body or with proteins in the blood.

albumin (*al bew min*) A protein found in the blood; serves to transport free fatty acids and can bind with some lipid-soluble drugs.

FIGURE 4.2 ■ Depot Binding with Blood Albumin Protein

(a) The drug begins to be absorbed from the stomach into the bloodstream, where it binds with albumin. (b) The albumin molecules are saturated with the drug and can hold no more. (c) Unbound molecules of the drug begin to enter the brain. Eventually, molecules of the drug will break away from the molecules of albumin and enter the brain.

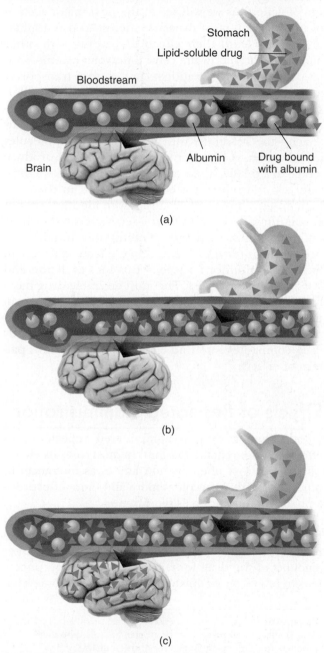

(a)

(b)

(c)

FIGURE 4.3 ■ A Dose-Response Curve

Increasingly stronger doses of the drug produce increasingly larger effects until the maximum effect is reached. After that point, increments in the dose do not produce any increments in the drug's effect. However, the risk of adverse side effects increases.

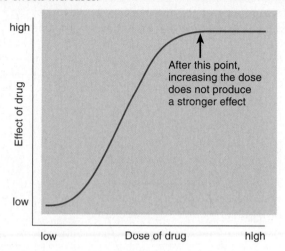

Drug Effectiveness

Drugs vary widely in their effectiveness. The effects of a small dose of a relatively effective drug can equal or exceed the effects of larger amounts of a relatively ineffective drug. The best way to measure the effectiveness of a drug is to plot a **dose-response curve.** To do this, subjects are given various doses of a drug, usually defined as milligrams of drug per kilogram of a subject's body weight, and the effects of the drug are plotted. Because the molecules of most drugs distribute themselves throughout the blood and then throughout the rest of the body, a heavier subject (human or laboratory animal) will require a larger quantity of a drug to achieve the same concentration as a smaller quantity will produce in a smaller subject. As Figure 4.3 shows, increasingly stronger doses of a drug cause increasingly larger effects until the point of maximum effect is reached. At this point, increasing the dose of the drug does not produce any more effect. (See *Figure 4.3.*)

Most drugs have more than one effect. Opiates such as morphine and codeine produce analgesia (reduced sensitivity to pain), but they also depress the activity of neurons in the medulla that control heart rate and respiration. A physician who prescribes an opiate to relieve a patient's pain wants to administer a dose that is large enough to produce analgesia but not large enough to depress heart rate and respiration—effects that could be fatal. Figure 4.4 shows two dose-response curves, one for the analgesic effects of a painkiller and one for the drug's depressant effects on respiration. The difference between these curves indicates the drug's margin of

dose-response curve A graph of the magnitude of an effect of a drug as a function of the amount of drug administered.

FIGURE 4.4 ■ Dose Response Curves for Morphine

The dose-response curve on the left shows the analgesic effect of morphine, and the curve on the right shows one of the drug's adverse side effects: its depressant effect on respiration. A drug's margin of safety is reflected by the difference between the dose-response curve for its therapeutic effects and that for its adverse side effects.

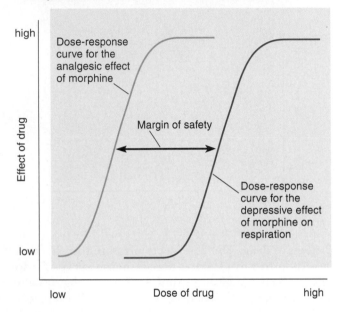

safety. Obviously, the most desirable drugs have a large margin of safety. (See *Figure 4.4.*)

One measure of a drug's margin of safety is its **therapeutic index.** This measure is obtained by administering varying doses of the drug to a group of laboratory animals such as mice. Two numbers are obtained: the dose that produces the desired effects in 50 percent of the animals and the dose that produces toxic effects in 50 percent of the animals. The therapeutic index is the ratio of these two numbers. For example, if the toxic dose is five times higher than the effective dose, then the therapeutic index is 5.0. The lower the therapeutic index, the more care must be taken in prescribing the drug. For example, barbiturates have relatively low therapeutic indexes—as low as 2 or 3. In contrast, tranquilizers such as Valium have therapeutic indexes of well over 100. As a consequence, an accidental overdose of a barbiturate is much more likely to have tragic effects than a similar overdose of or Valium.

Why do drugs vary in their effectiveness? There are two reasons. First, different drugs—even those with the same behavioral effects—may have different sites of action. For example, both morphine and aspirin have analgesic effects, but morphine suppresses the activity of neurons in the spinal cord and brain that are involved

in pain perception, whereas aspirin reduces the production of a chemical involved in transmitting information from damaged tissue to pain-sensitive neurons. Because the drugs act very differently, a given dose of morphine (expressed in terms of milligrams of drug per kilogram of body weight) produces much more pain reduction than the same dose of aspirin does.

The second reason that drugs vary in their effectiveness has to do with the affinity of the drug with its site of action. As we will see in the next major section of this chapter, most drugs of interest to psychopharmacologists exert their effects by binding with other molecules located in the central nervous system—with presynaptic or postsynaptic receptors, with transporter molecules, or with enzymes involved in the production or deactivation of neurotransmitters. Drugs vary widely in their **affinity** for the molecules to which they attach—the readiness with which the two molecules join together. A drug with a high affinity will produce effects at a relatively low concentration, whereas a drug with a low affinity must be administered in relatively high doses. Thus, even two drugs with identical sites of action can vary widely in their effectiveness if they have different affinities for their binding sites. In addition, because most drugs have multiple effects, a drug can have high affinities for some of its sites of action and low affinities for others. The most desirable drug has a high affinity for sites of action that produce therapeutic effects and a low affinity for sites of action that produce toxic side effects. One of the goals of research by drug companies is to find chemicals with just this pattern of effects.

Effects of Repeated Administration

Often, when a drug is administered repeatedly, its effects will not remain constant. In most cases its effects will diminish—a phenomenon known as **tolerance.** In other cases a drug becomes more and more effective—a phenomenon known as **sensitization.**

Let's consider tolerance first. Tolerance is seen in many drugs that are commonly abused. For example, a regular user of heroin must take larger and larger amounts of the drug for it to be effective. And once a person has taken an opiate regularly enough to develop

therapeutic index The ratio between the dose that produces the desired effect in 50 percent of the animals and the dose that produces toxic effects in 50 percent of the animals.

affinity The readiness with which two molecules join together.

tolerance A decrease in the effectiveness of a drug that is administered repeatedly.

sensitization An increase in the effectiveness of a drug that is administered repeatedly.

tolerance, that individual will suffer **withdrawal symptoms** if he or she suddenly stops taking the drug. Withdrawal symptoms are primarily the opposite of the effects of the drug itself. For example, heroin produces euphoria; withdrawal from it produces *dysphoria*—a feeling of anxious misery. (*Euphoria* and *dysphoria* mean "easy to bear" and "hard to bear," respectively.) Heroin produces constipation; withdrawal from it produces nausea and cramping. Heroin produces relaxation; withdrawal from it produces agitation.

Withdrawal symptoms are caused by the same mechanisms that are responsible for tolerance. Tolerance is the result of the body's attempt to compensate for the effects of the drug. That is, most systems of the body, including those controlled by the brain, are regulated so that they stay at an optimal value. When the effects of a drug alter these systems for a prolonged time, compensatory mechanisms begin to produce the opposite reaction, at least partially compensating for the disturbance from the optimal value. These mechanisms account for the fact that more and more of the drug must be taken to achieve a given level of effects. Then, when the person stops taking the drug, the compensatory mechanisms make themselves felt, unopposed by the action of the drug.

Research suggests that there are several types of compensatory mechanisms. As we will see, many drugs that affect the brain do so by binding with receptors and activating them. The first compensatory mechanism involves a decrease in the effectiveness of such binding. Either the receptors become less sensitive to the drug (that is, their affinity for the drug decreases) or the receptors decrease in number. The second compensatory mechanism involves the process that couples the receptors to ion channels in the membrane or to the production of second messengers. After prolonged stimulation of the receptors, one or more steps in the coupling process become less effective. (Of course, *both* effects can occur.) The details of these compensatory mechanisms are described in Chapter 18, which discusses the causes and effects of drug abuse.

As we saw, many drugs have several different sites of action and thus produce several different effects. This means that some of the effects of a drug may show tolerance but others may not. For example, barbiturates cause sedation and also depress neurons that control respiration. The sedative effects show tolerance, but the respiratory depression does not. This means that if larger and larger doses of a barbiturate are taken to achieve the same level of sedation, the person begins to run the risk of taking a dangerously large dose of the drug.

Sensitization is, of course, the exact opposite of tolerance: Repeated doses of a drug produce larger and larger effects. Because compensatory mechanisms tend to correct for deviations away from the optimal values of physiological processes, sensitization is less common than tolerance. And some of the effects of a drug may show sensitization while others show tolerance. For example, repeated injections of cocaine become more and more likely to produce movement disorders and convulsions, whereas the euphoric effects of the drug do not show sensitization—and may even show tolerance.

Placebo Effects

A **placebo** is an innocuous substance that has no specific physiological effect. The word comes from the Latin *placere,* "to please." A physician may sometimes give a placebo to anxious patients to placate them. (You can see that the word *placate* also has the same root.) But although placebos have no *specific* physiological effect, it is incorrect to say that they have *no* effect. If a person thinks that a placebo has a physiological effect, then administration of the placebo may actually produce that effect.

When experimenters want to investigate the behavioral effects of drugs in humans, they must use control groups whose members receive placebos, or they cannot be sure that the behavioral effects they observe were caused by specific effects of the drug. Studies with laboratory animals must also use placebos, even though we need not worry about the animals' "beliefs" about the effects of the drugs we give them. Consider what you must do to give a rat an intraperitoneal injection of a drug. You reach into the animal's cage, pick the animal up, hold it in such a way that its abdomen is exposed and its head is positioned to prevent it from biting you, insert a hypodermic needle through its abdominal wall, press the plunger of the syringe, and replace the animal in its cage, being sure to let go of it quickly so that it cannot turn and bite you. Even if the substance you inject is innocuous, the experience of receiving the injection would activate the animal's autonomic nervous system, cause the secretion of stress hormones, and have other physiological effects. If we want to know what the behavioral effects of a drug are, we must compare the drug-treated animals with other animals who receive a placebo, administered in exactly the same way as the drug. (By the way, a skilled and experienced researcher can handle a rat so gently that it shows very little reaction to a hypodermic injection.)

---

withdrawal symptom The appearance of symptoms opposite to those produced by a drug when the drug is administered repeatedly and then suddenly no longer taken.

placebo (*pla see boh*) An inert substance that is given to an organism in lieu of a physiologically active drug; used experimentally to control for the effects of mere administration of a drug.

Interim Summary

Principles of Psychopharmacology

Psychopharmacology is the study of the effects of drugs on the nervous system and behavior. Drugs are exogenous chemicals that are not necessary for normal cellular functioning that significantly alter the functions of certain cells of the body when taken in relatively low doses. Drugs have *effects,* physiological and behavioral, and they have *sites of action*—molecules with which they interact to produce these effects.

Pharmacokinetics is the fate of a drug as it is absorbed into the body, circulates throughout the body, and reaches its sites of action. Drugs may be administered by intravenous, intraperitoneal, intramuscular, and subcutaneous injection; they may be administered orally, sublingually, intrarectally, by inhalation, and topically (on skin or mucous membrane); and they may be injected intracerebrally or intracerebroventricularly. Lipid-soluble drugs easily pass through the blood–brain barrier, whereas others pass this barrier slowly or not at all.

The time courses of various routes of drug administration are different. And once molecules of a drug reach the blood, they may bind with albumin protein, and they may bind with storage depots in fat tissue, muscles, or bones. Eventually, drugs disappear from the body. Some are deactivated by enzymes, especially in the liver, and others are simply excreted.

The dose-response curve represents a drug's effectiveness; it relates the amount administered (usually in milligrams per kilogram of the subject's body weight) to the resulting effect. Most drugs have more than one site of action and thus more than one effect. The safety of a drug is measured by the difference between doses that produce desirable effects and those that produce toxic side effects. Drugs vary in their effectiveness because of the nature of their sites of actions and the affinity between molecules of the drug and these sites of action.

Repeated administration of a drug can cause either tolerance, often resulting in withdrawal symptoms, or sensitization. Tolerance can be caused by decreased affinity of a drug with its receptors, by decreased numbers of receptors, or by decreased coupling of receptors with the biochemical steps it controls. Some of the effects of a drug may show tolerance, while others may not—or may even show sensitization.

SITES OF DRUG ACTION

Throughout the history of our species, people have discovered that plants—and a few animals—produce chemicals that act on synapses. (Of course, the people who discovered these chemicals knew nothing about neurons and synapses.) Some of these chemicals have been used for their pleasurable effects; others have been used to treat illness, reduce pain, or poison other animals (or enemies). More recently, scientists have learned to produce completely artificial drugs, some with potencies far greater than those of the naturally occurring drugs. The traditional uses of drugs remain, but in addition, they can be used in research laboratories to investigate the operations of the nervous system. Most drugs that affect behavior do so by affecting synaptic transmission. Drugs that affect synaptic transmission are classified into two general categories. Those that block or inhibit the postsynaptic effects are called **antagonists.** Those that facilitate them are called **agonists.** (The Greek word *agon* means "contest." Thus, an *agonist* is one who takes part in the contest.)

This section will describe the basic effects of drugs on synaptic activity. Recall from Chapter 2 that the sequence of synaptic activity goes like this: Neurotransmitters are synthesized and stored in synaptic vesicles. The synaptic vesicles travel to the presynaptic membrane, where they become docked. When an axon fires, voltage-dependent calcium channels in the presynaptic membrane open, permitting the entry of calcium ions. The calcium ions interact with the docking proteins and initiate the release of the neurotransmitters into the synaptic cleft. Molecules of the neurotransmitter bind with postsynaptic receptors, causing particular ion channels to open, which produces excitatory or inhibitory postsynaptic potentials. The effects of the neurotransmitter are kept relatively brief by their reuptake by transporter molecules in the presynaptic membrane or by their destruction by enzymes. In addition, the stimulation of presynaptic autoreceptors regulates the synthesis and release of the neurotransmitter. The discussion of the effects of drugs in this section

antagonist A drug that opposes or inhibits the effects of a particular neurotransmitter on the postsynaptic cell.

agonist A drug that facilitates the effects of a particular neurotransmitter on the postsynaptic cell.

follows the same basic sequence. All of the effects I will describe are summarized in Figure 4.5, with some details shown in additional figures. I should warn you that some of the effects are complex, so the discussion that follows bears careful reading. I recommend that you study **MyPsychKit 4.1**, *Actions of Drugs*, which reviews this material.

mypsychkit Where learning comes to life!

Animation 4.1
Actions of Drugs

Effects on Production of Neurotransmitters

The first step is the synthesis of the neurotransmitter from its precursors. In some cases the rate of synthesis and release of a neurotransmitter is increased when a precursor is administered; in these cases the precursor itself serves as an agonist. (See step 1 in *Figure 4.5*.)

The steps in the synthesis of neurotransmitters are controlled by enzymes. Therefore, if a drug inactivates one of these enzymes, it will prevent the neurotransmitter from being produced. Such a drug serves as an antagonist. (See step 2 in *Figure 4.5*.)

Effects on Storage and Release of Neurotransmitters

Neurotransmitters are stored in synaptic vesicles, which are transported to the presynaptic membrane, where the chemicals are released. The storage of neurotransmitters in vesicles is accomplished by the same kind of transporter molecules that are responsible for reuptake of a neurotransmitter into a terminal button. The transporter molecules are located in the membrane of synaptic vesicles, and their action is to pump molecules of the neurotransmitter across the membrane, filling the vesicles. Some of the transporter molecules that fill synaptic vesicles are capable of being blocked by a drug. Molecules of the drug bind with a particular site on the transporter and inactivate it. Because the synaptic vesicles remain empty, nothing is released when the vesicles eventually rupture against the presynaptic membrane. The drug serves as an antagonist. (See step 3 in *Figure 4.5*.)

Some drugs act as antagonists by preventing the release of neurotransmitters from the terminal button. They do so by deactivating the proteins that cause docked synaptic vesicles to fuse with the presynaptic

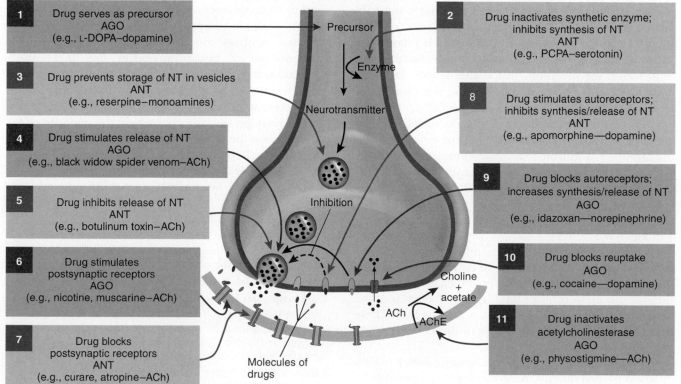

FIGURE 4.5 ■ Drug Affects on Synaptic Transmission

The figure summarizes the ways in which drugs can affect the synaptic transmission (AGO = agonist; ANT = antagonist; NT = neurotransmitter). Drugs that act as agonists are marked in blue; drugs that act as antagonists are marked in red.

1 Drug serves as precursor
AGO
(e.g., L-DOPA–dopamine)

2 Drug inactivates synthetic enzyme; inhibits synthesis of NT
ANT
(e.g., PCPA–serotonin)

3 Drug prevents storage of NT in vesicles
ANT
(e.g., reserpine–monoamines)

8 Drug stimulates autoreceptors; inhibits synthesis/release of NT
ANT
(e.g., apomorphine—dopamine)

4 Drug stimulates release of NT
AGO
(e.g., black widow spider venom–ACh)

9 Drug blocks autoreceptors; increases synthesis/release of NT
AGO
(e.g., idazoxan—norepinephrine)

5 Drug inhibits release of NT
ANT
(e.g., botulinum toxin–ACh)

10 Drug blocks reuptake
AGO
(e.g., cocaine—dopamine)

6 Drug stimulates postsynaptic receptors
AGO
(e.g., nicotine, muscarine–ACh)

11 Drug inactivates acetylcholinesterase
AGO
(e.g., physostigmine—ACh)

7 Drug blocks postsynaptic receptors
ANT
(e.g., curare, atropine–ACh)

Precursor
Enzyme
Neurotransmitter
Inhibition
Choline + acetate
ACh
AChE
Molecules of drugs

FIGURE 4.6 ■ Drug Actions at Binding Sites

(a) Competitive binding: Direct agonists and antagonists act directly on the neurotransmitter binding site. (b) Noncompetitive binding: Indirect agonists and antagonists act on an alternative binding site and modify the effects of the neurotransmitter on opening of the ion channel.

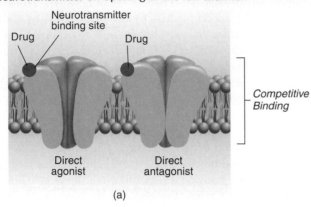

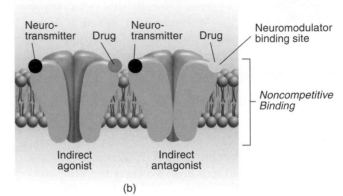

(a)   (b)

membrane and expel their contents into the synaptic cleft. Other drugs have just the opposite effect: They act as agonists by binding with these proteins and directly triggering release of the neurotransmitter. (See steps 4 and 5 in *Figure 4.5.*)

Effects on Receptors

The most important—and most complex—site of action of drugs in the nervous system is on receptors, both presynaptic and postsynaptic. Let's consider postsynaptic receptors first. (Here is where the careful reading should begin.) Once a neurotransmitter is released, it must stimulate the postsynaptic receptors. Some drugs bind with these receptors, just as the neurotransmitter does. Once a drug has bound with the receptor, it can serve as either an agonist or an antagonist.

A drug that mimics the effects of a neurotransmitter acts as a **direct agonist.** Molecules of the drug attach to the binding site to which the neurotransmitter normally attaches. This binding causes ion channels controlled by the receptor to open, just as they do when the neurotransmitter is present. Ions then pass through these channels and produce postsynaptic potentials. (See step 6 in *Figure 4.5.*)

Drugs that bind with postsynaptic receptors can also serve as antagonists. Molecules of such drugs bind with the receptors but do not open the ion channel. Because they occupy the receptor's binding site, they prevent the neurotransmitter from opening the ion channel. These drugs are called **receptor blockers** or **direct antagonists.** (See step 7 in *Figure 4.5.*)

Some receptors have multiple binding sites, to which different ligands can attach. Molecules of the neurotransmitter bind with one site, and other substances (such as neuromodulators and various drugs) bind with the others.

Binding of a molecule with one of these alternative sites is referred to as **noncompetitive binding,** because the molecule does not compete with molecules of the neurotransmitter for the same binding site. If a drug attaches to one of these alternative sites and prevents the ion channel from opening, the drug is said to be an **indirect antagonist.** The ultimate *effect* of an indirect antagonist is similar to that of a direct antagonist, but its site of action is different. If a drug attaches to one of the alternative sites and *facilitates* the opening of the ion channel, it is said to be an **indirect agonist.** (See *Figure 4.6.*)

As we saw in Chapter 2, the presynaptic membranes of some neurons contain autoreceptors that regulate the amount of neurotransmitter that is released. Because stimulation of these receptors causes less neurotransmitter to be released, drugs that selectively activate presynaptic receptors act as antagonists. Drugs that *block* presynaptic autoreceptors have the opposite effect: They *increase* the release of the neurotransmitter, acting as agonists. (Refer to steps 8 and 9 in *Figure 4.5.*)

We also saw in Chapter 2 that some terminal buttons form axoaxonic synapses—synapses of one terminal

---

direct agonist A drug that binds with and activates a receptor.

receptor blocker A drug that binds with a receptor but does not activate it; prevents the natural ligand from binding with the receptor.

direct antagonist A synonym for receptor blocker.

noncompetitive binding Binding of a drug to a site on a receptor; does not interfere with the binding site for the principal ligand.

indirect antagonist A drug that attaches to a binding site on a receptor and interferes with the action of the receptor; does not interfere with the binding site for the principal ligand.

indirect agonist A drug that attaches to a binding site on a receptor and facilitates the action of the receptor; does not interfere with the binding site for the principal ligand.

button with another. Activation of the first terminal button causes presynaptic inhibition or facilitation of the second one. The second terminal button contains **presynaptic heteroreceptors,** which are sensitive to the neurotransmitter released by the first one. (*Auto* means "self"; *hetero* means "other.") Presynaptic heteroreceptors that produce presynaptic inhibition do so by inhibiting the release of the neurotransmitter. Conversely, presynaptic heteroreceptors responsible for presynaptic facilitation *facilitate* the release of the neurotransmitter. So drugs can block or facilitate presynaptic inhibition or facilitation, depending on whether they block or activate presynaptic heteroreceptors. (See *Figure 4.7.*)

Finally (yes, this is the last site of action I will describe in this subsection), you will recall from Chapter 2 that autoreceptors are located in the membrane of dendrites of some neurons. When these neurons become active, their dendrites, as well as their terminal

FIGURE 4.8 ■ Dendritic Autoreceptors

The dendrites of certain neurons release some neurotransmitter when the cell is active. Activation of dendritic autoreceptors by the neurotransmitter (or by a drug that binds with these receptors) hyperpolarizes the membrane, reducing the neuron's rate of firing. Blocking of dendritic autoreceptors by a drug prevents this effect.

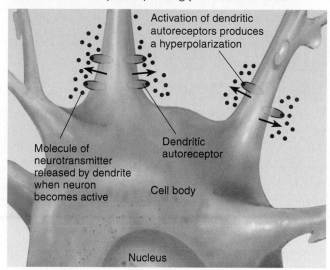

FIGURE 4.7 ■ Presynaptic Heteroreceptors

Presynaptic facilitation is caused by activation of receptors that facilitate the opening of calcium channels near the active zone of the postsynaptic terminal button, which promotes release of the neurotransmitter. Presynaptic inhibition is caused by activation of receptors that inhibit the opening of these calcium channels.

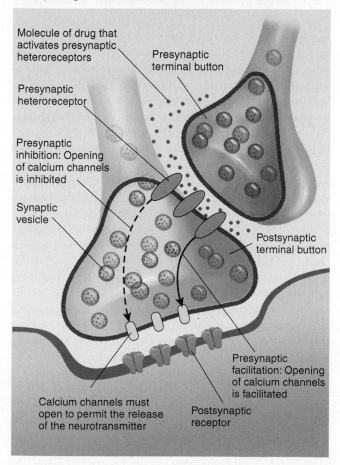

buttons, release neurotransmitter. The neurotransmitter released by the dendrites stimulates autoreceptors located on these same dendrites, which decrease neural firing by producing hyperpolarizations. This mechanism has a regulatory effect, serving to prevent these neurons from becoming too active. Thus, drugs that bind with and *activate* dendritic autoreceptors will serve as *antagonists*. Those that bind with and *block* dendritic autoreceptors will serve as *agonists*, because they will prevent the inhibitory hyperpolarizations. (See *Figure 4.8.*)

As you will surely realize, the effects of a particular drug that binds with a particular type of receptor can be very complex. The effects depend on where the receptor is located, what its normal effects are, and whether the drug activates the receptor or blocks its actions.

Effects on Reuptake or Destruction of Neurotransmitters

The next step after stimulation of the postsynaptic receptor is termination of the postsynaptic potential. Two processes accomplish that task: Molecules of the neurotransmitter are taken back into the terminal

---

presynaptic heteroreceptor A receptor located in the membrane of a terminal button that receives input from another terminal button by means of an axoaxonic synapse; binds with the neurotransmitter released by the presynaptic terminal button.

button through the process of reuptake, or they are destroyed by an enzyme. Drugs can interfere with either of these processes. In the first case, molecules of the drug attach to the transporter molecules responsible for reuptake and inactivate them, thus blocking reuptake. In the second case, molecules of the drug bind with the enzyme that normally destroys the neurotransmitter and prevents the enzymes from working. The most important example of such an enzyme is acetylcholinesterase, which destroys acetylcholine. Because both types of drugs prolong the presence of the neurotransmitter in the synaptic cleft (and hence in a location where they can stimulate postsynaptic receptors), they serve as *agonists*. (Refer to steps 10 and 11 in *Figure 4.5*.)

InterimSummary

Sites of Drug Action

The process of synaptic transmission entails the synthesis of the neurotransmitter, its storage in synaptic vesicles, its release into the synaptic cleft, its interaction with postsynaptic receptors, and the consequent opening of ion channels in the postsynaptic membrane. The effects of the neurotransmitter are then terminated by reuptake into the terminal button or, in the case of acetylcholine, by enzymatic deactivation.

Each of the steps necessary for synaptic transmission can be interfered with by drugs that serve as *antagonists,* and a few can be stimulated by drugs that serve as *agonists.* Thus, drugs can increase the pool of available precursor, block a biosynthetic enzyme, prevent the storage of neurotransmitter in synaptic vesicles, stimulate or block the release of the neurotransmitter, stimulate or block presynaptic or postsynaptic receptors, retard reuptake, or deactivate enzymes that destroy the neurotransmitter. A drug that activates postsynaptic receptors serves as an agonist, whereas one that activates presynaptic or dendritic autoreceptors or serves as an antagonist. A drug that blocks postsynaptic receptors serves as an antagonist, whereas one that blocks autoreceptors serves as an agonist. A drug that activates or blocks presynaptic heteroreceptors serves as an agonist or antagonist, depending on whether the heteroreceptors are responsible for presynaptic facilitation or inhibition.

NEUROTRANSMITTERS AND NEUROMODULATORS

Because neurotransmitters have two general effects on postsynaptic membranes—depolarization (EPSP) or hyperpolarization (IPSP)—one might expect that there would be two kinds of neurotransmitters, excitatory and inhibitory. Instead, there are many different kinds—several dozen at least. In the brain most synaptic communication is accomplished by two neurotransmitters: one with excitatory effects (glutamate) and one with inhibitory effects (GABA). (Another inhibitory neurotransmitter, glycine, is found in the spinal cord and lower brain stem.) Most of the activity of local circuits of neurons involves balances between the excitatory and inhibitory effects of these chemicals, which are responsible for most of the information transmitted from place to place within the brain. In fact, there are probably no neurons in the brain that do not receive excitatory input from glutamate-secreting terminal buttons and inhibitory input from neurons that secrete either GABA or glycine. And with the exception of neurons that detect painful stimuli, all sensory organs transmit information to the brain through axons whose terminals release glutamate. (Pain-detecting neurons secrete a peptide.)

What do all the other neurotransmitters do? In general, they have modulating effects rather than information-transmitting effects. That is, the release of neurotransmitters other than glutamate and GABA tends to activate or inhibit entire circuits of neurons that are involved in particular brain functions. For example, secretion of acetylcholine activates the cerebral cortex and facilitates learning, but the information that is learned and remembered is transmitted by neurons that secrete glutamate and GABA. Secretion of norepinephrine increases vigilance and enhances readiness to act when a signal is detected. Secretion of serotonin suppresses certain categories of species-typical behaviors and reduces the likelihood that the animal acts impulsively. Secretion of dopamine in some regions of the brain generally activates voluntary movements but does not specify which movements will occur. In other regions, secretion of dopamine reinforces ongoing behaviors and makes them more likely to occur at a later time. Because particular drugs can selectively affect neurons that secrete particular neurotransmitters, they can have specific effects on behavior.

This section introduces the most important neurotransmitters, discusses some of their behavioral functions, and describes the drugs that interact with them. As we saw in the previous section of this chapter, drugs have many different sites of action. Fortunately for your information-processing capacity (and perhaps your sanity), not all types of neurons are affected by all types of drugs. As you will see, that still leaves a good number of drugs to be mentioned by name. Obviously, some are more important than others. Those whose effects I describe in some detail are more important than those I mention in passing. If you want to learn more details about these drugs (and many others), you should consult one of the psychopharmacology texts listed in the suggested readings at the end of this chapter.

Acetylcholine

Acetylcholine is the primary neurotransmitter secreted by efferent axons of the peripheral nervous system. All muscular movement is accomplished by the release of acetylcholine, and ACh is also found in the ganglia of the autonomic nervous system and at the target organs of the parasympathetic branch of the ANS. Because ACh is found outside the central nervous system in locations that are easy to study, this neurotransmitter was the first to be discovered, and it has received much attention from neuroscientists. Some terminology: These synapses are said to be *acetylcholinergic. Ergon* is the Greek word for "work." Thus, *dopaminergic* synapses release dopamine, *serotonergic* synapses release serotonin, and so on. (The suffix *-ergic* is pronounced "*ur jik*".)

The axons and terminal buttons of acetylcholinergic neurons are distributed widely throughout the brain. Three systems have received the most attention from neuroscientists: those originating in the dorsolateral pons, the basal forebrain, and the medial septum. The effects of ACh release in the brain are generally facilitatory. The acetylcholinergic neurons located in the dorsolateral pons play a role in REM sleep (the phase of sleep during which dreaming occurs). Those located in the basal forebrain are involved in activating the cerebral cortex and facilitating learning, especially perceptual learning. Those located in the medial septum control the electrical rhythms of the hippocampus and modulate its functions, which include the formation of particular kinds of memories.

Figure 4.9 shows a schematic midsagittal view of a rat brain. On it are indicated the most important sites of acetylcholinergic cell bodies and the regions served by the branches of their axons. The figure illustrates a rat brain because most of the neuroanatomical tracing studies have been performed with rats. Presumably, the location and projections of acetylcholinergic neurons in the human brain resemble those found in the rat brain, but we cannot yet be certain. The methods used for tracing particular systems of neurons in the brain and the difficulty of doing such studies with the human brain are described in Chapter 5. (See *Figure 4.9.*)

FIGURE 4.9 ■ Acetylcholinergic Pathways in a Rat Brain

This schematic figure shows the locations of the most important groups of acetylcholinergic neurons and the distribution of their axons and terminal buttons.

(Adapted from Woolf, N. J. *Progress in Neurobiology*, 1991, *37*, 475–524.)

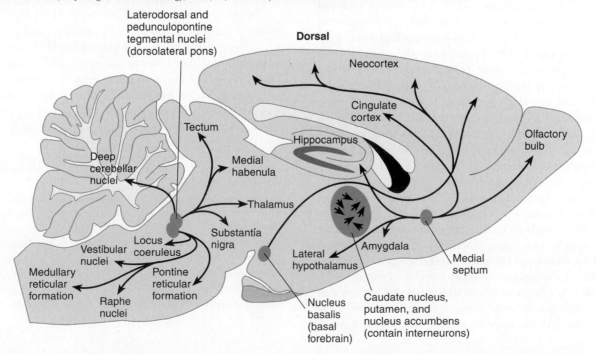

FIGURE 4.10 ■ Biosynthesis of Acetylcholine

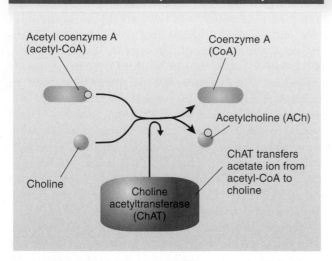

Acetylcholine is composed of two components: *choline,* a substance derived from the breakdown of lipids, and *acetate,* the anion found in vinegar, also called acetic acid. Acetate cannot be attached directly to choline; instead, it is transferred from a molecule of *acetyl-CoA.* CoA (coenzyme A) is a complex molecule, consisting in part of the vitamin pantothenic acid (one of the B vitamins). CoA is produced by the mitochondria, and it takes part in many reactions in the body. **Acetyl-CoA** is simply CoA with an acetate ion attached to it. ACh is produced by the following reaction: In the presence of the enzyme **choline acetyltransferase (ChAT),** the acetate ion is transferred from the acetyl-CoA molecule to the choline molecule, yielding a molecule of ACh and one of ordinary CoA. (See *Figure 4.10.*)

A simple analogy will illustrate the role of coenzymes in chemical reactions. Think of acetate as a hot dog and choline as a bun. The task of the person (enzyme) who operates the hot dog vending stand is to put a hot dog into the bun (make acetylcholine). To do so, the vendor needs a fork (coenzyme) to remove the hot dog from the boiling water. The vendor inserts the fork into the hot dog (attaches acetate to CoA) and transfers the hot dog from fork to bun.

Two drugs, botulinum toxin and the venom of the black widow spider, affect the release of acetylcholine. **Botulinum toxin** is produced by *clostridium botulinum,* a bacterium that can grow in improperly canned food. This drug prevents the release of ACh (step 5 of Figure 4.5). As we saw in this chapter's opening case, botulinum toxin drug is an extremely potent poison because the paralysis it can cause leads to suffocation. In contrast, **black widow spider venom** has the opposite effect: It stimulates the release of ACh (step 4 of Figure 4.5). Although the effects of black widow spider venom can also be fatal, the venom is much less toxic than botulinum toxin. In fact, most healthy adults would have to

receive several bites, but infants or frail, elderly people would be more susceptible.

You may have been wondering why double vision was the first symptom of botulism in the opening case. The answer is that the delicate balance among the muscles that move the eyes is upset by any interference with acetylcholinergic transmission. You undoubtedly know that *botox* treatment has become fashionable. A very dilute (obviously!) solution of botulinum toxin is injected into people's facial muscles to stop muscular contractions that are causing wrinkles in the skin. I'm not planning on getting a botox treatment, but if I did, I would want to be sure that the solution was sufficiently dilute.

You will recall from Chapter 2 that after being released by the terminal button, ACh is deactivated by the enzyme acetylcholinesterase (AChE), which is present in the postsynaptic membrane. The deactivation produces choline and acetate from ACh. Because the amount of choline that is picked up by the soma from the general circulation and then sent to the terminal buttons by means of axoplasmic flow is not sufficient to keep up with the loss of choline by an active synapse, choline must be recycled. After ACh is destroyed by the AChE in the postsynaptic membrane, the choline is returned to the terminal buttons by means of reuptake. There, it is converted back into ACh. This process has an efficiency of 50 percent; that is, half of the choline is retrieved and recycled. (See *Figure 4.11.*)

Reuptake of choline can be blocked by a drug called **hemicholinium.** Because this drug prevents the recycling of choline, the terminal button must rely solely on transport of this substance from the cell body. The result is the production (and release) of less acetylcholine, which means that hemicholinium serves as an acetylcholine antagonist. (See *Figure 4.11.*)

Drugs that deactivate AChE (step 11 of Figure 4.5) are used for several purposes. Some are used as insecticides. These drugs readily kill insects but not humans and other mammals, because our blood contains enzymes that destroy them. (Insects lack the enzyme.) Other AChE inhibitors are used medically. For example, a hereditary disorder called *myasthenia gravis* is caused by an attack of a person's immune system against acetylcholine receptors

acetyl-CoA (*a see tul*) A cofactor that supplies acetate for the synthesis of acetylcholine.

choline acetyltransferase (ChAT) (*koh leen a see tul **trans** fer ace*) The enzyme that transfers the acetate ion from acetyl coenzyme A to choline, producing the neurotransmitter acetylcholine.

botulinum toxin (*bot you **lin** um*) An acetylcholine antagonist; prevents release by terminal buttons.

black widow spider venom A poison produced by the black widow spider that triggers the release of acetylcholine.

hemicholinium (*hem ee koh **lin** um*) A drug that inhibits the uptake of choline.

FIGURE 4.11 ■ The Acetycholine–Choline Cycle

The figure illustrates the destruction of acetylcholine by acetylcholinesterase and the reuptake of choline. The drug hemicholinium blocks the reuptake of choline.

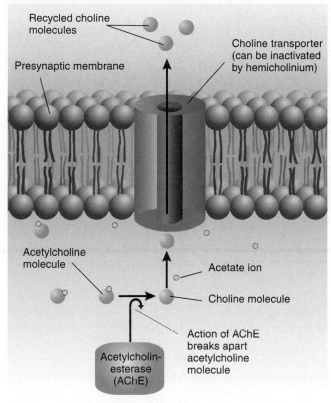

production of second messengers, their actions are slower and more prolonged than those of nicotinic receptors. The central nervous system contains both kinds of ACh receptors, but muscarinic receptors predominate. Some nicotinic receptors are found at axoaxonic synapses in the brain, where they produce presynaptic facilitation. Activation of these receptors is responsible for the addictive effect of the nicotine found in tobacco smoke.

Just as two different drugs stimulate the two classes of acetylcholine receptors, two different drugs *block* them (step 7 of Figure 4.5). Both drugs were discovered in nature long ago, and both are still used by modern medicine. The first, **atropine,** blocks muscarinic receptors. The drug is named after *Atropos,* the Greek fate who cut the thread of life (which a sufficient dose of atropine will certainly do). Atropine is one of several *belladonna alkaloids* extracted from a plant called the deadly nightshade, and therein lies a tale. Many years ago, women who wanted to increase their attractiveness to men put drops containing belladonna alkaloids into their eyes. In fact, *belladonna* means "pretty lady." Why was the drug used this way? One of the unconscious responses that occurs when we are interested in something is dilation of our pupils. By blocking the effects of acetylcholine on the pupil, belladonna alkaloids such as atropine make the pupils dilate. This change makes a woman appear more interested in a man when she looks at him, and, of course, this apparent sign of interest makes him regard her as more attractive.

Another drug, **curare,** blocks nicotinic receptors. Because these receptors are the ones found on muscles, curare, like botulinum toxin, causes paralysis. However, the effects of curare are much faster. The drug is extracted from several different species of plants found in South America, where it was discovered long ago by people who used it to coat the tips of arrows and darts. Within minutes of being struck by one of these points, an animal collapses, ceases breathing, and dies. Nowadays, curare (and other drugs with the same site of action) are used to paralyze patients who are to undergo surgery so that their muscles will relax completely and not contract when they are cut with a scalpel. An anesthetic must also be used, because a person who receives only curare will remain perfectly conscious and

located on skeletal muscles. The person becomes weaker and weaker as the muscles become less responsive to the neurotransmitter. If the person is given an AChE inhibitor such as **neostigmine,** the person will regain some strength because the acetylcholine that is released has a more prolonged effect on the remaining receptors. (Neostigmine cannot cross the blood–brain barrier, so it does not affect the AChE found in the central nervous system.)

There are two types of ACh receptors: one ionotropic and one metabotropic. These receptors were identified when investigators discovered that different drugs activated them (step 6 of Figure 4.5). The ionotropic ACh receptor is stimulated by nicotine, a drug found in tobacco leaves. (The Latin name of the plant is *Nicotiniana tabacum.*) The metabotropic ACh receptor is stimulated by muscarine, a drug found in the poison mushroom *Amanita muscaria.* Consequently, these two ACh receptors are referred to as **nicotinic receptors** and **muscarinic receptors,** respectively. Because muscle fibers must be able to contract rapidly, they contain the rapid, ionotropic nicotinic receptors.

Because muscarinic receptors are metabotropic in nature and thus control ion channels through the

neostigmine (*nee o stig meen*) A drug that inhibits the activity of acetylcholinesterase.

nicotinic receptor An ionotropic acetylcholine receptor that is stimulated by nicotine and blocked by curare.

muscarinic receptor (*muss ka rin ic*) A metabotropic acetylcholine receptor that is stimulated by muscarine and blocked by atropine.

atropine (*a tro peen*) A drug that blocks muscarinic acetylcholine receptors.

curare (*kew rahr ee*) A drug that blocks nicotinic acetylcholine receptors.

sensitive to pain, even though paralyzed. And, of course, a respirator must be used to supply air to the lungs.

The Monoamines

Dopamine, norepinephrine, epinephrine, and serotonin are four chemicals that belong to a family of compounds called **monoamines.** Because the molecular structures of these substances are similar, some drugs affect the activity of all of them to some degree. The first three—dopamine, norepinephrine, and epinephrine—belong to a subclass of monoamines called **catecholamines.** It is worthwhile learning the terms in Table 4.1, because they will be used many times throughout the rest of this book. (See *Table 4.1.*)

The monoamines are produced by several systems of neurons in the brain. Most of these systems consist of a relatively small number of cell bodies located in the brain stem, whose axons branch repeatedly and give rise to an enormous number of terminal buttons distributed throughout many regions of the brain. Monoaminergic neurons thus serve to modulate the function of widespread regions of the brain, increasing or decreasing the activities of particular brain functions.

Dopamine

The first catecholamine in Table 4.1, **dopamine (DA),** produces both excitatory and inhibitory postsynaptic potentials, depending on the postsynaptic receptor. Dopamine is one of the more interesting neurotransmitters because it has been implicated in several important functions, including movement, attention, learning, and the reinforcing effects of drugs that people tend to abuse; therefore, it is discussed in Chapters 8, 9, 13, and 18.

The synthesis of the catecholamines is somewhat more complicated than that of ACh, but each step is a simple one. The precursor molecule is modified slightly, step by step, until it achieves its final shape. Each step is controlled by a different enzyme, which causes a small part to be added or taken off. The precursor for the two major catecholamine neurotransmitters (dopamine and norepinephrine) is *tyrosine,* an essential amino acid that we must obtain from our diet. Tyrosine receives a

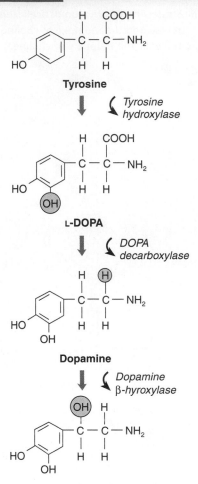

FIGURE 4.12 ■ **Biosynthesis of the Catecholamines**

hydroxyl group (OH—an oxygen atom and a hydrogen atom) and becomes L-**DOPA** (L-3,4-dihydroxyphenylalanine). The enzyme that adds the hydroxyl group is called *tyrosine hydroxylase.* L-DOPA then loses a carboxyl group (COOH—one carbon atom, two oxygen atoms, and one hydrogen atom) through the activity of the enzyme *DOPA decarboxylase* and becomes dopamine. Finally, the enzyme *dopamine β-hydroxylase* attaches a hydroxyl group to dopamine, which becomes norepinephrine. These reactions are shown in *Figure 4.12.*

TABLE 4.1 ■ Classification of the Monoamine Neurotransmitters

CATECHOLAMINES	INDOLAMINES
Dopamine	Serotonin
Norepinephrine	
Epinephrine	

monoamine (***mahn** o a meen*) A class of amines that includes indolamines, such as serotonin; and catecholamines, such as dopamine, norepinephrine, and epinephrine.

catecholamine (*cat a **kohl** a meen*) A class of amines that includes the neurotransmitters dopamine, norepinephrine, and epinephrine.

dopamine (DA) (***dope** a meen*) A neurotransmitter; one of the catecholamines.

L-**DOPA** (*ell **dope** a*) The levorotatory form of DOPA; the precursor of the catecholamines; often used to treat Parkinson's disease because of its effect as a dopamine agonist.

FIGURE 4.13 ■ Dopaminergic Pathways in a Rat Brain

This schematic figure shows the locations of the most important groups of dopaminergic neurons and the distribution of their axons and terminal buttons.

(Adapted from Fuxe, K., Agnati, L. F., Kalia, M., Goldstein, M., Andersson K., and Harfstrand A. in *Basic and Clinical Aspects of Neuroscience: The Dopaminergic System,* edited by E. Fluckinger, E. E. Muller, and M. O. Thomas. Berlin: Springer–Verlag, 1985.)

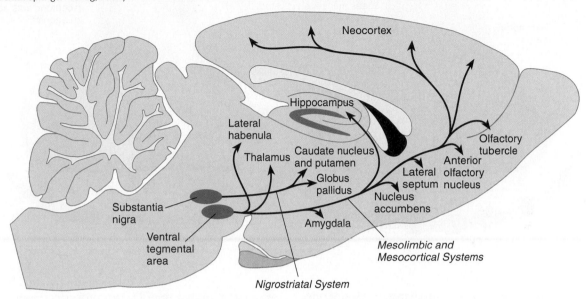

The brain contains several systems of dopaminergic neurons. The three most important of these originate in the midbrain: in the substantia nigra and in the ventral tegmental area. (The substantia nigra was shown in Figure 3.23; the ventral tegmental area is located just below this region.) The cell bodies of neurons of the **nigrostriatal system** are located in the substantia nigra and project their axons to the neostriatum: the caudate nucleus and the putamen. The neostriatum is an important part of the basal ganglia, which is involved in the control of movement. The cell bodies of neurons of the **mesolimbic system** are located in the ventral tegmental area and project their axons to several parts of the limbic system, including the nucleus accumbens, amygdala, and hippocampus. The nucleus accumbens plays an important role in the reinforcing (rewarding) effects of certain categories of stimuli, including those of drugs that people abuse. The cell bodies of neurons of the **mesocortical system** are also located in the ventral tegmental area. Their axons project to the prefrontal cortex. These neurons have an excitatory effect on the frontal cortex and affect such functions as formation of short-term memories, planning, and strategy preparation for problem solving. These three systems of dopaminergic neurons are shown in *Figure 4.13.*

Degeneration of dopaminergic neurons that connect the substantia nigra with the caudate nucleus causes **Parkinson's disease,** a movement disorder characterized by tremors, rigidity of the limbs, poor balance, and

difficulty in initiating movements. The cell bodies of these neurons are located in a region of the brain called the *substantia nigra* ("black substance"). This region is normally stained black with melanin, the substance that gives color to skin. This compound is produced by the breakdown of dopamine. (The brain damage that causes Parkinson's disease was discovered by pathologists who observed that the substantia nigra of a deceased person who had had this disorder was pale rather than black.) People with Parkinson's disease are given L-DOPA, the precursor to dopamine. Although dopamine cannot cross the blood–brain barrier, L-DOPA can. Once L-DOPA reaches the brain, it is taken up by dopaminergic neurons and is converted to dopamine (step 1 of Figure 4.5). The increased synthesis of dopamine causes more

nigrostriatal system (*nigh grow stry **ay** tul*) A system of neurons originating in the substantia nigra and terminating in the neostriatum (caudate nucleus and putamen).

mesolimbic system (*mee zo **lim** bik*) A system of dopaminergic neurons originating in the ventral tegmental area and terminating in the nucleus accumbens, amygdala, and hippocampus.

mesocortical system (*mee zo **kor** ti kul*) A system of dopaminergic neurons originating in the ventral tegmental area and terminating in the prefrontal cortex.

Parkinson's disease A neurological disease characterized by tremors, rigidity of the limbs, poor balance, and difficulty in initiating movements; caused by degeneration of the nigrostriatal system.

FIGURE 4.14 ■ Effects of Low and High Doses of Apomorphine

At low doses, apomorphine serves as a dopamine antagonist; at high doses, it serves as an agonist.

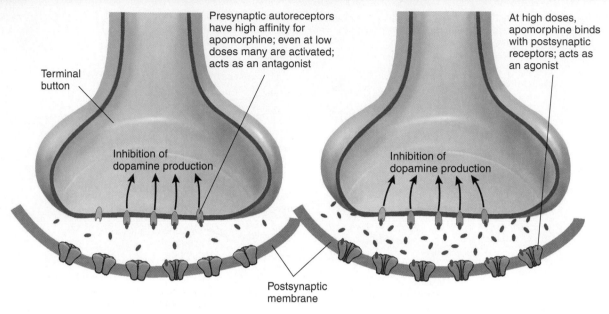

Terminal button

Presynaptic autoreceptors have high affinity for apomorphine; even at low doses many are activated; acts as an antagonist

Inhibition of dopamine production

At high doses, apomorphine binds with postsynaptic receptors; acts as an agonist

Inhibition of dopamine production

Postsynaptic membrane

dopamine to be released by the surviving dopaminergic neurons in patients with Parkinson's disease. As a consequence, the patients' symptoms are alleviated.

Another drug, **AMPT** (or α-methyl-*p*-tyrosine), inactivates tyrosine hydroxylase, the enzyme that converts tyrosine to L-DOPA (step 2 of Figure 4.5). Because this drug interferes with the synthesis of dopamine (and of norepinephrine as well), it serves as a catecholamine antagonist. The drug is not normally used medically, but it has been used as a research tool in laboratory animals.

The drug **reserpine** prevents the storage of monoamines in synaptic vesicles by blocking the transporters in the membrane of vesicles of monoaminergic neurons (step 3 of Figure 4.5). Because the synaptic vesicles remain empty, no neurotransmitter is released when an action potential reaches the terminal button. Reserpine, then, is a monoamine antagonist. The drug, which comes from the root of a shrub, was discovered over 3000 years ago in India, where it was found to be useful in treating snakebite and seemed to have a calming effect. Pieces of the root are still sold in markets in rural areas of India. In Western medicine, reserpine was previously used to treat high blood pressure, but it has been replaced by drugs with fewer side effects.

Several different types of dopamine receptors have been identified, all metabotropic. Of these, two are the most common: D_1 *receptors* and D_2 *receptors*. It appears that D_1 receptors are exclusively postsynaptic, whereas D_2 receptors are found both presynaptically and postsynaptically in the brain. Stimulation of D_1 receptors increases the production of the second messenger cyclic AMP,

whereas stimulation of D_2 receptors decreases it, as does stimulation of D_3 and D_4 receptors. Several drugs stimulate or block specific types of dopamine receptors.

Autoreceptors are found in the dendrites, soma, and terminal buttons of dopaminergic neurons. Activation of the autoreceptors in the dendritic and somatic membrane decreases neural firing by producing hyperpolarizations. The presynaptic autoreceptors located in the terminal buttons suppress the activity of the enzyme tyrosine hydroxylase and thus decrease the production of dopamine—and ultimately its release. Dopamine autoreceptors resemble D_2 receptors, but there seem to be some differences. For example, the drug **apomorphine** is a D_2 agonist, but it seems to have a greater affinity for presynaptic D_2 receptors than for postsynaptic D_2 receptors. A low dose of apomorphine acts as an antagonist, because it stimulates the presynaptic receptors and inhibits the production and release of dopamine. Higher doses begin to stimulate postsynaptic D_2 receptors, and the drug begins to act as a direct agonist. (See *Figure 4.14.*)

AMPT A drug that blocks the activity of tyrosine hydroxylase and thus interferes with the synthesis of the catecholamines.

reserpine (*ree sur peen*) A drug that interferes with the storage of monoamines in synaptic vesicles.

apomorphine (*ap o more feen*) A drug that blocks dopamine autoreceptors at low doses; at higher doses, blocks postsynaptic receptors as well.

FIGURE 4.15 ■ Role of Monoamine Oxidase

This schematic shows the role of monoamine oxidase in dopaminergic terminal buttons and the action of deprenyl.

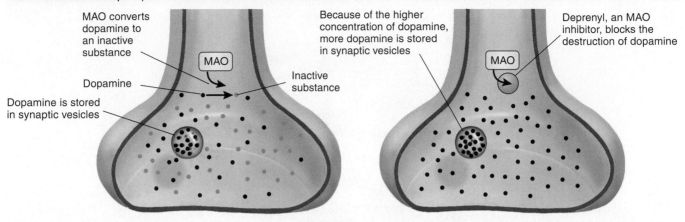

Several drugs inhibit the reuptake of dopamine, thus serving as potent dopamine agonists (step 10 of Figure 4.5). The best known of these drugs are amphetamine, cocaine, and methylphenidate. Amphetamine has an interesting effect: It causes the release of both dopamine and norepinephrine by causing the transporters for these neurotransmitters to run in reverse, propelling DA and NE into the synaptic cleft. Of course, this action also blocks reuptake of these neurotransmitters. Cocaine and **methylphenidate** simply block dopamine reuptake. Because cocaine also blocks voltage-dependent sodium channels, it is sometimes used as a topical anesthetic, especially in the form of eye drops for eye surgery. Methylphenidate (Ritalin) is used to treat children who have attention deficit disorder.

The production of the catecholamines is regulated by an enzyme called **monoamine oxidase (MAO).** This enzyme is found within monoaminergic terminal buttons, where it destroys excessive amounts of neurotransmitter. A drug called **deprenyl** destroys the particular form of monoamine oxidase (MAO-B) that is found in dopaminergic terminal buttons. Because deprenyl prevents the destruction of dopamine, more dopamine is released when an action potential reaches the terminal button. Thus, deprenyl serves as a dopamine agonist. (See *Figure 4.15.*)

MAO is also found in the blood, where it deactivates amines that are present in foods such as chocolate and cheese. Without such deactivation these amines could cause dangerous increases in blood pressure.

Dopamine has been implicated as a neurotransmitter that might be involved in schizophrenia, a serious mental disorder whose symptoms include hallucinations, delusions, and disruption of normal, logical thought processes. Drugs such as **chlorpromazine,** which block D_2 receptors, alleviate these symptoms (step 7 of Figure 4.5). Hence, investigators have speculated that schizophrenia is produced by overactivity of dopaminergic neurons. More recently discovered drugs—the so-called *atypical antipsychotics*—have more complicated actions, which are discussed in Chapter 16.

Norepinephrine

Because **norepinephrine (NE),** like ACh, is found in neurons in the autonomic nervous system, this neurotransmitter has received much experimental attention. I should note that the terms *Adrenalin* and *epinephrine* are synonymous, as are *noradrenalin* and *norepinephrine.* Let me explain why. **Epinephrine** is a hormone produced by the adrenal medulla, the central core of the adrenal glands, located just above the kidneys. Epinephrine also serves as a neurotransmitter in the brain, but it is of minor importance compared with norepinephrine.

---

methylphenidate (*meth ul **fen** i date*) A drug that inhibits the reuptake of dopamine.

monoamine oxidase (MAO) (*mahn o a meen*) A class of enzymes that destroy the monoamines: dopamine, norepinephrine, and serotonin.

deprenyl (*depp ra nil*) A drug that blocks the activity of MAO-B; acts as a dopamine agonist.

chlorpromazine (*klor **proh** ma zeen*) A drug that reduces the symptoms of schizophrenia by blocking dopamine D_2 receptors.

norepinephrine (NE) (*nor epp i **neff** rin*) One of the catecholamines; a neurotransmitter found in the brain and in the sympathetic division of the autonomic nervous system.

epinephrine (*epp i **neff** rin*) One of the catecholamines; a hormone secreted by the adrenal medulla; serves also as a neurotransmitter in the brain.

FIGURE 4.16 ■ Noradrenergic Pathways in a Rat Brain

This schematic figure shows the locations of the most important groups of noradrenergic neurons and the distribution of their axons and terminal buttons.

(Adapted from Cotman, C. W. and McGaugh, J. L. *Behavioral Neuroscience: An Introduction.* New York: Academic Press, 1980.)

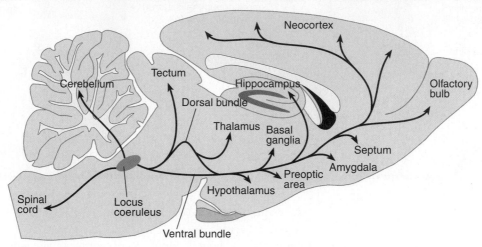

Ad renal is Latin for "toward the kidney." In Greek, one would say *epi nephron* ("upon the kidney"), hence the term *epinephrine.* The latter term has been adopted by pharmacologists, probably because the word *Adrenalin* was appropriated by a drug company as a proprietary name; therefore, to be consistent with general usage, I will refer to the neurotransmitter as *norepinephrine.* The accepted adjectival form is *noradrenergic;* I suppose that *norepinephrinergic* never caught on because it takes so long to pronounce.

We have already seen the biosynthetic pathway for norepinephrine in Figure 4.12. The drug AMPT, which prevents the conversion of tyrosine to L-DOPA, blocks the production of norepinephrine as well as dopamine (step 2 of Figure 4.5).

Most neurotransmitters are synthesized in the cytoplasm of the terminal button and then stored in newly formed synaptic vesicles. However, for norepinephrine the final step of synthesis occurs inside the vesicles themselves. The vesicles are first filled with dopamine. Then the dopamine is converted to norepinephrine through the action of the enzyme dopamine β-hydroxylase located within the vesicles. The drug **fusaric acid** inhibits the activity of the enzyme dopamine-β-hydroxylase and thus blocks the production of norepinephrine without affecting the production of dopamine.

Excess norepinephrine in the terminal buttons is destroyed by monoamine oxidase, type A. The drug **moclobemide** specifically blocks MAO-A and hence serves as a noradrenergic agonist.

Almost every region of the brain receives input from noradrenergic neurons. The cell bodies of most of these neurons are located in seven regions of the pons

and medulla and one region of the thalamus. The cell bodies of the most important noradrenergic system begin in the **locus coeruleus,** a nucleus located in the dorsal pons. The axons of these neurons project to the regions shown in Figure 4.16. As we will see later, the primary effect of activation of these neurons is an increase in vigilance—attentiveness to events in the environment. (See *Figure 4.16.*)

Most neurons that release norepinephrine do not do so through terminal buttons on the ends of axonal branches. Instead, they usually release them through **axonal varicosities,** beadlike swellings of the axonal branches. These varicosities give the axonal branches of catecholaminergic neurons the appearance of beaded chains.

There are several types of noradrenergic receptors, identified by their differing sensitivities to various drugs. Actually, these receptors are usually called *adrenergic* receptors rather than *noradrenergic* receptors, because they are sensitive to epinephrine (Adrenalin) as well as norepinephrine. Neurons in the central nervous system

fusaric acid (*few **sahr** ik*) A drug that inhibits the activity of the enzyme dopamine-β-hydroxylase and thus blocks the production of norepinephrine.

moclobemide (*mok low **bem** ide*) A drug that blocks the activity of MAO-A; acts as a noradrenergic agonist.

locus coeruleus (*sur oo lee us*) A dark-colored group of noradrenergic cell bodies located in the pons near the rostral end of the floor of the fourth ventricle.

axonal varicosity An enlarged region along the length of an axon that contains synaptic vesicles and releases a neurotransmitter or neuromodulator.

contain β_1- and β_2-*adrenergic receptors* and α_1- and α_2-*adrenergic receptors*. All four kinds of receptors are also found in various organs of the body besides the brain and are responsible for the effects of epinephrine and norepinephrine when they act as hormones outside the central nervous system. In the brain, all autoreceptors appear to be of the α_2 type. (The drug **idazoxan** blocks α_2 autoreceptors and hence acts as an agonist.) All adrenergic receptors are metabotropic, coupled to G proteins that control the production of second messengers.

Adrenergic receptors produce both excitatory and inhibitory effects. In general, the *behavioral* effects of the release of NE are excitatory. In the brain, α_1 receptors produce a slow depolarizing (excitatory) effect on the postsynaptic membrane, while α_2 receptors produce a slow hyperpolarization. Both types of β receptors increase the responsiveness of the postsynaptic neuron to its excitatory inputs, which presumably related to the role this neurotransmitter plays in vigilance. Noradrenergic neurons—in particular, α_2 receptors—are also involved in sexual behavior and in the control of appetite.

Serotonin

The third monoamine neurotransmitter, **serotonin** (also called **5-HT,** or 5-hydroxytryptamine), has also received much experimental attention. Its behavioral effects are complex. Serotonin plays a role in the regulation of mood; in the control of eating, sleep, and arousal; and in the regulation of pain. Serotonergic neurons are involved somehow in the control of dreaming.

The precursor for serotonin is the amino acid *tryptophan*. The enzyme *tryptophan hydroxylase* adds a hydroxyl group, producing *5-HTP* (5-hydroxytryptophan). The enzyme *5-HTP decarboxylase* removes a carboxyl group from 5-HTP, and the result is 5-HT (serotonin). (See *Figure 4.17.*) The drug **PCPA** (*p*-chlorophenylalanine) blocks the activity of tryptophan hydroxylase and thus serves as a serotonergic antagonist.

The cell bodies of serotonergic neurons are found in nine clusters, most of which are located in the raphe nuclei of the midbrain, pons, and medulla. Like norepinephrine, 5-HT is released from varicosities rather than terminal buttons. The two most important clusters of serotonergic cell bodies are found in the dorsal and medial raphe nuclei, and I will restrict my discussion to these clusters. The word *raphe* means "seam" or "crease" and refers to the fact that most of the raphe nuclei are found at or near the midline of the brain stem. Both the dorsal and median raphe nuclei project axons to the cerebral cortex. In addition, neurons in the dorsal raphe innervate the basal ganglia, and those in the median raphe innervate the dentate gyrus, a part of the hippocampal formation. These and other connections are shown in *Figure 4.18.*

FIGURE 4.17 ■ **Biosynthesis of Serotonin (5-Hydroxytryptamine, or 5-HT)**

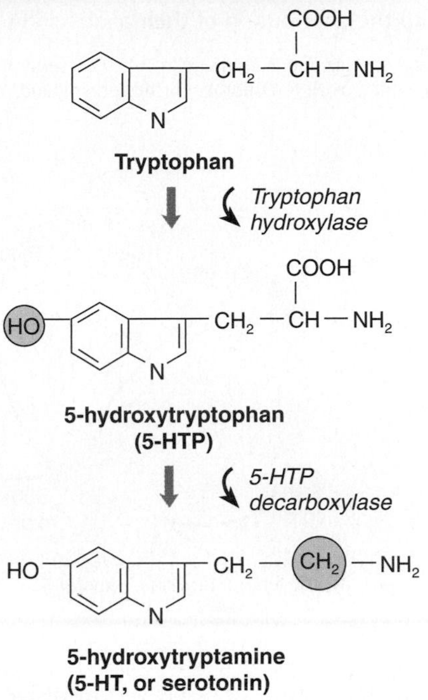

Investigators have identified at least nine different types of serotonin receptors: 5-HT_{1A-1B}, 5-HT_{1D-1F} 5-HT_{2A-2C}, and 5-HT_3. Of these the 5-HT_{1B} and 5-HT_{1D} receptors serve as presynaptic autoreceptors. In the dorsal and median raphe nuclei, 5-HT_{1A} receptors serve as autoreceptors in the membrane of dendrites and soma. All 5-HT receptors are metabotropic except for the 5-HT_3 receptor, which is ionotropic. The 5-HT_3 receptor controls a chloride channel, which means that it produces inhibitory postsynaptic potentials. These receptors appear to play a role in nausea and vomiting, because 5-HT_3 antagonists have been found to be useful in treating the side effects of chemotherapy and radiotherapy for the treatment of cancer. Pharmacologists have discovered drugs that serve as agonists or antagonists for some, but not all, of the types of 5-HT receptors.

Drugs that inhibit the reuptake of serotonin have found a very important place in the treatment of mental disorders. The best known of these, **fluoxetine** (Prozac), is used to treat depression, some forms of

idazoxan A drug that blocks presynaptic noradrenergic α_2 receptors and hence acts as an agonist, facilitating the synthesis and release of NE.

serotonin (5-HT) (*sair a **toe** nin*) An indolamine neurotransmitter; also called 5-hydroxytryptamine.

PCPA A drug that inhibits the activity of tryptophan hydroxylase and thus interferes with the synthesis of 5-HT.

fluoxetine (*floo **ox** i teen*) A drug that inhibits the reuptake of 5-HT.

FIGURE 4.18 ■ Serotonergic Pathways in a Rat Brain

This schematic figure shows the locations of the most important groups of serotonergic neurons and the distribution of their axons and terminal buttons.

(Adapted from Consolazione, A. and Cuello, A. C. CNS serotonin pathways. In *Biology of Serotonergic Transmission,* edited by N. N. Osborne. Chichester: England: Wiley & Sons, 1982.)

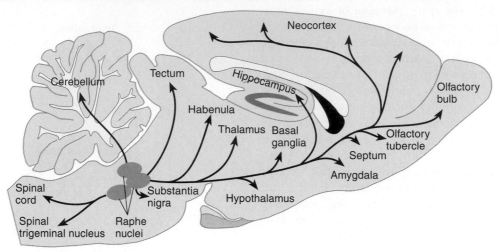

anxiety disorders, and obsessive-compulsive disorder. These disorders—and their treatment—are discussed in Chapters 16 and 17. Another drug, **fenfluramine,** which causes the release of serotonin as well as inhibits its reuptake, has been used as an appetite suppressant in the treatment of obesity. Chapter 12 discusses the topic of obesity and its control by means of drugs.

Several hallucinogenic drugs produce their effects by interacting with serotonergic transmission. **LSD** (lysergic acid diethylamide) produces distortions of visual perceptions that some people find awesome and fascinating but that simply frighten other people. This drug, which is effective in extremely small doses, is a direct agonist for postsynaptic 5-HT_{2A} receptors in the forebrain. Another drug, **MDMA** (methylenedioxymethamphetamine), is both a noradrenergic and a serotonergic agonist and has both excitatory and hallucinogenic effects. Like its relative amphetamine, MDMA (popularly called "ecstasy") causes noradrenergic transporters to run backwards, this causing the release of NE and inhibiting its reuptake. This site of action is apparently responsible for the drug's excitatory effect. MDMA also causes serotonergic transporters to run backwards, and this site of action is apparently responsible for the drug's hallucinogenic effects. Unfortunately, research indicates that MDMA can damage serotonergic neurons and cause cognitive deficits.

Amino Acids

So far, all of the neurotransmitters I have described are synthesized within neurons: acetylcholine from choline, the catecholamines from the amino acid tyrosine, and serotonin from the amino acid tryptophan. Some neurons secrete simple amino acids as neurotransmitters. Because amino acids are used for protein synthesis by all cells of the brain, it is difficult to prove that a particular amino acid is a neurotransmitter. However, investigators suspect that at least eight amino acids may serve as neurotransmitters in the mammalian central nervous system. As we saw in the introduction to this section, three of them are especially important because they are the most common neurotransmitters in the CNS: glutamate, gamma-aminobutyric acid (GABA), and glycine.

Glutamate

Because **glutamate** (also called *glutamic acid*) and GABA are found in very simple organisms, many investigators believe that these neurotransmitters were the first to have evolved. Besides producing postsynaptic potentials by activating postsynaptic receptors, they also have direct excitatory effects (glutamic acid) and inhibitory effects (GABA) on axons; they raise or lower the threshold of excitation, thus affecting the rate at which action potentials occur. These direct effects suggest that these substances had a general modulating role even before

fenfluramine (*fen* **fluor** i *meen*) A drug that stimulates the release of 5-HT.

LSD A drug that stimulates 5-HT_{2A} receptors.

MDMA A drug that serves as a noradrenergic and serotonergic agonist, also known as "ecstasy"; has excitatory and hallucinogenic effects.

glutamate An amino acid; the most important excitatory neurotransmitter in the brain.

the evolutionary development of specific receptor molecules.

Glutamate is the principal excitatory neurotransmitter in the brain and spinal cord. It is produced in abundance by the cells' metabolic processes. There is no effective way to prevent its synthesis without disrupting other activities of the cell.

Investigators have discovered four major types of glutamate receptors. Three of these receptors are ionotropic and are named after the artificial ligands that stimulate them: the **NMDA receptor,** the **AMPA receptor,** and the **kainate receptor.** The other glutamate receptor—the **metabotropic glutamate receptor**—is (obviously!) metabotropic. Actually, there appear to be at least eight subtypes of metabotropic glutamate receptors, but little is known about their functions except that some of them serve as presynaptic autoreceptors. The AMPA receptor is the most common glutamate receptor. It controls a sodium channel, so when glutamate attaches to the binding site, it produces EPSPs. The kainate receptor has similar effects.

The NMDA receptor has some special—and very important—characteristics. It contains at least six different binding sites: four located on the exterior of the receptor and two located deep within the ion channel. When it is open, the ion channel controlled by the NMDA receptor permits both sodium and calcium ions to enter the cell. The influx of both of these ions causes a depolarization, of course, but the entry of calcium (Ca^{2+}) is especially important. Calcium serves as a second messenger, binding with—and activating—various enzymes within the cell. These enzymes have profound effects on the biochemical and structural properties of the cell. As we shall see, one important result is alteration in the characteristics of the synapse that provide one of the building blocks of a newly formed memory. These effects of NMDA receptors will be discussed in much more detail in Chapter 13. The drug **AP5 (2-amino-5-phosphonopentanoate)** blocks the glutamate binding site on the NMDA receptor and impairs synaptic plasticity and certain forms of learning.

Figure 4.19 presents a schematic diagram of an NMDA receptor and its binding sites. Obviously, glutamate binds with one of these sites, or we would not call it a glutamate receptor. However, glutamate by itself cannot open the calcium channel. For that to happen, a molecule of glycine must be attached to the glycine binding site, located on the outside of the receptor. (We do not yet understand why glycine—which also serves as an inhibitory neurotransmitter in some parts of the central nervous system—is required for this ion channel to open.) (See *Figure 4.19.*)

An additional requirement for the opening of the calcium channel is that a magnesium ion *not* be attached to the magnesium binding site, located deep within the channel. Under normal conditions, when the

FIGURE 4.19 ■ NMDA Receptor

This schematic illustration of an NMDA receptor shows its binding sites.

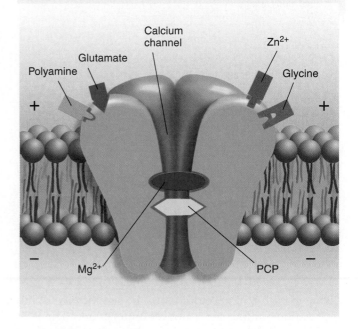

postsynaptic membrane is at the resting potential, a magnesium ion (Mg^{2+}) is attracted to the magnesium binding site and blocks the calcium channel. If a molecule of glutamate attaches to its binding site, the channel widens, but the magnesium ion still blocks it, so no calcium can enter the postsynaptic neuron. However, if the postsynaptic membrane is partially depolarized, the magnesium ion is repelled from its binding site. Thus, the NMDA receptor opens only if glutamate is present *and* the postsynaptic membrane is depolarized. The NMDA receptor, then, is a voltage- and neurotransmitter-dependent ion channel. (See *Figure 4.19.*)

What about the other three binding sites? If a zinc ion (Zn^{2+}) binds with the zinc binding site, the activity of the NMDA receptor is decreased. On the other hand, the polyamine site has a facilitatory effect. (Polyamines are chemicals that have been shown to be important for

---

NMDA receptor A specialized ionotropic glutamate receptor that controls a calcium channel that is normally blocked by Mg^{2+} ions; has several other binding sites.

AMPA receptor An ionotropic glutamate receptor that controls a sodium channel; stimulated by AMPA.

kainate receptor (*kay in ate*) An ionotropic glutamate receptor that controls a sodium channel; stimulated by kainic acid.

metabotropic glutamate receptor (*meh tab a troh pik*) A category of metabotropic receptors that are sensitive to glutamate.

AP5 (2-amino-5-phosphonopentanoate) A drug that blocks the glutamate binding site on NMDA receptors.

TABLE 4.2 ■ Behavioral Symptoms of Phencyclidine (PCP)

Altered body image

Feelings of isolation and aloneness

Cognitive disorganization

Drowsiness and apathy

Negativism and hostility

Feelings of euphoria and inebriation

Dreamlike states

Adapted from Feldman, Meyer, and Quenzer, 1997.

tissue growth and development. The significance of the polyamine binding site is not yet understood.) The PCP site, located deep within the ion channel near the magnesium binding site, binds with a hallucinogenic drug, **PCP** (phencyclidine, also known as "angel dust"). PCP serves as an indirect antagonist; when it attaches to its binding site, calcium ions cannot pass through the ion channel. PCP is a synthetic drug and is not produced by the brain. Thus, it is not the natural ligand of the PCP binding site. What that ligand is and what useful functions it serves are not yet known. The behavioral symptoms of PCP are listed in *Table 4.2.*

Several drugs affect glutamatergic synapses. As you already know, NMDA, AMPA, and kainate (more precisely, *kainic acid*) serve as direct agonists at the receptors named after them. In addition, one of the most common drugs—alcohol—serves as an antagonist of NMDA receptors. As we will see in Chapter 19, this effect is responsible for the seizures that can be provoked by sudden withdrawal from heavy long-term alcohol intake.

GABA

GABA (gamma-aminobutyric acid) is produced from glutamic acid by the action of an enzyme (glutamic acid decarboxylase, or GAD) that removes a carboxyl group. The drug **allylglycine** inactivates GAD and thus prevents the synthesis of GABA (step 2 of Figure 4.5). GABA is an inhibitory neurotransmitter, and it appears to have a widespread distribution throughout the brain and spinal cord. Two GABA receptors have been identified: $GABA_A$ and $GABA_B$. The $GABA_A$ receptor is ionotropic and controls a chloride channel; the $GABA_B$ receptor is metabotropic and controls a potassium channel.

As you know, neurons in the brain are greatly interconnected. Without the activity of inhibitory synapses these interconnections would make the brain unstable. That is, through excitatory synapses neurons would excite their neighbors, which would then excite *their* neighbors, which would then excite the originally active neurons, and so on, until most of the neurons in the brain would be firing uncontrollably. In fact, this event does sometimes occur, and we refer to it as a *seizure.* (*Epilepsy* is a neurological disorder characterized by the presence of seizures.) Normally, an inhibitory influence is supplied by GABA-secreting neurons, which are present in large numbers in the brain. Some investigators believe that one of the causes of epilepsy is an abnormality in the biochemistry of GABA-secreting neurons or in GABA receptors.

Like NMDA receptors, $GABA_A$ receptors are complex; they contain at least five different binding sites. The primary binding site is, of course, for GABA. The drug **muscimol** (derived from the ACh agonist muscarine) serves as a direct agonist for this site (step 6 of Figure 4.5). Another drug, **bicuculline,** blocks this GABA binding site, serving as a direct antagonist (step 7 of Figure 4.5). A second site on the $GABA_A$ receptor binds with a class of tranquilizing drugs called the **benzodiazepines.** These drugs include diazepam (Valium) and chlordiazepoxide (Librium), which are used to reduce anxiety, promote sleep, reduce seizure activity, and produce muscle relaxation. The third site binds with barbiturates. The fourth site binds with various steroids, including some steroids used to produce general anesthesia. The fifth site binds with picrotoxin, a poison found in an East Indian shrub. In addition, alcohol binds with one of these sites—probably the benzodiazepine binding site. (See *Figure 4.20.*)

Barbiturates, drugs that bind to the steroid site, and benzodiazepines all promote the activity of the $GABA_A$ receptor; thus, all these drugs serve as indirect agonists. The benzodiazepines are very effective **anxiolytics,** or

PCP Phencyclidine; a drug that binds with the PCP binding site of the NMDA receptor and serves as an indirect antagonist.

GABA An amino acid; the most important inhibitory neurotransmitter in the brain.

allylglycine A drug that inhibits the activity of GAD and thus blocks the synthesis of GABA.

muscimol (*musk i mawl*) A direct agonist for the GABA binding site on the $GABA_A$ receptor.

bicuculline (*by kew kew leen*) A direct antagonist for the GABA binding site on the $GABA_A$ receptor.

benzodiazepine (*ben zoe dy azz a peen*) A category of anxiolytic drugs; an indirect agonist for the $GABA_A$ receptor.

anxiolytic (*angz ee oh lit ik*) An anxiety-reducing effect.

FIGURE 4.20 ■ GABA$_A$ Receptor

This schematic illustration of a GABA$_A$ receptor shows its binding sites.

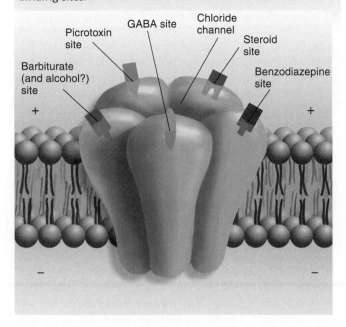

"anxiety-dissolving" drugs. They are often used to treat people with anxiety disorders. In addition, some benzodiazepines serve as effective sleep medications, and others are used to treat some types of seizure disorder.

In low doses, barbiturates have a calming effect. In progressively higher doses they produce difficulty in walking and talking, unconsciousness, coma, and death. Although veterinarians sometimes use barbiturates to produce anesthesia for surgery, the therapeutic index—the ratio between a dose that produces anesthesia and one that causes fatal depression of the respiratory centers of the brain—is small. As a consequence, these drugs are rarely used by themselves to produce surgical anesthesia in humans.

Picrotoxin has effects opposite to those of benzodiazepines and barbiturates: It *inhibits* the activity of the GABA$_A$ receptor, thus serving as an indirect antagonist. In high enough doses, this drug causes convulsions.

Various steroid hormones are normally produced in the body, and some hormones related to progesterone (the principal pregnancy hormone) act on the steroid binding site of the GABA$_A$ receptor, producing a relaxing, anxiolytic sedative effect. However, the brain does not produce Valium, barbiturates, or picrotoxin. What are the natural ligands for these binding sites? So far, most research has concentrated on the benzodiazepine binding site. These binding sites are more complex than the others. They can be activated by drugs such as the benzodiazepines, which promote the activity of the receptor and thus serve as indirect agonists. They can

also be activated by other drugs that have the opposite effect—that inhibit the activity of the receptor, thus serving as indirect antagonists. Presumably, the brain produces natural ligands that act as indirect agonists or antagonists at the benzodiazepine binding site, but so far, such a chemical has not been identified.

What about the GABA$_B$ receptor? This metabotropic receptor, coupled to a G protein, serves as both a postsynaptic receptor and a presynaptic autoreceptor. A GABA$_B$ agonist, baclofen, serves as a muscle relaxant. Another drug, CGP 335348, serves as an antagonist. The activation of GABA$_B$ receptors opens potassium channels, producing hyperpolarizing inhibitory postsynaptic potentials.

Glycine

The amino acid **glycine** appears to be the inhibitory neurotransmitter in the spinal cord and lower portions of the brain. Little is known about its biosynthetic pathway; there are several possible routes, but not enough is known to decide how neurons produce glycine. The bacteria that cause tetanus (lockjaw) release a chemical that prevents the release of glycine (and GABA as well); the removal of the inhibitory effect of these synapses causes muscles to contract continuously.

The glycine receptor is ionotropic, and it controls a chloride channel. Thus, when it is active, it produces inhibitory postsynaptic potentials. The drug **strychnine,** an alkaloid found in the seeds of the *Strychnos nux vomica,* a tree found in India, serves as a glycine antagonist. Strychnine is very toxic, and even relatively small doses cause convulsions and death. No drugs have yet been found that serve as specific glycine agonists.

Researchers have discovered that some terminal buttons in the brain release both glycine and GABA (Jonas, Bischofberger, and Sandkühler, 1998; Nicoll and Malenka, 1998). The apparent advantage for the co-release of these two inhibitory neurotransmitters is the production of rapid, long-lasting postsynaptic potentials: The glycine stimulates rapid ionotropic receptors, and the GABA stimulates long-lasting metabotropic receptors. Obviously, the postsynaptic membrane at these synapses contains both glycine and GABA receptors.

Peptides

Recent studies have discovered that the neurons of the central nervous system release a large variety of peptides. Peptides consist of two or more amino acids linked together by peptide bonds. All the peptides that

glycine (*gly* seen) An amino acid; an important inhibitory neurotransmitter in the lower brain stem and spinal cord.

strychnine (*strik* neen) A direct antagonist for the glycine receptor.

have been studied so far are produced from precursor molecules. These precursors are large polypeptides that are broken into pieces by special enzymes. Neurons manufacture both the polypeptides and the enzymes needed to break them apart in the right places. The appropriate sections of the polypeptides are retained, and the rest are destroyed. Because the synthesis of peptides takes place in the soma, vesicles containing these chemicals must be delivered to the terminal buttons by axoplasmic transport.

Peptides are released from all parts of the terminal button, not just from the active zone; thus, only a portion of the molecules are released into the synaptic cleft. The rest presumably act on receptors belonging to other cells in the vicinity. Once released, peptides are destroyed by enzymes. There is no mechanism for reuptake and recycling of peptides.

Several different peptides are released by neurons. Although most peptides appear to serve as neuromodulators, some act as neurotransmitters. One of the best known families of peptides is the **endogenous opioids.** (*Endogenous* means "produced from within"; *opioid* means "like opium.") Several years ago it became clear that opiates (drugs such as opium, morphine, and heroin) reduce pain because they have direct effects on the brain. (Please note that the term *opioid* refers to endogenous chemicals, and *opiate* refers to drugs.) Pert, Snowman, and Snyder (1974) discovered that neurons in a localized region of the brain contain specialized receptors that respond to opiates. Then, soon after the discovery of the opiate receptor, other neuroscientists discovered the natural ligands for these receptors (Terenius and Wahlström, 1975; Hughes et al., 1975), which they called **enkephalins** (from the Greek word *enkephalos,* "in the head"). We now know that the enkephalins are only two members of a family of endogenous opioids, all of which are synthesized from one of three large peptides that serve as precursors. In addition, we know that there are at least three different types of opiate receptors: μ (mu), δ (delta), and κ (kappa).

Several different neural systems are activated when opiate receptors are stimulated. One type produces analgesia, another inhibits species-typical defensive responses such as fleeing and hiding, and another stimulates a system of neurons involved in reinforcement ("reward"). The last effect explains why opiates are often abused. The situations that cause neurons to secrete endogenous opioids are discussed in Chapter 7, and the brain mechanisms of opiate addiction are discussed in Chapter 18.

So far, pharmacologists have developed only two types of drugs that affect neural communication by means of opioids: direct agonists and antagonists. Many synthetic opiates, including heroin (dihydromorphine) and Percodan (levorphanol), have been developed, and some are used clinically as analgesics (step 6 of Figure 4.5).

Several opiate receptors blockers have also been developed (step 7 of Figure 4.5). One of them, **naloxone,** is used clinically to reverse opiate intoxication. This drug has saved the lives of many drug abusers who would otherwise have died of an overdose of heroin.

As we saw in Chapter 2, many terminal buttons contain two different types of synaptic vesicles, each filled with a different substance. These terminal buttons release peptides in conjunction with a "classical" neurotransmitter (one of those I just described). One reason for the corelease of peptides is their ability to regulate the sensitivity of presynaptic or postsynaptic receptors to the neurotransmitter. For example, the terminal buttons of the salivary nerve of the cat (which control the secretion of saliva) release both acetylcholine and a peptide called VIP. When the axons fire at a low rate, only ACh is released and only a little saliva is secreted. At a higher rate, both ACh and VIP are secreted, and the VIP dramatically increases the sensitivity of the muscarinic receptors in the salivary gland to ACh; thus, much saliva is released.

Several peptide hormones released by endocrine glands are also found in the brain, where they serve as neuromodulators. In some cases the peripheral and central peptides perform related functions. For example, outside the nervous system the hormone angiotensin acts directly on the kidneys and blood vessels to produce effects that help the body cope with the loss of fluid, and inside the nervous system circuits of neurons that use angiotensin as a neurotransmitter perform complementary functions, including the activation of neural circuits that produce thirst. The existence of the blood–brain barrier keeps hormones in the general circulation separate from the extracellular fluid in the brain, which means that the same peptide molecule can have different effects in these two regions.

Many peptides produced in the brain have interesting behavioral effects, which will be discussed in subsequent chapters.

Lipids

Various substances derived from lipids can serve to transmit messages within or between cells. The best known, and probably the most important, are the two **endocannabinoids** ("endogenous cannabis-like substances")—natural ligands for the receptors that are

---

endogenous opioid (*en* **dodge** *en us* **oh** *pee oyd*) A class of peptides secreted by the brain that act as opiates.

enkephalin (*en* **keff** *a lin*) One of the endogenous opioids.

naloxone (*na* **lox** *own*) A drug that blocks opiate receptors.

endocannabinoid (*en do can* **ab** *in oid*) A lipid; an endogenous ligand for cannabinoid receptors, which also bind with THC, the active ingredient of marijuana.

FIGURE 4.21 ■ Cannaboid Receptors in a Rat Brain

In this autoradiogram the brain has been incubated in a solution containing a radioactive ligand for cannabinoid receptors. The receptors are indicated by dark areas. (Autoradiography is described in Chapter 5.) (Br St = brain stem, Cer = cerebellum, CP = caudate nucleus/putamen, Cx = cortex, EP = entopeduncular nucleus, GP = globus pallidus, Hipp = hippocampus, SNr = substantia nigra.)

(Courtesy of Miles Herkenham, National Institute of Mental Health, Bethesda, MD.)

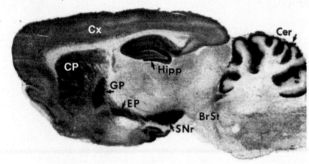

responsible for the physiological effects of the active ingredient in marijuana. Matsuda et al. (1990) discovered that **THC** (tetrahydrocannabinol, the active ingredient of marijuana) stimulates cannabinoid receptors located in specific regions of the brain. (See *Figure 4.21.*) Two types of cannabinoid receptors, CB_1 and CB_2, both metabotropic, have since been discovered. CB_1 receptors are found in the brain, especially in the frontal cortex, anterior cingulate cortex, basal ganglia, cerebellum, hypothalamus, and hippocampus. Very low levels of CB_1 receptors are found in the brain stem, which accounts for the low toxicity of THC. CB_2 receptors are found outside the brain, especially in cells of the immune system.

THC produces analgesia and sedation, stimulates appetite, reduces nausea caused by drugs used to treat cancer, relieves asthma attacks, decreases pressure within the eyes in patients with glaucoma, and reduces the symptoms of certain motor disorders. On the other hand, THC interferes with concentration and memory, alters visual and auditory perception, and distorts perceptions of the passage of time (Iversen, 2003). Devane et al. (1992) discovered the first natural ligand for the THC receptor: a lipidlike substance that they named **anandamide,** from the Sanskrit word *ananda,* or "bliss." A few years after the discovery of anandamide, Mechoulam et al. (1995) discovered another endocannabinoid, 2-arachidonyl glycerol (2-AG).

Anandamide seems to be synthesized on demand; that is, it is produced and released as it is needed and is not stored in synaptic vesicles. It is deactivated by an enzyme, **FAAH** (fatty acid amide hydrolase), which is present in anandamide-secreting neurons. Because the enzyme is found there, molecules of anandamide must be transported back into these neurons, which is accomplished by anandamide transporters. Besides THC, several drugs have been discovered that affect the actions of the endocannabinoids. CB_1 receptors are blocked by the drug **rimonabant,** the enzyme FAAH is inhibited by **MAFP,** and reuptake is inhibited by **AM1172.**

CB_1 receptors are found on terminal buttons of glutamatergic, GABAergic, acetylcholinergic, noradrenergic, dopaminergic, and serotonergic neurons, where they serve as presynaptic heteroreceptors, regulating neurotransmitter release (Iversen, 2003). When activated, the receptors open potassium channels in the terminal buttons, shortening the duration of action potentials there and decreasing the amount of neurotransmitter that is released. When neurons release cannabinoids, the chemicals diffuse a distance of approximately 20 μm in all directions, and their effects persist for several tens of seconds. The short-term memory impairment that accompanies marijuana use appears to be caused by the action of THC on CB_1 receptors in the hippocampus. Endocannabinoids also appear to play an essential role in the reinforcing effects of opiates: A targeted mutation that prevents the production of CB_1 receptors abolishes the reinforcing effects of morphine but not of cocaine, amphetamine, or nicotine (Cossu et al., 2001). These effects of cannabinoids are discussed further in Chapter 18.

I mentioned three paragraphs ago that THC (and, of course, the endocannabinoids) have an analgesic effect. Agarwal et al. (2007) found that THC exerts its analgesic effects by stimulating CB_1 receptors in the peripheral nervous system. In addition, a commonly used over-the-counter analgesic, *acetaminophen* (known as *paracetamol* in many countries), also acts on these receptors. Once it enters the blood, acetaminophen is converted into another compound that then joins with arachidonic acid, the precursor of anandamide. This compound binds with peripheral CB_1 receptors and activates them, reducing pain sensation. Administration of a CB_1 antagonist completely blocks the analgesic effect of acetaminophen (Bertolini et al., 2006).

THC The active ingredient in marijuana; activates CB_1 receptors in the brain.

anandamide (*a **nan** da mide*) The first cannabinoid to be discovered and probably the most important one.

FAAH Fatty acid amide hydrolase, the enzyme that destroys anandamide after it is brought back into the cell by anandamide transporters.

rimonabant A drug that blocks CB_1 receptors.

MAFP A drug that inhibits FAAH; prevents the breakdown of anandamide.

AM1172 A drug that inhibits the reuptake of anandamide.

Nucleosides

A nucleoside is a compound that consists of a sugar molecule bound with a purine or pyrimidine base. One of these compounds, **adenosine** (a combination of ribose and adenine), serves as a neuromodulator in the brain.

Adenosine is known to be released, apparently by glial cells as well as neurons, when cells are short of fuel or oxygen. The release of adenosine activates receptors on nearby blood vessels and causes them to dilate, increasing the flow of blood and helping to bring more of the needed substances to the region. Adenosine also acts as a neuromodulator, through its action on at least three different types of adenosine receptors. Adenosine receptors are coupled to G proteins, and their effect is to open potassium channels, producing inhibitory postsynaptic potentials. Because adenosine is present in all cells, investigators have not yet succeeded in distinguishing neurons that release this chemical as a neuromodulator. Thus, circuits of adenosinergic neurons have not yet been identified.

Because adenosine receptors suppress neural activity, adenosine and other adenosine receptor agonists have generally inhibitory effects on behavior. In fact, as we will see in Chapter 9, there is good evidence that adenosine receptors play an important role in the control of sleep. For example, the amount of adenosine in the brain increases during wakefulness and decreases during sleep. In fact, the accumulation of adenosine after prolonged wakefulness may be the most important cause of the sleepiness that ensues. A very common drug, **caffeine,** blocks adenosine receptors (step 7 of Figure 4.5) and hence produces excitatory effects. Caffeine is a bitter-tasting alkaloid found in coffee, tea, cocoa beans, and other plants. In much of the world a majority of the adult population ingests caffeine every day—fortunately, without apparent harm. (See *Table 4.3.*)

Prolonged use of caffeine leads to a moderate amount of tolerance, and people who suddenly stop taking caffeine complain of withdrawal symptoms, which include headaches, drowsiness, and difficulty in concentrating. If the person continues to abstain, the symptoms disappear within a few days. Caffeine does not produce the compulsive drug-taking behavior that is often seen in people who abuse amphetamine, cocaine, or the opiates. In addition, laboratory animals do not readily self-administer caffeine, as they do drugs that are commonly abused by humans.

Soluble Gases

Recently, investigators have discovered that neurons use at least two simple, soluble gases—nitric oxide and carbon monoxide—to communicate with one another. One of these, **nitric oxide (NO),** has received the most

TABLE 4.3 ■ Typical Caffeine Content of Chocolate and Several Beverages	
ITEM	CAFFEINE CONTENT
Chocolates	
Baking chocolate	35 mg/oz
Milk chocolate	6 mg/oz
Beverages	
Coffee	85 mg/5-oz cup
Decaffeinated coffee	3 mg/5-oz cup
Tea (brewed 3 minutes)	28 mg/5-oz cup
Cocoa or hot chocolate	30 mg/5-oz cup
Cola drink	30–46 mg/12-oz container

Based on data from Somani and Gupta, 1988.

attention. Nitric oxide (not to be confused with nitrous oxide, or laughing gas) is a soluble gas that is produced by the activity of an enzyme found in certain neurons. Researchers have found that NO is used as a messenger in many parts of the body; for example, it is involved in the control of the muscles in the wall of the intestines, it dilates blood vessels in regions of the brain that become metabolically active, and it stimulates the changes in blood vessels that produce penile erections (Culotta and Koshland, 1992). As we will see in Chapter 14, NO may also play a role in the establishment of neural changes that are produced by learning.

All of the neurotransmitters and neuromodulators discussed so far (with the exception of anandamide and perhaps adenosine) are stored in synaptic vesicles and released by terminal buttons. Nitric oxide is produced in several regions of a nerve cell—including dendrites—and is released as soon as it is produced. More accurately, it diffuses out of the cell as soon as it is produced.

adenosine (*a den oh seen*) A nucleoside; a combination of ribose and adenine; serves as a neuromodulator in the brain.

caffeine A drug that blocks adenosine receptors.

nitric oxide (NO) A gas produced by cells in the nervous system; used as a means of communication between cells.

It does not activate membrane-bound receptors but enters neighboring cells, where it activates an enzyme responsible for the production of a second messenger, cyclic GMP. Within a few seconds of being produced, nitric oxide is converted into biologically inactive compounds.

Nitric oxide is produced from arginine, an amino acid, by the activation of an enzyme known as **nitric oxide synthase.** This enzyme can be inactivated (step 2 of Figure 4.5) by a drug called L-NAME (nitro-L-arginine methyl ester).

You have undoubtedly heard of a drug called *sildenafil* (more commonly known as Viagra), which is used to treat men who have erectile dysfunction—difficulty maintaining a penile erection. As we just saw, nitric

oxide produces its physiological effects by stimulating the production of cyclic GMP. Although nitric oxide lasts only for a few seconds, cyclic GMP lasts somewhat longer but is ultimately destroyed by an enzyme. Molecules of sildenafil bind with this enzyme and thus cause cyclic GMP to be destroyed at a much slower rate. As a consequence, an erection is maintained for a longer time. (By the way, sildenafil has effects on other parts of the body and is used to treat altitude sickness and other vascular disorders.)

---

nitric oxide synthase   The enzyme responsible for the production of nitric oxide.

Interim Summary

Neurotransmitters and Neuromodulators

The nervous system contains a variety of neurotransmitters, each of which interacts with a specialized receptor. The neurotransmitters that have received the most study are acetylcholine and the monoamines: dopamine, norepinephrine, and 5-hydroxytryptamine (serotonin). The synthesis of these neurotransmitters is controlled by a series of enzymes. Several amino acids also serve as neurotransmitters, the most important of which are glutamate (glutamic acid), GABA, and glycine. Glutamate serves as an excitatory neurotransmitter; the others serve as inhibitory neurotransmitters.

Peptide neurotransmitters consist of chains of amino acids. Like proteins, peptides are synthesized at the ribosomes according to sequences coded for by the chromosomes. The best-known class of peptides in the nervous system includes the endogenous opioids, whose effects are mimicked by drugs such as opium and heroin. One lipid appears to serve as a chemical messenger: anandamide, the endogenous ligand for the CB_1 cannabinoid receptor. Adenosine, a nucleoside that has inhibitory effects on synaptic transmission, is released by neurons and glial cells in the brain. In addition, two soluble gases—nitric oxide and carbon monoxide—can diffuse out of the cell in which they are produced and trigger the production of a second messenger in adjacent cells.

This chapter has mentioned many drugs and their effects. They are summarized for your convenience in *Table 4.4.*

TABLE 4.4 ■ Drugs Mentioned in This Chapter			
NEUROTRANSMITTER	NAME OF DRUG	EFFECT OF DRUG	EFFECT ON SYNAPTIC TRANSMISSION
Acetylcholine (ACh)	Botulinum toxin	Blocks release of ACh	Antagonist
	Black widow spider venom	Stimulates release of ACh	Agonist
	Nicotine	Stimulates nicotinic receptors	Agonist
	Curare	Blocks nicotinic receptors	Antagonist
	Muscarine	Stimulates muscarinic receptors	Agonist
	Atropine	Blocks muscarinic receptors	Antagonist
	Neostigmine	Inhibits acetylcholinesterase	Agonist
	Hemicholinium	Inhibits reuptake of choline	Antagonist

(continued)

TABLE 4.4 ■ Drugs Mentioned in This Chapter *(continued)*

NEUROTRANSMITTER	NAME OF DRUG	EFFECT OF DRUG	EFFECT ON SYNAPTIC TRANSMISSION
Dopamine (DA)	L-DOPA	Facilitates synthesis of DA	Agonist
	AMPT	Inhibits synthesis of DA	Antagonist
	Reserpine	Inhibits storage of DA in synaptic vesicles	Antagonist
	Chlorpromazine	Blocks D_2 receptors	Antagonist
	Clozapine	Blocks D_4 receptors	Antagonist
	Cocaine, methylphenidate	Blocks DA reuptake	Agonist
	Amphetamine	Stimulates release of DA	Agonist
	Deprenyl	Blocks MAO-B	Agonist
Norepinephrine (NE)	Fusaric acid	Inhibits synthesis of NE	Antagonist
	Reserpine	Inhibits storage of NE in synaptic vesicles	Antagonist
	Idazoxan	Blocks α_2 autoreceptors	Agonist
	Desipramine	Inhibits reuptake of NE	Agonist
	Moclobemide	Inhibits MAO-A	Agonist
	MDMA, amphetamine	Stimulates release of NE	Agonist
Serotonin (5-HT)	PCPA	Inhibits synthesis of 5-HT	Antagonist
	Reserpine	Inhibits storage of 5-HT in synaptic vesicles	Antagonist
	Fenfluramine	Stimulates release of 5-HT	Agonist
	Fluoxetine	Inhibits reuptake of 5-HT	Agonist
	LSD	Stimulates 5-HT_{2A} receptors	Agonist
	MDMA	Stimulates release of 5-HT	Agonist
Glutamate	AMPA	Stimulates AMPA receptor	Agonist
	Kainic acid	Stimulates kainate receptor	Agonist
	NMDA	Stimulates NMDA receptor	Agonist
	AP5	Blocks NMDA receptor	Antagonist
GABA	Allylglycine	Inhibits synthesis of GABA	Antagonist
	Muscimol	Stimulates GABA receptors	Agonist
	Bicuculline	Blocks GABA receptors	Antagonist
	Benzodiazepines	Serve as indirect GABA agonist	Agonist
Glycine	Strychnine	Blocks glycine receptors	Antagonist
Opioids	Opiates (morphine, heroin, etc.)	Stimulates opiate receptors	Agonist
	Naloxone	Blocks opiate receptors	Antagonist
Anandamide	Rimonabant	Blocks cannabinoid CB_1 receptors	Antagonist
	THC	Stimulates cannabinoid CB_1 receptors	Agonist
	MAFP	Inhibits FAAH	Agonist
	AM1172	Blocks reuptake of anandamide	Agonist
Adenosine	Caffeine	Blocks adenosine receptors	Antagonist
Nitric oxide (NO)	L-NAME	Inhibits synthesis of NO	Antagonist
	Sildenafil	Inhibits destruction of cyclic GMPA	Agonist

SUGGESTED READINGS

Cooper, J. R., Bloom, F. E., and Roth, R. H. *The Biochemical Basis of Neuropharmacology,* 8th ed. New York: Oxford University Press, 2003.

Grilly, D. M. *Drugs and Human Behavior,* 5th ed. Boston: Allyn and Bacon, 2006.

Meyer, J. S., and Quenzer, L. F. *Psychopharmacology: Drugs, the Brain, and Behavior.* Sunderland, MA: Sinauer Associates, 2005.

ADDITIONAL RESOURCES

Visit www.mypsychkit.com for additional review and practice of the material covered in this chapter. Within MyPsychKit, you can take practice tests and receive a customized study plan to help you review. Dozens of animations, tutorials, and Web links are also available.

You can even review using the interactive electronic version of this textbook. You will need to register for MyPsychKit. See www.mypsychkit.com for complete details.

chapter
5

Methods and
Strategies of Research

outline

In July 1982, several young people began showing up at neurology clinics in northern California displaying dramatic, severe symptoms (Langston, Ballard, Tetrud, and Irwin, 1983). The most severely affected patients were almost totally paralyzed. They were unable to speak intelligibly, they drooled constantly, and their eyes were open with a fixed stare. Others, less severely affected, walked with a slow, shuffling gait and moved slowly and with great difficulty. The symptoms looked like those of Parkinson's disease, but that disorder has a very gradual onset. In addition, it rarely strikes people before late middle age, and the patients were all in their twenties or early thirties.

The common factor linking these patients was intravenous drug use; all of them had been taking a "new heroin," a synthetic opiate related to meperidine (Demerol). Some detective work revealed that the illicit drug was contaminated with a chemical that damaged dopaminergic neurons and caused the neurological symptoms. Because the symptoms looked like those of Parkinson's disease, the patients were given L-DOPA, the drug that is used to treat this disease, and they all showed significant improvement in their symptoms. Unfortunately, the improvement was temporary; the drug lost its effectiveness.

Fetal transplantation, an experimental neurosurgical method of treating parkinsonism has shown some promise. The rationale for the procedure is this: The symptoms of parkinsonism, whether from Parkinson's disease or the toxic effects of MPTP, are caused by the lack of dopamine in the caudate nucleus and putamen. There is at present no way to induce the brain to grow new dopaminergic neurons. However, if dopamine-secreting neurons can be introduced into the caudate nucleus and putamen and if they survive and secrete dopamine, then perhaps the parkinsonian symptoms will diminish. Because the implanted neurons must be healthy and vigorous and not trigger the recipient's immune system, the logical source for them is aborted human fetuses— or, perhaps some day, cultures of stem cells that have been induced to become dopamine-secreting neurons.

At least one of the people with MPTP poisoning received such a transplant. (Let's call him Mr. B.) Before the operation took place, Mr. B. was given an injection of radioactive L-DOPA, the precursor for dopamine. Then, one hour later, he was wheeled into a small room that housed a PET scanner. His head was positioned in the scanner, and for the next several minutes the machine gathered data from positrons that were being emitted as the radioactive particles in his head broke down.

A few weeks later, Mr. B. entered the hospital for his surgery. Technicians removed dopaminergic neurons from the substantia nigra of several aborted fetuses and prepared them for implantation into Mr. B.'s brain. Mr. B. was anesthetized, and the surgeon made cuts in his scalp to expose parts of his skull. He attached the frame of a stereotaxic apparatus to Mr. B's skull, made some measurements, and then drilled several holes. He used the stereotaxic apparatus to guide the injections of the fetal neurons into Mr. B.'s caudate nucleus and putamen. Once the injections were complete, the surgeon removed the stereotaxic frame and sutured the incisions he had made in the scalp.

The operation was quite successful; Mr. B. recovered much of his motor control. A little more than a year later, he was again given an injection of radioactive L-DOPA, and again his head was placed in the PET scanner. The results of the second scan showed what his recovery implied: The cells had survived and were secreting dopamine.

Study of the physiology of behavior involves the efforts of scientists in many disciplines, including physiology, neuroanatomy, biochemistry, psychology, endocrinology, and histology. Pursuing a research project in behavioral neuroscience requires competence in many experimental techniques. Because different procedures often produce contradictory results, investigators must be familiar with the advantages and limitations of the methods they employ. Scientific investigation entails a process of asking questions of nature. The method that is used frames the question. Often we receive a puzzling answer, only to realize later that we were not asking the question we thought we were. As we will see, the best conclusions about the physiology of behavior are made not by any single experiment, but by a program of research that enables us to compare the results of studies that approach the problem with different methods.

An enormous—and bewildering—array of research methods is available to the investigator. If I merely presented a catalog of them, it would not be surprising if you got lost—or simply lost interest. Instead, I will present only the most important and commonly used procedures, organized around a few problems that researchers have studied. This way, it should be easier to see the types of information provided by various research methods and to understand their advantages and disadvantages. It will also permit me to describe the strategies that researchers employ as they follow up the results of one experiment by designing and executing another one.

EXPERIMENTAL ABLATION

One of the most important research methods used to investigate brain functions involves destroying part of the brain and evaluating the animal's subsequent behavior. This method is called **experimental ablation** (from the Latin word *ablatus,* a "carrying away"). In most cases, experimental ablation does not involve the removal brain tissue; instead, the researcher destroys some tissue and leaves it in place. Experimental ablation is the oldest method used in neuroscience, and it remains in common use today.

Evaluating the Behavioral Effects of Brain Damage

A *lesion* is a wound or injury, and a researcher who destroys part of the brain usually refers to the damage as a *brain lesion.* Experiments in which part of the brain is damaged and the animal's behavior is subsequently observed are called **lesion studies.** The rationale for lesion studies is that the function of an area of the brain can be inferred from the behaviors that the animal can no longer perform after the area has been damaged. For example, if, after part of the brain has been destroyed, an animal can no longer perform tasks that require vision, we can conclude that the animal is blind—and that the damaged area plays some role in vision.

Just what can we learn from lesion studies? Our goal is to discover what functions are performed by different regions of the brain and then to understand how these functions are combined to accomplish particular behaviors. The distinction between *brain function* and *behavior* is an important one. Circuits within the brain perform functions, not behaviors. No one brain region or neural circuit is solely responsible for a behavior; each region performs a function (or set of functions) that contributes to performance of the behavior. For example, the act of reading involves functions required for controlling eye movements, focusing the lens of the eye, perceiving and recognizing words and letters, comprehending the meaning of the words, and so on. Some of these functions also participate in other behaviors; for example, controlling eye movement and focusing are required for any task that involves looking, and brain mechanisms used for comprehending the meanings of words also participate in comprehending speech. The researcher's task is to understand the functions that are required for performing a particular behavior and to determine what circuits of neurons in the brain are responsible for each of these functions.

The interpretation of lesion studies is complicated by the fact that all regions of the brain are interconnected. Suppose that we have a good understanding of the functions required for performance of a particular behavior. We find that damage to brain structure X impairs a particular behavior. Can we necessarily conclude that a function essential to this behavior is performed by circuits of neurons located in structure X? Unfortunately, we cannot. The functions we are interested in may actually be performed by neural circuits located elsewhere in the brain. Damage to structure X may simply interfere with the normal operation of the neural circuits in structure Y.

Producing Brain Lesions

How do we produce brain lesions? It is easy to destroy parts of the brain immediately beneath the skull; we anesthetize the animal, cut its scalp, remove part of its skull, and cut through the dura mater, bringing the cortex into view. Then we can use a suction device to aspirate the brain tissue. To accomplish this tissue removal, we place a glass pipette on the surface of the brain and suck away brain tissue with a vacuum pump attached to the pipette.

More often, we want to destroy regions that are hidden away in the depths of the brain. Brain lesions of subcortical regions (regions located beneath the cortex) are usually produced by passing electrical current through a stainless steel wire that is coated with an insulating varnish except for the very tip. We guide the wire stereotaxically so that its end reaches the appropriate location. (Stereotaxic surgery is described in the next subsection.) Then we turn on a lesion-making device, which produces radio frequency (RF) current—alternating current of a very high frequency. The passage of the current through the brain tissue produces heat that kills cells in the region surrounding the tip of the electrode. (See *Figure 5.1.*)

Lesions produced by these means destroy everything in the vicinity of the electrode tip, including neural cell bodies and the axons of neurons that pass through the region. A more selective method of producing brain lesions employs an excitatory amino acid such as *kainic acid,* which kills neurons by stimulating them to death. (As we saw in Chapter 4, kainic acid stimulates glutamate receptors.) Lesions produced in this way are referred to as **excitotoxic lesions.** When an excitatory amino acid is injected through a cannula into a region

---

experimental ablation   The removal or destruction of a portion of the brain of a laboratory animal; presumably, the functions that can no longer be performed are the ones the region previously controlled.

lesion study   A synonym for experimental ablation.

excitotoxic lesion (*ek sigh tow **tok** sik*)   A brain lesion produced by intracerebral injection of an excitatory amino acid, such as kainic acid.

FIGURE 5.1 ■ Radio Frequency Lesion

The arrows point to very small lesions produced by passing radio frequency current through the tips of stainless steel electrodes placed in the medial preoptic nucleus of a rat brain. The oblong hole in the middle of the photograph is the third ventricle. (Frontal section, cell-body stain.)

(From Turkenburg, J. L., Swaab, D. F., Endert, E., Louwerse, A. L., and van de Poll, N. E. *Brain Research Bulletin*, 1988, *21*, 215–224. Reprinted with permission.)

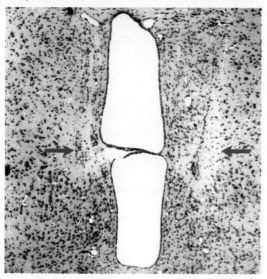

FIGURE 5.2 ■ Excitotoxic Lesion

(a) This section shows a normal hippocampus of a rat brain. (b) A lesion was produced by infusion of an excitatory amino acid in a region of the hippocampus. Arrowheads mark the ends of the region in which neurons have been destroyed.

(Courtesy of Benno Roozendaal, University of California, Irvine.)

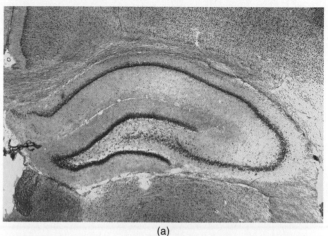

(a)

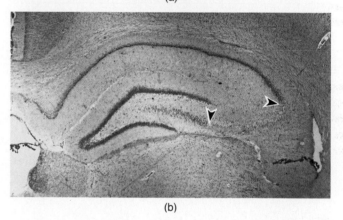

(b)

of the brain, the chemical destroys neural cell bodies in the vicinity but spares axons that belong to different neurons that happen to pass nearby. (See *Figure 5.2*.) This selectivity permits the investigator to determine whether the behavioral effects of destroying a particular brain structure are caused by the death of neurons located there or by the destruction of axons that pass nearby. For example, some researchers discovered that RF lesions of a particular region in the brain stem abolished REM sleep; therefore, they believed that this region was involved in the production of this stage of sleep. (REM sleep is the stage of sleep during which dreaming occurs.) But later studies showed that when kainic acid was used to destroy the neurons located there, the animals' sleep was *not* affected. Therefore, the RF lesions must have altered sleep by destroying the axons that pass through the area.

Even more specific methods of targeting and killing particular types of neurons are available. For example, molecular biologists have devised ways to attach toxic chemicals to antibodies that will bind with particular proteins found only on certain types of neurons in the brain. The antibodies target these proteins, and the toxic chemicals kill the cells to which the proteins are attached.

Note that when we produce subcortical lesions by passing RF current through an electrode or infusing a

chemical through a cannula, we always cause additional damage to the brain. When we pass an electrode or a cannula through the brain to get to our target, we inevitably cause a small amount of damage even before turning on the lesion maker or starting the infusion. Thus, we cannot simply compare the behavior of brain-lesioned animals with that of unoperated control animals; the incidental damage to the brain regions above the lesion may actually be responsible for some of the behavioral deficits we see. What we do is operate on a group of animals and produce **sham lesions.** To do so, we anesthetize each animal, put it in the stereotaxic apparatus (described below), cut open the scalp, drill the holes, insert the electrode or cannula, and lower it

sham lesion A placebo procedure that duplicates all the steps of producing a brain lesion except the one that actually causes the brain damage.

to the proper depth. In other words, we do everything we would do to produce the lesion except turn on the lesion maker or start the infusion. This group of animals serves as a control group; if the behavior of the animals with brain lesions is different from that of the sham-operated control animals, we can conclude that the lesions caused the behavioral deficits. (As you can see, a sham lesion serves the same purpose as a placebo does in a pharmacology study.)

Most of the time, investigators produce permanent brain lesions, but sometimes it is advantageous to disrupt the activity of a particular region of the brain temporarily. The easiest way to do so is to inject a local anesthetic or a drug called *muscimol* into the appropriate part of the brain. The anesthetic blocks action potentials in axons entering or leaving that region, thus effectively producing a temporary lesion (usually called a *reversible* brain lesion). Muscimol, a drug that stimulates GABA receptors, inactivates a region of the brain by inhibiting the neurons located there. (You will recall that GABA is the most important inhibitory neurotransmitter in the brain.)

Stereotaxic Surgery

So how do we get the tip of an electrode or cannula to a precise location in the depths of an animal's brain? The answer is **stereotaxic surgery.** *Stereotaxis* literally means "solid arrangement"; more specifically, it refers to the ability to locate objects in space. A *stereotaxic apparatus* contains a holder that fixes the animal's head in a standard position and a carrier that moves an electrode or a cannula through measured distances in all three axes of space. However, to perform stereotaxic surgery, one must first study a *stereotaxic atlas.*

The Stereotaxic Atlas

No two brains of animals of a given species are completely identical, but there is enough similarity among individuals to predict the location of particular brain structures relative to external features of the head. For instance, a subcortical nucleus of a rat might be so many millimeters ventral, anterior, and lateral to a point formed by the junction of several bones of the skull. Figure 5.3 shows two views of a rat skull: a drawing of the dorsal surface and, beneath it, a midsagittal view. (See *Figure 5.3.*) The skull is composed of several bones that grow together and form *sutures* (seams). The heads of newborn babies contain a soft spot at the junction of the coronal and sagittal sutures called the *fontanelle.* Once this gap closes, the junction is called **bregma,** from the Greek word meaning "front of head." We can find bregma on a rat's skull, too, and it serves as a convenient reference point. If the animal's skull is oriented as shown in the illustration, a particular region of the brain is found in a fairly constant position, relative to bregma.

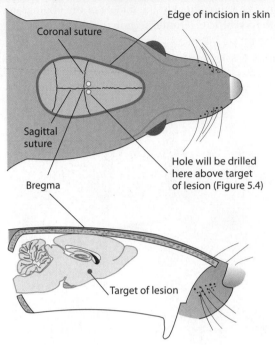

Edge of incision in skin
Coronal suture
Sagittal suture
Bregma
Hole will be drilled here above target of lesion (Figure 5.4)
Target of lesion

A **stereotaxic atlas** contains photographs or drawings that correspond to frontal sections taken at various distances rostral and caudal to bregma. For example, the page shown in Figure 5.4 is a drawing of a slice of the brain that contains a brain structure (shown in red) that we are interested in. If we wanted to place the tip of a wire in this structure (the fornix), we would have to drill a hole through the skull immediately above it. (See *Figure 5.4.*) Each page of the stereotaxic atlas is labeled according to the distance of the section anterior or posterior to bregma. The grid on each page indicates distances of brain structures ventral to the top of the skull and lateral to the midline. To place the tip of a wire in the fornix, we would drill a hole above the target and then lower the electrode through the hole until the tip was at the correct depth, relative to the skull height at bregma. (Compare *Figures 5.3* and *5.4.*) Thus, by finding a neural structure (which we cannot see in our

stereotaxic surgery (*stair ee oh tak sik*) Brain surgery using a stereotaxic apparatus to position an electrode or cannula in a specified position of the brain.

bregma The junction of the sagittal and coronal sutures of the skull; often used as a reference point for stereotaxic brain surgery.

stereotaxic atlas A collection of drawings of sections of the brain of a particular animal with measurements that provide coordinates for stereotaxic surgery.

FIGURE 5.4 ■ Stereotaxic Atlas

This sample page from a stereotaxic atlas shows a rat brain. The target (the fornix) is indicated in red. Labels have been removed for the sake of clarity.

(Adapted from Swanson, L. W. *Brain Maps: Structure of the Rat Brain.* New York: Elsevier, 1992.)

Target of lesion

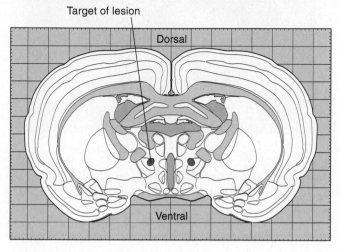

Dorsal

Ventral

FIGURE 5.5 ■ Stereotaxic Apparatus

This apparatus is used for performing brain surgery on rats.

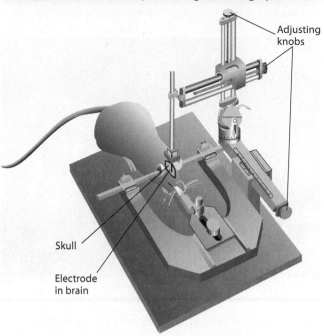

Adjusting knobs

Skull

Electrode in brain

---

animal) on one of the pages of a stereotaxic atlas, we can determine the structure's location relative to bregma (which we can see). Note that, because of variations in different strains and ages of animals, the atlas gives only an approximate location. We always have to try out a new set of coordinates, slice and stain the animal's brain, see the actual location of the lesion, correct the numbers, and try again. (Slicing and staining of brains are described later.)

The Stereotaxic Apparatus

A **stereotaxic apparatus** operates on simple principles. The device includes a head holder, which maintains the animal's skull in the proper orientation, a holder for the electrode, and a calibrated mechanism that moves the electrode holder in measured distances along the three axes: anterior–posterior, dorsal–ventral, and lateral–medial. Figure 5.5 illustrates a stereotaxic apparatus designed for small animals; various head holders can be used to outfit this device for such diverse species as rats, mice, hamsters, pigeons, and turtles. (See *Figure 5.5.*)

Once we obtain the coordinates from a stereotaxic atlas, we anesthetize the animal, place it in the apparatus, and cut the scalp open. We locate bregma, dial in the appropriate numbers on the stereotaxic apparatus, drill a hole through the skull, and lower the device into the brain by the correct amount. Now the tip of the cannula or electrode is where we want it to be, and we are ready to produce the lesion.

Of course, stereotaxic surgery may be used for purposes other than lesion production. Wires placed in the

brain can be used to stimulate neurons as well as to destroy them, and drugs can be injected that stimulate neurons or block specific receptors. We can attach cannulas or wires permanently by following a procedure that will be described later in this chapter. In all cases, once surgery is complete, the wound is sewn together, and the animal is taken out of the stereotaxic apparatus and allowed to recover from the anesthetic.

Stereotaxic apparatuses are made for humans, by the way. Sometimes a neurosurgeon produces subcortical lesions—for example, to reduce the symptoms of Parkinson's disease. Usually, the surgeon uses multiple landmarks and verifies the location of the wire (or other device) inserted into the brain by taking brain scans or recording the activity of the neurons in that region before producing a brain lesion. (See *Figure 5.6.*)

Histological Methods

After producing a brain lesion and observing its effects on an animal's behavior, we must slice and stain the brain so that we can observe it under the microscope and see the location of the lesion. Brain lesions often miss the mark, so we have to verify the precise location of the brain damage after testing the animal behaviorally. To do so, we must fix, slice, stain, and examine the brain.

---

stereotaxic apparatus A device that permits a surgeon to position an electrode or cannula into a specific part of the brain.

FIGURE 5.6 ■ Stereotaxic Surgery on a Human Patient

(Photograph courtesy of John W. Snell, University of Virginia Health System.)

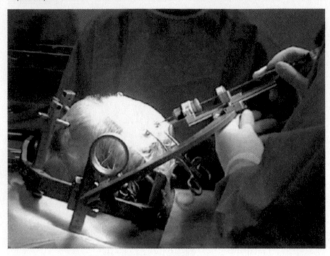

Together, these procedures are referred to as *histological methods*. (The prefix *histo-* refers to body tissue.)

Fixation and Sectioning

If we hope to study the tissue in the form it had at the time of the organism's death, we must destroy the autolytic enzymes (*autolytic* means "self-dissolving"), which will otherwise turn the tissue into mush. The tissue must also be preserved to prevent its decomposition by bacteria or molds. To achieve both of these objectives, we place the neural tissue in a **fixative.** The most commonly used fixative is **formalin,** an aqueous solution of formaldehyde, a gas. Formalin halts autolysis, hardens the very soft and fragile brain, and kills any microorganisms that might destroy it.

Before the brain is fixed (that is, put into a fixative solution), it is usually perfused. **Perfusion** of tissue (literally, "a pouring through") entails removal of the blood and its replacement with another fluid. The animal's brain is perfused because better histological results are obtained when no blood is present in the tissue. The animal whose brain is to be studied is humanely killed with an overdose of a general anesthetic. Blood vessels are opened so that the blood can be drained from them and replaced with a dilute salt solution. The brain is removed from the skull and placed in a container filled with the fixative.

Once the brain has been fixed, we must slice it into thin sections and stain various cellular structures to see anatomical details. Slicing is done with a **microtome** (literally, "that which slices small"). Slices prepared for examination under a light microscope are typically 10 to 80 μm in thickness; those prepared for the electron

FIGURE 5.7 ■ A Microtome

microscope are generally cut at less than 1 μm. (For some reason, slices of brain tissue are usually referred to as *sections*.)

A microtome contains three parts: a knife, a platform on which to mount the tissue, and a mechanism that advances the knife (or the platform) the correct amount after each slice so that another section can be cut. In most cases the platform includes an attachment that freezes the brain to make it hard enough to be cut into thin sections. Figure 5.7 shows a microtome. The knife holder slides forward on an oiled rail and takes a section off the top of the tissue mounted on the platform. The platform automatically rises by a predetermined amount as the knife and holder are pushed back so that the next forward movement of the knife takes off another section. (See *Figure 5.7*.)

After the tissue is cut, we attach the slices to glass microscope slides. We can then stain the tissue by putting the entire slide into various chemical solutions. Finally, we cover the stained sections with a small amount of a transparent liquid known as a *mounting medium* and place a very thin glass coverslip over the sections. The mounting medium keeps the coverslip in position. *MyPsychKit 5.1, Histological Methods,* shows these procedures.

mypsychkit™
Where learning comes to life!
Animation 5.1
Histological Methods

fixative A chemical such as formalin; used to prepare and preserve body tissue.

formalin (*for ma lin*) The aqueous solution of formaldehyde gas; the most commonly used tissue fixative.

perfusion (*per few zhun*) The process by which an animal's blood is replaced by a fluid such as a saline solution or a fixative in preparing the brain for histological examination.

microtome (*my krow tome*) An instrument that produces very thin slices of body tissues.

Staining

If you looked at an unstained section of brain tissue under a microscope, you would be able to see the outlines of some large cellular masses and the more prominent fiber bundles. However, no fine details would be revealed. For this reason the study of microscopic neuroanatomy requires special histological stains. Researchers have developed many different stains to identify specific substances within and outside of cells. For verifying the location of a brain lesion, we will use one of the simplest: a cell-body stain.

In the late nineteenth century Franz Nissl, a German neurologist, discovered that a dye known as methylene blue would stain the cell bodies of brain tissue. The material that takes up the dye, known as the *Nissl substance,* consists of RNA, DNA, and associated proteins located in the nucleus and scattered, in the form of granules, in the cytoplasm. Many dyes besides methylene blue can be used to stain cell bodies found in slices of the brain, but the most frequently used is cresyl violet. Incidentally, the dyes were not developed specifically for histological purposes; they were originally formulated for use in dyeing cloth.

The discovery of cell-body stains made it possible to identify nuclear masses in the brain. Figure 5.8 shows a frontal section of a cat brain stained with cresyl violet. Note that you can observe fiber bundles by their lighter appearance; they do not take up the stain. (See *Figure 5.8.*) The stain is not selective for *neural* cell bodies; all cells are stained, neurons and glia alike. It is up to the

FIGURE 5.8 ■ Frontal Section of a Cat Brain

The brain is stained with cresyl violet, a cell-body stain. The arrowheads point to *nuclei,* or groups of cell bodies.

(Histological material courtesy of Mary Carlson.)

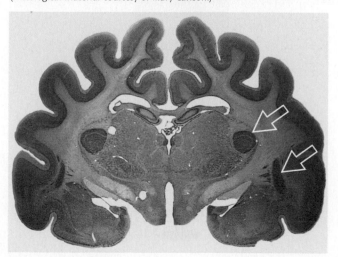

FIGURE 5.9 ■ Section Through an Axodendritic Synapse

On this electron photomicrograph of a section, two synaptic regions are indicated by arrows, and a circle indicates a region of pinocytosis in an adjacent terminal button, presumably representing recycling of vesicular membrane. T = terminal button; f = microfilaments; M = mitochondrion.

(From Rockel, A. J., and Jones, E. G. *Journal of Comparative Neurology,* 1973, *147,* 61–92. Reprinted with permission.)

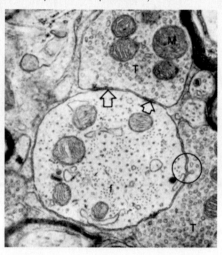

investigator to determine which is which—by size, shape, and location.

Electron Microscopy

The light microscope is limited in its ability to resolve extremely small details. Because of the nature of light itself, magnification of more than approximately 1500 times does not add any detail. To see such small anatomical structures as synaptic vesicles and details of cell organelles, investigators must use a **transmission electron microscope.** A beam of electrons is passed through a thin slice of the tissue to be examined. The beam of electrons casts a shadow of the tissue on a fluorescent screen, which can be photographed or scanned into a computer. Electron photomicrographs produced in this way can provide information about structural details on the order of a few tens of nanometers. (See *Figure 5.9.*)

A **scanning electron microscope** provides less magnification than a standard transmission electron microscope, which transmits the electron beam through the tissue. However, it shows objects in three dimensions.

transmission electron microscope A microscope that passes a focused beam of electrons through thin slices of tissue to reveal extremely small details.

scanning electron microscope A microscope that provides three-dimensional information about the shape of the surface of a small object by scanning the object with a thin beam of electrons.

FIGURE 5.10 ■ Neurons and Glia

A scanning electron micrograph was used to show neurons and glia.

(From *Tissues and Organs: A Text-Atlas of Scanning Electron Microscopy*, by Richard G. Kessel and Randy H. Kardon. Copyright © 1979 by W. H. Freeman and Co. Reprinted by permission of Barbara Kessel and Randy Kardon.)

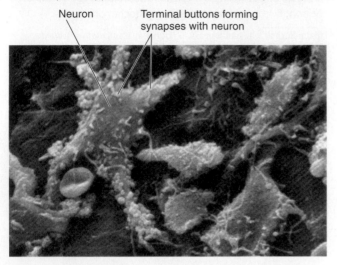

The microscope scans the tissue with a moving beam of electrons. The information from the reflection of the beam is received by a detector, and a computer produces a remarkably detailed three-dimensional view. (See *Figure 5.10.*)

Confocal Laser Scanning Microscopy

Conventional microscopy or transmission electron microscopy requires that the tissue be sliced into thin sections. The advent of the **confocal laser scanning microscope** makes it possible to see details inside thick sections of tissue or even in slabs of tissue maintained in tissue cultures or in the upper layers of tissue in the exposed living brain. The confocal microscope requires that the cells or parts of cells of interest be stained with a fluorescent dye. (This procedure, called

> **confocal laser scanning microscope** A microscope that provides high-resolution images of various depths of thick tissue that contains fluorescent molecules by scanning the tissue with light from a laser beam.

FIGURE 5.11 ■ Confocal Microscope

A laser scanning confocal microscope is shown in this simplified, schematic diagram.

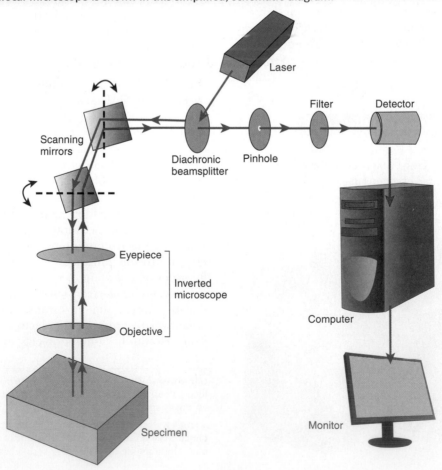

immunocytochemistry, is described in next section of this chapter.) For example, neurons that produce a particular peptide can be labeled with a fluorescent dye. A beam of light of a particular wavelength is produced by a laser and reflected off of a *dichroic* mirror—a special mirror that transmits light of certain wavelengths and reflects light of other wavelengths. Lenses in the microscope focus the laser light at a particular depth in the tissue. This light triggers fluorescence in the tissue, which passes through the lenses and is transmitted through the dichroic mirror to a pinhole aperture. This aperture blocks extraneous light caused by scattering within the tissue. The light that passes through the aperture is measured by a detector. Two moving mirrors cause the laser light to scan the tissue, which provides the computer with the information it needs to form an image of a slice of tissue located at a particular depth within the sample. If multiple scans are made while the location of the aperture is moved, a stack of images of slices through the tissue—remember, this can be living tissue—can be obtained. (See *Figure 5.11.*)

Figure 5.12 illustrates the use of confocal microscopy of the hippocampus of living anesthetized mice (Mizrahi et al., 2004). The skull was opened, and overlying cortex

FIGURE 5.12 ■ Branches of Dendrites

These photographs of branches of dendrites of hippocampal neurons of a living mouse were taken through a laser scanning confocal microscope. The images show that seizures caused by injection of drugs caused the disappearance of some dendritic spines.

(From Mizrahi, A., Crowley, J. C., Shtoyerman, E., and Katz, L. C. *Journal of Neuroscience*, 2004, *24*, 3147–3151. Copyright © 2004 by the Society for Neuroscience. Reprinted by permission of the Society for Neuroscience.)

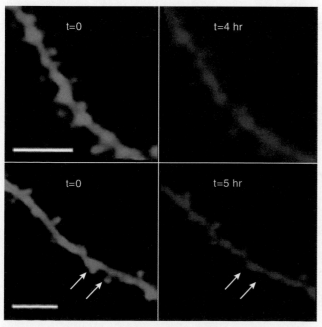

was dissected away so that images could be taken of the interior of the hippocampus. Molecular genetic methods had been used to insert a gene into the animals' DNA that produced a fluorescent protein dye in certain neurons in the hippocampus. Mizrahi and his colleagues obtained images of individual dendrites of these neurons before and after they induced seizures in the animals by administering excitatory drugs. Images made before the seizures are shown in green, and images made 4 to 5 hours after the seizures are shown in red. As you can see, the animals in which the seizures were induced showed a loss of dendritic spines (arrows), but there was no loss of spines in control animals. (See *Figure 5.12.*)

Tracing Neural Connections

Let's suppose that we were interested in discovering the neural mechanisms responsible for reproductive behavior. To start out, we wanted to study the physiology of sexual behavior of female rats. On the basis of some hints we received by reading reports of experiments by other researchers published in scientific journals, we performed stereotaxic surgery on two groups of female rats. We made a lesion in the ventromedial nucleus of the hypothalamus (VMH) of the rats in the experimental group and performed sham surgery on the rats in the control group. After a few days' recovery we placed the animals (individually, of course) with male rats. The females in the control group responded positively to the males' attention; they engaged in courting behavior followed by copulation. However, the females with the VMH lesions rejected the males' attention and refused to copulate with them. We confirmed with histology that the VMH was indeed destroyed in the brains of the experimental animals. (One experimental rat did copulate, but we discovered later that the lesion had missed the VMH in that animal, so we discarded the data from that subject.)

The results of our experiment indicate that neurons in the VMH appear to play a role in functions required for copulatory behavior in females. (By the way, it turns out that these lesions do not affect copulatory behavior in males.) So where do we go from here? What is the next step? In fact, there are many questions that we could pursue. One question concerns the system of brain structures that participate in female copulatory behavior. Certainly, the VMH does not stand alone; it receives inputs from other structures and sends outputs to still others. Copulation requires integration of visual, tactile, and olfactory perceptions and organization of patterns of movements in response to those of the partner. In addition, the entire network requires activation by the appropriate sex hormones. What is the precise role of the VMH in this complicated system?

Before we can hope to answer this question, we must know more about the connections of the VMH

FIGURE 5.13 ■ Tracing Neural Circuits

Once we know that a particular brain region is involved in a particular function, we may ask what structures provide inputs to the region and what structures receive outputs from it.

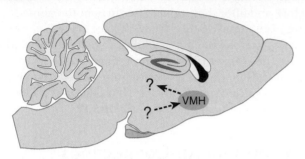

with the rest of the brain. What structures send their axons to the VMH, and to what structures does the VMH, in turn, send its axons? Once we know what the connections are, we can investigate the role of these structures and the nature of their interactions. (See *Figure 5.13.*)

How do we investigate the connections of the VMH? The question cannot be answered by means of histological procedures that stain all neurons, such as cell-body stains. If we look closely at a brain that has been prepared by these means, we see only a tangled mass of neurons. But in recent years, researchers have developed very precise methods that make specific neurons stand out from all of the others.

Tracing Efferent Axons

Eventually, the VMH must affect behavior. That is, neurons in the VMH must send axons to parts of the brain that contain neurons that are responsible for muscular movements. The pathway is probably not direct; more

likely, neurons in the VMH affect neurons in other structures, which influence those in yet other structures until, eventually, the appropriate motor neurons are stimulated. To discover this system, we want to be able to identify the paths followed by axons leaving the VMH. In other words, we want to trace the *efferent axons* of this structure.

We will use an **anterograde labeling method** to trace these axons. (*Anterograde* means "moving forward.") Anterograde labeling methods employ chemicals that are taken up by dendrites or cell bodies and are then transported through the axons toward the terminal buttons.

Over the years, neuroscientists have developed several different methods for tracing the pathways followed by efferent axons. For example, to discover the destination of the efferent axons of neurons located within the VMH, we could inject a minute quantity of **PHA-L** (a protein found in kidney beans) into that nucleus. (We would use a stereotaxic apparatus to do so, of course.) The molecules of PHA-L are taken up by dendrites and are transported through the soma to the axon, where they travel by means of fast axoplasmic transport to the terminal buttons. Within a few days the cells are filled in their entirety with molecules of PHA-L: dendrites, soma, axons and all their branches, and terminal buttons. Then we kill the animal, slice the brain, and mount the sections on microscope slides. A special *immunocytochemical* method is used to make the molecules of PHA-L visible, and the slides are examined under a microscope. (See *Figure 5.14.*)

anterograde labeling method (*ann ter oh grade*) A histological method that labels the axons and terminal buttons of neurons whose cell bodies are located in a particular region.

PHA-L Phaseolus vulgaris leukoagglutinin; a protein derived from kidney beans and used as an anterograde tracer; taken up by dendrites and cell bodies and carried to the ends of the axons.

FIGURE 5.14 ■ Using PHA-L to Trace Efferent Axons

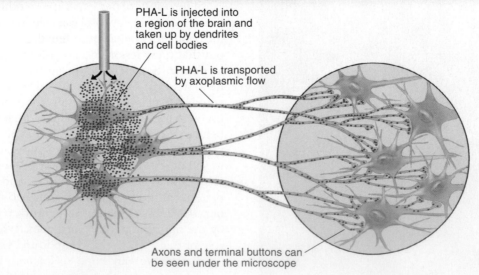

PHA-L is injected into a region of the brain and taken up by dendrites and cell bodies

PHA-L is transported by axoplasmic flow

Axons and terminal buttons can be seen under the microscope

Immunocytochemical methods take advantage of the immune reaction. The body's immune system has the ability to produce antibodies in response to antigens. *Antigens* are proteins (or peptides), such as those found on the surface of bacteria or viruses. *Antibodies,* which are also proteins, are produced by white blood cells to destroy invading microorganisms. Antibodies either are secreted by white blood cells or are located on their surface, in the way neurotransmitter receptors are located on the surface of neurons. When the antigens present on the surface of an invading microorganism come into contact with the antibodies that recognize them, the antibodies trigger an attack on the invader by the white blood cells.

Molecular biologists have developed methods for producing antibodies to any peptide or protein. The antibody molecules are attached to various types of dye molecules. Some of these dyes react with other chemicals and stain the tissue a brown color. Others are fluorescent; they glow when they are exposed to light of a particular wavelength. To determine where the peptide or protein (the antigen) is located in the brain, the investigator places fresh slices of brain tissue in a solution that contains the antibody/dye molecules. The antibodies attach themselves to their antigen. When the investigator examines the slices with a microscope (under light of a particular wavelength in the case of fluorescent dyes), he or she can see which parts of the brain—even which individual neurons—contain the antigen.

Figure 5.15 shows how PHA-L can be used to identify the efferents of a particular region of the brain. Molecules of this chemical were injected into the VMH.

Two days later, after the PHA-L had been taken up by the neurons in this region and transported to the ends of their axons, the animal was killed. Figure 5.15(a) shows the site of the injection; as you can see, the lectin fills nearby cell bodies and dendrites. (See *Figure 5.15a.*) Figure 5.15(b) shows a photomicrograph of the periaqueductal gray matter (PAG). This region contains some labeled axons and terminal buttons (gold color), which proves that some of the efferent axons of the VMH terminate in the PAG. (See *Figure 5.15b.*)

To continue our study of the role of the VMH in female sexual behavior, we would find the structures that receive information from neurons in the VMH (such as the PAG) and see what happens when each of them is destroyed. Let's suppose that damage to some of these structures also impairs female sexual behavior. We will inject these structures with PHA-L and see where *their* axons go. Eventually, we will discover the relevant pathways from the VMH to the motor neurons whose activity is necessary for copulatory behavior. (In fact, researchers have done so, and some of their results are presented in Chapter 10.)

Tracing Afferent Axons

Tracing efferent axons from the VMH will tell us only part of the story about the neural circuitry involved in female sexual behavior: the part between the VMH and the motor neurons. What about the circuits *before* the

immunocytochemical method A histological method that uses radioactive antibodies or antibodies bound with a dye molecule to indicate the presence of particular proteins of peptides.

FIGURE 5.15 ■ Anterograde Labeling Method

PHA-L was injected into the ventromedial nucleus of the hypothalamus, where it was taken up by dendrites and carried through the cells' axons to their terminal buttons. (a) The injection site. (b) Labeled axons and terminal buttons in the periaqueductal gray matter.

(Courtesy of Kirsten Nielsen Ricciardi and Jeffrey Blaustein, University of Massachusetts.)

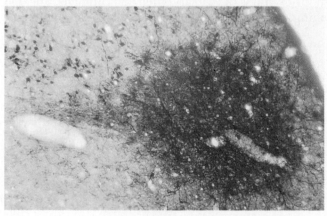

(a)

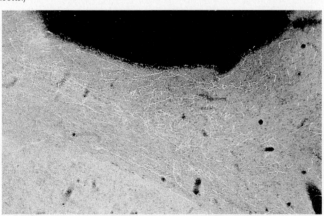

(b)

VMH? Is the VMH somehow involved in the analysis of sensory information (such as the sight, odor, or touch of the male)? Or perhaps the activating effect of a female's sex hormones on her behavior act through the VMH or through neurons whose axons form synapses there. To discover the parts of the brain that are involved in the "upstream" components of the neural circuitry, we need to find the inputs of the VMH—its afferent connections. To do so, we will employ a **retrograde labeling method.**

Retrograde means "moving backward." Retrograde labeling methods employ chemicals that are taken up by terminal buttons and carried back through the axons toward the cell bodies. The method for identifying the afferent inputs to a particular region of the brain is similar to the method used for identifying its efferents. First, we inject a small quantity of a chemical called **fluorogold** into the VMH. The chemical is taken up by terminal buttons and is transported back by means of retrograde axoplasmic transport to the cell bodies. A few days later we kill the animal, slice its brain, and examine the tissue under light of the appropriate wavelength. The molecules of fluorogold fluoresce under this light. We discover that the medial amygdala is one of the regions that provides input to the VMH. (See *Figure 5.16.*)

The anterograde and retrograde labeling methods that I have described identify a single link in a chain of neurons—neurons whose axons enter or leave a particular brain region. *Transneuronal* tracing methods identify a series of two, three, or more neurons that form serial synaptic connections with each other. The most effective retrograde transneuronal tracing method uses a **pseudorabies virus**—a weakened form of a pig herpes virus that was originally developed as a vaccine. For anterograde transneuronal tracing, a variety of the **herpes simplex virus,** similar to the one that causes cold sores, is used. The virus is injected directly into a brain region, is taken up by neurons there, and infects them. The virus spreads throughout the infected neurons and is eventually released, passing on the infection to other neurons that form synaptic connections with them.

The longer the experimenter waits after injecting the virus, the larger is the number of neurons that become infected. After the animal is killed and the brain is sliced, immunocytochemical methods are used to localize a protein produced by the virus. For example, Daniels, Miselis, and Flanagan-Cato (1999) injected pseudorabies virus in the muscles responsible for female rats' mating posture. After a few days, the rats were killed, and their brains were examined for evidence of viral infection. The study indicated that the virus found its way up the motor nerves to the motor neurons in the spinal cord, then to the reticular formation of the medulla, then to the periaqueductal gray matter, and finally to the VMH. These results confirm the results of the anterograde and retrograde labeling methods I just described. (Labeled neurons were found in other structures as well, but they are not relevant to this discussion.)

Together, anterograde and retrograde labeling methods—including transneuronal methods—enable us to discover circuits of interconnected neurons. Thus, these methods help to provide us with a "wiring diagram" of the brain. (See *Figure 5.17.*) Armed with other research methods (including some to be described later in this chapter), we can try to discover the functions of each component of this circuit.

Studying the Structure of the Living Human Brain

There are many good reasons to investigate the functions of brains of animals other than humans. For one thing, we can compare the results of studies made with

---

FIGURE 5.16 ■ A Retrograde Tracing Method

Fluorogold was injected in the VMH, where it was taken up by terminal buttons and transported back through the axons to their cell bodies. The photomicrograph shows these cell bodies, located in the medial amygdala.

(Courtesy of Yvon Delville, University of Massachusetts Medical School.)

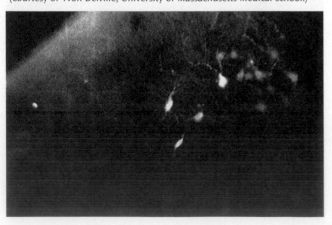

---

retrograde labeling method A histological method that labels cell bodies that give rise to the terminal buttons that form synapses with cells in a particular region.

fluorogold (*flew roh gold*) A dye that serves as a retrograde label; taken up by terminal buttons and carried back to the cell bodies.

pseudorabies virus A weakened form of a pig herpes virus used for retrograde transneuronal tracing, which labels a series of neurons that are interconnected synaptically.

herpes simplex virus A form of herpes virus used for anterograde transneuronal tracing, which labels a series of neurons that are interconnected synaptically.

FIGURE 5.17 ■ Results of Tracing Methods

The figure shows one of the inputs to the VMH and one of the outputs, as revealed by anterograde and retrograde labeling methods.

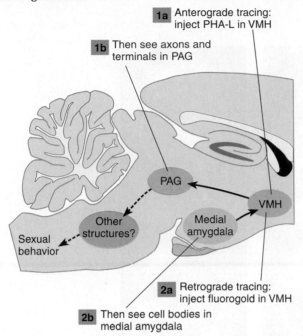

1a Anterograde tracing: inject PHA-L in VMH

1b Then see axons and terminals in PAG

PAG

Other structures?

Sexual behavior

Medial amygdala

VMH

2a Retrograde tracing: inject fluorogold in VMH

2b Then see cell bodies in medial amygdala

FIGURE 5.18 ■ Computerized Tomography (CT) Scanner

(© Larry Mulvihill/Rainbow.)

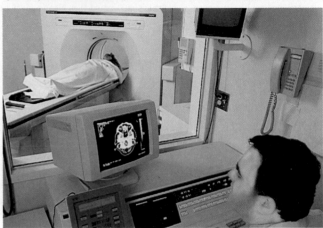

different species in order to make some inferences about the evolution of various neural systems. Even if our primary interest is in the functions of the human brain, we certainly cannot ask people to submit to brain surgery for the purposes of research. But diseases and accidents do occasionally damage the human brain, and if we know where the damage occurs, we can study the people's behavior and try to make the same sorts of inferences we make with deliberately produced brain lesions in laboratory animals. The problem is, where is the lesion?

In past years a researcher might have studied the behavior of a person with brain damage and never found out exactly where the lesion was located. The only way to be sure was to obtain the patient's brain when he or she died and examine slices of it under a microscope. But it was often impossible to do this. Sometimes the patient outlived the researcher. Sometimes the patient moved out of town. Sometimes (often, perhaps) the family refused permission for an autopsy. Because of these practical problems, study of the behavioral effects of damage to specific parts of the human brain made rather slow progress.

Recent advances in X-ray techniques and computers have led to the development of several methods for studying the anatomy of the living brain. These advances permit researchers to study the location and extent of brain damage while the patient is still living.

The first method that was developed is called **computerized tomography (CT)** (from the Greek for *tomos,* "cut," and *graphein,* "to write"). This procedure, usually referred to as a *CT scan,* works as follows: The patient's head is placed in a large doughnut-shaped ring. The ring contains an X-ray tube and, directly opposite it (on the other side of the patient's head), an X-ray detector. The X-ray beam passes through the patient's head, and the detector measures the amount of radioactivity that gets through it. The beam scans the head from all angles, and a computer translates the numbers it receives from the detector into pictures of the skull and its contents. (See *Figure 5.18.*)

Figure 5.19 shows a series of these CT scans taken through the head of a patient who sustained a stroke. The stroke damaged a part of the brain involved in bodily awareness and perception of space. The patient lost her awareness of the left side of her body and of items located on her left. You can see the damage as a white spot in the lower left corner of scan 5. (See *Figure 5.19.*)

An even more detailed, high-resolution picture of what is inside a person's head is provided by a process called **magnetic resonance imaging (MRI).** The MRI scanner resembles a CT scanner, but it does not use X-rays. Instead, it passes an extremely strong magnetic

computerized tomography (CT) The use of a device that employs a computer to analyze data obtained by a scanning beam of X-rays to produce a two-dimensional picture of a "slice" through the body.

magnetic resonance imaging (MRI) A technique whereby the interior of the body can be accurately imaged; involves the interaction between radio waves and a strong magnetic field.

FIGURE 5.19 ■ CT Scans of a Lesion

The patient has a lesion in the right occipital-parietal area (scan 5). The lesion appears white because it was accompanied by bleeding; blood absorbs more radiation than the surrounding brain tissue. Rostral is up, caudal is down; left and right are reversed. Scan 1 shows a section through the eyes and the base of the brain.

(Courtesy of J. McA. Jones, Good Samaritan Hospital, Portland, Oregon.)

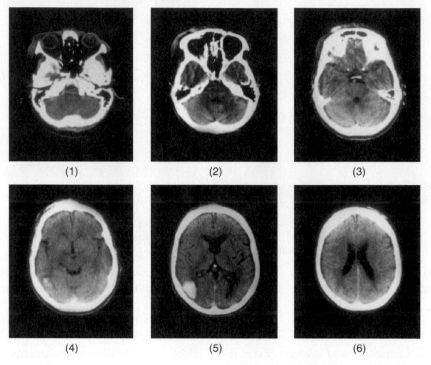

(1)   (2)   (3)

(4)   (5)   (6)

field through the patient's head. When a person's body is placed in a strong magnetic field, the nuclei of some atoms in molecules in the body spin with a particular orientation. If a radio frequency wave is then passed through the body, these nuclei emit radio waves of their own. Different molecules emit energy at different frequencies. The MRI scanner is tuned to detect the radiation from hydrogen atoms. Because these atoms are present in different concentrations in different tissues, the scanner can use the information to prepare pictures of slices of the brain. Unlike CT scans, which are generally limited to the horizontal plane, MRI scans can be taken in the sagittal or frontal planes as well. (See *Figure 5.20*.)

As you can see in Figure 5.20, MRI scans distinguish between regions of gray matter and white matter, so major fiber bundles (such as the corpus callosum) can be seen. However, small fiber bundles are not visible on these scans. A special modification of the MRI scanner permits the visualization of even small bundles of fibers and the tracing of fiber tracts. Above absolute zero, all molecules move in random directions because of thermal

FIGURE 5.20 ■ Midsagittal MRI Scan of a Human Brain

(Photo courtesy of Philips Medical Systems.)

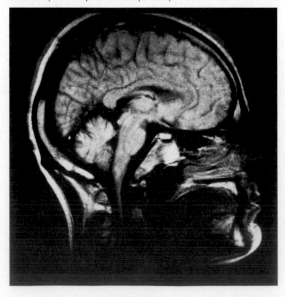

FIGURE 5.21 ■ Diffusion Tensor Imaging

A sagittal view of some of the axons that project from the thalamus to the cerebral cortex in the human brain is revealed by diffusion tensor imaging.

(From Wakana, S., Jian, H., Nagae-Poetscher, L. M., van Zijl, P. C. M., and Mori, S. *Radiology,* 2004, *230,* 77–87. Reprinted with permission.)

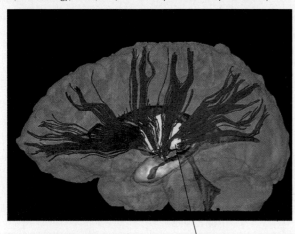

Thalamus

agitation: The higher the temperature, the faster the random movement. **Diffusion tensor imaging (DTI)** takes advantage of the fact that the movement of water molecules in bundles of white matter will not be random but will tend to be in a direction parallel to the axons that make up the bundles. The MRI scanner uses information about the movement of the water molecules to determine the location and orientation of bundles of axons in white matter. Figure 5.21 shows a sagittal view of some of the axons that project from the thalamus to the cerebral cortex in the human brain, as revealed by diffusion tensor imaging. The computer adds colors to distinguish different bundles of axons. (See *Figure 5.21.*)

diffusion tensor imaging (DTI) An imaging method that uses a modified MRI scanner to reveal bundles of myelinated axons in the living human brain.

InterimSummary

Experimental Ablation

The goal of research in behavioral neuroscience is to understand the brain functions required for the performance of a particular behavior and then to learn the location of the neural circuits that perform these functions. The lesion method is the oldest one employed in such research, and it remains one of the most useful. A subcortical lesion is made under the guidance of a stereotaxic apparatus. The coordinates are obtained from a stereotaxic atlas, and the tip of an electrode or cannula is placed at the target. A lesion is made by passing radio frequency current through the electrode or infusing an excitatory amino acid through the cannula, producing an excitotoxic lesion. The advantage of excitotoxic lesions is that they destroy only neural cell bodies; axons passing through the region are not damaged.

The location of a lesion must be determined after the animal's behavior has been observed. The animal is killed by humane means, the brain is perfused with a saline solution, and the brain is removed and placed in a fixative such as formalin. A microtome is used to slice the brain, which is usually frozen to make it hard enough to cut into thin sections. These sections are mounted on glass slides, stained with a cell-body stain, and examined under a microscope.

Light microscopes enable us to see cells and their larger organelles, but an electron microscope is needed to see small details, such as individual mitochondria and synaptic vesicles. Scanning electron microscopes provide a three-dimensional view of tissue but at a lower magnification than transmission electron microscopes. Confocal microscopes provide images of "slices" of tissues that can show the presence of particular molecules—even in living tissue.

The next step in a research program often requires the investigator to discover the afferent and efferent connections of the region of interest with the rest of the brain. Efferent connections (those that carry information from the region in question to other parts of the brain) are revealed with anterograde tracing methods, such as the one that uses PHA-L. Afferent connections (those that bring information to the region in question from other parts of the brain) are revealed with retrograde tracing methods, such as the one that uses fluorogold. Chains of neurons that form synaptic connections are revealed by transneuronal tracing method. The pseudorabies virus can be used as a retrograde tracer, and a variety of the herpes simplex virus can be used as an anterograde tracer.

Although brain lesions are not deliberately made in the human brain for the purposes of research, diseases and accidents can cause brain damage, and if we know where the damage is located, we can study people's behavior and make inferences about the location of the neural circuits that perform relevant functions. If the patient dies and the brain is available for examination, ordinary histological methods can be used. Otherwise, the living brain can be examined with CT scanners and MRI scanners. Diffusion tensor imaging (DTI) uses a modified MRI scanner to visualize bundles of myelinated axons in the living human brain.

Table 5.1 summarizes the research methods presented in this section.

TABLE 5.1 ■ Research Methods: Part I

GOAL OF METHOD	METHOD	REMARKS
Destroy or inactivate specific brain region	Radio frequency lesion	Destroys all brain tissue near tip of electrode
	Excitotoxic lesion; uses excitatory amino acid such as kainic acid	Destroys only cell bodies near tip of cannula; spares axons passing through the region
	6-HD lesion	Destroys catecholaminergic neurons near tip of cannula
	Infusion of local anesthetic; cryoloop	Temporarily inactivates specific brain region; animal can serve as its own control
Place electrode or cannula in specific region within brain	Stereotaxic surgery	Consult stereotaxic atlas for coordinates
Find location of lesion	Perfuse brain; fix brain; slice brain; stain sections	
Identify axons leaving a particular region and the terminal buttons of these axons	Anterograde tracing method, such as PHA-L	
Identify location of neurons whose axons terminate in a particular region	Retrograde tracing method, such as fluorogold	
Identify chain of neurons that are interconnected synaptically	Transneuronal tracing method; uses pseudorabies virus (for retrograde tracing) or herpes simplex virus (for anterograde tracing)	
Find location of lesion in living human brain	Computerized tomography (CT scanner)	Slows "slice" of brain; uses X-rays
	Magnetic resonance imaging (MRI scanner)	Shows "slice" of brain; better detail than CT scan; uses a magnetic field and radio waves
Find location of fiber bundles in living human brain	Diffusion tensor imaging (DTI)	Shows bundles of myelinated axons; uses an MRI scanner
Visualize details of cells in thick sections of tissue	Confocal laser scanning microscopy	Can be used to see "slices" of tissue in living brain; requires the presence of fluorescent molecules in the tissue

RECORDING AND STIMULATING NEURAL ACTIVITY

The first section of this chapter dealt with the anatomy of the brain and the effects of damage to particular regions. This section considers a different approach: studying the brain by recording or stimulating the activity of particular regions. Brain functions involve activity of circuits of neurons; thus, different perceptions and behavioral responses involve different patterns of activity in the brain. Researchers have devised methods to record these patterns of activity or artificially produce them.

Recording Neural Activity

Axons produce action potentials, and terminal buttons elicit postsynaptic potentials in the membrane of the cells with which they form synapses. These electrical events can be recorded (as we saw in Chapter 2), and changes in the electrical activity of a particular region can be used to determine whether that region plays a role in various behaviors. For example, recordings can be made during stimulus presentations, decision making, or motor activities.

Recordings can be made *chronically*, over an extended period of time after the animal recovers from surgery, or *acutely*, for a relatively short period of time during which the animal is kept anesthetized. Acute recordings, made while the animal is anesthetized, are usually restricted to studies of sensory pathways. Acute recordings seldom involve behavioral observations, since the behavioral capacity of an anesthetized animal is limited, to say the least.

Recordings with Microelectrodes

Drugs that affect serotonergic and noradrenergic neurons also affect REM sleep. Suppose that, knowing this fact, we wondered whether the activity of serotonergic and noradrenergic neurons would vary during different stages of sleep. To find out, we would record the activity of these neurons with microelectrodes. **Microelectrodes,** usually made of thin wires, have a very fine tip, small enough to record the electrical activity of individual neurons. This technique is usually called **single-unit recording** (a unit refers to an individual neuron).

Because we want to record the activity of single neurons over a long period of time in unanesthetized animals, we want more durable electrodes. We can purchase arrays of very fine wires, gathered together in a bundle, which can simultaneously record the activity of many different neurons. The wires are insulated so that only their tips are bare.

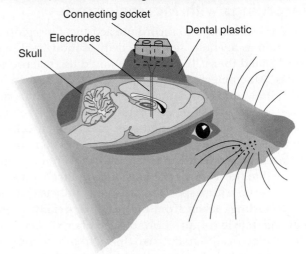

FIGURE 5.22 ■ Microelectrodes

This schematic shows a permanently attached set of electrodes, with a connecting socket cemented to the skull.

Connecting socket

Electrodes

Dental plastic

Skull

We implant the electrodes in the brains of animals through stereotaxic surgery. We attach the electrodes to miniaturized electrical sockets and bond the sockets to the animal's skull, using plastics that were originally developed for the dental profession. Then, after recovery from surgery, the animal can be "plugged in" to the recording system. Laboratory animals pay no heed to the electrical sockets on their skulls and behave quite normally. (See *Figure 5.22.*)

Researchers often attach rather complex devices to the animals' skulls when they implant microelectrodes. These devices include screw mechanisms that permit the experimenter to move the electrode—or array of electrodes—deeper into the brain so that they can record from several different parts of the brain during the course of their observations.

The electrical signals detected by microelectrodes are quite small and must be amplified. Amplifiers used for this purpose work just like the amplifiers in a stereo system, converting the weak signals recorded at the brain into stronger ones. These signals can be displayed on an oscilloscope and stored in the memory of a computer for analysis at a later time.

What about the results of our recordings from serotonergic and noradrenergic neurons? As you will learn in Chapter 9, if we record the activity of these neurons during various stages of sleep, we will find that their firing rates fall almost to zero during REM sleep. This

microelectrode A very fine electrode, generally used to record activity of individual neurons.

single-unit recording Recording of the electrical activity of a single neuron.

observation suggests that these neurons have an *inhibitory* effect on REM sleep. That is, REM sleep does not occur until these neurons stop firing.

Recordings with Macroelectrodes

Sometimes, we want to record the activity of a region of the brain as a whole, not the activity of individual neurons located there. To do this, we would use macroelectrodes. **Macroelectrodes** do not detect the activity of individual neurons; rather, the records that are obtained with these devices represent the postsynaptic potentials of many thousands—or millions—of cells in the area of the electrode. These electrodes can consist of unsharpened wires inserted into the brain, screws attached to the skull, or even metal disks attached to the human scalp with a special paste that conducts electricity. Recordings taken from the scalp, especially, represent the activity of an enormous number of neurons, whose electrical signals pass through the meninges, skull, and scalp before reaching the electrodes.

Occasionally, neurosurgeons implant macroelectrodes directly into the human brain. The reason for doing so is to detect the source of abnormal electrical activity that is giving rise to frequent seizures. Once the source has been determined, the surgeon can open the skull and remove the source of the seizures—usually scar tissue caused by brain damage that occurred earlier in life. Usually, the electrical activity of a human brain is recorded through electrodes attached to the scalp and displayed on a *polygraph.*

A polygraph contains a mechanism that moves a very long strip of paper past a series of pens. These pens are essentially the pointers of large voltmeters, moving up and down in response to the electrical signal sent to them by the biological amplifiers. (Often, the information is stored in a computer and displayed on a monitor rather than on strips of paper.) Figure 5.23 illustrates a record of electrical activity recorded from macroelectrodes attached to various locations on a person's scalp. (See *Figure 5.23.*) Such records are called **electroencephalograms (EEGs),** or "writings of electricity from the head." They can be used to diagnose epilepsy or study the stages of sleep and wakefulness, which are associated with characteristic patterns of electrical activity.

Another use of the EEG is to monitor the condition of the brain during procedures that could potentially damage it. I witnessed just such a procedure several years ago.

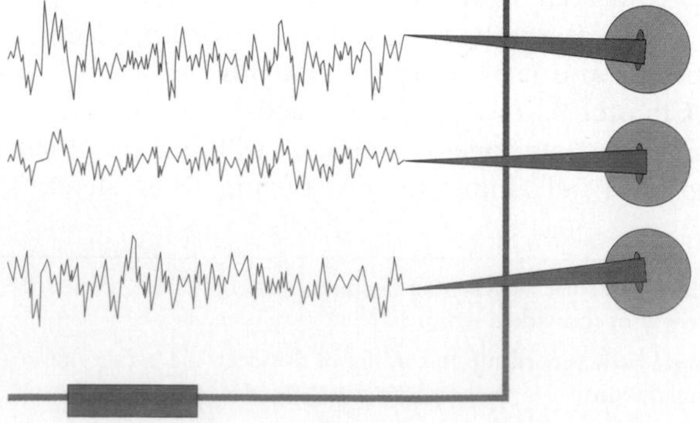

Mrs. F. had sustained one mild heart attack, and subsequent tests indicated a considerable amount of atherosclerosis, commonly referred to as "hardening of the arteries." Many of her arteries were narrowed by cholesterol-rich atherosclerotic plaque. A clot formed in a particularly narrow portion of one of her coronary arteries, which caused her heart attack. As the months passed after her heart attack, Mrs. F. had several *transient ischemic attacks,* brief episodes of neurological symptoms that appear to be caused by blood clots forming and then dissolving in cerebral blood vessels. In her case, they caused numbness in her right arm and difficulty in talking. Her physician referred her to a neurologist, who ordered an angiogram. This procedure revealed that her left carotid artery was almost totally blocked. The neurologist referred Mrs. F. to a neurosurgeon, who urged her to have an operation that would remove the plaque from part of her left carotid artery and increase the blood flow to the left side of her brain.

The procedure is called a *carotid endarterectomy.* I was chatting with Mrs. F.'s neurosurgeon after a conference, and he happened to mention that he would be performing the operation later that morning. I asked whether I could watch, and he agreed. When I entered the operating room, scrubbed and gowned, I found Mrs. F. already anesthetized, and the surgical nurse had prepared the left side of her neck for the incision. In addition, several EEG electrodes had been attached to her scalp, and I saw that Dr. L., a neurologist who

---

FIGURE 5.23 ■ **Record from a Polygraph**

Paper moves

---

macroelectrode An electrode used to record the electrical activity of large numbers of neurons in a particular region of the brain; much larger than a microelectrode.

electroencephalogram (EEG) An electrical brain potential recorded by placing electrodes on in the scalp.

specializes in clinical neurophysiology, was seated at his EEG machine.

The surgeon made an incision in Mrs. F.'s neck and exposed the carotid artery, at the point where the common carotid, coming from the heart, branched into the external and internal carotid arteries. He placed a plastic band around the common carotid artery and clamped it shut, stopping the flow of blood. "How does it look, Ken?" he asked Dr. L. "No good—I see some slowing. You'd better shunt."

The surgeon quickly removed the constricting band and asked the nurse for a shunt, a short length of plastic tubing a little thinner than the artery. He made two small incisions in the artery well above and well below the region that contained the plaque, and inserted the shunt. Now he could work on the artery without stopping the flow of blood to the brain. He made a longitudinal cut in the artery, exposing a yellowish mass that he dissected away and removed. He sewed up the incision, removed the shunt, and sutured the small cuts he had made to accommodate it. "Everything still okay?" he asked Dr. L. "Yes, her EEG is fine."

Most neurosurgeons prefer to do an endarterectomy by temporarily clamping the artery shut while they work on it. The work goes faster, and complications are less likely. Because the blood supply to the two hemispheres of the brain are interconnected (with special *communicating arteries*), it is often possible to shut down one of the carotid arteries for a few minutes without causing any damage. However, sometimes the blood flow from one side of the brain to the other is insufficient to keep the other side nourished with blood and oxygen. The only way the surgeon can know is to have the patient's EEG monitored. If the brain is not receiving a sufficient blood supply, the EEG will show the presence of characteristic "slow waves." That is what happened when Mrs. F.'s artery was clamped shut, and that is why the surgeon had to use a shunt tube. Without it, the procedure might have caused a stroke instead of preventing one.

By the way, Mrs. F. made a good recovery.

Magnetoencephalography

As you undoubtedly know, when electrical current flows through a conductor, it induces a magnetic field. This means that as action potentials pass down axons or as postsynaptic potentials pass down dendrites or sweep across the somatic membrane of a neuron, magnetic fields are also produced. These fields are exceedingly small, but engineers have developed superconducting detectors (called SQUIDs, or "superconducting quantum interference devices") that can detect magnetic fields that are approximately one-billionth of the size of the earth's magnetic field.

FIGURE 5.24 ■ Magnetoencephalography

The neuromagnetometer is shown on the monitor to the left. The regions of increased electrical activity are shown on the monitor to the right, superimposed on an image of the brain derived from an MRI scan.

(Courtesy of CTF Systems Inc.)

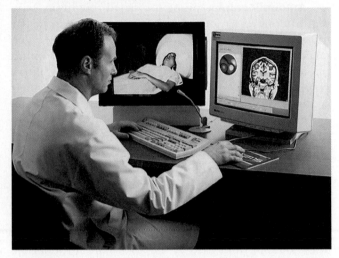

Magnetoencephalography is performed with *neuromagnetometers*, devices that contain an array of several SQUIDs, oriented so that a computer can examine their output and calculate the source of particular signals in the brain. The neuromagnetometer shown in Figure 5.2 contains 275 SQUIDs. These devices can be used clinically—for example, to find the sources of seizures so that they can be removed surgically. They can also be used in experiments to measure regional brain activity that accompanies the perception of various stimuli or the performance of various behaviors or cognitive tasks. (See *Figure 5.24*.)

Recording the Brain's Metabolic and Synaptic Activity

Electrical signals are not the only signs of neural activity. If the neural activity of a particular region of the brain increases, the metabolic rate of this region increases, too, largely as a result of increased operation of ion transporters in the membrane of the cells. This increased metabolic rate can be measured. The experimenter injects radioactive **2-deoxyglucose (2-DG)** into the animal's bloodstream. Because this chemical resembles

magnetoencephalography A procedure that detects groups of synchronously activated neurons by means of the magnetic field induced by their electrical activity; uses an array of superconducting quantum interference devices, or SQUIDs.

2-deoxyglucose (2-DG) (*dee ox ee gloo kohss*) A sugar that enters cells along with glucose but is not metabolized.

glucose (the principal food for the brain), it is taken into cells. Thus, the most active cells, which use glucose at the highest rate, will take up the highest concentrations of radioactive 2-DG. But unlike normal glucose, 2-DG cannot be metabolized, so it stays in the cell. The experimenter then kills the animal, removes the brain, slices it, and prepares it for *autoradiography*.

Autoradiography can be translated roughly as "writing with one's own radiation." Sections of the brain are mounted on microscope slides. The slides are then taken into a darkroom, where they are coated with a photographic emulsion (the substance found on photographic film). Several weeks later, the slides, with their coatings of emulsion, are developed, just like photographic film. The molecules of radioactive 2-DG show themselves as spots of silver grains in the developed emulsion because the radioactivity exposes the emulsion, just as X-rays or light will do.

The most active regions of the brain contain the most radioactivity, showing this radioactivity in the form of dark spots in the developed emulsion. Figure 5.25 shows an autoradiograph of a slice of a rat brain; the dark spots at the bottom (indicated by the arrow) are nuclei of the hypothalamus with an especially high metabolic rate. Chapter 9 describes the function of these nuclei. (See *Figure 5.25*.) *MyPsychKit 5.2, Autoradiography*, shows this procedure.

mypsychkit

Animation 5.2
Autoradiography

Another method of identifying active regions of the brain capitalizes on the fact that when neurons are activated (for example, by the terminal buttons that form synapses with them), particular genes in the nucleus called *immediate early genes* are turned on and particular proteins are produced. These proteins then bind with the chromosomes in the nucleus. The presence of these nuclear proteins indicates that the neuron has just been activated.

One of the nuclear proteins produced during neural activation is called **Fos.** You will remember that earlier in this chapter we began an imaginary research project on the neural circuitry involved in the sexual behavior of female rats. Suppose we want to use the Fos method in this project to see what neurons are activated during a female rat's sexual activity. We place female rats with males and permit the animals to copulate. Then we remove the rats' brains, slice them, and follow a procedure that stains Fos protein. Figure 5.26 shows the results: Neurons in the medial amygdala of a female rat that has just mated show the presence of dark spots, indicating the presence of Fos protein. Thus, these neurons appear to be activated by copulatory activity—perhaps by the physical stimulation of the genitals that occurs then. As you will recall, when we injected a retrograde tracer (fluorogold) into the VMH, we found that this region receives input from the medial amygdala. (See *Figure 5.26*.)

FIGURE 5.26 ■ Localization of Fos Protein

The photomicrograph shows a frontal section of the brain of a female rat, taken through the medial amygdala. The dark spots indicate the presence of Fos protein, localized by means of immunocytochemistry. The synthesis of Fos protein was stimulated by permitting the animal to engage in copulatory behavior.

(Courtesy of Marc Tetel, Wellesley College.)

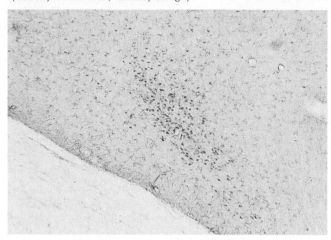

autoradiography A procedure that locates radioactive substances in a slice of tissue; the radiation exposes a photographic emulsion or a piece of film that covers the tissue.

Fos (*fahs*) A protein produced in the nucleus of a neuron in response to synaptic stimulation.

FIGURE 5.25 ■ 2-DG Autoradiogram

The frontal section of a rat brain, dorsal is at top, shows especially high regions of activity in the pair of nuclei in the hypothalamus, at the base of the brain.

(From Schwartz, W. J., and Gainer, H. *Science*, 1977, *197*, 1089–1091. Reprinted with permission.)

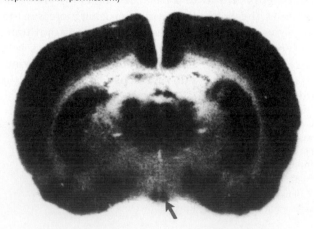

The metabolic activity of specific brain regions can be measured in human brains, too, by means of **functional imaging**—a computerized method of detecting metabolic or chemical changes within the brain. The first functional imaging method to be developed was **positron emission tomography (PET).** First, the person receives an injection of radioactive 2-DG. (Eventually, the chemical is broken down and leaves the cells. The dose given to humans is harmless.) The person's head is placed in a machine similar to a CT scanner. When the radioactive molecules of 2-DG decay, they emit subatomic particles called positrons, which are detected by the scanner. The computer determines which regions of the brain have taken up the radioactive substance, and it produces a picture of a slice of the brain, showing the activity level of various regions in that slice. (See *Figure 5.27.*)

One of the disadvantages of PET scanners is their operating cost. For reasons of safety the radioactive

chemicals that are administered have very short half-lives; that is, they decay and lose their radioactivity very quickly. For example, the half-life of radioactive 2-DG is 110 minutes; the half-life of radioactive water (also used for PET scans) is only 2 minutes. Because these chemicals decay so quickly, they must be produced on site, in an atomic particle accelerator called a *cyclotron*. Therefore, to the cost of the PET scanner must be added the cost of the cyclotron and the salaries of the personnel who operate it.

Another disadvantage of PET scans is the relatively poor spatial resolution (the blurriness) of the images. The temporal resolution is also relatively poor. The positrons being emitted from the brain must be sampled for a fairly long time, which means that rapid, short-lived events within the brain are likely to be missed. These disadvantages are not seen in functional MRI, described in the next paragraph. However, PET scanners can do something that functional MRI scanners cannot do: measure the concentration of particular chemicals in various parts of the brain. I will describe this procedure later in this chapter.

The brain imaging method with the best spatial and temporal resolution is known as **functional MRI (fMRI).** Engineers have devised modifications to existing MRI scanners and their software that permit the devices to acquire images that indicate regional metabolism. Brain activity is measured indirectly, by detecting levels of oxygen in the brain's blood vessels. Increased activity of a brain regions stimulates blood flow to that region, which increases the local blood oxygen level. The formal name of this type of imaging is *BOLD:* blood oxygen level–dependent signal. Functional MRI scans have a higher resolution than PET scans do, and they can be acquired much faster. Thus, they reveal more detailed information about the activity of particular brain regions. You will read about many functional imaging studies that employ fMRI scans in subsequent chapters of this book. (See *Figure 5.28.*)

Stimulating Neural Activity

So far, this section has been concerned with research methods that measure the activity of specific regions of the brain. But sometimes we may want to artificially

---

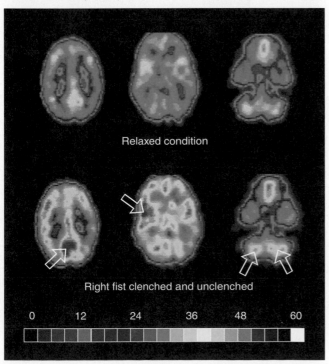

Relaxed condition

Right fist clenched and unclenched

| 0 | 12 | 24 | 36 | 48 | 60 |

---

functional imaging A computerized method of detecting metabolic or chemical changes in particular regions of the brain.

positron emission tomography (PET) A functional imaging method that reveals the localization of a radioactive tracer in a living brain.

functional MRI (fMRI) A functional imaging method; a modification of the MRI procedure that permits the measurement of regional metabolism in the brain, usually by detecting changes in blood oxygen level.

FIGURE 5.28 ■ Functional MRI Scans

These scans of human brains show localized average increases in neural activity of males (left) and females (right) while they were judging whether pairs of written words rhymed.

(From Shaywitz, B. A., et al., *Nature*, 1995, *373*, 607–609. Reprinted with permission.)

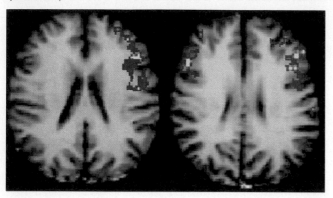

change the activity of these regions to see what effects these changes have on the animal's behavior. For example, female rats will copulate with males only if certain female sex hormones are present. If we remove the rats' ovaries, the loss of these hormones will abolish the rats' sexual behavior. We found in our earlier studies that VMH lesions disrupt this behavior. Perhaps if we *activate* the VMH, we will make up for the lack of female sex hormones and the rats will copulate again.

Electrical and Chemical Stimulation

How do we activate neurons? We can do so by electrical or chemical stimulation. Electrical stimulation simply involves passing an electrical current through a wire inserted into the brain, as you saw in Figure 5.22. Chemical stimulation is usually accomplished by injecting a small amount of an excitatory amino acid, such as kainic acid or glutamic acid, into the brain. As you learned in Chapter 4, the principal excitatory neurotransmitter in the brain is glutamic acid (glutamate), and both of these substances stimulate glutamate receptors, thus activating the neurons on which these receptors are located.

Injections of chemicals into the brain can be done through an apparatus that is permanently attached to the skull so that the animal's behavior can be observed several times. We place a metal cannula (a guide cannula) in an animal's brain and cement its top to the skull. At a later date we place a smaller cannula of measured length inside the guide cannula and then inject a chemical into the brain.

mypsychkit
Animation 5.3
Cannula Implantation

Because the animal is free to move about, we can observe the effects of the injection on its behavior. (See *Figure 5.29.*) *MyPsychKit 5.3, Cannula Implantation,* shows this surgical procedure.

The principal disadvantage of chemical stimulation is that it is slightly more complicated than electrical stimulation; chemical stimulation requires cannulas, tubes, special pumps or syringes, and sterile solutions of excitatory amino acids. However, it has a distinct advantage over

FIGURE 5.29 ■ Intracranial Cannula

(a) A guide cannula is permanently attached to the skull. (b) At a later time a thinner cannula can be inserted through the guide cannula into the brain. Chemicals can be infused into the brain through this device.

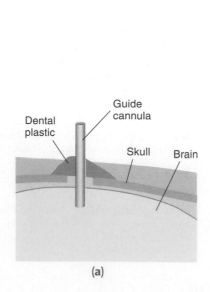

(a)

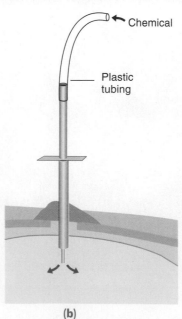

(b)

FIGURE 5.30 ■ Photostimulation

(a) Photosensitive proteins can be inserted into neural membranes by means of genetically modified viruses. Blue light causes ChR2 ion channels to depolarize the membrane, and yellow light causes NpHR ion transporters to hyperpolarize it. (b) The graph shows the effects on the membrane potential of different wavelengths of light acting on ChR2 or NpHR proteins. (c) Action potentials were elicited by pulses of blue light (blue arrows) and the inhibitory effects of the hyperpolarization were caused by yellow light.

(Part (a) adapted from Hausser, M., and Smith, S. L. *Nature,* 2007, *446,* 617–619, and parts (b) and (c) adapted from Zhang, F., Wang, L. P., Brauner, M., Liewald, J. F., Kay, K., Watzke, N., Wood, P. G., Bamberg, E., Nagel, G., Gottschalk, A., and Deisseroth, K. *Nature,* 2007, *446,* 633–639.)

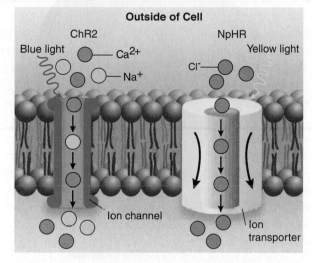

(a)

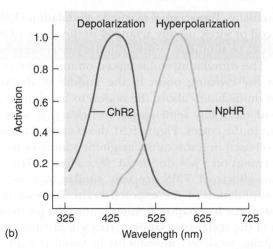

(b)

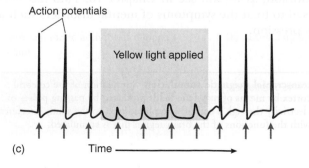

(c)

electrical stimulation: It activates cell bodies but not axons. Because only cell bodies (and their dendrites, of course) contain glutamate receptors, we can be assured that an injection of an excitatory amino acid into a particular region of the brain excites the cells there but not the axons of other neurons that happen to pass through the region. Thus, the effects of chemical stimulation are more localized than are the effects of electrical stimulation.

You might have noticed that I just said that kainic acid, which I described earlier as a neurotoxin, can be used to stimulate neurons. These two uses are not really contradictory. Kainic acid produces excitotoxic lesions by stimulating neurons to death. Whereas large doses of a concentrated solution kill neurons, small doses of a dilute solution simply stimulate them.

What about the results of our hypothetical experiment? In fact (as we shall see in Chapter 10), VMH stimulation *does* substitute for female sex hormones. Perhaps, then, the female sex hormones exert their effects in this nucleus. We will see how to test this hypothesis in the final section of this chapter.

When chemicals are injected into the brain through cannulas, molecules of the chemicals diffuse over a region that includes many different types of neurons: excitatory neurons, inhibitory neurons, interneurons that participate in local circuits, projection neurons that communicate with different regions of the brain, and neurons that release or respond to a wide variety of neurotransmitters and neuromodulators. Stimulating a particular brain region with electricity or an excitatory chemical affects all of these neurons, and the result is unlikely to resemble normal brain activity, which involves coordinated activation and inhibition of many different neurons. Ideally, we would like to be able to stimulate or inhibit selected populations of neurons in a given brain regions.

Photostimulation

Recent developments are providing the means to stimulate or inhibit particular types of neurons in particular brain regions (Boyden et al., 2005; Zhang et al., 2007). Photosensitive proteins have evolved in many organisms— even single-celled organisms such as algae and bacteria. Researchers have discovered that one of these proteins, *Channelrhodopsin-2 (ChR2),* found in green algae, controls an ion channel that, when open, permits the flow of sodium, potassium, and calcium ions. When blue light strikes a ChR2-ion channel, the channel opens, and the rush of positively charged sodium and calcium ions depolarizes the membrane. A second photosensitive protein, *Natronomonas pharaonis halorhodopsin (NpHR),* is found in a bacterium. This protein controls a transporter that moves chloride into the cell when activated by blue light. This influx of negatively charged ions hyperpolarizes the membrane. The action of both of these photosensitive proteins begins and ends very rapidly when light of the appropriate wavelength is turned on and off. (See *Figure 5.30.*)

FIGURE 5.34 ■ In Situ Hybridization

This simplified schematic shows the use of in situ hybridization to localize messenger RNA that is responsible for the synthesis of a particular protein or peptide.

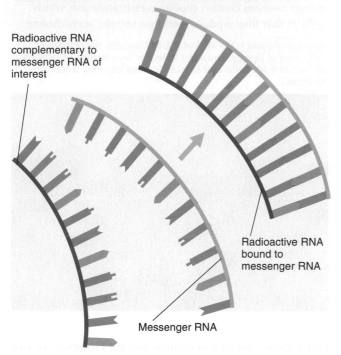

Radioactive RNA complementary to messenger RNA of interest

Radioactive RNA bound to messenger RNA

Messenger RNA

FIGURE 5.35 ■ In Situ Hybridization

The tissue was exposed to radioactive RNA that binds with the messenger RNA responsible for the synthesis of vasopressin, a peptide. The location of the radioactive RNA, revealed by means of autoradiography, shows up as white spots. The labeled neurons are located in a pair of nuclei in the hypothalamus.

(Courtesy of Geert DeVries, University of Massachusetts.)

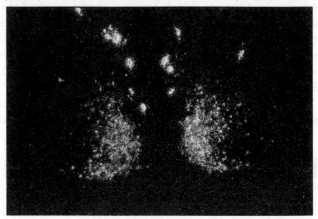

cases it is), molecular biologists can synthesize a piece of radioactive RNA that contains a sequence of nucleotides complementary to the sequence on the messenger RNA. We would expose slices of brain tissue to the radioactive RNA, which sticks to molecules of the appropriate messenger RNA. Then we would use autoradiographic methods (described in the second section of this chapter) to reveal the location of the messenger RNA and, by inference, the location of the cells producing the protein whose synthesis the RNA initiates.

Figure 5.34 explains the in situ hybridization method graphically, and Figure 5.35 shows the location of the messenger RNA responsible for the synthesis of a peptide, vasopressin, as revealed by this method. Side lighting of the microscope slide makes the silver grains in the photographic emulsion show up as white spots. (See *Figures 5.34* and *5.35*.)

Localizing Particular Receptors

As we saw in Chapter 2, neurotransmitters, neuromodulators, and hormones convey their messages to their target cells by binding with receptors on or in these cells. The location of these receptors can be determined by two different procedures.

The first procedure uses autoradiography. We expose slices of brain tissue to a solution containing a radioactive ligand for a particular receptor. Next, we rinse the slices so that the only radioactivity remaining in them is that of the molecules of the ligand bound to their receptors. Finally, we use autoradiographic methods to localize the radioactive ligand—and thus the receptors. Figure 5.36 shows an example of the results of this procedure. We see an autoradiogram of a slice of a rat's brain that was soaked in a solution that contained radioactive morphine, which bound with the brain's opiate receptors. (See *Figure 5.36*.)

The second procedure uses immunocytochemistry. Receptors are proteins; therefore, we can produce antibodies against them. We expose slices of brain tissue to the appropriate antibody (labeled with a fluorescent dye) and look at the slices with a microscope under light of a particular wavelength.

Let's apply the method for localizing receptors to the first line of investigation we considered in this chapter: the role of the ventromedial hypothalamus (VMH) in the sexual behavior of female rats. As we saw, lesions of the VMH abolish this behavior. We also saw that the behavior does not occur if the rat's ovaries are removed but that it can be activated by stimulation of the VMH with electricity or an excitatory amino acid. These results suggest that the sex hormones produced by the ovaries act on neurons in the VMH.

This hypothesis suggests two experiments. First, we could use the procedure shown in Figure 5.29 to place a small amount of the appropriate sex hormone directly into the VMH of female rats whose ovaries we had previously removed. As we shall see in Chapter 10, this

FIGURE 5.36 ■ Autoradiogram of a Rat Brain

In this horizontal section, rostral is at the top. The brain was incubated in a solution containing radioactive morphine, a ligand for opiate receptors. The receptors are indicated by white areas.

(From Herkenham, M. A., and Pert, C. B. *Journal of Neuroscience*, 1982, *2*, 1129–1149. Reprinted with permission.)

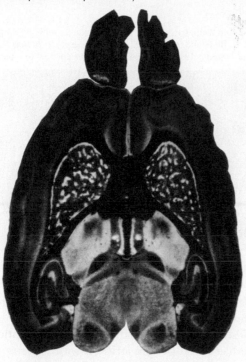

FIGURE 5.37 ■ Microdialysis

A dilute salt solution is slowly infused into the microdialysis tube, where it picks up molecules that diffuse in from the extracellular fluid. The contents of the fluid are then analyzed.

(Adapted from Hernandez, L., Stanley, B. G., and Hoebel, B. G. *Life Sciences*, 1986, *39*, 2629–2637.)

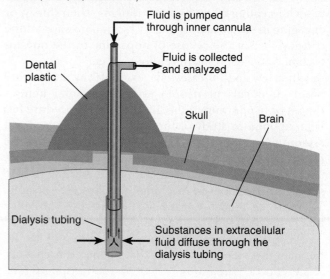

Fluid is pumped through inner cannula

Fluid is collected and analyzed

Dental plastic

Skull

Brain

Dialysis tubing

Substances in extracellular fluid diffuse through the dialysis tubing

procedure works; the hormone *does* reactivate the animals' sexual behavior. The second experiment would use autoradiography to look for the receptors for the sex hormone. We would expose slices of rat brain to the radioactive hormone, rinse them, and perform autoradiography. If we did so, we would indeed find radioactivity in the VMH. (And if we compared slices from the brains of female and male rats, we would find evidence of more hormone receptors in the females' brains.) We could also use immunocytochemistry to localize the hormone receptors, and we would obtain the same results.

Measuring Chemicals Secreted in the Brain

We know that cocaine—a particularly addictive drug—blocks the reuptake of dopamine, which suggests that the extracellular concentration of dopamine increases in some parts of the brain when a person takes cocaine. To measure the amount of dopamine in particular regions of the brain, we use a procedure called **microdialysis.**

Dialysis is a process in which substances are separated by means of an artificial membrane that is permeable to

some molecules but not others. A microdialysis probe consists of a small metal tube that introduces a solution into a section of dialysis tubing—a piece of artificial membrane shaped in the form of a cylinder, sealed at the bottom. Another small metal tube leads the solution away after it has circulated through the pouch. A drawing of such a probe is shown in *Figure 5.37.*

We use stereotaxic surgery to place a microdialysis probe in a rat's brain so that the tip of the probe is located in the region we are interested in. We pump a small amount of a solution similar to extracellular fluid through one of the small metal tubes into the dialysis tubing. The fluid circulates through the dialysis tubing and passes through the second metal tube, from which it is taken for analysis. As the fluid passes through the dialysis tubing, it collects molecules from the extracellular fluid of the brain, which are pushed across the membrane by the force of diffusion.

We analyze the contents of the fluid that has passed through the dialysis tubing by an extremely sensitive analytical method. This method is so sensitive that it can detect neurotransmitters (and their breakdown products) that have been released by the terminal buttons and have escaped from the synaptic cleft into the rest of the

microdialysis A procedure for analyzing chemicals present in the interstitial fluid through a small piece of tubing made of a semipermeable membrane that is implanted in the brain.

extracellular fluid. We find that the amount of dopamine present in the extracellular fluid of the nucleus accumbens, located in the basal forebrain, *does* increase when we give a rat an injection of cocaine. In fact, we find that the amount of dopamine in this region increases when we administer any addictive drug, such as heroin, nicotine, or alcohol. We even see increased dopamine secretion when the animal participates in a pleasurable activity such as eating when hungry, drinking when thirsty, or engaging in sexual activity. Such observations support the conclusion that the release of dopamine in the nucleus accumbens plays a role in reinforcement.

In a few special cases (for example, in monitoring brain chemicals of people with intracranial hemorrhages or head trauma) the microdialysis procedure has been applied to study of the human brain, but ethical reasons prevent us from doing so for research purposes. Fortunately, there is a noninvasive way to measure neurochemicals in the human brain. Although PET scanners are expensive machines, they are also versatile. They can be used to localize *any* radioactive substance that emits positrons.

Figure 5.38 shows PET scans of the brain of Mr. B., the man described in the case that opened this chapter. A stereotaxic apparatus was used to transplant fetal dopamine-secreting neurons into his basal ganglia. As we saw, a PET scan was taken of his brain before his surgery and a little more than a year afterward. He was given an injection of radioactive L-DOPA one hour before each scan was made. As you learned in Chapter 4, L-DOPA is taken up by the terminals of dopaminergic neurons, where it is converted to dopamine; thus, the radioactivity shown in the scans indicates the presence of dopamine-secreting terminals in the basal ganglia. The scans show the amount of radioactivity before (part a) and after (part b) he received the transplant. As you

FIGURE 5.38 ■ Pet Scans of Patient with Parkinsonian Symptoms

The scans show uptake of radioactive L-DOPA in the basal ganglia of a patient with parkinsonian symptoms induced by a toxic chemical before and after receiving a transplant of fetal dopaminergic neurons: (a) preoperative scan and (b) scan taken 13 months postoperatively. The increased uptake of L-DOPA indicates that the fetal transplant was secreting dopamine.

(Adapted from Widner, H., Tetrud, J., Rehncrona, S., Snow, B., Brundin, P., Gustavii, B., Björklund, A., Lindvall, O., and Langston, J. W. *New England Journal of Medicine*, 1992, *327*, 1556–1563. Scans reprinted with permission.)

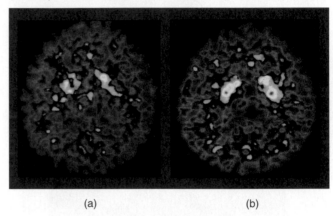

(a)   (b)

can see, the basal ganglia contained substantially more dopamine after the surgery. (See *Figure 5.38.*)

I wish I could say that the fetal transplantation procedure has cured people stricken with Parkinson's disease and those whose brains were damaged with the contaminated drug. Unfortunately, as we will see in Chapter 15, the therapeutic effects of the transplant are often temporary, and with time, serious side effects often emerge.

InterimSummary

Neurochemical Methods

Neurochemical methods can be used to determine the location of an enormous variety of substances in the brain. They can identify neurons that secrete a particular neurotransmitter or neuromodulator and those that possess receptors that respond to the presence of these substances. Peptides and proteins can be directly localized, through immunocytochemical methods; the tissue is exposed to an antibody that is linked to a molecule that fluoresces under light of a particular wavelength. Other substances can be detected by immunocytochemical localization of an enzyme that is required for their synthesis. Peptides and proteins can also be detected by in situ hybridization methods that reveal the presence of the messenger RNA that directs their synthesis.

Receptors for neurochemicals can be localized by two means. The first method uses autoradiography to reveal the distribution of a radioactive ligand to which the tissue has been exposed. The second method uses immunocytochemistry to detect the presence of the receptors themselves, which are proteins. Combined staining methods can localize neurons that possess a particular receptor or a particular peptide and also have connections with particular regions of the brain.

The secretions of neurotransmitters and neuromodulators can be measured by implanting the tip of a microdialysis probe in a particular region of the brain. A PET scanner can be used to perform similar observations of the human brain. People are given an injection of a radioactive tracer such as a drug that binds with a particular receptor or a chemical that is incorporated into a particular neurotransmitter, and a PET scanner reveals the location of the tracer in the brain.

Table 5.3 summarizes the research methods presented in this section.

TABLE 5.3 ■ Research Methods: Part III

GOAL OF METHOD	METHOD	REMARKS
Measure neurotransmitters and neuro-modulators released by neurons	Microdialysis	A wide variety of substances can be analyzed
Identify neurons producing a particular neurotransmitter or neuromodulator	Immunocytochemical localization of peptide or protein	Requires a specific antibody
	Immunocytochemical localization of enzyme responsible for synthesis of substance	Useful if substance is not a peptide or protein
Identify neurons that contain a particular type of receptor	Autoradiographic localization of radioactive ligand	
	Immunocytochemical localization of receptor	Requires a specific antibody
Genetic methods	Twin studies	Comparison of concordance rates of mono-zygotic and dizygotic twins estimates heritability of trait
	Adoption studies	Similarity of offspring and adoptive and biological parents estimates heritability of trait
	Targeted mutations	Inactivation, insertion, or increased expression of a gene
	Antisense oligonucleotides	Bind with messenger RNA; prevent synthesis of protein

GENETIC METHODS

All behavior is determined by interactions between an individual's brain and his or her environment. Many behavioral characteristics—such as talents, personality variables and mental disorders—seem to run in families. This fact suggests that genetic factors may play a role in the development of physiological differences that are ultimately responsible for these characteristics. In some cases the genetic link is very clear: A defective gene interferes with brain development, and a neurological abnormality causes behavioral deficits. In other cases the links between heredity and behavior are much more subtle, and special genetic methods must be used to reveal them.

Twin Studies

A powerful method for estimating the influence of heredity on a particular trait is to compare the *concordance rate* for this trait in pairs of monozygotic and dizygotic twins. Monozygotic twins (identical twins) have identical genotypes—that is, their chromosomes, and the genes they contain, are identical. In contrast, the genetic similarity between dizygotic twins (fraternal twins) is, on the average, 50 percent. Investigators study records to identify pairs of twins in which at least one member has the trait—for example, a diagnosis of a particular mental disorder. If both twins have been diagnosed with this disorder, they are said to be *concordant*. If only one has received this diagnosis, the twins are said to be *discordant*. Thus, if a disorder has a genetic basis, the percentage of monozygotic twins who are concordant for the diagnosis will be higher than that for dizygotic twins. For example, as we will see in Chapter 16, the concordance rate for schizophrenia in twins is at least four times higher for monozygotic twins than for dizygotic twins, a finding that provides strong evidence that schizophrenia is a heritable trait. Twin studies have found that many individual characteristics, including personality traits, prevalence of obesity, incidence of alcoholism, and a wide variety of mental disorders, are influenced by genetic factors.

Adoption Studies

Another method for estimating the heritability of a particular behavioral trait is to compare people who were adopted early in life with their biological and adoptive parents. All behavioral traits are affected to some degree by hereditary factors, environmental factors, and an interaction between hereditary and environmental factors. Environmental factors are both social and biological in nature. For example, the mother's health, nutrition, and drug-taking behavior during pregnancy are prenatal environmental factors, and the child's diet, medical care, and social environment (both inside and outside the home) are postnatal environmental factors. If a child is adopted soon after birth, the genetic factors will be associated with the biological parents, the prenatal environmental factors will be associated with the biological mother, and most of the postnatal environmental factors will be associated with the adoptive parents.

Adoption studies require that the investigator knows the identity of the parents of the people being studied and is able to measure the behavioral trait in the biological and adoptive parents. If the people being studied strongly resemble their biological parents, we conclude that the trait is probably influenced by genetic factors. To be certain, we will have to rule out possible differences in the prenatal environment of the adopted children. If, instead, the people resemble their adoptive parents, we conclude that the trait is influenced by environmental factors. (It would take further study to determine just what these environmental factors might be.) Of course, it is possible that both hereditary and environmental factors play roles, in which case the people being studied will resemble both their biological and adoptive parents.

Targeted Mutations

A genetic method developed by molecular biologists has put a powerful tool in the hands of neuroscientists. **Targeted mutations** are mutated genes produced in the laboratory and inserted into the chromosomes of mice. These mutated genes (also called knockout genes) are defective: They fail to produce a functional protein. In many cases the target of the mutation is an enzyme that controls a particular chemical reaction. For example, we will see in Chapter 13 that lack of a particular enzyme interferes with learning. This result suggests that the enzyme is partly responsible for changes in the structure of synapses required for learning to occur. In other cases the target of the mutation is a protein that itself serves useful functions in the cell. For example, we will see in Chapter 18 that a particular type of cannabinoid receptor is involved in the reinforcing and analgesic effects of opiates. Researchers can even produce *conditional knockouts* that cause the animal's genes to stop expressing a particular gene when the animal is given a particular drug. This permits the targeted gene to express itself normally during the animal's development and then be knocked out at a later time. Investigators can also use methods of genetic engineering to insert genes into the DNA of mice. These genes can cause increased production of proteins normally found in the host species, or they can produce entirely new proteins.

---

targeted mutation A mutated gene (also called a "knockout gene") produced in the laboratory and inserted into the chromosomes of mice; fails to produce a functional protein.

Antisense Oligonucleotides

Another genetic method involves the production of molecules that block the production of proteins encoded by particular genes by injecting **antisense oligonucleotides.** The most common type of antisense oligonucleotide are modified strands of RNA or DNA that will bind with specific molecules of messenger RNA and prevent them from producing their protein. Once the molecules of mRNA are trapped in this way, they are destroyed by enzymes present in the cell. The term

antisense refers to the fact that the synthetic oligonucleotides contain a sequence of bases complementary to those contained by a particular gene or molecule of mRNA. (As you can see, this method resembles the one used for in situ hybridization.)

antisense oligonucleotide (*oh li go new klee oh tide*) Modified strand of RNA or DNA that binds with a specific molecule of mRNA and prevents it from producing its protein.

InterimSummary

Genetic Methods

Because genes direct an organism's development, genetic methods are very useful in studies of the physiology of behavior. Twin studies compare the concordance rates of monozygotic (identical) and dizygotic (fraternal) twins for a particular trait. A higher concordance rate for monozygotic twins provides evidence that the trait is influenced by heredity. Adoption studies compare people who were adopted during infancy with their biological and adoptive parents. If the people resemble their biological parents, evidence is seen for genetic factors. If the people resemble their adoptive parents, evidence is seen for a role of factors in the family environment.

Targeted mutations permit neuroscientists to study the effects of a lack of a particular protein—for example, an enzyme, structural protein, or receptor—on an animal's

physiological and behavioral characteristics. Genes that cause the production of foreign proteins or increase production of native proteins can be inserted into the genome of strains of animals. Antisense oligonucleotides can be used to block the production of particular proteins.

Table 5.3 summarizes the research methods presented in this section.

Thought Questions

1. You have probably read news reports about studies of the genetics of human behavioral traits or seen them on television. What does it really mean when a laboratory reports the discovery of, say, a "gene for shyness"?
2. Most rats do not appear to like the taste of alcohol, but researchers have bred some rats that will drink alcohol in large quantities. Can you think of ways to use these animals to investigate the possible role of genetic factors in alcoholism in humans?

SUGGESTED READINGS

Laboratory Manual

Wellman, P. *Laboratory Exercises in Physiological Psychology.* Boston: Allyn and Bacon, 1994.

Stereotaxic Atlases

Paxinos, G., and Watson, C. *The Rat Brain in Stereotaxic Coordinates,* 4th ed. San Diego, CA: Academic Press, 1998.

Slotnick, B. M., and Leonard, C. M. *A Stereotaxic Atlas of the Albino Mouse Forebrain.* Rockville, MD: Public Health Service, 1975. (U.S. Government Printing Office Stock Number 017-024-00491-0)

Snider, R. S., and Niemer, W. T. *A Stereotaxic Atlas of the Cat Brain.* Chicago: University of Chicago Press, 1961.

Swanson, L. W. *Brain Maps: Structure of the Rat Brain.* Amsterdam: Elsevier, 1992.

Histological Methods

Heimer, L., and Záborsky, L. *Neuroanatomical Tract-Tracing Methods 2: Recent Progress.* New York: Plenum Press, 1989.

ADDITIONAL RESOURCES

Visit www.mypsychkit.com for additional review and practice of the material covered in this chapter. Within MyPsychKit, you can take practice tests and receive a customized study plan to help you review. Dozens of anima-

tions, tutorials, and Web links are also available. You can even review using the interactive electronic version of this textbook. You will need to register for MyPsychKit. See www.mypsychkit.com for complete details.

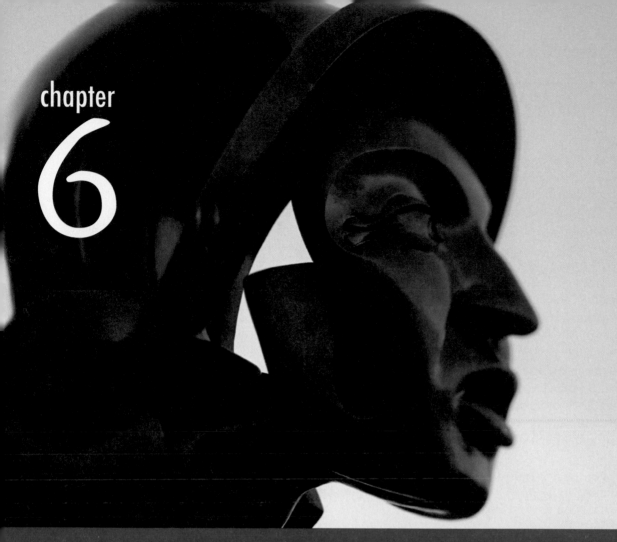

Dr. L., a young neuropsychologist, was presenting the case of Mrs. R. to a group of medical students doing a rotation in the neurology department at the medical center. The chief of the department had shown them Mrs. R.'s CT scans, and now Dr. L. was addressing the students. He told them that Mrs. R.'s stroke had not impaired her ability to talk or to move about, but it had affected her vision.

A nurse ushered Mrs. R. into the room and helped her find a seat at the end of the table.

"How are you, Mrs. R.?" asked Dr. L.

"I'm fine. I've been home for a month now, and I can do just about everything that I did before I had my stroke."

"Good. How is your vision?"

"Well, I'm afraid that's still a problem."

"What seems to give you the most trouble?"

"I just don't seem to be able to recognize things. When I'm working in my kitchen, I know what everything is as long as no one moves anything. A few times my husband tried to help me by putting things away, and I couldn't see them any more." She laughed. "Well, I could see them, but I just couldn't say what they were."

Dr. L. took some objects out of a paper bag and placed them on the table in front of her.

"Can you tell me what these are?" he asked. "No," he said, "please don't touch them."

Mrs. R. stared intently at the objects. "No, I can't rightly say what they are."

Dr. L. pointed to one of them, a wristwatch. "Tell me what you see here," he said.

Mrs. R. looked thoughtful, turning her head one way and then the other. "Well, I see something round, and it has two things attached to it, one on the top and one on the bottom." She continued to stare at it. "There are some things inside the circle, I think, but I can't make out what they are."

"Pick it up."

She did so, made a wry face, and said, "Oh. It's a wristwatch." At Dr. L.'s request, she picked up the rest of the objects, one by one, and identified each of them correctly.

"Do you have trouble recognizing people, too?" asked Dr. L.

"Oh, yes!" she sighed. "While I was still in the hospital, my husband and my son both came in to see me, and I couldn't tell who was who until my husband said something—then I could tell which direction his voice was coming from. Now I've trained myself to recognize my husband. I can usually see his glasses and his bald head, but I have to work at it. And I've been fooled a few times." She laughed. "One of our neighbors is bald and wears glasses, too, and one day when he and his wife were visiting us, I thought he was my husband, so I called him 'honey.' It was a little embarrassing at first, but everyone understood."

"What does a face look like to you?" asked Dr. L.

"Well, I know that it's a face, because I can usually see the eyes, and it's on top of a body. I can see a body pretty well, by how it moves." She paused a moment. "Oh, yes, I forgot, sometimes I can recognize a person by how he moves. You know, you can often recognize friends by the way they walk, even when they're far away. I can still do that. That's funny, isn't it? I can't see people's faces very well, but I can recognize the way they walk."

Dr. L. made some movements with his hands. "Can you tell what I'm pretending to do?" he asked.

"Yes, you're mixing something—like some cake batter."

He mimed the gestures of turning a key, writing, and dealing out playing cards, and Mrs. R. recognized them without any difficulty.

"Do you have any trouble reading?" he asked.

"Well, a little, but I don't do too badly."

Dr. L. handed her a magazine, and she began to read the article aloud—somewhat hesitantly but accurately. "Why is it," she asked, "that I can see the *words* all right but have so much trouble with *things* and with people's faces?"

As we saw in Chapter 3, the brain performs two major functions: It controls the movements of the muscles, producing useful behaviors, and it regulates the body's internal environment. To perform both these tasks, the brain must be informed about what is happening both in the external environment and within the body. Such information is received by the sensory systems. This chapter and the next are devoted to a discussion of the ways in which sensory organs detect changes in the environment and the ways in which the brain interprets neural signals from these organs.

We receive information about the environment from **sensory receptors**—specialized neurons that detect a variety of physical events. (Do not confuse sensory receptors with receptors for neurotransmitters, neuromodulators, and hormones. Sensory receptors are specialized neurons, and the other types of receptors are specialized proteins that bind with certain molecules.) Stimuli impinge on the receptors and, through various processes, alter their membrane potentials. This process is known as **sensory transduction** because sensory events are *transduced* ("transferred") into changes in the cells' membrane potential. These electrical changes are

sensory receptor A specialized neuron that detects a particular category of physical events.

sensory transduction The process by which sensory stimuli are transduced into slow, graded receptor potentials.

FIGURE 6.1 ■ The Electromagnetic Spectrum

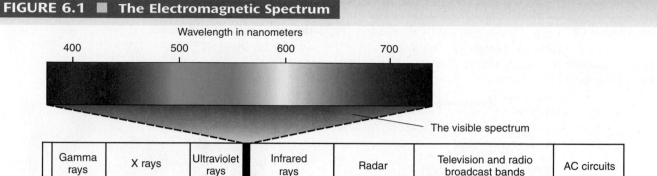

called **receptor potentials.** Most receptors lack axons; a portion of their somatic membrane forms synapses with the dendrites of other neurons. Receptor potentials affect the release of neurotransmitters and hence modify the pattern of firing in neurons with which these cells form synapses. Ultimately, the information reaches the brain.

People often say that we have five senses: sight, hearing, smell, taste, and touch. Actually, we have more than five, but even experts disagree about how the lines between the various categories should be drawn. Certainly, we should add the vestibular senses; as well as providing us with auditory information, the inner ear supplies information about head orientation and movement. The sense of touch (or, more accurately, *somatosensation*) detects changes in pressure, warmth, cold, vibration, limb position, and events that damage tissue (that is, produce pain). Everyone agrees that we can detect these stimuli; the issue is whether we should say that they are detected by separate senses.

This chapter considers vision, the sensory modality that receives the most attention from psychologists, anatomists, and physiologists. One reason for this attention derives from the fascinating complexity of the sensory organs of vision and the relatively large proportion of the brain that is devoted to the analysis of visual information. Approximately 20 percent of the cerebral cortex plays a direct role in the analysis of visual information (Wandell, Dumoulin, and Brewer, 2007). Another reason, I am sure, is that vision is so important to us as individuals. A natural fascination with such a rich source of information about the world leads to curiosity about how this sensory modality works. Chapter 7 deals with the other sensory modalities: audition, the vestibular senses, the somatosenses, gustation, and olfaction.

THE STIMULUS

As we all know, our eyes detect the presence of light. For humans, light is a narrow band of the spectrum of electromagnetic radiation. Electromagnetic radiation with a wavelength of between 380 and 760 nm (a nanometer, nm, is one-billionth of a meter) is visible to us. (See *Figure 6.1*.) Other animals can detect different ranges of electromagnetic radiation. For example, honeybees can detect differences in ultraviolet radiation reflected by flowers that appear white to us. The range of wavelengths we call *light* is not qualitatively different from the rest of the electromagnetic spectrum; it is simply the part of the continuum that we humans can see.

The perceived color of light is determined by three dimensions: *hue, saturation,* and *brightness*. Light travels at a constant speed of approximately 300,000 kilometers (186,000 miles) per second. Thus, if the frequency of oscillation of the wave varies, the distance between the peaks of the waves will vary similarly but in inverse fashion. Slower oscillations lead to longer wavelengths, and faster ones lead to shorter wavelengths. Wavelength determines the first of the three perceptual dimensions of light: **hue.** The visible spectrum displays the range of hues that our eyes can detect.

Light can also vary in intensity, which corresponds to the second perceptual dimension of light: **brightness.** If the intensity of the electromagnetic radiation is increased, the apparent brightness increases, too. The third dimension, **saturation,** refers to the relative purity of the light that is being perceived. If all the radiation is of one wavelength, the perceived color is pure, or fully saturated. Conversely, if the radiation contains all wavelengths, it produces no sensation of hue—it appears white. Colors with intermediate amounts of saturation consist of different mixtures of wavelengths. Figure 6.2 shows some color samples, all with the same hue but with different levels of brightness and saturation. (See *Figure 6.2*.)

receptor potential A slow, graded electrical potential produced by a receptor cell in response to a physical stimulus.

hue One of the perceptual dimensions of color; the dominant wavelength.

brightness One of the perceptual dimensions of color; intensity.

saturation One of the perceptual dimensions of color; purity.

FIGURE 6.2 ■ Color Wavelength and Saturation

This figure shows examples of colors with the same dominant wavelength (hue) but different levels of saturation or brightness.

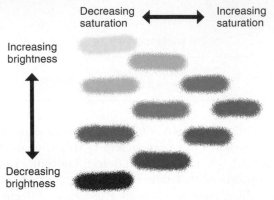

FIGURE 6.3 ■ The Extraocular Muscles, Which Move the Eyes

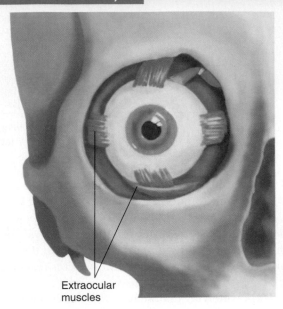

Extraocular muscles

ANATOMY OF THE VISUAL SYSTEM

For an individual to see, an image must be focused on the retina, the inner lining of the eye. This image causes changes in the electrical activity of millions of neurons in the retina, which results in messages being sent through the optic nerves to the rest of the brain. (I said "the rest" because the retina is actually part of the brain; it and the optic nerve are in the central—not peripheral—nervous system.) This section describes the anatomy of the eyes, the photoreceptors in the retina that detect the presence of light, and the connections between the retina and the brain.

The Eyes

The eyes are suspended in the *orbits*, bony pockets in the front of the skull. They are held in place and moved by six extraocular muscles attached to the tough, white outer coat of the eye called the *sclera*. (See ***Figure 6.3.***) Normally, we cannot look behind our eyeballs and see these muscles, because their attachments to the eyes are hidden by the *conjunctiva*. These mucous membranes line the eyelid and fold back to attach to the eye (thus preventing a contact lens that has slipped off the cornea from "falling behind the eye"). Figure 6.4 illustrates the anatomy of the eye. (See ***Figure 6.4.***)

The eyes make three types of movements: vergence movements, saccadic movements, and pursuit movements. **Vergence movements** are cooperative movements that keep both eyes fixed on the same target—or, more precisely, that keep the image of the target object on corresponding parts of the two retinas. If you hold up a finger in front of your face, look at it, and then bring your finger closer to your face, your eyes will make vergence movements toward your nose. If you then look at an object on the other side of the room, your eyes will rotate outward, and you will see two separate blurry images of your finger.

When you scan the scene in front of you, your gaze does not roam slowly and steadily across its features. Instead, your eyes make jerky **saccadic movements**—you shift your gaze abruptly from one point to another. (*Saccade* comes from the French word for "jerk.") When you read a line in this book, your eyes stop several times, moving very quickly between each stop. You cannot consciously control the speed of movement between stops; during each *saccade* the eyes move as fast as they can. Only by performing a **pursuit movement**—say, by looking at your finger while you move it around—can you make your eyes move more slowly.

The white outer layer of most of the eye, the sclera, is opaque and does not permit entry of light. However, the cornea, the outer layer at the front of the eye, is transparent and admits light. The amount of light that enters is regulated by the size of the pupil, which is an

vergence movement The cooperative movement of the eyes, which ensures that the image of an object falls on identical portions of both retinas.

saccadic movement (*suh **kad** ik*) The rapid, jerky movement of the eyes used in scanning a visual scene.

pursuit movement The movement that the eyes make to maintain an image of a moving object on the fovea.

FIGURE 6.4 ■ The Human Eye

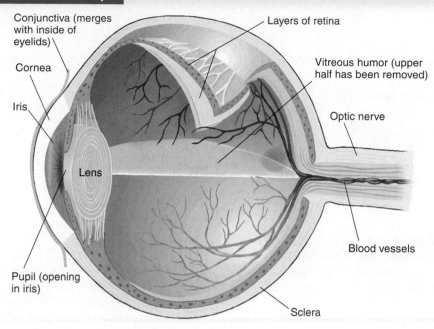

opening in the iris, the pigmented ring of muscles situated behind the cornea. The lens, situated immediately behind the iris, consists of a series of transparent, onionlike layers. Its shape can be altered by contraction of the *ciliary muscles*. These changes in shape permit the eye to focus images of near or distant objects on the retina—a process called **accommodation.**

After passing through the lens, light traverses the main part of the eye, which is filled with *vitreous humor* ("glassy liquid"), a clear, gelatinous substance. After passing through the vitreous humor, light falls on the **retina,** the interior lining of the back of the eye. In the retina are located the receptor cells, the **rods** and **cones** (named for their shapes), collectively known as **photoreceptors.**

The human retina contains approximately 120 million rods and 6 million cones. Although they are greatly outnumbered by rods, cones provide us with most of the information about our environment. In particular, they are responsible for our daytime vision. They provide us with information about small features in the environment and thus are the source of vision of the highest sharpness, or *acuity* (from the Latin *acus*, "needle"). The **fovea,** or central region of the retina, which mediates our most acute vision, contains only cones. Cones are also responsible for color vision—our ability to discriminate light of different wavelengths. Although rods do not detect different colors and provide vision of poor acuity, they are more sensitive to light. In a very dimly lighted environment we use our rod vision; therefore, in dim light we are color-blind and lack foveal vision. (See *Table 6.1.*)

Another feature of the retina is the **optic disk,** where the axons conveying visual information gather together and leave the eye through the optic nerve. The optic disk produces a *blind spot* because no receptors are located there. We do not normally perceive our blind spots, but their presence can be demonstrated. If you have not found yours, you may want to try the exercise described in *Figure 6.5.*

Close examination of the retina shows that it consists of several layers of neuron cell bodies, their axons and dendrites, and the photoreceptors. Figure 6.6 illustrates a cross section through the primate retina, which is divided into three main layers: the photoreceptive

accommodation Changes in the thickness of the lens of the eye, accomplished by the ciliary muscles, that focus images of near or distant objects on the retina.

retina The neural tissue and photoreceptive cells located on the inner surface of the posterior portion of the eye.

rod One of the receptor cells of the retina; sensitive to light of low intensity.

cone One of the receptor cells of the retina; maximally sensitive to one of three different wavelengths of light and hence encodes color vision.

photoreceptor One of the receptor cells of the retina; transduces photic energy into electrical potentials.

fovea (*foe vee a*) The region of the retina that mediates the most acute vision of birds and higher mammals. Color-sensitive cones constitute the only type of photoreceptor found in the fovea.

optic disk The location of the exit point from the retina of the fibers of the ganglion cells that form the optic nerve; responsible for the blind spot.

TABLE 6.1 ■ Locations and Response Characteristics of Photoreceptors

CONES	RODS
Most prevalent in the central retina; found in the fovea	Most prevalent in the peripheral retina; not found in the fovea
Sensitive to moderate to high levels of light	Sensitive to low levels of light
Provide information about hue	Provide only monochromatic information
Provide excellent acuity	Provide poor acuity

layer, the bipolar cell layer, and the ganglion cell layer. Note that the photoreceptors are at the *back* of the retina; light must pass through the overlying layers to get to them. Fortunately, these layers are transparent. (See *Figure 6.6.*)

The photoreceptors form synapses with **bipolar cells,** neurons whose two arms connect the shallowest and deepest layers of the retina. In turn, these neurons form synapses with the **ganglion cells,** neurons whose axons travel through the optic nerves (the second cranial nerves) and carry visual information into the rest of the brain. In addition, the retina contains **horizontal cells** and **amacrine cells,** both of which transmit information in a direction parallel to the surface of the retina and thus combine messages from adjacent photoreceptors. (See *Figure 6.6.*)

The primate retina contains approximately 55 different types of neurons: one type of rod, three types of cones, two types of horizontal cells, ten types of bipolar cells, 24–29 types of amacrine cells, and 10–15 types of ganglion cells (Masland, 2001).

Photoreceptors

Figure 6.7 shows a drawing of two rods and a cone. Note that each photoreceptor consists of an outer segment connected by a cilium to the inner segment, which contains the nucleus. (See *Figure 6.7.*) The outer segment contains several hundred **lamellae,** or thin plates of

bipolar cell A bipolar neuron located in the middle layer of the retina, conveying information from the photoreceptors to the ganglion cells.

ganglion cell A neuron located in the retina that receives visual information from bipolar cells; its axons give rise to the optic nerve.

horizontal cell A neuron in the retina that interconnects adjacent photoreceptors and the outer processes of the bipolar cells.

amacrine cell (*amm a krine*) A neuron in the retina that interconnects adjacent ganglion cells and the inner processes of the bipolar cells.

lamella A layer of membrane containing photopigments; found in rods and cones of the retina.

FIGURE 6.5 ■ A Test for the Blind Spot

With your left eye closed, look at the plus sign with your right eye, and move the page nearer to and farther from you. When the page is about 20 cm from your face, the green circle disappears because its image falls on the blind spot of your right eye.

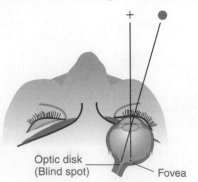

Optic disk
(Blind spot)   Fovea

FIGURE 6.6 ■ Details of Retinal Circuitry

(Adapted from Dowling, J. E., and Boycott, B. B. *Proceedings of the Royal Society of London, B*, 1966, *166*, 80–111.)

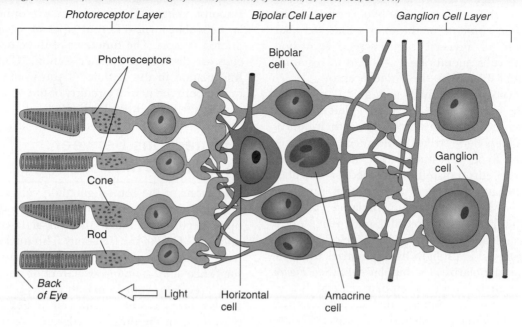

membrane. (*Lamella* is the diminutive form of *lamina,* "thin layer.")

Let's consider the nature of transduction of visual information. The first step in the chain of events that leads to visual perception involves a special chemical called a photopigment. **Photopigments** are special molecules embedded in the membrane of the lamellae; a single human rod contains approximately 10 million of them. The molecules consist of two parts: an **opsin** (a protein) and **retinal** (a lipid). There are several forms of opsin; for example, the photopigment of human rods, **rhodopsin,** consists of *rod opsin* plus retinal. (*Rhod-* refers to the Greek *rhodon,* "rose," not to *rod.* Before it is bleached by the action of light, rhodopsin has a pinkish hue.) Retinal is synthesized from vitamin A, which explains why carrots, which are rich in this vitamin, are said to be good for your eyesight.

When a molecule of rhodopsin is exposed to light, it breaks into its two constituents: rod opsin and retinal. When that happens, the rod opsin changes from its rosy color to a pale yellow; hence, we say that the light *bleaches* the photopigment. The splitting of the photopigment produces the receptor potential: hyperpolarization of the membrane of the photoreceptor.

In the vertebrate retina, photoreceptors provide input to both bipolar cells and horizontal cells. Figure 6.8 shows the neural circuitry from a photoreceptor to a ganglion cell. The circuitry is much simplified and omits the horizontal cells and amacrine cells. The first two types of cells in the circuit—photoreceptors and bipolar

FIGURE 6.7 ■ Photoreceptors

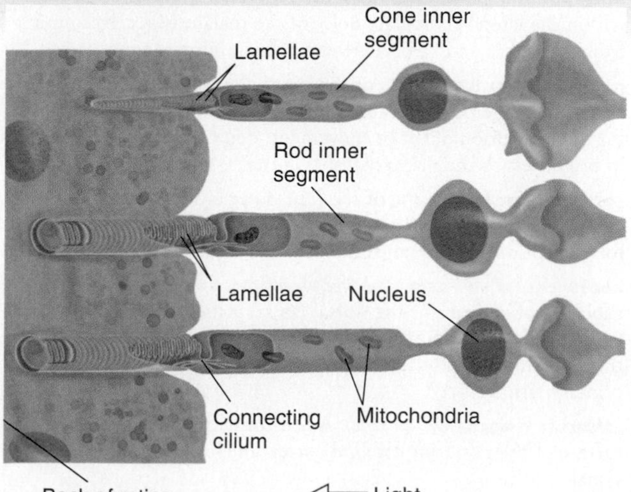

photopigment A protein dye bonded to retinal, a substance derived from vitamin A; responsible for transduction of visual information.

opsin (*opp sin*) A class of protein that, together with retinal, constitutes the photopigments.

retinal (*rett i nahl*) A chemical synthesized from vitamin A; joins with an opsin to form a photopigment.

rhodopsin (*roh dopp sin*) A particular opsin found in rods.

cells—do not produce action potentials. Instead, their release of the neurotransmitter (glutamate) is regulated by the value of their membrane potential; depolarizations increase the release, and hyperpolarizations decrease it. The contents of the circles indicate what would be seen on an oscilloscope screen recording changes in the cells' membrane potentials in response to a spot of light shining on the photoreceptor.

The hyperpolarizing effect of light on the membrane of a photoreceptor is shown in the left circle. In the dark, photoreceptors constantly release their neurotransmitter. When light strikes molecules of the photopigment, the hyperpolarization that ensues reduces the amount of neurotransmitter released by the photoreceptor. Because the neurotransmitter normally hyperpolarizes the dendrites of the bipolar cell by binding with inhibitory metabotropic alutamate receptors, a *reduction* in its release causes the membrane of the bipolar cell to *depolarize*. Thus, light hyperpolarizes the photoreceptor and depolarizes the bipolar cell. (See *Figure 6.8.*) The depolarization of the bipolar cell causes it to release more neurotransmitter, which depolarizes the membrane of the ganglion cell and raises this cell's rate of firing. Thus, light shining on the photoreceptor excites the ganglion cell and increases the rate of firing of its axon.

The circuit shown in Figure 6.8 illustrates a ganglion cell whose firing rate increases in response to light. As we will see, other ganglion cells *decrease* their firing rate in response to light. These neurons are connected to bipolar cells that form different types of synapses with the photoreceptors. The functions of these two types of circuits are discussed in a later section, "Coding of Visual Information in the Retina." If you would like to know more about the neural circuitry of the retina, you should consult the book by Rodieck (1998).

Connections Between Eye and Brain

The axons of the retinal ganglion cells bring information to the rest of the brain. They ascend through the optic nerves and reach the **dorsal lateral geniculate nucleus (LGN)** of the thalamus. This nucleus receives its name from its resemblance to a bent knee (*genu* is Latin for "knee"). It contains six layers of neurons, each of which receives input from only one eye. The neurons in the two inner layers contain cell bodies that are larger than those in the outer four layers. For this reason the inner two layers are called the **magnocellular layers,** and the outer four layers are called the **parvocellular layers** (*parvo-* refers to the small size of the cells). A third set of neurons in the **koniocellular sublayers** are found ventral to each of the magnocellular and parvocellular layers. (*Konis* is the Greek word for "dust.") As we will see later, these three sets of layers belong to different systems, which are responsible for the analysis of different types of visual information. They receive input from different types of retinal ganglion cells. (See *Figure 6.9.*)

The neurons in the LGN send their axons through a pathway known as the *optic radiations* to the primary visual cortex—the region surrounding the **calcarine fissure**

FIGURE 6.8 ■ Neural Circuitry in the Retina

Light striking a photoreceptor produces a hyperpolarization, so the photoreceptor releases *less* neurotransmitter. Because the neurotransmitter normally hyperpolarizes the membrane of the bipolar cell, the reduction causes a *depolarization*. This depolarization causes the bipolar cell to release *more* neurotransmitter, which excites the ganglion cell.

(Adapted from Dowling, J. E., in *The Neurosciences: Fourth Study Program*, edited by F. O. Schmitt and F. G. Worden. Cambridge, Mass.: MIT Press, 1979.)

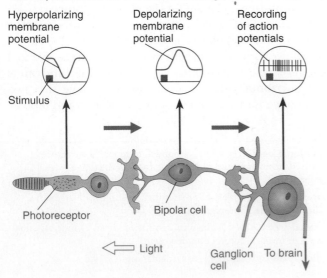

dorsal lateral geniculate nucleus (LGN) A group of cell bodies within the lateral geniculate body of the thalamus; receives inputs from the retina and projects to the primary visual cortex.

magnocellular layer One of the inner two layers of neurons in the dorsal lateral geniculate nucleus; transmits information necessary for the perception of form, movement, depth, and small differences in brightness to the primary visual cortex.

parvocellular layer One of the four outer layers of neurons in the dorsal lateral geniculate nucleus; transmits information necessary for perception of color and fine details to the primary visual cortex.

koniocellular sublayer (*koh nee oh **sell** yew lur*) One of the sublayers of neurons in the dorsal lateral geniculate nucleus found ventral to each of the magnocellular and parvocellular layers; transmits information from short-wavelength ("blue") cones to the primary visual cortex.

calcarine fissure (*kal ka rine*) A horizontal fissure on the inner surface of the posterior cerebral cortex; the location of the primary visual cortex.

FIGURE 6.9 ■ Lateral Geniculate Nucleus

The photomicrograph shows a section through the right lateral geniculate nucleus of a rhesus monkey (cresyl violet stain) Layers 1, 4, and 6 receive input from the contralateral (left) eye, and layers 2, 3, and 5 receive input from the ipsilateral (right) eye. Layers 1 and 2 are the magnocellular layers; layers 3–6 are the parvocellular layers. The koniocellular sublayers are found ventral to each of the parvocellular and magnocellular layers. The receptive fields of all six principal layers are in almost perfect registration; cells located along the line of the unlabeled arrow have receptive fields centered on the same point.

(Photomicrograph from Hubel, D. H., Wiesel, T. N., and Le Vay, S. *Philosophical Transactions of the Royal Society of London, B,* 1977, *278,* 131–163. Reprinted with permission.)

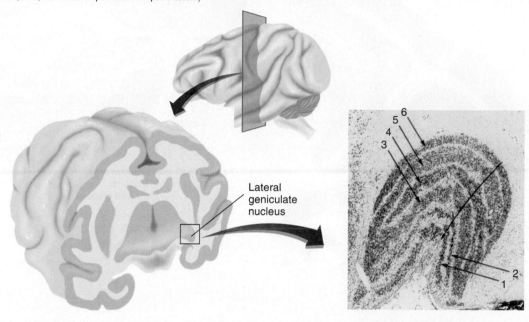

Lateral geniculate nucleus

(*calcarine* means "spur-shaped"), a horizontal fissure located in the medial and posterior occipital lobe. The primary visual cortex is often called the **striate cortex** because it contains a dark-staining layer (*striation*) of cells. (See *Figure 6.10.*)

Figure 6.11 shows a diagrammatical view of a horizontal section of the human brain. The optic nerves join together at the base of the brain to form the X-shaped **optic chiasm** (*khiasma* is the Greek for "cross"). There, axons from ganglion cells serving the inner halves of the retina (the nasal sides) cross through the chiasm and ascend to the LGN on the opposite side of the brain. The axons from the outer halves of the retina (the temporal sides) remain on the same side of the brain. (See *Figure 6.11.*) The lens inverts the image of the world projected on the retina (and similarly reverses left and right). Therefore, because the axons from the nasal halves of the retinas cross to the other side of the brain, each hemisphere receives information from the contralateral half (opposite side) of the visual scene. That is, if a person looks straight ahead, the right hemisphere receives information from the left half of the visual field,

and the left hemisphere receives information from the right. It is *not* correct to say that each hemisphere receives visual information solely from the contralateral eye. (See *Figure 6.11.*)

Besides the primary retino-geniculo-cortical pathway, fibers from the retina take several other pathways. For example, one pathway to the hypothalamus synchronizes an animal's activity cycles to the 24-hour rhythms of day and night. (We will study this system in Chapter 9.) Other pathways, especially those that travel to the optic tectum and the pretectal nuclei, coordinate eye movements, control the muscles of the iris (and thus the size of the pupil) and the ciliary muscles (which control the lens), and help to direct our attention to sudden movements in the periphery of our visual field.

---

striate cortex (***stry** ate*) The primary visual cortex.

optic chiasm A cross-shaped connection between the optic nerves, located below the base of the brain, just anterior to the pituitary gland.

FIGURE 6.10 ■ Striate Cortex

This photomicrograph is a cross section through the striate cortex of a rhesus macaque monkey. The ends of the striate cortex are shown by arrows.

(From Hubel, D. H., and Wiesel, T. N. *Proceedings of the Royal Society of London, B*, 1977, *198*, 1–59. Reprinted with permission.)

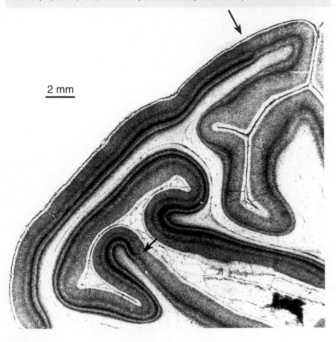

2 mm

FIGURE 6.11 ■ The Primary Visual Pathway

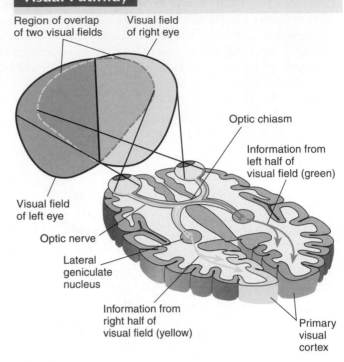

Region of overlap of two visual fields

Visual field of right eye

Optic chiasm

Information from left half of visual field (green)

Visual field of left eye

Optic nerve

Lateral geniculate nucleus

Information from right half of visual field (yellow)

Primary visual cortex

Interim Summary

The Stimulus and Anatomy of the Visual System

Light consists of electromagnetic radiation, similar to radio waves but of a different frequency and wavelength. Color can vary in three perceptual dimensions: hue, brightness, and saturation, which correspond to the physical dimensions of wavelength, intensity, and purity.

The photoreceptors in the retina—the rods and the cones—detect light. Muscles move the eyes so that images of particular parts of the environment fall on the retina. Accommodation is accomplished by the ciliary muscles, which change the shape of the lens. Photoreceptors communicate through synapses with bipolar cells, which communicate through synapses with ganglion cells. In addition, horizontal cells and amacrine cells combine messages from adjacent photoreceptors.

When light strikes a molecule of photopigment in a photoreceptor, the retinal molecule detaches from the opsin molecule. This detachment hyperpolarizes the membrane of the photoreceptor. As a result, the rate of firing of the ganglion cell changes, signaling the detection of light.

Visual information from the retina reaches the striate cortex surrounding the calcarine fissure after being relayed through the magnocellular, parvocellular, and koniocellular layers of the LGN. Several other regions of the brain, including the hypothalamus and the tectum, also receive visual information. These regions help to regulate activity during the day–night cycle, coordinate eye and head movements, control attention to visual stimuli, and regulate the size of the pupils.

Thought Question

People who try to see faint, distant lights at night are often advised to look just to the side of the location where they expect to see the lights. Can you explain the reason for this advice?

CODING OF VISUAL INFORMATION IN THE RETINA

This section describes the way in which cells of the retina encode information they receive from the photoreceptors.

Coding of Light and Dark

One of the most important methods for studying the physiology of the visual system is the use of microelectrodes to record the electrical activity of single neurons. As we saw in the previous section, some ganglion cells become excited when light falls on the photoreceptors that communicate with them. The **receptive field** of a neuron in the visual system is the part of the visual field that an individual neuron "sees"—that is, the place in which a visual stimulus must be located to produce a response in that neuron. Obviously, the location of the receptive field of a particular neuron depends on the location of the photoreceptors that provide it with visual information. If a neuron receives information from photoreceptors located in the fovea, its receptive field will be at the fixation point—the point at which the eye is looking. If the neuron receives information from photoreceptors located in the periphery of the retina, its receptive field will be located off to one side.

At the periphery of the retina many individual receptors converge on a single ganglion cell, bringing information from a relatively large area of the retina—and hence a relatively large area of the visual field. However, the fovea contains approximately equal numbers of ganglion cells and cones. These receptor-to-axon relationships explain the fact that our foveal (central) vision is very acute but our peripheral vision is much less precise. (See *Figure 6.12.*)

Over seventy years ago, Hartline (1938) discovered that the frog retina contained three types of ganglion cells. ON cells responded with an excitatory burst when the retina was illuminated, OFF cells responded when the light was turned off, and ON/OFF cells responded briefly when the light went on and again when it went off. Kuffler (1952, 1953), recording from ganglion cells in the retina of the cat, discovered that their receptive field consists of a roughly circular center, surrounded by a ring. Stimulation of the center or surrounding fields had contrary effects: ON cells were excited by light falling in the central field (*center*) and were inhibited by light falling in the surrounding field (*surround*), whereas OFF cells responded in the opposite manner. ON/OFF ganglion cells were briefly excited when light was turned on or off. In primates most of these ON/OFF cells project primarily to the superior colliculus, which is primarily involved in visual reflexes (Schiller and Malpeli, 1977); thus, these cells do not

FIGURE 6.12 ■ Central Versus Peripheral Acuity

Ganglion cells in the fovea receive input from a smaller number of photoreceptors than those in the periphery and hence provide more acute visual information.

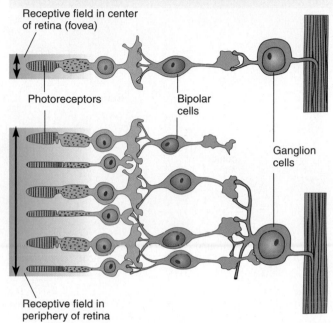

Receptive field in center of retina (fovea)

Photoreceptors   Bipolar cells   Ganglion cells

Receptive field in periphery of retina

appear to play a direct role in form perception. (See *Figure 6.13.*)

Figure 6.13 also illustrates a rebound effect that occurs when the light is turned off again. Neurons whose firing is inhibited while the light is on will show a brief burst of excitation when it is turned off. In contrast, neurons whose firing is increased will show a brief period of inhibition when the light is turned off. (See *Figure 6.13.*)

The two major categories of ganglion cells (ON and OFF) and the organization of their receptive fields into contrasting center and surround provide useful information to the rest of the visual system. Let us consider these two types of ganglion cells first. As Schiller (1992) notes, ganglion cells normally fire at a relatively low rate. Then, when the level of illumination in the center of their receptive field increases or decreases (for example, when an object moves or the eye makes a saccade), they signal the change. In particular, ON cells signal increases and OFF cells signal decreases, but both signal them by an increased rate of firing. Such a system is particularly efficient. Theoretically, a single type of ganglion

---

receptive field That portion of the visual field in which the presentation of visual stimuli will produce an alteration in the firing rate of a particular neuron.

FIGURE 6.13 ■ ON and OFF Ganglion Cells

The figure shows responses of ON and OFF ganglion cells to stimuli presented in the center or the surround of the receptive field.

(Adapted from Kuffler, S. W. *Cold Spring Harbor Symposium for Quantitative Biology*, 1952, *17*, 281–292.)

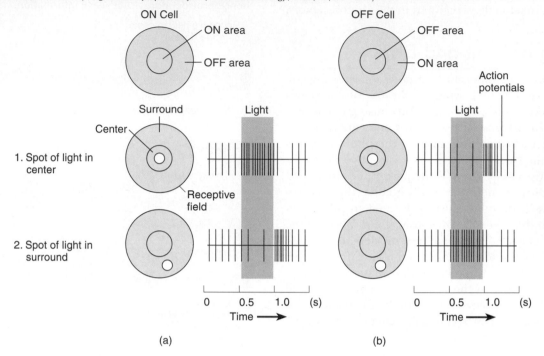

cell could fire at an intermediate rate and signal changes in the level of illumination by increases or decreases in rate of firing. However, in this case the average rate of firing of the one million axons in each optic nerve would have to be much higher.

Several studies have shown that ON cells and OFF cells do, indeed, signal different kinds of information. Schiller, Sandell, and Maunsell (1986) injected monkeys with APB (2-amino-4-phosphonobutyrate), a drug that selectively blocks synaptic transmission in ON bipolar cells. They found that the animals had difficulty detecting spots that were made brighter than the background but had no difficulty detecting spots that were slightly darker than the background. In addition, Dolan and Schiller (1989) found that an injection of APB completely blocked vision in very dim light, which is normally mediated by rods. Thus, rod bipolar cells must all be of the ON type. (If you think about it, that arrangement makes sense; in very dim light we are more likely to see brighter objects against a dark background than dark objects against a light background.)

The second characteristic of the receptive fields of ganglion cells—their center-surround organization—enhances our ability to detect the outlines of objects even when the contrast between the object and the background is low. Figure 6.14 illustrates this phenomenon.

This figure shows six gray squares arranged in order of brightness. The right side of each square looks lighter than the left side, which makes the borders between the squares stand out. But these exaggerated borders do not exist in the illustration; they are added by our visual system because of the center-surround organization of the receptive fields of the retinal ganglion cells. (See *Figure 6.14*.)

Figure 6.15 explains how this phenomenon works. We see the centers and surrounds of the receptive fields of several ganglion cells. (In reality these receptive fields would be overlapping, but the simplified arrangement is easier to understand. This example also includes only ON cells—again, for the sake of simplicity.) The image

FIGURE 6.14 ■ Enhancement of Contrast

Although each gray square is of uniform darkness, the right edge of each square looks somewhat lighter, and the left edge looks somewhat darker. This effect appears to be caused by the opponent center-surround arrangement of the receptive fields of the retinal ganglion cells.

FIGURE 6.15 ■ A Schematic Explanation of Phenomenon Shown in Figure 6.14

Only ON cells are shown; OFF cells are responsible for the darker appearance of the left side of the darker square.

All of the surrounds of the ON cells whose receptive fields fall within the lighter gray are evenly illuminated; this illumination partially inhibits the firing of these cells

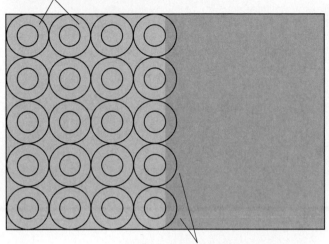

A portion of the inhibitory surrounds of the ON cells near the border receives less illumination; thus, these cells have the highest rate of firing

of the transition between lighter and darker regions falls across some of these receptive fields. The cells whose centers are located in the brighter region but whose surrounds are located at least partially in the darker region will have the highest rate of firing. (See *Figure 6.15.*)

Coding of Color

So far, we have been examining the monochromatic properties of ganglion cells—that is, their responses to light and dark. But, of course, objects in our environment selectively absorb some wavelengths of light and reflect others, which, to our eyes, gives them different colors. The retinas of humans, Old World monkeys, one species of New World monkey, and apes contain three different types of cones, which provides them (and us) with the most elaborate form of color vision (Jacobs, 1996; Hunt et al., 1998). Although monochromatic (black-and-white) vision is perfectly adequate for most purposes, color vision gave our primate ancestors the ability to distinguish ripe fruit from unripe fruit and made it more difficult for other animals to hide themselves by means of camouflage (Mollon, 1989). In fact, the photopigments of primates with three types of cones seem well suited for distinguishing red and yellow fruits against a background of green foliage (Regan et al., 2001).

Color Mixing

Various theories of color vision have been proposed for many years—long before it was possible to disprove or validate them by physiological means. In 1802, Thomas Young, a British physicist and physician, proposed that the eye detected different colors because it contained three types of receptors, each sensitive to a single hue. His theory was referred to as the *trichromatic* (three-color) *theory*. It was suggested by the fact that for a human observer any color can be reproduced by mixing various quantities of three colors judiciously selected from different points along the spectrum.

I must emphasize that *color mixing* is different from *pigment mixing*. If we combine yellow and blue pigments (as when we mix paints), the resulting mixture is green. Color mixing refers to the addition of two or more light sources. If we shine a beam of red light and a beam of bluish green light together on a white screen, we will see yellow light. If we mix yellow and blue light, we get white light. When white appears on a color television screen or computer monitor, it actually consists of tiny dots of red, blue, and green light. (See *Figure 6.16.*)

Another fact of color perception suggested to a German physiologist, Ewald Hering (1905/1965), that hue might be represented in the visual system as *opponent colors;* red versus green and yellow versus blue. People interested in color perception have long regarded yellow, blue, red, and green as primary colors—colors that seem unique and do not appear to be blends of other colors. (Black and white are primary, too, but we perceive them as colorless.) All other colors can be described as mixtures of these primary colors. The trichromatic system cannot explain why *yellow* is included in this group—why it is perceived as a pure color. In addition, some colors appear to blend, whereas others do not. For example, one can speak of a bluish green or a yellowish green, and orange appears to have both red and yellow qualities. Purple resembles both red and blue. But try to imagine a reddish green or a bluish yellow. It is impossible; these colors seem to be opposite to each other. Again, these facts are not explained by the trichromatic theory. As we shall see in the following section, the visual system uses both trichromatic and opponent-color systems to encode information related to color.

Photoreceptors: Trichromatic Coding

Physiological investigations of retinal photoreceptors in higher primates have found that Young was right: Three different types of photoreceptors (three different types of cones) are responsible for color vision. Investigators have studied the absorption characteristics of individual photoreceptors, determining the amount of light of different wavelengths that is absorbed by the photopigments. These characteristics are controlled by the

FIGURE 6.16 ■ Additive Color Mixing and Paint Mixing

When blue, red, and green lights of the proper intensity are all shone together, the result is white light. When red, blue, and yellow paints are mixed together, the result is a dark gray.

particular opsin a photoreceptor contains; different opsins absorb particular wavelengths more readily. Figure 6.17 shows the absorption characteristics of the four types of photoreceptors in the human retina: rods and the three types of cones. (See *Figure 6.17.*)

The peak sensitivities of the three types of cones are approximately 420 nm (blue-violet), 530 nm (green),

FIGURE 6.17 ■ Absorbance of Light by Rods and Cones

The graph shows the relative absorbance of light by rods and three types of cones in the human retina.

(From Dartnall, H. J. A., Bowmaker, J. K., and Mollon, J. D. Human visual pigments: Microspectrophotometric results from the eyes of seven persons. *Proceedings of the Royal Society of London, B,* 1983, *220,* 115–130. Reprinted with permission.)

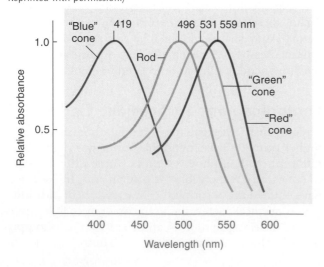

and 560 nm (yellow-green). The peak sensitivity of the short-wavelength cone is actually 440 nm in the intact eye because the lens absorbs some short-wavelength light. For convenience the short-, medium-, and long-wavelength cones are traditionally called "blue," "green," and "red" cones, respectively. The relative number of "red" and "green" cones varies considerably from person to person. Amazingly, even large differences in the relative numbers of these cones have no measurable effect on a person's color vision. The retina contains a much smaller number of "blue" cones—approximately 8 percent of the total. Even fewer of the bipolar cells in the primate retina—approximately 1.5 percent—transmit information from blue cones to the ganglion cells (Kouyama and Marshak, 1992).

Evidence suggests that the first cone opsin to evolve was most sensitive to long wavelengths of light. These "red" cones were supplemented by the evolution of "blue" cones, which provides the limited dichromatic vision found in most mammals (Haverkamp et al., 2005). The trichromatic color vision found in humans and Old World monkeys was made possible when the "red" opsin gene was duplicated and one of the copies mutated into the gene that produced the "green" opsin (Solomon and Lennie, 2007). (You will recall from Chapter 3 that an important factor in evolutionary development is genetic duplication, which permits the process of natural selection to "experiment" with mutations of the extra gene. In this case the gene for the old "red" opsin was retained, and the new gene for the "green" opsin produced a third category of color-sensitive cones.)

Genetic defects in color vision result from anomalies in one or more of the three types of cones (Wissinger and Sharpe, 1998; Nathans, 1999). The first two kinds of defective color vision described here involve genes on the X chromosome; thus, because males have only one X chromosome, they are much more likely to have this disorder. (Females are likely to have a normal gene on one of their X chromosomes, which compensates for the defective one.) People with **protanopia** ("first-color defect") confuse red and green. They see the world in shades of yellow and blue; both red and green look yellowish to them. Their visual acuity is normal, which suggests that their retinas do not lack "red" or "green" cones. This fact and their sensitivity to lights of different wavelengths suggest that their "red" cones are filled with "green" cone opsin. People with **deuteranopia** ("second-color defect") also confuse red and green and also have normal visual acuity. Their "green" cones appear to be filled with "red" cone opsin. (In other words, their vision is dichromatic, like that of our ancestors.)

Tritanopia ("third-color defect") is rare, affecting fewer than 1 in 10,000 people. This disorder involves a faulty gene that is not located on an X chromosome; thus, it is equally prevalent in males and females. People with tritanopia have difficulty with hues of short wavelengths and see the world in greens and reds. To them a clear blue sky is a bright green, and yellow looks pink. Their retinas lack "blue" cones. Because the retina contains so few of these cones, their absence does not noticeably affect visual acuity.

Retinal Ganglion Cells: Opponent-Process Coding

At the level of the retinal ganglion cell the three-color code gets translated into an opponent-color system. Daw (1968) and Gouras (1968) found that these neurons respond specifically to pairs of primary colors, with red opposing green and blue opposing yellow. Thus, the retina contains two kinds of color-sensitive ganglion cells: *red-green* and *yellow-blue*. Some color-sensitive ganglion cells respond in a center-surround fashion. For example, a cell might be excited by red and inhibited by green in the center of their receptive field while showing the opposite response in the surrounding ring. (See *Figure 6.18.*) Other ganglion cells that receive input from cones do not respond differentially to different wavelengths but simply encode relative brightness in the center and surround. These cells serve as "black-and-white detectors."

The response characteristics of retinal ganglion cells to light of different wavelengths are obviously determined by the particular circuits that connect the three types of cones with the two types of ganglion cells. These circuits involve different types of bipolar cells, amacrine cells, and horizontal cells.

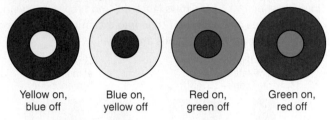

FIGURE 6.18 ■ Receptive Fields of Color-Sensitive Ganglion Cells

When a portion of the receptive field is illuminated with the color shown, the cell's rate of firing increases. When a portion is illuminated with the complementary color, the cell's rate of firing decreases.

Yellow on, blue off   Blue on, yellow off   Red on, green off   Green on, red off

Figure 6.19 helps to explain how particular hues are detected by the "red," "green," and "blue" cones and translated into excitation or inhibition of the red-green and yellow-blue ganglion cells. The diagram does not show the actual neural circuitry, which includes the retinal neurons that connect the cones with the ganglion cells. The arrows in Figure 6.19 refer merely to the *effects* of the light falling on the retina. The book by Rodieck (1998) describes the actual neural circuitry in considerable detail.

Detection and coding of pure red, green, or blue light is the easiest to understand. For example, red light excites "red" cones, which causes the excitation of red-green ganglion cells. (See *Figure 6.19a.*) Green light excites "green" cones, which causes the *inhibition* of red-green cells. (See *Figure 6.19b.*) But consider the effect of yellow light. Because the wavelength that produces the sensation of yellow is intermediate between the wavelengths that produce red and green, it will stimulate both "red" and "green" cones about equally. Yellow-blue ganglion cells are excited by both "red" and "green" cones, so their rate of firing increases. However, red-green ganglion cells are excited by red and inhibited by green, so their firing rate does not change. The brain detects an increased firing rate from the axons of yellow-blue ganglion cells, which it interprets as yellow. (See *Figure 6.19c.*) Blue light simply inhibits the activity of yellow-blue ganglion cells. (See *Figure 6.19d.*) By the way,

protanopia (*pro tan **owe** pee a*) An inherited form of defective color vision in which red and green hues are confused; "red" cones are filled with "green" cone opsin.

deuteranopia (*dew ter an **owe** pee a*) An inherited form of defective color vision in which red and green hues are confused; "green" cones are filled with "red" cone opsin.

tritanopia (*try tan **owe** pee a*) An inherited form of defective color vision in which hues with short wavelengths are confused; "blue" cones are either lacking or faulty.

FIGURE 6.19 ■ Color Coding in the Retina

(a) Red light stimulates a "red" cone, which causes excitation of a red-green ganglion cell.
(b) Green light stimulates a "green" cone, which causes inhibition of a red-green ganglion
cell. (c) Yellow light stimulates "red" and "green" cones equally but does not affect "blue"
cones. The stimulation of "red" and "green" cones causes excitation of a yellow-blue
ganglion cell. (d) Blue light stimulates a "blue" cone, which causes inhibition of a yellow-
blue ganglion cell. The arrows labeled E and I represent neural circuitry within the retina
that translates excitation of a cone into excitation or inhibition of a ganglion cell. For
clarity, only some of the circuits are shown.

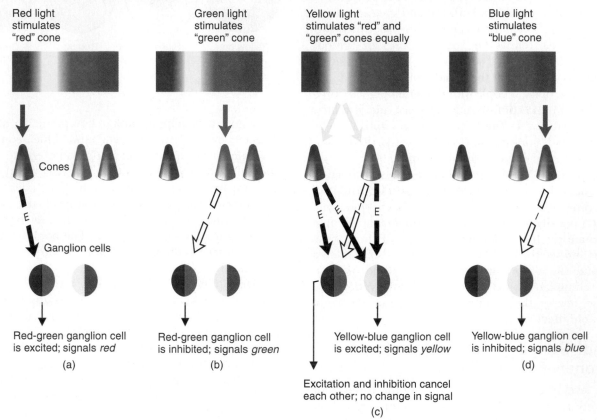

although "blue" cones constitute approximately 8 percent of the total number of cones in the primate retina,
"blue" bipolar cells constitute only about 1.5 percent of
the bipolar cells in the primate retina.

The opponent-color system employed by the ganglion cells explains why we cannot perceive a reddish
green or a bluish yellow: An axon that signals red or
green (or yellow or blue) can either increase or
decrease its rate of firing; it cannot do both at the same
time. A reddish green would have to be signaled by a
ganglion cell firing slowly and rapidly at the same time,
which is obviously impossible.

Adaptation: Negative Afterimages

Figure 6.20 demonstrates an interesting property of the
visual system: the formation of a **negative afterimage.**
Stare at the cross in the center of the image on the left
for approximately 30 seconds. Then quickly look at the

cross in the center of the white rectangle to the right.
You will have a fleeting experience of seeing the red and
green colors of a radish—colors that are complementary,
or opposite, to the ones on the left. (See *Figure 6.20.*)
Complementary items go together to make up a whole.
In this context **complementary colors** are those that
make white (or shades of gray)
when added together. (This
phenomenon is demonstrated
even more vividly in *MyPsychKit
6.1, Complementary colors.*)

mypsychkit
Where learning comes to life!
Animation 6.1
Complementary Colors

negative afterimage The image seen after a portion of the retina
is exposed to an intense visual stimulus; consists of colors
complementary to those of the physical stimulus.

complementary colors Colors that make white or gray when
mixed together.

FIGURE 6.20 ■ A Negative Afterimage

Stare for approximately 30 seconds at the plus sign in the center of the left figure; then quickly transfer your gaze to the plus sign in the center of the right figure. You will see colors that are complementary to the originals.

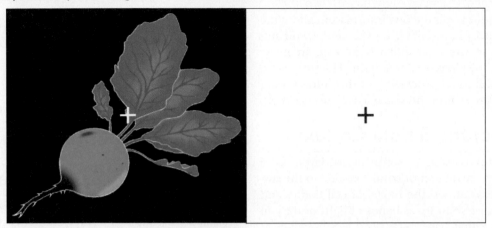

The most important cause of negative afterimages is adaptation in the rate of firing of retinal ganglion cells. When ganglion cells are excited or inhibited for a prolonged period of time, they later show a *rebound effect*, firing faster or slower than normal. For example, the green of the radish in Figure 6.20 inhibits some red-green ganglion cells. When this region of the retina is then stimulated with the neutral-colored light reflected off the white rectangle, the red-green ganglion cells—no longer inhibited by the green light—fire faster than normal. Thus, we see a red afterimage of the radish.

Interim Summary

Coding of Visual Information in the Retina

Recordings of the electrical activity of single neurons in the retina indicate that each ganglion cell receives information from photoreceptors—just one in the fovea and many more in the periphery. The receptive field of most retinal ganglion cells consists of two concentric circles, with the cells becoming excited when light falls in one region and becoming inhibited when it falls in the other. This arrangement enhances the ability of the nervous system to detect contrasts in brightness. ON cells are excited by light in the center, and OFF cells are excited by light in the surround. ON cells detect light objects against dark backgrounds; OFF cells detect dark objects against light backgrounds.

Color vision occurs as a result of information provided by three types of cones, each of which is sensitive to light of a certain wavelength: long, medium, or short. The absorption characteristics of the cones are determined by the particular opsin that their photopigment contains. Most forms of defective color vision appear to be caused by alterations in cone opsins. The "red" cones of people with protanopia are filled with "green" cone opsin, and the "green" cones of people with deuteranopia are filled with "red" cone opsin. The retinas of people with tritanopia appear to lack "blue" cones.

Most color-sensitive ganglion cells respond in an opposing center-surround fashion to the pairs of primary colors: red and green, and blue and yellow. The responses of these neurons is determined by the retinal circuitry connecting them with the photoreceptors. Negative afterimages produced by staring at a colored stimulus and then looking at a neutral background provide an image with colors complementary to the original stimulus. This phenomenon is caused by adaptation of retinal cells that show rebound activity in the opposite direction from that produced by sight of the original stimulus.

Thought Questions

Why is color vision useful? Birds, some fish, and some primates have full, three-cone color vision. Considering our own species, what other benefits (besides the ability to recognize ripe fruit, which I mentioned in the previous section) might come from the evolution of color vision?

ANALYSIS OF VISUAL INFORMATION: ROLE OF THE STRIATE CORTEX

The retinal ganglion cells encode information about the relative amounts of light falling on the center and surround regions of their receptive field and, in many cases, about the wavelength of that light. The striate cortex performs additional processing of this information, which it then transmits to the visual association cortex.

Anatomy of the Striate Cortex

The striate cortex consists of six principal layers (and several sublayers), arranged in bands parallel to the surface. These layers contain the nuclei of cell bodies and dendritic trees that show up as bands of light or dark in sections of tissue that have been dyed with a cell-body stain. (See *Figure 6.21*.)

If we consider the striate cortex of one hemisphere as a whole—if we imagine that we remove it and spread it out on a flat surface—we find that it contains a map of the contralateral half of the visual field. (Remember that each side of the brain sees the opposite side of the visual field.) The map is distorted; approximately 25 percent of the striate cortex is devoted to the analysis of information from the fovea, which represents a small part of the visual field. (The area of the visual field seen by the fovea is approximately the size of a large grape held at arm's length.)

The pioneering studies of David Hubel and Torsten Wiesel at Harvard University during the 1960s began a

revolution in the study of the physiology of visual perception (see Hubel and Wiesel, 1977, 1979). Hubel and Wiesel discovered that neurons in the visual cortex did not simply respond to spots of light; they selectively responded to specific *features* of the visual world. That is, the neural circuitry within the visual cortex combines information from several sources (for example, from axons carrying information received from several different ganglion cells) in such a way as to detect features that are larger than the receptive field of a single ganglion cell or a single cell in the LGN. The following subsections describe the visual characteristics that researchers have studied so far: orientation and movement, spatial frequency, retinal disparity, and color.

Orientation and Movement

Most neurons in the striate cortex are sensitive to *orientation*. That is, if a line or an edge (the border of a light and a dark region) is positioned in the cell's receptive field and rotated around its center, the cell will respond only when the line is in a particular position—a particular orientation. Some neurons respond best to a vertical line, some to a horizontal line, and some to a line oriented somewhere in between. Figure 6.22 shows

FIGURE 6.22 ■ Orientation Sensitivity

An orientation-sensitive neuron in the striate cortex will become active only when a line of a particular orientation appears within its receptive field. For example, the neuron depicted in this figure responds best to a bar that is vertically oriented.

(Adapted from Hubel, D. H., and Wiesel, T. N. *Journal of Physiology (London)*, 1959, *148*, 574–591.)

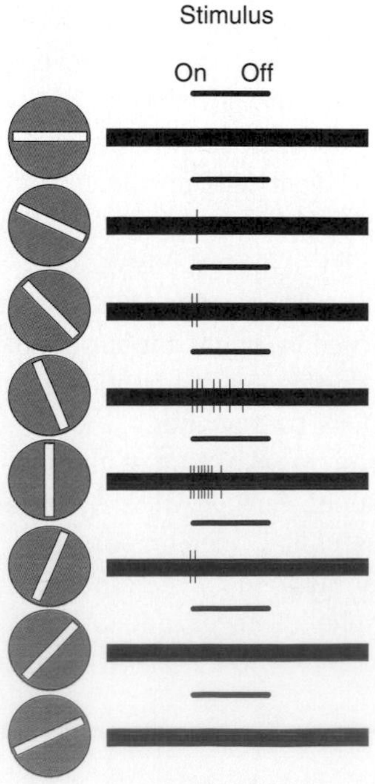

FIGURE 6.21 ■ Striate Cortex, Showing the Six Principal Layers

In this photomicrograph of a small section of striate cortex, the letter W refers to the white matter that underlies the visual cortex; beneath the white matter is layer VI of the striate cortex on the opposite side of the gyrus.

(From Hubel, D. H., and Wiesel, T. N. *Proceedings of the Royal Society of London, B*, 1977, *198*, 1–59. Reprinted with permission.)

the responses of a neuron in the striate cortex when lines were presented at various orientations. As you can see, this neuron responded best when a vertical line was presented in its receptive field. (See *Figure 6.22.*)

Some orientation-sensitive neurons have receptive fields organized in an opponent fashion. Hubel and Wiesel referred to them as **simple cells.** For example, a line of a particular orientation (say, a dark 45° line against a white background) might excite a cell if placed in the center of the receptive field but inhibit the cell if moved away from the center. (See *Figure 6.23a.*) Another type of neuron, which the researchers referred to as a **complex cell,** also responded best to a line of a particular orientation but did not show an inhibitory surround; that is, it continued to respond while the line was moved within the receptive field. In fact, many complex cells increased their rate of firing when the line was moved perpendicular to its angle of orientation—often only in one direction. Thus, these neurons also served as movement detectors. In addition, complex cells responded equally well to white lines against black backgrounds and

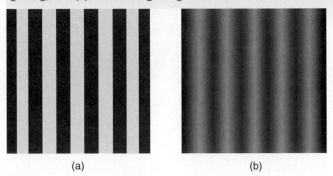

FIGURE 6.24 ■ Parallel Gratings

Two kinds of gratings are compared: (a) square-wave grating, and (b) sine-wave grating.

(a)   (b)

black lines against white backgrounds. (See *Figure 6.23b.*) Finally, **hypercomplex cells** responded to lines of a particular orientation but had an inhibitory region at the end (or ends) of the lines, which meant that the cells detected the location of *ends* of lines of a particular orientation. (See *Figure 6.23c.*)

Spatial Frequency

Although the early studies by Hubel and Wiesel suggested that neurons in the primary visual cortex detected lines and edges, subsequent research found that they actually responded best to sine-wave gratings (De Valois, Albrecht, and Thorell, 1978). Figure 6.24 compares a sine-wave grating with a more familiar square-wave grating. A square-wave grating consists of a simple set of rectangular bars that vary in brightness; the brightness along the length of a line perpendicular to them would vary in a stepwise (square-wave) fashion. (See *Figure 6.24a.*) A **sine-wave grating** looks like a series of fuzzy, unfocused parallel bars. Along any line perpendicular to the long axis of the grating, the brightness varies according to a sine-wave function. (See *Figure 6.24b.*)

A sine-wave grating is designated by its spatial frequency. We are accustomed to the expression of frequencies (for example, of sound waves or radio waves) in terms of time or distance (such as cycles per second

FIGURE 6.23 ■ Orientation-Sensitive Neurons

The response characteristics of neurons to orientation in the primary visual cortex are (a) simple cell, (b) complex cell, and (c) hypercomplex cell.

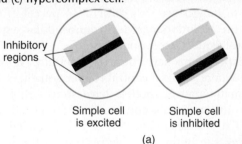

Inhibitory regions

Simple cell   Simple cell
is excited   is inhibited

(a)

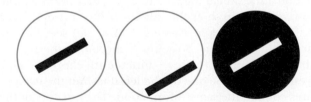

Complex cell is excited by all three stimuli

(b)

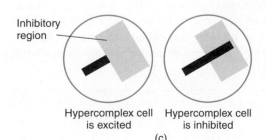

Inhibitory region

Hypercomplex cell   Hypercomplex cell
is excited   is inhibited

(c)

simple cell An orientation-sensitive neuron in the striate cortex whose receptive field is organized in an opponent fashion.

complex cell A neuron in the visual cortex that responds to the presence of a line segment with a particular orientation located within its receptive field, especially when the line moves perpendicularly to its orientation.

hypercomplex cell A neuron in the visual cortex that responds to the presence of a line segment with a particular orientation that ends at a particular point within the cell's receptive field.

sine-wave grating A series of straight parallel bands varying continuously in brightness according to a sine-wave function, along a line perpendicular to their lengths.

FIGURE 6.25 ■ Concepts of Visual Angle and Spatial Frequency

Angles are drawn between the sine waves, with the apex at the viewer's eye. The *visual angle* between adjacent sine waves is smaller when the waves are closer together.

or wavelength in cycles per meter). But because the image of a stimulus on the retina varies in size according to how close it is to the eye, the visual angle is generally used instead of the physical distance between adjacent cycles. Thus, the **spatial frequency** of a sine-wave grating is its variation in brightness measured in cycles per degree of visual angle. (See *Figure 6.25.*)

Most neurons in the striate cortex respond best when a sine-wave grating of a particular spatial frequency is placed in the appropriate part of the visual field. Different neurons detect different spatial frequencies. For orientation-sensitive neurons the grating must be aligned at the appropriate angle of orientation. Albrecht

FIGURE 6.26 ■ The Experiment by Albrecht, 1978

(a) The stimulus is presented to the animal. (b) The graph shows the response of a simple cell in the primary visual cortex.

(Adapted from De Valois, R. L., and De Valois, K. K. *Spatial Vision.* New York: Oxford University Press, 1988.)

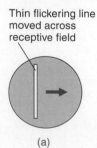

Thin flickering line moved across receptive field

(a)

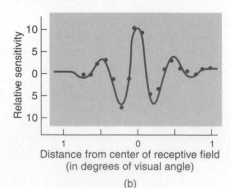

Distance from center of receptive field
(in degrees of visual angle)

(b)

(1978) mapped the shapes of receptive fields of simple cells by observing their response while moving a very thin flickering line of the appropriate orientation through their receptive fields. He found that many of them had multiple inhibitory and excitatory regions surrounding the center. The profile of the excitatory and inhibitory regions of the receptive fields of such neurons looked like a modulated sine wave—precisely what would be needed to detect a few cycles of a sine-wave grating. (See *Figure 6.26.*) In most cases a neuron's receptive field is large enough to include between 1.5 and 3.5 cycles of the grating (De Valois, Thorell, and Albrecht, 1985).

What is the point of having neural circuits that analyze spatial frequency? A complete answer requires some rather complicated mathematics, so I will give a simplified one here. (If you are interested, you can consult a classic book by De Valois and De Valois, 1988.) Consider the types of information provided by high and low spatial frequencies. Small objects, details within a large object, and large objects with sharp edges provide a signal rich in high frequencies, whereas large areas of light and dark are represented by low frequencies. An image that is deficient in high-frequency information looks fuzzy and out of focus, like the image seen by a nearsighted person who is not wearing corrective lenses. This image still provides much information about forms and objects in the environment; thus, the most important visual information is that contained in *low spatial frequencies.* When low-frequency information is removed, the shapes of images are very difficult to perceive. (As we will see, the evolutionary older magnocellular system provides low-frequency information.)

Many experiments have confirmed that the concept of spatial frequency plays a central role in visual perception, and mathematical models have shown that the information present in a scene can be represented very efficiently if it is first encoded in terms of spatial frequency. Thus, the brain probably represents the information in a similar way. Here I will describe just one example to help show the validity of the concept. Look at the two pictures in *Figure 6.27.* You can see that the picture on the right looks much more like the face of Abraham Lincoln, the nineteenth-century U.S. President, than the one on the left does. Yet the two pictures contain the same information. The creators of the pictures, Harmon and Julesz (1973), used a computer to construct the figure on the left, which consists of a series of squares, each representing the average brightness of a portion of a picture of Lincoln. The one on the right is simply a transformation of the first one in which high frequencies have been removed. Sharp edges contain

spatial frequency The relative width of the bands in a sine-wave grating, measured in cycles per degree of visual angle.

FIGURE 6.27 ■ Spatial Filtering

The two pictures contain the same amount of low-frequency information, but extraneous high-frequency information has been filtered from the picture on the right. If you look at the pictures from across the room, they look identical.

(From Harmon, L. D., and Julesz, B. *Science*, 1973, *180*, 1191–1197. Copyright 1973 by the American Association for the Advancement of Science. Reprinted with permission.)

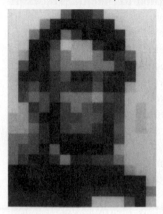

high spatial frequencies, so the transformation eliminates them. In the case of the picture on the left, these frequencies have nothing to do with the information contained in the original picture; thus, they can be seen as visual "noise." The filtration process (accomplished by a computer) removes this noise—and makes the image much clearer to the human visual system. Presumably, the high frequencies produced by the edges of the squares in the left figure stimulate neurons in the striate cortex that are tuned to high spatial frequencies. When the visual association cortex receives this noisy information, it has difficulty perceiving the underlying form.

If you want to watch the effect of filtering the extraneous high-frequency noise, try the following demonstration. Put the book down and look at the pictures in Figure 6.27 from across the room. The distance "erases" the high frequencies, because they exceed the resolving power of the eye, and the two pictures look identical. Now walk toward the book, focusing on the left figure. As you get closer, the higher frequencies reappear, and this picture looks less and less like the face of Lincoln. (See *Figure 6.27*.)

Retinal Disparity

We perceive depth by many means, most of which involve cues that can be detected monocularly, that is, by one eye alone. For example, perspective, relative retinal size, loss of detail through the effects of atmospheric haze, and relative apparent movement of retinal images as we move our heads all contribute to depth perception and do not require binocular vision. However, binocular vision provides a vivid perception of depth through the process of stereoscopic vision, or *stereopsis*. If you have used a stereoscope (such as a View-Master) or have seen a three-dimensional movie, you know what I mean. Stereopsis is particularly important in the visual guidance of fine movements of the hands and fingers, such as we use when we thread a needle.

Most neurons in the striate cortex are *binocular*—that is, they respond to visual stimulation of either eye. Many of these binocular cells, especially those found in a layer that receives information from the magnocellular system, have response patterns that appear to contribute to the perception of depth (Poggio and Poggio, 1984). In most cases the cells respond most vigorously when each eye sees a stimulus in a slightly *different* location. That is, the neurons respond to **retinal disparity,** a stimulus that produces images on slightly different parts of the retina of each eye. This is exactly the information that is needed for stereopsis; each eye sees a three-dimensional scene slightly differently, and the presence of retinal disparity indicates differences in the distance of objects from the observer.

Color

In the striate cortex, information from color-sensitive ganglion cells is transmitted, through the parvocellular and koniocellular layers of the LGN, to special cells grouped together in **cytochrome oxidase (CO) blobs.** CO blobs were discovered by Wong-Riley (1978), who found that a stain for cytochrome oxidase, an enzyme that is present in mitochondria, showed a patchy distribution. (The presence of high levels of cytochrome oxidase in a cell indicates that the cell normally has a high rate of metabolism.) Subsequent research with the stain (Horton and Hubel, 1980; Humphrey and Hendrickson, 1980) revealed the presence of a polka-dot pattern of dark columns extending through layers 2 and 3 and (more faintly) layers 5 and 6. The columns are oval in cross section, approximately 150 × 200 μm in diameter, and spaced at 0.5-mm intervals (Fitzpatrick, Itoh, and Diamond, 1983; Livingstone and Hubel, 1987).

Figure 6.28 shows a photomicrograph of a slice through the striate cortex (also called V1 because it is the first area of visual cortex) and an adjacent area of visual association cortex (area V2) of a macaque monkey. The visual cortex has been flattened out and stained

retinal disparity The fact that points on objects located at different distances from the observer will fall on slightly different locations on the two retinas; provides the basis for stereopsis.

cytochrome oxidase (CO) blob The central region of a module of the primary visual cortex, revealed by a stain for cytochrome oxidase; contains wavelength-sensitive neurons; part of the parvocellular system.

FIGURE 6.28 ■ Blobs and Stripes in Visual Cortex

A photomicrograph (actually, a montage of several different tissue sections) shows a slice through the primary visual cortex (area V1) and a region of visual association cortex (V2) of a macaque monkey, stained for cytochrome oxidase. Area V1 shows spots ("blobs"), and area V2 shows three types of stripes: thick, thin (both dark), and pale.

(Reprinted from Sincich, L. C., and Horton, J. C. *Annual Review of Neuroscience*, Volume 28 © 2005, 303–326 by Annual Reviews www.annualreviews.org)

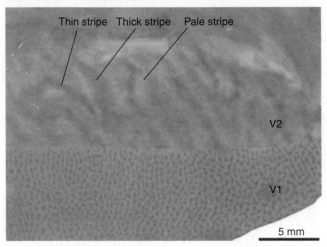

Thin stripe   Thick stripe   Pale stripe

V2

V1

5 mm

for the mitochondrial enzyme. You can clearly see the CO blobs within the striate cortex. The distribution of CO-rich neurons in area V2 consists of three kinds of stripes: *thick stripes, thin stripes,* and *pale stripes.* The thick and thin stripes stain heavily for cytochrome oxidase; the pale stripes do not. (See *Figure 6.28.*)

Until recently, researchers believed that the parvocellular system transmitted all information pertaining to color to the striate cortex. However, we now know that the parvocellular system receives information only from "red" and "green" cones; additional information from "blue" cones is transmitted through the koniocellular system (Hendry and Yoshioka, 1994; Chatterjee and Callaway, 2003).

To summarize, neurons in the striate cortex respond to several different features of a visual stimulus, including orientation, movement, spatial frequency, retinal disparity, and color. Now let us turn our attention to the way in which this information is organized within the striate cortex.

Modular Organization of the Striate Cortex

Most investigators believe that the brain is organized in modules, which probably range in size from a hundred thousand to a few million neurons. Each module

receives information from other modules, performs some calculations, and then passes the results to other modules. In recent years, investigators have been learning the characteristics of the modules that are found in the visual cortex.

The striate cortex is divided into approximately 2500 modules, each approximately 0.5 × 0.7 mm and containing approximately 150,000 neurons. The neurons in each module are devoted to the analysis of various features contained in one very small portion of the visual field. Collectively, these modules receive information from the entire visual field, the individual modules serving like the tiles in a mosaic mural. Input from the parvocellular, koniocellular, and magnocellular layers of the LGN is received by different sublayers of the striate cortex: The parvocellular input is received by layer 4Cβ, the magnocellular input is received by layer 4Cα, and the koniocellular input is received by layer 4A.

The modules actually consist of two segments, each surrounding a CO blob. Neurons located within the blobs have a special function: Most of them are sensitive to color, and all of them are sensitive to low spatial frequencies. They are relatively insensitive to other visual features: They do not respond selectively to different orientations and have relatively large receptive fields, which means that they do not provide information useful for form perception. In addition, their receptive fields are monocular—they receive visual information from only one eye (Kaas and Collins, 2001; Landisman and Ts'o, 2002).

Outside the CO blob, neurons show sensitivity to orientation, movement, spatial frequency, and binocular disparity, but most do not respond to color (Livingstone and Hubel, 1984; Born and Tootell, 1991; Edwards, Purpura, and Kaplan, 1995). Each half of the module receives input from only one eye, but the circuitry within the module combines the information from both eyes, which means that most of the neurons are binocular. Depending on their locations within the module, neurons receive varying percentages of input from each of the eyes.

If we record from neurons anywhere within a single module, we will find that their receptive fields overlap. Thus, all the neurons in a module analyze information from the same region of the visual field. Furthermore, if we insert a microelectrode straight down into an interblob region of the striate cortex (that is, in a location in a module outside one of the CO blobs), we will find both simple and complex cells, but all of the orientation-sensitive cells will respond to lines of the same orientation. In addition, they will all have the same **ocular dominance**—that is, the same percentage of input

ocular dominance The extent to which a particular neuron receives more input from one eye than from the other.

FIGURE 6.29 ■ **One Module of Primary Visual Cortex**

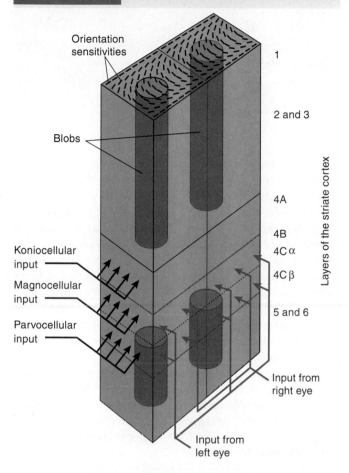

FIGURE 6.30 ■ **Organization of Spatial Frequency**

Optimal spatial frequency of neurons in striate cortex is shown as a function of the distance of the neuron from the center of the nearest cytochrome oxidase blob.

(Adapted from Edwards, D. P., Purpura, K. P., and Kaplan, E. *Vision Research*, 1995, *35*, 1501–1523.)

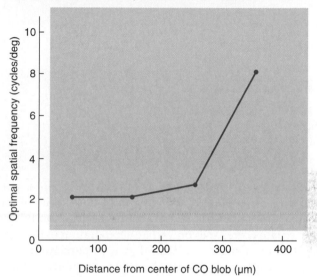

from each of the eyes. If we move our electrode around the module, we will find that these two characteristics—orientation sensitivity and ocular dominance—vary systematically and are arranged at right angles to each other. (See *Figure 6.29*.)

How does spatial frequency fit into this organization? Edwards, Purpura, and Kaplan (1995) found that neurons within the CO blobs responded to low spatial frequencies but were sensitive to small differences in brightness. Outside the blobs, sensitivity to spatial frequency varied with the distance from the center of the nearest blob. Higher frequencies were associated with greater distances. (See *Figure 6.30*.) However, neurons outside the blobs were less sensitive to contrast; the difference between the bright and dark areas of the sine-wave grating had to be greater for these neurons than for neurons within the blobs.

InterimSummary

Analysis of Visual Information: Role of the Striate Cortex

The striate cortex (area V1) consists of six layers and several sublayers. Visual information is received from the magnocellular, parvocellular, and koniocellular layers of the dorsal lateral geniculate nucleus (LGN). Information from V1 is sent to area V2, the first region of the visual association cortex. The magnocellular system is phylogenetically older, color-blind, and sensitive to movement, depth, and small differences in brightness. The parvocellular and koniocellular systems are of more recent origin. The parvocellular system receives information from "red" and "green" cones and is able to discriminate finer details. The koniocellular system provides additional information about color, received from "blue" cones.

The striate cortex (area V1) is organized into modules, each surrounding a pair of CO blobs, which are revealed by a stain for cytochrome oxidase, an enzyme found in mitochondria. Each half of a module receives information from one eye; but because information is shared, most of the neurons respond to information from both eyes. The neurons in the CO blobs are sensitive to color and to low-frequency sine-wave gratings, whereas those between the blobs are sensitive to sine-wave gratings of higher spatial frequencies, orientation, retinal disparity, and movement.

Thought Question

Look at the scene in front of you, and try to imagine how its features are encoded by neurons in your striate cortex. Try to picture how the objects you see can be specified by an analysis of orientation, spatial frequency, and color.

ANALYSIS OF VISUAL INFORMATION: ROLE OF THE VISUAL ASSOCIATION CORTEX

Although the striate cortex is necessary for visual perception, perception of objects and of the totality of the visual scene does not take place there. Each module of the striate cortex sees only what is happening in one tiny part of the visual field. Thus, for us to perceive objects and entire visual scenes, the information from these individual modules must be combined. That combination takes place in the visual association cortex.

Two Streams of Visual Analysis

Visual information received from the striate cortex is analyzed in the visual association cortex. Neurons in the striate cortex send axons to the **extrastriate cortex,** the region of the visual association cortex that surrounds the striate cortex. The primate extrastriate cortex consists of several regions, each of which contains one or more independent maps of the visual field. Each region is specialized, containing neurons that respond to particular features of visual information, such as orientation, movement, spatial frequency, retinal disparity, or color. So far, investigators have identified over two dozen distinct regions and subregions of the visual cortex of the rhesus monkey. These regions are arranged hierarchically, beginning with the striate cortex (Grill-Spector and Malach, 2004; Wandell, Dumoulin, and Brewer, 2007). Most of the information passes up the hierarchy; each region receives information from regions located beneath it in the hierarchy (closer to the striate cortex), analyzes the information, and passes the results on to "higher" regions for further analysis. Some information is also transmitted in the opposite direction, but axons that descend the hierarchy are much less numerous than those that ascend it.

The results of a functional-imaging study by Murray, Boyaci, and Kersten (2006) demonstrate a phenomenon

that owes its existence to information that follows pathways that travel up the hierarchy, from regions of the visual association cortex back to the striate cortex. First, try the following demonstration. Stare at an object (for example, an illuminated light bulb) that has enough contrast with the background to produce an afterimage. Then look at a nearby surface, such as the back of your hand. Before the afterimage fades away, look at a more distant surface, such as the far wall of the room (assuming that you are indoors). You will see that the afterimage looks much larger when it is seen against a distant background. The investigators presented subjects with stimuli like those shown in Figure 6.31: spheres positioned against a background in locations that made them look closer to or farther from the observer. Although the spheres were actually the same size, their location on the background made the one that was apparently farther away look larger than the other one. (See *Figure 6.31*.)

Murray and his colleagues used functional MRI (fMRI) to record activation of the striate cortex while the subjects looked at the spheres. They found that looking at the sphere that appeared to be larger activated a larger area of the striate cortex. We know that perception of apparent distance in a background like that shown in Figure 6.31 cannot take place in the striate cortex but requires neural circuitry found in the visual association cortex. This fact means that computations made in higher levels of the visual system can act back on the striate cortex and modify the activity taking place there.

Figure 6.32 shows the location of the striate cortex and several regions in the extrastriate cortex of the human brain. The views of brain in Figures 6.32(a) and 6.32(b) are nearly normal in appearance. Figures 6.32(c) and 6.32(d) show "inflated" cortical surfaces, enabling us to see regions that are normally hidden in the depths of sulci and fissures. The hidden regions are shown in dark gray, while regions that are normally

---

extrastriate cortex A region of visual association cortex; receives fibers from the striate cortex and from the superior colliculi and projects to the inferior temporal cortex.

FIGURE 6.31 ■ **Display Used by Murray, Boyaci, and Kersten (2006)**

The ball that appears to be farther away looks larger than the closer one, even though the images they cast on the retina are exactly the same size.

(From Sterzer, P., and Rees, G. *Nature Neuroscience*, 2006, *9*, 302–304. Reprinted with permission.)

visible (the surfaces of gyri) are shown in light gray. Figure 6.32(e) shows an unrolling of the cortical surface caudal to the dotted red line and green lines in Figure 6.32(c) and 6.32(d). (See *Figure 6.32.*)

The outputs of the striate cortex (area V1) are sent to area V2, a region of the extrastriate cortex just adjacent to V1. As we saw in Figure 6.28, a dye for cytochrome oxidase reveals blobs in V1 and three kinds of stripes in V2. Neurons in V1 blobs project to thin stripes, and neurons outside the blobs in V1 project to thick stripes and pale stripes. Thus, neurons in the thin stripes of V2 receive information concerning color, and those in the thick stripes and pale stripes receive information about orientation, spatial frequency, movement, and retinal disparity. (See *Figure 6.33.*)

The receptive fields of neurons in V2 are several times larger in diameter than those of neurons in the striate cortex, which suggests that V2 cells receive input from several V1 cells. Approximately 70 percent of orientation-sensitive neurons in V2 encode the presence of stimuli with same orientation throughout their receptive field. However, a significant minority respond to stimuli with one orientation in part of the receptive field and to those with a different orientation in the rest of the field (Anzai, Peng, and Van Essen, 2007). Presumably, these cells are able to recognize elements of more complex stimuli, such as their corners.

FIGURE 6.32 ■ **Striate Cortex and Regions of Extrastriate Cortex**

These views of a human brain show (a) a nearly normal lateral view, (b) a nearly normal midsagittal view, (c) an "inflated" lateral view, (d) an "inflated" midsagittal view, and (e) an unrolling of the cortical surface caudal to the dotted red line and green lines shown in (c) and (d).

(From Tootell, B. H., and Hadjikhani, N. *Cerebral Cortex*, 2001, *11*, 298–311. Reprinted with permission.)

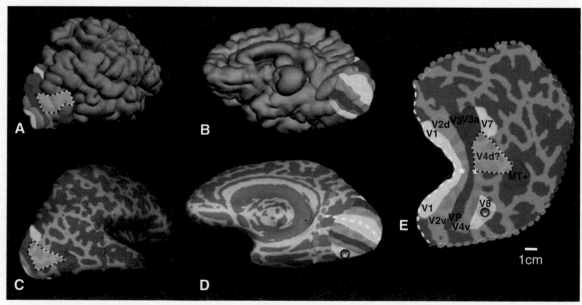

FIGURE 6.33 ■ Connections Between Areas V1 and V2

(Adapted from Sincich and Horton, *Annual Review of Neuroscience*, 2005, *28*, 303–326.)

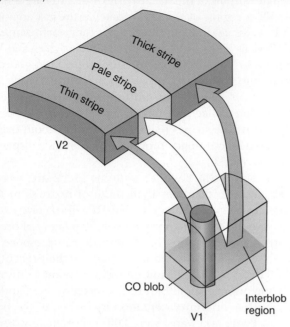

FIGURE 6.34 ■ Human Visual System

The figure shows the human visual system from the eye to the two streams of the visual association cortex.

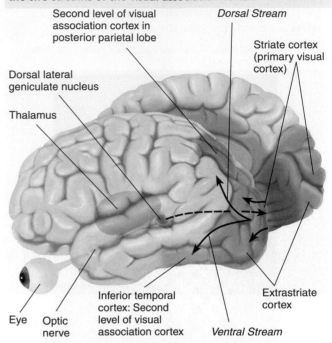

At this point, the visual association cortex divides into two pathways. On the basis of their own research and a review of the literature, Ungerleider and Mishkin (1982) concluded that the visual association cortex contains two streams of analysis: the **dorsal stream** and the **ventral stream.** Subsequent anatomical studies have confirmed this conclusion (Baizer, Ungerleider, and Desimone, 1991). One stream continues forward toward a series of regions that constitute the ventral stream, terminating in the **inferior temporal cortex.** The other stream ascends into regions of the dorsal stream, terminating in the **posterior parietal cortex.** Some axons conveying information received from the magnocellular system bypass area V2: They project from area V1 directly to area V5, a region of the extrastriate cortex devoted to the analysis of movement. The ventral stream recognizes *what* an object is and what color it has, and the dorsal stream recognizes *where* the object is located and, if it is moving, its speed and direction of movement. (See *Figure 6.34.*)

As we saw, the parvocellular, koniocellular, and magnocellular systems provide different kinds of information. The magnocellular system is found in all mammals, whereas the parvocellular and koniocellular systems are found only in primates. These systems receive information from different types of ganglion cells, which are connected to different types of bipolar cells and photoreceptors. Only the cells in the parvocellular and koniocellular system receive information about wavelength from cones; thus, these systems provide information concerning

color. Cells in the parvocellular system also show high spatial resolution and low temporal resolution; that is, they are able to detect very fine details, but their response is slow and prolonged. The koniocellular system, which receives information only from "blue" cones, which are much less numerous than "red" and "green" cones, does not provide information about fine details. In contrast, neurons in the magnocellular system are color-blind. They are not able to detect fine details, but they can detect smaller contrasts between light and dark. They are also especially sensitive to movement. (See *Table 6.2.*) The dorsal stream receives mostly magnocellular input, but the ventral stream receives approximately equal input from the magnocellular and the parvocellular/koniocellular systems.

dorsal stream A system of interconnected regions of visual cortex involved in the perception of spatial location, beginning with the striate cortex and ending with the posterior parietal cortex.

ventral stream A system of interconnected regions of visual cortex involved in the perception of form, beginning with the striate cortex and ending with the inferior temporal cortex.

inferior temporal cortex The highest level of the ventral stream of the visual association cortex; involved in perception of objects, including people's bodies and faces.

posterior parietal cortex The highest level of the dorsal stream of the visual association cortex; involved in perception of movement and spatial location.

TABLE 6.2 ■ Properties of the Magnocellular, Parvocellular, and Koniocellular Divisions of the Visual System			
PROPERTY	MAGNOCELLULAR DIVISION	PARVOCELLULAR DIVISION	KONIOCELLULAR DIVISION
Color	No	Yes (from "red" and "green" cones)	Yes (from "blue" cones)
Sensitivity to contrast	High	Low	Low
Spatial resolution (ability to detect fine details)	Low	High	Low
Temporal resolution	Fast (transient response)	Slow (sustained response)	Slow (sustained response)

Many neurons throughout almost all regions of the visual cortex are responsive to binocular disparity, which, as we saw earlier, serves as the basis for stereoscopic depth perception (Parker, 2007; Roe et al., 2007). The disparity-sensitive neurons found in the dorsal stream, which is involved in spatial perception, respond to large, extended visual surfaces, whereas those found in the ventral stream, which is involved in object perception, respond to the contours of three-dimensional objects.

Perception of Color

As we saw earlier, neurons within the CO blobs in the striate cortex respond differentially to colors. Like the ganglion cells in the retina (and the parvocellular and koniocellular neurons in the dorsal lateral geniculate nucleus), these neurons respond in opponent fashion. This information is analyzed by the regions of the visual association cortex that constitute the ventral stream.

Studies with Laboratory Animals

In the monkey brain, color-sensitive neurons in the CO blobs of the striate cortex send color-related information to the thin stripes in area V2. Neurons in V2 send information to an adjacent region of the extrastriate cortex called V4. Zeki (1980) found that neurons this region respond selectively to colors, but their response characteristics are much more complex than those of neurons in V1 or V2. Unlike the neurons we have encountered so far, these neurons respond to a variety of wavelengths, not just the wavelengths that correspond to red, green, yellow, and blue.

The appearance of the colors of objects remains much the same whether we observe them under artificial light, under an overcast sky, or at noon on a cloudless day. This phenomenon is known as **color constancy.** Our visual system does not simply respond according to the wavelength of the light reflected by objects in each

part of the visual field; instead, it compensates for the source of the light. This compensation appears to be made by simultaneously comparing the color composition of each point in the visual field with the average color of the entire scene. If the scene contains a particularly high level of long-wavelength light (as it would if an object were illuminated by the light of a setting sun), then some long-wavelength light is "subtracted out" of the perception of each point in the scene. This compensation helps us to see what is actually out there.

Schein and Desimone (1990) performed a careful study of the response characteristics of neurons in area V4 of the monkey extrastriate cortex. They found that these neurons responded to specific colors. Some also responded to colored bars of specific orientation; thus, area V4 seems to be involved in the analysis of form as well as color. The color-sensitive neurons had a rather unusual secondary receptive field: a large region surrounding the primary field. When stimuli were presented in the secondary receptive field, the neuron did not respond. However, stimuli presented there could suppress the neuron's response to a stimulus presented in the primary field. For example, if a cell would fire when a red spot was presented in the primary field, it would fire at a slower rate (or not at all) when an additional red stimulus was presented in the surrounding secondary field. In other words, these cells responded to particular wavelengths of light but subtracted out the amount of that wavelength that was present in the background. As Schein and Desimone point out, this subtraction could serve as the basis for color constancy.

Walsh et al. (1993) confirmed this prediction; damage to area V4 does disrupt color constancy. The investigators found that although monkeys could still

color constancy The relatively constant appearance of the colors of objects viewed under varying lighting conditions.

discriminate between different colors after area V4 had been damaged, their performance was impaired when the color of the overall illumination was changed. But the fact that the monkeys could still perform a color discrimination task under constant illumination means that some region besides area V4 must be involved in color vision. *MyPsychKit 6.2, Color Constancy,* illustrates the effects of the color of overall illumination on color perception.

Animation 6.2
Color Constancy

A study by Heywood, Gaffan, and Cowey (1995) suggested that a portion of the inferior temporal cortex just anterior to area V4—a region of the monkey brain that is usually referred to as area TEO—plays a critical role in visual discrimination. The investigators destroyed area TEO, leaving area V4 intact, and observed severe impairment in color discrimination. The monkeys had no difficulty discriminating shades of gray, so the deficit was restricted to impaired color perception.

A more recent study by Conway, Moeller, and Tsao (2007) performed a detailed analysis of the responsiveness of neurons in a large region of the visual association cortex in monkeys, including areas V4 and TEO. Using fMRI, the investigators identified color "hot spots"—small scattered regions that were strongly activated by changes in the color of visual stimuli. Next, they recorded the response characteristics of neurons inside and outside these spots, which they called *globs*. (I'm sure the similarity between the terms "blobs" and "globs" was intentional.) They found that glob neurons were indeed responsive to colors but also had some weak sensitivity to shapes. In contrast, interglob neurons (those located outside globs) did not respond to colors but were strongly selective to shape. Thus, within a large region of visual association cortex, patches of neurons were strongly sensitive to colors or to shape but not to both. The fact that color-sensitive globs are spread across a wide area of visual association cortex probably explains why only rather large brain lesions case severe disruptions in perception of color.

Studies with Humans

Lesions of a restricted region of the human extrastriate cortex can cause loss of color vision without disruption of visual acuity. The patients describe their vision as resembling a black-and-white film. In addition, they cannot even imagine colors or remember the colors of objects they saw before their brain damage occurred (Damasio et al., 1980; Heywood and Kentridge, 2003). The condition is known as **cerebral achromatopsia** ("vision without color"). If the brain damage is unilateral, people will lose color vision in only half of the visual field.

As we just saw, Heywood, Gaffan, and Cowey (1995) found a region of the inferior temporal cortex of the monkey brain whose damage disrupted the ability to make color discriminations. The analogous region appears to play a critical role in color perception in humans. An fMRI study by Hadjikhani et al. (1998) found a color-sensitive region that included the lingual and fusiform gyri, in a location corresponding to area TEO in the monkey's cortex, which they called area V8. An analysis of 92 cases of achromatopsia by Bouvier and Engel (2006) confirmed that damage to this region (which is adjacent to and partly overlaps the *fusiform face area*, discussed later in this chapter) disrupts color vision. (Refer to *Figure 6.32*.)

The function of our ability to perceive different colors is to help us perceive different objects in our environment. Thus, to perceive and understand what is in front of us, we must have information about color combined with other forms of information. Some people with brain damage lose the ability to perceive shapes but can still perceive colors. For example, Zeki et al. (1999) described a patient who could identify colors but was otherwise blind. Patient P. B. received an electrical shock that caused both cardiac and respiratory arrest. He was revived, but the period of anoxia caused extensive damage to his extrastriate cortex. As a result, he lost all form perception. However, even though he could not recognize objects presented on a video monitor, he could still identify their colors.

Perception of Form

The analysis of visual information that leads to the perception of form begins with neurons in the striate cortex that are sensitive to orientation and spatial frequency. These neurons send information to area V2 and then on to the subregions of the visual association cortex and constitute the ventral stream.

Studies with Laboratory Animals

In primates the recognition of visual patterns and identification of particular objects take place in the inferior temporal cortex, located on the ventral part of the temporal lobe. This region of visual association cortex is located at the end of the ventral stream. It is here that analyses of form and color are put together, and perceptions of three-dimensional objects and backgrounds are achieved. The inferior temporal cortex consists of two major regions: a posterior area (TEO) and an anterior area (TE). Damage to these regions causes severe deficits in visual discrimination (Mishkin, 1966; Gross, 1973; Dean, 1976).

cerebral achromatopsia (*ay krohm a **top** see a*) Inability to discriminate among different hues; caused by damage to area V8 of the visual association cortex.

As we saw earlier, the analysis of visual information is hierarchical: Area V1 is concerned with the analysis of elementary aspects of information in very small regions of the visual field, and successive regions analyze more complex characteristics. The size of the receptive fields also grows as the hierarchy is ascended. The receptive fields of neurons in area TEO are larger than those in area V4, and the receptive fields of neurons in area TE are the largest of all, often encompassing the entire contralateral half of the visual field (Boussaoud, Desimone, and Ungerleider, 1991). In general, these neurons respond best to three-dimensional objects (or photographs of them). They respond poorly to simple stimuli such as spots, lines, or sine-wave gratings. Most of them continue to respond even when complex stimuli are moved to different locations, are changed in size, are placed against a different background, or are partially occluded by other objects (Rolls and Baylis, 1986; Kovács, Vogels, and Orban, 1995). Thus, they appear to participate in the recognition of objects rather than the analysis of specific features.

The fact that neurons in the primate inferior temporal cortex respond to very specific complex shapes suggests that the development of the circuits responsible for detecting them must involve learning. Indeed, that seems to be the case. For example, several studies have found neurons in the inferior temporal cortex that respond specifically to objects that the monkeys have already seen many times but not to unfamiliar objects (Kobatake, Tanaka, and Tamori, 1992; Logothetis, Pauls, and Poggio, 1995; Baker, Behrman, and Olson, 2002). The role of the inferior temporal cortex in learning will be discussed in more detail in Chapter 13.

Studies with Humans

Study of people who have sustained brain damage to the visual association cortex has told us much about the organization of the human visual system. In recent years our knowledge has been greatly expanded by functional-imaging studies.

Visual Agnosia.

Damage to the human visual association cortex can cause a category of deficits known as **visual agnosia.** *Agnosia* ("failure to know") refers to an inability to perceive or identify a stimulus by means of a particular sensory modality, even though its details can be detected by means of that modality and the person retains relatively normal intellectual capacity.

Mrs. R., whose case was described in the opening of this chapter, had visual agnosia caused by damage to the ventral stream of her visual association cortex. As we saw, she could not identify common objects by sight, even though she had relatively normal visual acuity. However, she could still read, even small print, which indicates that reading involves different brain regions than object

perception does. (Chapter 14 discusses research that has identified brain regions involved in visual recognition of letters and words.) When she was permitted to hold an object that she could not recognize visually, she could immediately recognize it by touch and say what it is, which proves that she had not lost her memory for the object or simply forgotten how to say its name.

Analysis of Specific Categories of Visual Stimuli.

Visual agnosia is caused by damage to the parts of the visual association cortex that contribute to the ventral stream. In fact, damage to specific regions of the ventral stream can impair the ability to recognize specific categories of visual stimuli. Of course, even if specific regions of the visual association cortex are involved in analyzing specific categories of stimuli, the boundaries of brain lesions will seldom coincide with the boundaries of brain regions with particular functions.

With the advent of functional imaging, investigators have studied the responses of the normal human brain and have discovered several regions of the ventral stream that are activated by the sight of particular categories of visual stimuli. For example, researchers have identified regions of the inferior temporal and lateral occipital cortex that are specifically activated by categories such as animals, tools, cars, flowers, letters and letter strings, faces, bodies, and scenes. (See Tootell, Tsao, and Vanduffel, 2003, and Grill-Spector and Malach, 2004, for a review.) However, not all of these findings have been replicated, and, of course, general-purpose regions contain circuits that can learn to recognize shapes that do not fall into these categories. A relatively large region of the ventral stream of the visual association cortex, the **lateral occipital complex (LOC),** appears to respond to a wide variety of objects and shapes.

A functional-imaging study by Downing et al. (2006) suggests that there are few regions of the visual association cortex devoted to the analysis of specific categories of stimuli. The investigators presented images of objects from 19 different categories to normal subjects and found only three regions that showed the greatest activation to the sight of specific categories: faces, bodies, and scenes.

A common symptom of visual agnosia is **prosopagnosia,** inability to recognize particular faces (*prosopon* is Greek for "face"). That is, patients with this disorder can recognize that they are looking at a face,

visual agnosia (*ag no zha*) Deficits in visual perception in the absence of blindness; caused by brain damage.

lateral occipital complex (LOC) A region of the extrastriate cortex, involved in perception of objects other than people's bodies and faces.

prosopagnosia (*prah soh pag no zha*) Failure to recognize particular people by the sight of their faces.

but they cannot say whose face it is—even if it belongs to a relative or close friend. They see eyes, ears, a nose, and a mouth, but they cannot recognize the particular configuration of these features that identifies an individual face. They still remember who these people are and will usually recognize them when they hear the person's voice. As one patient said, "I have trouble recognizing people from just faces alone. I look at their hair color, listen to their voices . . . I use clothing, voice, and hair. I try to associate something with a person one way or another . . . what they wear, how their hair is worn" (Buxbaum, Glosser, and Coslett, 1999, p. 43).

Studies with brain-damaged people and functional-imaging studies suggest that these special face-recognizing circuits are found in the **fusiform face area (FFA),** located in the fusiform gyrus on the base of the temporal lobe. For example, Grill-Spector, Knouf, and Kanwisher (2004) obtained fMRI scans of the brains of people who looked at pictures of faces and several other categories of objects. Figure 6.35 shows the results, projected on an "inflated" ventral view of the cerebral cortex. The black outlines show the regions of the fusiform cortex that were activated by viewing faces, drawn on all images of the brain for comparison with the activation produced by other categories of objects. As you can see,

FIGURE 6.35 ■ Visual Stimuli

These functional MRI scans are of people looking at six categories of visual stimuli. Neural activity is shown on "inflated" ventral views of the cerebral cortex. The fusiform face area is shown as a black outline, derived from the responses to faces shown in the upper left scan.

(From Grill-Spector, K., Knouf, N., and Kanwisher, N. *Nature Neuroscience,* 2004, 7, 555–561. Reprinted with permission.)

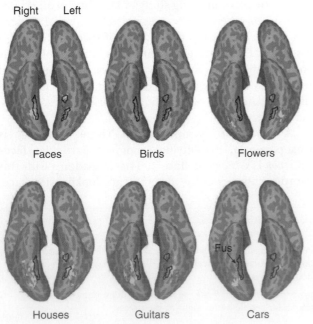

FIGURE 6.36 ■ Visual Object Agnosia Without Prosopagnosia

A patient could recognize the face in this painting but not the flowers, fruits, and vegetables that compose it.

(Giuseppe Arcimboldo. 1527–1593. *Vertumnus.* Erich Lessing/Art Resource, New York.)

images of faces activated the regions indicated by these outlines better than other categories of visual stimuli. (See *Figure 6.35.*)

Perhaps the strangest piece of evidence for a special face-recognition region comes from a report by Moscovitch, Winocur, and Behrmann (1997), who studied a man with a visual agnosia for objects but not for faces. For example, he recognized the face shown in Figure 6.36 but not the flowers, fruits, and vegetables that compose it. (See *Figure 6.36.*) Presumably, some regions of his visual association cortex were damaged, but the fusiform face region was not.

A functional-imaging study by Cox, Meyers, and Sinha (2004) found that visual cues correlated with faces can activate the fusiform face area. They found

fusiform face area (FFA) A region of the visual association cortex located in the inferior temporal lobe; involved in perception of faces and other complex objects that require expertise to recognize.

FIGURE 6.37 ■ Implied Faces

The fusiform face area was activated by actual faces (e) and by a blurry gray shape in the appropriate position that implied the presence of a face (a).

(From Cox, D., Meyers, E., and Sinha, P. *Science*, 2004, *304*, 115–117. Copyright © 2004 American Association for the Advancement of Science. Reprinted with permission.)

FIGURE 6.38 ■ Perception of Faces and Bodies

The fusiform face area (FFA) and extrastriate body area (EBA) were activated by images of faces, headless bodies, body parts, and assorted objects.

(Adapted from Schwarzlose, R. F., Baker, C. I., and Kanwisher, N. *Journal of Neuroscience*, 2005, *23*, 11055–11059.)

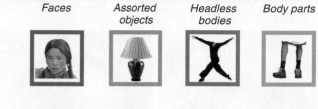
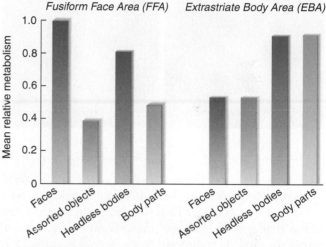

that photographs that implied the presence of a face (a blurry gray shape above a photograph of a man's torso) activated the FFA even though no facial features were present. This finding suggests not simply that the FFA is prewired to recognize facial features, but also that the activity of this region can be affected by previously learned information. (See *Figure 6.37.*)

Another interesting region of the ventral stream is the **extrastriate body area (EBA),** which is just posterior to the FFA and partly overlaps it. Downing et al. (2001) found that this region was specifically activated by photographs, silhouettes, or stick drawings of human bodies or body parts and not by control stimuli such as photographs or drawings of tools, scrambled silhouettes, or scrambled stick drawings of human bodies. Figure 6.38 shows the magnitude of the fMRI response in the nonoverlapping regions of the FFA and EBA to several categories of stimuli (Schwarzlose, Baker, and Kanwisher, 2005). As you can see, the FFA responded to faces more than any of the other categories, and the EBA showed the greatest response to headless bodies and body parts. (See *Figure 6.38.*)

Urgesi, Berlucchi, and Aglioti (2004) used transcranial magnetic stimulation to temporarily disrupt the normal neural activity of the EBA. (As we saw in Chapter 5, the TMS procedure applies a strong localized magnetic field to the brain by passing an electrical current through a coil of wire placed on the scalp.) The investigators

found that the disruption temporarily impaired people's ability to recognize photographs of body parts, but not parts of faces or motorcycles.

As we will see in Chapter 14, the hippocampus and nearby regions of the medial temporal cortex are involved in spatial perception and memory. Several studies have identified a **parahippocampal place area (PPA),** located in a region of limbic cortex bordering the ventromedial temporal lobe, that is activated by the sight of scenes and backgrounds. For example, Steeves et al. (2004) studied Patient D. F., a 47-year-old woman who had sustained brain damage caused by accidental carbon monoxide poisoning 14 years earlier. Bilateral damage to her lateral occipital cortex (an important part of the ventral stream) caused a profound visual agnosia for objects. However, she was able to recognize both natural

extrastriate body area (EBA) A region of the visual association cortex located in the lateral occipitotemporal cortex; involved in perception of the human body and body parts other than faces.

parahippocampal place area (PPA) A region of limbic cortex on the medial temporal lobe; involved in perception of particular places ("scenes").

FIGURE 6.39 ■ **Parahippocampal Place Area**

The scans show the activation of the parahippocampal cortex in (a) Patient D. F., a woman with a profound visual agnosia for objects, in response to viewing scenes and (b) similar responses in a control subject.

(From Steeves, J. K. E., Humphrey, G. K., Culham, J. C., Menon, R. A., Milner, A. D., and Goodale, M. A. *Journal of Cognitive Neuroscience*, 2004, *16*, 955–965. Reprinted by permission.)

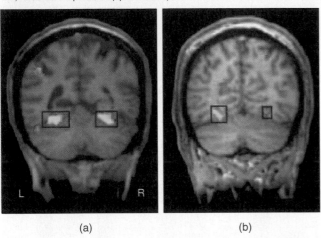

(a)   (b)

FIGURE 6.40 ■ **Greebles**

"Greebles" are computer-created objects from the study by Gauthier and Tarr (1997). They were categorized by family and gender, and different individuals each had their own particular shapes. Two greebles of the same gender and family would resemble each other more closely than any other two greebles.

(From Gauthier, I., and Tarr, M. J. *Vision Research*, 1997, *37*, 1673–1682. Copyright © 1997. Reprinted with permission of Elsevier Science.)

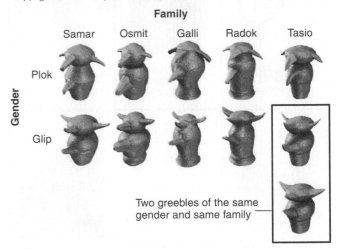

Two greebles of the same gender and same family

and human-made scenes (beaches, forests, deserts, cities, markets, and rooms). Functional imaging showed activation of her intact PPA. These results suggest that scene recognition does not depend on recognition of particular objects found within the scene, because D. F. was incapable of recognizing these objects. Figure 6.39 shows the activation in her brain and that of a control subject. (See *Figure 6.39.*)

Are Faces Special?   As we just saw, the ability to recognize faces by sight depends on a specific region of the fusiform gyrus. But must we conclude that the development of this region is a result of natural selection, and that the FFA comes prewired with circuits devoted to the analysis of faces? Several kinds of evidence suggest that the answer is no—that the face-recognition circuits develop as a result of the experience we have of seeing people's faces. Because of the extensive experience we have of looking at faces, we are all experts at recognizing them.

What about people who have become experts at recognizing other types of objects? It appears that recognition of specific complex stimuli by experts, too, is disrupted by lesions that cause prosopagnosia: inability of a farmer to recognize his cows, inability of a bird expert to recognize different species of birds, and inability of a driver to recognize his own car except by reading its license plate (Bornstein, Stroka, and Munitz, 1969; Damasio, Damasio, and Van Hoesen, 1982). Two functional-imaging studies (Gauthier et al., 2000; Xu, 2005) found that when bird or

car experts (but not nonexperts) viewed pictures of birds or cars, the fusiform face area was activated. Another study (Gauthier et al., 1999) found that when people had spent a long time becoming familiar with computer-generated objects they called "greebles," viewing the greebles activated the fusiform face area. (See *Figure 6.40.*) Tarr and Gauthier (2000) suggested we should relabel the FFA as the *flexible* fusiform area.

A functional-imaging study (Golby et al., 2001) found higher activation of the fusiform face area when people viewed pictures of faces of members of their own race (African Americans or European Americans). Indeed, the subjects in this study were able to recognize faces of people of their own race more accurately than faces of people of the other race. Presumably, this difference reflected the fact that people have more experience of seeing other members of their own race, which indicates that expertise does appear to play a role in face recognition.

There is no doubt that a region of the fusiform gyrus plays an essential role in the analysis of particular faces. In fact, a face-responsive area exists in a similar location in the monkey brain, and this area contains neurons that respond to the faces of both monkeys and humans (Tsao et al., 2006). Two issues are still disputed by investigators interested in the FFA. First, is analysis of faces the sole function of this region, or is it really a "flexible fusiform area" involved in visual analysis of categories of very similar stimuli that can be discriminated only by experts?

The activation of the FFA by greebles in the brains of greeble experts suggests that the FFA is an expertise area rather than an exclusively face area. However, according to Kanwisher and Yovel (2006), "since Greebles resemble faces (and/or bodies), they are a poor choice of stimulus to distinguish between the face-specificity and expertise hypotheses" (p. 2113). One study using high-resolution fMRI found evidence for small patches of the FFA that contained neurons that responded to nonface objects such as cars and animals (Grill-Spector, Sayres, and Ress, 2007), but an article by Baker, Hutchison, and Kanwisher (2007) disputes the results of this study for technical reasons. Perhaps a more important issue is the relative roles of genetic programming and experience in development of a brain region critically involved in face perception.

A functional-imaging study indicates that although the relative size of the LOC, which responds to objects other than faces and bodies, is the same in children and adults, the left FFA does not reach its eventual size until adulthood, and the ability to recognize faces is directly related to the expansion of the FFA (Golarai et al., 2007). These findings are consistent with the suggestion that the ability to recognize faces is a learned skill that grows with experience. Figure 6.41 shows the regions on the left and right fusiform cortex of an eight-year-old child and an adult. You can see the age-related size difference and also the difference between the size of this region in the left and right hemispheres. (See *Figure 6.41.*)

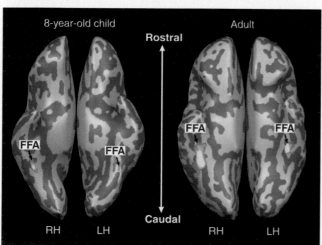

Evidence indicates that newborn babies prefer to look at stimuli that resemble faces, which suggests the presence of prewired circuits in the human brain that dispose babies to look at faces and hence learn to recognize them. Farroni et al. (2005) presented newborn babies (between 13 and 168 hours old) with pairs of stimuli and found that they preferred to look at the ones that bore the closest resemblance to faces viewed in their normal, upright orientation, with the lighting coming from above, as it normally does. Figure 6.42 illustrates the stimuli that Farroni and her colleagues used. An asterisk above a stimulus indicates that the babies spent more time looking at it than at the other member of the pair. If neither stimulus is marked with an asterisk, that means that the baby indicated no preference—and as you can see, these pairs of stimuli bore the least resemblance to a face illuminated from above. (See *Figure 6.42.*)

A review of the literature by Johnson (2005) suggests that a baby's preference for faces is controlled by a fast, low-spatial-frequency, subcortical pathway that is present in newborn infants. This circuit survives in many adults with prosopagnosia caused by cortical damage, who can realize that they are looking at a face even though they cannot recognize it and can even recognize facial expressions such as happiness, fear, or anger. (This phenomenon is discussed in more detail in Chapter 11, which deals with emotion.) The subcortical pathway guarantees that babies will look at faces, which increases social bonding with other humans as well as facilitating the development of face-sensitive circuits in the cerebral cortex.

A study by Le Grand et al. (2001) discovered that the experience of seeing faces very early in life plays a critical role in the development of the skills necessary for recognizing them later in life. The investigators tested the ability of people (aged 9–21 years) who had been born with congenital cataracts to recognize subtle differences between pairs of faces. These people had been unable to

see more than light and dark until they received eye surgery at 62–187 days of age that made normal vision possible. The early visual deprivation resulted in a severe deficit, compared with the performance of control subjects, in recognizing the facial differences.

A follow-up study by Le Grand et al. (2003) tested people who were born with cataracts in only one eye. Because of the immaturity of the newborn brain, visual information received by one eye is transmitted only to the contralateral visual cortex. (You may recall that I said earlier in this chapter that it is not correct to say that each hemisphere receives visual information solely from the contralateral eye. However, my admonition does not apply to newborn babies.) This means that the right hemisphere of a person born with a cataract in the left eye does not receive patterned visual information until the cataract is removed. Le Grand and his colleagues predicted that because the right fusiform gyrus is critical for facial recognition, people born with cataracts in their left eye would show a deficit in recognizing faces but that people born with cataracts in the right eye would show normal discrimination—and that is exactly what they found.

By the way, there are three basic ways in which we can recognize individual faces: differences in features (for example, the size and shape of the eyes, nose, and mouth), differences in contour (the overall shape of the face), and differences in configuration of features (for example, the spacing of the eyes, nose, and mouth). Figure 6.43 illustrates these differences in a series of composite faces from the study by Le Grand et al. (2003). (You can see that the face on the far left is the same in each of the rows.) The top row of faces contain different features: eyes and mouths from photos of different people. (The noses are all the same.) The middle row of faces are all of the same person, but the contours of the faces have different shapes. The bottom row contain different configurations of features from one individual. In these faces, the spacing between the eyes and between the eyes and the mouth have been altered. Differences in configuration are the most difficult to detect, and the people with early visual deprivation showed a deficit in configural recognition. (See *Figure 6.43.*)

As we will see in Chapter 17, people with autistic disorder fail to develop normal social relations with other people. Indeed, in severe cases they give no signs that they recognize that other people exist. Grelotti, Gauthier, and Schultz (2002) found that people with autistic disorder showed a deficit in the ability to recognize faces and that looking at faces failed to activate the fusiform gyrus. The authors speculate that the lack of interest in other people, caused by the brain abnormalities responsible for autism, resulted in a lack of motivation that normally promotes the acquisition of expertise in recognizing faces as a child grows up. Chapter 17 discusses autistic disorder in more detail.

FIGURE 6.43 ■ Composite Faces

The faces in the top row contain different features: eyes and mouths from photos of different people. The middle row of faces are all of the same person, but the contours of the faces have different shapes. The faces in the bottom row contain different configurations of features from one individual: The spacing between the eyes and between the eyes and the mouth have been altered.

(From Le Grand, R., Mondloch, C. J., Maurer, D., and Brent, H. P. *Nature Neuroscience*, 2003, 6, 1108–1112. Reprinted with permission.)

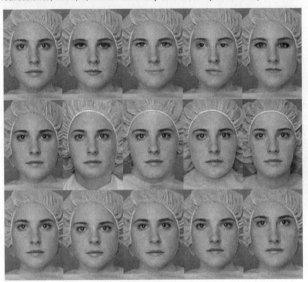

Perception of Movement

We need to know not only what things are, but also where they are and where they are going. Without the ability to perceive the direction and velocity of movement of objects, we would have no way to predict where they will be. We would be unable to catch the objects (or avoid letting them catch us). This section examines the perception of movement; the final section examines the perception of location.

Studies with Laboratory Animals

One of the regions of the extrastriate cortex—area V5, also known as area MT, for *medial temporal*—contains neurons that respond to movement. Damage to this region severely disrupts a monkey's ability to perceive moving stimuli (Siegel and Andersen, 1986). Area V5 receives input directly from the striate cortex and from several regions of the extrastriate cortex. It also receives input from the superior colliculus, which is involved in visual reflexes, including reflexive control of eye movements.

Accurately determining the velocity and direction of movement of an object is an important ability. That moving object could be a prey animal trying to run away, a predator trying to catch you, or a projectile you are trying to catch (or keep from hitting you). If we are to accurately track moving objects, the information received by V5 must be up to date. In fact, the axons that transmit information from the magnocellular system are thick and heavily myelinated, which increases the rate at which they conduct action potentials. Petersen, Miezin, and Allman (1988) recorded the responses of neurons in areas V4 and V5. As you can see in Figure 6.44, visual information reached the V5 neurons sooner than it reached those in area V4, whose neurons are involved in the analysis of form and color. (See *Figure 6.44.*)

The input from the superior colliculus contributes in some way to the movement sensitivity of neurons in area V5. Rodman, Gross, and Albright (1989, 1990) found that destruction of the striate cortex or the superior colliculus alone does not eliminate the movement sensitivity of V5 neurons, but destruction of both areas does. The roles played by these two sources of input are not yet known. Clearly, both inputs provide useful information; Seagraves et al. (1987) found that monkeys still could detect movement after lesions of the striate cortex but had difficulty estimating its rate.

A region adjacent to area V5, area MST, or *medial superior temporal*, receives information about movement from V5 and performs a further analysis. MST neurons respond to complex patterns of movement, including radial, circular, and spiral motion (see Vaina, 1998, for a review). One important function of this region—in

---

FIGURE 6.44 ■ Responses of Neurons in Areas V4 and V5

Note that neurons in the motion-sensitive area V5 responded sooner to stimuli presented in their receptive field, and their firing ceased sooner, than neurons in the form- and color-sensitive area V4. The faster, briefer response is what one would expect of neurons involved in perceiving a moving object's velocity and direction of movement.

(Adapted from Petersen, S., Miezin, F., and Allman, J. Transient and sustained responses in four extrastriate visual areas of the owl monkey. *Experimental Brain Research*, 1988, *70*, 55–60.

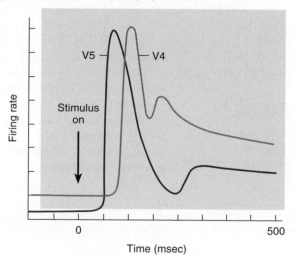

particular, the dorsolateral MST, or MSTd—appears to be analysis of **optic flow.** As we move around in our environment or as objects in our environment move in relation to us, the sizes, shapes, and locations of environmental features on our retinas change. Imagine the image seen by a video camera as you walk along a street, pointing the lens of the camera straight in front of you. Suppose your path will pass just to the right of a mailbox. The image of the mailbox will slowly get larger. Finally, as you pass the mailbox, its image will veer to the left and disappear. Points on the sidewalk will move downward, and branches of trees that you pass under will move upward. Analysis of the relative movement of the visual elements of your environment—the optic flow—will tell you where you are heading, how fast you are approaching different items in front of you, and whether you will pass to the left or right (or under or over) these items. The point toward which we are moving does not move, but all other points in the visual scene move away from it. Therefore, this point is called the *center of expansion*. If we keep moving in the same direction, we will eventually bump into an object that lies at the center of expansion. We can also use optic flow to determine whether an object approaching us will hit us or pass us by.

Bradley et al. (1996) recorded from single units in MSTd of monkeys and found that particular neurons responded selectively to expansion foci located in particular regions of the visual field. These neurons compensated for eye movements, which means that their activity identified the location in the environment toward which an animal was moving. (The ability of the visual system to compensate for eye movements is discussed in the next subsection of this chapter.) Britten and van Wezel (1998) found that electrical stimulation of MSTd disrupted monkeys' ability to perceive the apparent direction in which they were heading; thus, these neurons do indeed seem to play an essential role in heading estimation derived from optic flow.

Studies with Humans

Perception of Motion.
Functional-imaging studies suggest that a motion-sensitive area V5 (usually called MT/MST) is found within the inferior temporal sulcus of the human brain (Dukelow et al., 2001). However, a more recent study suggests that this region is located in the lateral occipital cortex, between the lateral and inferior occipital sulci (Annese, Gazzaniga, and Toga, 2004). Annese and his colleagues examined sections of the brains of deceased subjects that had been stained for

---

optic flow The complex motion of points in the visual field caused by relative movement between the observer and environment; provides information about the relative distance of objects from the observer and of the relative direction of movement.

FIGURE 6.45 ■ Location of Visual Area V5

The location of area V5 (also called MT/MST or MST+) in a human brain is identified by a stain that showed the presence of a dense projection of thick, heavily myelinated axons. (LOS = lateral occipital sulcus, IOS = inferior occipital sulcus.)

(From Annese, J., Gazzaniga, M. S., and Toga, A. W. *Cerebral Cortex*, 2005, 15, 1044–1053. Reprinted with permission.)

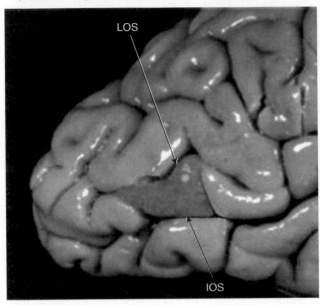

LOS

IOS

the presence of myelin. As we just saw, area V5 receives a dense projection of thick, heavily myelinated axons, and the location of this region was revealed by the myelin stain. (See *Figure 6.45.*)

Bilateral damage to the human brain that includes area V5 produces an inability to perceive movement—**akinetopsia.** For example, Zihl et al. (1991) reported the case of a woman with bilateral lesions of the lateral occipital cortex and area MT/MST.

Patient L. M. had an almost total loss of movement perception. She was unable to cross a street without traffic lights, because she could not judge the speed at which cars were moving. Although she could perceive movements, she found moving objects very unpleasant to look at. For example, while talking with another person, she avoided looking at the person's mouth because she found its movements very disturbing. When the investigators asked her to try to detect movements of a visual target in the laboratory, she said, "First the target is completely at rest. Then it suddenly jumps upwards and downwards" (Zihl et al., 1991, p. 2244). She was able to see that the target was constantly changing its position, but she was unaware of any sensation of movement.

Walsh et al. (1998) used transcranial magnetic stimulation (TMS) to temporarily inactivate area MT/MST in normal human subjects. The investigators found that during the stimulation, people were unable to detect which of several objects displayed on a computer screen was moving. When the current was off, the subjects had no trouble detecting the motion. The current had no effect on the subjects' ability to detect stimuli that varied in their form. (*MyPsychKit 6.3, Motion Aftereffects,* illustrates an interesting movement-related phenomenon.)

mypsychkit
Animation 6.3
Motion Aftereffects

Optic Flow.   As we saw in the previous subsection, neurons in area MSTd of the monkey brain respond to optic flow, an important source of information about the direction in which the animal is heading. A functional-imaging study by Peuskens et al. (2001) found that area MT/MST became active when people judged their heading while viewing a display showing optic flow. Vaina and her colleagues (Jornales et al., 1997; Vaina, 1998) found that people with lesions that included this region were able to perceive motion but could not perceive heading from optic flow.

Form from Motion.   Perception of movement can even help us to perceive three-dimensional forms—a phenomenon known as *form from motion.* Johansson (1973) demonstrated just how much information we can derive from movement. He dressed actors in black and attached small lights to several points on their bodies, such as their wrists, elbows, shoulders, hips, knees, and feet. He made movies of the actors in a darkened room while they were performing various behaviors, such as walking, running, jumping, limping, doing push-ups, and dancing with a partner who was also equipped with lights. Even though observers who watched the films could see only a pattern of moving lights against a dark background, they could readily perceive the pattern as belonging to a moving human and could identify the behavior the actor was performing. Subsequent studies (Kozlowski and Cutting, 1977; Barclay, Cutting, and Kozlowski, 1978) showed that people could even tell, with reasonable accuracy, the sex of the actor wearing the lights. The cues appeared to be supplied by the relative amounts of movement of the shoulders and hips as the person walked. (For a demonstration of this phenomenon, see *MyPsychKit 6.4, Form from Motion.*)

mypsychkit
Animation 6.4
Form from Motion

akinetopsia Inability to perceive movement, caused by damage to area V5 (also called MST) of the visual association cortex.

McCleod et al. (1996) suggest that the ability to perceive form from motion does not involve area V5. They reported that patient L. M. (studied by Zihl et al., 1991) could recognize people depicted solely by moving points of light *even though she could not perceive the movements themselves.* Vaina and her colleagues (reported by Vaina, 1998) found a patient with a lesion in the medial right occipital lobe who showed just the opposite deficits: Patient R. A. could perceive movement—even complex radial and circular optic flow—but could not perceive form from motion. Thus, perception of motion and perception of form from motion involve different regions of the visual association cortex.

A functional-imaging study by Grossman et al. (2000) found that when people viewed a video that showed form from motion, a small region on the ventral bank of the posterior end of the superior temporal sulcus became active. More activity was seen in the right hemisphere, whether the images were presented to the left or right visual field. Grossman and Blake (2001) found that this region became active even when people *imagined* that they were watching points of light representing form from motion. (See *Figure 6.46.*) Grossman, Battelli, and Pascual-Leone (2005) found that inactivation of this area with transcranial magnetic stimulation disrupted perception of form from motion.

Perception of form from motion might not seem like a phenomenon that has any importance outside the laboratory. However, this phenomenon does occur under natural circumstances, and it appears to involve brain mechanisms different from those involved in normal object perception. For example, as we saw in the prologue to this chapter, people with visual agnosia can often still perceive *actions* (such as someone pretending to stir something in a bowl or deal out some playing cards) even though they cannot recognize objects by sight. They may be able to recognize friends by the way the friends walk, even though they cannot recognize the friends' faces.

Lê et al. (2002) reported the case of patient S. B., a 30-year-old man whose ventral stream was damaged extensively bilaterally by encephalitis when he was three years old. As a result, he was unable to recognize objects, faces, textures, or colors. However, he could perceive movement and could even catch a ball that was thrown to him. Furthermore, he could recognize other people's arm and hand movements that mimed common activities such as cutting something with a knife or brushing one's teeth, and he could recognize people he knew by their gait.

Biological Motion. As we saw earlier in this chapter, neurons in the extrastriate body area (EBA) are activated by the sight of human body parts. A functional-imaging study by Pelphrey et al. (2005) showed subjects a computer-generated image of a person who made hand, eye, and mouth movements. (Note that the subjects were perceiving motion made by a human being, not form from the motion of individual points of light as described in the previous subsection.) The investigators found that movements of different body parts activated different locations just anterior to the EBA.

Compensation for Eye Movements. So far, this discussion has been confined to movement of objects in the visual field. But if a person moves his or her eyes, head, or whole body, the image on the retina will move even if everything within the person's visual field remains stable. Often, of course, *both* kinds of movements will occur at the same time. The problem for the visual system is to determine which of these images are produced by movements of objects in the environment and which are produced by the person's own eye, head, and body movements.

To illustrate this problem, think about how the page of this book looks as you read it. If we could make a videotape of one of your retinas, we would see that the image of the page projected there is in constant movement as your eyes make several saccades along a line and then snap back to the beginning of the next line. Yet the page seems perfectly still to you. On the other hand, if you look at a single point on the page (say, a period at the end of a sentence) and then move the

FIGURE 6.46 ■ Responses to Viewing Form from Motion

The figure shows horizontal and lateral views of neural activity while the subject was viewing videos of biological motion such as those shown in MyPsychKit 6.4. Maximum activity is seen in a small region on the ventral bank of the posterior end of the superior temporal sulcus, primarily in the right hemisphere.

(From Grossman, E. D., and Blake, R. *Vision Research*, 2001, *41*, 1475–1482. Reprinted with permission.)

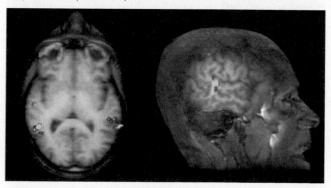

page around while following the period with your eyes, you perceive the book as moving, even though the image on your retina remains relatively stable. (Try it.) Then think about the images on your retina while you are driving in busy traffic, constantly moving your eyes around to keep track of your own location and that of other cars moving in different directions at different speeds. You are perceiving not only the simple movement of objects, but optic flow as well, which helps you keep track of the trajectories of the objects relative to each other and to yourself.

Haarmeier et al. (1997) reported the case of a patient with bilateral damage to the extrastriate cortex who could not compensate for image movement caused by head and eye movements. When the patient moved his eyes, it looked to him as if the world was moving in the opposite direction. Without the ability to compensate for head and eye movements, any movement of a retinal image was perceived as movement of the environment. On the basis of evidence from EEG and MEG (magnetoencephalography) studies in human subjects and single-unit recordings in monkeys, Thier et al. (2001) suggest that this compensation involves extrastriate cortex located at the junction of the temporal and parietal lobes near a region involved in the analysis of signals from the vestibular system. Indeed, the investigators note that when patients with damage to this region move their eyes, the lack of compensation for these movements makes them feel very dizzy.

Perception of Spatial Location

The parietal lobe is involved in spatial and somatosensory perception, and it receives visual, auditory, somatosensory, and vestibular information to perform these tasks. Damage to the parietal lobes disrupts performance on a variety of tasks that require perceiving and remembering the locations of objects and controlling movements of the eyes and the limbs. The dorsal stream of the visual association cortex terminates in the posterior parietal cortex.

The anatomy of the posterior parietal cortex is shown in Figure 6.47. We see an "inflated" dorsal view of the left hemisphere of a human brain. Five regions within the **intraparietal sulcus (IPS)** are of particular interest: AIP, LIP, VIP, CIP, and MIP (anterior, lateral, ventral, caudal, and medial IPS) are indicated. (See *Figure 6.47.*)

Single-unit studies with monkeys and functional-imaging studies with humans indicate that neurons in the IPS are involved in visual attention and control of saccadic eye movements (LIP and VIP), visual control of reaching and pointing (VIP and MIP), visual control of grasping and manipulating hand movements (AIP), and perception of depth from stereopsis (CIP) (Snyder, Batista, and

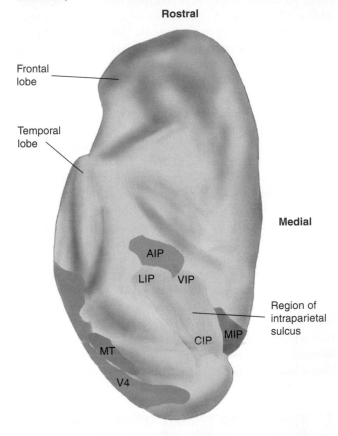

FIGURE 6.47 ■ Dorsal View of the Left Hemisphere

This "inflated" view of a human brain shows the anatomy of the posterior parietal cortex.

(Adapted from Astafiev, S. V., Shulman, G. L., Stanley, C. M., Snyder, A. Z., Van Essen, D. C., and Corbetta, M. *Journal of Neuroscience*, 2003, *23*, 4689–4699.)

Andersen, 2000; Culham and Kanwisher, 2001; Astafiev et al., 2003; Tsao et al., 2003; Frey et al., 2005).

Goodale and his colleagues (Goodale and Milner, 1992; Goodale et al., 1994; Goodale and Westwood, 2004) suggested that the primary function of the dorsal stream of the visual cortex is to guide actions rather than simply to perceive spatial locations. As Ungerleider and Mishkin (1982) originally put it, the ventral and dorsal streams tell us "what" and "where." Goodale and his colleagues suggested that the better terms are "what" and "how." First, they noted that the visual cortex of the posterior parietal lobe is extensively connected to regions of the frontal lobe involved in controlling eye movements, reaching movements of the limbs, and

intraparietal sulcus (IPS) The end of the dorsal stream of the visual association cortex; involved in perception of location, visual attention, and control of eye and hand movements.

grasping movements of the hands and fingers. Second, they noted that damage to the dorsal stream can produce deficits in visually guided movements. (Chapter 8 discusses in more detail the role of the posterior parietal cortex in control of movements.) They cited the case of a woman with damage to the dorsal stream who had no difficulty recognizing line drawings (that is, her ventral stream was intact) but who had trouble picking up objects (Jakobson et al., 1991). The patient could easily perceive the difference in size of wooden blocks that were set out before her, but she failed to adjust the distance between her thumb and forefinger to the size of the block she was about to pick up. In contrast, a patient with profound visual agnosia caused by damage to the ventral stream could not distinguish between wooden blocks of different sizes but *could* adjust the distance between her thumb and forefinger when she picked them up. She made this adjustment by means of vision, before she actually touched them (Milner et al., 1991; Goodale et al., 1994). A functional-imaging study of this patient (James et al., 2003) showed normal activity in the dorsal stream while she was picking up objects—especially in the anterior intraparietal sulcus (AIP), which is involved in manipulating and grasping.

The suggestion by Goodale and his colleagues seems a reasonable one. Certainly, the dorsal stream is involved in perception of the location of object's space—but then, if its primary role is to direct movements, it *must* be involved in location of these objects, or else how could it direct movements toward them? In addition, it must contain information about the size and shape of objects, or else how could it control the distance between thumb and forefinger?

Two functional-imaging studies provide further evidence that the dorsal stream is involved in visual control of movement. Valyear et al. (2006) presented photographs of pairs of elongated stimuli, one after the other, and noted which regions of the brain responded to the difference between the two stimuli. They found that a region of the ventral stream responded differentially to pairs of stimuli that differed in their form (for example, a fork versus a clarinet) but did not distinguish between the same object shown in different orientations (for example, one tipped 45 degrees to the right of vertical and the other tipped 45 degrees to the left). In contrast, a region of the dorsal stream distinguished between different orientations but ignored changes in the identify of the two objects. A follow-up study published the next year (Rice et al., 2007) showed subjects photographs of two different types of objects: graspable ones, such as forks and hammers, and non-graspable ones, such as tractors and pieces of furniture. The investigators found that, as before, the region of the dorsal stream ignored changes in the identity of the objects but distinguished between orientations.

However, the region distinguished between the orientations only of stimuli that a person could grasp. This region did *not* distinguish between the orientations of photos of stimuli that could not be picked up, such as tractors and pieces of furniture.

A fascinating (and delightful) study with young children demonstrates the importance of communication between the dorsal and ventral streams of the visual system (DeLoache, Uttal, and Rosengren, 2004). The experimenters let children play with large toys: an indoor slide that they could climb and slide down, a chair that they could sit on, and a toy car that they could enter. After the children played in and on the large toys, the children were taken out of the room, the large toys were replaced with identical miniature versions, and the children were then brought back into the room. When

FIGURE 6.48 ■ Ventral and Dorsal Streams of the Visual Cortex

The figure shows some major components of the ventral stream and some of the dorsal stream of the visual cortex. This view is similar to that seen in Figure 6.32(e).

(Adapted from Tootell, R. B. H., Tsao, D., and Vanduffel, W. *Journal of Neuroscience*, 2003, *23*, 3981–3989.)

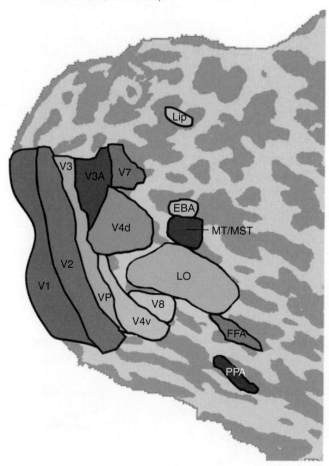

the children played with the miniature toys, they acted as if they were the large versions: They tried to climb onto the slide, climb into the car, and sit on the chair. MyPsychKit 6.5 shows a video of a 2-year-old child trying to climb into the toy car. He says "In!" several times and turns to his mother, apparently asking her to help him. The authors suggest that this behavior reflects incomplete maturation of connections between the dorsal and ventral streams. The ventral stream recognizes the identity of the objects, and the dorsal stream recognizes their size, but the information is not adequately shared between these two systems. (See *MyPsychKit 6.5, Dissociation of Perception and Action.*)

mypsychkit

Animation 6.5
Dissociation of Perception and Action

I realize that I have presented a large amount of information in this section—and I'm sure you do, too. The importance of the visual system is attested to by the fact that approximately 25 percent of our cerebral cortex is devoted to this sense modality and by the many discoveries being made in the laboratories that are busy discovering interesting things about vision. *Figure 6.48* shows the location of the regions that make up the ventral stream and some of the dorsal stream. (The rest of the dorsal stream lies in the intraparietal sulcus, which is illustrated in Figure 6.47.) *Table 6.3* lists these regions and summarizes their major functions.

TABLE 6.3 ■ Regions of the Human Visual Cortex and Their Functions

REGION OF HUMAN VISUAL CORTEX	NAME OF REGION (IF DIFFERENT)	FUNCTION
V1	Striate cortex	Small modules that analyze orientation, movement, spatial frequency, retinal disparity, and color
V2		Further analysis of information from V1
Ventral Stream		
V3 + VP		Further analysis of information from V2
V3A		Processing of visual information across entire visual field of contralateral eye
V4d/V4v	V4 dorsal/ventral	Analysis of form Processing of color constancy V4d = lower visual field, V4v = upper visual field
V8		Color perception
LO	Lateral occipital complex	Object recognition
FFA	Fusiform face area	Face recognition, object recognition by experts ("flexible fusiform area")
PPA	Parahippocampal place area	Recognition of particular places
EBA	Extrastriate body area	Perception of body parts other than face

TABLE 6.3 ■ Regions of the Human Visual Cortex and Their Functions *(continued)*

REGION OF HUMAN VISUAL CORTEX	NAME OF REGION (IF DIFFERENT)	FUNCTION
Dorsal Stream		
V7		Visual attention Control of eye movements
MT/MST	Medial temporal/medial superior temporal (named for locations in monkey brain)	Perception of motion Perception of biological motion and optic flow in specific subregions
LIP	Lateral intraparietal area	Visual attention Control of saccadic eye movements
VIP	Ventral intraparietal area	Control of visual attention to particular locations Control of eye movements Visual control of pointing
AIP	Anterior intraparietal area	Visual control of hand movements: grasping, manipulation
MIP	Middle intraparietal area Parietal reach region (monkeys)	Visual control of reaching
CIP	Caudal intraparietal area Caudal parietal disparity region	Perception of depth from stereopsis

Interim Summary

Analysis of Visual Information: Role of the Visual Association Cortex

The visual cortex consists of the striate cortex and two streams of visual association cortex. The ventral stream, which ends with the inferior temporal cortex, is involved with perception of objects. Lesions of this region disrupt visual object perception. Also, single neurons in the inferior temporal cortex respond best to complex stimuli and continue to do so even if the object is moved to a different location, changed in size, placed against a different background, or partially hidden. The dorsal stream, which ends with the posterior parietal cortex, is involved with perception of movement, location, visual attention, and control of eye and hand movements. There are at least two dozen different subregions of the visual cortex, arranged in a hierarchical fashion. Each region analyzes a particular characteristic of visual information and passes the results of this analysis to other regions in the hierarchy. However, some information from the association cortex is sent back to the striate cortex. Neurons in the thin stripes of V2 receive information concerning color from the blobs in the striate cortex (V1), and those in the thick stripes and pale stripes receive information about orientation, spatial frequency, movement, and retinal disparity from the interblob regions of V1.

Damage to area V4 abolishes color constancy (accurate perception of color under different lighting conditions), and damage to area V8 causes cerebral achromatopsia, a loss of color vision but not of form perception. A condition opposite to achromatopsia can also be seen: A patient with extensive damage to the extrastriate cortex was functionally blind but could still recognize colors. His brain damage apparently destroyed regions of the visual association cortex that are responsible for form perception but not those for color perception.

Functional-imaging studies indicate that specific regions of the cortex are involved in perception of form, movement, and color, and these studies are enabling us to discover the correspondences between the anatomy of the human visual system and that of laboratory animals. Humans who have sustained damage to the ventral stream of visual association cortex have difficulty recognizing objects by sight, even though fine details can often be detected—a disorder known as visual agnosia. Prosopagnosia—failure to recognize faces—is caused by damage to the fusiform face area (FFA), a region on the base of the right temporal lobe. The development of this region may be a result of extensive experience looking at faces; expertise with other complex stimuli such as artificial creatures (greebles) causes the development of circuits devoted to the perception of these stimuli as well.

The extrastriate body area (EBA), a region adjacent to the FFA, contains neurons that respond to the sight of bodies or body parts, and the parahippocampal place area (PPA) responds to scenes that depict particular places. Newborn babies prefer to look at facelike stimuli, a preference that may involve subcortical mechanisms. Babies deprived of visual input for the first few months of life because of congenital cataracts demonstrate impaired discrimination of faces later in life. The fusiform face area fails to develop in people with autism, presumably because of insufficient motivation to become expert in recognizing other people's faces.

Damage to area V5 (also called area MT) disrupts an animal's ability to perceive movement, and damage to the posterior parietal cortex disrupts perception of the spatial location of objects. Damage to the human visual association cortex corresponding to area V5 disrupts perception of movement, producing a disorder known as akinetopsia. In addition, transcranial magnetic stimulation of V5 causes a temporary disruption, and functional-imaging studies show that perception of moving stimuli activate this region. In both monkeys and humans, area MSTd, a region of extrastriate cortex that is adjacent to area V5, appears to be specialized for perceiving optic flow, one of the cues we use to perceive the direction in which we are heading.

The ability to perceive form from motion—recognition of complex movements of people indicated by lights attached to parts of their body—is probably related to the ability to recognize people by the way they walk. This ability apparently depends on a region of cerebral cortex on the ventral bank of the posterior end of the superior temporal sulcus. The visual association cortex receives information about eye movements from the motor system and information about movement of retinal images from the visual cortex and determines which movements are caused by head and eye movements and which are caused by movements in the environment. A patient with extrastriate damage was unable to compensate for eye movements; when he moved his eyes, he perceived movement in the environment. The location of the region responsible for this compensation appears to be in the extrastriate cortex at the junction of the temporal and parietal lobes.

Some people with visual agnosia caused by damage to the ventral stream can still perceive the meanings of mimed actions or recognize friends by the way the friends walk, which indicates that the dorsal stream of these people's visual cortex is largely intact. Most of the visual association cortex at the end of the dorsal stream is located in the intraparietal sulcus: LIP and VIP are involved in visual attention and control of saccadic eye movements, VIP and MIP are involved in visual control of reaching and pointing, AIP is involved in visual control of grasping and manipulating, and CIP is involved in perception of depth from stereopsis.

Goodale and his colleagues suggest that the primary function dorsal stream of visual association cortex is better characterized as "how" rather than "where"; the role of the posterior parietal cortex in control of reaching, grasping, and manipulation requires visually derived information of movement, depth, and location.

Thought Question

Some psychologists are interested in "top-down" processes in visual perception—that is, the effects of context on perceiving ambiguous stimuli. For example, if you are in a dimly lighted kitchen and see a shape that could be either a loaf of bread or a country mailbox, you will be more likely to perceive the object as a loaf of bread. Where in the brain might contextual information affect perception?

SUGGESTED READINGS

Gregory, R. L. *Eye and Brain: The Psychology of Seeing*, 5th ed. Princeton, NJ: Princeton University Press, 1997.

Grill-Spector, K., and Malach, R. The human visual cortex. *Annual Review of Neuroscience*, 2002, *27*, 649–677.

Oyster, C. W. *The Human Eye: Structure and Function*. Sunderland, MA: Sinauer Associates, 1999.

Purves, D., and Lotto, R. B. *Why We See What We Do: An Empirical Theory of Vision*. Sunderland, MA: Sinauer Associates, 2003.

Rodieck, R. W. *The First Steps in Seeing*. Sunderland, MA: Sinauer Associates, 1998.

Solomon, S. G., and Lennie, P. The machinery of colour vision. *Nature Reviews: Neuroscience*, 2007, 8, 276–286.

Wandell, B. A., Dumoulin, S. O., and Brewer, A. A. Visual field maps in human cortex. *Neuron*, 2007, *56*, 366–383.

ADDITIONAL RESOURCES

Visit www.mypsychkit.com for additional review and practice of the material covered in this chapter. Within MyPsychKit, you can take practice tests and receive a customized study plan to help you review. Dozens of animations, tutorials, and Web links are also available. You can even review using the interactive electronic version of this textbook. You will need to register for MyPsychKit. See www.mypsychkit.com for complete details.

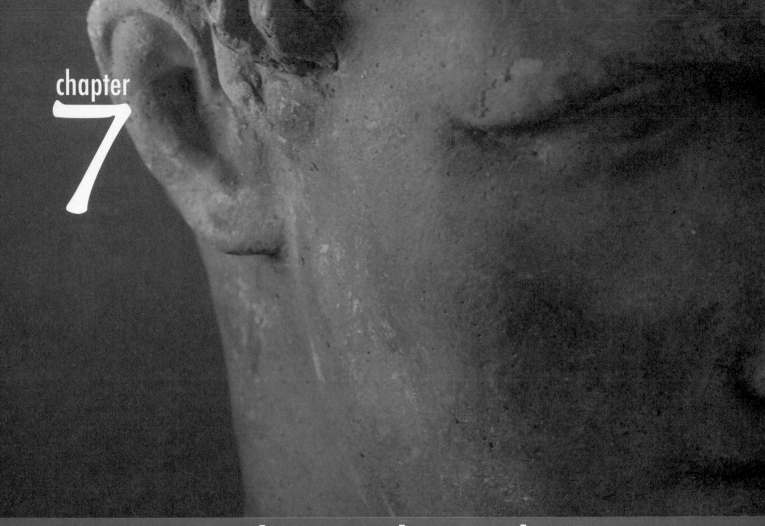

chapter
7

Audition, the Body Senses, and the Chemical Senses

outline

Nine-year-old Sara tried to think of something else, but the throbbing pain in her thumb was relentless. Earlier in the day, her brother had slammed the car door on it.

"Why does it have to hurt so much, Daddy?" she asked piteously.

"I wish I could help you, sweetheart," he answered. "Pain may be useful, but it sure isn't fun."

"What do you mean, useful?" she asked in astonishment. "You mean it's good for me?" She looked at her father reproachfully.

"Well, this probably isn't the time to tell you about the advantages of pain, because it's hard to appreciate them when you're suffering." A glimmer of interest began to grow in her eyes. For as long as she could remember, Sara loved to have her father explain things to her, even when his explanations got a little confusing.

"You know," he said, "there are some people who never feel any pain. They are born that way."

"Really?" Her eyes widened. "They're lucky!"

"No, they really aren't. Without the sense of pain, they keep injuring themselves. When they touch something hot, they don't know enough to let go, even when their hand is getting burned. If the water in the shower gets too hot, they don't realize they're getting scalded. If their shoes don't fit right, they get huge blisters without knowing what's happening. If they fall and sprain an ankle—or even break a bone—they don't feel that something bad has happened to them, and their injury will just get worse. Some people who have no ability to feel pain have died when their appendix burst because they didn't know that something bad was happening inside them."

Sara looked thoughtful. Her father's explanation seemed to be distracting her from her pain.

"Parents of children who can't feel pain say that it's difficult to teach them to avoid danger. When a child does something that causes pain, she quickly learns to avoid repeating her mistake. Remember when you were three years old and walked on the grill of the heater in the cabin floor? You had just gotten out of the shower, and you burned the bottoms of your feet."

"I *think* so," she said. "Yes, you bought me a bag of candy corn to make me forget how much it hurt."

"That's right. Your mom and I had told you that the grill was dangerous when the heater was on, but it took an actual experience to teach you to stay away. We feel pain when parts of our bodies are damaged. The injured cells make a chemical that's picked up by nerve endings, and the nerves send messages to the brain to warn it that something bad is happening. Our brains automatically try to get us away from whatever it is that hurts us—and we also learn to become afraid of it. After you burned your feet on the grill, you stayed away from it even when you were wearing shoes. Kids who can't feel pain can learn to stay away from dangerous things, but it's not an automatic, gut-level kind of learning. They have to pay attention all the time, and if they let down their guard, it's easy for them to injure themselves. Pain isn't fun, but it's hard to survive without it."

"I guess so," said Sara reluctantly. She looked at her bandaged thumb, and the sudden realization of how much it hurt brought tears to her eyes again. "But the pain could go away now, because it's already taught me everything I need to know."

One chapter was devoted to vision, but the rest of the sensory modalities must share a chapter. This unequal allocation of space reflects the relative importance of vision to our species and the relative amount of research that has been devoted to it. This chapter is divided into five major sections, which discuss audition, the vestibular system, the somatosenses, gustation, and olfaction.

AUDITION

For most people, audition is the second most important sense. The value of verbal communication makes audition even more important than vision in some respects; for example, a blind person can join others in conversation far more easily than a deaf person can. (Of course,

deaf people can use sign language to converse with each other.) Acoustic stimuli also provide information about things that are hidden from view, and our ears work just as well in the dark. This section describes the nature of the stimulus, the sensory receptors, the brain mechanisms devoted to audition, and some of the details of the physiology of auditory perception.

The Stimulus

We hear sounds, which are produced by objects that vibrate and set molecules of air into motion. When an object vibrates, its movements cause molecules of air surrounding it alternately to condense and rarefy (pull apart), producing waves that travel away from the object at approximately 700 miles per hour. If the vibration ranges between approximately 30 and 20,000 times per second, these waves will stimulate

FIGURE 7.1 ■ Sound Waves

Changes in air pressure from sound waves move the eardrum in and out. Air molecules are closer together in regions of higher pressure and farther apart in regions of lower pressure.

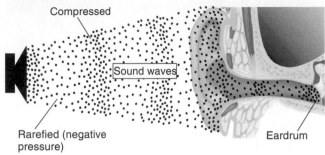

Compressed

Sound waves

Rarefied (negative pressure)

Eardrum

receptor cells in our ears and will be perceived as sounds. (See *Figure 7.1.*)

In Chapter 6 we saw that light has three perceptual dimensions—hue, brightness, and saturation—that correspond to three physical dimensions. Similarly, sounds vary in their pitch, loudness, and timbre. The perceived **pitch** of an auditory stimulus is determined by the frequency of vibration, which is measured in **hertz (Hz),** or cycles per second. (The term honors Heinrich Hertz, a nineteenth-century German physicist.) **Loudness** is a function of intensity—the degree to which the condensations and rarefactions of air differ from each other. More vigorous vibrations of an object produce more intense sound waves and hence louder ones. **Timbre** provides information about the nature of the particular sound—for example, the sound of an oboe or a train whistle. Most natural acoustic stimuli are complex, consisting of several different frequencies of vibration. The particular mixture determines the sound's timbre. (See *Figure 7.2.*)

The eye is a *synthetic* organ (literally, "a putting together"). When two different wavelengths of light are mixed, we perceive a single color. For example, when we see a mixture of red and bluish green light, we perceive pure yellow light and cannot detect either of the two constituents. In contrast, the ear is an *analytical* organ (from *analyein,* "to undo"). When two different frequencies of sound waves are mixed, we do not perceive an intermediate tone; instead, we hear both original tones. As we will see, the ability of our auditory system to detect the individual component frequencies of a complex tone gives us the capacity to identify the nature of particular sounds, such as those of different musical instruments.

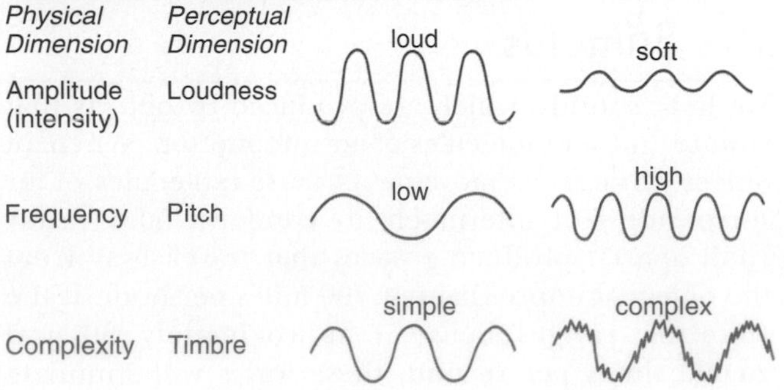

Anatomy of the Ear

Figure 7.3 shows a section through the ear and auditory canal and illustrates the apparatus of the middle and inner ear. (See *Figure 7.3.*) Sound is funneled via the *pinna* (external ear) through the ear canal to the **tympanic membrane** (eardrum), which vibrates with the sound.

The *middle ear* consists of a hollow region behind the tympanic membrane, approximately 2 ml in volume. It contains the bones of the middle ear, called the **ossicles,** which are set into vibration by the tympanic membrane. (As we saw in Chapter 1, two of these bones evolved from part of the reptilian jaw.) The **malleus** (hammer) connects with the tympanic membrane and transmits vibrations via the **incus** (anvil) and **stapes** (stirrup) to the **cochlea,** the structure that contains the receptors. The baseplate of the stapes presses against the membrane behind the **oval window,** the opening in the bony process surrounding the cochlea. (See *Figure 7.3.*)

The cochlea is part of the *inner ear.* It is filled with fluid; therefore, sounds transmitted through the air must be transferred into a liquid medium. This process normally is very inefficient—99.9 percent of the energy

FIGURE 7.2 ■ Physical and Perceptual Dimensions of Sound Waves

Physical Dimension	Perceptual Dimension		
Amplitude (intensity)	Loudness	loud	soft
Frequency	Pitch	low	high
Complexity	Timbre	simple	complex

pitch A perceptual dimension of sound; corresponds to the fundamental frequency.

hertz (Hz) Cycles per second.

loudness A perceptual dimension of sound; corresponds to intensity.

timbre (*tim ber* or *tamm ber*) A perceptual dimension of sound; corresponds to complexity.

tympanic membrane The eardrum.

ossicle (*ahss i kul*) One of the three bones of the middle ear.

malleus The "hammer"; the first of the three ossicles.

incus The "anvil"; the second of the three ossicles.

stapes (*stay peez*) The "stirrup"; the last of the three ossicles.

cochlea (*cock lee uh*) The snail-shaped structure of the inner ear that contains the auditory transducing mechanisms.

oval window An opening in the bone surrounding the cochlea that reveals a membrane, against which the baseplate of the stapes presses, transmitting sound vibrations into the fluid within the cochlea.

FIGURE 7.3 ■ The Auditory Apparatus

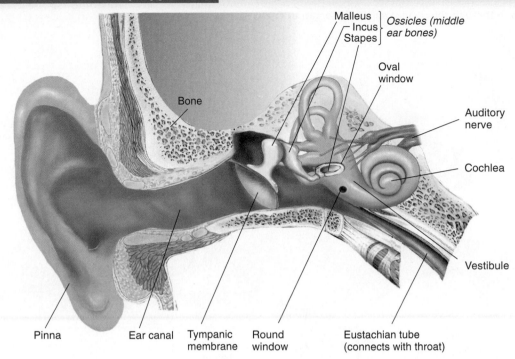

Ossicles (middle ear bones) — Malleus, Incus, Stapes

Oval window

Auditory nerve

Cochlea

Bone

Vestibule

Pinna

Ear canal

Tympanic membrane

Round window

Eustachian tube (connects with throat)

of airborne sound would be reflected away if the air impinged directly against the oval window of the cochlea. The chain of ossicles serves as an extremely efficient means of energy transmission. The bones provide a mechanical advantage, with the baseplate of the stapes making smaller but more forceful excursions against the oval window than the tympanic membrane makes against the malleus.

The name *cochlea* comes from the Greek word *kokhlos,* or "land snail." It is indeed snail-shaped, consisting of two-and-three-quarters turns of a gradually tapering cylinder, 35 mm (1.37 in.) long. The cochlea is divided longitudinally into three sections, the *scala vestibuli* ("vestibular stairway"), the *scala media* ("middle stairway"), and the *scala tympani* ("tympanic stairway"), as shown in *Figure 7.4.* The receptive organ, known as the **organ of Corti**, consists of the *basilar membrane,* the *hair cells,* and the *tectorial membrane.* The auditory receptor cells are called **hair cells,** and they are anchored, via rodlike **Deiters's cells,** to the **basilar membrane.** The cilia of the hair cells pass through the *reticular membrane,* and the ends of some of them attach to the fairly rigid **tectorial membrane,** which projects overhead like a shelf. (See *Figure 7.4.*) Sound waves cause the basilar membrane to move relative to the tectorial membrane, which bends the cilia of the hair cells. This bending produces receptor potentials.

Georg von Békésy—in a lifetime of brilliant studies on the cochleas of various animals, from human cadavers to elephants—found that the vibratory energy exerted

on the oval window causes the basilar membrane to bend (von Békésy, 1960). Because of the physical characteristics of the basilar membrane, the portion that bends the most is determined by the frequency of the sound: High-frequency sounds cause the end nearest the oval window to bend.

Figure 7.5 shows this process in a cochlea that has been partially straightened. If the cochlea were a closed system, no vibration would be transmitted through the oval window, because liquids are essentially incompressible. However, there is a membrane-covered opening, the **round window,** that allows the fluid inside the cochlea to move back and forth. The baseplate of the stapes vibrates against the membrane behind the oval window and introduces sound waves of high or low frequency into the

organ of Corti The sensory organ on the basilar membrane that contains the auditory hair cells.

hair cell The receptive cell of the auditory apparatus.

Deiters's cell (*dye terz*) A supporting cell found in the organ of Corti; sustains the auditory hair cells.

basilar membrane (*bazz i ler*) A membrane in the cochlea of the inner ear; contains the organ of Corti.

tectorial membrane (*tek torr ee ul*) A membrane located above the basilar membrane; serves as a shelf against which the cilia of the auditory hair cells move.

round window An opening in the bone surrounding the cochlea of the inner ear that permits vibrations to be transmitted, via the oval window, into the fluid in the cochlea.

FIGURE 7.4 ■ The Cochlea

This cross section through the cochlea shows the organ of Corti.

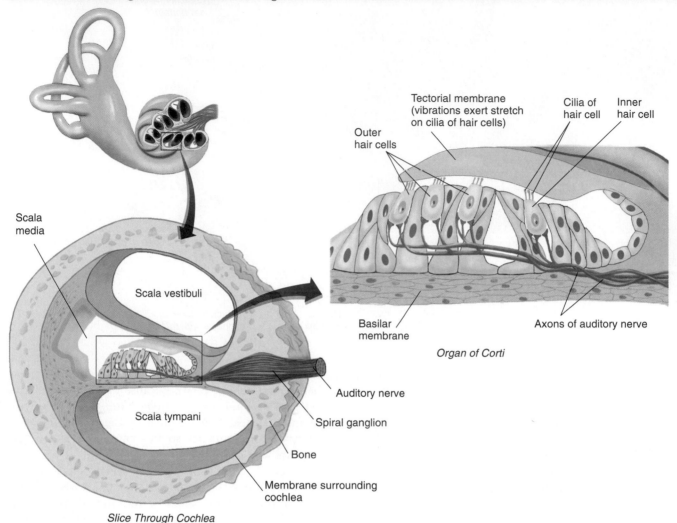

Organ of Corti

Slice Through Cochlea

cochlea. The vibrations cause part of the basilar membrane to flex back and forth. Pressure changes in the fluid underneath the basilar membrane are transmitted to the membrane of the round window, which moves in and out in a manner opposite to the movements of the oval window. That is, when the baseplate of the stapes pushes in, the membrane behind the round window bulges out. As we will see in a later subsection, different frequencies of sound vibrations cause different portions of the basilar membrane to flex. (See *Figure 7.5.*)

Some people suffer from a middle ear disease that causes the bone to grow over the round window. Because their basilar membrane cannot easily flex back and forth, these people have a severe hearing loss. However, their hearing can be restored by a surgical procedure called *fenestration* ("window making"), in which a tiny hole is drilled in the bone where the round window should be.

Auditory Hair Cells and the Transduction of Auditory Information

Two types of auditory receptors, *inner* and *outer* auditory hair cells, are located on the basilar membrane. Hair cells contain **cilia** ("eyelashes"), fine hairlike appendages, arranged in rows according to height. The human cochlea contains approximately 3500 inner hair cells and 12,000 outer hair cells. The hair cells form synapses with dendrites of bipolar neurons whose axons bring auditory information to the brain. Figure 7.6

cilium A hairlike appendage of a cell involved in movement or in transducing sensory information; found on the receptors in the auditory and vestibular system.

FIGURE 7.5 ■ Responses to Sound Waves

When the stapes pushes against the membrane behind the oval window, the membrane behind the round window bulges outward. Different high-frequency and medium-frequency sound vibrations cause flexing of different portions of the basilar membrane. In contrast, low-frequency sound vibrations cause the tip of the basilar membrane to flex in synchrony with the vibrations.

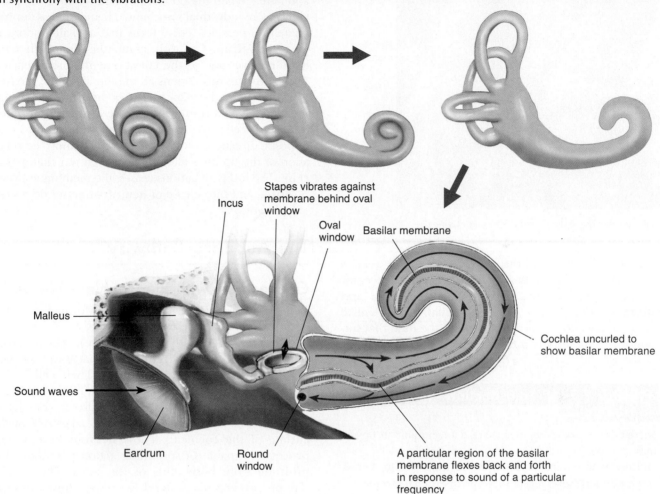

Incus

Stapes vibrates against membrane behind oval window

Oval window

Basilar membrane

Malleus

Cochlea uncurled to show basilar membrane

Sound waves

Eardrum

Round window

A particular region of the basilar membrane flexes back and forth in response to sound of a particular frequency

---

shows the appearance of the inner and outer hair cells and the reticular membrane in a photograph taken by means of a scanning electron microscope. Note the three rows of outer hair cells on the right and the single row of inner hair cells on the left. (See *Figure 7.6.*)

Sound waves cause both the basilar membrane and the tectorial membrane to flex up and down. These movements bend the cilia of the hair cells in one direction or the other. The tips of the cilia of outer hair cells are attached directly to the tectorial membrane. The cilia of the inner hair cells do not touch the overlying tectorial membrane, but the relative movement of the two membranes causes the fluid within the cochlea to flow past them, making them bend back and forth, too.

Cilia contain a core of actin filaments surrounded by myosin filaments, and these proteins make the cilia

stiff and rigid (Flock, 1977). Adjacent cilia are linked to each other by elastic filaments known as **tip links.** Each tip link is attached to the end of one cilium and to the side of an adjacent cilium. The points of attachment, known as **insertional plaques,** look dark under an electron microscope. As we will see, receptor potentials are triggered at the insertional plaques. (See *Figure 7.7.*)

Normally, tip links are slightly stretched, which means that they are under a small amount of tension. Thus, movement of the bundle of cilia in the direction of the tallest of them further stretches these linking fibers,

---

tip link An elastic filament that attaches the tip of one cilium to the side of the adjacent cilium.

insertional plaque The point of attachment of a tip link to a cilium.

FIGURE 7.6 ■ Organ of Corti

This scanning electron photomicrograph of a portion of the organ of Corti shows the cilia of the inner and outer hair cells.

(Photomicrograph courtesy of I. Hunter-Duvar, The Hospital for Sick Children, Toronto, Ontario.)

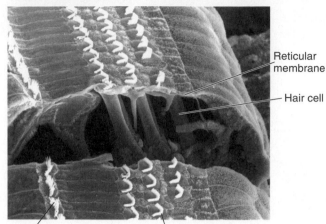

Reticular membrane

Hair cell

Cilia of inner hair cells   Cilia of outer hair cells

whereas movement in the opposite direction relaxes them. The bending of the bundle of cilia causes receptor potentials (Pickles and Corey, 1992; Hudspeth and Gillespie, 1994; Gillespie, 1995; Jaramillo, 1995). Unlike the fluid that surrounds most neurons, the fluid that surrounds the auditory hair cells is rich in potassium. Each

FIGURE 7.7 ■ Transduction Apparatus in Hair Cells

These electron micrographs show (a) a longitudinal section through three adjacent cilia: tip links, elastic filaments attached to insertional plaques, link adjacent cilia. (b) A cross section through several cilia shows an insertional plaque.

(From Hudspeth, A. J., and Gillespie, P. G. *Neuron*, 1994, *12*, 1–9. Copyright © 1994 Cell Press. Reprinted with permission.)

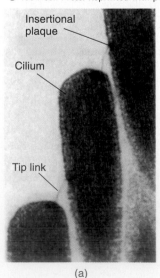

Insertional plaque

Cilium

Tip link

(a)

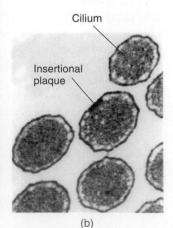

Cilium

Insertional plaque

(b)

insertional plaque contains a single cation channel, which Corey et al. (2004) identified as TRPA1, a member of the *transient receptor potential cation channel*, subfamily A, type 1. (I mention the TRP family of receptors because, as we shall see later in this chapter, this family includes receptors involved in perception of touch, temperature, and taste.) When the bundle of cilia is straight, the probability of an individual ion channel being open is approximately 10 percent. This means that a small amount of the cations K^+ and Ca^{2+} diffuses into the cilium. When the bundle moves toward the tallest one, the increased tension on the tip links opens all the ion channels, the flow of cations into the cilia increases, and the membrane depolarizes. As a result, the release of neurotransmitter by the hair cell increases. When the bundle moves in the opposite direction, toward the shortest cilium, the relaxation of the tip links allows the opened ion channels to close. The influx of cations ceases, the membrane hyperpolarizes, and the release of neurotransmitter decreases. (See *Figure 7.8*.)

The Auditory Pathway

Connections with the Cochlear Nerve

The organ of Corti sends auditory information to the brain by means of the **cochlear nerve,** a branch of the auditory nerve (eighth cranial nerve). The neurons that give rise to the afferent axons that travel through this nerve are of the bipolar type. Their cell bodies reside in the *cochlear nerve ganglion*. (This ganglion is also called the *spiral ganglion* because it consists of clumps of cell bodies arranged in a spiral caused by the curling of the cochlea.) These neurons have axonal processes, capable of sustaining action potentials that protrude from both ends of the soma. The end of one process acts like a dendrite, responding with excitatory postsynaptic potentials when the neurotransmitter is released by the auditory hair cells. The excitatory postsynaptic potentials trigger action potentials in the auditory nerve axons, which form synapses with neurons in the medulla. (Refer to *Figure 7.4*.)

Each cochlear nerve contains approximately 50,000 afferent axons. The dendrites of approximately 95 percent of these axons form synapses with the inner hair cells (Dallos, 1992). These axons are thick and myelinated. The other 5 percent of the sensory fibers in the cochlear nerve form synapses with the much more numerous outer hair cells, at a ratio of approximately one fiber per thirty outer hair cells. These axons are thin and unmyelinated. Thus, although the inner hair cells represent only 29 percent of the total number of receptor cells, their

cochlear nerve The branch of the auditory nerve that transmits auditory information from the cochlea to the brain.

FIGURE 7.8 ■ Transduction in Hair Cells of the Inner Ear

(a) The figure shows the appearance of the cilia of an auditory hair cell. (b) Movement of the bundle of cilia toward the tallest one increases the firing rate of the cochlear nerve axon attached to the hair cell, while movement away from the tallest one decreases it. (c) Movement toward the tallest cilium increases tension on the tip links, which opens the ion channels and increases the influx of K^+ and Ca^{2+} ions. Movement toward the shortest cilium removes tension from the tip links, which permits the ion channels to close, stopping the influx of cations.

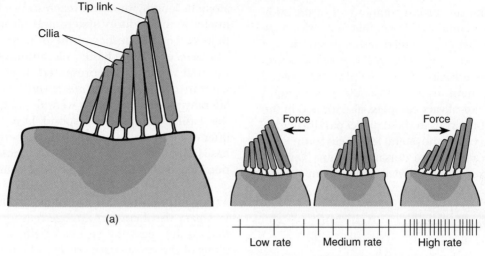

(a)

Action Potentials in Cochlear Nerve Axon

(b)

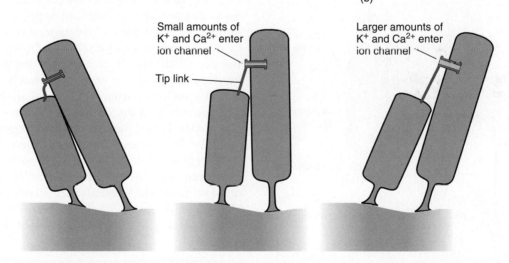

(c)

connections with auditory nerves suggest that they are of primary importance in the transmission of auditory information to the central nervous system (CNS).

Physiological and behavioral studies confirm the inferences made from the synaptic connections of the two types of hair cells: The inner hair cells are necessary for normal hearing. In fact, Deol and Gluecksohn-Waelsch (1979) found that a mutant strain of mice whose cochleas contain *only* outer hair cells apparently cannot hear at all. Subsequent research indicates that the outer hair cells are *effector* cells, involved in altering the mechanical characteristics of the basilar membrane and thus influencing the effects of sound vibrations on the inner hair cells. I will discuss the role of outer hair cells in the section on place coding of pitch.

The cochlear nerve contains efferent axons as well as afferent ones. The source of the efferent axons is the superior olivary complex, a group of nuclei in the medulla; thus, the efferent fibers constitute the

olivocochlear bundle. The fibers form synapses directly on outer hair cells and on the dendrites that serve the inner hair cells. The neurotransmitter at the afferent synapses is glutamate. The efferent terminal buttons secrete acetylcholine, which has an inhibitory effect on the hair cells.

The Central Auditory System

The anatomy of the subcortical components of the auditory system is more complicated than that of the visual system. Rather than giving a detailed verbal description of the pathways, I will refer you to *Figure 7.9.* Note that axons enter the **cochlear nucleus** of the medulla and synapse there. Most of the neurons in the cochlear nucleus send axons to the **superior olivary complex,** also located in the medulla. Axons of neurons in these nuclei pass through a large fiber bundle called the **lateral lemniscus** to the inferior colliculus, located in the dorsal midbrain. Neurons there send their axons to the medial geniculate nucleus of

the thalamus, which sends its axons to the auditory cortex of the temporal lobe. As you can see, there are many synapses along the way to complicate the story. Each hemisphere receives information from both ears but primarily from the contralateral one. Auditory information is relayed to the cerebellum and reticular formation as well.

If we unrolled the basilar membrane into a flat strip and followed afferent axons serving successive points along its length, we would reach successive points in the nuclei of the auditory system and ultimately successive points along the surface of the primary auditory cortex. The *basal* end of the basilar membrane (the end toward the oval window) is represented most medially in the auditory cortex, and the *apical* end is represented most laterally there. Because, as we will see, different parts of the basilar membrane respond best to different frequencies of sound, this relationship between cortex and basilar membrane is referred to as **tonotopic representation** (*tonos* means "tone," and *topos* means "place").

As we saw in Chapter 6, the visual cortex is arranged in a hierarchy. Modules in the striate cortex (primary visual cortex) analyze features of visual information and pass the results of this analysis to subregions of the extrastriate cortex, which perform further analyses and pass information on to other regions, culminating in the highest levels of visual association cortex in the parietal and inferior temporal lobes. The dorsal stream, which ends in the parietal cortex, is involved in perception of location ("where"), while the ventral stream, which ends in the inferior temporal cortex, is involved in perception of form ("what").

The auditory cortex seems to be similarly arranged. The primary auditory cortex lies hidden on the upper bank of the lateral fissure. The **core region,** which contains the primary auditory cortex, actually consists of three regions, each of which receives a separate tonotopic map of auditory information from the ventral division from the medial geniculate nucleus (Kaas, Hackett, and Tramo, 1999; Hackett, Preuss, and Kaas, 2001; Poremba et al., 2003; Petkov et al., 2006). The

FIGURE 7.9 ■ Pathways of the Auditory System

The major pathways are indicated by heavy arrows.

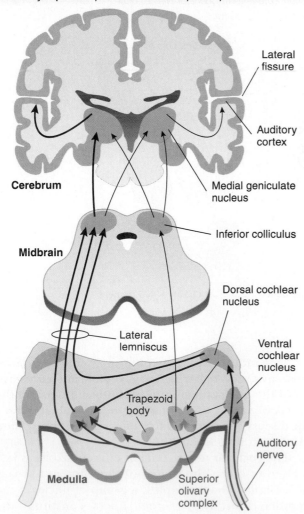

olivocochlear bundle   A bundle of efferent axons that travel from the olivary complex of the medulla to the auditory hair cells on the cochlea.

cochlear nucleus   One of a group of nuclei in the medulla that receive auditory information from the cochlea.

superior olivary complex   A group of nuclei in the medulla; involved with auditory functions, including localization of the source of sounds.

lateral lemniscus   A band of fibers running rostrally through the medulla and pons; carries fibers of the auditory system.

tonotopic representation   (*tonn oh top ik*)   A topographically organized mapping of different frequencies of sound that are represented in a particular region of the brain.

core region   The primary auditory cortex, located on a gyrus on the dorsal surface of the temporal lobe.

FIGURE 7.10 ■ **Auditory Cortex of the Rhesus Monkey**

The auditory cortex is situated in the temporal cortex on the bottom of the lateral fissure. The core region contains three auditory fields (A1, R, and RT), each of which contains a tonotopic map of auditory information. The belt region contains at least seven auditory fields (CM, RM, RTM, RTL, AL, ML, CL, and perhaps MM).

(Adapted from Petkov, C. I., Kayser, C., Augath, M., and Logothetis, N. K. *PLoS Biology*, 2006, *4*, 1213–1226.)

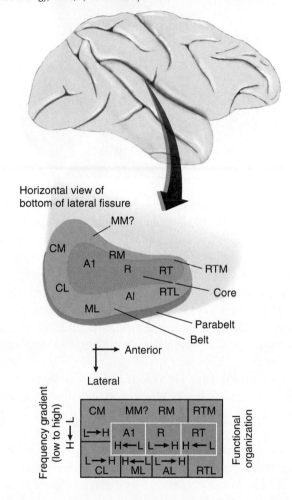

FIGURE 7.11 ■ **Anatomical Coding of Pitch**

Stimuli of different frequencies maximally deform different regions of the basilar membrane.

(Adapted from von Békésy, G. *Journal of the Acoustical Society of America*, 1949, *21*, 233–245.)

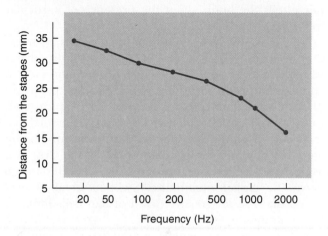

cortex, is involved with sound localization, and the ventral stream, which terminates in the parabelt region of the anterior temporal lobe, is involved with analysis of complex sounds (Rauschecker and Tian, 2000). Research on the functions of these streams is described later.

Perception of Pitch

As we have seen, the perceptual dimension of pitch corresponds to the physical dimension of frequency. The cochlea detects frequency by two means: moderate to high frequencies by place coding and low frequencies by rate coding. These two types of coding are described next.

Place Coding

The work of von Békésy has shown us that because of the mechanical construction of the cochlea and basilar membrane, acoustic stimuli of different frequencies cause different parts of the basilar membrane to flex back and forth. Figure 7.11 illustrates the amount of deformation along the length of the basilar membrane produced by stimulation with tones of various frequencies. Note that higher frequencies produce more displacement at the basal end of the membrane (the end closest to the stapes). (See *Figure 7.11.*)

These results suggest that at least some frequencies of sound waves are detected by means of a **place code.**

first level of auditory association cortex, the **belt region,** surrounds the primary auditory cortex, much as the extrastriate cortex surrounds the primary visual (striate) cortex. The belt region, which consists of at least seven divisions, receives information both from the primary auditory cortex and from the dorsal and medial divisions of the medial geniculate nucleus. The highest level of auditory association cortex, the **parabelt region,** receives information from the belt region and from the divisions of the medial geniculate nucleus that project to the belt region. (See *Figure 7.10.*)

Like the visual cortex, the auditory cortex is arranged in two streams, dorsal and ventral. The dorsal stream, which terminates in the posterior parietal

belt region The first level of auditory association cortex; surrounds the primary auditory cortex.

parabelt region The second level of auditory association cortex; surrounds the belt region.

place code The system by which information about different frequencies is coded by different locations on the basilar membrane.

In this context a code represents a means by which neurons can represent information. Thus, if neurons at one end of the basilar membrane are excited by higher frequencies and those at the other end are excited by lower frequencies, we can say that the frequency of the sound is *coded* by the particular neurons that are active. In turn, the firing of particular axons in the cochlear nerve tells the brain about the presence of particular frequencies of sound.

Evidence for place coding of pitch comes from several sources. High doses of the antibiotic drugs kanamycin and neomycin produce degeneration of the auditory hair cells. The damage begins at the basal end of the cochlea and progresses toward the apical end; this pattern can be verified by killing experimental animals after dosing them with the antibiotic for varying amounts of time. Longer exposures to the drug are associated with increased progress of hair cell damage down the basilar membrane. Stebbins et al. (1969) found that the progressive death of hair cells induced by an antibiotic closely parallels a progressive hearing loss: The highest frequencies are the first to go, and the lowest are the last.

Good evidence for place coding of pitch (at least, in humans) comes from the effectiveness of cochlear implants. **Cochlear implants** are devices that are used to restore hearing in people with deafness caused by damage to the hair cells. The external part of a cochlear implant consists of a microphone and a miniaturized electronic signal processor. The internal part contains a very thin, flexible array of 16–22 electrodes, which the surgeon carefully inserts into the cochlea in such a way that it follows the snaillike curl and ends up resting along the entire length of the basilar membrane. Each electrode in the array stimulates a different part of the basilar membrane. Information from the signal processor is passed to the electrodes by means of flat coils of wire, implanted under the skin. (See *Figure 7.12*.)

The primary purpose of a cochlear implant is to restore a person's ability to understand speech. Because most of the important acoustical information in speech is contained in frequencies that are too high to be accurately represented by a rate code, the multichannel electrode was developed in an attempt to duplicate the place coding of pitch on the basilar membrane (Copeland and Pillsbury, 2004). When different regions of the basilar membrane are stimulated, the person perceives sounds with different pitches. The signal processor in the external device analyzes the sounds detected by the microphone and sends separate signals to the appropriate portions of the basilar membrane. This device can work well; most people with cochlear implants can understand speech well enough to use a telephone (Shannon, 2007).

FIGURE 7.12 ■ A Child with a Cochlear Implant

The microphone and processor are worn over the ear, and the headpiece contains a coil that transmits signals to the implant.

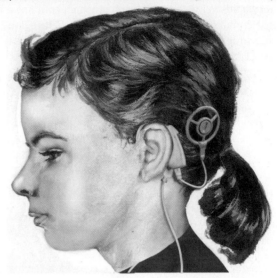

Many people in the Deaf community, who communicate with each other by means of signing, have negative feelings toward oral communication. The difficult task of deciphering lip movements makes them feel tense. They realize that their pronunciation is imperfect and that their voices sound strange to others. They feel at a disadvantage with respect to hearing people in a spoken conversation. In contrast, they feel relaxed and at ease when communicating with other deaf people.

Like other people who closely identify with their cultures, members of the Deaf community feel pride in their common heritage and react to perceived threats. Some deaf people say that if they were given the opportunity to hear, they would refuse it. Some deaf parents have expressed happiness when they learned that their children were born deaf too. They no longer needed to fear that their children would not be a part of their own Deaf culture.

Some members of the Deaf community perceive the cochlear implant as a serious threat to their culture. This device is most useful for two groups: people who became deaf in adulthood and very young children. Cochlear implants in postlingually deaf adults pose no threat to the Deaf community because these people never were members of the culture. But putting a cochlear implant in a young child means that the child's early education will be

---

cochlear implant An electronic device surgically implanted in the inner ear that can enable a deaf person to hear.

committed to the oralist approach. In addition, many deaf people resent the implication that deafness is something that needs to be repaired. They see themselves as different but not at all defective.

As we just saw, cochlear implants are most effective when they are implanted in very young children. One reason that they are less effective in older children is that the auditory cortex, lacking input from the ears, develops connections with other sensory modalities (especially vision) and becomes at least partially committed to serving functions other than audition (Doucet et al., 2006). In addition, the synaptic connections between terminals of auditory nerve neurons and neurons of the medulla are abnormal in congenitally deaf mammals, and some investigators have suggested that these abnormalities contributed to the decreased effectiveness of cochlear implants later in life. However, Ryugo, Kretzmer, and Niparko (2005) placed cochlear implants in the inner ears of congenitally deaf cats and found that three months of stimulation restored the synaptic connections to their normal state.

The work of von Békésy indicated that although the basilar membrane codes for frequency along its length, the coding was not very specific. His studies and those of investigators who followed him indicated that a given frequency causes a large region of the basilar membrane to be deformed. This finding contrasted with the observation that people can detect changes in frequency of only 2 or 3 Hz.

The reason for this discrepancy is now clear. Because of technical limitations, von Békésy had to observe the cochleas of animals that were no longer living or, at best, cochleas that had been damaged by the procedure necessary to make the measurements. More recently, investigators have used much more sensitive—and less damaging—procedures to observe movements of the basilar membrane in response to different frequencies of sound. It appears that the point of maximum vibration of the basilar membrane to a particular frequency is very precisely localized—but only when the cells in the organ of Corti are alive and healthy (Evans, 1992; Ruggero, 1992; Narayan et al., 1998).

The fact that the tuning characteristics of the basilar membrane change when the cells in the organ of Corti die suggested that these cells somehow affect the mechanical properties of the basilar membrane. We now know that the outer hair cells are responsible for this selective tuning and for amplification of the vibration of the basilar membrane produced by sound waves. As I mentioned earlier, outer hair cells are capable of motion. They contain contractile proteins, just as muscle fibers do. When these cells are exposed to an electrical current or when acetylcholine is placed on them, they contract by up to 10 percent of their length (Brownell et al., 1985; Zenner, Zimmermann, and Schmitt, 1985). Because the tips of their cilia are embedded in the tectorial membrane, contraction alters the mechanical characteristics of the basilar membrane—and consequently the response properties of the inner hair cells.

When the basilar membrane vibrates, movement of the cilia of the outer hair cells opens and closes ion channels, causing changes in the membrane potential. These changes cause movements of the contractile proteins, thus lengthening and shortening the cells. These changes in length amplify the vibrations of the basilar membrane. As a consequence, the signal that is received by inner hair cells is enhanced, which greatly increases the sensitivity of the inner ear to sound waves.

Figure 7.13 illustrates the importance of outer hair cells to the sensitivity and frequency selectivity of inner hair cells (Fettiplace and Hackney, 2006). The three V-shaped *tuning curves* indicate the sensitivity of individual inner hair cells, as shown by the response of individual afferent auditory nerve axons to pure tones. The low points of the three solid curves indicate that the hair cells will respond to a faint sound only if it is of a specific frequency—for these cells, either 0.5 kHz (red curve), 2.0 kHz (green curve), or 8.0 kHz (blue curve). If the sound is louder, the cells will respond to frequencies above and below their preferred frequencies. The dashed line indicates the response of the "blue" neuron after the outer hair cells have been destroyed. As you can see, this cell loses both sensitivity and selectivity: It will respond only to loud sounds but to a wide range of frequencies. (See *Figure 7.13.*)

Rate Coding

We have seen that the frequency of a sound can be detected by place coding. However, the lowest frequencies do not appear to be accounted for in this manner. Kiang (1965) was unable to find any cells that responded best to frequencies of less than 200 Hz. How, then, can animals distinguish low frequencies? It appears that lower frequencies are detected by neurons that fire in synchrony to the movements of the apical end of the basilar membrane. Thus, lower frequencies are detected by means of a **rate coding.**

The most convincing evidence of rate coding of pitch also comes from studies of people with cochlear implants. Pijl and Schwartz (1995a, 1995b) found that stimulation of a single electrode with pulses of electricity produced sensations of pitch that were proportional to the frequency of the stimulation. In fact, the subjects

---

rate coding The system by which information about different frequencies is coded by the rate of firing of neurons in the auditory system.

FIGURE 7.13 ■ Tuning Curves

The figure shows the responses of single axons in the cochlear nerve that receive information from inner hair cells on different locations of the basilar membrane. The cells are more frequency selective at lower sound intensities. The dashed line shows the loss of sensitivity and selectivity of the high-frequency neuron after destruction of the outer hair cells.

(Adapted from Fettiplace, R., and Hackney, C. M. *Nature Reviews: Neuroscience*, 2006, 7, 19–29.)

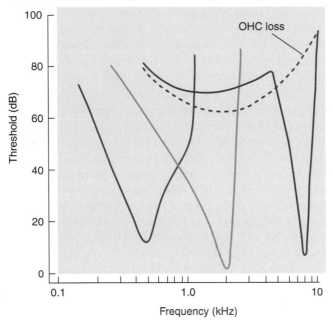

showing that the auditory system is very sensitive. Thus, in very quiet environments a young, healthy ear is limited in its ability to detect sounds in the air by the masking noise of blood rushing through the cranial blood vessels rather than by the sensitivity of the auditory system itself. More recent studies using modern instruments (reviewed by Hudspeth, 1983) have essentially confirmed Wilska's measurements. The softest sounds that can be detected appear to move the tip of the hair cells between 1 and 100 picometers (trillionths of a meter). They achieve their maximum response when the tips are moved 100 nanometers (Corwin and Warchol, 1991).

The axons of the cochlear nerve appear to inform the brain of the loudness of a stimulus by altering their rate of firing. Louder sounds produce more intense vibrations of the eardrum and ossicles, which produce a more intense shearing force on the cilia of the auditory hair cells. As a result, these cells release more neurotransmitter, producing a higher rate of firing by the cochlear nerve axons. This explanation seems simple for the axons involved in place coding of pitch; in this case, pitch is signaled by which neurons fire, and loudness is signaled by their rate of firing. However, the neurons in the apex of the basilar membrane that signal the lowest frequencies do so by their rate of firing. If they fire more frequently, they signal a higher pitch. Therefore, most investigators believe that the loudness of low-frequency sounds is signaled by the *number* of axons arising from these neurons that are active at a given time.

Perception of Timbre

Although laboratory investigations of the auditory system often employ pure sine waves as stimuli, these waves are seldom encountered outside the laboratory. Instead, we hear sounds with a rich mixture of frequencies—sounds of complex timbre. For example, consider the sound of a clarinet playing a particular note. If we hear it, we can easily say that it is a clarinet and not a flute or a violin. The reason we can do so is that these three instruments produce sounds of different timbre, which our auditory system can distinguish.

Figure 7.14 shows the waveform from a clarinet playing a steady note (*top*). The shape of the waveform repeats itself regularly at the **fundamental frequency,** which corresponds to the perceived pitch of the note. A Fourier analysis of the waveform shows that it actually consists of a series of sine waves that includes the fundamental frequency and many **overtones,** multiples of the fundamental frequency. Different instruments produce overtones with

could even recognize familiar tunes produced by modulating the pulse frequency. (The subjects had become deaf later in life, after they had already learned to recognize the tunes.) As we would expect, the subjects' perceptions were best when the tip of the basilar membrane was stimulated, and only low frequencies could be distinguished by this method. (See *MyPsychKit 7.1, Perception of Pitch.*)

Animation 7.1
Perception of Pitch

Perception of Loudness

The cochlea is an extremely sensitive organ. Wilska (1935) used an ingenious procedure to estimate the smallest vibration needed to produce a perceptible sound. He glued a small wooden rod to a volunteer's tympanic membrane (temporarily, of course) and made the rod vibrate longitudinally by means of an electromagnetic coil that could be energized with alternating current. He could vary the frequency and intensity of the current, which consequently changed the perceived pitch and loudness of the stimulus. He found that subjects could detect a sound even when the eardrum was vibrated over a distance less than the diameter of a hydrogen atom,

fundamental frequency The lowest, and usually most intense, frequency of a complex sound; most often perceived as the sound's basic pitch.

overtone The frequency of complex tones that occurs at multiples of the fundamental frequency.

FIGURE 7.14 ■ Sound Wave from Clarinet

Shown in the figure is the shape of a sound wave from a clarinet (top) and the individual frequencies into which it can be analyzed.

(Reprinted from *Stereo Review,* copyright © 1977 by Diamandis Communications Inc.)

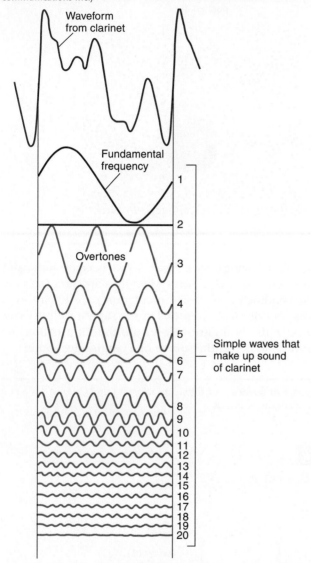

different intensities. (See *Figure 7.14.*) Electronic synthesizers simulate the sounds of real instruments by producing a series of overtones of the proper intensities, mixing them, and passing them through a loudspeaker.

When the basilar membrane is stimulated by the sound of a clarinet, different portions respond to each of the overtones. This response produces a unique anatomically coded pattern of activity in the cochlear nerve, which is subsequently identified by circuits in the auditory association cortex.

Actually, the recognition of complex sounds is not quite that simple. Figure 7.14 shows the analysis of a *sustained* sound of a clarinet. But most sounds (including those produced by a clarinet) are dynamic; that is, their beginning, middle, and end are different from each other. The beginning of a note played on a clarinet (the *attack*) contains frequencies that appear and disappear in a few milliseconds. And at the end of the note (the *decay*), some harmonics disappear before others. If we are to recognize different sounds, the auditory cortex must analyze a complex sequence of multiple frequencies that appear, change in amplitude, and disappear. And when you consider the fact that we can listen to an orchestra and identify several instruments that are playing simultaneously, you can appreciate the complexity of the analysis performed by the auditory system. We will revisit this process later in this chapter.

Perception of Spatial Location

So far, I have discussed coding of pitch, loudness, and timbre only (the last of which is actually a complex frequency analysis). The auditory system also responds to other qualities of acoustic stimuli. For example, our ears are very good at determining whether the source of a sound is to the right or left of us. Two separate physiological mechanisms detect the location of sound sources: We use phase differences for low frequencies (less than approximately 3000 Hz) and intensity differences for high frequencies. In addition, we use another mechanism—analysis of timbre—to determine the height of the source of a sound and whether it is in front of us or behind us.

Localization by Means of Arrival Time and Phase Differences

If we are blindfolded, we can still determine with rather good accuracy the location of a stimulus that emits a click. We are most accurate at judging the *azimuth*—that is, the horizontal (left or right) angle of the source of the sound relative to the midline of our body. Neurons in our auditory system respond selectively to different *arrival times* of the sound waves at the left and right ears. If the source of the click is to the right or left of the midline, the sound pressure wave will reach one ear sooner and initiate action potentials there first. Only if the stimulus is straight ahead will the ears be stimulated simultaneously. Many neurons in the auditory system respond to sounds presented to either ear. Some of these neurons, especially those in the superior olivary complex of the medulla, respond according to the difference in arrival times of sound waves produced by clicks presented *binaurally* (that is, to both ears). Their response rates reflect differences as small as a fraction of a millisecond.

Of course, we can hear continuous sounds as well as clicks, and we can also perceive the location of their source. We detect the source of continuous low-pitched sounds by means of phase differences.

FIGURE 7.15 ■ Sound Localization

This method localizes low- and medium-frequency sounds through phase differences.
(a) Source of a 1000-Hz tone to the right. The pressure waves on each eardrum are out of
phase; one eardrum is pushed in while the other is pushed out. (b) Source of a sound
directly in front. The vibrations of the eardrums are synchronized (in phase).

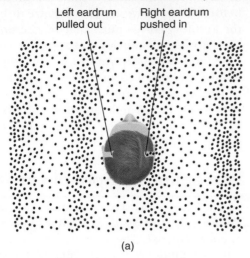

Left eardrum   Right eardrum
pulled out   pushed in

(a)

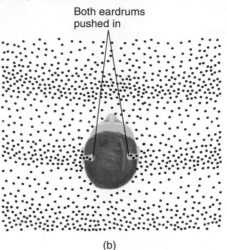

Both eardrums
pushed in

(b)

Phase differences refer to the simultaneous arrival, at each ear, of different portions (phases) of the oscillating sound wave. For example, if we assume that sound travels at 700 miles per hour through the air, adjacent cycles of a 1000-Hz tone are 12.3 inches apart. Thus, if the source of the sound is located to one side of the head, one eardrum is pulled out while the other is pushed in. The movement of the eardrums will reverse, or be 180° *out of phase.* If the source were located directly in front of the head, the movements would be perfectly in phase

(0° out of phase). (See *Figure 7.15.*) Because some auditory neurons respond only when the eardrums (and thus the bending of the basilar membrane) are at least somewhat out of phase, neurons in the superior olivary complex in the brain are able to use the information they provide to detect the source of a continuous sound.

---

phase difference   The difference in arrival times of sound waves at each of the eardrums.

FIGURE 7.16 ■ Model of a Coincidence Detector

This model detector can determine differences in arrival times at each ear of an auditory stimulus.

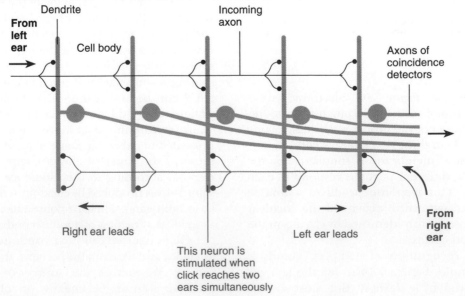

Dendrite   Incoming
                                axon
**From
left
ear**   Cell body

                                              Axons of
                                              coincidence
                                              detectors

Right ear leads   Left ear leads

                                              **From
                                              right
                                              ear**

This neuron is
stimulated when
click reaches two
ears simultaneously

A possible mechanism to explain the ability of the nervous system to detect very short delays in the arrival times of two signals was first proposed by Jeffress (1948). He suggested that neurons received information from two sets of axons coming from the two ears. Each neuron served as a *coincidence detector;* it responded only if it received signals simultaneously from synapses belonging to both sets of axons. If a signal reached the two ears simultaneously, neurons in the middle of the array would fire. However, if the signal reached one ear before the other, then neurons farther away from the "early" ear would be stimulated. (See *Figure 7.16.*)

In fact, that is exactly how the mechanism works. Carr and Konishi (1988, 1990) obtained anatomical evidence in support of Jeffress's hypothesis from the brain of the barn owl, a nocturnal bird that can very accurately detect the source of a sound (such as that made by an unfortunate mouse). Figure 7.17 shows a drawing of the distribution of the branches of two axons, one from each ear, projecting to the nucleus laminaris, the barn owl analog of the mammalian medial superior olive. As you can see, axons from the ipsilateral and contralateral ears penetrate the nucleus from opposite directions; therefore, dorsally located neurons within the nucleus are stimulated by sounds that first reach the contralateral ear. (Compare *Figures 7.16* and *7.17.*) Carr and Konishi recorded from single units within the nucleus and found that the response characteristics of the neurons located there were perfectly consistent with these anatomical facts. (See *MyPsychKit 7.2, Sound Localization.*)

mypsychkit
Animation 7.2
Sound of Localization

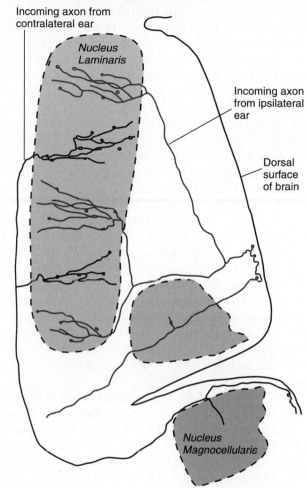

FIGURE 7.17 ■ Evidence for a Coincidence Detector in the Brain of a Barn Owl

Compare the branches of the axons with those of Figure 7.16. The drawing was prepared from microscopic examination of sections of stained tissue.

(Adapted from Carr, C. E., and Konishi, M. *Proceedings of the National Academy of Sciences, USA,* 1989, *85,* 8311–8315.)

Incoming axon from contralateral ear

Nucleus Laminaris

Incoming axon from ipsilateral ear

Dorsal surface of brain

Nucleus Magnocellularis

Localization by Means of Intensity Differences

The auditory system cannot readily detect binaural phase differences of high-frequency stimuli; the differences in phases of such rapid sine waves are just too short to be measured by the neurons. However, high-frequency stimuli that occur to the right or left of the midline stimulate the ears unequally. The head absorbs high frequencies, producing a "sonic shadow," so the ear closest to the source of the sound receives the most intense stimulation. Some neurons in the auditory system respond differentially to binaural stimuli of different intensity in each ear, which means that they provide information that can be used to detect the source of tones of high frequency.

The neurons that detect binaural differences in loudness are located in the superior olivary complex. But whereas neurons that detect binaural differences in phase or arrival time are located in the *medial* superior olivary complex, these neurons are located in the *lateral* superior olivary complex. Information from both sets of neurons is sent to other levels of the auditory system.

Localization by Means of Timbre

We just saw that left–right localization of the source of high- and low-frequency sounds is accomplished by two different mechanisms. But how can we determine the elevation of the source of a sound and perceive whether it is in front of us or behind us? One answer is that we can turn and tilt our heads, thus transforming the discrimination into a left–right decision. But we have another means by which we can determine elevation and distinguish front from back: analysis of timbre. This method involves a part of the auditory system that I have not said much about: the external ear (pinna). If you look at someone's external ear, you will see that it contains several folds and ridges. Most of the sound waves that we hear bounce off the folds and ridges of the pinna before they enter the ear canal.

This process changes the nature of the sounds that we hear. Depending on the angle at which the sound waves strike these folds and ridges, different frequencies will be enhanced or attenuated. In other words, the pattern of reflections will change with the location of the source of the sound, which will alter the timbre of the sound that is perceived. Sounds coming from behind the head will sound different from those coming from above the head or in front of it, and sounds coming from above will sound different from those coming from the level of our ears.

Figure 7.18 shows the effects of elevation on the intensity of sounds of various frequencies received at an ear (Oertel and Young, 2004). The experimenters placed a small microphone in a cat's ear and recorded the sound produced by an auditory stimulus presented at various elevations relative to the cat's head. They used a computer to plot the ear's transfer functions—a graph that compares the intensity of various frequencies of sound received by the ear to the intensity of these frequencies received by a microphone in open air. What is important in Figure 7.18 is not the shape of the transfer functions, but the fact that these functions varied with the elevation of the source of the sound. The transfer function for a sound directly in front of the cat (0° of elevation) is shown in green. This curve is shown at the 60°, 30°, and −30° positions as well so that they can be compared with the curves obtained with the sound source at these locations, too (red, orange, and blue, respectively). That sounds complicated, I know, but if you look at the figure, you will clearly see that the timbre of sounds that reaches the cat's ear changed along with elevation of the source of the sound. (See *Figure 7.18.*)

The first level of analysis of information about the elevation of a sound appears to take place in the dorsal cochlear nucleus. Destruction of the axons that connect this nucleus with the rest of the auditory system disrupt a cat's ability to orient to sources of sounds that differ in elevation (Sutherland, Masterton, and Glendenning, 1998). In addition, Kanold and Young (2001) found that the dorsal cochlear nucleus receives information from sensory receptors muscles that move a cat's ear. Presumably, this information can be used to correct for changes in the orientation of the cat's ear, just as the visual system can compensate for movements of the eye.

People's ears differ in shape; thus, the changes in the timbre of a sound coming from different locations will also differ from person to person. This means that each individual must learn to recognize the subtle changes in the timbre of sounds that originate in locations in front of the head, behind it, above it, or below it. The neural circuits that accomplish this task are not genetically programmed—they must be acquired as a result of experience.

FIGURE 7.18 ■ Changes in Timbre of Sounds with Changes in Elevation

The graphs are transfer functions, which compare the intensity of various frequencies of sound received by the ear to the intensity of these frequencies received by a microphone in open air. For ease of comparison, the 0° transfer function (green) is superimposed on the transfer functions obtained at 60° (red), 30° (orange), and −30° (blue). The differences in the transfer functions at various elevations provide cues that aid in perception of the location of a sound source.

(Adapted from Oertel, D., and Young, E. D. *Trends in Neuroscience,* 2004, *27,* 104–110.)

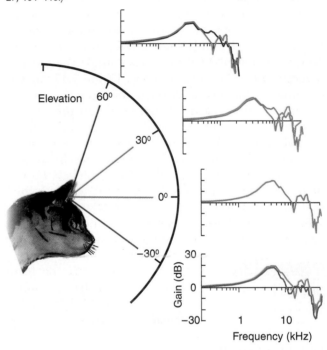

An experiment by Zwiers, Van Opstal, and Cruysberg (2001) found evidence for the role of experience in calibrating the sensitivity of the auditory system to changes in elevation. They found that blind people had more difficulty judging the elevation of sounds than sighted people did, especially if some noise was present. Presumably, the increased accuracy of sighted people reflected the fact that they had had the opportunity to calibrate the changes in timbre of sounds caused by changes in the height of their sources, which they could see. In contrast, the ability of blind people to perceive the horizontal location of the sources of sounds was as good as that of sighted people. After all, blind people have much experience navigating to and around the sources of sounds located at ground level (and objects that reflect sounds, such as that of a tapping cane). These perceptions can be calibrated by physical contact with these objects.

Perception of Complex Sounds

Hearing has three primary functions: to detect sounds, to determine the location of their sources, and to recognize the identity of these sources—and thus their meaning and relevance to us (Heffner and Heffner, 1990; Yost, 1991). Let us consider the third function: recognizing the identity of a sound source. Unless you are in a completely silent location, pay attention to what you can hear. Right now, I am sitting in an office and can hear the sound of a fan in a computer, the tapping of the keys as I write this, the footsteps of someone passing outside the door, and the voices of some people talking in the hallway. How can I recognize these sources? The axons in my cochlear nerve contain a constantly changing pattern of activity corresponding to the constantly changing mixtures of frequencies that strike my eardrums. Somehow, the auditory system of my brain recognizes particular patterns that belong to particular sources, and I perceive each of them as an independent entity.

Perception of Environmental Sounds and Their Location

The task of the auditory system in identifying sound sources, is one of *pattern recognition*. The auditory system must recognize that particular patterns of constantly changing activity belong to different sound sources. And as we saw, few patterns are simple mixtures of fixed frequencies. For example, notes of different pitches produce different patterns of activity in our cochlear nerve, yet we recognize each of the notes as belonging to a clarinet. In addition, the notes played on a clarinet have a characteristic attack and decay. And consider the complexity of sounds that occur in the environment: cars honking, birds chirping, people coughing, doors slamming, and so on. (I will discuss speech recognition—an even more complicated task—in Chapter 14.) Needless to say, we are far from understanding how pattern recognition of such complex sounds works.

Perception of complex sounds appears to be accomplished by circuits of neurons in the auditory cortex. However these sounds are recognized, it is clear that the circuits that perform the analysis must receive accurate information. Recognition of complex sounds requires that the timing of changes in the components of the sounds be preserved all the way to the auditory cortex. In fact, the neurons that convey information to the auditory cortex contain special features that permit them to conduct this information rapidly and accurately (Trussell, 1999). Their axons contain special low-threshold voltage-gated potassium channels that produce very short action potentials. Their terminal buttons are large and release large amounts of glutamate, and the postsynaptic membrane contains neurotransmitter-dependent ion channels that act unusually rapidly; thus, these synapses produce very strong EPSPs. The terminal buttons form synapses with the somatic membrane of the postsynaptic neurons, which minimizes the distance between the synapses and the axon—and the delay in conducting information to the axon of the postsynaptic neuron.

As I mentioned earlier in this chapter, the auditory cortex, like the visual cortex, is organized into two streams: a dorsal stream, involved in perception of location, and a ventral stream, involved in perception of form. In a single-unit recording study, Rauschecker and Tian (2000) found that neurons in the "what" system discriminated between different monkey calls, while neurons in the "where" system discriminated between different locations of loudspeakers presenting these calls.

Figure 7.19 compares the regions of the monkey brain that are devoted to the processing of visual and auditory information, as determined by 2-DG autoradiography. As you can see, the dorsal streams of both systems overlap in the parietal lobe. This overlap is undoubtedly related to the fact that monkeys (and humans too) can use the convergence of sight and sound to recognize which of several objects in the environment is making a noise. In addition, we can learn the association between the sight of an object and the sounds it makes. Information from both the visual and auditory systems is also projected to specific regions of the frontal lobes—again, with a region where both systems overlap. The role of the frontal lobes in learning and memory is discussed in Chapter 13. (See *Figure 7.19.*)

As we saw in Chapter 6, lesions of the visual association cortex can selectively impair various aspects of visual perception. Damage to the ventral stream can

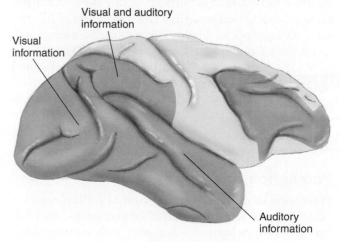

FIGURE 7.19 ■ Sensory Processing

The figure shows regions of the monkey brain that are devoted to the processing of visual and auditory information.

(Adapted from Poremba, A., Saunders, R. C., Crane, A. M., Cook, M., Sokoloff, L., and Mishkin, M. *Science,* 2003, *299,* 569–572.)

Visual and auditory information

Visual information

Auditory information

produce visual agnosias—the inability to recognize objects even though the visual acuity may be good, and lesions of the dorsal stream disrupt performance on a variety of tasks that require perceiving and remembering the locations of objects. Similarly, lesions of the auditory association cortex can produce auditory agnosias—impairment of various aspects of auditory perception, even though the individuals are not deaf.

Clarke et al. (2000) reported the cases of three patients with brain damage that affected different portions of the auditory cortex. The investigators tested the patients' ability to recognize environmental sounds, to identify the locations from which the sounds were coming, and to detect when the source of a sound was moving. Patient F. D. had difficulty recognizing environmental sounds but could identify sound location or movement. Patient C. Z. could recognize environmental sounds but could not identify sound location or movement. Finally, although patient M. A. was not deaf, she showed deficits in all three tasks: recognition, localization, and perception of movement. Although the lesions in these patients were too large for us to determine the exact locations of the brain regions responsible for perception of environmental sounds and the location of their sources, we can certainly conclude that different regions of the auditory cortex are involved in perceiving *what* and *where*.

A review of thirty-eight functional-imaging studies with human subjects (Arnott et al., 2004) reported a consistent result: Perception of the identity of sounds activated the ventral stream of the auditory cortex and perception of the location of sounds activated the dorsal stream. A functional MRI (fMRI) study by Alain, He, and Grady (2008) supports this conclusion. The investigators presented people with sounds of animals, humans, and musical instruments (for example, the bark of a dog, a cough, and the sound of a flute) in one of three locations: 90° to the left, straight ahead, or 90° to the right. On some blocks of trials the subjects were asked to press a button when they heard two sounds of any kind from the same location. On other blocks of trials they were asked to indicate when they heard the same kind of sound twice in a row, regardless of its location. As Figure 7.20 shows, judgments of location activated dorsal regions ("where"), and judgments of the nature of a sound activated ventral regions ("what"). (See *Figure 7.20.*) An interesting experiment is described in *MyPsychKit 7.3: Perception of Environmental Sounds.*

mypsychkit
Where learning comes to life!

Animation 7.3
Perception of Environmental Sounds

Perception of Music

Perception of music is a special form of auditory perception. Music consists of sounds of various pitches and timbres played in a particular sequence with an underlying

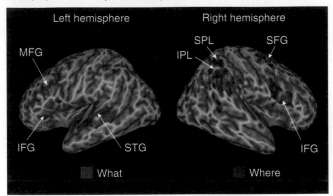

FIGURE 7.20 ■ "Where" versus "What"

This figure shows regional brain activity in response to judgments of category (blue) and location (red) of sounds. IFG = inferior frontal gyrus, IPL = inferior parietal lobule, MFG = middle frontal gyrus, SFG = superior frontal gyrus, SPL = superior parietal lobule, STG = superior temporal gyrus.

(From Alain, C., He, Y., and Grady, C. *Journal of Cognitive Neuroscience,* 2008, *20,* 285–295. Reprinted with permission.)

rhythm. Particular combinations of musical notes played simultaneously are perceived as consonant or dissonant, pleasant or unpleasant. The intervals between notes of musical scales follow specific rules, which may vary in the music of different cultures. In Western music, melodies played using notes that follow one set of rules (the major mode) usually sound happy, while those played using another set of rules (the minor mode) generally sound sad. In addition, a melody is recognized by the relative intervals between its notes, not by their absolute value. A melody is perceived as unchanging even when it is played in different keys—that is, when the pitches of all the notes are raised or lowered without changing the relative intervals between them. Thus, musical perception requires recognition of sequences of notes, their adherence to rules that govern permissible pitches, harmonic combinations of notes, and rhythmical structure. Because the duration of musical pieces is several seconds to many minutes, musical perception involves a substantial memory capacity. Thus, the neural mechanisms required for musical perception must obviously be complex.

The analysis of music in the brain, like that of all other acoustical stimuli, begins with the subcortical auditory pathways and the primary auditory cortex. Then more complex aspects of music are analyzed by regions of the auditory association cortex. For example, we saw that the sounds made by musical instruments are complex. Pitch is determined by the fundamental frequency, and timbre is determined by the mixture of overtones. Studies with monkeys and humans have found that the primary auditory cortex responds to pure tones of different frequencies but that recognition of the pitch of complex sounds is accomplished only by the auditory

FIGURE 7.21 ■ Human Auditory Cortex

The auditory cortex is located on the cortex on the bottom of the lateral fissure. A1 = primary auditory cortex, aSTG = anterior superior temporal gyrus, H = high, L = low, PT = planum temporale (region of auditory association cortex), R = area R (rostral auditory cortex).

(Adapted from Bendor, D., and Wang, X. *Current Opinion in Neurobiology*, 2006, *16*, 391–399.)

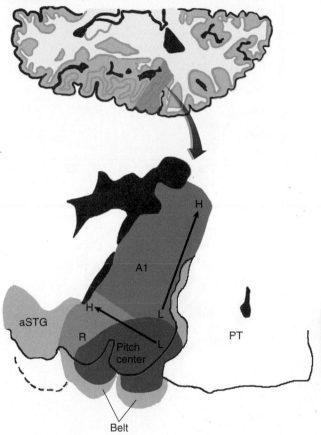

ganglia are involved in timing of musical rhythms, as they are in the timing of movements.

Everyone learns a language, but only some people become musicians. Musical training obviously makes changes in the brain—changes in motor systems involved in singing or playing an instrument, and changes in the auditory system involved in recognizing subtle complexities of harmony, rhythm, and other characteristics of musical structure. Here, I will consider aspects of musical expertise related to audition. Some of the effects of musical training can be seen in changes in the structure or activity of portions of the auditory system of the brain. For example, a study by Schneider et al. (2002) found that the volume of the primary auditory cortex of musicians was 130 percent larger than that of nonmusicians, and the neural response in this area to musical tones was 102 percent greater in musicians. Moreover, both of these measures were positively related to a person's musical aptitude.

Even differences in subcortical components of the auditory cortex are related to musical ability. Musacchia et al. (2007) found that EEG recordings of brain stem activity in response to musical stimuli had a higher amplitude and more faithfully represented the stimuli in musicians than in nonmusicians. In addition, the quality of the responses were strongly correlated with the length of the person's musical training.

Patient I. R., a right-handed woman in her early forties, sustained bilateral damage during surgical treatment of aneurysms located on her middle cerebral arteries. Aneurysms (discussed in more detail in Chapter 15), are balloonlike swellings on blood vessels that are sometimes subject to rupture, which can have fatal consequences. The surgery successfully clipped off the aneurysms but resulted in damage to most of the left superior temporal gyrus, some of the inferior frontal and parietal lobes bordering the lateral fissure. Damage to the right hemisphere was less severe but included the anterior third of the superior temporal gyrus and the right inferior and middle frontal gyri.

Ten years after the surgery, Peretz and her colleagues studied the effects of the patient's brain damage on her musical ability (Peretz, Gagnon, and Bouchard, 1998). Although Patient I. R. had normal hearing, could understand speech and converse normally, and could recognize environmental sounds, she showed a nearly complete **amusia**—loss of the ability to perceive or produce melodic or rhythmic aspects of music. She had been raised in a musical environment; both her grandmother and brother were professional musicians. After her surgery, she

association cortex (Bendor and Wang, 2006). Functional-imaging studies with humans indicate that pitch discrimination takes place in a region of the superior temporal gyrus rostral and lateral to the primary auditory cortex. (See *Figure 7.21*.)

Different regions of the brain are involved in different aspects of musical perception (Peretz and Zatorre, 2005). For example, the inferior frontal cortex appears to be involved in recognition of harmony, the right auditory cortex appears to be involved in perception of the underlying beat in music, and the left auditory cortex appears to be involved in perception of rhythmic patterns that are superimposed on the rhythmic beat. (Think of a drummer indicating the regular, underlying beat by operating the foot pedal of the bass drum and superimposing a more complex pattern of beats on smaller drums with the drumsticks.) In addition, the cerebellum and basal

amusia (*a mew zia*) Loss or impairment of musical abilities, produced by hereditary factors or brain damage.

lost the ability to recognize melodies that she had been familiar with previously, including simple pieces such as "Happy Birthday." She was no longer able to sing.

Remarkably, despite her inability to recognize melodic and rhythmic aspects of music, she insisted that she still enjoyed listening to music. Peretz and her colleagues discovered that I. R. was still able to recognize emotional aspects of music. Although she could not recognize pieces that the experimenters played for her, she recognized whether the music sounded happy or sad. She could also recognize happiness, sadness, fear, anger, surprise, and disgust in a person's tone of voice. The ability to recognize emotion in music contrasts with her inability to recognize dissonance in music—a quality that normal listeners find intensely unpleasant. Peretz and her colleagues (2001) discovered that I. R. was totally insensitive to changes in music that irritate normal listeners. Even four-month-old babies prefer consonant music to dissonant music, which shows that recognition of dissonance develops very early in life (Zentner and Kagan, 1998.) You can listen to music that varies in emotional content (happy, sad, peaceful, and scary) and dissonance in *MyPsychKit 7.4, Emotion and Dissonance in Music.* I think it's fascinating that Patient I. R. could not distinguish between the dissonant and consonant versions but could still identify happy and sad music.

mypsychkit
Animation 7.4
Emotion and Dissonance in Music

Approximately 4 percent of the population exhibits congenital amusia—a severe and persistent deficit in musical ability (but not in perception of speech or environmental sounds) that becomes apparent early in life. As we have seen, people with musical training show differences in the response of cortical and subcortical components of the auditory system in response to musical sounds. Studies have found that people with congenital amusia also show differences in the structure of their auditory system. Structural MRI studies by Hyde et al. (2006, 2007) found that the auditory cortex of the right superior temporal gyrus (STG) and the cortex of the right inferior frontal gyrus (IFG) was thicker in people with congenital amusia, and the white matter of the right IFG was thinner. The authors note that the increased thickness of the cortex is a probable indication of cortical malformations, perhaps caused by abnormal neuronal migration. In fact, Hyde and her colleagues found that the thickness of these two cortical regions was negatively correlated with musical ability. (See *Figure 7.22.*)

Musical ability in general and congenital amusia in particular appear to have a genetic basis. Drayna et al. (2001) had pairs of twins listen to simple popular melodies and determine which ones contained some wrong—and discordant—notes. They found that the

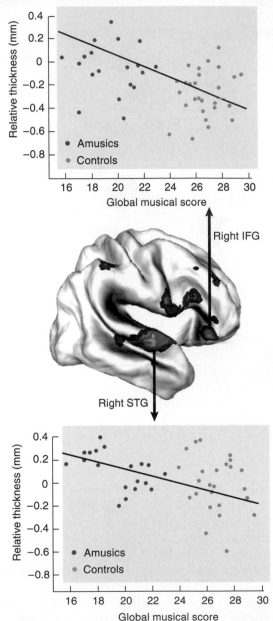

FIGURE 7.22 ■ Cortex and Musical Ability

The thickness of the cortex of the right inferior frontal gyrus (IFG) and right superior temporal gyrus (STG) and musical ability was found to be inversely related to musical ability.

(From Hyde, K. L., Lerch, J. P., Zatorre, R .J., Griffiths, T. D., Evans, A. C., and Peretz, I. *Journal of Neuroscience,* 2007, *27,* 13028–13032. Reprinted with permission.)

Right IFG

Right STG

correlation between the scores of the twin pairs was .67 for monozygotic twins but only .44 for dizygotic twins. These results indicate a heritability index in this kind of musical ability of approximately .75 (on a scale of 0–1.0). Peretz, Cummings, and Dubé (2007) found that 39 percent of first-degree relatives (siblings, parents, or children) of people with amusia also had amusia, compared with an incidence of only 3 percent in the first-degree relatives of people in control families.

InterimSummary

Audition

The receptive organ for audition is the organ of Corti, located on the basilar membrane. When sound strikes the tympanic membrane, it sets the ossicles into motion, and the baseplate of the stapes pushes against the membrane behind the oval window. Pressure changes thus applied to the fluid within the cochlea cause a portion of the basilar membrane to flex, causing the basilar membrane to move laterally with respect to the tectorial membrane that overhangs it. This movement pulls directly on the cilia of the outer hair cells and changes their membrane potential. This change causes contractions or relaxations of contractile proteins within the cell, which amplify movements of the basilar membrane and sharpen their focus. These events cause movements in the fluid within the cochlea, which, in turn, causes the cilia of the inner hair cells to wave back and forth. These mechanical forces open cation channels in the tips of the hair cells and thus produce receptor potentials.

The inner hair cells form synapses with the dendrites of the bipolar neurons whose axons give rise to the cochlear branch of the eighth cranial nerve. The central auditory system involves several brain stem nuclei, including the cochlear nuclei, superior olivary complexes, and inferior colliculi. The medial geniculate nucleus relays auditory information to the primary auditory cortex on the medial surface of the temporal lobe. The primary auditory cortex contains three separate tonotopic representations of auditory information and is surrounded by two levels of auditory association cortex: the belt region, which contains seven tonotopic maps, and the parabelt region. As we saw in Chapter 6, the visual association cortex is divided into two streams, one analyzing color and form and the other analyzing location and movement. Similarly, the auditory association cortex is organized into streams that analyze the nature of sounds and the and location of their sources.

Pitch is encoded by two means. High-frequency sounds cause the base of the basilar membrane (near the oval window) to flex; low-frequency sounds cause the apex (opposite end) to flex. Because high and low frequencies thus stimulate different groups of auditory hair cells, frequency is encoded anatomically. The lowest frequencies cause the apex of the basilar membrane to flex back and forth in time with the acoustic vibrations. The outer hair cells act as motive elements rather than as sensory transducers, contracting in response to activity of the efferent axons and modifying the mechanical properties of the basilar membrane.

The auditory system is analytical in its operation. That is, it can discriminate between sounds with different timbres by detecting the individual overtones that constitute the sounds and producing unique patterns of neural firing in the auditory system.

Left–right localization is performed by analyzing binaural differences in arrival time, in phase relations, and in intensity. The location of the azimuth of the sources of brief sounds (such as clicks) and sounds of frequencies below approximately 3000 Hz is detected by neurons in the medial superior olivary complex, which respond most vigorously when one ear receives the click first or when the phase of a sine wave received by one ear leads that received by the other. The location of the azimuth of the sources of high-frequency sounds is detected by neurons in the lateral superior olivary complex, which respond most vigorously when one organ of Corti is stimulated more intensely than the other. Localization of the elevation of the sources of sounds can be accomplished by turning the head or by perception of subtle differences in the timbre of sounds coming from different directions. The folds and ridges in the external ear (pinna) reflect different frequencies into the ear canal, changing the timbre of the sound according to the location of its source.

To recognize the source of sounds, the auditory system must recognize the constantly changing patterns of activity received from the axons in the cochlear nerve. Like the visual cortex, the auditory cortex is organized into two streams. Electrophysiological and functional-imaging studies indicate that the ventral stream is involved in the analysis of the sound, and the dorsal stream is involved in perception of its location. Localized lesions of the auditory association cortex can impair people's ability to recognize environmental sounds, sound location, or sound movement.

Perception of music requires recognition of sequences of notes, their adherence to rules governing permissible pitches, harmonic combinations of notes, and rhythmical structure. Perception of pitch activates regions of the superior temporal gyrus rostral and lateral to the primary auditory cortex. Other regions of the brain—especially in the right hemisphere—are involved in perception of the underlying beat of music and the specific rhythmic patterns of a particular piece. Musical training appears to increase the size and responsiveness of the primary auditory cortex. A case study indicates that recognition of emotion in music involves some brain mechanisms independent of those that recognize dissonance. Congenital amusia appears to be related to abnormal development of the right superior temporal gyrus and the right inferior frontal gyrus. Musical ability, and the occurrence of amusia, appear to have a genetic basis.

Thought Question

A naturalist once noted that when a male bird stakes out his territory, he sings with a very sharp, staccato song that says, in effect, "Here I am, and stay away!" In contrast, if a predator appears in the vicinity, many birds will emit alarm calls that consist of steady whistles that start and end slowly. Knowing what you do about the two means of localizing sounds, why do these two types of calls have different characteristics?

VESTIBULAR SYSTEM

The vestibular system has two components: the vestibular sacs and the semicircular canals. They represent the second and third components of the *labyrinths* of the inner ear. (We just studied the first component, the cochlea.) The **vestibular sacs** respond to the force of gravity and inform the brain about the head's orientation. The **semicircular canals** respond to angular acceleration—changes in the rotation of the head—but not to steady rotation. They also respond (but rather weakly) to changes in position or to linear acceleration.

The functions of the vestibular system include balance, maintenance of the head in an upright position, and adjustment of eye movement to compensate for head movements. Vestibular stimulation does not produce any readily definable sensation; certain low-frequency stimulation of the vestibular sacs can produce nausea, and stimulation of the semicircular canals can produce dizziness and rhythmic eye movements (*nystagmus*). However, we are not directly aware of the information received from these organs. This section describes the vestibular system: the vestibular apparatus, the receptor cells, and the vestibular pathway in the brain.

Anatomy of the Vestibular Apparatus

Figure 7.23 shows the labyrinths of the inner ear, which include the cochlea, the semicircular canals, and the two vestibular sacs: the **utricle** ("little pouch") and the **saccule** ("little sack"). (See *Figure 7.23.*) The semicircular canals approximate the three major planes of the head: sagittal, transverse, and horizontal. Receptors in each canal respond maximally to angular acceleration in one plane. The semicircular canal consists of a membranous canal floating within a bony one; the

membranous canal contains a fluid called *endolymph.* An enlargement called the **ampulla** contains the organ in which the sensory receptors reside. The sensory receptors are hair cells similar to those found in the cochlea. Their cilia are embedded in a gelatinous mass called the **cupula,** which blocks part of the ampulla. (See *Figure 7.23.*)

To explain the effects of angular acceleration on the semicircular canals, I will first describe an "experiment." If we place a glass of water on the exact center of a turntable and then start the turntable spinning, the water in the glass will, at first, remain stationary (the glass will move with respect to the water it contains). Eventually, however, the water will begin rotating with the container. If we then stop the turntable, the water will continue spinning for a while because of its inertia.

The semicircular canals operate on the same principle. The endolymph within these canals, like the water in the glass, resists movement when the head begins to rotate. This inertial resistance pushes the endolymph against the cupula, causing it to bend, until the fluid begins to move at the same speed as the head. If the head rotation is then stopped, the endolymph, still circulating through the canal, pushes the cupula the other way. Angular acceleration is thus translated into bending of the cupula,

vestibular sac One of a set of two receptor organs in each inner ear that detect changes in the tilt of the head.

semicircular canal One of the three ringlike structures of the vestibular apparatus that detect changes in head rotation.

utricle (*you trih kul*) One of the vestibular sacs.

saccule (*sak yule*) One of the vestibular sacs.

ampulla (*am pull uh*) An enlargement in a semicircular canal; contains the cupula and the crista.

cupula (*kew pew luh*) A gelatinous mass found in the ampulla of the semicircular canals; moves in response to the flow of the fluid in the canals.

FIGURE 7.23 ■ Receptive Organ of the Semicircular Canals

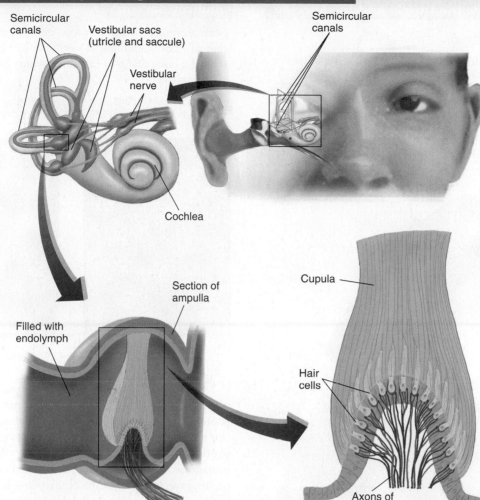

which exerts a shearing force on the cilia of the hair cells. (Of course, unlike the glass of water in my example, we do not normally spin around in circles; the semicircular canals measure very slight and very brief rotations of the head.)

The vestibular sacs (the utricle and saccule) work very differently. These organs are roughly circular, and each contains a patch of receptive tissue. The receptive tissue is located on the "floor" of the utricle and on the "wall" of the saccule when the head is in an upright position. The receptive tissue, like that of the semicircular canals and cochlea, contains hair cells. The cilia of these receptors are embedded in an overlying gelatinous mass, which contains something rather unusual: *otoconia,* which are small crystals of calcium carbonate. (See *Figure 7.24.*) The weight of the crystals causes the gelatinous mass to shift in position as the orientation of the head changes. Thus, movement produces a shearing force on the cilia of the receptive hair cells.

The Receptor Cells

The hair cells of the semicircular canal and vestibular sacs are similar in appearance. Each hair cell contains several cilia, graduated in length from short to long. These hair cells resemble the auditory hair cells found in the cochlea, and their transduction mechanism is also similar: A shearing force of the cilia opens ion channels, and the entry of potassium ions depolarizes the ciliary membrane. All three forms of hair cells employ the same receptor molecules: TRPA1, which I described earlier in this chapter. Figure 7.25 shows two views of a hair cell of a bullfrog saccule made by a scanning electron microscope. (See *Figure 7.25.*)

The Vestibular Pathway

The vestibular and cochlear nerves constitute the two branches of the eighth cranial nerve (auditory nerve). The bipolar cell bodies that give rise to the

FIGURE 7.24 ■ Receptive Tissue of the Vestibular Sacs: the Utricle and the Saccule

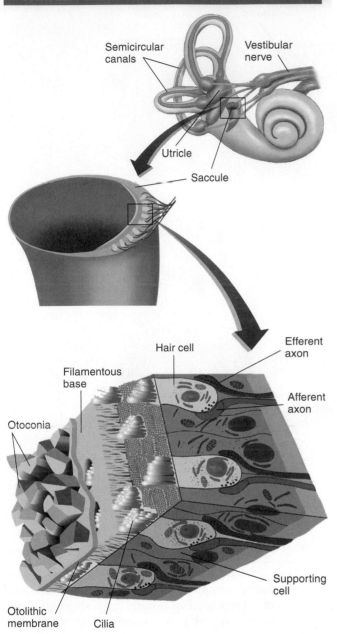

FIGURE 7.25 ■ Hair Cells of a Bullfrog Saccule

These scanning electron microscope views show (a) oblique view of a normal bundle of vestibular hair cells and (b) top view of a bundle of hair cells from which the longest has been detached.

(From Hudspeth, A. J., and Jacobs, R. *Proceedings of the National Academy of Sciences, USA*, 1979, *76*, 1506–1509. Reprinted with permission.)

(a)   (b)

mined. Most investigators believe that the cortical projections are responsible for feelings of dizziness; the activity of projections to the lower brain stem can produce the nausea and vomiting that accompany motion sickness. Projections to brain stem nuclei controlling neck muscles are clearly involved in maintaining an upright position of the head.

Perhaps the most interesting connections are those to the cranial nerve nuclei (third, fourth, and sixth) that control the eye muscles. As we walk or (especially) run, the head is jarred quite a bit. The vestibular system exerts direct control on eye movement to compensate for the sudden head movements. This process, called the *vestibulo-ocular reflex,* maintains a fairly steady retinal image. Test this reflex yourself: Look at a distant object and hit yourself (gently) on the side of the head. Note that your image of the world jumps a bit but not too much. People who have suffered vestibular damage and who lack the vestibulo-ocular reflex have difficulty seeing anything while walking or running. Everything becomes a blur of movement.

afferent axons of the vestibular nerve (a branch of the eighth cranial nerve) are located in the **vestibular ganglion,** which appears as a nodule on the vestibular nerve.

Most of the axons of the vestibular nerve synapse within the vestibular nuclei in the medulla, but some axons travel directly to the cerebellum. Neurons of the vestibular nuclei send their axons to the cerebellum, spinal cord, medulla, and pons. There also appear to be vestibular projections to the temporal cortex, but the precise pathways have not been deter-

vestibular ganglion A nodule on the vestibular nerve that contains the cell bodies of the bipolar neurons that convey vestibular information to the brain.

Interim**Summary**

Vestibular System

The semicircular canals are filled with fluid. When the head begins rotating or comes to rest after rotation, inertia causes the fluid to push the cupula to one side or the other. This movement exerts a shearing force on the cupula, the organ that contains the vestibular hair cells. The vestibular sacs contain a patch of receptive tissue that contains hair cells whose cilia are embedded in a gelatinous mass. The weight of the otoconia in the gelatinous mass shifts when the head tilts, causing a shearing force on some of the cilia of the hair cells.

Each hair cell contains one long cilium and several shorter ones. These cells form synapses with dendrites of bipolar neurons whose axons travel through the vestibular nerve. The receptors also receive efferent terminal buttons from neurons located in the cerebellum and medulla, but the function of these connections is not known. Vestibular information is received by the vestibular nuclei in the medulla, which relay it on to the cerebellum, spinal cord, medulla, pons, and temporal cortex. These pathways are responsible for control of posture, head movements, and eye movements and the puzzling phenomenon of motion sickness.

Thought Questions

Why can slow, repetitive vestibular stimulation cause nausea and vomiting? Obviously, there are connections between the vestibular system and the area postrema, which (as you learned in Chapter 2) controls vomiting. Can you think of any useful functions that might be served by these connections?

SOMATOSENSES

The somatosenses provide information about what is happening on the surface of our body and inside it. The **cutaneous senses** (skin senses) include several submodalities commonly referred to as *touch*. **Proprioception** and **kinesthesia** provide information about body position and movement. The muscle receptors are discussed in this section and in Chapter 8. The **organic senses** arise from receptors in and around the internal organs. Because the cutaneous senses are the most studied of the somatosenses, both perceptually and physiologically, I will devote most of my discussion to them.

The Stimuli

The cutaneous senses respond to several different types of stimuli: pressure, vibration, heating, cooling, and events that cause tissue damage (and hence pain). Feelings of pressure are caused by mechanical deformation of the skin. Vibration is produced in the laboratory or clinic by tuning forks or mechanical devices, but it more commonly occurs when we move our fingers across a rough surface. Thus, we use vibration sensitivity to judge an object's roughness. Obviously, sensations of warmth and coolness are produced by objects that change skin temperature from normal. Sensations of pain can be caused by many different types of stimuli, but it appears that most cause at least some tissue damage.

Kinesthesia is provided by stretch receptors in skeletal muscles that report changes in muscle length to the central nervous system and by stretch receptors in tendons that measure the force being exerted by the muscles. Receptors within joints between adjacent bones respond to the magnitude and direction of limb movement. Muscle length detectors, located within the muscles, do not give rise to conscious sensations; their information is used to control movement. These receptors will be discussed separately in Chapter 8.

We are aware of some of the information received by means of the organic senses, which can provide us with unpleasant sensations such as stomachaches or gallbladder attacks, or pleasurable ones such as those provided by a warm drink on a cold winter day. We are unaware of some information, such as that provided from receptors in the digestive system, kidneys, liver, heart, and blood vessels that are sensitive to nutrients and minerals. This information, which plays a role in the control of metabolism and water and mineral balance, is described in Chapter 12.

---

cutaneous sense (*kew tane ee us*) One of the somatosenses; includes sensitivity to stimuli that involve the skin.

proprioception Perception of the body's position and posture.

kinesthesia Perception of the body's own movements.

organic sense A sense modality that arises from receptors located within the inner organs of the body.

Anatomy of the Skin and Its Receptive Organs

The skin is a complex and vital organ of the body—one that we tend to take for granted. We cannot survive without it; extensive skin burns are fatal. Our cells, which must be bathed by a warm fluid, are protected from the hostile environment by the skin's outer layers. The skin participates in thermoregulation by producing sweat, thus cooling the body, or by restricting its circulation of blood, thus conserving heat. Its appearance varies widely across the body, from mucous membrane to hairy skin to the smooth, hairless skin of the palms and the soles of the feet.

Skin consists of subcutaneous tissue, dermis, and epidermis and contains various receptors scattered throughout these layers. Figure 7.26 shows cross sections through hairy and **glabrous skin** (hairless skin, found on our fingertips and palms and on the bottoms of our toes and feet). Hairy skin contains unencapsulated (free) nerve endings; **Ruffini corpuscles,** which respond to indentation of the skin; and **Pacinian corpuscles,** which respond to rapid vibrations. Pacinian corpuscles are the largest sensory end organs in the body. Their size, approximately 0.5 × 1.0 mm, makes them visible to the naked eye. They consist of up to seventy onionlike layers wrapped around the dendrite of a single myelinated axon. Free nerve endings, which detect painful stimuli and changes in temperature, are found just below the surface of the skin. Other free nerve endings are found in a basketwork around the base of hair follicles and around the emergence of hair shafts from the skin. These fibers detect movement of hairs. (See *Figure 7.26.*)

Glabrous skin contains a more complex mixture of free nerve endings and axons that terminate within specialized end organs (Iggo and Andres, 1982). The increased complexity reflects the fact that we use the palms of our hands and the inside surfaces of our fingers to explore the environment actively: We use them to hold and touch objects. In contrast, the rest of our body most often contacts the environment passively; that is, other things come into contact with it.

Glabrous skin, like hairy skin, contains free nerve endings, Ruffini corpuscles, and Pacinian corpuscles. (Pacinian corpuscles are also found in the joints and in various internal organs.) Glabrous skin also contains **Meissner's corpuscles,** which are found in *papillae* ("nipples"), small elevations of the dermis that project up into the epidermis. These end organs are innervated by between two and six axons. They respond to low-frequency vibration or to brief taps on the skin. **Merkel's disks,** which respond to indentation of the skin, are found at the base of the epidermis, in the same general location as Meissner's corpuscles, adjacent to sweat ducts. (See *Figure 7.26.*)

The *mechanoreceptors* in the skin (that is, those receptors that respond to mechanical stimulation) can be divided into four categories, depending on the size of their receptive field in the skin and the speed with which they adapt to a constant stimulus. (The process of adaptation is described in the next subsection.) Glabrous skin, with its increased cutaneous sensitivity, contains receptors with the smallest receptive fields: Meissner's corpuscles and Merkel's disks. (See *Table 7.1.*)

Perception of Cutaneous Stimulation

The three most important qualities of cutaneous stimulation are touch, temperature, and pain. These qualities, along with itch, are described in the sections that follow.

FIGURE 7.26 ■ Cutaneous Receptors

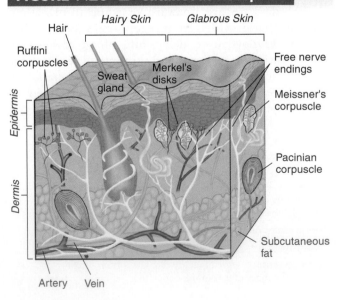

Hairy Skin   Glabrous Skin

Hair
Ruffini corpuscles
Epidermis
Dermis
Sweat gland
Merkel's disks
Free nerve endings
Meissner's corpuscle
Pacinian corpuscle
Subcutaneous fat
Artery   Vein

---

glabrous skin (*glab russ*) Skin that does not contain hair; found on the palms and the soles of the feet.

Ruffini corpuscle A vibration-sensitive organ located in hairy skin.

Pacinian corpuscle (*pa chin ee un*) A specialized, encapsulated somatosensory nerve ending that detects mechanical stimuli, especially vibrations.

Meissner's corpuscle The touch-sensitive end organs located in the papillae, small elevations of the dermis that project up into the epidermis.

Merkel's disk The touch-sensitive end organs found at the base of the epidermis, adjacent to sweat ducts.

TABLE 7.1 ■ Categories of Mechanoreceptors in Glabrous Skin

SPEED OF ADAPTATION	SIZE OF RECEPTIVE FIELD	IDENTITY OF RECEPTOR
Slow	Small, sharp borders	Merkel's disk
Slow	Large, diffuse borders	Ruffini corpuscles
Rapid	Small, sharp borders	Meissner's corpuscles
Rapid	Large, diffuse borders	Pacinian corpuscles

Touch

Sensitivity to pressure and vibration is caused by movement of the skin, which moves the dendrites of mechanoreceptors. Most investigators believe that the encapsulated nerve endings serve only to modify the physical stimulus transduced by the dendrites that reside within them. But what is the mechanism of transduction? How does movement of the dendrites of mechanoreceptors produce changes in membrane potentials? It appears that the movement causes ion channels to open, and the flow of ions into or out of the dendrite causes a change in the membrane potential. You will recall that TRPA1, a member of the TRP (transient receptor potential) family of receptor proteins is responsible for transduction of mechanical information in auditory and vestibular hair cells. Although evidence from a study by Maroto et al. (2005) suggested that another member of the same family—TRPC1—was responsible for mechanoreception in vertebrates, further research cast doubt on this conclusion (Gottlieb et al., 2008). At present, there appears to be no good candidate for a receptor for light, nonpainful, touch.

Most information about tactile sensation is precisely localized—that is, we can perceive the location on our skin where we are being touched. Until very recently, neuroscientists believed that in humans this information was transmitted to the central nervous system only by fast-conducting myelinated axons. However, a recent study discovered a new category of tactile sensation that is transmitted by small-diameter unmyelinated axons (Olausson et al., 2002).

At age 31, patient G. L., a 54-year-old woman, "suffered a permanent and specific loss of large myelinated afferents after episodes of acute polyradiculitis and polyneuropathy that affected her whole body below the nose. A sural nerve biopsy indicated a complete loss of large-diameter myelinated fibers. . . . Before the present study, she denied having any touch sensibility below the nose, and she lost the ability to perceive tickle when she became ill. She states that her perceptions of temperature, pain and itch are intact" (Olausson et al., 2002, pp. 902–903).

G. L. could indeed detect the stimuli that are normally attributed to small-diameter unmyelinated axons—temperature, pain, and itch—but she could not detect vibratory or normal tactile stimuli. But when the hairy skin on her forearm or the back of her hand was stroked with a soft brush, she reported a faint, pleasant sensation. However, she could not determine the direction of the stroking or its precise location. An fMRI analysis showed that this stimulation activated the insular cortex, a region that is known to be associated with emotional responses. The somatosensory cortex was not activated. When regions of hairy skin of control subjects were stimulated this way, fMRI showed activation of the primary and secondary somatosensory cortex as well as the insular cortex because the stimulation activated both large and small axons. The glabrous skin on the palm of the hand is served only by large-diameter, myelinated axons. When this region was stroked with a brush, G. L. reported no sensation at all, presumably because of the absence of these axons.

The investigators concluded that besides conveying information about noxious and thermal stimuli, small-diameter unmyelinated axons constitute a "system for limbic touch that may underlie emotional, hormonal and affiliative responses to caresslike, skin-to-skin contact between individuals" (Olausson et al., 2002, p. 900) And as we saw, G. L. could no longer perceive tickle. Olausson and his colleagues note that tickling sensations, which were previously believed to be transmitted by these small axons, are apparently transmitted by the large, myelinated axons that were destroyed in patient G. L.

Our cutaneous senses are used much more often to analyze shapes and textures of stimulus objects that are moving with respect to the surface of the skin. Sometimes, the object itself moves; but more often, we do the moving ourselves. If I placed an object in your palm and asked you to keep your hand still, you would have a great deal of difficulty recognizing the object by touch alone. If I said that you could now move your hand, you would manipulate the object, letting its surface slide across your palm and the pads of your fingers. You would be able to describe the object's three-dimensional shape, hardness, texture, slipperiness, and so on. Obviously, your motor system must cooperate, and you need kinesthetic sensation from your muscles and joints, besides the cutaneous information. If you squeeze the object and feel a lot of well-localized pressure in return, it is hard. If you feel a less intense, more diffuse pressure in return, it is soft. If it produces vibrations as it moves over the ridges on your fingers, it is rough. If very little effort is needed to move the object while pressing it against your skin, it is slippery. If it does not produce vibrations as it moves across your skin, but moves in a jerky fashion, and if it takes effort to remove your fingers from its surface, it is sticky. Thus, our somatosenses work dynamically with the motor system to provide useful information about the nature of objects that come into contact with our skin.

Studies of people who make especially precise use of their fingertips show changes in the regions of somatosensory cortex that receive information from this part of the body. For example, violinists must make very precise movements of the four fingers of the left hand, which are used to play notes by pressing the strings against the fingerboard. Tactile feedback and proprioceptive feedback are very important in accurately moving and positioning these fingers so that sounds of the proper pitch are produced. In contrast, placement of the thumb, which slides along the bottom of the neck of the violin, is less critical. In a study of violin players, Elbert et al. (1995) found that the portions of their right somatosensory cortex that receive information from the four fingers of their left hand were enlarged relative to the corresponding parts of the left somatosensory cortex. The amount of somatosensory cortex that receives information from the thumb was not enlarged. (The right hand holds the bow, and the violinist makes precise movements with the arm and wrist, but tactile information from the fingers of this hand is much less important.)

Temperature

Feelings of warmth and coolness are relative, not absolute, except at the extremes. There is a temperature level that, for a particular region of skin, will produce a sensation of temperature neutrality—neither warmth nor coolness. This neutral point is not an absolute value but depends on the prior history of thermal stimulation of that area. If the temperature of a region of skin is raised by a few degrees, the initial feeling of warmth is replaced by one of neutrality. If the skin temperature is lowered to its initial value, it now feels cool. Thus, increases in temperature lower the sensitivity of warmth receptors and raise the sensitivity of cold receptors. The converse holds for decreases in skin temperature. This adaptation to ambient temperature can be demonstrated easily by placing one hand in a bucket of warm water and the other in a bucket of cool water until some adaptation has taken place. If you then simultaneously immerse both hands in water at room temperature, it will feel warm to one hand and cool to the other.

There are two categories of thermal receptors: those that respond to warmth and those that respond to coolness. Cold sensors in the skin are located just beneath the epidermis, and warmth sensors are located more deeply in the skin. Information from cold sensors is conveyed to the CNS by thinly myelinated A δ fibers, and information from warmth sensors is conveyed by unmyelinated C fibers. We can detect thermal stimuli over a very wide range of temperatures, from less than 8° C (noxious cold) to over 52° C. (noxious heat). Investigators have long believed that no single receptor could detect such a range of temperatures, and recent research indicates that this belief was correct. At present we know of six mammalian thermoreceptors—all members of the TRP family (Bandell, Macpherson, and Patapoutian, 2007; Romanovsky, 2007). (See *Figure 7.27* and *Table 7.2*.)

Some of the thermal receptors respond to particular chemicals as well as to changes in temperature. For example, the *M* in TRPM8 stands for menthol, a compound found in the leaves of many members of the mint family. As you undoubtedly know, peppermint tastes cool in the mouth, and menthol is added to some cigarettes to make the smoke feel cooler (and perhaps to try to delude smokers into thinking that the smoke is less harsh and damaging to the lungs). Menthol provides a cooling sensation because it binds with and stimulates the TRPM8 receptor and produces neural activity that the brain interprets as coolness. As we will see in the next subsection, chemicals can produce the sensation of heat also.

Bautista et al. (2007) prepared mice with a targeted mutation (knockout) of the TRPM8 receptor gene.

FIGURE 7.27 ■ Activity of TRP Channels

The activity of cold-activated (blue) and heat-activated (orange) temperature-sensitive TRP channels are shown as a function of temperature.

(Adapted from Romanovsky, A. A. *American Journal of Physiology*, 2007, *292*, R37–R46.)

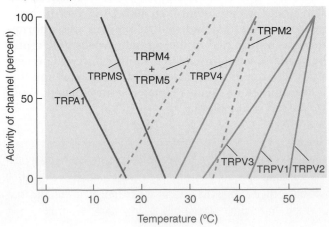

TABLE 7.2 ■ Categories of Mammalian Thermal Receptors

NAME OF RECEPTOR	TYPE OF STIMULUS	TEMPERATURE RANGE
TRPV2	Noxious heat	Above 52° C
TRPV1	Heat	Above 43° C
TRPV3	Warmth	Above 31° C
TRPV4	Warmth	Above 25° C
TRPM8	Coolness	Below 28° C
TRPA1	Cold [?] (also found in cilia of auditory and vestibular hair cells)	Below 18° C

They found that the mutation severely impaired the response of the mice to environmental cold. The investigators assessed sensitivity to cold by placing the mice in a box that contained two chambers connected by an opening through which the animals could pass. The floors of the chambers consisted of temperature-controlled metal plates. Normal mice preferred to spend time on a plate held at 30° C and avoided a 20° C plate. However, the mice without TRPM8 receptors showed no preference until the temperature on the cool plate dropped to 15° C. In additions, electrical recordings of cutaneous C fibers of the mutant mice showed no evidence of cold-sensitive neurons.

Pain

The story of pain is quite different from that of temperature and pressure; pain is a very complicated sensation. It is obvious that our awareness of pain and our emotional reaction to it are controlled by mechanisms within the brain. For example, we can have a tooth painlessly removed under hypnosis, which has no effect on the sensitivity of pain receptors. Stimuli that produce pain also tend to trigger species-typical escape and withdrawal responses. Subjectively, these stimuli *hurt,* and we try hard to avoid or escape from them. However, sometimes we are better off ignoring pain and getting on with important tasks. In fact, our brains possess mechanisms that can reduce pain, partly through the action of the endogenous opioids. These mechanisms are described in more detail in a later section of this chapter.

Pain reception, like thermoreception, is accomplished by the networks of free nerve endings in the skin. There appear to be at least three types of pain receptors (usually referred to as *nociceptors,* or "detectors of noxious stimuli"). High-threshold mechanoreceptors are free nerve endings that respond to intense pressure, which might be caused by something striking, stretching, or pinching the skin. A second type of free nerve ending appears to respond to extremes of heat, to acids, and to the presence of *capsaicin,* the active ingredient in chile peppers. (Note that we say that chile peppers make food taste "hot.") This type of fiber contains TRPV1 receptors (Kress and Zeilhofer, 1999). The *V* stands for *vanilloid*—a group of chemicals of which capsaicin is a member. Caterina et al. (2000) found that mice with a knockout of the gene for the TRPV1 receptor showed less sensitivity to painful high-temperature stimuli and would drink water to which capsaicin had been added. The mice responded normally to noxious mechanical stimuli. Presumably, the TRPV1 receptor is responsible for pain produced by burning of the skin. It also appears to play a role in regulation of body temperature. In addition, Ghilardi et al. (2005) found that a drug that blocks TRPV1 receptors reduced pain in patients with bone cancer, which is apparently caused by the production of acid by the tumors.

Another type of nociceptive fiber contains TRPA1 receptors, which, as we saw earlier in this chapter, are found in the cilia of auditory and vestibular hair cells. Although several studies have suggested that these receptors serve as detectors of extreme cold, more recent evidence calls this conclusion into question. TRPA1 receptors are sensitive to pungent irritants found in mustard oil, wintergreen oil, horseradish, and garlic and to a variety of environmental irritants, including those found in vehicle exhaust and tear gas (Bautista et al., 2006; Nilius et al., 2007). The primary function of this receptor appears to be to provide information about the presence of chemicals that produce inflammation.

Itch

Another noxious sensation, itch (or more formally, *pruritus*) is caused by skin irritation. Itch was defined by a seventeenth-century German physician as an "unpleasant sensation that elicits the desire or reflex to scratch" (Ikoma et al., 2006, p. 535.) If an adult sees a child scratching at an insect bite or other form of skin irritation, the adult is likely to say, "Stop that—it will only make it worse!" The scratching may indeed make the irritation worse, but the immediate effect of scratching is to reduce the itching. Scratching reduces itching because pain suppresses itching (and, ironically, itching reduces pain). Histamine and other chemicals released by skin irritation and allergic reactions are important sources of itching. Experiments have shown that painful stimuli such as heat and electrical shock can reduce sensations of itch produced by an injection of histamine into the skin, even when the painful stimuli are applied up to 10 cm from the site of irritation (Ward, Wright, and McMahon, 1996; Nilsson, Levinsson, and Schouenborg, 1997). On the other hand, the administration of an opiate into the epidural space around the spinal cord diminishes pain but often produces itching as an unwelcome side effect (Chaney, 1995). Naloxone, a drug that blocks opiate receptors, has been used to reduce *cholestatic pruritus,* a condition of itching that sometimes accompanies pregnancy (Bergasa, 2005).

Little is known about the receptors that are responsible for the sensation of itch, but at least two different types of neurons transmit itch-related information to the CNS. Johanek et al. (2007) produced itch in volunteers with intradermal injections of histamine and applications of *cowhage spicules*—tiny, needlelike plant fibers that contain an enzyme that breaks down proteins in the skin. Both treatments produce intense itch, but only histamine produces an area of vasodilation. Pretreatment of a patch of skin with a topical antihistamine prevented histamine from producing an itch at that spot but had no effect on the itch produced by cowhage. In contrast, pretreatment of a patch of skin with capsaicin prevented cowhage-induced itch but not histamine-induced itch.

The Somatosensory Pathways

Somatosensory axons from the skin, muscles, or internal organs enter the central nervous system via spinal nerves. Those located in the face and head primarily enter through the trigeminal nerve (fifth cranial nerve). The cell bodies of the unipolar neurons are located in the dorsal root ganglia and cranial nerve ganglia. Axons that convey precisely localized information, such as fine touch, ascend through the *dorsal columns* in the white matter of the spinal cord to nuclei in the lower medulla. From there, axons cross the brain and ascend through the *medial lemniscus* to the *ventral posterior nuclei of the thalamus,* the relay nuclei for somatosensation. Axons from the thalamus project to the primary somatosensory cortex, which in turn sends axons to the secondary somatosensory cortex. In contrast, axons that convey poorly localized information, such as pain or temperature, form synapses with other neurons as soon as they enter the spinal cord. The axons of these neurons cross to the other side of the spinal cord and ascend through the *spinothalamic tract* to the ventral posterior nuclei of the thalamus. (See *Figure 7.28.*)

Recall from Chapter 6 that the primary visual cortex contains columns of cells, each of which responds

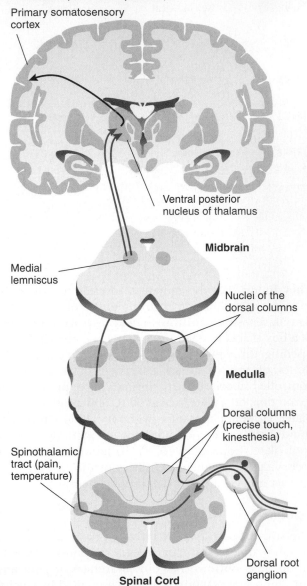

FIGURE 7.28 ■ Somatosensory Pathways

The figure shows the Somatosensory pathways from the spinal cord to the somatosensory cortex. Note that precisely localized information (such as fine touch) and imprecisely localized information (such as pain and temperature) are transmitted by different pathways.

Primary somatosensory cortex

Ventral posterior nucleus of thalamus

Midbrain

Medial lemniscus

Nuclei of the dorsal columns

Medulla

Dorsal columns (precise touch, kinesthesia)

Spinothalamic tract (pain, temperature)

Dorsal root ganglion

Spinal Cord

to particular features, such as orientation, ocular dominance, or spatial frequency. Within these columns are blobs that contain cells that respond to particular colors. The somatosensory cortex also has a columnar arrangement; in fact, cortical columns were discovered there by Mountcastle (1957) before they were found in the visual and auditory cortex. Within a column, neurons respond to a particular type of stimulus (for example, temperature or pressure) applied to a particular part of the body.

Dykes (1983) has reviewed research indicating that the primary and secondary somatosensory cortical areas are divided into at least five (and perhaps as many as ten) different maps of the body surface. Within each map, cells respond to a particular submodality of somatosensory receptors. Separate areas have been identified that respond to slowly adapting cutaneous receptors, rapidly adapting cutaneous receptors, receptors that detect changes in muscle length, receptors located in the joints, and Pacinian corpuscles.

As you learned in Chapter 6, the extrastriate cortex consists of several subareas, each of which contains an independent representation of the visual field. For example, one area responds specifically to color and form, and another responds to movement. The somatosensory cortex appears to follow a similar scheme: Each cortical map of the body contains neurons that respond to a specific submodality of stimulation.

As we saw in Chapter 6, damage to the visual association cortex can cause visual agnosia, and as we saw earlier in this chapter, damage to the auditory association cortex can cause auditory agnosia. You will not be surprised to learn that damage to the somatosensory association cortex can cause tactile agnosia. Reed, Caselli, and Farah (1996) described patient E. C., a woman with left parietal lobe damage who was unable to recognize common objects by touch. For example, the patient identified a pine cone as a brush, a ribbon as a rubber band, and a snail shell as a bottle cap. The deficit was not due to a simple loss of tactile sensitivity; the patient was still sensitive to light touch and to warm and cold objects, and she could easily discriminate objects by their size, weight, and roughness.

Nakamura et al. (1998) described patient M. T., who had a different type of tactile agnosia. Patient M. T. had bilateral lesions of the angular gyrus, a region of the parietal lobe surrounding the caudal end of the lateral fissure. This patient, like patient E. C., had normal tactile sensitivity, but he could not identify objects by touch. However, unlike patient E. C., he could *draw* objects that he touched even though he could not recognize what they were. (See *Figure 7.29*.) The fact that he could draw the objects means that his ability to perceive three-dimensional objects by touch must have been intact. However, the brain damage prevented the information analyzed by the somatosensory association cortex to be transmitted to parts of the brain responsible for control of language—and for consciousness.

As I mentioned earlier, recognition of objects by touch requires cooperation between the somatosensory and motor systems. When we attempt to identify objects by touch alone, we explore them with moving fingers. Valenza et al. (2001) reported the case of a patient with brain damage to the right hemisphere that produced a disorder they called *tactile apraxia*. As we will see in

FIGURE 7.29 ■ Tactile Agnosia

(a) Drawings of wrenches felt but not seen by M. T., a patient with associative tactile agnosia. Although the patient did not recognize the objects as wrenches, he was able to draw them accurately. (b) Drawings of objects felt but not seen by E. C., a patient with apperceptive tactile agnosia. The patient could neither recognize the objects by touch nor draw them accurately.

(From Nakamura, J., Endo, K., Sumida, T., and Hasegawa, T. *Cortex*, 1998, *34*, 375–388, and Reed, C. L., Caselli, R. J., and Farah, M. J. *Brain*, 1996, *119*, 875–888. Reprinted with permission.)

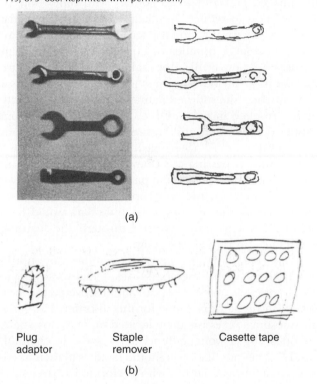

(a)

Plug adaptor   Staple remover   Casette tape

(b)

Chapter 8, *apraxia* refers to a difficulty in carrying out purposeful movements in the absence of paralysis or muscular weakness. When the experimenters gave the patient objects to identify by touch with her left hand, the patient explored the objects with her fingers in a disorganized fashion. (Exploration and identification using her right hand were normal.) If the experimenters guided the patient's fingers and explored an object the way people normally do, she was able to recognize the object's shape. Thus, her deficit was caused by a movement disorder, not by damage to brain mechanisms involved in tactile perception.

Perception of Pain

Pain is a curious phenomenon. It is more than a mere sensation; it can be defined only by some sort of withdrawal reaction or, in humans, by verbal report. Pain can be modified by opiates, by hypnosis, by the administration of pharmacologically inert sugar pills, by emotions, and even by other forms of stimulation, such as acupuncture. Recent research efforts have made remarkable progress in discovering the physiological bases of these phenomena.

We might reasonably ask *why* we experience pain. The answer is that in most cases pain serves a constructive role. For example, inflammation, which often accompanies injuries to skin or muscle, greatly increases sensitivity of the inflamed region to painful stimuli. This effect motivates the individual to minimize movement of the injured part and avoid contact with other objects. The effect is to reduce the likelihood of further injury. People who have congenital insensitivity to pain suffer an abnormally large number of injuries, such as cuts and burns. One woman did not make the shifts in posture that we normally do when our joints start to ache. As a consequence, she suffered damage to the spine that ultimately resulted in death. Other people have died from ruptured appendixes and ensuing abdominal infections that they did not feel (Sternbach, 1968). I am sure that a person who is passing a kidney stone would not find much comfort in the fact that pain does more good than ill; pain is, nevertheless, very important to our existence. (As we saw in the opening case, little Sara would have preferred not to have suffered through the learning experience provided by her pain.)

Cox et al. (2006) studied three families from northern Pakistan whose members included several people with a complete absence of pain and discovered the location of the gene responsible for this disorder. The gene, an autosomal recessive allele located on chromosome 2, encodes for a voltage-dependent sodium channel, $Na_X1.7$. The case that brought the families to their attention was a ten-year-old boy who performed a "street theater" during which he would thrust knives through his arms and walk on burning coals without feeling any pain. He died just before his fourteenth birthday after jumping off the roof of a house. All six of the affected people in the three families had injuries to their lips or tongues caused by self-inflicted bites. They all suffered from bruises and cuts, and many sustained bone fractures that they did not notice until the injuries impaired their mobility. Despite their total lack of pain from any type of noxious stimulus, they had normal sensations of touch, warmth, coolness, proprioception, tickle, and pressure.

Some environmental events diminish the perception of pain. For example, Beecher (1959) noted that many wounded American soldiers back from the battle at Anzio, Italy, during World War II reported that they felt no pain from their wounds. They did not even want medication. It would appear that their perception of pain was diminished by the relief they felt from surviving such a terrible ordeal. There are other instances in which people report the perception of pain but are not

bothered by it. Some tranquilizers have this effect, and damage to parts of the brain does too.

Pain appears to have three different perceptual and behavioral effects (Price, 2000). First is the sensory component—the pure perception of the intensity of a painful stimulus. The second component is the immediate emotional consequences of pain—the unpleasantness or degree to which the individual is bothered by the painful stimulus. It is this characteristic that was reduced in some of the soldiers at Anzio. The third component is the long-term emotional implications of chronic pain—the threat that such pain represents to one's future comfort and well-being.

These three components of pain appear to involve different brain mechanisms. The purely sensory component of pain is mediated by a pathway from the spinal cord to the ventral posterolateral thalamus to the primary and secondary somatosensory cortex. The immediate emotional component of pain appears to be mediated by pathways that reach the anterior cingulate cortex (ACC) and insular cortex. The long-term emotional component appears to be mediated by pathways that reach the prefrontal cortex. (See *Figure 7.30.*)

Let's look at some evidence for brain mechanisms involved in short-term and long-term emotional responses to pain. Several studies have found that painful stimuli

---

FIGURE 7.30 ■ The Three Components of Pain

A simplified, schematic diagram shows the brain mechanisms involved in the sensory component, the immediate emotional component, and the long-term emotional component of pain.

(Adapted from Price, D. B. *Science*, 2000, *288*, 1769–1772.)

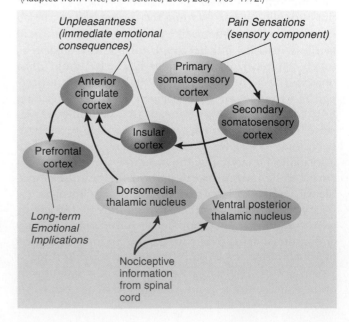

activate the insular cortex and the ACC. In addition, Ostrowsky et al. (2002) found that electrical stimulation of the insular cortex caused reports of painful burning and stinging sensations. Damage to this region decreases people's emotional response to pain (Berthier, Starkstein, and Leiguarda, 1988): They continue to feel the pain but do not seem to recognize that it is harmful. They do not withdraw from pain or the threat of pain.

Rainville et al. (1997) produced pain sensations in human subjects by having them put their arms in ice water. Under one condition the researchers used hypnosis to diminish the unpleasantness of the pain. The hypnosis worked; the subjects said that the pain was less unpleasant, even though it was still as intense. Meanwhile, the investigators used a PET scanner to measure regional activation of the brain. They found that the painful stimulus increased the activity of both the primary somatosensory cortex and the ACC. When the subjects were hypnotized and found the pain less unpleasant, the activity of the ACC decreased, but the activity of the primary somatosensory cortex remained high. Presumably, the primary somatosensory cortex is involved in the perception of pain, and the ACC is involved in its immediate emotional effects—its unpleasantness. (See *Figure 7.31.*)

In another study from the same laboratory, Hofbauer et al. (2001) produced the opposite effect. They presented subjects with a painful stimulus and used hypnotic suggestion to reduce the perceived intensity of the pain. They found that the suggestion reduced subject's ratings of pain and also decreased the activation of the somatosensory cortex. Thus, changes in perceived *intensity* of pain is reflected in changes in activation of the somatosensory cortex, whereas changes in perceived *unpleasantness* of pain is reflected in changes in activation of the ACC.

Several functional-imaging studies have shown that under certain conditions, stimuli associated with pain can activate the ACC even when no actual painful stimulus is applied. Osaka et al. (2004) found that the ACC was activated when subjects heard Japanese words that vividly denote various types of pain (for example, a throbbing pain, a splitting headache, or the pain caused by being stuck with thorns). In a test of romantically involved couples, Singer et al. (2004) found that when women received a painful electrical shock to the back of their hand, their ACC, anterior insular cortex, thalamus, and somatosensory cortex became active. When they saw their partners receive a painful shock but did not receive one themselves, the same regions (except for the somatosensory cortex) became active. Thus, the emotional component of pain—in this case, a vicarious experience of pain, provoked by empathy with the feelings of someone a person loved—caused responses in the brain similar to the ones caused by actual pain. Just as we saw in the study by Rainville et al.

FIGURE 7.31 ■ Sensory and Emotional Components of Pain

The PET scans show regions of the brain responding to pain. *Top:* Dorsal views of the brain. Activation of the primary somatosensory cortex (circled in red) by a painful stimulus was not affected by a hypnotically suggested reduction in unpleasantness of a painful stimulus, indicating that this region responded to the sensory component of pain. *Bottom:* Midsagittal views of the brain. The anterior cingulate cortex (circled in red) showed much less activation when the unpleasantness of the painful stimulus was reduced by hypnotic suggestion.

(From Rainville, P., Duncan, G. H., Price, D. D., Carrier, Benoit, and Bushnell, M. C. *Science,* 1997, *277,* 968–971. Copyright © 1997 American Association for the Advancement of Science. Reprinted with permission.)

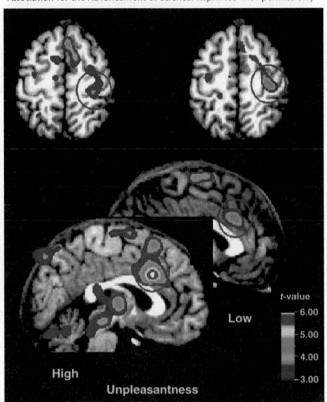

(1997), the somatosensory cortex is activated only by an actual noxious stimulus.

The final component of pain—the emotional consequences of chronic pain—appears to involve the prefrontal cortex. As we will see in Chapter 11, damage to the prefrontal cortex impairs people's ability to make plans for the future and to recognize the personal significance of situations in which they are involved. Along with the general lack of insight, people with prefrontal damage tend not to be concerned with the implications of chronic conditions—including chronic pain—for their future.

A particularly interesting form of pain sensation occurs after a limb has been amputated. After the limb is gone, up to 70 percent of amputees report that they feel as though the missing limb still exists and that it often hurts. This phenomenon is referred to as the **phantom limb** (Melzak, 1992; Ramachandran and Hirstein, 1998). People with phantom limbs report that the limb feels very real, and they often say that if they try to reach out with it, it feels as though it were responding. Sometimes, they perceive it as sticking out, and they may feel compelled to avoid knocking it against the side of a doorframe or sleeping in a position that would make it come between them and the mattress. People have reported all sorts of sensations in phantom limbs, including pain, pressure, warmth, cold, wetness, itching, sweatiness, and prickliness.

The classic explanation for phantom limbs has been activity of the sensory axons belonging to the amputated limb. Presumably, the nervous system interprets this activity as coming from the missing limb. When nerves are cut and connections cannot be reestablished between the proximal and distal portions, the cut ends of the proximal portions form nodules known as *neuromas*. The treatment for phantom pain has been to cut the nerves above these neuromas, to cut the dorsal roots that bring the afferent information from these nerves into the spinal cord, or to make lesions in somatosensory pathways in the spinal cord, thalamus, or cerebral cortex. Sometimes these procedures work for a while, but often the pain returns.

Melzak (1992) suggested that the phantom limb sensation is inherent in the organization of the parietal cortex. As we saw in the discussion of unilateral neglect in Chapter 1, the parietal cortex is involved in our awareness of our own bodies. Indeed, people with lesions of the parietal lobe (especially in the right hemisphere) have been known to push their own leg out of bed, believing that it belongs to someone else. Melzak reports that some people who were born with missing limbs nevertheless experience phantom limb sensations, which would suggest that our brains are genetically programmed to provide sensations for all four limbs.

Endogenous Modification of Pain Sensitivity

For many years, investigators have known that perception of pain can be modified by environmental stimuli. Recent work, beginning in the 1970s, has revealed the existence of neural circuits whose activity can produce analgesia. A variety of environmental stimuli can activate these analgesia-producing circuits. Most of these stimuli cause the release of the endogenous opioids, which were described in Chapter 4.

Electrical stimulation of particular locations within the brain can cause analgesia, which can even be profound enough to serve as an anesthetic for surgery in rats (Reynolds, 1969). The most effective locations appear to be within the periaqueductal gray matter and in the

rostroventral medulla. For example, Mayer and Liebeskind (1974) reported that electrical stimulation of the periaqueductal gray matter produced analgesia in rats equivalent to that produced by at least 10 milligrams (mg) of morphine per kilogram of body weight, which is a large dose. The technique has even found an application in reducing severe, chronic pain in humans: Fine wires are surgically implanted in parts of the central nervous system and attached to a radio-controlled device that permits the patient to administer electrical stimulation when necessary (Kumar, Wyant, and Nath, 1990).

Analgesic brain stimulation apparently triggers the neural mechanisms that reduce pain, primarily by causing endogenous opioids to be released. Basbaum and Fields (1978, 1984), who summarized their work and that of others, proposed a neural circuit that mediates opiate-induced analgesia. Basically, they proposed the following: Endogenous opioids (released by environmental stimuli or administered as a drug) stimulate opiate receptors on neurons in the periaqueductal gray matter. Because the effect of opiates appears to be inhibitory (Nicoll, Alger, and Jahr, 1980), Basbaum and Fields proposed that the neurons that contain opiate receptors are themselves inhibitory interneurons. Thus, the administration of opiates activates the neurons on which these interneurons synapse. (See *Figure 7.32.*)

Neurons in the periaqueductal gray matter send axons to the **nucleus raphe magnus,** located in the medulla. The neurons in this nucleus send axons to the dorsal horn of the spinal cord gray matter; destruction of these axons eliminates analgesia induced by an injection of morphine. The inhibitory effects of these neurons apparently involve one or two interneurons in the spinal cord. (See *Figure 7.32.*)

Pain sensitivity can be regulated by direct neural connections, as well as by secretion of the endogenous opioids. The periaqueductal gray matter receives inputs from the frontal cortex, amygdala, and hypothalamus (Beitz, 1982; Mantyh, 1983). These inputs permit learning and emotional reactions to affect an animal's responsiveness to pain even without the secretion of opioids.

Biological Significance of Analgesia

It appears that a considerable amount of neural circuitry is devoted to reducing the intensity of pain. What functions do these circuits perform? When an animal encounters a noxious stimulus, the animal

phantom limb Sensations that appear to originate in a limb that has been amputated.

nucleus raphe magnus A nucleus of the raphe that contains serotonin-secreting neurons that project to the dorsal gray matter of the spinal cord and is involved in analgesia produced by opiates.

FIGURE 7.32 ■ Opiate-Induced Analgesia

The schematic shows the neural circuit that mediates opiate-induced analgesia, as hypothesized by Basbaum and Fields (1978).

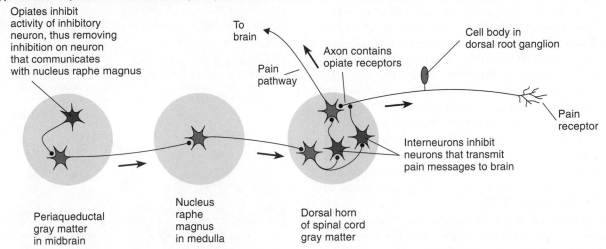

usually stops what it is doing and engages in withdrawal or escape behaviors. Obviously, these responses are quite appropriate. However, they are sometimes counterproductive. For example, males fighting for access to females during mating season will fail to pass on their genes if pain elicits withdrawal responses that interfere with fighting. In fact, fighting and sexual activity both stimulate brain mechanisms of analgesia.

Komisaruk and Larsson (1971) found that gentle probing of a rat's vagina with a glass rod produced analgesia. Such probing also increases the activity of neurons in the periaqueductal gray matter and decreases the responsiveness of neurons in the ventrobasal thalamus to painful stimulation (Komisaruk and Steinman, 1987). The phenomenon also occurs in humans; Whipple and Komisaruk (1988) found that self-administered vaginal stimulation reduces sensitivity to painful stimuli but not to neutral tactile stimuli. Presumably, copulation triggers analgesic mechanisms. The adaptive significance of this phenomenon is clear: Painful stimuli encountered during the course of copulation are less likely to cause the behavior to be interrupted; thus, the chances of pregnancy are increased.

Pain can also be reduced, at least in some people, by administering a pharmacologically inert placebo. When some people take a medication that they believe will reduce pain, it triggers the release of endogenous opioids and actually does so. This effect is eliminated if the people are given an injection of naloxone, a drug that blocks opiate receptors (Benedetti, Arduino, and Amanzio, 1999). Thus, for some people a placebo is not pharmacologically inert—it has a physiological effect. The placebo effect may be mediated through connections of the frontal cortex with the periaqueductal gray matter. A functional-imaging study by Zubieta et al.

(2005) found that placebo-induced analgesia did indeed cause the release of endogenous opiates. They used a PET scanner to detect the presence of μ-opioid neurotransmission in the brains of people who responded to the effects of a placebo. As Figure 7.33 shows, several regions of the brain, including the anterior cingulate cortex and insular cortex, showed evidence of increased endogenous opioid activity. (See *Figure 7.33*.)

FIGURE 7.33 ■ Effects of a Placebo on μ-Opioid Neurotransmission

The figure shows the brains of people who responded to the effects of analgesic placebo. ACC = anterior cingulate cortex, DLPFC = dorsolateral prefrontal cortex, NAC = nucleus accumbens.

(From Zubieta, J.-K., Bueller, J. A., Jackson, L. R., Scott, D .J., Xu, Y., Koeppe, R. A., Nichols, T. E., and Stohler, C. S. *Journal of Neuroscience*, 2005, *25*, 7754–7762. Reprinted with permission.)

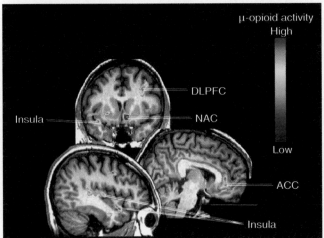

A functional-imaging study by Wager et al. (2004) supports the suggestion that the prefrontal cortex plays a role in placebo analgesia. They administered painful stimuli (heat or electrical shocks) to the skin with or without the application of an "analgesic" skin cream that was actually an unmedicated placebo. They observed a placebo effect—reports of less intense pain and decreased activity in the primary pain-reactive regions of the brain, including the thalamus, ACC, and insular cortex. They also observed *increased* activity in the prefrontal cortex and the periaqueductal gray matter of the midbrain. Presumably, the expectation of decreased sensitivity to pain caused the increased activity of the prefrontal cortex, and connections of this region with the periaqueductal gray matter activated endogenous mechanisms of analgesia. (See *Figure 7.34*.)

An interesting study by deCharms et al. (2005) found that reduced activity of the anterior cingulate cortex decreased perceptions of pain. The investigators used feedback from fMRI to teach people to suppress the activity of their own ACC. Previous studies have shown that people can change their blood pressure, heart rate, EEG rhythms, and other psychophysical characteristics if they are given feedback about these characteristics (usually called *biofeedback*). The subjects in the present study were given feedback that indicated the level of ACC activation while their brains were being scanned. During alternating 60-second periods, they were asked to try to increase and then decrease the activation. They soon learned to do so. Their pain thresholds were then assessed during periods of high and low ACC activation by applying heat to the palm of their left hand. The results showed that the subjects could tolerate more

FIGURE 7.34 ■ The Placebo Effect

Functional MRI scans show increased activity in the dorsolateral prefrontal cortex and the periaqueductal gray matter of the midbrain of subjects who showed decreased sensitivity to pain in response to administration of a placebo.

(From Wager, T. D., Rilling, J. K., Smith, E. E., Sokolik, A., Casey, K. L., Davidson, R. J., Kosslyn, S. M., Rose, R. M., and Cohen, J. D. *Science*, 2004, *303*, 1162–1166. Copyright © 2004 American Association for the Advancement of Science. Reprinted with permission.)

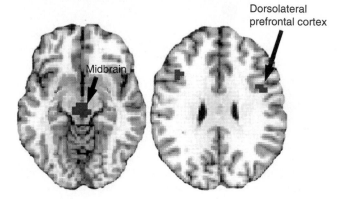

pain when the ACC activation was decreased and were more sensitive to pain when it was increased. Control conditions showed that false feedback about ACC activation had no effect on pain sensitivity, nor did training subjects to change the activity of the posterior cingulate cortex, a brain region unrelated to pain perception. deCharms and his colleagues even found that patients who suffered from chronic pain could reduce their pain after learning to control the activity of their ACC by these means.

Interim Summary

Somatosenses

Cutaneous sensory information is provided by specialized receptors in the skin. Pacinian corpuscles provide information about vibration. Ruffini corpuscles, similar to Pacinian corpuscles but considerably smaller, respond to indentation of the skin. Meissner's corpuscles, found in papillae and innervated by several axons, respond to low-frequency vibration or to brief taps on the skin. Merkel's disks, also found in papillae, consist of single, flattened dendritic endings next to specialized epithelial cells. These receptors respond to pressure. Painful stimuli and changes in temperature are detected by free nerve endings.

When the dendrites of mechanoreceptors bend, ion channels open, producing a receptor potential. Although most tactile information is transmitted to the CNS via fast-conducting myelinated axons, gentle stroking produces a pleasant sensation mediated by small, unmyelinated axons. This information is received by the insular cortex, a region associated with emotional responses.

Unless the skin is moving, tactile sensation provides little information about the nature of objects we touch. Movement and manipulation provide information about the shape, mass, texture, and other physical characteristics of objects we feel. Tactile experience, such as that gained by musicians, increases the portion of the somatosensory cortex devoted to the fingers involved in this experience.

Temperature receptors adapt to the ambient temperature; moderate changes in skin temperature are soon perceived as neutral, and deviations above or below this temperature are perceived as warmth or coolness. Transduction of different ranges of temperatures is accomplished by six members of the TRP (transient receptor potential) family of receptors. One of the coolness receptors, TRPM8, also responds to menthol and is involved in responsiveness to environmental cold. There are at least three different types of pain receptors: high-threshold mechanoreceptors; fibers with capsaicin receptors (TRPV1 receptors), which detect extremes of heat, acids, and the presence of capsaicin; and fibers with TRPA1 receptors, which are sensitive to chemical irritants and inflammation. Itch is an unpleasant sensation conveyed by unknown receptors. Pain and itch are mutually inhibitory.

Precise, well-localized somatosensory information is conveyed by a pathway through the dorsal columns and their nuclei and the medial lemniscus, connecting the dorsal column nuclei with the ventral posterior nuclei of the thalamus. Information about pain and temperature ascends the spinal cord through the spinothalamic system. Organic sensibility reaches the central nervous system by means of axons that travel through nerves of the autonomic nervous systems.

The neurons in the primary somatosensory cortex are topographically arranged, according to the part of the body from which they receive sensory information (somatotopic representation). Columns within the somatosensory cortex respond to a particular type of stimulus from a particular region of the body. Different types of somatosensory receptors send their information to separate areas of the somatosensory cortex. Damage to the somatosensory association cortex can cause tactile agnosia, inability to recognize common objects by means of touch.

A particular voltage-dependent sodium channel, $Na_x1.7$, plays an essential role in pain sensation. Mutations of the gene for this protein produce total insensitivity to pain. Pain perception is not a simple function of stimulation of pain receptors; it is a complex phenomenon with sensory and emotional components that can be modified by experience and the immediate environment. The sensory component is mediated by the primary and secondary somatosensory cortex, the immediate emotional component appears to be mediated by the anterior cingulate cortex and the insular cortex, and the long-term emotional component appears to be mediated by the prefrontal cortex. Functional-imaging studies using hypnotic suggestion found that a decrease in the sensory component of pain reduced activation of the somatosensory cortex and that reduction of the unpleasantness of pain reduced the activation of the anterior cingulate cortex. The phantom limb phenomenon, which often is accompanied by phantom pain, appears to be inherent in the organization of the parietal lobe.

Just as we have mechanisms to perceive pain, we have mechanisms to reduce it—to produce analgesia. In the appropriate circumstances, neurons in the periaqueductal gray matter are stimulated through synaptic connections with the frontal cortex, amygdala, and hypothalamus. In addition, some neurosecretory cells in the brain release enkephalins, a class of endogenous opioids. Connections from the periaqueductal gray matter to the nucleus raphe magnus of the medulla activate serotonergic neurons located there. These neurons send axons to the dorsal horn of the spinal cord gray matter, where they inhibit the transmission of pain information to the brain. In humans, chronic pain is sometimes treated by implanting electrodes in the periaqueductal gray matter or the thalamus and permitting the patients to stimulate the brain through these electrodes when the pain becomes severe.

Analgesia occurs when it is important for an animal to continue a behavior that would tend to be inhibited by pain—for example, mating or fighting. The administration of a placebo can also produce analgesia. Because this effect is blocked by naloxone, it must involve the release of endogenous opioids. Functional-imaging studies suggest that the placebo effect may be caused by increased activity of the prefrontal cortex, which activates the periaqueductal gray matter and inhibits the activity of the anterior cingulate cortex and insular cortex, inducing analgesia. Learning to increase or decrease the activity of one's own anterior cingulate cortex using feedback from fMRI increases or decreases a person's sensitivity to pain.

Thought Question

Our fingertips and our lips are the most sensitive parts of our bodies; relatively large amounts of the primary somatosensory cortex are devoted to analyzing information from these parts of the body. It is easy to understand why fingertips are so sensitive: We use them to explore object by touch. But why are our lips so sensitive? Does it have something to do with eating?

GUSTATION

The stimuli that we have encountered so far produce receptor potentials by imparting physical energy: thermal, photic (involving light), or kinetic. However, the stimuli received by the last two senses to be studied—gustation and olfaction—interact with their receptors chemically. This section discusses the first of them: gustation.

The Stimuli

Gustation is clearly related to eating; this sense modality helps us to determine the nature of things we put in our mouths. For a substance to be tasted, molecules of it must dissolve in the saliva and stimulate the taste receptors on the tongue. Tastes of different substances vary, though much less than we generally realize. There are only five qualities of taste: *bitterness, sourness, sweetness, saltiness,* and *umami.* You are familiar with the first four qualities, and I will explain the fifth one later. Flavor, as opposed to taste, is a composite of olfaction and gustation. Much of the flavor of food depends on its odor; *anosmic* people (who lack the sense of smell) or people whose nostrils are stopped up have difficulty distinguishing between different foods by taste alone.

Most vertebrates possess gustatory systems that respond to all five taste qualities. (An exception is the cat family; lions, tigers, leopards, and house cats do not detect sweetness—but then, none of the food they normally eat is sweet.) Clearly, sweetness receptors are food detectors. Most sweet-tasting foods, such as fruits and some vegetables, are safe to eat (Ramirez, 1990). Saltiness receptors detect the presence of sodium chloride. In some environments, inadequate amounts of this mineral are obtained from the usual source of food, so sodium chloride detectors help the animal to detect its presence. Injuries that cause bleeding deplete an organism of its supply of sodium rapidly, so the ability to find it quickly can be critical. In recent years, researchers have recognized the existence of a fifth taste quality: *umami.* **Umami,** a Japanese word that means "good taste," refers to the taste of monosodium glutamate (MSG), a substance that is often used as a flavor enhancer in Asian cuisine (Kurihara, 1987; Scott and Plata-Salaman, 1991). The umami receptor detects the presence of glutamate, an amino acid found in proteins. Presumably, the umami receptor provides the ability to taste proteins, an important nutrient.

Most species of animals will readily ingest substances that taste sweet or somewhat salty. Similarly, they are attracted to foods that are rich in amino acids, which explains the use of MSG as a flavor enhancer. However, they will tend to avoid substances that taste sour or bitter. Because of bacterial activity, many foods become acidic when they spoil. In addition, most unripe fruits are acidic. Acidity tastes sour and causes an avoidance reaction. (Of course, we have learned to make highly preferred mixtures of sweet and sour, such as lemonade.) Bitterness is almost universally avoided and cannot easily be improved by adding some sweetness. Many plants produce poisonous alkaloids, which protect them from being eaten by animals. Alkaloids taste bitter; thus, the bitterness receptor undoubtedly serves to warn animals away from these chemicals.

Anatomy of the Taste Buds and Gustatory Cells

The tongue, palate, pharynx, and larynx contain approximately 10,000 taste buds. Most of these receptive organs are arranged around *papillae,* small protuberances of the tongue. *Fungiform papillae,* located on the anterior two-thirds of the tongue, contain up to eight taste buds, along with receptors for pressure, touch, and temperature. *Foliate papillae* consist of up to eight parallel folds along each edge of the back of the tongue. Approximately 1300 taste buds are located in these folds. *Circumvallate papillae,* arranged in an inverted V on the posterior third of the tongue, contain approximately 250 taste buds. They are shaped like little plateaus surrounded by moatlike trenches. Taste buds consist of groups of twenty to fifty receptor cells, specialized neurons arranged somewhat like the segments of an orange. Cilia are located at the end of each cell and project through the opening of the taste bud (the pore) into the saliva that coats the tongue. Tight junctions between adjacent taste cells prevent substances in the saliva from diffusing freely into the taste bud itself. Figure 7.35 shows the appearance of a circumvallate papilla; a cross section through the surrounding trench contains a taste bud. (See *Figure 7.35.*)

Taste receptor cells form synapses with dendrites of bipolar neurons whose axons convey gustatory information to the brain through the seventh, ninth, and tenth cranial nerves. The neurotransmitter released by the receptor cells is adenosine triphosphate (ATP), the molecule produced by mitochondria that stores energy within cells (Finger et al., 2005). The receptor cells have a life span of only ten days. They quickly wear out, being directly exposed to a rather hostile environment. As they degenerate, they are replaced by newly developed cells; the dendrite of the bipolar neuron is passed on to the new cell (Beidler, 1970).

---

umami *(oo mah mee)* The taste sensation produced by glutamate

FIGURE 7.35 ■ The Tongue

The figure shows (a) papillae on the surface of the tongue and (b) taste buds.

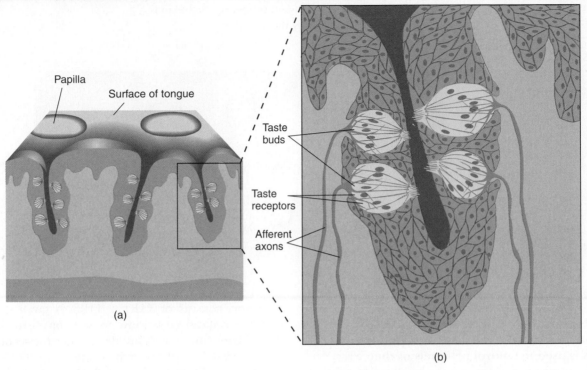

Papilla

Surface of tongue

Taste buds

Taste receptors

Afferent axons

(a)

(b)

Perception of Gustatory Information

Transduction of taste is similar to the chemical transmission that takes place at synapses: The tasted molecule binds with the receptor and produces changes in membrane permeability that cause receptor potentials. Different substances bind with different types of receptors, producing different taste sensations. In this section I will describe what we know about the nature of the molecules with particular tastes and the receptors that detect their presence.

To taste salty, a substance must ionize. Although the best stimulus for saltiness receptors is sodium chloride (NaCl), a variety of salts containing metallic cations (such as Na^+, K^+, and Li^+) with a small anion (such as Cl^-, Br^-, SO_4^{2-}, or NO_3^-) taste salty. The receptor for saltiness seems to be a simple sodium channel. When present in the saliva, sodium enters the taste cell and depolarizes it, triggering action potentials that cause the cell to release neurotransmitter (Avenet and Lindemann, 1989; Kinnamon and Cummings, 1992). The best evidence that sodium channels are involved is the fact that amiloride, a drug that is known to block sodium channels, prevents sodium chloride from activating taste cells and decreases sensations of saltiness. However, the drug does not completely block these sensations in humans, so most investigators believe that more than one type of

receptor is involved (Schiffman, Lockhead, and Maes, 1983; Ossebaard, Polet, and Smith, 1997).

Sourness receptors appear to respond to the hydrogen ions present in acidic solutions. However, because the sourness of a particular acid is not simply a function of the concentration of hydrogen ions, the anions must have an effect as well. The reason for this anion effect is not yet known. Huang et al. (2006) reported the discovery of the sourness receptor: a transient receptor potential ion channel known as PKD2L1. (The unfortunate name for this channel is *polycystic-kidney-disease-like ion channel.*) Figure 7.36 shows activity recorded from nerves that serve the gustatory receptors of the tongues of normal mice and mice with a knockout of the gene for the PKD2L1 ion channel. As you can see, afferent taste fibers of normal mice showed responses to all five categories of taste stimuli, but PKD2L1 knockout mice showed no response to the acids. However, the responses of these mice to the other four categories of tastes were normal. (See *Figure 7.36.*)

Huang and her colleagues also found evidence that PKD2L1 sourness receptors also detect pH levels (levels of acidity versus alkalinity) in the cerebrospinal fluid. Neurons that expressed these receptors were found at the border of the central canal of the spinal cord, and processes of these neurons extended past the border into the CSF itself. The investigators found that the firing rate of these

FIGURE 7.36 ■ The Sourness Receptor

Responses were recorded from nerves that serve the gustatory receptors of the tongues of normal mice and mice with a knockout (KO) of the PKD2L1 transient receptor potential ion channel. Only sensitivity to sour-tasting substances is affected by the mutation.

(Adapted from Huang, A. L., Chen, X., Hoon, M.A., Chandrashekar, J., Guo, W., Tränker, D., Ryba, N. J. P., and Zuker, C. S. *Nature*, 2006, *442*, 934–938.)

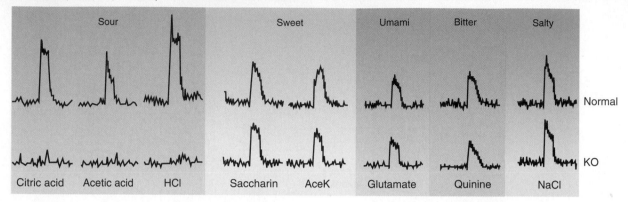

neurons, but not of adjacent neurons that did not express PKD2L1 receptors, was related to pH levels: Low pH levels (more acidity) produced higher levels of activity. It seems likely that information from these neurons is used to control pH levels of the CSF.

The stimulus for sourness—the presence of acids—is clear. However, bitter and sweet substances are more difficult to characterize. The typical stimulus for bitterness is a plant alkaloid such as quinine; for sweetness it is a sugar such as glucose or fructose. The fact that some molecules elicit both sensations suggested to early researchers that bitterness and sweetness receptors may be similar. For example, the Seville orange rind contains a glycoside (complex sugar) that tastes extremely bitter; the addition of a hydrogen ion to the molecule makes it taste intensely sweet (Horowitz and Gentili, 1974). Some amino acids taste sweet. Indeed, the commercial sweetener aspartame consists of just two amino acids: aspartate and phenylalanine.

Receptors sensitive to bitterness and sweetness are linked to a G protein known as *gustducin,* which is very similar in structure to *transducin,* the G protein involved in transduction of photic information in the retina. Receptors sensitive to umami are linked to both gustducin and transducin (McLaughlin et al., 1993; Wong, Gannon, and Margolskee, 1996; He et al., 2004). When a bitter molecule binds with the receptor, the G protein activates an enzyme that begins a cycle of chemical reactions that causes the release of ATP, the neurotransmitter of taste receptor cells.

Recent evidence has discovered two families of receptors responsible for detecting sweet, bitter, and umami tastes, (see Scott, 2004, for a review). The first family, T1R, has three members: T1R1, T1R2, and T1R3, produced by three different genes. The sweet receptor

consists of two components: T1R2 + T1R3. The umami receptors consists of T1R1 + T1R3. A given gustatory receptor cell can be sensitive to sweet or umami but not both. Bitter compounds are detected by the second family of receptors, T2R, of which there are 30 members (Matsunami, Montmayeur, and Buck, 2000). A given gustatory receptor cell sensitive to bitterness contains one of the many different varieties of T2R, which indicates that each cell can detect the presence of many different bitter-tasting molecules. As we saw, many compounds found in nature that taste bitter to us are poisonous. Rather than entrusting detection of these compounds to a single receptor, the process of evolution has given us the ability to detect a wide variety of compounds with different molecular shapes. (See *Figure 7.37.*)

I mentioned earlier that cats are insensitive to sweet tastes. Li et al. (2005) discovered the reason for the absence of sweet sensitivity: The DNA of members of the cat family (the investigators tested domestic cats, tigers, and cheetahs) lacks functional genes that produce T1R2 proteins, which are necessary for sweet receptors. (See *Figure 7.37.*) The investigators concluded that this mutation was probably an important event in the evolution of cats' carnivorous behavior.

For many years, researchers have known that many species of animals (including our own) show a distinct preference for high-fat foods. Because there is not a distinct taste that is associated with the presence of fat, most investigators concluded that we detected fat by its odor and texture ("mouth feel"). However, Fukuwatari et al. (2003) found that rats whose olfactory sense was destroyed continued to show a preference for a liquid diet containing a long-chain fatty acid, one of the breakdown products of fat. Laugerette et al. (2005) reported that a fatty acid transporter found in the papillae of the tongue,

FIGURE 7.37 ■ Structure of Receptors

This schematic drawing shows the structure of receptors responsible for detection of sweet, bitter, and umami tastes.

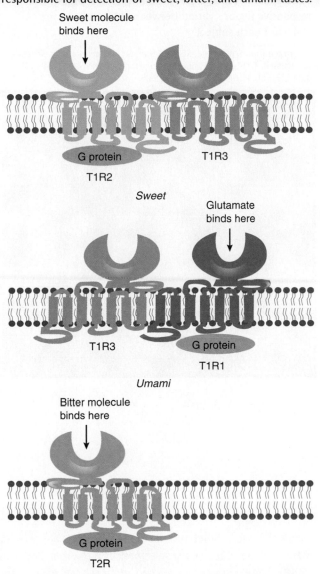

Sweet molecule binds here

G protein

T1R3

T1R2

Sweet

Glutamate binds here

T1R3

G protein

T1R1

Umami

Bitter molecule binds here

G protein

T2R

Bitter

CD36, appear to serve as a fatty-acid detector. The investigators found that in contrast to normal mice, mice with a knockout of the CD36 gene showed no preference for fatty acids. When fats reach the tongue, some of these molecules are broken down by an enzyme called *lingual lipase,* which is found in the vicinity of taste buds. The activity of lingual lipase ensures that fatty acid detectors are stimulated when food containing fat enters the mouth.

The Gustatory Pathway

Gustatory information is transmitted through cranial nerves 7, 9, and 10. Information from the anterior part of the tongue travels through the **chorda tympani,** a branch of the seventh cranial nerve (facial nerve). Taste receptors in the posterior part of the tongue send information through the lingual (tongue) branch of the ninth cranial nerve (glossopharyngeal nerve); the tenth cranial nerve (vagus nerve) carries information from receptors of the palate and epiglottis. The chorda tympani gets its name because it passes through the middle ear just beneath the tympanic membrane. Because of its convenient location, it is accessible to a recording or stimulating electrode. Investigators have even recorded from this nerve during the course of human ear operations.

The first relay station for taste is the **nucleus of the solitary tract,** located in the medulla. In primates the taste-sensitive neurons of this nucleus send their axons to the ventral posteromedial thalamic nucleus, a nucleus that also receives somatosensory information received from the trigeminal nerve (Beckstead, Morse, and Norgren, 1980). Thalamic taste-sensitive neurons send their axons to the primary gustatory cortex, which is located in the base of the frontal cortex and in the insular cortex (Pritchard et al., 1986). Neurons in this region project to the secondary gustatory cortex, located in the caudolateral orbitofrontal cortex (Rolls, Yaxley, and Sienkiewicz, 1990). Unlike most other sense modalities, taste is ipsilaterally represented in the brain; that is, the right side of the tongue projects to the right side of the brain, and the left side projects to the left. (See *Figure 7.38.*)

In a functional-imaging study, Schoenfeld et al. (2004) had people sip water that was flavored with sweet, sour, bitter, and umami tastes. The investigators found that tasting each flavor activated different regions in the primary gustatory area of the insular cortex. Although the locations of the taste-responsive regions differed from subject to subject, the same pattern was seen when a given subject was tested on different occasions. Thus, the representation of tastes in the gustatory cortex is idiosyncratic but stable. (See *Figure 7.39.*)

Gustatory information also reaches the amygdala and the hypothalamus and adjacent basal forebrain (Nauta, 1964; Russchen, Amaral, and Price, 1986). Many investigators believe that the hypothalamic pathway plays a role in mediating the reinforcing effects of sweet, umami, and slightly salty tastes. In fact, some neurons in the hypothalamus respond to sweet stimuli only when the animal is hungry (Rolls et al., 1986).

chorda tympani A branch of the facial nerve that passes beneath the eardrum; conveys taste information from the anterior part of the tongue and controls the secretion of some salivary glands.

nucleus of the solitary tract A nucleus of the medulla that receives information from visceral organs and from the gustatory system.

FIGURE 7.38 ■ Neural Pathways of the Gustatory System

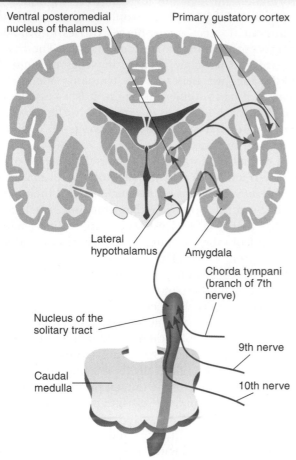

FIGURE 7.39 ■ Activation in the Primary Gustatory Cortex

Functional MRI images of six subjects show that the responsive regions varied between subjects but were stable for each subject.

(Reprinted from Schoenfeld, M. A., Neuer, G., Tempelmann, C., Schüssler, K., Noesselt, T., Hopf, J.-M., and Heinze, H.-J. *Neuroscience*, 2004, *127*, 347–353 with permission from Elsevier.)

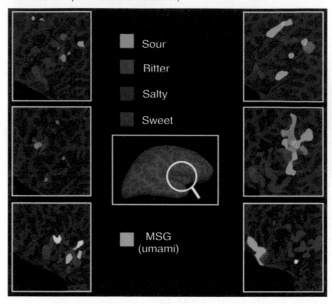

Interim Summary

Gustation

Taste receptors detect only five sensory qualities: bitterness, sourness, sweetness, saltiness, and umami (umaminess?). Bitter foods often contain plant alkaloids, many of which are poisonous. Sour foods have usually undergone bacterial fermentation, which can produce toxins. On the other hand, sweet foods (such as fruits) are usually nutritious and safe to eat, and salty foods contain an essential cation: sodium. The fact that people in affluent cultures today tend to ingest excessive amounts of sweet and salty foods suggests that these taste qualities are naturally reinforcing. Umami, the taste of glutamate, identifies proteins.

Saltiness receptors appear to be simple sodium channels. Sourness receptors appear to detect the presence of hydrogen ions, which activate a transient receptor potential ion channel known as PKD2L1. Bitter, sweet, and umami tastes are detected by two families of receptors: sweetness by a receptor consisting of T1R2 + T1R3, umami by one consisting of T1R1 + T1R3, and bitterness by thirty different members of the T2R family. A fatty-acid transporter, CD36, found in the papillae of the tongue, detects molecules of fatty acids produced when an enzyme, lingual lipase, breaks down some molecules of fat in the mouth.

Gustatory information from the anterior part of the tongue travels through the chorda tympani, a branch of the facial nerve that passes beneath the eardrum on its way to the brain. The posterior part of the tongue sends gustatory information through the glossopharyngeal nerve, and the palate and epiglottis send gustatory information through the vagus nerve. Gustatory information is received by the nucleus of the solitary tract (located in the medulla) and is relayed by the ventral posteromedial thalamus to the primary gustatory cortex in the basal frontal and insular areas. Different tastes activate different regions of the primary gustatory cortex. The caudolateral orbitofrontal cortex contains the secondary gustatory cortex. Gustatory information is also sent to the amygdala, hypothalamus, and basal forebrain.

Thought Question
Bees and birds can taste sweet substances, but cats and alligators cannot. Obviously, the ability to taste particular

substances is related to the range of foods a species eats. If, through the process of evolution, a species develops a greater range of foods, what do you think comes first: the food or the receptor? Would a species start eating something with a new taste (say, something sweet) and later develop the appropriate taste receptors, or would the taste receptors evolve first and then lead the animal to a new taste?

OLFACTION

Olfaction, the second chemical sense, helps us to identify food and avoid food that has spoiled and is unfit to eat. It helps the members of many species to track prey or detect predators and to identify friends, foes, and receptive mates. Although many other mammals, such as dogs, have more sensitive olfactory systems than humans do, we should not underrate our own. The olfactory system is second only to the visual system in the number of sensory receptor cells, with an estimated 10 million cells. We can smell some substances at lower concentrations than the most sensitive laboratory instruments can detect.

For years I have told my students that one reason for the difference in sensitivity between our olfactory system and those of other mammals is that other

mammals put their noses where odors are the strongest—just above the ground. For example, a dog following an odor trail sniffs along the ground, where the odors of a passing animal may have clung. Even a bloodhound's nose would not be very useful if it were located five or six feet above the ground, as ours is. I was gratified to learn that a scientific study established the fact that when people sniff the ground as dogs do, their olfactory system works much better. Porter et al. (2007) prepared a scent trail—a string moistened with essential oil of chocolate and laid down in a grassy field. The subjects were blindfolded and wore earmuffs, kneepads, and gloves, which prevented them from using anything other than their noses to follow the scent trail. They did quite well, and they adopted the same zigzag strategy used by dogs. (See *Figure 7.40*.) As the authors wrote, these findings "suggest that the poor reputation of human olfaction may reflect, in part, behavioral demands rather than ultimate abilities" (Porter et al., 2007, p. 27).

The Stimulus

The stimulus for odor (known formally as *odorants*) consists of volatile substances having a molecular weight in the range of approximately 15 to 300. Almost all odorous compounds are lipid soluble and of organic origin. However, many substances that meet these criteria have no odor at all, so we still have much to learn about the nature of odorants.

Anatomy of the Olfactory Apparatus

Our 6 million olfactory receptor cells reside within two patches of mucous membrane (the **olfactory epithelium**), each having an area of about 1 square inch.

FIGURE 7.40 ■ Scent-Tracking Behavior
The path followed by a dog and a human during the scent tracking is shown in red.

(From Porter, J., Craven, B., Khan, R. M., Chang, S.-J., Kang, I Judkewitz, B., Volpe, J., Settles, G., and Sobel, N. *Nature Neuroscience*, 2007, *10*, 27–29. Reprinted with permission.)

---

olfactory epithelium The epithelial tissue of the nasal sinus that covers the cribriform plate; contains the cilia of the olfactory receptors.

FIGURE 7.41 ■ The Olfactory System

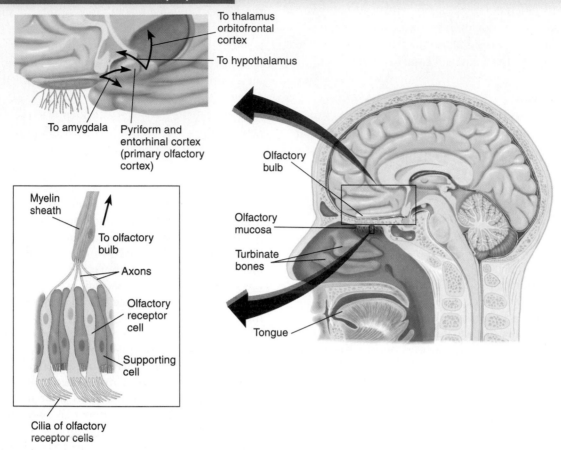

The olfactory epithelium is located at the top of the nasal cavity, as shown in *Figure 7.41.* Less than 10 percent of the air that enters the nostrils reaches the olfactory epithelium; a sniff is needed to sweep air upward into the nasal cavity so that it reaches the olfactory receptors.

The inset in Figure 7.41 illustrates a group of olfactory receptor cells, along with their supporting cells. (See *inset, Figure 7.41.*) Olfactory receptor cells are bipolar neurons whose cell bodies lie within the olfactory mucosa that lines the *cribriform plate,* a bone at the base of the rostral part of the brain. There is a constant production of new olfactory receptor cells, but their life is considerably longer than those of gustatory receptor cells. Supporting cells contain enzymes that destroy odorant molecules and thus help to prevent them from damaging the olfactory receptor cells.

Olfactory receptor cells send a process toward the surface of the mucosa, which divides into ten to twenty cilia that penetrate the layer of mucus. Odorous molecules must dissolve in the mucus and stimulate receptor molecules on the olfactory cilia. Approximately thirty-five bundles of axons, ensheathed by glial cells, enter the skull through small holes in the cribriform ("perforated") plate. The olfactory mucosa also contains free

nerve endings of trigeminal nerve axons; these nerve endings presumably mediate sensations of pain that can be produced by sniffing some irritating chemicals, such as ammonia.

The **olfactory bulbs** lie at the base of the brain on the ends of the stalklike olfactory tracts. Each olfactory receptor cell sends a single axon into an olfactory bulb, where it forms synapses with dendrites of **mitral cells** (named for their resemblance to a bishop's miter, or ceremonial headgear). These synapses take place in the complex axonal and dendritic arborizations called **olfactory glomeruli** (from *glomus,* "ball"). There are approximately 10,000 glomeruli, each of which receives input from a bundle of approximately 2000 axons. The axons of the mitral cells travel to the rest of the brain

olfactory bulb The protrusion at the end of the olfactory tract; receives input from the olfactory receptors.

mitral cell A neuron located in the olfactory bulb that receives information from olfactory receptors; axons of mitral cells bring information to the rest of the brain.

olfactory glomerulus (*glow mare you luss*) A bundle of dendrites of mitral cells and the associated terminal buttons of the axons of olfactory receptors.

through the olfactory tracts. Some of these axons terminate in other regions of the ipsilateral forebrain; others cross the brain and terminate in the contralateral olfactory bulb.

Olfactory tract axons project directly to the amygdala and to two regions of the limbic cortex: the piriform cortex and the entorhinal cortex. (See *Figure 7.41.*) The amygdala sends olfactory information to the hypothalamus, the entorhinal cortex sends it to the hippocampus, and the piriform cortex sends it to the hypothalamus and to the orbitofrontal cortex via the dorsomedial nucleus of the thalamus (Buck, 1996; Shipley and Ennis, 1996). As you may recall, the orbitofrontal cortex also receives gustatory information; thus, it may be involved in the combining of taste and olfaction into flavor. The hypothalamus also receives a considerable amount of olfactory information, which is probably important for the acceptance or rejection of food and for the olfactory control of reproductive processes seen in many species of mammals.

Most mammals have another organ that responds to chemicals in the environment: the *vomeronasal organ*. Because it plays an important role in animals' responses to pheromones, chemicals produced by other animals that affect reproductive physiology and behavior, its structure and function are described in Chapter 10.

Efferent fibers from several locations in the brain enter the olfactory bulbs. These include acetylcholinergic, noradrenergic, dopaminergic, and serotonergic inputs (Shipley and Ennis, 1996).

Transduction of Olfactory Information

For many years, researchers have recognized that olfactory cilia contain receptors that are stimulated by molecules of odorants, but the nature of the receptors was unknown. Jones and Reed (1989) identified a particular G protein, which they called G_{olf}. This protein is able to activate an enzyme that catalyzes the synthesis of cyclic AMP, which, in turn, can open sodium channels and depolarize the membrane of the olfactory cell (Nakamura and Gold, 1987; Firestein, Zufall, and Shepherd, 1991; Menco et al., 1992).

As we saw in Chapter 2, G proteins serve as the link between metabotropic receptors and ion channels: When a ligand binds with a metabotropic receptor, the G protein either opens ion channels directly or does so indirectly, by triggering the production of a second messenger. The discovery of G_{olf} suggested that olfactory cilia contained odorant receptors linked to this G protein. Indeed, Buck and Axel (1991) used molecular genetics techniques and discovered a family of genes that code for a family of olfactory receptor proteins (and in 2004 won a Nobel Prize for doing so). So far, olfactory receptor genes have been isolated in more than twelve species of vertebrates, including mammals, birds, and amphibians (Mombaerts, 1999). Humans have 339 different olfactory receptor genes, and mice have 913 (Godfrey, Malnic, and Buck, 2004; Malnic, Godfrey, and Buck, 2004). Molecules of odorant bind with olfactory receptors, and the G proteins coupled to these receptors open sodium channels and produce depolarizing receptor potentials.

Perception of Specific Odors

For many years, recognition of specific odors has been an enigma. Humans can recognize up to ten thousand different odorants, and other animals can probably recognize even more (Shepherd, 1994). Even with 339 different olfactory receptors, that leaves many odors unaccounted for. And every year, chemists synthesize new chemicals, many with odors unlike those that anyone has previously detected. How can we use a relatively small number of receptors to detect so many different odorants?

Before I answer this question, we should look more closely at the relationship among receptors, olfactory neurons, and the glomeruli to which the axons of these neurons project. First, the cilia of each olfactory neuron contain only one type of receptor (Nef et al., 1992; Vassar, Ngai, and Axel, 1993). As we saw, each glomerulus receives information from approximately two thousand different olfactory receptor cells. Using in situ hybridization methods to identify particular receptor proteins in individual cells, Ressler, Sullivan, and Buck (1994) discovered that although a given glomerulus receives information from approximately two thousand different olfactory receptor cells, each of these cells contains the same type of receptor molecule. According to Serizawa et al. (2003), early in development, one olfactory receptor gene randomly turns on in a given olfactory receptor cell, and as soon as it does so, it inhibits the expression of all other olfactory genes. Thus, there are as many types of glomeruli as there are types of receptor molecules. Furthermore, the location of particular types of glomeruli (defined by the type of receptor that sends information to them) appears to be the same in each of the olfactory bulbs in a given animal and may even be the same from one animal to another. (See *Figure 7.42.*)

An ingenious study by Zou et al. (2001) investigated the specificity of olfactory information in the pathway from olfactory receptors to olfactory glomeruli to the olfactory cortex. To accomplish this feat, they inserted a gene for a transneuronal tracer

FIGURE 7.42 ■ Connections of Olfactory Receptor Cells with Glomeruli

Each glomerulus of the olfactory bulb receives information from only one type of receptor cell. Olfactory receptor cells of different colors contain different types of receptor molecules.

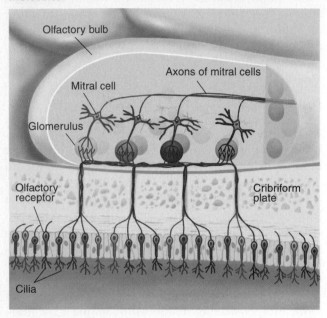

protein (barley lectin) into the DNA of mice adjacent to two different olfactory receptor genes. Because of the location of this gene, it was turned on only in olfactory receptor cells that produced one of the two selected receptor genes. Thus, two different types of olfactory receptor cells expressed barley lectin. This protein is transmitted from one neuron to others, with which it forms synapses. Thus, it was carried to glomeruli and from there to a third set of neurons in the olfactory cortex. The investigators found that just as retinotopic information is maintained in the visual system and tonotopic information is maintained in the auditory system, "olfactotopic" information is maintained in the olfactory system. That is, the particular glomeruli that receive information from particular olfactory receptors send this information to specific regions of olfactory cortex. These regions appeared to occur in identical locations in different mice.

Now let's get back to the question I just posed: How can we use a relatively small number of receptors to detect so many different odorants? The answer is that a particular odorant binds to more than one receptor. Thus, because a given glomerulus receives information from only one type of receptor, different odorants produce different *patterns* of activity in different glomeruli. Recognizing a particular odor, then, is a matter of recognizing a particular pattern of

activity in the glomeruli. The task of chemical recognition is transformed into a task of spatial recognition.

Figure 7.43 illustrates this process (Malnic et al., 1999). The left side of the figure shows the shapes of eight hypothetical odorants. The right side shows four hypothetical odorant receptor molecules. If a portion of the odorant molecule fits the binding site of the receptor molecule, it will activate it and stimulate the olfactory neuron. As you can see, each odorant molecule fits the binding site of at least one of the receptors and in most cases fits more than one of them. Notice also that the *pattern* of receptors activated by each of the eight odorants is different, which means that if we know which pattern of receptors is activated, we know which odorant is present. Of course, even though a particular odorant might bind with several different types of receptor molecules, it might not bind equally well with each of them. For example, it might bind very well with one receptor molecule, moderately well with another, weakly with another, and so on. (See *Figure 7.43*.)

FIGURE 7.43 ■ Coding of Olfactory Information

A hypothetical explanation of the coding shows that different odorant molecules attach to different combinations of receptor molecules. (Activated receptor molecules are shown in blue.) Unique patterns of activation represent particular odorants.

(Adapted from Malnic, B., Hirono, J, Sato, T., and Buck, L. B. *Cell*, 1999, *96*, 713–723.)

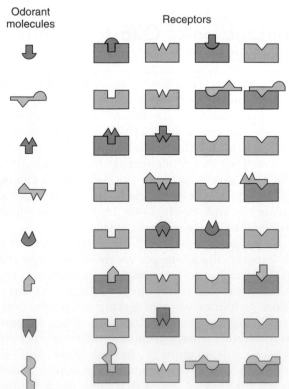

As we just saw, the spatial pattern of "olfactotopic" information is maintained in the olfactory cortex. Presumably, the olfactory cortex recognizes particular odors by recognizing different patterns of activation there.

Igarashi and Mori (2004) obtained evidence to support this model. They presented a variety of odorants to rats and recorded the regions of activation on the surface of an exposed olfactory bulb. They found that different features of the molecules activated different regions of the olfactory bulb. Figure 7.44 shows the patterns of activity. The common features of the molecules that activated glomeruli in different regions of the olfactory bulb are outlined in red or green. (See *Figure 7.44*.)

Several studies have found that interactions can take place between glomeruli within the olfactory bulb. For example, some odors have the ability to mask others. (The existence of the deodorant and air-freshener industries depends on this fact.) Cooks in various cultures have long known that as long as it is not too strong, the unpleasant, rancid off-flavor of spoiled food can be masked by the spices fennel and clove. Takahashi, Nagayama, and Mori (2004) found

that evidence for this masking can be seen in the responses of olfactory glomeruli. The investigators used a dental drill to thin the skull of anesthetized rats above an olfactory bulb and moistened it with mineral oil so that it became transparent. Examination of the olfactory bulbs with a microscope showed local areas of increased blood flow when different odorants were presented, which indicated regions of increased neural activity. Takahashi and his colleagues mapped the regions of the olfactory bulb that responded to bad-smelling odorants (alkylamines and aliphatic aldehydes) and to the odors of fennel and clove. They found that responses to the bad odors were suppressed by the presence of the spice odors, indicating that the masking took place in the olfactory bulbs. Presumably, the glomeruli that responded to the spice odors inhibited those that responded to the rancid ones.

Interactions can take place in the olfactory cortex as well. Zou and Buck (2006) found that some neurons in the mouse olfactory cortex responded to the combinations of two odorants but not to either one of them in isolation. A functional-imaging study with humans by Gottfried, Winston, and Dolan (2006) found that the

FIGURE 7.44 ■ Clusters and Zones in the Rodent Olfactory Bulb

Analysis of information in an unrolled map of the rodent olfactory bulb. The olfactory bulb can be subdivided into nine clusters in four zones. Each cluster responds to particular features of odorant molecules, indicated by green or red shading. A = anterior, D = dorsal, P = posterior, M = medial, V = ventral.

(From Mori, K., Takahashi, Y. K., Igarashi, K. M., and Yamaguchi, M. *Physiological Reviews*, 2006, *86*, 400–433. Reprinted with permission.)

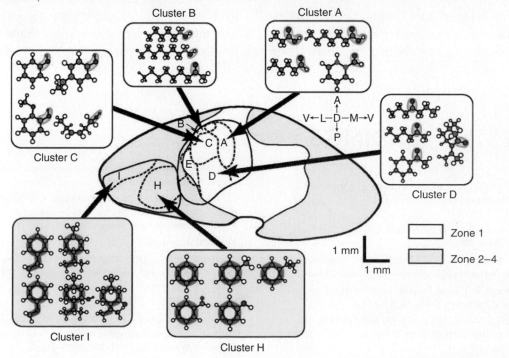

coding of olfactory information differed in two regions of the piriform cortex. Groups of neurons in the anterior region represent the chemical structures of odorants, whereas groups of neurons in the posterior region represent the qualities of odorants. The investigators asked the subjects to sniff eight different odorants from two categories of chemical structure (alcohols or aldehydes) and two categories of perceptual quality (lemonlike or vegetablelike). They found that the alcohols and the aldehydes activated two different clusters of neurons in the anterior piriform cortex but that the lemonlike and vegetablelike odorants activated two different clusters of neurons in the posterior piriform cortex. The anterior region of the piriform cortex received information directly from the olfactory bulb, and the posterior region receives the information indirectly, from the anterior region. As the investigators note, we do not yet know how maps of chemical structure are combined to form maps of perceptual quality. Presumably, learning plays some role in this process.

Interim Summary

Olfaction

The olfactory receptors consist of bipolar neurons located in the olfactory epithelium that lines the roof of the nasal sinuses, on the bone that underlies the frontal lobes. The receptors send processes toward the surface of the mucosa, which divide into cilia. The membranes of these cilia contain receptors that detect aromatic molecules dissolved in the air that sweeps past the olfactory mucosa. The axons of the olfactory receptors pass through the perforations of the cribriform plate into the olfactory bulbs, where they form synapses in the glomeruli with the dendrites of the mitral cells. These neurons send axons through the olfactory tracts to the brain, principally to the amygdala, the piriform cortex, and the entorhinal cortex. The hippocampus, hypothalamus, and orbitofrontal cortex receive olfactory information indirectly.

Aromatic molecules produce membrane potentials by interacting with a newly discovered family of receptor molecules, which appears to contain 339 members. These receptor molecules are coupled to a special G protein, G_{olf}. When an odorant molecule bind with and stimulates one of these receptors, G_{olf} catalyzes the synthesis of cyclic AMP, which opens sodium channels and depolarizes the membrane. Each glomerulus receives information from only one type of olfactory receptor, and "olfactotopic" coding is maintained all the way to the olfactory cortex. This means that the task of detecting different odors is a spatial one; the brain recognizes odors by means of the patterns of activity created in the olfactory cortex. The anterior piriform cortex appears to code odor information according to the structure of the odorant molecules, and the posterior piriform cortex codes the information it receives from the anterior region according to the odorants' perceptual categories.

Thought Question

Have you ever encountered an odor that you knew was somehow familiar, but you couldn't say exactly why? Can you think of any explanations? Might this phenomenon have something to do with the fact that the sense of olfaction developed very early in our evolutionary history?

SUGGESTED READINGS

Audition

Copeland, B. J., and Pillsbury, H. C. Cochlear implantation for the treatment of deafness. *Annual Review of Medicine,* 2004, *55,* 157–167.

Ehret, G., and Romand, R. *The Central Auditory System.* New York: Oxford University Press, 1997.

Fettiplace, R., and Hackney, C. M. The sensory and motor roles of auditory hair cells. *Nature Reviews: Neuroscience,* 2006, *7,* 19–29.

Peretz, I., and Zatorre, R. J. Brain organization for music processing. *Annual Review of Psychology,* 2005, *56,* 89–114.

Stewart, L., von Kriegstein, K., Warren, J. D., and Griffiths, T. D. Music and the brain: Disorders of musical listening. *Brain,* 2006, *129,* 2533–2553.

Yost, W. A. *Fundamentals of Hearing: An Introduction,* 4th ed. San Diego: Academic Press, 2000.

Zatorre, R. J., Chen, J. L., and Penhune, V. B. When the brain plays music: Auditory–motor interactions in music perception and production. *Nature Reviews: Neuroscience,* 2007, *8,* 547–558.

Vestibular System

Cohen, B., Tomko, D. L., and Guedry, F. E. *Sensing and Controlling Motion: Vestibular and Sensorimotor Function.* New York: New York Academy of Sciences, 1992.

Somatosenses

Benedetti, F., Mayberg, H. S., Wager, T. D., Stohler, C. S., and Zubieta, J.-K. Neurobiological mechanisms of the placebo effect. *Journal of Neuroscience,* 2005, *25,* 10390–10402.

Dhaka, A., Viswanath, V., and Patapoutian, A. TRP ion channels and temperature sensation. *Annual Review of Neuroscience,* 2006, *29,* 135–161.

García-Añoveros, J., and Corey, D. P. The molecules of mechanosensation. *Annual Review of Neuroscience,* 1997, *20,* 567–594.

Ikoma, A., Steinhoff, M., Ständer, S., Yosipovitch, G., and Schmelz, M. The neurobiology of itch. *Nature Reviews: Neuroscience,* 2006, *7,* 535–547.

Kruger, L. *Pain and Touch: Handbook of Perception and Cognition,* 2nd ed. San Diego: Academic Press, 1996.

Melzak, R. Phantom limbs. *Scientific American,* 1992, *266*(4), 120–126.

Olfaction and Gustation

Lledo, P.-M., Gheusi, G., and Vincert, J.-D. Information processing in the mammalian olfactory system. *Physiological Reviews,* 2005, *85,* 281–317.

Simon, S. A., de Araujo, I. E., Gutierrez, R., and Nicolelis, M. A. L. The neural mechanisms of gustation: A distributed processing code. *Nature Reviews: Neuroscience,* 2006, *7,* 890–901.

Wilson, R. L., and Mainen, Z. F. Early events in olfactory processing. *Annual Review of Neuroscience,* 2006, *29,* 163–201.

ADDITIONAL RESOURCES

Visit www.mypsychkit.com for additional review and practice of the material covered in this chapter. Within MyPsychKit, you can take practice tests and receive a customized study plan to help you review. Dozens of animations, tutorials, and Web links are also available. You can even review using the interactive electronic version of this textbook. You will need to register for MyPsychKit. See www.mypsychkit.com for complete details.

FIGURE 8.3 ■ Action Potentials and Contractions of a Muscle Fiber

A rapid succession of action potentials can cause a muscle fiber to produce a sustained contraction. Each dot represents an individual action potential.

(Adapted from Devanandan, M. S., Eccles, R. M., and Westerman, R. A. *Journal of Physiology (London)*, 1965, *178*, 359–367.)

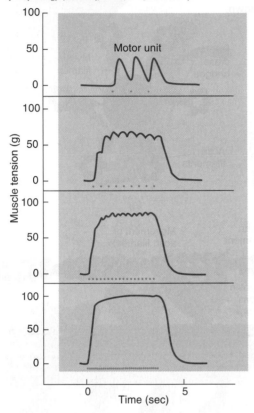

the twitches of the constituent muscle fibers. Obviously, the strength of a muscular contraction is determined by the average rate of firing of the various motor units. If, at a given moment, many units are firing, the contraction will be forceful. If few are firing, the contraction will be weak.

Sensory Feedback from Muscles

As we saw, the intrafusal muscle fibers contain sensory endings that are sensitive to stretch. The intrafusal muscle fibers are arranged in parallel with the extrafusal muscle fibers. Therefore, they are stretched when the muscle lengthens and are relaxed when it shortens. Thus, even though these afferent neurons are *stretch receptors*, they serve as *muscle length detectors*. This distinction is important. Stretch receptors are also located within the tendons, in the **Golgi tendon organ.** These receptors detect the total amount of stretch exerted by the muscle, through its tendons, on the bones to which the muscle is attached. The stretch receptors of the Golgi

tendon organ encode the degree of stretch by the rate of firing. They respond not to a muscle's length but to how hard it is pulling. In contrast, the receptors on intrafusal muscle fibers detect muscle length, not tension.

Figure 8.4 shows the response of afferent axons of the muscle spindles and Golgi tendon organ to various types of movements. Figure 8.4(a) shows the effects of passive lengthening of muscles, the kind of movement that would be seen if your forearm, held in a completely relaxed fashion, were slowly lowered by someone who was supporting it. The rate of firing of one type of muscle spindle afferent neuron (MS_1) increases, while the activity of the afferent of the Golgi tendon organ remains unchanged. (See *Figure 8.4a.*) Figure 8.4(b) shows the results when the arm is dropped quickly; note that this time the second type of muscle spindle afferent neuron (MS_2) fires a rapid burst of impulses. This fiber, then, signals rapid changes in muscle length. (See *Figure 8.4b.*) Figure 8.4(c) shows what would happen if a weight were suddenly dropped into your hand while your forearm was held parallel to the ground. Neurons MS_1 and MS_2 (especially MS_2, which responds to rapid changes in muscle length) briefly fire, because your arm lowers briefly and then comes back to the original position. The Golgi tendon organ, monitoring the strength of contraction, fires in proportion to the stress on the muscle, so it increases its rate of firing as soon as the weight is added. (See *Figure 8.4c.*)

Smooth Muscle

Our bodies contain two types of **smooth muscle,** both of which are controlled by the autonomic nervous system. *Multiunit smooth muscles* are found in large arteries, around hair follicles (where they produce *piloerection*, or fluffing of fur), and in the eye (controlling lens adjustment and pupillary dilation). This type of smooth muscle is normally inactive, but it will contract in response to neural stimulation or to certain hormones. In contrast, *single-unit smooth muscles* normally contract in a rhythmical fashion. Some of these cells spontaneously produce *pacemaker potentials*, which we can regard as self-initiated excitatory postsynaptic potentials. These slow potentials elicit action potentials, which are propagated by adjacent smooth muscle fibers, causing a wave of muscular contraction. The efferent nerve supply (and various hormones) can modulate the rhythmical rate, increasing or decreasing it. Single-unit smooth muscles

Golgi tendon organ The receptor organ at the junction of the tendon and muscle that is sensitive to stretch.

smooth muscle Nonstriated muscle innervated by the autonomic nervous system, found in the walls of blood vessels, in the reproductive tracts, in sphincters, within the eye, in the digestive system, and around hair follicles.

FIGURE 8.4 ■ Responses of Muscle and Tendon Receptors

The figure shows the effects of arm movements on the firing of muscle and tendon afferent axons: (a) slow, passive extension of the arm, (b) rapid extension of the arm, and (c) addition of a weight to an arm held in a horizontal position. MS1 and MS2 are two types of muscle spindles; GTO is an afferent fiber from the Golgi tendon organ.

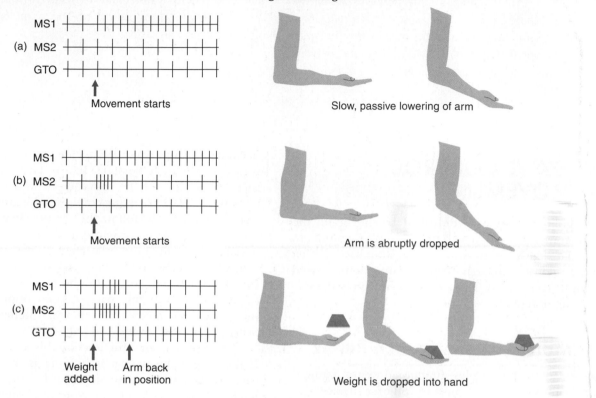

Slow, passive lowering of arm

Arm is abruptly dropped

Weight is dropped into hand

are found chiefly in the gastrointestinal system, uterus, and small blood vessels.

Cardiac Muscle

As its name implies, **cardiac muscle** is found in the heart. This type of muscle looks somewhat like striated muscle but acts like single-unit smooth muscle. The heart beats regularly, even if it is denervated. Neural activity and certain hormones (especially the cate-cholamines) serve to modulate the heart rate. A group of cells in the *pacemaker* of the heart are rhythmically active and initiate the contractions of cardiac muscle that constitute the heartbeat.

---

cardiac muscle The muscle responsible for the contraction of the heart.

InterimSummary

Muscles

Our bodies possess skeletal muscle, smooth muscle, and cardiac muscle. Skeletal muscles contain extrafusal muscle fibers, which provide the force of contraction. The alpha motor neurons form synapses with the extrafusal muscle fibers and control their contraction. Skeletal muscles also contain intrafusal muscle fibers, which detect changes in muscle length. The length of the intrafusal muscle fiber, and hence its sensitivity to increases in muscle length, is controlled by the gamma motor neuron. Besides the intrafusal muscle fibers, the muscles contain stretch receptors in the Golgi tendon organs, located at the ends of the muscles.

The force of muscular contraction is provided by long protein molecules called actin and myosin, arranged in overlapping parallel arrays. When an action potential, initiated by the synapse at the motor endplate, causes calcium ions to enter the muscle fiber, the myofibrils extract energy from ATP and cause a twitch of the muscle fiber, producing a ratchetlike "rowing" movement of the myosin cross bridges.

Smooth muscle is controlled by the autonomic nervous system through direct neural connections and indirectly through the endocrine system. Multiunit smooth muscles contract only in response to neural or hormonal stimulation. In contrast, single-unit smooth muscles normally contract rhythmically, but their rate is controlled by the autonomic nervous system. Cardiac muscle also contracts spontaneously, and its rate of contraction, too, is influenced by the autonomic nervous system.

REFLEXIVE CONTROL OF MOVEMENT

Although behaviors are controlled by the brain, the spinal cord possesses a certain degree of autonomy. Particular kinds of somatosensory stimuli can elicit rapid responses through neural connections located within the spinal cord. These reflexes constitute the simplest level of motor integration.

The Monosynaptic Stretch Reflex

The activity of the simplest functional neural pathway in the body is easy to demonstrate. Sit on a surface high enough to allow your legs to dangle freely, and have someone lightly tap your patellar tendon, just below the kneecap. This stimulus briefly stretches your quadriceps muscle, on the top of your thigh. The stretch causes the muscle to contract, which makes your leg kick forward. (I am sure few of you will bother with this demonstration, because you are already familiar with it; physical examinations often include a test of this reflex.) The time interval between the tendon tap and the start of the leg extension is about 50 milliseconds. That interval is too short for the involvement of the brain; it would take considerably longer for sensory information to be relayed to the brain and for motor information to be relayed back. For example, suppose a person is asked to move his or her leg as quickly as possible after being *touched* on the knee. This response would not be reflexive but would involve sensory and motor mechanisms of the brain. In this case the interval between the stimulus and the start of the response would be several times greater than the time required for the patellar reflex.

Obviously, the patellar reflex as such has no utility; no selective advantage is bestowed on animals that kick a limb when a tendon is tapped. However, if a more natural stimulus is applied, the utility of this mechanism becomes apparent. Figure 8.5 shows the effects of placing a weight in a person's hand. This time I have included a piece of the spinal cord, with its roots, to show the neural circuit that composes the **monosynaptic stretch reflex.** First, follow the circuit: Starting at the muscle spindle, afferent impulses are conducted to terminal buttons in the gray matter of the spinal cord. These terminal buttons synapse on an alpha motor neuron that innervates the extrafusal muscle fibers of the same muscle. Only one synapse is encountered along the route from receptor to effector—hence the term *monosynaptic*. (See *Figure 8.5a.*)

Now consider a useful function this reflex performs. If the weight the person is holding is increased, the forearm begins to move downward. This movement lengthens the muscle and increases the firing rate of the muscle spindle afferent neurons, whose terminal buttons then stimulate the alpha motor neurons, increasing their rate of firing. Consequently, the strength of the muscular contraction increases, and the arm pulls the weight up. (See *Figure 8.5b.*)

Another important role played by the monosynaptic stretch reflex is control of posture. To stand, we must keep our center of gravity above our feet, or we will fall. As we stand, we tend to oscillate forward and back and from side to side. Our vestibular sacs and our visual system play important roles in the maintenance of posture. However, these systems are aided by the activity of the monosynaptic stretch reflex. For example, consider what happens when a person begins to lean forward. The large calf muscle (gastrocnemius) is stretched, and this stretching elicits compensatory muscular contraction that pushes the toes downward, thus restoring upright posture. (See *Figure 8.6.*)

---

monosynaptic stretch reflex A reflex in which a muscle contracts in response to its being quickly stretched; involves a sensory neuron and a motor neuron, with one synapse between them.

FIGURE 8.5 ■ The Monosynaptic Stretch Reflex

(a) Neural circuit. (b) A useful function.

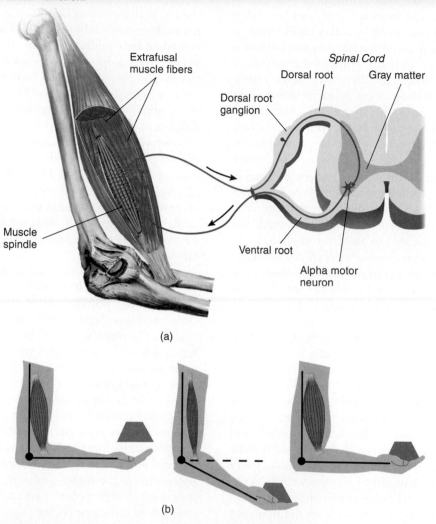

Extrafusal
muscle fibers

Spinal Cord

Dorsal root   Gray matter

Dorsal root
ganglion

Muscle
spindle

Ventral root

Alpha motor
neuron

(a)

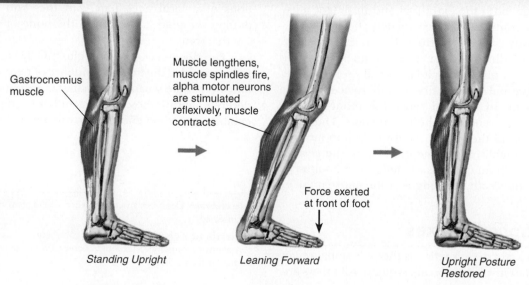

(b)

FIGURE 8.6 ■ The Role of the Monosynaptic Stretch Reflex in Postural Control

Gastrocnemius
muscle

Muscle lengthens,
muscle spindles fire,
alpha motor neurons
are stimulated
reflexively, muscle
contracts

Force exerted
at front of foot

Standing Upright   *Leaning Forward*   *Upright Posture
Restored*

The Gamma Motor System

The muscle spindles are very sensitive to changes in muscle length; they will increase their rate of firing when the muscle is lengthened by a very small amount. The interesting thing is that this detection mechanism is adjustable. Remember that the ends of the intrafusal muscle fibers can be contracted by activity of the associated efferent axons of the gamma motor neurons; their rate of firing determines the degree of contraction. When the muscle spindles are relaxed, they are relatively insensitive to stretch. However, when the gamma motor neurons are active, they become shorter and hence become much more sensitive to changes in muscle length. This property of adjustable sensitivity simplifies the role of the brain in controlling movement. The more control that can occur in the spinal cord, the fewer messages must be sent to and from the brain.

We have already seen that the afferent axons of the muscle spindle help to maintain limb position even when the load carried by the limb is altered. Efferent control of the muscle spindles permits these muscle length detectors to assist in changes in limb position as well. Consider a single muscle spindle. When its efferent axon is completely silent, the spindle is completely relaxed and extended. As the firing rate of the efferent axon increases, the spindle gets shorter and shorter. If, simultaneously, the rest of the entire muscle also gets shorter, there will be no stretch on the central region that contains the sensory endings, and the afferent axon will not respond. However, if the muscle spindle contracts faster than does the muscle as a whole, there will be a considerable amount of afferent activity.

The motor system makes use of this phenomenon in the following way: When commands from the brain are issued to move a limb, both the alpha motor neurons and the gamma motor neurons are activated. The alpha motor neurons start the muscle contracting. If there is little resistance, both the extrafusal and intrafusal muscle fibers will contract at approximately the same rate, and little activity will be seen from the afferent axons of the muscle spindle. However, if the limb meets with resistance, the intrafusal muscle fibers will shorten more than the extrafusal muscle fibers, and hence sensory axons will begin to fire and cause the monosynaptic stretch reflex to strengthen the contraction. Thus, the brain makes use of the gamma motor system in moving the limbs. By establishing a rate of firing in the *gamma motor system*, the brain controls the length of the muscle spindles and, indirectly, the length of the entire muscle.

Polysynaptic Reflexes

The monosynaptic stretch reflex is the only spinal reflex we know of that involves only one synapse. All others are *polysynaptic*. Examples include relatively simple ones, such as limb withdrawal in response to noxious stimulation, and relatively complex ones, such as the ejaculation of semen. Spinal reflexes do not exist in isolation; they are normally controlled by the brain. For example, Chapter 2 described how inhibition from the brain can prevent a person from dropping a hot casserole dish, even though the painful stimuli received by the fingers cause reflexive extension of the fingers. This section will describe some general principles by which polysynaptic spinal reflexes operate.

Before I begin the discussion, I should mention that the simple circuit diagrams used here (including the one you just looked at in Figure 8.6) are much too simple. Reflex circuits are typically shown as a single chain of neurons, but in reality most reflexes involve thousands of neurons. Each axon usually synapses on many neurons, and each neuron receives synapses from many different axons.

As we saw previously, the afferent axons from the Golgi tendon organ serve as detectors of muscle stretch. There are two populations of afferent axons from the Golgi tendon organ, with different sensitivities to stretch. The more sensitive afferent axons tell the brain how hard the muscle is pulling. The less sensitive ones have an additional function. Their terminal buttons synapse on spinal cord interneurons—neurons that reside entirely within the gray matter of the spinal cord and serve to interconnect other spinal neurons. These interneurons synapse on the alpha motor neurons serving the same muscle. The terminal buttons liberate glycine and hence produce inhibitory postsynaptic potentials on the motor neurons. (See *Figure 8.7.*) The function of this reflex pathway is to decrease the strength of muscular contraction when there is danger of damage to the tendons or bones to which the muscles are attached. Weight lifters can lift heavier weights if their Golgi tendon organs are deactivated with injections of a local anesthetic, but they run the risk of pulling the tendon away from the bone or even breaking the bone.

The discovery of the inhibitory Golgi tendon organ reflex provided the first real evidence of neural inhibition, long before the synaptic mechanisms were understood. A **decerebrate** cat, whose brain stem has been cut through, exhibits a phenomenon known as **decerebrate rigidity.**

---

decerebrate Describes an animal whose brain stem has been transected.

decerebrate rigidity Simultaneous contraction of agonist and antagonistic muscles; caused by decerebration or damage to the reticular formation.

FIGURE 8.7 ■ Polysynaptic Inhibitory Reflex

Input from the Golgi tendon organ can cause inhibitory postsynaptic potentials to occur on the alpha motor neuron.

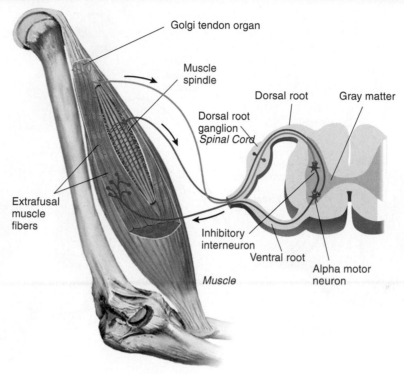

The animal's back is arched, and its legs are extended stiffly from its body. This rigidity results from excitation originating in the caudal reticular formation, which greatly facilitates all stretch reflexes, especially of extensor muscles, by increasing the activity of the gamma motor system. Rostral to the brain stem transection is an inhibitory region of the reticular formation, which normally counterbalances the excitatory one. The transection removes the inhibitory influence, leaving only the excitatory one. If you attempt to flex the outstretched leg of a decerebrate cat, you will meet with increasing resistance, which will suddenly melt away, allowing the limb to flex. It almost feels as though you were closing the blade of a pocketknife—hence the term **clasp-knife reflex.** The sudden release is, of course, mediated by activation of the Golgi tendon organ reflex.

Even the monosynaptic stretch reflex serves as the basis of polysynaptic reflexes. Muscles are arranged in opposing pairs. The **agonist** moves the limb in the direction being studied, and because muscles cannot push back, the **antagonist** muscle must move the limb back in the opposite direction. Consider this finding: When a stretch reflex is elicited in the agonist, it contracts quickly, thus causing the antagonist to lengthen. It would appear, then, that the antagonist is presented with a stimulus that should elicit *its* stretch reflex. Yet the antagonist relaxes instead. Let us see why.

Afferent axons of the muscle spindles, besides sending terminal buttons to the alpha motor neuron and to the brain, also synapse on inhibitory interneurons. The terminal buttons of these interneurons synapse on the alpha motor neurons that innervate the antagonistic muscle. (See *Figure 8.8.*) Thus, a stretch reflex excites the agonist and *inhibits the antagonist* so that the limb can move in the direction controlled by the stimulated muscle.

clasp-knife reflex A reflex that occurs when force is applied to flex or extend the limb of an animal showing decerebrate rigidity; resistance is replaced by sudden relaxation.

agonist A muscle whose contraction produces or facilitates a particular movement.

antagonist A muscle whose contraction resists or reverses a particular movement.

FIGURE 8.8 ■ Secondary Reflexes

Firing of the muscle spindle causes excitation on the alpha motor neuron of the agonist and inhibition on the antagonist.

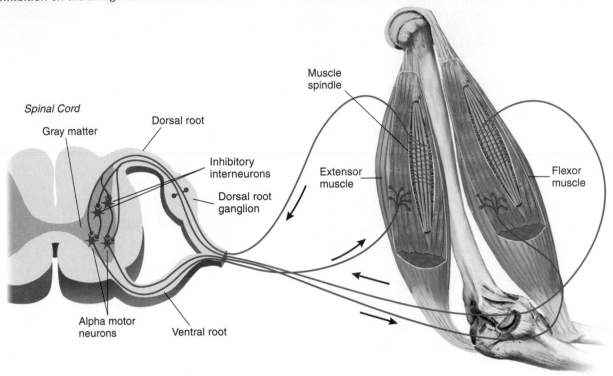

Spinal Cord

Gray matter

Dorsal root

Inhibitory interneurons

Dorsal root ganglion

Alpha motor neurons

Ventral root

Muscle spindle

Extensor muscle

Flexor muscle

Interim Summary

Reflexive Control of Movement

Reflexes are simple circuits of sensory neurons, interneurons (usually), and efferent neurons that control simple responses to particular stimuli. In the monosynaptic stretch reflex the terminal buttons of axons that receive sensory information from the intrafusal muscle fibers synapse with alpha motor neurons that innervate the same muscle. Thus, a sudden lengthening of the muscle causes the muscle to contract. By setting the length of the intrafusal muscle fibers, and hence their sensitivity to increases in muscle length, the motor system of the brain can control limb position. Changes in a weight being held that cause the limb to move will be quickly compensated for by means of the monosynaptic stretch reflex.

Polysynaptic reflexes contain at least one interneuron between the sensory neuron and the motor neuron. For example, when a strong muscular contraction threatens to damage muscles or limbs, the increased rate of firing of the afferent axons of Golgi tendon organs stimulates inhibitory interneurons, which inhibit the alpha motor neurons of those muscles. And when the afferent axons of intrafusal muscle fibers fire, they excite inhibitory interneurons that slow the rate of firing of the alpha motor neurons that serve the antagonistic muscles, causing the antagonist to relax and the agonist to contract.

CONTROL OF MOVEMENT BY THE BRAIN

Movements can be initiated by several means. For example, rapid stretch of a muscle triggers the monosynaptic stretch reflex, a stumble triggers righting reflexes, and the rapid approach of an object toward the face causes a startle response, a complex reflex consisting of movements of several muscle groups. Other stimuli initiate sequences of movements that we have previously learned. For example, the presence of food causes eating, and the sight of a loved one evokes a hug and a kiss. Because there is no single cause of behavior, we cannot

FIGURE 8.9 ■ Motor Cortex and the Motor Homunculus

Stimulation of various regions of the primary motor cortex causes movement in muscles of various parts of the body.

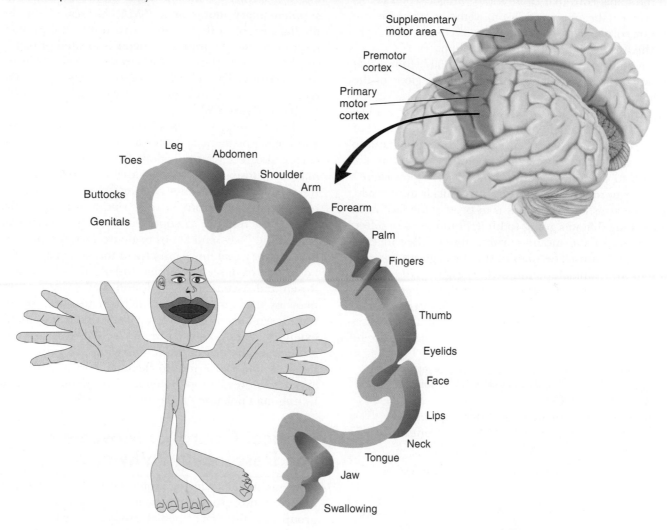

find a single starting point in our search for the neural mechanisms that control movement.

The brain and spinal cord include several different motor systems, each of which can simultaneously control particular kinds of movements. For example, a person can walk and talk with a friend simultaneously. While doing so, the person can gesture with the hands to emphasize a point, scratch an itch, brush away a fly, wipe sweat off his or her forehead, and so on. Walking, postural adjustments, talking, movement of the arms, and movements of the fingers all involve different specialized motor systems.

Organization of the Motor Cortex

The primary motor cortex lies on the precentral gyrus, just rostral to the central sulcus. Stimulation studies (including those in awake humans) have shown that the activation of neurons located in particular parts of the

primary motor cortex causes movements of particular parts of the body. In other words, the primary motor cortex shows **somatotopic organization** (from *soma,* "body," and *topos,* "place"). Figure 8.9 shows a *motor homunculus* based on the observations of Penfield and Rasmussen (1950). Note that a disproportionate amount of cortical area is devoted to movements of the fingers and the muscles used for speech. (See *Figure 8.9.*)

It is important to recognize that the primary motor cortex is organized in terms of particular *movements* of particular parts of the body. Each movement may be accomplished by the contraction of several muscles. For example, when the arm is extended in a particular direction, many muscles in the shoulder, upper arm,

somatotopic organization A topographically organized mapping of parts of the body that are represented in a particular region of the brain.

and forearm must contract. This fact means that complex neural circuitry is located between individual neurons in the primary motor cortex and the motor neurons in the spinal cord that cause motor units to contract. As we will see, the commands for movement initiated in the motor cortex are assisted and modified—most notably, by the basal ganglia and the cerebellum.

A study by Graziano and Aflalo (2007) found that although brief stimulation of particular regions of the primary motor cortex of monkeys caused brief movements of various parts of the body, prolonged stimulation produced much more complex movements. For example, stimulation of one region caused the hand to close and then approach the mouth and then the mouth to open. Stimulation of another region caused the face to squint, the head to turn quickly to one side, and the arms to fling up, as if to protect the face from something that was going to hit it. Stimulation of different zones of the motor cortex caused different categories of actions. The map of these categories was consistent from animal to animal. (See *Figure 8.10*.)

The principal cortical input to the primary motor cortex is the frontal association cortex, located rostral to it. Two regions immediately adjacent to the primary motor cortex—the *supplementary motor area* and the

FIGURE 8.10 ■ Stimulation of the Motor Cortex

Categories of movements were elicited by prolonged stimulation of specific regions of the motor cortex of monkeys.

(Adapted from Graziano, M. S. A., and Aflalo, T. N. *Neuron*, 2007, *56*, 239–251.)

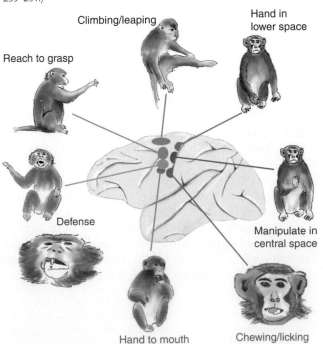

Climbing/leaping

Hand in lower space

Reach to grasp

Defense

Manipulate in central space

Hand to mouth

Chewing/licking

premotor cortex—are especially important in the control of movement. Both regions receive sensory information from the parietal and temporal lobes, and both send efferent axons to the primary motor cortex. The **supplementary motor area (SMA)** is located on the medial surface of the brain, just rostral to the primary motor cortex. The **premotor cortex** is located primarily on the lateral surface, also just rostral to the primary motor cortex. The roles that these regions play in the control of movement is discussed later in this chapter. (Refer to *Figure 8.9*.)

Besides receiving input from the premotor cortex and the supplementary motor area, the primary motor cortex also receives projections from the adjacent primary somatosensory cortex, located just across the central sulcus. The connections between these two areas are quite specific: Neurons in the primary somatosensory cortex that respond to stimuli applied to a particular part of the body send axons to neurons in the primary motor cortex that move muscles in the same part of the body. For example, Asanuma and Rosén (1972) and Rosén and Asanuma (1972) found that somatosensory neurons that respond to a touch on the back of the thumb send axons to motor neurons that cause thumb extension, and somatosensory neurons that respond to a touch on the ball of the thumb send axons to motor neurons that cause thumb flexion. This organization appears to provide rapid feedback to the motor system during manipulation of objects.

Cortical Control of Movement: The Descending Pathways

Neurons in the primary motor cortex control movements by two groups of descending tracts, the **lateral group** and the **ventromedial group,** named for their locations in the white matter of the spinal cord. The lateral group consists of the *corticospinal tract,* the *corticobulbar tract,* and the *rubrospinal tract.* This system is primarily involved in control of independent limb movements, particularly movements of the hands and fingers. *Independent* limb movements mean that the right and left limbs make different movements or one limb moves while the other remains still. These movements

supplementary motor area (SMA) A region of motor association cortex of the dorsal and dorsomedial frontal lobe, rostral to the primary motor cortex.

premotor cortex A region of motor association cortex of the lateral frontal lobe, rostral to the primary motor cortex.

lateral group The corticospinal tract, the corticobulbar tract, and the rubrospinal tract.

ventromedial group The vestibulospinal tract, the tectospinal tract, the reticulospinal tract, and the ventral corticospinal tract.

FIGURE 8.11 ■ Lateral Group of Descending Motor Tracts

The figure shows the lateral corticospinal tract (light blue lines), corticobulbar tract (green lines), and rubrospinal tract (red lines). The ventral corticospinal tract (dark blue lines) is part of the ventromedial group.

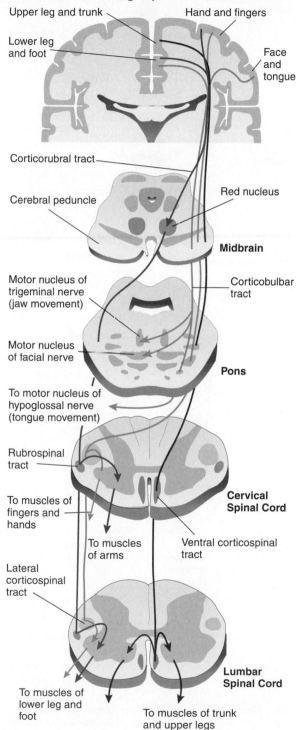

Upper leg and trunk

Lower leg and foot

Hand and fingers

Face and tongue

Corticorubral tract

Cerebral peduncle

Red nucleus

Midbrain

Motor nucleus of trigeminal nerve (jaw movement)

Corticobulbar tract

Motor nucleus of facial nerve

Pons

To motor nucleus of hypoglossal nerve (tongue movement)

Rubrospinal tract

To muscles of fingers and hands

To muscles of arms

Cervical Spinal Cord

Ventral corticospinal tract

Lateral corticospinal tract

To muscles of lower leg and foot

Lumbar Spinal Cord

To muscles of trunk and upper legs

contrast with coordinated limb movements, such as those involved in locomotion. The ventromedial group consists of the *vestibulospinal tract,* the *tectospinal tract,* the *reticulospinal tract,* and the *ventral corticospinal tract.* These tracts control more automatic movements: gross movements of the muscles of the trunk and coordinated trunk and limb movements involved in posture and locomotion.

Let's first consider the lateral group of descending tracts. The **corticospinal tract** consists of axons of cortical neurons that terminate in the gray matter of the spinal cord. The largest concentration of cell bodies responsible for these axons is located in the primary motor cortex, but neurons in the parietal and temporal lobes also send axons through the corticospinal pathway. The axons leave the cortex and travel through subcortical white matter to the ventral midbrain, where they enter the cerebral peduncles. They leave the peduncles in the medulla and form the **pyramidal tracts,** so called because of their shape. At the level of the caudal medulla, most of the fibers decussate (cross over) and descend through the contralateral spinal cord, forming the **lateral corticospinal tract.** The rest of the fibers descend through the ipsilateral spinal cord, forming the **ventral corticospinal tract.** Because of its location and function, the ventral corticospinal tract is actually part of the ventromedial group. (See the light and dark blue lines in *Figure 8.11.*)

Most of the axons in the lateral corticospinal tract originate in the regions of the primary motor cortex and supplementary motor area that control the distal parts of the limbs: the arms, hands, and fingers and the lower legs, feet, and toes. They form synapses, directly or via interneurons, with motor neurons in the gray matter of the spinal cord—in the lateral part of the ventral horn. These motor neurons control muscles of the distal limbs, including those that move the arms, hands, and fingers. (See the light blue lines in *Figure 8.11.*)

The axons in the ventral corticospinal tract originate in the upper leg and trunk regions of the primary motor cortex. They descend to the appropriate region of the spinal cord and divide, sending terminal buttons

corticospinal tract The system of axons that originates in the motor cortex and terminates in the ventral gray matter of the spinal cord.

pyramidal tract The portion of the corticospinal tract on the ventral border of the medulla.

lateral corticospinal tract The system of axons that originates in the motor cortex and terminates in the contralateral ventral gray matter of the spinal cord; controls movements of the distal limbs.

ventral corticospinal tract The system of axons that originates in the motor cortex and terminates in the ipsilateral ventral gray matter of the spinal cord; controls movements of the upper legs and trunk.

into both sides of the gray matter. They control motor neurons that move the muscles of the upper legs and trunk. (See the dark blue lines in *Figure 8.11*.)

The corticospinal pathway controls hand and finger movements and is indispensable for moving the fingers independently when reaching and manipulating. Postural adjustments of the trunk and use of the limbs for reaching and locomotion are unaffected; therefore, these types of movements are controlled by other systems. Because the monkeys had difficulty releasing their grasp when they picked up objects but had no trouble doing so when climbing the walls of the cage, we can conclude that the same behavior (opening the hand) is controlled by different brain mechanisms in different contexts.

The second of the lateral group of descending pathways, the **corticobulbar tract,** projects to the medulla (sometimes called the *bulb*). This pathway is similar to the corticospinal pathway, except that it terminates in the motor nuclei of the fifth, seventh, ninth, tenth, eleventh, and twelfth cranial nerves (the trigeminal, facial, glossopharyngeal, vagus, spinal accessory, and hypoglossal nerves). These nerves control movements of the face, neck, and tongue and parts of the extraocular eye muscles. (See the green lines in *Figure 8.11*.)

The third member of the lateral group is the **rubrospinal tract.** This tract originates in the red nucleus (*nucleus ruber*) of the midbrain. The red nucleus receives its most important inputs from the motor cortex via the **corticorubral tract** and (as we shall see later) from the cerebellum. Axons of the rubrospinal tracts terminate on motor neurons in the spinal cord that control independent movements of the forearms and hands—that is, movements that are independent of trunk movements. (They do not control the muscles that move the fingers.) (See the red lines in *Figure 8.11*.)

Now let's consider the second set of pathways originating in the brain stem: the ventromedial group. This group includes the **vestibulospinal tracts,** the **tectospinal tracts,** and the **reticulospinal tracts,** as well as the ventral corticospinal tract (already described). These tracts control motor neurons in the ventromedial part of the spinal cord gray matter. Neurons of all these tracts receive input from the portions of the primary

FIGURE 8.12 ■ Ventromedial Group of Descending Motor Tracts

The figure shows the tectospinal tract (blue lines), lateral reticulospinal tract (purple lines), medial reticulospinal tract (orange lines), and vestibulospinal tract (green lines).

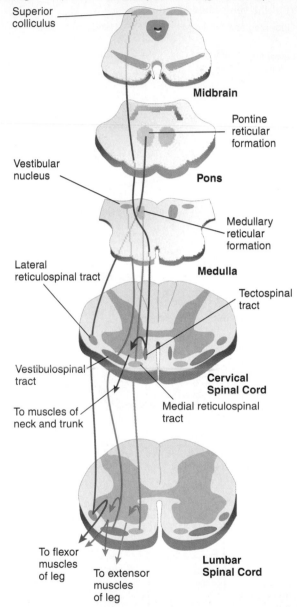

corticobulbar tract A bundle of axons from the motor cortex to the fifth, seventh, ninth, tenth, eleventh, and twelfth cranial nerves; controls movements of the face, neck, tongue, and parts of the extraocular eye muscles.

rubrospinal tract The system of axons that travels from the red nucleus to the spinal cord; controls independent limb movements.

corticorubral tract The system of axons that travels from the motor cortex to the red nucleus.

vestibulospinal tract A bundle of axons that travels from the vestibular nuclei to the gray matter of the spinal cord; controls

postural movements in response to information from the vestibular system.

tectospinal tract A bundle of axons that travels from the tectum to the spinal cord; coordinates head and trunk movements with eye movements.

reticulospinal tract A bundle of axons that travels from the reticular formation to the gray matter of the spinal cord; controls the muscles responsible for postural movements.

TABLE 8.1 ■ Major Motor Pathways

	ORIGIN	TERMINATION	MUSCLE GROUP	FUNCTION
Lateral group				
Lateral corticospinal tract	Finger, hand, and arm region of motor cortex	Spinal cord	Fingers, hands, and arms	Grasping and manipulating objects
Rubrospinal tract	Red nucleus	Spinal cord	Hands (not fingers), lower arms, feet, and lower legs	Movement of forearms and hands independent from that of the trunk
Corticobulbar tract	Face region of motor cortex	Cranial nerve nuclei: 5, 7, 9, 10, 11, and 12	Face and tongue	Face and tongue movements
Ventromedial group				
Vestibulospinal tract	Vestibular nuclei	Spinal cord	Trunk and legs	Posture
Tectospinal tract	Superior colliculi	Spinal cord	Neck and trunk	Coordination of eye movements with those of trunk and head
Lateral reticulospinal tract	Medullary reticular formation	Spinal cord	Flexor muscles of legs	Walking
Medial reticulospinal tract	Pontine reticular formation	Spinal cord	Extensor muscles of legs	Walking
Ventral corticospinal tract	Trunk and upper leg region of motor cortex	Spinal cord	Hands (not fingers), lower arms, feet, and lower legs	Locomotion and posture

motor cortex that control movements of the trunk and proximal muscles (that is, the muscles located on the parts of the limbs close to the body). In addition, the reticular formation receives a considerable amount of input from the premotor cortex and from several subcortical regions, including the amygdala, hypothalamus, and basal ganglia. The cell bodies of neurons of the vestibulospinal tracts are located in the vestibular nuclei. As you might expect, this system plays a role in the control of posture. The cell bodies of neurons in the tectospinal tracts are located in the superior colliculus and are involved in coordinating head and trunk movements with eye movements. The cell bodies of neurons of the reticulospinal tracts are located in many nuclei in the brain stem and midbrain reticular formation. These neurons control several automatic functions, such as muscle tonus, respiration, coughing, and sneezing; but they are also involved in behaviors that are under direct neocortical control, such as walking. (See *Figure 8.12*.)

Table 8.1 summarizes the names of these pathways, their locations, and the muscle groups they control. (See *Table 8.1*.)

Planning and Initiating Movements: Role of the Motor Association Cortex

The supplementary motor area and the premotor cortex are involved in the planning of movements, and they execute these plans through their connections with the primary motor cortex. Functional-imaging studies show that when people execute sequences of movements—or even imagine them—these regions become activated (Roth et al., 1996). More recent evidence indicates that the motor association cortex is also involved in imitating the actions of other people (an ability that makes it possible to learn new behaviors

FIGURE 8.13 ■ Cortical Control of Movement

The posterior association cortex is involved with perceptions and memories, and the frontal association cortex is involved with plans for movement.

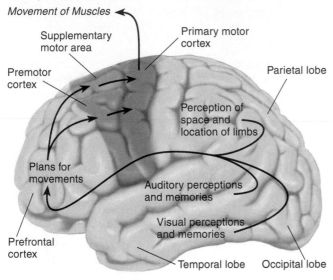

Movement of Muscles ←

Supplementary motor area

Primary motor cortex

Premotor cortex

Parietal lobe

Perception of space and location of limbs

Plans for movements

Auditory perceptions and memories

Visual perceptions and memories

Prefrontal cortex

Temporal lobe   Occipital lobe

general, the supplementary motor cortex is involved in learning and performing behaviors that consist of sequences of movements. A nearby region appears to be involved in initiating spontaneous movements. The premotor cortex is involved in imitating responses of other people and in understanding and predicting these actions.

The Supplementary Motor Area

The supplementary motor area plays a critical role in behavioral sequences. Damage to this region disrupts the ability to execute well learned sequences of responses in which the performance of one response serves as the signal that the next response must be made. Chen et al. (1995) found that lesions of the supplementary motor area severely impaired monkeys' ability to perform a simple sequence of two responses: pushing a lever in and then turning it to the left, receiving a peanut after each response. (See *Figure 8.14.*)

A single-unit recording study came to similar conclusions. Mushiake, Inase, and Tanji (1991) trained monkeys to perform a memorized series of responses, pressing

from them) and even in understanding the functions of other people's behavior.

The supplementary motor area and the premotor cortex receive information from association areas of the parietal and temporal cortex. As we saw in Chapter 6, the visual association cortex is organized in two streams: dorsal and ventral. The ventral stream, which terminates in the inferior temporal cortex, is involved in perceiving and recognizing particular objects—the "what" of visual perception. The dorsal stream, which terminates in the posterior parietal lobe, is involved in perception of location—the "where" of visual perception. In addition, the parietal lobes are involved in organizing visually guided movements—the "how" of visual perception. Besides receiving visual information about space, the parietal lobe receives information about spatial location from the somatosensory, vestibular, and auditory systems and integrates this information with visual information. Thus, the regions of the frontal cortex that are involved in planning movements receive the information they need about what is happening and where it is happening from the temporal and parietal lobes. Because the parietal lobes contain spatial information, the pathway from them to the frontal lobes is especially important in controlling both locomotion and arm and hand movements. After all, meaningful locomotion requires us to know where we are, and meaningful movements of our arms and hands require us to know where objects are located in space. (See *Figure 8.13.*)

Let's look at the functions of the supplementary motor area and the premotor cortex in more detail. In

FIGURE 8.14 ■ The Task Used in the Experiment by Chen et al.

The monkey was required to (1) push the handle in and then (2) turn it to the left, receiving a piece of food in the door above the lever after each component of the sequence.

(Adapted from Chen, Y.-C., Thaler, D., Nixon, P. D., Stern, C. E., and Passingham, R. E. *Experimental Brain Research*, 1995, *102*, 461–473.)

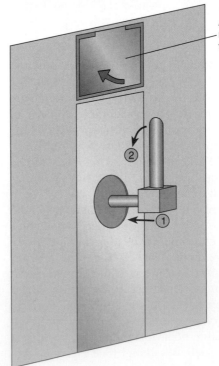

After correct response, monkey pushes panel to receive food

each of three buttons in a specific sequence. While the monkeys were performing this task, more than half of the neurons in the supplementary motor area became activated. However, when the sequence was cued by visual stimuli—the monkeys simply had to press the button that was illuminated—these neurons showed little activity.

Shima and Tanji (2000) taught monkeys six sequences of three motor responses. For example, one of the sequences was push, then pull, then turn. They recorded from neurons in the supplementary motor area and found neurons whose activity appeared to encode elements of these sequences. For example, some neurons responded just before a particular sequence of three movements occurred; some neurons responded between two particular responses; and some neurons responded as the monkey was preparing the make the last response of the sequence. Presumably, these neurons were members of circuits that encoded the information necessary to perform the six sequences. Figure 8.15 shows the response of a neuron that responded during a pulling movement, but only if it was to be followed by a pushing movement. (See *Figure 8.15*.)

Shima and Tanji (1998) temporarily inactivated the supplementary motor area in monkeys with injections of muscimol, a drug that stimulates GABA receptors and thus inhibits neural activity. They found that after inactivation of this region, monkeys could still reach for objects or make particular movements in response to visual cues but could no longer make a sequence of three movements they had previously learned.

Studies with human subjects have obtained results similar to those obtained with monkeys. For example, a functional-imaging study by Hikosaka et al. (1996) observed increased activity in the posterior SMA during performance of a learned sequence of button presses. Gerloff et al. (1997) taught people to make a sequence of sixteen finger presses on an electronic piano. When the experimenters disrupted the activity of the SMA with transcranial magnetic stimulation, the performance of the sequence was disrupted. However, the disruption was not immediate: The subjects continued the sequence for approximately one second and then stopped, saying that they "did not know anymore which series of keys to press next." Apparently, the SMA is involved in planning the elements *yet to come* in sequences of movements. The actual execution of the movements appears to be controlled elsewhere—presumably by the primary motor cortex.

A region just anterior to the supplementary motor area, the *pre-SMA*, appears to be involved in control of spontaneous movements—or at least in the perception of control. It has long been known that although electrical stimulation of the motor cortex causes movements, it does not produce the *desire* to move. The movement is perceived as automatic and involuntary. In contrast, electrical stimulation of the medial surface of the frontal

FIGURE 8.15 ■ Firing Patterns of a Supplementary Motor Area Neuron

The figure shows the firing patterns of a single neuron in the supplementary motor area of a monkey. The animal performed three sequences of movements. The neuron responded only during a pulling response that was to be followed by a pushing response. Black hash marks indicate action potentials during each trial, and blue histograms indicate the total number of action potentials summed across all trials.

(From Shima, K., and Tanji, J. *Journal of Neurophysiology*, 2000, *84*, 2148–2160. Reprinted with permission of the American Physiological Society.)

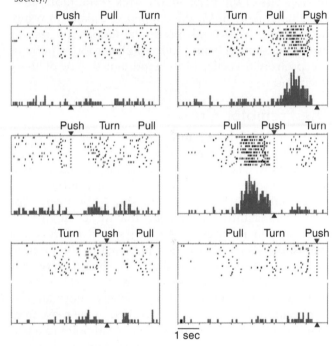

lobes (including the SMA and pre-SMA) often provokes the urge to make a movement or at least the anticipation that a movement is about to occur (Fried et al., 1991).

A functional-imaging study by Lau et al. (2004) found that the pre-SMA became active just before people performed spontaneous movements. The experimenters asked the subjects to make a finger movement from time to time, whenever they felt like doing so. The subjects watched a red light that moved around a clock face at about 2.5 seconds per revolution. They were asked to pay attention to the instant when they decided to make the movement and report the position of the red dot at that time. The decision appeared to occur approximately 0.2 sec before the movement began. However, fMRI showed that the activity of the pre-SMA actually began to increase approximately 2–3 seconds earlier than that, which suggests that the neural activity responsible for the decision to move begins before a person is even aware of making that decision.

The most important input to the supplementary motor area comes from the parietal lobes. Sirigu et al. (2004) used a task similar to the one in the study by Lau et al. to investigate decision making in people with lesions of the parietal cortex. They found that people with parietal lesions could accurately report when they started the movement, but they were not aware of an intention to move prior to making the movement. These results suggest that information received from the parietal lobes permits the pre-SMA to detect that a decision to move has been made. The location of the neural circuits actually responsible for the decision are not known, although Sirigu and her colleagues note that lesions of the prefrontal cortex (even more anterior than the pre-SMA) disrupt people's plans for voluntary action. People with prefrontal lesions will *react* to events but show deficits in *initiating* behavior, so perhaps the prefrontal cortex is an important source of these decisions.

The Premotor Cortex

The premotor cortex is involved in learning and executing complex movements that are guided by sensory information. The results of several studies suggest that the premotor cortex is involved in using arbitrary stimuli to indicate what movement should be made. For example, reaching for an object that we see in a particular location involves *nonarbitrary* spatial information; that is, the visual information provided by the location of the object specifies just where we should target our reaching movement. But we also have the ability to learn to make movements based on *arbitrary* information— information that is not directly related to the movement that it signals. For example, a person can point to a particular object when someone says its name, or a dancer can make a particular movement when asked to do so by a choreographer. Different languages use different sounds to indicate the names of objects, and different choreographers could invent different names for movements used in their dances. Or a person could be told to "wave your left hand when you hear the buzz and touch your nose when you hear the bell." The associations between these stimuli and the movements they designate are arbitrary and must be learned.

Kurata and Hoffman (1994) trained monkeys to move their hands toward the right or left in response to either a spatial or a nonspatial signal. The spatial signal required a monkey to move in the direction indicated by signal lights located to the right and left of its hand. The nonspatial signal consisted of a pair of lights, one red and one green, located in the middle of the display. The red light signaled a movement to the left, and the green light signaled a movement to the right. The investigators temporarily inactivated the premotor cortex with injections of muscimol. When this region was inactivated, the monkeys could still move their hands toward a signal light located to the left or right (a nonarbitrary signal), but they could no longer make the appropriate movements when the red or green signal lights were illuminated.

Similar results are seen in people with damage to the premotor cortex. Halsband and Freund (1990) found that patients with these lesions could learn to make six different movements in response to spatial cues but not to arbitrary visual cues. That is, they could learn to point to one of six locations in which they had just seen a visual stimulus, but they could not learn to use a set of visual, auditory, and tactile cues to make particular movements.

Imitating and Comprehending Movements: Role of the Mirror Neuron System

Rizzolatti and his colleagues (Gallese et al., 1996; Rizzolatti et al. 2001) performed an interesting study that has changed the way we think about imitating and comprehending the behavior of others. The investigators found that neurons in an area of the rostral part of the ventral premotor cortex in the monkey brain (area F5) became active when monkeys saw people or other monkeys perform various grasping, holding, or manipulating movements or when they performed these movements themselves. Thus, the neurons responded to either the sight or the execution of particular movements. The investigators named these cells **mirror neurons.** The location of these neurons, the ventral premotor cortex, is reciprocally connected with the *inferior parietal lobule,* a region of the posterior parietal lobe, and further investigation found that this region also contains mirror neurons. Given the characteristics of mirror neurons, we might expect that they play a role in a monkey's ability to imitate the movements of other monkeys—and Rizzolatti and his colleagues found that this inference was correct.

Figure 8.16 shows the anatomy of the major regions of the parietal lobe of the human brain that I will discuss in the next several subsections of this chapter. For example, you can see the inferior parietal lobule (IPL), where mirror neurons were first discovered. (See *Figure 8.16.*)

Several functional-imaging studies have shown that the human brain also contains a circuit of mirror neurons in the inferior parietal lobule and the ventral

---

mirror neurons Neurons located in the ventral premotor cortex and inferior parietal lobule that respond when the individual makes a particular movement or sees another individual making that movement.

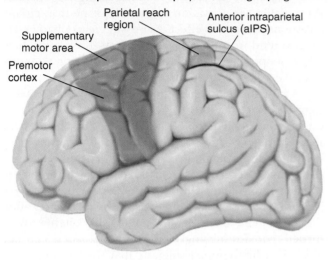

In the human brain, the inferior parietal lobule and the ventral premotor cortex constitute the primary mirror neuron circuit. The parietal reach region plays a role in reaching, and the anterior intraparietal sulcus plays a role in grasping.

Parietal reach region

Anterior intraparietal sulcus (aIPS)

Supplementary motor area

Premotor cortex

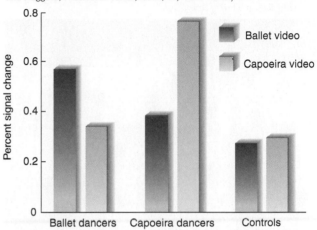

FIGURE 8.17 ■ Mirror Neurons and Special Skills

The graph shows activation in the dorsal premotor cortex of ballet dancers, capoeira dancers, and control subjects watching videos of ballet dancing or capoeira dancing.

(Adapted from Calvo-Merino, B., Glaser, D. E., Grèzes, J., Passingham, R. E., and Haggard, P. *Cerebral Cortex*, 2005, *15*, 1243–1249.)

Ballet video

Capoeira video

Percent signal change

Ballet dancers   Capoeira dancers   Controls

premotor area. For example, in a functional-imaging study, Buccino et al. (2004) asked nonmusicians to watch and then imitate video clips of an expert guitarist placing his fingers on the neck of a guitar to play a chord. The investigators found that both watching and imitating the movements activated the mirror neuron circuit. Several studies have found that the mirror neuron system is activated most strongly when one is watching a behavior in which one is already competent. For example, Calvo-Merino et al. (2005) had professional ballet dancers, professional capoeira dancers, and nonexperts watch videos of dancers performing ballet and capoeira movements. (*Capoeira* is a Brazilian form of dance/game/martial art.) All subjects showed activation of the mirror neuron system. The professional dancers showed more activation than the nonexperts did, and the two groups of dancers showed greater activation when they watched the kind of dance in which they were proficient. (See *Figure 8.17.*)

Mirror neurons are activated not only by the performance of an action or the sight of someone else performing that action, but also by sounds that indicate the occurrence of a familiar action. For example, Kohler et al. (2002) found that mirror neurons in the ventral prefrontal cortex of monkeys became active when the animals heard sounds they recognized, such as a peanut breaking, a piece of paper being ripped, or a stick being dropped. Individual neurons—the researchers called them *audiovisual neurons*—responded to the sounds of particular actions and to the sight of those actions. Presumably, activation of these neurons by these familiar

sounds reminds the animals of the actions the sounds represent. Lahav, Saltzman, and Schlaug (2007) found that the connections of audiovisual neurons could be established very quickly. The investigators taught nonmusicians to play a simple tune on a piano. Next, they obtained fMRI scans from the subjects while they listened to the tune they had learned and, as control conditions, listened to familiar tunes that they had not learned to play and to the same notes they had played but in a different order. Although the subjects rested quietly in the scanner without moving, their frontoparietal mirror neuron network was activated when they heard the tune they had learned to play.

Haslinger et al. (2005) found that the interaction between audition and vision worked in the other direction as well. The investigators showed professional pianists silent videos of a hand playing the piano or making meaningless finger movements above a piano keyboard. (See *Figure 8.18.*) Functional imaging showed that when the subjects watched actual piano playing, the mirror neuron system and visual cortex were activated, as would be expected, but the auditory cortex was activated as well. Presumably, the musicians imagined what it was like to make the meaningful hand and finger movements, activating the mirror neuron system, but also imagined what the piano would sound like when the keys were pressed, activating the auditory cortex.

Rizzolatti, Fogassi, and Gallese (2001) suggest that the mirror neuron circuit helps us to understand the actions of others. "[A]n action is understood when its

FIGURE 8.18 ■ Mirror Neurons in Musicians

Videos of piano playing (a), but not meaningless finger movements (b), activated the mirror neuron system and also the auditory cortex of professional pianists.

(From Haslinger, B., Erhard, P., Altenmüller, E., Schroeder, U., Boecker, H., and Ceballos-Baumann, A. O. *Journal of Cognitive Neuroscience,* 2005, *17,* 282–293. Reprinted with permission.)

(a)

(b)

observation causes the motor system of the observer to 'resonate.' So, when we observe a hand grasping an apple, the same population of neurons that control the execution of grasping movements becomes active in the observer's motor areas. . . . In other words, we understand an action because the motor representation of that action is activated in our brain" (p. 661). By "resonation," Rizzolatti and his colleagues mean that the neural circuits responsible for performing a particular action are activated when we see someone else beginning to perform that action or even when we hear the characteristic sounds produced by that action.

Feedback from the activation of these circuits gives rise to the recognition of the action.

The next time you intently watch someone executing a skilled action—say, pitching a baseball, swinging a bat, kicking a football, or jumping a hurdle—see whether you don't find yourself tensing the muscles that you would use if you were performing the action. Presumably, the activation of the mirror neuron circuit is responsible for this effect. As we will see in Chapter 11, we also tend to copy facial expressions of emotion that other people make, and feedback from doing so tends to evoke a similar emotional state in us.

A functional-imaging study by Iacoboni et al. (2005) suggests that the mirror neuron system helps us to understand other people's intentions. The researchers showed subjects video clips of an arm and hand reaching for and grasping a drinking mug. The actions were shown in isolation or in the context of objects set out for a snack (mug, teapot, milk pitcher, sugar bowl, sealed jam jar, plate of cookies, etc.) or the same objects after the snack had been eaten (mug, milk pitcher overturned, cookies missing from the plate, open jam jar, etc.). The first context suggests that the intent of the action is that of drinking, and the second suggests that the intent is that of cleaning up. The investigators found that watching the reaching action activated the mirror neuron system of the ventral premotor cortex, but there were differences in the activation when the action occurred in the two different contexts. (There were no differences in the activation caused by simply looking at the contexts.) The authors concluded that the mirror neuron system encodes not only an action but the intent of that action. (See *Figure 8.19.*)

FIGURE 8.19 ■ Understanding Intentions

The actions and contexts were presented to the subjects in the experiment by Iacoboni et al. (2005).

(From Iacoboni, M., Molnar-Szakacs, I., Gallese, V., Buccino, G., Mazziotta, J. C., and Rizzolatti, G. *PLoS Biology,* 2005, *3,* e79. Reprinted under open access license.)

Control of Reaching and Grasping

Much of our behavior involves interacting with objects in our environment. Many of these interactions involve reaching for something and then doing something with it, such as picking it up, moving it, or otherwise manipulating it. Researchers investigating these interactions classify them into two major categories: reaching and grasping. It turns out that different brain mechanisms are involved in these two activities.

Most reaching behavior is controlled by vision. As we saw in Chapter 6, the dorsal stream of the visual system is involved in determining the location of objects and, if they are moving, the direction and speed of their movement. You will not be surprised to learn that connections between the parietal lobe (the endpoint of the dorsal stream of the visual association cortex) and the frontal lobe play a critical role in reaching. As we saw in Chapter 6, several regions of the visual association cortex are named for particular types of objects that we perceive, for example, fusiform face area, extrastriate body area, and parahippocampal place area. One region of the medial posterior parietal cortex has been named the **parietal reach region.** Connolly, Andersen, and Goodale (2003) found that when people were about to make a pointing or reaching movement to a particular location this region became active. Presumably, the parietal cortex determines the location of the target and supplies information about this location to motor mechanisms in the frontal cortex. (See *Figure 8.20* and refer to *Figure 8.16*.)

Another region of the posterior parietal cortex, the anterior part of the intraparietal sulcus (aIPS), is involved in controlling hand and finger movements involved in grasping the target object. A functional-imaging study by Frey et al. (2005) had people reach for objects of

FIGURE 8.20 ■ The Parietal Reach Region

An inflated left cerebral hemisphere shows fMRI activation of the parietal reach region (PRR) just as people were about to make a pointing or reaching movement. POS = parieto-occipital sulcus.

(From Connolly, J.D., Andersen, R. A., and Goodale, M.A. *Experimental Brain Research,* 2003, *153,* 140–145. Reprinted with permission.)

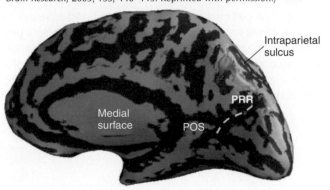

FIGURE 8.21 ■ Activation of the Anterior Intraparietal Sulcus

The activation is produced by grasping movements made while reaching for objects with different shapes. Activity made by reaching for and simply touching the objects was subtracted from activity made by reaching and grasping, leaving only the grasping component of fMRI activation.

(From Frey, S. H., Vinton, D., Norlund, R., and Grafton, S. T. *Cognitive Brain Research,* 2005, *23,* 397–405. Reprinted with permission.)

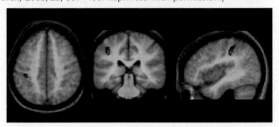

different shapes, which required them to make a variety of hand and finger movements to hold onto the objects. The brain activity directly related to grasping movements was determined by subtracting the activity produced by reaching for and simply touching the objects from the activity produced by reaching for and grasping the objects. The grasping activity activated the aIPS. See *Figure 8.21.*)

An experiment by Tunik, Frey, and Grafton (2005) confirmed the importance of the aIPS to grasping. The investigators had subjects reach for and grasp a rectangular object that was oriented with its long side in a vertical or horizontal position. On some trials ("perturbed trials") the object rotated during the subjects' reaching movements, which required the subjects to adjust the position of their hand or fingers before they reached the object. On some of these perturbed trials the investigators applied transcranial magnetic stimulation that disrupted the activity of the aIPS. When the disruptive stimulation occurred within 65 ms after the rotation of the object, the subjects' ability to accurately change grip posture was disrupted. Stimulation of the hand area of the primary motor cortex or other parts of the parietal lobe had no effect.

The visual input to the aIPS is, of course, from the dorsal stream of the visual system. In a functional-imaging study by Shmuelof and Zohary (2005), subjects watched brief videos of a hand reaching out to grasp a variety of objects. Sometimes the hand appeared in the left visual field and the object appeared in the right visual field; sometimes the locations for the hand and

parietal reach region A region in the medial posterior parietal cortex that plays a critical role in control of pointing or reaching with the hands.

the object were reversed. (The subjects focused their gaze on a fixation point located between the hand and the object.) This procedure means that for a particular trial, visual information about an object was transmitted to one side of the brain, and visual information about a hand shaped to grasp the object was transmitted to the other side of the brain. Analysis of the brain activation showed that information about the object activated the ventral stream of the visual system and information about the shape of the hand activated the aIPS—part of the dorsal stream. The results suggest that the aIPS is involved in recognition of grasping movements as well as their execution.

Deficits of Skilled Movements: The Apraxias

Damage to the frontal or parietal cortex on the left side of the brain can produce a category of deficits called **apraxia.** Literally, the term means "without action," but apraxia differs from paralysis or weakness that occurs when motor structures such as the precentral gyrus, basal ganglia, brain stem, or spinal cord are damaged. Apraxia refers to the inability to imitate movements or produce them in response to verbal instructions or inability to demonstrate the movements that would be made in using a familiar tool or utensil (Leiguarda and Marsden, 2000). Neuropsychological studies of the apraxias have provided information about the way skilled behaviors are organized and initiated.

There are four major types of apraxia, two of which I will discuss in this chapter. *Limb apraxia* refers to problems with movements of the arms, hands, and fingers. *Oral apraxia* refers to problems with movements of the muscles used in speech. *Apraxic agraphia* refers to a particular type of writing deficit. *Constructional apraxia* refers to difficulty in drawing or constructing objects. Because of their relation to language, I will describe oral apraxia and the various forms of agraphia in Chapter 14.

Limb Apraxia

Limb apraxia is characterized by movement of the wrong part of the limb, incorrect movement of the correct part, or correct movements but in the incorrect sequence. It is assessed by asking patients to perform movements—for example, imitating hand gestures made by the examiner. The most difficult movements involve pantomiming particular acts without the presence of the objects that are normally acted upon. For example, the examiner might say to the patient, "Pretend you have a key in your hand and open a door with it." In response, a patient with limb apraxia might wave his wrist back and forth rather than rotating it or might rotate his wrist first and then pretend to insert the key. Or if asked to pretend that she is brushing her teeth, a patient might use her finger as though it

were a toothbrush rather than pretending to hold a toothbrush in her hand.

To perform behaviors on verbal command without having a real object to manipulate, a person must comprehend the command and be able to imagine the missing article as well as to make the proper movements; therefore, these requests are the most difficult to carry out. Somewhat easier are tasks that involve imitating behaviors performed by the experimenter. Sometimes, a patient who cannot mime the use of a key can copy the examiner's hand movements. The easiest tasks involve the actual use of objects. For example, the examiner might give the patient a door key and ask him or her to demonstrate its use. If the brain lesion makes it impossible for the patient to understand speech, then the examiner cannot assess the ability to perform behaviors on verbal command. In this case the examiner can only measure the patient's ability to imitate movements or use actual objects. (See Heilman, Rothi, and Kertesz, 1983, for a review.)

Why does damage to the left parietal hemisphere, but usually not the right, cause an apraxia of both hands? The answer seems to be that the right hemisphere is involved with extrapersonal space and the left hemisphere is involved with one's own body. A functional-imaging study by Chaminade, Meltzoff, and Decety (2005) supports this explanation. The investigators asked subjects to watch another person perform hand and arm gestures and then either imitate the gestures or make different ones with the same arm or the other arm. On the basis of the activity seen by fMRI scans, the authors concluded that posterior regions of the right hemisphere tracked the movements of the model in space, while the left parietal lobe organized the movements that would be made in response.

Although the frontal and parietal lobes are both involved in the imitating hand gestures made by other people, the frontal cortex appears to play a more important role in recognizing the meaning of these gestures. Pazzaglia et al. (2008) tested thirty-three patients with damage to the left hemisphere and eight patients with damage to the right hemisphere and found that twenty-one of them (all with left hemisphere damage to the frontal or parietal lobes) had limb apraxia. They tested the apraxic patients for recognition of hand gestures by having them watch video clips in which a person performed the gestures correctly or incorrectly. For example, incorrect gestures included playing a broom as if it were a guitar or pretending to hitchhike by extending the little finger instead of the thumb. Apraxic patients

---

apraxia Difficulty in carrying out purposeful movements, in the absence of paralysis or muscular weakness.

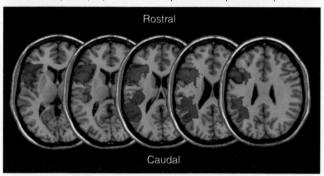

FIGURE 8.22 ■ Lesions Causing Limb Apraxia

Left hemisphere lesions in the frontal and parietal lobes cause limb apraxia. Lesions in red regions interfere with the ability of patients to comprehend the gestures made by other people.

(From Pazzaglia, M., Smania, N., Corato, E., and Aglioti, S. M. *Journal of Neuroscience,* 2008, *28,* 3030–3041. Reprinted with permission.)

Rostral

Caudal

with damage to the inferior frontal gyrus, but not to the parietal cortex, showed deficits in comprehension of the gestures. (See *Figure 8.22.*)

Constructional Apraxia

Constructional apraxia is caused by lesions of the right hemisphere, particularly the right parietal lobe. People with this disorder do not have difficulty making most types of skilled movements with their arms and hands. They have no trouble using objects properly, imitating their use, or pretending to use them. However, they have trouble drawing pictures or assembling objects from elements such as toy building blocks.

The primary deficit in constructional apraxia appears to involve the ability to perceive and imagine geometrical relations. Because of this deficit, a person cannot draw a picture, say, of a cube, because he or she cannot imagine what the lines and angles of a cube look like, not because of difficulty controlling the movements of his or her arm and hand. (See *Figure 8.23.*) Besides being unable to draw accurately, a person with constructional apraxia invariably has trouble with other tasks involving spatial perception, such as following a map.

The Basal Ganglia

Anatomy and Function

The basal ganglia constitute an important component of the motor system. We know that they are important because their destruction by disease or injury causes severe motor deficits. The motor nuclei of the basal ganglia include the caudate nucleus, putamen, and globus

FIGURE 8.23 ■ Constructional Apraxia

A patient with constructional apraxia caused by a lesion of the right parietal lobe attempted to copy a cube.

(From *Fundamentals of Human Neuropsychology,* by B. Kolb and I. Q. Whishaw. W. H. Freeman and Company. Copyright © 1980. Reprinted with permission.)

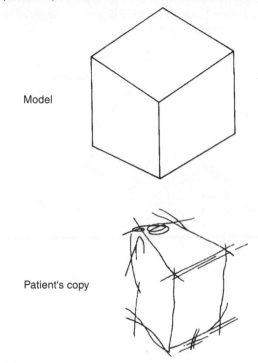

Model

Patient's copy

pallidus. The basal ganglia receive most of their input from all regions of the cerebral cortex (but especially the primary motor cortex and primary somatosensory cortex) and the substantia nigra. They have two primary outputs: the primary motor cortex, supplementary motor area, and premotor cortex (via the thalamus) and motor nuclei of the brain stem that contribute to the ventromedial pathways. Through these connections the basal ganglia influence movements under the control of the primary motor cortex and exert some direct control over the ventromedial system.

constructional apraxia Difficulty in drawing pictures or diagrams or in making geometrical constructions of elements such as building blocks or sticks; caused by damage to the right parietal lobe.

caudate nucleus A telencephalic nucleus, one of the input nuclei of basal ganglia; involved with control of voluntary movement.

putamen A telencephalic nucleus; one of the input nuclei of the basal ganglia; involved with control of voluntary movement.

globus pallidus A telencephalic nucleus; the primary output nucleus of the basal ganglia; involved with control of voluntary movement.

FIGURE 8.24 ■ Basal Ganglia

(a) The locations of the components of the basal ganglia and associated structures. (b) The major connections of the basal ganglia and associated structures. Excitatory connections are shown as black lines; inhibitory connections are shown as red lines. The indirect pathway is indicated by arrows with broken lines. Many connections, such as the inputs to the substantia nigra, are omitted for clarity. The internal division of the globus pallidus, the primary output of the basal ganglia and the target of stereotaxic surgery for Parkinson's disease, is outlined in gray.

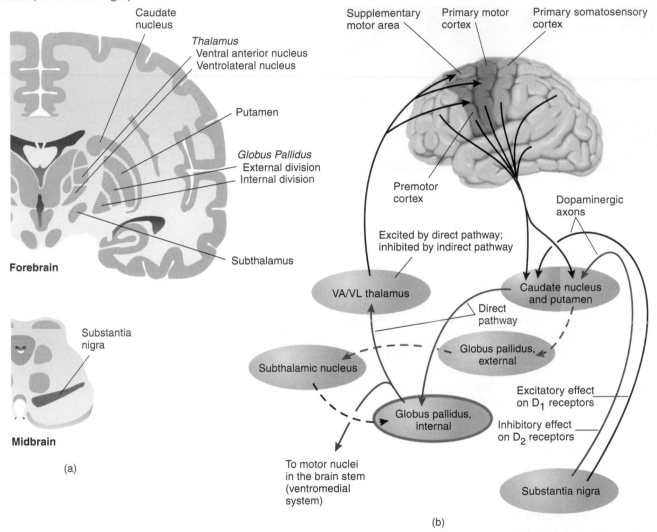

(a)

(b)

Figure 8.24(a) illustrates the components of the basal ganglia: the **caudate nucleus,** the **putamen,** and the **globus pallidus.** It also shows some nuclei associated with the basal ganglia: the **ventral anterior nucleus** and **ventrolateral nucleus** of the thalamus and the substantia nigra of the ventral midbrain. (See *Figure 8.24a.*)

Figure 8.24(b) shows some of the more important connections of the basal ganglia and helps to explain the role these structures play in the control of movement. For the sake of clarity this figure leaves out many connections, including inputs to the substantia nigra from the

basal ganglia and other structures. First, let's take a quick look at the loop formed between the cortex and the basal ganglia. The frontal, parietal, and temporal cortex

ventral anterior nucleus (of thalamus) A thalamic nucleus that receives projections from the basal ganglia and sends projections to the motor cortex.

ventrolateral nucleus (of thalamus) A thalamic nucleus that receives projections from the basal ganglia and sends projections to the motor cortex.

send axons to the caudate nucleus and the putamen, which then connect with the globus pallidus. The globus pallidus sends information back to the motor cortex via the ventral anterior and ventrolateral nuclei of the thalamus, completing the loop. Thus, the basal ganglia can monitor somatosensory information and are informed of movements being planned and executed by the motor cortex. Using this information (and other information they receive from other parts of the brain), they can then influence the movements controlled by the motor cortex. Throughout this circuit, information is represented somatotopically. That is, projections from neurons in the motor cortex that cause movements in particular parts of the body project to particular parts of the putamen, and this segregation is maintained all the way back to the motor cortex. (See *Figure 8.24b.*)

Another important input to the basal ganglia comes from the substantia nigra of the midbrain. We saw in Chapter 4 that degeneration of the nigrostriatal bundle, the dopaminergic pathway from the substantia nigra to the caudate nucleus and putamen (the *neostriatum*), causes Parkinson's disease. (I will say more about the neural circuits involved in this disorder in the next subsection.) (See *Figure 8.24b.*)

Now let's consider some of the complexities of the cortical–basal ganglia loop. The links in the loop are made by both excitatory (glutamate-secreting) neurons and inhibitory (GABA-secreting) neurons. The caudate nucleus and putamen receive excitatory input from the cerebral cortex. They send inhibitory axons to the external and internal divisions of the globus pallidus (the GP_i and the GP_e, respectively). The pathway that includes the GP_i is known as the **direct pathway** (arrows with solid lines). Neurons in GP_i sends inhibitory axons to the ventral anterior and ventrolateral thalamus (VA/VL thalamus), which send excitatory projections to the motor cortex. The net effect of the loop is excitatory because it contains two inhibitory links. Each inhibitory link (red arrow) reverses the sign of the input to that link. Thus, excitatory input to the caudate nucleus and putamen causes the these structures to *inhibit* neurons in the GP_i. This inhibition *removes* the inhibitory effect of the connections between the GP_i on the VA/VL thalamus; in other words, neurons in the VA/VL thalamus become more excited. This excitation is passed on to the motor cortex. (See *Figure 8.24b.*)

The pathway that includes the GP_e is known as the **indirect pathway** (arrows with broken lines). Neurons in GP_e send inhibitory input to the subthalamic nucleus, which sends excitatory input to the GP_i. From there on, the circuit is identical to the one we just examined— except that the ultimate effect of this loop on the thalamus and frontal cortex is *inhibitory*. The globus pallidus also sends axons to various motor nuclei in the brain stem that contribute to the ventromedial system. (See *Figure 8.24b.*)

Parkinson's Disease

Now that you understand the cortical–basal ganglia loop, you can understand the symptoms and treatment of two important neurological disorders: Parkinson's disease and Huntington's disease. The primary symptoms of Parkinson's disease are muscular rigidity, slowness of movement, a resting tremor, and postural instability. For example, once a person with Parkinson's disease is seated, he or she finds it difficult to arise. Once the person begins walking, he or she has difficulty stopping. Thus, a person with Parkinson's disease cannot easily pace back and forth across a room. Reaching for an object can be accurate, but the movement usually begins only after a considerable delay, and the individual components of the movement (a series of trunk, arm, hand, and finger movements) are poorly coordinated (Poizner et al., 2000). Writing is slow and labored, and as it progresses, the letters get smaller and smaller. Postural movements are impaired. A normal person who is bumped while standing will quickly move to restore balance—for example, by taking a step in the direction of the impending fall or by reaching out with the arms to grasp a piece of furniture. However, a person with Parkinson's disease fails to do so and simply falls. A person with this disorder is even unlikely to put out his or her arms to break the fall.

Parkinson's disease also produces a resting tremor—vibratory movements of the arms and hands that diminish somewhat when the individual makes purposeful movements. The tremor is accompanied by rigidity; the joints appear stiff. However, the tremor and rigidity are not the cause of the slow movements. In fact, some patients with Parkinson's disease show extreme slowness of movements but little or no tremor.

Let's look at Figure 8.24(b) again to see why damage to the nigrostriatal bundle causes slowness of movements and disrupts postural adjustments. Normal movements require an appropriate balance between the direct (excitatory) and indirect (inhibitory) pathways. The caudate nucleus and putamen consist of two different zones, both of which receive input from dopaminergic neurons of the substantia nigra. One of these zones contains D_1 dopamine receptors, which produce excitatory effects. Neurons in this zone send their axons to the

---

direct pathway (in basal ganglia) The pathway that includes the caudate nucleus and putamen, the internal division of the globus pallidus, and the ventral anterior/ventrolateral thalamic nuclei; has an excitatory effect on movement.

indirect pathway (in basal ganglia) The pathway that includes the caudate nucleus and putamen, the external division of the globus pallidus, the subthalamic nucleus, the internal division of the globus pallidus, and the ventral anterior/ventrolateral thalamic nuclei; has an inhibitory effect on movement.

FIGURE 8.25 ■ Huntington's Disease

(a) A slice through a normal human brain shows the normal appearance of the caudate nuclei and putamen (arrowheads) and lateral ventricles. (b) A slice through the brain of a person who had Huntington's disease. The arrowheads indicate the location of the caudate nuclei and putamen, which are severely degenerated. As a consequence of the degeneration, the lateral ventricles (open spaces in the middle of the slice) have enlarged.

(Courtesy of Harvard Medical School/Betty G. Martindale and Anthony D'Agostino, Good Samaritan Hospital, Portland, Oregon.)

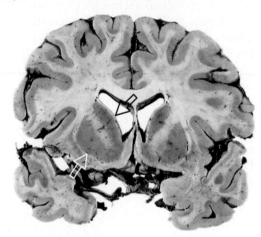

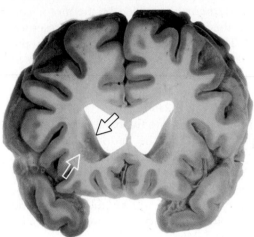

(a) Normal   (b) Huntington's disease

GP$_i$. Neurons in the other zone contain D$_2$ receptors, which produce inhibitory effects. These neurons send their axons to the GP$_e$. (See *Figure 8.24b.*) The first of these circuits, beginning with the black arrow from the substantia nigra, goes through two inhibitory synapses (red arrows) before it reaches the VA/VL thalamus; thus, this circuit has an excitatory effect on behavior. The second of these circuits begins with an inhibitory input to the caudate nucleus and putamen, but it goes through *four* inhibitory synapses in the following pathway: substantia nigra → caudate/putamen → GP$_e$ → subthalamic nucleus → GP$_i$ → VA/VL thalamus. Thus, the effect of this pathway, too, is excitatory; thus, dopaminergic input to the caudate nucleus and putamen facilitate movements. Note that the GP$_i$ also sends axons to the ventromedial system. A decrease in this inhibitory output is probably responsible for the muscular rigidity and poor control of posture seen in Parkinson's disease. (See *Figure 8.24b.*)

As we saw in Chapter 4, the standard treatment for Parkinson's disease is L-DOPA, the precursor of dopamine. When an increased amount of L-DOPA is present, the remaining nigrostriatal dopaminergic neurons in a patient with Parkinson's disease will produce and release more dopamine. But this compensation often produces *dyskinesias* and *dystonias*—involuntary movements and postures that are presumably caused by too much stimulation of dopamine receptors in the basal ganglia. In addition, L-DOPA does not work indefinitely;

eventually, the number of nigrostriatal dopaminergic neurons declines to such a low level that the symptoms become worse. Some patients—especially those whose symptoms began when they were relatively young—eventually become bedridden, scarcely able to move.

In recent years, clinicians have worked developing new ways to treat Parkinson's disease, and much research has been done on discovering the causes of the brain damage. I will describe these efforts in Chapter 15.

Huntington's Disease

Another basal ganglia disease, **Huntington's disease,** is caused by degeneration of the caudate nucleus and putamen, especially of GABAergic and acetylcholinergic neurons. (See *Figure 8.25.*) Whereas Parkinson's disease causes a poverty of movements, Huntington's disease, formerly called *Huntington's chorea*, causes uncontrollable ones, especially jerky limb movements. (The term *chorea* derives from the Greek *khoros*, meaning "dance.") The movements of Huntington's disease look like fragments of purposeful movements but occur involuntarily. This disease is progressive and eventually causes death.

Huntington's disease A fatal inherited disorder that causes degeneration of the caudate nucleus and putamen; characterized by uncontrollable jerking movements, writhing movements, and dementia.

The symptoms of Huntington's disease usually begin in the patient's thirties or forties but can sometimes begin in the early twenties. The first signs of neural degeneration occur in the caudate nucleus and the putamen—specifically, in the medium-sized spiny inhibitory neurons whose axons travel to the external division of the globus pallidus. The loss of inhibition provided by these GABA-secreting neurons increases the activity of the GP_e, which then inhibits the subthalamic nucleus. As a consequence, the activity level of the GP_i decreases, and excessive movements occur. (Refer to *Figure 8.24b*.) As the disease progresses, the caudate nucleus and putamen degenerate until almost all of their neurons disappear. The patient dies from complications of immobility. Unfortunately, there is at present no effective treatment for this disorder.

Huntington's disease is a hereditary disorder, caused by a dominant gene on chromosome 4. In fact, the gene has been located, and its defect has been identified as a repeated sequence of bases that code for the amino acid glutamine (Collaborative Research Group, 1993). This repeated sequence causes the gene product—a protein called *huntingtin*—to contain an elongated stretch of glutamine. Longer stretches of glutamine are associated with patients whose symptoms began at a younger age, which strongly suggests that this abnormal portion of the huntingtin molecule is responsible for the disease. Research on the role that abnormal huntingtin plays in death of basal ganglia neurons is described in Chapter 15.

The Cerebellum

The cerebellum is an important part of the motor system. It contains about 50 billion neurons, compared to the approximately 22 billion neurons in the cerebral cortex (Robinson, 1995). Its outputs project to every major motor structure of the brain. When it is damaged, people's movements become jerky, erratic, and uncoordinated. The cerebellum consists of two hemispheres that contain several deep nuclei situated beneath the wrinkled and folded cerebellar cortex. Thus, the cerebellum resembles the cerebrum in miniature. The medial part of the cerebellum is phylogenetically older than the lateral part, and it participates in control of the ventromedial system. The **flocculonodular lobe,** located at the caudal end of the cerebellum, receives input from the vestibular system and projects axons to the vestibular nucleus. You will not be surprised to learn that this system is involved in postural reflexes. (See the green lines in *Figure 8.26*.) The **vermis** ("worm"), located on the midline, receives auditory and visual information from the tectum and cutaneous and kinesthetic information from the spinal cord. It sends its outputs to the **fastigial nucleus** (one of the set of deep cerebellar nuclei). Neurons in the fastigial nucleus send axons to

the vestibular nucleus and to motor nuclei in the reticular formation. Thus, these neurons influence behavior through the vestibulospinal and reticulospinal tracts, two of the three ventromedial pathways. (See the blue lines in *Figure 8.26*.)

The rest of the cerebellar cortex receives most of its input from the cerebral cortex, including the primary motor cortex and association cortex. This input is relayed to the cerebellar cortex through the pontine tegmental reticular nucleus. The intermediate zone of the cerebellar cortex projects to the **interposed nuclei,** which in turn project to the red nucleus. Thus, the intermediate zone influences the control of the rubrospinal system over movements of the arms and legs. The interposed nuclei also send outputs to the ventrolateral thalamic nucleus, which projects to the motor cortex. (See the red lines in *Figure 8.26*.)

The lateral zone of the cerebellum is involved in the control of independent limb movements, especially rapid, skilled movements. Such movements are initiated by neurons in the frontal association cortex, which control neurons in the primary motor cortex. But although the frontal cortex can plan and initiate movements, it does not contain the neural circuitry needed to calculate the complex, closely timed sequences of muscular contractions that are needed for rapid, skilled movements. That task falls to the lateral zone of the cerebellum.

Both the frontal association cortex and the primary motor cortex send information about intended movements to the lateral zone of the cerebellum via the **pontine nucleus.** The lateral zone also receives information from the somatosensory system, which informs it about the current position and rate of movement of the limbs—information that is necessary for computing the details of a movement. When the cerebellum receives information that the motor cortex has begun to initiate a movement, it computes the contribution that various muscles will have to make to perform that movement. The results of this computation are sent to the **dentate nucleus,** another of

flocculonodular lobe A region of the cerebellum; involved in control of postural reflexes.

vermis The portion of the cerebellum located at the midline; receives somatosensory information and helps to control the vestibulospinal and reticulospinal tracts through its connections with the fastigial nucleus.

fastigial nucleus A deep cerebellar nucleus; involved in the control of movement by the reticulospinal and vestibulospinal tracts.

interposed nuclei A set of deep cerebellar nuclei; involved in the control of the rubrospinal system.

pontine nucleus A large nucleus in the pons that serves as an important source of input to the cerebellum.

dentate nucleus A deep cerebellar nucleus; involved in the control of rapid, skilled movements by the corticospinal and rubrospinal systems.

FIGURE 8.26 ■ Inputs and Outputs of the Cerebellum

The figure shows the inputs and outputs of three systems: the flocculonodular lobe (green lines), the vermis (blue lines), and the intermediate zone of the cerebellar cortex (red lines).

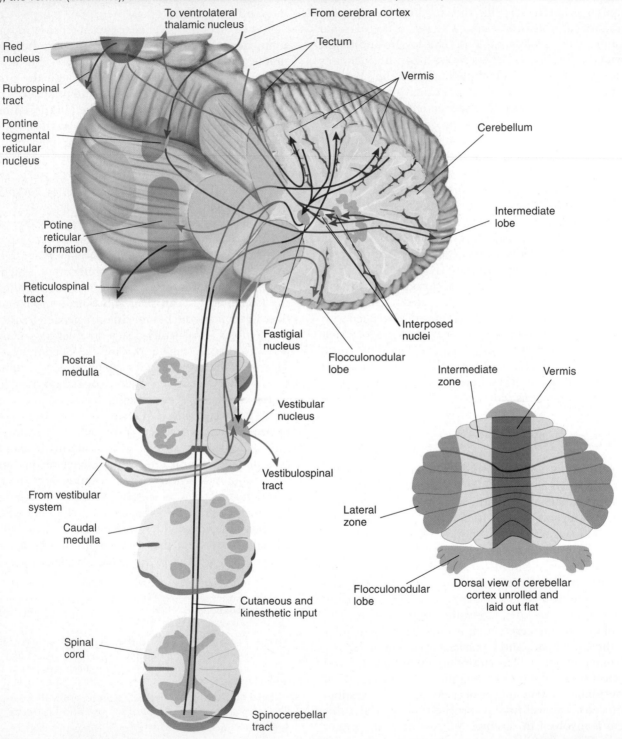

the deep cerebellar nuclei. Neurons in the dentate nucleus pass the information on to the ventrolateral thalamus, which projects to the primary motor cortex. The projection from the ventrolateral thalamus to the primary motor cortex enables the cerebellum to modify the ongoing movement that was initiated by the frontal cortex. The lateral zone of the cerebellum also sends efferents to the red nucleus (again, via the dentate nucleus); thus, it helps to control independent limb movements through this system as well. (See *Figure 8.27*.)

In humans, lesions of different regions of the cerebellum produce different symptoms. Damage to the flocculonodular lobe or the vermis causes disturbances in posture and balance. Damage to the intermediate zone produces deficits in movements controlled by the rubrospinal system; the principal symptom of this damage is limb rigidity. Damage to the lateral zone causes weakness and *decomposition of movement*. For example, a person with this kind of damage who is attempting to bring the hand to the mouth will make separate movements of the joints of the shoulder, elbow, and wrist instead of performing simultaneous smooth movements.

Lesions of the lateral zone of the cerebellar cortex also appear to impair the timing of rapid *ballistic* movements. Ballistic (literally, "throwing") movements occur too fast to be modified by feedback. The sequence of muscular movements must then be programmed in advance, and the individual muscles must be activated at the proper times. You might like to try this common neurological test: Have a friend place his or her finger in front of your face, about three-quarters of an arm's length away. While your friend slowly moves his or her finger around to serve as a moving target, alternately touch your nose and your friend's finger as rapidly as you can. If your cerebellum is normal, you can successfully hit your nose and your friend's finger without too much trouble. People with lateral cerebellar damage have great difficulty; they tend to miss the examiner's hand and poke themselves in the eye. (I have often wondered why neurologists do not adopt a less dangerous test.)

When making rapid, aimed movements, we cannot rely on feedback to stop the movement when we reach the target. By the time we perceive that our finger has reached the proper place, it is too late to stop the movement, and we will overshoot the target if we try to stop it then. Instead of relying on feedback, the movement appears to be timed. We estimate the distance between our hand and the target, and our cerebellum calculates the amount of time that the muscles will have to be turned on. After the proper amount of time, the cerebellum briefly turns on antagonistic muscles to stop the movement. In fact, Kornhuber (1974) suggested that one of the primary functions of the cerebellum is timing the duration of rapid movements. Obviously, learning must play a role in controlling such movements.

Timmann, Watts, and Hore (1999) reported an interesting example of the role the cerebellum plays in timing sequences of muscular contractions. When tossing a ball at a target using an overarm throw, a person raises his or her hand above the shoulder, rotates the arm forward, and then releases the ball by extending the fingers—moving them apart. The timing of the release is critical: too soon and the ball goes too high, too late and it goes too low. The researchers found that normal subjects released the ball within an 11-msec window 95 percent of the time. Patients with cerebellar lesions did five times worse: Their window was 55 msec wide.

The cerebellum also appears to integrate successive *sequences* of movements that must be performed one after the other. For example, Holmes (1939) reported that one of his patients said, "The movements of my left arm are done subconsciously, but I have to think out each movement of the right [affected] arm. I come to a dead stop in turning and have to think before I start again."

FIGURE 8.27 ■ Inputs and Outputs of the Cerebellar Cortex

The lateral zone receives information about impending movements from the frontal lobes and helps to smooth and integrate the movements through its connections to the primary motor cortex and red nucleus through the dentate nucleus and ventral thalamus.

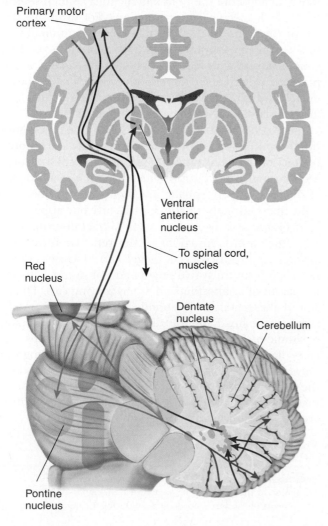

Primary motor cortex

Ventral anterior nucleus

To spinal cord, muscles

Red nucleus

Dentate nucleus

Cerebellum

Pontine nucleus

Thach (1978) obtained experimental evidence that corroborates this role. He found that many neurons in the dentate nuclei (which receive inputs from the lateral zone of the cerebellar cortex) showed response patterns that predicted the *next* movement in a sequence rather than the one that was currently taking place. Presumably, the cerebellum was planning these movements.

Dr. S., a professor of neurology at the medical school, stood on the stage of the auditorium as he presented a case to a group of physicians and students. He discussed the symptoms and possible causes of cerebellar–brain stem degeneration. "Now I'd like to present Mr. P.," he said, as a set of MRI scans appeared on the screen. "As you can see, Mr. P.'s cerebellum shows substantial degeneration, but we can't see evidence of any damage to the brain stem."

Dr. S. left the stage and returned, pushing Mr. P. onstage in a wheelchair.

"Mr. P., how are you feeling today?"

"I'm fine," he replied. "Of course, I'd feel better if I could have walked out here myself."

"Of course."

Dr. S. chatted with Mr. P. for a few minutes, getting him to talk enough so that we could see that his mental condition was lucid and that he had no obvious speech or memory problems.

"Okay, Mr. P., I'd like you to make some movements." He faced Mr. P. and said, "Please stretch your hands out and hold them like this." Dr. S. suddenly raised his arms from his sides and held them out straight in front of him, palms down, fingers pointing forward.

Mr. P. did not respond immediately. He looked as if he were considering what to do. Suddenly, his arms straightened out and lifted from the armrests of the wheelchair. Instead of stopping when they were pointed straight ahead of him, they continued upward. Mr. P. grunted, and his arms began flailing around—up, down, left, and right—until he finally managed to hold them outstretched in front of him. He was panting with the effort to hold them there.

"Thank you, Mr. P. Please put your arms down again. Now try this." Dr. S. very slowly raised his arms from his sides until they were straight out in front of them. Mr. P. did the same, and this time there was no overshoot.

After a few more demonstrations, Dr. S. thanked Mr. P. and wheeled him offstage. When he returned, he reviewed what we had seen.

"When Mr. P. tried to quickly raise his arms in front of him, his primary motor cortex sent messages to the appropriate muscles, and his arms straightened out and began to rise.

Normally, the cerebellum is informed about the movement and, through its connections back to the motor cortex, begins to contract the antagonistic muscles at the appropriate time, bringing the arms to rest in the intended position. Mr. P. could get the movement started just fine, but the damage to his cerebellum eliminated the help this structure gives to rapid movements, and he couldn't stop his arms in time. When he tried to move slowly, he could use visual and kinesthetic feedback from the position of his arms to control the movement. Your cerebellum isn't nearly as important in the control of simple, slow movements. For that you need your basal ganglia, but that's another story."

The Reticular Formation

The reticular formation consists of a large number of nuclei located in the core of the medulla, pons, and midbrain. The reticular formation controls the activity of the gamma motor system and hence regulates muscle tonus. In addition, the pons and medulla contain several nuclei with specific motor functions. For example, different locations in the medulla control automatic or semiautomatic responses such as respiration, sneezing, coughing, and vomiting. As we saw, the ventromedial pathways originate in the superior colliculi, vestibular nuclei, and reticular formation. Thus, the reticular formation plays a role in the control of posture.

The reticular formation also plays a role in locomotion. Stimulation of the **mesencephalic locomotor region,** located ventral to the inferior colliculus, causes a cat to make pacing movements (Shik and Orlovsky, 1976). The mesencephalic locomotor region does not send fibers directly to the spinal cord but apparently controls the activity of reticulospinal tract neurons.

Other motor functions of the reticular formation are also being discovered. Siegel and McGinty (1977) recorded from thirty-five single neurons in the reticular formation of unanesthetized, freely moving cats. Thirty-two of these neurons responded during *specific* movements of the head, tongue, facial muscles, ears, forepaw, or shoulder. The specific nature of the relationships suggests that the neurons play some role in controlling the movements. For example, one neuron responded when the tongue moved out and to the left. The functions of these neurons and the range of movements they control are not yet known.

---

mesencephalic locomotor region A region of the reticular formation of the midbrain whose stimulation causes alternating movements of the limbs normally seen during locomotion.

Interim Summary

Control of Movement by the Brain

The motor systems of the brain are complex. (Having read this section, you do not need me to tell you that.) A good way to review the systems is through an example. Suppose you see, out of the corner of your eye, that something is moving. You quickly turn your head and eyes toward the source of the movement and discover that a vase of flowers on a table someone has just bumped is ready to fall. You quickly reach forward, grab it, and restore it to a stable, upright position. (For simplicity's sake I will assume that you are right-handed.)

The rapid movement of your head and eyes is controlled by mechanisms that involve the superior colliculi and nearby nuclei. The head movement and corresponding movement of the trunk are mediated by the tectospinal tract. You perceive the tipping vase because of the activity of neurons in your visual association cortex. The dorsal stream of your visual association cortex also contributes spatial information to the parietal reaching region in your left hemisphere, which calculates the reaching movement you must make and transmits this information to the motor association cortex in your left frontal lobe. During your reaching movement the cortex located in your anterior intraparietal sulcus sends information to your motor association cortex that moves your hand and fingers so that you will be ready to grasp the falling vase. Because the movement will have to be very rapid, your cerebellum controls its timing on the basis of information it receives from the association cortex of the frontal and parietal lobes. Your hand stops just as it touches the vase, and connections between the somatosensory cortex and the primary motor cortex initiate a reflex that closes your hand around the vase.

The muscles of your arm and hand are controlled through a cooperation between the corticospinal, rubrospinal, and ventromedial pathways. Even before your hand moves, the ventral corticospinal tract and the ventromedial pathways (vestibulospinal and reticulospinal system, largely under the influence of the basal ganglia) begin adjusting your posture so that you will not fall forward when you suddenly reach in front of you. Depending on how far forward you will have to reach, the reticulospinal tract may even cause one leg to step forward to take your weight. The rubrospinal tract controls the muscles of your upper arm, and the lateral corticospinal tract controls your finger and hand movements. Perhaps you say, triumphantly, "I got it!" The corticobulbar pathway, under the control of speech mechanisms in the left hemisphere, causes the muscles of your vocal apparatus to say these words.

The supplementary motor area (SMA) and the premotor cortex receive information from the parietal lobe and help to initiate movements through their connections with the primary motor cortex. The SMA is involved in well-learned behavioral sequences. Neurons there fire at particular points in behavioral sequences, and disruption or damage impairs the ability to perform these sequences. The pre-SMA is involved in awareness of our decisions to make spontaneous movements. The premotor cortex is involved in learning and executing complex movements that are guided by arbitrary sensory information, such as verbal instructions. This region and the inferior parietal lobule constitute a mirror neuron system that plays an important role in imitation and understanding the actions and intensions of others.

A person with apraxia will have difficulty making controlled movements of the limb in response to a verbal request or an attempt to imitate another person's action. Most cases of apraxia are produced by lesions of the left frontal or parietal cortex. The left parietal cortex directly controls movement of the right limb by activating neurons in the left primary motor cortex and indirectly controls movement of the left limb by sending information to the right frontal association cortex.

The basal ganglia are part of a circuit that includes the cerebral cortex, the subthalamic nucleus, thalamic motor nuclei, and the substantia nigra. This system is involved in coordinating and timing of movements that are slower than those controlled by the cerebellum. Parkinson's disease is caused by degeneration of dopamine-secreting neurons of the substantia nigra that send axons to the basal ganglia. Huntington's disease, a fatal disease caused by a mutation that caused production of abnormal huntingtin protein, causes degeneration of the caudate nucleus and putamen. Although identification of the faulty protein provides hope for understanding the causes of the neural degeneration, there is still no treatment for this disorder.

SUGGESTED READINGS

Graziano, M. S. A. *The Intelligent Movement Machine: An Ethological Perspective on the Primate Motor System.* New York: Oxford University Press, 2009.

Kandel, E. R., Schwartz, J. H., and Jessell, T. M. *Principles of Neural Science,* 4th ed. New York: McGraw-Hill, 2000.

Kolb, B., and Whishaw, I. Q. *Fundamentals of Human Neuropsychology,* 4th ed. New York: W. H. Freeman, 1996.

Shumway-Cook, A., and Woollacott, M. *Motor Control: Translating Research into Clinical Practice,* 3rd ed. Philadelphia: Lippincott Williams & Wilkins, 2007.

ADDITIONAL RESOURCES

Visit www.mypsychkit.com for additional review and practice of the material covered in this chapter. Within MyPsychKit, you can take practice tests and receive a customized study plan to help you review. Dozens of animations, tutorials, and Web links are also available. You can even review using the interactive electronic version of this textbook. You will need to register for MyPsychKit. See www.mypsychkit.com for complete details.

Sleep and Biological Rhythms

Lately, Michael felt almost afraid of going to bed because of the unpleasant experiences he had been having. His dreams seemed to have become more intense in a rather disturbing way. Several times in the past few months, he felt as if he were paralyzed as he lay in bed, waiting for sleep to come. It was a strange feeling; was he *really* paralyzed, or was he just not trying hard enough to move? He always fell asleep before he was able to decide. A couple of times he woke up just before it was time for his alarm to go off and felt unable to move. Then the alarm would ring, and he would quickly shut it off. That meant that he really wasn't paralyzed, didn't it? Was he going crazy?

One night brought the worst experience of all. As Michael was falling asleep, he felt again as if he were paralyzed. Then he saw his former roommate enter his bedroom. But that wasn't possible! Since the time he graduated from college, he had lived alone, and he always locked the door. He tried to say something, but he couldn't. His roommate was holding a hammer. He walked up to the bed, stood over Michael, and suddenly raised the hammer, as if to smash in his forehead. When Michael awoke in the morning, he shuddered with the remembrance. It had seemed so real! It must have been a dream, but he didn't think he was asleep. He was in bed. Can a person really dream that he is lying in bed, not yet asleep?

That day at the office, he had trouble concentrating on his work. He forced himself to review his notes, because he had to present the details of the new project to the board of directors. This was his big chance; if the project were accepted, he would certainly be chosen to lead it, and that would mean a promotion and a substantial raise. Naturally, with so much at stake, he felt nervous when he entered the boardroom. His boss introduced Michael and asked him to begin. Michael glanced at his notes and opened his mouth to talk. Suddenly, he felt his knees buckle. All his strength seemed to slip away. He fell heavily to the floor. He could hear people running over and asking what had happened. He couldn't move anything except his eyes. His boss got down on his knees, looked into Michael's face, and asked, "Michael, are you all right?" Michael looked at his boss and tried to answer, but he couldn't say a thing. A few seconds later, he felt his strength coming back. He opened his mouth and said, "I'm okay." He struggled to his knees and then sat in a chair, feeling weak and frightened.

"You undoubtedly have a condition known as narcolepsy," said the doctor whom Michael visited. "It's a problem that concerns the way your brain controls sleep. I'll have you spend a night in the sleep clinic and get some recordings done to confirm my diagnosis, but I'm sure that I'll be proved correct. You told me that lately you've been taking short naps during the day. What were these naps like? Were you suddenly struck by an urge to sleep?" Michael nodded. "I just had to put my head on the desk, even though I was afraid that my boss might see me. But I don't think I slept more than five minutes or so." "Did you still feel sleepy when you woke?" "No," he replied, "I felt fine again." The doctor nodded. "All the symptoms you have reported—the sleep attacks, the paralysis you experienced before sleeping and after waking up, the spell you had today—they all fit together. Fortunately, we can usually control narcolepsy with medication. In fact, we have a new one that does an excellent job. I'm sure we'll have you back to normal, and there is no reason why you can't continue with your job. If you'd like, I can talk with your boss and reassure him, too."

Why do we sleep? Why do we spend at least one-third of our lives doing something that provides most of us with only a few fleeting memories? I will attempt to answer this question in several ways. In the first two parts of this chapter I will describe what is known about the phenomenon of sleep and its disorders, including insomnia, narcolepsy, sleepwalking, and other sleep-related disorders. In the third part I will discuss research on the functions performed by sleep. In the fourth part I will describe the search for the chemicals and the neural circuits that control sleep and wakefulness. In the final part of the chapter I will discuss the brain's biological clock—the mechanism that controls daily rhythms of sleep and wakefulness.

A PHYSIOLOGICAL AND BEHAVIORAL DESCRIPTION OF SLEEP

Sleep is a behavior. That statement might seem peculiar, because we usually think of behaviors as activities that involve movements, such as walking or talking. Except for the rapid eye movements that accompany a particular stage, sleep is not distinguished by movement. What characterizes sleep is that the insistent urge of sleepiness forces us to seek out a quiet, warm, comfortable place; lie down; and remain there for several hours. Because we remember very little about what happens while we sleep, we tend to think of sleep more as a state of consciousness than as a behavior. The change in consciousness is undeniable, but it should not prevent us from noticing the behavioral changes.

Stages of Sleep

The best research on human sleep is conducted in a sleep laboratory. A sleep laboratory, usually located at a university or medical center, consists of one or several small bedrooms adjacent to an observation room, where the experimenter spends the night (trying to stay awake). The experimenter prepares the sleeper for electrophysiological measurements by attaching electrodes to the scalp to monitor the electroencephalogram (EEG) and to the chin to monitor muscle activity, recorded as the **electromyogram (EMG).** Electrodes attached around the eyes monitor eye movements, recorded as the **electro-oculogram (EOG).** In addition, other electrodes and transducing devices can be used to monitor autonomic measures such as heart rate, respiration, and changes in the ability of the skin to conduct electricity. (See *Figure 9.1.*)

During wakefulness the EEG of a normal person shows two basic patterns of activity: *alpha activity* and *beta activity.* **Alpha activity** consists of regular, medium-frequency waves of 8–12 Hz. The brain produces this activity when a person is resting quietly, not particularly aroused or excited and not engaged in strenuous mental activity (such as problem solving). Although alpha waves

FIGURE 9.1 ■ A Subject Prepared for a Night's Sleep in a Sleep Laboratory

(Philippe Platilly/Science Photo Library/Photo Researchers Inc.)

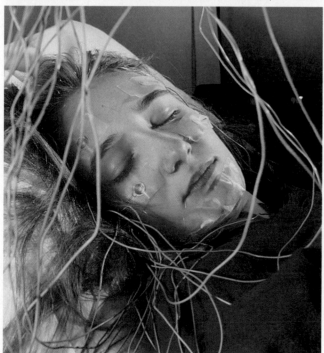

FIGURE 9.2 ■ EEG Recording of the Stages of Sleep

(From Horne, J. A. *Why We Sleep: The Functions of Sleep in Humans and Other Mammals.* Oxford, England: Oxford University Press, 1988. Reprinted with permission.)

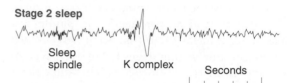

Awake

Alpha activity   Beta activity

Stage 1 sleep

Theta activity

Stage 2 sleep

Sleep spindle   K complex   Seconds

Stage 3 sleep

Delta activity

Stage 4 sleep

Delta activity

REM sleep

Theta activity   Beta activity

sometimes occur when a person's eyes are open, they are much more prevalent when the eyes are closed. The other type of waking EEG pattern, **beta activity,** consists of irregular, mostly low-amplitude waves of 13–30 Hz. Beta activity shows *desynchrony;* it reflects the fact that many different neural circuits in the brain are actively processing information. Desynchronized activity occurs when a person is alert and attentive to events in the environment or is thinking actively. (See *Figure 9.2.*)

electromyogram (EMG) *(my oh gram)* An electrical potential recorded from an electrode placed on or in a muscle.

electro-oculogram (EOG) *(ah kew loh gram)* An electrical potential from the eyes, recorded by means of electrodes placed on the skin around them; detects eye movements.

alpha activity Smooth electrical activity of 8–12 Hz recorded from the brain; generally associated with a state of relaxation.

beta activity Irregular electrical activity of 13–30 Hz recorded from the brain; generally associated with a state of arousal.

Let us look at a typical night's sleep of a female college student in a sleep laboratory. (Of course, we would obtain similar results from a male, with one exception, which is noted later.) The experimenter attaches the electrodes, turns the lights off, and closes the door. Our subject becomes drowsy and soon enters stage 1 sleep, marked by the presence of some **theta activity** (3.5–7.5 Hz), which indicates that the firing of neurons in the neocortex is becoming more synchronized. This stage is actually a transition between sleep and wakefulness; if we watch our volunteer's eyelids, we will see that from time to time they slowly open and close and that her eyes roll upward and downward. (See *Figure 9.2.*) About 10 minutes later she enters stage 2 sleep. The EEG during this stage is generally irregular but contains periods of theta activity, *sleep spindles,* and *K complexes.* Sleep spindles are short bursts of waves of 12–14 Hz that occur between two and five times a minute during stages 1–4 of sleep. K complexes are sudden, sharp waveforms, which, unlike sleep spindles, are usually found only during stage 2 sleep. They spontaneously occur at the rate of approximately one per minute but often can be triggered by noises—especially unexpected noises. A functional MRI study by Czisch et al. (2004) indicated that K complexes, triggered by an auditory stimulus, represents an inhibitory mechanism that presumably protects the sleeper from awakening. As we will see, K complexes appear to be the forerunner of delta waves, which appear in the deepest levels of sleep. (See *Figure 9.2.*)

The subject is sleeping soundly now; but if awakened, she might report that she has not been asleep. This phenomenon often is reported by nurses who awaken loudly snoring patients early in the night (perhaps to give them a sleeping pill) and find that the patients insist that they were lying there awake all the time. About 15 minutes later the subject enters stage 3 sleep, signaled by the occurrence of high-amplitude **delta activity** (less than 3.5 Hz). (See *Figure 9.2.*) The distinction between stage 3 and stage 4 is not clear-cut; stage 3 contains 20–50 percent delta activity, and stage 4 contains more than 50 percent. Because slow-wave EEG activity predominates during sleep stages 3 and 4, these stages are collectively referred to as **slow-wave sleep.** (See *Figure 9.2.*)

In recent years, researchers have begun studying the details of the EEG activity that occurs during slow-wave sleep and the brain mechanisms responsible for this activity (Steriade, 2003, 2006). It turns out that the most important feature of slow-wave activity during sleep are slow oscillations of less than 1 Hz. Each oscillation consists of a single high-amplitude biphasic (down and up) wave of slightly less than 1 Hz. The first part of the wave indicates a **down state**—a period of inhibition during which neurons in the neocortex are absolutely silent. Presumably, it is during this down state that neocortical neurons are able to rest. The second part indicates an **up state**—a period of excitation

FIGURE 9.3 ■ EEG and Single-Cell Activity

Recordings of neocortical EEG and single-cell activity were made during the slow-wave sleep of a cat. The right side of the figure shows three slow oscillations, each of which includes a sleep spindle. During the descending phase of the slow oscillation (down state), the neuron is hyperpolarized and does not fire. During the ascending phase, the neuron fires.

(From Contreras, D., and Steriade, M. *Journal of Neuroscience,* 1995, *15,* 604–632. Reprinted with permission.)

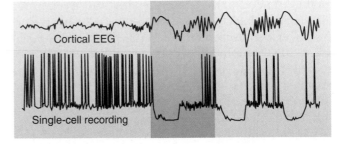

during which these neurons briefly fire at a high rate. Other components of slow-wave sleep, including K complexes, sleep spindles, and delta waves, are synchronized with these slow oscillations. Figure 9.3 shows the EEG recording and a single-cell recording from the neocortex of a sleeping cat. Three slow oscillations are shown at the right of the figure. Each oscillation consists of an inhibitory hyperpolarizing silent phase (down state) followed by an excitatory depolarizing phase during which the neuron fires at a high rate (up state). (See *Figure 9.3.*) As we will see in Chapter 13, the slow oscillations that occur during sleep play an important role in learning and memory.

About 90 minutes after the beginning of sleep (and about 45 minutes after the onset of stage 4 sleep), we notice an abrupt change in a number of physiological measures recorded from our subject. The EEG suddenly becomes mostly desynchronized, with a sprinkling of theta waves, very similar to the record obtained during stage 1 sleep. (Refer to *Figure 9.2.*) We also note that her eyes are rapidly darting back and forth beneath her closed eyelids. We can see this activity in the EOG, recorded from electrodes attached to the skin around her eyes, or we can observe the eye movements

theta activity EEG activity of 3.5–7.5 Hz that occurs intermittently during early stages of slow-wave sleep and REM sleep.

delta activity Regular, synchronous electrical activity of less than 4 Hz recorded from the brain; occurs during the deepest stages of slow-wave sleep.

slow-wave sleep Non-REM sleep, characterized by synchronized EEG activity during its deeper stages.

down state A period of inhibition during a slow oscillation during slow-wave sleep; neurons in the neocortex are silent and resting.

up state A period of excitation during a slow oscillation during slow-wave sleep; neurons in the neocortex briefly fire at a high rate.

directly—the cornea produces a bulge in the closed eyelids that can be seen to move about. We also see that the EMG becomes silent; there is a profound loss of muscle tone. In fact, physiological studies have shown that, aside from occasional twitching, a person actually becomes paralyzed during REM sleep. This peculiar stage of sleep is quite distinct from the quiet sleep we saw earlier. It is usually referred to as **REM sleep** (for the **r**apid **e**ye **m**ovements that characterize it).

By most criteria, stage 4 is the deepest stage of sleep; only loud noises will cause a person to awaken, and when awakened, the person acts groggy and confused. During REM sleep a person might not react to noises, but he or she is easily aroused by meaningful stimuli, such as the sound of his or her name. Also, when awakened from REM sleep, a person appears alert and attentive.

If we arouse our volunteer during REM sleep and ask her what was going on, she will almost certainly report that she had been dreaming. The dreams of REM sleep tend to be narrative in form, with a storylike progression of events. If we wake her during slow-wave sleep and ask, "Were you dreaming?" she will most likely say, "No." However, if we question her more carefully, she might report the presence of a thought, an image, or some emotion.

During the rest of the night our subject's sleep alternates between periods of REM and **non-REM sleep.** Each cycle is approximately 90 minutes long, containing a 20- to 30-minute bout of REM sleep. Thus, an 8-hour sleep will contain four or five periods of REM sleep. Figure 9.4 shows a graph of a typical night's sleep. The vertical axis indicates the EEG activity that is being recorded; thus, REM sleep and stage 1 sleep are placed on the same line because similar patterns of EEG activity occur at these times. Note that most slow-wave sleep (stages 3 and 4) occurs during the first half of night. Subsequent bouts of non-REM sleep contain more and more stage 2 sleep, and bouts of REM sleep (indicated by the horizontal bars) become more prolonged. (See *Figure 9.4.*)

The fact that REM sleep occurs at regular 90-minute intervals suggests that a brain mechanism alternately causes REM and slow-wave sleep. Normally, a period of slow-wave sleep must precede REM sleep. In addition, there seems to be a minimum amount of time after each occurrence of REM sleep during which time REM sleep cannot take place again. In fact, the cyclical nature of REM sleep appears to be controlled by a "clock" in the brain that also controls an activity cycle that continues through waking.

As we saw, during REM sleep we become paralyzed; most of our spinal and cranial motor neurons are strongly inhibited. (Obviously, the ones that control respiration and eye movements are spared.) At the same time the brain is very active. Cerebral blood flow

FIGURE 9.4 ■ Sleep Stages During a Single Night

In this typical pattern of sleep stages, the dark blue shading indicates REM sleep.

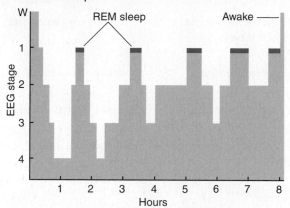

and oxygen consumption are accelerated. In addition, during most periods of REM sleep a male's penis will become at least partially erect, and a female's vaginal secretions will increase. However, Fisher, Gross, and Zuch (1965) found that in males, genital changes do not signify that the person is experiencing a dream with sexual content. (Of course, people can have dreams with frank sexual content. In males some dreams culminate in ejaculation—the so-called nocturnal emissions, or "wet dreams." Females, too, sometimes experience orgasm during sleep.)

The fact that penile erections occur during REM sleep, independent of sexual arousal, has been used clinically to assess the causes of impotence (Karacan, Salis, and Williams, 1978; Singer and Weiner, 1996). A subject sleeps in the laboratory with a device attached to his penis that measures its circumference. If penile enlargement occurs during REM sleep, then his failure to obtain an erection during attempts at intercourse is not caused by physiological problems such as nerve damage or a circulatory disorder. (A neurologist told me that there is a less expensive way to gather the same data: The patient obtains a strip of postage stamps, moistens them, and applies them around his penis before going to bed. In the morning he checks to see whether the perforations are broken.)

The important differences between REM sleep and slow-wave sleep are listed in *Table 9.1.*

REM sleep A period of desynchronized EEG activity during sleep, at which time dreaming, rapid eye movements, and muscular paralysis occur; also called *paradoxical sleep.*

non-REM sleep All stages of sleep except REM sleep.

TABLE 9.1 ■ Principal Characteristics of REM and Slow-Wave Sleep	
REM SLEEP	**SLOW-WAVE SLEEP**
EEG desynchrony (rapid, irregular waves)	EEG synchrony (slow waves)
Lack of muscle tonus	Moderate muscle tonus
Rapid eye movements	Slow or absent eye movements
Penile erection or vaginal secretion	Lack of genital activity
Dreams	

Mental Activity During Sleep

Although sleep is a period during which we do not respond very much to the environment, it is incorrect to refer to sleep as a state of unconsciousness. Consciousness during sleep certainly differs from waking consciousness, but we *are* conscious then. In the morning we usually forget what we experienced while asleep, so in retrospect we conclude that we were unconscious. However, when experimenters wake sleeping subjects, the reports that the subjects give make it clear that they were conscious.

Researchers have found that the rate of cerebral blood flow in the human brain during REM sleep is high in the visual association cortex but low in the primary visual cortex and the prefrontal cortex (Madsen et al., 1991; Braun et al., 1998). The lack of activity in the primary visual cortex reflects the fact that the eyes are not receiving visual input; the high level of activity in the visual association cortex undoubtedly reflects the visual hallucinations that occur during dreams. As we shall see in Chapter 13, the prefrontal cortex is involved in making plans, keeping track of the organization of events in time, and distinguishing illusion from reality. As Madsen and his colleagues noted, dreams are characterized by good visual images but are poorly organized with respect to time; for example, past, present, and future are often interchanged (Hobson, 1988). And as Melges (1982) put it, "the dreamer often has no feeling of striving for long-term goals but rather is carried along by the flow of time by circumstances that crop up in an unpredictable way." This quote could just as well be describing the daily life of a person whose prefrontal cortex has been damaged.

Several investigators have suggested that the eye movements made during REM sleep are related to the visual imagery that occurs while we dream. Roffwarg et al. (1962) recorded the eye movements of subjects during REM sleep and then awakened the subjects and asked them to describe what had been happening in their dreams. They found that the eye movements were similar to what would have been expected if the subjects had actually been watching these events. In addition, evidence indicates that the particular brain mechanisms that become active during a dream are those that would become active if the events in the dream were actually occurring. For example, cortical and subcortical motor mechanisms become active during a dream that contains movement, as if the person were actually moving (McCarley and Hobson, 1979). In addition, if a dream involves talking and listening, regions of the dreamer's brain that are involved in speaking and listening become especially active (Hong et al., 1996). (Brain mechanisms of verbal communication are discussed in Chapter 15.)

Although narrative, storylike dreaming most often occurs during REM sleep, mental activity can also accompany slow-wave sleep. Some of the most terrifying nightmares occur during slow-wave sleep, especially stage 4 sleep (Fisher et al., 1970). In French the word for nightmare is *cauchemar,* or "pressing devil." Figure 9.5 shows a victim of a nightmare (undoubtedly in the throes of stage 4 slow-wave sleep) being squashed by an *incubus* (from the Latin *incubare,* "to lie upon"). (See *Figure 9.5.*)

FIGURE 9.5 ■ The Nightmare, 1781, by Henry Fuseli, Swiss, 1741–1825

(Gift of Mr. and Mrs. Bert L. Smokler and Mr. and Mrs. Lawrence A. Fleischman, Acc. No. 55.5. Courtesy of the Detroit Institute of Arts.)

InterimSummary

A Physiological and Behavioral Description of Sleep

Sleep is generally regarded as a state, but it is nevertheless a behavior. The stages of non-REM sleep, stages 1–4, are defined by EEG activity. Slow-wave sleep (stages 3 and 4) includes the two deepest stages. Alertness consists of desynchronized beta activity (13–30 Hz); relaxation and drowsiness consist of alpha activity (8–12 Hz); stage 1 sleep consists of alternating periods of alpha activity, irregular fast activity, and theta activity (3.5–7.5 Hz); the EEG of stage 2 sleep lacks alpha activity but contains sleep spindles (short periods of 12–14 Hz activity) and occasional K complexes; stage 3 sleep consists of 20–50 percent delta activity (less than 3.5 Hz); and stage 4 sleep consists of more than 50 percent delta activity. Waves of delta activity are organized around the sharp slow waves of less than 1 Hz that begin with a silent down state followed by an active up state. About 90 minutes after the beginning of sleep, people enter REM sleep. Cycles of REM and slow-wave sleep alternate in periods of approximately 90 minutes.

REM sleep consists of rapid eye movements, a desynchronized EEG, sensitivity to external stimulation, muscular paralysis, genital activity, and dreaming. Mental activity can accompany slow-wave sleep too, but most narrative dreams occur during REM sleep.

Thought Questions

1. Have you ever been resting quietly and suddenly heard someone tell you that you had obviously been sleeping because you were snoring? Did you believe them, or were you certain that you were really awake? Do you think it was likely that you had actually entered stage 1 sleep?

2. What is accomplished by dreaming? Some researchers believe that the subject matter of a dream does not matter; it is the REM sleep itself that is important. Others believe that the subject matter *does* count. Some researchers believe that if we remember a dream, then the dream failed to accomplish all of its functions; others say that remembering dreams is useful, because it can give us some insights into our problems. What do you think of these controversies?

3. Some people report that they are "in control" of some of their dreams—that they feel as if they determine what comes next and are not simply swept along passively. Have you ever had this experience? Have you ever had a "lucid dream," in which you aware of the fact that you were dreaming?

DISORDERS OF SLEEP

Because we spend about one-third of our lives sleeping, sleep disorders can have a significant impact on our quality of life. They can also affect the way we feel while we are awake.

Insomnia

Insomnia is a problem that is said to affect approximately 25 percent of the population occasionally and 9 percent regularly (Ancoli-Israel and Roth, 1999). But we need to define *insomnia* carefully. First, there is no single definition of insomnia that can apply to all people. The amount of sleep that individuals require is quite variable. A short sleeper may feel fine with 5 hours; a long sleeper may still feel unrefreshed after 10 hours of sleep. Insomnia must be defined in relation to a person's particular sleep needs.

The second consideration in defining insomnia is the unreliability of self-reports. Most patients who receive a prescription for a sleeping medication are given one on the basis of their own description of their symptoms. That is, they tell their physician that they sleep very little at night, and the drug is prescribed on the basis of this testimony. Very few patients are observed during a night's sleep in a sleep laboratory; thus, insomnia is one of the few medical problems that physicians treat without having direct clinical evidence for its existence. But studies on the sleep of people who complain of insomnia show that most of them grossly underestimate the amount of time they actually sleep. In fact, Rosa and Bonnet (2000) evaluated in a sleep laboratory the sleep of people who complained of insomnia and people who did not and found no differences between the two groups. They *did*, however, find personality differences, which could account for the complaints.

For many years the goal of sleeping medication was to help people fall asleep, and when drug companies evaluated potential medications, they concentrated on that property. However, if we think about the ultimate goal of sleeping medication, it is to make the person feel more refreshed the next day. If a medication puts people to sleep right away but produces a hangover of grogginess and difficulty concentrating the next day, it is worse than useless. In fact, many drugs that are traditionally used to treat insomnia had just this effect. More recently, researchers have recognized that the true evaluation of a sleeping medication must be made during wakefulness the following day, and "hangover-free" drugs are finally being developed (Hajak et al., 1995; Ramakrishnan and Scheid, 2007).

A particular form of insomnia is caused by an inability to sleep and breathe at the same time. Patients with this disorder, called **sleep apnea,** fall asleep and then cease to breathe. (Nearly all people, especially people who snore, have occasional episodes of sleep apnea, but not to the extent that it interferes with sleep.) During a period of sleep apnea the level of carbon dioxide in the blood stimulates chemoreceptors (neurons that detect the presence of certain chemicals), and the person wakes up, gasping for air. The oxygen level of the blood returns to normal, the person falls asleep, and the whole cycle begins again. Because sleep is disrupted, people with this disorder typically feel sleepy and groggy during the day. Fortunately, many cases of sleep apnea are caused by an obstruction of the airway that can be corrected surgically or relieved by a device that attaches to the sleeper's face and provides pressurized air that keeps the airway open (Sher, 1990; Piccirillo, Duntley, and Schotland, 2000).

Research on the effects of severe sleep apnea suggest that diagnosis and treatment of this disorder should begin as soon as possible. Although patients who have been successfully treated for obstructive sleep apnea normally feel less sleepy during the day, many of them report that some sleepiness remains. The findings of a study by Zhu et al. (2007) suggest that the periods of anoxia that occur during sleep apnea may damage neurons that play an important role in wakefulness and alertness. Zhu and his colleagues found that exposing mice to brief, intermittent periods of low atmospheric oxygen caused degeneration of noradrenergic and dopaminergic neurons in the midbrain and pons.

Narcolepsy

Narcolepsy (*narke* means "numbness," and *lepsis* means "seizure") is a neurological disorder characterized by sleep (or some of its components) at inappropriate times (Nishino, 2007). The symptoms can be described in terms of what we know about the phenomena of sleep. The primary symptom of narcolepsy is the **sleep attack.** The narcoleptic sleep attack is an overwhelming urge to sleep that can happen at any time but occurs most often under monotonous, boring conditions. Sleep (which appears to be entirely normal) generally

lasts for 2–5 minutes. The person usually wakes up feeling refreshed.

Another symptom of narcolepsy—in fact, the most striking one—is **cataplexy** (from *kata*, "down," and *plexis*, "stroke"). During a cataplectic attack, a person will sustain varying amounts of muscle weakness. In some cases the person will become completely paralyzed and slump down to the floor. The person will lie there, *fully conscious,* for a few seconds to several minutes. What apparently happens is that one of the phenomena of REM sleep—muscular paralysis—occurs at an inappropriate time. As we saw, this loss of tonus is caused by massive inhibition of motor neurons in the spinal cord. When this happens during waking, the victim of a cataplectic attack loses control of his or her muscles. (As in REM sleep, the person continues to breathe and is able to control eye movements.)

Cataplexy is quite different from a narcoleptic sleep attack; cataplexy is usually precipitated by strong emotional reactions or by sudden physical effort, especially if the patient is caught unawares. Laughter, anger, or an effort to catch a suddenly thrown object can trigger a cataplectic attack. In fact, as Guilleminault, Wilson, and Dement (1974) noted, even people who do not have cataplexy sometimes lose muscle strength after a bout of intense laughter. (Perhaps that is why we say a person can become "weak from laughter.") Common situations that bring on cataplexy are attempting to discipline one's children and making love (an unfortunate time to become paralyzed!). Michael, the man described in the opener to this chapter, had his first cataplectic attack when he was addressing the board of directors of the company he worked for. Wise (2004) notes that patients with narcolepsy often try to avoid thoughts and situations that are likely to evoke strong emotions because they know that these emotions are likely to trigger cataplectic attacks.

REM sleep paralysis sometimes intrudes into waking at a time that does not present any physical danger—just before or just after normal sleep, when a person is already lying down. This symptom of narcolepsy is referred to as **sleep paralysis,** an inability to move just before the onset of sleep or upon waking in the morning. A person can be snapped out of sleep paralysis by being touched or by hearing someone call his or her

---

sleep apnea (*app nee a*) Cessation of breathing while sleeping.

narcolepsy (*nahr ko lep see*) A sleep disorder characterized by periods of irresistible sleep, attacks of cataplexy, sleep paralysis, and hypnagogic hallucinations.

sleep attack A symptom of narcolepsy; an irresistible urge to sleep during the day, after which the person awakens feeling refreshed.

cataplexy (*kat a plex ee*) A symptom of narcolepsy; complete paralysis that occurs during waking.

sleep paralysis A symptom of narcolepsy; paralysis occurring just before a person falls asleep.

FIGURE 9.6 ■ A Dog Undergoing a Cataplectic Attack

The attack was triggered by the dog's excitement at finding some food on the floor.
(a) The dog sniffs the food. (b) Muscles begin to relax. (c) The dog is temporarily paralyzed,
as it would be during REM sleep.

(Photos courtesy of the Sleep Disorders Foundation, Stanford University.)

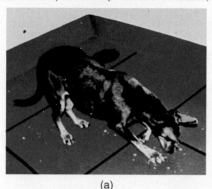
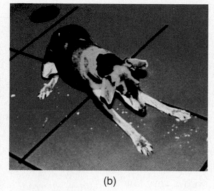
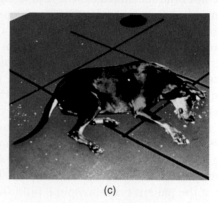

(a)   (b)   (c)

name. Sometimes, the mental components of REM sleep intrude into sleep paralysis; that is, the person dreams while lying awake, paralyzed. These episodes, called **hypnagogic hallucinations,** are often alarming or even terrifying. (The term *hypnagogic* comes from the Greek words *hupnos,* "sleep," and *agogos,* "leading.") During a hypnagogic hallucination, Michael thought that his former roommate was trying to attack him with a hammer.

Narcolepsy is produced by a brain abnormality that disrupts the neural mechanisms that control various aspects of sleep and arousal. As we saw, narcoleptic patients have difficulty staying awake, and aspects of REM sleep intrude into the waking state. In addition, they often skip the slow-wave sleep that normally begins a night's sleep and go directly into REM sleep from waking. Finally, their sleep is often fragmented—disrupted by periods of wakefulness.

Fortunately, human narcolepsy is relatively rare, with an incidence of approximately one in 2000 people. This hereditary disorder appears to involve a gene found on chromosome 6, but it is strongly influenced by unknown environmental factors (Mignot, 1998; Mahowald and Schenck, 2005; Nishino, 2007). Years ago, researchers began a program to maintain breeds of dogs that are afflicted with narcolepsy, with the hopes that discovery of the causes of canine narcolepsy would further our understanding of the causes of human narcolepsy. (See *Figure 9.6.*) Eventually, this research paid off. Lin et al. (1999) discovered that a mutation of a specific gene is responsible for canine narcolepsy. The product of this gene is a receptor for a peptide neurotransmitter called *hypocretin* by some researchers and *orexin* by others. The name "hypocretin" comes from the fact that the lateral *hypo*thalamus contains the cell bodies of all of the neurons that se*crete* this peptide. The

name "orexin" comes from the role this peptide plays in the control of eating and metabolism, which are discussed in more detail in Chapter 12. (*Orexis* means "appetite" in Greek.) Most researchers appear to have settled on the word **orexin,** so I will use this term also. There are two orexin receptors. Lin and his colleagues discovered that the mutation responsible for canine narcolepsy involves the orexin B receptor.

Chemelli et al. (1999) prepared a targeted mutation in mice against the orexin gene and found that the animals showed symptoms of narcolepsy. Like human patients with narcolepsy, they went directly into REM sleep from waking and showed periods of cataplexy while they were awake. (Videos of narcoleptic dogs, mice, and people are shown in *MyPsychKit 9.1, Narcolepsy.*) Gerashchenko et al. (2001, 2003) prepared a toxin that attacked only orexinergic neurons and administered it to rats. The destruction of the orexin system produced the symptoms of narcolepsy.

mypsychkit
Where learning comes to life!
Animation 9.1
Narcolepsy

Loss of orexinergic neurons is the cause of narcolepsy in humans as well. Nishino et al. (2000) performed an analysis of the cerebrospinal fluid of normal subjects and patients with narcolepsy. They found a complete absence of orexin in seven of the nine narcoleptic

---

hypnagogic hallucination *(hip na **gah** jik)* A symptom of narcolepsy; vivid dreams that occur just before a person falls asleep; accompanied by sleep paralysis.

orexin A peptide, also known as *hypocretin,* produced by neurons whose cell bodies are located in the hypothalamus; their destruction causes narcolepsy.

patients. They hypothesized that the cause of narcolepsy in these seven patients was a hereditary disorder that caused the immune system to attack and destroy orexin-secreting neurons. Most patients with narcolepsy are born with orexinergic neurons, but during adolescence the immune system attacks these neurons, and the symptoms of narcolepsy begin. The narcolepsy that is seen in two patients with high levels of orexin may have been caused by a mutation of a gene responsible for production of the orexin B receptor—the same mutation that causes canine narcolepsy. Peyron et al. (2002) reported the case of one patient with early onset narcolepsy (before age 2) with a different genetic defect: mutation of the gene responsible for the production of orexin. (See *MyPsychKit 9.1, Narcolepsy.*) Scammell et al. (2001) reported the case of a patient who developed narcolepsy after a stroke that damaged the hypothalamus. An analysis of the patient's cerebrospinal fluid showed a very low level of orexin; apparently, the stroke had damaged the patient's orexinergic neurons.

The symptoms of narcolepsy can be treated with drugs. Sleep attacks can be diminished by stimulants such a methylphenidate (Ritalin), a catecholamine agonist (Vgontzas and Kales, 1999). The REM sleep phenomena (cataplexy, sleep paralysis, and hypnagogic hallucinations) can be alleviated by antidepressant drugs, which facilitate both serotonergic and noradrenergic activity (Mitler, 1994; Hublin, 1996). As we will see in Chapter 16, abnormalities in patterns of REM sleep are seen in people suffering from depression. The fact that drugs that reduce depression also suppress the phenomena of REM sleep is probably not coincidental.

More recently, modafinil, a stimulant drug whose precise site of action is still unknown, has been used to treat narcolepsy (Fry, 1998; Nishino, 2007). (Michael, the man discussed in the opener to this chapter, is now taking this drug.) A study by Scammell et al. (2000) suggests that modafinil acts, directly or indirectly, on orexinergic neurons. The investigators found that administration of modafinil increased the expression of Fos protein in orexinergic neurons, which indicates that the neurons had been activated.

The connections of orexinergic neurons with other regions of the brain are discussed later in this chapter.

REM Sleep Behavior Disorder

Several years ago, Schenck et al. (1986) reported the existence of an interesting disorder: **REM sleep behavior disorder.** As you now know, REM sleep is accompanied by paralysis. Although the motor cortex and subcortical motor systems are extremely active during REM sleep (McCarley and Hobson, 1979), people are unable to move at this time.

The fact that people are paralyzed while they dream suggests the possibility that but for the paralysis, they would act out their dreams. Indeed, they would. The behavior of people who exhibit REM sleep behavior disorder corresponds with the contents of their dreams. Consider the following case:

> I was a halfback playing football, and after the quarterback received the ball from the center he lateraled it sideways to me and I'm supposed to go around end and cut back over tackle and—this is very vivid—as I cut back over tackle there is this big 280-pound tackle waiting, so I, according to football rules, was to give him my shoulder and bounce him out of the way . . . when I came to I was standing in front of our dresser and I had [gotten up out of bed and run and] knocked lamps, mirrors and everything off the dresser, hit my head against the wall and my knee against the dresser. (Schenck et al., 1986, p. 294)

Like narcolepsy, REM sleep behavior disorder appears to be a neurodegenerative disorder with at least some genetic component (Schenck et al., 1993). It is often associated with better-known neurodegenerative disorders such as Parkinson's disease and multiple system atrophy (Boeve et al., 2007). These disorders are called *α-synucleinopathies* because they involve the inclusion of α-synuclein protein in degenerating neurons. In addition, REM sleep behavior disorder can be caused by brain damage—in some cases to the neural circuits in the brain stem that control the phenomena of REM sleep, which will be discussed later in this chapter (Culebras and Moore, 1989). The symptoms of REM sleep behavior disorder are the opposite of those of cataplexy; that is, rather than exhibiting paralysis outside REM sleep, patients with REM sleep behavior disorder *fail* to exhibit paralysis *during* REM sleep. As you might expect, the drugs that are used to treat the symptoms of cataplexy will aggravate the symptoms of REM sleep behavior disorder (Schenck and Mahowald, 1992). REM sleep behavior disorder is usually treated by clonazepam, a benzodiazepine (Schenck, Hurwitz, and Mahowald, 1993).

Problems Associated with Slow-Wave Sleep

Some maladaptive behaviors occur during slow-wave sleep, especially during its deepest phase, stage 4. These behaviors include bedwetting (*nocturnal enuresis*), sleepwalking (*somnambulism*), and night terrors (*pavor nocturnus*). All three events occur most frequently in

---

REM sleep behavior disorder A neurological disorder in which the person does not become paralyzed during REM sleep and thus acts out dreams.

children. Often bedwetting can be cured by training methods, such as having a special electronic circuit ring a bell when the first few drops of urine are detected in the bed sheet (a few drops usually precede the ensuing flood). Night terrors consist of anguished screams, trembling, a rapid pulse, and usually no memory of what caused the terror. Night terrors and somnambulism usually cure themselves as the child gets older. Neither of these phenomena is related to REM sleep; a sleepwalking person is *not* acting out a dream. Especially when it occurs in adulthood, sleepwalking appears to have a genetic component (Hublin et al., 1997).

Sometimes, people can engage in complex behaviors while sleepwalking. Consider the following cases:

> One evening Ed Weber got up from a nap on the sofa, polished off a half-gallon of chocolate chip ice cream, then dozed off again. He woke up an hour later and went looking for the ice cream, summoning his wife to the kitchen and insisting, to her astonishment, that someone else must have eaten it.
>
> [T]elevision talk show host Montel Williams . . . told viewers he had removed raw foods from his refrigerator because "I wake up in the morning and there's a pack of chicken and there's a bite missing out of it . . . . I can take a whole pound of ham or bologna . . . and then wake up in the morning and not realize that I had [eaten] it and ask, 'Who ate my lunch meat?'" (Boodman, 2004, p. HE01)

Schenck et al. (1991) reported nineteen cases of people with histories of eating during the night while they were asleep, which they labeled **sleep-related eating disorder.** Almost half of the patients had become overweight from night eating. Once patients realize that they are eating in their sleep, they often employ stratagems such as keeping their food under lock and key or setting alarms that will awaken them when they try to open the refrigerator.

Sleep-related eating disorder usually responds well to dopaminergic agonists, anxiolytic drugs, or antianxiety drugs. An increased incidence of nocturnal eating in family members of people with this disorder suggests that heredity may play a role (De Ocampo et al., 2002). Preliminary evidence also suggests that nocturnal eating may also be associated with the use of some sleeping medications used to treat insomnia (Morgenthaler and Silber, 2002; Najjar, 2007).

sleep-related eating disorder A disorder in which the person leaves his or her bed and seeks out and eats food while sleepwalking, usually without a memory for the episode the next day.

Interim**Summary**

Disorders of Sleep

Although many people believe that they have insomnia—that they do not obtain as much sleep as they would like—insomnia is not a disease. Insomnia can be caused by depression, pain, illness, or even excited anticipation of a pleasurable event. Only recently have sleeping medications been developed that do not cause a hangover of grogginess and difficulty concentrating the next day. Sometimes, insomnia is caused by sleep apnea, which can often be corrected surgically or treated by wearing a mask that delivers pressurized air.

Narcolepsy is characterized by four symptoms. *Sleep attacks* consist of overwhelming urges to sleep for a few minutes. *Cataplexy* is sudden paralysis, during which the person remains conscious. *Sleep paralysis* is similar to cataplexy, but it occurs just before sleep or on waking. *Hypnagogic hallucinations* are dreams that occur during periods of sleep paralysis, just before a night's sleep. Sleep attacks are treated with stimulants such as amphetamine, and the other symptoms are treated with serotonin agonists. Studies with narcoleptic dogs and humans indicate that this disorder is caused by pathologies in a system of neurons that secrete a neuropeptide known as orexin (also known as hypocretin). REM sleep behavior disorder is a neurodegenerative disease that damages brain mechanisms that produce paralysis during REM sleep. As a result, the patient acts out his or her dreams.

During slow-wave sleep, especially during stage 4, some people are afflicted by bedwetting (nocturnal enuresis), sleepwalking (somnambulism), or night terrors (pavor nocturnus). These problems are most common in children, who usually outgrow them. People with sleep-related eating disorder seek and consume food while sleepwalking.

Thought Question

Suppose you were spending the night at a friend's house and, hearing a strange noise during the night, got out of bed and found your friend walking around, still asleep. How would you tell whether your friend was sleepwalking or had REM sleep behavior disorder?

WHY DO WE SLEEP?

We all know how insistent the urge to sleep can be and how uncomfortable we feel when we have to resist it and stay awake. With the exception of the effects of severe pain and the need to breathe, sleepiness is probably the most insistent drive that we can experience. People can commit suicide by refusing to eat or drink, but even the most stoical person cannot indefinitely defy the urge to sleep. Sleep will come, sooner or later, no matter how hard a person tries to stay awake. Although the issue is not yet settled, most researchers believe that the primary function of slow-wave sleep is to permit the brain to rest. In addition, slow-wave sleep and REM sleep promote different types of learning, and REM sleep appears to promote brain development.

Functions of Slow-Wave Sleep

Sleep is a universal phenomenon among vertebrates. As far as we know, all mammals and birds sleep (Durie, 1981). Reptiles also sleep, and fish and amphibians enter periods of quiescence that probably can be called sleep. However, only warm-blooded vertebrates (mammals and birds) exhibit unequivocal REM sleep, with muscular paralysis, EEG signs of desynchrony, and rapid eye movements. Obviously, birds such as flamingos, which sleep while perched on one leg, do not lose tone in the muscles they use to remain standing. Also, animals such as moles, which move their eyes very little while awake, show few signs of eye movement while asleep. The functions of REM sleep will be discussed separately, in a later section.

Sleep appears to be essential to survival. Evidence for this assertion comes from the fact that sleep is found in some species of mammals that would seem to be better off without it. For example, the Indus dolphin (*Platanista indi*) lives in the muddy waters of the Indus estuary in Pakistan (Pilleri, 1979). Through the process of evolution it has become blind, presumably because vision is not useful in the animal's environment. (It has an excellent sonar system, which it uses to navigate and find prey.) However, despite the dangers caused by sleeping, sleep has not disappeared in this species. The Indus dolphin never stops swimming; doing so would result in injury, because of the dangerous currents and the vast quantities of debris carried by the river during the monsoon season. Pilleri captured two dolphins and studied their habits. He found that they slept a total of 7 hours a day, in brief naps of 4–60 seconds each. If sleep were simply an adaptive response, why was it not eliminated (as vision was) through the process of natural selection?

Some other species of marine mammals have developed an extraordinary pattern of sleep: The cerebral hemispheres take turns sleeping, presumably because that strategy always permits at least one hemisphere to be

FIGURE 9.7　■　Sleep in a Dolphin

The two hemispheres sleep independently, presumably so that the animal remains behaviorally alert.

(Adapted from Mukhametov, L. M., in *Sleep Mechanisms,* edited by A. A. Borbély and J. L. Valatx. Munich: Springer-Verlag, 1984.)

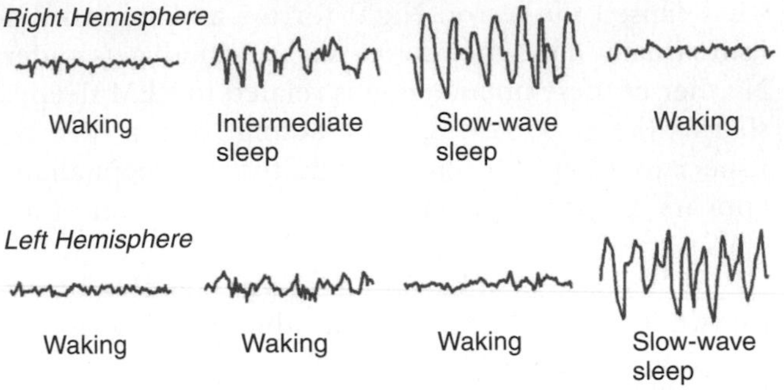

alert. In addition, the eye contralateral to the active hemisphere remains open. The bottlenose dolphin (*Tursiops truncatus*) and the porpoise (*Phocoena phocoena*) both sleep this way (Mukhametov, 1984). Figure 9.7 shows the EEG recordings from the two hemispheres; note that slow-wave sleep occurs independently in the left and right hemispheres. (See *Figure 9.7.*)

Effects of Sleep Deprivation

When we are forced to miss a night's sleep, we become very sleepy. The fact that sleepiness is so motivating suggests that sleep is a necessity of life. If so, it should be possible to deprive people or laboratory animals of sleep and see what functions are disrupted. We should then be able to infer the role that sleep plays. However, the results of sleep deprivation studies have not revealed as much as investigators had originally hoped.

Studies with Humans. Deprivation studies with human subjects have not obtained persuasive evidence that sleep is needed to keep the body functioning normally. Horne (1978) reviewed over fifty experiments in which people had been deprived of sleep. He reported that most of them found that sleep deprivation did not interfere with people's ability to perform physical exercise. In addition, the studies found no evidence of a physiological stress response to sleep deprivation. Thus, the primary role of sleep does not seem to be rest and recuperation of the *body*. However, people's cognitive abilities were affected; some people reported perceptual distortions or even hallucinations and had trouble concentrating on mental tasks. Perhaps sleep provides the opportunity for the brain to rest.

What happens to sleep-deprived subjects after they are permitted to sleep again? Most of them sleep longer the next night or two, but they never regain all of the sleep they lost. In one remarkable case a seventeen-year-old boy stayed awake for 264 hours so that he could obtain a place in the *Guinness Book of World Records* (Gulevich, Dement, and Johnson, 1966). After his ordeal the boy slept for a little less than 15 hours and awoke feeling fine. He slept slightly more than 10 hours the second night and just under 9 hours the third. Almost 67 hours were never made up. However, percentages of recovery were not equal for all stages of sleep. Only 7 percent of stages 1 and 2 were made up, but 68 percent of stage 4 slow-wave sleep and 53 percent of REM sleep were made up. Other studies (for example, Kales et al., 1970) have found similar results, suggesting that stage 4 sleep and REM sleep are more important than the other stages.

What do we know about the possible functions of slow-wave sleep? What happens then that is so important? Both cerebral metabolic rate and cerebral blood flow decline during slow-wave sleep, falling to about 75 percent of the waking level during stage 4 sleep (Sakai et al., 1979; Buchsbaum et al., 1989; Maquet, 1995). In particular, the regions that have the highest levels of activity during waking show the highest levels of delta waves—and the lowest levels of metabolic activity—during slow-wave sleep. Thus, the presence of slow-wave activity in a particular region of the brain appears to indicate that that region is resting. As we know from behavioral observation, people are unreactive to all but intense stimuli during slow-wave sleep and, if awakened, act groggy and confused, as if their cerebral cortex has been shut down and has not yet resumed its functioning. In addition, several studies have shown that missing a single night's sleep impairs people's cognitive abilities; presumably, the brain needs sleep to function at peak efficiency (Harrison and Horne, 1998, 1999) These observations suggest that during stage 4 sleep the brain is indeed resting.

The available evidence does indeed suggest that the brain needs to rest periodically to recuperate from adverse side effects of its waking activity. But what is the nature of these adverse effects? Siegel (2005) suggests that one of the waste products produced by the high metabolic rate associated with waking activity of the brain are free radicals, chemicals that contain at least one unpaired electron. Free radicals are highly reactive oxidizing agents; they can bind with electrons from other molecules and damage the cells in which they are found, a process known as *oxidative stress*. During slow-wave sleep the lowered rate of metabolism permits restorative mechanisms in the cells to destroy the free radicals and prevent their damaging effects. Indeed, Ramanathan et al. (2002) found evidence that prolonged sleep deprivation caused an increase in free radicals in the brains of rats and caused oxidative stress.

An inherited neurological disorder called **fatal familial insomnia** results in damage to portions of the thalamus (Sforza et al., 1995; Gallassi et al., 1996; Montagna et al., 2003). The symptoms of this disease, which is related to Creuzfeldt-Jacob disease and bovine spongiform encephalopathy ("mad cow disease"), include deficits in attention and memory, followed by a dreamlike, confused state; loss of control of the autonomic nervous system and the endocrine system; increased body temperature; and insomnia. The first signs of sleep disturbances are reductions in sleep spindles and K complexes. As the disease progresses, slow-wave sleep completely disappears, and only brief episodes of REM sleep (without the accompanying paralysis) remain. As the name indicates, the disease is fatal. Whether the insomnia, caused by the brain damage, contributes to the other symptoms and to the patient's death is not known. However, as we shall see in the next subsection, when laboratory animals are kept awake indefinitely, they too will die.

Schenkein and Montagna (2006a, 2006b) describe the case of a man diagnosed with a form of fatal familial insomnia that usually causes death within 12 months. Because several relatives had died of this disorder, the man knew what to expect, and he enlisted the aid of several physicians to administer drugs and treatments designed to help him sleep. For several months the treatments did help him to sleep, and the man survived about a year longer than would have been expected. Further studies will be needed to determine whether his increased survival time was a direct result of the increased sleep. In any event, his quality of life during most of the period of his illness was much improved.

Studies with Laboratory Animals. Rechtschaffen and his colleagues (Rechtschaffen et al., 1983, 1989; Rechtschaffen and Bergmann, 1995, 2002) devised a procedure to keep rats from sleeping without forcing them to exercise continuously. (A group of control animals exercised just as much as the experimental subjects but were able to engage in normal amounts of sleep.)

Sleep deprivation had serious effects. The control animals remained in perfect health. However, the experimental animals looked sick and stopped grooming their fur. They became weak and uncoordinated and lost their ability to regulate their body temperature. Although they began eating much more food than normal, their metabolic rates became so high that they continued to lose weight. Eventually, the rats died. The cause of death is still not understood. The rats' brains appeared to be normal, and there were no obvious signs of inflammation or damage to other internal organs. The animals' levels of stress hormones were not unusually high, so the deaths could not be attributed to simple stress. If they were given a high-calorie diet to compensate for their increased metabolic rate, the rats lived longer, but eventually they succumbed (Everson and Wehr, 1993). As we

fatal familial insomnia A fatal inherited disorder characterized by progressive insomnia.

will see later in this chapter, damage to parts of the brain cause insomnia, and if the animals do not recover from their sleeplessness, they die.

As we just saw, the symptoms of the neurodegenerative disorder called fatal familial insomnia resembles the effects of forced sleep deprivation in rats. Budka et al. (1998) reported another similarity. The investigators studied five people with fatal familial insomnia who, along with insomnia, memory loss, and autonomic dysfunction, exhibited prominent weight loss.

Effects of Exercise on Slow-Wave Sleep

Sleep deprivation studies with humans suggest that the brain may need slow-wave sleep in order to recover from the day's activities. Another way to determine whether sleep is needed for restoration of physiological functioning is to look at the effects of daytime activity on nighttime sleep. If the function of sleep is to repair the effects on the body of physical activity during waking hours, then we should expect that sleep and exercise are related. That is, we should sleep more after a day of vigorous exercise than after a day spent quietly at an office desk.

However, the relationship between sleep and exercise is not very compelling. For example, Ryback and Lewis (1971) found no changes in slow-wave or REM sleep of healthy subjects who spent six weeks resting in bed. If sleep repairs wear and tear, we would expect these people to sleep less. Adey, Bors, and Porter (1968) studied the sleep of almost completely immobile quadriplegics and paraplegics and found only a small decrease in slow-wave sleep as compared with uninjured people. Thus, although sleep certainly provides the body with rest, its primary function appears to be something else.

Effects of Brain Activity on Slow-Wave Sleep

If the primary function of slow-wave sleep is to permit the brain to rest and recover from its daily activity, then we might expect that a person would spend more time in slow-wave sleep after a day of intense cerebral activity. Tasks that demand alertness and mental activity *do* increase glucose metabolism in the brain (Roland, 1984). The most significant increases are seen in the frontal lobes, where slow-wave activity is most intense during non-REM sleep. In an experiment that supports this interpretation, Huber et al. (2004) had people perform a motor learning task just before going to sleep. The task required to subjects to make hand movements whose directions were indicated by a visual display. During sleep the subjects showed increased slow-wave activity in the region of the neocortex that became active while they were performing the task. Presumably, the increased activity of these cortical neurons called for more rest during the following night's sleep. A follow-up study by

Huber et al. (2006) found that immobilizing a person's arm for 12 hours produced the opposite result: During sleep the person showed *less* slow-wave activity in the regions of the neocortex that received somatosensory information from that arm and controlled its movements.

In an ingenious study, Horne and Minard (1985) found a way to increase mental activity without affecting physical activity and without causing stress. The investigators told subjects to show up for an experiment in which they were supposed to take some tests designed to measure reading skills. When the subjects turned up, however, they were told that the plans had been changed. They were invited for a day out, at the expense of the experimenters. (Not surprisingly, the subjects willingly accepted.) They spent the day visiting an art exhibition, a shopping center, a museum, an amusement park, a zoo, and an interesting mansion. After a scenic drive through the countryside they watched a movie in a local theater. They were driven from place to place and certainly did not become overheated by exercise. After the movie they returned to the sleep laboratory. They said they were tired, and they readily fell asleep. Their sleep duration was normal, and they awoke feeling refreshed. However, their slow-wave sleep—particularly stage 4 sleep—was increased. After all that mental exercise, the brain appears to have needed more rest than usual.

Functions of REM Sleep

Clearly, REM sleep is a time of intense physiological activity. The eyes dart about rapidly, the heart rate shows sudden accelerations and decelerations, breathing becomes irregular, and the brain becomes more active. It would be unreasonable to expect that REM sleep has the same functions as slow-wave sleep. An early report on the effects of REM sleep deprivation (Dement, 1960) observed that as the deprivation progressed, subjects had to be awakened from REM sleep more frequently; the "pressure" to enter REM sleep built up. Furthermore, after several days of REM sleep deprivation, subjects would show a **rebound phenomenon** when permitted to sleep normally; they spent a much greater-than-normal percentage of the recovery night in REM sleep. This rebound suggests that there is a need for a certain amount of REM sleep—that REM sleep is controlled by a regulatory mechanism. If selective deprivation causes a deficiency in REM sleep, the deficiency is made up later, when uninterrupted sleep is permitted.

Researchers have long been struck by the fact that the highest proportion of REM sleep is seen during the

rebound phenomenon The increased frequency or intensity of a phenomenon after it has been temporarily suppressed; for example, the increase in REM sleep seen after a period of REM sleep deprivation.

most active phase of brain development. Perhaps, then, REM sleep plays a role in this process (Siegel, 2005). Infant animals born with well-developed brains spend less time in REM sleep than do animals born with immature brain. For example, guinea pigs, which are born with teeth, claws, and fur and are able to walk within an hour of birth, spend around 1 hour in REM sleep each day (Jouvet-Mounier, Astic, and Lacote, 1970). In contrast, ferrets, which are born with less-developed brains, spend around 6 hours in REM sleep each day (Jha, Coleman, and Frank, 2006). Humans, too, are born with immature brains. Studies of human fetuses and infants born prematurely indicate that REM sleep begins to appear 30 weeks after conception and peaks at around 40 weeks (Roffwarg, Muzio, and Dement, 1966; Petre-Quadens and De Lee, 1974; Inoue et al., 1986). Approximately 70 percent of a newborn infant's sleep is REM sleep. By six months of age this proportion has declined to approximately 30 percent. By eight years of age it has fallen to approximately 22 percent, and by late adulthood it is less than 15 percent. Clearly, there is a relation between brain development and REM sleep.

But if the function of REM sleep is to promote brain development, why do adults have REM sleep? One possibility is that REM sleep facilitates the massive changes in the brain that occur during development but also some of the more modest changes responsible for learning that occur later in life. As we will see in the next subsection, evidence does suggest that REM sleep facilitates learning—but so does slow-wave sleep.

Sleep and Learning

Research with both humans and laboratory animals indicates that sleep does more than allow the brain to rest: It also aids in the consolidation of long-term memories (Marshall and Born, 2007). In fact, slow-wave sleep and REM sleep play different roles in memory consolidation.

As we will see in Chapter 13, there are two major categories of long-term memory: *declarative memory* (also called *explicit memory*) and *nondeclarative memory* (also called *implicit memory*). Declarative memories include those that people can talk about, such as memories of past episodes in their lives. They also include memories of the relationships between stimuli or events, such as the spatial relationships between landmarks that permit us to navigate around our environment. Nondeclarative memories include those gained through experience and practice that do not necessarily involve an attempt to "memorize" information, such as learning to drive a car, throw and catch a ball, or recognize a person's face. Research has found that slow-wave sleep and REM sleep play different roles in the consolidation of declarative and nondeclarative memories.

Before I tell you about the results of this research, let's review the consciousness of a person engaged in each of these stages of sleep. During REM sleep, people normally have a high level of consciousness. If we awaken people during REM sleep, they will be alert and clear-headed and will almost always be able to describe the details of a dream that they were having. However, if we awaken people during slow-wave sleep, they will be groggy and confused, and will usually tell us that nothing was happening. So which stages of sleep do you think aid in the consolidation of declarative and nondeclarative memories?

I would have thought that REM sleep would be associated with declarative memories and slow-wave sleep with nondeclarative memories. However, just the opposite is true. Let's look at evidence from two studies that looked at the effects of a nap on memory consolidation. Mednick, Nakayama, and Stickgold (2003) had subjects learn a nondeclarative visual discrimination task at 9:00 A.M. The subjects' ability to perform the task was tested 10 hours later, at 7:00 P.M. Some, but not all, of the subjects took a 90-minute nap during the day between training and testing. The investigators recorded the EEGs of the sleeping subjects to determine which of them engaged in REM sleep and which of them did not. (Obviously, all of them engaged in slow-wave sleep, because this stage of sleep always comes first in healthy people.) The investigators found that the performance of subjects who did not take a nap was worse when they were tested at 7:00 P.M. than it had been at the end of training. The subjects who engaged only in slow-wave sleep did about the same during testing as they had done at the end of training. However, the subjects who engaged in REM sleep performed significantly better. Thus, REM sleep strongly facilitated the consolidation of a nondeclarative memory. (See *Figure 9.8.*)

In the second study, Tucker et al. (2006) trained subjects on two tasks: a declarative task (learning a list of paired words) and a nondeclarative task (learning to trace a pencil-and-paper design while looking at the paper in a mirror). Afterward, some of subjects were permitted to take a nap lasting for about 1 hour. Their EEGs were recorded, and they were awakened before they could engage in REM sleep. The subjects' performance on the two tasks was then tested 6 hours after the original training. The investigators found that compared with subjects who stayed awake, a nap consisting of just slow-wave sleep increased the subjects' performance on the declarative task but had no effect on performance of the nondeclarative task. (See *Figure 9.9.*) So these two experiments (and many others I have not described) indicate that REM sleep facilitates consolidation of nondeclarative memories and that slow-wave sleep facilitates consolidation of declarative memories.

FIGURE 9.8 ■ REM Sleep and Learning

Only after a 90-minute nap that included both slow-wave sleep and REM sleep did the subjects' performance on a nondeclarative visual discrimination task improve.

(Adapted from Mednick, S., Nakayama, K., and Stickgold, R. *Nature Neuroscience*, 2003, *6*, 697–698.)

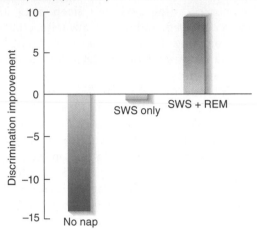

FIGURE 9.9 ■ Slow-Wave Sleep in Learning

Subjects learned a declarative learning task (list of paired words) and a nondeclarative learning task (mirror tracing). After a nap that included just slow-wave sleep, only subjects who learned the declarative learning task showed improved performance, compared with subjects who stayed awake.

(Adapted from Tucker, M. A., Hirota, Y., Wamsley, E. J., Lau, H., Chaklader, A., and Fishbein, W. *Neurobiology of Learning and Memory*, 2006, *86*, 241–247.)

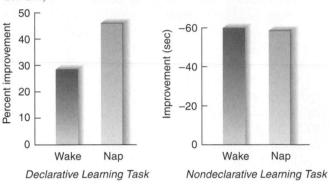

One last experiment on this topic: Peigneux et al. (2004) had human subjects learn their way around a computerized virtual-reality town. This task is very similar to what people do when they learn their way around a real town. They must learn the relative locations of landmarks and streets that connect them so that they can find particular locations when the experimenter "placed" them at various starting points. As we will see in Chapter 13, the hippocampus plays an essential role in learning of this kind. Peigneux and his colleagues used functional brain imaging to measure regional brain activity and found that the same regions of the hippocampus were activated during route learning and during slow-wave sleep the following night. These patterns were *not* seen during REM sleep. Thus, although people who are awakened during slow-wave sleep seldom report that they had been dreaming, the sleeping brain rehearses information that was acquired during the previous day.

Many studies with laboratory animals have directly recorded the activity of individual neurons in the animals' brains. These studies, too, indicate that the brain appears to rehearse newly learned information during slow-wave sleep. Chapter 13 reviews this evidence and describes research on the relevant brain mechanisms.

Interim Summary

Why Do We Sleep?

The two principal explanations for sleep are that sleep serves as an adaptive response or that it provides a period of restoration. The fact that all vertebrates sleep, including some that would seem to be better off without it, suggests that sleep is more than an adaptive response.

In humans the effects of several days of sleep deprivation include perceptual distortions and (sometimes) mild hallucinations and difficulty performing tasks that require prolonged concentration. These effects suggest that sleep deprivation impairs cerebral functioning. Deep slow-wave sleep appears to be the most important stage, and perhaps its function is to permit the brain to recuperate. Fatal familial insomnia is an inherited disease that results in degeneration of parts of the thalamus, deficits in attention and memory, a dreamlike state, loss of control of the autonomic nervous system and the endocrine system, insomnia, and death. Animals that are sleep-deprived eventually die. Their symptoms include increased body temperature and metabolic rate, voracious eating and weight loss but no obvious signs of a stress response.

The primary function of sleep does not seem to be to provide an opportunity for the body to repair the wear and tear that occurs during waking hours. Changes in a person's level of exercise do not significantly alter the amount of sleep the person needs the following night. Instead, the most important function of slow-wave sleep seem to be to lower the brain's metabolism and permit it to rest. In support of this hypothesis, research has shown that slow-wave sleep does indeed reduce the brain's metabolic rate and that increased mental activity (the surprise treat experiment) can cause an increase in slow-wave sleep the next night.

The functions of REM sleep are even less understood than those of slow-wave sleep. REM sleep may promote brain development. Both REM sleep and slow-wave sleep promote learning: REM sleep facilitates nondeclarative learning, and slow-wave sleep facilitates declarative learning.

Thought Question

The evidence presented in this section suggests that the primary function of sleep is to permit the brain to rest. But could sleep also have some other functions? For example, could sleep serve as an adaptive response, keeping animals out of harm's way *as well as* providing some cerebral repose? Sleep researcher William Dement pointed out that one of the functions of the lungs is communication. Obviously, the *primary* function of our lungs is to provide oxygen and rid the body of carbon dioxide, and this function explains the evolution of the respiratory system. But we can also use our lungs to vibrate our vocal cords and provide sounds used to talk, so they play a role in communication, too. Other functions of our lungs are to warm our cold hands (by breathing on them), to kindle fires by blowing on hot coals, or to blow out candles. With this perspective in mind, can you think of some other useful functions of sleep?

PHYSIOLOGICAL MECHANISMS OF SLEEP AND WAKING

So far, I have discussed the nature of sleep, problems associated with it, and its functions. Now it is time to examine what researchers have discovered about the physiological mechanisms that are responsible for the behavior of sleep and for its counterpart, alert wakefulness.

Chemical Control of Sleep

As we have seen, sleep is *regulated*; that is, if an organism is deprived of slow-wave sleep or REM sleep, the organism will make up at least part of the missed sleep when permitted to do so. In addition, the amount of slow-wave sleep that a person obtains during a daytime nap is deducted from the amount of slow-wave sleep he or she obtains the next night (Karacan et al., 1970). These facts suggest that some physiological mechanism monitors the amount of sleep that an organism needs—in other words, keeps track of the sleep debt we incur during hours of wakefulness. What might this mechanism be?

The simplest explanation would be that the body produces a sleep-promoting substance during that accumulates during wakefulness and is destroyed during sleep. The longer someone is awake, the longer he or she has to sleep to deactivate this substance. And because REM sleep deprivation produces an independent REM sleep debt, there might have to be two substances, one for each stage of sleep.

Where might such substances be produced? They do not appear to be found in the general circulation of the body. As we saw earlier, the cerebral hemispheres of the bottlenose dolphin sleep at different times (Mukhametov, 1984). If sleep were controlled by chemicals in the blood, the hemispheres should sleep at the same time. This observation suggests that if sleep is controlled by chemicals, these chemicals are produced within the brain and act there. In support of this suggestion Oleksenko et al. (1992) obtained evidence that indicates that each hemisphere of the brain incurs its own sleep debt. The researchers deprived a bottlenose dolphin of sleep in only one hemisphere. When they allowed the animal to sleep normally, they saw a rebound of slow-wave sleep only in the deprived hemisphere.

Benington, Kodali, and Heller (1995) suggested that **adenosine,** a nucleoside neuromodulator, might play a primary role in the control of sleep, and subsequent studies have supported this suggestion. As we saw in Chapter 2, astrocytes maintain a small stock of nutrients in the form of glycogen, an insoluble carbohydrate that is also stocked by the liver and the muscles. In times of increased brain activity this glycogen is converted into fuel for neurons; thus, prolonged wakefulness causes a decrease in the level of glycogen in the brain (Kong et al., 2002). A fall in the level of glycogen causes an increase in the level of extracellular adenosine, which has an inhibitory effect on neural activity. This accumulation of adenosine serves as a sleep-promoting substance. During slow-wave sleep, neurons in the brain rest, and the astrocytes renew their stock of glycogen

---

adenosine A neuromodulator that is released by neurons engaging in high levels of metabolic activity, may play a primary role in the initiation of sleep.

(Basheer et al., 2004; Wigren et al., 2007). If wakefulness is prolonged, even more adenosine accumulates, which inhibits neural activity and produces the cognitive and emotional effects that are seen during sleep deprivation. (As we saw in Chapter 4, caffeine blocks adenosine receptors. I don't need to tell you the effect that caffeine has on sleepiness.)

The role of adenosine as a sleep-promoting factor is discussed in more detail later in this chapter, in a section devoted to the neural control of sleep.

Neural Control of Arousal

As we have seen, sleep is not a unitary condition but consists of several different stages with very different characteristics. The waking state, too, is nonuniform; sometimes we are alert and attentive, and sometimes we fail to notice much about what is happening around us. Of course, sleepiness has an effect on wakefulness; if we are fighting to stay awake, the struggle might impair our ability to concentrate on other things. But everyday observations suggest that even when we are not sleepy, our alertness can vary. For example, when we observe something very interesting (or frightening or simply surprising), we become more alert and aware of our surroundings.

Circuits of neurons that secrete at least five different neurotransmitters play a role in some aspect of an animal's level of alertness and wakefulness—what is commonly called arousal: acetylcholine, norepinephrine, serotonin, histamine, and orexin.

Acetylcholine

One of the most important neurotransmitters involved in arousal—especially of the cerebral cortex—is acetylcholine. Two groups of acetylcholinergic neurons, one in the pons and one located in the basal forebrain, produce activation and cortical desynchrony when they are stimulated (Jones, 1990; Steriade, 1996). A third group of acetylcholinergic neurons, located in the medial septum, controls the activity of the hippocampus.

Researchers have long known that acetylcholinergic agonists increase EEG signs of cortical arousal and that acetylcholinergic antagonists decrease them (Vanderwolf, 1992). Marrosu et al. (1995) used microdialysis probes to measure the release of acetylcholine in the hippocampus and neocortex, two regions whose activity is closely related to an animal's alertness and behavioral arousal. They found that the levels of ACh in these regions were high during both waking and REM sleep—periods during which the EEG displayed desynchronized activity—but low during slow-wave sleep. (See *Figure 9.10.*)

Rasmusson, Clow, and Szerb (1994) electrically stimulated a region of the dorsal pons and found that the stimulation activated the cerebral cortex and increased the release of acetylcholine there by 350 percent (as

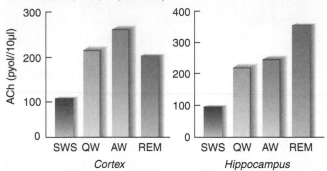

FIGURE 9.10 ■ Acetylcholine and the Sleep–Waking Cycle

The graphs show the release of acetylcholine from the cortex and hippocampus during the sleep-wake cycle. SWS = slow-wave sleep, QW = quiet waking, AW = active, and REM sleep.

(Adapted from Marrosu, F., Portas, C., Mascia, M. S., Casu, M. A., et al. *Brain Research*, 1995, *671*, 329–332.)

measured by microdialysis probes). A group of acetylcholinergic neurons located in the basal forebrain forms an essential part of the pathway that is responsible for this effect. If these neurons were deactivated by infusing a local anesthetic or drugs that blocked synaptic transmission, the activating effects of the pontine stimulation were abolished. Lee et al. (2004) found that most neurons in the basal forebrain showed a high rate of firing during both waking and REM sleep and a low rate of firing during slow-wave sleep.

Norepinephrine

Investigators have long known that catecholamine agonists such as amphetamine produce arousal and sleeplessness. These effects appear to be mediated primarily by the noradrenergic system of the **locus coeruleus (LC),** located in the dorsal pons. Neurons of the locus coeruleus give rise to axons that branch widely, releasing norepinephrine (from axonal varicosities) throughout the neocortex, hippocampus, thalamus, cerebellar cortex, pons, and medulla; thus, they potentially affect widespread and important regions of the brain.

Aston-Jones and Bloom (1981) recorded the activity of noradrenergic neurons of the LC across the sleep-waking cycle in unrestrained rats. They found that this activity was closely related to behavioral arousal: The firing rate of these neurons was high during wakefulness, low during slow-wave sleep, and almost zero during REM sleep. Within a few seconds of awakening, the rate of firing increased dramatically. (See *Figure 9.11.*)

---

locus coeruleus (LC) *(sa roo lee us)* A dark-colored group of noradrenergic cell bodies located in the pons near the rostral end of the floor of the fourth ventricle; involved in arousal and vigilance.

FIGURE 9.11 ■ Activity of Noradrenergic Neurons

This figure shows the activity of noradrenergic neurons in the locus coeruleus of freely moving rats during various stages of sleep and waking.

(From Aston-Jones, G., and Bloom, F. E. *The Journal of Neuroscience*, 1981, *1*, 876–886. Copyright © 1981, The Society for Neuroscience.)

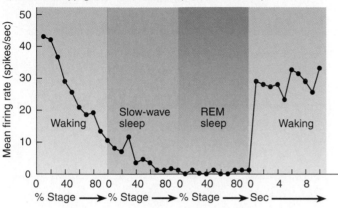

FIGURE 9.12 ■ Activity of Serotonergic Neurons

This figure shows the activity of serotonergic (5-HT-secreting) neurons in the dorsal raphe nuclei of freely moving cats during various stages of sleep and waking.

(Adapted from Trulson, M. E., and Jacobs, B. L. *Brain Research*, 1979, *163*, 135–150. Redrawn with permission.)

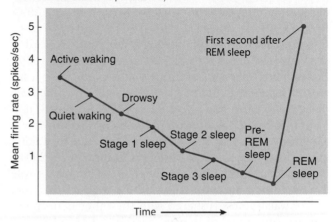

Most investigators believe that activity of noradrenergic LC neurons increases an animal's vigilance—its ability to pay attention to stimuli in the environment. For example, Aston-Jones et al. (1994) recorded the electrical activity of noradrenergic LC neurons in monkeys performing a task that required them to watch for a particular stimulus that would appear on a video display. The investigators observed that the monkeys performed best when the rate of firing of the LC neurons was high. After the monkeys worked for a long time at the task, the neurons' rate of firing fell, and so did the monkeys' performance. These results support the conclusion that the activation of LC neurons (and their release of norepinephrine) increases vigilance.

Serotonin

A third neurotransmitter, serotonin (5-HT) also appears to play a role in activating behavior. Almost all of the brain's serotonergic neurons are found in the **raphe nuclei,** which are located in the medullary and pontine regions of the reticular formation. The axons of these neurons project to many parts of the brain, including the thalamus, hypothalamus, basal ganglia, hippocampus, and neocortex. Stimulation of the raphe nuclei causes locomotion and cortical arousal (as measured by the EEG), whereas PCPA, a drug that prevents the synthesis of serotonin, reduces cortical arousal (Peck and Vanderwolf, 1991).

Jacobs and Fornal (1999) suggested that one specific contribution of serotonergic neurons to activation is facilitation of continuous, automatic movements, such as pacing, chewing, and grooming. On the other hand, when animals engage in orienting responses to novel stimuli, the activity of serotonergic neurons decreases.

Perhaps serotonergic neurons are involved in facilitating ongoing activities and suppressing the processing of sensory information, preventing reactions that might disrupt the ongoing activities.

Figure 9.12 shows the activity of serotonergic neurons, recorded by Trulson and Jacobs (1979). As you can see, these neurons, like the noradrenergic neurons studied by Aston-Jones and Bloom (1981), were most active during waking. Their firing rate declined during slow-wave sleep and became virtually zero during REM sleep. However, once the period of REM sleep ended, the neurons temporarily became very active again. (See *Figure 9.12.*)

Histamine

The fourth neurotransmitter implicated in the control of wakefulness and arousal is histamine, a compound synthesized from histidine, an amino acid. You are undoubtedly aware that antihistamines, which are used to treat allergies, can cause drowsiness. They do so by blocking histamine H_1 receptors in the brain. More modern antihistamines cannot cross the blood–brain barrier, so they do not cause drowsiness.

The cell bodies of histaminergic neurons are located in the **tuberomammillary nucleus (TMN)** of the

raphe nuclei (*ruh fay*) A group of nuclei located in the reticular formation of the medulla, pons, and midbrain, situated along the midline; contain serotonergic neurons.

tuberomammillary nucleus (TMN) A nucleus in the ventral posterior hypothalamus, just rostral to the mammillary bodies; contains histaminergic neurons involved in cortical activation and behavioral arousal.

hypothalamus, located at the base of the brain just rostral to the mammillary bodies. The axons of these neurons project primarily to the cerebral cortex, thalamus, basal ganglia, basal forebrain, and hypothalamus. The projections to the cerebral cortex directly increase cortical activation and arousal, and projections to acetylcholinergic neurons of the basal forebrain and dorsal pons do so indirectly, by increasing the release of acetylcholine in the cerebral cortex (Khateb et al., 1995; Brown, Stevens, and Haas, 2001). The activity of histaminergic neurons is high during waking but low during slow-wave sleep and REM sleep (Steininger et al., 1996). In addition, injections of drugs that prevent the synthesis of histamine or block histamine H_1 receptors decrease waking and increase sleep (Lin, Sakai, and Jouvet, 1998). Also, infusion of histamine into the basal forebrain region of rats causes an increase in waking and a decrease in non-REM sleep (Ramesh et al., 2004).

Although histamine clearly plays an important role in wakefulness and arousal, evidence suggests that control of wakefulness is shared with the other neurotransmitters discussed in this section. For example, Parmentier et al. (2002) found that mice with a targeted mutation that blocks the synthesis of histamine showed normal amounts of spontaneous wakefulness. However, the animals showed less arousal in response to environmental stimuli. For example, normal mice placed in a novel environment remain awake for 2–3 hours, but the histamine-deprived mice fell asleep within a few minutes. Similarly, Takahashi, Lin, and Sakai (2006) found that histaminergic neurons did not respond to environmental stimuli unless the stimuli elicited a state of overt attention. Gerashchenko et al. (2004) suggest that the brain's arousal systems promote wakefulness at different times or in different situations and that no one of these systems plays a critical role under all conditions.

Orexin

As we saw in the section on sleep disorders, the cause of narcolepsy is degeneration of orexinergic neurons in humans and a hereditary absence of orexin-B receptors in dogs. The cell bodies of neurons that secrete orexin (also called hypocretin, as we saw) are located in the lateral hypothalamus. Although there are only about 7000 orexinergic neurons in the human brain, the axons of these neurons project to almost every part of the brain, including the cerebral cortex and all of the regions involved in arousal and wakefulness, including the locus coeruleus, raphe nuclei, tuberomammillary nucleus, and acetylcholinergic neurons in the dorsal pons and basal forebrain (Sakurai, 2007). Orexin has an excitatory effect in all of these regions.

Mileykovskiy, Kiyashchenko, and Siegel (2005) recorded the activity of single orexinergic neurons in unanesthetized rats and found that the neurons fired at a high rate during alert or active waking and at a low

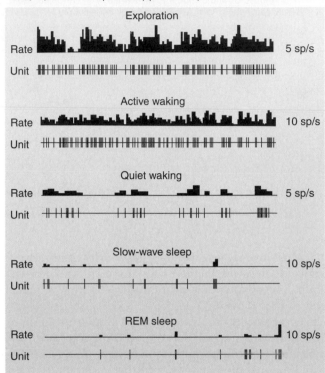

FIGURE 9.13 ■ Activity of Single Hypocretinergic Neurons

This figure shows neuron activity during phases of sleep and waking.

(From Mileykovskiy, B. Y., Kiyashchenko, L. I., and Siegel, J. M. *Neuron*, 2005, *46*, 787–798. Reprinted by permission.)

rate during quiet waking, slow-wave sleep, and REM sleep. The highest rate of firing was seen when the rats were engaged in exploratory activity. (See *Figure 9.13*.)

Neural Control of Slow-Wave Sleep

When we are awake and alert, most of the neurons in our brain—especially those of the forebrain—are active, which enables us to pay attention to sensory information and process this information, to think about what we are perceiving, to retrieve and think about memories, and to engage in a variety of behaviors that we are called on to perform during the day. The level of brain activity is largely controlled by the five sets of arousal neurons described in the previous section. A high level of activity of these neurons keeps us awake, and a low level puts us to sleep.

But what controls the activity of the arousal neurons? What causes this activity to fall, thus putting us to sleep? The first hint at an answer to this question was suggested in the early twentieth century by careful observations of a Viennese neurologist, Constantin von Economo, who noticed that patients afflicted by a new type of encephalitis that was sweeping through Europe and North America showed severe disturbance in sleep and waking (Triarhou, 2006). Most patients slept excessively, waking only to eat and drink. According to von Economo, these patients had brain damage at the junction of the brain stem and forebrain, at a location that would destroy the axons of the arousal neurons entering the forebrain. Some patients, however, showed just the opposite symptoms: They slept only a few hours each day. Although they were tired, they had difficulty falling asleep and usually awakened shortly thereafter. Von Economo reported that patients who displayed insomnia had damage to the region of the anterior hypothalamus. We now know that this region, usually referred to as the *preoptic area*, is the one most involved in control of sleep. The preoptic area contains neurons whose axons form inhibitory synaptic connections with the brain's arousal neurons. When our preoptic neurons (let's call them *sleep neurons*) become active, they suppress the activity of our arousal neurons, and we fall asleep (Saper, Scammell, and Lu, 2005).

Nauta (1946) found that destruction of the preoptic area produced total insomnia in rats. The animals subsequently fell into a coma and died; the average survival time was only three days. The effects of effects of electrical stimulation of the preoptic area are just the opposite. Sterman and Clemente (1962a, 1962b) found that electrical stimulation of this region produced signs of drowsiness in the behavior and the EEG of unanesthetized, freely moving cats. The average latency period between the stimulation and the changes in the EEG was 30 seconds, but sometimes the effect was immediate. The animals often subsequently fell asleep.

The majority of the sleep neurons are located in the **ventrolateral preoptic area (vlPOA).** In addition, some are located in the *median preoptic nucleus (MnPN)*. Damage to vlPOA neurons suppresses sleep (Lu et al., 2000), and the activity of these neurons, measured by their levels of Fos protein, increases during sleep. Anatomical and histochemical studies indicate that the sleep neurons secrete the inhibitory neurotransmitter GABA, and that they send their axons to the orexinergic, histaminergic, acetylcholinergic, serotonergic, and noradrenergic neurons of the lateral hypothalamus, tuberomammillary nucleus, dorsal pons, raphe nuclei, and locus coeruleus (Sherin et al., 1998; Gvilia et al., 2006; Suntsova et al., 2007). As we saw in the previous section, activity of neurons in these five regions causes cortical activation and behavioral arousal. Inhibition of these regions, then, is a necessary condition for sleep.

The sleep neurons in the preoptic area receive inhibitory inputs from some of the same regions they inhibit, including the tuberomammillary nucleus, raphe nuclei, and locus coeruleus (Chou et al., 2002); thus, they are inhibited by histamine, serotonin, and norepinephrine. As Saper et al. (2001) suggest, this mutual inhibition may provide the basis for establishing

FIGURE 9.14 ■ The Sleep/Waking Flip-Flop

According to Saper et al. (2001), the major sleep-promoting region (the vlPOA) and the major wakefulness-promoting regions (the basal forebrain and pontine regions that contain acetylcholinergic neurons; the locus coeruleus, which contains noradrenergic neurons; the raphe nuclei, which contain serotonergic neurons; and the tuberomammillary nucleus of the hypothalamus, which contains histaminergic neurons) are reciprocally connected by inhibitory GABAergic neurons. (a) When the flip-flop is in the "wake" state, the arousal systems are active and the vlPOA is inhibited, and the animal is awake. (b) When the flip-flop is in the "sleep" state, the vlPOA is active and the arousal systems are inhibited, and the animal is asleep.

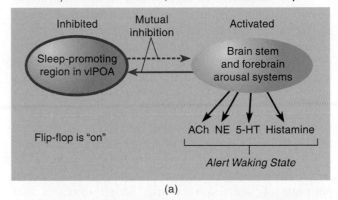
(a)

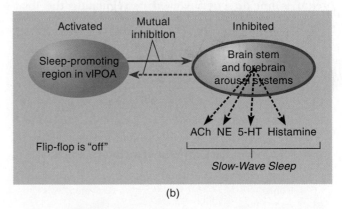
(b)

periods of sleep and waking. They note that reciprocal inhibition also characterizes an electronic circuit known as a *flip-flop*. A flip-flop can assume one of two states, usually referred to as on or off—or 0 or 1 in computer applications. Thus, either the sleep neurons are active and inhibit the wakefulness neurons or the wakefulness neurons are active and inhibit the sleep neurons. Because these regions are mutually inhibitory, it is impossible for neurons in both sets of regions to be active at the same time. (See *Figure 9.14.*)

ventrolateral preoptic area (vlPOA) A group of GABAergic neurons in the preoptic area whose activity suppresses alertness and behavioral arousal and promotes sleep.

A flip-flop has an important advantage: When it switches from one state to another, it does so quickly. Clearly, it is most advantageous to be either asleep or awake; a state that has some of the characteristics of both sleep and wakefulness would be maladaptive. However, there is one problem with flip-flops: They can be unstable. In fact, people with narcolepsy and animals with damage to the orexinergic system of neurons exhibit just this characteristic. They have great difficulty remaining awake when nothing interesting is happening, and they have trouble remaining asleep for an extended amount of time. (They also show intrusions of the characteristics of REM sleep at inappropriate time. I will discuss this phenomenon in the next section.)

Saper et al. (2001) suggest that an important function of orexinergic neurons is to help stabilize the sleep/waking flip-flop through their excitatory connections to the wakefulness neurons . Activity of this system of neurons tips the activity of the flip-flop toward the waking state, thus promoting wakefulness and inhibiting sleep. Perhaps your success at staying awake in a boring lecture depends on maintaining a high rate of firing of your orexinergic neurons, which would keep the flip-flop in the waking state. (See *Figure 9.15*.)

Mochizuki et al. (2004) found that mice with a targeted mutation against the orexin gene showed normal amounts of sleep and waking, which indicates that orexin is not directly involved in regulating the total amount of time spent in these behavioral states. However, the animals' bouts of wakefulness and slow-wave sleep were very brief, showing many more transitions between sleep and waking. (You will recognize that these symptoms describe those of people with narcolepsy, which is caused by a loss of orexinergic neurons.) Brisbare-Roch et al. (2007) developed a drug that blocks orexin receptors. The drug, which has been tested in rats, dogs, and humans, promotes sleep, which provides further evidence that orexin promotes wakefulness.

The evidence I have reviewed so far in this section concerns the brain mechanisms that are promote wakefulness and sleep. As we all know, sleepiness is controlled by two factors: time of day and length of time our brains have been awake and active. The final section of this chapter describes the internal clock, located in the hypothalamus, that controls daily rhythms of sleep and waking. But what is responsible for the sleepiness that accumulates as a result of prolonged waking and mental activity?

As we saw earlier in this chapter, adenosine is produced when neurons are metabolically active, and the accumulation of adenosine produces drowsiness and sleep. Porkka-Heiskanen, Strecker, and McCarley (2000) used microdialysis to measure adenosine levels in several regions of the brain. They found that the level of adenosine increased during wakefulness and slowly decreased during sleep, especially in the basal forebrain. (See *Figure 9.16*.)

As we all know, caffeine, a drug that blocks adenosine receptors, helps to decrease drowsiness and promote alertness. Huang et al. (2005) found that mice with a targeted mutation against the gene responsible for the production of adenosine 2A receptors blocked the arousing effect of caffeine. Scammell et al. (2001) found that infusion of an adenosine agonist into the vlPOA activated neurons there, decreased the activity of histaminergic neurons of the tuberomammillary nucleus, and increased slow-wave sleep. However, adenosine receptors are found on neurons in many regions of the brain, including the orexinergic neurons of the lateral hypothalamus (Thakkar, Winston, and McCarley, 2002). Thus, it is unlikely that all of the sleep-promoting effects of adenosine involve the neurons of the vlPOA.

Aging has detrimental effects on the quality of sleep: Sleep becomes fragmented, time spent waking increases, and the amount of delta activity (a measure of depth of slow-wave sleep) decreases. A study by Murillo-Rodriguez et al. (2004) suggests one cause of this phenomenon. They observed no differences in the number of vlPOA neurons in young and old rats but found that the administration of adenosine or an adenosine agonist had a smaller effect on sleep in older animals. The authors suggested that a reduced number

---

FIGURE 9.15 ■ Role of Orexinergic Neurons in Sleep

The schematic diagram shows the effect of activation of the orexinergic system of neurons of the lateral hypothalamus on the sleep/waking flip-flop. Motivation to remain awake or events that disturb sleep activate the orexinergic neurons.

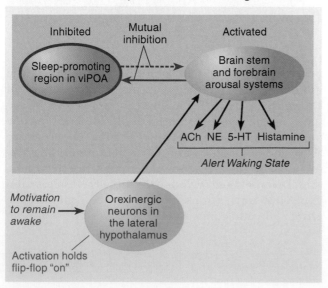

FIGURE 9.16 ■ Adenosine and Sleep

Extracellular adenosine in the cat basal forebrain region during 6 hours of prolonged waking and 3 hours of recovery sleep was measured by microdialysis.

(From Porkka-Heiskanen, T., Strecker, R. E., Thakkar, M., Bjorkum, A. A., Greene, R. W., and McCarley, R. W. *Science*, 1997, *276*, 1265–1268. Copyright © 1997 by the American Association for the Advancement of Science. Reprinted with permission.)

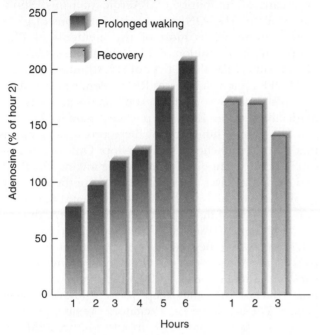

FIGURE 9.17 ■ Adenosine, Time of Day, and Hunger

This figure shows the role of adenosine, time of day, and hunger and satiety signals on the sleep/waking flip-flop.

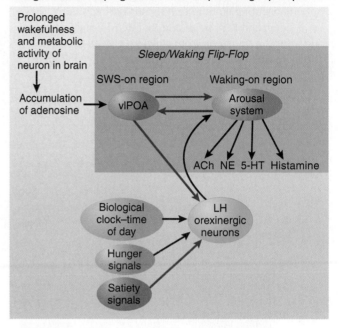

of adenosine receptors in the vlPOA might be the cause of this decreased sensitivity.

Seeing that orexinergic neurons help to hold the sleep/waking flip-flop in the waking state, the obvious question to ask is, what factors control the activity of orexinergic neurons? During the waking part of the day/night cycle, orexinergic neurons receive an excitatory signal from the biological clock that controls rhythms of sleep and waking. These neurons also receive an signals from brain mechanisms that monitor the animal's nutritional state: Hunger-related signals activate orexinergic neurons, and satiety-related signals inhibit them. Thus, orexinergic neurons maintain arousal during the times when an animal should search for food. In fact, if normal mice (but not mice with a targeted mutation against orexin receptors) are given less food than they would normally eat, they stay awake longer each day (Yamanaka et al., 2003, Sakurai, 2007). Finally, orexinergic neurons receive inhibitory input from the vlPOA, which means that sleep signals that arise from the accumulation of adenosine can eventually overcome excitatory input to orexinergic neurons, and sleep can occur. (See *Figure 9.17.*)

Neural Control of REM Sleep

As we saw earlier in this chapter, REM sleep consists of desynchronized EEG activity, muscular paralysis, rapid eye movements, and increased genital activity. The rate of cerebral metabolism during REM sleep is as high as it is during waking (Maquet et al., 1990), and were it not for the state of paralysis, the level of *physical* activity would also be high.

As we shall see, REM sleep is controlled by a flip-flop similar to the one that controls cycles of sleep and waking. The sleep/waking flip-flop determines when we wake and when we sleep; and once we fall asleep, the REM flip-flop controls our cycles of REM sleep and slow-wave sleep.

Acetylcholinergic neurons play an important role in cerebral activation during alert wakefulness. Researchers have also found that they are involved in the neocortical activation that accompanies REM sleep. Several studies (for example, El Mansari, Sakai, and Jouvet, 1989; Steriade et al., 1990; Kayama, Ohta, and Jodo, 1992) have shown that ACh neurons in the dorsal pons fire at a high rate during both REM sleep and active wakefulness or during REM sleep alone. Figure 9.18 shows the activity of a so-called *REM-ON* cell, which fires at a high rate only during REM sleep. As you can see, this neuron increased its activity approximately 80 sec before the onset of REM sleep.

FIGURE 9.18 ■ Firing Pattern of a REM-ON Cell

The actycholinergic REM-ON cell is located in the peribrachial area of the pons. The figure shows: (a) action potentials during 60-minute intervals during waking, slow-wave sleep, and REM sleep and (b) rate of firing just before and after the transition from slow-wave sleep to REM sleep. The increase in activity begins approximately 80 seconds before the onset of REM sleep.

(Adapted from El Mansari, M., Sakai, K., and Jouvet, M. *Experimental Brain Research*, 1989, *76*, 519–529.)

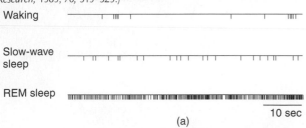

(a)

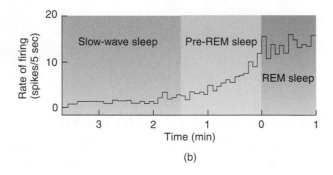

(b)

(See *Figure 9.18.*) Findings such as these suggested that the ACh neurons of the dorsal pons served as the trigger mechanism that initiated a period of REM sleep. However, more recent research suggests that REM sleep is controlled by the activity of a flip-flop whose elements do not include acetylcholinergic neurons.

Reviews by Lu et al. (2006), Fuller, Saper, and Lu (2007), and Luppi et al. (2007) summarize the evidence for the REM flip-flop. A region of the dorsal pons, just ventral to the locus coeruleus, contains REM-ON neurons. In rats this region is known as the **sublaterodorsal nucleus (SLD).** A region of the dorsal midbrain, the **ventrolateral periaqueductal gray matter (vlPAG),** contains REM-OFF neurons. For simplicity I will simply refer to the *REM-ON* and *REM-OFF* regions. The REM-ON and REM-OFF regions are interconnected by means of inhibitory GABAergic neurons. Stimulation of the REM-ON region with infusions of glutamate agonists elicits most of the elements of REM sleep, whereas inhibition of this region with GABA agonists disrupts REM sleep. In contrast, stimulation of the REM-OFF region suppresses REM sleep, whereas damage to this region or infusions of GABA agonists dramatically increases REM sleep. (See *Figure 9.19.*)

The mutual inhibition of these two regions means that that they function like a flip-flop: Only one region can be active at any given time. During waking, the REM-OFF region receives excitatory input from the orexinergic neurons of the lateral hypothalamus, and this activation tips the REM flip-flop into the off state. (Additional excitatory input to the REM-OFF region is received from two sets of wakefulness neurons: the noradrenergic neurons of the locus coeruleus and the serotonergic neurons of the raphe nuclei.) When the sleep/waking flip-flop switches into the sleep phase, slow-wave sleep begins. The activity of the excitatory orexinergic, noradrenergic, and serotonergic inputs to the REM-OFF region begins to decrease. Eventually, the REM-sleep flip-flop switches to the on state, and REM sleep begins. Figure 9.20 shows the control of the REM-sleep flip-flop by the sleep-waking flip-flop. (See *Figure 9.20.*)

sublaterodorsal nucleus (SLD) A region of the dorsal pons, just ventral to the locus coeruleus, that contains REM-ON neurons; part of the REM flip-flop.

ventrolateral periaqueductal gray matter (vlPAG) A region of the dorsal midbrain, that contains REM-OFF neurons; part of the REM flip-flop.

FIGURE 9.19 ■ The REM-Sleep Flip-Flop

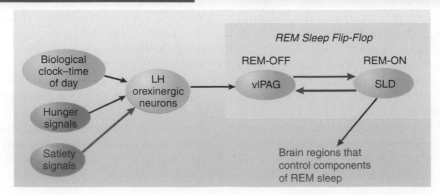

FIGURE 9.20 ■ REM Sleep

This schematic shows the interaction between the sleep/waking flip-flop and the REM sleep flip-flop.

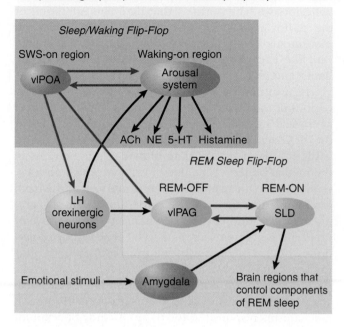

FIGURE 9.21 ■ Humor and Narcolepsy

The graph shows the activation of the hypothalamus and amygdala in normal subjects and patients with narcolepsy who are watching neutral and humorous sequences of photos.

(From Schwartz, S., Ponz, A., Poryazova, R., Werth, E., Boesiger, P., Khatami, R., and Bassetti, C. L. *Brain*, 2008, *131*, 514–522. Reprinted by permission.)

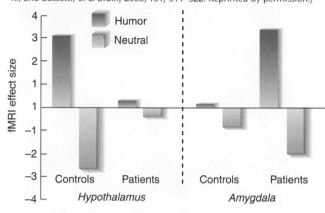

Once sleep begins, the activity of orexinergic neurons ceases, which removes one source of the excitatory input to the REM-OFF region. As we saw in Figures 9.11 and 9.12, as sleep progresses, the activity of noradrenergic and serotonergic neurons gradually decreases. As a consequence, more of the excitatory input to the REM-OFF region is removed. The REM flip-flop tips to the on state, and REM sleep begins. Presumably, an internal clock—perhaps located in the pons—controls the alternating periods of REM sleep and slow-wave sleep that follow.

We can see now why degeneration of orexinergic neurons causes narcolepsy. The daytime sleepiness and the fragmented sleep occur because without the influence of orexin, the sleep/waking flip-flop becomes unstable. The release of orexin in the REM-OFF region normally keeps the REM flip-flop in the off state. With the loss of orexinergic neurons, emotional episodes such as laughter or anger, which activate the amygdala, tip the REM flip-flop into the on state, and the result is an attack of cataplexy. (See *Figure 9.20*.) In fact, a functional-imaging study by Schwartz et al. (2008) found that when people with cataplexy watched humorous sequences of photographs, the hypothalamus was activated less, and the amygdala was activated more, than the same structures in control subjects. The investigators suggest that the loss of hypocretinergic neurons removed an inhibitory influence of the hypothalamus on the amygdala. The increased amygdala activity could account at least in part for the increased activity of

REM-ON neurons that occurs even during waking in people with cataplexy. (See *Figure 9.21*.)

As we saw earlier, REM sleep has several behavioral components, including rapid eye movements, genital activity, and muscular paralysis. The last of these phenomena, muscular paralysis, is particularly interesting. We also saw that patients with REM sleep behavior disorder fail to become paralyzed during REM sleep and therefore act out their dreams. The same thing happens to cats when a lesion is placed in a particular region of the midbrain. Jouvet (1972) described this phenomenon:

> To a naive observer, the cat, which is standing, looks awake since it may attack unknown enemies, play with an absent mouse, or display flight behavior. There are orienting movements of the head or eyes toward imaginary stimuli, although the animal does not respond to visual or auditory stimuli. These extraordinary episodes . . . are a good argument that "dreaming" occurs during [REM sleep] in the cat. (Jouvet, 1972, pp. 236–237)

Jouvet's lesions destroyed a set of neurons that are responsible for the muscular paralysis that occurs during REM sleep. These neurons are located just ventral to the area we now know to be part of the REM-ON region. Some of the axons that leave this region travel to the spinal cord, where they excite inhibitory interneurons whose axons form synapses with motor neurons. This means that when the REM flip-flop tips to the on state, motor neurons in the spinal cord become inhibited and cannot respond to the signals arising from the motor cortex in the course of a dream. Damage to the

FIGURE 9.22 ■ Control of REM Sleep

Components of REM sleep are controlled by the REM-ON region.

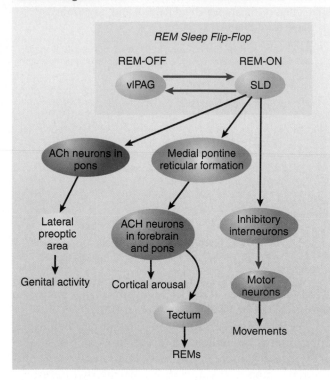

REM-ON region removes this inhibition, and the person (or one of Jouvet's cats) acts out his or her dreams. (See *Figure 9.22.*)

The fact that our brains contain an elaborate mechanism whose sole function is to keep us paralyzed while we dream—that is, to prevent us from acting out our dreams—suggests that the motor components of dreams are as important as the sensory components. Perhaps the practice our motor system gets during REM sleep helps us to improve our performance of behaviors we have learned that day. The inhibition of the motor neurons in the spinal cord prevents the movements being practiced from actually occurring, with the exception of a few harmless twitches of the hands and feet.

Neurons in the REM-ON region also send axons to regions of the thalamus that are involved in the control of cortical arousal, which may at least partly account for the EEG activation that is seen during REM sleep. In addition, they send axons to glutamatergic neurons in the medial pontine reticular formation, which, in turn, send axons to the acetylcholinergic neurons of the basal forebrain. Activation of these forebrain neurons produces arousal and cortical desynchrony. The control of rapid eye movements appears to be achieved by projections from acetylcholinergic neurons in the dorsal pons to the tectum (Webster and Jones, 1988).

Little is known about the function of genital activity that occurs during REM sleep or about the neural mechanisms responsible for them. A study by Schmidt et al. (2000) found that lesions of the lateral preoptic area in rats suppressed penile erections during REM sleep but had no effect on erections during waking. Salas et al. (2007) found that penile erections could be triggered by electrical stimulation of acetylcholinergic neurons in the pons that become active during REM sleep. The investigators note that evidence suggests that these pontine neurons may be directly connected with neurons in the lateral preoptic area and thus may be responsible for the erections. (See *Figure 9.22.*)

Interim Summary

Physiological Mechanisms of Sleep and Waking

The fact that the amount of sleep is regulated suggests that sleep-promoting substances (produced during wakefulness) or wakefulness-promoting substances (produced during sleep) may exist. The sleeping pattern of the dolphin brain suggests that such substances do not accumulate in the blood. Evidence suggests that adenosine, released when neurons are obliged to utilize the supply of glycogen stored in astrocytes, serves as the link between increased brain metabolism and the necessity of sleep.

Five systems of neurons appear to be important for alert, active wakefulness: the acetylcholinergic system of the peribrachial area of the pons and the basal forebrain, involved in cortical activation; the noradrenergic system of the locus coeruleus, involved in vigilance; the serotonergic system of the raphe nuclei, involved in activation of automatic behaviors such as locomotion and grooming; the histaminergic neurons of the tuberomammillary nucleus, involved in maintaining wakefulness; and the orexinergic system of the lateral hypothalamus, also involved in maintaining wakefulness.

Slow-wave sleep occurs when neurons in the ventrolateral preoptic area (vlPOA) become active. These neurons inhibit the systems of neurons that promote wakefulness. In turn, the vlPOA is inhibited by these same wakefulness-promoting regions, thus forming a kind of flip-flop that keeps us either awake or asleep. The accumulation of adenosine promotes sleep by inhibiting wakefulness-promoting regions and activating the sleep-promoting neurons of the vlPOA. Activity of the orexinergic neurons of the lateral hypothalamus keeps the flip-flop that controls sleep and waking in the "waking" state.

REM sleep is controlled by another flip-flop. The sublaterodorsal nucleus (SLD) serves as the REM-ON region, and the ventrolateral periaqueductal gray region (vlPAG) serves as the REM-OFF region. This flip-flop is controlled by the sleep/waking flip-flop; only when the sleep/waking flip-flop is in the "sleeping" state can the REM flip-flop switch to the "REM" state. The muscular paralysis that prevents our acting out our dreams is produced by connections between neurons in the SLD that excite inhibitory interneurons in the spinal cord. Penile erections during REM sleep (but not during waking) are abolished by lesions of the lateral preoptic area. Rapid eye movements are produced by indirect connections between the SLD and the tectum, through the medial pontine reticular formation and acetylcholinergic neurons in the pons.

Thought Questions

Have you ever been lying in bed, almost asleep, when you suddenly thought of something important you had forgotten to do? Did you then suddenly become fully awake and alert? If so, neurons in your brain's arousal systems undoubtedly became active, which aroused your cerebral cortex. What do you think the source of this activation was? What activated the arousal systems? How would you go about answering this question? What research methods described in Chapter 5 would be helpful?

BIOLOGICAL CLOCKS

Much of our behavior follows regular rhythms. For example, we saw that the stages of sleep are organized around a 90-minute cycle of REM and slow-wave sleep. And, of course, our daily pattern of sleep and waking follows a 24-hour cycle. Finally, many animals display seasonal breeding rhythms in which reproductive behaviors and hormone levels show yearly fluctuations. In recent years investigators have learned much about the neural mechanisms that are responsible for these rhythms.

Circadian Rhythms and Zeitgebers

Daily rhythms in behavior and physiological processes are found throughout the plant and animal world. These cycles are generally called **circadian rhythms.** (*Circa* means "about," and *dies* means "day"; therefore, a circadian rhythm is one with a cycle of approximately 24 hours.) Some of these rhythms are passive responses to changes in illumination. However, other rhythms are controlled by mechanisms within the organism—by "internal clocks." For example, Figure 9.23 shows the activity of a rat during various conditions of illumination. Each horizontal line represents 24 hours. Vertical tick marks represent the animal's activity in a running wheel. The upper portion of the figure shows the activity of the rat during a normal day–night cycle, with alternating 12-hour periods of light and dark. Notice that the animal is active during the night, which is normal for a rat. (See *Figure 9.23.*)

Next, the dark–light cycle was shifted by 6 hours; the animal's activity cycle quickly followed the change. (See *Figure 9.23.*) Finally, dim lights were left on continuously. The cyclical pattern in the rat's activity

FIGURE 9.23 ■ Wheel-Running Activity of a Rat

Note that the animal's activity occurs at "night" (that is, during the 12 hours the light is off) and that the active period is reset when the light period is changed. When the animal is maintained in constant dim illumination, it displays a free-running activity cycle of approximately 25 hours.

(From Groblewski, T. A., Nuñez, A., and Gold, R. M. Paper presented at the meeting of the Eastern Psychological Association, April 1980. Reprinted with permission.)

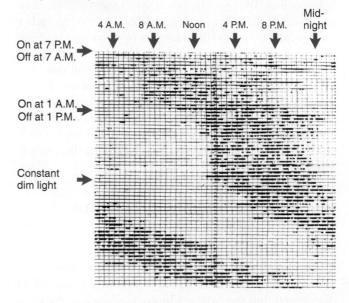

circadian rhythm (*sur kay dee un or sur ka dee un*) A daily rhythmical change in behavior or physiological process.

remained. Because there were no cycles of light and dark in the rat's environment, the source of rhythmicity must be located within the animal; that is, the animal must possess an internal, biological clock. You can see that the rat's clock was not set precisely to 24 hours; when the illumination was held constant, the clock ran a bit slow. The animal began its bout of activity almost 1 hour later each day. (See *Figure 9.23*.)

The phenomenon illustrated in Figure 9.23 is typical of the circadian rhythms shown by many species. A free-running clock, with a cycle of approximately 24 hours, controls some biological functions—in this case, motor activity. Regular daily variation in the level of illumination (that is, sunlight and darkness) normally keeps the clock adjusted to 24 hours. Light serves as a **zeitgeber** (German for "time giver"); it synchronizes the endogenous rhythm. Studies with many species of animals have shown that if they are maintained in constant darkness (or constant dim light), a brief period of bright light will reset their internal clock, advancing or retarding it depending upon when the light flash occurs (Aschoff, 1979). For example, if an animal is exposed to bright light soon after dusk, the biological clock is set back to an earlier time—as if dusk had not yet arrived. On the other hand, if the light occurs late at night, the biological clock is set ahead to a later time—as if dawn had already come.

Like other animals, humans exhibit circadian rhythms. Our normal period of inactivity begins several hours after the start of the dark portion of the day–night cycle and persists for a variable amount of time into the light portion. Without the benefits of modern civilization we would probably go to sleep earlier and get up earlier than we do; we use artificial lights to delay our bedtime and window shades to extend our time for sleep. Under constant illumination our biological clocks will run free, gaining or losing time like a watch that runs too slow or too fast. Different people have different cycle lengths, but most people in that situation will begin to live a "day" that is approximately 25 hours long. This works out quite well, because the morning light, acting as a zeitgeber, simply resets the clock.

The Suprachiasmatic Nucleus

Role in Circadian Rhythms

Researchers working independently in two laboratories (Moore and Eichler, 1972; Stephan and Zucker, 1972) discovered that the primary biological clock of the rat is located in the **suprachiasmatic nucleus (SCN)** of the hypothalamus; they found that lesions disrupted circadian rhythms of wheel running, drinking, and hormonal secretion. The SCN also provides the primary control over the timing of sleep cycles. Rats are nocturnal

FIGURE 9.24 ■ The SCN

The figure shows the location and appearance of the suprachiasmatic nuclei in a rat. Cresyl violet stain was used to color the nuclei in this cross section of a rat brain.

(Courtesy of Geert DeVries, University of Massachusetts.)

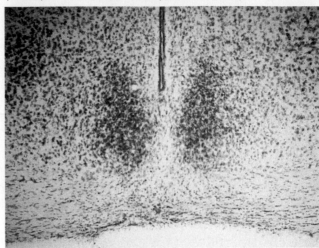

animals; they sleep during the day and forage and feed at night. Lesions of the SCN abolish this pattern; sleep occurs in bouts randomly dispersed throughout both day and night (Ibuka and Kawamura, 1975; Stephan and Nuñez, 1977). However, rats with SCN lesions still obtain the same amount of sleep that normal animals do. The lesions disrupt the circadian pattern but do not affect the total amount of sleep.

Figure 9.24 shows the suprachiasmatic nuclei in a cross section through the hypothalamus of a mouse; they appear as two clusters of dark-staining neurons at the base of the brain, just above the optic chiasm. (See *Figure 9.24*.) The suprachiasmatic nuclei of the rat consist of approximately 8,600 small neurons, tightly packed into a volume of 0.036 mm^3 (Moore, Speh, and Leak, 2002).

Because light is the primary zeitgeber for most mammals' activity cycles, we would expect that the SCN receives fibers from the visual system. Indeed, anatomical studies have revealed a direct projection of fibers from the retina to the SCN: the *retinohypothalamic pathway* (Hendrickson, Wagoner, and Cowan, 1972; Aronson et al., 1993). If you look carefully at Figure 9.24, you can see small dark spots within the optic chiasm, just ventral and medial to the base of the SCN; these are cell bodies of oligodendroglia that serve axons that enter the SCN and provide information from the retina. (See *Figure 9.24*.)

zeitgeber (*tsite gay ber*) A stimulus (usually the light of dawn) that resets the biological clock that is responsible for circadian rhythms.

suprachiasmatic nucleus (SCN) (*soo pra ky az **mat** ik*) A nucleus situated atop the optic chiasm. It contains a biological clock that is responsible for organizing many of the body's circadian rhythms.

The photoreceptors in the retina that provide photic information to the SCN are neither rods nor cones—the cells that provide us with the information used for visual perception. Freedman et al. (1999) found that targeted mutations against genes necessary for production of both rods and cones did not disrupt the synchronizing effects of light. However, when they removed the mice's eyes, these effects *were* disrupted. These results suggest that there is a special photoreceptor that provides information about the ambient level of light that synchronizes circadian rhythms. Provencio et al. (2000) found the photochemical responsible for this effect, which they named **melanopsin.**

Unlike the other retinal photopigments, which are found in rods and cones, melanopsin is present in ganglion cells—the neurons whose axons transmit information from the eyes to the rest of the brain. Melanopsin-containing ganglion cells are sensitive to light, and their axons terminate in the SCN, the vlPOA, the thalamus, and the olivary pretectal nuclei (Berson, Dunn, and Takao, 2002; Hattar et al., 2002; Gooley et al., 2003). As we saw in the previous section of this chapter, the vlPOA is involved in control of slow-wave sleep. The pretectal nuclei are involved in the pupillary light response. As you know, our pupils dilate in dim light and constrict in bright light. Apparently, melanopsin-containing ganglion cells, and not rods and cones, are involved in this response. (See *Figure 9.25.*)

FIGURE 9.25 ■ Melanopsin-Containing Ganglion Cells in the Retina

The axons of the ganglion cells form the retinohypothalamic tract. These neurons detect the light of dawn that resets the biological clock in the SCN.

(From Hattar, S., Liao, H.-W., Takao, M., Berson, D. M., and Yau, K.-W. *Science,* 2002, *295,* 1065–1070. Copyright © 2002 The American Association for the Advancement of Science. Reprinted with permission.)

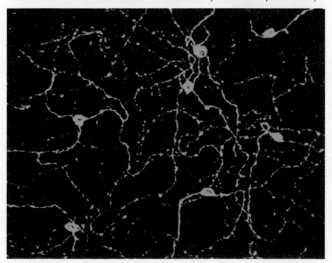

How does the SCN control cycles of sleep and waking? Efferent axons of the SCN responsible for organizing cycles of sleep and waking terminate in the *subparaventricular zone (SPZ),* a region just dorsal to the SCN (Deurveilher and Semba, 2005). Lu et al. (2001) found that excitotoxic lesions of the ventral part of the SPZ disrupted circadian rhythms of sleep and waking. The ventral SPZ projects to the *dorsomedial nucleus of the hypothalamus (DMH),* which in turn projects to several brain regions, including two that play critical roles in the control of sleep and waking: the vlPOA and the orexinergic neurons of the lateral hypothalamus. The projections to the vlPOA are inhibitory and thus inhibit sleep, whereas the projections to the orexinergic neurons are excitatory and thus promote wakefulness (Saper, Scammell, and Lu, 2005). Of course, the activity of these connections varies across the day/night cycle. In diurnal animals (such as ourselves) the activity of these connections is high during the day and low during the night. (See *Figure 9.26.*)

Although connections of SCN neurons with the SPZ appear to play a critical role in circadian control of sleep and waking, several experiments suggest that the SCN can also control these rhythms by secretion of chemicals that diffuse through the brain's extracellular fluid. Lehman et al. (1987) destroyed the SCN and then transplanted in its place a new set of suprachiasmatic nuclei obtained from donor animals. The grafts succeeded in reestablishing circadian rhythms, even though very few synaptic connections were observed between the graft and the recipient's brain.

The most convincing evidence for chemical communication between the SCN and other parts of the brain comes from a transplantation study by Silver et al. (1996). Silver and her colleagues first destroyed the SCN in a group of hamsters, abolishing their circadian rhythms. Then, a few weeks later, they removed SCN tissue from donor animals and placed it in tiny semipermeable capsules, which they then implanted in the animals' third ventricles. Nutrients and other chemicals could pass through the walls of the capsules, keeping the SCN tissue alive, but the neurons inside the capsules were not able to establish synaptic connections with the surrounding tissue. Nevertheless, the transplants re-established circadian rhythms in the recipient animals. The identity of the chemical signal is not yet known, but two candidates have been proposed: *transforming growth factor-α (TGF-α)* and *prokineticin 2* (Cheng et al., 2002; Kramer et al., 2005). Presumably, the chemicals secreted by cells in the SCN affect

melanopsin *(mell a **nop** sin)* A photopigment present in ganglion cells in the retina whose axons transmit information to the SCN, the thalamus, and the olivary pretectal nuclei.

FIGURE 9.26 ■ Control of Circadian Rhythms

The SCN controls circadian rhythms in sleep and waking. During the day cycle, the DMH inhibits the vlPOA and excites the brain stem and forebrain arousal systems, thus stimulating wakefulness.

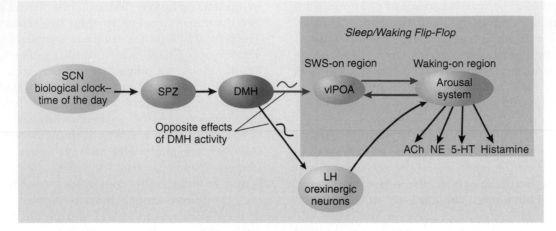

rhythms of sleep and waking by diffusing into the SPZ and binding with receptors on neurons located there.

The Nature of the Clock

All clocks must have a time base. Mechanical clocks use flywheels or pendulums; electronic clocks use quartz crystals. The SCN, too, must contain a physiological mechanism that parses time into units. After years of research, investigators are finally beginning to discover the nature of the biological clock in the SCN.

Several studies have demonstrated daily activity rhythms in the SCN, which indicates that the circadian clock is indeed located there. A study by Schwartz and Gainer (1977) nicely demonstrated day–night fluctuations in the activity of the SCN. The investigators injected some rats with radioactive 2-DG during the day and injected others at night. The animals were then killed, and autoradiographs of cross sections through the brain were prepared. (2-DG autoradiography was described in Chapter 5.) Figure 9.27 shows photographs of two of these cross sections. Note the evidence of radioactivity (and hence a high metabolic rate) in the SCN of the brain that was injected during the day (*left*). (See *Figure 9.27*.)

Schwartz and his colleagues (Schwartz et al., 1983) found a similar pattern of activity in the SCN of squirrel monkeys, which are diurnal animals (active during the day). These results suggest that it is not differences in the SCN that determine whether an animal is nocturnal or diurnal but differences elsewhere in the brain. The SCN keeps track of day and night, but it is up to mechanisms located elsewhere to determine when the animal is to be awake or asleep at these times. As we saw, the dorsomedial nucleus of the thalamus transmits information from the SCN to

brain regions involved in the control of sleep and waking. This nucleus seems to be the most likely place that the sign (excitatory or inhibitory) of circadian signals could be reversed so that the same clock could control sleep/waking rhythms in both nocturnal and diurnal animals.

The "ticking" of the biological clock within the SCN could involve interactions of circuits of neurons, or it could be intrinsic to individual neurons themselves. Evidence suggests the latter—that each neuron contains a clock. Several studies have succeeded in keeping individual SCN neurons alive in a culture medium. For example, Welsh et al. (1995) removed tissue from the rat SCN and dissolved the connections between the cells with papain, an enzyme that is

FIGURE 9.27 ■ Circadian Rhythms in the SCN

The autoradiographs show cross sections of brains of rats injected with radioactive 1_DG during the day (left) and the night (right). The dark region at the base of the brain (arrows) indicates increased metabolic activity of the suprachiasmatic nuclei.

(From Schwartz, W. J., and Gainer, H. *Science*, 1977, *197*, 1089–1091. Copyright © 1977 The American Association for the Advancement of Science. Reprinted with permission.)

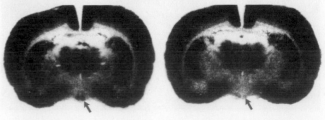

FIGURE 9.28 ■ Firing Rate of SCN Neurons

This figure shows the firing rate of individual neurons in a tissue culture. Color bars have been added to emphasize the daily peaks. Note that although each neuron has a period of approximately 1 day, the activity cycles of the neurons are not synchronized.

(From Welsh, D. K., Logothetis, D. E., Meister, M., and Reppert, S. M. *Neuron*, 1995, *14*, 697–706. Copyright © 1995 Cell Press. Reprinted with permission.)

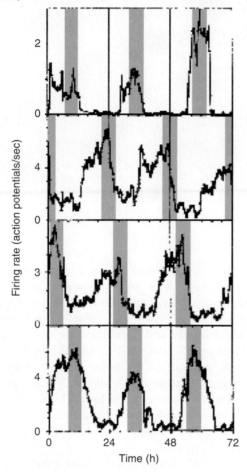

FIGURE 9.29 ■ Circadian Rhythms in the SCN

This schematic is a simplified explanation of the molecular control of the "ticking" of neurons of the SCN.

The protein enters the nucleus, suppressing the gene responsible for its production. No more messenger RNA is made.

The level of the protein falls, so the gene becomes active again.

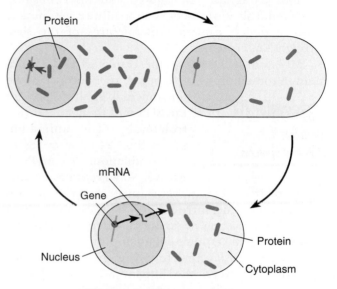

The gene is active; messenger RNA leaves the nucleus and causes the production of the protein.

sometimes used as a meat tenderizer. The cells were placed on top of an array of microelectrodes so that their electrical activity could be measured. Unlike neurons in the intact SCN, whose rhythms are synchronized, the cultured neurons displayed individual, independent, circadian rhythms in activity. Figure 9.28 shows the activity cycles of four neurons. As you can see, all showed circadian rhythms, but their periods of peak activity occurred at different times of day. (See *Figure 9.28*.)

What causes intracellular ticking? For many years, investigators have believed that circadian rhythms were produced by the production of a protein that, when it reached a certain level in the cell, inhibited its own production. As a result, the levels of the protein would begin to decline, which would remove the inhibition, starting the production cycle again. (See *Figure 9.29*.)

Just such a mechanism was discovered in *Drosophila melanogaster,* the common fruit fly. Subsequent research with mammals discovered a similar system (Shearman et al., 2000; Reppert and Weaver, 2001; Van Gelder et al., 2003). The system involves at least seven genes and their proteins and two interlocking feedback loops. When one of the proteins produced by the first loop reaches a sufficient level, it starts the second loop, which eventually inhibits the production of proteins in the first loop, and the cycle begins again. Thus, the intracellular ticking is regulated by the time it takes to produce and degrade a set of proteins. Yan and Silver (2004) found that if animals that are kept in the dark are exposed to a brief period of bright light, the levels of these proteins change. Furthermore, pulses of light presented during different phases of the animal's circadian rhythm produce different effects on the production of these proteins. As we saw earlier, exposure to light at different times of day had different effects on the animals' behavior, setting their biological clocks either forward or back (Aschoff, 1979).

A study by Yamaguchi et al. (2003) devised a procedure that measured the cycles of protein production in individual neurons in the SCN. They inserted a *luciferase reporter gene* into a strain of mice. This gene, which produces the protein responsible for the light emitted by fireflies, was linked to the gene responsible for the production of *per1,* one of the proteins involved in the feedback loops I mentioned in the previous paragraph. Yamaguchi and his colleagues preserved a slice of the SCN in a culture medium. A sensitive digital camera took pictures of the slice through a microscope.

MyPsychKit 9.2 shows the results of this experiment. Production of *per1* began in neurons in the dorsomedial shell region of the SCN and then spread ventrolaterally. The animation shows four 24-hour cycles of *per1* production. (See *MyPsych Kit 9.2, SCN Protein Synthesis.*) When a drug that prevents protein synthesis was added to the fluid surrounding the slice, the cycling stopped, but cycling began again when the drug was rinsed out of the culture chamber.

mypsychkit
Where learning comes to life!
Animation 9.2
SCN Protein Synthesis

Genetic studies have found further evidence for similarities between the human SCN and that of laboratory animals. Toh et al. (2001) found that a mutation on chromosome 2 of a gene for one of the proteins involved in the feedback loops I mentioned earlier in this subsection (*per2*) is responsible for the **advanced sleep phase syndrome.** This syndrome causes a 4-hour advance in rhythms of sleep and temperature cycles. People with this syndrome fall asleep around 7:30 P.M. and awaken around 4:30 A.M. The mutation appears to change the relationship between the zeitgeber of morning light and the phase of the circadian clock that operates in the cells of the SCN. Ebisawa et al. (2001) found evidence that the opposite disorder, the **delayed sleep phase syndrome,** may be caused by mutations of the *per3* gene, found on chromosome 1. This syndrome consists of a 4-hour delay in sleep/waking rhythms. People with this disorder are typically unable to fall asleep before 2:00 A.M. and have great difficulty waking before mid-morning.

Control of Seasonal Rhythms: The Pineal Gland and Melatonin

Although the SCN has an intrinsic rhythm of approximately 24 hours, it plays a role in much longer rhythms. (We could say that it is involved in a biological calendar as well as a biological clock.) For example, male hamsters show annual rhythms of testosterone secretion,

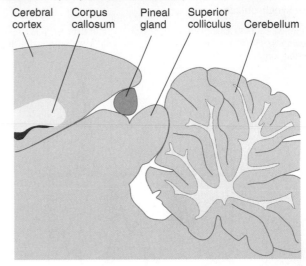

FIGURE 9.30   ■   The Pineal Gland

The pineal gland is located on the dorsal surface of the midbrain.

(Adapted from Paxinos, G., and Watson, C. *The Rat Brain in Stereotaxic Coordinates.* Sydney: Academic Press, 1982.)

Cerebral cortex   Corpus callosum   Pineal gland   Superior colliculus   Cerebellum

which appear to be based on the amount of light that occurs each day. Their breeding season begins as the day length increases and ends when it decreases. Lesions of the SCN abolish these annual breeding cycles; the animals' testes then secrete testosterone all year (Rusak and Morin, 1976). Possibly, the lesions disrupt these annual cycles because they destroy the 24-hour clock against which the daily light period is measured to determine the season. That is, if the light period is considerably shorter than 12 hours, the season is winter; if it is considerably longer than 12 hours, the season is summer.

The control of seasonal rhythms involves another part of the brain: the **pineal gland** (Bartness et al., 1993). This structure sits on top of the midbrain, just in front of the cerebellum. (See *Figure 9.30.*) The

advanced sleep phase syndrome A 4-hour advance in rhythms of sleep and temperature cycles, apparently caused by a mutation of a gene (*per2*) involved in the rhythmicity of neurons of the SCN.

delayed sleep phase syndrome A 4-hour delay in rhythms of sleep and temperature cycles, possibly caused by a mutation of a gene (*per3*) involved in the rhythmicity of neurons of the SCN.

pineal gland (*py* **nee** *ul*) A gland attached to the dorsal tectum; produces melatonin and plays a role in circadian and seasonal rhythms.

pineal gland secretes a hormone called **melatonin**, so named because it has the ability in certain animals (primarily fish, reptiles, and amphibians) to turn the skin temporarily dark. (The dark color is produced by a chemical known as *melanin*.) In mammals, melatonin controls seasonal rhythms. Neurons in the SCN make indirect connections with neurons in the *paraventricular nucleus of the hypothalamus* (the PVN). The axons of these neurons travel all the way to the spinal cord, where they form synapses with preganglionic neurons of the sympathetic nervous system. The postganglionic neurons innervate the pineal gland and control the secretion of melatonin.

In response to input from the SCN, the pineal gland secretes melatonin during the night. This melatonin acts back on various structures in the brain (including the SCN, whose cells contain melatonin receptors) and controls hormones, physiological processes, and behaviors that show seasonal variations. During long nights a large amount of melatonin is secreted, and the animals go into the winter phase of their cycle. Lesions of the SCN, the paraventricular nucleus (PVN), or the pineal gland disrupt seasonal rhythms that are controlled by day length—and so do knife cuts that interrupt the neural connection between the SCN and the PVN, which indicates that this is one function of the SCN that is mediated through its neural connections with another structure. Furthermore, although transplants of fetal suprachiasmatic nuclei will restore circadian rhythms, they will not restore seasonal rhythms, because the transplanted tissue does not establish neural connections with the PVN (Ralph and Lehman, 1991).

Changes in Circadian Rhythms: Shift Work and Jet Lag

When people abruptly change their daily rhythms of activity, their internal circadian rhythms, controlled by the SCN, become desynchronized with those in the external environment. For example, if a person who normally works on the day shift begins working on a night shift or if someone travels east or west across several time zones, his or her SCN will signal the rest of the brain that it is time to sleep during the work shift (or the middle of the day in the case of jet travel). This disparity between internal rhythms and the external environment results in sleep disturbances and mood changes and interferes with people's ability to function during waking hours. Problems such as ulcers, depression, and accidents related to sleepiness are more common in people whose work schedules shift frequently (Drake et al., 2004).

Jet lag is a temporary phenomenon; after several days, people who have crossed several time zones find it easier to fall asleep at the appropriate time, and their daytime alertness improves. Shift work can present a more enduring problem when people are required to change shifts often. Obviously, the solution to jet lag and to the problems caused by shift work is to get the internal clock synchronized with the external environment as quickly as possible. The most obvious way to start is to try to provide strong zeitgebers at the appropriate time. If a person is exposed to bright light before the low point in the daily rhythm of body temperature (which occurs an hour or two before the person usually awakens), the person's circadian rhythm is delayed. If the exposure to bright light occurs after the low point, the circadian rhythm is advanced (Dijk et al., 1995). In fact, several studies have shown that exposure to bright lights at the appropriate time helps to ease the transition (Boulos et al., 1995). Similarly, people adapt to shift work more rapidly if artificial light is kept at a brighter level in the workplace and if the bedroom is kept as dark as possible (Horowitz et al., 2001).

As we saw in the previous subsection, the role of melatonin in seasonal rhythms is well established. Studies in recent years suggest that melatonin may also be involved in circadian rhythms. As we saw, melatonin is secreted during the night, which, for diurnal mammals such as ourselves, is the period during which we sleep. But although our species lacks strong seasonal rhythms, the daily rhythm of melatonin secretion persists. Thus, melatonin must have some functions besides regulation of seasonal rhythms.

Studies have found that melatonin, acting on receptors in the SCN, can affect the sensitivity of SCN neurons to zeitgebers and can itself alter circadian rhythms (Gillette and McArthur, 1995; Starkey et al., 1995). Researchers do not yet understand exactly what role melatonin plays in the control of circadian rhythms, but they have already discovered practical applications. Melatonin secretion normally reaches its highest levels early in the night, at around bedtime. Investigators have found that the administration of melatonin at the appropriate time (in most cases, just before going to bed) significantly reduces the adverse effects of both jet lag and shifts in work schedules (Arendt et al., 1995; Deacon and Arendt, 1996). Bedtime melatonin has even helped to synchronize circadian rhythms and has improved the sleep of blind people for whom light cannot serve as a zeitgeber (Skene, Lockley, and Arendt, 1999).

---

melatonin *(mell a **tone** in)* A hormone secreted during the night by the pineal body; plays a role in circadian and seasonal rhythms.

InterimSummary

Biological Clocks

Our daily lives are characterized by cycles in physical activity, sleep, body temperature, secretion of hormones, and many other physiological changes. Circadian rhythms—those with a period of approximately 1 day—are controlled by biological clocks in the brain. The principal biological clock appears to be located in the suprachiasmatic nuclei of the hypothalamus; lesions of these nuclei disrupt most circadian rhythms, and the activity of neurons located there correlates with the day–night cycle. Light, detected by special retinal ganglion cells that contain a photopigment called melanopsin, serves as a zeitgeber for most circadian rhythms. The human biological clock tends to run a bit slow, with a period of approximately 25 hours. The sight of sunlight in the morning is conveyed from the retina to the core region of the SCN. The shell region, which contains neurons that show circadian rhythms of activity, receives a signal from neurons in the core region that reset the clock to the start of a new cycle.

Individual neurons, rather than circuits of neurons, are responsible for the "ticking." Each tick, approximately 24 hours long, consists of the production and breakdown of a series of proteins in two interlocking loops that act back on the genes responsible for their own production.

The SCN and the pineal gland control annual rhythms. During the night the SCN signals the pineal gland to secrete melatonin. Prolonged melatonin secretion, which occurs during winter, causes the animals to enter the winter phase of their annual cycle. Melatonin also appears to be involved in synchronizing circadian rhythms: The hormone can help people to adjust to the effects of shift work or jet lag and even synchronize the daily rhythms of blind people for whom light cannot serve as a zeitgeber.

Thought Question

Until recently (in terms of the evolution of our species), our ancestors tended to go to sleep when the sun set and wake up when it rose. Once our ancestors learned how to control fire, they undoubtedly stayed up somewhat later, sitting in front of a fire. But it was only with the development of cheap, effective lighting that many members of our species adopted the habit of staying up late and waking several hours after sunrise. Considering that our biological clock and the neural mechanisms it controls evolved long ago, do you think the changes in our daily rhythms impair any of our physical and intellectual abilities?

SUGGESTED READINGS

Hobson, J. A. *Dreaming: An Introduction to the Science of Sleep.* Oxford, England: Oxford University Press, 2004.

Jouvet, M. *The Paradox of Sleep: The Story of Dreaming.* Cambridge, MA: The MIT Press, 2001.

Kryger, M. H., Roth, T., and Dement, W. C. *Principles and Practice of Sleep Medicine,* 3rd ed. Philadelphia: Saunders, 2000.

Luppi, P.-H. *Sleep: Circuits and Function.* Boca Raton, FL: CRC Press, 2005.

Pace-Schott, E. F., Solms, M., Blagrove, M., and Harnad, S. *Sleep and Dreaming: Scientific Advances and Reconsiderations.* Cambridge, England: Cambridge University Press, 2003.

Vu, T. T., Desseilles, M., Petit, D., Mazza, S., Montplaisir, J. Maquet, P. Neuroimaging in sleep medicine. *Sleep Medicine,* 2007, *8,* 350–373.

ADDITIONAL RESOURCES

Visit www.mypsychkit.com for additional review and practice of the material covered in this chapter. Within MyPsychKit, you can take practice tests and receive a customized study plan to help you review. Dozens of animations, tutorials, and Web links are also available. You can even review using the interactive electronic version of this textbook. You will need to register for MyPsychKit. See www.mypsychkit.com for complete details.

Reproductive Behavior

outline

At first a tragic surgical accident appeared to suggest that people's sexual identity and sexual orientation were not under the strong control of biological factors and that these characteristics could be shaped by the way a child was raised (Money and Ehrhardt, 1972). Identical twin boys were raised normally until seven months of age, at which time the penis of one of the boys was accidentally removed during circumcision. The cautery (a device that cuts tissue by means of electric current) was adjusted too high, and instead of removing the foreskin, the current burned off the entire penis. After a period of agonized indecision the parents decided to raise the child as a girl. John became Joan.

Joan's parents started dressing her in girl's clothing and treating her like a little girl. Surgeons performed a sex change operation, removing the testes and creating a vagina. At first, psychologists who studied Joan reported that she was a normal, happy girl and concluded that it was a child's upbringing that determined his or her sexual identity. Many writers saw this case as a triumph of socialization over biology.

Unfortunately, this conclusion was premature (Diamond and Sigmundson, 1997). It turned out that although Joan did not know she had been born as a boy, she was unhappy as a girl. She felt that she really was a boy and even tried to stand to urinate. When, as an unhappy adolescent, she threatened to commit suicide, her family and physicians agreed to a sex change. The estrogen treatment she had been receiving was terminated, she started taking androgens, she had a mastectomy, and surgeons began creating a phallus. Joan became John again. His father finally told him that he was born a boy, a revelation that John received with great relief. John later married and adopted his wife's children.

We now know this person's real identity—actually, Bruce became Brenda, who then chose the name David after deciding to become a boy again. A book has told his story (Colapinto, 2000), and a 2002 television documentary (*NOVA*'s "Sex: Unknown") presented interviews with David, his mother, Dr. Diamond, and others involved in this unfortunate case. Sadly, David subsequently lost his job, he and his wife separated, and in May 2004, at the age of 38, he committed suicide.

Reproductive behaviors constitute the most important category of social behaviors, because without them, most species would not survive. These behaviors—which include courting, mating, parental behavior, and most forms of aggressive behaviors—are the most striking categories of **sexually dimorphic behaviors,** that is, behaviors that differ in males and females (*di* + *morphous*, "two forms"). As you will see, hormones that are present both before and after birth play a very special role in the development and control of sexually dimorphic behaviors.

This chapter describes male and female sexual development and then discusses the neural and hormonal control of two sexually dimorphic behaviors that are most important to reproduction: sexual behavior and parental behavior.

SEXUAL DEVELOPMENT

A person's chromosomal sex is determined at the time of fertilization. However, this event is merely the first in a series of steps that culminate in the development of a male or female. This section considers the major features of sexual development.

Production of Gametes and Fertilization

All cells of the human body (other than sperms or ova) contain twenty-three pairs of chromosomes. The genetic information that programs the development of a human is contained in the DNA that constitutes these chromosomes. We pride ourselves on our ability to miniaturize computer circuits on silicon chips, but that accomplishment looks primitive when we consider that the blueprint for a human being is too small to be seen by the naked eye.

The production of **gametes** (ova and sperms; *gamein* means "to marry") entails a special form of cell division. This process produces cells that contain one member of each of the twenty-three pairs of chromosomes. The development of a human begins at the time of fertilization, when a single sperm and ovum join, sharing their twenty-three single chromosomes to reconstitute the twenty-three pairs.

sexually dimorphic behavior A behavior that has different forms or that occurs with different probabilities or under different circumstances in males and females.

gamete (*gamm eet*) A mature reproductive cell; a sperm or ovum.

FIGURE 10.1 ■ Determination of Gender

The gender of the offspring depends on whether the sperm cell that fertilizes the ovum carries an X or a Y chromosome.

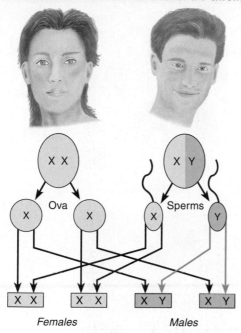

A person's genetic sex is determined at the time of fertilization of the ovum by the father's sperm. Twenty-two of the twenty-three pairs of chromosomes determine the organism's physical development independent of its sex. The last pair consists of two **sex chromosomes,** which determine whether the offspring will be a boy or a girl.

There are two types of sex chromosomes: X chromosomes and Y chromosomes. Females have two X chromosomes (XX); thus, all the ova that a woman produces will contain an X chromosome. Males have an X and a Y chromosome (XY). When a man's sex chromosomes divide, half the sperms contain an X chromosome, and the other half contain a Y chromosome. A Y-bearing sperm produces an XY-fertilized ovum and therefore a male. An X-bearing sperm produces an XX-fertilized ovum and therefore a female. (See *Figure 10.1.*)

Development of the Sex Organs

Men and women differ in many ways: Their bodies are different, parts of their brains are different, and their reproductive behaviors are different. Are all these differences directly encoded on the tiny Y chromosome, the sole piece of genetic material that distinguishes males from females? The answer is no. The X chromosome and the twenty-two nonsex chromosomes found in the cells of both males and females contain all the information needed to develop the bodies of either sex. Exposure to sex hormones, both before and after birth,

is responsible for our sexual dimorphism. What the Y chromosome does control is the development of the glands that produce the male sex hormones.

Gonads

There are three general categories of sex organs: the gonads, the internal sex organs, and the external genitalia. The **gonads**—testes or ovaries—are the first to develop. Gonads (from the Greek *gonos,* "procreation") have a dual function: They produce ova or sperms, and they secrete hormones. Through the sixth week of prenatal development, male and female fetuses are identical. Both sexes have a pair of identical undifferentiated gonads, which have the potential of developing into either testes or ovaries. The factor that controls their development appears to be a single gene on the Y chromosome called *Sry* (sex-determining region Y). This gene produces a protein that binds to the DNA of cells in the undifferentiated gonads and causes them to become testes. (Testes are also known as *testicles,* Latin for "little testes.") Believe it or not, the words "testis" and "testify" have the same root, meaning "witness." Legend has it that ancient Romans placed their right hand over their genitals while swearing that they would tell the truth in court. (Only men were permitted to testify.) If the Sry gene is not present, the undifferentiated gonads become ovaries (Sinclair et al., 1990; Smith, 1994; Koopman, 2001). In fact, a few cases of XX males have been reported. This anomaly can occur when the Sry gene becomes translocated from the Y chromosome to the X chromosome during production of the father's sperms (Warne and Zajac, 1998). (A test based on a molecular probe for Sry was used to ensure that potential competitors for the women's Olympic events in Atlanta in 1996 had no Sry gene.) Although the Sry gene begins the process of gonadal differentiation, at least two other genes are also necessary for completion of this process (Nikolova and Vilain, 2006).

Once the gonads have developed, a series of events is set into action that determines the individual's gender. These events are directed by hormones, which affect sexual development in two ways. During prenatal development these hormones have **organizational effects,** which influence the development of a person's

---

sex chromosome The X and Y chromosomes, which determine an organism's gender. Normally, XX individuals are female, and XY individuals are male.

gonad (rhymes with *moan* ad) An ovary or testis.

Sry The gene on the Y chromosome whose product instructs the undifferentiated fetal gonads to develop into testes.

organizational effect (of hormone) The effect of a hormone on tissue differentiation and development.

FIGURE 10.2 ■ Development of Internal Sex Organs

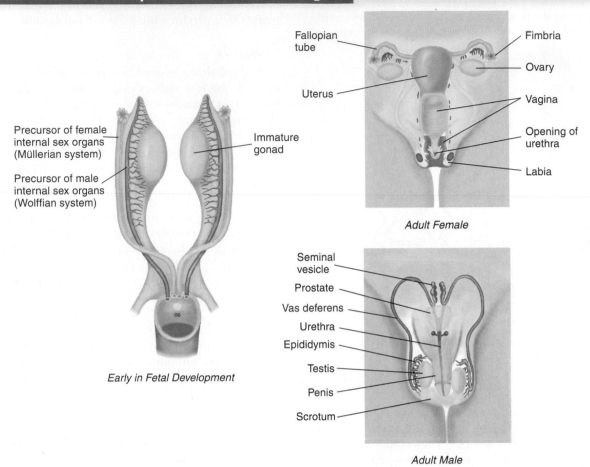

Early in Fetal Development

Adult Female

Adult Male

sex organs and brain. These effects are permanent; once a particular path is followed in the course of development, there is no going back. The second role of sex hormones is their **activational effect.** These effects occur later in life, after the sex organs have developed. For example, hormones activate the production of sperms, make erection and ejaculation possible, and induce ovulation. Because the bodies of adult males and females have been organized differently, sex hormones will have different activational effects in the two sexes.

Internal Sex Organs

Early in embryonic development the internal sex organs are *bisexual*; that is, all embryos contain the precursors for both female and male sex organs. However, during the third month of gestation, only one of these precursors develops; the other withers away. The precursor of the internal female sex organs, which develops into the *fimbriae* and *Fallopian tubes,* the *uterus,* and the *inner two-thirds of the vagina,* is called the **Müllerian system.** The precursor of the internal male sex organs, which develops into the *epididymis, vas deferens,* and

seminal vesicles, is called the **Wolffian system.** (These systems were named after their discoverers, Müller and Wolff. See *Figure 10.2.*)

The gender of the internal sex organs of a fetus is determined by the presence or absence of hormones secreted by the testes. If these hormones are present, the Wolffian system develops. If they are not, the Müllerian system develops. The Müllerian (female) system needs no hormonal stimulus from the gonads to develop; it just normally does so. (Turner's syndrome, a disorder of sexual development that I will discuss later, provides the evidence for this assertion.) In contrast, the cells of the Wolffian (male) system do not develop unless they are stimulated to do so by a

activational effect (of hormone) The effect of a hormone that occurs in the fully developed organism; may depend on the organism's prior exposure to the organizational effects of hormones.

Müllerian system The embryonic precursors of the female internal sex organs.

Wolffian system The embryonic precursors of the male internal sex organs.

hormone. Thus, testes secrete two types of hormones. The first, a peptide hormone called **anti-Müllerian hormone,** does exactly what its name says: It prevents the Müllerian (female) system from developing. It therefore has a **defeminizing effect.** The second, a set of steroid hormones called **androgens,** stimulates the development of the Wolffian system. (This class of hormone is also aptly named: *Andros* means "man," and *gennan* means "to produce.") Androgens have a **masculinizing effect.**

Two different androgens are responsible for masculinization. The first, **testosterone,** is secreted by the testes and gets its name from these glands. An enzyme called *5α reductase* converts some of the testosterone into another androgen, known as **dihydrotestosterone.**

As you will recall from Chapter 2, hormones exert their effects on target cells by stimulating the appropriate hormone receptor. Thus, the precursor of the male internal sex organs—the Wolffian system—contains androgen receptors that are coupled to cellular mechanisms that promote growth and division. When molecules of androgens bind with these receptors, the epididymis, vas deferens, and seminal vesicles develop and grow. In contrast, the cells of the Müllerian system contain receptors for anti-Müllerian hormone that *prevent* growth and division. Thus, the presence of anti-Müllerian hormone prevents the development of the female internal sex organs.

The fact that the internal sex organs of the human embryo are bisexual and could potentially develop as either male or female is dramatically illustrated by two genetic disorders: *androgen insensitivity syndrome* and *persistent Müllerian duct syndrome.* Some people are insensitive to androgens; they have **androgen insensitivity syndrome,** one of the more aptly named disorders (Money and Ehrhardt, 1972; MacLean, Warne, and Zajac, 1995). The cause of androgen insensitivity syndrome is a genetic mutation that prevents the formation of functioning androgen receptors. (The gene for the androgen receptor is located on the X chromosome.) The primitive gonads of a genetic male fetus with androgen insensitivity syndrome become testes and secrete both anti-Müllerian hormone and androgens. The lack of androgen receptors prevents the androgens from having a masculinizing effect; thus, the epididymis, vas deferens, seminal vesicles, and prostate fail to develop. However, the anti-Müllerian hormone still has its defeminizing effect, preventing the female internal sex organs from developing. The uterus, fimbriae, and Fallopian tubes fail to develop, and the vagina is shallow. The external genitalia are female, and at puberty the person develops a woman's body. Of course, lacking a uterus and ovaries, the person cannot have children. (See *Figure 10.3.*)

FIGURE 10.3 ■ XY Female Displaying Androgen Insensitivity Syndrome

(From Money, J., and Ehrhardt, A. A. *Man & Woman, Boy & Girl.* Copyright © 1973 by The Johns Hopkins University Press, Baltimore, Maryland. Reprinted with permission.)

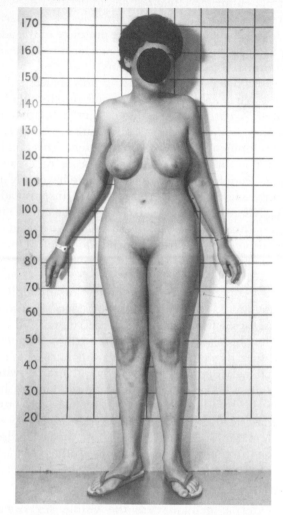

anti-Müllerian hormone A peptide secreted by the fetal testes that inhibits the development of the Müllerian system, which would otherwise become the female internal sex organs.

defeminizing effect An effect of a hormone present early in development that reduces or prevents the later development of anatomical or behavioral characteristics typical of females.

androgen (an dro jen) A male sex steroid hormone. Testosterone is the principal mammalian androgen.

masculinizing effect An effect of a hormone present early in development that promotes the later development of anatomical or behavioral characteristics typical of males.

testosterone (tess **tahss** ter own) The principal androgen found in males.

dihydrotestosterone (dy hy dro tess **tahss** ter own) An androgen, produced from testosterone through the action of the enzyme 5α reductase.

androgen insensitivity syndrome A condition caused by a congenital lack of functioning androgen receptors; in a person with XY sex chromosomes, causes the development of a female with testes but no internal sex organs.

The second genetic disorder, **persistent Müllerian duct syndrome,** has two causes: either a failure to produce anti-Müllerian hormone or the absence of receptors for this hormone (Warne and Zajac, 1998). When this syndrome occurs in genetic males, androgens have their masculinizing effect but defeminization does not occur. Thus, the person is born with *both* sets of internal sex organs, male and female. The presence of the additional female sex organs usually interferes with normal functioning of the male sex organs.

So far, I have been discussing only male sex hormones. What about prenatal sexual development in females? A chromosomal anomaly indicates that the hormones produced by female sex organs are not needed for development of the Müllerian system. This fact has led to the dictum "Nature's impulse is to create a female." People with **Turner's syndrome** have only one sex chromosome: an X chromosome. (Thus, instead of having XX cells, they have X0 cells—0 indicating a missing sex chromosome.) In most cases the existing X chromosome comes from the mother, which means that the cause of the disorder lies with a defective sperm (Knebelmann et al., 1991). Because a Y chromosome is not present, testes do not develop. In addition, because two X chromosomes are needed to produce ovaries, these glands are not produced either. But even though people with Turner's syndrome have no gonads at all, they develop into females, with normal female internal sex organs and external genitalia—which proves that fetuses do not need ovaries or the hormones they produce to develop as females. Of course, they must be given estrogen pills to induce puberty and sexual maturation. And they cannot bear children, because without ovaries they cannot produce ova.

External Genitalia

The external genitalia are the visible sex organs, including the penis and scrotum in males and the labia, clitoris, and outer part of the vagina in females. (See *Figure 10.4.*) As we just saw, the external genitalia do not need to be stimulated by female sex hormones to become female; they just naturally develop that way. In the presence of dihydrotestosterone the external genitalia will become male. Thus, the gender of a person's external genitalia is determined by the presence or absence of an androgen, which explains why people with Turner's syndrome have female external genitalia even though they lack ovaries. People with androgen insensitivity syndrome have female external genitalia too, because without androgen receptors their cells cannot respond to the androgens produced by their testes.

Figure 10.5 summarizes the factors that control the development of the gonads, internal sex organs, and genitalia. (See *Figure 10.5.*)

FIGURE 10.4 ■ Development of External Genitalia

(Adapted from Spaulding, M. H., in *Contributions to Embryology*, Vol. 13. Washington, DC: Carnegie Institute of Washington, 1921.)

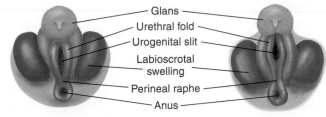
Indifferent Stage

Seventh to Eighth Week

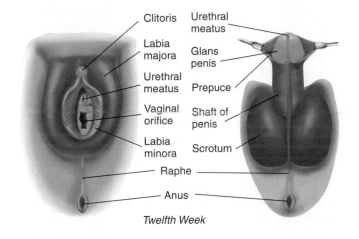
Twelfth Week

Sexual Maturation

The *primary* sex characteristics include the gonads, internal sex organs, and external genitalia. These organs are present at birth. The *secondary* sex characteristics, such as enlarged breasts and widened hips or a beard and deep voice, do not appear until puberty. Without seeing genitals, we must guess the sex of a prepubescent child from his or her haircut and clothing; the bodies of young boys

persistent Müllerian duct syndrome A condition caused by a congenital lack of anti-Müllerian hormone or receptors for this hormone; in a male, causes development of both male and female internal sex organs.

Turner's syndrome The presence of only one sex chromosome (an X chromosome); characterized by lack of ovaries but otherwise normal female sex organs and genitalia.

FIGURE 10.5 ■ Hormonal Control of Development

This schematic shows the hormonal control of internal sex organ development.

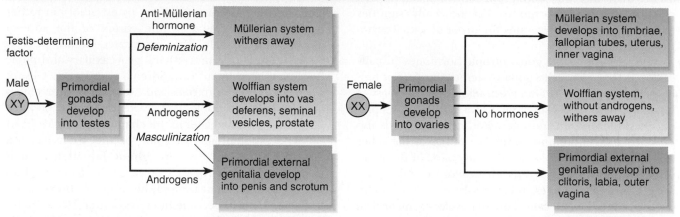

and girls are rather similar. However, at puberty the gonads are stimulated to produce their hormones, and these hormones cause the person to mature sexually. The onset of puberty occurs when cells in the hypothalamus secrete **gonadotropin-releasing hormones (GnRH),** which stimulate the production and release of two **gonadotropic hormones** by the anterior pituitary gland. The gonadotropic ("gonad-turning") hormones stimulate the gonads to produce *their* hormones, which are ultimately responsible for sexual maturation. (See *Figure 10.6.*)

The two gonadotropic hormones are **follicle-stimulating hormone (FSH)** and **luteinizing hormone (LH),**

named for the effects they produce in the female (production of a *follicle* and its subsequent *luteinization,* to be described in the next section of this chapter). However, the same hormones are produced in the male, where they stimulate the testes to produce sperms and to secrete testosterone. If male and female pituitary glands are exchanged in rats, the ovaries and testes respond perfectly to the hormones secreted by the new glands (Harris and Jacobsohn, 1951–1952).

For over a century the age at which children (particularly girls) reach puberty has been diminishing in developed countries, presumably because of improved nutrition (Foster and Nagatani, 1999). Girls who remain unusually thin through exercise and diet tend to reach puberty later than normal (think of a female Olympic gymnast), while obese girls tend to reach puberty sooner (Frisch, 1990). As we will see in Chapter 12, *leptin,* a peptide hormone secreted by well-nourished fat cells, provides an important signal to the brain concerning the amount of fat tissue in the body. If body fat increases, the level of leptin in the blood increase and signals the brain to suppress appetite. This hormone also appears to play a role in determining the onset of puberty in females. Chehab et al. (1997) gave juvenile female mice daily injections of leptin. Although the appetite-suppressing

FIGURE 10.6 ■ Sexual Maturation

Puberty is initiated when the hypothalamus secretes gonadotropin-releasing hormones, which activate the secretion of gonadotropic hormones by the anterior pituitary gland.

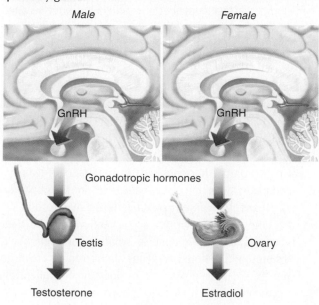

gonadotropin-releasing hormone (GnRH) (go *nad oh* **trow** *pin*) A hypothalamic hormone that stimulates the anterior pituitary gland to secrete gonadotropic hormone.

gonadotropic hormone A hormone of the anterior pituitary gland that has a stimulating effect on cells of the gonads.

follicle-stimulating hormone (FSH) The hormone of the anterior pituitary gland that causes development of an ovarian follicle and the maturation of an ovum.

luteinizing hormone (LH) (**lew** *tee a nize ing*) A hormone of the anterior pituitary gland that causes ovulation and development of the ovarian follicle into a corpus luteum.

effects of the injections caused the animals' body weight to decrease, these mice nevertheless entered puberty earlier than mice that were given control injections of a placebo. Thus, a hormone that normally signifies increased body fat accelerates the onset of sexual maturity, at least in females.

In response to the gonadotropic hormones (usually called *gonadotropins*) the gonads secrete steroid sex hormones. The ovaries produce **estradiol,** one of a class of hormones known as **estrogens.** As we saw, the testes produce testosterone, an androgen. Both types of glands also produce a small amount of the hormones of the other sex. The gonadal steroids affect many parts of the body. Both estradiol and androgens initiate closure of the growing portions of the bones and thus halt skeletal growth. In females, estradiol also causes breast development, growth of the lining of the uterus, changes in the deposition of body fat, and maturation of the female genitalia. In males, androgens stimulate growth of facial, axillary (underarm), and pubic hair; lower the voice; alter the hairline on the head (often causing baldness later in life);

stimulate muscular development; and cause genital growth. This description leaves out two of the female secondary characteristics: axillary and pubic hair. These characteristics are produced not by estradiol but rather by androgens secreted by the cortex of the adrenal glands. Even a male who is castrated before puberty (whose testes are removed) will grow axillary and pubic hair, stimulated by his own adrenal androgens. A list of the principal sex hormones and examples of their effects are presented in Table 10.1. Note that some of these effects are discussed later in this chapter. (See *Table 10.1.*)

The bipotentiality of some of the secondary sex characteristics remains throughout life. If a man is treated with an estrogen (for example, to control an androgen-dependent tumor), he will grow breasts, and his facial hair will become finer and softer. However, his voice will remain low, because the enlargement of the larynx is permanent. Conversely, a woman who receives high levels of an androgen (usually from a tumor that secretes androgens) will grow a beard, and her voice will become lower.

Interim**Summary**

Sexual Development

Gender is determined by the sex chromosomes: XX produces a female, and XY produces a male. Males are produced by the action of the Sry gene on the Y chromosome, which contains the code for the production of a protein that in turn causes the primitive gonads to become testes. The testes secrete two kinds of hormones that cause a male to develop. Testosterone and dihydrotestosterone (androgens) stimulate the development of the Wolffian system (masculinization), and anti-Müllerian hormone suppresses the development of the Müllerian system (defeminization). Androgen insensitivity syndrome results from a hereditary defect in androgen receptors, and persistent Müllerian duct syndrome results from a hereditary defect in production of anti-Müllerian hormone or its receptors.

By default the body is female ("Nature's impulse is to create a female"); only by the actions of testicular hormones does it become male. Masculinization and defeminization are referred to as *organizational* effects of hormones; *activational* effects occur after development is complete. A person with Turner's syndrome (X0) fails to develop gonads but nevertheless develops female internal sex organs and external genitalia. The external genitalia develop from common precursors. In the absence of gonadal hormones the precursors develop the female form; in the presence of

androgens (primarily dihydrotestosterone, which derives from testosterone through the action of 5α reductase) they develop the male form (masculinization).

Sexual maturity occurs when the hypothalamus begins secreting gonadotropin-releasing hormone, which stimulates the secretion of follicle-stimulating hormone and luteinizing hormone by the anterior pituitary gland. These hormones stimulate the gonads to secrete their hormones, thus causing the genitals to mature and the body to develop the secondary sex characteristics (activational effects). Leptin, a hormone secreted by well-nourished fat tissue, appears to be one of the signals that stimulates the onset of puberty, at least in females.

Thought Questions

1. Suppose that parents could determine the sex of their child, say, by having one of the would-be parents take a drug before conceiving the baby. What would the consequences be?

2. With appropriate hormonal treatment the uterus of a postmenopausal woman can be made ready for the implantation of another woman's ovum, fertilized in vitro, and she can become a mother. In fact, several women in their fifties and sixties have done so. What do you think about this procedure? Should decisions about using it be left to couples and their physicians, or does the rest of society (represented by their legislators) have an interest?

TABLE 10.1 ■ Classification of Sex Steroid Hormones		
CLASS	**PRINCIPAL HORMONE IN HUMANS (WHERE PRODUCED)**	**EXAMPLES OF EFFECTS**
Androgens	Testosterone (testes)	Development of Wolffian system; production of sperms; growth of facial, pubic, and axillary hair; muscular development; enlargement of larynx; inhibition of bone growth; sex drive in men (and women?)
	Dihydrotestosterone (produced from testosterone by action of 5α reductase)	Maturation of male external genitalia
	Androstenedione (adrenal glands)	In women, growth of pubic and axillary hair; less important than testosterone and dihydrotestosterone in men
Estrogens	Estradiol (ovaries)	Maturation of female genitalia; growth of breasts; alterations in fat deposits; growth of uterine lining; inhibition of bone growth; sex drive in women (?)
Gestagens	Progesterone (ovaries)	Maintenance of uterine lining
Hypothalamic hormones	Gonadotropin-releasing hormone (hypothalamus)	Secretion of gonadotropins
Gonadotropins	Follicle-stimulating hormone (anterior pituitary)	Development of ovarian follicle
	Luteinizing hormone (anterior pituitary)	Ovulation; development of corpus luteum
Other hormones	Prolactin (anterior pituitary)	Milk production; male refractory period (?)
	Oxytocin (posterior pituitary)	Milk ejection; orgasm; pair bonding (especially females); bonding with infants
	Vasopressin (posterior pituitary)	Pair bonding (especially males)

HORMONAL CONTROL OF SEXUAL BEHAVIOR

We have seen that hormones are responsible for sexual dimorphism in the structure of the body and its organs. Hormones have organizational and activational effects on the internal sex organs, genitals, and secondary sex characteristics. Naturally, all of these effects influence a person's behavior. Simply having the physique and genitals of a man or a woman exerts a powerful effect.

But hormones do more than give us masculine or feminine bodies; they also affect behavior by interacting directly with the nervous system. Androgens that are present during prenatal development affect the

---

estradiol (*ess tra **dye** ahl*) The principal estrogen of many mammals, including humans.

estrogen (*ess trow jen*) A class of sex hormones that cause maturation of the female genitalia, growth of breast tissue, and development of other physical features characteristic of females.

development of the nervous system. In addition, both male and female sex hormones have activational effects on the adult nervous system that influence both physiological processes and behavior. This section considers some of these hormonal effects.

Hormonal Control of Female Reproductive Cycles

The reproductive cycle of female primates is called a **menstrual cycle** (from *mensis,* meaning "month"). Females of other species of mammals also have reproductive cycles, called **estrous cycles.** *Estrus* means "gadfly"; when a female rat is in estrus, her hormonal condition goads her to act differently than she does at other times. (For that matter, her condition goads male rats to act differently too.) The primary feature that distinguishes menstrual cycles from estrous cycles is the monthly growth and loss of the lining of the uterus. The other features are approximately the same, except that the estrous cycle of rats takes four days. Also, the sexual behavior of female mammals with estrous cycles is linked to ovulation, whereas most female primates can mate at any time during their menstrual cycle.

Menstrual cycles and estrous cycles consist of a sequence of events that are controlled by hormonal secretions of the pituitary gland and ovaries. These glands interact, the secretions of one affecting those of the other. A cycle begins with the secretion of gonadotropins by the anterior pituitary gland. These hormones (especially FSH) stimulate the growth of **ovarian follicles,** small spheres of epithelial cells surrounding each ovum. Women normally produce one ovarian follicle each month; if two are produced and fertilized, dizygotic (fraternal) twins will develop. As ovarian follicles mature, they secrete estradiol, which causes growth of the lining of the uterus in preparation for implantation of the ovum, should it be fertilized by a sperm. Feedback from the increasing level of estradiol eventually triggers the release of a surge of LH by the anterior pituitary gland. (See *Figure 10.7* and *MyPsychKit 10.1, The Menstrual Cycle.*)

mypsychkit
Animation 10.1
The Menstrual Cycle

The LH surge causes *ovulation:* The ovarian follicle ruptures, releasing the ovum. Under the continued influence of LH, the ruptured ovarian follicle becomes a **corpus luteum** ("yellow body"), which produces estradiol and **progesterone.** (See *Figure 10.7.*) The latter hormone promotes pregnancy (*gestation*). It maintains the lining of the uterus, and it inhibits the ovaries from producing another follicle. Meanwhile, the ovum enters one of the Fallopian tubes and begins its progress toward the uterus. If it meets sperm cells during its travel down the Fallopian

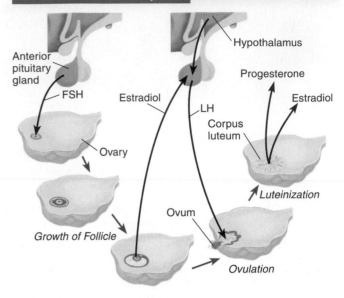

FIGURE 10.7 ■ **Neuroendocrine Control of the Menstrual Cycle**

tube and becomes fertilized, it begins to divide, and several days later it attaches itself to the uterine wall.

If the ovum is not fertilized or if it is fertilized too late to develop sufficiently by the time it gets to the uterus, the corpus luteum will stop producing estradiol and progesterone, and then the lining of the walls of the uterus will slough off. At this point, menstruation will commence.

Hormonal Control of Sexual Behavior of Laboratory Animals

The interactions between sex hormones and the human brain are difficult to study. We must turn to two sources of information: experiments with animals and various developmental disorders in humans, which serve as nature's own "experiments." Let us first consider the evidence gathered from research with laboratory animals.

menstrual cycle (*men strew al*) The female reproductive cycle of most primates, including humans; characterized by growth of the lining of the uterus, ovulation, development of a corpus luteum, and (if pregnancy does not occur), menstruation.

estrous cycle The female reproductive cycle of mammals other than primates.

ovarian follicle A cluster of epithelial cells surrounding an oocyte, which develops into an ovum.

corpus luteum (*lew tee um*) A cluster of cells that develops from the ovarian follicle after ovulation; secretes estradiol and progesterone.

progesterone (*pro jess ter own*) A steroid hormone produced by the ovary that maintains the endometrial lining of the uterus during the later part of the menstrual cycle and during pregnancy; along with estradiol it promotes receptivity in female mammals with estrous cycles.

Males

Male sexual behavior is quite varied, although the essential features of *intromission* (entry of the penis into the female's vagina), *pelvic thrusting* (rhythmic movement of the hindquarters, causing genital friction), and *ejaculation* (discharge of semen) are characteristic of all male mammals. Humans, of course, have invented all kinds of copulatory and noncopulatory sexual behavior. For example, the pelvic movements leading to ejaculation may be performed by the woman, and sex play can lead to orgasm without intromission.

The sexual behavior of rats has been studied more than that of any other laboratory animal (Hull and Dominguez, 2007). Male rats reach sexual maturity between 45 and 75 days of age. When an adult male rat encounters a receptive female, he will spend some time nuzzling her and sniffing her face and genitals, mount her, and engage in several rapid, shallow pelvic thrusts. If he detects her vagina, he will make a deeper thrust, achieve intromission, and then dismount. He will mount her several times, achieving intromission on most of the mountings. After eight to fifteen intromissions approximately 1 minute apart (each lasting only about one-quarter of a second), the male will ejaculate.

After ejaculating, the male refrains from sexual activity for a period of time (minutes, in the rat). Most mammals will return to copulate several times, finally showing a longer pause, called a **refractory period.** (The term comes from the Latin *refringere,* "to break off.") An interesting phenomenon occurs in some mammals. If a male, after finally becoming "exhausted" by repeated copulation with the same female, is presented with a new female, he begins to respond quickly—often as fast as he did in his initial contact with the first female. Successive introductions of new females can keep his performance up for prolonged periods of time. This phenomenon is undoubtedly important in species in which a single male inseminates all the females in his harem. Species with approximately equal numbers of reproductively active males and females are less likely to act this way.

In one of the most unusual studies I have read about, Beamer, Bermant, and Clegg (1969) tested the ability of a ram (male sheep) to recognize ewes with which he had mated. A ram that is given a new ewe each time will quickly begin copulating and will ejaculate within 2 minutes. (In one study, a ram kept up this performance with twelve ewes. The experimenters finally got tired of shuffling sheep around; the ram was still ready to go.) Beamer and his colleagues tried to fool rams by putting trench coats and Halloween face masks on females with which the rams had mated. (No, I'm not making this up.) The males were not fooled by the disguise; they apparently recognized their former partners by their odor and were no longer interested in them.

The rejuvenating effect of a new female, also seen in roosters, is usually called the **Coolidge effect.** The following story is reputed to be true, but I cannot vouch for that. (If it is not true, it ought to be.) The former U.S. president Calvin Coolidge and his wife were touring a farm when Mrs. Coolidge asked the farmer whether the continuous and vigorous sexual activity among the flock of hens was the work of just one rooster. The reply was yes. She smiled and said, "You might point that out to Mr. Coolidge." The president looked thoughtfully at the birds and then asked the farmer whether a different hen was involved each time. The answer, again, was yes. "You might point that out to Mrs. Coolidge," he said.

Sexual behavior of male rodents depends on testosterone, a fact that has long been recognized (Bermant and Davidson, 1974). If a male rat is castrated (that is, if his testes are removed), his sexual activity eventually ceases. However, the behavior can be reinstated by injections of testosterone. I will describe the neural basis of this activational effect later in this chapter.

Other hormones play a role in male sexual behavior. **Oxytocin** is a hormone produced by the posterior pituitary gland that contracts the milk ducts and thus causes milk ejection in lactating females. It is also produced in males, where it obviously plays no role in lactation. Oxytocin is released at the time of orgasm in both males and females and appears to contribute to the contractions of the smooth muscle in the male ejaculatory system and of the vagina and uterus (Carmichael et al., 1987; Carter, 1992). The effects of this hormonal release can easily be seen in lactating women, who often eject some milk at the time of orgasm. Oxytocin plays a role in establishment of pair bonding, a phenomenon that will be discussed later in this chapter.

Females

The mammalian female has been described as the passive participant in copulation. It is true that in some species the female's role during the act of copulation is

refractory period (*ree frak to ree*) A period of time after a particular action (for example, an ejaculation by a male) during which that action cannot occur again.

Coolidge effect The restorative effect of introducing a new female sex partner to a male that has apparently become "exhausted" by sexual activity.

oxytocin (*ox ee tow sin*) A hormone secreted by the posterior pituitary gland; causes contraction of the smooth muscle of the milk ducts, the uterus, and the male ejaculatory system; also serves as a neurotransmitter in the brain.

merely to assume a posture that exposes her genitals to the male. This behavior is called the **lordosis** response (from the Greek *lordos*, meaning "bent backward"). The female will also move her tail away (if she has one) and stand rigidly enough to support the weight of the male. However, the behavior of a female rodent in *initiating* copulation is often very active. Certainly, if a male attempts to copulate with a nonestrous rodent, the female will either actively flee or rebuff him. But when the female is in a receptive state, she will often approach the male, nuzzle him, sniff his genitals, and show behaviors characteristic of her species. For example, a female rat will exhibit quick, short, hopping movements and rapid ear wiggling, which most male rats find irresistible (McClintock and Adler, 1978).

Sexual behavior of female rodents depends on the gonadal hormones present during estrus: estradiol and progesterone. In rats, estradiol increases about 40 hours before the female becomes receptive; just before receptivity occurs, the corpus luteum begins secreting large quantities of progesterone (Feder, 1981). Ovariectomized rats (rats whose ovaries have been removed) are not sexually receptive. Although sexual receptivity can be produced in ovariectomized rodents by administering large doses of estradiol alone, the most effective treatment duplicates the normal sequence of hormones: a small amount of estradiol, followed by progesterone. Progesterone alone is ineffective; thus, the estradiol "primes" its effectiveness. Priming with estradiol takes about 16–24 hours, after which an injection of progesterone produces receptive behaviors within an hour (Takahashi, 1990). The neural mechanisms that are responsible for these effects will be described later in this chapter.

Studies with targeted mutations confirm the importance of estradiol and progesterone on sexual behavior in female rodents. Rissman et al. (1997) found that female mice without estrogen receptors were unreceptive to males even after treatment with estradiol and progesterone, and Lydon et al. (1995) observed similar effects in female mice without progesterone receptors.

The sequence of estradiol followed by progesterone has three effects on female rats: It increases their receptivity, their proceptivity, and their attractiveness. *Receptivity* refers to their ability and willingness to copulate—to accept the advances of a male by holding still and displaying lordosis when he attempts to mount her. *Proceptivity* refers to a female's eagerness to copulate, as shown by the fact that she seeks out a male and engages in behaviors that tend to arouse his sexual interest. *Attractiveness* refers to physiological and behavioral changes that affect the male. The male rat (along with many other male mammals) is most responsive to females who are in estrus ("in heat"). Males will ignore a female whose ovaries have been removed, but injections of estradiol and progesterone will restore her attractiveness (and also change her behavior toward the male). The stimuli that arouse a male rat's sexual interest include her odor and her behavior. In some species, visible changes, such as the swollen sex skin in the genital region of a female monkey, also affect sex appeal.

Even though women do not show obvious physical changes during the fertile period of the menstrual cycle, some subtle changes do occur. Roberts et al. (2004) took photos of women's faces during fertile and nonfertile periods of their menstrual cycles and found that both men and women judged the photos taken during the fertile period to be more attractive than those taken during a nonfertile period. (The women whose pictures were taken were not told the object of the study until afterward, to prevent them from unknowingly changing their facial expressions in a way that might bias the results.)

Organizational Effects of Androgens on Behavior: Masculinization and Defeminization

The dictum "Nature's impulse is to create a female" applies to sexual behavior as well as to sex organs. That is, if a rodent's brain is *not* exposed to androgens during a critical period of development, the animal will engage in female sexual behavior as an adult (if the animal is then given estradiol and progesterone). Fortunately for experimenters this critical time comes shortly after birth for rats and for several other species of rodents that are born in a rather immature condition. Thus, if a male rat is castrated immediately after birth, permitted to grow to adulthood, and then given injections of estradiol and progesterone, it will respond to the presence of another male by arching its back and presenting its hindquarters. In other words, it will act as if it were a female (Blaustein and Olster, 1989).

In contrast, if a rodent brain is exposed to androgens during development, two phenomena occur: behavioral defeminization and behavioral masculinization. *Behavioral defeminization* refers to the organizational effect of androgens that prevents the animal from displaying female sexual behavior in adulthood. As we shall see later, this effect is accomplished by suppressing the development of neural circuits controlling female sexual behavior. For example, if a

---

lordosis   A spinal sexual reflex seen in many four-legged female mammals; arching of the back in response to approach of a male or to touching the flanks, which elevates the hindquarters.

FIGURE 10.8 ■ Organizational Effects of Testosterone

Around the time of birth, testosterone masculinizes and defeminizes rodents' sexual behavior.

Hormone Treatment

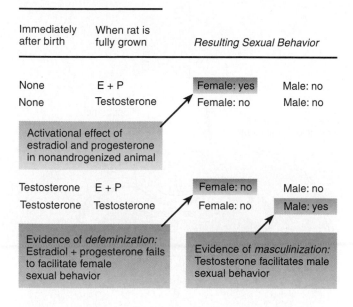

Immediately after birth	When rat is fully grown	Resulting Sexual Behavior	
None	E + P	Female: yes	Male: no
None	Testosterone	Female: no	Male: no

Activational effect of estradiol and progesterone in nonandrogenized animal

Testosterone	E + P	Female: no	Male: no
Testosterone	Testosterone	Female: no	Male: yes

Evidence of *defeminization:* Estradiol + progesterone fails to facilitate female sexual behavior

Evidence of *masculinization:* Testosterone facilitates male sexual behavior

female rodent is ovariectomized and given an injection of testosterone immediately after birth, she will *not* respond to a male rat when, as an adult, she is given injections of estradiol and progesterone. *Behavioral masculinization* refers to the organizational effect of androgens that enables animals to engage in male sexual behavior in adulthood. This effect is accomplished by stimulating the development of neural circuits controlling male sexual behavior. For example, if the female rodent in my previous example is given testosterone in adulthood rather than estradiol and progesterone, she will mount and attempt to copulate with a receptive female. (See Breedlove, 1992, and Carter, 1992, for references to specific studies.) (See *Figure 10.8.*)

Effects of Pheromones

Hormones transmit messages from one part of the body (the secreting gland) to another (the target tissue). Another class of chemicals, called **pheromones,** carries messages from one animal to another. Some of these chemicals, like hormones, affect reproductive behavior. Karlson and Luscher (1959) coined the term, from the Greek *pherein,* "to carry," and *horman,* "to excite." Pheromones are released by one animal and directly affect the physiology or behavior of another. In mammalian species most pheromones are detected by means of olfaction.

Pheromones can affect reproductive physiology or behavior. First, let us consider the effects on reproductive physiology. When groups of female mice are housed together, their estrous cycles slow down and eventually stop. This phenomenon is known as the **Lee-Boot effect** (van der Lee and Boot, 1955). If groups of females are exposed to the odor of a male (or of his urine), they begin cycling again, and their cycles tend to be synchronized. This phenomenon is known as the **Whitten effect** (Whitten, 1959). The **Vandenbergh effect** (Vandenbergh, Whitsett, and Lombardi, 1975) is the acceleration of the onset of puberty in a female rodent caused by the odor of a male. Both the Whitten effect and the Vandenbergh effect are caused by a group of compounds that are present only in the urine of intact adult males (Ma, Miao, and Novotny, 1999; Novotny et al., 1999); the urine of a juvenile or castrated male has no effect. Thus, the production of the pheromone by a male requires the presence of testosterone.

The **Bruce effect** (Bruce, 1960a, 1960b) is a particularly interesting phenomenon: When a recently impregnated female mouse encounters a normal male mouse other than the one with which she mated, the pregnancy is very likely to fail. This effect, too, is caused by a substance secreted in the urine of intact males— but not of males that have been castrated. Thus, a male mouse that encounters a pregnant female is able to prevent the birth of infants carrying another male's genes and subsequently impregnate the female himself. This phenomenon is advantageous even from the female's point of view. The fact that the new male has managed to take over the old male's territory indicates that he is probably healthier and more vigorous, and therefore, his genes will contribute to the formation of offspring that are more likely to survive.

As you learned in Chapter 7, detection of odors is accomplished by the olfactory bulbs, which constitute the primary olfactory system. However, some of the effects that pheromones have on reproductive cycles

pheromone (*fair oh moan*) A chemical released by one animal that affects the behavior or physiology of another animal; usually smelled or tasted.

Lee-Boot effect The slowing and eventual cessation of estrous cycles in groups of female animals that are housed together; caused by a pheromone in the animals' urine; first observed in mice.

Whitten effect The synchronization of the menstrual or estrous cycles of a group of females, which occurs only in the presence of a pheromone in a male's urine.

Vandenbergh effect The earlier onset of puberty seen in female animals that are housed with males; caused by a pheromone in the male's urine; first observed in mice.

Bruce effect Termination of pregnancy caused by the odor of a pheromone in the urine of a male other than the one that impregnated the female; first identified in mice.

FIGURE 10.9 ■ The Rodent Accessory Olfactory System

(Adapted from Wysocki, C. J. *Neuroscience & Biobehavioral Reviews*, 1979, 3, 301–341.)

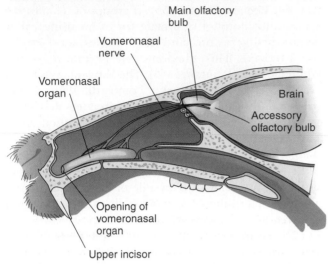

FIGURE 10.10 ■ The Amygdala

This schematic of cross section through a rat brain shows the location of the amygdala.

(Adapted from Swanson, L. W. *Brain Maps: Structure of the Rat Brain.* New York: Elsevier, 1992.)

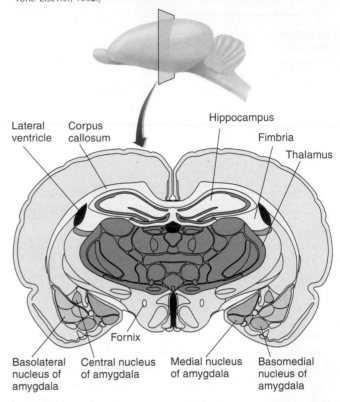

appear to be mediated by another sensory organ—the **vomeronasal organ (VNO)**—which consists of a small group of sensory receptors arranged around a pouch connected by a duct to the nasal passage. The vomeronasal organ, which is present in all orders of mammals except for cetaceans (whales and dolphins), projects to the **accessory olfactory bulb,** located immediately behind the olfactory bulb (Wysocki, 1979). (See *Figure 10.9.*) The VNO contains over 200 G-protein-linked receptor molecules that detect many of the chemicals that serve as pheromones (Dulac and Axel, 1995; Ryba and Tirindelli, 1997; Stowers and Marton, 2005). These receptor molecules are only distantly related to the ones present in the olfactory epithelium.

Removal of the accessory olfactory bulb disrupts the Lee-Boot effect, the Whitten effect, the Vandenbergh effect, and the Bruce effect; thus, the vomeronasal system is essential for these phenomena (Halpern, 1987). The accessory olfactory bulb sends axons to the **medial nucleus of the amygdala,** which in turn projects to the preoptic area and anterior hypothalamus and to the ventromedial nucleus of the hypothalamus. (As you learned in Chapter 7, so does the main olfactory bulb.) Thus, the neural circuit responsible for the effects of these pheromones appears to involve these regions. As we shall see, the preoptic area, the medial amygdala, the ventromedial nucleus of the hypothalamus, and the medial preoptic area all play important roles in reproductive behavior. (See *Figure 10.10.*)

Although the vomeronasal organ can respond to some airborne molecules, it is primarily sensitive to nonvolatile compounds found in urine or other substances (Brennan and Keverne, 2004). In fact, stimulation

of a nerve that serves the nasal region of the hamster causes fluid to be pumped into the vomeronasal organ, which exposes the receptors to any substances that may be present (Meredith and O'Connell, 1979). This pump is activated whenever the animal encounters a novel stimulus (Meredith, 1994).

Luo, Fee, and Katz (2003) implanted microelectrodes into the accessory olfactory bulb and recorded the activity of single neurons in freely moving mice while they investigated lightly anesthetized mice introduced into the test cage. They found that neurons of the vomeronasal system responded only when the mice were actively investigating the other animal's mouth or

vomeronasal organ (VNO) (*voah mer oh **nay** zul*) A sensory organ that detects the presence of certain chemicals, especially when a liquid is actively sniffed; mediates the effects of some pheromones.

accessory olfactory bulb A neural structure located in the main olfactory bulb that receives information from the vomeronasal organ.

medial nucleus of the amygdala (*a **mig** da la*) A nucleus that receives olfactory information from the olfactory bulb and accessory olfactory bulb; involved in the effects of odors and pheromones on reproductive behavior.

anogenital region. In addition, the neurons showed sharply tuned response characteristics, distinguishing between different strains of mice and between male and female mice. (See *MyPsychKit 10.2, VNO Responses.*)

mypsychkit
Animation 10.2
VNO Responses

He et al. (2008) used molecular genetic techniques to prepare a strain of mice that expressed a fluorescent dye in neurons in their main and accessory olfactory systems that would glow when the neurons were stimulated. They then placed slices of the VNO in a culture medium and examined the cells as they exposed the tissue to diluted mouse urine. Although many neurons were activated by mouse urine, only two or three responded uniquely to male urine, and approximately eight responded uniquely to female urine. It seems, then, that a small number of receptors are tuned specifically to chemicals secreted by males or females. Urine from different *individuals* produced different patterns of activity of large numbers of neurons, undoubtedly reflecting the existence of different concentrations of large numbers of chemicals in the animals' urine. In fact, as He et al. note, mouse urine contains many hundreds—perhaps thousands—of different compounds.

The vomeronasal organ is essential for the ability of a rodent to identify the sex of another individual. If transduction of chemical information in the vomeronasal organ is prevented by a knockout of a gene required for this process (TRPC2), mice can no longer distinguish between males and females (Stowers et al., 2002). In fact, male mice with this genetic knockout will attempt to mate with both males and females, and they will not attack a strange male that invades their territory, as a normal mouse will do. But the fact that they can successfully engage in sexual behavior with female mice—and in fact can impregnate them—indicates that input from the vomeronasal system is not essential for this behavior.

Normally, when a male mouse smells another mouse, he approaches and sniffs the other mouse's face and anogenital region. Urine contains many nonvolatile compounds, and the secretions of lachrymal glands (tears) contain at least one nonvolatile pheromone that is detected by the vomeronasal system (Kimoto et al., 2005). This investigatory behavior permits the animal to detect nonvolatile chemicals secreted by the other animal. If the other animal is an estrous female, the male courts and mates with it. If it is a strange male, he attacks it. If it is a familiar male (say, one of its littermates), he will usually tolerate its presence. So the main olfactory system stimulates investigatory behavior when the presence of another mouse is detected, and information provided by the vomeronasal system determines the gender, estrous condition, and identity of the other animal. Without the information from the VNO, the animal indiscriminately displays sexual behavior. In contrast, if a male mouse is made anosmic (incapable of detecting odors) by removal of the olfactory bulbs, by a genetic knockout that prevents transduction of olfactory information in the main olfactory epithelium, or by application of a chemical that damages olfactory receptors, he will not approach and sniff another mouse and, consequently, will not attack or attempt to mate with it (Mandiyan, Coats, and Shah, 2005; Wang et al., 2006).

The males of some species produce sex-attractant pheromones that affect the behavior of females. For example, a pheromone present in the saliva of boars (male pigs) elicits sexual behavior in sows. This response persists even after the sow's VNO is destroyed, which indicates that the main olfactory system can detect the pheromone and elicit the behavior (Dorries, Adkins, and Halpern, 1997). Some male pheromones that attract females are detected by the main olfactory system. For example, Mak et al. (2007) found that the odor of soiled bedding taken from the cage of a male mouse activated neurons in the main olfactory system and hippocampus of female mice. The odor even stimulated neurogenesis (production of new neurons); Mak and her colleagues found new neurons in the olfactory bulb and hippocampus. Moreover, bedding from cages of dominant males stimulated neurogenesis more effectively than did bedding from subordinate males.

It appears that at least some pheromone-related phenomena occur in humans. McClintock (1971) studied the menstrual cycles of women attending an all-female college. She found that women who spent a large amount of time together tended to have synchronized cycles: Their menstrual periods began within a day or two of one another. In addition, women who regularly spent some time in the presence of men tended to have shorter cycles than those who rarely spent time with (smelled?) men.

Russell, Switz, and Thompson (1980) obtained direct evidence that pheromones can synchronize women's menstrual cycles. The investigators collected daily samples of a woman's underarm sweat. They dissolved the samples in alcohol and swabbed them on the upper lips of a group of women three times each week, in the order in which they were originally taken. The cycles of the women who received the extract (but not those of control subjects whose lips were swabbed with pure alcohol) began to synchronize with the cycle of the odor donor. These results were confirmed by a similar study by Stern and McClintock (1998), who found that compounds from the armpits of women that were taken around the time of ovulation lengthened other women's menstrual cycles, and compounds taken late in the cycle shortened them. Preti et al. (2003) performed a similar experiment but exposed women to extracts of sweat collected from men. They found that the extract (but not a placebo) advanced the onset of the next pulse of the women's LH secretion, reduced tension, and increased relaxation.

Several studies have found that two compounds present in human sweat have different effects in men and women. Singh and Bronstad (2001) had men smell T-shirts that had been worn by women for several days. The men reported that shirts worn by women during the fertile phase of the menstrual cycle smelled more pleasant and more sexy than those worn during the nonfertile phase. Jacob and McClintock (2000) found that the androgenic chemical *androstadienone* (*AND*) increases alertness and positive mood in women but decreased positive mood in men. Wyart et al. (2007) found that women who smelled AND showed higher levels of cortisol (an adrenal hormone involved in a variety of emotional behaviors) as well as reporting a more positive mood and an increase in sexual arousal. A functional-imaging study by Savic et al. (2001) found that AND activated the preoptic area and ventromedial hypothalamus in women but not in men, whereas the estrogenic chemical *estratetraene* (*EST*) activated the paraventricular nucleus and dorsomedial hypothalamus in men but not in women.

Whether or not pheromones play a role in sexual attraction in humans, the familiar odor of a sex partner probably has a positive effect on sexual arousal—just like the sight of a sex partner or the sound of his or her voice. We are not generally conscious of the fact, but we can identify other people on the basis of olfactory cues. For example, a study by Russell (1976) found that people were able to distinguish by odor between T-shirts that they had worn and those previously worn by other people. They could also tell whether the unknown owner of a T-shirt was male or female. Thus, it is likely that men and women can learn to be attracted by their partners' characteristic odors, just as they can learn to be attracted by the sound of their voice. In an instance like this, the odors are serving simply as sensory cues, not as pheromones.

What sensory organ detects the presence of human pheromones? Although humans have a small vomeronasal organ located along the nasal septum (bridge of tissue between the nostrils) approximately 2 cm from the opening of the nostril (Garcia-Velasco and Mondragon, 1991), the human VNO to be a vestigial, nonfunctional, organ. The density of neurons in the VNO is very sparse, and investigators have not found any neural connections from this organ to the brain (Doty, 2001). Evidence clearly shows that human reproductive physiology is affected by pheromones, but it appears that these chemical signals are detected by the "standard" olfactory system—the receptor cells in the olfactory epithelium—and not by cells in the VNO.

Human Sexual Behavior

Human sexual behavior, like that of other mammals, is influenced by activational effects of gonadal hormones and, almost certainly, by organizational effects as well.

If hormones have organizational effects on human sexual behavior, they must exert these effects by altering the development of the brain. Although there is good evidence that prenatal exposure to androgens affects development of the human brain, we cannot yet be certain that this exposure has long-lasting behavioral effects. The evidence pertaining to these issues is discussed later, in a section on sexual orientation.

Activational Effects of Sex Hormones in Women

As we saw, the sexual behavior of most female mammals other than higher primates is controlled by the ovarian hormones estradiol and progesterone. (In some species, such as cats and rabbits, only estradiol is necessary.) As Wallen (1990) points out, the ovarian hormones control not only the *willingness* (or even eagerness) of an estrous female to mate but also her *ability* to mate. That is, a male rat cannot copulate with a female rat that is not in estrus. Even if he would overpower her and mount her, her lordosis response would not occur, and he would be unable to achieve intromission. Thus, the evolutionary process in rats seems to have selected animals that mate only at a time when the female is able to become pregnant. (The neural control of the lordosis response and the effects of ovarian hormones on it are described later in this chapter.)

In higher primates (including our own species) the ability to mate is not controlled by ovarian hormones. There are no physical barriers to sexual intercourse during any part of the menstrual cycle. If a woman or other female primate consents to sexual activity at any time (or is forced to submit by a male), intercourse can certainly take place.

Although ovarian hormones do not *control* women's sexual activity, they may still have an influence on women's sexual interest. Early studies reported that fluctuations in the level of ovarian hormones had only a minor effect on women's sexual interest (Adams, Gold, and Burt, 1978; Morris et al., 1987). However, as Wallen (1990) notes, these studies have almost all involved married women who live with their husbands. In stable, monogamous relationships in which the partners are together on a daily basis, sexual activity can be instigated by either of them. Normally, a husband does not force his wife to have intercourse with him, but even if she is not interested in engaging in sexual activity at that moment, she may find that she wants to do so because of her affection for him. Thus, changes in sexual interest and arousability might not always be reflected in changes in sexual behavior. In fact, a study of lesbian couples (whose menstrual cycles are likely to be synchronized) found a significant increase in sexual interest and activity during the middle portions of the women's cycles (Matteo and Rissman, 1984), which suggests that ovarian hormones *do* influence women's sexual interest.

A study by Van Goozen et al. (1997) supports this suggestion. The investigators found that the sexual activity initiated by men and women showed very different relations to the woman's menstrual cycle (and hence to her level of ovarian hormones). Men initiated sexual activity at about the same rate throughout the woman's cycle, whereas sexual activity initiated by women showed a distinct peak around the time of ovulation, when estradiol levels are highest. (See *Figure 10.11.*) Bullivant et al. (2004) found that women were more likely to initiate sexual activity and were more likely to engage in sexual fantasies just before and during the surge in luteinizing hormone that stimulates ovulation.

Wallen (2001) points out that although ovarian hormones may affect a woman's sexual interest, her behavior can be influenced by other factors as well. For example, if a woman does not want to become pregnant and does not have absolute confidence in her method of birth control, she may avoid sexual intercourse at mid-cycle, around the time of ovulation— even if her potential sexual interest is at a peak. In fact, Harvey (1987) found that women were more likely to engage in autosexual activity at this time. On the other hand, women who *want* to become pregnant are more likely to initiate sexual intercourse during the time when they are most likely to conceive.

Several studies suggest that women's sexual interest can be stimulated by androgens. There are two primary sources of androgens in the female body: the ovaries and the adrenal glands. The primary ovarian sex steroids are, of course, estradiol and progesterone, but these glands also produce testosterone. The adrenal glands produce another androgen, androstenedione, along with other adrenocortical steroids. However, the available evidence indicates that androgens by themselves (in the absence of estradiol) do not directly stimulate women's sexual interest but appear to amplify the effects of estradiol. For example, Shifren et al. (2000) studied ovariectomized women aged 31–56 years who were receiving estrogen-replacement therapy. The women were given, in addition to the estrogen, either a placebo or one of two different doses of testosterone, delivered through transdermal patches. Although the placebo produced a positive effect, the testosterone produced an even greater increase in sexual activity and rate of orgasm. At the higher dose, the percentage of women who had sex fantasies, masturbated, and had intercourse increased two to three times over baseline levels, and these women reported higher levels of well-being.

Activational Effects of Sex Hormones in Men

Although women and mammals with estrous cycles differ in their behavioral responsiveness to sex hormones, men resemble other mammals in their behavioral responsiveness to testosterone. With normal levels they can be potent and fertile; without testosterone, sperm production ceases, and sooner or later, so does sexual potency. In a double-blind study, Bagatell et al. (1994) gave a placebo or a gonadotropin-releasing hormone (GnRH) antagonist to young male volunteers to suppress secretion of testicular androgens. Within two weeks, the subjects who received the GnRH antagonist reported a decrease in sexual interest, sexual fantasy, and intercourse. Men who received replacement doses of testosterone along with the antagonist did not show these changes.

The decline of sexual activity after castration is quite variable. As reported by Money and Ehrhardt (1972), some men lose potency immediately, whereas others show a slow, gradual decline over several years. Perhaps at least some of the variability is a function of prior experience; practice not only may "make perfect," but also may forestall a decline in function. Although there is no direct evidence with respect to this possibility in humans, Wallen and his colleagues (Wallen et al., 1991; Wallen,

FIGURE 10.11 ■ Sexual Activity of Heterosexual Couples

This graph shows the distribution of sexual activity initiated by the man and by the woman.

(Adapted from Wallen, K. *Hormones and Behavior*, 2001, *40*, 339–357. After data from Van Goozen et al., 1997.)

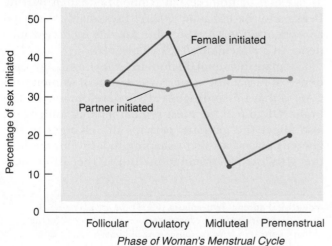

Phase of Woman's Menstrual Cycle

2001) injected a GnRH antagonist in seven adult male rhesus monkeys that were part of a larger group. The injection suppressed testosterone secretion, and sexual behavior declined after one week. However, the decline was related to the animal's social rank and sexual experience: More sexually experienced, high-ranking males continued to copulate. In fact, the highest-ranking male continued to copulate and ejaculate at the same rate as before, even though his testosterone secretion was suppressed for almost eight weeks. The mounting behavior of the lowest-ranking monkey completely ceased and did not resume until testosterone secretion recovered from the anti-GnRH treatment.

Testosterone not only affects sexual activity but also is affected by it—or even by thinking about it. A scientist stationed on a remote island (Anonymous, 1970) removed his beard with an electric shaver each day and weighed the clippings. Just before he left for visits to the mainland (and to the company of a female companion), his beard began growing faster. Because rate of beard growth is related to androgen levels, the effect indicates that his anticipation of sexual activity stimulated testosterone production. Confirming these results, Hellhammer, Hubert, and Schurmeyer (1985) found that watching an erotic film increased men's testosterone level.

Sexual Orientation

What controls a person's sexual orientation, that is, the gender of the preferred sex partner? Some investigators believe that sexual orientation is determined by childhood experiences, especially interactions between the child and parents. A large-scale study of several hundred male and female homosexuals reported by Bell, Weinberg, and Hammersmith (1981) attempted to assess the effects of these factors. The researchers found no evidence that homosexuals had been raised by domineering mothers or submissive fathers, as some clinicians had suggested. The best predictor of adult homosexuality was a self-report of homosexual feelings, which usually preceded homosexual activity by three years. The investigators concluded that their data did not support social explanations for homosexuality but were consistent with the possibility that homosexuality is at least partly biologically determined.

If homosexuality does have a physiological cause, it certainly is not variations in the levels of sex hormones during adulthood. Many studies have examined the levels of sex steroids in male homosexuals (Meyer-Bahlburg, 1984), and the vast majority found these levels to be similar to those of heterosexuals. A few studies suggest that about 30 percent of female homosexuals have elevated levels of testosterone (but still lower than those found in men). Whether these differences are related to a biological cause of female homosexuality or whether differences in lifestyles may increase the secretion of testosterone is not yet known.

A more likely biological cause of homosexuality is a subtle difference in brain structure caused by differences in the amount of prenatal exposure to androgens. Perhaps, then, the brains of male homosexuals are neither masculinized nor defeminized, those of female homosexuals are masculinized and defeminized, and those of bisexuals are masculinized but not defeminized. Of course, these are *hypotheses,* not *conclusions.* They should be regarded as suggestions to guide future research.

Prenatal Androgenization of Genetic Females

Evidence suggests that prenatal androgens can affect human social behavior and sexual orientation, as well as anatomy. In a disorder known as **congenital adrenal hyperplasia (CAH),** the adrenal glands secrete abnormal amounts of androgens. (*Hyperplasia* means "excessive formation.") The secretion of androgens begins prenatally; thus, the syndrome causes prenatal masculinization. Boys born with CAH develop normally; the extra androgen does not seem to have significant effects. However, a girl with CAH will be born with an enlarged clitoris, and her labia may be partly fused together. (As Figure 10.4 shows, the scrotum and labia develop from the same tissue in the fetus.) If masculinization of the genitals is pronounced, surgery is sometimes performed to correct them. In any event, once the syndrome has been identified, the person will be given a synthetic hormone that suppresses the abnormal secretion of androgens.

As a group, females with CAH have an increased likelihood of becoming sexually attracted to other women; approximately one-third describe themselves as bisexual or homosexual (Cohen-Bendahan, van de Beek, and Berenbaum, 2005). Presumably, prenatal androgenization is responsible for this increased incidence of a masculinized sexual orientation.

A plausible explanation for the increased incidence of masculine sexual orientation of women with CAH is that the androgens affect development of the brain. Of course, we must remember that androgens also affect the genitals; perhaps the changes in the genitals played a role in shaping the development of the girls' sexual orientation. But if the differences

---

congenital adrenal hyperplasia (CAH) (*hy per* **play** *zha*) A condition characterized by hypersecretion of androgens by the adrenal cortex; in females, causes masculinization of the external genitalia.

seen in sexual orientation *are* caused by effects of prenatal androgens on brain development, then we could reasonably conclude that prenatal androgens play a role in establishing a sexual orientation toward females, too. That is, these results support the hypothesis that male sexual orientation is at least partly determined by masculinizing (and defeminizing) effects of androgens on the human brain.

A study by Iijima et al. (2001) found that prenatal androgenization may be responsible for other sexually dimorphic behaviors as well. The investigators asked young children, including girls with CAH, to draw pictures. Typically, boys are more likely to make drawings that use dark or cold colors and to feature moving objects such as cars, trucks, trains, and airplanes, whereas girls are more likely to use light and warm colors and to include people, flowers, and butterflies. The investigators found that masculine motifs were much more likely to appear in the drawings of girls with CAH. (See *Figure 10.12*.)

Children typically show sex differences in toy preferences (Alexander, 2003). Boys generally prefer toys that can be used actively, especially those that move or can be propelled by the child. Girls generally prefer toys that provide the opportunity for nurturance. Of course, it is an undeniable fact that both caregivers and peers often encourage "sex-typical" toy choices. However, evidence suggests that biology may play a role in the nature of these choices. For example, even at one day of age, baby boys prefer to watch a moving mobile and baby girls prefer to look at a female face (Connellan et al., 2000). Alexander and Hines (2002) found that young vervet monkeys showed the same sexually dimorphic preferences in choice of toys: Males chose to play with a car and a ball, whereas females preferred to play with a doll and a pot. (See *Figure 10.13*.)

Pasterski et al. (2005) found that girls with CAH were more likely to choose male toys than were their non-CAH sisters or female cousins. The girls' parents reported that they made a special effort to encourage their CAH daughters to play with "girls' toys" but that this encouragement appeared not to succeed. Thus, the girls' tendencies to make male toy choices did not seem to be a result of parental pressure.

Because of the masculinizing effects of CAH, some clinicians have suggested that parents of strongly androgenized girls with greatly enlarged clitorises raise them as boys and not subject them to surgical procedures that feminize their genitalia. The rationale for this strategy is that the social behavior and sexual orientation of strongly androgenized girls is likely to be masculine anyway, and the enlarged clitoris could serve for sexual intercourse with women. The disadvantages include the fact that the girls' ovaries would have to be removed, which would eliminate the possibility of natural

FIGURE 10.12 ■ Prenatal Androgens and Children's Drawings

This series of drawings show: (a) drawing by a 5-year-old girl, (b) drawing by a 5-year-old boy, and (c) drawing by a 5-year-old girl with congenital adrenal hyperplasia.

(From Iijima, M., Arisaka, O., Minamoto, F., and Arai, Y. *Hormones and Behavior*, 2001, *20*, 99–104. Reprinted with permission.)

(a)

(b)

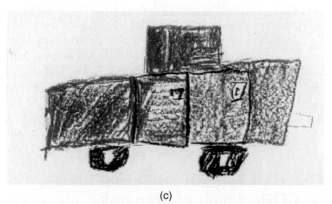

(c)

parenthood. (Most women with CAH are able to conceive and bear children.) A review by Meyer-Bahlburg (2001) provides a thoughtful and sensitive discussion of these issues.

FIGURE 10.13 ■ Sex-typical Toy Choices

Infant vervet monkeys show sex-typical toy choices: (a) a female playing with a doll and (b) a male playing with a toy car.

(Photograph courtesy of Gerianne M. Alexander, Texas A & M University)

(a)   (b)

Failure of Androgenization of Genetic Males

As we saw, genetic males with androgen insensitivity syndrome develop as females, with female external genitalia—but also with testes and without uterus or Fallopian tubes. If an individual with this syndrome is raised as a girl, all is well. Normally, the testes are removed because they often become cancerous; but if they are not removed, the body will mature into that of a woman at the time of puberty through the effects of the small amounts of estradiol produced by the testes. (If the testes are removed, the person will be given estradiol to accomplish the same result.) At adulthood the individual will function sexually as a woman, although surgical lengthening of the vagina may be necessary. Women with this syndrome report average sex drives, including normal frequency of orgasm in intercourse. Most marry and lead normal sex lives.

There are no reports of bisexuality or homosexuality (sexual orientation toward women) of XY women with androgen insensitivity syndrome. Thus, the lack of androgen receptors appears to prevent both the masculinizing and defeminizing effects of androgens on a person's sexual interest. Of course, it is also possible that rearing an XY child with androgen insensitivity syndrome as a girl plays a role in that person's sexual orientation.

Effects of Rearing on Sexual Identity and Orientation of Prenatally Androgenized Genetic Males

The case presented in the opener to this chapter (Bruce/Brenda/David) suggests that people's sexual identity and sexual orientation are strongly influenced by

biological factors and cannot easily be changed by the way a child is raised. Presumably, the exposure of Bruce's brain to androgens prenatally and during the first few months of life affected his neural development, favoring the emergence of male sexual identity and an orientation toward women as romantic and sexual partners. Fortunately, cases of penile ablation are rare. However, a developmental abnormality known as *cloacal exstrophy* results in the birth of a boy with normal testes but urogenital abnormalities, often including the lack of a penis. In the past, many boys born with this condition were raised as females, primarily because it is relatively easy to surgically construct a vagina that can function in intercourse but very difficult to construct a functioning penis. However, studies have shown that approximately 50 percent of such people later expressed dissatisfaction with their gender assignment and began living as men, often undergoing sex-change procedures (Meyer-Bahlburg, 2005; Reiner, 2005; Gooren, 2006). Such people are almost always sexually oriented toward females. If we consider the social and parental pressure against someone who has been raised as a girl subsequently adopting a male sex role, 50 percent is an impressively large number. Meyer-Bahlburg (2005) reports the case of an exstrophy patient raised as a female who underwent a gender change at the age of 52, after both parents had died. Presumably, fear of parental disapproval had prevented this person from making the change earlier. In contrast, there appear to be no reported cases of boys with cloacal exstrophy raised as boys who later became dissatisfied with their gender assignment. Reiner (2005) flatly concludes that "genetic males with male-typical prenatal androgen effects should be reared male" (p. 549).

Sexual Orientation and the Brain

The human brain is a sexually dimorphic organ. This fact has long been suspected, even before confirmation was received from anatomical studies and studies of regional cerebral metabolism using PET and functional MRI (fMRI). For example, neurologists discovered that the two hemispheres of a woman's brain appear to share functions more than those of a man's brain do. If a man sustains a stroke that damages the left side of the brain, he is more likely to show impairments in language than will a woman with similar damage. Presumably, the woman's right hemisphere shares language functions with the left, so damage to one hemisphere is less devastating in women than it is in men. Also, men's brains are, on average, somewhat larger—apparently because men's bodies are generally larger than those of women. In addition, the sizes of some specific regions of the telencephalon and diencephalon are different in males and females, and the shape of the corpus callosum may also be sexually dimorphic. (For specific references, see Breedlove, 1994; Swaab, Gooren, and Hofman, 1995; and Goldstein et al., 2001.)

Most investigators believe that the sexual dimorphism of the human brain is a result of differential exposure to androgens prenatally and during early postnatal life. Of course, additional changes could occur at the time of puberty, when another surge in androgens occurs. Sexual dimorphism in the human brain could even be a result of differences in the social environments of males and females. We cannot manipulate the hormone levels of humans before and after birth as we can with laboratory animals, so it might be a long time before enough evidence is gathered to permit us to make definite conclusions.

Several studies have examined the brains of deceased heterosexual and homosexual men and heterosexual women. So far, these studies have found differences in the size of three different subregions of the brain: the suprachiasmatic nucleus (SCN), a sexually dimorphic nucleus of the hypothalamus, and the anterior commissure (Swaab and Hofman, 1990; LeVay, 1991; Allen and Gorski, 1992). You are already familiar, from Chapter 9, with the suprachiasmatic nucleus; the anterior commissure is a fiber bundle that interconnects parts of the left and right temporal lobes. However, from what we know about brain functions, there is no reason to expect that differences in the SCN or the corpus callosum would play a role in sexual orientation. Also, a follow-up study confirmed the existence of a sexually dimorphic nucleus in the hypothalamus but failed to find a relationship between its size and sexual orientation in men (Byne et al., 2001.). At this point, there is no good evidence for differences in brain structure that might account for differences in sexual orientation.

Approximately 8 percent of domestic rams (male sheep) show a sexual preference for other males. These animals do not show female-typical behavior; they show typical male mounting behavior but direct this behavior toward other males in preference to females (Price et al., 1988). A study by Roselli et al. (2004) discovered a sexually dimorphic nucleus in the medial preoptic/anterior hypothalamic area that was significantly larger in males than in females. They also found that this nucleus was twice as large in female-oriented (heterosexual) rams as in male-oriented (homosexual) rams. (See *Figure 10.14.*)

As we saw earlier, a functional-imaging study by Savic et al. (2001) found that the brains of heterosexual men and women reacted differently to the odors of AND and EST, two chemicals that may serve as human pheromones. Savic, Berglund, and Lindström (2005) investigated the patterns of brain activation in heterosexual women and homosexual and heterosexual men in response to the odors of these chemicals. They found the same sex differences in the responses of heterosexual men and women that were seen in the earlier study. They also found that the response of homosexual

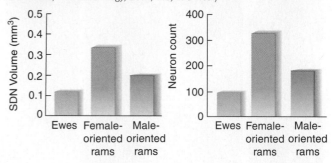

FIGURE 10.14   ■   Sexual Orientation and the SDN

The graphs show sexual dimorphism and the role of sexual orientation in the volume and number of neurons in the sexually dimorphic nucleus (SDN) in sheep.

(Adapted from Roselli, C. E., Larkin, K., Resko, J. A., Stellflug, J. N., and Stromshak, F. *Endocrinology,* 2004, *145,* 478–483.)

men was similar to that of the heterosexual women, which suggests that the response pattern was affected by the person's sexual orientation. People with an orientation to women (heterosexual men) showed brain activation in the paraventricular and dorsomedial nuclei of the hypothalamus when they smelled EST, whereas people with an orientation to men (heterosexual women and homosexual men) showed brain activation in the preoptic area and ventromedial hypothalamus when they smelled AND.

Although this section has been considering sexual orientation—the sex of those to whom an individual is sexually and romantically attracted—another sexual characteristic is related to structural differences in the brain. Zhou et al. (1995) found that the size of a particular region of the forebrain, the central subdivision of the *bed nucleus of the stria terminalis (BNST),* is larger in males than in females. They also found that in male-to-female transsexuals this nucleus is as small as it is in normal females. The size of this nucleus was as large in male homosexuals as in male heterosexuals. Thus, its size was related to sexual *identity,* not to sexual *orientation.* Kruijver et al. (2000) replicated these results and found that the size of this region in female-to-male transsexuals was within the range of sizes seen in normal males. (See *Figure 10.15.*)

Male transsexuals are men who regard themselves as females trapped in male bodies. Some go so far as to seek medical assistance to obtain female sex hormones and sex-change operations. (Most male homosexuals have male sexual identities; although they are romantically and sexually oriented toward other men, they do not regard themselves as women, nor do they wish to be.) Whether the BNST actually plays a role in a person's sexual identity will have to be determined by further research.

FIGURE 10.15 ■ The Human BNST

These photomicrographs of slices of the human brain contain the central subdivision of the bed nucleus of the stria terminalis (BNST).

(From Zhou, J.-N., Hofman, M. A., Gooren, L. J. G., and Swaab, D. F. *Nature,* 1995, *378,* 68–70. Reprinted with permission.)

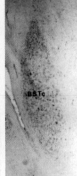

| Heterosexual man | Heterosexual woman | Homosexual man | Transexual male-to-female |

We cannot necessarily conclude that any of the brain regions I mentioned in this section are directly involved in people's sexual orientation (or sexual identity). For example, Martin and Nguyen (2004) found differences in the relative lengths of bones in the arms and legs of heterosexual and homosexual men and women. (See *Figure 10.16.*) Of course, no one would say that differences in relative bone lengths are the causes of differences in sexual orientation. Instead, the bone differences might simply reflect differences in the people's exposure

FIGURE 10.16 ■ Arm Length and Sexual Orientation

This graph shows sexual dimorphism and the role of sexual orientation in the ratio of arm length to stature.

(Adapted from Martin, J. T., and Nguyen, D. H. *Hormones and Behavior,* 2004, *45,* 31–39.)

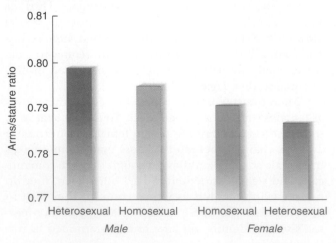

to sex hormones during prenatal or early postnatal development. (Bone growth is strongly affected by both androgens and estrogens.) Similarly, we should be cautious about concluding that differences in brain structure seen so far are the causes of differences in sexual orientation. The *real* differences—if indeed sexual orientation is determined by prenatal exposure to androgens—may lie elsewhere in the brain, in some regions as yet unexplored by researchers. However, the observation that differences in body or brain structure are related to sexual orientation does suggest that exposure to prenatal hormones has an effect on the nature of a person's sexuality.

Possible Causes of Differences in Brain Development

If sexual orientation is, indeed, affected by differences in exposure of the developing brain to androgens, what factors might cause this exposure to vary? Presumably, something must decrease the prenatal androgen levels to which male homosexuals are exposed and increase the levels to which female homosexuals are exposed. As we saw, congenital adrenal hyperplasia exposes the developing fetus to increased levels of androgens, but most homosexual women do not have CAH. So far, no other plausible sources of high levels of prenatal androgens have been proposed.

Studies performed with laboratory animals suggest an event that could potentially interfere with prenatal androgenization of males: maternal stress. Ward (1972) subjected pregnant rats to periods of stress by confining them and exposing them to a bright light, which suppresses androgen production in male fetuses. The male rats born to the stressed mothers were less likely than control subjects to display male sexual behavior and were more likely to display female sexual behavior when they were given injections of estradiol and progesterone. Another study (Ward and Stehm, 1991) found that the play behavior of juvenile male rats whose mothers were stressed while pregnant resembled that of females more than that of males—that is, the animals showed less rough-and-tumble play. Thus, the behavioral effects caused by prenatal stress are not restricted to changes in sexual behavior.

Other studies with laboratory animals have shown that besides having behavioral effects, prenatal stress reduces the size of a sexually dimorphic nucleus of the preoptic area, which normally is larger in males than in females and which (as we will see in a later section) plays an important role in male sex behavior (Anderson et al., 1986). Although we cannot assume that prenatal stress in humans and laboratory animals has similar effects on the brain and behavior, the results of these studies are consistent with the hypothesis that male homosexuality may be related to events that reduce exposure to prenatal androgens.

Studies by Blanchard and his colleagues (Blanchard, 2001) and by Bogaert (2006) suggest another factor that can influence sexual differentiation of the brain. In these studies, investigators found that homosexual men tend to have more older brothers—but not more older sisters or younger brothers or sisters—than heterosexual men. In contrast, the numbers of brothers or sisters (younger or older) of homosexual and heterosexual women did not differ, nor did the age of the mother or father or the interval between births. The presence of older brothers and sisters had no effect on women's sexual orientation. The data obtained by Blanchard and his colleagues suggest that the odds of a boy becoming homosexual increased by approximately 3.3 percent for each older brother. Assuming a 2 percent rate of homosexuality in boys without older brothers, the predicted rate would be 3.6 percent for a boy with two older brothers and 6.3 percent for one with four older brothers. Thus, the odds are still strongly against the incidence of homosexuality even in a family with several boys.

The authors suggest that when mothers are exposed to several male fetuses, their immune system may become sensitized to proteins that only males possess. As a result, the response of the mother's immune system may affect the prenatal brain development of later male fetuses. Of course, most men who have several older brothers are heterosexual, so even if this hypothesis is correct, it appears that only some women become sensitized to a protein produced by their male fetuses.

Heredity and Sexual Orientation

Another factor that may play a role in sexual orientation is heredity. Twin studies take advantage of the fact that identical twins have identical genes, whereas the genetic similarity between fraternal twins is, on the average, 50 percent. Bailey and Pillard (1991) studied pairs of male twins in which at least one member identified himself as homosexual. If both twins are homosexual, they are said to be *concordant* for this trait. If only one is homosexual, the twins are said to be *discordant*. Thus, if homosexuality has a genetic basis, the percentage of monozygotic twins who are concordant for homosexuality should be higher than that for dizygotic twins. This is exactly what Bailey and Pillard found: The concordance rate was 52 percent for identical twins and only 22 percent for fraternal twins—a difference of 30 percent. Other studies have shown differences of up to 60 percent (Gooren, 2006).

Genetic factors also appear to affect female homosexuality. Bailey et al. (1993) found that the concordance of female monozygotic twins for homosexuality was 48 percent, while that of dizygotic twins was 16 percent. Another study, by Pattatucci and Hamer (1995), found an increased incidence of homosexuality and bisexuality in sisters, daughters, nieces, and female cousins (through a paternal uncle) of homosexual women.

For several years, investigators have been puzzled by an apparent paradox. On average, male homosexuals have approximately 80 percent fewer children than male heterosexuals do (Bell and Weinberg, 2978). This reduced fecundity should exert strong selective pressure against any genes that predispose men to become homosexual. Some investigators have suggested that homosexuals may play a supportive role in their families, thus increasing the fecundity of their heterosexual brothers and sisters, who share some of their genes (Wilson, 1975). However, more recent studies (Bobrow and Bailey, 2001; Rahman and Hull, 2005) have found that homosexuals do not provide more financial or emotional support to their siblings than heterosexuals do. A study by Camperio-Ciani, Corna, and Capiluppi (2004) suggests a possible explanation. They found that the female maternal relatives (for example, maternal aunts and grandmothers) of male homosexuals had higher fecundity rates did than female maternal relatives of male heterosexuals. No differences were found in the female paternal relatives of homosexuals and heterosexuals. Because men are likely to share an X chromosome with female maternal relatives but not with female paternal relatives, the investigators suggested that a gene or genes on the X chromosome that increase a male's likelihood of becoming homosexual also increase a female's fecundity.

To summarize, evidence suggests that two biological factors—prenatal hormonal exposure and heredity—may affect a person's sexual orientation. These research findings certainly contradict the suggestion that a person's sexual orientation is a moral issue. It appears that homosexuals are no more responsible for their sexual orientation than heterosexuals are. Morris et al. (2004) pointed out the unlikelihood of a person's sexual orientation being a simple matter of choice. It is difficult to imagine someone saying to himself, "Let's see, I'll have gym at school today, so I'll wear white socks and tennis shoes. Gosh, as long as I'm making decisions I guess I better be attracted to girls for the rest of my life, too" (Morris et al., 2004, p. 475). Ernulf, Innala, and Whitam (1989) found that people who believed that homosexuals were "born that way" expressed more positive attitudes toward homosexuals than did people who believed that they "chose to be" or "learned to be" that way. Thus, we can hope that research on the origins of homosexuality will reduce prejudice based on a person's sexual orientation. The question "Why does someone become homosexual?" will probably be answered when we find out why someone becomes *heterosexual*.

Interim Summary

Hormonal Control of Sexual Behavior

Sexual behaviors are controlled by the organizational and activational effects of hormones. The female reproductive cycle (menstrual cycle or estrous cycle) begins with the maturation of one or more ovarian follicles, which occurs in response to the secretion of FSH by the anterior pituitary gland. As the ovarian follicle matures, it secretes estradiol, which causes the lining of the uterus to develop. When estradiol reaches a critical level, it causes the pituitary gland to secrete a surge of LH, triggering ovulation. The empty ovarian follicle becomes a corpus luteum, under the continued influence of LH, and secretes estradiol and progesterone. If pregnancy does not occur, the corpus luteum dies and stops producing hormones, and menstruation begins.

The sexual behavior of males of all mammalian species appears to depend on the presence of androgens. Oxytocin has a facilitatory effect on erection and ejaculation. The proceptivity, receptivity, and attractiveness of female mammals other than primates depend primarily on estradiol and progesterone. In particular, estradiol has a priming effect on the subsequent appearance of progesterone.

In most mammals, female sexual behavior is the norm, just as the female body and female sex organs are the norm. In other words, unless prenatal androgens masculinize and defeminize the animal's brain, its sexual behavior will be feminine. Behavioral masculinization refers to the androgen-stimulated development of neural circuits that respond to testosterone in adulthood, producing male sexual behavior. Behavioral defeminization refers to the inhibitory effects of androgens on the development of neural circuits that respond to estradiol and progesterone in adulthood, producing female sexual behavior.

Pheromones can affect sexual physiology and behavior. Odorants present in the urine of female mice affect their estrous cycles, lengthening and eventually stopping them (the Lee-Boot effect). Odorants present in the urine of male mice abolish these effects and cause the females' cycles to become synchronized (the Whitten effect). (Phenomena similar to the Lee-Boot effect and the Whitten effect also occur in women.) Odorants can also accelerate the onset of puberty in females (the Vandenbergh effect). In addition, the odor of the urine from a male other than the one that impregnated a female mouse will cause her to abort (the Bruce effect). Connections between the olfactory system and the amygdala appear to be important in stimulating male sexual behavior.

The main olfactory system detects volatile chemicals that signal the presence of another animal, and the vomeronasal organ determines the other animal's sex, estrous condition, and identity.

Pheromones present in the underarm sweat of both men and women affect women's menstrual cycles, and substances present in male sweat improve women's moods. Because the human vomeronasal organ does not appear to have sensory functions, these effects must be mediated by the main olfactory bulb. The search for sex attractant pheromones in humans has so far been fruitless, although we might well recognize our sex partners by their odors.

Testosterone has an activational effect on the sexual behavior of men, just as it does on the behavior of other male mammals. Women do not require estradiol or progesterone to experience sexual interest or to engage in sexual behavior. These hormones may affect the quality and intensity of their sex drive, and studies comparing the sexual behavior of female monkeys housed in small groups with that of females housed in large groups in large cages suggest that the sexual proceptivity may be related to ovarian hormones, even in higher primates. Studies with women suggest that variations in levels of ovarian hormones across the menstrual cycle do affect sexual interest but that other factors (such as initiation of sexual activity by partners or a desire to avoid or attain pregnancy) can also affect sexual behavior. In addition, the presence of androgens may facilitate the effect of estradiol on women's sexual interest.

Sexual orientation (that is, heterosexuality or homosexuality) may be influenced by prenatal exposure to androgens. Studies of prenatally androgenized girls suggest that organizational effects of androgens influence the development of sexual orientation; androgenization appears to enhance interest in activities and toys usually preferred by boys and to increase the likelihood of a sexual orientation toward women. If androgens cannot act (as they cannot in cases of androgen insensitivity syndrome), the person's anatomy and behavior are feminine. So far, evidence concerning brain structures and sexual orientation is inconclusive. The size of a region of the forebrain has been found to be related to sexual identity. The case of a twin boy whose penis was accidentally destroyed during infancy suggests that the behavioral effects of prenatal androgenization are not easily reversed by the way a child is reared, and studies of genetic males with cloacal exstrophy, who are born without a penis, support the conclusion that prenatal exposure to androgens promotes a male sexual identity and an orientation toward females. The fact that male homosexuals tend to have more older brothers than male heterosexuals do has led to the suggestion that a woman's immune system may become sensitized to a protein that is expressed only in male fetuses. Finally, twin studies suggest that heredity may play a role in sexual orientation in both men and women.

Thought Question

Whatever the relative roles played by biological and environmental factors may be, most investigators believe that a person's sexual orientation is not a matter of choice. Why do you think so many people consider sexual orientation to be a moral issue?

NEURAL CONTROL OF SEXUAL BEHAVIOR

The control of sexual behavior—at least in laboratory animals—involves different brain mechanisms in males and females. This section describes these mechanisms.

Males

Spinal Mechanisms

Some sexual responses are controlled by neural circuits contained within the spinal cord. Men with injuries that totally transect the spinal cord have become fathers when their wives have been artificially inseminated with semen obtained by mechanical stimulation (Hart, 1978). Brackett et al. (1998) found that vibratory stimulation of the penis successfully elicited ejaculation in most men with complete spinal cord transection above the tenth thoracic segment, which suggests that the circuits in the spinal cord that control the ejaculatory response are located below this region. Because the spinal damage prevents sensory information from reaching the brain, these men are unable to feel the stimulation and do not experience an orgasm.

Ejaculation occurs after sufficient amount of tactile stimulation of an erect penis causes activation of the spinal ejaculation generator. Coolen and her colleagues (Coolen et al., 2004) have identified a group of neurons in the lumbar region of the rat spinal cord that appear to constitute a critical part of the spinal ejaculation generator. These neurons project to a nucleus in a specific part of the posterior intralaminar thalamus, so the investigators refer to them as *lumbar spinothalamic (LSt) cells.* Ejaculation, but not mounts or intromissions without ejaculation, activates these neurons, as shown by the presence of Fos protein. Specific destruction of these neurons with a toxin completely abolishes ejaculation but does not impair the animal's ability to mount an estrous female and achieve intromission.

Brain Mechanisms

As we just saw, erection and ejaculation are controlled by circuits of neurons that reside in the spinal cord. However, brain mechanisms have both excitatory and inhibitory control of these circuits. Although tactile stimulation of a man's genitals can stimulate erection and ejaculation, these responses can be inhibited by the context. For example, the outcomes of tactile stimulation of a man's penis will have different outcomes when his physician is carrying out physical examination or when his partner touches him while they are lying in bed. In addition, a man's penis can become erect when he sees his partner or has erotic thoughts—even when his penis is not being touched. Thus, there must be brain mechanisms that can activate or suppress the spinal mechanisms that control genital reflexes.

As we saw earlier in this chapter, the primary and accessory olfactory systems play important roles in reproductive behavior. Both of these systems send fibers to the medial nucleus of the amygdala. Several studies (for example, Lehman and Winans, 1982; Heeb and Yahr, 2000) have found that lesions of the medial amygdala abolished the sexual behavior of male rodents. Thus, the amygdala is part of the system that mediates the effects of pheromones on male sexual behavior.

The **medial preoptic area (MPA),** located just rostral to the hypothalamus, is the forebrain region most critical for male sexual behavior. (As we will see later in this chapter, it is also critical for other sexually dimorphic behavior, including maternal behavior.) Electrical stimulation of this region elicits male copulatory behavior (Malsbury, 1971), and sexual activity increases the firing rate of single neurons in the MPA (Shimura, Yamamoto, and Shimokochi, 1994; Mas, 1995). In addition, the act of copulation increases the metabolic activity of the MPA and induces the production of Fos protein (Oaknin et al., 1989; Robertson et al., 1991; Wood and Newman, 1993). (The significance of the Fos protein as an indicator of neural activation was discussed in

---

medial preoptic area (MPA) An area of cell bodies just rostral to the hypothalamus; plays an essential role in male sexual behavior.

FIGURE 10.17 ■ Preoptic Area of the Rat Brain

These photomicrographs of sections through the preoptic area of a rat brain show
(a) normal male, (b) normal female, and (c) androgenized female. SDN-POA = sexually
dimorphic nucleus of the preoptic area; OC = optic chiasm; V = third ventricle;
SCN = suprachiasmatic nucleus; AC = anterior commissure.

(From Gorski, R. A., in *Neuroendocrine Perspectives*, Vol. 2, edited by E. E. Müller and R. M. MacLeod. Amsterdam:
Elsevier-North Holland, 1983. Reprinted with permission.)

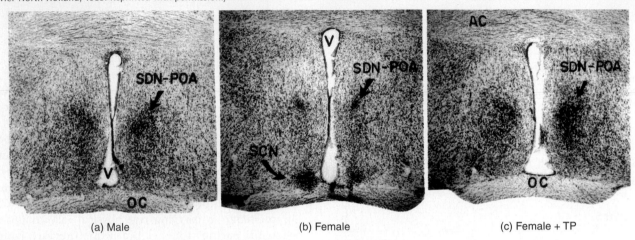

(a) Male   (b) Female   (c) Female + TP

Chapter 5.) Dominguez, Gil, and Hull (2006) found that mating increased the release of glutamate in the MPA and that infusion of glutamate into the MPA increased the frequency of ejaculation. Finally, destruction of the MPA abolishes male sexual behavior (Heimer and Larsson, 1966/1967).

The organizational effects of androgens are responsible for sexual dimorphisms in brain structure. Gorski et al. (1978) discovered a nucleus within the MPA of the rat that is three to seven times larger in males than in females. This area is called (appropriately enough) the **sexually dimorphic nucleus (SDN)** of the preoptic area. The size of this nucleus is controlled by the amount of androgens present during fetal development. According to Rhees, Shryne, and Gorski (1990a, 1990b), the critical period for masculinization of the SDN appears to start on the eighteenth day of gestation and end once the animals are five days old. (Normally, rats are born on the twenty-second day of gestation.) De Jonge et al. (1989) found that lesions of the SDN decrease masculine sexual behavior. (See *Figure 10.17*.)

The medial amygdala, like the medial preoptic area, is sexually dimorphic: One region within this structure (which contains an especially high concentration of androgen receptors) is 85 percent larger in male rats than in female rats (Hines, Allen, and Gorski, 1992). In addition, destruction of the medial amygdala disrupts the sexual behavior of male rats. De Jonge et al. (1992) found that the rats with these lesions took longer to mount receptive females and to

ejaculate. Wood and Newman (1993) observed that mating increased the production of Fos protein in the medial amygdala.

The MPA receives chemosensory input from the vomeronasal organ and the main olfactory system through connections with the medial amygdala and the BNST. (You will recall that in humans the BNST is sexually dimorphic and that it is smaller in transsexual males.) The MPA also receives somatosensory information from the genitals through connections with the central tegmental field of the midbrain and the medial amygdala. The act of copulation induces the production of Fos protein in both of these regions (Gréco et al., 1998). (See *Figure 10.18*.)

Androgens exert their activational effects on neurons in the MPA and associated brain regions. If a male rodent is castrated in adulthood, its sexual behavior will cease. However, the behavior can be reinstated by implanting a small amount of testosterone directly into the MPA or in two regions whose axons project to the MPA: the central tegmental field and the medial amygdala (Sipos and Nyby, 1996; Coolen and Wood, 1999). Both of these regions contain a high concentration of androgen receptors in the male rat brain (Cottingham and Pfaff, 1986).

---

sexually dimorphic nucleus A nucleus in the preoptic area that is much larger in males than in females; first observed in rats; plays a role in male sexual behavior.

FIGURE 10.18 ■ Brain Regions Involved in Sexual Behavior

In this schematic, cross sections through the rat brain show the location of the medial preoptic area, the medial amygdala, the bed nucleus of the stria terminalis, and the central tegmental field of the midbrain, which are involved in sexual behavior.

(Adapted from Swanson, L. W. *Brain Maps: Structure of the Rat Brain.* New York: Elsevier, 1992.)

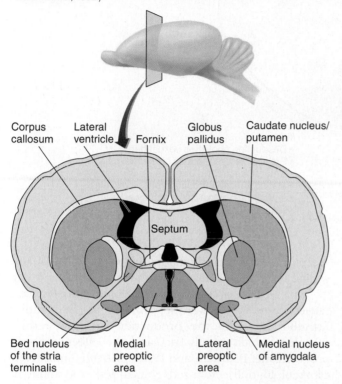

Corpus callosum · Lateral ventricle · Fornix · Globus pallidus · Caudate nucleus/putamen

Septum

Bed nucleus of the stria terminalis · Medial preoptic area · Lateral preoptic area · Medial nucleus of amygdala

As we saw in the previous subsection, the lumbar region of the spinal cord contains a group of neurons that play a critical role in ejaculation (Coolen et al., 2004). Anatomical tracing studies suggest that the most important connections between the MPA and the ejaculation generator in the spinal cord are accomplished through the **periaqueductal gray matter (PAG)** of the midbrain and the **nucleus paragigantocellularis (nPGi)** of the medulla. (Marson and McKenna, 1996; Normandin and Murphy, 2008). The nPGi has inhibitory effects on spinal cord sexual reflexes, so one of the tasks of the pathway originating in the MPA is to suppress this inhibition. The MPA suppresses the nPGi directly through an inhibitory pathway and does so indirectly by inhibiting the activity of the PAG, which normally excites the nPGi.

The inhibitory connections between neurons of the nPGi and those of the lumbar spinothalamic neurons are serotonergic. As Marson and McKenna (1992) showed, application of serotonin (5-HT) to the spinal cord suppresses ejaculation. This connection may

explain a well-known side effect of specific serotonin reuptake inhibitors (SSRIs). Men who take SSRIs as a treatment for depression often report that they have no trouble attaining an erection but have difficulty achieving an ejaculation. Presumably, the action of the drug as an agonist at the serotonergic synapses in the spinal cord increases the inhibitory influence of nPGi on the spinal ejaculation generator.

Figure 10.19 summarizes the evidence I have presented so far in this section. (See *Figure 10.19.*)

A functional-imaging study by Holstege et al. (2003b) examined the pattern of brain activation during ejaculation in men that was elicited by the men's female partners by means of manual stimulation. Ejaculation was accompanied by neural activity in many brain regions, including the junction between the midbrain and the diencephalon, which includes the ventral tegmental area (probably involved in the pleasurable, reinforcing effects of orgasm), other midbrain regions, several thalamic nuclei, the lateral putamen (part of the basal ganglia), and the cerebellum. *Decreased* activity was seen in the amygdala and the nearby entorhinal cortex. As we will see in Chapter 11, the amygdala is involved in defensive behavior and in negative emotions such as fear and anxiety. Decreased activation is also seen in this structure when people who are deeply in love see pictures of their loved ones (Bartels and Zeki, 2000, 2004).

Females

Just as the MPA plays an essential role in male sex behavior, another region in the ventral forebrain plays a similar role in female sexual behavior: the **ventromedial nucleus of the hypothalamus (VMH).** A female rat with bilateral lesions of the ventromedial nuclei will not display lordosis, even if she is treated with estradiol and progesterone. Conversely, electrical stimulation of the ventromedial nucleus facilitates female sexual behavior (Pfaff and Sakuma, 1979). (See *Figure 10.20.*)

As we saw in the previous section, the medial amygdala of males receives chemosensory information from the vomeronasal system and somatosensory information from the genitals, and it sends efferent axons to the medial preoptic area. These connections are found

periaqueductal gray matter (PAG) The region of the midbrain that surrounds the cerebral aqueduct; plays an essential role in various species-typical behaviors, including female sexual behavior.

nucleus paragigantocellularis (nPGi) A nucleus of the medulla that receives input from the medial preoptic area and contains neurons whose axons form synapses with motor neurons in the spinal cord that participate in sexual reflexes in males.

ventromedial nucleus of the hypothalamus (VMH) A large nucleus of the hypothalamus located near the walls of the third ventricle; plays an essential role in female sexual behavior.

FIGURE 10.19 ■ Male Sexual Behavior

This schematic shows a possible explanation of the interacting excitatory effects of pheromones, genital stimulation, and testosterone on male sexual behavior.

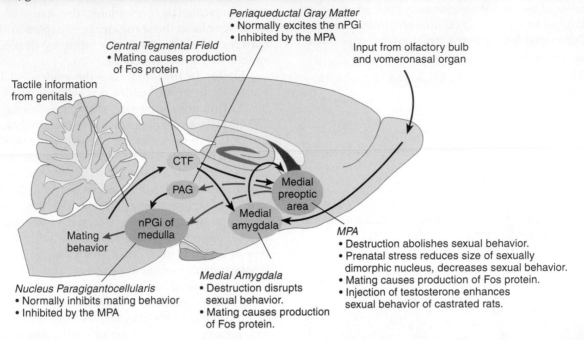

in females as well. In addition, neurons in the medial amygdala also send efferent axons to the VMH. In fact, copulation or mechanical stimulation of the genitals or flanks of a female rat increases the production of Fos protein in both the medial amygdala and the VMH (Pfaus et al., 1993; Tetel, Getzinger, and Blaustein, 1993).

As we saw earlier, sexual behavior of female rats is activated by a priming dose of estradiol, followed by progesterone. The estrogen sets the stage, so to speak, and the progesterone stimulates the sexual behavior. Injections of these hormones directly into the VMH will stimulate sexual behavior even in females whose ovaries have been removed (Rubin and Barfield, 1980; Pleim and Barfield, 1988). And if a chemical that blocks the production of progesterone receptors is injected into the VMH, the animal's sexual behavior is disrupted (Ogawa et al., 1994). Thus, estradiol and progesterone exert their effects on female sexual behavior by activating neurons in this nucleus.

Rose (1990) recorded from single neurons in the ventromedial hypothalamus of freely moving female hamsters and found that injections of progesterone (following estradiol pretreatment) increased the activity level of these neurons, particularly when the animals were displaying lordosis. Tetel, Celentano, and Blaustein (1994) found that neurons in both the VMH and the medial amygdala that showed increased Fos production when the animal's genitals were stimulated also contained estrogen

receptors. Thus, the stimulating effects of estradiol and genital stimulation converge on the same neurons.

The mechanism by which estradiol primes a female's sensitivity to progesterone appears to be simple: Estradiol increases the production of progesterone receptors, which greatly increases the effectiveness of progesterone. Blaustein and Feder (1979) administered estradiol to ovariectomized guinea pigs and found a 150 percent increase in the number of progesterone receptors in the hypothalamus. Presumably, the estradiol activates genetic mechanisms in the nucleus that are responsible for the production of progesterone receptors.

Figure 10.21 shows two slices through the hypothalamus of ovariectomized guinea pigs, stained for progesterone receptors. One of the animals had previously received a priming dose of estradiol; the other had not. As this figure shows, the estradiol dramatically increased the number of cells containing progesterone receptors. See *Figure 10.21*.)

The neurons of the ventromedial nucleus send axons to the periaqueductal gray matter. This region, too, has been implicated in female sexual behavior; Sakuma and Pfaff (1979a, 1979b) found that electrical stimulation of the PAG facilitates lordosis in female rats and that lesions there disrupt it. In addition, Hennessey et al. (1990) found that lesions that disconnect the VMH from the PAG abolish female sexual behavior. Finally, Sakuma and Pfaff (1980a, 1980b) found that estradiol treatment or electrical stimulation of the

FIGURE 10.20 ■ The Ventromedial Nucleus of the Hypothalamus

This cross section through the rat brain shows the location of the ventromedial nucleus of the hypothalamus.

(Adapted from Swanson, L. W. *Brain Maps: Structure of the Rat Brain.* New York: Elsevier, 1992.)

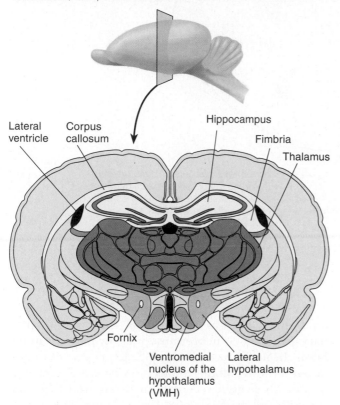

ventromedial nuclei increased the firing rate of neurons in the PAG. (The PAG contains both estrogen and progesterone receptors.)

Daniels, Miselis, and Flanagan-Cato (1999) injected a transneuronal retrograde tracer, pseudorabies virus, in the muscles responsible for the lordosis response in female rats. They found that the pathway innervating these muscles was as previous studies predicted: VMH → PAG → nPGi → motor neurons in the ventral horn of the lumbar region of the spinal cord.

As we saw in the previous subsection, the brain regions that control male genital reflexes include the MPA, PAG, and nPGi. Anatomical tracing studies (Marson, 1995; Marson and Murphy, 2006) injected pseudorabies virus into the clitorises and vaginas of female rats and found heavy retrograde labeling in these three brain structures (and some others as well). Thus, it seems likely that erections of the penis and clitoris are controlled by similar brain mechanisms. This finding is not surprising, because these organs derive from the same embryonic tissue.

Figure 10.22 summarizes the evidence I have presented so far in this section. (See *Figure 10.22.*)

A functional-imaging study by Holstege et al. (2003a) investigated the neural activation that accompanies orgasm in women, elicited by manual clitoral stimulation supplied by the women's male partners. They observed activation in the junction between the midbrain and diencephalon, the lateral putamen, and the cerebellum, just as was observed in men (Holstege et al., 2003b). They also saw activation in the PAG, a critical region for copulatory behavior in female laboratory animals.

FIGURE 10.21 ■ Progesterone Receptors

These photomicrographs of sections through the hypothalamus of ovariectomized guinea pigs were stained for progesterone receptors: (a) no priming, and (b) after receiving a priming dose of estradiol.

(Courtesy of Joanne Turcotte and Jeffrey Blaustein, University of Massachusetts.)

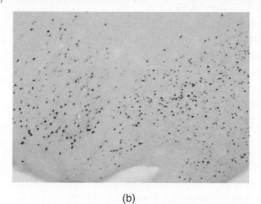

(a)   (b)

FIGURE 10.22 ■ Female Sexual Behavior

This schematic shows a possible explanation of the interacting excitatory effects of pheromones, genital stimulation, and estradiol and progesterone on female sexual behavior.

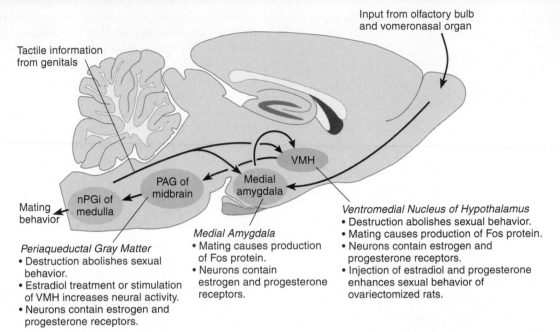

Tactile information from genitals

Input from olfactory bulb and vomeronasal organ

VMH

PAG of midbrain

Medial amygdala

nPGi of medulla

Mating behavior

Periaqueductal Gray Matter
• Destruction abolishes sexual behavior.
• Estradiol treatment or stimulation of VMH increases neural activity.
• Neurons contain estrogen and progesterone receptors.

Medial Amygdala
• Mating causes production of Fos protein.
• Neurons contain estrogen and progesterone receptors.

Ventromedial Nucleus of Hypothalamus
• Destruction abolishes sexual behavior.
• Mating causes production of Fos protein.
• Neurons contain estrogen and progesterone receptors.
• Injection of estradiol and progesterone enhances sexual behavior of ovariectomized rats.

Formation of Pair Bonds

In approximately 5 percent of mammalian species, heterosexual couples form monogamous, long-lasting bonds. In humans, such bonds can be formed between members of homosexual couples as well. As naturalists and anthropologists have pointed out, monogamy is not always exclusive: In many species of animals, humans included, individuals sometimes cheat on their partners. In addition, some people display serial monogamy—intense relationships that last for a period of time, only to be replaced with similarly intense relationships with new partners. And, of course, some cultures condone (or even encourage) polygamy. But there is no doubt that pair bonding occurs in some species, including our own.

Several laboratories have investigated pair bonding in some closely related species of voles (small rodents that are often mistaken for mice). Prairie voles (*Microtus ochrogaster*) are monogamous; males and females form pair bonds after mating, and the fathers help to care for the pups. In the wild, most prairie voles whose mates die never take another partner (Getz and Carter, 1996). Meadow voles (*Microtus pennsylvanicus*) are promiscuous; after mating, the male leaves, and the mother cares for the pups by herself.

Several studies have revealed a relationship between monogamy and the levels of two peptides in the brain: vasopressin and oxytocin. These compounds are both released as hormones by the posterior pituitary gland and as neurotransmitters by neurons in the brain. In males, vasopressin appears to play the more important role. Monogamous voles have a higher level of V1a vasopressin receptors in the ventral forebrain than do polygamous voles (Insel, Wang, and Ferris, 1994). This difference appears to be responsible for the presence or absence of monogamy. Lim and Young (2004) found that mating induced the production of Fos protein in the ventral forebrain of male prairie voles and that an injection into this region of a drug that blocks V1a receptors disrupted the formation of pair bonds. Lim et al. (2004) performed an even more convincing experiment. They injected a genetically modified virus that contained the gene for the V1a receptor into the ventral forebrain of normally polygamous male meadow voles. This manipulation increased the synthesis of the V1a receptor in this brain region and transformed the animals from polygamy to monogamy. Unlike prairie voles, which spend much time in physical contact with their partners after mating, meadow voles spend little time with them. Lim and her colleagues found that male meadow voles with artificially increased levels of V1a receptors spent much more time huddling side by side with their mates.

In female voles, oxytocin appears to play a major role in pair bonding. Mating stimulates the release of oxytocin, and peripheral injection of oxytocin or injection into the cerebral ventricles facilitates pair bonding in female prairie voles (Williams et al., 1994). In contrast, a drug that blocks oxytocin receptor disrupts formation of pair bonds (Cho et al., 1999).

Many investigators believe that oxytocin and vasopressin may play a role in the formation of pair bonding in humans. For example, after intercourse, at a time when blood levels of oxytocin are increased, people report feelings of calmness and well-being, which are certainly compatible with the formation of bonds with one's partner. However, it is difficult to envision ways to perform definitive research on this topic. Experimenters can study the effects of these hormones or their antagonists on pair bonding in laboratory animals, but they certainly cannot do so with humans. However, Heinrichs et al. (2003) found that injections of oxytocin caused relaxation and a reduction of anxiety in human subjects. In addition, Kosfeld

et al. (2005) found that oxytocin increased trust. The investigators had subjects play a "trust" game, in which they were given money, some or all of which they could give to another player (the trustee) to invest. If the trustee made money with his investment (all subjects were male), he could share the winnings with the first player—or he could be selfish and keep everything for himself. Fifty minutes before the start of the game, the subjects received a nasal spray containing oxytocin or a placebo. The oxytocin appeared to increase people's trust in other people: The subjects who received the oxytocin gave 18 percent more money to the trustees, and they were twice as likely to give them all of their money to invest. A second study found that oxytocin did not simply make the subjects feel more confident. The were no more likely to risk their money in an investment game that did not involve other people.

Interim Summary

Neural Control of Sexual Behavior

Sexual reflexes such as sexual posturing, erection, and ejaculation are organized in the spinal cord. Vibratory stimulation of the penis can elicit ejaculation in men with complete transection of the spinal cord, as long as the damage is located above the tenth thoracic segment. LSt cells, a group of neurons in the lumbar region of the rat spinal cord play a critical role in triggering an ejaculation.

In laboratory animals, different brain mechanisms control male and female sexual behavior. The medial preoptic area is the forebrain region that is most critical for male sexual behavior. Stimulating this area produces copulatory behavior; destroying it permanently abolishes the behavior. The sexually dimorphic nucleus, located in the medial preoptic area, develops only if an animal is exposed to androgens early in life. A sexually dimorphic nucleus is found in humans as well. Destruction of the SDN (part of the MPA) in laboratory animals impairs mating behavior.

Neurons in the MPA contain testosterone receptors. Copulatory activity causes an increase in the activity of neurons in this region. Implantation of testosterone directly into the MPA reinstates copulatory behavior that was previously abolished by castration in adulthood. Neurons in the MPA are part of a circuit that includes the periaqueductal gray matter, the nucleus paragigantocellularis of the medulla, and motor neurons that control genital reflexes in the spinal cord. Connections of the nPGi with the spinal cord are inhibitory.

Ejaculation in men is accompanied by increased behavior in the brain's reinforcement mechanisms, several thalamic nuclei, the lateral putamen, and the cerebellum. Activity of the amygdala decreases.

The most important forebrain region for female sexual behavior is the ventromedial nucleus of the hypothalamus (VMH). Its destruction abolishes copulatory behavior, and its stimulation facilitates this behavior. Both estradiol and progesterone exert their facilitating effects on female sexual behavior in this region, and studies have confirmed the existence of progesterone and estrogen receptors there. The priming effect of estradiol is caused by an increase in progesterone receptors in the VMH. The steroid-sensitive neurons of the VMH send axons to the periaqueductal gray matter (PAG) of the midbrain; these neurons, through their connections with the medullary reticular formation, control the particular responses that constitute female sexual behavior. Orgasm in women is accompanied by increased activity in regions similar to those activated during ejaculation in men and, in addition, in the periaqueductal gray matter.

Vasopressin and oxytocin, peptides that serve as hormones and as neurotransmitters in the brain, appear to facilitate pair bonding. In fact, the insertion of the gene for vasopressin receptors in the basal forebrain of polygamous male voles induces monogamous behavior. Vasopressin plays the most important role in males, and oxytocin plays the most important role in females. In humans, oxytocin appears to increase trust in other people.

PARENTAL BEHAVIOR

In most mammalian species, reproductive behavior takes place after the offspring are born as well as at the time they are conceived. This section examines the role of hormones in the initiation and maintenance of maternal behavior and the role of the neural circuits that are responsible for their expression. Most of the research has involved rodents; less is known about the neural and endocrine bases of maternal behavior in primates.

Although most research on the physiology of parental behavior has focused on maternal behavior, some researchers are now studying paternal behavior shown by the males of some species of rodents. It goes without saying that the human paternal behavior is very important for the offspring of our species, but the physiological basis of this behavior has not yet been studied.

Maternal Behavior of Rodents

The final test of the fitness of an animal's genes is the number of offspring that survive to a reproductive age. Just as the process of natural selection favors reproductively competent animals, it favors those that care adequately for their young, if their young in fact require care. Rat and mouse pups certainly do; they cannot survive without a mother to attend to their needs.

At birth, rats and mice resemble fetuses. The infants are blind (their eyes are still shut), and they can only wriggle helplessly. They are poikilothermous ("cold-blooded"); their brain is not yet developed enough to regulate body temperature. They even lack the ability to release their own urine and feces spontaneously and must be helped to do so by their mother. As we will see shortly, this phenomenon actually serves a useful function.

During gestation, female rats and mice build nests. The form this structure takes depends on the material available for its construction. In the laboratory the animals are usually given strips of paper or lengths of rope or twine. A good *brood nest,* as it is called, is shown in Figure 10.23. This nest is made of hemp rope; a piece of the rope is shown below the nest. The mouse laboriously shredded the rope and then wove an enclosed nest, with a small hole for access to the interior. (See *Figure 10.23.*)

At the time of **parturition** (delivery of offspring) the female begins to groom and lick the area around her vagina. As a pup begins to emerge, she assists the uterine contractions by pulling the pup out with her teeth. She then eats the placenta and umbilical cord and cleans off the fetal membranes—a quite delicate operation. (A newborn pup looks as though it is sealed in very thin plastic wrap.) After all the pups have been born and cleaned up, the mother will probably nurse them. Milk is usually present in the mammary glands very near the time of birth.

FIGURE 10.23 ■ A Mouse's Brood Nest

Beside the nest is a length of the kind of rope the mouse used to construct it.

Periodically, the mother licks the pups' anogenital region, stimulating reflexive urination and defecation. Friedman and Bruno (1976) have shown the utility of this mechanism. They noted that a lactating female rat produces approximately 48 grams (g) of milk on the tenth day of lactation. This milk contains approximately 35 milliliters (ml) of water. The experimenters injected some of the pups with tritiated (radioactive) water and later found radioactivity in the mother and in the littermates. They calculated that a lactating rat normally consumes 21 ml of water in the urine of her young, thus recycling approximately two-thirds of the water she gives to the pups in the form of milk. The water, traded back and forth between mother and young, serves as a vehicle for the nutrients—fats, protein, and sugar—contained in milk. Because each day the milk production of a lactating rat is approximately 14 percent of her body weight (for a human weighing 120 pounds, that would be around 2 gallons), the recycling is extremely useful, especially when the availability of water is a problem.

Besides cleaning, nursing, and purging her offspring, a female rodent will retrieve pups if they leave or are removed from the nest. The mother will even construct another nest in a new location and move her litter there, should the conditions at the old site become unfavorable (for example, when an inconsiderate

---

parturition (*par tew ri shun*) The act of giving birth.

FIGURE 10.24 ■ A Female Mouse Carrying a Pup

experimenter puts a heat lamp over it). The way a female rodent picks up her pup is quite consistent: She gingerly grasps the animal, managing not to injure it with her very sharp teeth. (I can personally attest to the sharpness of a mouse's teeth and the strength of its jaw muscles.) She then carries the pup with a characteristic prancing walk, her head held high. (See *Figure 10.24.*) The pup is brought back to the nest and is left there. The female then leaves the nest again to search for another pup. She continues to retrieve pups until she finds no more; she does not count her pups and then stop retrieving when she has them all. A mouse or rat will usually accept all the pups she is offered, if they are young enough. I once observed two lactating female mice with nests in corners of the same cage, diagonally opposite each other. I disturbed their nests, which triggered a long bout of retrieving, during which each mother stole youngsters from the other's nest. The mothers kept up their exchange for a long time, passing each other in the middle of the cage.

Under normal conditions, one of the stimuli that induce a female rat to begin taking care of pups is the act of parturition. Female rodents normally begin taking care of their pups as soon as they are born. Some of this effect is caused by prenatal hormones, but the passage of the pups through the birth canal also stimulates maternal behavior: Artificially distending the birth canal in nonpregnant females stimulates maternal behavior, whereas cutting the sensory nerves that innervate the birth canal retards the appearance of maternal behavior (Graber and Kristal, 1977; Yeo and Keverne, 1986).

Hormonal Control of Maternal Behavior

As we saw earlier in this chapter, most sexually dimorphic behaviors are controlled by the organizational and activational effects of sex hormones. Maternal behavior is somewhat different in this respect. First, there is no evidence that organizational effects of hormones play a role; as we will see, under the proper conditions, even males will take care of infants. (Obviously, they cannot provide the infants with milk.) Second, although maternal behavior is affected by hormones, it is not *controlled* by them. Most virgin female rats will begin to retrieve and care for young pups after having infants placed with them for several days (Wiesner and Sheard, 1933). And once the rats are sensitized, they will thereafter take care of pups as soon as they encounter them; sensitization lasts for a lifetime.

Although hormones are not essential for the activation of maternal behavior, many aspects of maternal behavior are facilitated by hormones. Nest-building behavior is facilitated by progesterone, the principal hormone of pregnancy (Lisk, Pretlow, and Friedman, 1969). After parturition, mothers continue to maintain their nests, and they construct new nests if necessary, even though their blood level of progesterone is very low then.

Although pregnant female rats will not immediately care for foster pups that are given to them during pregnancy, they will do so as soon as their own pups are born. The hormones that influence a female rodent's responsiveness to her offspring are the ones that are present shortly before parturition. Figure 10.25 shows the levels of the three hormones that have been implicated in maternal behavior: progesterone, estradiol, and **prolactin.** Note that just before parturition the level of estradiol begins rising, then the level of progesterone falls dramatically, followed by a sharp increase in prolactin, the hormone produced by the anterior pituitary gland that is responsible for milk production. (See *Figure 10.25.*) If ovariectomized virgin female rats are given progesterone, estradiol, and prolactin in a pattern that duplicates this sequence, the time it takes to sensitize their maternal behavior is drastically reduced (Bridges et al., 1985.)

As we saw in the previous section, pair bonding involves vasopressin and oxytocin. In at least some species, oxytocin also appears to be involved in formation of a bond between mother and offspring. In rats the administration of oxytocin facilitates the establishment of maternal behavior (Insel, 1997). Van Leengoed, Kerker, and Swanson (1987) injected an

prolactin A hormone of the anterior pituitary gland, necessary for production of milk; also facilitates maternal behavior.

hypothesis that facial expression of emotion uses an innate, species-typical repertoire of movements of facial muscles (Darwin, 1872/1965). For example, Ekman and Friesen (1971) studied the ability of members of an isolated tribe in New Guinea to recognize facial expressions of emotion produced by Westerners. They had no trouble doing so and themselves produced facial expressions that Westerners readily recognized. Figure 11.17 shows four photographs taken from videotapes of a man from this tribe reacting to stories designed to evoke facial expressions of happiness, sadness, anger, and disgust. I am sure that you will have no trouble recognizing which is which. (See *Figure 11.17*.)

Because the same facial expressions were used by people who had not previously been exposed to each other, Ekman and Friesen concluded that the expressions were unlearned behavior patterns. In contrast, different cultures use different words to express particular concepts; production of these words does not involve innate responses but must be learned.

Other researchers have compared the facial expressions of blind and normally sighted children. They reasoned that if the facial expressions of the two groups are similar, then the expressions are natural for our species and do not require learning by imitation. (Studies of blind adults would not be conclusive because adults would probably have heard enough descriptions of facial expressions to be able to pose them.) In fact, the facial expressions of young blind and sighted children are very similar (Woodworth and Schlosberg, 1954; Izard, 1971). Thus, both the cross-cultural studies and the investigations with blind children confirm the naturalness of these expressions.

Neural Basis of the Communication of Emotions: Recognition

Effective communication is a two-way process. That is, the ability to display one's emotional state by changes in expression is useful only if other people are able to recognize them. In fact, Kraut and Johnston (1979) unobtrusively observed people in circumstances that would be likely to make them happy. They found that happy situations (such as making a strike while bowling, seeing the home team score, or experiencing a beautiful day) produced only small signs of happiness when the people were alone. However, when the people were interacting socially with other people, they were much more likely to smile. For example, bowlers who made a strike usually did not smile when the ball hit the pins, but when they turned around to face their companions, they often smiled. Jones et al. (1991) found that even 10-month-old children showed this tendency. (No, I'm not suggesting that infants have been observed while bowling.)

Recognition of another person's facial expression of emotions is generally automatic, rapid, and accurate. Tracy and Robbins (2008) found that observers quickly recognized brief expressions of a variety of emotions. If they were given more time to think about the expression they had seen, they showed very little improvement.

Laterality of Emotional Recognition

We recognize other people's feelings by means of vision and audition—seeing their facial expressions and hearing their tone of voice and choice of words. Many studies have found that the right hemisphere plays a more important role than the left hemisphere in comprehension of emotion. For example, Bowers et al. (1991) found that patients with right hemisphere damage had difficulty producing or describing mental images of facial expressions of emotions. Subjects were asked to imagine the face of someone who was very happy (or very sad, angry, or afraid). Then they were asked questions about the facial expression—for example, *Do the eyes look twinkly? Is the brow raised? Are the corners of the lips raised up?* People with right hemisphere damage had trouble answering these questions but could easily answer questions about nonemotional images, such as

FIGURE 11.17 ■ Facial Expressions in a New Guinea Tribesman

The tribesman made faces when he heard the following stories: (a) "Your friend has come and you are happy." (b) "Your child had died." (c) "You are angry and about to fight." (d) "You see a dead pig that has been lying there a long time."

(From Ekman, P., *The Face of Man: Expressions of Universal Emotions in a New Guinea Village.* New York: Garland STPM Press, 1980. Reprinted with permission.)

(a)

(b)

(c)

(d)

FIGURE 11.18 ■ Perception of Emotions

The PET scans indicate brain regions activated by listening to emotions expressed by tone of voice (green) or meanings of words (red).

(From George, M. S., Parekh, P. I., Rosinsky, N., Ketter, T. A., Kimbrell, T. A., Heilman, K. M., Herscovitch, P., and Post, R. M., *Archives of Neurology*, 1996, *53*, 665–670. Reprinted with permission.)

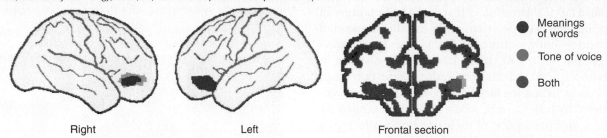

Right   Left   Frontal section

● Meanings of words
● Tone of voice
● Both

What's higher off the ground: a horse's knee or the top of its tail? or *What number from one to nine does a peanut look like?*

Several functional-imaging studies have confirmed these results. For example, George et al. (1996) had subjects listen to some sentences and identify their emotional content. In one condition the subjects listened to the meaning of the words and said whether they described a situation in which someone would be happy, sad, angry, or neutral. In another condition the subjects judged the emotional state from the tone of the voice. The investigators found that comprehension of emotion from word meaning increased the activity of the prefrontal cortex bilaterally, the left more than the right. Comprehension of emotion from tone of voice increased the activity of only the right prefrontal cortex. (See *Figure 11.18*.)

Heilman, Watson, and Bowers (1983) recorded a particularly interesting case of a man with a disorder called *pure word deafness,* caused by damage to the left temporal cortex. (This syndrome is described in Chapter 14). The man could not comprehend the meaning of speech but had no difficulty identifying the emotion being expressed by its intonation. This case, like the functional-imaging study by George et al. (1996), indicates that comprehension of words and recognition of tone of voice are independent functions.

Role of the Amygdala

As we saw in the previous section, the amygdala plays a special role in emotional responses. It plays a role in emotional recognition as well. For example, several studies have found that lesions of the amygdala (the result of degenerative diseases or surgery for severe seizure disorders) impair people's ability to recognize facial expressions of emotion, especially expressions of fear (Adolphs et al., 1994, 1995; Young et al., 1995; Calder et al., 1996; Adolphs et al., 1999). In addition, functional-imaging studies (Morris et al., 1996; Whalen et al., 1998) have found large increases in the activity of the amygdala when people view photographs of faces expressing fear

but only small increases (or even decreases) when they look at photographs of happy faces. However, although amygdala lesions impair visual recognition of facial expressions of emotion, several studies have found that they do not appear to affect people's ability to recognize emotions in tone of voice (Anderson and Phelps, 1998; Adolphs and Tranel, 1999).

Several studies suggest that the amygdala receives visual information that we use to recognize facial expressions of emotion directly from the thalamus and not from the visual association cortex. Adolphs (2002) notes that the amygdala receives visual input from two sources: subcortical and cortical. The subcortical input (from the superior colliculus and the *pulvinar,* a large nucleus in the posterior thalamus) appears to provide the most important information for this task. In fact, some people with blindness caused by damage to the visual cortex can recognize facial expressions of emotion *even though they have no conscious awareness of looking at a person's face,* a phenomenon known as **affective blindsight** (de Gelder et al., 1999, Anders et al., 2004). Morris et al. (2001) performed a functional-imaging study with one such patient and discovered that when he viewed faces with fearful expressions (of which he had no conscious perception), the superior colliculus, posterior thalamus, and amygdala became active. Presumably, this subcortical pathway provides the visual information to the amygdala and other brain regions involved in emotional perception.

People can express emotions through their body language, as well as through muscular movements of their face (de Gelder, 2006). For example, a clenched fist might accompany an angry facial expression, and a fearful person might run away. The sight of photographs of

affective blindsight The ability of a person who cannot see objects in his or her blind field to accurately identify facial expressions of emotion while remaining unconscious of perceiving them; caused by damage to the visual cortex.

bodies posed in gestures of fear activates the amygdala, just as the sight of fearful faces does (Hadjikhani and de Gelder, 2003). Meeren, van Heijnsbergen, and de Gelder (2005) prepared computer-modified photographs of people showing facial expressions of emotions that were either congruent with the person's body posture (for example, a facial expression of fear and a body posture of fear) or incongruent (for example, a facial expression of anger and a body posture of fear). The investigators asked people to identify the facial expressions shown in the photos and found that the ratings were faster and more accurate when the facial and body expressions were congruent. In other words, when we look at other people's faces, our perception of their emotion is affected by their body posture as well as their facial expressions.

As we saw in Chapter 6, the visual cortex receives information from two systems of neurons. The *magnocellular system* (named for layers of large cells in the lateral geniculate nucleus of the thalamus that relay visual information from the eye to the visual cortex) provides information about movement, depth, and very subtle differences in brightness in the scene before our eyes. This system appeared early in evolution of the mammalian brain and provides most mammals (dogs and cats, for example) with a monochromatic, somewhat fuzzy image of the world. The *parvocellular system* (named for layers of small cells in the lateral geniculate nucleus) is found only in some primates, including humans. This system provides us with color vision and detection of fine details. The part of the visual association cortex responsible for recognition of faces, the *fusiform face area*, receives information primarily (but not solely) from the parvocellular system. The information that the amygdala receives from the superior colliculus and the pulvinar has its origin in the more primitive magnocellular system.

An ingenious functional-imaging study by Vuilleumier, Armony, and Dolan (2003) presented people with pictures of faces showing neutral or fearful expressions. Some of the pictures were normal, some had been filtered with a computer program so that they showed only high spatial frequencies, and some had been filtered to show only low spatial frequencies. (Chapter 6 described the concept of spatial frequencies.) As Figure 11.19 shows, high spatial frequencies show fine details of transitions between light and dark, and low spatial frequencies show fuzzy images. As you may have deduced, these photos primarily stimulate the parvocellular and magnocellular systems, respectively. (See *Figure 11.19*.)

Vuilleumier and his colleagues found that the fusiform face area was better at recognizing individual faces and primarily used high spatial frequency (parvocellular) information to do so. In contrast, the amygdala (and the superior colliculus and pulvinar, which provide it with visual information) was able to recognize an expression of fear based on low spatial frequency

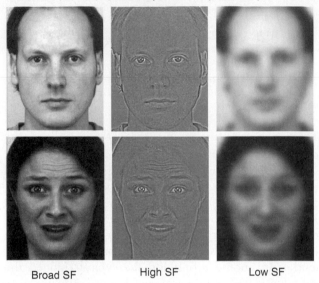

FIGURE 11.19 ■ Functional-Imaging Study by Vuilleumier et al. (2003)

The figure shows the stimuli used by Vuilleumier et al. (2003). The more primitive magnocellular system is sensitive to low spatial frequencies (SF), and the more recently evolved parvocellular system is sensitive to high spatial frequencies.

(From Vuilleumier, P., Armony, J. L., Driver, J., and Dolan, R. J. *Nature Neuroscience*, 2003, 6, 624–631. Reprinted with permission.)

Broad SF   High SF   Low SF

(magnocellular) information but not on high spatial frequency information.

Krolak-Salmon et al. (2004) recorded electrical potentials from the amygdala and visual association cortex through electrodes that had been implanted in people who were being evaluated for neurosurgery to alleviate a seizure disorder. They presented the people with photographs of faces showing neutral expressions or expressions of fear, happiness, or disgust. The found that fearful faces produced the largest response and that the amygdala showed activity before the visual cortex did. The rapid response supports the conclusion that the amygdala receives visual information from the magnocellular system (which conducts information very rapidly) that permits it to recognize facial expressions of fear.

So far, the evidence suggests that the amygdala plays an indispensable role in recognition of facial expressions of fear. However, a study by Adolphs et al. (2005) suggests that under the appropriate conditions, other regions of the brain can perform this task. Adolphs and his colleagues discovered that S. M., a woman with bilateral amygdala damage, failed to look at the eyes when she examined photographs of faces. Spezio et al. (2007) conducted a similar study, but this one measured S. M.'s eye movements while she was actually conversing with another person. Like the study by Adolphs et al., this one found that S. M. failed to

FIGURE 11.20 ■ Eye Fixations After Amygdala Damage

The figure shows the numbers of fixations on a person's face made by a patient with bilateral amygdala damage (Patient S. M.) and a normal subject. Warmer colors indicate increasing numbers of fixations. Note that Patient S. M. does not look at the other person's eyes.

(From Spezio, M. L., Huang, P.-Y. S., Castelli, F., and Adolphs, R. *Journal of Neuroscience*, 2007, *27*, 3994–3997. Copyright © 2007, The Society for Neuroscience. Reprinted with permission.)

Patient S. M.   Normal subject

FIGURE 11.21 ■ Functional-Imaging Study by Whalen et al. (2004)

The stimuli used in the study show that the whites of the eyes alone can convey the impression of a fearful expression.

(From Whalen, P. J., Kagan, J., Cook, R. G., Davis, F. C., Kim, H., Polis, S., McLaren, D. G., Somerville, L. H., McLean, A. A., Maxwell, J. S., and Johnstone, T. *Science*, 2004, *306*, 2061. Copyright © 2004 American Association for the Advancement of Science. Reprinted with permission.)

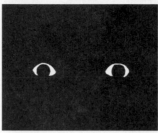

Fear   Happiness

direct her gaze to the other person's eyes. In contrast, she spent an abnormally large amount of time looking at the other person's mouth. (See *Figure 11.20.*)

By themselves, eyes are able to convey a fearful expression. (See *Figure 11.21.*) A functional-imaging study by Whalen et al. (2004) found that the viewing the fearful eyes shown in Figure 11.21 activated the ventral amygdala, the region that receives the majority of the cortical and subcortical inputs to the amygdala. So the fact that S. M. did not look at eyes suggests a cause for her failure to detect only this emotion. In fact, when Adolphs et al. (2005) instructed S. M. to look at the eyes of the face she was examining, she was able to recognize an expression of fear. However, unless she was reminded to do so, she soon stopped looking at eyes, and her ability to recognize a fearful expression disappeared again. It will be interesting to learn whether other people with amygdala damage can also recognize expressions of fear if they are instructed to look at eyes.

Perception of Direction of Gaze

Perrett and his colleagues (see Perrett et al., 1992) have discovered an interesting brain function that may be related to recognition of emotional expression. They found that neurons in the monkey's superior temporal sulcus (STS) are involved in recognition of the direction

of another monkey's gaze—or even that of a human. They found that some neurons in this region responded when the monkey looked at photographs of a monkey's face or a human face but only when the gaze of the face in the photograph was oriented in a particular direction. For example, Figure 11.22 shows the activity level of a neuron that responded when a human face was looking upward. (See *Figure 11.22.*)

Why is gaze important in recognition of emotions? First, it is important to know whether an emotional expression is directed toward you or toward someone else. For example, an angry expression directed toward you means something very different from a similar expression directed toward someone else. And if someone else shows signs of fear, the expression can serve as a useful warning, but only if you can figure out what the person is looking at. In fact, Adams and Kleck (2005) found that people more readily recognized anger if the eyes of another person were directed toward the observer and fear if they were directed somewhere else. As Blair (2008) notes, an angry expression directed toward the observer means that the other person wants the observer to stop what he or she is doing.

The neocortex that lines the STS seems to provide such information. Lesions there disrupt monkeys' ability to discriminate the direction of another animal's gaze, but they do not impair the monkeys' ability to recognize other animals' faces (Campbell et al., 1990; Heywood and Cowey, 1992). As we saw in Chapter 6, the posterior parietal cortex—the endpoint of the dorsal stream of visual analysis—is concerned with perceiving the location of objects in space. A functional-imaging study by Pelphrey et al. (2003) had people watch an animated cartoon of a face. When the direction of gaze changed, increased activity was seen in the right STS and posterior parietal cortex. Presumably, the connections between

FIGURE 11.22 ■ A Gaze-Direction Cell

The graph shows the responses of a single neuron in the cortex lining of the superior temporal sulcus of a monkey's brain. The cell fired most vigorously when the monkey was presented a photograph of a face looking upward.

(From Perrett, D. I., Harries, M. H., Mistlin, A. J., Hietanen, J. K., Benson, P. J., Bevan, R., Thomas, S., Oram, M. W., Ortega, J., and Brierley, K., *International Journal of Comparative Psychology*, 1990, *4*, 25–55. Reprinted with permission.)

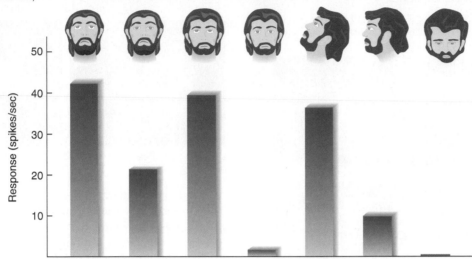

neurons in the STS and the parietal cortex enable the orientation of another person's gaze to direct one's attention to a particular location in space.

Role of Imitation in Recognition of Emotional Expressions: The Mirror Neuron System

Adolphs et al. (2000) discovered a possible link between somatosensation and emotional recognition. They compiled computerized information about the locations of brain damage in 108 patients with localized brain lesions and correlated this information with the patients' ability to recognize and identify facial expressions of emotions. They found that the most severe damage to this ability was caused by damage to the somatosensory cortex of the right hemisphere. (See *Figure 11.23.*) They suggest that when we see a facial expression of an emotion, we unconsciously imagine ourselves making that expression. Often, we do more than imagine making the expressions—we actually imitate what we see. Adolphs et al. suggest that the somatosensory representation of what it feels like to make the perceived expression provides cues that we use to recognize the emotion being expressed in the face we are viewing. In support of this hypothesis, Adolphs and his colleagues report that the ability of patients with right hemisphere lesions to recognize facial expressions of emotions is correlated with their ability to perceive somatosensory stimuli. That is, patients with somatosensory impairments (caused by

right-hemisphere lesions) also had impairments in recognition of emotions.

We are beginning to understand the neural circuit that provides this form of feedback. In Chapter 8 I described the role of *mirror neurons* in the control of movement. Mirror neurons are activated when an animal performs a particular behavior or when it sees another animal performing that behavior. Presumably, these neurons are involved in learning to imitate the actions of others. These neurons, which are located in the ventral

FIGURE 11.23 ■ Brain Damage and Recognition of Facial Expressions of Emotion

In this computer-generated representation of performance of subjects with localized brain damage on recognition of facial expressions of emotion, the colored areas outline the site of the lesions. Good performance is shown in shades of blue; poor performance is shown in red and yellow.

(From Adolphs, R., Damasio, H., Tranel, D., Cooper, G., and Damasio, A. R. *The Journal of Neuroscience*, 2000, *20*, 2683–2690. Copyright © 2000 by the Society for Neuroscience. Reprinted with permission.)

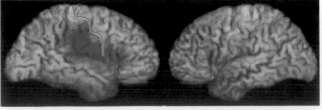

Right hemisphere   Left hemisphere

premotor area of the frontal lobe, receive input from the superior temporal sulcus and the posterior parietal cortex. As we saw in Chapter 8, this circuit is activated when we see another person perform a goal-directed action, and feedback from this activity helps us to understand what the person is trying to accomplish. Carr et al. (2003) suggest that the mirror neuron system, which is activated when we observe facial movements of other people, provides feedback that helps us to understand how other people feel. In other words, the mirror neuron system may be involved in our ability to empathize with the emotions of other people. (I will have more to say about empathy in the last section of this chapter.)

A neurological disorder known as Moebius syndrome provides further support for this hypothesis. Moebius syndrome is a congenital condition that involves defective development of the sixth (abducens) and seventh (facial) cranial nerves and results in facial paralysis and inability to make lateral eye movements. Because of this paralysis, people affected with Moebius syndrome cannot make facial expressions of emotion. In addition, they have difficulty recognizing the emotional expressions of other people (Cole, 2001). Perhaps their inability to produce facial expressions of emotions makes it impossible for them to imitate the expressions of other people, and the lack of internal feedback from the motor system to the somatosensory cortex may make the task of recognition more difficult.

In Chapter 8 I described research on *audiovisual neurons*—neurons that respond to the *sounds* of particular actions and to the sight of those actions. Warren et al. (2006) obtained evidence that audiovisual neurons play a role in communication of emotions, too. The investigators asked volunteers to make emotional sounds in response to written scenarios that presented situations expected to evoke triumph, amusement, fear, and disgust. The volunteers were asked not to make verbal responses such as "yuck" or "yippee" but to restrict themselves to nonverbal vocal responses. These sounds were presented to subjects while they underwent fMRI scanning. The scans showed that hearing the emotional vocalizations activated the same regions of the brain that were activated by facial expressions of these emotions. In other words, when we hear other people make nonverbal emotional sounds, our mirror neuron system is activated, and the feedback from this activation may contribute to our recognition of the emotions being expressed by these sounds.

Disgust

And now for something completely different. Several studies have found that damage to the insular cortex and basal ganglia impairs people's ability to recognize facial expressions of disgust (Sprengelmeyer et al., 1996, 1997; Calder et al., 2000). In addition, a functional-imaging study by Wicker et al. (2003) found that both smelling a disgusting odor and seeing a face of a person showing an expression of disgust activate the insular cortex. Disgust (literally, "bad taste") is an emotion provoked by something that tastes or smells bad—or by an action that we consider to be in bad taste (figuratively, not literally). Disgust produces a very characteristic facial expression; if you want to see a good example, refer to Figure 11.17d. As we saw in Chapter 7, the insula contains the primary gustatory cortex, so perhaps it is not a coincidence that this region is also involved in recognition of "bad taste."

A functional-imaging study by Thielscher and Pessoa (2007) asked subjects to press one of two levers to indicate whether the facial expression they saw was one of disgust or fear. The expressions varied in intensity, and one of them was actually neutral, indicating neither disgust nor fear. Nevertheless, the subjects were asked to press one of the levers on every trial, indicating disgust or fear. When the subjects saw faces expressing disgust, the insular cortex and part of the basal ganglia were activated. What was particularly interesting was that even when the subjects were watching a neutral expression, if they pressed the "disgust" lever, the "disgust" regions of the brain were activated.

The results of an online survey presented on the British Broadcasting Corporation Science web site suggests that the emotion of disgust has its origins in avoidance of disease. The survey presented pairs of photos and asked people to indicate which photos were more disgusting. The people who responded indicated that the one that appeared to hold a potential threat of disease was more disgusting. For example, a yellow liquid that has soaked a tissue looks more like a body fluid than a blue liquid does. (See *Figure 11.24*.)

Neural Basis of the Communication of Emotions: Expression

Facial expressions of emotion are automatic and involuntary (although, as we saw, they can be modified by display rules). It is not easy to produce a realistic facial expression of emotion when we do not really feel that way. In fact, Ekman and Davidson have confirmed an early observation by a nineteenth-century neurologist, Guillaume-Benjamin Duchenne de Boulogne, that genuinely happy smiles, as opposed to false smiles or social smiles people make when they greet someone else, involve contraction of a muscle near the eyes, the lateral part of the orbicularis oculi—now sometimes referred to as Duchenne's muscle (Ekman, 1992; Ekman and Davidson, 1993). As Duchenne put it, "The first [zygomatic major muscle] obeys the will but the second [orbicularis oculi] is only put in play by the sweet

FIGURE 11.24 ■ Disease and Disgust

This figure shows pairs of photographs with high and low relation to the threat of disease that were used in the online survey presented on the BBC Science web site. The numbers in red or green indicate the mean ratings (range = 1–5) made by people who completed the survey.

(From Curtis, V., Aunger, R., and Rabie, T. *Biology Letters,* 2004, *271,* S131–S133. Reprinted with permission.)

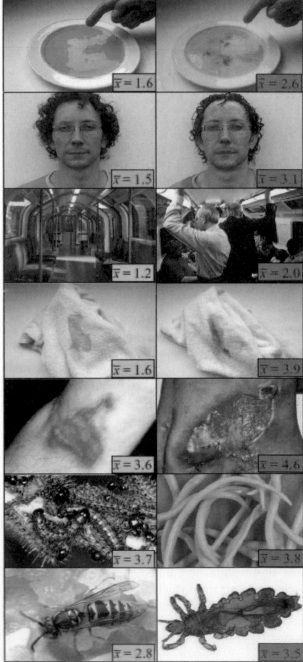

Disease irrelevant   Disease relevant

FIGURE 11.25 ■ An Artificial Smile

The photograph shows Dr. Duchenne electrically stimulating muscles in the face of a volunteer, causing contraction of muscles around the mouth that become active during a smile. As Duchenne discovered, however, a true smile also involves muscles around the eyes.

Corbis

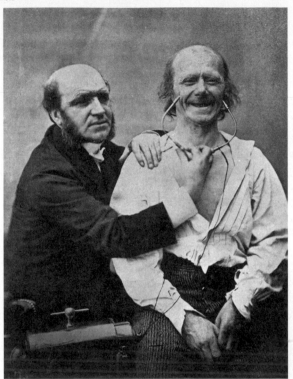

emotions of the soul; the . . . fake joy, the deceitful laugh, cannot provoke the contraction of this latter muscle" (Duchenne, 1862/1990, p. 72). (See *Figure 11.25.*) The difficulty actors have in voluntarily producing a convincing facial expression of emotion is one of the reasons that led Constantin Stanislavsky to develop his system of *method acting,* in which actors attempt to imagine themselves in a situation that would lead to the desired emotion. Once the emotion is evoked, the facial expressions follow naturally (Stanislavsky, 1936).

This observation is confirmed by two neurological disorders with complementary symptoms (Hopf et al., 1992; Topper et al., 1995; Urban et al., 1998; Michel et al., 2008). The first, **volitional facial paresis,** is caused by damage to the face region of the primary motor cortex or to the fibers connecting this region with the motor

volitional facial paresis Difficulty in moving the facial muscles voluntarily; caused by damage to the face region of the primary motor cortex or its subcortical connections.

nucleus of the facial nerve, which controls the muscles responsible for movement of the facial muscles. (*Paresis,* from the Greek "to let go," refers to a partial paralysis.) The interesting thing about volitional facial paresis is that the patient cannot voluntarily move the facial muscles but will express a genuine emotion with those muscles. For example, Figure 11.26(a) shows a woman trying to pull her lips apart and show her teeth. Because of the lesion in the face region of her right primary motor cortex, she could not move the left side of her face. However, when she laughed (Figure 11.26b), both sides of her face moved normally. (See *Figure 11.26a* and *11.26b.*) In contrast, **emotional facial paresis** is caused by damage

FIGURE 11.26 ■ Emotional and Volitional Paresis

(a) A woman with volitional facial paresis caused by a right hemisphere lesion trys to pull her lips apart and show her teeth. Only the right side of her face responds. (b) The same woman shows a genuine smile. (c) A man with emotional facial paresis caused by a left-hemisphere lesion shows his teeth. (d) The same man is smiling. Only the left side of his face responds.

(From Hopf, H. C., Mueller-Forell, W., and Hopf, N. J., *Neurology,* 1992, *42,* 1918–1923. Reprinted with permission.)

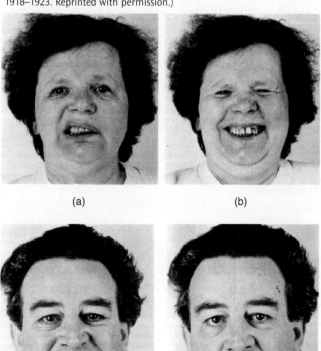

(a)   (b)

(c)   (d)

to the insular region of the prefrontal cortex, to the white matter of the frontal lobe, or to parts of the thalamus. This system joins the system responsible for voluntary movements of the facial muscles in the medulla or caudal pons. People with this disorder can move their face muscles voluntarily but do not express emotions on the affected side of the face. Figure 11.26(c) shows a man pulling his lips apart to show his teeth, which he had no trouble doing. Figure 11.26(d) shows him smiling; as you can see, only the left side of his mouth is raised. He had a stroke that damaged the white matter of the left frontal lobe. (See *Figure 11.26c* and *11.26d.*) These two syndromes clearly indicate that different brain mechanisms are responsible for voluntary movements of the facial muscles and automatic, involuntary expression of emotions involving the same muscles.

Several studies have investigated the brain mechanisms involved in laughter, an expression of emotion more intense than smiling. Arroyo et al. (1993) reported the case of a patient who had seizures that were accompanied by mirthless laughter—that is, the patient laughed but was neither happy nor amused. Recordings made with depth electrodes indicated that the seizure began in the left anterior cingulate gyrus. Removal of a noncancerous tumor located nearby ended both the seizures and the mirthless laughter. The authors suggest that anterior cingulate cortex may be involved in the muscular movements that produce laughter. Shammi and Stuss (1999) found that damage to the right vmPFC impaired people's ability to understand—and be amused by—jokes. For example, consider the following joke:

> Mr. Smith's neighbor approaches him and asks, "Say, are you using your lawnmower this afternoon?" "Yes, I am," replies Mr. Smith.
>
> Which alternative below finishes the joke?
>
> a. "Oops!" as he steps on a rake that barely misses his face.
>
> b. "Fine, then you won't be wanting your golf clubs—I'll just borrow them."
>
> c. "Oh well, can I borrow it when you're done, then?"
>
> d. "The birds are always eating my grass seed."

emotional facial paresis Lack of movement of facial muscles in response to emotions in people who have no difficulty moving these muscles voluntarily; caused by damage to the insular prefrontal cortex, subcortical white matter of the frontal lobe, or parts of the thalamus.

The funny alternative is, of course, (b). But people with ventromedial prefrontal damage usually chose (a), presumably because its slapstick aspect reminded them of humor that they had seen in the past. Clearly, they did not get the point of the joke.

A functional-imaging study by Goel and Dolan (2001) found that different types of jokes activated different regions of the brain, but all of them activated one region: the right ventromedial prefrontal cortex. Another functional-imaging study by the same authors (Goel and Dolan, 2007) presented subjects with socially appropriate and socially inappropriate cartoon jokes. (The inappropriate jokes had strong sexual content that some subjects found offensive.) The investigators found that increasingly funny jokes produced increasing activation of several regions, including the nucleus accumbens (a region involved in reinforcement and reward) and the right vmPFC, while jokes with increasing violation of social norms produced increasing activation of several regions, including the right amygdala and the left orbital frontal cortex. (See *Figure 11.27*.)

As we saw in the previous subsection, the right hemisphere plays a more significant role in recognizing emotions in the voice or facial expressions of other people—especially negative emotions. The same hemispheric specialization appears to be true for expressing emotions. When people show emotions with their facial muscles, the left side of the face usually makes a more intense expression. For example, Sackeim and Gur (1978) cut photographs of people who were expressing emotions into right and left halves, prepared mirror images of each of them, and pasted them together, producing so-called *chimerical faces* (from the mythical Chimera, a fire-breathing monster, part goat, part lion, and part serpent). They found that the left

FIGURE 11.28 ■ Chimerical Faces

(a) Original photo. (b) Composite of the right side of the man's face. (c) Composite of the left side of the man's face.

(Reprinted from *Neuropsychologia, 16,* H. A. Sackeim and R. C. Gur, Lateral asymmetry in intensity of emotional expression, 1978, with kind permission from Pergamon Press, Ltd., Headington Hill Hall, Oxford OX3 0BW, UK.)

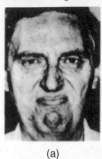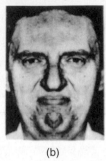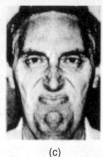

(a)   (b)   (c)

halves were more expressive than the right ones. (See *Figure 11.28.*) Because motor control is contralateral, the results suggest that the right hemisphere is more expressive than the left.

Moscovitch and Olds (1982) made more natural observations of people in restaurants and parks and found that the left side of the face appears to make stronger expressions of emotions. They confirmed these results in the laboratory by analyzing videotapes of people telling sad or humorous stories. A review of the literature by Borod et al. (1998) found forty-eight other studies that obtained similar results.

Using the chimerical faces technique, Hauser (1993) found that rhesus monkeys, like humans, express emotions more strongly in the left sides of their faces. Analysis of videotapes further showed that emotional expressions also begin sooner in the left side of the face. These findings suggest that

FIGURE 11.27 ■ Humor and Violation of Social Norms

The graphs show activation of the right ventromedial prefrontal cortex and the left orbitofrontal cortex, as measured by fMRI, by exposure to humorous cartoons with increasing funniness and increasing violation of social norms.

(Data from Goel, V., and Dolan, R. J. *Journal of Cognitive Neuroscience,* 2007, *19,* 1574–1580.)

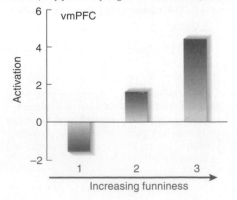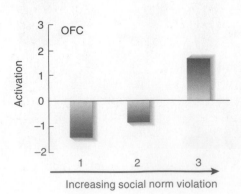

FIGURE 11.29 ■ Emotional Expression and the Right Hemisphere

Successive frames from a videotape of a rhesus monkey show a fear grimace in response to an interaction with a more dominant monkey. The movement begins in the left side of the face, controlled by the right hemisphere.

(From Hauser, M. D., *Science*, 1993, *261*, 475–477. Copyright © 1993, American Association for the Advancement of Science. Reprinted with permission.)

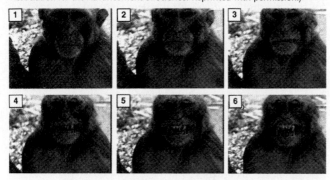

hemispherical specialization for emotional expression appeared before the emergence of our own species. Figure 11.29 shows six videotape frames of a monkey's fear grimace expressed during the course of an interaction with a more dominant monkey. (See *Figure 11.29*.)

Left hemisphere lesions do not usually impair vocal expressions of emotion. For example, people with Wernicke's aphasia (described in Chapter 14) usually modulate their voice according to mood, even though the words they say make no sense. In contrast, right-hemisphere lesions do impair expression of emotion, both facially and by tone of voice.

We saw in the previous subsection that the amygdala is involved in the recognition of facial expression of emotions. Research indicates that it is not involved in emotional *expression*. Anderson and Phelps (2000) reported the case of S. P., a 54-year-old woman whose right amygdala was removed to treat a serious seizure disorder. Because of a preexisting lesion of the left amygdala, the surgery resulted in a bilateral amygdalectomy. After the surgery, S. P. lost the ability to recognize facial expressions of fear, but she had no difficulty recognizing individual faces, and she could easily identify male and female faces and accurately judge their ages. What is particularly interesting is that the amygdala lesions did not impair S. P.'s ability to produce her own facial expressions of fear. She had no difficulty accurately expressing fear, anger, happiness, sadness, disgust, and surprise. However, when she saw a photograph of herself showing fear, she could not tell what emotion her face had been expressing.

Interim Summary

Communication of Emotions

We (and members of other species) communicate our emotions primarily through facial gestures. Darwin believed that such expressions of emotion were innate—that these muscular movements were inherited behavioral patterns. Ekman and his colleagues performed cross-cultural studies with members of an isolated tribe in New Guinea. Their results supported Darwin's hypothesis.

Recognition of other people's emotional expressions involves the right hemisphere more than the left. Studies with normal people have shown that people can judge facial expressions or tone of voice better when the information is presented to the right hemisphere than when it is presented to the left hemisphere. Functional imaging indicates that when people judge the emotions of voices, the right hemisphere is activated more than the left. Studies of people with left- or right-hemisphere brain damage corroborate these findings. In addition, studies show that recognition of particular faces involves neural circuits different from those needed

to recognize facial expressions of emotions. Finally, the amygdala plays a role in recognition of facial expressions of fearfulness; lesions of the amygdala disrupt this ability, and functional-imaging studies show increased activity of the amygdala while the subject is engaging in this task. The ability to judge emotions by a person's tone of voice is not affected.

The amygdala receives magnocellular (primitive) visual information from the superior colliculus and pulvinar, and this information is used in making judgments about fearful expressions. Because of this input, people with damage to the visual cortex that leads to blindness in part of the visual field can nevertheless recognize facial expressions of emotions presented there, a phenomenon called affective blindsight. We can also recognize emotions expressed in a person's body posture or movement, and the amygdala receives and processes this input as well. One of the reasons that bilateral amygdala damage impairs recognition of fearful facial expressions appears to be failure to look at people's eyes.

The direction of the gaze of a person expressing an emotion has information value. Neurons in the superior temporal sulcus are sensitive to direction of gaze and transmit this information to other parts of the brain, including the amygdala. Mirror neurons in the ventral premotor cortex receive visual information concerning the facial expression of other people that activate the neural circuits responsible for these expressions. Feedback from this activity, which may be transmitted to the somatosensory cortex, helps us to comprehend the emotional intentions of other people. Damage to the basal ganglia and insular cortex disrupts recognition of facial expressions of disgust, and functional-imaging studies show increased activity in the insular cortex (which contains the primary gustatory cortex) when people smell disgusting odors or see faces displaying disgust.

Facial expression of emotions (and other stereotypical behaviors such as laughing and crying) are almost impossible to simulate. For example, only a genuine smile of pleasure causes the contraction of the lateral part of the orbicularis oculi (Duchenne's muscle). The anterior cingulate gyrus appears to play a role in the motor aspects of laughter, while the appreciation of humor appears to involve the right ventromedial prefrontal cortex. Genuine expressions of emotion are controlled by special neural circuits. The best evidence for this assertion comes from the complementary syndromes of emotional and volitional facial paresis. People with emotional facial paresis can move their facial muscles voluntarily but not in response to an emotion, whereas people with volitional facial paresis show the opposite symptoms. In addition, the left halves of people's faces—and the faces of monkeys—tend to be more expressive than the right halves.

FEELINGS OF EMOTIONS

So far, we have examined two aspects of emotions: the organization of patterns of responses that deal with the situation that provokes the emotion and the communication of emotional states with other members of the species. The final aspect of emotion to be examined in this chapter is the subjective component: feelings of emotion.

The James-Lange Theory

William James (1842–1910), an American psychologist, and Carl Lange (1834–1900), a Danish physiologist, independently suggested similar explanations for emotion, which most people refer to collectively as the **James-Lange theory** (James, 1884; Lange, 1887). Basically, the theory states that emotion-producing situations elicit an appropriate set of physiological responses, such as trembling, sweating, and increased heart rate. The situations also elicit behaviors, such as clenching of the fists or fighting. The brain receives sensory feedback from the muscles and from the organs that produce these responses, and it is this feedback that constitutes our feeling of emotion.

James said that our own emotional feelings are based on what we find ourselves doing and on the sensory feedback we receive from the activity of our muscles and internal organs. For example, when we find ourselves trembling and feel queasy, we experience fear. Where feelings of emotions are concerned, we are self-observers. Thus, the two aspects of emotions reported in the first two sections of this chapter (patterns of emotional responses and expressions of emotions) give rise to the third: feelings. (See *Figure 11.30.*)

FIGURE 11.30 ■ The James-Lange Theory of Emotion

This schematic shows that an event in the environment triggers behavioral, autonomic, and endocrine responses. Feedback from these responses produces feelings of emotions.

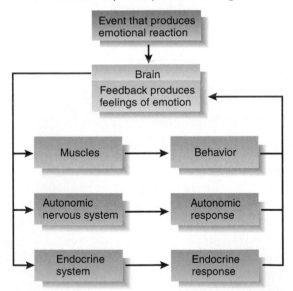

James's description of the process of emotion might strike you as being at odds with your own experience. Many people think that they experience emotions directly, internally. They consider the outward manifestations of emotions to be secondary events. But have you ever

James-Lange theory A theory of emotion that suggests that behaviors and physiological responses are directly elicited by situations and that feelings of emotions are produced by feedback from these behaviors and responses.

found yourself in an unpleasant confrontation with someone else and discovered that you were trembling, even though you did not think that you were so bothered by the encounter? Or did you ever find yourself blushing in response to some public remark that was made about you? Or did you ever find tears coming to your eyes while you watched a film that you did not think was affecting you? What would you conclude about your emotional states in situations like these? Would you ignore the evidence from your own physiological reactions?

A well-known physiologist, Walter Cannon, criticized James's theory. He said that the internal organs were relatively insensitive and that they could not respond very quickly, so feedback from them could not account for our feelings of emotion. In addition, he observed that cutting the nerves that provide feedback from the internal organs to the brain did not alter emotional behavior (Cannon, 1927). However, subsequent research indicated that Cannon's criticisms are not relevant. For example, although the viscera are not sensitive to some kinds of stimuli, such as cutting and burning, they provide much better feedback than Cannon suspected. Moreover, many changes in the viscera can occur rapidly enough that they could be the causes of feelings of emotion.

Cannon cited the fact that cutting the sensory nerves between the internal organs and the central nervous system does not abolish emotional behavior in laboratory animals. However, this observation misses the point. It does not prove that feelings of emotion survive this surgical disruption—only that emotional *behaviors* do. We do not know how the animals feel; we know only that they will snarl and attempt to bite if threatened. In any case, James did not attribute all feelings of emotion to the internal organs; he also said that feedback from muscles was important. The threat might make the animal snarl and bite, and the feedback from the facial and neck muscles might constitute a "feeling" of anger, even if feedback from the internal organs was cut off. But we have no way to ask the animal how it felt.

James's theory is difficult to verify experimentally because it attempts to explain *feelings* of emotion, not the causes of emotional responses, and feelings are private events. Some anecdotal evidence supports the theory. For example, Sweet (1966) reported the case of a man in whom some sympathetic nerves were severed on one side of the body to treat a cardiovascular disorder. The man—a music lover—reported that the shivering sensation he felt while listening to music now occurred only on the unoperated side of his body. He still enjoyed listening to music, but the surgery had altered his emotional reaction.

In one of the few tests of James's theory, Hohman (1966) collected data from people with spinal cord damage. He asked these people about the intensity of their emotional feelings. If feedback is important, one would expect that emotional feelings would be less intense if the injury were high (that is, close to the brain) than if it were low, because a high spinal cord injury would make the person become insensitive to a larger part of the body. In fact, this result is precisely what Hohman found: The higher the injury, the less intense the feeling was. As one of Hohman's subjects said:

> I sit around and build things up in my mind, and I worry a lot, but it's not much but the power of thought. I was at home alone in bed one day and dropped a cigarette where I couldn't reach it. I finally managed to scrounge around and put it out. I could have burned up right there, but the funny thing is, I didn't get all shook up about it. I just didn't feel afraid at all, like you would suppose. (Hohman, 1966, p. 150)

Another subject showed that angry behavior (an emotional response) does not appear to depend on *feelings* of emotion. Instead, the behavior is evoked by the situation (and by the person's evaluation of it) even if the spinal cord damage has reduced the intensity of the person's emotional feelings.

> Now, I don't get a feeling of physical animation, it's sort of cold anger. Sometimes I act angry when I see some injustice. I yell and cuss and raise hell, because if you don't do it sometimes, I've learned people will take advantage of you, but it doesn't have the heat to it that it used to. It's a mental kind of anger. (Hohman, 1966, p. 151)

Feedback from Simulated Emotions

James stressed the importance of two aspects of emotional responses: emotional behaviors and autonomic responses. As we saw earlier in this chapter, a particular set of muscles—those of the face—helps us to communicate our emotional state to other people. Several experiments suggest that feedback from the contraction of facial muscles can affect people's moods and even alter the activity of the autonomic nervous system.

Ekman and his colleagues (Ekman, Levenson, and Friesen, 1983; Levenson, Ekman, and Friesen, 1990) asked subjects to move particular facial muscles to simulate the emotional expressions of fear, anger, surprise, disgust, sadness, and happiness. They did not tell the subjects what emotion they were trying to make them produce, but only what movements they should make. For example, to simulate fear, they told the subjects, "Raise your brows. While holding them raised, pull your brows together. Now raise your upper eyelids and tighten the lower eyelids. Now stretch your lips horizontally." (These movements produce a facial expression of fear.)

While the subjects made the expressions, the investigators monitored several physiological responses controlled by the autonomic nervous system.

The simulated expressions *did* alter the activity of the autonomic nervous system. In fact, different facial expressions produced somewhat different patterns of activity. For example, anger increased heart rate and skin temperature, fear increased heart rate but decreased skin temperature, and happiness decreased heart rate without affecting skin temperature.

Why should a particular pattern of movements of the facial muscles cause changes in mood or in the activity of the autonomic nervous system? Perhaps the connection is a result of experience; in other words, perhaps the occurrence of particular facial movements along with changes in the autonomic nervous system leads to classical conditioning, so that feedback from the facial movements becomes capable of eliciting the autonomic response—and a change in perceived emotion. Or perhaps the connection is innate. As we saw earlier, the adaptive value of emotional expressions is that they communicate feelings and intentions to others. The research presented earlier in this chapter on the role of mirror neurons and the somatosensory cortex suggests that one of the ways we communicate feelings is through unconscious imitation.

A functional-imaging study by Damasio et al. (2000) asked people to recall and try to re-experience past episodes from their lives that evoked feelings of sadness, happiness, anger, and fear. The investigators found that recalling these emotions activated the subjects' somatosensory cortex and upper brain stem nuclei involved in control of internal organs and detection of sensations received from them. These responses are certainly compatible with James's theory. As Damasio et al. put it,

> [Emotions are part of a neural mechanism] based on structures that regulate the organism's current state by executing specific actions via the musculoskeletal system, ranging from facial and postural expressions to complex behaviors, and by producing chemical and neural responses aimed at the internal milieu, viscera and telencephalic neural circuits. The consequences of such responses are represented in both subcortical regulatory structures . . . and in cerebral cortex . . . , and those representations constitute a critical aspect of the neural basis of feelings (p. 1049).

I suspect that if James were still alive, he would approve of these words.

The tendency to imitate the expressions of other people appears to be innate. Field et al. (1982) had adults make facial expressions in front of infants. The infants' own facial expressions were videotaped and were subsequently rated by people who did not know what expressions the adults were displaying. Field and her colleagues found that even newborn babies (with an average age of 36 hours) tended to imitate the expressions they saw. Clearly, the effect occurs too early in life to be a result of learning. Figure 11.31 shows three photographs of the adult expressions and the expressions they elicited in a baby. Can you look at them yourself without changing your own expression, at least a little? (See *Figure 11.31*.)

FIGURE 11.31 ■ Imitation in an Infant

The photographs show happy, sad, and surprised faces posed by an adult and the responses made by the infant.

(From Field, T., in *Development of Nonverbal Behavior in Children*, edited by R. S. Feldman. New York: Springer-Verlag, 1982. Reprinted with kind permission of Springer Science and Business Media.)

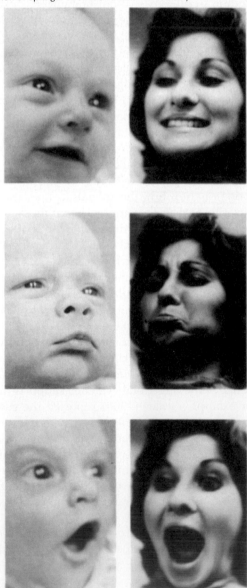

Perhaps imitation provides one of the channels by which organisms communicate their emotions—and evoke feelings of empathy. For example, if we see someone looking sad, we tend to assume a sad expression ourselves. The feedback from our own expression helps to put us in the other person's place and makes us more likely to respond with solace or assistance. And perhaps one of the reasons we derive pleasure from making someone else smile is that their smile makes *us* smile and feel happy. In fact, a functional-imaging study by Pfeifer et al. (2008) found that when normal 10-year-old children watched and imitated emotional expressions, increased activity was seen in the frontal mirror neuron system. In addition, the level of neural activation was positively correlated with measures of the children's empathetic behavior and their interpersonal skills.

Interim Summary

Feelings of Emotions

From the earliest times, people recognized that emotions were accompanied by feelings that seemed to come from inside the body, which probably provided the impetus for developing physiological theories of emotion. James and Lange suggested that emotions were primarily responses to situations. Feedback from the physiological and behavioral reactions to emotion-producing situations gave rise to the feelings of emotion; thus, feelings are the *results*, not the *causes*, of emotional reactions. Hohman's study of people with spinal cord damage supported the James-Lange theory; people who could no longer feel the reactions from most of their body reported that they no longer experienced intense emotional states.

Ekman and his colleagues have shown that even simulating an emotional expression causes changes in the activity of the autonomic nervous system. Perhaps feedback from these changes explains why an emotion can be "contagious": We see someone smile with pleasure, we ourselves imitate the smile, and the internal feedback makes us feel at least somewhat happier. The tendency to mimic the facial expression of others appears to be a consequence of activity in the brain's system of mirror neurons.

SUGGESTED READINGS

Damasio, A. R. *Looking for Spinoza: Joy, Sorrow, and the Feeling Brain.* New York: Harcourt, 2003.

Lane, R. D., and Nadel, L. (eds.) *Cognitive Neuroscience of Emotion.* New York: Oxford University Press, 2000.

LeDoux, J. E. Emotional circuits in the brain. *Annual Review of Neuroscience,* 2000, *23,* 155–184.

Moll, J., Zahn, R., de Oliveira-Souza, R., Krueger, F., and Grafman, J. The neural basis of human moral cognition. *Nature Reviews: Neuroscience,* 2005, *6,* 799–809.

Nelson, R. J., and Trainor, B. C. Neural mechanisms of aggression. *Nature Reviews: Neuroscience,* 2007, *8,* 536–546.

Pessoa, L. On the relationship between emotion and cognition. *Nature Reviews: Neuroscience,* 2008, *9,* 148–158.

Popova, N. K. From genes to aggressive behavior: The role of serotonergic system. *BioEssays,* 2006, *28,* 495–503.

Stoff, D. M., and Susman, E. J. (eds.). *Developmental Psychobiology of Aggression.* New York: Cambridge University Press, 2005.

ADDITIONAL RESOURCES

Visit www.mypsychkit.com for additional review and practice of the material covered in this chapter. Within MyPsychKit, you can take practice tests and receive a customized study plan to help you review. Dozens of animations, tutorials, and Web links are also available. You can even review using the interactive electronic version of this textbook. You will need to register for MyPsychKit. See www.mypsychkit.com for complete details.

FIGURE 12.1 ■ Example of a Regulatory System

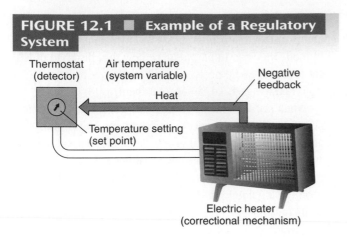

(the optimal value of the system variable), a **detector** that monitors the value of the system variable, and a **correctional mechanism** that restores the system variable to the set point.

An example of a regulatory system is a room whose temperature is regulated by a thermostatically controlled heater. The system variable is the air temperature of the room, and the detector for this variable is a thermostat. This device can be adjusted so that contacts of a switch will be closed when the temperature falls below a preset value (the set point). Closure of the contacts turns on the correctional mechanism—the coils of the heater. (See *Figure 12.1.*) If the room cools below the set point of the thermostat, the thermostat turns the heater on, and the heater warms the room. The rise in room temperature causes the thermostat to turn the heater off. Because the activity of the correctional mechanism (heat production) feeds back to the thermostat and causes it to turn the heater off, this process is called **negative feedback.** Negative feedback is an essential characteristic of all regulatory systems.

This chapter considers regulatory systems that involve ingestive behaviors: drinking and eating. These behaviors are correctional mechanisms that replenish the body's depleted stores of water or nutrients. Because of the delay between ingestion and replenishment of the depleted stores, ingestive behaviors are controlled by **satiety mechanisms** as well as by detectors that monitor the system variables. Satiety mechanisms are required because of the physiology of our digestive system. For example, suppose you spend some time in a hot, dry environment and lose body water. The loss of water causes internal detectors to initiate the correctional mechanism: drinking. You quickly drink a glass or two of water and then stop. What stops your ingestive behavior? The water is still in your digestive system, not yet in the fluid surrounding your cells, where it is needed. Therefore, although drinking was initiated by detectors that measure your body's need for water, *it was stopped by other means.* There must be a satiety mechanism that says, in effect, "Stop—this water, when absorbed by the digestive system into the blood, will eventually replenish the body's need." Satiety mechanisms monitor the activity of the correctional mechanism (in this case, drinking), not the system variables themselves. When a sufficient amount of drinking occurs, the satiety mechanisms stop further drinking *in anticipation* of the replenishment that will occur later. (See *Figure 12.2.*)

DRINKING

To maintain our internal milieu at its optimal state, we have to drink some water from time to time. This section describes the control of this form of ingestive behavior.

detector In a regulatory process, a mechanism that signals when the system variable deviates from its set point.

correctional mechanism In a regulatory process, the mechanism that is capable of changing the value of the system variable.

negative feedback A process whereby the effect produced by an action serves to diminish or terminate that action; a characteristic of regulatory systems.

satiety mechanism A brain mechanism that causes cessation of hunger or thirst, produced by adequate and available supplies of nutrients or water.

FIGURE 12.2 ■ Outline of the System Controlling Drinking

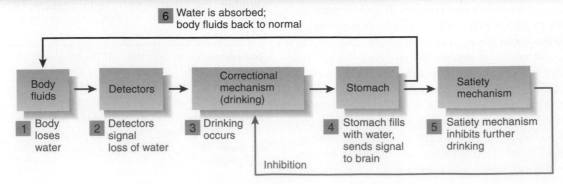

Some Facts About Fluid Balance

Before you can understand the physiological control of drinking, you must know something about the fluid compartments of the body and their relationships with each other. The body contains four major fluid compartments: one compartment of intracellular fluid and three compartments of extracellular fluid. Approximately two-thirds of the body's water is contained in the **intracellular fluid,** the fluid portion of the cytoplasm of cells. The rest is **extracellular fluid,** which includes the **intravascular fluid** (the blood plasma), the cerebrospinal fluid, and the **interstitial fluid.** *Interstitial* means "standing between"; indeed, the interstitial fluid stands between our cells—it is the "seawater" that bathes them. For the purposes of this chapter we will ignore the cerebrospinal fluid and concentrate on the other three compartments. (See *Figure 12.3.*)

Two of the fluid compartments of the body must be kept within precise limits: the intracellular fluid and the intravascular fluid. The intracellular fluid is controlled by the concentration of solutes in the interstitial fluid. (*Solutes* are the substances dissolved in a solution.) Normally, the interstitial fluid is **isotonic** (from *isos,* "equal," and *tonos,* "tension") with the intracellular fluid. That is, the concentration of solutes in the cells and in the interstitial fluid that bathes them is balanced, so water does not tend to move into or out of the cells. If the interstitial fluid loses water (becomes more concentrated, or **hypertonic**), water will be pulled out of the cells. On the other hand, if the interstitial fluid gains water (becomes more dilute, or **hypotonic**), water will move into the cells. Either condition endangers cells; a loss of water deprives them of the ability to perform many chemical reactions,

FIGURE 12.4 ■ Solute Concentration

The figure shows the effects of differences in solute concentration on the movement of water molecules.

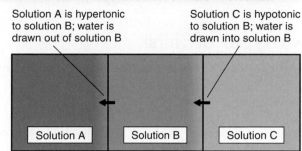

and a gain of water can cause their membranes to rupture. Therefore, the concentration of the interstitial fluid must be closely regulated. (See *Figure 12.4.*)

The volume of the blood plasma must be closely regulated because of the mechanics of the operation of the heart. If the blood volume falls too low, the heart can no longer pump the blood effectively; if the volume is not restored, heart failure will result. This condition is called **hypovolemia,** literally "low volume of the blood" (*-emia* comes from the Greek *haima,* "blood"). The vascular system of the body can make some adjustments for loss of blood volume by contracting the muscles in smaller veins and arteries, thereby presenting a smaller space for the blood to fill, but this correctional mechanism has definite limits.

The two important characteristics of the body fluids—the solute concentration of the intracellular fluid and the volume of the blood—are monitored by two different sets of receptors. A single set of receptors would not work because it is possible for one of these fluid compartments to be changed without affecting the other. For example, a loss of blood obviously reduces the volume of the intravascular fluid, but it has no effect on the volume of the intracellular fluid. On the other hand, a salty meal will increase

FIGURE 12.3 ■ Relative Size of the Body's Fluid Compartments

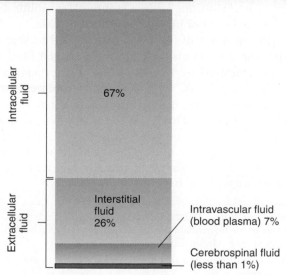

intracellular fluid The fluid contained within cells.

extracellular fluid All body fluids outside cells: interstitial fluid, blood plasma, and cerebrospinal fluid.

intravascular fluid The fluid found within the blood vessels.

interstitial fluid The fluid that bathes the cells, filling the space between the cells of the body (the "interstices").

isotonic Equal in osmotic pressure to the contents of a cell. A cell placed in an isotonic solution neither gains nor loses water.

hypertonic The characteristic of a solution that contains enough solute that it will draw water out of a cell placed in it, through the process of osmosis.

hypotonic The characteristic of a solution that contains so little solute that a cell placed in it will absorb water, through the process of osmosis.

hypovolemia (*hy poh voh lee mee a*) Reduction in the volume of the intravascular fluid.

the solute concentration of the interstitial fluid, drawing water out of the cells, but it will not cause hypovolemia. Thus, the body needs two sets of receptors, one measuring blood volume and another measuring cell volume.

Two Types of Thirst

As we just saw, for our bodies to function properly, the volume of two fluid compartments—intracellular and intravascular—must be regulated. Most of the time, we ingest more water and sodium than we need, and the kidneys excrete the excess. However, if the levels of water or sodium fall too low, correctional mechanisms—drinking water or ingesting sodium—are activated. Everyone is familiar with the sensation of thirst, which occurs when we need to ingest water. However, a salt appetite is much more rare because it is difficult for people *not* to get enough sodium in their diet, even if they do not put extra salt on their food. Nevertheless, the mechanisms to increase sodium intake exist, even though they are seldom called upon in members of our species.

Because loss of water from either the intracellular or intravascular fluid compartments stimulates drinking, researchers have adopted the terms *osmometric thirst* and *volumetric thirst* to describe them. The term *volumetric* is clear; it refers to the metering (measuring) of the volume of the blood plasma. The term *osmometric* requires more explanation, which will be provided in the next section. The term *thirst* means different things in different circumstances. Its original definition referred to a sensation that people say they have when they are dehydrated. Here I use it in a descriptive sense. Because we do not know how other animals feel, *thirst* simply means a tendency to seek water and to ingest it.

Our bodies lose water continuously, primarily through evaporation. Each breath exposes the moist inner surfaces of the respiratory system to the air; thus, each breath causes the loss of a small amount of water. In addition, our skin is not completely waterproof; some water finds its way through the layers of the skin and evaporates from the surface. The moisture that is lost through evaporation is, of course, pure distilled water. (The body

FIGURE 12.5 ■ Water Loss Through Evaporation

Skin

1 Water is lost through evaporation

2 Concentration of interstitial fluid increases

3a Capillaries lose water by osmosis

3b Cells lose water by osmosis

FIGURE 12.6 ■ An Osmoreceptor

The figure shows a hypothetical explanation of the workings of an osmoreceptor.

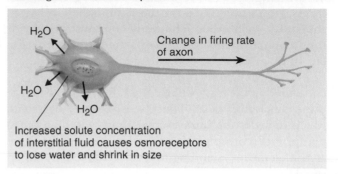

H_2O

H_2O

H_2O

Change in firing rate of axon

Increased solute concentration of interstitial fluid causes osmoreceptors to lose water and shrink in size

loses water through sweating, too; but because it loses salt along with the water, sweating produces a need for sodium as well as for water.) Figure 12.5 illustrates how the loss of water through evaporation depletes both the intracellular fluid and intravascular fluid compartments. For the sake of simplicity, only a few cells are shown, and the volume of the interstitial fluid is greatly exaggerated. Water is lost directly from the interstitial fluid, which becomes slightly more concentrated than either the intracellular or the intravascular fluid. This change draws water from both the cells and the blood plasma. Eventually, the loss of water from the cells and the blood plasma will be great enough that both osmometric thirst and volumetric thirst will be produced. (See *Figure 12.5.*)

Osmometric Thirst

Osmometric thirst occurs when the solute concentration of the interstitial fluid increases. This increase draws water out of the cells, and they shrink in volume. The term *osmometric* refers to the fact that the detectors are actually responding to (metering) changes in the concentration of the interstitial fluid that surrounds them. *Osmosis* is the movement of water through a semipermeable membrane from a region of low solute concentration to one of high solute concentration.

The existence of neurons that respond to changes in the solute concentration of the interstitial fluid was first hypothesized by Verney (1947). Verney suggested that these detectors, which he called **osmoreceptors,** were neurons whose firing rate was affected by their level of hydration. That is, if the interstitial fluid surrounding them became more concentrated, they would lose water through osmosis. The shrinkage would cause them to alter their firing rate, which would send signals to other parts of the brain. (See *Figure 12.6.*)

osmometric thirst Thirst produced by an increase in the osmotic pressure of the interstitial fluid relative to the intracellular fluid, thus producing cellular dehydration.

osmoreceptor A neuron that detects changes in the solute concentration of the interstitial fluid that surrounds it.

FIGURE 12.7 ■ The Circumventricular Organs

This sagittal section of the rat brain shows the location of the circumventricular organs. *Inset:* A hypothetical circuit connecting the subfornical organ with the median preoptic nucleus.

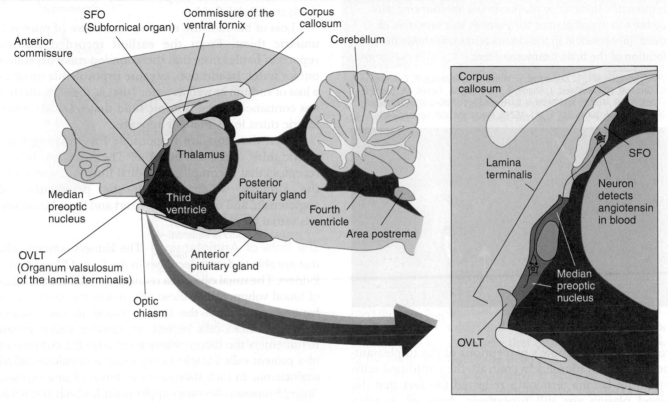

When we eat a salty meal, we incur a pure osmometric thirst. The salt is absorbed from the digestive system into the blood plasma; hence, the blood plasma becomes hypertonic. This condition draws water from the interstitial fluid, which makes this compartment become hypertonic too and thus causes water to leave the cells. The osmoreceptors responsible for osmometric thirst are located in a region known as the *lamina terminalis,* found just rostral to the ventral portion of the third ventricle. The lamina terminalis contains two specialized *circumventricular organs:* the *OVLT* and the *SFO.* The brain contains several circumventricular organs, specialized regions with rich blood supplies located along the ventricular system. You are already familiar with one of these: the area postrema, discussed in Chapter 2. (See **Figure 12.7.**)

The **OVLT** (if you really want to know, that stands for the *organum vasculosum of the lamina terminalis*) and the **subfornical organ (SFO),** like the other circumventricular organs, are located outside the blood–brain barrier. That means that substances dissolved in the blood pass easily into the interstitial fluid within these organs. Evidence suggests that most of the osmoreceptors responsible for osmometric thirst are located in the OVLT, but some are also located in the SFO (McKinley et al., 2004).

A functional-imaging study by Egan et al. (2003) found that the human lamina terminalis also appears to contain osmoreceptors. The investigators administered intravenous injections of hypertonic saline to normal subjects while their brain were being scanned. They observed strong activation of several brain regions, including the lamina terminalis and the anterior cingulate cortex. When the subjects were permitted to drink water, they did so and almost immediately reported that their thirst had been satisfied. Simultaneously, the activity in the anterior cingulate cortex returned to baseline values. However, the activity in the lamina terminalis remained high. These results suggest that the activity of the anterior cingulate cortex reflected the subjects' thirst, which was immediately relieved by a drink of water. (As we saw in Chapter 7, activity of this region is

OVLT (organum vasculosum of the lamina terminalis) A circumventricular organ located anterior to the anteroventral portion of the third ventricle; served by fenestrated capillaries and thus lacks a blood–brain barrier.

subfornical organ (SFO) A small organ located in the confluence of the lateral ventricles, attached to the underside of the fornix; contains neurons that detect the presence of angiotensin in the blood and excite neural circuits that initiate drinking.

circumventricular organs. In fact, research indicates that the subfornical organ is the site at which blood angiotensin acts to produce thirst. This structure gets its name from its location, just below the commissure of the ventral fornix. (Refer to *Figure 12.7.*)

Simpson, Epstein, and Camardo (1978) found that very low doses of angiotensin injected directly into the SFO caused drinking and that destruction of the SFO or injection of a drug that blocks angiotensin receptors abolished the drinking that normally occurs when angiotensin is injected into the blood. In addition, Phillips and Felix (1976) found that injections of minute quantities of angiotensin into the SFO increased the firing rate of single neurons located there; evidently, these neurons contain angiotensin receptors.

Neurons in the subfornical organ send their axons to another part of the lamina terminalis: the **median preoptic nucleus** (not to be confused with the *medial* preoptic nucleus), a small nucleus wrapped around the front of the anterior commissure, a fiber bundle that connects the amygdala and anterior temporal lobe. (Refer to the inset in *Figure 12.7.*)

On the basis of these findings, Thrasher and his colleagues (see Thrasher, 1989) suggested that the median preoptic nucleus acts as an integrating system for most or all of the stimuli for osmometric and volumetric thirst. As you just saw, the median preoptic nucleus receives information from angiotensin-sensitive neurons in the SFO. In addition, this nucleus receives information from the OVLT (which contains osmoreceptors) and from the nucleus of the solitary tract (which receives information from the atrial baroreceptors). Excitotoxic lesions of the median preoptic nucleus, which destroy cell bodies but spare axons passing through the structure, cause severe deficits in osmometric thirst (Cunningham et al., 1992). According to Thrasher and his colleagues, the median preoptic nucleus integrates the information it receives and, through its efferent connections with other parts of the brain, controls drinking. (See *Figure 12.10.*)

The region of the lamina terminalis seems to play a critical role in fluid regulation in humans as well. As we saw earlier, functional imaging indicates that osmometric

FIGURE 12.10 ■ Neural Circuitry Concerned with the Control of Drinking

In this schematic not all connections are shown, and some connections may be indirect. Although most of the osmoreceptors are located in the OVLT (organum vasculosum of the lamina terminalis), some are also located in the subfornical organ.

(Adapted from Thrasher, T. N. *Acta Physiologica Scandinavica*, 1989, *136*, 141–150.)

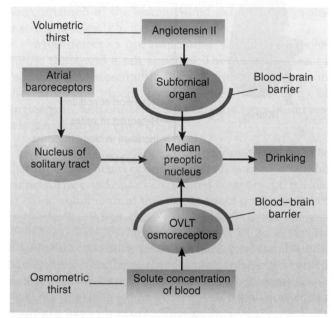

thirst increases the activity of this region. In addition, McIver et al. (1991) reported that brain damage that includes this region can cause *adipsia*—lack of drinking. The patients report no sensation of thirst even after they are given an injection of hypertonic saline. To survive, they must deliberately drink water at regular intervals each day, even though they feel no need to do so.

median preoptic nucleus A small nucleus situated around the decussation of the anterior commissure; plays a role in thirst stimulated by angiotensin.

Interim Summary

Physiological Regulatory Mechanisms and Drinking

A regulatory system contains four features: a system variable (the variable that is regulated), a set point (the optimal value of the system variable), a detector to measure the system variable, and a correctional mechanism to change it.

Physiological regulatory systems, such as control of body fluids and nutrients, require a satiety mechanism to anticipate the effects of the correctional mechanism, because the changes brought about by eating and drinking occur only after a considerable period of time.

The body contains three major fluid compartments: intracellular, interstitial, and intravascular. Sodium and water can easily pass between the intravascular fluid and the interstitial fluid, but sodium cannot penetrate the cell membrane. The solute concentration of the interstitial fluid must be closely regulated. If it becomes hypertonic, cells lose water; if it becomes hypotonic, they gain water. The volume of the intravascular fluid (blood plasma) must also be kept within bounds.

Osmometric thirst occurs when the interstitial fluid becomes hypertonic, drawing water out of cells. This event, which can be caused by evaporation of water from the body or by ingestion of a salty meal, is detected by osmoreceptors in the OVLT and the SFO, circumventricular organs located in the lamina terminalis. Activation of these osmoreceptors stimulates drinking.

Volumetric thirst occurs along with osmometric thirst when the body loses fluid through evaporation. Pure volumetric thirst is caused by blood loss, vomiting, and diarrhea. One stimulus for volumetric thirst is provided by a fall in blood flow to the kidneys. This event triggers the secretion of renin, which converts plasma angiotensinogen to angiotensin I. Angiotensin I is subsequently converted to its active form, angiotensin II, which acts on neurons in the SFO and stimulates thirst. The hormone also increases blood pressure and stimulates the secretion of pituitary and adrenal hormones that inhibit the secretion of water and sodium by the kidneys and induce a sodium appetite. (Sodium is needed to help restore the plasma volume.) Volumetric drinking can also be stimulated by a set of baroreceptors in the atria of the heart that detect decreased blood volume and send this information to the brain.

A nucleus in the lamina terminalis, the median preoptic nucleus, receives and integrates osmometric and volumetric information. Information about hypovolemia conveyed by angiotensin reaches the median preoptic nucleus directly from the SFO. Information about hypovolemia conveyed by the atrial stretch receptor system reaches this nucleus by means of a relay in the nucleus of the solitary tract. Osmotic information reaches this nucleus from osmoreceptors in both the OVLT and the SFO. Neurons in the median preoptic nucleus stimulate drinking through their connections with other parts of the brain. The anterior cingulate cortex also receives information from the OVLT via the dorsal midline nucleus of the thalamus, and this region may be involved in sensations of thirst.

Thought Question

How do we know that we are thirsty? What does thirst feel like? It cannot simply be a dry mouth or throat, because a real thirst is not quenched by taking a small sip of water, which moistens the mouth and throat as well as a big drink does.

EATING: SOME FACTS ABOUT METABOLISM

Clearly, eating is one of the most important things we do, and it can also be one of the most pleasurable. Much of what an animal learns to do is motivated by the constant struggle to obtain food; therefore, the need to ingest undoubtedly shaped the evolutionary development of our own species. After having read the first part of this chapter, in which you saw that the signals that cause thirst are well understood, you might be surprised to learn that researchers are only now discovering what the system variables for hunger are. Control of ingestive behavior is even more complicated than the control of drinking and sodium intake. We can achieve water balance by the intake of two ingredients: water and sodium chloride. When we eat, we must obtain adequate amounts of carbohydrates, fats, amino acids, vitamins, and minerals other than sodium. Thus, our food-ingestive behaviors are more complex, as are the physiological mechanisms that control them.

The rest of this chapter describes research on the control of eating: metabolism, regulation of body weight, the environmental and physiological factors that begin and stop a meal, and the neural mechanisms that monitor the nutritional state of our bodies and control our ingestive behavior. It also describes the most serious eating disorders: obesity and anorexia nervosa. Despite all the effort that has gone into understanding the physiology of ingestive behavior, these disorders are still difficult to treat. Our best hope of finding effective treatments is to achieve a better understanding of the physiology of metabolism and ingestive behavior.

As you saw in the discussion of the physiology of drinking, you must know something about the fluid compartments of the body and the functions of the kidney to understand the physiology of drinking. Therefore, you will not be surprised that this part of the chapter begins with a discussion of metabolism. Your first inclination might be to skip over this section; but if you do so, you will find that you will not understand experiments that are described later. For example, the system variables that cause an animal to seek food and eat it are obviously related to the animal's metabolism. This section will discuss only as much about this subject as you will need to understand these experiments.

When we eat, we incorporate into our own bodies molecules that were once part of other living organisms, plant and animal. We ingest these molecules for two reasons: to construct and maintain our own organs and to obtain energy for muscular movements and for keeping our bodies warm. In other words, we need both building blocks and fuel. Although food used for building blocks is essential, I will discuss only the food used for fuel, because most of the molecules we eat get "burned" to provide energy for movement and heating.

To stay alive, our cells must be supplied with fuel and oxygen. Obviously, fuel comes from the digestive tract, and its presence there is a result of eating. But the digestive tract is sometimes empty; in fact, most of us wake up in the morning in that condition. So there has to be a reservoir that stores nutrients to keep the cells of the body nourished when the gut is empty. Indeed, there are two reservoirs: one short-term and the other long-term. The short-term reservoir stores carbohydrates, and the long-term reservoir stores fats.

The short-term reservoir is located in the cells of the liver and the muscles, and it is filled with a complex, insoluble carbohydrate called **glycogen.** For simplicity I will consider only one of these locations: the liver. Cells in the liver convert glucose (a simple, soluble carbohydrate) into glycogen and store the glycogen. They are stimulated to do so by the presence of **insulin,** a peptide hormone produced by the pancreas. Thus, when glucose and insulin are present in the blood, some of the glucose is used as a fuel, and some of it is stored as glycogen. Later, when all of the food has been absorbed from the digestive tract, the level of glucose in the blood begins to fall.

The fall in glucose is detected by cells in the pancreas and in the brain. The pancreas responds by stopping its secretion of insulin and starting to secrete a different peptide hormone: **glucagon.** The effect of glucagon is opposite that of insulin: It stimulates the conversion of glycogen into glucose. (Unfortunately, the terms *glucose, glycogen,* and *glucagon* are similar enough that it is easy to confuse them. Even worse, you will soon encounter another one: *glycerol.*) (See *Figure 12.11.*) Thus, the liver soaks up excess glucose and stores it as glycogen when plenty of glucose is available, and the liver releases glucose from its reservoir when the digestive tract becomes empty and the level of glucose in the blood begins to fall.

The carbohydrate reservoir in the liver is reserved primarily for the central nervous system (CNS). When you wake in the morning, your brain is being fed by your liver, which is in the process of converting glycogen to glucose and releasing it into the blood. The glucose reaches the CNS, where it is absorbed and metabolized by the neurons and the glia. This process can continue for a few hours, until all of the carbohydrate reservoir in the liver is used up. (The average liver holds approximately 300

calories of carbohydrate.) Usually, we eat some food before this reservoir gets depleted, which permits us to refill it. But if we do not eat, the CNS has to start living on the products of the long-term reservoir.

Our long-term reservoir consists of adipose tissue (fat tissue). This reservoir is filled with fats or, more precisely, with **triglycerides.** Triglycerides are complex molecules that contain **glycerol** (a soluble carbohydrate, also called *glycerine*) combined with three **fatty acids** (stearic acid, oleic acid, and palmitic acid). Adipose tissue is found beneath the skin and in various locations in the abdominal cavity. It consists of cells that are capable of absorbing nutrients from the blood, converting them to triglycerides, and storing them. These cells can expand enormously in size; in fact, the primary physical difference between an obese person and a person of normal weight is the size of their fat cells, which is determined by the amount of triglycerides that these cells contain.

The long-term fat reservoir is obviously what keeps us alive when we are fasting. As we begin to use the contents of our short-term carbohydrate reservoir, fat cells start converting triglycerides into fuels that the cells can use and releasing these fuels into the bloodstream. As we just saw, when we wake in the morning with an empty digestive tract, our brain (in fact, all of the CNS) is living on glucose released by the liver. But what about the other cells of the body? They are living on fatty acids, sparing the glucose for the brain. As you will recall from Chapter 3, the sympathetic nervous system is primarily involved in the breakdown and utilization of stored nutrients. When the digestive system is empty, there is an increase in the activity of the sympathetic axons that innervate adipose tissue, the pancreas, and the adrenal medulla. All three effects (direct neural stimulation, secretion of glucagon, and secretion of catecholamines) cause triglycerides in the long-term fat reservoir to be broken down into glycerol and fatty acids. The fatty acids can be directly metabolized by cells in all of the body *except the*

---

glycogen (*gly ko jen*) A polysaccharide often referred to as *animal starch*; stored in liver and muscle; constitutes the short-term store of nutrients.

insulin A pancreatic hormone that facilitates entry of glucose and amino acids into the cell, conversion of glucose into glycogen, and transport of fats into adipose tissue.

glucagon (*gloo ka gahn*) A pancreatic hormone that promotes the conversion of liver glycogen into glucose.

triglyceride (*try gliss er ide*) The form of fat storage in adipose cells; consists of a molecule of glycerol joined with three fatty acids.

glycerol (*gliss er all*) A substance (also called glycerine) derived from the breakdown of triglycerides, along with fatty acids; can be converted by the liver into glucose.

fatty acid A substance derived from the breakdown of triglycerides, along with glycerol; can be metabolized by most cells of the body except for the brain.

FIGURE 12.11 ■ Effects of Insulin and Glucagon on Glucose and Glycogen

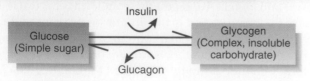

FIGURE 12.12 ■ Metabolic Pathways During the Fasting Phase and Absorptive Phase of Metabolism

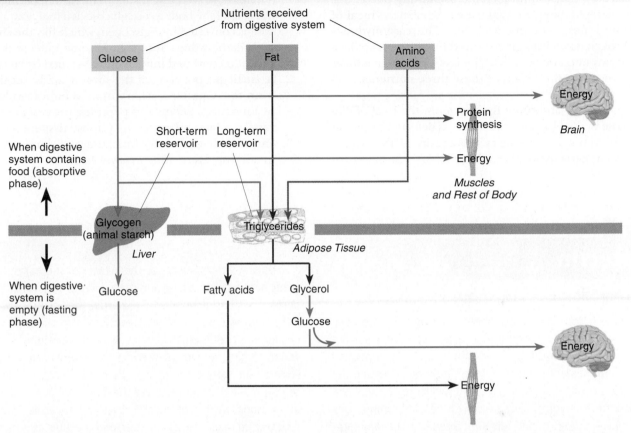

brain, which needs glucose. That leaves glycerol. The liver takes up glycerol and converts it to glucose. That glucose, too, is available to the brain.

You may be asking why the cells of the rest of the body treat the brain so kindly, letting it consume almost all the glucose that the liver releases from its carbohydrate reservoir and constructs from glycerol. The answer is simple: Insulin has several other functions besides causing glucose to be converted to glycogen. One of these functions is controlling the entry of glucose into cells. Glucose easily dissolves in water, but it will not dissolve in fats. Cell membranes are made of lipids (fatlike substances); thus, glucose cannot directly pass through them. To be taken into a cell, glucose must be transported there by *glucose transporters*—protein molecules that are situated in the membrane and are similar to those responsible for the reuptake of transmitter substances. Glucose transporters contain insulin receptors, which control their activity; only when insulin binds with these receptors can glucose be transported into the cell. But the cells of the nervous system are an exception to this rule. Their glucose transporters do not contain insulin receptors; thus, these cells can absorb glucose *even when insulin is not present.*

Figure 12.12 reviews what I have said so far about the metabolism that takes place while the digestive tract is empty, which physiologists refer to as the **fasting phase** of metabolism. A fall in the blood glucose level causes the pancreas to stop secreting insulin and to start secreting glucagon. The absence of insulin means that most of the cells of the body can no longer use glucose; thus, all the glucose present in the blood is reserved for the CNS. The presence of glucagon and the absence of insulin instruct the liver to start drawing on the short-term carbohydrate reservoir—to start converting its glycogen into glucose. The presence of glucagon and the absence of insulin, along with increased activity of the sympathetic nervous system, also instruct fat cells to start drawing on the long-term fat reservoir—to start breaking down triglycerides into fatty acids and glycerol. Most of the body lives on the fatty acids, and the glycerol, which is converted into glucose by the liver, gets used by the brain. If fasting is prolonged, proteins (especially protein found in muscle) will be broken down to amino acids, which can be metabolized by all of the body except the CNS. (See *Figure 12.12.*)

---

fasting phase The phase of metabolism during which nutrients are not available from the digestive system; glucose, amino acids, and fatty acids are derived from glycogen, protein, and adipose tissue during this phase.

The phase of metabolism that occurs when food is present in the digestive tract is called the **absorptive phase.** Now that you understand the fasting phase, this one is simple. Suppose that we eat a balanced meal of carbohydrates, proteins, and fats. The carbohydrates are broken down into glucose, and the proteins are broken down into amino acids. The fats basically remain as fats. Let us consider each of these three nutrients.

1. As we start absorbing the nutrients, the level of glucose in the blood rises. This rise is detected by cells in the brain, which causes the activity of the sympathetic nervous system to decrease and the activity of the parasympathetic nervous system to increase. This change tells the pancreas to stop secreting glucagon and to begin secreting insulin. The insulin permits all the cells of the body to use glucose as a fuel. Extra glucose is converted into glycogen, which fills the short-term carbohydrate reservoir. If some glucose is left over, it is converted into fat and absorbed by fat cells.

2. A small proportion of the amino acids received from the digestive tract are used as building blocks to construct proteins and peptides; the rest are converted to fats and stored in adipose tissue.

3. Fats are not used at this time; they are simply stored in adipose tissue. (See *Figure 12.12.*)

Interim Summary

Eating: Some Facts About Metabolism

Metabolism consists of two phases. During the absorptive phase we receive glucose, amino acids, and fats from the intestines. The blood level of insulin is high, which permits all cells to metabolize glucose. In addition, the liver and the muscles convert glucose to glycogen, which replenishes the short-term reservoir. Excess carbohydrates and amino acids are converted to fats, and fats are placed into the long-term reservoir in the adipose tissue.

During the fasting phase the activity of the parasympathetic nervous system falls, and the activity of the sympathetic nervous system increases. In response, the level of insulin falls, and the levels of glucagon and the adrenal catecholamines rise. These events cause liver glycogen to be converted to glucose and triglycerides to be broken down into glycerol and fatty acids. In the absence of insulin, only the central nervous system can use the glucose that is available in the blood; the rest of the body lives on fatty acids. Glycerol is converted to glucose by the liver, and the glucose is metabolized by the brain.

WHAT STARTS A MEAL?

Regulation of body weight requires a balance between food intake and energy expenditure. If we ingest more calories than we burn, we will gain weight; carbohydrates and proteins contain about 4 kcal/gm, and fats contain about 9 kcal/gm. (The "calorie" you might see printed on a food label is actually a kilocalorie—1000 calories—or enough energy to raise the temperature of a liter of water by 1° C.) In this section I will discuss the factors that control when and how much we eat. I will have more to say about control of energy expenditure later in this chapter.

Assuming that our energy expenditure is constant, we need two mechanisms to maintain a relatively constant body weight. One mechanism must increase our motivation to eat if our long-term nutrient reservoir is becoming depleted, and another must restrain our food intake if we begin to take in more calories than we need.

Signals from the Environment

The environment of our ancestors shaped the evolution of these regulatory mechanisms. In the past, starvation was a much greater threat to survival than overeating was. In fact, a tendency to overeat in times of plenty provided a reserve that could be drawn upon if food became scarce again—which it often did. A feast-or-famine environment favored the evolution of mechanisms that were quick to detect losses from the long-term reservoir and provide a strong signal to seek and eat food. Natural selection for mechanisms that detected weight gain and suppressed overeating was much less significant.

The answer to the question posed by the title of this section, "What starts a meal?" is not simple. Most people,

---

absorptive phase The phase of metabolism during which nutrients are absorbed from the digestive system; glucose and amino acids constitute the principal source of energy for cells during this phase, and excess nutrients are stored in adipose tissue in the form of triglycerides.

if they were asked why they eat, would say that they do so because they get hungry. By that, they probably mean that something happens inside the body that provides a sensation that makes them want to eat. But if this is true, just what is happening inside our bodies? The factors that motivate us to eat when food is readily available are very different from those that motivate us when food is scarce. When food is plentiful, we tend to eat when the stomach and upper intestine are empty. This emptiness provides a hunger signal—a message to our brain that indicates that we should begin to eat. The time it takes for food to leave the stomach would seem to encourage the establishment of a pattern of eating three meals a day. In addition, our ancestors undoubtedly found it most practical to prepare food for a group of people and have everyone eat at the same time. Most modern-day work schedules follow this routine as well.

Although an empty stomach is an important signal, many factors start a meal, including the sight of a plate of food, the smell of food cooking in the kitchen, the presence of other people sitting around the table, or the words "It's time to eat!" I am writing this in the late afternoon. I am anticipating a tasty meal this evening and look forward to eating it. I don't feel particularly hungry, but I like good food and expect to enjoy my dinner. My short-term and long-term nutrient reservoirs are well stocked, so my motivation to eat will not be based upon a physiological need for nourishment.

Signals from the Stomach

As we just saw, an empty stomach and upper intestine provide an important signal to the brain that it is time to start thinking about finding something to eat. Recently, researchers discovered one of the ways this signal may be communicated to the brain. The gastrointestinal system (especially the stomach) releases a peptide hormone called **ghrelin** (Kojima et al., 1999). The name *ghrelin* is a contraction of *GH releasin,* which reflects the fact that this peptide is involved in controlling the release of growth hormone, usually abbreviated as *GH.* Studies with laboratory animals have found that blood levels of ghrelin increase with fasting and are reduced after a meal and that ghrelin antibodies or ghrelin receptor antagonists inhibit eating. Also, subcutaneous injections or infusions of ghrelin into the cerebral ventricles cause weight gain by increasing food intake and decreasing the metabolism of fats. In humans, blood levels of ghrelin increase shortly before each meal, which suggests that this peptide is involved in the initiation of a meal and that feelings of hunger are correlated with blood levels of ghrelin (Tschöp, Smiley, and Heiman, 2000; Ariyasu et al., 2001; Wren et al., 2001; Bagnasco et al., 2003). Figure 12.13 shows the relationship between blood levels of ghrelin and food intake during the course of a day. (See *Figure 12.13.*)

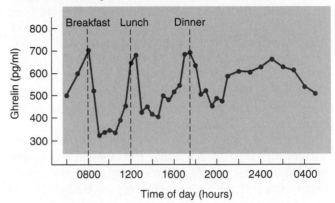

FIGURE 12.13 ■ Levels of Ghrelin in Human Blood Plasma

The graph shows that a rise in the level of this peptide preceded each meal.

(Adapted from Cummings, D. E., Purnell, J. Q., Frayo, R. S., Schmidova, K., Wisse, B. E., and Weigle, D. S. *Diabetes,* 2001, *50,* 1714–1719.)

Ghrelin is a potent stimulator of food intake, and it even stimulates thoughts about food. Schmid et al. (2005) found that a single intravenous injection of ghrelin not only enhanced appetite in normal subjects, it also elicited vivid images of foods that the subjects liked to eat. As we saw in the case at the beginning of this chapter, one of the most obvious symptoms of Prader-Willi syndrome is obesity caused by almost continuous eating. The probable cause of the overeating is a chronic elevation in the blood level of ghrelin—a level that remains high even after a meal (DelParigi et al., 2002). We do not yet know the reason for the increased ghrelin secretion. In addition, as we will see later in this chapter, the most successful form of obesity surgery—a particular type of gastric bypass operation—suppresses ghrelin secretion and probably owes at least some of its success to this fact.

Ghrelin secretion is suppressed when an animal eats or when an experimenter infuses food into the animal's stomach. Injections of nutrients into the blood do *not* suppress ghrelin secretion, so the release of the hormone is controlled by the contents of the digestive system and not by the availability of nutrients in the blood (Schaller et al., 2003). Williams et al. (2003) operated on rats and installed devices similar to miniature blood-pressure cuffs around the *pylorus,* the junction between the stomach and the *duodenum.* The **duodenum** is the uppermost part of the small intestine. (The original Greek name for this part of the gut was *dodekadaktulon,*

---

ghrelin (*grell in*) A peptide hormone released by the stomach that increases eating; also produced by neurons in the brain.

duodenum The first portion of the small intestine, attached directly to the stomach.

or "twelve fingers long." In fact, the duodenum is twelve finger *widths* long.) When a cuff was inflated, it would compress the pylorus and prevent the contents of the stomach from leaving. Surprisingly, Williams and her colleagues found that an infusion of food into the stomach would not suppress the secretion of ghrelin when the cuff was inflated. However, when the pylorus was open, an infusion of food into the stomach, which could then reach the small intestine, *did* suppress ghrelin secretion. Subsequently, Overduin et al. (2004) found that infusion of food directly into the small intestine, bypassing the stomach, suppressed ghrelin secretion. Thus, although the stomach secretes ghrelin, its secretion appears to be controlled by receptors present in the upper part of the small intestine, not in the stomach itself.

Although ghrelin is an important short-term hunger signal, it clearly cannot be the only one. For example, people who have undergone successful gastric bypass surgery have almost negligible levels of ghrelin in the blood. Although they eat less and lose weight, they certainly do not stop eating. In addition, mice with a targeted mutation against the ghrelin gene or the ghrelin receptor have normal food intake and body weight (Sun, Ahmed, and Smith, 2003; Sun et al., 2004). However, Zigman et al. (2005) found that this mutation protected mice from overeating and gaining weight when they were fed a tasty, high-fat diet that induces obesity in normal mice. Thus, alternative mechanisms can stimulate feeding, which, given the vital importance of food, is not surprising. In fact, one of the factors that complicates research on ingestive behavior is the presence of redundant systems. (But we are undoubtedly better off with these complicating factors.)

Metabolic Signals

Most of the time, we begin a meal a few hours after the previous meal, so our nutrient reservoirs are seldom in serious need of replenishment. But if we skip several meals, we get hungrier and hungrier, presumably because of physiological signals indicating that we have been withdrawing nutrients from our long-term reservoir. What happens to the level of nutrients in our body as time passes after a meal? As you learned earlier in this chapter, during the absorptive phase of metabolism we live on food that is being absorbed from the digestive tract. After that we start drawing on our nutrient reservoirs: The brain lives on glucose, and the rest of the body lives on fatty acids. Although the metabolic needs of the cells of the body are being met, we are taking fuel out of our long-term reservoir—making withdrawals rather than deposits. Clearly, this is the time to start thinking about our next meal.

A fall in blood glucose level (a condition known as *hypoglycemia*) is a potent stimulus for hunger. Hypoglycemia can be produced experimentally by giving an animal a large injection of insulin, which causes cells in the liver, muscles, and adipose tissue to take up glucose and store it away. We can also deprive cells of glucose by injecting an animal with 2-deoxyglucose (2-DG). You are already familiar with this chemical, because I described several experiments in previous chapters that used radioactive 2-DG in conjunction with PET scanners or autoradiography to study the metabolic rate of different parts of the brain. When (nonradioactive) 2-DG is given in large doses, it interferes with glucose metabolism by competing with glucose for access to the mechanism that transports glucose through the cell membrane and for access to the enzymes that metabolize glucose. Both hypoglycemia and 2-DG cause **glucoprivation;** that is, they deprive cells of glucose. And glucoprivation, whatever its cause, stimulates eating.

Hunger can also be produced by causing **lipoprivation**—depriving cells of lipids. More precisely, they are deprived of the ability to metabolize fatty acids through injection of a drug such as *mercaptoacetate*.

What is the nature of the detectors that monitor the level of metabolic fuels, and where are they located? The evidence that has been gathered so far indicates that there are two sets of detectors: one set located in the brain and another set located in the liver. The detectors in the brain monitor the nutrients that are available on their side of the blood–brain barrier, and the detectors in the liver monitor the nutrients that are available to the rest of the body. Because the brain can use only glucose, its detectors are sensitive to glucoprivation, and because the rest of the body can use both glucose and fatty acids, the detectors in the liver are sensitive to both glucoprivation and lipoprivation.

Let's first review the evidence for the detectors in the liver. A study by Novin, VanderWeele, and Rezek (1973) suggested that receptors in the liver can stimulate glucoprivic hunger; when these neurons are deprived of nutrients, they cause eating. The investigators infused 2-DG into the **hepatic portal vein.** This vein brings blood from the intestines to the liver; thus, an injection of a drug into this vein (an *intraportal* infusion) delivers it directly to the liver. (See *Figure 12.14.*) The investigators found that the intraportal infusions of 2-DG caused immediate eating. When they cut the vagus nerve, which connects the liver with the brain, the infusions no longer stimulated eating. Thus, the brain receives the hunger signal through the vagus nerve.

Now let's look at some of the evidence that indicates that the brain has its own nutrient detectors. Because

glucoprivation A dramatic fall in the level of glucose available to cells; can be caused by a fall in the blood level of glucose or by drugs that inhibit glucose metabolism.

lipoprivation A dramatic fall in the level of fatty acids available to cells; usually caused by drugs that inhibit fatty acid metabolism.

hepatic portal vein The vein that transports blood from the digestive system to the liver.

FIGURE 12.14 ■ The Hepatic Portal Blood Supply

The liver receives water, minerals, and nutrients from the digestive system through this blood supply.

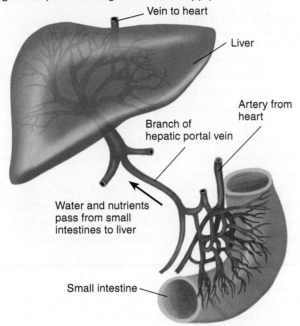

- Vein to heart
- Liver
- Artery from heart
- Branch of hepatic portal vein
- Water and nutrients pass from small intestines to liver
- Small intestine

FIGURE 12.15 ■ Nutrient Receptors

The figure shows probable location of nutrient receptors responsible for hunger signals.

The brain cannot metabolize fatty acids; receptors detect only glucose levels.

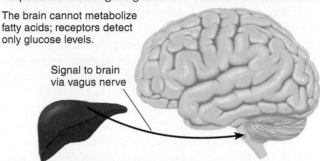

Signal to brain via vagus nerve

The liver can metabolize glucose and fatty acids; receptors detect levels of both nutrients.

the brain can use only glucose, it would make sense that these detectors respond to glucoprivation—and, indeed, they do. Ritter, Dinh, and Zhang (2000) found that injections of 5-TG into two regions of the hindbrain—the dorsomedial and ventrolateral medulla—induced eating. (5-TG, like 2-DG, produces glucoprivation.) The role of these regions in control of food intake and metabolism is discussed later in this chapter.

Liproprivic hunger appears to be stimulated by receptors in the liver. Ritter and Taylor (1990) induced liproprivic hunger with an injection of mercaptoacetate and found that cutting the vagus nerve abolished this hunger. Thus, the liver appears to contain receptors that detect

low availability of glucose or fatty acids (glucoprivation or lipoprivation) and send this information to the brain through the vagus nerve (Friedman, Horn, and Ji, 2005).

To summarize: The brain contains detectors that monitor the availability of glucose (its only fuel) inside the blood–brain barrier, and the liver contains detectors the monitor the availability of nutrients (glucose and fatty acids) outside the blood–brain barrier. (See *Figure 12.15.*)

Before closing this section, I should emphasize that no single set of receptors is solely responsible for the information the brain uses to control eating. As we saw, mice with a knockout of the ghrelin gene eat almost normally. In addition, Tordoff, Hopfenbeck, and Novin (1982) found that cutting the hepatic branch of the vagus nerve, which prevents hunger signals originating in the liver from reaching the brain, had little effect on an animal's day-to-day eating. Finally, lesions of the medulla that abolish both glucoprivic and lipoprivic signals do not lead to long-term disturbances in the control of feeding (Ritter et al., 1992). Apparently, the control of metabolism and ingestive behavior is just too important to entrust to one mechanism.

InterimSummary

What Starts a Meal?

Many stimuli, environmental and physiological, can initiate a meal. Natural selection has endowed us with strong mechanisms to encourage eating but weaker ones to prevent overeating and weight gain. Stimuli associated with eating—such as clocks indicating lunchtime or dinnertime, the smell or sight of food, or an empty stomach—increase appetite. Ghrelin, a peptide hormone released by the stomach when it and the upper intestine are empty, is a potent stimulator of food intake.

Studies with inhibitors of the metabolism of glucose and fatty acids indicate that low levels of both of these nutrients are involved in hunger; that is, animals will eat in response to both glucoprivation and lipoprivation. These signals are normally present only after several meals have been missed. Receptors in the liver detect both glucoprivation and lipoprivation and transmit this information to the brain through sensory axons of the vagus nerve. Glucoprivic eating can also be stimulated by interfering with glucose metabolism in the medulla; thus, the brain stem contains its own glucose-sensitive detectors.

WHAT STOPS A MEAL?

There are two primary sources of satiety signals—the signals that stop a meal. Short-term satiety signals come from the immediate effects of eating a particular meal, which begin long before the food is digested. To search for these signals, we will follow the pathway traveled by ingested food: the eyes, nose, and mouth; the stomach; the small intestine; and the liver. Each of these locations can potentially provide a signal to the brain that indicates that food has been ingested and is progressing on the way toward absorption. Long-term satiety signals arise in the adipose tissue, which contains the long-term nutrient reservoir. These signals do not control the beginning and end of a particular meal, but they do, in the long run, control the intake of calories by modulating the sensitivity of brain mechanisms to the hunger and satiety signals that they receive.

As I mentioned earlier, because the consequences of starvation are much more serious than the consequences of overeating, the process of natural selection has given us strong mechanisms to start eating and weaker ones to stop eating. The relatively weak inhibitory control of eating is shown by the fact that many environmental factors can increase meal size. If there were strong constraints on food intake, then it should be difficult to get a person to eat more than a normal amount. De Castro and his colleagues (reviewed by de Castro, 2004) found that people ate a larger meal if the food was especially tasty, if it was eaten in the company of other people, if the portions were larger, and if the meal began later in the day. People ate a smaller meal if their stomachs still contained food or if the food was unpalatable. (However, palatability is not normally an important factor, because people usually choose what they eat.)

Head Factors

The term *head factors* refers to several sets of receptors located in the head: the eyes, the nose, the tongue, and the throat. Information about the appearance, odor, taste, texture, and temperature of food has some automatic effects on food intake, but most of the effects involve learning. The mere act of eating does not produce long-lasting satiety; an animal with a **gastric fistula** (a tube that drains food out of the stomach before it can be digested) will eat indefinitely.

Undoubtedly, the most important role of head factors in satiety is the fact that taste and odor of food can serve as stimuli that permit animals to learn about the caloric contents of different foods. For example, Cecil, Francis, and Read (1998) found that people became more satiated when they ate a bowl of high-fat soup than when the experimenters infused an equal amount of soup into their stomachs with a flexible tube. Apparently, the act of tasting and swallowing the soup contributed to the feeling of fullness caused by the presence of the soup in the stomach.

Gastric Factors

The stomach apparently contains receptors that can detect the presence of nutrients. Davis and Campbell (1973) allowed rats to eat their fill, and shortly thereafter, they removed food from the rats' stomachs through an implanted tube. When the rats were permitted to eat again, they ate almost exactly the same amount of food that had been taken out. This finding suggests that animals are able to monitor the amount of food in their stomachs.

Deutsch and Gonzalez (1980) confirmed and extended these findings. They found that when they removed food from the stomach of a rat that had just eaten all it wanted, the animal would immediately eat just enough food to replace what had been removed—even if the experimenters replaced the food with a nonnutritive saline solution. Obviously, the rats did not simply measure the volume of the food in their stomachs, because they were not fooled by the infusion of a saline solution. Of course, this study indicates only that the stomach contains nutrient receptors; it does not prove that there are not detectors in the intestines as well.

Intestinal Factors

Indeed, the intestines do contain nutrient detectors. Studies with rats have shown that afferent axons arising from the duodenum are sensitive to the presence of glucose, amino acids, and fatty acids (Ritter et al., 1992). In fact, some of the chemoreceptors found in the duodenum are also found in the tongue. These axons may transmit a satiety signal to the brain.

A study by Feinle, Grundy, and Read (1997) found evidence for intestinal satiety factors in humans. The investigators had people swallow an inflatable bag attached to the end of a thin, flexible tube. When the stomach and duodenum were empty, the subjects reported that they simply felt bloated when the bag was inflated, filling the stomach. However, when fats or carbohydrates were infused into the duodenum while the bag was being inflated, the people reported sensations of fullness like those experienced after eating a meal. Thus, stomach and intestinal satiety factors can interact. That's not surprising, given the fact that by the time we finish a normal meal, our stomachs are full and a small quantity of nutrients has been received by the duodenum.

gastric fistula A tube that drains out the contents of the stomach.

After food reaches the stomach, it is mixed with hydrochloric acid and pepsin, an enzyme that breaks proteins into their constituent amino acids. As digestion proceeds, food is gradually introduced from the stomach into the duodenum. There, the food is mixed with bile and pancreatic enzymes, which continue the digestive process. The duodenum controls the rate of stomach emptying by secreting a peptide hormone called **cholecystokinin (CCK).** This hormone receives its name from the fact that it causes the gallbladder (cholecyst) to contract, injecting bile into the duodenum. (Bile breaks fats down into small particles so that they can be absorbed from the intestines.) CCK is secreted in response to the presence of fats, which are detected by receptors in the walls of the duodenum. In addition to stimulating contraction of the gallbladder, CCK causes the pylorus to constrict and inhibits gastric contractions, thus keeping the stomach from giving the duodenum more food.

Obviously, the blood level of CCK must be related to the amount of nutrients (particularly fats) that the duodenum receives from the stomach. Thus, this hormone could potentially provide a satiety signal to the brain, telling it that the duodenum was receiving food from the stomach. Many studies have indeed found that injections of CCK suppress eating (Gibbs, Young, and Smith, 1973; Smith, Gibbs, and Kulkosky, 1982). In addition, a strain of rats with a genetic mutation that prevents the production of CCK receptors become obese, apparently because of a disruption in normal satiety (Moran et al., 1998). CCK does not act directly on the brain; instead, it acts on receptors located in the junction between the stomach and the duodenum (Moran et al., 1989). South and Ritter (1988) found that the appetite-suppressing effect of CCK was abolished by the application of *capsaicin* to the vagus nerve. Capsaicin, a drug extracted from chile peppers, destroys sensory axons in the vagus nerve. This finding indicates that signals from CCK receptors are transmitted to the brain via the vagus nerve.

As we saw earlier in this chapter, secretion of ghrelin by the stomach provides a signal that increases hunger. Once a meal begins and food starts to enter the duodenum, the secretion of ghrelin is suppressed. The suppression of a hunger signal could well be another factor that brings a meal to its end.

Investigators have discovered a chemical produced by cells in the gastrointestinal tract that appears to serve as an additional satiety signal. This chemical, **peptide YY$_{3-36}$** (let's just call it **PYY**) is released after a meal in amounts proportional to the calories that were just ingested (Pedersen-Bjergaard et al., 1996). Only nutrients caused PYY to be secreted; a large drink of water had no effect. Injections of PYY significantly decreased the size of meals eaten by members of several species, including rats and humans (Batterham et al., 2002, 2007). (See *Figure 12.16.*)

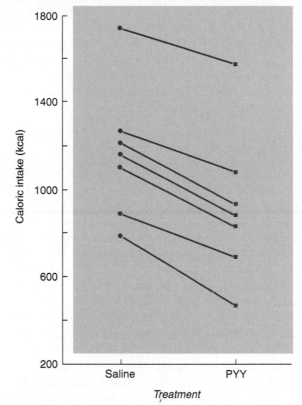

FIGURE 12.16 ■ Effects of PYY on Hunger

The graph shows the amount of food (in kilocalories) eaten at a buffet meal thirty minutes after people received a ninety-minute intravenous infusion of saline or PYY. Data points from each subject are connected by straight lines.

(Data from Batterham, R. L., ffytche, D. H., Rosenthal, J. M., Zelaya, R. O., Barker, G. J., Withers, D. J., and Williams, S. C. R. *Nature*, 2007, *450*, 106–109.)

Liver Factors

Satiety produced by gastric factors and duodenal factors is anticipatory; that is, these factors predict that the food in the digestive system will, when absorbed, eventually restore the system variables that cause hunger. Not until nutrients are absorbed from the intestines can they be used to nourish the cells of the body and replenish the body's nutrient reservoirs. The last stage of satiety appears to occur in the liver, which is the first organ to learn that food is finally being received from the intestines.

Evidence that nutrient detectors in the liver play a role in satiety comes from several sources. For example,

cholecystokinin (CCK) (*coal i sis toe ky nin*) A hormone secreted by the duodenum that regulates gastric motility and causes the gallbladder (cholecyst) to contract; appears to provide a satiety signal transmitted to the brain through the vagus nerve.

peptide YY$_{3-36}$ (PYY) A peptide released by the gastrointestinal system after a meal in amounts proportional to the size of the meal.

Tordoff and Friedman (1988) infused small amounts of two nutrients, glucose and fructose, into the hepatic portal vein. The amounts they used were similar to those that are produced when a meal is being digested. The infusions "fooled" the liver; both nutrients reduced the amount of food that the rats ate. Fructose cannot cross the blood–brain barrier and is metabolized very poorly by cells in the rest of the body, but it can readily be metabolized by the liver. Therefore, the signal from this nutrient must have originated in the liver. These results strongly suggest that when the liver receives nutrients from the intestines, it sends a signal to the brain that produces satiety. More accurately, the signal *continues* the satiety that was already started by signals arising from the stomach and upper intestine.

Insulin

As you will recall, the absorptive phase of metabolism is accompanied by an increased level of insulin in the blood. Insulin permits organs other than the brain to metabolize glucose, and it promotes the entry of nutrients into fat cells, where they are converted into triglycerides. You will also recall that cells in the brain do not need insulin to metabolize glucose. Nevertheless, the brain contains insulin receptors (Unger et al., 1989). What purpose do these insulin receptors serve? The answer is that they appear to detect insulin present in the blood, which tells the brain that the body is probably in the absorptive phase of metabolism. Thus, insulin may serve as a satiety signal (Woods et al., 2006).

Insulin is a peptide and would not normally be admitted to the brain. However, a transport mechanism delivers it through the blood–brain barrier, and it reaches neurons in the hypothalamus that are involved in regulation of hunger and satiety. Infusion of insulin into the third ventricle inhibits eating and causes a loss of body weight (Woods et al., 1979). In addition, Brüning et al. (2000) prepared a mutation in mice that prevented the synthesis of insulin receptors in the brain without affecting their production elsewhere in the body. The mice became obese, especially when they were fed a tasty, high-fat diet, which would be expected if one of the factors that promotes satiety was absent.

Long-Term Satiety: Signals from Adipose Tissue

So far, I have discussed short-term satiety factors—those arising from a single meal. But in most people, body weight appears to be regulated over the long term. If an animal is force-fed so that it becomes fatter than normal, it will reduce its food intake once it is permitted to choose how much to eat (Wilson et al., 1990). (See *Figure 12.17.*) Similar studies have shown

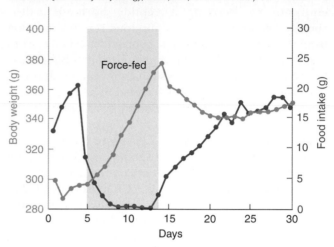

FIGURE 12.17 ■ Effects of Force Feeding

After rats were fed an excess of their normal food intake, their subsequent food intake fell and recovered only when their body weights returned to normal.

(Adapted from Wilson, B. E., Meyer, G. E., Cleveland, J. C., and Weigle, D. S. *American Journal of Physiology*, 1990, *259*, R1148–R1155.)

that an animal will adjust its food intake appropriately if it is given a high-calorie or low-calorie diet. And if an animal is put on a diet that reduces its body weight, short-term satiety factors become much less effective (Cabanac and Lafrance, 1991). Thus, signals arising from the long-term nutrient reservoir may alter the sensitivity of the brain to hunger signals or short-term satiety signals.

What exactly is the system variable that permits the body weight of most organisms to remain relatively stable? It seems highly unlikely that body *weight* itself is regulated; this variable would have to be measured by detectors in the soles of our feet or (for those of us who are sedentary) in the skin of our buttocks. What is more likely is that some variable related to body fat is regulated. The basic difference between obese and nonobese people is the amount of fat stored in their adipose tissue. Perhaps fat tissue provides a signal to the brain that indicates how much of it there is. If so, the signal is almost certainly some sort of chemical, because cutting the nerves that serve the fat tissue in an animal's body does not affect its total body fat.

The discovery of a long-term satiety signal from fat tissue came after years of study with a strain of genetically obese mice. The **ob mouse** (as this strain is called) has a low metabolism, overeats, and gets exceedingly fat. It also develops diabetes in adulthood, just as many obese people do. Researchers in

ob mouse A strain of mice whose obesity and low metabolic rate are caused by a mutation that prevents the production of leptin.

several laboratories reported the discovery of the cause of the obesity (Campfield et al., 1995; Halaas et al., 1995; Pelleymounter et al., 1995). A particular gene, called OB, normally produces a peptide that has been given the name **leptin** (from the Greek word *leptos,* "thin"). Leptin is normally secreted by well-nourished fat cells. Because of a genetic mutation, the fat cells of ob mice are unable to produce leptin.

Leptin has profound effects on metabolism and eating, acting as an antiobesity hormone. If ob mice are given daily injections of leptin, their metabolic rate increases, their body temperature rises, they become more active, and they eat less. As a result, their weight returns to normal. Figure 12.18 shows a picture of an untreated ob mouse and an ob mouse that has received injections of leptin. (See *Figure 12.18.*)

The discovery of leptin has stimulated much interest among researchers who are interested in finding ways to treat human obesity. Because it is a natural hormone, it appeared to provide a way to help people to lose weight without the use of drugs that have potentially harmful effects. Unfortunately, as we shall see later in this chapter, leptin is not a useful treatment for obesity.

FIGURE 12.18 ■ Effects of Leptin on Obesity in Mice

The photograph shows an ob (obese) strain mouse on the left that is untreated; the mouse on the right received daily injections of leptin.

(Photo courtesy of Dr. J. Sholtis, The Rockefeller University. Copyright © 1995 Amgen, Inc.)

leptin A hormone secreted by adipose tissue; decreases food intake and increases metabolic rate, primarily by inhibiting NPY-secreting neurons in the arcuate nucleus.

InterimSummary

What Stops a Meal?

Short-term satiety signals control the size of a meal. These signals include feedback from the nose and mouth about the nutritive value of the food eaten, from gastric factors that are activated by the entry of food into the stomach, from intestinal factors that are activated by the passage of food from the stomach into the duodenum, and from liver factors that are activated by the presence of newly digested nutrients in the blood carried by the hepatic portal artery.

The signals from the stomach include information about the volume and chemical nature of the food it contains. A fall in ghrelin secretion also provides a satiety signal to the brain. Another satiety signal from the intestine is provided by CCK, which is secreted by the duodenum when it receives food from the stomach. Information about the secretion of CCK is transmitted to the brain through the afferent axons of the vagus nerve. PYY, a peptide secreted after a meal by the intestines, also acts as a satiety signal. A satiety signal also comes from the liver, which detects nutrients being received from the intestines through the hepatic portal vein. Finally, although a high level of insulin in the blood causes glucoprivic eating by driving down blood levels of glucose, moderately high levels, associated with the absorptive phase of metabolism, provide a satiety signal to the brain.

Signals arising from fat tissue affect food intake on a long-term basis, apparently by modulating the effectiveness of short-term hunger and satiety signals. Force-feeding facilitates satiety, and starvation inhibits it. Studies of the ob mouse led to the discovery of leptin, a peptide hormone secreted by well-nourished adipose tissue that increases an animal's metabolic rate and decreases food intake.

Thought Questions

1. Do you find hunger unpleasant? I find that when I'm looking forward to a meal I particularly like, I don't mind being hungry, knowing that I'll enjoy the meal that much more. But then, I've never gone without eating for several days.

2. The drive-reduction hypothesis of motivation and reinforcement says that drives are aversive and satiety is pleasurable. Clearly, *satisfying* hunger is pleasurable, but what about *satiety*? Which do you prefer, eating a meal while you are hungry or feeling full afterward?

BRAIN MECHANISMS

Although hunger and satiety signals originate in the digestive system and in the body's nutrient reservoirs, the target of these signals is the brain. This section looks at some of the research on the brain mechanisms that control food intake and metabolism.

Brain Stem

Ingestive behaviors are phylogenetically ancient; obviously, all our ancestors ate and drank or died. Therefore, we should expect that the basic ingestive behaviors of chewing and swallowing are programmed by phylogenetically ancient brain circuits. Indeed, studies have shown that these behaviors can be performed by decerebrate rats, whose brains were transected between the diencephalon and the midbrain (Norgren and Grill, 1982; Flynn and Grill, 1983; Grill and Kaplan, 1990). **Decerebration** disconnects the motor neurons of the brain stem and spinal cord from the neural circuits of the cerebral hemispheres (such as the cerebral cortex and basal ganglia) that normally control them. The only behaviors that decerebrate animals can display are those that are directly controlled by neural circuits located within the brain stem. (See *Figure 12.19*.)

Decerebrate rats cannot approach and eat food; the experimenters must place food, in liquid form, into their mouths. Decerebrate rats can distinguish between different tastes; they drink and swallow sweet or slightly salty liquids and spit out bitter ones. They even respond to

FIGURE 12.19 ■ Decerebration

The operation disconnects the forebrain from the hindbrain so that the muscles involved in ingestive behavior are controlled solely by hindbrain mechanisms.

Decerebration is accomplished by transecting the brain stem.

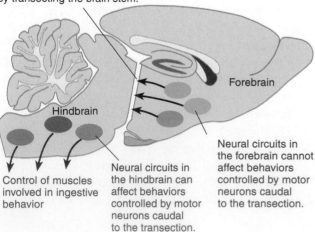

Hindbrain

Forebrain

Control of muscles involved in ingestive behavior

Neural circuits in the hindbrain can affect behaviors controlled by motor neurons caudal to the transection.

Neural circuits in the forebrain cannot affect behaviors controlled by motor neurons caudal to the transection.

FIGURE 12.20 ■ Role of the Brain Stem in Hunger

Lesions of the nucleus of the solitary tract and adjacent area postrema abolished both lipoprivic hunger (MA treatment) and glucoprivic hunger (2-DG treatment).

(Based on data from Ritter, S., and Taylor, J. S. *American Journal of Physiology*, 1990, *258*, R1395–R1401.)

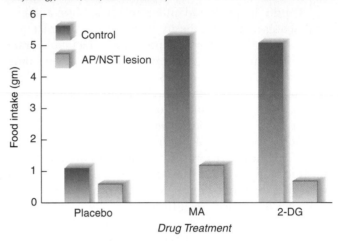

hunger and satiety signals. They drink more sucrose after having been deprived of food for 24 hours, and they drink less of it if some sucrose is first injected directly into their stomachs. They also eat in response to glucoprivation. These studies indicate that the brain stem contains neural circuits that can detect hunger and satiety signals and control at least some aspects of food intake.

The area postrema and the nucleus of the solitary tract (henceforth referred to as the AP/NST) receive taste information from the tongue and a variety of sensory information from the internal organs, including signals from detectors in the stomach, duodenum, and liver. In addition, we saw that this region contains a set of detectors that are sensitive to the brain's own fuel: glucose. All this information is transmitted to regions of the forebrain that are more directly involved in control of eating and metabolism. Evidence indicates that events that produce hunger increase the activity of neurons in the AP/NST. In addition, lesions of this region abolish both glucoprivic and lipoprivic feeding (Ritter and Taylor, 1990; Ritter, Dinh, and Friedman, 1994). (See *Figure 12.20*.)

Hypothalamus

Discoveries made in the 1940s and 1950s focused the attention of researchers interested in ingestive behavior on two regions of the hypothalamus: the lateral

decerebration A surgical procedure that severs the brain stem, disconnecting the hindbrain from the forebrain.

FIGURE 12.21 ■ Regions of the Hypothalamus Involved in Eating and Metabolism

(Adapted from Swanson, L. W. *Brain Maps: Structure of the Rat Brain.* New York: Elsevier, 1992.)

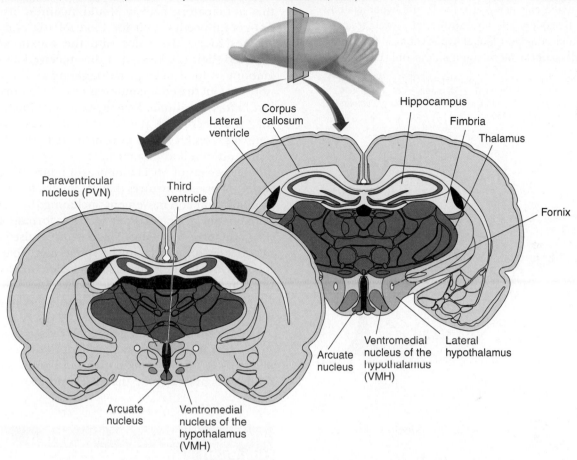

area and the ventromedial area. For many years, investigators believed that these two regions controlled hunger and satiety, respectively; one was the accelerator, and the other was the brake. The basic findings were these: After the lateral hypothalamus was destroyed, animals stopped eating or drinking (Anand and Brobeck, 1951; Teitelbaum and Stellar, 1954). Electrical stimulation of the same region would produce eating, drinking, or both behaviors. Conversely, lesions of the ventromedial hypothalamus produced overeating that led to gross obesity, whereas electrical stimulation suppressed eating (Hetherington and Ranson, 1942). (See *Figure 12.21.*)

Role in Hunger

Researchers have discovered several peptides produced by neurons in the hypothalamus that play a special role in the control of feeding and metabolism (Arora and Anubhuti, 2006). Two of these peptides, **melanin-concentrating hormone (MCH)** and **orexin,** produced by

neurons in the lateral hypothalamus, stimulate hunger and decrease metabolic rate, thus increasing and preserving the body's energy stores. (See *Figure 12.22.*)

Melanin-concentrating hormone received its name from its role in regulating changes in skin pigmentation in fish and other nonmammalian vertebrates (Kawauchi et al., 1983). In mammals it serves as a neurotransmitter. Orexin (from the Greek word *orexin,* "appetite") was discovered by Sakurai et al. (1998). (This peptide is also known as *hypocretin.*) As we saw in Chapter 9, degeneration of neurons that secrete orexin is responsible for narcolepsy. Evidence reviewed there suggests that it

melanin-concentrating hormone (MCH) A peptide neurotransmitter found in a system of lateral hypothalamic neurons that stimulate appetite and reduce metabolic rate.

orexin A peptide neurotransmitter found in a system of lateral hypothalamic neurons that stimulate appetite and reduce metabolic rate.

FIGURE 12.22 ■ Peptides in the Hypothalamus

The diagram shows melanin-concentrating hormone (MCH) neurons and orexin neurons of the lateral hypothalamus. Abbreviations: ic = internal capsule, ot = optic tract, ZI = zona incerta, LH = lateral hypothalamus, fx = fornix, 3v = third ventricle, Mt = mammillothalamic tract.

(Adapted from Elias, C. F., Saper, C. B., Maratos-Flier, E., Tritos, N. A., Lee, C., Kelly, J., Tatro, J. B., Hoffman, G. E., Ollmann, M. M., Barsh, G. S., Sakurai, T., Yanagisawa, M., and Elmquist, J. K. *Journal of Comparative Neurology*, 1998, *402*, 442–459.)

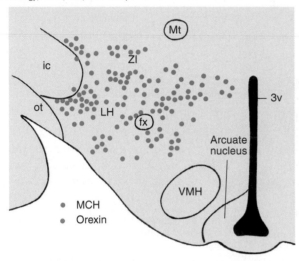

plays a role in keeping the brain's sleep–waking switch in the "waking" position.

Researchers refer to MCH and orexin as *orexigens*, "appetite-inducing chemicals." Injections of either of these peptides into the lateral ventricles or various regions of the brain induce eating. If rats are deprived of food, messenger RNA levels for MCH and orexin in the lateral hypothalamus increase (Qu et al., 1996; Sakurai et al., 1998; Dube, Kalra, and Kalra, 1999). Of these two orexigenic hypothalamic peptides, MCH appears to play the more important role in stimulating feeding. Mice with a targeted mutation against the MCH gene eat less than normal mice and are consequently underweight (Shimada et al., 1998). However, although mice with a targeted mutation against orexin eat somewhat less than normal mice, they eventually become obese in late adulthood (Hara et al., 2001). Finally, genetically engineered mice with increased production of MCH in the hypothalamus overeat and gain weight (Ludwig et al., 2001).

Besides playing a role in sleep, orexin may also play a role in the relationship between eating and sleep. Mieda et al. (2004) studied mice with a targeted mutation against the orexin gene. Normal mice that are fed one meal at the same time each day show increased locomotor activity shortly before mealtime. They also show increased levels of fos protein in orexin-secreting neurons at this time. Presumably, in the wild this anticipatory activity would manifest itself in a hunger-induced search for food. Mieda and his colleagues found that mice with the orexin knockout adjusted their intake so that they received a sufficient amount of food during the single meal, but they did not show increased wakefulness or anticipatory activity just before mealtime. Yamanaka et al. (2003) suggest that the decreased activity seen in orexin-secreting neurons after feeding may contribute to the sleepiness that is often felt after a meal.

The axons of MCH and orexin neurons travel to a variety of brain structures that are known to be involved in motivation and movement, including the neocortex, periaqueductal gray matter, reticular formation, thalamus, and locus coeruleus. These neurons also have connections with neurons in the spinal cord that control the autonomic nervous system, which explains how they can affect the body's metabolic rate (Sawchenko, 1998; Nambu et al., 1999). These connections are shown in *Figure 12.23*.

As we saw earlier, hunger signals caused by an empty stomach or by glucoprivation or lipoprivation arise from detectors in the abdominal cavity and brain stem. How do these signals activate the MCH and orexin neurons

FIGURE 12.23 ■ Feeding Circuits in the Brain

The schematic shows connections of the MCH neurons and orexin neurons of the lateral hypothalamus.

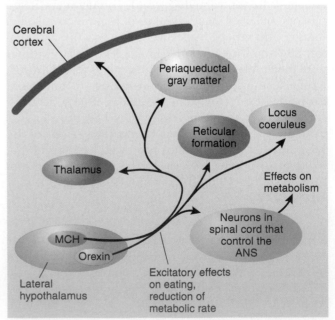

of the lateral hypothalamus? Part of the pathway involves a system of neurons that secrete a neurotransmitter called **neuropeptide Y (NPY),** an extremely potent stimulator of food intake (Clark et al., 1984). Infusion of NPY into the hypothalamus produces ravenous, almost frantic eating. Rats that receive an infusion of this peptide into the brain will work very hard, pressing a lever many times for each morsel of food; they will eat food made bitter with quinine; and they will continue to drink milk even when doing so means that they receive an electric shock to the tongue (Flood and Morley, 1991; Jewett et al., 1992).

The cell bodies of most of the neurons that secrete NPY are found in the **arcuate nucleus,** located in the hypothalamus at the base of the third ventricle. The arcuate nucleus also contains neurosecretory cells whose hormones control the secretions of the anterior pituitary gland. (Refer to *Figure 12.21.*)

Neurons that secrete NPY are affected by hunger and satiety signals; Sahu, Kalra, and Kalra (1988) found that hypothalamic levels of NPY are increased by food deprivation and lowered by eating. In addition, Myers et al. (1995) found that hypothalamic injections of a drug that blocks neuropeptide Y receptors suppress eating caused by food deprivation. Specific hunger signals have been shown to activate NPY neurons. For example, Sindelar et al. (2004) found that in normal mice, glucoprivation produced by injections of 2-DG caused a 240 percent increase in NPY messenger RNA in the hypothalamus. They also found that mice with a targeted mutation against the gene for NPY showed a feeding deficit in response to glucoprivation.

As we saw earlier, injections of 5-TG (a drug similar to 2-DG) into two regions of the medulla activates glucose-sensitive neurons and produces eating. One of these regions, located in the ventrolateral medulla, contains neurons that release NPY. The terminals of these neurons ascend to the forebrain, where they form synapses with NPY neurons in the arcuate nucleus. Li and Ritter (2004) found that glucoprivation increased the production of NPY in these neurons. The investigators also injected a toxin into the arcuate nucleus that was taken up by the NPY terminals and carried back to the cell bodies in the medulla, which subsequently died. The death of these neurons abolished glucoprivic eating. Thus, the signal for glucoprivic eating is carried by the axons of NPY neurons in the medulla to NPY neurons in the arcuate nucleus. As we saw earlier, glucoprivation is detected in the liver and in the medulla. Presumably, the neurons destroyed by Li and Ritter also carry signals from detectors in the liver to the hypothalamus.

As we saw earlier, ghrelin, released by the stomach, provides a potent hunger signal to the brain. Shuto et al. (2002) found that rats with a genetic alteration that

prevents ghrelin receptors from being produced in the hypothalamus ate less and gained weight more slowly than normal rats did. Evidence indicates that ghrelin exerts its effects on appetite and metabolism by stimulating receptors located on NPY neurons (Willesen, Kristensen, and Romer, 1999; Nakazato et al., 2001; Van den Top et al., 2004). Thus, two important hunger signals—glucoprivation and ghrelin—activate the orexigenic NPY neurons.

A study by Abizaid et al. (2007) found that ghrelin also activates neurons in the mesolimbic motivation and reinforcement system. Dopaminergic neurons in the ventral tegmental area (VTA) contain ghrelin receptors, and the investigators found that intraperitoneal injections of ghrelin or infusion of ghrelin directly into the VTA elicited eating in rats and mice. The administration of ghrelin also increased the activity of DA neurons and triggered the release of activity in the nucleus accumbens, the major target of VTA dopaminergic neurons.

NPY neurons of the arcuate nucleus send a projection directly to MCH and orexin neurons in the lateral hypothalamus (Broberger et al., 1998; Elias et al., 1998a). These connections appear to be primarily responsible for the feeding elicited by activation of NPY neurons. In addition, NPY neurons send a projection of axons to the **paraventricular nucleus (PVN),** a region of the hypothalamus where infusions of NPY affect metabolic functions, including the secretion of insulin (Bai et al., 1985).

The terminals of hypothalamic NPY neurons release another orexigenic peptide in addition to neuropeptide Y: **agouti-related peptide,** otherwise known as **AGRP** (Hahn et al., 1998). AGRP, like NPY, is a potent and extremely long-lasting orexigen. Infusion of a very small amount of this peptide into the third ventricle of rats produces an increase in food intake that lasts for six days (Lu et al., 2001).

I should briefly mention one other category of orexigenic compounds: the endocannabinoids. (See

---

neuropeptide Y (NPY) A peptide neurotransmitter found in a system of neurons of the arcuate nucleus that stimulate feeding, insulin and glucocorticoid secretion, decrease the breakdown of triglycerides, and decrease body temperature.

arcuate nucleus A nucleus in the base of the hypothalamus that controls secretions of the anterior pituitary gland; contains NPY-secreting neurons involved in feeding and control of metabolism.

paraventricular nucleus (PVN) A nucleus of the hypothalamus located adjacent to the dorsal third ventricle; contains neurons involved in control of the autonomic nervous system and the posterior pituitary gland.

agouti-related protein (AGRP) A neuropeptide that acts as an antagonist at MC-4 receptors and increases eating.

Di Marzo and Matias, 2005, for a review of the evidence cited in this paragraph.) One of the effects of the THC contained in marijuana is an increase in appetite—especially for highly palatable foods. The endocannabinoids, whose actions are mimicked by THC, stimulate eating, apparently by increasing the release of MCH and orexin. (As you will recall from Chapter 4, cannabinoid receptors are found on terminal buttons, where they regulate the release of other neurotransmitters.) Levels of endocannabinoids are highest during fasting and lowest during feeding. A genetic mutation that disrupts the production of FAAH, the enzyme that destroys the endocannabinoids after they have been released, causes overweight and obesity. Cannabinoid agonists have been used to increase the appetite of cancer patients, and cannabinoid antagonists have been used as an aid to weight reduction. (I will discuss this use later in this chapter, in the section on obesity.)

In summary, activity of MCH and orexin neurons of the lateral hypothalamus increases food intake and decreases metabolic rate. These neurons are activated by NPY/AGRP-secreting neurons of the arcuate nucleus, which are sensitive to ghrelin and which receive excitatory input from NPY neurons in the medulla that are sensitive to glucoprivation. The NPY/AGRP neurons of the arcuate nucleus also project to the paraventricular nucleus, which plays a role in control of insulin secretion and metabolism. The endocannabinoids stimulate appetite by increasing the release of MCH and orexin. (See *Figure 12.24*.)

Role in Satiety

As we saw earlier in this chapter, leptin, a hormone secreted by well-fed adipose tissue, suppresses eating and raises the animal's metabolic rate. The interactions of this long-term satiety signal with neural circuits involved in hunger are now being discovered. Leptin produces its behavioral and metabolic effects by binding with receptors in the brain—in particular, on neurons that secrete the orexigenic peptides NPY and AGRP.

Activation of leptin receptors on NPY/AGRP-secreting neurons in the arcuate nucleus has an inhibitory effect on these neurons (Glaum et al., 1996; Jobst, Enriori, and Cowley, 2004). Because NPY/AGRP neurons normally activate orexin and MCH neurons, the presence of leptin in the arcuate nucleus decreases the release of these orexigens.

When leptin was discovered, researchers hoped that this naturally occurring peptide could be used to treat obesity. In fact, a drug company paid a large sum of money for the rights to develop this compound. However, it turns out that most obese people already

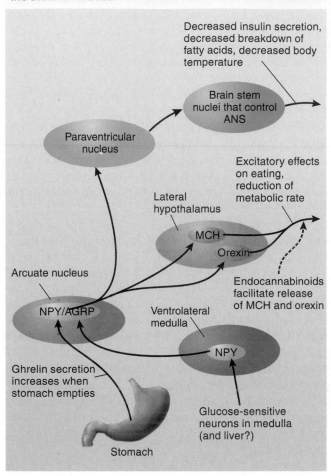

FIGURE 12.24 ■ Action of Hunger Signals on Feeding Circuits in the Brain

The diagram shows connections of the NPY neurons of the arcuate nucleus.

have a high blood level of leptin, and increasing this level with injections of the peptide has little or no effect on their food intake. I will further discuss this issue in the section on obesity later in this chapter.

The arcuate nucleus contains another system of neurons that secrete two peptides, both of which serve as *anorexigens*, or "appetite-suppressing chemicals." Douglass, McKinzie, and Couceyro (1995) discovered the first of these peptides, which is now called **CART** (for *cocaine- and amphetamine-regulated transcript*). When cocaine or amphetamine is administered to an animal,

CART Cocaine- and amphetamine-regulated transcript; a peptide neurotransmitter found in a system of neurons of the arcuate nucleus that inhibit feeding.

levels of this peptide increase, which may have something to do with the fact that these drugs suppress appetite. CART neurons appear to play an important role in satiety. If animals are deprived of food, levels of CART decrease. CART is almost totally absent in ob mice, which lack leptin, but injection of leptin into their cerebral ventricles stimulates the production of CART. Injection of CART into the cerebral ventricles inhibits feeding, including the feeding stimulated by NPY. Finally, infusion of a CART antibody increases feeding (Kristensen et al., 1998).

CART neurons are located in the arcuate nucleus and send their axons to a variety of locations, including several other hypothalamic nuclei, the periaqueductal gray matter, and regions of the spinal cord that control the autonomic nervous system (Koylu et al., 1998). Activity of CART neurons appears to suppress eating by inhibiting MCH and orexin neurons and to increase metabolic rate through connections of these neurons with neurons in the paraventricular nucleus. CART neurons contain leptin receptors that have an *excitatory* effect; thus, activation of CART-secreting neurons appears to be at least partly responsible for the satiating effect of leptin (Elias et al., 1998b).

A second anorexigen, α-**melanocyte-stimulating hormone (α-MSH),** is also released by CART neurons. This peptide is an agonist of the **melanocortin-4 receptor (MC-4R)**; it binds with the receptor and inhibits feeding. You will recall that NPY neurons also release AGRP, which stimulates eating. Both α-MSH and AGRP bind with the MC-4R. However, whereas AGRP binds with MC4 receptors and causes feeding (as we saw in the previous subsection), α-MSH binds with MC4 receptors and *inhibits* eating. CART/α-MSH neurons are activated by leptin, and NPY/AGRP neurons are inhibited by leptin. (See Elmquist, Elias, and Saper, 1999, and Wynne, Stanley, and McGowan, 2005, for specific references.) So leptin stimulates the release of the anorexigens CART and α-MSH and inhibits the release of the orexigens NPY and AGRP.

The peptide α-MSH and the MC4 receptor appear to play an important role in the control of eating. Although the MC4 receptor has two natural ligands, one with an orexigenic effect and the other with an anorexigenic effect, its primary function seems to be to suppress appetite: Huszar et al. (1997) found that mice with a targeted mutation against the gene for the MC4 receptor became obese. Agonists for the MC4 receptor such as α-MSH increase metabolic rate as well as suppress eating; thus, activation of this receptor reduces body weight by its effects on metabolism as well as on behavior (Hwa et al., 2001).

We saw that ghrelin activates NPY/AGRP neurons, which increases appetite. Ghrelin also inhibits CART/α-MSH neurons, which decreases the anorexic effect

of these two peptides (Cowley et al., 2003). Another appetite-stimulating peptide, orexin, also inhibits CART/α-MSH neurons (Ma et al., 2007). Thus, two important orexigenic peptides inhibit the activity of anorexigenic peptides.

Earlier in this chapter, I mentioned an anorexigenic peptide, PYY, which is produced by cells in the gastrointestinal tract in amounts proportional to the calories that were just ingested. PYY binds with the Y2 receptor, an inhibitory autoreceptor found on NPY/AGRP neurons in the arcuate nucleus of the hypothalamus. When PYY binds with the Y2 receptor, it suppresses the release of NPY and AGRP. Both peripheral injection of PYY and infusion directly into the arcuate nucleus of the hypothalamus suppresses food intake (Batterham et al., 2002).

Batterham et al. (2007) note that in most industrialized societies today, food intake is largely determined by nonhomeostatic factors, including cognitive and emotional factors such as the signals from the environment that I described earlier. In a functional-imaging study, Batterham and her colleagues scanned the brains of hungry subjects while they received intravenous infusions of saline or PYY. Thirty minutes later, the subjects were offered a large buffet meal. The investigators found that when the subjects were being infused with saline (the placebo), the level of activation of the hypothalamus predicted the amount of food they ate later: Increased hypothalamic activation predicted larger meals. When the subjects were being infused with PYY, the level of activation of the orbitofrontal cortex predicted the amount of food eaten: Increased orbitofrontal activation predicted *smaller* meals. (See *Figure 12.25.*) Presumably, the activation of the hypothalamus during the saline infusions reflected the activity of circuits involved in hunger. Infusion of PYY reduced the activity of these circuits, and nonhomeostatic cognitive and emotional factors took control of the amount of food that the subjects ate.

In summary, leptin appears to exert at least some of its satiating effects by binding with leptin receptors on neurons in the arcuate nucleus. Leptin inhibits NPY/AGRP neurons, which suppresses the feeding that these peptides stimulate and prevents the decrease in metabolic rate that they provoke. Leptin activates CART/α-MSH neurons, which then

---

α-**melanocyte-stimulating hormone (α-MSH)** A neuropeptide that acts as an agonist at MC-4 receptors and inhibits eating.

melanocortin-4 receptor (MC-4R) A receptor found in the brain that binds with α-MSH and agouti-related protein; plays a role in control of appetite.

FIGURE 12.25 ■ Effects of PYY on Activity of the Hypothalamus and OFC

The graphs show correlation between activation of the hypothalamus and orbitofrontal cortex (OFC) and the amount that subjects ate after receiving intravenous infusions of PYY or saline. On saline days, high activity of the hypothalamus predicted increased food intake. On PYY days, high activity of the OFC predicted decreased food intake.

(Data from Batterham, R. L., ffytche, D. H., Rosenthal, J. M., Zelaya, R. O., Barker, G. J., Withers, D. J., and Williams, S. C. R. *Nature*, 2007, *450*, 106–109.)

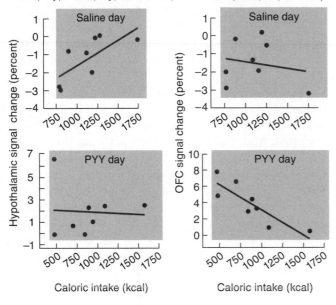

FIGURE 12.26 ■ Action of Satiety Signals on Hypothalamic Neurons Involved in Control of Hunger and Satiety

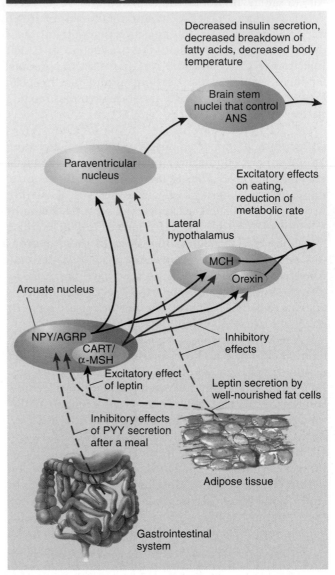

inhibit MCH and orexin neurons in the lateral hypothalamus and prevent their stimulatory effect on appetite. PYY, released by the gastrointestinal tract just after a meal, inhibits orexigenic NPY/AGRP neurons. (See *Figure 12.26*.)

Interim Summary

Brain Mechanisms

The brain stem contains neural circuits that are able to control acceptance or rejection of sweet or bitter foods and can even be modulated by satiation or physiological hunger signals, such as a decrease in glucose metabolism or the presence of food in the digestive system. The area postrema and nucleus of the solitary tract and (AP/NST) receive signals from the tongue, stomach, small intestine, and liver and send the

information on to many regions of the forebrain. These signals interact and help to control food intake. Lesions of the AP/NST disrupt both glucoprivic and lipoprivic eating.

The lateral hypothalamus contains two sets of neurons whose activity increases eating and decreases metabolic rate. These neurons secrete the peptides orexin and MCH (melanin-concentrating hormone). Food deprivation increases the level of these peptides, and mice with a targeted mutation against MCH undereat. Secretion of orexin also keeps animals from sleeping through the time for a meal if food is available only intermittently. The axons of these neurons project to regions of the brain involved in motivation, movement, and metabolism.

The release of neuropeptide Y in the lateral hypothalamus induces ravenous eating, an effect that is produced by excitatory connections of NPY-secreting neurons with the orexin and MCH neurons. NPY neurons in the arcuate nucleus of the hypothalamus receive input from glucose-sensitive neurons in the medulla. NPY neurons are the primary target of ghrelin in the hypothalamus. Ghrelin also activates the mesolimbic reinforcement system by stimulating dopaminergic neurons in the ventral tegmental area, which increases the release of DA in the nucleus accumbens. When NPY is infused in the paraventricular nucleus, it decreases metabolic rate. Levels of NPY increase when an animal is deprived of food and fall again when the animal eats. A drug that blocks NPY receptors suppresses eating. NPY neurons also release a peptide called AGRP. This peptide serves as an antagonist at MC4 receptors and, like NPY, stimulates eating. Endocannabinoids, whose action is mimicked by THC, the active ingredient in marijuana, also stimulate eating, apparently by increasing the release of MCH and orexin.

Leptin, the long-term satiety hormone secreted by well-stocked adipose tissue, desensitizes the brain to hunger signals. It binds with receptors in the arcuate nucleus of the hypothalamus, where it inhibits NPY-secreting neurons, increasing metabolic rate and suppressing eating. The arcuate nucleus also contains neurons that secrete CART (cocaine- and amphetamine-regulated transcript), a peptide that suppresses eating. These neurons, which are *activated* by leptin, have inhibitory connections with MCH and orexin neurons in the lateral hypothalamus. CART neurons also secrete a peptide called α-MSH, which serves as an agonist at MC4 receptors and inhibits eating. Ghrelin, which activates NPY/AGRP neurons and stimulates hunger, also inhibits CART/α-MSH neurons and suppresses the satiating effect of the peptides secreted by these neurons. The anorexigenic peptide, PYY, which is released by the gastrointestinal system, inhibits NPY neurons.

OBESITY

Obesity is a widespread problem that can have serious medical consequences. In the United States, approximately 67 percent of men and 62 percent of women are overweight, defined as having a body mass index (BMI) of over 25. In the past twenty years, the incidence of obesity, defined as a BMI of over 30, has doubled in the population as a whole and has tripled for adolescents. Obesity is also increasing in developing countries as household incomes rise. For example, over a ten-year period, the incidence of obesity in young urban children in China increased by a factor of eight (Ogden, Carroll, and Flegal, 2003; Zorrilla et al., 2006). The known health hazards of obesity include cardiovascular disease, diabetes, stroke, arthritis, and some forms of cancer. One hundred years ago, type 2 diabetes was almost never seen in people before the age of 40. However, because of the increased incidence of obesity in children, this disorder is seen even in 10-year-old children. (See *Figure 12.27*.)

Possible Causes

What causes obesity? As we will see, genetic differences—and their effects on development of the endocrine system and brain mechanisms that control food intake and metabolism—appear to be responsible for the overwhelming majority of people with extreme obesity. But as we just saw, the problem of obesity has been growing over recent years. Clearly, changes in the gene pool cannot account for this increase; instead, we must look

FIGURE 12.27 ■ Obesity and Type 2 Diabetes

The maps show the prevalence of obesity and type 2 ("adult onset") diabetes in the United States.

(Based on data from the Centers for Disease Control and Prevention.)

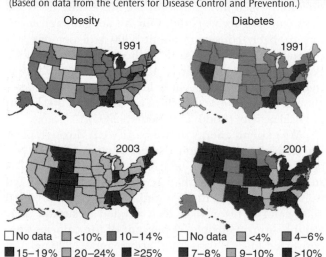

to environmental causes that have produced changes in people's behavior.

Body weight is the result of the difference between two factors: calories consumed and energy expended. If we consume more calories than we expend as heat and work, we gain weight. If we expend more than we consume, we lose weight. In modern industrialized societies, inexpensive, convenient, good-tasting, high-fat food is readily available, which promotes an increase in intake. In most places, fast-food restaurants are close at hand, parking is convenient (or even unnecessary at restaurants with drive-up windows), and the size of the portions they serve has increased in recent years. People eat out more often than they used to, and most often they do so at inexpensive fast-food restaurants.

Of course, fast-food restaurants are not the only environmental factor responsible for the increased incidence of obesity. Snack foods are available in convenience stores and vending machines, and even school cafeterias make high-calorie, high-fat foods and sweetened beverages available to their young students. In fact, school administrators often welcome the installation of vending machines because of the income they provide. As Bray, Nielsen, and Popkin (2004) point out, intake of high-fructose corn syrup, which is found in many prepared foods, including soft drinks, fruit drinks, flavored yogurts, and baked goods, may contribute to obesity. Fructose, unlike glucose, does not stimulate insulin secretion or enhance leptin production, so this form of sugar is less likely to activate the brain's satiety mechanisms. A 1994–1996 survey reported that the average daily consumption of fructose for an American over 2 years of age is approximately 318 kcal. Almost certainly, this figure is larger today.

Another modern trend that contributes to the obesity epidemic involves changes in people's expenditure of energy. The proportion of people employed in jobs that require a high level of physical activity has decreased considerably, which means that on the average we need less food than our forbears did. Our hunter-gatherer ancestors probably consumed about 3000 kcal per day and expended about 1000 kcal in their everyday activities. People with sedentary occupations in today's industrialized societies consume a little less than their ancestors—about 2400 kcal—but they burn up only about 300 kcal in physical activity (Booth and Neufer, 2005).

We expend energy in two basic ways: through physical activity and through the production of heat. Not all physical activity can be categorized as "exercise." A study by Levine, Eberhardt, and Jensen (1999) fed nonobese people a diet for eight weeks that contained 1000 calories more than they needed to sustain their weight. Approximately 39 percent of the calories were converted into fat tissue, and approximately 26 percent

went into lean tissue, increased resting metabolic rate, and the energy required to digest the extra food. The rest, approximately 33 percent, went into an increase in involuntary activity: muscle tone, postural changes, and fidgeting. Levine and his colleagues referred to this phenomenon as "nonexercise activity thermogenesis," or NEAT. The amount of fat tissue that a person gained was inversely related to his or her level of NEAT. Levine et al. (2005) measured NEAT levels in a group of people with sedentary lifestyles that included both lean and moderately obese individuals. The investigators found that the overweight people remained seated 2.5 hours per day more than the lean people. The difference in energy expenditure amounted to 350 kcal per day.

One biological factor that may control NEAT levels is orexin, the brain peptide that enhances wakefulness and activity as well as eating. Kiwaki et al. (2004) measured NEAT levels in rats. The investigators found a dose-related increase in energy expenditure when orexin was infused into the paraventricular nucleus of the hypothalamus through chronically implanted intracranial cannulas. (See *Figure 12.28*.)

Differences in body weight—perhaps reflecting physiological differences in metabolism, activity levels (including NEAT), or appetite—have a strong hereditary basis. Twin studies suggest that between 40 percent and 85 percent of the variability in body fat is due to genetic differences (Price and Gottesman, 1991; Allison et al., 1996; Comuzzie and Allison, 1998). Twin studies have found strong genetic effects on the amount of weight that people gain or lose when they are placed on high- or low-calorie diets (Bouchard et al., 1990; Hainer

FIGURE 12.28 ■ Orexin and NEAT

The graph shows the effect of infusion of orexin into the paraventricular nucleus of the hypothalamus of rats on nonexercise activity thermogenesis (NEAT).

(Adapted from Levine, J. A. *Journal of Physiology, Endocrinology, and Metabolism*, 2004, *286*, E675–E685.)

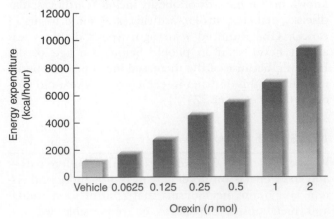

et al., 2001). Thus, heredity appears to affect people's metabolic efficiency.

Just as cars differ in their fuel efficiency, so do living organisms, and hereditary factors can affect the level of efficiency. For example, farmers have bred cattle, pigs, and chickens for their efficiency in converting feed into muscle tissue, and researchers have done the same with rats (Pomp and Nielsen, 1999). As we just saw, people differ in this form of efficiency too. Those with an efficient metabolism have calories left over to deposit in the long-term nutrient reservoir; thus, they have difficulty keeping this reservoir from growing. Researchers have referred to this condition as a "thrifty phenotype." In contrast, people with an inefficient metabolism (a "spendthrift phenotype") can eat large meals without getting fat. A fuel-efficient automobile is desirable, but a fuel-efficient body runs the risk of becoming obese—at least in an environment where food is cheap and plentiful.

Why are there genetic differences in metabolic efficiency? As we saw earlier in this chapter, natural selection for mechanisms that helped our ancestors to avoid starvation was much stronger than natural selection for mechanisms that helped them to avoid becoming obese. Perhaps individual differences in metabolic efficiency reflect the nature of the environment experienced by their ancestors. Perhaps people whose ancestors lived in regions where food was scarce or subject to periods of famine are more likely to have inherited efficient metabolisms.

This hypothesis has received support from epidemiological studies. Ravussin et al. (1994) studied two groups of Pima Indians, who live in the southwestern United States and northwestern Mexico. Members of the two groups appear to have the same genetic background; they speak the same language and have common historical traditions. The two groups separated 700–1000 years ago and now live under very different environmental conditions. The Pima Indians in the southwestern United States eat a high-fat American diet and weigh an average of 90 kg (198 lb), men and women combined. In contrast, the lifestyle of the Mexican Pimas is probably similar to that of their ancestors. They spend long hours working at subsistence farming and eat a low-fat diet—and weigh an average of 64 kg (141 lb). The cholesterol level of the American Pimas is much higher than that of the Mexican Pimas, and the American Pimas' rate of diabetes is more than five times higher. These findings show that genes that promote an efficient metabolism are of benefit to people who must work hard for their calories but that these same genes turn into a liability when people live in an environment where the physical demands are low and high-calorie food is cheap and plentiful.

As we saw earlier, study of the ob mouse led to the discovery of leptin, the hormone secreted by well-nourished adipose tissue. So far, researchers have found several cases of familial obesity caused by the absence of leptin produced by the mutation of the gene responsible for leptin production or the production of the leptin receptor (Farooqi and O'Rahilly, 2005). Treatment of leptin-deficient people with injections of leptin has dramatic effects on the people's body weights. (See *Figure 12.29.*) Unfortunately, leptin has no effect on people who lack leptin receptors. In any case, mutations of the genes for leptin or leptin receptors are very rare, so they do not explain the vast majority of cases of obesity.

FIGURE 12.29 ■ Hereditary Leptin Deficiency

The photographs show three patients with hereditary leptin deficiency before (a) and after (b) treatment with leptin for 18 months. The faces of the patients are obscured for privacy. Two normal-weight nurses are shown for comparison purposes.

(From Licinto, J., Caglayan, S., Ozata, M., Yildiz, B. O., de Miranda, P. B., O'Kirwan, F., Whitby, R., Liang, L., Cohen, P., Bhasin, S., Krauss, R. M., Veldhuios, J. D., Wagner, A. J., DePaoli, A. M., McCann, S. M., and Wong, M.-L. *Proceedings of the National Academy of Science, USA*, 2004, *101*, 4531–4536. Reprinted with permission.)

As I mentioned earlier in this chapter, leptin has not proved to provide a useful treatment for obesity. Indeed, obese people already have elevated blood levels of leptin, and additional leptin has no effect on their food intake or body weight. In other words, obese people show *leptin resistance*. Leptin is a peptide, and peptides normally cannot cross the blood–brain barrier. However, an active mechanism transports molecules of leptin across this barrier so that it can exert its behavioral and metabolic effects (Banks et al., 1996; Golden, MacCagnan, and Pardridge, 1997). Caro et al. (1996) suggested that differences in the effectiveness of this transport system may be one cause of obesity. If not much leptin gets across the blood–brain barrier, the leptin signal in the brain will be weaker than it should be. Caro and his colleagues found that although the level of leptin in the blood was 318 percent higher in obese people, it was only 30 percent higher in their cerebrospinal fluid (which is presumably related to the concentration of the hormone in the brain). Thus, leptin resistance could be caused by reduced transport of leptin molecules into the brain.

Although the role of leptin as a long-term satiety signal is well established, several investigators have suggested that a *fall* in blood levels of leptin should be regarded as a *hunger* signal. Starvation decreases the blood level of leptin, which removes an inhibitory influence on NPY/AGRP neurons and an excitatory influence on CART/α-MSH neurons. That is, a low level of leptin increases the release of orexigenic peptides and decreases the release of anorexigenic peptides. And as Flier (1998) suggests, people with a thrifty metabolism should show resistance to a high level of leptin, which would permit weight gain in times of plenty. People with a spendthrift metabolism should not show leptin resistance and should eat less as their level of leptin rises.

As you may have noticed, many people gain weight as they grow older. There are undoubtedly several causes for this tendency, including a decrease in levels of physical activity. But some evidence suggests that there can be age-related changes in sensitivity to leptin. Scarpace, Matheny, and Tümer (2001) found that hypothalamic neurons in aged obese rats showed a smaller response to leptin than those in rats of normal weight. They also observed a 50 percent reduction in the number of leptin receptors, which may account for this difference.

As we saw earlier in this chapter, the MC4 receptor plays a role in the control of eating and metabolism. Several groups of researchers have found families with severe obesity caused by mutations of the MC4 receptor gene (Farooqi and O'Rahilly, 2005). It appears that mutation of the MC4 receptor is the most common simple genetic cause of severe obesity. Approximately 4 percent of such people have a mutation of the gene for these receptors.

The final physiological factor that I will mention in this section is a chemical known as **uncoupling protein (UCP).** This protein is found in mitochondria and may be one of the factors that determine the rate at which an animal burns off its calories. In other words, it may be a factor in determining metabolic efficiency. Three different uncoupling proteins exist. One of them, UCP3, is found in muscles, and this form probably plays the most important role in metabolic efficiency. Several metabolic signals—including leptin—increase the expression of UCP3, thus increasing metabolic rate and burning off calories (Scarpace et al., 2000). Using methods of molecular genetics, Clapham et al. (2000) prepared a strain of mice that produced an abnormally high level of UCP3 in their skeletal muscles. These animals ate more than normal mice did but were lean and had a much lower level of body fat. In addition, Schrauwen et al. (1999) found that levels of UCP3 in Pima Indians were negatively correlated with body mass index and positively correlated with metabolic rate. In other words, Pima Indians with high levels of UCP3 had spendthrift phenotypes that helped to protect them from developing obesity.

Treatment

Obesity is extremely difficult to treat; the enormous financial success of diet books, health spas, and weight reduction programs attests to the trouble people have in losing weight. More precisely, many programs help people to lose weight initially, but then the weight is quickly regained. Kramer et al. (1989) reported that four to five years after participating in a fifteen-week behavioral weight-loss program, fewer than 3 percent of the participants managed to maintain the weight loss they had achieved during the program. Some experts have suggested that given the extremely low long-term success rate, perhaps we should stop treating people for obesity until our treatments are more successful. As Wooley and Garner (1994) said,

> We should stop offering ineffective treatments aimed at weight loss. Researchers who think they have invented a better mousetrap should test it in controlled research before setting out their bait for the entire population. Only by admitting that our treatments do not work—and showing that we mean it by refraining from offering them—can we begin to undo a century of recruiting fat people for failure. (p. 656)

---

uncoupling protein (UCP) A mitochondrial protein that facilitates the conversion of nutrients into heat.

Whatever the cause of obesity, the metabolic fact of life is this: If calories in exceed calories out, then body fat will increase. Because it is difficult to increase the "calories out" side of the equation enough to bring an obese person's weight back to normal, most treatments for obesity attempt to reduce the "calories in." The extraordinary difficulty that obese people have in reducing caloric intake for a sustained period of time (that is, for the rest of their lives) has led to the development of some extraordinary means. In this section I shall describe some surgical, pharmacological, and behavioral methods that have been devised to make obese people eat less.

Surgeons have become involved in trying to help obese people lose weight. The procedures they have developed (called *bariatric surgery*, from the Greek *barys*, "heavy," and *iatrikos*, "medical") are designed to reduce the amount of food that can be eaten during a meal or interfere with absorption of calories from the intestines. Bariatric surgery has been aimed at the stomach, the small intestine, or both.

The most effective form of bariatric surgery is a special form of gastric bypass called the *Roux-en-Y gastric bypass*, or *RYGB*. This procedure produces a small pouch in the upper end of the stomach. The jejunum (the second part of the small intestine, immediately "downstream" from the duodenum) is cut, and the upper end is attached to the stomach pouch. The effect is to produce a small stomach whose contents enter the jejunum, bypassing the duodenum. Digestive enzymes that are secreted into the duodenum pass through the upper intestine and meet up with the meal that has just been received from the stomach pouch. (See *Figure 12.30*.)

The RYGB procedure appears to work well, although it often causes an iron and vitamin B_{12} deficiency. Brolin (2002) reported that the average post-surgical loss of excessive weight of obese patients was 65–75 percent, or about 35 percent of their initial weight. Even patients who sustained smaller weight losses showed improved health, including reductions in hypertension and diabetes. A meta-analysis of 147 studies by Maggard et al. (2005) reported an average weight loss of 43.5 kg (approximately 95 lb) one year after RYGB surgery and 41.5 kg after three years. Unfortunately, a recent study of 16,155 patients in the United States who underwent bariatric surgery found a higher rate of mortality than had previously been reported (Flum et al., 2005). The thirty-day, ninety-day, and one-year mortality rates were 2.0 percent, 2.8 percent, and 4.6 percent, respectively. Men were more likely than women to die, and the mortality rate increased with age. The mortality rate was also higher for patients of surgeons with less than average experience performing the procedures. Clearly, the

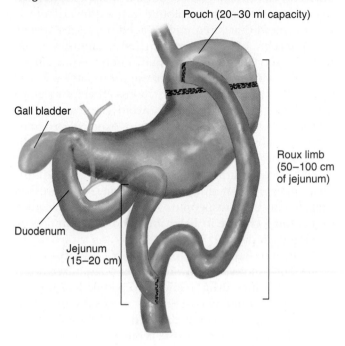

FIGURE 12.30 ■ Roux-en-Y Gastric Bypass (RYGB) Surgery

This procedure almost totally suppresses the secretion of ghrelin.

Pouch (20–30 ml capacity)

Gall bladder

Roux limb (50–100 cm of jejunum)

Duodenum

Jejunum (15–20 cm)

decision to undergo bariatric surgery should not be taken lightly.

One important reason for the success of the RYGB procedure appears to be that it disrupts the secretion of ghrelin. The procedure also increases blood levels of PYY (Chan et al., 2006; Reinehr et al., 2006). Both of these changes should decrease food intake: A decrease in ghrelin should reduce hunger, and an increase in PYY should increase satiety. A plausible explanation for the decreased secretion of ghrelin could be disruption of communication between the upper intestine and the stomach; as you will recall, although ghrelin is secreted by the stomach, the upper intestine controls this secretion. Presumably, because the surgery decreases the speed at which food moves through the small intestine, more PYY is secreted. Suzuki et al. (2005) prepared an animal model of the RYGB procedure. They performed bariatric surgery on a genetically obese strain of rats and monitored the animals' food intake, weight, ghrelin, and PYY levels. They found that the rats that sustained a RYGB procedure (but not rats that received sham surgery) ate less, lost weight, and showed decreased ghrelin levels and increased PYY levels. These results support the conclusion that changes in the secretion of these two peptides contribute to the success of this procedure.

A less drastic form of therapy for obesity—exercise—has significant benefits. As I mentioned earlier, decreased physical activity is an important cause of the increased number of overweight people. Exercise burns off calories, of course, but it also appears to have beneficial effects on metabolic rate. Bunyard et al. (1998) found that when middle-aged men participated in an aerobic exercise program for six months, their body fat decreased and their daily energy requirement increased—by 5 percent for obese men and by 8 percent for lean men. (Remember, a less efficient metabolism means that it is easier to avoid gaining weight.) Gurin et al. (1999) found that an exercise program helped obese children to lose fat and had the additional benefit of increasing bone density. Hill et al. (2003) calculated that an increased energy expenditure through exercise of only 100 kcal per day could prevent weight gain in most people. The effort would require only a small change in behavior—about 14 minutes of walking each day.

Another type of therapy for obesity—drug treatment—shows some promise. There are three possible ways in which drugs could help people lose weight: reduce the amount of food they eat, prevent some of the food they eat from being digested, and increase their metabolic rate (that is, provide them with a "spendthrift phenotype").

Some serotonergic agonists suppress eating. A review by Bray (1992) concluded that these drugs can be of benefit in weight-loss programs. However, one of the drugs most commonly used for this purpose, fenfluramine, was found to have hazardous side effects, including pulmonary hypertension and damage to the valves of the heart, so the drug was withdrawn from the market in the United States (Blundell and Halford, 1998). Fenfluramine acts by stimulating the release of 5-HT. Fortunately, another drug, sibutramine, has similar therapeutic effects and has not yet been associated with serious side effects. Sibutramine—which also acts as an antidepressant drug—is an inhibitor of the three monoamines: serotonin, norepinephrine, and dopamine. The mechanism by which sibutramine reduces weight is not clear. Some other antidepressant drugs that block the reuptake of serotonin, such as paroxetine (Paxil), actually increase food intake and lead to a weight gain (Deshmukh and Franco, 2003).

Another drug, orlistat, interferes with the absorption of fats by the small intestine. As a result, some of the fat in the person's diet passes through the digestive system and is excreted with the feces. Possible side effects include a deficiency of fat-soluble vitamins and anal leaking of undigested fat. A double-blind, placebo-controlled study by Hill et al. (1999) found that orlistat helped people to maintain weight loss they had achieved by participating in a conventional weight-loss program. People who received the placebo were much more likely to regain the weight they had lost.

As I mentioned earlier, the fact that marijuana often elicits a craving for highly palatable foods led to the discovery that the endocannabinoids have an orexigenic effect. The drug rimonabant, which blocks CB1 cannabinoid receptors, was found to suppress appetite. Two large phase III clinical trials of rimonabant, with a total of 5580 patients, were carried out in North America and Europe. The results showed a significant weight loss, lower blood levels of triglycerides and insulin, increased blood levels of HDL ("good" cholesterol), and minimal adverse side effects (Di Marzo and Matias, 2005). Rimonabant is currently approved for the treatment of obesity in 42 countries, but the U.S. Food and Drug Administration has not yet approved it (Isoldi and Arrone, 2008). As we will see in Chapter 19, rimonabant has also been shown to help people stop smoking, which suggests that craving for nicotine also involves the release of endocannabinoids in the brain.

As we have seen, appetite can be stimulated by activation of NPY, MCH, orexin, and ghrelin receptors, and it can be suppressed by the activation of leptin, CCK, CART, and MC4 receptors. Appetite can also be suppressed by activation of inhibitory presynaptic Y2 autoreceptors by PYY. Most of these orexigenic and anorexigenic chemicals also affect metabolism: Orexigenic chemicals tend to decrease metabolic rate, and anorexigenic chemicals tend to increase it. In addition, uncoupling protein causes nutrients to be "burned"—converted into heat instead of adipose tissue. Do these discoveries hold any promise for the treatment of obesity? Is there any possibility that researchers will find drugs that will stimulate or block these receptors, thus decreasing people's appetite and increasing the rate at which they burn rather than store their calories? Drug companies certainly hope so, and they are working hard on developing medications that will do so, because they know that there will be a very large number of people willing to pay for them.

The variety of methods—surgical, behavioral, and pharmacological—that therapists and surgeons have developed to treat obesity attests to the tenacity of the problem. The basic difficulty, beyond that caused by having an efficient metabolism, is that eating is pleasurable and satiety signals are easy to ignore or override. Despite the fact that relatively little success has been seen until now, I am personally optimistic about what the future may hold. I think that if we learn more about the physiology of hunger signals, satiety signals, and the reinforcement provided by eating, we will be able to develop safe and effective drugs that attenuate the signals that encourage us to eat and strengthen those that encourage us to stop eating.

InterimSummary

Obesity

Obesity presents serious health problems. As we saw earlier, natural selection has given us strong hunger mechanisms and weaker mechanisms of long-term satiety. Obesity is strongly affected by hereditary. Some people have inherited a thrifty metabolism, which makes it difficult for them to lose weight. One of the manifestations of a thrifty metabolism is a low level of nonexercise activity thermogenesis, or NEAT. A high percentage of Pima Indians who live in the United States and consume a high-fat diet become obese and, as a consequence, develop diabetes. In contrast, Mexican Pima Indians, who work hard at subsistence farming and eat a low-fat diet, remain thin and have a low incidence of obesity.

Obesity in humans is related to a hereditary absence of leptin or leptin receptors only in a few families. In general, obese people have very high levels of leptin in their blood. However, they show resistance to the effects of this peptide, apparently because the transport of leptin through the blood–brain barrier is reduced. The most significant simple genetic cause of severe obesity is mutation of the MC4 receptor, which responds to the orexigen AGRP and the anorexigen α-MSH. Genetic variations in uncoupling protein, which controls the conversion of nutrients into heat by the mitochondria, may also be involved in people's metabolic efficiency.

Researchers have tried many behavioral, surgical, and pharmacological treatments for obesity, but no panacea has yet been found. The RYGB procedure, a special form of gastric bypass operation, is the most successful form of bariatric surgery. The effectiveness of this operation is probably due primarily to its suppression of ghrelin secretion and stimulation of PYY secretion. The best hope for the future probably comes from drugs. One drug, rimonabant, blocks cannabinoid receptors and suppresses appetite. At present many pharmaceutical companies are trying to apply the results of the discoveries of orexigens and anorexigens described in this chapter to the development of antiobesity drugs.

This section and the previous one introduced several neuropeptides and peripheral peptides that play a role in control of eating and metabolism. Table 12.1 summarizes information about these compounds. (See *Table 12.1.*)

Thought Question

One of the last prejudices that people admit to publicly is a dislike of fat people. Is this fair, given that genetic differences in metabolism are such an important cause of obesity?

TABLE 12.1 ■ Neuropeptides and Peripheral Peptides Involved in Control of Food Intake and Metabolism

NEUROPEPTIDES

Name	Location of Cell Bodies	Location of Terminals	Interaction with Other Peptides	Physiological or Behavioral Effects
Melanin-concentrating hormone (MCH)	Lateral hypothalamus	Neocortex, periaqueductal gray matter, reticular formation, thalamus, locus coeruleus, neurons in spinal cord that control the sympathetic nervous system	Activated by NPY/AGRP; inhibited by leptin and CART/α-MSH	Eating, decreased metabolic rate
Orexin	Lateral hypothalamus	Similar to those of MCH neurons	Activated by NPY/AGRP; inhibited by leptin and CART/α-MSH	Eating, decreased metabolic rate
Neuropeptide Y (NPY)	Arcuate nucleus of hypothalamus	Paraventricular nucleus, MCH and orexin neurons of the lateral hypothalamus	Activated by ghrelin; inhibited by leptin	Eating, decreased metabolic rate

(continued)

TABLE 12.1 ■ Neuropeptides and Peripheral Peptides Involved in Control of Food Intake and Metabolism *(continued)*

Agouti-related protein (AGRP)	Arcuate nucleus of hypothalamus (colocalized with NPY)	Same regions as NPY neurons	Inhibited by leptin	Eating, decreased metabolic rate; acts as antagonist at MC4 receptors
Cocaine- and amphetamine-regulated transcript (CART)	Arcuate nucleus of hypothalamus	Paraventricular nucleus, lateral hypothalamus, periaqueductal gray matter, neurons in spinal cord that control the sympathetic nervous system	Activated by leptin	Suppression of eating, increased metabolic rate
α-melanocyte stimulating hormone (α-MSH)	Arcuate nucleus of hypothalamus (colocalized with CART)	Same regions as CART neurons	Activated by leptin	Suppression of eating, increased metabolic rate; acts as agonist at MC4 receptors

PERIPHERAL PEPTIDE

Name	*Where Produced*	*Site of Actions*	*Physiological or Behavioral Effects*
Leptin	Fat tissue	Inhibits NPY/AGRP neurons; excites CART/α-MSH neurons	Suppression of eating, increased metabolic rate
Insulin	Pancreas	Similar to leptin	Similar to leptin
Ghrelin	Gastrointestinal system	Activates NPY/AGRP neurons	Eating
Cholecystokinin (CCK)	Duodenum	Neurons in pylorus	Suppression of eating
Peptide YY$_{3-36}$ (PYY)	Gastrointestinal system	Inhibits NPY/AGRP neurons	Suppression of eating

ANOREXIA NERVOSA/ BULIMIA NERVOSA

Most people, if they have an eating problem, tend to overeat. However, some people, especially among adolescent women, have the opposite problem: They eat too little, even to the point of starvation. This disorder is called **anorexia nervosa.** Another eating disorder, **bulimia nervosa,** is characterized by a loss of control of food intake. (The term *bulimia* comes from the Greek *bous,* "ox," and *limos,* "hunger.") People with bulimia nervosa periodically gorge themselves with food, especially dessert or snack food and especially in the afternoon or evening. These binges are usually followed by self-induced vomiting or the use of laxatives, along with feelings of depression and guilt. With this combination of binging and purging, the net nutrient intake (and consequently, the body weight) of bulimics can vary; Weltzin et al. (1991) reported that 19 percent of bulimics undereat, 37 percent eat a normal amount, and 44 percent overeat. Episodes of bulimia are seen in some patients with anorexia nervosa. The incidence of anorexia nervosa is estimated at 0.5–2 percent; that of bulimia nervosa at 1–3 percent. Women are ten to twenty times more likely than men to develop anorexia

anorexia nervosa A disorder that most frequently afflicts young women; exaggerated concern with being overweight that leads to excessive dieting and often compulsive exercising; can lead to starvation.

bulimia nervosa Bouts of excessive hunger and eating, often followed by forced vomiting or purging with laxatives; sometimes seen in people with anorexia nervosa.

nervosa and approximately ten times more likely to develop bulimia nervosa. (See Klein and Walsh, 2004.)

Possible Causes

The literal meaning of the word *anorexia* suggests a loss of appetite, but people with this disorder are usually interested in—even preoccupied with—food. They may enjoy preparing meals for others to consume, collect recipes, and even hoard food that they do not eat. Although anorexics might not be oblivious to the effects of food, they express an intense fear of becoming obese, which continues even if they become dangerously thin. Many exercise by cycling, running, or almost constant walking and pacing.

Anorexia is a serious disorder. Five to 10 percent of people with anorexia die of complications of the disease or of suicide. Many anorexics suffer from osteoporosis, and bone fractures are common. When the weight loss becomes severe enough, anorexic women cease menstruating. Some disturbing reports (Artmann et al., 1985; Herholz, 1996; Kingston et al., 1996; Katzman et al., 2001) indicate the presence of enlarged ventricles and widened sulci in the brains of anorexic patients, which indicate shrinkage of brain tissue.

Many researchers and clinicians have concluded that anorexia nervosa and bulimia nervosa are symptoms of an underlying mental disorder. However, evidence suggests just the opposite: that the symptoms of eating disorders are actually symptoms of starvation. A famous study carried out at the University of Minnesota by Keyes and his colleagues (Keyes et al., 1950) recruited thirty-six physically and psychologically healthy young men to observe the effects of semistarvation. For six months, the men ate approximately 50 percent of what they had been eating previously and, as a result, lost approximately 25 percent of their original body weight. As the volunteers lost weight, they began displaying disturbing symptoms, including preoccupation with food and eating, ritualistic eating, erratic mood, impaired cognitive performance, and physiological changes such as decreased body temperature. They began hoarding food and nonfood objects and were unable to explain (even to themselves) why they bothered to accumulate objects for which they had no use. At first, they were gregarious, but as time went on, they became withdrawn and isolated. They lost interest in sex, and many even "welcomed the freedom from sexual tensions and frustrations normally present in young adult men" (Keyes et al., p. 840).

The obsessions with food and weight loss and the compulsive rituals that people with anorexia nervosa develop suggest a possible linkage with obsessive-compulsive disorder (described in more detail in Chapter 18). However, the fact that these obsessions and compulsions were seen in the subjects of the Minnesota study—none of whom showed these symptoms previously—suggests that they are effects rather than causes of the eating disorder.

Both anorexia and semistarvation include symptoms such as mood swings, depression, and insomnia. Even hair loss is seen in both conditions. The suicide rate in patients with anorexia is higher than that of the rest of the population (Pompili et al., 2004). None of the volunteers in the Minnesota study committed suicide, but one cut off three of his fingers. This volunteer said, "I have been more depressed than ever in my life. . . I thought that there was only one thing that would pull me out of the doldrums, that is release from [the experiment]. I decided to get rid of some fingers . . . It was premeditated" (Keyes et al., 1950, pp. 894–895).

Although binge eating is a symptom of anorexia, eating very slowly is, too. Patients with anorexia tend to dawdle over a meal, and so did the volunteers in the Minnesota study. "Toward the end of starvation some of the men would dawdle for almost two hours over a meal, which previously they would have consumed in a matter of minutes" (Keyes et al., p. 833).

As we saw, excessive exercising is a prominent symptom of anorexia (Zandian et al., 2007). In fact, Manley, O'Brien, and Samuels (2008) found that many fitness instructors recognize that some of their clients may have an eating disorder and have expressed concern about ethical or liability issues in permitting such clients to participate in their classes or facilities.

Studies with animals suggest that the increased activity may actually be a result of the fasting. When rats are allowed access to food for one hour each day, they will spend more and more time running in a wheel if one is available and will lose weight and eventually die of emaciation (Smith, 1989). Nergårdh et al. (2007) placed rats in individual cages. Some of the cages were equipped with running wheels so that the rats' running activity levels could be measured. After adaptation to the cages, the animals were given access to food once a day for varying amounts of time between one and twenty-four hours (no food restriction). The rats in cages with running wheels that received food on restricted schedules began to spend more time running. In fact, the rats with the most restrictive feeding schedules ran the most. Clearly, the increased running was counterproductive, because these animals lost much more weight than did the animals housed in cages without running wheels. (See *Figure 12.31*.)

One explanation for the increased activity of rats on a semistarvation diet is that it reflects an innate tendency to seek food when it becomes scarce. Normally, rats would expend their activity by exploring the environment and searching for food, but because of their confinement, the tendency to explore results only in futile wheel running. Another possible explanation is the low body temperature that accompanies a semistarvation

FIGURE 12.31 ■ Activity, Food Restriction, and Weight Loss

The graph shows changes in body weight of rats that were permitted one-hour or twenty-four-hour access to food each day. Rats with access to a running wheel spent time running and lost weight, especially those with only one hour of access to food.

(Data from Nergårdh, R., Ammar, A., Brodin, U., Bergström, Scheurink, A., and Södersten, P. *Psychoneuroendocrinology*, 2007, *32*, 493–502.)

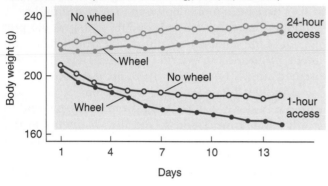

diet. (In fact, patients with anorexia complain that they feel cold.) The increased activity may simply reflect an attempt to keep warm. Whichever explanation is correct, the fact that starving rats increase their activity suggests that the excessive activity of anorexic patients may be a symptom of starvation, not simply a weight-loss strategy.

Blood levels of NPY are elevated in patients with anorexia. As we saw earlier in this chapter, NPY normally stimulates eating. Nergårdh et al. (2007) found that intracerebroventricular infusions of NPY further increased the time spent running in rats on a restricted feeding schedule. Normally, NPY stimulates eating (as it did in rats with unlimited access to food), but under conditions of starvation it stimulates wheel-running activity instead.

By now, you are probably wondering why anorexia gets started in the first place. Even if the symptoms of anorexia are largely those of starvation, what begins the behavior that leads to starvation? The simple answer is that we still do not know. One possibility is a genetic predisposition for this behavior. There is good evidence, primarily from twin studies, that hereditary factors play an important role in the development of anorexia nervosa (Russell and Treasure, 1989; Walters and Kendler, 1995; Kortegaard et al., 2001). In fact, between 58 and 76 percent of the variability in the occurrence of anorexia nervosa appears to be under control of genetic factors (Klein and Walsh, 2004). In addition, the incidence of anorexia nervosa is higher in girls who were born prematurely or who had sustained birth trauma during complicated deliveries (Cnattingius et al., 1999), which suggests that biological factors independent of heredity

may play a role. Perhaps some young women (and a small number of young men) go on a diet to bring their body weights closer to what they perceive as ideal. Once they get set on this course and begin losing weight, physiological and endocrinological changes bring about the symptoms of starvation outlined above, and the vicious cycle begins.

The fact that anorexia nervosa is seen primarily in young women has prompted both biological and social explanations. Most psychologists favor the latter, concluding that the emphasis that most modern industrialized societies place on slimness—especially in women—is responsible for this disorder. Another possible cause could be the changes in hormones that accompany puberty. Whatever the cause, young men and women differ in their responses to even a short period of fasting. Södersten, Bergh, and Zandian (2006) had high school students visit their laboratory at noon one day, where they were given all the food that wanted to eat for lunch. Seven days later, they returned to the laboratory again. This time, they had been fasting since lunch on the previous day. The men ate more food than they had the first time. However, the women actually ate *less* than they had before. (See *Figure 12.32.*) Apparently, women have difficulty compensating for a period of food deprivation by eating more food. As the authors note, "dieting may be dangerous in women and in particular in those who are physically active and therefore need to eat more food, [such as] athletes" (p. 575).

Treatment

Anorexia is very difficult to treat successfully. Cognitive behavior therapy, considered by many clinicians to be the most effective approach, has a success rate of less than 50 percent and a relapse rate of 22 percent during a one-year treatment period (Pike et al., 2003). A meta-analysis by Steinhausen (2002) indicates that the success rate in treating anorexia has not improved in the last fifty years. As Ben-Tovim (2003) notes, "Much of the literature on the treatment and outcome of eating disorders lacks methodological robustness and ignores basic epidemiological principles. The absence of authoritative evidence for treatment effectiveness makes it increasingly hard to protect resource intensive treatments in anorexia and bulimia nervosa, and existing theories of the causation of the disorders are too nonspecific to generate effective programs of prevention. New models are urgently required" (p. 65).

Researchers have tried to treat anorexia nervosa with many drugs that increase appetite in laboratory animals or in people without eating disorders—for example, antipsychotic medications, drugs that stimulate adrenergic α_2 receptors, L-DOPA, and THC (the

FIGURE 12.32 ■ **Reactions of Young Men and Women on Fasting**

The graph shows food intake and eating rate during a buffet lunch after a 24-hour period of fasting or after a period during which they ate meals at their normal times.

(Data from Södersten, P., Bergh, C., and Zandian, M. *Hormones and Behavior*, 2006, *50*, 572–578.)

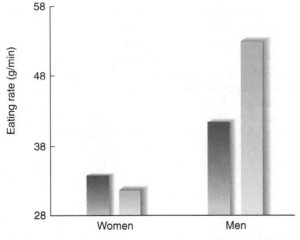

active ingredient in marijuana). Unfortunately, none of these drugs has been shown to be helpful (Mitchell, 1989). In any event the fact that anorexics are usually obsessed with food (and show high levels of neuropeptide Y and ghrelin) suggests that the disorder is not caused by the absence of hunger. Researchers have had better luck with bulimia nervosa; several studies suggest that serotonin agonists such as fluoxetine (an antidepressant drug that is best known as Prozac) may aid in the treatment of this disorder (Advokat and Kutlesic, 1995; Kaye et al., 2001). However, fluoxetine does not help anorexic patients (Attia et al., 1998).

Bergh, Södersten, and their colleagues (Zandian et al., 2007; Court, Bergh, and Södersten, 2008) have devised a novel and apparently effective treatment protocol for anorexia. The patients are taught to eat faster by placing a plate of food on an electronic scale attached to a computer that displays the time course of their actual and ideal intake. After the meal the patients are kept in a warm room, which reduces their anxiety and their activity level. A study with rats in cages with running wheels (Hillebrand et al., 2005) put the animals on a restricted feeding schedule and observed the symptoms I described earlier: hyperactivity, reduced food intake, severe body weight loss, and decreased body temperature. If a warm metal plate was placed in the cage, the rats with restricted access to food spent less time running in the activity wheel and more time on the warm plate. Rats that had unrestricted access to food ignored the presence of the warm plate.

Anorexia nervosa and bulimia nervosa are serious conditions; understanding their causes is more than an academic matter. We can hope that research on the biological and social control of feeding and metabolism and the causes of compulsive behaviors will help us to understand this puzzling and dangerous disorder.

Interim Summary

Anorexia Nervosa/Bulimia Nervosa

Anorexia nervosa is a serious—even life-threatening—disorder. Although anorexic patients avoid eating, they are often preoccupied with food. Bulimia nervosa consists of periodic binging and purging and a low body weight. Anorexia nervosa has a strong hereditary component and is seen primarily in young women.

Some researchers believe that the symptoms of anorexia—preoccupation with food and eating, ritualistic eating, erratic mood, excessive exercising, impaired cognitive performance, and physiological changes such as decreased body temperature—are symptoms of starvation and not the underlying causes of anorexia. A study carried out over fifty years ago found that several months of semi-starvation caused similar symptoms to emerge in previously healthy people. If rats are allowed access to food for

Patient H. M. has a relatively pure amnesia. His intellectual ability and his immediate verbal memory appear to be normal. He can repeat seven numbers forward and five numbers backward, and he can carry on conversations, rephrase sentences, and perform mental arithmetic. He is unable to remember events that occurred during several years preceding his brain surgery, but he can recall older memories very well. He showed no personality change after the operation, and he appears to be generally polite and good-natured.

However, since the operation, H. M. has been unable to learn anything new. He cannot identify by name people he has met since the operation (performed in 1953, when he was twenty-seven years old). His family moved to a new house after his operation, and he never learned how to get around in the new neighborhood. (He now lives in a nursing home, where he can be cared for.) He is aware of his disorder and often says something like this:

Every day is alone in itself, whatever enjoyment I've had, and whatever sorrow I've had. . . . Right now, I'm wondering. Have I done or said anything amiss? You see, at this moment everything looks clear to me, but what happened just before? That's what worries me. It's like waking from a dream; I just don't remember. (Milner, 1970, p. 37)

H. M. is capable of remembering a small amount of verbal information as long as he is not distracted; constant rehearsal can keep information in his immediate memory for a long time. However, rehearsal does not appear to have any long-term effects. If he is distracted for a moment, he will completely forget whatever he had been rehearsing. He works very well at repetitive tasks. Indeed, because he so quickly forgets what previously happened, he does not easily become bored. He can endlessly reread the same magazine or laugh at the same jokes, finding them fresh and new each time. His time is typically spent solving crossword puzzles and watching television.

Experiences change us; encounters with our environment alter our behavior by modifying our nervous system. As many investigators have said, an understanding of the physiology of memory is the ultimate challenge to neuroscience research. The brain is complex, and so are learning and remembering. However, despite the difficulties, the long years of work finally seem to be paying off. New approaches and new methods have evolved from old ones, and real progress has been made in understanding the anatomy and physiology of learning and remembering.

THE NATURE OF LEARNING

Learning refers to the process by which experiences change our nervous system and hence our behavior. We refer to these changes as *memories*. Although it is convenient to describe memories as if they were notes placed in filing cabinets, this is certainly not the way experiences are reflected within the brain. Experiences are not "stored"; rather, they change the way we perceive, perform, think, and plan. They do so by physically changing the structure of the nervous system, altering neural circuits that participate in perceiving, performing, thinking, and planning.

Learning can take at least four basic forms: perceptual learning, stimulus-response learning, motor learning, and relational learning. **Perceptual learning** is the ability to learn to recognize stimuli that have been perceived before. The primary function of this type of learning is the ability to identify and categorize objects (including other members of our own species) and situations. Unless we have learned to recognize something,

we cannot learn how we should behave with respect to it—we will not profit from our experiences with it, and profiting from experience is what learning is all about.

Each of our sensory systems is capable of perceptual learning. We can learn to recognize objects by their visual appearance, the sounds they make, how they feel, or how they smell. We can recognize people by the shape of their faces, the movements they make when they walk, or the sound of their voices. When we hear people talk, we can recognize the words they are saying and, perhaps, their emotional state. As we shall see, perceptual learning appears to be accomplished primarily by changes in the sensory association cortex. That is, learning to recognize complex visual stimuli involves changes in the visual association cortex, learning to recognize complex auditory stimuli involves changes in the auditory association cortex, and so on.

Stimulus-response learning is the ability to learn to perform a particular behavior when a particular stimulus is present. Thus, it involves the establishment of connections between circuits involved in perception and those involved in movement. The behavior could be an automatic response such as a defensive reflex, or it could be a complicated sequence of movements. Stimulus-response learning includes two major categories of learning that psychologists have studied extensively: *classical conditioning* and *instrumental conditioning*.

---

perceptual learning Learning to recognize a particular stimulus.

stimulus-response learning Learning to automatically make a particular response in the presence of a particular stimulus; includes classical and instrumental conditioning.

FIGURE 13.1 ■ A Simple Neural Model of Classical Conditioning

When the 1000-Hz tone is presented just before the puff of air to the eye, synapse T is strengthened.

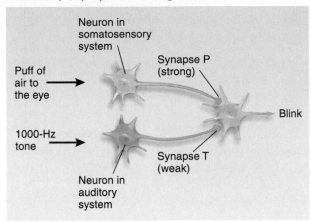

Classical conditioning is a form of learning in which an unimportant stimulus acquires the properties of an important one. It involves an *association between two stimuli.* A stimulus that previously had little effect on behavior becomes able to evoke a reflexive, species-typical behavior. For example, a defensive eyeblink response can be conditioned to a tone. If we direct a brief puff of air toward a rabbit's eye, the eye will automatically blink. The response is called an **unconditional response (UR)** because it occurs unconditionally, without any special training. The stimulus that produces it (the puff of air) is called an **unconditional stimulus (US).** Now we begin the training. We present a series of brief 1000-Hz tones, each followed 500 ms later by a puff of air. After several trials the rabbit's eye begins to close even before the puff of air occurs. Classical conditioning has occurred; the **conditional stimulus (CS**—the 1000-Hz tone) now elicits the **conditional response (CR**—the eyeblink). (See *Figure 13.1.*)

When classical conditioning takes place, what kinds of changes occur in the brain? Figure 13.1. shows a simplified neural circuit that could account for this type of learning. For the sake of simplicity we will assume that the US (the puff of air) is detected by a single neuron in the somatosensory system and that the CS (the 1000-Hz tone) is detected by a single neuron in the auditory system. We will also assume that the response—the eyeblink—is controlled by a single neuron in the motor system. Of course, learning actually involves many thousands of neurons—sensory neurons, interneurons, and motor neurons—but the basic principle of synaptic change can be represented by this simple figure. (See *Figure 13.1.*)

Let's us see how this circuit works. If we present a 1000-Hz tone, we find that the animal makes no reaction because the synapse connecting the tone-sensitive neuron

with the neuron in the motor system is weak. That is, when an action potential reaches the terminal button of synapse T (tone), the excitatory postsynaptic potential (EPSP) that it produces in the dendrite of the motor neuron is too small to make that neuron fire. However, if we present a puff of air to the eye, the eye blinks. This reaction occurs because nature has provided the animal with a strong synapse between the somatosensory neuron and the motor neuron that causes a blink (synapse P, for "puff"). To establish classical conditioning, we first present the 1000-Hz tone and then quickly follow it with a puff of air. After we repeat these pairs of stimuli several times, we find that we can dispense with the air puff; the 1000-Hz tone produces the blink all by itself.

Over fifty years ago, Donald Hebb proposed a rule that might explain how neurons are changed by experience in a way that would cause changes in behavior (Hebb, 1949). The **Hebb rule** says that if a synapse repeatedly becomes active at about the same time that the postsynaptic neuron fires, changes will take place in the structure or chemistry of the synapse that will strengthen it. How would the Hebb rule apply to our circuit? If the 1000-Hz tone is presented first, then weak synapse T (for "tone") becomes active. If the puff is presented immediately afterward, then strong synapse P becomes active and makes the motor neuron fire. The act of firing then strengthens any synapse with the motor neuron *that has just been active.* Of course, this means synapse T. After several pairings of the two stimuli and after several increments of strengthening, synapse T becomes strong enough to cause the motor neuron to fire by itself. Learning has occurred. (See *Figure 13.1.*)

When Hebb formulated his rule, he was unable to determine whether it was true or false. Now, finally, enough progress has been made in laboratory techniques that the strength of individual synapses can be determined, and investigators are studying the physiological bases of learning. We will see the results of some of these approaches in the next section of this chapter.

The second major class of stimulus-response learning is **instrumental conditioning** (also called *operant*

---

classical conditioning A learning procedure; when a stimulus that initially produces no particular response is followed several times by an **unconditional stimulus (US)** that produces a defensive or appetitive response (the **unconditional response—UR**), the first stimulus (now called a **conditional stimulus—CS**) itself evokes the response (now called a **conditional response—CR**).

Hebb rule The hypothesis proposed by Donald Hebb that the cellular basis of learning involves strengthening of a synapse that is repeatedly active when the postsynaptic neuron fires.

instrumental conditioning A learning procedure whereby the effects of a particular behavior in a particular situation increase (reinforce) or decrease (punish) the probability of the behavior; also called *operant conditioning.*

FIGURE 13.2 ■ A Simple Neural Model of Instrumental Conditioning

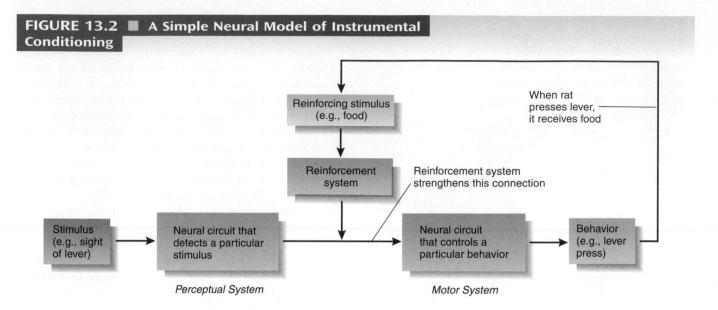

Perceptual System   *Motor System*

conditioning). Whereas classical conditioning involves automatic, species-typical responses, instrumental conditioning involves behaviors that have been learned. And whereas classical conditioning involves an association between two stimuli, instrumental conditioning involves an *association between a response and a stimulus.* Instrumental conditioning is a more flexible form of learning. It permits an organism to adjust its behavior according to the consequences of that behavior. That is, when a behavior is followed by favorable consequences, the behavior tends to occur more frequently; when it is followed by unfavorable consequences, it tends to occur less frequently. Collectively, "favorable consequences" are referred to as **reinforcing stimuli,** and "unfavorable consequences" are referred to as **punishing stimuli.** For example, a response that enables a hungry organism to find food will be reinforced, and a response that causes pain will be punished. (Psychologists often refer to these terms as *reinforcers* and *punishers.*)

Let's consider the process of reinforcement. Briefly stated, reinforcement causes changes in an animal's nervous system that increase the likelihood that a particular stimulus will elicit a particular response. For example, when a hungry rat is first put in an operant chamber (a "Skinner box"), it is not very likely to press the lever mounted on a wall. However, if it does press the lever and if it receives a piece of food immediately afterward, the likelihood of its pressing the lever increases. Put another way, reinforcement causes the sight of the lever to serve as the stimulus that elicits the lever-pressing response. It is not accurate to say simply that a particular behavior becomes more frequent. If no lever is present, a rat that has learned to press one will not wave its paw around in the air. The *sight of a lever* is needed to produce the response. Thus, the process of reinforcement strengthens

a connection between neural circuits involved in perception (the sight of the lever) and those involved in movement (the act of lever pressing). As we will see later in this chapter, the brain contains reinforcement mechanisms that control this process. (See *Figure 13.2.*)

The third major category of learning, **motor learning,** is actually a component of stimulus-response learning. For simplicity's sake we can think of perceptual learning as the establishment of changes within the sensory systems of the brain, stimulus-response learning as the establishment of connections between sensory systems and motor systems, and motor learning as the establishment of changes within motor systems. But, in fact, motor learning cannot occur without sensory guidance from the environment. For example, most skilled movements involve interactions with objects: bicycles, pinball machines, tennis racquets, knitting needles, and so on. Even skilled movements that we make by ourselves, such as solitary dance steps, involve feedback from the joints, muscles, vestibular apparatus, eyes, and contact between the feet and the floor. Motor learning differs from other forms of learning primarily in the degree to which new forms of behavior are learned; the more novel the behavior, the more the neural circuits in the motor systems of the brain must be modified. (See *Figure 13.3.*)

reinforcing stimulus An appetitive stimulus that follows a particular behavior and thus makes the behavior become more frequent.

punishing stimulus An aversive stimulus that follows a particular behavior and thus makes the behavior become less frequent.

motor learning Learning to make a new response.

FIGURE 13.3 ■ An Overview of Perceptual, Stimulus-Response (S-R), and Motor Learning

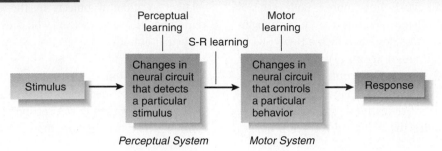

A particular learning situation can involve varying amounts of all three types of learning that I have described so far: perceptual, stimulus-response, and motor. For example, if we teach an animal to make a new response whenever we present a stimulus it has never seen before, the animal must learn to recognize the stimulus (perceptual learning) and make the response (motor learning), and a connection must be established between these two new memories (stimulus-response learning). If we teach the animal to make a response that it has already learned whenever we present a new stimulus, only perceptual learning and stimulus-response learning will take place.

The three forms of learning I have described so far consist primarily of changes in one sensory system, between one sensory system and the motor system, or in the motor system. But obviously, learning is usually more complex than that. The fourth form of learning involves learning the *relationships* among individual stimuli. For example, a somewhat more complex form of perceptual learning involves connections between different areas of the association cortex. When we hear the sound of a cat meowing in the dark, we can imagine what a cat looks like and what it would feel like if we stroked its fur. Thus, the neural circuits in the auditory

association cortex that recognize the meow are somehow connected to the appropriate circuits in the visual association cortex and the somatosensory association cortex. These interconnections, too, are accomplished as a result of learning.

Perception of spatial location—*spatial learning*— also involves learning about the relationships among many stimuli. For example, consider what we must learn to become familiar with the contents of a room. First, we must learn to recognize each of the objects. In addition, we must learn the relative locations of the objects with respect to each other. As a result, when we find ourselves in a particular place in the room, our perceptions of these objects and their locations relative to us tell us exactly where we are.

Other types of relational learning are even more complex. *Episodic learning*—remembering sequences of events (episodes) that we witness—requires us to keep track of and remember not only individual events but also the order in which they occur. As we will see in the last section of this chapter, a special system that involves the hippocampus and associated structures appears to perform coordinating functions required for many types of learning that go beyond simple perceptual, stimulus-response, or motor learning.

Interim Summary

The Nature of Learning

Learning produces changes in the way we perceive, act, think, and feel. It does so by producing changes in the nervous system in the circuits responsible for perception, in those responsible for the control of movement, and in connections between the two.

Perceptual learning consists primarily of changes in perceptual systems that make it possible for us to recognize stimuli so that we can respond to them appropriately.

Stimulus-response learning consists of connections between perceptual and motor systems. The most important forms are classical and instrumental conditioning. Classical conditioning occurs when a neutral stimulus is followed by an unconditional stimulus (US) that naturally elicits an unconditional response (UR). After this pairing, the neutral stimulus becomes a conditional stimulus (CS); it now elicits the response by itself, which we refer to as the conditional response (CR).

Instrumental conditioning occurs when a response is followed by a reinforcing stimulus, such as a drink of water for a thirsty animal. The reinforcing stimulus increases the likelihood that the other stimuli that were present when the response was made will evoke the response. Both forms of stimulus-response learning may occur as a result of strengthened synaptic connections, as described by the Hebb rule.

Motor learning, although it may primarily involve changes within neural circuits that control movement, is guided by sensory stimuli; thus, it is actually a form of stimulus-response learning. Relational learning, the most complex form of learning, includes the ability to recognize objects through more than one sensory modality, to recognize the relative location of objects in the environment, and to remember the sequence in which events occurred during particular episodes.

Thought Question

Can you think of specific examples of each of the categories of learning described in this section? Can you think of some examples that include more than one category?

SYNAPTIC PLASTICITY: LONG-TERM POTENTIATION AND LONG-TERM DEPRESSION

On theoretical considerations alone, it would appear that learning must involve synaptic plasticity: changes in the structure or biochemistry of synapses that alter their effects on postsynaptic neurons. Recent years have seen an explosion of research on this topic, largely stimulated by the development of methods that permit researchers to observe structural and biochemical changes in microscopically small structures: the presynaptic and postsynaptic components of synapses.

Induction of Long-Term Potentiation

Electrical stimulation of circuits within the hippocampal formation can lead to long-term synaptic changes that seem to be among those responsible for learning. Lømo (1966) discovered that intense electrical stimulation of axons leading from the entorhinal cortex to the dentate gyrus caused a long-term increase in the magnitude of excitatory postsynaptic potentials in the postsynaptic neurons; this increase has come to be called **long-term potentiation (LTP).** (The word *potentiate* means "to strengthen, to make more potent.")

First, let's review some anatomy. The **hippocampal formation** is a specialized region of the limbic cortex located in the temporal lobe. (Its location in a human brain is shown in Figure 3.19.) Because the hippocampal formation is folded in one dimension and then curved in another, it has a complex, three-dimensional shape. Therefore, it is difficult to show what it looks like with a diagram on a two-dimensional sheet of paper. Fortunately, the structure of the hippocampal formation is orderly; a slice taken anywhere perpendicular to its curving long axis contains the same set of circuits.

Figure 13.4 shows a slice of the hippocampal formation, illustrating a typical procedure for producing long-term potentiation. The primary input to the hippocampal formation comes from the *entorhinal cortex*. The axons of neurons in the entorhinal cortex pass through the *perforant path* and form synapses with the granule cells of the *dentate gyrus*. A stimulating electrode is placed in the perforant path, and a recording electrode is placed in the dentate gyrus, near the granule cells. (See *Figure 13.4b*.) First, a single pulse of electrical stimulation is delivered to the perforant path, and then the resulting population EPSP is recorded in the dentate gyrus. The **population EPSP** is an extracellular measurement of the excitatory postsynaptic potentials (EPSP) produced by the synapses of the perforant path axons with the dentate granule cells. The size of the first population EPSP indicates the strength of the synaptic connections before long-term potentiation has taken place. Long-term potentiation can be induced by stimulating the axons in the perforant path with a burst of approximately one hundred pulses of electrical stimulation, delivered within a few seconds. Evidence that long-term potentiation has occurred is obtained by periodically delivering single pulses to the perforant path and recording the response in the dentate gyrus. If the response is greater than it was before the burst of pulses was delivered, long-term potentiation has occurred. (See *Figure 13.5*.)

Long-term potentiation can be produced in other regions of the hippocampal formation and in many other places in the brain. It can last for several months

long-term potentiation (LTP) A long-term increase in the excitability of a neuron to a particular synaptic input caused by repeated high-frequency activity of that input.

hippocampal formation A forebrain structure of the temporal lobe, constituting an important part of the limbic system; includes the hippocampus proper (Ammon's horn), dentate gyrus, and subiculum.

population EPSP An evoked potential that represents the EPSPs of a population of neurons.

FIGURE 13.4 ■ The Hippocampal Formation and Long-Term Potentiation

The schematic shows the connections of the components of the hippocampal formation and the procedure for producing long-term potentiation.

(Photograph from Swanson, L. W., Köhler, C., and Björklund, A., in *Handbook of Chemical Neuroanatomy. Vol. 5: Integrated Systems of the CNS, Part I.* Amsterdam: Elsevier Science Publishers, 1987. Reprinted with permission.)

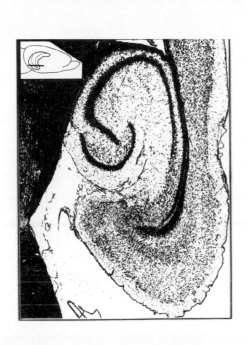

(a)

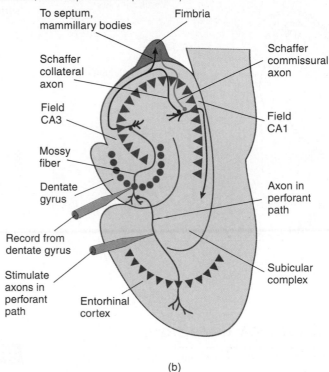

(b)

FIGURE 13.5 ■ Long-Term Potentiation

Population EPSPs were recorded from the dentate gyrus before and after electrical stimulation that led to long-term potentiation.

(From Berger, T. W. *Science*, 1984, *224*, 627–630. Copyright © 1984 by the American Association for the Advancement of Science. Reprinted with permission.)

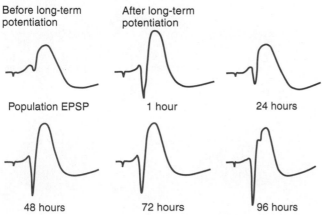

(Bliss and Lømo, 1973). It can be produced in isolated slices of the hippocampal formation as well as in the brains of living animals, which allows researchers to stimulate and record from individual neurons and to analyze biochemical changes. The brain is removed from the skull, the hippocampal complex is dissected, and slices are placed in a temperature-controlled chamber filled with liquid that resembles interstitial fluid. Under optimal conditions a slice remains alive for up to forty hours.

Many experiments have demonstrated that long-term potentiation in hippocampal slices can follow the Hebb rule. That is, when weak and strong synapses to a single neuron are stimulated at approximately the same time, the weak synapse becomes strengthened. This phenomenon is called **associative long-term potentiation,** because it is produced by the association (in time) between the activity of the two sets of synapses. (See *Figure 13.6.*)

associative long-term potentiation A long-term potentiation in which concurrent stimulation of weak and strong synapses to a given neuron strengthens the weak ones.

FIGURE 13.6 ■ Associative Long-Term Potentiation

If the weak stimulus and strong stimulus are applied at the same time, the synapses activated by the weak stimulus will be strengthened.

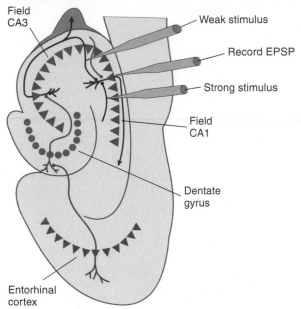

FIGURE 13.7 ■ The Role of Summation in Long-Term Potentiation

If axons are stimulated rapidly, the EPSPs produced by the terminal buttons will summate, and the postsynaptic membrane will depolarize enough for long-term potentiation to occur. If axons are stimulated slowly, the EPSPs will not summate, and long-term potentiation will not occur.

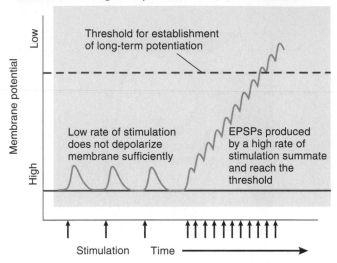

Role of NMDA Receptors

Nonassociative long-term potentiation requires some sort of additive effect. That is, a series of pulses delivered at a high rate all in one burst will produce LTP, but the same number of pulses given at a slow rate will not. (In fact, as we shall see, low-frequency stimulation can lead to the opposite phenomenon: long-term *depression*.) The reason for this phenomenon is now clear. A rapid rate of stimulation causes the excitatory postsynaptic potentials to summate, because each successive EPSP occurs before the previous one has dissipated. This means that rapid stimulation depolarizes the postsynaptic membrane much more than slow stimulation does. (See *Figure 13.7*.)

Several experiments have shown that synaptic strengthening occurs when molecules of the neurotransmitter bind with postsynaptic receptors located in a dendritic spine that is already depolarized. Kelso, Ganong, and Brown (1986) found that if they used a microelectrode to artificially depolarize a neuron in field CA1 and then stimulated the axons that formed synapses with this neuron, the synapses became stronger. However, if the stimulation of the synapses and the depolarization of the neuron occurred at different times, no effect was seen; therefore, the two events had to occur together. (See *Figure 13.8*.)

Experiments such as the ones I just described indicate that LTP requires two events: activation of synapses and depolarization of the postsynaptic neuron. The

FIGURE 13.8 ■ Long-Term Potentiation

Synaptic strengthening occurs when synapses are active while the membrane of the postsynaptic cell is depolarized.

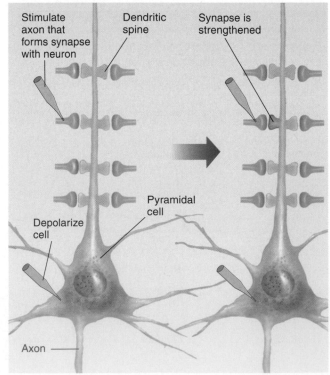

FIGURE 13.9 ■ The NMDA Receptor

The NMDA receptor is a neurotransmitter- and voltage-dependent ion channel. (a) When the postsynaptic membrane is at the resting potential, Mg^{2+} blocks the ion channel, preventing Ca^{2+} from entering. (b) When the membrane is depolarized, the magnesium ion is evicted. Thus, the attachment of glutamate to the binding site causes the ion channel to open, allowing calcium ions to enter the dendritic spine.

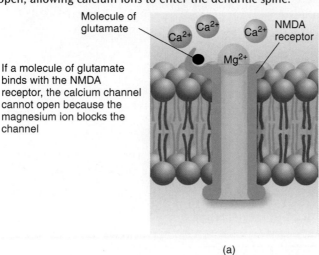

If a molecule of glutamate binds with the NMDA receptor, the calcium channel cannot open because the magnesium ion blocks the channel

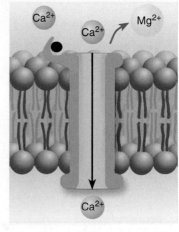

Depolarization of the membrane evicts the magnesium ion and unblocks the channel. Now glutamate can open the ion channel and permit the entry of calcium ions.

Depolarization

(a)

(b)

explanation for this phenomenon, at least in some parts of the brain, lies in the characteristics of a very special type of glutamate receptor. The **NMDA receptor** has some unusual properties. It is found in the hippocampal formation, especially in field CA1. It gets its name from a drug that specifically activates it: *N*-methyl-D-aspartate. The NMDA receptor controls a calcium ion channel. This channel is normally blocked by a magnesium ion (Mg^{2+}), which prevents calcium ions from entering the cell even when the receptor is stimulated by glutamate. But if the postsynaptic membrane is depolarized, the Mg^{2+} is ejected from the ion channel, and the channel is free to admit Ca^{2+} ions. Thus, calcium ions enter the cells through the channels controlled by NMDA receptors only when glutamate is present *and* when the postsynaptic membrane is depolarized. This means that the ion channel controlled by the NMDA receptor is a neurotransmitter- *and* voltage-dependent ion channel. (See *Figure 13.9.* and *MyPsychKit 13.1, The NMDA Receptor.*)

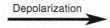
Animation 13.1
The NMDA Receptor

Cell biologists have discovered that the calcium ion is used by many cells as a second messenger that activates various enzymes and triggers biochemical processes. The entry of calcium ions through the ion channels controlled by NMDA receptors is an essential step in long-term potentiation (Lynch et al., 1984). **AP5** (2-amino-5-phosphonopentanoate), a drug that blocks NMDA receptors, prevents calcium ions from entering the dendritic spines and thus blocks the establishment of LTP

(Brown et al., 1989). These results indicate that the activation of NMDA receptors is necessary for the first step in the process events that establishes LTP: the entry of calcium ions into dendritic spines.

In Chapter 2 you learned that only axons are capable of producing action potentials. Actually, they can also occur in dendrites of some types of pyramidal cells, including those in field CA1 of the hippocampal formation. The threshold of excitation for **dendritic spikes** (as these action potentials are called) is rather high. As far as we know, they occur only when an action potential is triggered in the axon of the pyramidal cell. The backwash of depolarization across the cell body triggers a dendritic spike, which is propagated up the trunk of the dendrite. This means that whenever the axon of a pyramidal cell fires, all of its dendritic spines become depolarized for a brief time.

A study by Magee and Johnston (1997) proved that the simultaneous occurrence of synaptic activation and a dendritic spike strengthens the active synapse. The investigators injected individual CA1 pyramidal cells in hippocampal slices with calcium-green-1, a fluorescent

NMDA receptor A specialized ionotropic glutamate receptor that controls a calcium channel that is normally blocked by Mg^{2+} ions; involved in long-term potentiation.

AP5 2-Amino-5-phosphonopentanoate, a drug that blocks NMDA receptors.

dendritic spike An action potential that occurs in the dendrite of some types of pyramidal cells.

FIGURE 13.10 ■ Associative Long-Term Potentiation

If the activity of strong synapses is sufficient to trigger an action potential in the neuron, the dendritic spike will depolarize the membrane of dendritic spines, priming NMDA receptors so that any weak synapses active at that time will become strengthened.

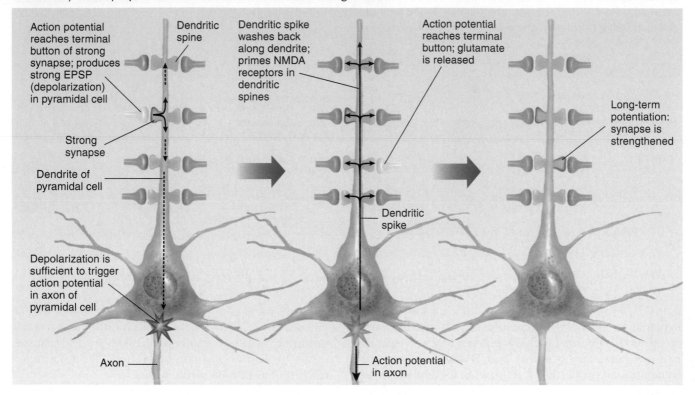

dye that permitted them to observe the influx of calcium. They found that when individual synapses became active at the same time that a dendritic spike had been triggered, calcium "hot spots" occurred near the activated synapses. Moreover, the size of the excitatory postsynaptic potential produced by these activated synapses became larger. In other words, these synapses became strengthened. To confirm that the dendritic spikes were necessary for the synaptic potentiation to take place, the investigators infused a small amount of tetrodotoxin (TTX) onto the base of the dendrite just before triggering an action potential. The TTX prevented the formation of dendritic spikes by blocking voltage-dependent sodium channels. Under these conditions, long-term potentiation did not occur.

I think that considering what you already know about associative LTP, you can anticipate the role that NMDA receptors play in this phenomenon. If weak synapses are active by themselves, nothing happens because the membrane of the dendritic spine does not depolarize sufficiently for the calcium channels controlled by the NMDA receptors to open. (Remember that for these channels to open, the postsynaptic membrane must first depolarize and displace the magnesium ions that normally block them.) However, if the activity

of strong synapses located elsewhere on the postsynaptic cell has caused the cell to fire, then a dendritic spike will depolarize the postsynaptic membrane enough to eject the magnesium ions from the calcium channels of the NMDA receptors in the dendritic spines. If some weak synapses then become active, calcium will enter the dendritic spines and cause the synapses to become strengthened. Thus, the special properties of NMDA receptors account not only for the existence of long-term potentiation but also for its associative nature. (See *Figure 13.10* and *MyPsychKit 13.2, Associative LTP.*)

Animation 13.2

Associative LTP

Mechanisms of Synaptic Plasticity

What is responsible for the increases in synaptic strength that occur during long-term potentiation? Dendritic spines on CA1 pyramidal cells contain two types of glutamate receptors: NMDA receptors and **AMPA receptors.** Research indicates that strengthening

---

AMPA receptor An ionotropic glutamate receptor that controls a sodium channel; when open, it produces EPSPs.

of an individual synapse appears to be accomplished by insertion of additional AMPA receptors into the postsynaptic membrane of the dendritic spine. AMPA receptors control sodium channels; thus, when they are activated by glutamate, they produce EPSPs in the membrane of the dendritic spine. Therefore, with more AMPA receptors present, the release of glutamate by the terminal button causes a larger excitatory postsynaptic potential. In other words, the synapse becomes stronger.

Where do these new AMPA receptors come from? Shi et al. (1999) used a harmless virus to insert a gene for a subunit of the AMPA receptor into rat hippocampal neurons maintained in a tissue culture. The AMPA receptors produced by the gene had a fluorescent dye molecule attached to them, which permitted the investigators to use a two-photon laser scanning microscope to see the exact location of AMPA receptors in dendritic spines of CA1 neurons. The investigators induced LTP by stimulating axons that form synapses with these dendrites. Before LTP was induced, they saw AMPA receptors clustered at the base of the dendritic spines. Fifteen minutes after the induction of LTP, the AMPA receptors flooded into the spines and

---

FIGURE 13.11 ■ Role of AMPA Receptors in Long-Term Potentiation

Two-photon laser scanning microscopy of the CA1 region of living hippocampal slices shows delivery of AMPA receptors into dendritic spines after long-term potentiation. The AMPA receptors were tagged with a fluorescent dye molecule. The two photographs at the bottom are higher magnifications of the ones above. The arrows labeled *a* and *b* point to dendritic spines that became filled with AMPA receptors after the induction of long-term potentiation.

(From Shi, S.-H., Hayashi, Y., Petralia, R. S., Zaman, S. H., Wenthold, R. J., Svoboda, K., and Malinow, R. *Science,* 1999, *284,* 1811–1816. Copyright © 1999 by the American Association for the Advancement of Science. Reprinted with permission.)

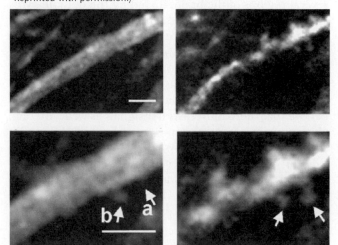

Before LTP   After LTP

---

moved to their tips—the location of the postsynaptic membrane. This movement of AMPA receptors did not occur when AP5, the drug that blocks NMDA receptors, was added to the culture medium. (See *Figure 13.11.*)

How does the entry of calcium ions into the dendritic spine cause AMPA receptors to move into the postsynaptic membrane? This process appears to involve several enzymes, including **CaM-KII** (type II calcium-calmodulin kinase), an enzyme found in dendritic spines. CaM-KII is a *calcium-dependent* enzyme, which is inactive until a calcium ion binds with it and activates it. Many studies have shown that CaM-KII plays a critical role in long-term potentiation. For example, Silva et al. (1992a) produced a targeted mutation against the gene responsible for the production of CaM-KII in mice. The mice had no obvious neuroanatomical defects, and the responses of their NMDA receptors were normal. However, the investigators were unable to produce LTP in field CA1 of hippocampal slices taken from these animals. Lledo et al. (1995) found that injection of activated CaM-KII directly into CA1 pyramidal cells mimicked the effects of LTP: It strengthened synaptic transmission in those cells.

As we saw in Chapter 3, when synapses are examined under an electron microscope, a dark band is seen just inside the postsynaptic membrane. This band, known as the *postsynaptic density,* contains a variety of proteins: receptors, enzymes, messenger proteins, and scaffolding proteins—structural proteins that anchor the receptors, enzymes, and messengers in place (Allison et al., 2000). Shen and Meyer (1999) used a harmless virus to insert a gene for a fluorescent dye molecule attached to CaM-KII into cultured hippocampal neurons. They found that after LTP was induced, CaM-KII molecules became concentrated in the postsynaptic densities of dendritic spines, where the postsynaptic receptors are located. (See *Figure 13.12.*)

Two other changes that accompany LTP are alteration of synaptic structure and production of new synapses. Many studies have found that the establishment of LTP includes changes in the size and shape of dendritic spines. For example, Bourne and Harris (2007) suggest that LTP causes the enlargement of thin spines into fatter, mushroom-shaped spines. Figure 13.13. shows the variety of shapes that dendritic spines and their associated postsynaptic density can take. (See *Figure 13.13.*) Nägerl et al. (2007) found that the establishment of LTP caused the growth of new dendritic spines. After about fifteen to nineteen hours, the new spines formed synaptic connections with terminals of nearby axons. (See *Figure 13.14.*)

CaM-KII Type II calcium-calmodulin kinase, an enzyme that must be activated by calcium; may play a role in the establishment of long-term potentiation.

FIGURE 13.12 ■ Role of CaM-KII in Long-Term Potentiation

CaM-KII molecules migrate into the postsynaptic densities of dendritic spines after long-term potentiation. (a) A single hippocampal pyramidal neuron is stained for the presence of CaM-KII, before NMDA receptor stimulation. (b) The same neuron after NMDA receptor stimulation. (c) An enlargement of the area in (a) is marked by a white rectangle. The presence of CaM-KII is shown in green. (d) An enlargement of the area in (b) is marked by a white rectangle. The presence of CaM-KII that has moved into dendritic spines is shown in red.

(From Shen, K., and Meyer, T. *Science*, 1999, *284*, 162–166. Copyright © 1999 by the American Association for the Advancement of Science. Reprinted with permission.)

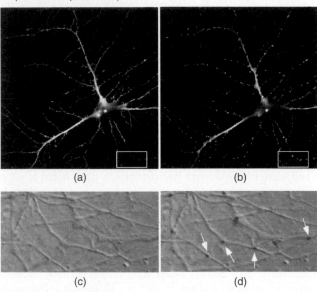

(a)   (b)

(c)   (d)

FIGURE 13.13 ■ Dendritic Spines in Field CA1

According to Bourne and Harris (2007), long-term potentiation may convert thin spines into mushroom-shaped spines. (a) Colorized photomicrograph: Dendrite shafts are yellow, spine necks are blue, spine heads are green, and presynaptic terminals are orange. (b) Three-dimensional reconstruction of a portion of a dendrite (yellow) shows the variation I size and shape of postsynaptic densities (red).

(From Bourne, J., and Harris, K. M. *Current Opinion in Neurobiology*, 2007, *17*, 381–386. Reprinted with permission.)

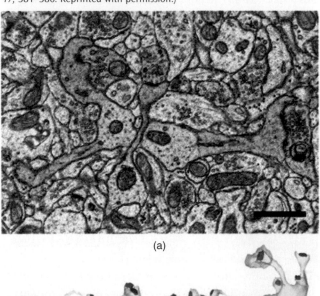

(a)

(b)

Researchers believe that LTP may also involve *presynaptic* changes in existing synapses, such as an increase in the amount of glutamate that is released by the terminal button. But how could a process that begins postsynaptically, in the dendritic spines, cause presynaptic changes? A possible answer comes from the discovery that a simple molecule, nitric oxide, can communicate messages from one cell to another. As we saw in Chapter 4, nitric oxide is a soluble gas produced from the amino acid arginine by the activity of an enzyme known as **nitric oxide synthase.** Once produced, NO lasts only a short time before it is destroyed. Thus, if it were produced in dendritic spines in the hippocampal formation, it could diffuse only as far as the nearby terminal buttons, where it might produce changes related to the induction of LTP.

Several experiments suggest that NO may indeed be a retrograde messenger involved in LTP. (*Retrograde* means "moving backward"; in this context it refers to messages sent from the dendritic spine back to the terminal button.) Several studies have shown that drugs that block nitric oxide synthase prevent the establishment of LTP in field CA1 (Haley, Wilcox, and Chapman,

1992). In addition, Endoh, Maiese, and Wagner (1994) found that a calcium-activated NO synthase is found in several regions of the brain, including the dentate gyrus and fields CA1 and CA3 of the hippocampus. Arancio et al. (1995) obtained evidence that NO acts by stimulating the production of cyclic GMP, a second messenger, in presynaptic terminals. Although there is good evidence that NO is one of the signals the dendritic spine uses to communicate with the terminal button, most investigators believe that there must be other signals as well. After all, alterations in synapses require coordinated changes in both presynaptic and postsynaptic elements.

For several years after its discovery, researchers believed that LTP involved a single process. Since then it has become clear that LTP consists of several stages.

nitric oxide synthase An enzyme responsible for the production of nitric oxide.

FIGURE 13.14 ■ Growth of Dendritic Spines After Long-Term Potentiation

Two-photon microscopic images show a segment of a dendrite of a CA1 pyramidal neuron before and after electrical stimulation that established long-term potentiation. Numbers in each box indicate the time before or after the stimulation.

(From Nägerl, U. V., Köstinger, G., Anderson, J. C., Martin, K. A. C., and Bonhoeffer, T. *Journal of Neuroscience,* 2007, *27,* 8149–8156. Reprinted with permission.)

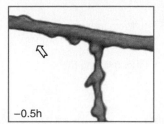
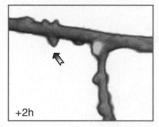
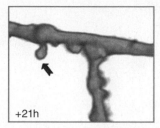

Long-lasting LTP—that is, LTP that lasts more than a few hours—requires protein synthesis. Frey et al. (1988) found that drugs that block protein synthesis prevented the establishment of long-lasting LTP in field CA1. If the drug was administered before, during, or immediately after a prolonged burst of stimulation was delivered, LTP occurred, but it disappeared a few hours later. However, if the drug was administered one hour after the synapses had been stimulated, the LTP persisted. Apparently, the protein synthesis necessary for establishing the later phase of long-lasting LTP is accomplished within an hour of stimulation.

According to Raymond (2007), there are actually three types of LTP. The first type, LTP1, involves almost immediate changes in synaptic strength caused by insertion of AMPA receptors. This form of LTP lasts for an hour or two. The second type, LTP2, involves local protein synthesis. Dendrites contain messenger RNAs that can be translated into proteins. These RNAs include codes for various enzymes, components of receptors, and structural proteins (Martin and Zukin, 2006). The most durable type of long-term potentiation, LTP3, involved production of mRNA in the nucleus that is then transported to the dendrites, where protein synthesis takes place. The long-lasting form of LTP also requires the presence of dopamine, which stimulates D1 receptors present on the dendrites. The importance of dopamine in the establishment of long-term memories is discussed later in this chapter.

For several years, investigators were puzzled about the mechanism that controlled the location of the protein synthesis initiated by production of mRNA in the nucleus. As we saw, LTP involves individual synapses: Only the synapses that are activated when the postsynaptic membrane is depolarized are strengthened. What mechanism delivers proteins produced in the cell body by translation of newly produced mRNA to the appropriate dendritic spines?

Evidence suggests that LTP initiates two processes: the production of plasticity-related proteins through normal synthesis of messenger RNA in the nucleus of the cell and the production of a chemical "tag" in the dendritic spines where the LTP has taken place. The new proteins then diffuse throughout the dendrites of the cell and are captured by the tags and used to stabilize temporary synaptic changes and establish the longest-lasting LTP (U. Frey and Morris, 1997; Frey and Frey, 2008). (See *Figure 13.15.*)

Figure 13.16 summarizes the biochemistry discussed in this subsection. I suspect that you might feel overwhelmed by all the new terms I have introduced here, and I hope that the figure will help to clarify things. The evidence we have seen so far indicates that activation of a terminal button releases glutamate, which binds with NMDA receptors in the postsynaptic membrane of the dendritic spine. If this membrane was depolarized by a dendritic spike, then calcium ions will enter through channels controlled by the NMDA receptors and activate CaM-KII, a calcium-dependent protein kinase. CaM-KII travels to the postsynaptic density of dendritic spines, where it causes the insertion of AMPA receptors into the postsynaptic density. In addition, LTP initiates rapid changes in synaptic structure and the production of new synapses. (See *Figure 13.16.*) The entry of calcium also activates a calcium-dependent NO synthase, and the newly produced NO then presumably diffuses out of the dendritic spine, back to the terminal button. There, it may trigger unknown chemical reactions that increase the release of glutamate. (See *Figure 13.16.*) Finally, long-lasting LTP (LTP2 and LTP3) requires the presence of dopamine and local and remote synthesis of new proteins that stabilize the changes made in the structure of the potentiated synapse. (See *MyPsychKit 13.3, Chemistry of LTP.*)

mypsychkit

Animation 13.3

Chemistry of LTP

FIGURE 13.15 ■ The "Tag" Hypothesis of Frey and Morris (1998)

This hypothesis suggests how proteins, whose synthesis is initiated by synapses that are undergoing long-term potentiation, can be directed to the locations where they are needed to sustain long-lasting long-term potentiation.

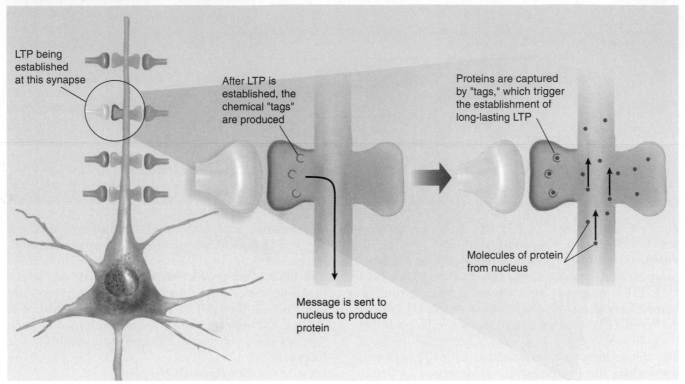

LTP being established at this synapse

After LTP is established, the chemical "tags" are produced

Proteins are captured by "tags," which trigger the establishment of long-lasting LTP

Molecules of protein from nucleus

Message is sent to nucleus to produce protein

Long-Term Depression

I mentioned earlier that low-frequency stimulation of the synaptic inputs to a cell can *decrease* rather than increase their strength. This phenomenon, known as **long-term depression (LTD),** also plays a role in learning. Apparently, neural circuits that contain memories are established by strengthening some synapses and weakening others. Dudek and Bear (1992) stimulated Schaffer collateral inputs to CA1 neurons in hippocampal slices with 900 pulses of electrical current, delivered at rates ranging from 1 to 50 Hz. They found that frequencies above 10 Hz caused long-term potentiation, whereas those below 10 Hz caused long-term depression. Both of these effects were blocked by application of AP5, the NMDA receptor blocker; thus, both effects require the activation of NMDA receptors. (See *Figure 13.17.*)

Several studies have demonstrated *associative* long-term depression, which is produced when synaptic inputs are activated at the same time that the postsynaptic membrane is either weakly depolarized or hyperpolarized (Debanne, Gähwiler, and Thompson, 1994; Thiels et al., 1996).

As we saw, the most commonly studied form of long-term potentiation involves an increase in the number of AMPA receptors in the postsynaptic membrane of dendritic spines. Long-term depression appears to involve the opposite: a *decrease* in the number of AMPA receptors in these spines (Carroll et al., 1999). And just as AMPA receptors are inserted into dendritic spines during LTP, they are removed from the spines in vesicles during LTD (Lüscher et al., 1999).

In field CA1, long-term depression, like long-term potentiation, involves the activation of NMDA receptors, and its establishment is disrupted by AP5. How can activation of the same receptor produce opposite effects? An answer was suggested by Lisman (1989), who noted that sustained, low-frequency stimulation of synapses on pyramidal cells in this region that produces LTD would cause a modest but prolonged increase in

long-term depression (LTD) A long-term decrease in the excitability of a neuron to a particular synaptic input caused by stimulation of the terminal button while the postsynaptic membrane is hyperpolarized or only slightly depolarized.

FIGURE 13.16 ■ Chemistry of Long-Term Potentiation

These chemical reactions appear to be triggered by the entry of an adequate amount of calcium into the dendritic spine.

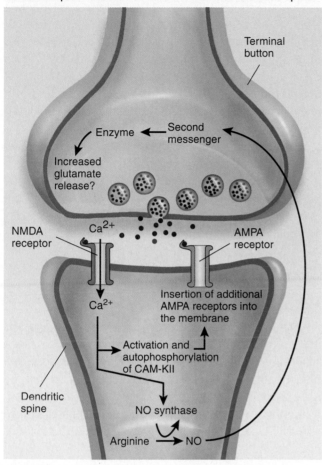

FIGURE 13.17 ■ Long-Term Potentiation and Long-Term Depression

The graph shows changes in the sensitivity of synapses of Schaffer collateral axons with CA1 pyramidal cells after electrical stimulation at various frequencies.

(Adapted from Dudek, S. M., and Bear, M. F. *Proceedings of the National Academy of Sciences,* 1992, *89,* 4363–4367.)

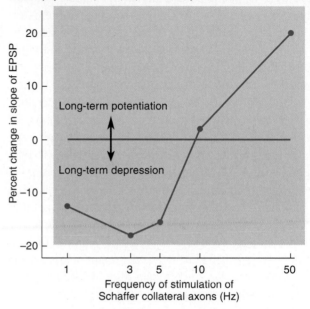

buildup of a modest but prolonged increase in intracellular calcium.

Other Forms of Long-Term Potentiation

Long-term potentiation was discovered in the hippocampal formation and has been studied more in this region than in others, but it also occurs in many other regions of the brain. Later in this chapter we will see the role of LTP in particular forms of learning. In some but not all of these regions, LTP is initiated by stimulation of NMDA receptors. For example, in the hippocampal formation, NMDA receptors are present in highest concentrations in field CA1 and in the dentate gyrus. However, very few NMDA receptors are found in the region of field CA3 that receives mossy fiber input from the dentate gyrus (Monaghan and Cotman, 1985). High-frequency stimulation of the mossy fibers produces LTP that gradually decays over a period of several hours (Lynch et al., 1991). AP5, the drug that blocks NMDA receptors and prevents the establishment of LTP in CA1 neurons, has no effect on LTP in field CA3. In addition, long-term potentiation in field CA3 appears to involve only presynaptic changes; no alterations are seen in the structure of dendritic spines after LTP has taken place (Reid et al., 2004).

intracellular Ca^{2+}, whereas the intense, high-frequency stimulation that produces LTP would cause a much greater increase in Ca^{2+}. Perhaps small and large increases in intracellular calcium ions trigger different mechanisms.

Evidence in favor of this hypothesis was obtained by a study by Liu et al. (2004). NMDA receptors come in at least two forms. One form contains one type of subunit, and the other contains a different type of subunit. Liu and his colleagues found that LTP was prevented by a drug that blocked one type of NMDA receptor and that LTD was prevented by a drug that blocked the other type of NMDA receptor. Receptors that produce LTP permit an influx of large amounts of Ca^{2+} if they are stimulated repeatedly in a short amount of time. In contrast, receptors that produce LTD permit less calcium to enter the cell, but if they are stimulated slowly over a long period of time, they permit the

InterimSummary

Synaptic Plasticity: Long-Term Potentiation and Long-Term Depression

The study of long-term potentiation in the hippocampal formation has suggested a mechanism that might be responsible for at least some of the synaptic changes that occur during learning. A circuit of neurons passes from the entorhinal cortex through the hippocampal formation. High-frequency stimulation of the axons in this circuit strengthens synapses; it leads to an increase in the size of the EPSPs in the dendritic spines of the postsynaptic neurons. Associative long-term potentiation can also occur, in which weak synapses are strengthened by the action of strong ones. In fact, the only requirement for LTP is that the postsynaptic membrane be depolarized at the same time that the synapses are active.

In field CA1, in the dentate gyrus, and in several other parts of the brain, NMDA receptors play a special role in LTP. These receptors, sensitive to glutamate, control calcium channels but can open them only if the membrane is already depolarized. Thus, the combination of membrane depolarization (for example, from a dendritic spike produced by the activity of strong synapses) and activation of an NMDA receptor causes the entry of calcium ions. The increase in calcium activates several calcium-dependent enzymes, including CaM-KII. CaM-KII causes the insertion of AMPA receptors into the membrane of the dendritic spine, increasing their sensitivity to glutamate released by the terminal button. This change is accompanied by structural alterations in the shape of the dendritic spine and by the growth of new spines, which establish new synapses. LTP may also involve presynaptic changes, through the activation of NO synthase, an enzyme responsible for the production of nitric oxide. This soluble gas may diffuse into nearby terminal buttons, where it facilitates the release of glutamate. Long-lasting LTP requires protein synthesis. The presence of "tag" molecules in potentiated dendritic spines may capture proteins produced in the soma and incorporate them into the synapse.

Long-term depression occurs when a synapse is activated at the time that the postsynaptic membrane is hyperpolarized or only slightly depolarized. In field CA1, LTP and LTD are established by slightly different forms of NMDA receptors. If LTP and LTD occurred only in the hippocampal formation, their discovery would still be an interesting finding, but the fact that they also occur in several other regions of the brain suggests that they play an important role in many forms of learning.

Thought Question

The brain is the most complex organ in the body, and it is also the most malleable. Every experience leaves at least a small trace, in the form of altered synapses. When we tell someone something or participate in an encounter that the other person will remember, we are (literally) changing connections in the person's brain. How many synapses change each day? What prevents individual memories from becoming confused?

PERCEPTUAL LEARNING

Learning enables us to adapt to our environment and to respond to changes in it. In particular, it provides us with the ability to perform an appropriate behavior in an appropriate situation. Situations can be as simple as the sound of a buzzer or as complex as the social interactions of a group of people. The first part of learning involves learning to perceive particular stimuli.

Perceptual learning involves learning to *recognize* things, not *what to do* when they are present. (Learning what to do is discussed in the next three sections of this chapter.) Perceptual learning can involve learning to recognize entirely new stimuli, or it can involve learning to recognize changes or variations in familiar stimuli. For example, if a friend gets a new hairstyle or replaces glasses with contact lenses, our visual memory of that person changes. We also learn that particular stimuli are found in particular locations or contexts or in the presence of other stimuli. We can even learn and remember particular *episodes:* sequences of events taking place at a particular time and place. The more complex forms of perceptual learning will be discussed in the last section of this chapter, which is devoted to relational learning.

Learning to Recognize Stimuli

In mammals with large and complex brains, objects are recognized visually by circuits of neurons in the visual association cortex. Visual learning can take place very rapidly, and the number of items that can be remembered is enormous. In fact, Standing (1973) showed people 10,000 color slides and found that they could recognize most of the slides weeks later. Other primates are capable of remembering items that they have seen for just a few seconds, and the experience changes the responses of neurons in their visual association cortex (Rolls, 1995b).

As we saw in Chapter 6, the primary visual cortex receives information from the lateral geniculate nucleus of the thalamus. After the first level of analysis the information is sent to the extrastriate cortex, which surrounds the primary visual cortex (striate cortex). After analyzing particular attributes of the visual scene, such as form, color, and movement, subregions of the extrastriate cortex send the results of their analysis to the next level of the visual association cortex, which is divided into two "streams." The *ventral stream,* which is involved with object recognition, continues ventrally into the inferior temporal cortex. The *dorsal stream,* which is involved with perception of the location of objects, continues dorsally into the posterior parietal cortex. As some investigators have said, the ventral stream is involved with the *what* of visual perception, and the dorsal stream is involved with the *where.* (See *Figure 13.18.*)

Many studies have shown that lesions that damage the inferior temporal cortex—part of the ventral stream—disrupt the ability to discriminate between different visual stimuli. These lesions impair the ability to perceive (and thus to learn to recognize) particular kinds of visual information. As we saw in Chapter 6, people with damage to the inferior temporal cortex may have excellent vision but be unable to recognize familiar, everyday objects such as scissors, clothespins, or light bulbs—and faces of friends and relatives.

Perceptual learning clearly involves changes in synaptic connections in the visual association cortex that establish new neural circuits—changes such as the ones described in the previous section of this chapter. At a later time, when the same stimulus is seen again and the same pattern of activity is transmitted to the cortex, these circuits become active again. This activity constitutes the recognition of the stimulus—the readout of the visual memory, so to speak. For example, Yang and Maunsell (2004) trained monkeys to detect small differences in visual stimuli whose images were projected onto a specific region of the retina. After the training was complete, the monkeys were able to detect differences much smaller than those they could detect when the training first started. However, they were unable to detect these differences when the patterns were projected onto other regions of the retina. Recordings of single neurons in the visual association cortex showed that the response properties of neurons that received information from the "trained" region of the retina—but not from other regions—had become sensitive to small differences in the stimuli. Clearly, neural circuits in that region alone had been modified by the training.

Let's look at some evidence from studies with humans that supports the conclusion that activation of neural circuits in the sensory association cortex constitutes the "readout" of a perceptual memory. Many years ago, Penfield and Perot (1963) discovered that when they stimulated the visual and auditory association cortex as patients were undergoing seizure surgery, the patients reported memories of images or sounds—for example, images of a familiar street or the sound of the patient's mother's voice. (You will recall from the opening case in Chapter 3 that seizure surgery is performed under a local anesthetic so that the surgeons can test the effects of brain stimulation on the patients' cognitive functions.)

Damage to regions of the brain involved in visual perception not only impair the ability to recognize visual stimuli but also disrupt people's memory of the visual properties of familiar stimuli. For example, Vandenbulcke et al. (2006) found that Patient J. A., who had sustained damage to the right fusiform gyrus, performed poorly on tasks that required her to draw or describe visual features of various animals, fruits, vegetables, tools, vehicles, or pieces of furniture. Her other cognitive abilities, including the ability to describe nonvisual attributes of objects, were normal. In addition, an fMRI study found that when normal control subjects were asked to perform the visual tasks that she performed poorly, activation was seen in the region of their brains that corresponded to J. A.'s lesion.

Kourtzi and Kanwisher (2000) found that specific kinds of visual information can activate very specific regions of visual association cortex. As we saw in Chapter 6, a region of the visual association cortex, MT/MST, plays an essential role in perception of movement. The investigators presented subjects with photographs that implied motion—for example, an athlete getting ready to throw a ball. They found that photographs like these, but not photographs of people remaining still, activated area MT/MST. Obviously, the

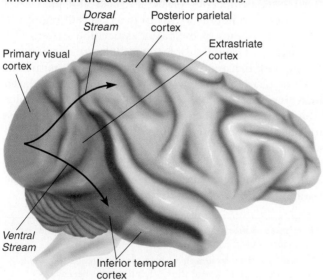

FIGURE 13.18 ■ The Major Divisions of the Visual Cortex of the Rhesus Monkey

The arrows indicate the primary direction of the flow of information in the dorsal and ventral streams.

Dorsal Stream

Posterior parietal cortex

Extrastriate cortex

Primary visual cortex

Ventral Stream

Inferior temporal cortex

FIGURE 13.19 ■ **Evidence of Retrieval of Visual Memories of Movement**

The bars represent the level of activation, measured by fMRI, of MT/MST, a region of the visual association cortex that responds to movement. Subjects looked at photographs of static scenes or scenes that implied motion similar to the ones shown here.

(Adapted from Kourtzi, A. and Kanwisher, N. *Journal of Cognitive Neuroscience*, 2000, *12*, 48–55.)

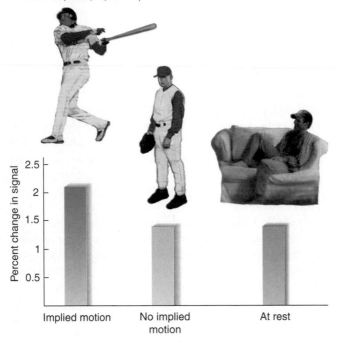

photographs did not move, but presumably, the subjects' memories contained information about movements they had previously seen. (See *Figure 13.19*.)

A functional-imaging study by Goldberg, Perfetti, and Schneider (2006) asked people questions that involved visual, auditory, tactile, and gustatory information. The researchers found that answering the questions activated the regions of association cortex involved in perception of the relevant sensory information. For example, questions about flavor activated the gustatory cortex, questions about tactile information activated the somatosensory cortex, and questions about visual and auditory information activated the visual and auditory association cortex.

Perceptual Short-Term Memory

So far, all the studies I have mentioned involved recognition of stimuli, either particular objects or their locations. Often, recognition is all that is necessary: We see a stimulus and immediately make the appropriate response. But sometimes the situation demands that we make the appropriate response after a delay, even after the stimulus is no longer visible. For example, suppose that we have driven into a large parking lot, and because

we will have to carry a heavy package, we want to park as near as possible to the entrance of a store located just in front of us. We look to the left and see a space about 100 feet away. We then look to the right and see a space about 50 feet away. Mentally comparing the distances, we turn to the right. Because we could not look in both directions simultaneously, we had to compare the distance to the second space with our memory of the distance to the first one. In other words, we had to compare a perception with a short-term memory of something else we had just perceived. A **short-term memory** is the memory of a stimulus or an event that lasts for a short while—usually on the order of a few seconds.

As we just saw, learning to recognize a stimulus involves synaptic changes in the appropriate regions of the sensory association cortex that establish new circuits of neurons. *Recognition* of a stimulus occurs when sensory input activates these established sets of neural circuits. Short-term memory of a stimulus involves activity of these circuits—or other circuits that are activated by them—that continues even after the stimulus disappears. For example, *learning* to recognize a friend's face produces changes in synaptic strengths in neural circuits in the fusiform face region of our visual association cortex, *recognizing* that she is present involves activation of the circuits that are established by these changes, and *remembering* that she is still in the room even when we look elsewhere involves continued activity of these circuits (or other circuits connected to them).

Functional-imaging studies have shown that retention of specific types of short-term visual memories involves activity of specific regions of the visual association cortex. One region of the ventral stream, the *fusiform face area,* is involved in recognition of faces, and another region, the *parahippocampal place area,* is involved in recognition of places. A functional-imaging study by Ranganath, DeGutis, and D'Esposito (2004) found evidence that short-term memory for particular faces and places was associated with neural activity in two different regions of the ventral stream of the visual association cortex. The investigators trained people on a delayed matching-to-sample task that required them to remember particular faces or places for a short period of time. In a **delayed matching-to-sample task,** a subject is shown a stimulus (the sample), and then, after a delay, the subject must indicate which of several alternatives is the same as the sample. Ranganath and his colleagues found that short-term memories of faces activated the fusiform face area and that short-term memories of places activated the parahippocampal place area. (See *Figure 13.20.*)

short-term memory Memory of a stimulus or an event that lasts for a short while.

delayed matching-to-sample task A task that requires the subject to indicate which of several stimuli has just been perceived.

FIGURE 13.20 ■ Short-Term Perceptual Memory

The fusiform face area and parahippocampal place area are activated by information about faces or places in short-term memory during cue and delay periods of a delayed matching-to-sample task.

(Adapted from Ranganath, C., DeGutis, J., and D'Esposito, M. *Cognitive Brain Research*, 2004, *20*, 37–45.)

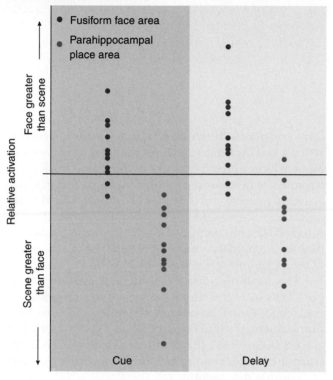

As we saw in Chapter 6, transcranial magnetic stimulation (TMS) of the visual association cortex interferes with visual perception. TMS induces a weak electrical current in the brain that disrupts neural activity and thus interferes with the normal functions of the stimulated region. Oliveri et al. (2001) trained people on a delayed matching-to-sample task that required them to remember either abstract figures or the locations of a white square on a video screen. On some trials the investigators applied TMS to the association cortex of either the ventral stream or the dorsal stream during the delay interval, after the sample stimuli had been turned off. They found that stimulating the ventral stream interfered with short-term memory for visual patterns and stimulating the dorsal stream interfered with short-term memory for location.

Although the neural circuits responsible for learning to recognize particular stimuli appear to reside in the sensory association cortex, perceptual short-term memories involve other brain regions as well—especially the prefrontal cortex. Miyashita (2004) suggests that the role of the prefrontal cortex in short-term memory is to "manipulate and organize to-be-remembered information, devise strategies for retrieval, and also monitor the outcome" of these processes.

An example of this role was seen in a functional-imaging study by Blumenfeld and Ranganarh (2006). The investigators presented subjects with groups of three words arranged vertically. The words were names of animals or tangible objects, such as *owl, pillow,* and *skunk.* Above each set of three words was a heading that said REHEARSE or REORDER. In the REHEARSE condition the subjects attempted to remember the words by simply rehearsing them subvocally—silently saying the words to themselves. In the REORDER condition the subjects were told to rearrange the three words according to the relative weights of the items they denoted. For example, if the three words were "spider, tank, jar," they should remember them as "spider, jar, tank." After a delay, one of the words that had just been seen was presented along with a number, and the subjects had to indicate whether or not the number indicated the location of the word in the sequence. For example, "tank" would be in position 2 after the words "spider, tank, jar" had been rearranged according to weight.

Blumenfeld and Ranganarh found that the dorsolateral prefrontal cortex was activated during REORDER trials. In fact, when the subjects were tested later, after they left the scanner, they were most likely to remember words from REORDER trials that were accompanied by the greatest amount of activity in this brain region.

Interim Summary

Perceptual Learning

Perceptual learning occurs as a result of changes in synaptic connections within the sensory association cortex. Damage to the inferior temporal cortex—the highest level of the ventral stream of the visual association cortex—disrupts visual perceptual learning. Functional-imaging studies with humans have shown that retrieval of memories of pictures, sounds, movements, or spatial locations activates the appropriate regions of the sensory association cortex.

Perceptual short-term memory involves sustained activity of neurons in the sensory association cortex. Functional-imaging studies have shown that retention of specific types of short-term visual memories involves activity of specific regions of the visual association cortex. Transcranial magnetic stimulation of various regions of the human sensory association cortex disrupt short-term perceptual memories. The prefrontal cortex is also involved in short-term memory. This region encodes information pertaining to the stimulus that must be remembered and is involved in manipulating and organizing information in short-term memory.

Thought Questions

1. How many perceptual memories does your brain hold? How many images, sounds, and odors can you recognize, and how many objects and surfaces can you recognize by touch? Is there any way we could estimate these quantities?

2. Can you think of times when you saw something that you needed to remember and did so by keeping in mind a response you would need to make rather than an image of the stimulus you just perceived?

CLASSICAL CONDITIONING

Neuroscientists have studied the anatomy and physiology of classical conditioning using many models, such as the gill withdrawal reflex in *Aplysia* (a marine invertebrate) and the eyeblink reflex in the rabbit (Carew, 1989; Lavond, Kim, and Thompson, 1993). I have chosen to describe a simple mammalian model of classical conditioning—the conditioned emotional response—to illustrate the results of such investigations.

The amygdala is part of an important system involved in a particular form of stimulus-response learning: classically conditioned emotional responses. An aversive stimulus such as a painful foot shock produces a variety of behavioral, autonomic, and hormonal responses: freezing, increased blood pressure, secretion of adrenal stress hormones, and so on. A classically conditioned emotional response is established by pairing a neutral stimulus (such as a tone of a particular frequency) with an aversive stimulus (such as a brief foot shock). As we saw in Chapter 11, after these stimuli are paired, the tone becomes a CS; when it is presented by itself, it elicits the same type of responses as the unconditional stimulus does.

A conditioned emotional response can occur in the absence of the auditory cortex (LeDoux et al., 1984); thus, I will confine my discussion to the subcortical components of this process. Information about the CS (the tone) reaches the lateral nucleus of the amygdala. This nucleus also receives information about the US (the foot shock) from the somatosensory system. Thus, these two sources of information converge in the lateral nucleus, which means that synaptic changes responsible for learning could take place in this location.

A hypothetical neural circuit is shown in Figure 13.21. The lateral nucleus of the amygdala contains neurons whose axons project to the central nucleus. Terminal buttons from neurons that transmit auditory and somatosensory information to the lateral nucleus

form synapses with dendritic spines on these neurons. When a rat encounters a painful stimulus, somatosensory input activates strong synapses in the lateral nucleus. As a result, the neurons in this nucleus begin firing, which activates neurons in the central nucleus, evoking an unlearned (unconditional) emotional response. If a tone is paired with the painful stimulus, the weak synapses in the lateral amygdala are strengthened through the action of the Hebb rule. (See *Figure 13.21*.)

This hypothesis has a considerable amount of support. Lesions of the lateral nucleus of the amygdala disrupt conditioned emotional responses that involve a simple auditory stimulus as a CS and a shock to the feet as a US (Kapp et al., 1979; Nader et al., 2001). Thus, the synaptic changes responsible for this learning may take place within this circuit.

FIGURE 13.21 ■ Conditioned Emotional Responses

The figure shows the probable location of the changes in synaptic strength produced by the classically conditioned emotional response that results from pairing a tone with a foot shock.

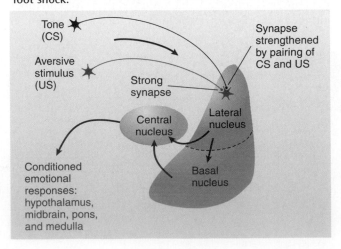

FIGURE 13.22 ■ Classical Conditioning in the Lateral Amygdala

The graph shows the change in rate of firing of neurons in the lateral amygdala in response to the tone, relative to baseline levels.

(Adapted from Quirk, G. J., Repa, J. C., and LeDoux, J. E. *Neuron*, 1995, *15*, 1029–1039.)

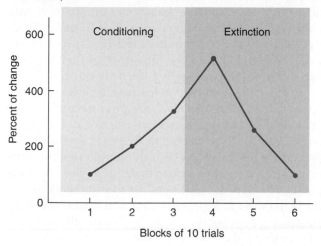

Quirk, Repa, and LeDoux (1995) found evidence for synaptic changes in the lateral nucleus of the amygdala. They recorded the activity of neurons in this nucleus in freely moving rats before, during, and after pairing of a tone with a foot shock. Within a few trials the neurons became more responsive to the tone, and many neurons that had not previously responded to the tone began doing so. When they repeatedly presented the tone without the foot shock, the response extinguished, and the rate of firing of the neurons in the lateral nucleus returned to baseline levels. (See *Figure 13.22*.) Maren (2000) confirmed these results and also found that the magnitude of the increased firing rate of neurons in the lateral nucleus correlated with the magnitude of the conditioned emotional response.

The evidence from many studies indicates that the changes in the lateral amygdala responsible for acquisition of a conditioned emotional response involve LTP. LTP in many parts of the brain—including the amygdala—is accomplished through the activation of NMDA receptors. Rodrigues, Schafe, and LeDoux (2001) used a drug that blocks the NR2B subunit of the NMDA receptor. The investigators found that infusion of this drug into the lateral amygdala prevented the acquisition of a conditioned emotional response. Injections of drugs that block LTP into the amygdala prevent the establishment of conditioned emotional responses.

Rumpel et al. (2005) used a harmless virus to insert a gene for a fluorescent dye coupled to a subunit of the AMPA receptor into the lateral amygdala of rats. They paired a tone with a shock and established a conditioned emotional response. They found that the learning experience caused AMPA receptors to be driven into dendritic spines of synapses between lateral amygdala neurons and axons that provide auditory input. The investigators also inserted a gene for a fluorescent dye coupled with a defective subunit of the AMPA receptor into the lateral amygdala. The defective subunit prevented AMPA receptors from being driven into the dendritic spines. As a result, conditioning did not take place. In fact, infusion of a wide variety of drugs into the lateral amygdala that prevent long-term potentiation in this nucleus disrupt acquisition of a conditioned emotional response (Rodrigues, Schafe, and LeDoux, 2004; Schafe et al., 2005; Schafe, Doyère, and LeDoux, 2005). The results of these studies support the conclusion that LTP in the lateral amygdala, mediated by NMDA receptors, plays a critical role in the establishment of conditioned emotional responses.

InterimSummary

Classical Conditioning

You have already encountered the conditioned emotional response in Chapter 11 and in the previous section of this chapter, in which I discussed perceptual learning. When an auditory stimulus (CS) is paired with a foot shock (US), the two types of information converge in the lateral nucleus of the amygdala. This nucleus is connected, directly and via the basal nucleus and accessory basal nucleus, with the central nucleus, which is connected with brain regions that control various components of the emotional response. Lesions anywhere in this circuit disrupt the response.

Recordings of single neurons in the lateral nucleus of the amygdala indicate that classical conditioning changes the response of neurons to the CS. The mechanism of synaptic plasticity in this system appears to be NMDA-mediated long-term potentiation. Infusion of drugs that block LTP into the lateral nucleus blocks establishment of conditioned emotional responses.

INSTRUMENTAL CONDITIONING

Instrumental (operant) conditioning is the means by which we (and other animals) profit from experience. If, in a particular situation, we make a response that has favorable outcomes, we will tend to make the response again. This section first describes the neural pathways involved in instrumental conditioning and then discusses the neural basis of reinforcement.

Basal Ganglia

As we saw earlier in this chapter, instrumental conditioning entails the strengthening of connections between neural circuits that detect a particular stimulus and neural circuits that produce a particular response. Clearly, the circuits that are responsible for instrumental conditioning begin in various regions of the sensory association cortex, where perception takes place, and end in the motor association cortex of the frontal lobe, which controls movements. But what pathways are responsible for these connections, and where do the synaptic changes responsible for the learning take place?

There are two major pathways between the sensory association cortex and the motor association cortex: direct transcortical connections (connections from one area of the cerebral cortex to another) and connections via the basal ganglia and thalamus. (A third pathway, involving the cerebellum and thalamus, also exists, but the role of this pathway in instrumental conditioning has until very recently received little attention from neuroscientists.) Both of these pathways appear to be involved in instrumental conditioning, but they play different roles.

In conjunction with the hippocampal formation, the transcortical connections are involved in the acquisition of episodic memories—complex perceptual memories of sequences of events that we witness or that are described to us. (The acquisition of these types of memories is discussed in the last section of this chapter.) The transcortical connections are also involved in the acquisition of complex behaviors that involve deliberation or instruction. For example, a person learning to drive a car with a manual transmission might say, "Let's see, push in the clutch, move the shift lever to the left and then away from me—there, it's in gear—now let the clutch come up—oh! It died—I should have given it more gas. Let's see, clutch down, turn the key. . . ." A memorized set of rules (or an instructor sitting next to us) provides a script for us to follow. Of course, this process does not have to be audible or even involve actual movements of the speech muscles; a person can think in words with neural activity that does not result in overt behavior. (Animals that cannot communicate by means of language can acquire complex responses by observing and imitating the behavior of other animals.)

At first, performing a behavior through observation or by following a set of rules is slow and awkward. And because so much of the brain's resources are involved in recalling the rules and applying them to our behavior, we cannot respond to other stimuli in the environment—we must ignore events that might distract us. But then, with practice, the behavior becomes much more fluid. Eventually, we perform it without thinking and can easily do other things at the same time, such as carrying on a conversation with passengers as we drive our car.

Evidence suggests that as learned behaviors become automatic and routine, they are "transferred" to the basal ganglia. The process seems to work like this: As we deliberately perform a complex behavior, the basal ganglia receive information about the stimuli that are present and the responses we are making. At first the basal ganglia are passive "observers" of the situation, but as the behaviors are repeated again and again, the basal ganglia begin to learn what to do. Eventually, they take over most of the details of the process, leaving the transcortical circuits free to do something else. We need no longer think about what we are doing.

The neostriatum—the caudate nucleus and the putamen—receives sensory information from all regions of the cerebral cortex. It also receives information from the frontal lobes about movements that are planned or are actually in progress. (So as you can see, the basal ganglia have all the information they need to monitor the progress of someone learning to drive a car.) The outputs of the caudate nucleus and the putamen are sent to another part of the basal ganglia: the globus pallidus. The outputs of this structure are sent to the frontal cortex: to the premotor and supplementary motor cortex, where plans for movements are made, and to the primary motor cortex, where they are executed. (See *Figure 13.23*.)

Studies with laboratory animals have found that lesions of the basal ganglia disrupt instrumental conditioning but do not affect other forms of learning. For example, Fernandez-Ruiz et al. (2001) destroyed the portions of the caudate nucleus and putamen that receive visual information from the ventral stream. They found that although the lesions did not disrupt visual perceptual learning, they impaired the monkeys' ability to learn to make a visually guided operant response.

Williams and Eskandar (2006) trained monkeys to move a joystick in a particular direction (left, right, forward, or backward) when they saw a particular visual stimulus. Correct responses were reinforced with a sip of fruit juice. As the monkeys learned the task, the rate of firing of single neurons in the caudate nucleus increased. In fact, the activity of caudate neurons was correlated with the animals' rate of learning. When the

stimulation of :
the mesolimbi
ment systems.

Functions c

A reinforceme
detect the pre
recognize that
strengthen the
detect the disc
lever) and the
response (a lev

Assuming
several questio
gic neurons in
tons to release
dopamine play
Where do thes
that suggests s
tions is discusse

Detecting R
occurs when ne
and cause the a
ventral tegmen
lus is not a simp
forcer on one o
example, the p
ior of a hungry
just eaten. Thu
matically activa
its activation als

Studies by S
activity of dopa
bens, have disc
appears to be a
For example,
taught monkeys
make a respons
During training
responded rapid
liquid) was deliv
the task, the VT
tory stimulus wa
stimulus was del
ulus does not o
dopaminergic n
2007). A functic
found similar re
that when a sma
in people's mou
bens was activate
predictable, no s

Schultz and
the dopaminerg

FIGURE 13.23 ■ The Basal Ganglia and Their Connections

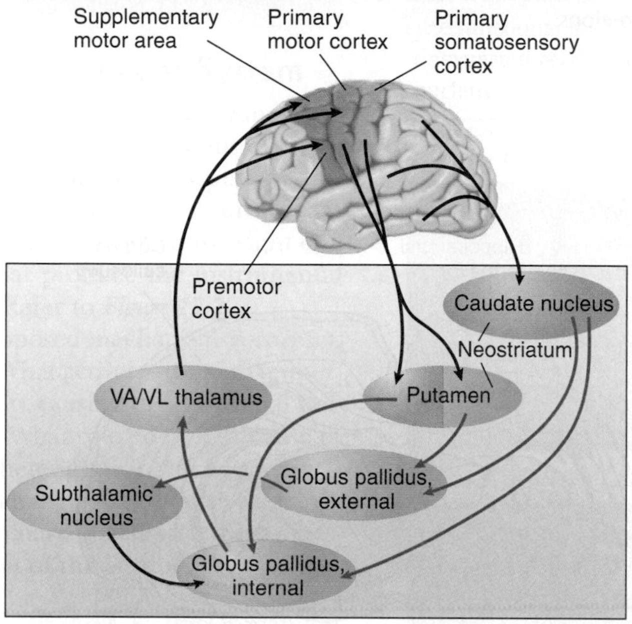

investigators increased the activation of caudate neurons through low-intensity, high-frequency electrical stimulation during the reinforcement period, they monkeys learned a particular stimulus-response association more quickly. These results provide further evidence for the role of the basal ganglia in instrumental conditioning.

As we saw in the previous section, long-term potentiation appears to play a critical role in classical conditioning. This form of synaptic plasticity appears to be involved in instrumental conditioning, as well. Packard and Teather (1997) found that blocking NMDA receptors in the basal ganglia with an injection of AP5 disrupted learning guided by a simple visual cue.

Reinforcement

Learning provides a means for us to profit from experience—to make responses that provide favorable outcomes. When good things happen (that is, when reinforcing stimuli occur), reinforcement mechanisms in the brain become active, and the establishment of synaptic changes is facilitated. The discovery of the existence of such reinforcement mechanisms occurred by accident.

Neural Circuits Involved in Reinforcement

In 1954, James Olds, a young assistant professor, and Peter Milner, a graduate student, attempted to determine whether electrical stimulation of the reticular formation would facilitate maze learning in rats. They planned to turn on the stimulator briefly each time an animal reached a choice point in the maze. First, however, they had to be certain that the stimulation was not aversive, because an aversive stimulus would undoubtedly interfere with learning. As Olds reported,

> I applied a brief train of 60-cycle sine-wave electrical current whenever the animal entered one corner of the enclosure. The animal did not stay away from that corner, but rather came back quickly after a brief sortie which followed the first stimulation and came back even more quickly after a briefer sortie which followed the second stimulation. By the time the third electrical stimulus had been applied the animal seemed indubitably to be "coming back for more." (Olds, 1973, p. 81)

Realizing that they were on to something big, Olds and Milner decided to drop their original experiment and study the phenomenon they had discovered. Subsequent research discovered that although there are several different reinforcement mechanisms, the activity of dopaminergic neurons plays a particularly important role in reinforcement. As we saw in Chapter 4, the mesolimbic system of dopaminergic neurons begins in the **ventral tegmental area (VTA)** of the midbrain and projects rostrally to several forebrain regions, including the amygdala, hippocampus, and **nucleus accumbens (NAC).** This nucleus is located in the basal forebrain rostral to the preoptic area and immediately adjacent to the septum. (In fact, the full name of this region is the *nucleus accumbens septi,* or "nucleus leaning against the septum.") (See *Figure 13.24.*) Neurons in the NAC project to the ventral part of the basal ganglia, which, as we just saw, are involved in learning. The mesocortical system also plays a role in reinforcement. This system also begins in the ventral tegmental area but projects to the prefrontal cortex, the limbic cortex, and the hippocampus.

Chapter 5 described a research technique called *microdialysis,* which enables an investigator to analyze the contents of the interstitial fluid within a specific region of the brain. Researchers using this method have shown that reinforcing electrical stimulation of the medial forebrain bundle or the ventral tegmental area or the administration of cocaine or amphetamine causes the release of dopamine in the nucleus accumbens (Moghaddam and Bunney, 1989; Nakahara et al., 1989; Phillips et al., 1992). (The medial forebrain bundle

ventral tegmental area (VTA) A group of dopaminergic neurons in the ventral midbrain whose axons form the mesolimbic and mesocortical systems; plays a critical role in reinforcement.

nucleus accumbens A nucleus of the basal forebrain near the septum; receives dopamine-secreting terminal buttons from neurons of the ventral tegmental area and is thought to be involved in reinforcement and attention.

FIGURE 13.2
Nucleus Accu

Diagrams of section

(Adapted from Swanson

Sep
area

Anterior
commissure

connects the ven
accumbens. See *F*
also found that th
as water, food, or
dopamine in the
of reinforcing br
many ways to tho

Although mi
brain of humans
imaging studies h
vate the human
Knutson et al. (20
became more ac
being released th
stimuli that ind
money. Aharon e
sexual men woul
of beautiful wom
when they saw th
accumbens incre

I should note
that aversive stim
cause the release
brain, including
1992). Thus, it is
function of do
appear to be inv

(the opportunity to win some money) increased the activation of the ventral tegmentum and some of its projection regions (including the nucleus accumbens) in humans. The investigators found that the subjects were more likely to remember pictures that they had seen while they were anticipating the chance to win some money.

As we have seen, the prefrontal cortex provides an important input to the ventral tegmental area. The terminal buttons of the axons connecting these two areas secrete glutamate, an excitatory neurotransmitter, and the activity of these synapses makes dopaminergic neurons in the ventral tegmental area fire in a bursting pattern, which greatly increases the amount of dopamine they secrete in the nucleus accumbens (Gariano and Groves, 1988). The prefrontal cortex is generally involved in devising strategies, making plans, evaluating progress made toward goals, and judging the appropriateness of one's own behavior. Perhaps the prefrontal cortex turns on the reinforcement mechanism when it determines that the ongoing behavior is bringing the organism nearer to its goals—that the present strategy is working.

Even private behaviors such as thinking and planning may be subject to reinforcement. For example, recall the last time you were thinking about a problem and suddenly had an idea that might help you to solve it. Did you suddenly feel excited and happy? It would be interesting if we could record the activity of the axons leading from your frontal cortex to your ventral tegmental area at times like that.

Strengthening Neural Connections: Dopamine and Neural Plasticity.

Like classical conditioning, instrumental conditioning involves strengthening of synapses located on neurons that have just been active. However, instrumental conditioning involves three elements: a discriminative stimulus, a response, and a reinforcing stimulus. How are the neural manifestations of these three elements combined?

Let's consider a hungry rat learning to press a lever and obtain food. As in classical conditioning, one element (the discriminative stimulus—in this case the sight of the lever) activates only weak synapses on motor neurons responsible for a movement that causes a lever press. The second element—the particular circumstance that happened to induce the animal to press the lever—activates strong synapses, making the neurons fire. The third element comes into play only if the response is followed by a reinforcing stimulus. If it is, the reinforcement mechanism triggers the secretion of a neurotransmitter or neuromodulator throughout the region in which the synaptic changes take place. This chemical is the third element; only if it is present can weak synapses be strengthened. Dopamine serves such a role. Several studies have shown that long-term potentiation

is essential for instrumental conditioning and that dopamine is an essential ingredient in long-lasting long-term potentiation.

Smith-Roe and Kelley (2000) found that the presence of dopamine and the activation of NMDA receptors in the nucleus accumbens both appear to be necessary for instrumental conditioning to take place. They found that a low dose of a dopamine D1 receptor antagonist or a low dose of AP5 into the nucleus accumbens had no effect on rats' ability to learn a lever-pressing task. However, simultaneous infusion of the same doses of the two drugs severely impaired the animals' ability to learn this task. Knecht et al. (2004) taught people a vocabulary of artificial words. The learning took place gradually, during five daily sessions. In a double-blind procedure, some subjects were given L-DOPA 90 minutes before each session, and others were given a placebo. (As you know, L-DOPA is the precursor for dopamine, and administration of this drug increases the release of dopamine in the brain.) The subjects who received the L-DOPA learned the artificial vocabulary faster and remembered it better than those who received the placebo.

As I mentioned earlier, the prefrontal cortex may activate the reinforcement system when it detects that the animal's behavior is resulting in progress toward a goal. But the prefrontal cortex is a *target* of dopaminergic neurons as well as a source of their control. For example, Stein and Belluzzi (1989) found that rats will press a lever that produces an injection of a dopamine agonist into this region. Duvauchelle and Ettenberg (1991) found that if a rat's prefrontal cortex is electrically stimulated while the animal is in a particular location, the animal will learn to prefer that location to others where the stimulation did not take place. This learning appears to involve the release of dopamine, because it is prevented by injections of a drug that blocks dopamine receptors. And in a microdialysis study, Hernandez and Hoebel (1990) found that when rats were performing a food-reinforced lever-pressing task, the levels of dopamine in the prefrontal cortex increased.

Dopamine modulates LTP in the prefrontal cortex as well as in the nucleus accumbens. Gurden, Tassin, and Jay (1999) found that stimulation of the VTA enhanced LTP in the prefrontal cortex produced by electrical stimulation of the hippocampus. Gurden, Takita, and Jay (2000) found that infusion of D1 receptor agonists into the prefrontal cortex did so as well but that D1 antagonists impaired LTP. A study by Bissière, Humeau, and Luthi (2003) found that dopamine facilitates LTP in the lateral amygdala as well. These experiments provide further evidence that dopamine plays a modulating role in synaptic plasticity in parts of the brain that are involved in learning.

InterimSummary

Instrumental Conditioning

Instrumental conditioning entails the strengthening of connections between neural circuits that detect stimuli and neural circuits that produce responses. One of the locations of these changes appears to be the basal ganglia, especially the changes responsible for learning of automated and routine behaviors. The basal ganglia receive sensory information and information about plans for movement from the neocortex. Instrumental conditioning activates the basal ganglia, and damage to the basal ganglia or infusion of a drug that blocks NMDA receptors there disrupts instrumental conditioning.

Olds and Milner discovered that rats would perform a response that caused electrical current to be delivered through an electrode placed in the brain; thus, the stimulation was reinforcing. Subsequent studies found that stimulation of many locations had reinforcing effects but that the medial forebrain bundle produced the strongest and most reliable ones.

Although several neurotransmitters may play a role in reinforcement, one is particularly important: dopamine. The cell bodies of the most important system of dopaminergic neurons are located in the ventral tegmental area, and their axons project to the nucleus accumbens, prefrontal cortex, and amygdala.

Microdialysis studies have also shown that natural and artificial reinforcers stimulate the release of dopamine in the nucleus accumbens, and functional-imaging studies have shown that reinforcing stimuli activate the nucleus accumbens in humans. The dopaminergic reinforcement system appears to be activated by unexpected reinforcers or by stimuli that predict the occurrence of a reinforcer. Conditions such as novelty or the expectation of a reinforcing stimulus facilitate learning. The prefrontal cortex may play a role in reinforcement that occurs when our own behavior brings us nearer to a goal.

Dopamine induces synaptic plasticity by facilitating associative long-term potentiation. Evidence indicates that dopamine can facilitate long-term potentiation in the nucleus accumbens, amygdala, and prefrontal cortex.

Thought Question

Have you ever been working hard on a problem and suddenly thought of a possible solution? Did the thought make you feel excited and happy? What would we find if we had a microdialysis probe in your nucleus accumbens?

RELATIONAL LEARNING

So far, this chapter has discussed relatively simple forms of learning, which can be understood as changes in circuits of neurons that detect the presence of particular stimuli or as strengthened connections between neurons that analyze sensory information and those that produce responses. But most forms of learning are more complex; most memories of real objects and events are related to other memories. Seeing a photograph of an old friend may remind you of the sound of the person's name and of the movements you have to make to pronounce it. You may also be reminded of things you have done with your friend: places you have visited, conversations you have had, experiences you have shared. Each of these memories can contain a series of events, complete with sights and sounds, that you will be able to recall in the proper sequence. Obviously, the neural circuits in the visual association cortex that recognize your friend's face are connected to circuits in many other parts of the brain, and these circuits are connected to many others. This section discusses research on relational learning, which includes the establishment and retrieval of memories of events, episodes, and places.

Human Anterograde Amnesia

One of the most dramatic and intriguing phenomena caused by brain damage is *anterograde amnesia*, which, at first glance, appears to be the inability to learn new information. However, when we examine the phenomenon more carefully, we find that the basic abilities of perceptual learning, stimulus-response learning, and motor learning are intact but that complex relational learning, of the type I just described, is gone. This section discusses the nature of anterograde amnesia in humans and its anatomical basis. The section that follows discusses related research with laboratory animals.

The term **anterograde amnesia** refers to difficulty in learning new information. A person with pure anterograde amnesia can remember events that occurred in the past, from the time before the brain damage occurred, but cannot retain information encountered

anterograde amnesia Amnesia for events that occur after some disturbance to the brain, such as head injury or certain degenerative brain diseases.

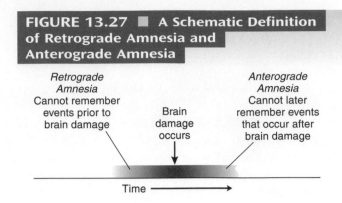

FIGURE 13.27 ■ A Schematic Definition of Retrograde Amnesia and Anterograde Amnesia

after the damage. In contrast, **retrograde amnesia** refers to the inability to remember events that happened *before* the brain damage occurred. (See *Figure 13.27.*) As we will see, pure anterograde amnesia is rare; usually, there is also a retrograde amnesia for events that occurred for a period of time before the brain damage occurred.

In 1889, Sergei Korsakoff, a Russian physician, first described a severe memory impairment caused by brain damage, and the disorder was given his name. The most profound symptom of **Korsakoff's syndrome** is a severe anterograde amnesia: The patients appear to be unable to form new memories, although they can still remember old ones. They can converse normally and can remember events that happened long before their brain damage occurred, but they cannot remember events that happened afterward. As we will see in Chapter 15, the brain damage that causes Korsakoff's syndrome is usually (but not always) a result of chronic alcohol abuse.

Anterograde amnesia can also be caused by damage to the temporal lobes. Scoville and Milner (1957) reported that bilateral removal of the medial temporal lobe produced a memory impairment in humans that was apparently identical to that seen in Korsakoff's syndrome. H. M., the man described in the case that opened this chapter, received the surgery in an attempt to treat his severe epilepsy, which could not be controlled even by high doses of anticonvulsant medication. The epilepsy appears to have been caused by a head injury he received when he was struck by a bicycle at age nine (Corkin et al., 1997).

The surgery successfully treated H. M.'s seizure disorder, but it became apparent that the operation had produced a serious memory impairment. Further investigation revealed that the critical site of damage was the hippocampus. Once it was known that bilateral medial temporal lobectomy causes anterograde amnesia, neurosurgeons stopped performing this operation and are now careful to operate on only one temporal lobe.

H. M.'s history and memory deficits were described in the introduction to this chapter (Milner, Corkin, and Teuber, 1968; Milner, 1970; Corkin et al., 1981). Because of his relatively pure amnesia, he has been extensively studied. Milner and her colleagues based the following conclusions on his pattern of deficits:

1. *The hippocampus is not the location of long-term memories; nor is it necessary for the retrieval of long-term memories.* If it were, H. M. would not have been able to remember events from early in his life, he would not know how to talk, he would not know how to dress himself, and so on.
2. *The hippocampus is not the location of immediate (short-term) memories.* If it were, H. M. would not be able to carry on a conversation, because he would not remember what the other person said long enough to think of a reply.
3. *The hippocampus is involved in converting immediate (short-term) memories into long-term memories.* This conclusion is based on a particular hypothesis of memory function: that our immediate memory of an event is retained by neural activity and that long-term memories consist of relatively permanent biochemical or structural changes in neurons. The conclusion seems a reasonable explanation for the fact that when presented with new information, H. M. seems to understand it and remember it as long as he thinks about it but that a permanent record of the information is just never made.

As we will see, these three conclusions are too simple. Subsequent research on patients with anterograde amnesia indicates that the facts are more complicated—and more interesting—than they first appeared to be. But to appreciate the significance of the findings of more recent research, we must understand these three conclusions and remember the facts that led to them.

As we saw earlier in this chapter, most psychologists believe that learning consists of at least two stages: short-term memory and long-term memory. They conceive of short-term memory as a means of storing a limited amount of information temporarily and long-term memory as a means of storing an unlimited amount (or at least an enormously large amount) of information permanently. We can remember a new item of information (such as a telephone number) for as long as we want by engaging in a particular behavior: rehearsal. However, once we stop rehearsing the information, we might or might not be able to remember it later; that is, the information might or might not get stored in long-term memory.

retrograde amnesia Amnesia for events that preceded some disturbance to the brain, such as a head injury or electroconvulsive shock.

Korsakoff's syndrome Permanent anterograde amnesia caused by brain damage resulting from chronic alcoholism or malnutrition.

FIGURE 13.28 ■ **A Simple Model of the Learning Process**

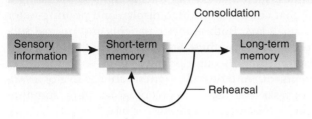

The simplest model of the memory process says that sensory information enters short-term memory, rehearsal keeps it there, and eventually, the information makes its way into long-term memory, where it is permanently stored. The conversion of short-term memories into long-term memories has been called **consolidation,** because the memories are "made solid," so to speak. (See *Figure 13.28.*)

Now you can understand the original conclusions of Milner and her colleagues: If H. M.'s short-term memory is intact and if he can remember events from before his operation, then the problem must be that consolidation does not take place. Thus, the role of the hippocampal formation in memory is consolidation—converting short-term memories to long-term memories.

Spared Learning Abilities

H. M.'s memory deficit is striking and dramatic. However, when he and other patients with anterograde amnesia are studied more carefully, it becomes apparent that the amnesia does not represent a total failure in learning ability. When the patients are appropriately trained and tested, we find that they are capable of three of the four major types of learning described earlier in this chapter: perceptual learning, stimulus-response learning, and motor learning. A review by Spiers, Maguire, and Burgess (2001) summarized 147 cases of anterograde amnesia that are consistent with the description that follows.

First, let us consider perceptual learning. Figure 13.29. shows two sample items from a test of the ability to recognize broken drawings; note how the drawings are successively more complete. (See *Figure 13.29.*) Subjects are first shown the least complete set (set I) of each of twenty different drawings. If they do not recognize a figure (and most people do not recognize set I), they are shown more complete sets until they identify it. One hour later, the subjects are tested again for retention, starting with set I. When H. M. was given this test and was retested an hour later, he showed considerable improvement (Milner, 1970). When he was retested four months later, he *still* showed this improvement. His performance was not

FIGURE 13.29 ■ **Examples of Broken Drawings**

(Reprinted with permission of author and publisher from Gollin, E. S. Developmental studies of visual recognition of incomplete objects. *Perceptual and Motor Skills,* 1960, 11, 289–298.)

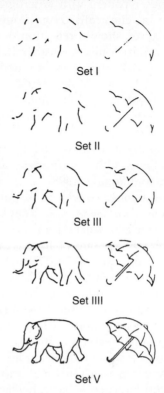

as good as that of normal control subjects, but he showed unmistakable evidence of long-term retention. (You can try the broken drawing task and some other tasks that people with anterograde amnesia can successfully learn by running *MyPsychKit 13.4, Implicit Memory Tasks.*)

mypsychkit
Animation 13.4
Implicit Memory Tasks

Johnson, Kim, and Risse (1985) found that patients with anterograde amnesia could learn to recognize faces. The researchers played unfamiliar melodies from Korean songs to amnesic patients and found that when they were tested later, the patients preferred these melodies to ones they had not heard before. The experimenters also presented photographs of two men along with stories of their lives: One man was said to be dishonest, mean, and vicious; the other was said to be nice enough to invite home to dinner. (Half of the patients heard that one of the men was the bad one, and the other half heard that the other man was.) Twenty days later, the amnesic patients

consolidation The process by which short-term memories are converted into long-term memories.

said they liked the picture of the "nice" man better than that of the "nasty" one.

Investigators have also succeeded in demonstrating stimulus-response learning by H. M. and other amnesic subjects. For example, Woodruff-Pak (1993) found that H. M. and another patient with anterograde amnesia could acquire a classically conditioned eyeblink response. H. M. even showed retention of the task two years later: He acquired the response again in one-tenth the number of trials that were needed previously. Sidman, Stoddard, and Mohr (1968) successfully trained patient H. M. on an instrumental conditioning task—a visual discrimination task in which pennies were given for correct responses.

Finally, several studies have demonstrated motor learning in patients with anterograde amnesia. For example, Reber and Squire (1998) found that subjects with anterograde amnesia could learn a sequence of button presses in a *serial reaction time task*. They sat in front of a computer screen and watched an asterisk appear—apparently randomly—in one of four locations. Their task was to press the one of four buttons that corresponded to the location of the asterisk. As soon as they did so, the asterisk moved to a new location, and they pressed the corresponding button. (See *Figure 13.30*.)

Although experimenters did not say so, the sequence of button presses specified by the moving asterisk was not random. For example, it might be DBCACBDCBA, a ten-item sequence that is repeated continuously. With practice, subjects become faster and faster at this task. It is clear that their rate increases because they have learned the sequence, because if the sequence is changed, their performance decreases. The amnesic subjects learned this task just as well as normal subjects did.

FIGURE 13.30 ■ The Serial Reaction Time Task

In the procedure of the study by Reber and Squire (1998), subjects pressed the button in a sequence indicated by movement of the asterisk on the computer screen.

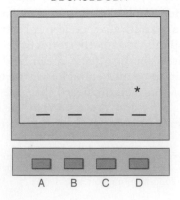

DBCACBDCBA

A study by Cavaco et al. (2004) tested amnesic patients on a variety of tasks modeled on real-world activities, such as weaving, tracing figures, operating a stick that controlled a video display, and pouring water into small jars. Both amnesic patients and normal subjects did poorly on these tasks at first, but their performance improved through practice. Thus, as you can see, patients with anterograde amnesia are capable of a variety of tasks that require perceptual learning, stimulus-response learning, and motor learning.

Declarative and Nondeclarative Memories

If amnesic patients can learn tasks like these, you might ask, why do we call them *amnesic*? The answer is this: Although the patients can learn to perform these tasks, they do not remember anything about having learned them. They do not remember the experimenters, the room in which the training took place, the apparatus that was used, or any events that occurred during the training. Although H. M. learned to recognize the broken drawings, he denied that he had ever seen them before. Although the amnesic patients in the study by Johnson, Kim, and Risse learned to like some of the Korean melodies better, they did not recognize that they had heard them before; nor did they remember having seen the pictures of the two young men. Although H. M. successfully acquired a classically conditioned eyeblink response, he did not remember the experimenter, the apparatus, or the headband he wore that held the device that delivered a puff of air to his eye.

In the experiment by Sidman, Stoddard, and Mohr, although H. M. learned to make the correct response (press a panel with a picture of a circle on it), he was unable to recall having done so. In fact, once H. M. had learned the task, the experimenters interrupted him, had him count his pennies (to distract him for a little while), and then asked him to say what he was supposed to do. He seemed puzzled by the question; he had absolutely no idea. But when they turned on the stimuli again, he immediately made the correct response. Finally, although the amnesic subjects in Reber and Squire's study obviously learned the sequence of button presses, they were completely unaware that there was, in fact, a sequence; they thought that the movement of the asterisk was random.

The distinction between what people with anterograde amnesia can and cannot learn is obviously important because it reflects the basic organization of the learning process. Clearly, there are at least two major categories of memories. Psychologists have given them several different names. For example, some investigators (Eichenbaum, Otto, and Cohen, 1992;

Squire, 1992) suggest that patients with anterograde amnesia are unable to form **declarative memories,** which have been defined as those that are "explicitly available to conscious recollection as facts, events, or specific stimuli" (Squire, Shimamura, and Amaral, 1989, p. 218). The term *declarative* obviously comes from *declare,* which means "to proclaim; to announce." The term reflects the fact that patients with anterograde amnesia cannot talk about experiences that they have had since the time of their brain damage. Thus, according to Squire and his colleagues, declarative memory is memory of events and facts that we can think and talk about.

Declarative memories are not simply verbal memories. For example, think about some event in your life, such as your last birthday. Think about where you were, when the event occurred, what other people were present, what events occurred, and so on. Although you could describe ("declare") this episode in words, the memory itself would not be verbal. In fact, it would probably be more like a video clip running in your head, one whose starting and stopping points—and fast forwards and rewinds—you could control.

The other category of memories, often called **nondeclarative memories,** includes instances of perceptual, stimulus-response, and motor learning that we are not necessarily conscious of. (Some psychologists refer to these two categories as *explicit* and *implicit* memories, respectively.) Nondeclarative memories appear to operate automatically. They do not require deliberate attempts on the part of the learner to memorize something. They do not seem to include facts or experiences; instead, they control behaviors. For example, think about when you learned to ride a bicycle. You did so quite consciously and developed declarative memories about your attempts: who helped you learn, where you rode, how you felt, how many times you fell, and so on. But you also formed nondeclarative stimulus-response and motor memories; *you learned to ride.* You learned to make automatic adjustments with your hands and body that kept your center of gravity above the wheels.

The acquisition of specific behaviors and skills is probably the most important form of implicit memory. Driving a car, turning the pages of a book, playing a musical instrument, dancing, throwing and catching a ball, sliding a chair backward as we get up from the dinner table—all of these skills involve coordination of movements with sensory information received from the environment and from our own moving body parts. We do not need to be able to describe these activities in order to perform them. We may not even be aware of all the movements we make while we are performing them.

Patient E. P. developed a profound anterograde amnesia when he was stricken with a case of viral encephalitis that destroyed much of his medial temporal lobe. Bayley, Frascino, and Squire (2005) taught patient E. P. to point to a particular member of each of a series of eight pairs of objects. He eventually learned to do so, but he had no explicit memory of which objects were correct. When asked why he chose a particular object, he said, "It just seems that's the one. It's here (pointing to head) somehow or another and the hand goes for it. . . . I can't say memory. I just feel this is the one. . . . It's just jumping out at me. 'I'm the one. I'm the one'" (Bayley, Frascino, and Squire, 2005, p. 551). Clearly, he learned a nondeclarative stimulus-response task without at the same time acquiring any declarative memories about what he had learned.

What brain regions are responsible for the acquisition of nondeclarative memories? As we saw earlier in this chapter, perceptual memories involve the sensory regions of the cerebral cortex. The basal ganglia appear to play an essential role in stimulus-response and motor learning. Several experiments have shown that people with diseases of the basal ganglia have deficits that can be attributed to difficulty in learning automatic responses. For example, Owen et al. (1992) found that patients with Parkinson's disease were impaired on learning a visually cued instrumental conditioning task, and Willingham and Koroshetz (1993) found that patients with Huntington's disease failed to learn a sequence of button presses. (Parkinson's disease and Huntington's disease are both degenerative diseases of the basal ganglia.)

Table 13.1 lists the declarative and nondeclarative memory tasks that I have described so far. (See *Table 13.1.*)

Anatomy of Anterograde Amnesia

The phenomenon of anterograde amnesia—and its implications for the nature of relational learning—has led investigators to study this phenomenon in laboratory animals. But before I review this research (which has provided some very interesting results), we should examine the brain damage that produces anterograde amnesia. One fact is clear: Damage to the hippocampus or to regions of the brain that supply its inputs and receive its outputs causes anterograde amnesia.

declarative memory Memory that can be verbally expressed, such as memory for events in a person's past.

nondeclarative memory Memory whose formation does not depend on the hippocampal formation; a collective term for perceptual, stimulus-response, and motor memory.

TABLE 13.1 ■ Examples of Declarative and Nondeclarative Memory Tasks

DECLARATIVE MEMORY TASKS	
Remembering past experiences	
Finding one's way in new environment	

NONDECLARATIVE MEMORY TASKS	TYPE OF LEARNING
Learning to recognize broken drawings	Perceptual
Learning to recognize pictures and objects	Perceptual
Learning to recognize faces	Perceptual (and stimulus-response?)
Learning to recognize melodies	Perceptual
Classical conditioning (eyeblink)	Stimulus-response
Instrumental conditioning (choose circle)	Stimulus-response
Learning sequence of button presses	Motor

As we saw earlier in this chapter, the hippocampal formation consists of the dentate gyrus, the CA fields of the hippocampus itself, and the subiculum (and its subregions). The most important input to the hippocampal formation is the entorhinal cortex; neurons there have axons that terminate in the dentate gyrus, CA3, and CA1. The entorhinal cortex receives its inputs from the amygdala, various regions of the limbic cortex, and all association regions of the neocortex, either directly or via two adjacent regions of limbic cortex: the **perirhinal cortex** and the **parahippocampal cortex.** Collectively, these three regions constitute the *limbic cortex of the medial temporal lobe.* (See *Figure 13.31.*)

The outputs of the hippocampal system come primarily from field CA1 and the subiculum. Most of these outputs are relayed back through the entorhinal, perirhinal, and parahippocampal cortex to the same regions of association cortex that provide inputs.

The hippocampal formation also receives input from subcortical regions via the fornix. These inputs select and modulate the functions of the hippocampal formation. The fornix carries dopaminergic axons from the ventral tegmental area, noradrenergic axons from the locus coeruleus, serotonergic axons from the raphe nuclei, and acetylcholinergic axons from the medial septum. The fornix also connects the hippocampal formation with the mammillary bodies, located in the posterior hypothalamus. The most prominent brain damage seen in cases of Korsakoff's syndrome—and presumably the cause of the anterograde amnesia—is degeneration of the mammillary bodies. (See *Figure 13.32.*)

The clearest evidence that damage restricted to the hippocampal formation produces anterograde amnesia came from a case studied by Zola-Morgan, Squire, and Amaral (1986). Patient R. B., a 52-year-old man with a history of heart trouble, sustained a cardiac arrest. Although his heart was successfully restarted, the period of anoxia caused by the temporary halt in blood flow resulted in brain damage. The primary symptom of this brain damage was permanent anterograde amnesia, which Zola-Morgan and his colleagues carefully documented. Five years after

perirhinal cortex A region of limbic cortex adjacent to the hippocampal formation that, along with the parahippocampal cortex, relays information between the entorhinal cortex and other regions of the brain.

parahippocampal cortex A region of limbic cortex adjacent to the hippocampal formation that, along with the perirhinal cortex, relays information between the entorhinal cortex and other regions of the brain.

FIGURE 13.31 ■ Cortical Connections of the Hippocampal Formation

The figure shows (a) a view of the base of a monkey's brain and (b) connections with the cerebral cortex.

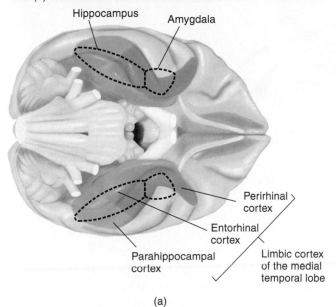

Hippocampus Amygdala

Perirhinal cortex

Entorhinal cortex

Parahippocampal cortex

Limbic cortex of the medial temporal lobe

(a)

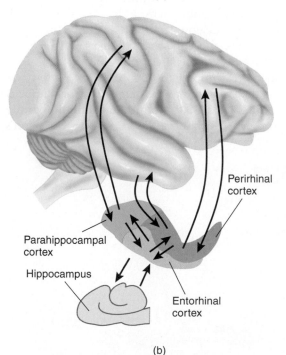

Perirhinal cortex

Parahippocampal cortex

Hippocampus

Entorhinal cortex

(b)

the onset of the amnesia, R. B. died of heart failure. His family gave permission for histological examination of his brain.

The investigators discovered that field CA1 of the hippocampal formation was gone; its neurons had completely degenerated. Subsequent studies reported other patients with anterograde amnesia caused by CA1 damage

(Victor and Agamanolis, 1990; Kartsounis, Rudge, and Stevens, 1995; Rempel-Clower et al., 1996). (See *Figure 13.33.*) In addition, several studies have found that a period of anoxia causes damage to field CA1 in monkeys and in rats and that the damage causes anterograde amnesia in these species too (Auer, Jensen, and Whishaw, 1989; Zola-Morgan et al., 1992).

Why is field CA1 of the hippocampus so sensitive to anoxia? The answer appears to lie in the fact that this region is especially rich in NMDA receptors. For some reason, metabolic disturbances of various kinds, including seizures, anoxia, or hypoglycemia, cause glutamatergic terminal buttons to release glutamate at abnormally high levels. The effect of this glutamate release is to stimulate NMDA receptors, which permit the entry of calcium. Within a few minutes, excessive amounts of intracellular calcium begin to destroy the neurons. If animals are pretreated with drugs that block NMDA receptors, a period of anoxia is much less likely to produce brain damage (Rothman and Olney, 1987). CA1 neurons contain many NMDA receptors, so long-term potentiation can quickly become established there. This flexibility undoubtedly contributes to our ability to learn as quickly as we do. But it also renders these neurons particularly susceptible to damage by metabolic disturbances.

Role of the Hippocampal Formation in Consolidation of Declarative Memories

As we saw earlier in this chapter, the hippocampus is not the location of either short-term or long-term memories; after all, patients with damage to the hippocampal formation can remember events that happened before the brain became damaged, and their short-term memory is relatively normal. But the hippocampal formation clearly plays a role in the process through which declarative memories are formed. Most researchers believe that the process works something like this: The hippocampus receives information about what is going on from sensory and motor association cortex and from some subcortical regions, such as the basal ganglia and amygdala. It processes this information and then, through its *efferent* connections with these regions, modifies the memories that are being consolidated there, linking them together in ways that will permit us to remember the relationships among the elements of the memories—for example, the order in which events occurred, the context in which we perceived a particular item, and so on. Without the hippocampal formation we would be left with individual, isolated memories without the linkage that makes it possible to remember—and think about—episodes and contexts.

FIGURE 13.32 ■ The Major Subcortical Connections of the Hippocampal Formation

A midsagittal view of a rat brain shows these connections.

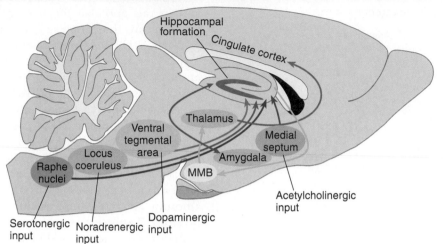

If the hippocampus does modify memories as they are being formed, then experiences that lead to declarative memories should activate the hippocampal formation. In fact, several studies have found this prediction to be true. In general, pictorial or spatial information activates the right hippocampal formation, and verbal information activates the left hippocampal formation. For example, Brewer et al. (1998) had normal subjects look at a series of complex color photos and later tested their ability to say whether they remembered them. (As we saw, people with anterograde amnesia are capable of perceptual learning, but they cannot *say* whether they have seen a particular item.) While the subjects were studying the pictures the first time, the experimenters recorded regional brain activity by functional MRI. Brewer and his colleagues found that the pictures that the subjects were most likely to remember later were those that caused the most activation of the right hippocampal region, suggesting that this region was involved in the encoding phase of memory formation. A study by Alkire et al. (1998) found that activation of the *left* hippocampal formation was related to a person's

FIGURE 13.33 ■ Damage to Field CA1 Caused by Anoxia

The scans show (a) section through a normal hippocampus and (b) section through the hippocampus of patient G. D. The pyramidal cells of field CA1 (between the two arrowheads) have degenerated. (DG = dentate gyrus, gl, ml, pl = layers of the dentate gyrus, PaS = parasubiculum, PrS = presubiculum, S = subiculum.)

(From Rempel-Clower, N. L., Zola, S. M., Squire, L. R., and Amaral, D. G. *Journal of Neuroscience*, 1996, *16*, 5233–5255. Reprinted with permission.)

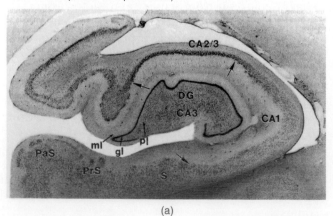

(a)

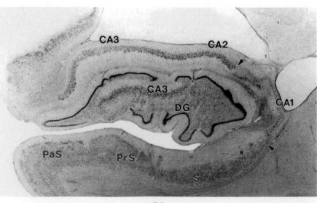

(b)

ability to remember a list of words: The subjects with the greatest amount of activation showed the best memory for the words. (See *Figure 13.34.*)

As we saw, anterograde amnesia is usually accompanied by retrograde amnesia—the inability to remember events that occurred for a period of time before the

FIGURE 13.34 ■ The Hippocampal Formation and Encoding of Declarative Memories

(a) The scan shows regions whose metabolic activity during learning correlated with likelihood of recall later. "Hot" colors reflect positive correlations; "cool" colors reflect negative correlations. The arrow points to the hippocampal formation. (b) The graph shows the percentage correct during free recall as a function of relative metabolic rate of the left hippocampal formation of the nine subjects in the study.

(Adapted from Alkire, M. T., Haier, R. J., Fallon, J. H., and Cahill, L. *Proceedings of the National Academy of Sciences, USA,* 1998, 95, 14506–14510.)

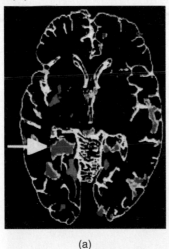

(a)

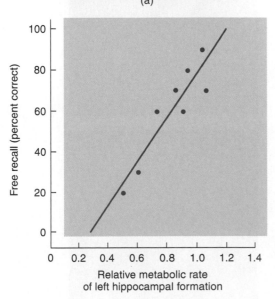

(b)

brain damage occurred. The duration of the retrograde amnesia appears to be related to the amount of damage to the medial temporal lobe (Squire and Bayley, 2007; Kirwan et al., 2008). Damage limited to the hippocampus (including the dentate gyrus and subiculum) results in a retrograde amnesia lasting a few years. Additional damage to the entorhinal cortex produces a retrograde amnesia of one to two decades. Damage that involves the hippocampus and much of the medial temporal lobe produces a retrograde amnesia that spares only those memories from early life. The memories that are spared in all these cases include semantic memories acquired early in life, memories of personal episodes when the patient was younger, and the ability to navigate in or describe the early home neighborhood.

The following examples illustrate retrieval of early memories by a patient with a profound anterograde amnesia.

Patient E. P. made the following response when he was asked to describe an incident from the period before he attended school.

> When I was 5 years old, we moved from Oakland to the country. I was very excited and looked forward to the change. I remember the truck that dad rented. It was hardly full because we didn't have much furniture. When it was time to leave, mom got in the car and followed behind the truck. I rode in the truck with dad. (Reed and Squire, 1998, p. 3951)

Patient E. P. is also able to find his way around the neighborhood where he grew up but is completely lost in the neighborhood to which he moved after he became amnesic (Teng and Squire, 1999).

The fact that retrograde amnesia extends back for a limited period of time suggests that a gradual process controlled by the hippocampal formation transforms memories located elsewhere. Before this transformation is complete, the hippocampal formation is required for the retrieval of these memories. Later, retrieval of these memories can be accomplished even if the hippocampal formation has been damaged. A functional-imaging study by Takashima et al. (2006) supports this hypothesis. The investigators had normal subjects look at 320 different photographs of landscapes for 5.5 seconds each. The subjects were encouraged to try to memorize the photographs. For example, the investigators gave the subjects specific examples of learning strategies, such as "'Where on the picture would you like to be most?', 'Where do you think the place is?', and 'Look for very special, distinct objects on the picture'" (p. 759). Later that day, one day later, one month later, and three months later, the investigators presented photographs that included a mixture of new photographs and a sample of the photographs the subjects had previously seen and asked the subjects to identify which ones were familiar to them. A different sample of previously seen photographs was presented at each session, which meant that the memories for the initial set of 320 photographs got progressively older. The subjects brains were scanned during each memory-testing session.

Takashima and her colleagues found that initially, the degree of hippocampal activation correlated with the subjects' memory of the photographs they had previously seen. However, as time went on, the hippocampal activation decreased, and the activation of the prefrontal cortex showed a correlation with correct identification. (See *Figure 13.35.*) The investigators concluded that the hippocampus played a role in retrieval of early memories but that this task was transferred to the prefrontal cortex as time went on. They

suggest that it is unlikely that the memories for the photographs were stored in the prefrontal cortex but hypothesized that this region, with its rich connections with other regions of the cerebral cortex, might be involved in organizing and linking information stored elsewhere.

You might wonder why the hippocampus would be involved in a perceptual memory in the first place. After all, we saw earlier that people with hippocampal damage can learn to recognize visual stimuli. The answer is that when people with anterograde amnesia are shown images that they had previously seen (but after the onset of their amnesia), they will deny having seen them before. However, if they are given a forced choice between an old image and a new one, they will point to the one they had previously seen, without showing any signs of real recognition. You will recall that patient E. P. said, "I can't say memory. I just feel this is the one. . . . It's just jumping out at me" (Bayley, Frascino, and Squire, 2005, p. 551). This nondeclarative perceptual memory is different from the declarative memory that the subjects in the study by Takashima et al., who

---

FIGURE 13.35 ■ Changing Roles of Hippocampus and Prefrontal Cortex in Memory

The role of the ventromedial prefrontal cortex (top) increased over time, and the role of the hippocampus (bottom) decreased over time.

(From Takashima, A., Petersson, K. M., Rutters, F., Tendolkar, I., Jensen, O., Zwarts, M. J., McNaughton, B. L., and Fernández, G. *Proceedings of the National Academy of Sciences, USA,* 2006, *103,* 756–761. Reprinted with permission.)

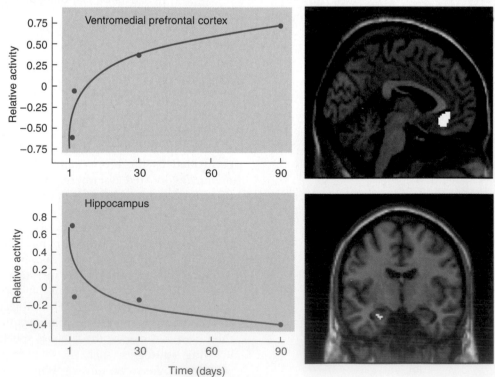

deliberately encouraged their subjects to think about the photographs and try to remember them.

You might also wonder why the role of the hippocampus in maintaining access to a memory appears to end in less than three months, whereas retrograde amnesia caused by hippocampal damage lasts for at least several years. The most likely explanation is that when investigators test for the extent of a patient's retrograde amnesia, they ask questions about more complex memories, such as autobiographical episodes, which involve sequences of many individual memories. Retrieval of such complex sets of memories may require the participation of the hippocampus for a much longer time.

Episodic and Semantic Memories

Evidence suggests that semantic and episodic memories are distinct forms of declarative memory. **Episodic memories** involve context; they include information about when and under what conditions a particular episode occurred and the order in which the events in the episode took place. Episodic memories are specific to a particular time and place, because a given episode—by definition—occurs only once. **Semantic memories** involve facts, but they do not include information about the context in which the facts were learned. In other words, semantic memories are less specific than episodic memories. For example, knowing that the sun is a star involves a less specific memory than being able to remember when, where, and from whom you learned this fact. Semantic memories can be acquired gradually, over time. Episodic memories must be learned all at once.

Acquisition of both major categories of declarative memories—episodic and semantic—appears to require the participation of the hippocampus. Manns, Hopkins, and Squire (2003) found that five patients with damage limited to the hippocampal formation showed an anterograde amnesia for semantic as well as episodic information.

As we saw earlier in this chapter, perceptual memories appear to be located in the sensory association cortex, the regions where the perceptions take place. Presumably, episodic memories, which consist of an integrated sequence of perceptual memories, are also located there. What about semantic memories—memories for factual information? Knowing that the sun is a star certainly involves memories different from knowing what the sun looks like. Thus, semantic memories are not simply perceptual memories. A degenerative neurological disorder known as **semantic dementia** suggests that the temporal lobe plays an important role in storing semantic information. Semantic dementia is caused by degeneration of the neocortex of the anterolateral temporal lobe (Lambon Ralph and Patterson, 2008). At least in the early stages of the degenerative process the hippocampal

formation and the rest of the medial temporal lobe are not affected. Murre, Graham, and Hodges (2001) describe the case of patient A. M., born in 1930 and studied by the investigators between 1994 and 1997.

A. M. was an active, intelligent man who had received an undergraduate degree in engineering and a master's degree in science. He worked for an internationally renowned company, where he was responsible for managing over 450 employees. His neurological symptoms began with progressive difficulty in understanding the speech of others and finding appropriate words of his own. By the time Murre and his colleagues met A. M., his speech was fluent and grammatical but contained little meaning.

Examiner: Can you tell me about a time you were in hospital?

A. M.: Well one of the best places was in April last year here (ha ha) and then April, May, June, July, August, September and then October, and then April today.

Examiner: Can you remember April last year?

A. M.: April last year, that was the first time, and ch, on the Monday, for example, they were checking all my whatsit, and that was the first time, when my brain was, eh, shown, you know, you know that bar of the brain (indicates left), not the, the other one was okay, but that was lousy, so they did that and then doing everything like that, like this and probably a bit better than I am just now (indicates scanning by moving his hands over his head). (Murre, Graham, and Hodges, 2001, p. 651)

Patient A. M.'s loss of semantic information had a profound effect on his everyday activities. He seemed not to understand functions of commonplace objects. For example, he held a closed umbrella horizontally over his head during a rainstorm and brought his wife a lawnmower when she had asked for a stepladder. He put sugar into a glass of wine and put yogurt on a raw defrosting salmon steak and ate it. He nevertheless showed some surprisingly complex behaviors. Because he could not be trusted to drive a car, his wife surreptitiously removed the car keys from his key ring. He

episodic memory Memory of a collection of perceptions of events organized in time and identified by a particular context.

semantic memory A memory of facts and general information.

semantic dementia Loss of semantic memories caused by progressive degeneration of the neocortex of the lateral temporal lobes.

noticed their absence, and rather than complaining to her (presumably, he realized that would be fruitless), he surreptitiously removed the car keys from her key ring, went to a locksmith, and had a duplicate set made.

Although his semantic memory was severely damaged, his episodic memory was surprisingly good. The investigators reported that even when his dementia had progressed to the point at which he was scoring at chance levels on a test of semantic information, he answered a phone call that was meant for his wife, who was out of the house. When she returned later, he remembered to tell her about the call.

As you can see, the symptoms of semantic dementia are quite different from those of anterograde amnesia. Semantic information is lost, but episodic memory for recent events can be spared. The hippocampal formation and the limbic cortex of the medial temporal lobe appear to be involved in the consolidation and retrieval of declarative memories, both episodic and semantic, but the semantic memories themselves appear to be stored in the neocortex—in particular, in the neocortex of the anterolateral temporal lobe. Pobric, Jefferies, and Lambon Ralph (2007) found that transcranial magnetic stimulation of the left anterior temporal lobe, which disrupted the normal neural activity of this region, produced the symptoms of semantic dementia. The subjects had difficulty naming pictures of objects and understanding the meanings of words, but they had no trouble performing other, nonsemantic, tasks such as naming six-digit numbers and matching large numbers according to their approximate size. Also, a functional-imaging study by Rogers et al. (2006) recorded activation of the anterolateral temporal lobes when people performed a picture-naming task.

Spatial Memory

I mentioned earlier in this chapter that patient H. M. has not been able to find his way around his present environment. Although spatial information need not be declared (we can demonstrate our topographical memories by successfully getting from place to place), people with anterograde amnesia are unable to consolidate information about the location of rooms, corridors, buildings, roads, and other important items in their environment.

Bilateral medial temporal lobe lesions produce the most profound impairment in spatial memory, but significant deficits can be produced by damage that is limited to the right hemisphere. For example, Luzzi et al. (2000) reported the case of a man with a lesion of the right parahippocampal gyrus who lost his ability to find

his way around a new environment. The only way he could find his room was by counting doorways from the end of the hall or by seeing a red napkin that was located on top of his bedside table.

Functional-imaging studies have shown that the right hippocampal formation becomes active when a person is remembering or performing a navigational task. For example Maguire, Frackowiak, and Frith (1997) had London taxi drivers describe the routes they would take in driving from one location to another. Functional imaging that was performed during their description of the route showed activation of the right hippocampal formation. London taxi drivers undergo extensive training to learn how to navigate efficiently in that city; in fact, this training takes about two years, and the drivers receive their license only after passing a rigorous set of tests. We would expect that this topographical learning would produce some changes in various parts of their brains, including their hippocampal formation. In fact, Maguire et al. (2000) found that the volume of the posterior hippocampus of London taxi drivers was larger than that of control subjects. Furthermore, the longer an individual taxi driver had spent in this occupation, the larger was the volume of the right posterior hippocampus. As we will see later in this chapter, the dorsal hippocampus of rats (which corresponds to the posterior hippocampus of humans) contains *place cells*—neurons that are directly involved in navigation in space.

Other experiments provides further evidence for the role of the hippocampus in spatial memory. Hartley et al. (2003) trained subjects to find their way in a computerized virtual-reality town. Some subjects became acquainted with the town by exploring it, giving them the opportunity to learn where various landmarks (shops, cafés, etc.) were located with respect to each other. Other subjects were trained to follow a specific pathway from one landmark to the next, making a sequence of turns to get from a particular starting point to another. The investigators hypothesized that the first task, which involved spatial learning, would require the participation of the hippocampus, while the second task, which involved learning a set of specific responses to a set of specific stimuli, would require the participation of the basal ganglia. The results were as predicted: Functional MRI revealed that the spatial task activated the hippocampus and the response task activated the caudate nucleus (a component of the basal ganglia).

Iaria et al. (2003) used a similar task that permitted subjects to learn a maze either through distant spatial cues or through a series of turns. About half of the subjects spontaneously used spatial cues, and the other half spontaneously learned to make a sequence of responses at specific locations. Again, fMRI showed the hippocampus was activated in subjects who followed the *spatial strategy* and the caudate nucleus was activated in subjects

FIGURE 13.36 ■ Spatial and Response Strategies

The figure shows the relation between volume of gray matter of the hippocampus (top) and caudate nucleus (bottom) and errors made on test trials in a virtual maze that could be performed only by using a response strategy. Increased density of the caudate nucleus was associated with better performance, and increased density of the hippocampus was associated with poorer performance.

(From Bohbot, V. D., Lerch, J., Thorndycraft, B., Iaria, G., and Zijdenbos, A. *Journal of Neuroscience,* 2007, *27,* 10078–10083. Reprinted with permission.)

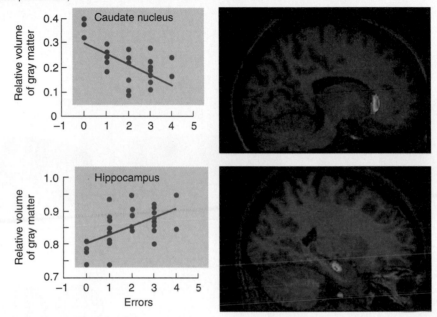

who followed the *response strategy*. In addition, a structural MRI study by Bohbot et al. (2007) found that people who tended to follow a spatial strategy in a virtual maze had a larger-than-average hippocampus, and people who tended to follow a response strategy had a larger-than-average caudate nucleus. (You will recall that the caudate nucleus, part of the basal ganglia, plays a role in stimulus-response learning.) Figure 13.36. shows the relationship between performance on test trials that could be performed only by using a response strategy. As you can see, the larger a person's caudate nucleus is (and the smaller a person's hippocampus is), the fewer errors that person made. (See *Figure 13.36.*)

Relational Learning in Laboratory Animals

The discovery that hippocampal lesions produced anterograde amnesia in humans stimulated interest in the exact role that this structure plays in the learning process. To pursue this interest, researchers have developed tasks that require relational learning, and laboratory animals with hippocampal lesions show memory deficits on such tasks, just as humans do.

Spatial Perception and Learning

As we saw, hippocampal lesions disrupt the ability to keep track of and remember spatial locations. For example, H. M. never learned to find his way home when his parents moved after his surgery. Laboratory animals show similar problems in navigation. Morris et al. (1982) developed a task that other researchers have adopted as a standard test of rodents' spatial abilities. The task requires rats to find a particular location in space solely by means of visual cues external to the apparatus. The "maze" consists of a circular pool, 1.3 meters in diameter, filled with a mixture of water and something to increase the opacity of the water, such as powdered milk. The water mixture hides the location of a small platform, situated just beneath the surface of the liquid. The experimenters put the rats into the water and let them swim until they encountered the hidden platform and climbed onto it. They released the rats from a new position on each trial. After a few trials, normal rats learned to swim directly to the hidden platform from wherever they were released.

The Morris water maze requires relational learning; to navigate around the maze, the animals get their bearings from the relative locations of stimuli located outside the maze—furniture, windows, doors, and so on. But the

FIGURE 13.37 ■ **The Morris Water Maze**

(a) Environmental cues present in the room provide information that permits the animals to orient themselves in space. (b) According to the task, start positions are variable or fixed. Normally, rats are released from a different position on each trial. If they are released from the same position every time, the rats can learn to find the hidden platform through stimulus-response learning. (c) The graphs show the performance of normal rats and rats with hippocampal lesions using variable or fixed start positions. Hippocampal lesions impair acquisition of the relational task. (d) Representative samples show the paths followed by normal rats and rats with hippocampal lesions on the relational task (variable start positions).

(Adapted from Eichenbaum, H. *Nature Reviews: Neuroscience*, 2000, *1*, 41–50. Data from Eichenbaum et al., 1990.)

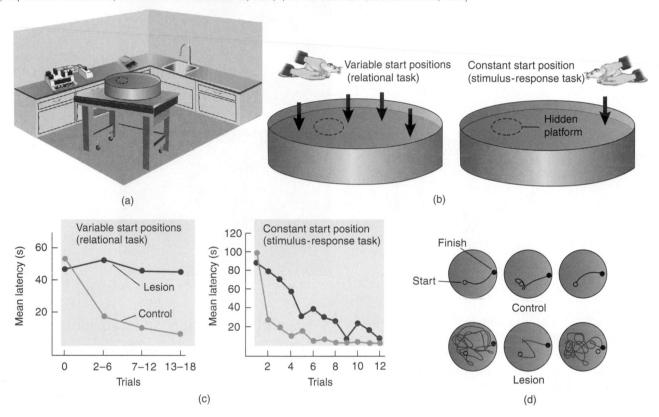

maze can be used for nonrelational, stimulus-response learning too. If the animals are always released at the same place, they learn to head in a particular direction—say, toward a particular landmark they can see above the wall of the maze (Eichenbaum, Stewart, and Morris, 1990).

If rats with hippocampal lesions are always released from the same place, they learn this nonrelational, stimulus-response task about as well as normal rats do. However, if they are released from a new position on each trial, they swim in what appears to be an aimless fashion until they finally encounter the platform. (See *Figure 13.37*.)

Many different types of studies have confirmed the importance of the hippocampus in spatial learning. For example, Gagliardo, Ioalé, and Bingman (1999) found that hippocampal lesions disrupted navigation in homing pigeons. The lesions did not disrupt the birds' ability to use the position of the sun at a particular time of day as a compass pointing toward their home roost. Instead,

the lesions disrupted their ability to keep track of where they were when they got near the end of their flight—at a time when the birds begin to use familiar landmarks to determine where they are. In a review of the literature, Sherry, Jacobs, and Gaulin (1992) reported that the hippocampal formation of species of birds and rodents that normally store seeds in hidden caches and later retrieve them (and that have excellent memories for spatial locations) is larger than that of animals without this ability.

Place Cells in the Hippocampal Formation

One of the most intriguing discoveries about the hippocampal formation was made by O'Keefe and Dostrovsky (1971), who recorded the activity of individual pyramidal cells in the hippocampus as an animal moved around the environment. The experimenters found that some neurons fired at a high rate only when the rat was in a particular location. Different neurons had different *spatial receptive fields;* that is, they responded when the animals were in different locations. A particular neuron might fire twenty times per second when the animal was in a particular location but only a few times per hour when the animal was located elsewhere. For obvious reasons these neurons were named **place cells.**

When a rat is placed in a symmetrical chamber, where there are few cues to distinguish one part of the apparatus from another, the animal must keep track of its location from objects it sees (or hears) in the environment outside the maze. Changes in these items affect the firing of the rats' place cells as well as their navigational ability. When experimenters move the stimuli as a group, maintaining their relative positions, the animals simply reorient their responses accordingly. However, when the experimenters interchange the stimuli so that they are arranged in a new order, the animals' performance (and the firing of their place cells) is disrupted. (Imagine how disoriented you might be if you entered a familiar room and found that the windows, doors, and furniture were in new positions.)

The fact that neurons in the hippocampal formation have spatial receptive fields does not mean that each neuron encodes a particular location. Instead, this information is undoubtedly represented by particular *patterns* of activity in circuits of large numbers of neurons within the hippocampal formation. In rodents most hippocampal place cells are found in the dorsal hippocampus, which corresponds to the posterior hippocampus in humans (Best, White, and Minai, 2001).

Evidence indicates that firing of hippocampal place cells appears to reflect the location where an animal "thinks" it is. Skaggs and McNaughton (1998) constructed an apparatus that contained two nearly identical chambers connected by a corridor. Each day, rats were placed in one of the chambers, and a cluster of electrodes in the animals' brains recorded the activity of hippocampal place cells. Each rat was always placed in the same chamber each day. Some of the place cells showed similar patterns of activity in each of the chambers, and some showed different patterns, which suggests that the hippocampus "realized" that there were two different compartments but also "recognized" the similarities between them. Then, on the last day of the experiment, the investigators placed the rats in the other chamber of the apparatus. For example, if a rat was usually placed in the north chamber, it was placed in the south chamber. The firing pattern of the place cells in at least half of the rats indicated that the hippocampus "thought" it was in the usual chamber—the one to the north. However, once the rat left the chamber and entered the corridor, it saw that it had to turn to the left to get to the other chamber and not to the right. The animal apparently realized its mistake, because for the rest of that session

FIGURE 13.38 ■ Apparatus Used by Skaggs and McNaughton (1998)

Place cells reflect the location where the animal "thinks" it is. Because the rat was normally placed in the north chamber, its hippocampal place cells responded as if it were there when it was placed in the south chamber one day. However, once it stuck its head into the corridor, it saw that the other chamber was located to its right, so it "realized" that it had just been in the south chamber. From then on, the pattern of firing of the hippocampal place cells accurately reflected the chamber in which the animal was located.

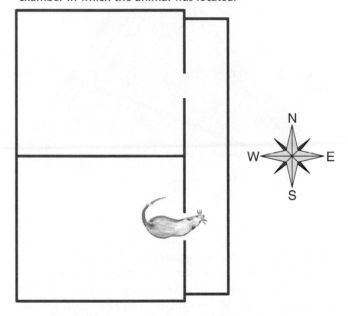

the neurons fired appropriately. They displayed the "north" pattern in the north chamber and the "south" pattern in the south chamber. (See *Figure 13.38.*)

The hippocampus appears to receive its spatial information from the parietal lobes by means of the entorhinal cortex. Sato et al. (2006) found that neurons in the medial parietal cortex of monkeys showed activity associated with specific movements at specific locations as the animals navigated a virtual environment with a joystick. (Yes, monkeys, too, can learn to play computer games.) When the investigators suppressed activity in the parietal cortex by infusing muscimol, the animals became lost. Quirk et al. (1992) found that neurons in the entorhinal cortex have spatial receptive fields, although these fields are not nearly as clear-cut as those of hippocampal pyramidal cells. Damage to the entorhinal cortex disrupts the spatial receptive fields of place cells in the hippocampus and impairs the animals' ability to navigate in spatial tasks (Miller and Best, 1980).

place cell A neuron that becomes active when the animal is in a particular location in the environment; most typically found in the hippocampal formation.

FIGURE 13.39 ■ Apparatus Used by Wood et al.

The rats were trained to turn right and turn left at the end of the stem of the T-maze on alternate trials. The firing patterns of hippocampal place cells with spatial receptive fields in the stem of the maze were different on trials during which the animals turned left or right.

(Adapted from Wood, E. R., Dudchenko, P. A., Robitsek, R. J., and Eichenbaum, H. *Neuron,* 2000, *27,* 623–633.)

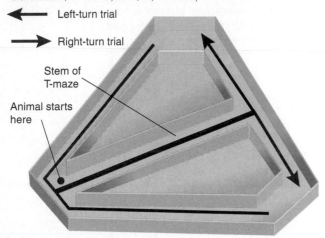

The activity of circuits of hippocampal place cells provide information about more than space. Wood et al. (2000) trained rats on a spatial alternation task in a T-maze. The task required the rats to enter the left and the right arms on alternate trials; when they did so, they received a piece of food in goal boxes located at the ends of the arms of the T. Corridors connected the goal boxes led back to the stem of the T-maze, where the next trial began. (See *Figure 13.39.*) Wood and her colleagues recorded from field CA1 pyramidal cells and, as expected, found that different cells fired when the rat was in different parts of the maze. However, two-thirds of the neurons fired differentially in the stem of the T on left-turn and right-turn trials. In other words, the cells not only encoded the rat's location in the maze, but also signaled whether the rat was going to turn right or turn left after it got to the choice point. Thus, pyramidal cells in CA1 encode both the current location and the intended destination.

Role of the Hippocampal Formation in Memory Consolidation

We have already seen evidence from functional-imaging studies and the effects of brain damage in humans that indicates that the hippocampal formation plays a critical role in consolidation of relational memories. Studies with laboratory animals support this conclusion. For example, Bontempi et al. (1999) trained mice in a spatial learning task. Five days later, they used a 2-DG imaging procedure to measure regional brain activation while they tested the

animals' memory for the task. The activity of the hippocampus was elevated and was positively correlated with the animal's performance—the higher the activity, the better the performance. At twenty-five days, hippocampal activity was down by 15–20 percent, and the correlation between activity and performance was gone. However, the activity of several regions of the cerebral cortex was elevated while the animals were being tested. The investigators conclude that these findings support the hypothesis that the hippocampus is involved in consolidation of spatial memories for a limited time, and the result of this activity is to help establish the memories in the cerebral cortex.

Maviel et al. (2004) trained mice in a Morris water maze and tested later for their memory of the location of the platform. Just before testing the animal's performance, the investigators temporarily deactivated specific regions of the animals' brains with intracerebral infusions of lidocaine, a local anesthetic. If the hippocampus was deactivated one day after training, the mice showed no memory of the task. However, if the hippocampus was deactivated thirty days after training, their performance was normal. In contrast, inactivation of several regions of the cerebral cortex impaired memory retrieval thirty days after training, but not one day after training. These findings indicate that the hippocampus is required for newly learned spatial information but not for information learned thirty days previously. The findings also suggest that sometime during these thirty days the cerebral cortex takes on a role in retention of this information. (See *Figure 13.40.*)

As we saw in Chapter 9, slow-wave sleep facilitates the consolidation of declarative memories in human subjects, while REM sleep facilitates the consolidation of nondeclarative memories. One advantage of recording place cells in the hippocampus while animals perform a spatial task is that the investigators can detect different patterns of activity in these cells that changes as the animals move through different environments. Lee and Wilson (2002)

FIGURE 13.40 ■ A Schematic Description of the Experiment by Maviel et al. (2004)

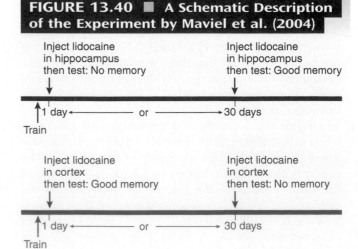

implanted an array of microelectrodes in field CA1 of rats and were able to record from 24 to 57 different neurons simultaneously in each animal. The rats ran through straight or U-shaped tracks, at the ends of which they found a piece of chocolate. The investigators recorded the sequences of place cell activity in field CA1 as the animals ran. They also recorded the activity of these cells while the animals slept. They found that particular cells had particular spatial receptive fields, so as the animals ran through the tracks, particular sequences of cell firing were seen. Recordings made after training showed the same patterns of activity while the animals engaged in slow-wave sleep. Presumably, these patterns indicate a replay of the animals' behavior as they moved through their environment and obtained the food, and the patterns facilitate consolidation of the memories of these episodes.

Reconsolidation of Memories

What happens to memories of events as time goes on? Clearly, if we learn something new about a particular subject, our memories pertaining to that subject must somehow be modified. For example, as I mentioned earlier in this chapter, if a friend gets a new hairstyle or replaces glasses with contact lenses, our visual memory of that person will change accordingly. And if you learn more about something—for example, the layout of a previously unfamiliar neighborhood—you will acquire a larger and larger number of interconnected memories. These examples indicate that memories can be altered or connected to newer memories. In recent years, researchers have been investigating a phenomenon known as **reconsolidation,** which appears to involve modification of long-term memories.

As we will see in Chapter 16, one of the side effects of a procedure known as electroconvulsive therapy is a period of retrograde amnesia. The procedure, used to treat cases of severe depression, involves the application of electricity through electrodes placed on a person's scalp. The current excites so many neurons in the brain that it produces a seizure. Presumably, the seizure erases short-term memories present at the time and thus prevents consolidation of these memories.

Misanin, Miller, and Lewis (1968) found that long-term memories, which are normally not affected by seizures, were vulnerable to disruption by electroconvulsive shock (ECS) if a reminder of the original learning experience was first presented. The investigators found that ECS given right after a learning experience prevented consolidation, but ECS given a day later did not. Apparently, the seizure given right after training disrupted the brain activity initiated by the training session and consequently interfered with consolidation. The seizure given the next day had no effect, because the memory had already been consolidated. However, if animals were given a "reminder" stimulus one day after training,

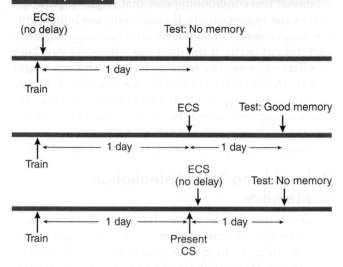

FIGURE 13.41 ■ A Schematic Description of the Experiment by Misanin, Miller, and Lewis (1968)

which presumably reactivated the memory, an ECS treatment administered immediately afterward caused amnesia for the task when the animals were tested the following day. Reactivation of the memory made it susceptible to disruption. (See *Figure 13.41*.)

A study by Ben Mamou, Gamache, and Nader (2006) found that the process of reconsolidation requires long-term potentiation. The investigators found that injection of *anisomycin*, a drug that prevents protein synthesis and thus interferes with memory consolidation, would disrupt memory of a previously learned avoidance task only if a reminder stimulus was presented. However, if an injection of an NMDA receptor antagonist was first infused into the amygdala (the region involved in learning this task), anisomycin had no effect on memory even if a reminder stimulus was presented. These results indicate that when synaptic plasticity is prevented, reconsolidation cannot occur. Thus, reconsolidation requires long-term potentiation.

The study by Misanin, Mamon, and their colleagues involved stimulus-response learning. More recent studies have found that long-term, well-consolidated relational memories are also susceptible to disruption. Presumably, the process of reconsolidation, which involves neural events similar to those responsible for the original consolidation, makes it possible for established memories to be altered or attached to new information (Nader, 2003). (Remember when I mentioned that seeing your friend with a new hairstyle would alter your visual memory of that person?) Events that interfere with consolidation

---

reconsolidation A process of consolidation of a memory that occurs subsequent to the original consolidation that can be triggered by a reminder of the original stimulus; thought to provide the means for modifying existing memories.

released from the same place in the maze, which turns the task into one of stimulus-response learning. The hippocampal formation contains place cells—neurons that respond when the animal is in a particular location, which implies that the hippocampus contains neural networks that keep track of the relationships among stimuli in the environment that define the animal's location. Neurons in the hippocampal formation reflect where an animal "thinks" it is. Topographical information reaches field CA1 of hippocampus from the parietal lobe by means of the entorhinal cortex. Place cells encode more than space; they can include information about the response that the animal will perform next.

Research has shown that the hippocampal formation plays a role in memory consolidation. A 2-DG imaging study found that the hippocampal activity correlates with animals' ability to remember a spatial learning task a few days after the original learning but that the correlation disappears after a few weeks. Similarly, deactivation of the dorsal hippocampus prevents consolidation if it occurs one day after the animal learns a Morris water maze task but has no effect if it occurs thirty days later. In contrast, deactivation of regions of the cerebral cortex thirty days after training disrupt performance if it occurs thirty days after training but has no effect if it occurs one day after training. Slow-wave sleep facilitates the consolidation of declarative memories, and REM sleep facilitates the consolidation of nondeclarative memories. During slow-wave sleep, place cells in field CA1 of rats replay the sequence of activity that they showed while navigating in an environment in the laboratory.

Memories can be altered or connected to newer memories—a process known as reconsolidation. When a long-term memory is reactivated by stimuli that provide a "reminder" of the original experience, the memories become susceptible to events that interfere with consolidation, such as electroconvulsive shock treatment, interference with long-term potentiation, or the administration of a drug that inhibits protein synthesis.

Learning involves long-term potentiation. When rats are trained in a maze, synaptic connections in the hippocampus are strengthened. A targeted mutation against the NMDA receptor gene that affects only field CA1 disrupts long-term potentiation and the ability to learn the Morris water maze.

The dentate gyrus is one of the two places in the brain where adult stem cells can divide and give rise to new neurons. These neurons establish connections with neurons in field CA3 and appear to participate in learning. Their ability to undergo long-term potentiation more easily than older neurons suggests that they facilitate the formation of new memories.

Thought Question

Although we can live only in the present, our memories are an important aspect of our identities. What do you think it would be like to have a memory deficit like H. M.'s? Imagine having no recollection of over thirty years of experiences. Imagine being surprised every time you see yourself in the mirror and discover someone who is more than thirty years older than you believe yourself to be.

SUGGESTED READINGS

Frey, S., and Frey, J. U. "Synaptic tagging" and "cross-tagging" and related associative reinforcement processes of functional plasticity as the cellular basis for memory formation. *Progress in Brain Research*, 2008, *169*, 117–143.

Patterson, K., Nestor, P. J., and Rogers, T. T. Where do you know what you know?: The representation of semantic knowledge in the human brain. *Nature Reviews: Neuroscience*, 2007, *8*, 976–987.

Schultz, W. Behavioral theories and the neurophysiology of reward. *Annual Review of Psychology*, 2006, *57*, 87–115.

Sigurdsson, T., Doyére, V., Cain, C. K., and LeDoux, J. E. Long-term potentiation in the amygdala: A cellular mechanism of fear learning and memory. *Neuropharmacology*, 2007, *52*, 215–227.

Spiers, H. J., and Maguire, E. A. The neuroscience of remote spatial memory: A tale of two cities. *Neuroscience*, 2007, *149*, 7–27.

Squire, L. R., Stark, C. E., and Clark, R. E. The medial temporal lobe. *Annual Review of Neuroscience*, 2004, *27*, 279–306.

Tronson, N. C., and Taylor, J. R. Molecular mechanisms of memory reconsolidation. *Nature Reviews: Neuroscience*, 2007, *8*, 262–275.

ADDITIONAL RESOURCES

Visit www.mypsychkit.com for additional review and practice of the material covered in this chapter. Within MyPsychKit, you can take practice tests and receive a customized study plan to help you review. Dozens of animations, tutorials, and Web links are also available. You can even review using the interactive electronic version of this textbook. You will need to register for MyPsychKit. See **www.mypsychkit.com** for complete details.

chapter

14

Human Communication

outline

While driving her car to visit some friends, R. F., a 39-year-old woman, was broadsided by an intoxicated driver who had ignored (or was too drunk to see) a stop sign. The left side of R. F.'s head was fractured, and the bone fragments caused considerable damage to her brain. A neurosurgeon repaired the damage as best he could, but R. F. remained in a coma for several weeks. By the time my colleagues and I met her, she had shown considerable recovery. However, she had difficulty remembering the names of even the most common objects, and she could no longer read.

Although R. F. could not read, she could match words with pictures, which indicated that she could still *perceive* words. This fact was made especially apparent one day when she was trying (without success) to read some words that I had typed. Suddenly, she said, "Hey! You spelled this one wrong." I looked at the word and realized that she was right; I had. But although she saw that the word was misspelled, she still could not say what it was, even when she tried very hard to sound it out. That evening I made up a list of eighty pairs of words, one spelled correctly and the other incorrectly. The next day I gave her a pencil and asked her to cross out the misspelled words. She was able to go through the list quickly and easily, correctly identifying 95 percent of the misspelled words. She was able to *read* only five of them.

Verbal behaviors constitute one of the most important classes of human social behavior. Our cultural evolution has been possible because we can talk and listen, write and read. Language enables our discoveries to be cumulative; knowledge gained by one generation can be passed on to the next.

The basic function of verbal communication is seen in its effects on other people. When we talk to someone, we almost always expect our speech to induce the person to engage in some sort of behavior. Sometimes, the behavior is of obvious advantage to us, as when we ask for an object or for help in performing a task. At other times we are simply asking for a social exchange: some attention and perhaps some conversation. Even "idle" conversation is not idle, because it causes another person to look at us and say something in return.

This chapter discusses the neural basis of verbal behavior: talking, understanding speech, reading, and writing.

SPEECH PRODUCTION AND COMPREHENSION: BRAIN MECHANISMS

Our knowledge of the physiology of language has been obtained primarily by observing the effects of brain lesions on people's verbal behavior. Although investigators have studied people who have undergone brain surgery or who have sustained head injuries, brain tumors, or infections, most of the observations have been made on people who have suffered strokes, or **cerebrovascular accidents.** The most common type of cerebrovascular accident is caused by obstruction of a blood vessel. The interruption in blood flow deprives a region of the brain of its blood supply, which causes cells in that region to die.

Another source of information about the physiology of language comes from studies using functional-imaging devices. In recent years, researchers have used these devices to gather information about language processes from normal subjects. In general, these studies have confirmed or complemented what we have learned by studying patients with brain damage.

The most important category of speech disorders is **aphasia,** a primary disturbance in the comprehension or production of speech, caused by brain damage. Not all speech disturbances are aphasias; to receive a diagnosis of aphasia, a patient must have difficulty comprehending, repeating, or producing meaningful speech, and this difficulty must not be caused by simple sensory or motor deficits or by lack of motivation. For example, inability to speak caused by deafness or paralysis of the speech muscles is not considered to be aphasia. In addition, the deficit must be relatively isolated; that is, the patient must appear to be aware of what is happening in his or her environment and to comprehend that others are attempting to communicate.

Lateralization

Verbal behavior is a *lateralized* function; most language disturbances occur after damage to the left side of the brain, whether people are left-handed or right-handed. Using an ultrasonic procedure to measure changes in cerebral blood flow while people performed a verbal task, Knecht et al. (2000) assessed the relationship between handedness and lateralization of speech mechanisms in people without any known brain damage. They found that right-hemisphere speech dominance was seen in only 4 percent of right-handed people, in 15 percent of ambidextrous people, and in 27 percent of

cerebrovascular accident A "stroke"; brain damage caused by occlusion or rupture of a blood vessel in the brain.

aphasia Difficulty in producing or comprehending speech not produced by deafness or a simple motor deficit; caused by brain damage.

left-handed people. If the left hemisphere is malformed or damaged early in life, then language dominance is very likely to pass to the right hemisphere (Vikingstad et al., 2000). Because the left hemisphere of approximately 90 percent of the total population is dominant for speech, you can assume that the brain damage described in this chapter is located in the left (speech-dominant) hemisphere unless I say otherwise.

Although the circuits that are *primarily* involved in speech comprehension and production are located in one hemisphere (almost always, the left hemisphere), it would be a mistake to conclude that the other hemisphere plays no role in speech. When we hear and understand words and when we talk about or think about our own perceptions or memories, we are using neural circuits besides those directly involved in speech. Thus, these circuits, too, play a role in verbal behavior. For example, damage to the right hemisphere makes it difficult for a person to read maps, perceive spatial relations, and recognize complex geometrical forms. People with such damage also have trouble talking about things like maps and complex geometrical forms or understanding what other people have to say about them. The right hemisphere also appears to be involved in organizing a narrative—selecting and assembling the elements of what we want to say (Gardner et al., 1983). As we saw in Chapter 11, the right hemisphere is involved in the expression and recognition of emotion in the tone of voice. And as we shall see in this chapter, it is also involved in control of *prosody*—the normal rhythm and stress found in speech. Therefore, both hemispheres of the brain have a contribution to make to our language abilities.

Speech Production

Being able to talk—that is, to produce meaningful speech—requires several abilities. First, the person must have something to talk about. Let us consider what this means. We can talk about something that is currently happening or something that happened in the past. In the first case we are talking about our perceptions: things we are seeing, hearing, feeling, smelling, and so on. In the second case we are talking about our memories of what happened in the past. Both perceptions of current events and memories of events that occurred in the past involve brain mechanisms in the posterior part of the cerebral hemispheres (the occipital, temporal, and parietal lobes). Thus, this region is largely responsible for our having something to say.

Of course, we can also talk about something that *did not* happen. That is, we can use our imagination to make up a story (or to tell a lie). We know very little about the neural mechanisms that are responsible for imagination, but it seems likely that they involve the mechanisms responsible for perceptions and memories;

after all, when we make up a story, we must base it on knowledge that we originally acquired through perception and have retained in our memory.

Given that a person has something to say, actually doing so requires some additional brain functions. As we shall see in this section, the conversion of perceptions, memories, and thoughts into speech makes use of neural mechanisms located in the frontal lobes.

Damage to a region of the inferior left frontal lobe (Broca's area) disrupts the ability to speak: It causes **Broca's aphasia.** This disorder is characterized by slow, laborious, and nonfluent speech. When trying to talk with patients who have Broca's aphasia, most people find it hard to resist supplying the words the patients are obviously groping for. But although they often mispronounce words, the ones the patients manage to come out with are usually meaningful. The posterior part of the cerebral hemispheres has something to say, but the damage to the frontal lobe makes it difficult for the patients to express these thoughts.

People with Broca's aphasia find it easier to say some types of words than others. They have great difficulty saying the little words with grammatical meaning, such as *a, the, some, in,* or *about.* These words are called **function words,** because they have important grammatical functions. The words that they do manage to say are almost entirely **content words**—words that convey meaning, including nouns, verbs, adjectives, and adverbs, such as *apple, house, throw,* or *heavy.* Here is a sample of speech from a man with Broca's aphasia, who is trying to describe the scene shown in *Figure 14.1.* As you will see, his words are meaningful, but what he says is certainly not grammatical. The dots indicate long pauses.

kid. . . . kk . . . can . . . candy . . . cookie . . . candy . . . well I don't know but it's writ . . . easy does it . . . slam . . . early . . . fall . . . men . . . many no . . . girl. Dishes . . . soap . . . soap . . . water . . . water . . . falling pah that's all . . . dish . . . that's all.

Cookies . . . can . . . candy . . . cookies cookies . . . he . . . down . . . That's all. Girl . . . slipping water . . . water . . . and it hurts . . . much to do . . . Her . . . clean up . . . Dishes . . . up there . . . I think that's doing it. (Obler and Gjerlow, 1999, p. 41)

Broca's aphasia A form of aphasia characterized by agrammatism, anomia, and extreme difficulty in speech articulation.

function word A preposition, article, or other word that conveys little of the meaning of a sentence but is important in specifying its grammatical structure.

content word A noun, verb, adjective, or adverb that conveys meaning.

FIGURE 14.1 ■ Assessment of Aphasia

The drawing of the kitchen story is part of the Boston Diagnostic Aphasia Test.

People with Broca's aphasia can comprehend speech much better than they can produce it. In fact, some observers have said that their comprehension is unimpaired, but as we will see, this is not quite true. Broca (1861) suggested that this form of aphasia is produced by a lesion of the frontal association cortex, just anterior to the face region of the primary motor cortex. Subsequent research proved him to be essentially correct, and we now call the region **Broca's area.** (See *Figure 14.2.*)

Lesions that produce Broca's aphasia are certainly centered in the vicinity of Broca's area. However, damage that is restricted to the cortex of Broca's area does not appear to produce Broca's aphasia; the damage

FIGURE 14.2 ■ Speech Areas

The figure shows the location of the primary speech areas of the brain. (Wernicke's area will be described later.)

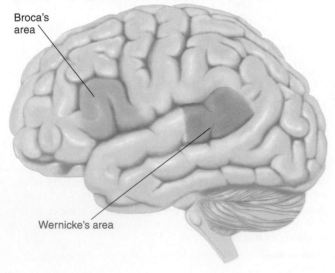

Broca's area

Wernicke's area

must extend to surrounding regions of the frontal lobe and to the underlying subcortical white matter (H. Damasio, 1989; Naeser et al., 1989). In addition, there is evidence that lesions of the basal ganglia—especially the head of the caudate nucleus—can also produce a Broca-like aphasia (Damasio, Eslinger, and Adams, 1984).

Watkins et al. (2002a, 2002b) studied three generations of the KE family, half of whose members are affected by a severe speech and language disorder caused by the mutation of a single gene found on chromosome 7. The primary deficit appears to involve the ability to perform the sequential movements necessary for speech, but the affected people also have difficulty repeating sounds they hear and forming the past tense of verbs. The mutation causes abnormal development of the caudate nucleus and the left inferior frontal cortex, including Broca's area.

What do the neural circuits in and around Broca's area do? Wernicke (1874) suggested that Broca's area contains motor memories—in particular, *memories of the sequences of muscular movements that are needed to articulate words.* Talking involves rapid movements of the tongue, lips, and jaw, and these movements must be coordinated with each other and with those of the vocal cords; thus, talking requires some very sophisticated motor control mechanisms. Obviously, circuits of neurons somewhere in our brain will, when properly activated, cause these sequences of movements to be executed. Because damage to the inferior caudal left frontal lobe (including Broca's area) disrupts the ability to articulate words, this region is a likely candidate for the location of these "programs." The fact that this region is directly connected to the part of the primary motor cortex that controls the muscles used for speech certainly supports this conclusion.

But the speech functions of the left frontal lobe include more than programming the movements used to speak. Broca's aphasia is much more than a deficit in pronouncing words. In general, three major speech deficits are produced by lesions in and around Broca's area: *agrammatism, anomia,* and *articulation difficulties.* Although most patients with Broca's aphasia will have all of these deficits to some degree, their severity can vary considerably from person to person—presumably, because their brain lesions differ. You can also hear the voice of an agrammatic patient and one with articulation difficulties in *MyPsychKit 14.1, Voices of Aphasia: Broca's Aphasia.*

mypsychkit

Animation 14.1
Voices of Aphasia:
Broca's Aphasia

Broca's area A region of frontal cortex, located just rostral to the base of the left primary motor cortex, that is necessary for normal speech production.

Agrammatism refers to a patient's difficulty in using grammatical constructions. This disorder can appear all by itself, without any difficulty in pronouncing words (Nadeau, 1988). As we saw, people with Broca's aphasia rarely use function words. In addition, they rarely use grammatical markers such as *-ed* or auxiliaries such as *have* (as in *I have gone*). For some reason they *do* often use *-ing*, perhaps because this ending converts a verb into a noun. A study by Saffran, Schwartz, and Marin (1980) illustrates this difficulty. The following quotations are from agrammatic patients attempting to describe pictures:

Picture of a boy being hit in the head by a baseball

The boy is catch . . . the boy is hitch . . . the boy is hit the ball. (Saffran, Schwartz, and Marin, 1980, p. 229)

Picture of a girl giving flowers to her teacher

Girl . . . wants to . . . flowers . . . flowers and wants to. . . . The woman . . . wants to. . . . The girl wants to . . . the flowers and the woman. (Saffran, Schwartz, and Marin, 1980, p. 234)

So far, I have described Broca's aphasia as a disorder in speech *production*. In an ordinary conversation, Broca's aphasics seem to understand everything that is said to them. They appear to be irritated and annoyed by their inability to express their thoughts well, and they often make gestures to supplement their scanty speech. The striking disparity between their speech and their comprehension often leads people to assume that their comprehension is normal. But it is not. Schwartz, Saffran, and Marin (1980) showed Broca's aphasics pairs of pictures in which agents and objects of the action were reversed: for example, a horse kicking a cow and a cow kicking a horse, a truck pulling a car and a car pulling a truck, and a dancer applauding a clown and a clown applauding a dancer. As they showed each pair of pictures, they read the subject a sentence, for example, *The horse kicks the cow*. The subjects' task was to point to the appropriate picture, indicating whether they understood the grammatical construction of the sentence. (See *Figure 14.3.*) They performed very poorly.

The correct picture in the study by Schwartz and her colleagues was specified by a particular aspect of grammar: word order. The agrammatism that accompanies Broca's aphasia appears to disrupt patients' ability to use grammatical information, including word order, to decode the meaning of a sentence. Thus, their deficit in comprehension parallels their deficit in production. If they heard a sentence such as *The man swats the mosquito*, they would understand that it concerns a man and a mosquito and the action of swatting. They would have no trouble figuring out who is doing what to whom. But

FIGURE 14.3 ■ Assessment of Grammatical Ability

This is an example of the stimuli used in the experiment by Schwartz, Saffran, and Marin (1980).

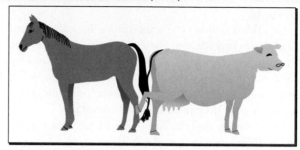

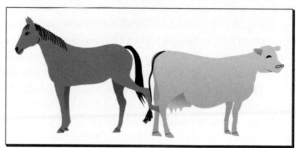

a sentence such as *The horse kicks the cow* does not provide any extra cues; if the grammar is not understood, neither is the meaning of the sentence.

Functional-imaging studies by Opitz and Friederici (2003, 2007) found that Broca's area was activated when people were taught an artificial grammar, which supports the conclusion that this region is involved in learning grammatical rules—especially complex ones. Sakai et al. (2002) had subjects read sentences that were correct, grammatically incorrect, or semantically incorrect (that is, did not make sense). While the subjects were judging the grammatical or semantic correctness of the sentences, the investigators applied transcranial magnetic stimulation to Broca's area. The parameters of stimulation were chosen to activate Broca's area, not disrupt its functioning. The investigators found that the stimulation facilitated grammatical judgments but not semantic judgments.

The second major speech deficit seen in Broca's aphasia is **anomia** ("without name"). Anomia refers to a word-finding difficulty; and because all aphasics omit words or use inappropriate ones, anomia is

---

agrammatism One of the usual symptoms of Broca's aphasia; a difficulty in comprehending or properly employing grammatical devices, such as verb endings and word order.

anomia Difficulty in finding (remembering) the appropriate word to describe an object, action, or attribute; one of the symptoms of aphasia.

actually a primary symptom of *all* forms of aphasia. However, because the speech of Broca's aphasics lacks fluency, their anomia is especially apparent; their facial expression and frequent use of sounds such as "uh" make it obvious that they are groping for the correct words.

The third major characteristic of Broca's aphasia is *difficulty with articulation*. Patients mispronounce words, often altering the sequence of sounds. For example, *lipstick* might be pronounced "likstip." People with Broca's aphasia recognize that their pronunciation is erroneous, and they usually try to correct it.

These three deficits are seen in various combinations in different patients, depending on the exact location of the lesion and, to a certain extent, on their stage of recovery. We can think of these deficits as constituting a hierarchy. On the lowest, most elementary level is control of the sequence of movements of the muscles of speech; damage to this ability leads to articulation difficulties. The next higher level is selection of the particular "programs" for individual words; damage to this ability leads to anomia. Finally, the highest level is selection of grammatical structure, including word order, use of function words, and word endings; damage to this ability leads to agrammatism.

We might expect that the direct control of articulation would involve the face area of the primary motor cortex and portions of the basal ganglia, while the selection of words, word order, and grammatical markers would involve Broca's area and adjacent regions of the frontal association cortex. Some studies indicate that

FIGURE 14.4 ■ The Insular Cortex

The cortex is normally hidden behind the rostral temporal lobe.

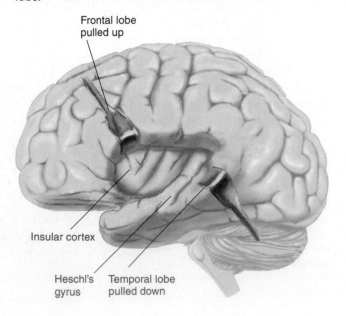

Frontal lobe pulled up

Insular cortex

Heschl's gyrus

Temporal lobe pulled down

FIGURE 14.5 ■ Involvement of the Insular Cortex in Speech Articulation

Evidence for involvement is shown by the percentage overlap in the lesions of twenty-five patients (a) with apraxia of speech and (b) without apraxia of speech. The only region common to all lesions that produced apraxia of speech was the precentral gyrus of the insular cortex.

(From Dronkers, N. F. *Nature*, 1996, *384*, 159–161. Reprinted with permission.)

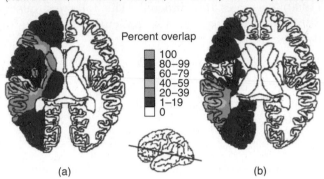

Percent overlap

100
80–99
60–79
40–59
20–39
1–19
0

(a)   (b)

different categories of symptoms of Broca's aphasia do indeed involve different brain regions. Dronkers (1996) appears to have found a critical location for control of speech articulation: the left precentral gyrus of the insula. The insular cortex is located on the lateral wall of the cerebral hemisphere behind the anterior temporal lobe. Normally, this region is hidden and can be seen only when the temporal lobe is dissected away. (See *Figure 14.4.*) Dronkers discovered the apparent role of this region by plotting the lesions of patients with and without apraxia of speech who had strokes that damaged the same general area of the brain. (**Apraxia of speech** is an impairment in the ability to program movements of the tongue, lips, and throat that are required to produce the proper sequence of speech sounds.) Figure 14.5(a) shows the overlap of the lesions of twenty-five patients with apraxia of speech. As you can see, a region of 100 percent overlap, shown in yellow, falls on the left precentral gyrus of the insula. (See *Figure 14.5a.*) In contrast, *none* of the lesions of nineteen patients who did not show apraxia of speech included damage to this region. (See *Figure 14.5b.*)

At least two functional-imaging studies support Dronkers's conclusion. Kuriki, Mori, and Hirata (1999) and Wise et al. (1999) found that pronunciation of words caused activation of the left anterior insula. However, other studies suggest that Broca's area is also involved in articulation (Nestor et al., 2003; Hillis et al.,

apraxia of speech Impairment in the ability to program movements of the tongue, lips, and throat required to produce the proper sequence of speech sounds.

2004). Stewart et al. (2001) used transcranial magnetic stimulation (TMS) to activate neurons in Broca's area or the adjacent area of primary motor cortex, which controls the muscles used for speech. The subjects reported that stimulation of the motor cortex made them feel as though they had lost control of their facial muscles. In contrast, stimulation of Broca's area made them feel as if they were unable to "get the word out."

Most of us have, at one time or other, had difficulty getting a word out even though the word was one that we knew well. This phenomenon has been called the "tip of the tongue phenomenon," or TOT. Shafto et al. (2007) found that people who often had difficulty thinking of the correct word to say but were sure that they knew it (that is, often had a TOT experience) showed loss of gray matter in the left insular cortex. These findings, too, support the role of this region in articulation.

Speech Comprehension

Comprehension of speech obviously begins in the auditory system, which detects and analyzes sounds. But *recognizing* words is one thing; *comprehending* them—understanding their meaning—is another. Recognizing a spoken word is a complex perceptual task that relies on memories of sequences of sounds. This task appears to be accomplished by neural circuits in the middle and posterior portion of the superior temporal gyrus of the left hemisphere, a region that has come to be known as **Wernicke's area.** (Refer to *Figure 14.2.*)

Wernicke's Aphasia: Description

The primary characteristics of **Wernicke's aphasia** are poor speech comprehension and production of meaningless speech. Unlike Broca's aphasia, Wernicke's aphasia is fluent and unlabored; the person does not strain to articulate words and does not appear to be searching for them. The patient maintains a melodic line, with the voice rising and falling normally. When you listen to the speech of a person with Wernicke's aphasia, it appears to be grammatical. That is, the person uses function words such as *the* and *but* and employs complex verb tenses and subordinate clauses. However, the person uses few content words, and the words that he or she strings together just do not make sense. In the extreme, speech deteriorates into a meaningless jumble, illustrated by the following quotation:

Examiner: What kind of work did you do before you came into the hospital?

Patient: Never, now mista oyge I wanna tell you this happened when happened when he rent. His—his kell come down here and is—he got ren something. It happened. In thesse ropiers

were with him for hi—is friend—like was. And it just happened so I don't know, he did not bring around anything. And he did not pay it. And he roden all o these arranjen from the pedis on from iss pescid. In these floors now and so. He hadn't had em round here. (Kertesz, 1981, p. 73)

Because of the speech deficit of people with Wernicke's aphasia, when we try to assess their ability to comprehend speech, we must ask them to use nonverbal responses. That is, we cannot assume that they do not understand what other people say to them just because they do not give the proper answer. A commonly used test of comprehension assesses their ability to understand questions by pointing to objects on a table in front of them. For example, they are asked to "Point to the one with ink." If they point to an object other than the pen, they have not understood the request. When tested this way, people with severe Wernicke's aphasia do indeed show poor comprehension.

A remarkable fact about people with Wernicke's aphasia is that they often seem unaware of their deficit. That is, they do not appear to recognize that their speech is faulty, nor do they recognize that they cannot understand the speech of others. They do not look puzzled when someone tells them something, even though they obviously cannot understand what they hear. Perhaps their comprehension deficit prevents them from realizing that what they say and hear makes no sense. They still follow social conventions, taking turns in conversation with the examiner, even though they do not understand what the examiner says and what they say in return makes little sense. They remain sensitive to the other person's facial expression and tone of voice and begin talking when he or she asks a question and pauses for an answer. One patient with Wernicke's aphasia made the following responses when asked to name ten common objects.

> *toothbrush* → "stoktery"
> *cigarette* → "cigarette"
> *pen* → "tankt"
> *knife* → "nike"
> *fork* → "fahk"
> *quarter* → "minkt"
> *pen* → "spentee"
> *matches* → "senktr"
> *key* → "seek"
> *comb* → "sahk"

Wernicke's area A region of auditory association cortex on the left temporal lobe of humans, which is important in the comprehension of words and the production of meaningful speech.

Wernicke's aphasia A form of aphasia characterized by poor speech comprehension and fluent but meaningless speech.

He acted sure of himself and gave no indication that he recognized that most of his responses were meaningless. The responses that he made were not simply new words that he had invented; he was asked several times to name the objects and gave different responses each time (except for *cigarette*, which he always named correctly). You can hear the speech of people with Wernicke's aphasia in *MyPsychKit 14.1, Voices of Aphasia: Wernicke's Aphasia.*

mypsychkit

Animation 14.1
Voices of Aphasia:
Wernicke's Aphasia

Wernicke's Aphasia: Analysis

Because the superior temporal gyrus is a region of auditory association cortex, and because a comprehension deficit is so prominent in Wernicke's aphasia, this disorder has been characterized as a *receptive* aphasia. Wernicke suggested that the region that now bears his name is the location of *memories of the sequences of sounds that constitute words.* This hypothesis is reasonable; it suggests that the auditory association cortex of the superior temporal gyrus recognizes the sounds of words, just as the visual association cortex of the inferior temporal gyrus recognizes the sight of objects.

But why should damage to an area that is responsible for the ability to recognize spoken words disrupt people's ability to speak? In fact, it does not; Wernicke's aphasia, like Broca's aphasia, actually appears to consist of several deficits. The abilities that are disrupted include *recognition of spoken words, comprehension of the meaning of words,* and the *ability to convert thoughts into words.* Let us consider each of these abilities in turn.

Recognition: Pure Word Deafness.   As I said in the introduction to this section, *recognizing* a word is not the same as *comprehending* it. If you hear a foreign word several times, you will learn to recognize it; but unless someone tells you what it means, you will not comprehend it. Recognition is a perceptual task; comprehension involves retrieval of additional information from memory.

Damage to the left temporal lobe can produce a disorder of auditory word recognition, uncontaminated by other problems. This syndrome is called **pure word deafness.** (See *Figure 14.6.*) Although people with pure word deafness are not deaf, they cannot understand speech. As one patient put it, "I can hear you talking, I just can't understand what you're saying." Another said, "It's as if there were a bypass somewhere, and my ears were not connected to my voice" (Saffran, Marin, and Yeni-Komshian, 1976, p. 211). These patients can recognize nonspeech sounds such as the barking of a dog, the sound of a doorbell, and the honking of a horn. Often, they can recognize the emotion expressed by the intonation of speech even though they cannot understand what is being said. More significantly, their own speech

FIGURE 14.6 ■ Pure Word Deafness

An MRI scan shows the damage to the superior temporal lobe of a patient with pure word deafness (arrow).

(From Stefanatos, G. A., Gershkoff, A., and Madigan, S. *Journal of the International Neuropsychological Society*, 2005, *11*, 456–470. Reprinted with permission.)

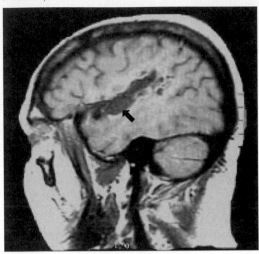

is excellent. They can often understand what other people are saying by reading their lips. They can also read and write, and they sometimes ask people to communicate with them in writing. Clearly, pure word deafness is not an inability to comprehend the meaning of words; if it were, people with this disorder would not be able to read people's lips or read words written on paper.

Functional-imaging studies confirm that perception of speech sounds activates neurons in the auditory association cortex of the superior temporal gyrus. For example, Scott et al. (2000) identified a region of the left anterior superior temporal gyrus was specifically activated by intelligible speech. (See *Figure 14.7.*) Sharp, Scott, and Wise (2004) found that deficits in speech comprehension were produced by lesions of the superior temporal lobe that damaged the region that is activated when people hear intelligible speech. Figure 14.8 shows a computer-generated depiction of the overlap of lesions of patients with brain damage that interfered with speech perception. Compare the regions of greatest overlap, shown in yellow and green, in *Figure 14.8* with the region shown in *Figure 14.7.*

What is involved in the analysis of speech sounds? Just what tasks does the auditory system have to accomplish? And what are the differences in the functions of the auditory association cortex of the left and right hemispheres? Most researchers believe that the left

pure word deafness   The ability to hear, to speak, and (usually) to read and write without being able to comprehend the meaning of speech; caused by damage to Wernicke's area or disruption of auditory input to this region.

FIGURE 14.7 ■ Responses to Speech Sounds

Results of PET scans indicate the regions of the superior temporal lobe that respond to speech sounds. *Red:* These regions responded to phonetic information (normal speech sounds or a computerized transformation speech that preserved the complexity of the speech sounds but rendered it unintelligible). *Orange:* This region responded to intelligible speech (normal speech sounds or a computerized transformation that removed most normal frequencies but preserved intelligibility).

(Adapted from Scott, S. K., Blank, E. C., Rosen, S., and Wise, R. J. S. *Brain,* 2000, *123,* 2400–2406.)

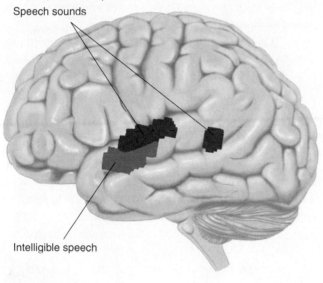

Speech sounds

Intelligible speech

FIGURE 14.8 ■ Speech Comprehension

The scan shows the overlap in the lesions of nine patients with deficits of speech comprehension. Note the similarity of the region of greatest overlap (yellow and green) and the regions that responded to speech sounds in Figure 14.7.

(From Sharp, D. J., Scott, S. K., and Wise, R. J. S. *Annals of Neurology,* 2004, *56,* 836–846. Reprinted with permission of John Wiley & Sons, Inc.)

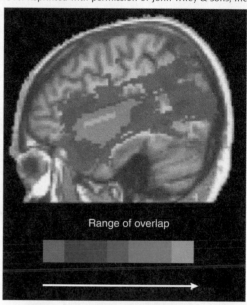

Range of overlap

hemisphere is primarily involved in judging the timing of the components of rapidly changing complex sounds, whereas the right hemisphere is primarily involved in judging more slowly changing components, including melody. Evidence suggests that the most crucial aspect of speech sounds is timing, not pitch. We can recognize words whether they are conveyed by the low pitch of a man or the high pitch of a woman or child. In fact, as you can hear in MyPsychKit 14.2, we can understand speech from which almost all tonal information has been removed, leaving only some noise modulated by the rapid stops and starts that characterize human speech sounds. (Listen to *MyPsychKit 14.2, Speech Perception.*)

mypsychkit
Where learning comes to life!
Animation 14.2
Speech Perception

Apparently, two types of brain injury can cause pure word deafness: disruption of auditory input to Wernicke's area or damage to Wernicke's area itself. Disruption of auditory input can be produced by bilateral damage to the primary auditory cortex, or it can be caused by damage to the white matter in the left temporal lobes that cuts axons bringing auditory information from the

primary auditory cortex to Wernicke's area (Poeppel, 2001; Stefanatos, Gershkoff, and Madigan, 2005). Either type of damage—disruption of auditory input or damage to Wernicke's area—disturbs the analysis of the sounds of words and hence prevents people from recognizing other people's speech. (See *Figure 14.9.*)

FIGURE 14.9 Brain Damage That Causes Pure Word Deafness

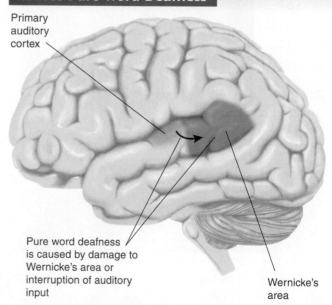

Primary auditory cortex

Pure word deafness is caused by damage to Wernicke's area or interruption of auditory input

Wernicke's area

As we saw in Chapter 8, our brains contain circuits of *mirror neurons*—neurons activated either when we perform an action or see the action performed by someone else. Feedback from these neurons may help us to understand the intent of the actions of others. Although speech recognition is clearly an auditory event, research indicates that hearing words automatically engages brain mechanisms that control speech. In other words, these mechanisms appear also to contain mirror neurons that are activated by the sounds of words. Several investigators have suggested that feedback from subvocal articulation (very slight movements of the muscles involved in speech that do not actually cause obvious movement) facilitate speech recognition. Fadiga et al. (2002) had Italian-speaking subjects listen to real words and pronounceable pseudowords that did or did not involve strong movement of the tongue. For example, the word *birra* ("beer") and the pseudoword *berro* require tongue movements, but the word *buffo* ("funny") and the pseudoword *biffo* do not. The investigators found that the excitability of the subjects' tongue muscles was increased only when they heard words that involved tongue movements. (See *Figure 14.10.*)

FIGURE 14.10 ■ Mirror Neurons and Speech

The graph shows excitability of tongue muscles when listening to syllables containing a sound produced by the tongue ("rr") or not produced by the tongue ("ff").

(Adapted from Fadiga, L., Craighero, L., Buccino, G., and Rizzolatti, G. *European Journal of Neuroscience*, 2002, *15*, 399–402.)

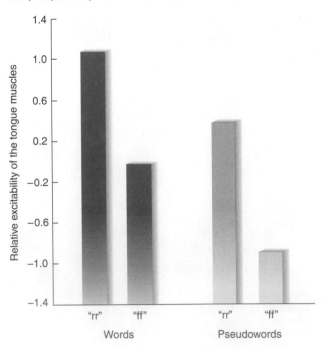

My experience with a patient I met several years ago suggests that monitoring of one's own speech plays an important role in the production of accurate and fluent speech.

Dr. D. introduced Mr. S., a patient with pure word deafness.

"Mr. S., will you tell us how you are feeling?" asked Dr. D.

The patient turned his head at the sound of his voice and said, "Sorry, I can't understand you."

"*How are you feeling?*" Dr. D. asked in a loud voice.

"Oh, I can hear you all right, I just can't understand you. Here," said Mr. S., handing Dr. D. a pencil and a small pad of paper.

Dr. D. took the pencil and paper and wrote something. He handed them back to Mr. S., who looked at it and said, "Fine. I'm just fine."

"Will you tell us about what you have been doing lately?" asked Dr. D. Mr. S. smiled, shook his head, and handed the paper and pencil to Dr. D. again.

"Oh sure," he said after reading the new question, and he proceeded to tell us about his garden and his other hobbies. "I don't get much from television unless there are a lot of close-ups, where I can read their lips. I like to listen to music on the radio, but, of course, the lyrics don't mean too much to me!" He laughed at his own joke, which had probably already seen some mileage.

"You mean that you can read lips?" someone asked.

Mr. S. immediately turned toward the sound of the voice and said, "What did you say? Say it slow, so I can try to read your lips." We all laughed, and Mr. S. joined us when the question was repeated slowly enough for him to decode. Another person tried to ask him a question, but apparently his Spanish accent made it impossible for Mr. S. to read his lips.

Suddenly, the phone rang. We all, including Mr. S., looked up at the wall where it was hanging. "Someone else had better get that," he said. "I'm not much good on the phone."

After Mr. S. had left the room, someone observed that although Mr. S.'s speech was easy to understand, it seemed a bit strange. "Yes," said a speech therapist, "he almost sounds like a deaf person who has learned to talk but doesn't get the pronunciation of the words just right."

Dr. D. nodded and played a tape for us. "This recording was made a few months after his strokes, ten years ago." We heard the same voice, but this time it sounded absolutely normal.

"Oh," said the speech therapist. "He has lost the ability to monitor his own speech, and over the years he has forgotten some of the details of how various words are pronounced."

"Exactly," said Dr. D. "The change has been a gradual one."

FIGURE 14.11 ■ Transcortical Sensory Aphasia and Wernicke's Aphasia

This schematic shows the location and interconnections of the posterior language area and an explanation of its role in transcortical sensory aphasia and Wernicke's aphasia.

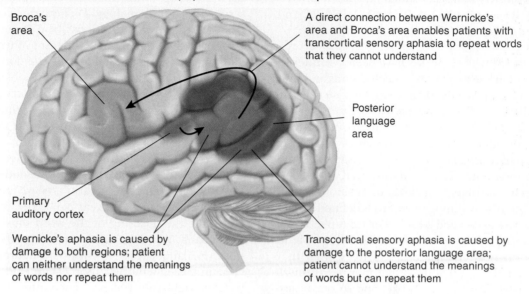

Broca's area

A direct connection between Wernicke's area and Broca's area enables patients with transcortical sensory aphasia to repeat words that they cannot understand

Posterior language area

Primary auditory cortex

Wernicke's aphasia is caused by damage to both regions; patient can neither understand the meanings of words nor repeat them

Transcortical sensory aphasia is caused by damage to the posterior language area; patient cannot understand the meanings of words but can repeat them

Comprehension: Transcortical Sensory Aphasia.

The other symptoms of Wernicke's aphasia—failure to comprehend the meaning of words and inability to express thoughts in meaningful speech—appear to be produced by damage that extends beyond Wernicke's area into the region that surrounds the posterior part of the lateral fissure, near the junction of the temporal, occipital, and parietal lobes. For want of a better term, I will refer to this region as the *posterior language area.* (See *Figure 14.11.*) The posterior language area appears to serve as a place for interchanging information between the auditory representation of words and the meanings of these words, stored as memories in the rest of the sensory association cortex.

Damage to the posterior language area alone, which isolates Wernicke's area from the rest of the posterior language area, produces a disorder known as **transcortical sensory aphasia.** (See *Figure 14.11.*) The difference between transcortical sensory aphasia and Wernicke's aphasia is that patients with transcortical sensory aphasia *can repeat what other people say to them;* therefore, they can recognize words. However, *they cannot comprehend the meaning of what they hear and repeat; nor can they produce meaningful speech of their own.* How can these people repeat what they hear? Because the posterior language area is damaged, repetition does not involve this part of the brain. Obviously, there must be a direct connection between Wernicke's area and Broca's area that bypasses the posterior language area. (See *Figure 14.11.*)

A woman sustained extensive brain damage from carbon monoxide produced by a faulty water heater. She spent several years in the hospital before she died, without ever saying anything meaningful on her own. She did not follow verbal commands or otherwise give signs of understanding them. However, she often repeated what was said to her. For example, if an examiner said "Please raise your right hand," she would reply, "Please raise your right hand." The repetition was not parrotlike; she did not imitate accents different from her own, and if someone made a grammatical error while saying something to her, she sometimes repeated the sentence correctly, without the error. She could also recite poems if someone started them. For example, when an examiner said, "Roses are red, violets are blue," she continued with "Sugar is sweet and so are you." She could sing and would do so when someone started singing a song she knew. She even learned new songs from the radio while in the hospital. Remember, though, that she gave *no signs of understanding anything she heard or said.* This disorder, transcortical sensory aphasia, along with pure word deafness, clearly confirms the conclusion that *recognizing* spoken words and *comprehending* them are different processes and involve different brain mechanisms (Geschwind, Quadfasel, and Segarra, 1968).

transcortical sensory aphasia A speech disorder in which a person has difficulty comprehending speech and producing meaningful spontaneous speech but can repeat speech; caused by damage to the region of the brain posterior to Wernicke's area.

In conclusion, transcortical sensory aphasia can be seen as Wernicke's aphasia without a repetition deficit. To put it another way, the symptoms of Wernicke's aphasia consist of those of pure word deafness plus those of transcortical sensory aphasia. As I tell my students, WA = TSA + PWD. By simple algebra, TSA = WA − PWD, and so on. (Refer to *Figure 14.11*.)

What Is Meaning?   As we have seen, Wernicke's area is involved in the analysis of speech sounds and thus in the recognition of words. Damage to the posterior language area does not disrupt people's ability to recognize words, but it does disrupt their ability to understand words or to produce meaningful speech of their own. But what, exactly, do we mean by the word *meaning*? And what types of brain mechanisms are involved?

Words refer to objects, actions, or relationships in the world. Thus, the meaning of a word is defined by particular memories associated with it. For example, knowing the meaning of the word *tree* means being able to imagine the physical characteristics of trees: what they look like, what the wind sounds like blowing through their leaves, what the bark feels like, and so on. It also means knowing facts about trees: about their roots, buds, flowers, nuts, and wood and the chlorophyll in their leaves. These memories are stored not in the primary speech areas but in other parts of the brain, especially regions of the association cortex. Different categories of memories may be stored in particular regions of the brain, but they are somehow tied together, so hearing the word *tree* activates all of them. (As we saw in Chapter 14,

the hippocampal formation is involved in this process of tying related memories together.)

In thinking about the brain's verbal mechanisms involved in recognizing words and comprehending their meaning, I find that the concept of a dictionary serves as a useful analogy. Dictionaries contain entries (the words) and definitions (the meanings of the words). In the brain we have at least two types of entries: auditory and visual. That is, we can look up a word according to how it sounds or how it looks (in writing). Let us just consider just one type of entry: the sound of a word. (I will discuss reading and writing later in this chapter.) We hear a familiar word and understand its meaning. How do we do so?

First, we must recognize the sequence of sounds that constitute the word—we find the auditory entry for the word in our "dictionary." As we saw, this entry appears in Wernicke's area. Next, the memories that constitute the meaning of the word must be activated. Presumably, Wernicke's area is connected—through the posterior language area—with the neural circuits that contain these memories. (See *Figure 14.12*.)

The Hebb rule, which we encountered in Chapter 13, can be invoked to explain the acquisition of words and their meanings. Recall that the Hebb rule says that when interconnected neurons are repeatedly active at the same time, the synaptic connections between them are strengthened. Thus, when we hear a word several times, a particular set of neurons in the superior temporal lobe become active, and their interconnections eventually become strengthened. (We could also hear the word

FIGURE 14.12 ■ The "Dictionary" in the Brain

Wernicke's area contains the auditory entries of words; the meanings are contained as memories in the sensory association areas. Black arrows represent comprehension of words—the activation of memories that correspond to a word's meaning. Red arrows represent translation of thoughts or perceptions into words.

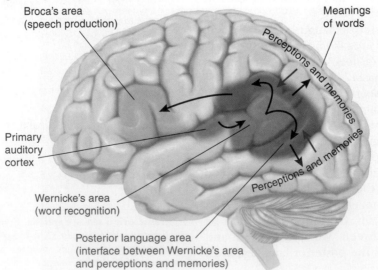

Broca's area
(speech production)

Meanings
of words

Perceptions and memories

Primary
auditory
cortex

Wernicke's area
(word recognition)

Perceptions and memories

Posterior language area
(interface between Wernicke's area
and perceptions and memories)

once and repeat it to ourselves, thus activating these neurons enough to strengthen their interconnections.) As Hebb put it, the coactivated neurons became a *cell assembly*—an assembly of interconnected neurons—in this case, an assembly that recognizes the sound of a particular word.

What evidence do we have that meanings of words are represented by cell assemblies located in various regions of the association cortex? The best evidence comes from the fact that damage to particular regions of the sensory association cortex can damage particular kinds of information and thus abolish particular kinds of meanings.

I met a patient who had recently had a stroke that damaged a part of her right parietal lobe that played a role in spatial perception. She was alert and intelligent and showed no signs of aphasia. However, she was confused about directions and other spatial relationships. When asked to, she could point to the ceiling and the floor, but she could not say which was *over* the other. Her perception of other people appeared to be entirely normal, but she could not say whether a person's head was at the *top* or *bottom* of the body.

I wrote a set of multiple-choice questions to test her ability to use words denoting spatial relations. The results of the test indicated that she did not know the meaning of words such as *up, down,* and *under* when they referred to spatial relationships but that she could use these words normally when they referred to nonspatial relations. For example, here are some of her incorrect responses when the words referred to spatial relations:

A tree's branches are *under* its roots.

The sky is *down.*

The ceiling is *under* the floor.

She made only ten correct responses on the sixteen-item test. In contrast, she got all eight items correct when the words referred to nonspatial relationships such as the following:

After exchanging pleasantries, they got *down* to business.

He got sick and threw *up.*

Damage to part of the association cortex of the *left* parietal lobe can produce an inability to name the body parts. The disorder is called **autotopagnosia,** or "poor knowledge of one's own topography." (A better name would have been *autotopanomia,* "poor naming of one's own topography.") People who can otherwise converse normally cannot reliably point to their elbow, knee, or cheek when asked to do so and cannot name body parts when the examiner points to them. However, they have no difficulty understanding the meaning of other words.

As we saw in Chapter 13, lesions of the anterolateral temporal lobe result in *semantic dementia*—loss of semantic

FIGURE 14.13 ■ Evaluating Metaphors

These images of neural activity were produced by evaluating the meaning of metaphors.

(From Sotillo, M., Carretié, L., Hinojosa, J. A., Tapia, M., Mercado, F., López-Martín, S., and Albert, J. *Neuroscience Letters,* 2005, 373, 5–9. Copyright © 2005, with permission from Elsevier.)

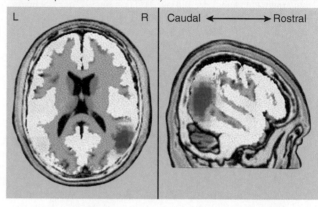

memories, including the names and even the functions of everyday objects. But speech also conveys abstract concepts, some of them quite subtle. Studies of brain-damaged patients (Brownell et al., 1983, 1990) suggest that comprehension of the more subtle, figurative aspects of speech involves the right hemisphere in particular—for example, understanding the meaning behind proverbs such as "People who live in glass houses shouldn't throw stones" or the moral of stories such as the one about the race between the tortoise and the hare.

Functional-imaging studies confirm these observations. Nichelli et al. (1995) found that judging the moral of Aesop's fables (in contrast to judging more superficial aspects of the stories) also activated regions of the right hemisphere. Sotillo et al. (2005) found that a task that required comprehension of metaphors such as "green lung of the city" (that is, a park) activated the right superior temporal cortex. (See *Figure 14.13.*)

Pobric et al. (2008) found that a temporary disruption of the right superior temporal cortex by means of transcranial magnetic stimulation impaired people's ability to understand novel metaphors, such as a "conscience storm." The stimulation had no effect on the subjects' ability to understand conventional metaphors, such as "sweet voice," that they have undoubtedly already heard or literal expressions such as "snow storm." (Expressions such as "sweet voice" are so common that many people do not realize that they are metaphorical. Sugar or honey are sweet, but a voice cannot be tasted.)

Repetition: Conduction Aphasia.   As we saw earlier in this section, the fact that people with transcortical

---

autotopagnosia Inability to name body parts or to identify body parts that another person names.

FIGURE 14.14 ■ Conduction Aphasia

MRI scans show the subcortical damage responsible for a case of conduction aphasia. This lesion damaged the arcuate fasciculus, a fiber bundle connecting Wernicke's area and Broca's area.

(From Arnett, P. A., Rao, S. M., Hussain, M., Swanson, S. J., and Hammeke, T. A. *Neurology*, 1996, *47*, 576–578. Reprinted with permission.)

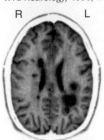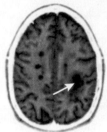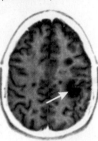

sensory aphasia can repeat what they hear suggests that there is a direct connection between Wernicke's area and Broca's area—and there is: the **arcuate fasciculus** ("arch-shaped bundle"). This bundle of axons appears to convey information about the *sounds* of words but not their *meanings*. The best evidence for this conclusion comes from a syndrome known as conduction aphasia, which is produced by damage to the inferior parietal lobe that extends into the subcortical white matter and damages the arcuate fasciculus (Damasio and Damasio, 1980). (See *Figure 14.14*.)

Conduction aphasia is characterized by meaningful, fluent speech and relatively good comprehension but very poor repetition. For example, the spontaneous speech of patient L. B. (observed by Margolin and Walker, 1981) was excellent; he made very few errors and had no difficulty naming objects. But let us see how patient L. B. performed when he was asked to repeat words. (You can hear this person's voice in *MyPsychKit 14.1, Voices of Aphasia: Conduction Aphasia.*)

mypsychkit
Animation 14.1
Voices of Aphasia:
Conduction Aphasia

Examiner: bicycle

Patient: bicycle

Examiner: hippopotamus

Patient: hippopotamus

Examiner: blaynge

Patient: I didn't get it.

Examiner: Okay, some of these won't be real words, they'll just be sounds. Blaynge.

Patient: I'm not . . .

Examiner: blanch

Patient: blanch

Examiner: north

Patient: north

Examiner: rilld

Patient: Nope, I can't say.

You will notice that the patient can repeat individual words (all nouns, in this case) but utterly fails to repeat nonwords. And as you can hear in the animation, he can repeat a meaningful three-word phrase but not three unrelated words. People with conduction aphasia can repeat speech sounds that they hear *only if these sounds have meaning.*

Sometimes, when a person with conduction aphasia is asked to repeat a word, he or she says a word with the same meaning—or at least one that is related. For example, if the examiner says *house*, the patient may say *home*. If the examiner says *chair*, the patient may say *sit*. One patient made the following response when asked to repeat an entire sentence:

Examiner: The auto's leaking gas tank soiled the roadway.

Patient: The car's tank leaked and made a mess on the street.

The symptoms that are seen in transcortical sensory aphasia and conduction aphasia lead to the conclusion that there are pathways connecting the speech mechanisms of the temporal lobe with those of the frontal lobe. The direct pathway through the arcuate fasciculus simply conveys speech sounds from Wernicke's area to Broca's area. We use this pathway to repeat unfamiliar words—for example, when we are learning a foreign language or a new word in our own language or when we are trying to repeat a nonword such as *blaynge*. The second pathway, between the posterior language area and Broca's area, is indirect and is based on the *meaning* of words, not on the sounds they make. When patients with conduction aphasia hear a word or a sentence, the meaning of what they hear evokes some sort of image related to that meaning. (The patient in the second example presumably imagined the sight of an automobile leaking fuel onto the pavement.) They are then able to describe that image, just as they would put their own thoughts into words. Of course, the words they choose might not be the

arcuate fasciculus A bundle of axons that connects Wernicke's area with Broca's area; damage causes conduction aphasia.

conduction aphasia An aphasia characterized by inability to repeat words that are heard but the ability to speak normally and comprehend the speech of others.

FIGURE 14.15 ■ A Hypothetical Explanation of Conduction Aphasia

A lesion that damages the arcuate fasciculus disrupts the transmission of auditory information, but not information related to meaning, to the frontal lobe.

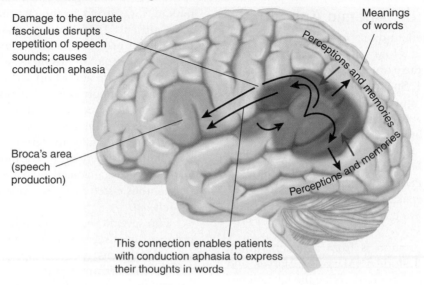

Damage to the arcuate fasciculus disrupts repetition of speech sounds; causes conduction aphasia

Meanings of words

Perceptions and memories

Broca's area (speech production)

Perceptions and memories

This connection enables patients with conduction aphasia to express their thoughts in words

same as the ones used by the person who spoke to them. (See *Figure 14.15.*)

A study by Catani, Jones, and ffytche (2005) provides the first anatomical evidence for the existence of the two pathways between Wernicke's area and Broca's area that I presented in Figure 14.15. The investigators used a special modification of a magnetic resonance imaging (MRI) apparatus that provides the ability to trace large fiber bundles in the brain—a procedure known as *diffusion tensor imaging.* The authors found one deep pathway that directly connects these two regions and a shallower pathway that consists of two segments. The anterior segment connects Broca's area with the inferior parietal cortex, and the posterior segment connects Wernicke's area with the inferior parietal cortex. Damage to the direct pathway would be expected to produce conduction aphasia, whereas damage to the indirect pathway would be expected to spare the ability to repeat speech but would impair comprehension. (See *Figure 14.16.*)

The symptoms of conduction aphasia suggest that the connection between Wernicke's area and Broca's area plays an important role in short-term memory of words and speech sounds that have just been heard. Presumably, rehearsal of such information can be accomplished by "talking to ourselves" inside our head without actually having to say anything aloud. Imagining ourselves saying the word activates the region of Broca's area, whereas imagining that we are hearing it activates the auditory association area of the temporal lobe. These two regions, connected by means of the arcuate fasciculus (which contains axons traveling in *both* directions)

circulate information back and forth, keeping the short-term memory alive. Baddeley (1993) refers to this circuit as the *phonological loop.*

Aziz-Zadeh et al. (2005) obtained evidence that we do use Broca's area when we talk to ourselves. The investigators applied TMS to Broca's area while people were silently counting the number of syllables in words presented on a screen. The parameters of stimulation

FIGURE 14.16 ■ Components of the Arcuate Fasciculus

A computer-generated reconstruction of the components of the arcuate fasciculus was obtained through diffusion tensor imaging.

(From Catani, M., Jones, D. K., and ffytche, D. H. *Annals of Neurology,* 2005, *57,* 8–16. Reprinted with permission.)

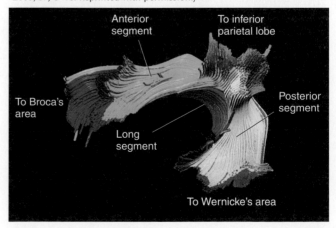

Anterior segment

To inferior parietal lobe

To Broca's area

Posterior segment

Long segment

To Wernicke's area

that the investigators used disrupts overt (actual) speech. They found that it disrupted covert speech as well; the subjects took longer to count the syllables when the TMS was on.

Memory of Words: Anomic Aphasia

As I have already noted, anomia, in one form or other, is a hallmark of aphasia. However, one category of aphasia consists of almost pure anomia, the other symptoms being inconsequential. Speech of patients with anomic aphasia is fluent and grammatical, and their comprehension is excellent, but they have difficulty finding the appropriate words. They often employ **circumlocutions** (literally, "speaking in a roundabout way") to get around missing words. Anomic aphasia is different from Wernicke's aphasia. People with anomic aphasia can understand what others say, and what these people say makes perfect sense, even if they often choose roundabout ways to say it.

The following quotation is from a patient whom some colleagues and I studied (Margolin, Marcel, and Carlson, 1985). We asked her to describe the kitchen picture shown earlier, in *Figure 14.1*. Her pauses, which are marked with three dots, indicate word-finding difficulties. In some cases, when she could not find a word, she supplied a definition instead (a form of circumlocution) or went off on a new track. I have added the words in brackets that I think she intended to use. (You can hear this person's voice on *MyPsychKit 14.1, Voices of Aphasia: Anomic Aphasia.*)

mypsychkit™
Where learning comes to life!

Animation 14.1
Voices of Aphasia:
Anomic Aphasia

Examiner: Tell us about that picture.

Patient: It's a woman who has two children, a son and a daughter, and her son is to get into the . . . cupboard in the kitchen to get out [*take*] some . . . cookies out of the [*cookie jar*] . . . that she possibly had made, and consequently he's slipping [*falling*] . . . the wrong direction [*backward*] . . . on the . . . what he's standing on [*stool*], heading to the . . . the cupboard [*floor*] and if he falls backwards he could have some problems [*get hurt*], because that [*the stool*] is off balance.

Anomia has been described as a partial amnesia for words. It can be produced by lesions in either the anterior or posterior regions of the brain, but only posterior lesions produce a *fluent* anomia. The most likely location of lesions that produce anomia without the other symptoms of aphasia, such as comprehension deficits, agrammatism, or difficulties in articulation, is the left temporal or parietal lobe, usually sparing Wernicke's area. In the case of the woman described

above, the damage included the left middle and inferior temporal gyri, which includes an important region of the visual association cortex. Wernicke's area was not damaged.

When my colleagues and I were studying the anomic patient, I was struck by the fact that she seemed to have more difficulty finding nouns than other types of words. I informally tested her ability to name actions by asking her what people shown in a series of pictures were doing. She made almost no errors in finding verbs. For example, although she could not say what a boy was holding in his hand, she had no trouble saying that he was *throwing* it. Similarly, she knew that a girl was *climbing* something but could not tell me the name of what she was climbing (a fence). In addition, she had no trouble finding nonvisual adjectives; for example, she could say that lemons tasted *sour,* that ice was *cold,* and that a cat's fur felt *soft.*

For several years I thought that our patient was unique. But other researchers have reported similar patterns of deficits. For example, Semenza and Zettin (1989) and Manning and Campbell (1992) described patients who had difficulty naming objects but not actions. Several studies have found that anomia for verbs (more correctly called *averbia*) is caused by damage to the frontal cortex, in and around Broca's area (Damasio and Tranel, 1993; Daniele et al., 1994; Bak et al., 2001). If you think about it, that makes sense. The frontal lobes are devoted to planning, organizing, and executing actions, so it should not surprise us that they are involved in the task of remembering the names of actions.

Shapiro, Shelton, and Caramazza (2000) reported the case of J. R., a man with a fluent anomic aphasia who, like our patient, had more difficulty with nouns than verbs. For example, J. R. had no difficulty completing the following sentence: "These people *sail*; this person _____" (*sails*). However, he had difficulty with sentences like this: "These are *sails*; this is a _____" (*sail*). Note that J. R. had difficulty with the word "sail" only when it was a noun.

Several functional-imaging studies have confirmed the importance of Broca's area and the region surrounding it in the production of verbs. For example, Hauk, Johnsrued, and Pulvermüller (2004) had subjects read verbs that related to movements of different parts of the body. For example, *bite, slap,* and *kick* involve movements of the face, arm, and leg, respectively. The investigators found that when the subjects read a verb, they saw activation in the regions of the motor cortex

circumlocution A strategy by which people with anomia find alternative ways to say something when they are unable to think of the most appropriate word.

FIGURE 14.17 ■ Verbs and Movements

The figure shows the relative activation of regions of the motor cortex that control movements of the face, arm, and leg when people read verbs that described movements of these regions, such as *bite, slap,* and *kick*.

(Adapted from From Hauk, O., Johnsrude, I., and Pulvermüller, F. *Neuron,* 2004, *41*, 301–307.)

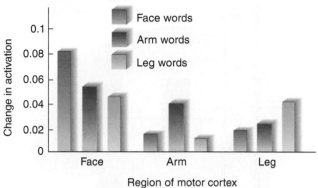

that controlled the relevant part of the body. (See *Figure 14.17*.) A similar study by Buccino et al. (2005) found that hearing sentences that involved hand movements (for example, *He turned the key*) activated the hand region of the motor cortex and that hearing sentences that involved foot movements (for example, *He stepped on the grass*) activated the foot region. Presumably, thinking about particular actions activated regions that control these actions.

The picture I have drawn so far suggests that comprehension of speech includes a flow of information from Wernicke's area to the posterior language area to various regions of sensory and motor association cortex, which contain memories that provide meanings to words. Production of spontaneous speech involves the flow of information concerning perceptions and memories from the sensory and motor association cortex to the posterior language area to Broca's area. This model is certainly an oversimplification, but it is a useful starting point in conceptualizing basic mental processes. For example, thinking in words probably involves two-way communication between the speech areas and surrounding association cortex (and regions such as the hippocampus and medial temporal lobe, of course).

Aphasia in Deaf People

So far, I have restricted my discussion to brain mechanisms of spoken and written language. But communication among members of the Deaf community involves another medium: sign language. Sign language is expressed manually, by movements of the hands. Sign language is *not* English; nor is it French, Spanish, or Chinese. The most common sign language in North America is ASL—American Sign Language. ASL is a full-fledged language, having signs for nouns, verbs, adjectives, adverbs, and all the other parts of speech contained in oral languages. People can converse rapidly and efficiently by means of sign language, can tell jokes, and can even make puns based on the similarity between signs. They can also use their language ability to think in words.

Some researchers believe that in the history of our species, sign language preceded spoken language—that our ancestors began using gestures to communicate before they switched to speech. As I mentioned earlier in this chapter, mirror neurons become active when we see or perform particular grasping, holding, or manipulating movements. Some of these neurons are found in Broca's area. Presumably, these neurons play an important role in learning to mimic other people's hand movements. Indeed, they might have been involved in the development of hand gestures used for communication in our ancestors, and they undoubtedly are used by deaf people when they communicate by sign language. A functional-imaging study by Iacoboni et al. (1999) found that Broca's area was activated when people observed and imitated finger movements. (See *Figure 14.18*.)

Several studies have found a linkage between speech and hand movements, which supports the suggestion that the spoken language of present-day humans evolved from hand gestures. For example, Gentilucci (2003) had subjects speak the syllables *ba* or *ga* while they were watching him grasp objects of different sizes. When the experimenter grasped a large object, the subjects opened their mouths more and said the syllable more loudly than when he grasped a small one. These results suggest that the region of the brain that controls grasping is also involved in controlling speech movements. (See *Figure 14.19*.)

The grammar of ASL is based on its visual, spatial nature. For example, if a person makes the sign for *John* in one place and later makes the sign for *Mary* in another place, she can place her hand in the *John* location and move it toward the *Mary* location while making the sign for *love*. As you undoubtedly figured out for yourself, she is saying, "John loves Mary." Signers can also modify the meaning of signs through facial expressions or the speed and vigor with which they make a sign. Thus, many of the prepositions, adjectives, and adverbs found in spoken languages do not require specific words in ASL. The fact that signed languages are based on three-dimensional hand and arm movements accompanied by facial expressions means that their grammars are very different from those of spoken languages. Therefore, a word-for-word translation from a spoken language to a signed language (or vice versa) is impossible.

The fact that the grammar of ASL is spatial suggests that aphasic disorders in deaf people who use sign

FIGURE 14.24 ■ Pure Alexia

This letter was written to Dr. Elizabeth Warrington by a patient with pure alexia. The letter reads as follows: "Dear Dr. Warrington, Thank you for your letter of September the 16th. I shall be pleased to be at your office between 10–10:30 am on Friday 17th October. I still find it very odd to be able to write this letter but not to be able to read it back a few minutes later. I much appreciate the opportunity to see you. Yours sincerely, Harry X."

(From McCarthy, R. A., and Warrington, E. K. *Cognitive Neuropsychology: A Clinical Introduction.* San Diego: Academic Press, 1990. Reprinted with permission.)

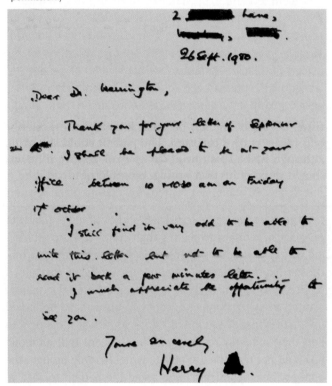

mypsychkit

Animation 14.3
Pure Alexia

patient could not read. (*MyPsychKit 14.3, Pure Alexia,* also illustrates the brain damage responsible for this disorder.) The first diagram shows the pathway that visual information would take if a person had damage *only to the left primary visual cortex.* In this case the person's right visual field would be blind; he or she would see nothing to the right of the fixation point. But people with this disorder can read. Their only problem is that they must look to the right of each word so that they can see all of it, which means that they read somewhat more slowly than someone with full vision. (See *Figure 14.25a.*)

Let us trace the flow of visual information for a person with this brain damage. Information from the left side of the visual field is transmitted to the right striate cortex (primary visual cortex) and then to regions of right visual association cortex. From there the information crosses the posterior corpus callosum and is transmitted to the left visual association cortex, where it is analyzed further. (I will describe these regions later.) The information is then transmitted to speech mechanisms located in the left frontal lobe. Thus, the person can read the words aloud. (See *Figure 14.25a.*)

The second diagram shows Dejerine's patient. Notice how the additional lesion of the corpus callosum prevents visual information concerning written text from reaching the posterior left hemisphere. Without this information the patient cannot read. (See *Figure 14.25b.*)

Mao-Draayer and Panitch (2004) reported the case of a man with multiple sclerosis who displayed the symptoms of pure alexia after sustaining a lesion that damaged both the subcortical white matter of the left occipital lobe and the posterior corpus callosum. As you can see in Figure 14.26, the lesions are in precisely the locations that Dejerine predicted would cause this syndrome. (See *Figure 14.26.*)

I must note that the diagrams shown in Figure 14.26 are as simple and schematic as possible. They illustrate only the pathway involved in seeing a word and pronouncing it, and they ignore neural structures that would be involved in understanding its meaning. As we will see later in this chapter, evidence from patients with brain lesions indicates that seeing and pronouncing words can take place independently of understanding them. Thus, although the diagrams are simplified, they are not unreasonable, given what we know about the neural components of the reading process.

You will recall that writing is not the only form of visible language; deaf people can communicate by means of sign language just as well as hearing people can communicate by means of spoken language. Hickok et al. (1995) reported on a case of "sign blindness" caused by damage similar to that which causes pure alexia. The patient, a right-handed deaf woman, sustained a stroke that damaged her left occipital lobe and the posterior corpus callosum. The lesion did not impair her ability to sign in coherent sentences, so she did not have a Wernicke-like aphasia. However, she could no longer understand other people's sign language, and she lost her ability to read. She had some ability to comprehend single signs (corresponding to single words), but she could not comprehend signed sentences.

Toward an Understanding of Reading

Reading involves at least two different processes: direct recognition of the word as a whole and sounding it out letter by letter. When we see a familiar word, we normally

FIGURE 14.25 ■ Pure Alexia

In this schematic, red arrows indicate the flow of information that has been interrupted by brain damage. (a) The route followed by information as a person with damage to the left primary visual cortex reads aloud. (b) Additional damage to the posterior corpus callosum interrupts the flow of information and produces pure alexia.

Damage to left primary visual cortex causes blindness in right visual field

Broca's area

Lateral geniculate nucleus

Wernicke's area

Extrastriate cortex receives information from left visual field through corpus callosum

Left primary visual cortex is destroyed

Information from left visual field

Damage to posterior corpus callosum prevents information from right extrastriate cortex from reaching left hemisphere

(a)   (b)

FIGURE 14.26 ■ Pure Alexia in a Patient with Multiple Sclerosis

The damage corresponds to that shown in Figure 14.25.

(From Mao-Draayer, Y., and Panitch, H. *Multiple Sclerosis,* 2004, *10,* 705–707. Reprinted with permission.)

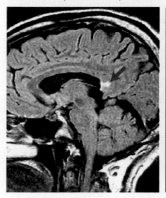
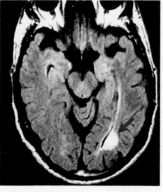

recognize it and pronounce it—a process known as **whole-word reading.** (With very long words we might instead perceive segments of several letters each.) The second method, which we use for unfamiliar words, requires recognition of individual letters and knowledge of the sounds they make. This process is known as **phonetic reading.**

Evidence for our ability to sound out words is easy to obtain. In fact, you can prove to yourself that phonetic reading exists by trying to read the following words:

glab trisk chint

whole-word reading Reading by recognizing a word as a whole; "sight reading."

phonetic reading Reading by decoding the phonetic significance of letter strings; "sound reading."

Well, as you could see, they are not really words, but I doubt that you had trouble pronouncing them. Obviously, you did not *recognize* them, because you probably never saw them before. Therefore, you had to use what you know about the sounds that are represented by particular letters (or groups of letters, such as *ch*) to figure out how to pronounce the words.

The best evidence that proves that people can read words without sounding them out, using the whole-word method, comes from studies of patients with acquired dyslexias. *Dyslexia* means "faulty reading." *Acquired* dyslexias are those caused by damage to the brains of people who already know how to read. In contrast, *developmental* dyslexias refer to reading difficulties that become apparent when children are learning to read. Developmental dyslexias, which may involve anomalies in brain circuitry, are discussed in a later section.

Figure 14.27 illustrates some elements of the reading processes. The diagram is an oversimplification of a very complex process, but it helps to organize some of the facts that investigators have obtained. It considers only reading and pronouncing single words, not understanding the meaning of text. When we see a familiar word, we normally recognize it as a whole and pronounce it. If we see an unfamiliar word or a pronounceable nonword, we must try to read it phonetically. (See *Figure 14.27.*)

Investigators have reported several types of acquired dyslexias, and I will describe three of them in this section: surface dyslexia, phonological dyslexia, and direct dyslexia. **Surface dyslexia** is a deficit in whole-word

FIGURE 14.27 ■ Model of the Reading Process

In this simplified model, whole-word reading is used for most familiar words; phonetic reading is used for unfamiliar words and for nonwords such as *glab, trisk,* or *chint.*

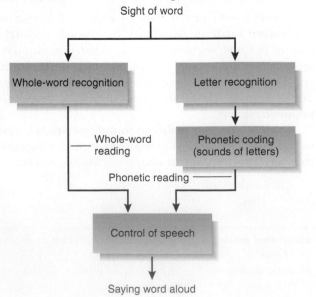

FIGURE 14.28 ■ Surface Dyslexia

In this hypothetical example, whole-word reading is damaged; only phonetic reading remains.

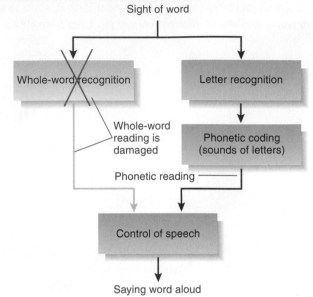

reading. The term *surface* reflects the fact that people with this disorder make errors related to the visual appearance of the words and to pronunciation rules, not to the meaning of the words, which is metaphorically "deeper" than the appearance.

Because patients with surface dyslexia have difficulty recognizing words as a whole, they are obliged to sound them out. Thus, they can easily read words with regular spelling, such as *hand, table,* or *chin.* However, they have difficulty reading words with irregular spelling, such as *sew, pint,* and *yacht.* In fact, they may read these words as *sue, pinnt,* and *yatchet.* They have no difficulty reading pronounceable nonwords, such as *glab, trisk,* and *chint.* Because people with surface dyslexia cannot recognize whole words by their appearance, they must, in effect, listen to their own pronunciation to understand what they are reading. If they read the word *pint* and pronounce it *pinnt,* they will say that it is not an English word (which it is not, pronounced that way). If the word is one member of a homophone, it will be impossible to understand it unless it is read in the context of a sentence. For example, if you hear the single word "pair" without additional information, you cannot know whether the speaker is referring to *pair, pear,* or *pare.* Thus, a patient with surface dyslexia who reads the word *pair* might say, ". . . it could be two of a kind, apples and . . . or what you do with your fingernails" (Gurd and Marshall, 1993, p. 594). (See *Figure 14.28.*)

surface dyslexia A reading disorder in which a person can read words phonetically but has difficulty reading irregularly spelled words by the whole-word method.

The symptoms of **phonological dyslexia** are opposite those of surface dyslexia: People with this disorder can read by the whole-word method but cannot sound words out. Thus, they can read words that they are already familiar with but have great difficulty figuring out how to read unfamiliar words or pronounceable nonwords (Beauvois and Dérouesné, 1979; Dérouesné and Beauvois, 1979). (In this context, *phonology*—loosely translated as "laws of sound"—refers to the relationship between letters and the sounds they represent.) People with phonological dyslexia may be excellent readers if they had already acquired a good reading vocabulary before their brain damage occurred.

Phonological dyslexia provides further evidence that whole-word reading and phonological reading involve different brain mechanisms. Phonological reading, which is the only way we can read nonwords or words we have not yet learned, entails some sort of letter-to-sound decoding. Obviously, phonological reading of English requires more than decoding of the sounds produced by single letters, because, for example, some sounds are transcribed as two-letter sequences (such as *th* or *sh*) and the addition of the letter *e* to the end of a word lengthens an internal vowel (*can* becomes *cane*). (See *Figure 14.29*.)

The Japanese language provides a particularly interesting distinction between phonetic and whole-word reading. The Japanese language makes use of two kinds of written symbols. *Kanji* symbols are pictographs, adopted from the Chinese language (although they are pronounced as Japanese words). Thus, they represent concepts by means of visual symbols but do not provide a guide to their pronunciation. Reading words expressed in kanji symbols is analogous, then, to whole-word reading. *Kana* symbols are phonetic representations of syllables; thus, they encode acoustical information. These symbols are used primarily to represent foreign words or Japanese words that the average reader would be unlikely to recognize if they were represented by their kanji symbols. Reading words expressed in kana symbols is obviously phonetic.

Studies of Japanese people with localized brain damage have shown that the reading of kana and kanji symbols involves different brain mechanisms (Iwata, 1984; Sakurai et al., 1994; Sakurai, Ichikawa, and Mannen, 2001). Difficulty reading kanji symbols is a form of surface dyslexia, whereas difficulty reading kana symbols is a form of phonological dyslexia. What regions are involved in these two kinds of reading?

Evidence from lesion and functional-imaging studies with readers of English, Chinese, and Japanese suggest that the process of whole-word reading follows the ventral stream of the visual system to the fusiform gyrus, located on the base of the temporal lobe. For example, functional-imaging studies by Thuy et al. (2004) and Liu et al. (2008) found that the reading of kanji words or Chinese characters (whole-word reading) activated the left fusiform gyrus. This region has come to be known as the **visual word-form area (VWFA).** As we saw in Chapter 6, this region is also involved in the perception of faces and other shapes that require expertise to distinguish—and certainly, recognizing whole words or kanji symbols requires expertise. Phonological reading appears to follow the dorsal stream to the region around the junction of the inferior parietal lobe and the superior temporal lobe (the temporoparietal cortex) and then follows a fiber bundle from this region to the inferior frontal cortex, which includes Broca's area (Sakurai et al., 2000; Jobard, Crivello, and Tzourio-Mazoyer, 2003; Thuy et al., 2004; Tan et al., 2005). The fact that phonological reading involves Broca's area suggests that it may actually involve articulation—that we sound out words not so much by "hearing" them in our heads as by feeling ourselves pronounce them silently to ourselves. Once words have been identified—by either means—their meaning must be accessed, which means that the two pathways converge on regions of the brain involved in recognition of word meaning, grammatical structure, and semantics. (See *Figure 14.30*.)

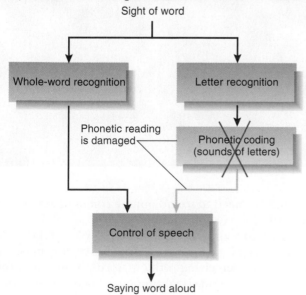

FIGURE 14.29 ■ Phonological Dyslexia

In this hypothetical example, phonetic reading is damaged; only whole-word reading remains.

Sight of word

Whole-word recognition

Letter recognition

Phonetic reading is damaged

Phonetic coding (sounds of letters)

Control of speech

Saying word aloud

phonological dyslexia A reading disorder in which a person can read familiar words but has difficulty reading unfamiliar words or pronounceable nonwords.

visual word-form area (VWFA) A region of the vusiform gyrus on the base of the temporal lobe that plays a critical role in whole-word recognition.

FIGURE 14.30 ■ Phonological and Whole-Word Reading

(a) Phonological reading. (b) Whole-word reading.

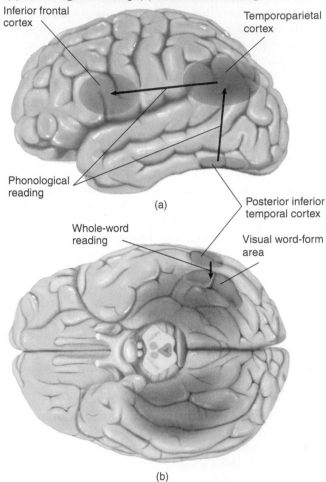

(a)

(b)

FIGURE 14.31 ■ Hearing and Reading Words

Progression of regional activity was measured by magnetoencephalography after hearing or reading a word.

(From Marinkovic, K., Dhond, R. P., Dale, A. M., Glessner, M., Carr, V., and Halgren, E. *Neuron*, 2003, *38*, 487–497. Reprinted with permission from Elsevier.)

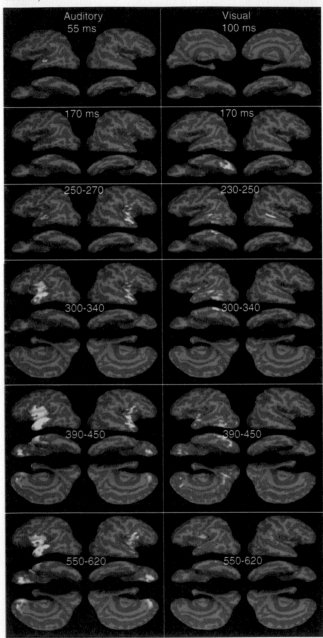

In fact, the neural circuits involved in written and auditory information must also eventually converge, because both must have access to the same linguistic and semantic information that identify words and their meaning. An interesting study by Marinkovic et al. (2003) used magnetoencephalography to trace regional brain activation as people heard or read individual words. As Figure 14.31 shows, neural activation responsible for the analysis of a spoken word begins in the auditory cortex of the temporal lobe and spreads to the auditory association cortex on the superior temporal lobe (including Wernicke's area) and then to the inferior frontal cortex (including Broca's area). The neural activation responsible for the analysis of a printed word begins in the visual cortex and spreads to the inferior temporal cortex and basal temporal lobe (including the fusiform gyrus) and then to the inferior frontal cortex. The temporoparietal cortex receives little activation, presumably because the subjects were fluent readers

who did not need to sound out the common words they were asked to read. (See *Figure 14.31.*)

Let's consider the role of the VWFA. Obviously, some parts of the visual association cortex must be involved in perceiving written words. You will recall from Chapter 6 that visual agnosia is a perceptual deficit in which people with bilateral damage to the visual association cortex cannot recognize objects by sight. However, people with visual agnosia can still read, which means that the perceptual analysis of objects and words involves at least some different brain mechanisms. This fact is both interesting and puzzling. Certainly, the ability to read cannot have shaped the evolution of the human brain, because the invention of writing is only a few thousand years old, and until very recently, the vast majority of the world's population was illiterate. Thus, reading and object recognition use brain mechanisms that undoubtedly existed long before the invention of writing. However, just as experience seeing faces affects the development of the fusiform face area in the right hemisphere, experience learning to read words undoubtedly affects the development of the visual word-form area—which, probably not coincidentally, is found in the fusiform cortex of the left hemisphere (McCandliss, Cohen, and Dehaene, 2003).

At least two written languages were invented by specific individuals. Hangul, the written form of the Korean language, was invented by King Sejong (and his scholars) in the fourteenth century. The characters of the Hangul alphabet are designed to look like the shapes the mouth makes when they are pronounced. In the early nineteenth century, Sequoyah, a Cherokee living in what is now the state of North Carolina, recognized the value of the "talking leaves" that European settlers used to record information and send messages to each other. He spent twelve years developing a written version of his language. At first, he tried to develop pictograms to represent individual words, but he abandoned that attempt when its complexity became obvious. He then analyzed the sounds of his language and selected eighty-five symbols—from English and Greek letters he found in books and some additional ones that he invented. He did not know the sounds that English and Greek letters represented, so the sounds he assigned to them bore no relationship to those of the languages they came from. Within a few months of the introduction of Sequoyah's alphabet, thousands of Cherokees learned to read and write.

The fusiform face area has the ability to quickly recognize unique configurations of people's eyes, noses, lips, and other features of their faces even when the differences between two people's faces are very similar. For example, parents and close friends of identical twins can see at a glance which twin they are looking at. Similarly, our VWFA can recognize a word even if it closely resembles another one. (See *Figure 14.32*.) It can also quickly recognize words written in different *typestyles*, fonts, or CASES. This means that the VWFA can recognize whole words with different shapes; certainly, chair and CHAIR do not look the same. It takes an experienced reader the same amount of time to read equally familiar three-letter

words and six-letter words (Nazir et al., 1998), which means that the whole-word reading process does not have to identify the letters one at a time, just as the face-recognition process in the right fusiform cortex does not have to identify each feature of a face individually before the face is recognized. Instead, we recognize several letters and their locations relative to each other.

A functional-imaging study by Vinckier et al. (2007) investigated the means by which the brain recognizes whole words. First, I need to provide some definitions. A *bigram* is a sequence of two letters (*bi*, "two"; *gram*, "something written"). *Frequent* bigrams are two-letter sequences that are often encountered in a particular language. For example, the bigram *SH* often occurs in English. In contrast, *LQ* is an *infrequent* bigram. *Quadrigrams* are strings of four letters and can be classified as frequent or infrequent. Now let's go to the study. Vinckier and his colleagues had adult readers look at the following stimuli: (1) strings of false fonts (nonsensical letterlike symbols), (2) strings of infrequent letters, (3) strings that contained only infrequent bigrams, (4) strings that contained frequent bigrams, (5) strings that contained frequent quadrigrams, and (6) real words. (See *Figure 14.33* for examples of these stimuli.)

FIGURE 14.32 ■ Subtle Differences in Written words

Unless you can read Arabic, Hindi, or Mandarin, you will probably have to examine these words carefully to find the small differences. However, as a reader of English, you will immediately recognize the words *cars* and *ears*.

(Adapted from Devlin, J. T., Jamison, H. L., Gonnerman, L. M., and Matthews, P. M. *Journal of Cognitive Neuroscience*, 2006, *18*, 911–922.)

English	Arabic	Hindi	Mandarin
cars ears	زمان رمان pomegranate time/era	आज आजा today come	夫 天 man sky

---

FIGURE 14.33 ■ Word Recognition in the VWFA

The diagram shows the structure of stimulus strings used in the experiment by Vinckier et al. (2007) and the frequency of their components. *Mouton* is the French word for sheep. (The experiment took place in France.)

Types of stimuli

False font	Infrequent letters	Frequent letters	Frequent bigrams	Frequent quadrigrams	Words
٦Ƅ⋂┼⊓Γ	JZWYWK	QOADTQ	QUMBSS	AVONIL	MOUTON

Examples

FIGURE 14.34 ■ Word Recognition in the VWFA

The scan shows regions of the brain that selectively responded to letterlike symbols, infrequent letters, frequent letters, bigrams, quadrigrams, and words. This range is indicated by colors that range from red to violet. Response gradients were seen in the VWFA (visual word-form area) and in Broca's area.

(From Vinckier, F., Dehaene, S., Jobert, A., Dubus, J.P., Sigman, M., and Cohen, L. *Neuron*, 2007, *55*, 143–156. Reprinted with permission from Elsevier.)

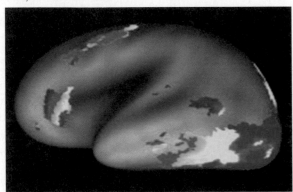

FIGURE 14.35 ■ Effects of VWFA Lesion

The scans show responses of brain regions to words, faces, houses, and tools before and after surgical removal of a small region of the VWFA. Note that the response to words (dark blue) is lost but responses to faces, houses, and tools remain. The lesion is indicated by the green arrowheads.

(From Gaillard, R., Naccache, L., Pinel, P., Clémenceau, S., Volle, E., Hasboun, D., Dupont, S., Baulac, M., Dehaene, S., Adam, C., and Cohen, L. *Neuron*, 2006, *50*, 191–204. Reprinted with permission from Elsevier.)

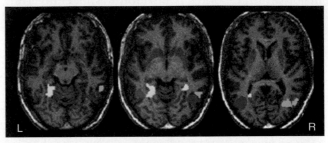

Before surgery

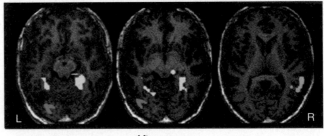

After surgery

Houses   Words   Faces   Tools   Lesion

Functional imaging showed that some brain regions were activated by all of the visual stimuli, including letterlike symbols, some were activated by letters but not symbols, and so on, up to regions that were activated by real words. The most selective region included the left anterior fusiform cortex, was activated only by actual words. In fact, as Figure 14.34 shows, the scans revealed a posterior-to-anterior gradient of selectivity, from symbol to whole word, along the base of the left occipital and temporal lobes. A second, smaller, gradient was seen in Broca's area. Presumably, this gradient represented phonetic reading—decoding of the sounds represented by the stimuli that the subjects viewed. Note that very little activity was produced by letterlike symbols (shown in red) in Broca's area. This makes sense, because there is no way to pronounce these symbols. (See *Figure 14.34.*)

Many studies have found that damage to the VWFA produces surface dyslexia—that is, impairment of whole-word reading. A study by Gaillard et al. (2006) combined fMRI and lesion evidence from a single subject that provides evidence that the left fusiform cortex does, indeed, contain this region. A patient with a severe seizure disorder became a candidate for surgery aimed at removal of a seizure focus. Before the surgery was performed, the patient viewed printed words and pictures of faces, houses, and tools while his brain was being scanned. He was warned that the seizure focus was located in a region that played a critical role in reading, but his symptoms were so severe that he elected to undergo the surgery. As expected, the surgery produced a deficit in whole-word reading. A combination of structural and functional imaging revealed that the lesion—a very small one—was located in the fusiform gyrus, the location of the VWFA. (See *Figure 14.35.*)

What about phonological reading? I mentioned earlier that Thuy et al. (2004) found that phonological reading activated the left temporoparietal cortex and Broca's area. A meta-analysis of thirty-five neuroimaging studies by Joberd, Crivello, and Tzourio-Mazoyer (2003) found that phonological reading activates the left temporoparietal region and Broca's area.

As we saw earlier in this chapter, recognizing a spoken word is different from understanding it. For example, patients with transcortical sensory aphasia can repeat what is said to them even though they show no signs of understanding what they hear or say. The same is true for reading. **Direct dyslexia** resembles transcortical sensory

direct dyslexia A language disorder caused by brain damage in which the person can read words aloud without understanding them.

aphasia, except that the words in question are written, not spoken (Schwartz, Marin, and Saffran, 1979; Lytton and Brust, 1989; Gerhand, 2001). Patients with direct dyslexia are able to read aloud *even though they cannot understand the words they are saying.* After sustaining a stroke that damaged his left frontal and temporal lobes, Lytton and Brust's patient lost the ability to communicate verbally; his speech was meaningless, and he was unable to comprehend what other people said to him. However, he could read words with which he was already familiar. He could *not* read pronounceable nonwords; therefore, he had lost the ability to read phonetically. His comprehension deficit seemed complete; when the investigators presented him with a word and several pictures, one of which corresponded to the word, he read the word correctly but had no idea what picture went with it. Gerhand's patient showed a similar pattern of deficits except that she was able to read phonetically: She could sound out pronounceable nonwords. These findings indicate that the brain regions responsible for phonetic reading and whole-word reading are each directly connected with brain regions responsible for speech.

Toward an Understanding of Writing

Writing depends on knowledge of the words that are to be used, along with the proper grammatical structure of the sentences they are to form. Therefore, if a patient is unable to express himself or herself by speech, we should not be surprised to see a writing disturbance (*dysgraphia*) as well. In addition, most cases of dyslexia are accompanied by dysgraphia.

One type of writing disorder involves difficulties in motor control—in directing the movements of a pen or pencil to form letters and words. Investigators have reported surprisingly specific types of writing disorders that fall into this category. For example, some patients can write numbers but not letters, some can write uppercase letters but not lowercase letters, some can write consonants but not vowels, some can write cursively but cannot print uppercase letters, and others can write letters normally but have difficulty placing them in an orderly fashion on the page (Cubelli, 1991; Alexander et al., 1992; Margolin and Goodman-Schulman, 1992; Silveri, 1996).

Many regions of the brain are involved in writing. For example, damage that produces various forms of aphasia will produce impairments in writing that are similar to those seen in speech. Organization of the motor aspects of writing involves the dorsal parietal lobe and the premotor cortex. These regions (and the primary motor cortex, of course) become activated when people engage in writing, and damage to these regions impairs writing (Otsuki et al., 1999; Katanoda, Yoshikawa, and Sugishita, 2001; Menon and Desmond, 2001). A functional-imaging study

FIGURE 14.36 ■ Writing and the Ventral Premotor Cortex

Subjects viewing letters activated the ventral premotor cortex in the hemisphere used for writing: the left hemisphere in right-handed subjects (yellow) and the right hemisphere in left-handed subjects (red).

(From Longcamp, M., Anton, J.-L., Roth, M., and Velay, J.-L. *Neuropsychologia*, 2005, *43*, 1801–1809. Reprinted with permission.)

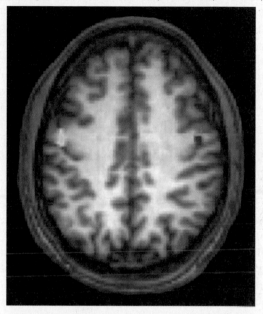

by Rijntjes et al. (1999) had people sign their names with either an index finger or a big toe. In both cases, doing so activated the premotor cortex that controlled movements of the hand. This finding suggests that when we learn to make a complex series of movements, the relevant information is stored in regions of the motor association cortex that control the part of the body that is being used but that this information can be used to control similar movements in other parts of the body. Longcamp et al. (2005) found that simply looking at alphabetical characters activated the premotor cortex: on the left side in right-handed people and on the right side in left-handed people. (See *Figure 14.36.*)

A more basic type of writing disorder involves problems in the ability to spell words, as opposed to problems with making accurate movements of the fingers. I will devote the rest of this section to this type of disorder. Like reading, writing (or, more specifically, spelling) involves more than one method. The first is related to audition. When children acquire language skills, they first learn the sounds of words, then learn to say them, then learn to read, and then learn to write. Undoubtedly, reading and writing depend heavily on the skills that are learned earlier. For example, to write most words, we must be able to "sound them out in our heads," that is, to hear them and to articulate them subvocally. If you want to demonstrate this to yourself, try to write a long word such as *antidisestablishmentarianism* from memory and see whether you can do it without saying the word to yourself. If you recite a poem or sing a song to yourself under your breath at the same time, you will see that the writing comes to a halt.

A second way of writing involves transcribing an image of what a particular word looks like—copying a visual mental image. Have you ever looked off into the distance to picture a word so that you could remember how to spell it? Some people are not very good at phonological spelling and have to write some words down to see whether they look correct. This method obviously involves *visual* memories, not acoustical ones.

A third way of writing involves memorization of letter sequences. We learn these sequences the way we learn poems or the lyrics to a song. For example, many Americans learned to spell *Mississippi* with a singsong chant that goes like this: **M**-i-s-s-**i**-s-s-**i**-p-p-**i**, emphasizing the boldfaced letters. (Similarly, most speakers of English say the alphabet with the rhythm of a nursery song that is commonly used to teach it.) This method involves memorizing sequences of letter names, not translating sounds into the corresponding letters.

Finally, the fourth way of writing involves motor memories. We undoubtedly memorize motor sequences for very familiar words, such as our own names. Most of us need not sound out our names to ourselves when we write our signature, nor need we say the sequence of letters to ourselves, nor need we imagine what our signature looks like.

Writing normally involves holding a pen or pencil and moving its point across a piece of paper. But we can create visual records with the keyboard of a typewriter or a computer. The first three methods of writing (sounding out the letters of a word, visualizing it, or reciting a memorized sequence of letters) apply as well to typing as they do to writing. However, the movements that we make with our hands and fingers are different when we write or type. Skilled typists learn automatic sequences of movements that produce frequently used words, but these are different from the movements we would make when we write these words. Otsuki et al. (2002) reported the case of a man who lost the ability to type after a stroke that damaged the ventral left frontal lobe. His ability to speak and understand speech, his ability to read, and his ability to write were not affected, and he showed no other obvious motor impairments besides his *dystypia,* as the investigators named it.

Neurological evidence supports at least the first three of these speculations. Brain damage can impair the first of these methods: phonetic writing. This deficit is called **phonological dysgraphia** (Shallice, 1981). (*Dysgraphia* refers to a writing deficit, just as *dyslexia* refers to a reading deficit.) People with this disorder are unable to sound out words and write them phonetically.

Thus, they cannot write unfamiliar words or pronounceable nonwords, such as the ones I presented in the section on reading. They can, however, visually imagine familiar words and then write them.

Phonological dysgraphia appears to be caused by damage to regions of the brain involved in phonological processing and articulation. Damage to Broca's area, the ventral precentral gyrus, and the insula cause this disorder, and phonological spelling tasks activate these regions (Omura et al., 2004; Henry et al., 2007).

Orthographic dysgraphia is just the opposite of phonological dysgraphia: It is a disorder of visually based writing. People with orthographic dysgraphia can *only* sound words out; thus, they can spell regular words such as *care* or *tree,* and they can write pronounceable nonsense words. However, they have difficulty spelling irregular words such as *half* or *busy* (Beauvois and Dérouesné, 1981); they may write *haff* or *bizzy.* Orthographic dysgraphia (impaired phonological writing), like surface dyslexia, is caused by damage to the VWFA on the base of the temporal lobe (Henry et al., 2007).

Both lesions studies and functional-imaging studies implicate the posterior inferior temporal cortex in writing of irregularly spelled English words or kanji symbols (Nakamura et al., 2000; Rapscak and Beeson, 2004). This region appears to be involved not in the motor aspects of writing but in knowledge of how irregular words are spelled or what strokes make up a kanji character.

Figure 14.37 shows the brain damage that causes phonological and orthographic dysgraphia. (See *Figure 14.37.*)

The third method of spelling depends on a person's having memorized sequences of letters that spell particular words. Cipolotti and Warrington (1996) reported the case of a patient who lacked this ability. The patient sustained a left hemisphere stroke that severely disrupted his ability to spell words orally and impaired his ability to recognize words that the examiners would spell aloud. Presumably, his ability to spell written words depended on the first two methods of writing: auditory and visual. The examiners noted that when they spelled out words to him, he would make writing movements with hand on top of his knee. When they asked him to clasp his hands together so that he could not make these writing movements, his ability to recognize four-letter words being spelled aloud dropped from 66 percent to 14 percent. It appears that he was using feedback from hand movements to recognize the words he was "writing" on his knee.

As we saw in the section on reading, some patients (those with direct dyslexia) can read aloud without being

phonological dysgraphia A writing disorder in which the person cannot sound out words and write them phonetically.

orthographic dysgraphia A writing disorder in which the person can spell regularly spelled words but not irregularly spelled ones.

FIGURE 14.37 ■ Phonological Dysgraphia and Orthographic Dysgraphia

The scans show the overlap in the lesions of (a) thirteen patients with phonological dysgraphia and (b) eight patients with orthographic dysgraphia. The highest degree of overlap is indicated in red, and the lowest degree is indicated in purple.

(From Henry, M. L., Beeson, P. M., Stark, A. J., and Rapcsak, S. Z. *Brain and Language*, 2007, *100*, 44–52. Reprinted with permission.)

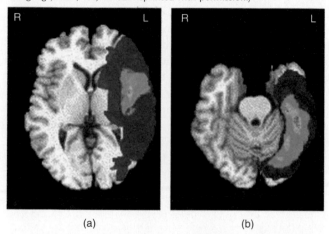

(a)   (b)

able to understand what they are reading. Similarly, some patients can write words that are dictated to them even though they cannot understand these words (Roeltgen, Rothi, and Heilman, 1986; Lesser, 1989). Of course, they cannot communicate by means of writing, because they cannot translate their thoughts into words. (In fact, because most of these patients have sustained extensive brain damage, their thought processes themselves are severely disturbed.) Some of these patients can even spell pronounceable nonwords, which indicates that their ability to spell phonetically is intact. Roeltgen et al. (1986) referred to this disorder as *semantic agraphia*, but perhaps the term *direct dysgraphia* would be more appropriate, because of the parallel with direct dyslexia.

Developmental Dyslexias

Some children have great difficulty learning to read and never become fluent readers, even though they are otherwise intelligent. Specific language learning disorders, called **developmental dyslexias,** tend to occur in families, a finding that suggests a genetic (and hence biological) component. The concordance rate of monozygotic twins ranges from 84 percent to 100 percent, and that of dizygotic twins ranges from 20 percent to 35 percent (Démonet, Taylor, and Chaix, 2004). Linkage studies suggest that the chromosomes 1, 2, 3, 6, 15, and 18 may contain genes responsible for different components of this disorder (Deffenbacher et al., 2004).

As we saw earlier, the fact that written language is a recent invention means that natural selection could not

have given us brain mechanisms whose only role is to interpret written language. Therefore, we should not expect that developmental dyslexia involves only deficits in reading. Indeed, researchers have found a variety of language deficits that do not directly involve reading. One common deficit is deficient phonological awareness. That is, people with developmental dyslexia have difficulty blending or rearranging the sounds of words that they hear (Eden and Zeffiro, 1998). For example, they have difficulty recognizing that if we remove the first sound from "cat," we are left with the word "at." They also have difficulty distinguishing the order of sequences of sounds (Helenius, Uutela, and Hari, 1999). Problems such as these might be expected to impair the ability to read phonetically. Dyslexic children also tend to have great difficulty in writing: They make spelling errors, they show poor spatial arrangements of letters, they omit letters, and their writing tends to have weak grammatical development (Habib, 2000).

Developmental dyslexia is a heterogeneous and complex trait; therefore, it undoubtedly has more than one cause. However, most studies that have closely examined the nature of the impairments seen in people with developmental dyslexia have found phonological impairments to be most common. For example, a study of sixteen dyslexics by Ramus et al. (2003) found that all had phonological deficits. Ten of the people also had auditory deficits, four also had a motor deficit, and two also had a visual deficit. These deficits—especially auditory deficits—aggravated the people's difficulty in reading but did not appear to be primarily responsible for the difficulty. Five of the people had only phonological deficits, and these deficits were sufficient to interfere with their ability to read.

Some evidence has been obtained from functional-imaging studies that suggests that the brains of dyslexics process written information differently than do the brains of proficient readers. For example, Shaywitz et al. (2002) had seventy dyslexic children and seventy-four nondyslexic children read words and pronounceable nonwords. The researchers found significantly different patterns of brain activation in the two groups. A child's reading skill was positively correlated with activation of the left occipitotemporal cortex. Hoeft et al. (2007) found that dyslexics showed decreased activation in the left temporoparietal cortex (dorsal to the region identified by Shaywitz et al.) and in the visual word-form area of the fusiform cortex. They also saw *hyperactivation* of the left inferior frontal cortex, including Broca's area. Presumably, the activation of Broca's area reflected an effort to decode the phonology of the

developmental dyslexia A reading difficulty in a person of normal intelligence and perceptual ability; of genetic origin or caused by prenatal or perinatal factors.

incomplete information being received from the poorly functioning regions of the more posterior brain regions involved in reading.

Most languages—including English—contain many irregular words. For example, consider *cough, rough, bough,* and *through.* Because there is no phonetic rule that describes how these words are to be pronounced, readers of English are obliged to memorize them. In fact, the forty sounds that distinguish English words can be spelled in up to 1120 different ways. In contrast, Italian is much more regular; this language contains twenty-five different sounds that can be spelled in only thirty-three combinations of letters (Helmuth, 2001). Paulesu et al. (2001) found that developmental dyslexia is rare among people who speak Italian and is much more common among speakers of English and French (another language with many irregular words). Paulesu and his colleagues identified college students with a history of dyslexia from Italy, France, and Great Britain. The Italian dyslexics were much harder to find, and their disorders were much less severe than those of their

English-speaking and French-speaking counterparts. However, functional imaging revealed that when all three groups were asked to read, their scans all showed the same pattern of activation: a decrease in the activity of the left occipitotemporal cortex—the same general region that Shaywitz et al. (2002) identified.

Paulesu and his colleagues (2001) concluded that the brain anomalies that cause dyslexia are similar in the three countries they studied but that the regularity of Italian spelling made it much easier for potential dyslexics in Italy to learn to read. By the way, other "dyslexia-friendly" languages include Spanish, Finnish, Czech, and Japanese. One of the authors of this study, Chris D. Frith, cites the case of an Australian boy who lived in Japan. He learned to read Japanese normally but was dyslexic in English (Recer, 2001). If the spelling of words in the English language were regularized (for example, *frend* instead of *friend, frate* instead of *freight, coff* instead of *cough*), many children who develop dyslexia under the present system would develop into much better readers. Somehow, I don't foresee that happening in the near future.

Interim Summary

Disorders of Reading and Writing

Brain damage can produce reading and writing disorders. With few exceptions, aphasias are accompanied by writing deficits that parallel the speech production deficits and by reading deficits that parallel the speech comprehension deficits. Pure alexia is caused by lesions that produce blindness in the right visual field and that destroy fibers of the posterior corpus callosum.

Research in the past few decades has discovered that acquired reading disorders (dyslexias) can fall into one of several categories, and the study of these disorders has provided neuropsychologists and cognitive psychologists with thought-provoking information that has helped them to understand the brain mechanisms involved in reading. Analysis of written words appears to begin in the left posterior inferior temporal cortex. Phonological information is then analyzed by the temporoparietal cortex and Broca's area, whereas word-form information is analyzed by the visual word-form area, located in the fusiform cortex. Surface dyslexia is a loss of whole-word reading ability. Phonological dyslexia is loss of the ability to read phonetically. Reading of kana (phonetic) and kanji (pictographic) symbols by Japanese people is equivalent to phonetic and whole-word reading, and damage to different parts of the brain interfere with these two forms of reading.

Direct dyslexia is analogous to transcortical sensory aphasia; the patients can read words aloud but cannot

understand what they are reading. Some can read both real words and pronounceable nonwords, so both phonetic and whole-word reading can be preserved.

Brain damage can disrupt writing ability by impairing people's ability to form letters—or even specific types of letters, such as uppercase or lowercase letters or vowels. The dorsal parietal cortex appears to be the most critical region for knowledge of the movements that produce letters. Other deficits involve the ability to spell words. We normally use at least four different strategies to spell words: phonetic (sounding the word out), visual (remembering how it looks on paper), sequential (recalling memorized sequences of letters), and motor (recalling memorized hand movements in writing very familiar words). Two types of dysgraphia—phonological and orthographic—represent difficulties in implementing phonetic and visual strategies, respectively. The existence of these two disorders indicates that several different brain mechanisms are involved in the process of writing. One case of dystypia—a specific deficit in the ability to type without other reading or writing disorders—has been reported. In addition, some patients have a deficit parallel to direct dyslexia: They can write words they can no longer understand.

Developmental dyslexia is a hereditary condition that may involve abnormal development of parts of the brain that play a role in language. Most developmental dyslexics have difficulty with phonological processing—of spoken words as well as written ones. Functional-imaging studies report decreased activation of a region of the left occipitotemporal and temporoparietal cortex and hyperactivation of Broca's area may be involved in developmental dyslexia. Children who learn to read languages that have writing with regular correspondence between spelling and pronunciation (such as Italian) are much less likely to become dyslexic than are those who learn to read languages with irregular spelling (such as English or French). A better understanding of the components of reading and writing may help us to develop effective teaching methods that will permit people with dyslexia to take advantage of the abilities that they do have.

Table 14.2 summarizes the disorders that were described in this section.

Thought Question

Suppose someone close to you suffered a head injury that caused phonological dyslexia. What would you do to try to help this person read better? (It would probably be best to build on the person's remaining abilities.) Suppose this person needed to learn to read some words that he or she had never seen before. How would you help the person to do so?

TABLE 14.2 ■ Reading and Writing Disorders Produced by Brain Damage

READING DISORDER	WHOLE-WORD READING	PHONETIC READING	REMARKS
Pure alexia	Poor	Poor	Can write
Surface dyslexia	Poor	Good	
Phonological dyslexia	Good	Poor	
Direct dyslexia	Good	Good	Cannot comprehend words
WRITING DISORDER	**WHOLE-WORD WRITING**	**PHONETIC WRITING**	
Phonological dysgraphia	Good	Poor	
Orthographic dysgraphia	Poor	Good	

SUGGESTED READINGS

Davis, G. A. *Aphasiology: Disorders and Clinical Practice.* Boston: Allyn and Bacon, 2000.

Démonet, J.-F., Taylor, M. J., and Chaix, Y. Developmental dyslexia. *Lancet,* 2004, *363,* 1451–1460.

Démonet, J.-F., Thierry, G., and Cardebat, D. Renewal of the neurophysiology of language: Functional neuroimaging. *Physiological Review,* 2005, *85,* 49–95.

Obler, L. K., and Gjerlow, K. *Language and the Brain.* Cambridge, England: Cambridge University Press, 1999.

Sarno, M. T. *Acquired Aphasia* (3rd ed.). New York: Academic Press, 1998.

Shaywitz, S.E., and Shaywitz, B.A. Dyslexia (specific reading disability). *Biological Psychiatry,* 2005, *57,* 1301–1309.

ADDITIONAL RESOURCES

Visit www.mypsychkit.com for additional review and practice of the material covered in this chapter. Within MyPsychKit, you can take practice tests and receive a customized study plan to help you review. Dozens of animations, tutorials, and Web links are also available. You can even review using the interactive electronic version of this textbook. You will need to register for MyPsychKit. See www.mypsychkit.com for complete details.

TABLE 15.2 ■ The Classification of Seizure Disorders

I. Generalized seizures (with no apparent local onset)

    A. Tonic-clonic (grand mal)

    B. Absence (petit mal)

    C. Atonic (loss of muscle tone; temporary paralysis)

II. Partial seizures (starting from a focus)

    A. Simple (no major change in consciousness)

        1. Localized motor seizure

        2. Motor seizure, with progression of movements as seizure spreads along the primary motor cortex

        3. Sensory (somatosensory, visual, auditory, olfactory, vestibular)

        4. Psychic (forced thinking, fear, anger, etc.)

        5. Autonomic (e.g., sweating, salivating, etc.)

    B. Complex (with altered consciousness)

        Includes 1–5, as above

III. Partial seizures (simple or complex) evolving to generalized cortical seizure: Starts as IIA or IIB, then becomes a grand mal seizure

---

changes in consciousness but do not cause *loss* of consciousness. In contrast, because of their particular location and severity, **complex partial seizures** lead to loss of consciousness. (See *Table 15.2.*)

The most severe form of seizure is often referred to as **grand mal.** This seizure is generalized, and because it includes the motor systems of the brain, it is accompanied by convulsions. Often, before having a grand mal seizure, a person has warning symptoms, such as changes in mood or perhaps a few sudden jerks of muscular activity upon awakening. (Almost everyone sometimes experiences these jolts while falling asleep.) A few seconds before the seizure occurs, the person often experiences an **aura,** which is presumably caused by excitation of neurons surrounding a seizure focus. This excitation has effects similar to those that would be produced by electrical stimulation of the region. Obviously, the nature of an aura varies according to the location of the focus. For example, because structures in the temporal lobe are involved in the control of emotional behaviors, seizures that originate from a focus located there often begin with feelings of fear and dread or, occasionally, euphoria.

The beginning of a grand mal seizure is called the **tonic phase.** All the patient's muscles contract forcefully. The arms are rigidly outstretched, and the person may make an involuntary cry as the tense muscles force air out of the lungs. (At this point the patient is completely unconscious.) The patient holds a rigid posture for about fifteen seconds, and then the **clonic phase** begins. (*Clonic* means "agitated.") The muscles begin trembling, then start jerking convulsively—quickly at first, then more and more slowly. Meanwhile, the eyes roll, the

patient's face is contorted with violent grimaces, and the tongue may be bitten. Intense activity of the autonomic nervous system manifests itself in sweating and salivation. After about thirty seconds, the patient's muscles relax; only then does breathing begin again. The patient falls into a stuporous, unresponsive sleep, which lasts for about fifteen minutes. After that the patient may awaken briefly but usually falls back into an exhausted sleep that may last for a few hours.

Recordings made during grand mal seizures from electrodes implanted into patients' brains show that neural firing first begins in the focus at the time of the aura; it then spreads to other regions of the brain (Adams and Victor, 1981). The activity spreads to regions surrounding the focus and then to the contralateral cortex (through the corpus callosum), the basal ganglia, the thalamus, and various nuclei of the brain stem reticular formation. At this point the symptoms begin. The excited subcortical regions feed back more excitation to the cortex, amplifying the activity there. Neurons in the motor cortex begin firing

complex partial seizure A partial seizure, starting from a focus and remaining localized, that produces loss of consciousness.

grand mal seizure A generalized, tonic-clonic seizure, which results in a convulsion.

aura A sensation that precedes a seizure; its exact nature depends on the location of the seizure focus.

tonic phase The first phase of a grand mal seizure, in which all of the patient's skeletal muscles are contracted.

clonic phase The phase of a grand mal seizure in which the patient shows rhythmic jerking movements.

FIGURE 15.6 ■ Primary Motor Cortex and Seizures

Mrs. R.'s seizure began in the foot region of the primary motor cortex, and as the seizure spread, more and more parts of her body became involved.

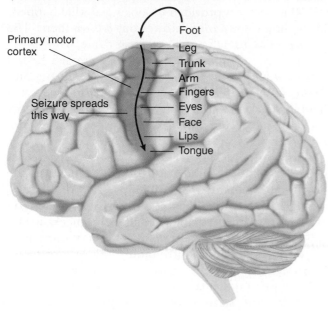

Primary motor cortex

Seizure spreads this way

Foot
Leg
Trunk
Arm
Fingers
Eyes
Face
Lips
Tongue

continuously, producing the tonic phase. Next, diencephalic structures begin quenching the seizure by sending inhibitory messages to the cortex. At first the inhibition comes in brief bursts; this causes the jerking movements of the clonic phase, as the muscles repeatedly relax and then contract again. Then the bursts of inhibition become more and more prolonged, and the jerks occur more and more slowly. Finally, the inhibition wins, and the patient's muscles relax.

Other types of seizures are far less dramatic. Partial seizures involve relatively small portions of the brain. The symptoms can include sensory changes, motor activity, or both. For example, a simple partial seizure that begins in or near the motor cortex can involve jerking movements that begin in one place and spread throughout the body as the excitation spreads along the precentral gyrus. In the case described at the beginning of the chapter I described such a progression, caused by a seizure triggered by a meningioma. The tumor was pressing against the "foot" region of the left primary motor cortex. When the seizure began, it involved the foot; and as it spread, it began involving the other parts of the body. (See *Figure 15.6.*) Mrs. R.'s first spell was a simple partial seizure, but her second one—much more severe—would be classed as a complex partial seizure, because she lost consciousness. A seizure that begins in the occipital lobe may produce visual symptoms such as spots of color, flashes of light, or temporary blindness; one originating in the parietal

lobe can evoke somatosensations, such as feelings of pins and needles or heat and cold. Seizures in the temporal lobes may cause hallucinations that include old memories; presumably, neural circuits involved in these memories are activated by the spreading excitation. Depending on the location and extent of the seizure, the patient may or may not lose consciousness.

Children are especially susceptible to seizure disorders. Many of them do not have grand mal episodes but instead have very brief seizures that are referred to as spells of **absence.** During an absence seizure, which is a *generalized* seizure disorder, they stop what they are doing and stare off into the distance for a few seconds, often blinking their eyes repeatedly. (These spells are also sometimes referred to as *petit mal* seizures.) During this time the children are unresponsive, and they usually do not notice their attacks. Because absence seizures can occur up to several hundred times each day, they can disrupt a child's performance in school. Unfortunately, many of these children are considered to be inattentive and unmotivated unless the disorder is diagnosed.

Seizures can have serious consequences: They can cause brain damage. Approximately 50 percent of patients with seizure disorders show evidence of damage to the hippocampus. The amount of damage is correlated with the number and severity of seizures the patient has had. Significant hippocampal damage can be caused by a single episode of **status epilepticus,** a condition in which the patient undergoes a series of seizures without regaining consciousness. The damage appears to be caused by an excessive release of glutamate during the seizure (Thompson et al., 1996).

Seizures have many causes. The most common cause is scarring, which may be produced by an injury, a stroke, a developmental abnormality, or the irritating effect of a growing tumor. For injuries the development of seizures may take a considerable amount of time. Often, a person who receives a head injury from an automobile accident will not start having seizures until several months later.

Various drugs and infections that cause a high fever can also produce seizures. High fevers are most common in children, and approximately 3 percent of children under the age of 5 years sustain seizures associated with fevers (Berkovic et al., 2006). In addition, seizures are commonly seen in alcohol or barbiturate addicts who suddenly stop taking the drug; the sudden release from

absence A type of seizure disorder often seen in children; characterized by periods of inattention, which are not subsequently remembered; also called *petit mal* seizure.

status epilepticus A condition in which a patient undergoes a series of seizures without regaining consciousness.

the inhibiting effects of the alcohol or barbiturate leaves the brain in a hyperexcitable condition. In fact, this condition is a medical emergency because it can be fatal.

Evidence suggests that NMDA receptors may be involved in the seizures caused by alcohol withdrawal. As you saw in Chapter 12, NMDA receptors are specialized glutamate receptors that control calcium channels. These channels open only when glutamate binds with the receptor *and* the membrane is depolarized. This double contingency is what seems to be responsible for at least one kind of synaptic modification involved in learning. Several studies have shown that alcohol blocks NMDA receptors (Gonzales, 1990). Perhaps, then, long-term suppression of NMDA receptors caused by chronic alcohol intake results in supersensitivity or "up-regulation," a compensatory mechanism produced by long-term inhibition of the receptors. When an alcoholic suddenly stops drinking, the NMDA receptors, which have been suppressed for so long, suddenly rebound. The increased activity causes seizures.

Genetic factors contribute to the incidence of seizure disorders (Berkovic et al., 2006). Nearly all of the genes that have been identified as playing a role in seizure disorders control the production of ion channels, which is not surprising, considering the fact that ion channels control the excitability of the neural membrane and are responsible for the propagation of action potentials. However, most seizure disorders are caused by nongenetic factors. In the past, many cases were considered to be *idiopathic* (of unknown causes, or literally "one's own suffering"). However, the development of MRIs with more and more resolution and sensitivity has meant that small brain abnormalities responsible for triggering seizures are more likely to be seen.

Seizure disorders are treated with anticonvulsant drugs, many of which work by increasing the effectiveness of inhibitory synapses. Most disorders respond well enough that the patient can lead a normal life. In a few instances, drugs provide little or no help. Sometimes, seizure foci remain so irritable that despite drug treatment, brain surgery is required, as we saw in the opening case of Chapter 3. The surgeon removes the region of the brain surrounding the focus (usually located in the medial temporal lobe). Most patients recover well, with their seizures eliminated or greatly reduced in frequency. Mrs. R.'s treatment, described in the opening case of this chapter, was a different matter; in her case the removal of a meningioma eliminated the source of the irritation and ended her seizures. No healthy brain tissue was removed.

Because seizure surgery often involves the removal of a substantial amount of brain tissue (usually from one of the temporal lobes), we might expect it to cause behavioral deficits. But in most cases the reverse is true; people's performance on tests of neuropsychological functioning usually *improves*. How can the removal of brain tissue improve a person's performance?

The answer is provided by looking at what happens in the brain not *during* seizures but *between* them. The seizure focus, usually a region of scar tissue, irritates the brain tissue surrounding it, causing increased neural activity that tends to spread to adjacent regions. Between seizures this increased excitatory activity is held in check by a compensatory increase in inhibitory activity. That is, inhibitory neurons in the region surrounding the seizure focus become more active. (This phenomenon is known as *interictal inhibition*; *ictus* means "stroke" in Latin.) A seizure occurs when the excitation overcomes the inhibition.

The problem is that the compensatory inhibition does more than hold the excitation in check; it also suppresses the normal functions of a rather large region of brain tissue surrounding the seizure focus. Thus, even though the focus may be small, its effects are felt over a much larger area even between seizures. Removing the seizure focus and some surrounding brain tissue eliminates the source of the irritation and makes the compensatory inhibition unnecessary. Freed from interictal inhibition, the brain tissue located near the site of the former seizure focus can now function normally, and the patient's neuropsychological abilities will show an improvement.

Many patients with seizure disorders obtain relief from seizures by following a *ketogenic diet* (Sinha and Kossoff, 2005). Most of the calories on such a diet come from fats, with a moderate amount from proteins and a very low amount from carbohydrates. This diet leads to the production of *ketones*—compounds that are produced when fats are broken down by the liver and the blood level of glucose is low. In such a condition the brain is nourished primarily by ketones. The benefits of a ketogenic diet have been known for at least eighty years, but only recently have researchers begun to investigate how it works (Rho, 2008). A study with rats by Garriga-Canut et al. (2006) administered daily electrical stimulation of the perforant path, the major fiber bundle that brings information to the hippocampus. This treatment eventually establishes seizures, presumably in much the same way as head injuries often establish seizure disorders several weeks or months later. The investigators then administered 2-DG, a drug that interferes with glucose metabolism. They found changes in the levels of several neurochemicals, which may provide clues to aid in the search for more effective antiseizure drugs.

FIGURE 15.7 ■ Strokes

The figure shows (a) the formation of thrombi and emboli and (b) an intracerebral hemorrhage.

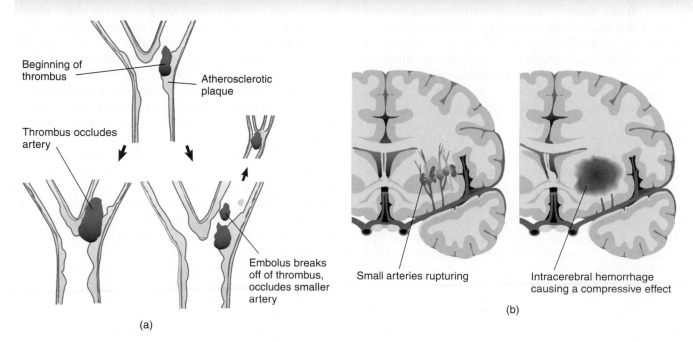

(a)

(b)

Small arteries rupturing

Intracerebral hemorrhage causing a compressive effect

CEREBROVASCULAR ACCIDENTS

You have already learned about the *effects* of cerebrovascular accidents, or *strokes*, in earlier chapters. For example, we saw that strokes can produce impairments in perception, emotional recognition and expression, memory, and language. This section will describe only their causes and treatments.

The incidence of strokes in the United States is approximately 600,000 per year. The likelihood of having a stroke is related to age; the probability doubles each decade after forty-five years of age and reaches 1 to 2 percent per year by age seventy-five (Wolfe et al., 1992). The two major types of strokes are *hemorrhagic* and *obstructive*. **Hemorrhagic strokes** are caused by bleeding within the brain, usually from a malformed blood vessel or from one that has been weakened by high blood pressure. The blood that seeps out of the defective blood vessel accumulates within the brain, putting pressure on the surrounding brain tissue and damaging it. **Obstructive strokes**—those that plug up a blood vessel and prevent the flow of blood—can be caused by thrombi or emboli. (Loss of blood flow to a region is called **ischemia,** from the Greek *ischein,* "to hold back," and *haima,* "blood.") A **thrombus** is a blood clot that forms in blood vessels, especially in places where their walls are already damaged. Sometimes, thrombi become so large

that blood cannot flow through the vessel, causing a stroke. People who are susceptible to the formation of thrombi are often advised to take a drug such as aspirin, which helps to prevent clot formation. An **embolus** is a piece of material that forms in one part of the vascular system, breaks off, and is carried through the bloodstream until it reaches an artery too small to pass through. It lodges there, damming the flow of blood through the rest of the vascular tree (the "branches" and "twigs" arising from the artery). Emboli can consist of a variety of materials, including bacterial debris from an infection in the lining of the heart or pieces broken off from a blood clot. As we will see in a later section, emboli can introduce a bacterial infection into the brain. (See *Figure 15.7.*)

Strokes produce permanent brain damage, but depending on the size of the affected blood vessel, the amount of damage can vary from negligible to massive. If

hemorrhagic stroke A cerebrovascular accident caused by the rupture of a cerebral blood vessel.

obstructive stroke A cerebrovascular accident caused by occlusion of a blood vessel.

ischemia (*is kee mee uh*) The interruption of the blood supply to a region of the body.

thrombus A blood clot that forms within a blood vessel, which may occlude it.

embolus (*em bo lus*) A piece of matter (such as a blood clot, fat, or bacterial debris) that dislodges from its site of origin and occludes an artery; in the brain an embolus can lead to a stroke.

a hemorrhagic stroke is caused by high blood pressure, medication is given to reduce it. If one is caused by weak and malformed blood vessels, brain surgery may be used to seal off the faulty vessels to prevent another hemorrhage. If a thrombus was responsible for the stroke, anticoagulant drugs will be given to make the blood less likely to clot, reducing the likelihood of another stroke. If an embolus broke away from a bacterial infection, antibiotics will be given to suppress the infection.

What, exactly, causes the death of neurons when the blood supply to a region of the brain is interrupted? We might expect that the neurons simply starve to death because they lose their supply of glucose and of oxygen to metabolize it. However, research indicates that the immediate cause of neuron death is the presence of excessive amounts of glutamate. In other words, the damage produced by loss of blood flow to a region of the brain is actually an excitotoxic lesion, just like one produced in a laboratory animal by the injection of a chemical such as kainic acid. (See Koroshetz and Moskowitz, 1996, for a review.)

When the blood supply to a region of the brain is interrupted, the oxygen and glucose in that region are quickly depleted. As a consequence, the sodium-potassium transporters, which regulate the balance of ions inside and outside the cell, stop functioning. Neural membranes become depolarized, which causes the release of glutamate. The activation of glutamate receptors further increases the inflow of sodium ions and causes cells to absorb excessive amounts of calcium through NMDA channels. The presence of excessive amounts of sodium and calcium within cells is toxic. The intracellular sodium causes the cells to absorb water and swell. The inflammation attracts microglia and activates them, causing them to become phagocytic. The phagocytic microglia begin destroying injured cells. Inflammation also attracts white blood cells, which can adhere to the walls of capillaries near the ischemic region and obstruct them. The presence of excessive amounts of calcium in the cells activates a variety of calcium-dependent enzymes, many of which destroy molecules that are vital for normal cell functioning. Finally, damaged mitochondria produce **free radicals**—molecules with unpaired electrons that act as powerful oxidizing agents. Free radicals are extremely toxic; they destroy nucleic acids, proteins, and fatty acids.

Researchers have sought ways to minimize the amount of brain damage caused by strokes. One approach has been to administer drugs that dissolve blood clots in an attempt to reestablish circulation to an ischemic brain region. This approach has met with some success. Administration of a clot-dissolving drug called tPA (tissue plasminogen activator) after the onset of a stroke has clear benefits if it is given within three hours (NINDS, 1995). tPA is an enzyme that converts the

plasminogen, a protein present in the blood, into *plasmin,* an enzyme that dissolves *fibrin,* a protein involved in clot formation. tPA can be synthesized and released by neurons and glia in the central nervous system, and it plays a role in cell migration and neural development.

More recent research indicates that although tPA helps to dissolve blood clots and restore cerebral circulation, it also has toxic effects in the central nervous system. Both tPA and plasmin are potentially neurotoxic if they are able to cross the blood–brain barrier and reach the interstitial fluid. Evidence suggests that in cases of severe stroke, in which the blood–brain barrier is damaged, tPA increases excitotoxicity, further damages the blood–brain barrier, and may even cause cerebral hemorrhage (Benchenane et al., 2004; Klaur et al., 2004). In cases in which tPA quickly restores blood flow, the blood–brain barrier is less likely to be damaged, and the enzyme will remain in the vascular system, where it will do no harm.

As you undoubtedly know, vampire bats live on the blood of other warm-blooded animals. They make a small incision in a sleeping animal's skin with their sharp teeth and lap up the blood with their tongues. One compound in their saliva acts as a local anesthetic and keeps the animal from awakening. Another compound (and this is the one we are interested in) acts as an anticoagulant, preventing the blood from clotting. The name of this enzyme is *Desmodus rotundus plasminogen activator* (DSPA), otherwise known as *desmoteplase.* (*Desmodus rotundus* is the Latin name for the vampire bat.) Research with laboratory animals indicate that unlike tPA, desmoteplase causes no excitotoxic injury when injected directly into the brain (Reddrop et al., 2005). A phase II placebo-controlled, double-blind clinical trial of desmoteplase (Hacke et al., 2005) found that desmoteplase restored blood flow and reduced clinical symptoms in a majority of patients if given up to nine hours after the occurrence of a stroke. (See *Figure 15.8.*)

How can strokes be prevented? Risk factors that can be reduced by medication or changes in lifestyle include high blood pressure, cigarette smoking, diabetes, and high blood levels of cholesterol. The actions we can take to reduce these risk factors are well known, so I need not describe them here. *Atherosclerosis,* a process in which the linings of arteries develop a layer of plaque, which consists of deposits of cholesterol, fats, calcium, and cellular waste products, is a precursor to heart attacks (myocardial infarction) and obstructive stroke, caused by clots that form around atherosclerotic plaques in cerebral and cardiac blood vessels.

---

free radical A molecule with unpaired electrons; acts as a powerful oxidizing agent; toxic to cells.

FIGURE 15.8 ■ Desmoteplase in Treatment of Strokes

The graph shows the effects of desmoteplase and a placebo on restoration of cerebral blood flow to affected area (*reperfusion*) and favorable clinical outcome.

(Adapted from Hacke, W., Albers, G., Al-Rawi, Y., Bogousslavsky, J., Davalos, A., Eliasziw, M., Fischer, M., Furlan, A., Kaste, M., Lees, K. R., Soehngen, M., Warach, S., and the DIAS Study Group. *Stroke*, 2005, *36*, 66–73.)

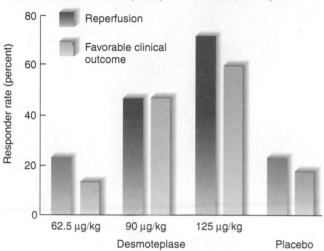

FIGURE 15.9 ■ Atherosclerotic Plaque

An angiogram shows an obstruction in the internal carotid artery caused by an atherosclerotic plaque.

(From Stapf, C., and Mohr, J. P. *Annual Review of Medicine*, 2002, *53*, 453–475. Reprinted with permission.)

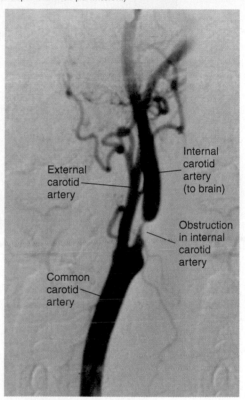

Atherosclerotic plaques often form in the internal carotid artery—the artery that supplies most of the blood flow to the cerebral hemispheres. These plaques can cause severe narrowing of the interior of the artery, greatly increasing the risk of a massive stroke. This narrowing can be visualized in an angiogram, produced by injecting a radiopaque dye into the blood and examining the artery with a computerized X-ray machine. (See *Figure 15.9.*) If the narrowing is severe, a *carotid endarterectomy* can be performed. The surgeon makes an incision in the neck that exposes the carotid artery, inserts a shunt in the artery, cuts the artery open, removes the plaque, and sews the artery back again (and the neck too, of course). Endarterectomy has been shown to reduce the risk of stroke by 50 percent in people under seventy-five years of age.

An even more effective—and possibly safer—surgical treatment involves the placement of a stent in a seriously narrowed carotid artery (Yadav et al., 2004). An arterial stent is an implantable device made of a metal mesh that is used to expand and hold open a partially occluded artery. The stent consists of a mesh tube made of springy metal collapsed inside a catheter—a flexible plastic tube. The surgeon cuts open a large artery in the groin and passes the catheter through large arteries up to the neck until the stent reaches the occlusion in the carotid artery. The end of the catheter holds a filter shaped like a collapsed parachute. When the catheter is retracted, the stent expands, opening the narrowed artery. The filter also springs open, catching any debris that is dislodged from the plaque, which would otherwise travel through the bloodstream and get trapped in a small artery, causing an infarct. The filter is then withdrawn, and the cannula is removed, leaving the expanded stent in place to keep the artery open. (See *Figure 15.10.*)

Depending on the location of the brain damage, people who have strokes will receive physical therapy and perhaps speech therapy to help them recover from their disability. Several studies have shown that exercise and sensory stimulation can facilitate recovery from the effects of brain damage (Cotman, Berchtold, and Christie, 2007). For example, Taub et al. (2006) studied patients with strokes that impaired their ability to use one arm and hand. The researchers put the *unaffected* arm in a sling for fourteen days and gave the patients training sessions during which the patients were forced to use the impaired arm. A placebo group received cognitive, relaxation, and physical fitness exercises for the same amount of time. This procedure (which is called *constraint-induced movement therapy*) produced long-term

FIGURE 15.10 ■ An Arterial Stent

The figure shows the placement of a stent in an obstructed internal carotid artery.

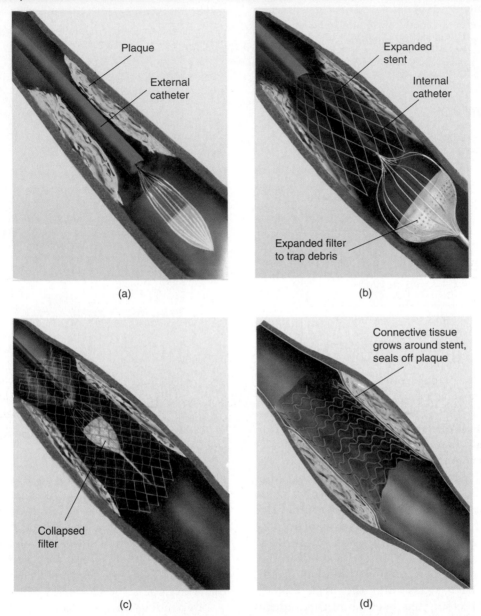

(a)

(b)

(c)

(d)

improvement in the patients' ability to use the affected arm. (See *Figure 15.11*.)

A study by Liepert et al. (2000) found that constraint-induced movement therapy caused changes in the connections of the primary motor cortex. The investigators used transcranial magnetic stimulation to map the area of the contralateral motor cortex that was involved in control of the impaired arm before and after treatment. Besides improving the patients' use of the impaired arm, the treatment caused an expansion of this region—apparently, into adjacent areas of the motor cortex—that was still present when the patients were tested six months later.

You will recall from Chapter 8 that mirror neurons in the parietal lobe and ventral premotor cortex become active when a person performs an action or sees someone else performing it. Ertelt et al. (2007) enrolled chronic stroke patients in a course of therapy that combined repetitive practice of hand and arm movements used in daily life with the watching of videos of actors performing the same movements. The patients' motor functions showed long-term improvement relative to those of patients in a control group who performed the exercises but watched videos of sequences of geometric symbols. Moreover, functional imaging showed increased activity in brain regions

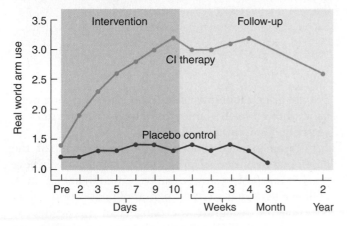

FIGURE 15.11 ■ Constraint-induced Movement Therapy

The graph shows the effects of constraint-induced (CI) therapy and placebo therapy on use of a limb whose movement was impaired by a stroke.

(Adapted from Taub, E., Uswatte, F., King, D. K., Morris, D., Crago, J. E., and Chatterjee, A. *Stroke,* 2006, *37,* 1045–1049.)

involved in movement, including the ventral premotor cortex and the supplementary motor area.

In some cases of brain damage or spinal cord damage, patients are unable to perform useful limb movements, even after intensive therapy. In such cases, investigators have attempted to devise brain–computer interfaces that permit the patient to control electronic and mechanical devices to perform useful actions. Developers of such interfaces have implanted arrays of microelectrodes directly into the patient's motor cortex and have applied surface electrodes to measure changes in EEG activity transmitted through the skull and scalp. These devices, while still experimental, permit patients to move prosthetic hands, perform actions with multijointed robotic arms, and move the cursor of a computer display and operate the computer (Wolpaw and McFarland, 2004; Hochberg et al., 2006).

InterimSummary

Tumors, Seizure Disorders, and Cerebrovascular Accidents

Neurological disorders have many causes. Because we have learned much about the functions of the human brain from studying the behavior of people with various neurological disorders, you have already learned about many of them in previous chapters of this book. Brain tumors are caused by the uncontrolled growth of various types of cells *other than neurons.* They can be benign or malignant. Benign tumors are encapsulated and thus have a distinct border; when one is surgically removed, the surgeon has a good chance of getting all of it. Tumors produce brain damage by compression and, in the case of malignant tumors, infiltration.

Seizures are periodic episodes of abnormal electrical activity of the brain. Partial seizures are localized, beginning with a focus—usually, some scar tissue caused by previous damage or a tumor. When they begin, they often produce an aura, consisting of particular sensations or changes in mood. Simple partial seizures do not produce profound changes in consciousness; complex partial seizures do. Generalized seizures may or may not originate at a single focus, but they involve most of the brain. Some seizures involve motor activity; the most serious are the grand mal convulsions that accompany generalized seizures. The convulsions are caused by involvement of the

brain's motor systems; the patient first shows a tonic phase, consisting of a few seconds of rigidity, and then a clonic phase, consisting of rhythmic jerking. Absence seizures, also called petit mal seizures, are common in children. These generalized seizures are characterized by periods of inattention and temporary loss of awareness. Seizures produced by abstinence after prolonged heavy intake of alcohol appear to be produced by supersensitivity (up-regulation) of NMDA receptors. Seizures are treated with anticonvulsant drugs and, in the case of intractable seizure disorders caused by an abnormal focus, seizure surgery, which usually involves the medial temporal lobe. A ketogenic diet, which is high in fat, moderate in protein, and low in carbohydrate, also relieves the symptoms of seizure disorders in some patients.

Cerebrovascular accidents damage parts of the brain through rupture of a blood vessel or occlusion (obstruction) of a blood vessel by a thrombus or embolus. A thrombus is a blood clot that forms within a blood vessel. An embolus is a piece of debris that is carried through the bloodstream and lodges in an artery. Emboli can arise from infections within the chambers of the heart or can consist of pieces of thrombi. The lack of blood flow appears to damage neurons primarily by stimulating a massive release of glutamate, which causes inflammation, phagocytosis by activated microglia, the production of free radicals, and activation of calcium-dependent enzymes. The best current treatment for stroke is administration of a drug that dissolves clots. Tissue plasminogen activator (tPA) must be given within three hours of the onset of the stroke and in some cases appears to cause brain damage on its own. Desmoteplase, an enzyme secreted in the saliva of vampire bats, is effective up to nine hours after a stroke and does not appear to cause damage. Carotid endarterectomy or insertion of a carotid stent can reduce the likelihood of a stroke in people with atherosclerotic plaque that obstructs the carotid arteries. After a stroke has occurred, physical therapy can facilitate recovery and minimize a patient's deficits. Constraint-induced movement therapy has been shown to be especially useful in restoring useful movement of limbs following unilateral damage to the motor cortex. Movement therapy combined with watching the movements being performed has beneficial effects, perhaps because of stimulation of the mirror neuron system.

DISORDERS OF DEVELOPMENT

As you will see in this section, brain development can be affected adversely by the presence of toxic chemicals during pregnancy and by genetic abnormalities, both hereditary and nonhereditary. In some instances the result is mental retardation.

Toxic Chemicals

A common cause of mental retardation is the presence of toxins that impair fetal development during pregnancy. For example, if a woman contracts rubella (German measles) early in pregnancy, the toxic chemicals released by the virus interfere with the chemical signals that control normal development of the brain. Most women who receive good health care will be immunized for rubella to prevent them from contracting it during pregnancy.

In addition to the toxins produced by viruses, various drugs can adversely affect fetal development. For example, mental retardation can be caused by the ingestion of alcohol during pregnancy, especially during the third to fourth week (Sulik, 2005). Babies born to alcoholic women are typically smaller than average and develop more slowly. Many of them exhibit **fetal alcohol syndrome,** which is characterized by abnormal facial development and deficient brain development. Figure 15.12 shows photographs of the faces of a child with fetal alcohol syndrome, of a mouse fetus whose mother was fed alcohol during pregnancy, and of a normal mouse fetus. As you can see, alcohol produces similar abnormalities in the offspring of both species. The facial abnormalities are relatively unimportant, of course. Much more serious are the abnormalities in the development of the brain. (See *Figure 15.12.*)

Research suggests that alcohol disrupts normal brain development by interfering with a **neural adhesion protein**—a protein that helps to guide the growth of neurons in the developing brain (Braun, 1996; Abrevalo, 2008). Prenatal exposure to alcohol even appears to have direct effects on neural plasticity. Sutherland, McDonald, and Savage (1997) found that the offspring of female rats that are given moderate amounts of alcohol during pregnancy showed smaller amounts of long-term potentiation (described in Chapter 12).

A woman need not be an alcoholic to impair the development of her offspring; some investigators believe that fetal alcohol syndrome can be caused by a single alcoholic binge during a critical period of fetal development. Now that we recognize the dangers of this syndrome, pregnant women are advised to abstain from alcohol (and from other drugs not specifically prescribed by their physicians) while their bodies are engaged in the task of sustaining the development of another human being.

Inherited Metabolic Disorders

Several inherited "errors of metabolism" can cause brain damage or impair brain development. Normal functioning of cells requires intricate interactions among countless biochemical systems. As you know, these systems depend on enzymes, which are responsible for constructing or breaking down particular chemical compounds. Enzymes are proteins and therefore are produced by mechanisms involving the chromosomes, which contain the recipes for their synthesis. "Errors of metabolism" refer to genetic abnormalities in which the recipe for a particular enzyme is in error, so the enzyme cannot be synthesized. If the enzyme is a critical one, the results can be very serious.

There are at least a hundred different inherited metabolic disorders that can affect the development of the brain. The most common and best-known is called **phenylketonuria (PKU).** This disease is caused by an

fetal alcohol syndrome A birth defect caused by ingestion of alcohol by a pregnant woman; includes characteristic facial anomalies and faulty brain development.

neural adhesion protein A protein that plays a role in brain development; helps to guide the growth of neurons.

phenylketonuria (PKU) (*fee nul kee ta **new** ree uh*) A hereditary disorder caused by the absence of an enzyme that converts the amino acid phenylalanine to tyrosine; the accumulation of phenylalanine causes brain damage unless a special diet is implemented soon after birth.

FIGURE 15.12 ■ Facial Malformations in Fetal Alcohol Syndrome

The photographs show a child with fetal alcohol syndrome, along with magnified views of mouse fetuses. (a) Mouse fetus whose mother received alcohol during pregnancy. (b) Normal mouse fetus.

(Photographs courtesy of Katherine K. Sulik.)

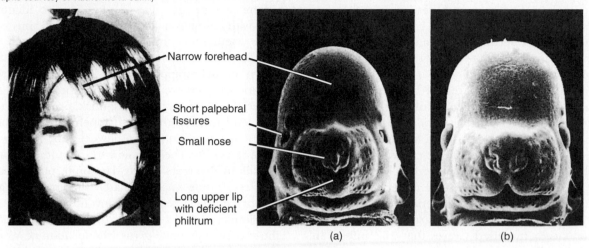

Narrow forehead

Short palpebral fissures

Small nose

Long upper lip with deficient philtrum

(a)   (b)

inherited lack of an enzyme that converts phenylalanine (an amino acid) into tyrosine (another amino acid). Excessive amounts of phenylalanine in the blood interfere with the myelinization of neurons in the central nervous system. Much of the myelinization of the cerebral hemispheres takes place after birth. Thus, when an infant born with PKU receives foods containing phenylalanine, the amino acid accumulates, and the brain fails to develop normally. The result is severe mental retardation, with an average IQ of approximately 20 by six years of age.

Fortunately, PKU can be treated by putting the infant on a low-phenylalanine diet. The diet keeps the blood level of phenylalanine low, and myelinization of the central nervous system takes place normally. Once myelinization is complete, the dietary restraints can be relaxed somewhat, because a high level of phenylalanine no longer threatens brain development. During prenatal development a fetus is protected by its mother's normal metabolism, which removes the phenylalanine from its circulation. However, if the *mother* has PKU, she must follow a strict diet during pregnancy or her infant will be born with brain damage. If she eats a normal diet, rich in phenylalanine, the high blood level of this compound will not damage her brain, but it will damage that of her fetus.

Diagnosing PKU immediately after birth is imperative so that the infant's brain is never exposed to high levels of phenylalanine. Consequently, many governments have passed laws that mandate a PKU test for all newborn babies. The test is inexpensive and accurate, and it has prevented many cases of mental retardation.

Other genetic errors of metabolism can be treated in similar fashion. For example, untreated **pyridoxine dependency** results in damage to cerebral white matter, to the thalamus, and to the cerebellum. It is treated by

large doses of vitamin B_6. Another error of metabolism, **galactosemia,** is an inability to metabolize galactose, a sugar found in milk. If it is not treated, it, too, causes damage to cerebral white matter and to the cerebellum. The treatment is use of a milk substitute that does not contain galactose. (Galactosemia should not be confused with *lactose intolerance,* which is caused by an insufficient production of lactase, the digestive enzyme that breaks down lactose. Lactose intolerance leads to digestive disturbance, not brain damage.)

Some other inherited metabolic disorders cannot yet be treated successfully. For example, **Tay-Sachs disease,** which occurs mainly in children of Eastern European Jewish descent, causes the brain to swell and damage itself against the inside of the skull and against the folds of the dura mater than encase it. The neurological symptoms begin by four months of age and include an exaggerated startle response to sounds, listlessness, irritability, spasticity, seizures, dementia, and finally death.

Tay-Sachs disease is one of several metabolic "storage" disorders. All cells contain sacs of material encased in membrane, called lysosomes ("dissolving bodies"). These sacs constitute the cell's rubbish-removal system; they contain enzymes that break down waste substances

pyridoxine dependency (*peer i dox een*) A metabolic disorder in which an infant requires larger-than-normal amounts of pyridoxine (vitamin B_6) to avoid neurological symptoms.

galactosemia (*ga lak tow see mee uh*) An inherited metabolic disorder in which galactose (milk sugar) cannot easily be metabolized.

Tay-Sachs disease A heritable, fatal, metabolic storage disorder; lack of enzymes in lysosomes causes accumulation of waste products and swelling of cells of the brain.

that cells produce in the course of their normal activities. The broken-down waste products are then recycled (used by the cells again) or excreted. Metabolic storage disorders are genetic errors of metabolism in which one or more vital enzymes are missing. Particular kinds of waste products cannot be destroyed by the lysosomes, so they accumulate. The lysosomes get larger and larger, the cells get larger and larger, and eventually the brain begins to swell and become damaged.

Researchers investigating hereditary errors of metabolism hope to prevent or treat these disorders in several ways. Some will be treated like PKU or galactosemia, by avoiding a constituent of the diet that cannot be tolerated. Others, such as pyridoxine dependency, will be treated by administering a substance that the body requires. Still others may be cured some day by the techniques of genetic engineering. Viruses infect cells by inserting their own genetic material into them and thus taking over the cells' genetic machinery, using it to reproduce themselves. Researchers hope to develop genetically modified viruses that will "infect" an infant's cells with genetic information that is needed to produce the enzymes that the cells lack, leaving the rest of the cells' functions intact.

Down Syndrome

Down syndrome is a congenital disorder that results in abnormal development of the brain, producing mental retardation in varying degrees. *Congenital* does not necessarily mean *hereditary*; it simply refers to a disorder that one is born with. Down syndrome is caused not by the inheritance of a faulty gene but by the possession of an extra twenty-first chromosome. The syndrome is closely associated with the mother's age; in most cases, something goes wrong with some of her ova, resulting in the presence of two (rather than one) twenty-first chromosomes. When fertilization occurs, the addition of the father's twenty-first chromosome makes three, rather than two. The extra chromosome presumably causes biochemical changes that impair normal brain development. The development of *amniocentesis,* a procedure whereby some fluid is withdrawn from a pregnant woman's uterus through a hypodermic syringe, has allowed physicians to identify fetal cells with chromosomal abnormalities and thus to determine whether the fetus carries Down syndrome.

Down syndrome, described in 1866 by John Langdon Down, occurs in approximately 1 out of 700 births. An experienced observer can recognize people with this disorder; they have round heads; thick, protruding tongues that tend to keep the mouth open much of the time; stubby hands; short stature; low-set ears; and somewhat slanting eyelids. They are slow to learn to talk, but most do talk by five years of age. The brain of a person with Down syndrome is approximately 10 percent lighter than that of a normal person, the convolutions (gyri and sulci) are simpler and smaller, the frontal lobes are small, and the superior temporal gyrus (the location of Wernicke's area) is thin. After age thirty the brain develops abnormal microscopic structures and begins to degenerate. Because this degeneration resembles that of Alzheimer's disease, it will be discussed in the next section.

A study by Fernandez et al. (2007) found that repeated low-dose injections of picrotoxin or pentylenetetrazole, drugs that serve as GABA antagonists, increased both long-term potentiation and performance on declarative learning tasks of a strain of mice that serve as a genetic model for Down syndrome. The drugs appeared to improve the animals' cognitive performance by suppressing excessive inhibition that is seen in their brains.

Down syndrome A disorder caused by the presence of an extra twenty-first chromosome, characterized by moderate-to-severe mental retardation and often by physical abnormalities.

Interim Summary

Disorders of Development

Developmental disorders can result in brain damage serious enough to cause mental retardation. During pregnancy the fetus is especially sensitive to toxins, such as alcohol or chemicals produced by some viruses. Several inherited metabolic disorders can also impair brain development. For example, phenylketonuria is caused by the lack of an enzyme that converts phenylalanine into tyrosine. Brain damage can be averted by feeding the infant a diet low in phenylalanine, so early diagnosis is essential. Other inherited metabolic disorders include pyridoxine dependency, which can be treated by vitamin B_6, and galactosemia, which can be treated with a diet that does not contain milk sugar. Storage disorders, such as Tay-Sachs disease, are caused by the inability of cells to destroy waste products within the lysosomes, which causes the cells to swell and eventually die. So far, these disorders cannot be treated. Down syndrome is produced by the presence of an extra twenty-first chromosome. The brain development of people with Down syndrome is abnormal, and after age 30 their brains develop features similar to those of people with Alzheimer's disease. A study with an animal model of Down syndrome suggests that administration of GABA antagonists might be useful.

DEGENERATIVE DISORDERS

Many disease processes cause degeneration of the cells of the brain. Some of these conditions injure particular kinds of cells, a fact that provides the hope that research will uncover the causes of the damage and find a way to halt it and prevent it from occurring in other people.

Transmissible Spongiform Encephalopathies

The outbreak of bovine spongiform encephalopathy (BSE, or "mad cow disease") in Great Britain in the late 1980s and early 1990s brought a peculiar form of brain disease to public attention. BSE is a **transmissible spongiform encephalopathy (TSE)**—a fatal contagious brain disease ("encephalopathy") whose degenerative process gives the brain a spongelike (or Swiss cheese–like) appearance. Besides BSE, these diseases include Creutzfeldt-Jakob disease, fatal familial insomnia, and kuru, which affect humans, and scrapie, which primarily affects sheep. Although scrapie cannot be transmitted to humans, BSE can, and it produces a variant of Creutzfeldt-Jakob disease. (See *Figure 15.13.*)

Unlike other transmissible diseases, TSEs are caused not by microorganisms but by simple proteins, which have been called **prions,** or "protein infectious agents" (Prusiner, 1982). Prion proteins are found primarily in the membrane of neurons, where they are believed to play a role in synaptic function. They are resistant to proteolytic enzymes—enzymes that are able to destroy proteins by breaking the peptide bonds that hold a protein's amino acids together. Prion proteins are also resistant to levels of heat that denature normal proteins, which explains why cooking meat from cattle with BSE does not destroy the infectious agent. The sequence of amino acids of normal prion protein (PrPc) and infectious prion (PrPSc) are identical. How, then, can two proteins with the same amino acid sequences have such different effects? The answer is that the functions of proteins are determined largely by their three-dimensional shapes. The only difference between PrPc and PrPSc is the way the protein is folded. Once misfolded PrPSc is introduced into a cell, it causes normal PrPc to become misfolded too, and the process of this transformation ultimately kills them. (See Hetz et al., 2003, for a review.)

A familial form of Creutzfeldt-Jakob disease is transmitted as a dominant trait, caused by a mutation of the *PRNP* gene located on the short arm of chromosome 20, which codes for the human prion protein gene. However, most cases of this disease are **sporadic.** That is, they occur in people without a family history of prion protein disease. Prion protein diseases are unique not only because they can be transmitted by means of a sim-

ple protein, but also because they can also be genetic or sporadic—and the genetic and sporadic forms can be transmitted to others. The most common form of transmission of Creutzfeldt-Jakob disease in humans is through transplantation of tissues such as dura mater or corneas, harvested from cadavers of people who were infected with a prion disease. One form of human prion protein disease, kuru, was transmitted through cannibalism: Out of respect for their recently departed relatives, members of a South Pacific tribe ate their brains and sometimes thus contracted the disease. This practice has since been abandoned.

Whatever role normal PrPc plays, it does not seem to be essential for the life of a cell. Bueler et al. (1993) found that the cells of mice with a targeted mutation of the prion protein gene produced absolutely no prion protein and did not develop mouse scrapie when they were inoculated with the misfolded prions that cause this disease. Normal mice inoculated with these prions died within six months.

FIGURE 15.13 ■ **Bovine Spongiform Encephalopathy and Creutzfeldt-Jakob Disease**

The graph shows the number of cases of BSE in cattle and variant Creutzfeldt-Jakob (vCJD) disease in humans in Great Britain between 1988 and March 31, 2008.

(Data from OIE–World Organization for Animal Health and the CJD Surveillance Unit.)

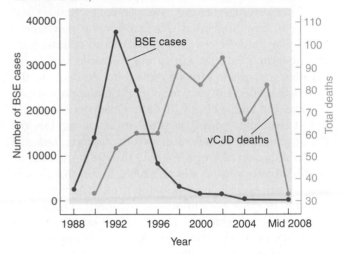

---

transmissible spongiform encephalopathy A contagious brain disease whose degenerative process gives the brain a spongelike appearance; caused by accumulation of misfolded prion protein.

prion (*pree on*) A protein that can exist in two forms that differ only in their three-dimensional shape; accumulation of misfolded prion protein is responsible for transmissible spongiform encephalopathies.

sporadic disease A disease that occurs rarely and is not obviously caused by heredity or an infectious agent.

A study by Steele et al. (2006) suggests that normal prion protein plays a role in neural development and differentiation in fetuses and neurogenesis in adults. The investigators produced a genetically engineered strain of mice that produced increased amounts of PrPc and found increased numbers of proliferating cells in the subventricular zone and more neurons in the dentate gyrus, compared with normal mice. Mice with a targeted mutation of the prion protein gene had fewer proliferating cells.

Some investigators (for example, Bailey, Kandel, and Si, 2004) have suggested that a prionlike mechanism could play a role in the establishment and maintenance of long-term memories. Long-term memories can last for decades, and prion proteins, which are resistant to the destructive effects of enzymes, might maintain synaptic changes for long periods of time. Criado et al. (2005) found that mice with a targeted mutation against the *PRNP* gene showed deficits in a spatial learning task and in establishment of long-term potentiation in the dentate gyrus. Papassotiropoulos et al. (2005) found that people with a particular allele of the prion protein gene remembered 17 percent more information 24 hours after a word list–learning task than did people with a different allele. (Both alleles are considered normal and are not associated with a prion protein disease.)

Mallucci et al. (2003) created a genetically modified mouse strain whose neurons produced an enzyme at twelve weeks of age that destroyed normal prion protein. When the animals were a few weeks of age, the experimenters infected them with misfolded mouse scrapie prions. Soon thereafter, the animals began to develop spongy holes in their brains, indicating that they were infected with mouse scrapie. Then, at twelve weeks, the enzyme became active and started destroying normal PrPc. Although analysis showed that glial cells in the brain still contained misfolded PrPSc, the disease process stopped. Neurons stopped making normal PrPc, which could no longer be converted into PrPSc, so the mice went on to live normal lives. The disease process continued to progress in mice without the special enzyme, and these animals soon died. The authors concluded that the process of conversion of PrPc to PrPSc is what kills cells. The mere presence of PrPSc in the brain (found in non-neuronal cells) does not cause the disease. Figure 15.14 shows the development of spongiform degeneration and its disappearance after the PrPc-destroying enzyme became active at twelve weeks of age. (See *Figure 15.14*.)

How might misfolded prion protein kill neurons? As we will see later in this chapter, the brains of people with several other degenerative diseases, including Parkinson's disease, Alzheimer's disease, amyotrophic lateral sclerosis, and Huntington's disease contain aggregations of misfolded proteins (Soto, 2003). As we saw in Chapter 3, cells contain the means by which they can commit suicide—a process known as *apoptosis*. Apoptosis can be triggered either externally, by a chemical signal telling the cell that it is no longer needed (for example, during development), or internally, by evidence that

---

FIGURE 15.14 ■ Experimental Treatment of a Prion Protein Infection

Neural death was prevented and early spongiosis was reversed in scrapie-infected mice after a genetically engineered enzyme began to destroy PrPc at twelve weeks of age. Arrows point to degenerating neurons in mice without the prion-destroying enzyme. Spongiosis is seen as holes in the brain tissue.

(From Mallucci, G., Dickinson, A., Linehan, J., Klöhn, P. C., Brandner, S., and Collinge, J. *Science*, 2003, *302*, 871–874. Copyright © 2003 by the American Association for the Advancement of Science. Reprinted with permission.)

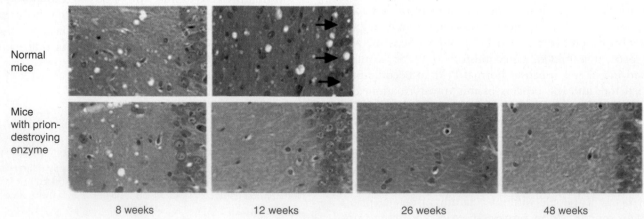

Time After Infection with Scrapies Prion

biochemical processes in the cell have become disrupted so that the cell is no longer functioning properly. Perhaps the accumulation of misfolded, abnormal proteins provides such a signal. Apoptosis involves production of "killer enzymes" called **caspases.** Mallucci et al. (2003) suggest that inactivation of caspase-12, the enzyme that appears to be responsible for the death of neurons infected with PrPSc, may provide a treatment that could arrest the progress of transmissible spongiform encephalopathies. Let's hope they are right.

Parkinson's Disease

One of the most common degenerative neurological disorders, Parkinson's disease, is caused by degeneration of the nigrostriatal system—the dopamine-secreting neurons of the substantia nigra that send axons to the basal ganglia. Parkinson's disease is seen in approximately 1 percent of people over sixty-five years of age. The primary symptoms of Parkinson's disease are muscular rigidity, slowness of movement, a resting tremor, and postural instability. For example, once a person with Parkinson's disease is seated, he or she finds it difficult to rise. Once the person begins walking, he or she has difficulty stopping. Thus, a person with Parkinson's disease cannot easily pace back and forth across a room. Reaching for an object can be accurate, but the movement usually begins only after a considerable delay. Writing is slow and labored, and as it progresses, the letters get smaller and smaller. Postural movements are impaired. A normal person who is bumped while standing will quickly move to restore balance—for example, by taking a step in the direction of the impending fall or by reaching out with the arms to grasp onto a piece of furniture. However, a person with Parkinson's disease fails to do so and simply falls. A person with this disorder is even unlikely to put out his or her arms to break the fall.

Parkinson's disease also produces a resting tremor—vibratory movements of the arms and hands that diminish somewhat when the individual makes purposeful movements. The tremor is accompanied by rigidity; the joints appear stiff. However, the tremor and rigidity are not the cause of the slow movements. In fact, some patients with Parkinson's disease show extreme slowness of movements but little or no tremor.

Examination of the brains of patients who had Parkinson's disease shows, of course, the near-disappearance of nigrostriatal dopaminergic neurons. Many surviving dopaminergic neurons show **Lewy bodies,** abnormal circular structures found within the cytoplasm. Lewy bodies have a dense protein core, surrounded by a halo of radiating fibers (Forno, 1996). (See *Figure 15.15.*) Although most cases of Parkinson's disease do not appear to have genetic origins, researchers have discovered that

FIGURE 15.15 ■ **Lewy Bodies**

A photomicrograph of the substantia nigra of a patient with Parkinson's disease shows a Lewy body, indicated by the arrow.

(Photograph courtesy of Dr. Don Born, University of Washington.)

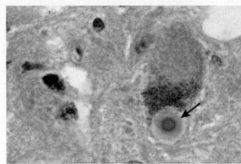

the mutation of a particular gene located on chromosome 4 will produce this disorder (Polymeropoulos et al., 1996). This gene produces a protein known as **α-synuclein,** which is normally found in the presynaptic terminals and is thought to be involved in synaptic transmission in dopaminergic neurons (Moore et al., 2005). The mutation produces what is known as a **toxic gain of function** because it produces a protein that results in effects that are toxic to the cell. Mutations that cause toxic gain of function are normally dominant because the toxic substance is produced whether one or both members of the pair of chromosomes contains the mutated gene. Abnormal α-synuclein becomes misfolded and forms aggregations, especially in dopaminergic neurons (Goedert, 2001). The dense core of Lewy bodies consists primarily of these aggregations, along with neurofilaments and synaptic vesicle proteins.

Another hereditary form of Parkinson's disease is caused by mutation of a gene on chromosome 6 that produces a gene that has been named **parkin** (Kitada et al., 1998). This mutation causes a **loss of function,**

caspase A "killer enzyme" that plays a role in apoptosis, or programmed cell death.

Lewy body Abnormal circular structures with a dense core consisting of α-synuclein protein; found in the cytoplasm of nigrostriatal neurons in people with Parkinson's disease.

α-synuclein A protein normally found in the presynaptic membrane, where it is apparently involved in synaptic plasticity. Abnormal accumulations are apparently the cause of neural degeneration in Parkinson's disease.

toxic gain of function Said of a genetic disorder caused by a dominant mutation that involves a faulty gene that produces a protein with toxic effects.

parkin A protein that plays a role in ferrying defective or misfolded proteins to the proteasomes; mutated parkin is a cause of familial Parkinson's disease.

loss of function Said of a genetic disorder caused by a recessive gene that fails to produce a protein that is necessary for good health.

which makes it a recessive disorder. If a person carries a mutated parkin gene on only one chromosome, the normal allele on the other chromosome can produce a sufficient amount of normal parkin for normal cellular functioning. Normal parkin plays a role in ferrying defective or misfolded proteins to the **proteasomes**—organelles responsible for destroying these proteins (Moore et al., 2005). This mutation permits high levels of defective protein to accumulate in dopaminergic neurons and ultimately damage them. Figure 15.16 illustrates the role of parkin in the action of proteasomes. Parkin assists in the tagging of abnormal or misfolded proteins with numerous molecules of **ubiquitin,** a small, compact globular protein. Ubiquitination (as this process is called) targets the abnormal proteins for destruction by the proteasomes, which break them down into their constituent amino acids. Defective parkin fails to ubiquinate abnormal proteins, and they accumulate in the cell, eventually killing it. For some reason, dopaminergic neurons are especially sensitive to this accumulation. (See *Figure 15.16.*)

Several other mutations have been discovered that produce Parkinson's disease. UCH-L1 is involved in the ubiquitin-proteasome system, DJ-1 plays a role in stabilizing messenger RNA and modulating its expression, and PINK1 is somehow involved with mitochondria (Vila and Przedborski, 2004). In addition, an epidemiological study found the existence of a mutation in mitochondrial DNA that caused Parkinson's disease, which was transmitted from mother to child (Swerdlow et al., 1998). (Sperm cells pass no mitochondria into a fertilized egg; all mitochondrial DNA are inherited from the mother.)

The overwhelming majority of the cases of Parkinson's disease (approximately 95 percent) are sporadic. That is, they occur in people without a family history of Parkinson's disease. What, then, triggers the accumulation of α-synuclein and the destruction of dopaminergic neurons? Research suggests that Parkinson's disease may be caused by toxins present in the environment, by faulty metabolism, or by unrecognized infectious disorders. For example, the insecticides rotenone and paraquat can also cause Parkinson's disease, and presumably, so can other unidentified toxins. All of these chemicals inhibit mitochondrial functions, which leads to the aggregation of misfolded α-synuclein, especially in dopaminergic neurons. These accumulated proteins eventually kill the cells (Dawson and Dawson, 2003).

As we saw in Chapter 4, the standard treatment for Parkinson's disease is L-DOPA, the precursor of dopamine. An increased level of L-DOPA in the brain causes a patient's remaining dopaminergic neurons to produce and secrete more dopamine and, for a

FIGURE 15.16 ■ The Role of Parkin in Parkinson's Disease

Parkin is involved in the destruction of abnormal or misfolded proteins by the ubiquitin-proteasome system. If parkin is defective because of a mutation, abnormal or misfolded proteins cannot be destroyed, so they accumulate in the cell. If α-synuclein is defective because of a mutation, parkin is unable to tag it with ubiquitin, and it accumulates in the cell.

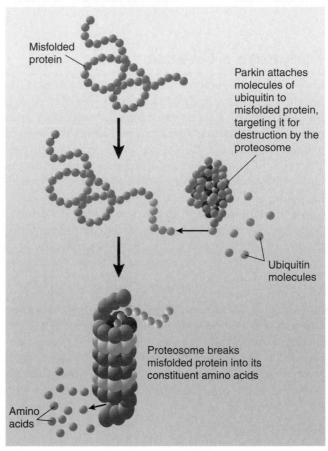

Misfolded protein

Parkin attaches molecules of ubiquitin to misfolded protein, targeting it for destruction by the proteosome

Ubiquitin molecules

Proteosome breaks misfolded protein into its constituent amino acids

Amino acids

time, alleviates the symptoms of the disease. But this compensation does not work indefinitely; eventually, the number of nigrostriatal dopaminergic neurons declines to such a low level that the symptoms become worse. In addition, high levels of L-DOPA produce side effects by acting on dopaminergic systems other than the nigrostriatal system. Some patients—especially those whose symptoms began

proteasome An organelle responsible for destroying defective or degraded proteins within the cell.

ubiquitin A protein that attaches itself to faulty or misfolded proteins and thus targets them for destruction by proteasomes.

when they were relatively young—become bedridden, scarcely able to move.

Another drug, deprenyl, is often given to patients with Parkinson's disease, usually in conjunction with L-DOPA. As we saw in Chapter 4, several people acquired the symptoms of Parkinson's disease after taking an illicit drug contaminated with MPTP. Subsequent studies with laboratory animals revealed that the toxic effects of this drug could be prevented by administration of deprenyl, a drug that inhibits the activity of the enzyme MAO-B. The original rationale for administering deprenyl to patients with Parkinson's disease was that it might prevent unknown toxins from producing further damage to dopaminergic neurons. In addition, Kumar and Andersen (2004) note that there is an age-related increase in MAO-B activity that might increase the level of oxidative stress in dopaminergic neurons. The intracellular breakdown of dopamine by MAO-B causes the formation of hydrogen peroxide, which can damage cells. Thus, a beneficial effect of MAO-B inhibitors might be to decrease normal, age-related oxidative stress. In addition, Czerniczyniec et al. (2007) found that deprenyl increased mitochondrial functions in the brains of mice. Interestingly, cigarette smokers have a lower incidence of Parkinson's disease, perhaps because compounds in tobacco inhibit MAO-B activity (Fowler et al., 2003). Of course, the increased incidence of lung cancer, emphysema, and other smoking-related diseases far outweighs any potential benefits effects on the incidence of Parkinson's disease.

What are the effects of the loss of dopaminergic neurons on normal brain functioning? Functional-imaging studies have shown that *akinesia* (difficulty in initiating movements) was associated with decreased activation of the supplementary motor area and that tremors are associated with abnormalities of a neural system involving the pons, midbrain, cerebellum, and thalamus (Grafton, 2004). A functional-imaging study by Buhmann et al. (2003) studied drug-naïve patients with akinetic hemiparkinsonism—difficulty in initiating movements on one side of the body. (Parkinson's disease often affects one side of the body more than the other, especially early in the course of the disease.) The investigators found decreased activation of the supplementary motor area and the primary motor cortex contralateral to the affected side while the patients performed a task that required them to touch a finger to their thumb. When the patients were given a dose of L-DOPA, the activation of these regions increased, and their motor performance improved. In fact, the improvements in motor performance were positively correlated with the increased brain activation.

Neurosurgeons have been developing three stereotaxic procedures designed to alleviate the symptoms of Parkinson's disease that no longer respond to treatment with L-DOPA. The first one, transplantation of fetal tissue, attempts to reestablish the secretion of dopamine in the neostriatum. The tissue is obtained from the substantia nigra of aborted human fetuses and implanted into the caudate nucleus and putamen by means of stereotaxically guided needles. As we saw in Chapter 5, PET scans have shown that dopaminergic fetal cells are able to grow in their new host and secrete dopamine, reducing the patient's symptoms—at lease, initially.

In a study of thirty-two patients with fetal tissue transplants, Freed (2002) found that those whose symptoms had previously responded to L-DOPA were most likely to benefit from the surgery. Presumably, these patients had a sufficient number of basal ganglia neurons with receptors that could be stimulated by the dopamine secreted by either the medication or the transplanted tissue. Unfortunately, many transplant patient later developed severe, persistent *dyskinesias*—troublesome and often painful involuntary movements. As a result, fetal transplants of dopaminergic fetal cells are no longer recommended (Olanow et al., 2003).

One potential source of dopaminergic neurons could come from cultures of neural stem cells—undifferentiated cells that have the ability, if appropriately stimulated, to develop into a variety of types of cells, including dopaminergic neurons (Snyder and Olanow, 2005). A significant advantage of human stem cells is that large numbers of cells could be transplanted, thus increasing the numbers of surviving cells in the patients' brains. Redmond et al. (2007) produced Parkinsonian symptoms in monkeys through injections of MPTP that destroyed most of the animals' nigrostriatal dopaminergic neurons. The investigators then implanted neural stem cells into the caudate nucleus and found that the stem cells differentiated not just into dopamine-secreting neurons, but also into astrocytes and other cells that protect and repair neurons. The implants also had a functional impact: The monkeys' motor behavior improved.

Another procedure has a long history, but only recently have technological developments in imaging methods and electrophysiological techniques led to an increase in its popularity. The principal output of the basal ganglia comes from the **internal division of the globus pallidus (GP$_i$).** (The caudate nucleus,

internal division of the globus pallidus (GP$_i$) A division of the globus pallidus that provides inhibitory input to the motor cortex via the thalamus; sometimes stereotaxically lesioned to treat the symptoms of Parkinson's disease.

putamen, and globus pallidus are the three major components of the basal ganglia.) This output, which is directed through the thalamus to the motor cortex, is inhibitory. Furthermore, a decrease in the activity of the dopaminergic input to the caudate nucleus and putamen causes an *increase* in the activity of the GP$_i$. Thus, damage to the GP$_i$ might be expected to relieve the symptoms of Parkinson's disease. (See *Figure 15.17.*)

In the 1950s, Leksell and his colleagues performed pallidotomies (surgical destruction of the internal division of the globus pallidus) in patients with severe Parkinson's disease (Svennilson et al., 1960; Laitinen, Bergenheim, and Hariz, 1992). The surgery often reduced the rigidity and enhanced the patient's ability to move. Unfortunately, the surgery occasionally made the patient's symptoms worse and

sometimes resulted in partial blindness. (The optic tract is located next to the GP$_i$.)

With the development of L-DOPA therapy in the late 1960s, pallidotomies were abandoned. However, it eventually became evident that L-DOPA worked for a limited time and that the symptoms of Parkinson's disease would eventually return. For that reason, in the 1990s, neurosurgeons again began experimenting with pallidotomies, first with laboratory animals and then with humans (Graybiel, 1996; Lai et al., 2000). This time, they used MRI scans to find the location of the GP$_i$ and then inserted an electrode into the target region. They could then pass low-intensity, high-frequency stimulation through the electrode, thus temporarily disabling the region around its tip. If the patient's rigidity disappeared (obviously, the patient is awake during the surgery), the electrode was in the right

FIGURE 15.17 ■ Connections of the Basal Ganglia

This schematic shows the major connections of the basal ganglia and associated structures. Excitatory connections are shown as black lines; inhibitory connections are shown as red lines. Many connections, such as the inputs to the substantia nigra, are omitted for clarity. Two regions that have been targets of stereotaxic surgery for Parkinson's disease—the internal division of the globus pallidus and the subthalamus—are outlined in gray. Damage to these regions reduces inhibitory input to the thalamus and facilitates movement.

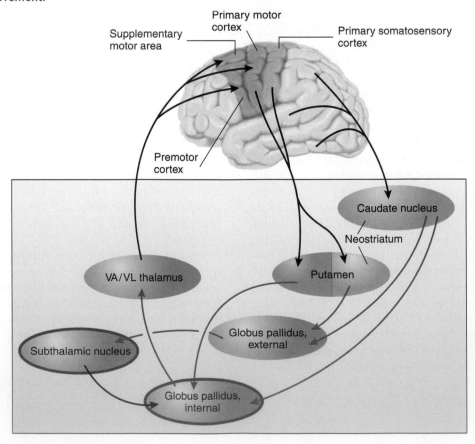

place. To make the lesion, the surgeon would then pass radiofrequency current of sufficient strength to heat and destroy the brain tissue. The results of this procedure have been so promising that several neurological teams have begun promoting its use in the treatment of relatively young patients whose symptoms no longer respond to L-DOPA. PET studies have found that after pallidotomy the metabolic activity in the premotor and supplementary motor areas of the frontal lobes, normally depressed in patients with Parkinson's disease, returns to normal levels (Grafton et al., 1995), a result indicating that lesions of the GP$_i$ do indeed release the motor cortex from inhibition.

Neurosurgeons have also targeted the subthalamus in patients with advanced Parkinson's disease. As Figure 15.17 shows, the subthalamus has an excitatory effect on the GP$_i$; therefore, damage to the subthalamus decreases the activity of this region and removes some of the inhibition on motor output. (See *Figure 15.17.*) Normally, damage to the subthalamus causes involuntary jerking and twitching movements. However, in people with Parkinson's disease, damage to this region brings motor activity, which is normally depressed, back to normal (Guridi and Obeso, 2001).

The third stereotaxic procedure aimed at relieving the symptoms of Parkinson's disease involves implanting electrodes in the subthalamic nucleus and attaching a device that permits the patient to electrically stimulate the brain through the electrodes. (See *Figure 15.18.*) According to some studies, deep brain stimulation is as effective as brain lesions in suppressing tremors and has fewer adverse side effects (Simuni et al., 2002; Speelman et al., 2002). In addition, a three-year follow-up study found no evidence of cognitive deterioration in patients who received implants for deep brain stimulation (Funkiewiez et al., 2004). The fact that either lesions or stimulation alleviates tremors suggests that the stimulation has an inhibitory effect on subthalamic neurons, but this hypothesis has not yet been confirmed.

Kaplitt et al. (2007) developed a remarkable procedure that might provide an alternative (or supplement) to deep brain stimulation. As we saw in Figure 15.17, the excitatory effect of the subthalamic nucleus on the GP$_i$ has an inhibitory effect on the thalamus and therefore on behavior. In a clinical trial designed to assess the safety of the new procedure, Kaplitt and his colleagues injected a genetically modified virus into the subthalamic nucleus of patients with Parkinson's disease that delivered a gene for GAD, the enzyme responsible for the biosynthesis of the major inhibitory neurotransmitter, GABA. The production of GAD turned some of the excitatory, glutamate-producing neurons in the subthalamic nucleus into inhibitory, GABA-producing neurons. As a result,

FIGURE 15.18 ■ Deep Brain Stimulation

Electrodes are implanted in the patient's brain, and wires are run under the skin to stimulation devices implanted near the collarbone.

(Illustration used with permission by Medtronic, Inc.)

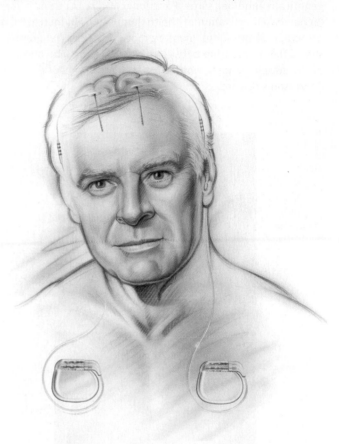

the activity of the GP$_i$ decreased, the activity of the supplementary motor area increased, and the symptoms of the patients improved. (See *Figure 15.19.*)

Huntington's Disease

Another basal ganglia disease, **Huntington's disease,** is caused by degeneration of the caudate nucleus and putamen. Whereas Parkinson's disease causes a poverty of movements, Huntington's disease causes uncontrollable ones, especially jerky limb movements. The movements of Huntington's disease look like fragments of purposeful movements but occur involuntarily. This disease is progressive, includes

Huntington's disease An inherited disorder that causes degeneration of the basal ganglia; characterized by progressively more severe uncontrollable jerking movements, writhing movements, dementia, and finally death.

FIGURE 15.19 ■ **Gene Therapy of Parkinson's Disease**

The gene for GAD, the enzyme responsible for biosynthesis of GABA, was delivered to cells in the subthalamic nucleus of patients with Parkinson's disease by means of a genetically modified virus. Functional MRI scans show (a) decreased activation of the thalamus and (b) increased activation of the supplementary motor area. (c) The graph shows the relationship between changes in the activation of the supplementary motor area and symptoms of Parkinson's disease.

(From Kaplitt, M. G., et al. *Lancet*, 2007, *369*, 2097–2105. Reprinted with permission.)

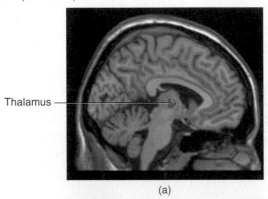

(a)

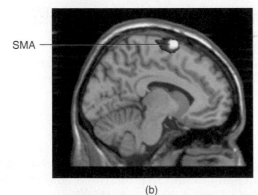

(b)

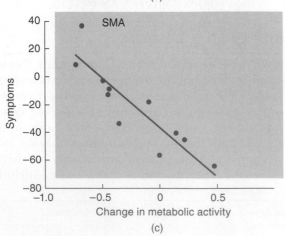

(c)

cognitive and emotional changes, and eventually causes death, usually within ten to fifteen years after the symptoms begin.

The symptoms of Huntington's disease usually begin in the person's thirties and forties but can sometimes begin in the early twenties. The first signs of neural degeneration occur in the putamen, in a specific group of inhibitory neurons: GABAergic medium spiny neurons. Damage to these neurons removes some inhibitory control exerted on the premotor and supplementary motor areas of the frontal cortex. Loss of this control leads to involuntary movements. As the disease progresses, neural degeneration is seen in many other regions of the brain, including the cerebral cortex.

Huntington's disease is a hereditary disorder, caused by a dominant gene on chromosome 4. In fact, the gene has been located, and its defect has been identified as a repeated sequence of bases that code for the amino acid glutamine (Collaborative Research Group, 1993). This repeated sequence causes the gene product—a protein called **huntingtin (htt)**—to contain an elongated stretch of glutamine. Abnormal htt becomes misfolded and forms aggregations that accumulate in the nucleus. Longer stretches of glutamine are associated with patients whose symptoms began at a younger age, a finding that indicates that this abnormal portion of the huntingtin molecule is responsible for the disease. These facts suggest that the mutation causes the disease through a toxic gain of function—that abnormal htt causes harm. In fact, the cause of death of neurons in Huntington's disease is apoptosis (cell "suicide"). Li et al. (2000) found that HD mice lived longer if they were given a caspase inhibitor, which suppresses apoptosis. Abnormal htt may trigger apoptosis by impairing the function of the ubiquitin-protease system, which activates caspase, one of the enzymes involved in apoptosis (Hague, Klaffke, and Bandmann, 2005).

Normal htt is found in cells throughout the body, but it occurs in especially high levels in neurons and in cells of the testes. The protein plays a critical role in development: O'Kusky et al. (1999) found that mice with a knockout of the gene that codes for huntingtin die before embryonic day 8.5. Heterozygous knockout mice, with one good htt gene, survive to adulthood, but because of the decreased level of htt, they show excessive motor activity and degeneration of neurons in the basal ganglia and subthalamic nuclei. These findings

huntingtin (htt) A protein that may serve to facilitate the production and transport of brain-derived neurotrophic factor. Abnormal huntingtin is the cause of Huntington's disease.

suggest that the mutation responsible for Huntington's disease may also cause brain damage through a loss of function. One of the most important functions of normal htt in adulthood appears to be to be facilitation of the production and transport of *brain-derived neurotrophic factor (BDNF)*. BDNF is a chemical that is necessary for survival of neurons in the caudate nucleus and putamen. This chemical is produced in the cerebral cortex and transported through axons to the basal ganglia. Abnormal htt interferes with BDNF activity in the caudate nucleus and putamen in two ways. First, its presence inhibits the expression of the BDNF gene (Zuccato et al., 2001, 2003). Second, it interferes with the transport of BDNF from the cerebral cortex to the basal ganglia (Gauthier et al., 2004).

Researchers have debated the role played by the accumulations of misfolded htt in the nucleus (known as *inclusion bodies*) in development of the disease. These inclusions could cause neural degeneration, they could have a protective role, or they could play no role at all. A study by Arrasate et al. (2004) strongly suggests that inclusion bodies actually protect neurons. The investigators prepared tissue cultures from rat striatal neurons that they infected with genes that expressed fragments of abnormal htt. Some of the neurons that produced the mutant htt formed inclusion bodies; others did not. Arrasate and her colleagues used a robotic microscope to see what happened to the cells over a period of almost ten days. They found that the inclusion bodies appeared to have a protective function. Neurons that contained inclusion bodies had lower levels of mutant htt elsewhere in the cell, and these neurons lived longer than those without these accumulations. (See *Figure 15.20.*)

At present there is no treatment for Huntington's disease. However, a study by DiFiglia et al. (2007) suggests a possible approach. The investigators used a genetically modified virus to deliver a mutant human htt gene into the striatum and overlying cortex of mice, which caused the development of neuropathology and motor deficits. DiFiglia and her colleagues then injected a small interfering RNA (siRNA) into the striatum that blocked the transcription of the htt genes—and hence the production of mutant htt—in this region. The treatment decreased the size of inclusion bodies in striatal neurons, prolonged the life of the striatal neurons, and reduced the animals' motor symptoms.

Alzheimer's Disease

Several neurological disorders result in **dementia,** a deterioration of intellectual abilities resulting from an organic brain disorder. A common form of dementia is called **Alzheimer's disease,** which occurs in approximately

FIGURE 15.20 ■ Infection with Abnormal Huntingtin

The photomicrograph shows two neurons that have been infected with genes that express fragments of abnormal huntingtin. The lower neuron shows an inclusion body (orange), and the upper one does not. Arresate et al. (2004) found that neurons with inclusion bodies survived longer than those without inclusion bodies. Blue ovals are the nuclei of uninfected neurons.

(Photo courtesy of Steven Finkbeiner, Gladstone Institute of Neurological Disease and the University of California, San Francisco.)

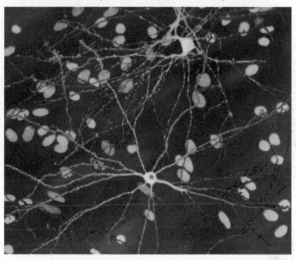

10 percent of the population above the age of sixty-five and almost 50 percent of people older than eighty-five years. It is characterized by progressive loss of memory and other mental functions. At first, people may have difficulty remembering appointments and sometimes fail to think of words or other people's names. As time passes, they show increasing confusion and increasing difficulty with tasks such as balancing a checkbook. The memory deficit most critically involves recent events, and thus it resembles the anterograde amnesia of Korsakoff's syndrome. If people with Alzheimer's disease venture outside alone, they are likely to get lost. They eventually become bedridden, then become completely helpless, and finally succumb (Terry and Davies, 1980).

Alzheimer's disease produces severe degeneration of the hippocampus, entorhinal cortex, neocortex (especially the association cortex of the frontal and temporal lobes), nucleus basalis, locus coeruleus, and raphe nuclei.

dementia (*da men sha*) A loss of cognitive abilities such as memory, perception, verbal ability, and judgment; common causes are multiple strokes and Alzheimer's disease.

Alzheimer's disease A degenerative brain disorder of unknown origin; causes progressive memory loss, motor deficits, and eventual death.

FIGURE 15.21 ■ Alzheimer's Disease

(a) This photograph shows a lateral view of the right side of the brain of a person with Alzheimer's disease. (Rostral is to the right; dorsal is up.) Note that the sulci of the temporal lobe and parietal lobe are especially wide, indicating degeneration of the neocortex (*arrowheads*). (b) This photograph shows a lateral view of the right side of a normal brain.

(Photo of diseased brain courtesy of A. D'Agostino, Good Samaritan Hospital, Portland, Oregon; photo of normal brain © Dan McCoy/Rainbow.)

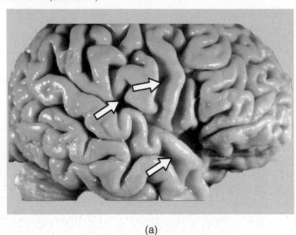

(a)

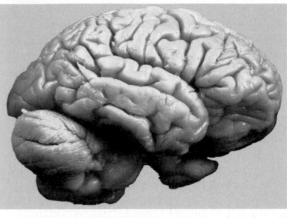

(b)

Figure 15.21 shows photographs of the brain of a patient with Alzheimer's disease and of a normal brain. You can see how much wider the sulci are in the patient's brain, especially in the frontal and temporal lobes, indicating substantial loss of cortical tissue. (See *Figure 15.21.*)

Earlier, I mentioned that the brains of patients with Down syndrome usually develop abnormal structures that are also seen in patients with Alzheimer's disease: *amyloid plaques* and *neurofibrillary tangles*. **Amyloid plaques** are extracellular deposits that consist of a dense core of a protein known as **β-amyloid,** surrounded by degenerating axons and dendrites, along with activated microglia and reactive astrocytes, cells that are involved in destruction of damaged cells. Eventually, the phagocytic glial cells destroy the degenerating axons and dendrites, leaving only a core of β-amyloid (usually referred to as Aβ).

Neurofibrillary tangles consist of dying neurons that contain intracellular accumulations of twisted filaments of hyperphosphorylated **tau protein.** Normal tau protein serves as a component of microtubules, which provide the cells' transport mechanism. During the progression of Alzheimer's disease, excessive amounts of phosphate ions become attached to strands of tau protein, thus changing its molecular structure. Abnormal filaments are seen in the soma and proximal dendrites of pyramidal cells in the cerebral cortex, which disrupt transport of substances within the cell, and the cell dies, leaving behind a tangle of protein filaments. (See *Figure 15.22.*)

Formation of amyloid plaques is caused by the production of a defective form of Aβ. The production of Aβ takes several steps. First, a gene encodes the production of the **β-amyloid precursor protein (APP),** a chain of approximately 700 amino acids. APP is then cut apart in

FIGURE 15.22 ■ Alzheimer's Disease

The photomicrographs from deceased patients with Alzheimer's disease show (a) an amyloid plaque, filled with β-amyloid protein, and (b) neurofibrillary tangles.

(Photos courtesy of D. J. Selkoe, Brigham and Women's Hospital, Boston.)

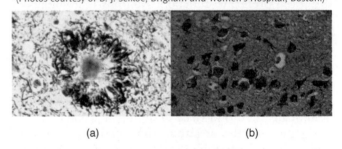

(a)   (b)

amyloid plaque (*amm i loyd*) An extracellular deposit containing a dense core of β-amyloid protein surrounded by degenerating axons and dendrites and activated microglia and reactive astrocytes.

β-amyloid (Aβ) A protein found in excessive amounts in the brains of patients with Alzheimer's disease.

neurofibrillary tangle (*new row fib ri lair y*) A dying neuron containing intracellular accumulations of abnormally phosphorylated tau-protein filaments that formerly served as the cell's internal skeleton.

tau protein A protein that normally serves as a component of microtubules, which provide the cell's transport mechanism and cytoskeleton.

β-amyloid precursor protein (APP) A protein produced and secreted by cells that serves as the precursor for β-amyloid protein.

two places by enzymes known as **secretases** to produce Aβ. The first, β-secretase, cuts the "tail" off of an APP molecule. The second, γ-secretase (gamma-secretase), cuts the "head" off. The result is a molecule of Aβ that contains either forty or forty-two amino acids. (See *Figure 15.23.*)

The location of the second cut of the APP molecule by γ-secretase determines which form is produced. In normal brains, 90–95 percent of the Aβ molecules are of the short form; the other 5–10 percent are of the long form. In patients with Alzheimer's disease the proportion of long Aβ rises to as much as 40 percent of the total. High concentrations of the long form have a tendency to fold themselves improperly and form aggregations, which have toxic effects on the cell. (As we saw earlier in this chapter, abnormally folded prions and α-synuclein proteins form aggregations that cause brain degeneration.) Small amounts of long Aβ can easily be cleared from the cell. The molecules are given a ubiquitin tag that marks them for destruction, and they are transported to the proteasomes, where they are rendered harmless. However, this system cannot keep up with abnormally high levels of production of long Aβ.

FIGURE 15.23 ■ β-Amyloid Protein

The schematic shows the production of β-amyloid protein (Aβ) from the amyloid precursor protein.

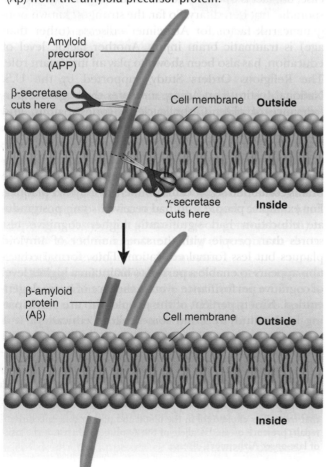

FIGURE 15.24 ■ Detection of β-Amyloid Protein

The PET scans show the accumulation of β-amyloid protein (Aβ) in the brains of a patient with Alzheimer's disease. AD = Alzheimer's disease, MR = structural magnetic resonance image, [C-11]PIB PET = PET scan of brains after an injection of a radioactive ligand for Aβ.

(Courtesy of William Klunk, Western Psychiatric Institute and Clinic, Pittsburgh, PA.)

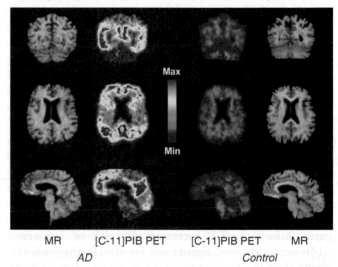

Acetylcholinergic neurons in the basal forebrain are among the first cells to be affected in Alzheimer's disease. Aβ serves as a ligand for the *p75 neurotrophin receptor*, a receptor that normally responds to stress signals and stimulates apoptosis (Sotthibundhu et al., 2008). Basal forebrain ACh neurons contain high levels of this receptors; thus, once the level of long-form Aβ reach a sufficiently high lever, these neurons begin to die.

Figure 15.24 shows the abnormal accumulation of Aβ in the brain of a person with Alzheimer's disease. Klunk and his colleagues (Klunk et al., 2003; Mathis et al., 2005) developed a chemical that binds with Aβ and readily crosses the blood–brain barrier. They gave the patient and a healthy control subject an injection of a radioactive form of this chemical and examined their brains with a PET scanner. You can see the accumulation of the protein in the patient's cerebral cortex. (See *Figure 15.24.*) The ability to measure the levels of Aβ in the brains of Alzheimer's patients will enable researchers to evaluate the effectiveness of potential treatments for the disease. If such a treatment is devised, the ability to identify the accumulation of Aβ early in the development of the disease will make it possible to begin a patient's treatment before significant degeneration—and the accompanying decline in cognitive abilities—has occurred.

secretase (*see cre tayss*) A class of enzymes that cut the β-amyloid precursor protein into smaller fragments, including β-amyloid.

general circulation and sensitizing the immune system to it, or perhaps the virus attaches itself to myelin. In addition, people born during the late winter and early spring are at higher risk, which suggests that infections contracted by a pregnant woman (for example, a viral disease contracted during the winter) may also increase susceptibility to this disease. In any event, the process is a long-lived one, lasting for many decades.

Only two treatments for multiple sclerosis have shown any promise. The first is *interferon β,* a protein that modulates the responsiveness of the immune system. Administration of interferon β has been shown to reduce the frequency and severity of attacks and to slow the progression of neurological disabilities in some patients with multiple sclerosis (Arnason, 1999). However, the treatment is only partially effective. Another partially effective treatment is *glatiramer acetate* (also known as copaxone or copolymer-1). Glatiramer acetate is a mixture of synthetic peptides composed from random sequences of the amino acids tyrosine, glutamate, alanine, and lysine. This compound was first produced in an attempt to induce the symptoms of multiple sclerosis in laboratory animals.

An experimentally induced demyelinating disease known as *experimental allergic encephalitis (EAE)* can be produced in laboratory animals by injecting them with protein found in myelin. The immune system then becomes sensitized to myelin protein and attacks the animal's own myelin sheaths. Glatiramer acetate turned out to do just the opposite; rather than *causing* EAE, it prevented its occurrence, apparently by stimulating certain cells of the immune system to secrete anti-inflammatory chemicals such as *interleukin 4,* which suppress the activity of immune cells that would otherwise attack the patient's myelin (Farina et al., 2005). As you might expect, researchers tested glatiramer acetate in people with MS and found that the drug reduced the symptoms of patients who showed the relapsing-remitting form of the disease: periodic occurrences of neurological symptoms followed by partial remissions. The drug is now approved for treatment of this disorder. A structural MRI study by Sormani et al. (2005) found a reduction of 20–54 percent in white matter lesions in 95 percent of patients treated with glatiramer acetate.

Although interferon β and glatiramer acetate provide some relief, neither treatment halts the progression of MS. We still need better forms of therapy. Because the symptoms of MS are often episodic—new or worsening symptoms followed by partial recovery—patients and their families often attribute the changes in the symptoms to whatever has happened recently. For example, if the patient has taken a new medication or gone on a new diet and the symptoms get worse, the patient will blame the symptoms on the medication or diet.

Conversely, if the patient gets better, he or she will credit the medication or diet.

Korsakoff's Syndrome

The last degenerative disorder I will discuss, Korsakoff's syndrome, is neither hereditary nor contagious. It is caused by environmental factors—usually (but not always) involving chronic alcoholism. The disorder actually results from a thiamine (vitamin B$_1$) deficiency caused by the alcoholism (Adams, 1969; Haas, 1988). Because alcoholics receive a substantial number of calories from the alcohol they ingest, they usually eat a poor diet, and their vitamin intake is consequently low. Furthermore, alcohol interferes with intestinal absorption of thiamine. The ensuing deficiency produces brain damage. Thiamine is essential for a step in metabolism: the carboxylation of pyruvate, an intermediate product in the breakdown of carbohydrates, fats, and amino acids. Korsakoff's syndrome sometimes occurs in people who have been severely malnourished and have then received intravenous infusions of glucose; the sudden availability of glucose to the cells of the brain without adequate thiamine with which to metabolize it damages the cells, probably because they accumulate pyruvate. Hence, standard medical practice is to administer thiamine along with intravenous glucose to severely malnourished patients.

As we saw in Chapter 13, the brain damage incurred in Korsakoff's syndrome causes anterograde amnesia. Although degeneration is seen in many parts of the brain, the damage that characterizes this disorder occurs in the mammillary bodies, located at the base of the brain, in the posterior hypothalamus. (See *Figure 15.29.*)

FIGURE 15.29 ■ Korsakoff's Syndrome

This brain slice shows the degeneration of the mammillary bodies in a patient with Korsakoff's syndrome.

(Courtesy of A. D'Agostino, Good Samaritan Hospital, Portland, Oregon.)

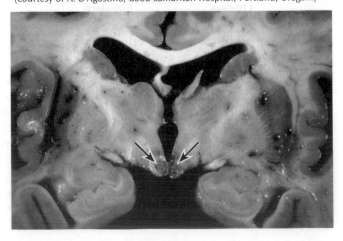

InterimSummary

Degenerative Disorders

Transmissible spongiform encephalopathies such as Creutzfeldt-Jakob disease, scrapie, and bovine spongiform encephalopathy ("mad cow disease") are unique among contagious diseases: They are produced by a simple protein molecule, not by a virus or microbe. The sequence of amino acids of normal prion protein (PrPc) and infectious prion protein (PrPSc) are identical, but their three-dimensional shapes differ in the way that they are folded. Somehow, the presence of a misfolded prion protein in a neuron causes normal prion proteins to become misfolded, and a chain reaction ensues. The transformation of PrPc into PrPSc kills the cell, apparently by triggering apoptosis. Creutzfeldt-Jakob disease is heritable as well as transmissible, but the most common form is sporadic—of unknown origin. Normal prion protein may play a role in neural development and neurogenesis, which may in turn affect the establishment and maintenance of long-term memories.

Parkinson's disease is caused by degeneration of dopamine-secreting neurons of the substantia nigra that send axons to the basal ganglia. Study of rare hereditary forms of Parkinson's disease reveals that the death of these neurons is caused by the aggregation of misfolded protein, α-synuclein. One mutation produces defective α-synuclein, and another produces defective parkin, a protein that assists in the tagging of abnormal proteins for destruction by the proteasomes. The accumulation of α-synuclein can also be triggered by some toxins, which suggests that nonhereditary forms of the disease may be caused by toxic substances present in the environment. Treatment of Parkinson's disease includes administration of L-DOPA, implantation of fetal dopaminergic neurons in the basal ganglia, stereotaxic destruction of a portion of the globus pallidus or subthalamus, and implantation of electrodes that enable the patient to electrically stimulate the subthalamus. Fetal transplants of dopaminergic neurons have turned out to be less successful than they had initially appeared to be, but transplants of neural stem cells shows more promise. A trial of gene therapy designed to reduce excitation in the subthalamic nucleus obtained promising results.

Huntington's disease, an autosomal dominant hereditary disorder, produces degeneration of the caudate nucleus and putamen. Mutated huntingtin misfolds and forms aggregations that accumulate in the nucleus of GABAergic neurons in the putamen. Although the primary effect of mutated huntingtin is gain of toxic function, the disease also appears to involve a loss of function; normal huntingtin facilitates the production and transport of BDNF, a protein necessary for survival of neurons in the striatum. Evidence also suggests that inclusion bodies have a protective function and that damage is done by mutated huntingtin dispersed throughout the cell. An animal study that transferred small interfering RNA targeted against the htt gene suggested that this approach might be fruitful.

Alzheimer's disease, another degenerative disorder, involves much more of the brain; the disease process eventually destroys most of the hippocampus and cortical gray matter. The brains of affected individuals contain many amyloid plaques, which contain a core of misfolded long-form Aβ protein surrounded by degenerating axons and dendrites, and neurofibrillary tangles, composed of dying neurons that contain intracellular accumulations of twisted filaments of tau protein. Hereditary forms of Alzheimer's disease involve defective genes for the amyloid precursor protein (APP), for the secretases that cut APP into smaller pieces, or for apolipoprotein E (ApoE), a glycoprotein involved in transport of cholesterol and the repair of cell membranes. A promising treatment is vaccination against Aβ, and transcutaneous administration of the antigen may provide a way to avoid triggering an inflammatory reaction. Temporary reduction of symptoms is seen in some patients who are treated with anticholinergic drugs or drugs that serve as NMDA antagonists. Exercise and intellectual stimulation appear to delay the onset of Alzheimer's disease, and obesity, high cholesterol levels, and diabetes are significant risk factors.

Amyotrophic lateral sclerosis is a degenerative disorder that attacks motor neurons. Ten percent of the cases are hereditary, caused by a mutation of the gene for SOD1; the other 90 percent are sporadic. The primary cause of sporadic ALS appears to be an abnormality in RNA editing, which results in the production of AMPA receptor subunits that permit the entry of excessive amounts of calcium into the cells. The only pharmacological treatment is riluzole, a drug that reduces glutamate-induced excitotoxicity. A virally introduced gene for IGF-1 has shown promise in an animal model of ALS.

Multiple sclerosis, a demyelinating disease, is characterized by periodic attacks of neurological symptoms, usually with partial remission between attacks. The damage appears to be caused by the body's immune system, which attacks the protein contained in myelin. Most investigators believe that a viral infection early in life somehow sensitizes the immune system to myelin protein. The only effective treatments for MS are interferon β and glatiramer acetate, a mixture of synthetic peptides that appears to stimulate certain immune cells to secrete anti-inflammatory chemicals.

Korsakoff's syndrome is usually a result of chronic alcohol abuse, but it can also be caused by malnutrition that results in a thiamine deficiency. The most obvious location of brain damage is the mammillary bodies, but damage also occurs in many other parts of the brain.

DISORDERS CAUSED BY INFECTIOUS DISEASES

Several neurological disorders can be caused by infectious diseases, transmitted by bacteria, fungi or other parasites, or viruses. The most common are encephalitis and meningitis. **Encephalitis** is an infection that invades the entire brain. The most common cause of encephalitis is a virus that is transmitted by mosquitoes, which pick up the infectious agent from horses, birds, or rodents. The symptoms of acute encephalitis include fever, irritability, and nausea, often followed by convulsions, delirium, and signs of brain damage, such as aphasia or paralysis. Unfortunately, there is no specific treatment besides supportive care, and between 5 and 20 percent of the cases are fatal; 20 percent of the survivors show some residual neurological symptoms.

Encephalitis can also be caused by the **herpes simplex virus,** which is the cause of cold sores (or "fever blisters") that most people develop in and around their mouth from time to time. Normally, the viruses live quietly in the *trigeminal nerve ganglia* nodules on the fifth cranial nerve that contain the cell bodies of somatosensory neurons that serve the face. The viruses proliferate periodically, traveling down to the ends of nerve fibers, where they cause sores to develop in mucous membrane. Unfortunately, they occasionally (but rarely) go the other way into the brain. Herpes encephalitis is a serious disease; the virus attacks the frontal and temporal lobes in particular and can severely damage them.

Two other forms of viral encephalitis are probably already familiar to you: polio and rabies. **Acute anterior poliomyelitis** ("polio") is fortunately very rare in developed countries since the development of vaccines that immunize people against the disease. The virus causes specific damage to motor neurons of the brain and spinal cord: neurons in the primary motor cortex; in the motor nuclei of the thalamus, hypothalamus, and brain stem; in the cerebellum; and in the ventral horns of the gray matter of the spinal cord. Undoubtedly, these motor neurons contain some chemical substance that either attracts the virus or in some way makes the virus become lethal to them.

Rabies is caused by a virus that is passed from the saliva of an infected mammal directly into a person's flesh by means of a bite wound. The virus travels through peripheral nerves to the central nervous system and there causes severe damage. It also travels to peripheral organs, such as the salivary glands, which makes it possible for the virus to find its way to another host. The symptoms include a short period of fever and headache, followed by anxiety, excessive movement and talking, difficulty in swallowing, movement disorders, difficulty in speaking, seizures, confusion, and, finally, death within two to seven days of the onset of the symptoms. The virus has a special affinity for cells in the cerebellum and hippocampus, and damage to the hippocampus probably accounts for the emotional changes that are seen in the early symptoms.

Fortunately, the incubation period for rabies lasts up to several months while the virus climbs through the peripheral nerves. (If the bite is received in the face or neck, the incubation time will be much shorter because the virus has a smaller distance to travel before it reaches the brain.) During the incubation period a person can receive a vaccine that will confer an immunity to the disease; the person's own immune system will destroy the virus before it reaches the brain.

Several infectious diseases cause brain damage even though they are not primarily diseases of the central nervous system. One such disease is acquired immune deficiency syndrome (AIDS). Records of autopsies have revealed that at least 75 percent of people who died of AIDS show evidence of brain damage (Levy and Bredesen, 1989). The brain damage often results in a syndrome called *AIDS dementia complex (ADC),* which is characterized by damage to synapses and death of neurons in the hippocampus, cerebral cortex, and basal ganglia (Mattson, Haughey, and Nath, 2005). The brain damage leads to a loss of cognitive and motor functions and is the leading cause of cognitive decline in people under 40 years of age. At first the patients may become forgetful, they may think and reason more slowly, and they may have word-finding difficulties (anomia). Eventually, they may become almost mute. Motor deficits may begin with tremor and difficulty in making complex movement but then may progress so much that the patient becomes bedridden (Maj, 1990).

For several years, researchers have been puzzled by the fact that although AIDS certainly causes neural damage, neurons are not themselves infected by the HIV virus (the organism responsible for the disease). Instead, ADC is caused by the glycoprotein *gp120* envelope that coats the RNA that is responsible for the AIDS infection. The gp120 binds with other proteins that trigger apoptosis—cell suicide (Mattson, Haughey, and Nath, 2005; Alirezaei et al., 2007).

Another category of infectious diseases of the brain actually involves inflammation of the meninges, the layers of connective tissue that surround the central nervous system. **Meningitis** can be caused by viruses or

encephalitis (*en seff a lye tis*) An inflammation of the brain; caused by bacteria, viruses, or toxic chemicals.

herpes simplex virus (*her peez*) A virus that normally causes cold sores near the lips but that can also cause brain damage.

acute anterior poliomyelitis (*poh lee oh my a lye tis*) A viral disease that destroys motor neurons of the brain and spinal cord.

rabies A fatal viral disease that causes brain damage; usually transmitted through the bite of an infected animal.

meningitis (*men in jy tis*) An inflammation of the meninges; can be caused by viruses or bacteria.

bacteria. The symptoms of all forms include headache, a stiff neck, and, depending on the severity of the disorder, convulsions, confusion or loss of consciousness, and sometimes death. The stiff neck is one of the most important symptoms. Neck movements cause the meninges to stretch; because they are inflamed, the stretch causes severe pain. Thus, the patient resists having his or her neck moved.

The most common form of viral meningitis usually does not cause significant brain damage. However, various forms of bacterial meningitis do. The usual cause is spread of a middle-ear infection into the brain, introduction of an infection into the brain from a head injury, or the presence of emboli that have dislodged

from a bacterial infection present in the chambers of the heart. Such an infection is often caused by unclean hypodermic needles; therefore, drug addicts are at particular risk for meningitis (as well as many other diseases). The inflammation of the meninges can damage the brain by interfering with circulation of blood or by blocking the flow of cerebrospinal fluid through the subarachnoid space, causing hydrocephalus. In addition, the cranial nerves are susceptible to damage. Fortunately, bacterial meningitis can usually be treated effectively with antibiotics. Of course, early diagnosis and prompt treatment are essential, because neither antibiotics nor any other known treatment can repair a damaged brain.

Interim**Summary**

Disorders Caused by Infectious Diseases

Infectious diseases can damage the brain. Encephalitis, usually caused by a virus, affects the entire brain. One form is caused by the herpes simplex virus, which infects the trigeminal nerve ganglia of most of the population. This virus tends to attack the frontal and temporal lobes. The polio virus attacks motor neurons in the brain and spinal cord, resulting in motor deficits or

even paralysis. The rabies virus, acquired by an animal bite, travels through peripheral nerves and attacks the brain, particularly the cerebellum and hippocampus. An AIDS infection also produces brain damage when the gp120 protein envelope of the HIV virus binds with other proteins that trigger apoptosis. Meningitis is an infection of the meninges, caused by viruses or bacteria. The bacterial form, which is usually more serious, is generally caused by an ear infection, a head injury, or an embolus from a heart infection.

SUGGESTED READINGS

Bossy-Wetzel, E., Schwarzenbacher, R., and Lipton, S. A. Molecular pathways to neurodegeneration. *Nature Medicine*, 2004, *10*, S2–S9.

Hardy, J. Toward Alzheimer therapies based on genetic knowledge. *Annual Review of Medicine*, 2004, *55*, 15–25.

Moore, D. J., West, A. B., Dawson, V. L., and Dawson, T. M. Molecular pathophysiology of Parkinson's disease. *Annual Review of Neuroscience*, 2005, *28*, 57–87.

Ropper, A. H., and Brown, R. H. *Adams and Victor's Principles of Neurology*, 8th ed. New York: McGraw-Hill Professional, 2005.

Weissmann, C., and Aguzzi, A. Approaches to therapy of prion diseases. *Annual Review of Medicine*, 2005, *56*, 321–344.

ADDITIONAL RESOURCES

Visit www.mypsychkit.com for additional review and practice of the material covered in this chapter. Within MyPsychKit, you can take practice tests and receive a customized study plan to help you review. Dozens of

animations, tutorials, and Web links are also available. You can even review using the interactive electronic version of this textbook. You will need to register for MyPsychKit. See www.mypsychkit.com for complete details.

MAJOR AFFECTIVE DISORDERS

Affect, as a noun, refers to feelings or emotions. Just as the primary symptom of schizophrenia is disordered thoughts, the major affective disorders (also called *mood disorders*) are characterized by disordered feelings.

Description

Feelings and emotions are essential parts of human existence; they represent our evaluation of the events in our lives. In a very real sense, feelings and emotions are what human life is all about. The emotional state of most of us reflects what is happening to us: Our feelings are tied to events in the real world and are usually the result of reasonable assessments of the importance these events have for our lives. But for some people, affect becomes divorced from reality. These people have feelings of extreme elation (*mania*) or despair (*depression*) that are not justified by events in their lives. For example, depression that accompanies the loss of a loved one is normal, but depression that becomes a way of life—and will not respond to the sympathetic effort of friends and relatives or even to psychotherapy—is pathological. Depression has a prevalence of approximately 3 percent in men and 7 percent in women, which makes it the fourth leading cause of disability (Kessler et al., 2003).

There are two principal types of major affective disorders. The first type is characterized by alternating periods of mania and depression—a condition called **bipolar disorder.** This disorder afflicts men and women in approximately equal numbers. Episodes of mania can last a few days or several months, but they usually take a few weeks to run their course. The episodes of depression that follow generally last three times as long as the mania. The second type is **major depressive disorder (MDD),** characterized by depression without mania. This depression may be continuous and unremitting or, more typically, may come in episodes. Mania without periods of depression sometimes occurs, but it is rare.

Severely depressed people usually feel extremely unworthy and have strong feelings of guilt. The affective disorders are dangerous; a person who suffers from a major affective disorder runs a considerable risk of death by suicide. According to Chen and Dilsaver (1996), 15.9 percent of people with MDD and 29.2 percent of people with bipolar disorder attempt to commit suicide. Schneider, Muller, and Philipp (2001) found that the rate of death by unnatural causes (not all suicides are diagnosed as such) for people with affective disorders was 28.8 times higher than expected for people of the same age in the general population.

Depressed people have very little energy, and they move and talk slowly, sometimes becoming almost torpid. At other times, they may pace around restlessly and aimlessly. They may cry a lot. They are unable to experience pleasure and lose their appetite for food and sex. Their sleep is disturbed; they usually have difficulty falling asleep and awaken early and find it difficult to get to sleep again. Even their body functions become depressed; they often become constipated, and their secretion of saliva decreases.

> [A psychiatrist] asked me if I was suicidal, and I reluctantly told him yes. I did not particularize—since there seemed no need to—did not tell him that in truth many of the artifacts of my house had become potential devices for my own destruction: the attic rafters (and an outside maple or two) a means to hang myself, the garage a place to inhale carbon monoxide, the bathtub a vessel to receive the flow from my opened arteries. The kitchen knives in their drawers had but one purpose for me. Death by heart attack seemed particularly inviting, absolving me as it would of active responsibility, and I had toyed with the idea of self-induced pneumonia—a long frigid, shirt-sleeved hike through the rainy woods. Nor had I overlooked an ostensible accident . . . by walking in front of a truck on the highway nearby. . . . Such hideous fantasies, which cause well people to shudder, are to the deeply depressed mind what lascivious daydreams are to persons of robust sexuality. (Styron, 1990, pp. 52–53)

Episodes of mania are characterized by a sense of euphoria that does not seem to be justified by circumstances. The diagnosis of mania is partly a matter of degree; one would not call exuberance and a zest for life pathological. People with mania usually exhibit nonstop speech and motor activity. They flit from topic to topic and often have delusions, but they lack the severe disorganization that is seen in schizophrenia. They are usually full of their own importance and often become angry or defensive if they are contradicted. Frequently, they go for long periods without sleep, working furiously on projects that are often unrealistic. (Sometimes, their work is fruitful; George Frideric Handel wrote *Messiah,* one of the masterpieces of choral music, during one of his periods of mania.)

---

bipolar disorder A serious mood disorder characterized by cyclical periods of mania and depression.

major depressive disorder (MDD) A serious mood disorder that consists of unremitting depression or periods of depression that do not alternate with periods of mania.

Heritability

Evidence indicates that a tendency to develop an affective disorder is a heritable characteristic. (See Hamet and Tremblay, 2005, for a review.) For example, Rosenthal (1971) found that close relatives of people who suffer from affective psychoses are ten times more likely to develop these disorders than are people without afflicted relatives. Gershon et al. (1976) found that if one member of a set of monozygotic twins was afflicted with an affective disorder, the likelihood that the other twin was similarly afflicted was 69 percent. In contrast, the concordance rate for dizygotic twins was only 13 percent. The heritability of the affective disorders implies that they have a physiological basis.

Genetic studies have found evidence that genes on several chromosomes may be implicated in the development of the affective disorders, but the findings of most linkage studies have not been replicated (Hamet and Tremblay, 2005). As we will see later in this chapter, the strongest candidate is the gene for the serotonin transporter, which plays an important role in brain development. This gene is found on chromosome 17.

Biological Treatments

There are several established and experimental biological treatments for major depressive disorder: monoamine oxidase (MAO) inhibitors, drugs that inhibit the reuptake of norepinephrine or serotonin or interfere with NMDA receptors, electroconvulsive therapy, transcranial magnetic stimulation, deep brain stimulation, vagus nerve stimulation, bright-light therapy (phototherapy), and sleep deprivation. (Phototherapy and sleep deprivation are discussed in a later section of this chapter.) Bipolar disorder can be treated by lithium and some anticonvulsant drugs. The fact that these disorders often respond to biological treatment provides additional evidence that they have a physiological basis. Furthermore, the fact that lithium is effective in treating bipolar affective disorder but not major depressive disorder suggests that there is a fundamental difference between these two illnesses (Soares and Gershon, 1998).

Before the 1950s there was no effective drug treatment for depression. In the late 1940s clinicians noticed that some drugs used for treating tuberculosis seemed to elevate the patient's mood. Researchers subsequently found that a derivative of these drugs, iproniazid, reduced symptoms of psychotic depression (Crane, 1957). Iproniazid inhibits the activity of MAO, which destroys excess monoamine transmitter substances within terminal buttons. Thus, the drug increases the release of dopamine, norepinephrine, and serotonin. Other MAO inhibitors were soon discovered. Unfortunately, MAO inhibitors can have harmful side effects, so they must be used with caution.

Fortunately, another class of antidepressant drugs was soon discovered that did not have these side effects: the **tricyclic antidepressants.** These drugs were found to inhibit the reuptake of 5-HT and norepinephrine by terminal buttons. By retarding reuptake, the drugs keep the neurotransmitter in contact with the postsynaptic receptors, thus prolonging the postsynaptic potentials. Thus, both the MAO inhibitors and the tricyclic antidepressant drugs are monoaminergic agonists.

Since the discovery of the tricyclic antidepressants, other drugs have been discovered that have similar effects. The most important of these are the **specific serotonin reuptake inhibitors (SSRI),** whose action is described by their name. These drugs (for example, fluoxetine (Prozac), citalopram (Celexa), and paroxetine (Paxil) are widely prescribed for their antidepressant properties and for their ability to reduce the symptoms of obsessive-compulsive disorder and social phobia (described in Chapter 17). Recently, another class of antidepressant drugs have been developed, the **serotonin and norepinephrine reuptake inhibitors (SNRI),** which do what their name indicates. These drugs have fewer nonspecific actions, and therefore fewer side effects, than the tricyclic antidepressants, which also have effects on both norepinephrine and serotonin reuptake (Stahl et al., 2005). The category of SNRIs includes milnacipran, duloxetine, and venlafaxine, with relative effects on 5-HT and noradrenergic transporters of 1:1, 1:10, and 1:30, respectively.

Another biological treatment for depression has an interesting history. Early in the twentieth century, a physician named von Meduna noted that psychotic patients who were also subject to epileptic seizures showed improvement immediately after each attack. He reasoned that the violent storm of neural activity in the brain that constitutes an epileptic seizure somehow improved the patients' mental condition. He developed a way to produce seizures by administering a drug, but the procedure was dangerous to the patient. In 1937, Ugo Cerletti, an Italian psychiatrist, developed a less dangerous method for producing seizures (Cerletti and Bini, 1938). He had previously learned that the local slaughterhouse applied a jolt of electricity to animals' heads to stun them before killing them. The electricity appeared to produce a

---

tricyclic antidepressant A class of drugs used to treat depression; inhibits the reuptake of norepinephrine and serotonin but also affects other neurotransmitters; named for the molecular structure.

specific serotonin reuptake inhibitor (SSRI) An antidepressant drug that specifically inhibits the reuptake of serotonin without affecting the reuptake of other neurotransmitters.

serotonin and norepinephrine reuptake inhibitor (SNRI) An antidepressant drug that specifically inhibits the reuptake of norepinephrine and serotonin without affecting the reuptake of other neurotransmitters.

seizure that resembled an epileptic attack. He decided to attempt to use electricity to more safely induce a seizure.

Cerletti tried the procedure on dogs and found that an electrical shock to the skull did produce a seizure and that the animals recovered with no apparent ill effects. He then used the procedure on humans and found it to be safer than the chemical treatment that was previously used. As a result of Cerletti's experiments, **electroconvulsive therapy (ECT)** became a common treatment for mental illness. Before a person receives ECT, he or she is anesthetized and is given a drug similar to curare, which paralyzes the muscles, preventing injuries that might be produced by a convulsion. (Of course, the patient is attached to a respirator until the effects of this drug wear off.) Electrodes are placed on the patient's scalp (most often to the non-speech-dominant hemisphere, to avoid damaging verbal memories), and a jolt of electricity triggers a seizure. Usually, a patient receives three treatments per week until maximum improvement is seen, which usually involves six to twelve treatments. The effectiveness of ECT has been established by placebo studies, in which some patients are anesthetized but not given shocks (Weiner and Krystal, 1994). Although ECT was originally used for a variety of disorders, including schizophrenia, we now know that its usefulness is limited to treatment of mania and depression. (See *Figure 16.16.*)

Antidepressant drug treatment has some adverse side-effects, including nausea, anxiety, sexual dysfunction, and weight gain. However, the major problem is that in a substantial percentage of patients, the drug fails to relieve their depression. Between 20 and 40 percent of patients with major depressive disorder do not show a significant response to initial treatment with an antidepressant drug. When patients do not respond, physicians will try different drugs. Some of these patients do eventually respond, but others do not and exhibit **treatment-resistant depression.** The reason that the list of biological treatments presented in the first paragraph of this section is so long is because no single treatment works for all patients—and for some patients, no treatment works at all. The existence of so many patients with treatment-resistant depression has motivated researchers to try to develop ways to alleviate the symptoms of patients who continue to suffer.

Even when depressed patients respond to treatment with antidepressant drugs, they do not do so immediately; improvement in symptoms is not usually seen before two to three weeks of drug treatment. In contrast, the effects of ECT are more rapid. A few seizures induced by ECT can often snap a person out of a deep depression within a few days. Although prolonged and excessive use of ECT causes brain damage, resulting in long-lasting impairments in memory (Squire, 1974), the judicious use of ECT during the interim period before

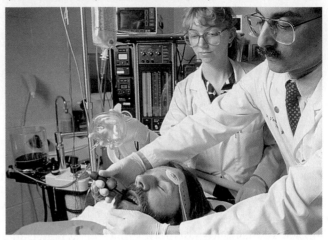

FIGURE 16.16 ■ **A Patient Prepared for Electroconvulsive Therapy**

(Will and Deni McIntyre/Photo Researchers, Inc.)

antidepressant drugs become effective has undoubtedly saved the lives of some suicidal patients.

How does ECT exert its antidepressant effect? It has been known for a long time that seizures have an anticonvulsant effect: ECT decreases brain activity and raises the seizure threshold of the brain, making it less likely for another seizure to occur (Sackeim et al., 1983; Nobler et al., 2001). The increased seizure threshold appears to be caused by an increased release of GABA and neuropeptide Y (Bolwig, Woldbye, and Mikkelsen, 1999; Sanacora et al., 2003). These changes may also be responsible for reducing the symptoms of depression.

Researchers have investigated another procedure designed to provide some of the benefits of ECT without introducing the risk of cognitive impairments or memory loss. As we saw in Chapter 5, transcranial magnetic stimulation (TMS) is accomplished by applying a strong localized magnetic field into the brain by passing an electrical current through a coil of wire placed on the scalp. The magnetic field induces an electrical current in the brain. Several studies suggested that TMS applied to the prefrontal cortex reduces the symptoms of depression without producing any apparent negative side effects (Padberg and Moller, 2003; Fitzgerald, 2004). However, a review of the literature by Mitchell and Loo (2006) concluded that although studies have obtained statistically significant results, the size of the effect is generally to small to be clinically significant.

---

electroconvulsive therapy (ECT) A brief electrical shock, applied to the head, that results in an electrical seizure; used therapeutically to alleviate severe depression.

treatment-resistant depression A major depressive disorder whose symptoms are not relieved after trials of several different treatments.

Preliminary research suggests that direct electrical stimulation of the brain (deep brain stimulation) may also be a useful therapy for treatment-resistant depression (Mayberg et al., 2005; Lozano et al., 2008). The investigators implanted stimulating electrodes just below the **subgenual anterior cingulate cortex (subgenual ACC),** a region of the medial prefrontal cortex. If you look at a sagittal view of the corpus callosum, you will see that the front of this structure looks like a bent knee—*genu,* in Latin. The subgenual ACC is located below the "knee" at the front of the corpus callosum. Response to the stimulation began soon, and it increased with time. One month after surgery, 35 percent of the patients showed an improvement in symptoms, and 10 percent showed a complete remission. Six months after surgery, 60 percent showed improvement, and 35 percent showed remission.

Another experimental treatment for depression, electrical stimulation of the vagus nerve, shows some promise of reducing the symptoms of depression (Groves and Brown, 2005). Vagus nerve stimulation provides an indirect form of brain stimulation. It is painless and does not elicit seizures—in fact, the procedure was originally developed as a treatment to prevent seizures in patients with seizure disorders. The stimulation is accomplished by means of an implanted device similar to the one used for deep brain stimulation, described in the section on Parkinson's disease in Chapter 15, except that the stimulating electrodes are attached to the vagus nerve. Approximately 80 percent of the axons in the vagus nerve are afferent, so electrical stimulation of the vagus nerve activates several regions of the brain stem. A review of the literature by Daban et al. (2008) concluded that the procedure showed promise in treatment of patients with treatment-resistant depression but that further double-blind clinical trials are needed to confirm its efficacy.

Functional-imaging studies have shown that both vagus nerve stimulation and deep brain stimulation cause progressive changes in the activity of several brain regions, including the subgenual ACC (Lozano et al., 2008; Pardo et al., 2008). As we will see later in this chapter, other research implicates this region in the development of depression. (See *Figure 16.17.*)

As I mentioned earlier, most antidepressant drugs currently in use act as noradrenergic or serotonergic agonists by inhibiting the reuptake of these neurotransmitters. Preliminary evidence suggest that an NMDA antagonist, ketamine, may alleviate the symptoms of treatment-resistant depression. Research with laboratory animals found that injections of ketamine reduced behaviors similar to those seen in depressed humans, and imaging studies with humans suggested that depressed patients showed increased brain levels of glutamate, which suggests that interference with gluta-

FIGURE 16.17 ■ Effects of Deep-Brain Stimulation and Vagus Nerve Stimulation

The scans show the effects of three months (a) and six months (b) of deep brain stimulation and the effects of six months (c) and one year (d) of vagus nerve stimulation of patients with treatment-resistant depression. "Warm" colors indicate increased activity, and "cool" colors indicate decreased activity in (a) and (b).

([a] and [b] from Lozano, A. M., Mayberg, H. S., Giacobbe, P., Hamani, C., Craddock, R. C., and Kennedy, S. H. *Biological Psychiatry*, 2008, *64*, 461–467; and [c] and [d] from Pardo, J. V., Sheikh, S. A., Schwindt, G. C., Lee, J. T., Kuskowski, M. A., Surerus, C., Lewis, S. M., Abuzzahab, F. S., Adson, D. E., and Rittberg, B. R. *NeuroImage*, 2008, *42*, 879–889.)

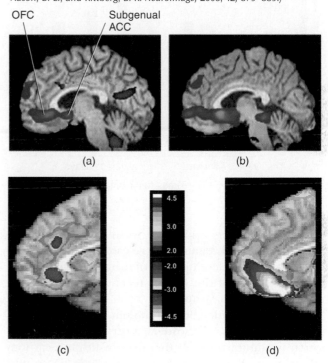

matergic transmission might have therapeutic effects (Yilmaz et al., 2002; Sanacora et al., 2004). Zarate et al., (2006) administered injections of ketamine or placebo to patients with treatment-resistant depression. In less than two hours after the ketamine injections, 71 percent of the patients showed an improvement in their symptoms, and 29 percent showed a remission of their symptoms. This positive response persisted for at least one week. (See *Figure 16.18.*) You will recall from the discussion of schizophrenia earlier in this chapter that chronic administration of ketamine or PCP, another NMDA antagonist, produces the symptoms of schizophrenia.

subgenual anterior cingulate cortex (subgenual ACC) A region of the medial prefrontal cortex located below the "knee" at the front of the corpus callosum; plays a role in the symptoms of depression.

FIGURE 16.18 ■ Treatment of Depression with Ketamine

The graph shows the effects of ketamine on symptoms of depression.

(Adapted from Zarate, C. A., Jaskaran, B. S., Carlson, P. J., Brutsche, N. E., Ameli, R., Luckenbaugh, D.A ., Charney, D. S., and Manji, H. K. *Archives of General Psychiatry*, 2006, *63*, 856–864.)

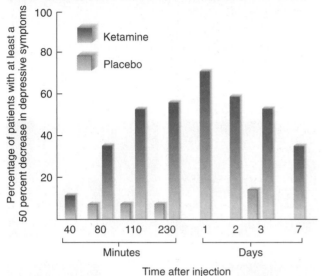

Clearly, depression cannot be treated with long-term administration of ketamine, but researchers are currently investigating the development of drugs that act on particular subunits of the NMDA receptor and hence do not elicit psychotic symptoms.

The therapeutic effect of **lithium,** the drug used to treat bipolar affective disorders, is very rapid. This drug, which is administered in the form of lithium carbonate, is most effective in treating the manic phase of a bipolar affective disorder; once the mania is eliminated, depression usually does not follow (Gerbino, Oleshansky, and Gershon, 1978; Soares and Gershon, 1998). Some clinicians and investigators have referred to lithium as psychiatry's wonder drug: It does not suppress normal feelings of emotions, but it leaves patients able to feel and express joy and sadness in response to events in their lives. Similarly, it does not impair intellectual processes; many patients have received the drug continuously for years without any apparent ill effects (Fieve, 1979). Between 70 and 80 percent of patients with bipolar disorder show a positive response to lithium within a week or two (Price and Heninger, 1994).

Lithium does have adverse side effects. The therapeutic index (the difference between an effective dose and an overdose) is low. Side effects include hand tremors, weight gain, excessive urine production, and thirst. Toxic doses produce nausea, diarrhea, motor incoordination, confusion, and coma. Because of the

low therapeutic index, patients' blood levels of lithium must be tested regularly to be certain that they do not receive an overdose. Unfortunately, some patients are not able to tolerate the side effects of lithium.

Researchers have found that lithium has many physiological effects, but they have not yet discovered the pharmacological effects of lithium that are responsible for its ability to eliminate mania (Phiel and Klein, 2001). Some researchers suggest that the drug stabilizes the population of certain classes of neurotransmitter receptors in the brain (especially serotonin receptors), thus preventing wide shifts in neural sensitivity (Jope et al., 1996) Others have shown that lithium may increase the production of neuroprotective proteins that help to prevent cell death (Manji, Moore, and Chen, 2001). In fact, Moore et al. (2000) found that four weeks of lithium treatment for bipolar disorder increased the volume of cerebral gray matter in the patients' brains, a finding that suggests that lithium facilitates neural or glial growth.

Because some patients cannot tolerate the side effects of lithium and because of the potential danger of overdose, researchers have been searching for alternative medications for bipolar disorder. The most promising results have come from the use of anticonvulsants, such as lamotrigine, valproate, and carbamazepine (Grunze, 2005).

The Monoamine Hypothesis

The fact that depression can be treated with MAO inhibitors and drugs that inhibit the reuptake of monoamines suggested the **monoamine hypothesis:** Depression is caused by insufficient activity of monoaminergic neurons. Because the symptoms of depression are not relieved by potent dopamine agonists such as amphetamine or cocaine, most investigators have focused their research efforts on the other two monoamines: norepinephrine and serotonin.

As we saw earlier in this chapter, the dopamine hypothesis of schizophrenia was suggested by the fact that dopamine agonists can produce the symptoms of schizophrenia and dopamine antagonists can reduce them. Similarly, the monoamine hypothesis of depression was suggested by the fact that monoamine antagonists can produce the symptoms of depression and monoamine agonists can reduce them. As you will recall from Chapter 4, the drug *reserpine* blocks the activity of transporters that fill synaptic vesicles in monoaminergic terminals with the

lithium A chemical element; lithium carbonate is used to treat bipolar disorder.

monoamine hypothesis A hypothesis that states that depression is caused by a low level of activity of one or more monoaminergic synapses.

neurotransmitter. Reserpine was previously used to lower blood pressure by blocking the release of norepinephrine in muscles in the walls of blood vessels, which causes these muscles to relax. However, reserpine has a serious side effect: By interfering with the release of serotonin and norepinephrine in the brain, it can cause depression. In fact, in the early years of its use as a hypotensive agent, up to 15 percent of the people who received it became depressed (Sachar and Baron, 1979). As we can see, a monoamine antagonist produces the symptoms of depression, and monoamine agonists alleviate them.

Delgado et al. (1990) developed an ingenious approach to study the role of serotonin in depression: the **tryptophan depletion procedure.** They studied depressed patients who were receiving antidepressant medication and were currently feeling well. For one day they had the patients follow a low-tryptophan diet (for example, salad, corn, cream cheese, and a gelatin dessert). Then the next day, the patients drank an amino acid "cocktail" that contained no tryptophan. The uptake of amino acids through the blood–brain barrier is accomplished by amino acid transporters. Because the patients' blood level of tryptophan was very low and that of the other amino acids was high, very little tryptophan found its way into the brain, and the level of tryptophan in the brain fell drastically. As you will recall, tryptophan is the precursor of 5-HT, or serotonin. Thus, the treatment lowered the level of serotonin in the brain.

Delgado and his colleagues found that the tryptophan depletion caused most of the patients to relapse back into depression. Then when they began eating a normal diet again, they recovered. These results strongly suggest that the therapeutic effect of at least some antidepressant drugs depends on the availability of serotonin in the brain. Subsequent studies have confirmed these results. These studies also indicate that tryptophan depletion has little or no effect on the mood of healthy subjects, but it does lower the mood of people with a personal history or family history of affective disorders (Young and Leyton, 2002; Neumeister et al., 2004).

Most investigators believe that the simple monoamine hypothesis—that depression is caused by low levels of norepinephrine or serotonin—is just that: too simple. The effects of tryptophan depletion certainly suggest that serotonin plays a role in depression, but depletion causes depression only in people with a personal or family history of depression. An acute decrease in serotonergic activity in healthy people with no family history of depression has no effect on mood. Thus, there appear to be physiological differences in the brains of the vulnerable people. Also, although SSRIs and SNRIs change the increase the level of 5-HT or norepinephrine in the brain very rapidly, the drugs do not relieve the symptoms of depression until they have been taken for several weeks. This fact suggests that some-

thing other than a simple increase in monoaminergic activity in responsible for the normalization of mood. Many investigators believe that the increased extracellular levels of monoamines produced by administration of antidepressant drugs begin a chain of events that eventually produce changes in the brain that are ultimately responsible for antidepressant effect.

The Amygdala and the Prefrontal Cortex: Role of the 5-HT Transporter

In a review of the relevant literature, Drevets (2001) suggests that the amygdala and several regions of the prefrontal cortex play special roles in the development of depression. As we saw in Chapter 11, the amygdala is critically involved in the expression of negative emotions. Functional-imaging studies indicate a 50–75 percent increase in blood flow and metabolism in the amygdala of depressed patients (Drevets et al., 1992; Links et al., 1996). A study by Abercrombie et al. (1998) found that the activity of the amygdala of depressed patients was correlated with the severity of their depression. In addition, the metabolic activity of the amygdala increases in normal subjects when they look at pictures of faces with expressions of sadness, and it also increases when depressed subjects remember episodes in their lives that made them sad (Drevets, 2000b; Liotti et al., 2002).

Another region of the medial prefrontal cortex, the subgenual ACC, shows a *lower* level of activation in depressed patients (Drevets et al., 1997). Figure 16.19 shows the decreased activity of the subgenual ACC in depressed patients. As the bar graph shows, the activity of this region is *increased* during a manic episode in patients with bipolar disorder (Drevets et al., 1997). Thus, the activity of this region decreases during times of negative mood and increases during times of positive mood. (See *Figure 16.19.*)

I mentioned earlier that two experimental therapies for treatment-resistant depression, deep-brain stimulation of the subgenual ACC and vagus nerve stimulation, decreased the activity of the subgenual ACC. These results appear to contradict the finding that the activity level of this region is *lower* in depressed people. So far, I have found no suggestions in the research literature that can explain the discrepancy, but at least all of the experiments point to the importance of the subgenual ACC. One other study suggests that this region plays a

---

tryptophan depletion procedure A procedure involving a low-tryptophan diet and a tryptophan-free amino acid "cocktail" that lowers brain tryptophan and consequently decreases the synthesis of 5-HT.

FIGURE 16.19 ■ Metabolic Rate of the Subgenual ACC in Mania and Depression

(a) Composite functional MRI images show decreased metabolic activity of this region in depressed patients. (b) The graph shows the mean relative metabolic rate of the subgenual prefrontal cortex in normal controls and depressed and manic patients.

(From Drevets, W. C., *Current Opinion in Neurobiology*, 2001, *11*, 240–249. Reprinted with permission.)

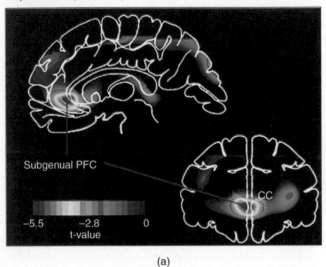

(a)

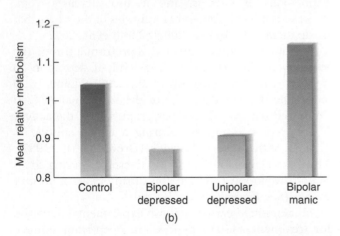

(b)

role in depression. A functional-imaging study by Siegel, Carter, and Thase (2006) found that depressed patients who initially exhibited a low response to emotional stimuli in the subgenual ACC and a high response in the amygdala responded best to cognitive behavior therapy. (Research has found that cognitive behavior therapy is the most effective form of psychotherapy for depression.)

In recent years, evidence has been accumulating that implicates the serotonin transporter in depression. A portion of the gene—the *promoter region*—for the 5-HT transporter (5-HTT) comes in two forms, short and long. (Most investigators refer to this promoter region as the 5-HTTLPR, although some call it SLC6A4. I'm telling

you this just in case you want to look up descriptions of future research on this topic.) As we saw in Chapter 11, people with one or two short alleles show greater activation of the amygdala when they look at photographs of faces expressing anger or fear. A longitudinal study by Caspi et al. (2003) followed 847 people over a period of more than twenty years, starting at 3 years of age, and recorded the occurrence of stressful events in their lives, including abuse during childhood, romantic disasters, bereavements, illnesses, and job crises. The investigators found that the probability of major depression and suicidality increased with the number of stressful life events the people had experienced. Moreover, the increase was much greater for people with one or two copies of the short alleles for the 5-HTT promoter. This study shows clear evidence of an interaction between environment and genetics. (See *Figure 16.20.*)

Other studies have confirmed the importance of the 5-HTT promoter to the development of depression. For example, Rausch et al. (2002) found that depressed people with two long alleles for this gene were more likely to respond to treatment with an antidepressant drug than were those with one or two short alleles. In fact, people with two long alleles were even more likely to respond to the placebo. A study by Lee et al. (2004) found that depressed people with two long alleles who were treated with antidepressant drugs had a much better long-term outcome (up to three years) than did people with one or two short alleles. Neumeister et al. (2002) found that tryptophan depletion was more likely to produce symptoms of depression in people with one or two short alleles. Finally, a functional-imaging study by Rhodes et al. (2007) found that people with higher levels of the 5-HT transporter in the amygdala showed less activation of the amygdala when the people looked at emotional faces.

What are the physiological effects of the long and short alleles of the 5-HTT promoter? The presence of short alleles means that fewer 5-HT transporters are produced which, in turn, means a slower rate of 5-HT reuptake and an increased amount of 5-HT in the extracellular fluid in the brain. But that SSRIs also raise the extracellular levels of 5-HT, so we would predict that long alleles, not short ones, would be associated with an increased risk of depression. This time, there is evidence that might explain the apparent contradiction. It turns out that serotonin has important roles in prenatal development as well as in postnatal brain functions (Gaspar, Cases, and Maroteaux, 2003). In fact, some glutamatergic neurons in the hippocampus and anterior cingulate cortex take up serotonin during a brief period during development. Presumably, this uptake has effects on the development of these brain regions.

The presence of high levels of serotonin in the brain during brain development has a very different effect than it does in adulthood. For example, when

FIGURE 16.20 ■ Stressful Life Events, 5-HTT, and Depression

The graphs show the probability of major depression and suicide ideation or attempts as a function of number of previous stressful life events of people with two long alleles (L/L), one short allele (S/L), or two short alleles (S/S) of the promoter region of the 5-HT transporter gene.

(Adapted from Caspi, A., Sugden, K., Moffitt, T. E., Taylor, A., Craig, I. W., Harrington, H., McClay, J., Mill, J., Martin, J., Braithwaite, A., and Poulton, R. *Science*, 2003, *301*, 386–389.)

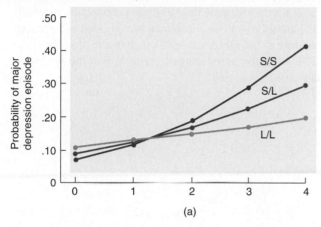

(a)

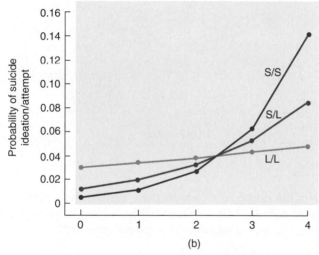

(b)

Number of stressful life events

neonatal mice are treated with a potent SSRI, they show signs of depression in adulthood such as abnormal sleep patterns and decreased motivation. However, when the drug is given only in adulthood, no such effects are seen (Popa et al., 2008).

A structural and functional-imaging study by Pezawas et al. (2005) suggests a mechanism that might explain the interaction between serotonin transporters and brain regions involved in mood. The investigators studied healthy people with no history of depression. They found that people with one or more short alleles had a 15 percent reduction in the volume in the amyg-

dala and a 25 percent reduction in the gray matter of the region around the genu of the corpus callosum. The largest reduction was in the subgenual ACC. Figure 16.21 shows this finding. Possession of a short allele on one or both chromosomes results in production of significantly decreased amounts of the 5-HT transporter. As we just saw, 5-HT is involved in prenatal development of the anterior cingulate cortex. Presumably, the possession of one or two short alleles of the 5-HTT promoter increases extracellular levels of 5-HT, affects prenatal development, and causes a reduction in the size of the subgenual ACC. (See *Figure 16.21*.)

Pezawas and his colleagues also showed people pictures of faces expressing fear or anger, which are known to activate the amygdala. The researchers studied moment-to-moment changes in neural activity and found a strong positive correlation between the activity of the subgenual ACC and the amygdala in people with two long alleles, which indicates that these two regions are functionally connected. They also found a strong *negative* correlation between the activity of the dorsal ACC and the amygdala: When the activity of the dorsal ACC went up, the activity of the amygdala went down, and when the activity of the dorsal ACC went down, the activity of the amygdala went up. Both of these correlations (especially the one between the amygdala and the subgenual ACC) were significantly lower in people with one or two short

FIGURE 16.21 ■ Gray Matter of the Amygdala and the Subgenual ACC

The scan shows reduction in the gray matter volume of the amygdala and subgenual anterior cingulate cortex (ACC) of normal subjects with one or two short alleles of the promoter region of the 5-HT transporter gene relative to subjects with two long alleles. "Hotter" colors indicate lower volume of brain tissue.

(From Pezawas, L., Meyer-Linderberg, A., Drabant, E. M., Verchinski, B. A., Munoz, K. E., Kolachana, B. S., Egan, M. F., Mattay, V. S., Hariri, A. R., and Weinberger, D. R. *Nature Neuroscience*, 2005, *8*, 828–834. Reprinted with permission.)

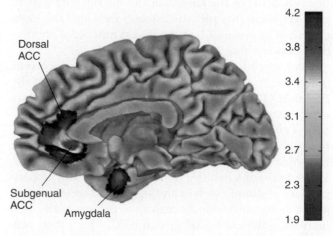

FIGURE 16.22 ■ Interactions Between the Amygdala and the Dorsal and Subgenual ACC

The scans show correlations between the activity of the dorsal and subgenual ACC and that of the amygdala, as measured by functional MRI while subjects were looking at pictures of faces displaying fear or anger. (a) Subjects with two long alleles of the promoter region of the 5-HT transporter gene. (b) Subjects with one or two short alleles of the same gene. Note that the scales differ for the two groups. Positive and negative numbers indicate positive and negative correlations. Warm colors indicate positive correlations; cool colors indicate negative correlations.

(From Pezawas, L., Meyer-Linderberg, A., Drabant, E. M., Verchinski, B. A., Munoz, K. E., Kolachana, B. S., Egan, M. F., Mattay, V. S., Hariri, A. R., and Weinberger, D. R. *Nature Neuroscience,* 2005, *8,* 828–834. Reprinted with permission.)

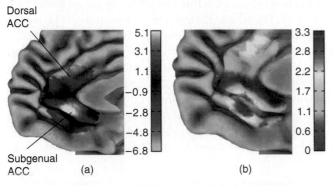

Dorsal ACC

5.1 / 3.1 / 1.1 / −0.9 / −2.8 / −4.8 / −6.8

3.3 / 2.8 / 2.2 / 1.7 / 1.1 / 0.6 / 0

Subgenual ACC (a)　(b)

FIGURE 16.23 ■ Neural Circuits Involved in Depression

The diagram shows a hypothetical feedback look, based on the results of the studies by Pezawas et al. (2005). Stressful or fearful information reaches the amygdala, which activates the subgenual anterior cingulate cortex (ACC). This region activates the dorsal ACC, which has inhibitory connections with the amygdala. The relative size of the arrows indicates the relative strength of the functional connections. People with one or two short alleles of the promoter region of the 5-HT transporter gene have a weaker connection between the amygdala and the subgenual ACC, as well as a reduced volume of the subgenual ACC, which means that the circuit is less effective in reducing the activation of the amygdala.

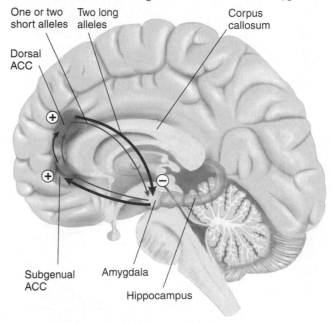

One or two short alleles　Two long alleles　Corpus callosum

Dorsal ACC

Subgenual ACC　Amygdala

Hippocampus

alleles than in those with two long alleles, which indicates that the functional connections were weaker. Figure 16.22 shows the correlations between the activity of the dorsal and subgenual ACC with that of the amygdala in people with both genotypes. (See *Figure 16.22.*)

Pezawas et al. administered a standardized psychological test to the subjects that measures a trait that has been shown to indicate a susceptibility to developing depression. They found that scores on this trait were negatively correlated with the degree of functional connectivity between the amygdala and the subgenual ACC. In other words, people whose scores indicated the highest risk of depression had the lowest functional connectivity between these brain regions.

Pezawas and his colleagues found a positive correlation between the activity of the subgenual ACC and the dorsal ACC, which indicates that they, too, are functionally connected. These findings suggest that changes in a neural circuit from the amygdala to the subgenual ACC to the dorsal ACC and back to the amygdala play a role in people's susceptibility to depression (Hamann, 2005). In the brain of a healthy, nondepressed person, anxiety-producing stimuli such as the sight of a face displaying anger or fear increase the activity of the amygdala. This increase activates the subgenual ACC, which then acti-

vates the dorsal ACC, which then inhibits the activity of the amygdala. This negative feedback loop helps to regulate the activity of amygdala. (See *Figure 16.23.*)

Role of Neurogenesis

As we saw in Chapters 3 and 13, neurogenesis can take place in the dentate gyrus—a region of the hippocampal formation—in the adult brain. Several studies with laboratory animals have shown that stressful experiences that produce the symptoms of depression suppress hippocampal neurogenesis, and the administration of antidepressant treatments, including MAO inhibitors, tricyclic antidepressants, SSRIs, ECT, and lithium, increases neurogenesis. In addition, the delay in the action of antidepressant treatments is about the same length as the time it takes for newborn neurons to mature. Moreover, if neurogenesis is suppressed by a low-level dose of X-radiation, antidepressant drugs lose their effectiveness. (See Sahay and Hen, 2007, for a review.) It is tempting to conclude that decreased hippocampal neurogenesis is the cause (or one of the causes) of depression. However, Sapolsky (2004) points out that it is difficult to find a link between the known functions of the hippocampus and the possible causes of depression. For example, disorders of memory, not of affect, are seen in people with hippocampal damage. We do not yet have enough evidence to decide whether neurogenesis plays a role in depression or whether the relationship between the two is coincidental.

There is currently no way to measure the rate of neurogenesis in the human brain. So far, all the evidence about human neurogenesis has been by extrapolation from studies with laboratory animals. However, a study by Pereira et al. (2007) used an MRI procedure that permitted them to estimate the blood volume of particular regions of the hippocampal formation in both mice and humans. They found that exercise (running wheels for the mice, an aerobic exercise regimen for humans) increased the blood volume of the dentate gyrus—the region where neurogenesis takes place—in both species. (As we will see in the next section of this chapter, exercise is an effective treatment for depression.) Histological procedures verified that increased neurogenesis in the mouse brain correlated with the increased blood volume, which strongly supports the conclusion that the exercise induces neurogenesis in the human brain, as well. (See *Figure 16.24.*)

Role of Circadian Rhythms

One of the most prominent symptoms of depression is disordered sleep. The sleep of people with depression tends to be shallow; slow-wave delta sleep (stages 3 and 4) is reduced, and stage 1 is increased. Sleep is fragmented; people tend to awaken frequently, especially toward the morning. In addition, REM sleep occurs earlier, the first half of the night contains a higher proportion of REM periods, and REM sleep contains an increased number of rapid eye movements (Kupfer, 1976; Vogel et al., 1980). (See *Figure 16.25.*)

REM Sleep Deprivation

One of the most effective antidepressant treatments is sleep deprivation, either total or selective. Selective deprivation of REM sleep, accomplished by monitoring people's EEG and awakening them whenever they show signs of REM sleep, alleviates depression (Vogel et al., 1975; Vogel et al., 1990). The therapeutic effect, like that of the antidepressant medications, occurs slowly, over the course of several weeks. Some patients show long-term improvement even after the deprivation is discontinued; thus, it is a practical as well as an effective treatment. In addition, regardless of their specific pharmacological effects, other treatments for depression

FIGURE 16.24 ■ Exercise and Neurogenesis

The scans show the effect of a program of aerobic exercise on the blood volume of regions of the human hippocampal formation. This measure serves as an indirect measure of neurogenesis. (a) Subregions of the hippocampus. EC = entorhinal cortex, DG = dentate gyrus, SUB = subiculum (b) Regional blood volume. "Hotter" colors indicate increased blood volume.

(From Pereira, A. C., Huddleston, D. E., Brickman, A. M., Sosunov, A. A., Hen, R., McKhann, G. M., Sloan, R., Gage, F. H., Brown, T. R., and Small, S. A. *Proceedings of the National Academy of Sciences, USA,* 2007, *104,* 5638–5643. Reprinted with permission.)

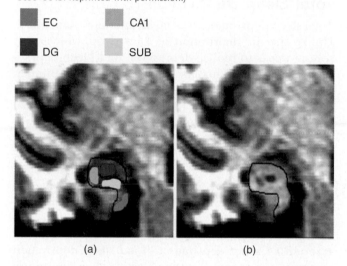

(a)   (b)

FIGURE 16.25 ■ Sleep and Depression

The diagram illustrates patterns of the stages of sleep of a normal subject and of a patient with major depression. Note the reduced sleep latency, reduced REM latency, reduction in slow-wave sleep (stages 3 and 4), and general fragmentation of sleep (arrows) in the depressed patient.

(From Gillin, J. C., and Borbély, A. A. *Trends in Neurosciences,* 1985, *8,* 537–542. Reprinted with permission.)

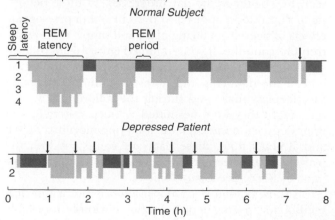

suppress REM sleep, delaying its onset and decreasing its duration (Scherschlicht et al., 1982; Vogel et al., 1990; Grunhaus et al., 1997; Thase, 2000). These facts suggest that REM sleep and mood might somehow be causally related. These results suggest that an important effect of successful antidepressant treatment may be to suppress REM sleep, and the changes in mood may be a result of this suppression. However, at least one antidepressant drug has been shown in a double-blind, placebo-controlled study *not* to suppress REM sleep (Mayers and Baldwin, 2005). Thus, suppression of REM sleep cannot be the *only* way in which antidepressant drugs work.

Total Sleep Deprivation

Total sleep deprivation also has an antidepressant effect. Unlike specific deprivation of REM sleep, which takes several weeks to reduce depression, total sleep deprivation produces immediate effects (Wu and Bunney, 1990). Typically, the depression is lifted by the sleep deprivation but returns the next day, after a normal night's sleep. Wu and Bunney suggest that during sleep, the brain produces a chemical that has a *depressogenic* effect in susceptible people. During waking, this substance is gradually metabolized and hence inactivated. Some of the evidence for this hypothesis is presented in Figure 16.26. The data are taken from eight different studies (cited by Wu and Bunney, 1990) and show self-ratings of depression of people who did and did not respond to sleep deprivation. Total sleep deprivation improves the mood of patients with major depression approximately two-thirds of the time. (See *Figure 16.26.*)

Why do only some people profit from sleep deprivation? This question has not yet been answered, but several studies have shown that it is possible to predict who will profit and who will not (Riemann, Wiegand, and Berger, 1991; Haug, 1992; Wirz-Justice and Van den Hoofdakker, 1999). In general, depressed patients whose mood remains stable will probably not benefit from sleep depression, whereas those whose mood fluctuates probably will. The patients who are most likely to respond are those who feel depressed in the morning but then gradually feel better as the day progresses. In these people, sleep deprivation appears to prevent the depressogenic effects of sleep from taking place and simply permits the trend to continue. If you examine Figure 16.26, you can see that the responders were already feeling better by the end of the day. This improvement continued through the sleepless night and during the following day. The next night they were permitted to sleep normally, and their depression was back the following morning. As Wu and Bunney note, these data are consistent with the hypothesis that sleep produces a substance with a depressogenic effect. (See *Figure 16.26.*)

Although total sleep deprivation is not a practical method for treating depression (it is obviously impossible

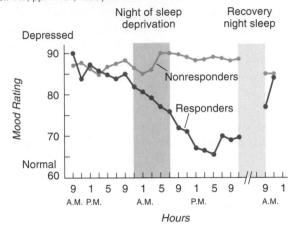

FIGURE 16.26 ■ Antidepressant Effects of Sleep Deprivation

The graph shows the mean mood rating of responding and nonresponding patients deprived of one night's sleep as a function of the time of day.

(Adapted from Wu, J. C., and Bunney, W. E. *American Journal of Psychiatry,* Vol. 147, pp. 14–21, 1990.)

to keep people awake indefinitely), several studies suggest that *partial* sleep deprivation can hasten the beneficial effects of antidepressant drugs (Szuba, Baxter, and Fairbanks, 1991; Leibenluft and Wehr, 1992). Some investigators have found that *intermittent* total sleep deprivation (say, twice a week for four weeks) can have beneficial results (Papadimitriou et al., 1993).

Role of Zeitgebers

Yet another phenomenon relates depression to sleep and waking—or, more specifically, to the mechanisms that are responsible for circadian rhythms. Some people become depressed during the winter season, when days are short and nights are long (Rosenthal et al., 1984). The symptoms of this form of depression, called **seasonal affective disorder (SAD),** are somewhat different from those of major depression; both forms include lethargy and sleep disturbances, but seasonal depression includes a craving for carbohydrates and an accompanying weight gain. (As you will recall, people with major depression tend to lose their appetite.)

SAD, like MDD and bipolar disorder, appears to have a genetic basis. In a study of 6439 adult twins, Madden et al. (1996) found that SAD ran in families, and they estimated that at least 29 percent of the variance in seasonal mood disorders could be attributed to genetic factors.

seasonal affective disorder (SAD) A mood disorder characterized by depression, lethargy, sleep disturbances, and craving for carbohydrates during the winter season when days are short.

SAD can be treated by **phototherapy:** exposing people to bright light for several hours a day (Rosenthal et al., 1985; Stinson and Thompson, 1990). As you will recall, circadian rhythms of sleep and wakefulness are controlled by the activity of the suprachiasmatic nucleus of the hypothalamus. Light serves as a *zeitgeber;* that is, it synchronizes the activity of the biological clock to the day–night cycle. One possibility is that people with SAD require a stronger-than-normal zeitgeber to reset their biological clock. According to Lewy et al. (2006), SAD is caused by a mismatch between cycles of sleep and cycles of melatonin secretion. Normally, secretion of melatonin begins in the evening, before people go to sleep. In fact, the time between the onset of melatonin secretion and the midpoint of sleep (halfway between falling asleep and waking up in the morning) is approximately six hours. People with SAD most often show a *phase delay* between cycles of melatonin and sleep; that is, the time interval between the onset of melatonin secretion and the midpoint of sleep is more than six hours. Exposure to bright light in the morning or administration of melatonin late in the afternoon (or, preferably, both treatments) advances the circadian cycle controlled by the biological clock in the suprachiasmatic nucleus. (These cycles were discussed in Chapter 9.) Those people with SAD who show a *phase advance* in their cycles can best be treated with exposure to bright light in the evening and administration of melatonin in the morning. (See *Figure 16.27.*) By the way, phototherapy has been found to help patients with major depressive disorder, especially in conjunction with administration of antidepressant drugs (Terman, 2007).

Phototherapy is safe and effective treatment for SAD. According to a study by Wirz-Justice et al. (1996), a special apparatus is not even needed. The authors found that a one-hour walk outside each morning reduced the symptoms of SAD. They noted that even on

FIGURE 16.27 ■ Cycles of Sleep and Melatonin Secretion

Normally, melatonin secretion begins in the evening, approximately six hours before the midpoint of sleep. Most people with seasonal affective disorder begin secreting melatonin earlier, showing a phase delay between cycles of melatonin and sleep. A few people with this disorder show a phase advance, with melatonin secretion beginning at a later time.

(Adapted from Lewy, A. K., Lefler, B. J., Emens, J. S., and Bauer, V. K. *Proceedings of the National Academy of Sciences, USA,* 2006, *103,* 7414–7419.)

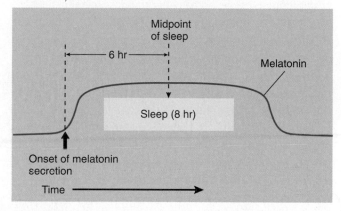

an overcast winter day, the early morning sky provides considerably more illumination than normal indoor artificial lighting, so a walk outside increases a person's exposure to light. The exercise helps, too. Many studies (for example, Dunn et al., 2005) have shown that a program of exercise improves the symptoms of depression.

---

phototherapy Treatment of seasonal affective disorder by daily exposure to bright light.

Interim Summary

Major Affective Disorders

The major affective disorders include bipolar disorder, with its cyclical episodes of mania and depression, and major depressive disorder. Heritability studies suggest that genetic anomalies are at least partly responsible for these disorders. MDD has been treated by several established or experimental biological treatments: MAO inhibitors, drugs that block the reuptake of norepinephrine and serotonin (tricyclic antidepressants, SSRIs, and SNRIs), ECT, TMS, deep brain stimulation, vagus nerve stimulation, and sleep deprivation. Bipolar

disorder can be successfully treated by lithium salts and anticonvulsant drugs. Lithium appears to stabilize neural transmission, especially in serotonin-secreting neurons. It also appears to protect neurons from damage and perhaps facilitate their repair.

The therapeutic effect of noradrenergic and serotonergic agonists and the depressant effect of reserpine, a monoaminergic antagonist, suggested the monoamine hypothesis of depression: that depression is caused by insufficient activity of monoaminergic neurons. Depletion of tryptophan (the precursor of 5-HT) in the brain causes a recurrence of depressive symptoms in depressed patients who are in remission, which lends further support to the conclusion that 5-HT plays a role in mood. However, although SSRIs have an immediate effect on serotonergic transmission in the brain, they do not relieve the symptoms of depression for several weeks, so the simple monoamine hypothesis appears not to be correct.

Functional-imaging studies found a increased activity in the amygdala and decreased activity in the subgenual ACC. Stressful life experiences increase the likelihood of depression in people with one or two short alleles of the 5-HT transporter promoter gene, and a better response to antidepressant treatment is seen in depressed people with two long alleles. Structural and functional-imaging studies have found a decrease in the volume of the amygdala and subgenual ACC and evidence for a weakened negative feedback loop from the amygdala to the subgenual ACC to the dorsal ACC back to the amygdala. Presumably, these changes occur because increased serotonergic activity associated with the presence of short alleles for the 5-HTT promoter affects prenatal brain development. Stressful experiences suppress hippocampal neurogenesis, and antidepressant treatments increase it. In addition, the effects of antidepressant treatments are abolished by suppression of neurogenesis.

Sleep disturbances are characteristic of affective disorders. In fact, total sleep deprivation rapidly (but temporarily) reduces depression in many people, and selective deprivation of REM sleep does so slowly (but more lastingly). In addition, almost all effective antidepressant treatments suppress REM sleep. A specific form of depression, seasonal affective disorder, can be treated by exposure to bright light. Clearly, the mood disorders are somehow linked to biological rhythms.

Thought Question

A television commentator, talking in particular about the suicide of a young pop star and in general about unhappy youth, asked with exasperation, "What would all these young people be doing if they had real problems like a Depression, World War II, or Vietnam?" People with severe depression often try to hide their pain because they fear that others will scoff at them and say that they have nothing to feel unhappy about. If depression is caused by abnormal brain functioning, are these remarks justified? How would you feel if you were severely depressed and people close to you berated you for feeling so sad and told you to snap out of it and quit feeling sorry for yourself? Do you think the expression of attitudes like this would decrease the likelihood of a depressed person committing suicide?

SUGGESTED READINGS

Barch, D. M. The cognitive neuroscience of schizophrenia. *Annual Review of Clinical Psychology*, 2005, *1*, 321–353.

Etkin, A., Pittenger, C., Polan, H. J., and Kandel, E. R. Toward a neurobiology of psychotherapy: Basic science and clinical applications. *Journal of Neuropsychiatry and Clinical Neurosciences*, 2005, *17*, 145–158.

Goodwin, D. W., and Guze, S. B. *Psychiatric Diagnosis*, 6th ed. New York: Oxford University Press, 1996.

Hasler, G., Drevets, W. C., Manji, H. K., and Charney, D. S. Discovering endophenotypes for major depression. *Neuropsychopharmacology*, 2004, *29*, 1765–1781.

Mueser, K. T., and McGurk, S. R. Schizophrenia. *The Lancet*, 2004, *363*, 2063–2072.

Southwick, S. M., Vythilingam, M., and Charney, D. S. The psychobiology of depression and resilience to stress: Implications for prevention and treatment. *Annual Review of Clinical Psychology*, 2005, *1*, 255–291.

Tsai, G., and Coyle, J. T. Glutamatergic mechanisms in schizophrenia. *Annual Review of Pharmacology and Toxicology*, 2002, *42*, 165–179.

Walker, E., Kestler, L., Bollini, A., and Hochman, K. M. Schizophrenia: Etiology and Course. *Annual Review of Psychology*, 2004, *55*, 401–430.

ADDITIONAL RESOURCES

Visit www.mypsychkit.com for additional review and practice of the material covered in this chapter. Within MyPsychKit, you can take practice tests and receive a customized study plan to help you review. Dozens of animations, tutorials, and Web links are also available. You can even review using the interactive electronic version of this textbook. You will need to register for MyPsychKit. See www.mypsychkit.com for complete details.

Anxiety Disorders, Autistic Disorder, Attention-Deficit/Hyperactivity Disorder, and Stress Disorders

In 1935 the report of an experiment with a chimpanzee triggered events whose repercussions are still being felt today. Jacobsen, Wolfe, and Jackson (1935) tested some chimpanzees on a behavioral task that requires the animal to remain quiet and remember the location of food that the experimenter has placed behind a screen. One animal, Becky, displayed a violent emotional reaction whenever she made an error while performing this task. "[When] the experimenter lowered . . . the opaque door to exclude the animal's view of the cups, she immediately flew into a temper tantrum, rolled on the floor, defecated, and urinated. After a few such reactions during the training period, the animal would make no further responses." After the chimpanzee's frontal lobes were removed, she became a model of good comportment. The chimpanzee "offered its usual friendly greeting, and eagerly ran from its living quarters to the transfer cage, and in turn went properly to the experimental cage. . . . If the animal made a mistake, it showed no evidence of emotional disturbance but quietly awaited the loading of the cups for the next trial" (Jacobsen, Wolfe, and Jackson, 1935, pp. 9–10).

These findings were reported at a scientific meeting in 1935, which was attended by Egas Moniz, a Portuguese neuropsychiatrist. He heard the report by Jacobsen and his colleagues and also one by Brickner (1936), which indicated that radical removal of the frontal lobes in a human patient (performed because of a tumor) did not appear to produce intellectual impairment; therefore, people could presumably get along without their frontal lobes. These two reports suggested to Moniz that "if frontal-lobe removal . . . eliminates frustrational behavior, why would it not be feasible to relieve anxiety states in man by surgical means?" (Fulton, 1949, pp. 63–64). In fact, Moniz persuaded a neurosurgeon to do so, and approximately one hundred operations were eventually performed under his supervision. (In 1949, Moniz received the Nobel Prize for the development of this procedure.)

I wrote that the repercussions of the 1935 meeting are still being felt today. Since that time, tens of thousands of people have received prefrontal lobotomies, primarily to reduce symptoms of emotional distress, and many of these people are still alive. At first, the medical community welcomed the procedure because it provided their patients with relief from emotional anguish. Only after many years were careful studies performed on the side effects of the procedure. These studies showed that although patients did perform well on standard tests of intellectual ability, they showed serious changes in personality, becoming irresponsible and childish. They also lost the ability to carry out plans, and most were unemployable. And although pathological emotional reactions were eliminated, so were normal ones. Because of these findings and because of the discovery of drugs and therapeutic methods that relieve the patients' symptoms without producing such drastic side effects, neurosurgeons eventually abandoned the prefrontal lobotomy procedure (Valenstein, 1986).

I should point out that the prefrontal lobotomies that were performed under Moniz's supervision, and by the neurosurgeons who followed, were not as drastic as the surgery performed by Jacobsen and his colleagues on Becky, the chimpanzee. In fact, no brain tissue was removed. Instead, the surgeons introduced various kinds of cutting devices into the frontal lobes and severed white matter (bundles of axons). One rather gruesome procedure did not even require an operating room; it could be performed in a physician's office. A *transorbital leucotome,* shaped like an ice pick, was introduced into the brain by passing it beneath the upper eyelid until the point reached the orbital bone above the eye. The instrument was hit with a mallet, driving it through the bone into the brain. The end was then swept back and forth so that it cut through the white matter. The patient often left the office within an hour.

Many physicians objected to the "ice pick" procedure because it was done blind (that is, the surgeon could not see just where the blade of the leucotome was located) and because it produced more damage than was necessary. Also, the fact that it was so easy and left no external signs other than a pair of black eyes may have tempted its practitioners to perform it too casually. In fact, at least 2,500 patients underwent this form of surgery (Valenstein, 1986).

What we know today about the effects of prefrontal lobotomy—whether done transorbitally or by more conventional means—tells us that such radical surgery should never have been performed. For too long the harmful side effects were ignored. (As we will see later in this chapter, neurosurgeons have developed a much restricted version of this surgery to treat intractable obsessive-compulsive disorder; it reduces the symptoms without producing such harmful side effects.)

Not too many years ago, the first three topics discussed in this chapter—the anxiety disorders, autism, and attention-deficit/hyperactivity disorder—would not have been covered in a book concerned with the physiology of behavior. (The importance of physiology to the fourth topic, stress, has long been recognized.) The anxiety disorders, autism, and attention-deficit/hyperactivity disorder were believed to be learned, primarily from parents who did a bad job raising their children. Although there was always at least some support for the suggestion that serious psychoses such as schizophrenia had a biological basis, other mental disorders were almost universally believed to be psychogenic in origin—that is, produced by "psychological" factors.

The tide has turned (or the pendulum has swung back, if you prefer that metaphor). Certainly, a person's family environment, social class, economic status, and similar factors affect the likelihood that he or she will develop a mental disorder and may help or hinder recovery. But physiological factors, including inherited ones and those that adversely affect development or damage the brain, play an important role too.

ANXIETY DISORDERS

As we saw in Chapter 16, the affective disorders are characterized by unrealistic extremes of emotion: depression or elation (mania). The **anxiety disorders** are characterized by unrealistic, unfounded fear and anxiety. This section describes three of the anxiety disorders that appear to have biological causes: panic disorder, generalized anxiety disorder, and social anxiety disorder. Although obsessive-compulsive disorder has traditionally been classified as an anxiety disorder, it has different symptoms from the other three disorders and involves different brain regions, so it is discussed separately.

Panic Disorder, Generalized Anxiety Disorder, and Social Anxiety Disorder

Description

People with **panic disorder** suffer from episodic attacks of acute anxiety—periods of acute and unremitting terror that grip them for variable lengths of time, from a few seconds to a few hours. The prevalence of this disorder is just under 2 percent (Kessler et al., 2005). Women appear to be a little more than twice as likely as men to suffer from panic disorder (Eaton et al., 1994). (See *Figure 17.1.*)

Panic attacks include many physical symptoms, such as shortness of breath, clammy sweat, irregularities in heartbeat, dizziness, faintness, and feelings of unreality. The victim of a panic attack often feels that he or she is going to die, and often seeks help in a hospital emergency room. Between panic attacks many people with panic disorder suffer from **anticipatory anxiety**—the fear that another panic attack will strike them. This anticipatory anxiety often leads to the development of a serious phobic disorder: **agoraphobia** (*agora* means "open space"). Agoraphobia can be severely disabling; some people with this disorder have stayed inside their homes

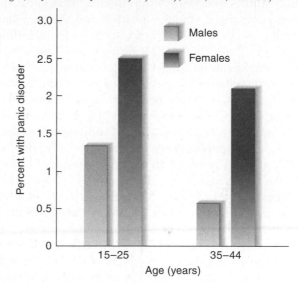

FIGURE 17.1 ■ **Prevalence of Panic Disorder**

The graph shows the percentages of men and women who receive a diagnosis of panic disorder earlier and later in life.

(Based on data from Eaton, W. W., Kessler, R. C., Wittchen, H. U., and Magee, W. J. *American Journal of Psychiatry*, 1994, *151*, 413–420.)

for years, afraid to venture outside where they might have a panic attack in public.

The primary characteristics of **generalized anxiety disorder** are excessive anxiety and worry, difficulty in controlling these symptoms, and clinically significant signs of distress and disruption to one's life. The prevalence of generalized anxiety disorder is approximately 3 percent, and the incidence is approximately two times greater in women than in men.

Social anxiety disorder (also called *social phobia*) is a persistent, excessive fear of being exposed to the scrutiny of other people that leads to avoidance of social situations in which the person is called on to perform (such as speaking or performing in public). If such situations

---

anxiety disorder A psychological disorder characterized by tension, overactivity of the autonomic nervous system, expectation of an impending disaster, and continuous vigilance for danger.

panic disorder A disorder characterized by episodic periods of symptoms such as shortness of breath, irregularities in heartbeat, and other autonomic symptoms, accompanied by intense fear.

anticipatory anxiety A fear of having a panic attack; may lead to the development of agoraphobia.

agoraphobia A fear of being away from home or other protected places.

generalized anxiety disorder A disorder characterized by excessive anxiety and worry serious enough to cause disruption to one's life.

social anxiety disorder A disorder characterized by excessive fear of being exposed to the scrutiny of other people that leads to avoidance of social situations in which the person is called on to perform.

are unavoidable, the person experiences intense anxiety and distress. The prevalence of social anxiety disorder, which is equally likely in men and women, is approximately 5 percent.

Possible Causes

Family studies and twin studies indicate that panic disorder, generalized anxiety disorder, and social anxiety disorder all have a hereditary component (Hettma, Neale, and Kendler, 2001; Merikangas and Low, 2005). Panic attacks can be triggered in people with a history of panic disorder by a variety of treatments that activate the autonomic nervous system, such as injections of lactic acid (a by-product of muscular activity), yohimbine (an α_2 adrenoreceptor antagonist), or doxapram (a drug used by anesthesiologists to increase breathing rate) or breathing air containing an elevated amount of carbon dioxide (Stein and Uhde, 1995). Lactic acid and carbon dioxide both increase heart rate and rate of respiration, just as exercise does; yohimbine has direct pharmacological effects on the nervous system.

As we saw in Chapters 11 and 16, the presence of one or two short alleles of the promoter region of the serotonin transporter (5-HTT) gene is associated with increased emotionality and susceptibility to depression, apparently because of differences in the structure and activity of the amygdala and particular regions of the prefrontal cortex. Evidence indicates that the presence of the short allele is also associated with higher levels of anxiety. Auerbach et al. (1999) found higher levels of negative emotionality in two-month-old infants with two short alleles. Furmark et al. (2004) found that people with social anxiety disorder whose chromosomes contained one or two short alleles showed increased levels of anxiety compared with people with two long alleles while they were anticipating speaking in front of an audience. In addition, functional imaging found that their amygdalas showed increased activations. (See *Figure 17.2.*) A study by Domschke et al. (2008) found that blushing (a symptom of social anxiety disorder) is more prevalent in people with social anxiety disorder whose chromosomes contain one or two short alleles of the 5-HTT promoter gene.

Functional-imaging studies suggest that the amygdala and the cingulate, prefrontal, and insular cortices are involved in anxiety disorders. Fischer et al. (1998) witnessed an unexpected panic attack in a subject while her regional cerebral blood flow was being measured by a PET scanner. They observed decreased activity in the right orbitofrontal cortex and anterior cingulate cortex. Pfleiderer et al. (2007) also observed a panic attack in a subject undergoing fMRI scanning and saw increased activity in the amygdala. Phan et al. (2005) found that people with social anxiety disorder showed increases in the activation of the amygdala when they looked at pictures of faces with angry, disgusted, or fearful expressions.

FIGURE 17.2 ■ **Role of Short or Long Alleles of the 5-HTT in Anxiety**

The graph shows the percentage change in regional blood flow (a measure of local brain activation) of the right amygdala of patients with social phobia when they anticipated speaking in front of an audience. Open circles represent subjects with two short alleles.

(Adapted from Furmark, T., Tillfors, M., Garpenstrand, H., Marteinsdottir, I., Långström, B., Oreland, L., and Fredrikson, M. *Neuroscience Letters*, 2004, *362*, 189–192.)

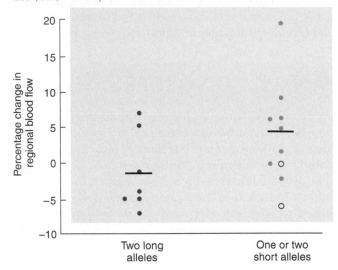

In addition, the activation of the amygdala was positively correlated with the severity of the people's symptoms. Monk et al. (2008) found that adolescents with generalized anxiety disorder showed increased activation of the amygdala and decreased activation of the ventrolateral prefrontal cortex while looking at angry faces. They also found evidence that activation of the ventromedial prefrontal cortex suppressed amygdala activation in healthy control subjects but not in those with anxiety disorder. Stein et al. (2007) found that college students with a high level of anxiety (but without a diagnosis of one of the anxiety disorders) showed increased activation of the amygdala and the insular cortex, both of which correlated positively with students' anxiety measures.

Treatment

Anxiety disorders are sometimes treated with benzodiazepines. As we just saw, increased activity of the amygdala is a common feature of the anxiety disorders. The amygdala contains a high concentration of $GABA_A$ receptors, which are the target of the benzodiazepines. Paulus et al. (2005) found that administration of a benzodiazepine (lorazepam) decreased the activation of both the amygdala and the insula of subjects looking at emotional faces. Administration of flumazenil, a benzodiazepine antagonist (having an action opposite that of the benzodiazepine tranquilizers), produces panic in

FIGURE 17.3 ■ Fluvoxamine and Panic Disorder

The graph shows the effects of fluvoxamine (an SSRI) on the severity of panic disorder.

(Adapted from Asnis, G. M., Hameedi, F. A., Goddard, A. W., Potkin, S. G., Black, D., Jameel, M., Desagani, K., and Woods, S. W. *Psychiatry Research*, 2001, *103*, 1–14.)

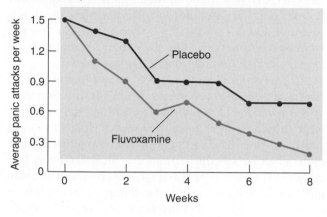

FIGURE 17.4 ■ D-Cycloserine and Social Anxiety Disorder

The graph shows the effects of D-cycloserine and a placebo on social anxiety scores of people with social anxiety disorder.

(Adapted from Hofmann, S. G., Meuret, A. E., Smits, J. A., Simon, N. M., et al. *Archives of General Psychiatry*, 2006, *63*, 298–304.)

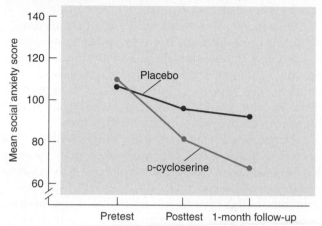

patients with panic disorder but not in control subjects (Nutt et al., 1990).

As we saw in Chapter 16, serotonin appears to play a role in depression. Much evidence suggests that serotonin plays a role in anxiety disorders too. Even though the symptoms of the anxiety disorders discussed in this subsection are very different from those of obsessive-compulsive disorder (described in the next subsection), specific serotonin reuptake inhibitors (SSRIs), which serve as potent serotonin agonists (such as fluoxetine), have become the first-line medications for treating all of these disorders—preferably in combination with cognitive behavior therapy (Asnis et al., 2001; Ressler and Mayberg, 2007). Figure 17.3 shows the effect of fluvoxamine, an SSRI, on the number of panic attacks in patients with panic disorder. (See *Figure 17.3.*)

As we also saw in Chapter 16, administration of indirect agonists of the NMDA receptor that attach to the glycine binding site have been used experimentally to successfully treat the symptoms of schizophrenia. Preliminary research suggests that the same may be true for anxiety disorder. Several studies have successfully used D-cycloserine in conjunction with cognitive behavior therapy to treat patients with anxiety disorder. For example, Ressler et al. (2004) found that the drug facilitated treatment of acrophobia (fear of heights), and Hofmann et al. (2006) found that it facilitated treatment of social anxiety disorder. (See *Figure 17.4.*)

In the treatment of anxiety disorders, cognitive behavior therapy often uses procedures that desensitize patients to the objects of situations they fear. For example, Ressler et al. used a computer program to expose their patients to a virtual glass elevator, gradually bringing them higher and higher from the ground. This procedure

appears to work by extinguishing a conditioned emotional response. In fact, a study by Walker et al. (2002) found that injections of D-cycloserine facilitated the extinction of a conditioned emotional response (CER) in rats. The drug had no effect performance of a CER unless it was administered along with extinction training. Injections of the drug by itself had no effect. Presumably, D-cycloserine also augments the ability of cognitive behavior therapy to extinguish fear responses.

Obsessive-Compulsive Disorder

Description

As the name implies, people with an **obsessive-compulsive disorder (OCD)** suffer from **obsessions**—thoughts that will not leave them—and **compulsions**—behaviors that they cannot keep from performing. Obsessions include concern or disgust with bodily secretions, dirt, germs, etc., fear that something terrible might happen, and a need for symmetry, order, or exactness. Most compulsions fall into one of four categories: *counting, checking, cleaning,* and *avoidance*. For example, people might repeatedly check burners on the stove to see that they are off and windows and locks to be sure that they are locked. Some

obsessive-compulsive disorder (OCD) A mental disorder characterized by obsessions and compulsions.

obsession An unwanted thought or idea with which a person is preoccupied.

compulsion The feeling that one is obliged to perform a behavior, even if one prefers not to do so.

people will wash their hands hundreds of times a day, even if their hands become covered with painful sores. Other people meticulously clean their house or endlessly wash, dry, and fold their clothes. Some become afraid to leave home because they fear contamination and refuse to touch other members of the family. If they do accidentally become "contaminated," they usually have lengthy purification rituals.

Obsessions are seen in a variety of mental disorders, including schizophrenia. However, unlike schizophrenics, people with obsessive-compulsive disorder recognize that their thoughts and behaviors are senseless and desperately wish that they would go away. Compulsions often become more and more demanding until they interfere with people's careers and daily lives.

The incidence of obsessive-compulsive disorder is 1–2 percent. Females are slightly more likely than males to have this diagnosis. OCD most commonly begins in young adulthood (Robbins et al., 1984). People with severe symptoms of this disorder are unlikely to marry, perhaps because of the common obsessional fear of dirt and contamination or because of the shame associated with the rituals they are compelled to perform, which causes them to avoid social contacts (Turner, Beidel, and Nathan, 1985).

Some investigators believe that the compulsive behaviors seen in OCD are forms of species-typical behaviors—for example, grooming, cleaning, and attention to sources of potential danger—that are released from normal control mechanisms by a brain dysfunction (Wise and Rapoport, 1988). Fiske and Haslam (1997) suggest that the behaviors seen in obsessive-compulsive disorder are simply pathological examples of a natural behavioral tendency to develop and practice social rituals. For example, people perform cultural rituals to mark transitions or changes in social status, to diagnose or treat illnesses, to restore relationships with deities, or to ensure the success of hunting or planting. Consider the following scenario (from Fiske and Haslam, 1997):

> Imagine that you are traveling in an unfamiliar country. Going out for a walk, you observe a man dressed in red, standing on a red mat in a red-painted gateway. . . . He utters the same prayer six times. He brings out six basins of water and meticulously arranges them in a symmetrical configuration in front of the gateway. Then he washes his hands six times in each of the six basins, using precisely the same motions each time. As he does this, he repeats the same phrase, occasionally tapping his right finger on his earlobe. Through your interpreter, you ask him what he is doing. He replies that there are dangerous polluting substances in the ground, . . . [and that] he must purify himself or something terrible will happen. He seems eager to tell you about his concerns. (p. 211)

Why is the man acting this way? Is he a priest following a sacred ritual, or does he have obsessive-compulsive disorder? Without knowing more about the spiritual rituals followed by the man's culture, we cannot say. Fiske and Haslam compared the features of OCD and other psychological disorders in descriptions of rituals, work, or other activities in fifty-two cultures. They found that the features of OCD were found in rituals in these cultures. The features of other psychological disorders were much less common. On the whole, the evidence suggests that the symptoms of obsessive-compulsive disorder represent an exaggeration of natural human tendencies.

Zhong and Liljenquist (2006) found that even well-educated people in an industrialized country (students at Northwestern University, in the United States) apparently unknowingly consider cleansing rituals to "wash away their sins." The investigators had the subjects recall in detail either an ethical or an unethical deed they had committed in the past. Later, they were asked to complete some word fragments by filling in letters where blanks occurred. Some word fragments could be made into words that did or did not pertain to cleansing. For example, W _ _ H, SH _ _ ER, and S _ _ P could be *wash*, *shower*, and *soap*, or they could be *wish*, *shaker*, and *step*. The subjects who had told about a misdeed were much more likely to think of cleansing-related words. And when offered a free gift—either a pencil or an antiseptic wipe—subjects who had told about a misdeed were more likely to chose the antiseptic wipe.

Possible Causes

Evidence indicates that obsessive-compulsive disorder is at least partly caused by hereditary factors. Several studies have found a greater concordance for obsessions and compulsions in monozygotic twins than in dizygotic twins (Hettema, Neale, and Kendler, 2001). Family studies have found that OCD is associated with a neurological disorder that appears during childhood (Pauls and Leckman, 1986; Pauls et al., 1986). This disorder, **Tourette's syndrome,** is characterized by muscular and vocal tics: facial grimaces, squatting, pacing, twirling, barking, sniffing, coughing, grunting, or repeating specific words (especially vulgarities). Leonard et al. (1992a, 1992c) found that many patients with obsessive-compulsive disorder had tics and that many patients with Tourette's syndrome showed obsessions and compulsions. Grados et al. (2001) found a family association between OCD and tic disorders (a broad category that includes Tourette's syndrome). Both groups of investigators believe that the two disorders are produced by the

Tourette's syndrome A neurological disorder characterized by tics and involuntary vocalizations and sometimes by compulsive uttering of obscenities and repetition of the utterances of others.

same underlying genotype. It is not clear why some people with this genotype gene develop Tourette's syndrome and others develop obsessive-compulsive disorder.

As with schizophrenia, not all cases of OCD have a genetic origin; the disorder sometimes occurs after brain damage caused by various means, such as birth trauma, encephalitis, and head trauma (Hollander et al., 1990; Berthier et al., 1996). In particular, the symptoms appear to be associated with damage to or dysfunction of the basal ganglia, cingulate gyrus, and prefrontal cortex (Giedd et al., 1995; Robinson et al., 1995).

Tic disorders (including OCD) can be caused by a group A β-hemolytic streptococcal infection. This infection can trigger several autoimmune diseases, in which the patient's immune system attacks and damages certain tissues of the body, including the valves of the heart, the kidneys, and—in this case—parts of the brain. Figure 17.5 shows the parallel course of a child's symptoms and the level of antistreptococcal DNA-B in her blood, which indicates the presence of an active infection. (See *Figure 17.5*.)

The symptoms of OCD appear to be produced by damage to the basal ganglia. Bodner, Morshed, and Peterson (2001) report the case of a 25-year-old man whose untreated sore throat (he lived in a religious group that prohibited antibiotics) developed into an autoimmune disease that produced obsessions and compulsions. The investigators found antibodies to type A β-hemolytic streptococcus, and MRI scans indicated abnormalities in the basal ganglia. An MRI study of thirty-four children with streptococcus-associated tics or OCD by Giedd et al. (2000) found an increase in the size of the basal ganglia that they attributed to an autoimmune inflammation of this region.

Several functional-imaging studies have found evidence of increased activity in the frontal lobes and caudate nucleus in patients with OCD. A review by Whiteside, Port, and Abramowitz (2004) found that functional-imaging studies consistently found increased activity of the caudate nucleus and the orbitofrontal cortex. Guehl et al. (2008) inserted microelectrodes into the caudate nuclei of three patients with OCD who were being evaluated for neurosurgery. They found that two of the patients, who reported the presence of obsessive thoughts during the surgery, showed increased activity in neurons in the caudate nucleus. The third patient, who did not report obsessive thoughts, showed a lower rate of neural activity.

A review by Saxena et al. (1998) described several studies that measured regional brain activity of OCD patients before and after successful treatment with drugs or cognitive behavior therapy. In general, the improvement in a patient's symptoms was correlated with a reduction in the activity of the caudate nucleus and orbitofrontal cortex. The fact that cognitive behavior therapy and drug therapy produced similar results is especially remarkable: It indicates that very different procedures may be bringing about physiological changes that alleviate a serious mental disorder.

Treatment

As we saw in the prologue to this chapter, clinicians developed procedures that damaged the prefrontal cortex or disconnected it from other parts of the brain to treat people with emotional reactions. Some patients with severe OCD have been successfully treated with *cingulotomy*—surgical destruction of specific fiber bundles in the subcortical frontal lobe, including the cingulum bundle (which connects the prefrontal and cingulate cortex with the limbic cortex of the temporal lobe) and a region that contains fibers that connect the basal ganglia with the prefrontal cortex (Ballantine et al., 1987; Mindus, Rasmussen, and Lindquist, 1994). These operations have a reasonably good success rate (Dougherty et al., 2002). Another reasonably successful surgical procedure, *capsulotomy*, destroys a region of a fiber bundle (the *internal capsule*) that connects the caudate nucleus with the medial prefrontal cortex (Rück et al., 2008). Of course, neurosurgery cannot be undone, so such procedures must be considered only as a last resort. As Rück and his colleagues report, some patients suffer from adverse side effects after surgery, such as problems of planning, apathy, or difficulty inhibiting socially inappropriate behavior.

FIGURE 17.5 ■ OCD and Streptococcal Hemolytic Infection

The graph shows the parallel course of a child's symptoms and the level of antistreptococcal DNA-B in her blood, which indicates the presence of an active infection. This relationship provides evidence that a group A β-hemolytic streptococcal infection can produce tics and the symptoms of OCD, presumably by affecting the basal ganglia.

(Adapted from Perlmutter, S. J., Garvey, M. A., Castellanos, X., Mittleman, B. B., Giedd, J., Rapoport, J. L., and Swedo, S. E. *American Journal of Psychiatry*, 1998, *155*, 1592–1598.)

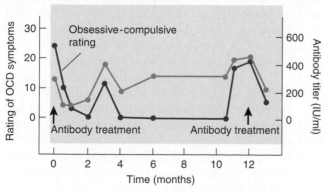

In one extraordinary case a patient performed his own psychosurgery. Solyom, Turnbull, and Wilensky (1987) reported the case of a young man with a serious obsessive-compulsive disorder whose ritual hand washing and other behaviors made it impossible for him to continue his schooling or lead a normal life. Finding that his life was no longer worthwhile, he decided to end it. He placed the muzzle of a .22-caliber rifle in his mouth and pulled the trigger. The bullet entered the base of the brain and damaged the frontal lobes. He survived, and he was amazed to find that his compulsions were gone. Fortunately, the damage did not disrupt his ability to make or execute plans; he went back to school and completed his education and now has a job. His IQ was unchanged. Ordinary surgery would have been less hazardous and messy, but it could hardly have been more successful.

As we saw in Chapter 15, deep brain stimulation (DBS) has been found to be useful in treating the symptoms of Parkinson's disease. Because OCD, like Parkinson's disease, appears to involve abnormalities in the basal ganglia, several clinics have tried to use DBS of the basal ganglia or fiber tracts connected with them to treat this disorder. This form of therapy appears to reduce the symptoms of OCD in some patients (Abelson et al., 2005; Larson, 2008). Fontaine et al. (2004) reported the case of a man with both Parkinson's disease and severe OCD. DBS of the subthalamic nucleus relieved both his motor disability and his obsessive-compulsive symptoms. A significant benefit of this procedure is that unlike psychosurgical procedures that destroy brain tissue, it is reversible: If no benefit is obtained from the stimulation, the electrodes can be removed.

As we saw in Chapters 8 and 14, the caudate nucleus and putamen receive information from the cerebral cortex. As this information is processed by the basal ganglia, it flows through two pathways before it passes to the thalamus and is sent back to the cortex. The *direct pathway* is excitatory, and the *indirect pathway* is inhibitory. (Refer to *Figure 8.24*.) Saxena et al. (1998) suggest that the symptoms of OCD may be a result of overactivity of the direct pathway. They propose that one of the functions of this pathway is control of previously learned behavior sequences that have become automatic so that they can be executed rapidly. The orbitofrontal cortex, which is involved in recognizing situations that have personal significance, can activate this pathway and the behaviors that it controls. The inhibitory indirect pathway is involved in suppressing these automatic behaviors, permitting the person to switch to other, more adaptive behaviors. Thus, obsessive-compulsive behavior could be a result of an imbalance between the direct and indirect pathways.

Three drugs are regularly used to treat the symptoms of OCD: clomipramine, fluoxetine, and fluvoxamine. These effective antiobsessional drugs are specific blockers of 5-HT reuptake; thus, they are serotonergic agonists. In general, serotonin has an inhibitory effect on species-typical behaviors, which has tempted several investigators to speculate that these drugs alleviate the symptoms of obsessive-compulsive disorder by reducing the strength of innate tendencies for counting, checking, cleaning, and avoidance behaviors that may underlie this disorder. Brain regions that have been implicated in OCD, including the orbitofrontal cortex and the basal ganglia, receive input from serotonergic terminals (Lavoie and Parent, 1990; El Mansari and Blier, 1997).

The importance of serotonergic activity in inhibiting compulsive behaviors is underscored by three interesting compulsions: trichotillomania, onychophagia, and acral lick dermatitis. *Trichotillomania* is compulsive hair pulling. People with this disorder (almost always females) often spend hours each night pulling hairs out one by one, sometimes eating them (Rapoport, 1991). *Onychophagia* is compulsive nail biting, which in its extreme can cause severe damage to the ends of the fingers. (For those who are sufficiently agile, toenail biting is not uncommon.) Double-blind studies have shown that both of these disorders can be treated successfully by clomipramine, the drug of choice for obsessive-compulsive disorder (Leonard et al., 1992a).

Acral lick dermatitis is a disease of dogs, not humans. Some dogs will continuously lick at a part of the body, especially a wrist or an ankle (called the *carpus* and the *hock*). The licking removes the hair and often erodes away the skin as well. The disorder seems to be genetic; it is seen almost exclusively in large breeds such as Great Danes, Labrador retrievers, and German shepherds, and it runs in families. A double-blind study found that clomipramine reduces this compulsive behavior (Rapoport, Ryland, and Kriete, 1992). At first, when I read the term "double-blind" in the report by Rapoport and her colleagues, I was amused to think that the investigators were careful not to let the dogs learn whether they were receiving clomipramine or a placebo. Then I realized that, of course, it was the dogs' owners who had to be kept in the dark.

We saw in the previous subsection that an NMDA receptor agonist, D-cycloserine, appears to be useful in treating the symptoms of anxiety disorders. This drug also appears to help in the treatment of the symptoms of OCD as well. A double-blind study by Kushner et al. (2007) found that compared with patients who received a placebo, patients who received D-cycloserine along with sessions of cognitive behavior therapy showed a more rapid decrease in their obsessive symptoms and were less likely to drop out of the treatment program. Presumably, the drug facilitated the extinction of the maladaptive thoughts and behaviors, just as it facilitates the extinction of conditioned emotional responses in patients with anxiety disorders.

InterimSummary

Anxiety Disorders

The anxiety disorders severely disrupt some people's lives. People with panic disorder periodically have panic attacks, during which they experience intense symptoms of autonomic activity and often feel as if they are going to die. Frequently, panic attacks lead to the development of agoraphobia, an avoidance of being away from a safe place, such as home. Family and twin studies have shown that panic disorder is at least partly heritable, which suggests that it has biological causes.

Panic attacks can be triggered in many susceptible people by conditions that activate the autonomic nervous system, such as an caffeine, yohimbine, injection of lactate, or inhalation of air containing an elevated amount of carbon dioxide. Panic attacks can be alleviated by the administration of a benzodiazepine, and a benzodiazepine antagonist can trigger a panic attack. Nowadays, the first choice of medical treatment for panic attacks is an SSRI. In addition, the presence of one or two short alleles of the promoter of the 5-HT transporter gene are associated with increased activation of the amygdala and higher levels of anxiety. Functional-imaging studies suggest that the amygdala and the cingulate, prefrontal, and insular cortices are involved in anxiety disorders.

Obsessive-compulsive disorder (OCD) is characterized by obsessions—unwanted thoughts—and compulsions—uncontrollable behaviors, especially those involving cleanliness and attention to danger. Some investigators believe that these behaviors represent overactivity of species-typical behavioral tendencies.

OCD has a heritable basis and is related to Tourette's syndrome, a neurological disorder characterized by tics and strange verbalizations. It can also be caused by brain damage at birth, encephalitis, and head injuries, especially when the basal ganglia are involved. A type A β-hemolytic streptococcus infection can stimulate an autoimmune attack—presumably on the basal ganglia—that produces the symptoms of OCD.

Functional imaging indicates that people with obsessive-compulsive disorder tend to show increased activity in the orbitofrontal cortex, cingulate cortex, and caudate nucleus. Drug treatment or behavior therapy that successfully reduces the symptoms of OCD generally reduces the activity of the orbitofrontal cortex and caudate nucleus. In severe cases of OCD that do not respond to other treatments, surgical procedures such as cingulotomy and capsulotomy may provide relief. Deep brain stimulation with implanted electrodes has been shown to be effective in some patients and, unlike cingulotomy and capsulotomy, has the benefits of being reversible. The most effective drugs are SSRIs such as clomipramine. Some investigators believe that clomipramine and related drugs alleviate the symptoms of OCD by increasing the activity of serotonergic pathways that play an inhibitory role on species-typical behaviors. Three other compulsions—hair pulling, nail biting, and (in dogs) acral lick syndrome—are also suppressed by clomipramine. In conjunction with cognitive behavior therapy, D-cycloserine, which acts as an indirect agonist at NMDA receptors, also appears to reduce the symptoms of OCD.

Thought Question

Most reasonable people would agree that a person with mental disorders cannot be blamed for his or her thoughts and behaviors. Most of us would sympathize with someone whose life is disrupted by panic attacks or obsessions and compulsions, and we would not see their plight as a failure of will power. After all, whether these disorders are caused by traumatic experiences or brain abnormalities (or both), the afflicted person has not chosen to be the way he or she is. But what about less dramatic examples: Should we blame people for their shyness or hostility or other maladaptive personality traits? If, as many psychologists believe, people's personality characteristics are largely determined by their heredity (and thus by the structure and chemistry of their brains), what are the implications for our concepts of "blame" and "personal responsibility"?

AUTISTIC DISORDER

Description

When a baby is born, the parents normally expect to love and cherish the child and to be loved and cherished in return. Unfortunately, some infants are born with a disorder that impairs their ability to return their parents' affection. The symptoms of **autistic disorder** (often simply referred to as *autism*) include a failure to develop normal social relations with other people, impaired

---

autistic disorder A chronic disorder whose symptoms include failure to develop normal social relations with other people, impaired development of communicative ability, lack of imaginative ability, and repetitive, stereotyped movements.

development of communicative ability, and the presence of repetitive, stereotyped behavior. Most people with autistic disorder display cognitive impairments. The syndrome was named and characterized by Kanner (1943), who chose the term (*auto*, "self," *-ism*, "condition") to refer to the child's apparent self-absorption.

According to a review by Fombonne (2005), the incidence of autistic disorder is approximately 13 in 10,000. The disorder is four times more common in males than in females. However, if only cases of autism with mental retardation are considered, the ratio falls to 2:1, and if only cases of high-functioning autism are considered (those with average or above-average intelligence and reasonably good communicative ability), the ratio rises to approximately 7:1. These data suggest that the social impairments are much more common in males but the cognitive and communicative impairments are more evenly shared by males and females. At one time, clinicians believed that autism was more prevalent in families with higher socioeconomic status, but more recent studies have found that the frequency of autism is the same in all social classes. The reported incidence of autism has increased in the past two decades, but according to Fombonne, evidence indicates that the apparent increase is a result of heightened awareness of the disorder and broadening of the diagnostic criteria. By the way, studies have failed to find evidence that autism is linked to childhood immunization (Andrews et al., 2004; Chen et al., 2004).

Autistic disorder is one of several pervasive developmental disorders that have similar symptoms. *Asperger's disorder* is generally less severe, and its symptoms do not include a delay in language development or the presence of important cognitive deficits. The primary symptoms are deficient or absent social interactions and repetitive and stereotyped behaviors along with obsessional interest in narrow subjects. *Rett's disorder* is a genetic neurological syndrome seen in girls that accompanies an arrest of normal brain development that occurs during infancy. Children with *childhood disintegrative disorder* show normal intellectual and social development and then, sometime between the ages of two and ten years, show a severe regression into autism. The prevalence of all forms of pervasive developmental disorders is approximately 60 in 10,000 (Fombonne, 2005).

According to the DSM-IV, a diagnosis of autistic disorder requires the presence of three categories of symptoms: impaired social interactions, absent or deficient communicative abilities, and the presence of stereotyped behaviors. Social impairments are the first symptoms to emerge. Infants with autistic disorder do not seem to care whether they are held, or they may arch their backs when picked up, as if they do not want to be held. They do not look or smile at their caregivers. If they are ill, hurt, or tired, they will not look to someone else for comfort.

As they get older, they do not enter into social relationships with other children and avoid eye contact with them. In severe cases, autistic people do not even seem to recognize the existence of other people.

Frith, Morton, and Leslie (1991) suggest that some of the symptoms of autism stem from abnormalities in the brain that prevent autistic people from forming a "theory of mind." That is, they are unable "to predict and explain the behavior of other humans in terms of their mental states" (p. 434). They cannot infer the thoughts, feelings, and intentions of other people from their emotional expressions, tone of voice, and behavior. As one autistic man complained, comparing his own social abilities with those of others, "Other people seem to have a special sense by which they can read other people's thoughts" (Rutter, 1983).

The language development of people with autism is abnormal or even nonexistent. They often echo what is said to them, and they may refer to themselves as others do—in the second or third person. For example, they may say, "You want some milk?" to mean "I want some milk." They may learn words and phrases by rote, but they fail to use them productively and creatively. Those who do acquire reasonably good language skills talk about their own preoccupations without regard for other people's interests. They usually interpret other people's speech literally. For example, when an autistic person is asked, "Can you pass the salt?," he might simply say "Yes"—and not because he is trying to be funny or sarcastic.

Autistic people generally show abnormal interests and behaviors. For example, they may show stereotyped movements, such as flapping their hand back and forth or rocking back and forth. They may become obsessed with investigating objects, sniffing them, feeling their texture, or moving them back and forth. They may become attached to a particular object and insist on carrying it around with them. They may become preoccupied in lining up objects or in forming patterns with them, oblivious to everything else that is going on around them. They often insist on following precise routines and may become violently upset when they are hindered from doing so. They show no make-believe play and are not interested in stories that involve fantasy. Although most autistic people are mentally retarded, not all are; and unlike most retarded people, they may be physically adept and graceful. Some have isolated skills, such as the ability to multiply two four-digit numbers very quickly, without apparent effort.

Possible Causes

When Kanner first described autism, he suggested that it was of biological origin; but not long afterward, influential clinicians argued that autism was learned. More precisely, it was taught—by cold, insensitive, distant, demanding, introverted parents. Bettelheim (1967) believed that autism was similar to the apathetic, withdrawn, and hopeless behavior seen in some of the survivors of the German concentration camps of World War II. You can imagine the guilt felt by parents who were told by a mental health professional that they were to blame for their child's pitiful condition. Some professionals saw the existence of autism as evidence for child abuse and advocated that autistic children be removed from their families and placed with foster parents.

Nowadays, researchers and mental health professionals are convinced that autism is caused by biological factors and that parents should be given help and sympathy, not blame. Careful studies have shown that the parents of autistic children are just as warm, sociable, and responsive as other parents (Cox et al., 1975). In addition, parents with one autistic child often raise one or more normal children. If the parents were at fault, we should expect *all* of their offspring to be autistic.

Heritability

Like all the mental disorders I have described so far, at least some forms of autism appear to be heritable. The best evidence for genetic factors comes from twin studies. These studies indicate that the concordance rate for autism in monozygotic twins is approximately 70 percent, while the rate in dizygotic twins studied so far is approximately 5 percent. The concordance rate for the more broadly defined *autistic spectrum disorders,* is 90 percent for monozygotic twins and 10 percent for dizygotic twins (Sebat et al., 2007). Genetic studies indicate that autistic disorder can be caused by a wide variety of mutations, especially those that interfere with neural development and communication (Autism Genome Project Consortium, 2007; Morrow et al., 2008).

Brain Pathology

The fact that autism is highly heritable is presumptive evidence that the disorder is a result of structural or biochemical abnormalities in the brain. In addition, a variety of medical disorders—especially those that occur during prenatal development—can produce the symptoms of autism. Evidence suggests that approximately 10 percent of all cases of autism have definable biological causes, such as rubella (German measles) during pregnancy; prenatal thalidomide; encephalitis caused by the herpes virus; and tuberous sclerosis, a genetic disorder that causes the formation of benign tumors in many organs, including the brain (DeLong, 1999; Rapin, 1999; Fombonne, 2005).

Evidence obtained in recent years indicates significant abnormalities in the development of the brains of autistic children. Courchesne et al. (2005, 2007) note that although the autistic brain is, on average, slightly smaller at birth, it begins to grow abnormally quickly, and by two to three years of age it is about 10 percent larger than a normal brain. Following this early spurt, the growth of the autistic brain slows down, so by adolescence it is only about 1–2 percent larger than normal.

Not all parts of the autistic brain show the same pattern of growth. The regions that appear to be most involved in the functions that are impaired in autism show the greatest growth early in life and the slowest growth between early childhood and adolescence. For example, the frontal cortex and temporal cortex of the autistic brain grow quickly during the first two years of life but then show little or no increase in size during the next four years, whereas these two regions grow by 20 percent and 17 percent, respectively, in normal brains. However, the growth pattern of "lower-order" regions of the cerebral cortex, such as the primary visual cortex and extrastriate cortex, are relatively normal in the autistic brain. The amygdala also shows an abnormal pattern of growth during development. By four years of age it is larger in autistic children. By the time of early adulthood it is the same size as the amygdala of nonautistic people but contains fewer neurons (Schumann and Amaral, 2006).

Autistic brains also show abnormalities in white matter. Herbert et al. (2004) found that in the autistic brain, the volume of white matter containing short-range axons was increased but that the volume of white matter containing long-range axons that connect distant regions of the brain was not. Courchesne et al. (2005, 2007) suggest that the production of excessive numbers of neurons early in development may cause the development of such a large number of short-range axons that the development of long-range axons is inhibited. The apparent hyperconnectivity of local regions of the cerebral cortex might possibly account for the exceptional isolated talents and skills shown by some autistics.

Researchers have employed structural- and functional-imaging methods to investigate the neural basis of the three categories of autistic symptoms. For example, Castelli et al. (2002) showed normal subjects and high-functioning people with autism or Asperger's disorder animations that depicted two triangles interacting in various goal-directed ways (for example, simply chasing or fighting) or in a way that suggested that one triangle was trying to trick or coax the other. For example, one normal subject described an animation in this way: "Triangles cuddling inside the house. Big wanted to persuade little to get out. He didn't want to . . . cuddling again" (p. 1843). People in the autism group were able to accurately describe the goal-directed interactions of

FIGURE 17.6 ■ Theory of Mind

The graph shows relative activation of specific brain regions of autistic adults and normal control subjects viewing a "theory of mind" animation of two triangles moving interactively with implied intentions. STS = superior temporal sulcus.

(Adapted from Castelli, F., Frith, C., Happé, F., and Frith, U. *Brain*, 2002, *125*, 1839–1849.)

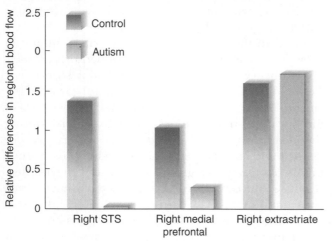

Regions with Significant Increase in Cerebral Blood Flow

FIGURE 17.7 ■ Fusiform Face Area and Autism

The scans show activation of the fusiform face area of control subjects but not of autistic subjects while looking at pictures of human faces.

(From Schultz, R. T. *International Journal of Developmental Neuroscience*, 2005, *23*, 125–141. Reprinted with permission.)

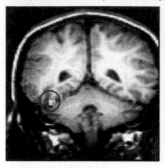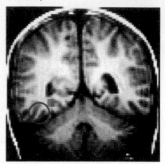

Control　　　　　　　　　Autistic

the triangles, but they had difficulty accurately describing the "intentions" of a triangle trying to trick or coax the other. In other words, they had difficulty forming a theory of mind. Functional imaging during presentation of the animations showed normal activation of early levels of the visual association cortex (the extrastriate cortex), but activation of the superior temporal sulcus (STS) and the medial prefrontal cortex was much lower in members of the autism group. (See *Figure 17.6* and *MyPsychKit 17.1, Inferring Causation.*) Previous research has shown that the STS plays an important role in detection of stimuli that indicate the actions of another individual (Allison, Puce, and McCarthy, 2000).

mypsychkit

Animation 17.1
Inferring Causation

The lack of interest in or understanding of other people is reflected in the response of the autistic brain to the sight of the human face. As we saw in Chapter 6, the fusiform face area (FFA), located on a region of visual association cortex on the base of the brain, is involved in the recognition of individual faces. A functional-imaging study by Schultz (2005) found little or no activity in the fusiform face area of autistic adults looking at pictures of human faces. (See *Figure 17.7.*) Autistics are poor at recognizing facial expressions of emotion or the direction of another person's gaze and have low rates of eye contact with other people. It seems likely that the FFA of autistics fails to respond to the sight of the human

face because these people spend very little time studying other people's faces and hence do not develop the expertise the rest of us acquire through normal interpersonal interactions. Grelotti et al. (2005) reported the case of an autistic boy who had a consuming interest in "Digimon" cartoon characters. Functional imaging showed no activation of the FFA when the boy viewed photos of faces, but photos of Digimon characters evoked strong activation of this region. This case supports the conclusion that the failure of the sight of faces to activate the FFA in people with autism is caused by a of lack of interest in faces, not by abnormalities in the FFA.

A study by Pelphrey et al. (2002) found that autistic people who were asked to identify the emotions shown in photographs of faces failed to look at other people's eyes, which are informative in making judgments of emotion. This tendency undoubtedly contributes to their impairment in analyzing social information. We saw in Chapter 11 that people with damage to the amygdala also fail to look at other people's eyes. The abnormal development of the amygdala in people with autism may be at least partly responsible for the low rates of eye contact with other people and their difficulty in assessing other people's emotional states.

In Chapters 8 and 11 I described the role of *mirror neurons* in the perception of emotions and behavioral intentions. These neurons, which are located in the ventral premotor area of the frontal lobe, receive input from the superior temporal sulcus and the posterior parietal cortex. This circuit is activated when we see another person produce an expression of emotion or perform a goal-directed action, and feedback from this activity helps us to understand what the person feels or is trying to accomplish. In other words, the mirror neuron system may be involved in our ability to understand what people are trying to do and to empathize with their emotions.

Iacoboni and Dapretto (2006) suggest that the social deficits seen in autism may be a result of abnormal development of the mirror neuron system. In fact, a functional-imaging study by Dapretto et al. (2006) observed deficient activation in the mirror motor neuron system of autistic children, and structural MRI study by Hadjikhani et al. (2006) found that the cerebral cortex in the mirror neuron system was thinner in autistics. A study by Senju et al. (2007) even found that children with autism failed to yawn when they saw a video of other people yawning. Control subjects showed an increased rate of yawning during or immediately after seeing videos that depicted yawning but not those that depicted other kinds of mouth movements. Presumably, the mirror neuron system is involved in this type of imitation.

Little is known about the nature of the brain abnormalities responsible for communicative deficits in autism, but several investigators suggest that regions of the temporal lobe involved in language abilities may be involved. A functional-imaging study by Gervais et al. (2004) found that the auditory cortex of autistic adults responded to sounds but did not show differential activation when speech sounds, as opposed to environmental nonspeech sounds, were presented.

Many investigators have noted that the presence of repetitive, stereotyped behavior and obsessive preoccupations with particular subjects resemble the symptoms of obsessive-compulsive disorder. As we saw earlier in this chapter, the symptoms of OCD appear to be related

FIGURE 17.8 ■ Caudate Nucleus and Stereotyped Behavior in Autism

The graph shows repetitive behavior scores of people with autistic spectrum disorders as a function of the volume of the right caudate nucleus. Larger volumes are associated with higher scores.

(Adapted from Hollander, E., Anagnostou, E., Chaplin, W., Esposito, K., Haznedar, M .M., Licalzi, E., Wasserman, S., Soorya, L., and Buchsbaum, M. *Biological Psychiatry*, 2005, *58*, 226–232.)

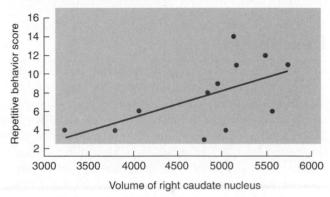

to increased activity of the caudate nucleus. Research suggests that the same may be true for the behavioral symptoms of autism. Several studies have observed increased volume of the caudate nucleus in autism (Sears et al., 1999; Langen et al., 2007). In fact, Hollander et al. (2005) found that the volume of the right caudate nucleus was positively correlated with ratings of repetitive behavior in patients with autism and Asperger's disorder. (See *Figure 17.8*.)

InterimSummary

Autistic Disorder

Autistic disorder occurs in approximately 13 of 10,000 infants. It is characterized by poor or absent social relations and communicative abilities and the presence of repetitive, stereotyped movements. Although autistics are usually, but not always, mentally retarded, they may have a particular, isolated talent. Autistic people have difficulty predicting the behavior of other people or forming a theory of mind to explain why they are acting as they do. They tend not to pay attention to other people's faces, as reflected in the lack of activation of the fusiform face area when they do so, and their ability to perceive emotional expressions on other people's faces is impaired.

In the past, clinicians blamed parents for autism, but now it is generally accepted as a disorder with biological roots. Twin studies have shown that autism is highly heritable but that many different genes are responsible for its development. Autism can also be caused by events that interfere with prenatal development, such as prenatal thalidomide or maternal infection with rubella. MRI studies indicate that the brains of babies who become autistic show abnormally rapid growth until the age of 2–3 years of age and then grow more slowly than the brains of unaffected children. The amygdala follows a similar pattern of development. Regions of the brain involved in higher-order processes such as communicative functions and interpretation of social stimuli develop more quickly in the autistic brain but then fail to continue to develop normally. In addition, long-distance communication between higher-order regions of the brain appears to be impaired. Abnormalities in the caudate nucleus may be associated with the repetitive, stereotyped behavior seen in autism.

ATTENTION-DEFICIT/ HYPERACTIVITY DISORDER

Some children have difficulty concentrating, remaining still, and working on a task. At one time or other, most children exhibit these characteristics. But children with **attention-deficit/hyperactivity disorder (ADHD)** display these symptoms so often that they interfere with the children's ability to learn.

Description

ADHD is the most common behavior disorder that shows itself in childhood. It is usually first discovered in the classroom, where children are expected to sit quietly and pay attention to the teacher or work steadily on a project. Some children's inability to meet these expectations then becomes evident. They have difficulty withholding a response, act without reflecting, often show reckless and impetuous behavior, and let interfering activities intrude into ongoing tasks.

According to the DSM-IV, the diagnosis of ADHD requires the presence of six or more of nine symptoms of inattention and six or more of nine symptoms of hyperactivity and impulsivity that have persisted for at least six months. Symptoms of inattention include such things as "often had difficulty sustaining attention in tasks of play activities" or "is often easily distracted by extraneous stimuli," and symptoms of hyperactivity and impulsivity include such things as "often runs about or climbs excessively in situations in which it is inappropriate" or "often interrupts or intrudes on others (e. g., butts into conversations or games)" (American Psychiatric Association, 1994, pp. 64–65).

ADHD can be very disruptive of a child's education and that of other children in the same classroom. It is seen in 4–5 percent of grade school children. Boys are about ten times more likely than girls to receive a diagnosis of ADHD, but in adulthood the ratio is approximately 2 to 1, which suggests that many girls with this disorder fail to be diagnosed. Because the symptoms can vary—some children's symptoms are primarily those of inattention, some are those of hyperactivity, and some show mixed symptoms—most investigators believe that this disorder has more than one cause. Diagnosis is often difficult, because the symptoms are not well defined. ADHD is often associated with aggression, conduct disorder, learning disabilities, depression, anxiety, and low self-esteem. Approximately 60 percent of children with ADHD continue to display symptoms of this disorder into adulthood, at which time a disproportionate number develop antisocial personality disorder and substance abuse disorder (Ernst et al., 1998). Adults with ADHD are also more likely to show cognitive impairments and lower occupational attainment than would be predicted by their education (Seidman et al., 1998). The most common treatment for ADHD is administration of methylphenidate (Ritalin), a drug that inhibits the reuptake of dopamine. Amphetamine, another dopamine agonist, also reduces the symptoms of ADHD, but this drug is used much less often.

Possible Causes

There is strong evidence from both family studies and twin studies for hereditary factors play an important role in determining a person's likelihood of developing ADHD. The estimated heritability of ADHD is high, ranging from 75 to 91 percent (Thapar, O'Donovan, and Owen, 2005).

According to Sagvolden and his colleagues (Sagvolden and Sargent, 1998; Sagvolden et al., 2005), the impulsive and hyperactive behaviors that are seen in children with ADHD are the result of a *delay of reinforcement gradient* that is steeper than normal. As we saw in Chapter 13, the occurrence of an appetitive stimulus can reinforce the behavior that just preceded it. For example, a piece of food can reinforce the lever press that a rat just made, and a smile can reinforce a person's attempts at conversation. Reinforcing stimuli are most effective if they immediately follow a behavior: The longer the delay, the less effective the reinforcement. Sagvolden and Sergeant suggest that deficiencies in dopaminergic transmission in the brains of people with ADHD increase the steepness of their delay of reinforcement gradient, which means that immediate reinforcement is even more effective in these children, but even slightly delayed reinforcement loses its potency. (See *Figure 17.9.*)

Why would a steeper delay of reinforcement gradient produce the symptoms of ADHD? According to Sagvolden and his colleagues, for people with a steep gradient, reinforcement with a short delay will be even more effective, thus producing overactivity. On the other hand, these people will be less likely to engage in behaviors that are followed by delayed reinforcement, as many of our behaviors (especially classroom activities) are. In support of this hypothesis, Sagvolden et al. (1998) trained normal boys and boys with ADHD on an instrumental conditioning task. When a signal was present, responses would be reinforced every 30 seconds with coins or trinkets. When the signal was not present, responses were never reinforced. The normal boys learned to respond only when the signal was present. When the signal was off, they waited patiently until it came on again. In contrast, the

---

attention-deficit/hyperactivity disorder (ADHD) A disorder characterized by uninhibited responses, lack of sustained attention, and hyperactivity; first shows itself in childhood.

FIGURE 17.9 ■ Delay of Reinforcement Gradients in ADHD

The graph illustrates different delay of reinforcement gradients as a function of time. Sagvolden and Sergeant (1998) hypothesize that a steeper gradient is responsible for the impulsive behavior of children with ADHD.

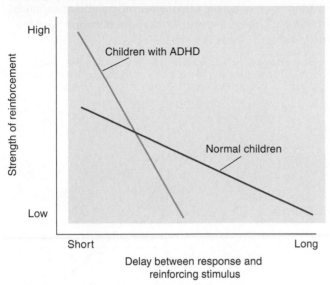

boys with ADHD showed impulsive behavior—intermittent bursts of rapid responses whether the signal was present or not. According to the investigators, this pattern of responding was what would be expected by a steep delay of reinforcement gradient.

The symptoms of ADHD resemble those produced by damage to the prefrontal cortex: distractibility, forgetfulness, impulsivity, poor planning, and hyperactivity (Aron, Robbins, and Poldrack, 2004). As we saw in Chapter 13, the prefrontal cortex plays a critical role in short-term memory. We use short-term memory to remember what we have just perceived, to remember information that we have just recalled from long-term memory, and to process ("work on") all of this information. For this reason, short-term memory is often referred to as *working memory*. The prefrontal cortex uses working memory to guide thoughts and behavior, regulate attention, monitor the effects of our actions, and organize plans for future actions (Arnsten, 2008). Damage or abnormalities in the neural circuits that perform these functions give rise to the symptoms of ADHD.

As we saw in Chapter 16, the fact that dopamine antagonists were discovered to reduce the positive symptoms of schizophrenia suggested the hypothesis that schizophrenia is caused by overactivity of dopaminergic transmission. Similarly, the fact that methylphenidate, a dopamine *agonist*, alleviates the symptoms of ADHD has suggested the hypothesis that this disorder is caused by *underactivity* of dopaminergic transmission. As we saw in Chapter 13, normal functioning of the prefrontal cortex

is impaired by low levels of dopamine receptor stimulation in this region, so the suggestion that abnormalities in dopaminergic transmission play a role in ADHD seem reasonable.

Berridge et al. (2006) administered methylphenidate to rats and established a moderate dose that improved their performance on tasks that required attention and working memory—tasks that involve the participation of the prefrontal cortex. They use microdialysis to measure the release of dopamine and norepinephrine and found that the drug increased the levels of both of these neurotransmitters in the prefrontal cortex but not in other brain regions. A follow-up study by Devilbiss and Berridge (2008) found that a moderate dose of methylphenidate increased the responsiveness of neurons in the prefrontal cortex. A high dose of methylphenidate profoundly *suppressed* neural activity.

Many studies have shown that the effect of dopamine levels in the prefrontal cortex on the functions of this region follow an inverted U-shaped curve. (See *Figure 17.10*.) Graphs of many behavioral functions have an inverted U shape. For example, moderate levels of motivation increase performance on most tasks, but very low levels fail to induce a person to perform, and very high levels tend to make people nervous and interfere with their performance. The dose-response curve for the effects of methylphenidate also follow an inverted U-shaped function, which is why Berridge and his colleagues tested different doses of the drug to find

FIGURE 17.10 ■ An Inverted U Curve

The graph illustrates an inverted U-curve function, in which low and high values of the variable on the horizontal axis are associated with low values of the variable on the vertical axis and moderate values are associated with high values. Presumably, the relationship between brain dopamine levels and the symptoms of ADHD follow a function like this one.

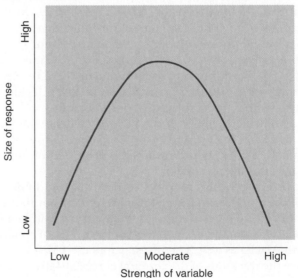

a dose that optimized the animals' performance. Clinicians have found the same to be true for the treatment of ADHD: Doses that are too low are ineffective, and doses that are too high produce increases in activity level that disrupt children's attention and cognition.

Good evidence that the levels of dopamine in the human prefrontal cortex have effects on behavior comes from studies of people with two different variants of the gene for an enzyme that affects dopamine levels in the brain. COMT (catechol-O-methyltransferase) is an enzyme that breaks down catecholamines (including dopamine and norepinephrine) in the extracellular fluid. Although reuptake is the primary means of removing catecholamines from the synapse, COMT also plays a role in deactivating these neurotransmitters after they are released. Mattay et al. (2003) noted that the clinical effects of amphetamine (which are similar to those of methylphenidate) are variable. In some people, amphetamine increases positive mood and facilitates performance on cognitive tasks, but in other people it has the opposite effect. Mattay et al. tested the effect of amphetamine on the tasks that made demands on working memory in people with two different variants of the COMT gene. They found that people with the *val-val* variant, who have lower brain levels of catecholamines, performed better when they were given low doses of amphetamine. In contrast, administration of amphetamine to people with the *met-met* variant, who have higher brain levels of catecholamines, actually impaired their performance. Presumably, the first group was pushed up the U-shaped curve, and the second group, already around the top of the curve, was pushed down the other side. (See *Figure 17.11*.)

I mentioned a few paragraphs ago that Berridge et al. (2006) found that methylphenidate increased the level of both dopamine and norepinephrine in the prefrontal cortex. It appears that both of these effects improve the symptoms of ADHD. Drugs that block α_2 receptors (one of the families of receptors that respond to norepinephrine) impair performance of monkeys on working-memory tasks and produce the symptoms of ADHD. Conversely, drugs that stimulate these receptors improve performance (Arnsten and Li, 2005). Evidence suggests that optimal levels of both dopamine and norepinephrine in the prefrontal cortex facilitate the functions of this region, and the effects of methylphenidate on both of these neurotransmitters is responsible for the drug's therapeutic effects.

We saw in the previous section that the brains of children with autism develop differently from those of unaffected children. A study by Shaw et al. (2007) found differences in the development of the brains of children with ADHD as well. The investigators used MRI to measure the thickness of the cerebral cortex in 446 children,

FIGURE 17.11 ■ Interactions Between Amphetamine and COMT Alleles on Working Memory

The graph shows the differential effects of amphetamine on the performance on a working memory task of people with two different variants of the gene for the COMT enzyme. The performance of people with the *val-val* variant was enhanced by amphetamine, and the performance of people with the *met-met* variant was reduced.

(Adapted from Mattay, V. S., Goldberg, T. E., Fera, F., Hariri, A. R., Tessitore, A., Egan, M. F., Kolachana, B., Callicott, J. H., and Weinberger, D. R. *Proceedings of the National Academy of Sciences, USA*, 2003, *100*, 6186–6191.)

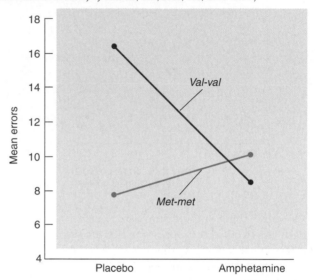

half of them with ADHD. Many of the children were scanned several times, which permitted the construction of detailed maps of the growth of the cerebral cortex of both groups. The investigators found that cortical growth was delayed in children with ADHD. In fact, the cortical thickness of the brains of children with ADHD at age 10.5 years was about the same as that of the brains of unaffected children at 7.5 years. Ultimately, the growth of the brains of the children with ADHD caught up with those of unaffected children.

Most investigators believe that ADHD is caused by abnormalities in a network of brain regions that involves the striatum (caudate nucleus and putamen) as well as the prefrontal cortex, which has reciprocal connections with the striatum. Functional-imaging studies lend support to this hypothesis. Studies have reported decreased activation of the caudate nucleus (Rubia et al., 1999; Durston et al., 2003; Vaidya et al., 2005) or medial prefrontal cortex (Rubia et al., 1999; Tamm et al., 2004) while subjects with ADHD were performing tasks that required careful attention and the ability to inhibit a response. Given the importance of dopaminergic innervation of both regions, abnormalities in dopaminergic transmission may be responsible for the alterations in brain functions.

InterimSummary

Attention-Deficit/Hyperactivity Disorder

Attention-deficit/hyperactivity disorder is the most common behavior disorder that first appears in childhood. Children with ADHD show symptoms of inattention, hyperactivity, and impulsivity. The most common medical treatment is methylphenidate, a dopamine agonist.

Family and twin studies indicate a heritable component in this disorder. Evidence suggests that a steeper delay of reinforcement gradient may account for impulsiveness and hyperactivity. Molecular genetic studies have found an association between ADHD and different alleles for COMT, an enzyme that deactivates monoamines.

Growth of the brains of children with ADHD follows that of the brains of unaffected children, but the rate of growth is slower. Most investigators believe that ADHD is caused by abnormalities in a network of brain regions that involves the striatum and the prefrontal cortex. Functional-imaging studies have shown hypoactivation of these structures in the brains of people with ADHD while they are performing tasks that require careful attention and the ability to inhibit a response.

STRESS DISORDERS

Aversive stimuli can harm people's health. Many of these harmful effects are produced not by the stimuli themselves but by our reactions to them. Walter Cannon, the physiologist who criticized the James-Lange theory described in Chapter 11, introduced the term **stress** to refer to the physiological reaction caused by the perception of aversive or threatening situations.

The word *stress* was borrowed from engineering, in which it refers to the action of physical forces of mechanical structures. The word can be a noun or a verb; and the noun can refer to situations or the individual's response to them. When we say that someone was subjected to stress, we really mean that someone was exposed to a situation that elicited a particular reaction in that person: a **stress response.**

The physiological responses that accompany the negative emotions prepare us to threaten rivals or fight them or to run away from dangerous situations. Walter Cannon introduced the phrase **fight-or-flight response** to refer to the physiological reactions that prepare us for the strenuous efforts required by fighting or running away. Normally, once we have bluffed or fought with an adversary or run away from a dangerous situation, the threat is over, and our physiological condition can return to normal. The fact that the physiological responses may have adverse long-term effects on our health is unimportant as long as the responses are brief. But sometimes, the threatening situations are continuous rather than episodic, producing a more or less continuous stress response. And as we will see in the section on posttraumatic stress disorder, sometimes threatening situations are so severe that they trigger responses that can last for months or years.

Physiology of the Stress Response

As we saw in Chapter 11, emotions consist of behavioral, autonomic, and endocrine responses. The latter two components, the autonomic and endocrine responses, are the ones that can have adverse effects on health. (Well, I guess the behavioral components can too, if a person rashly gets into a fight with someone who is much bigger and stronger.) Because threatening situations generally call for vigorous activity, the autonomic and endocrine responses that accompany them are catabolic; that is, they help to mobilize the body's energy resources. The sympathetic branch of the autonomic nervous system is active, and the adrenal glands secrete epinephrine, norepinephrine, and steroid stress hormones. Because the effects of sympathetic activity are similar to those of the adrenal hormones, I will limit my discussion to the hormonal responses.

Epinephrine affects glucose metabolism, causing the nutrients stored in muscles to become available to provide energy for strenuous exercise. Along with norepinephrine, the hormone also increases blood flow to the muscles by increasing the output of the heart. In doing so, it also increases blood pressure, which, over the long term, contributes to cardiovascular disease.

Besides serving as a stress hormone, norepinephrine is (as you know) secreted in the brain as a neurotransmitter. Some of the behavioral and physiological

---

stress A general, imprecise term that can refer either to a stress response or to a situation that elicits a stress response.

stress response A physiological reaction caused by the perception of aversive or threatening situations.

fight-or-flight response A species-typical response preparatory to fighting or fleeing; thought to be responsible for some of the deleterious effects of stressful situations on health.

responses produced by aversive stimuli appear to be mediated by noradrenergic neurons. For example, microdialysis studies have found that stressful situations increase the release of norepinephrine in the hypothalamus, frontal cortex, and lateral basal forebrain (Yokoo et al. 1990; Cenci et al., 1992). Montero, Fuentes, and Fernandez-Tome (1990) found that destruction of the noradrenergic axons that ascend from the brain stem to the forebrain prevented the rise in blood pressure that is normally produced by social isolation stress. The stress-induced release of norepinephrine in the brain is controlled by a pathway from the central nucleus of the amygdala to the locus coeruleus, the nucleus of the brain stem that contains norepinephrine-secreting neurons (Van Bockstaele et al., 2001).

The other stress-related hormone is *cortisol*, a steroid secreted by the adrenal cortex. Cortisol is called a **glucocorticoid** because it has profound effects on glucose metabolism. In addition, glucocorticoids help to break down protein and convert it to glucose, help to make fats available for energy, increase blood flow, and stimulate behavioral responsiveness, presumably by affecting the brain. They decrease the sensitivity of the gonads to luteinizing hormone (LH), which suppresses the secretion of the sex steroid hormones. In fact, Singer and Zumoff (1992) found that the blood level of testosterone in male hospital residents (doctors, not patients) was severely depressed, presumably because of the stressful work schedule they were obliged to follow. Glucocorticoids have other physiological effects, too, some of which are only poorly understood. Almost every cell in the body contains glucocorticoid receptors, which means that few of them are unaffected by these hormones.

The secretion of glucocorticoids is controlled by neurons in the paraventricular nucleus of the hypothalamus (PVN), whose axons terminate in the median eminence, where the hypothalamic capillaries of the portal blood supply to the anterior pituitary gland are located. (The pituitary portal blood supply was described in Chapter 3.) The neurons of the PVN secrete a peptide called **corticotropin-releasing hormone (CRH),** which stimulates the anterior pituitary gland to secrete **adrenocorticotropic hormone (ACTH).** ACTH enters the general circulation and stimulates the adrenal cortex to secrete glucocorticoids. (See *Figure 17.12.*)

CRH (previously called CRF, or corticotropin-releasing factor) is also secreted within the brain, where it serves as a neuromodulator/neurotransmitter, especially in regions of the limbic system that are involved in emotional responses, such as the periaqueductal gray matter, the locus coeruleus, and the central nucleus of the amygdala. The behavioral effects produced by an injection of CRH into the brain are similar to those produced by aversive situations; thus, some elements of the

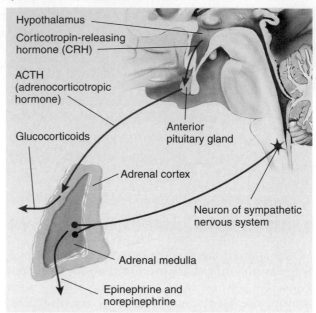

FIGURE 17.12 ■ **Control of Secretion of Stress Hormones**

The diagram illustrates control of the secretion of glucocorticoids by the adrenal cortex and of catecholamines by the adrenal medulla.

Hypothalamus
Corticotropin-releasing hormone (CRH)
ACTH (adrenocorticotropic hormone)
Glucocorticoids
Anterior pituitary gland
Adrenal cortex
Neuron of sympathetic nervous system
Adrenal medulla
Epinephrine and norepinephrine

stress response appear to be produced by the release of CRH by neurons in the brain. For example, intracerebroventricular injection of CRH decreases the amount of time a rat spends in the center of a large open chamber (Britton et al., 1982), enhances the acquisition of a classically conditioned fear response (Cole and Koob, 1988), and increases the startle response elicited by a sudden loud noise (Swerdlow et al., 1986). On the other hand, intracerebroventricular injection of a CRH antagonist *reduces* the anxiety caused by a variety of stressful situations (Kalin, Sherman, and Takahaski, 1988; Heinrichs, et al., 1994; Skutella et al., 1994).

The secretion of glucocorticoids does more than help an animal react to a stressful situation: It helps the animal to survive. If a rat's adrenal glands are removed, the rat becomes much more susceptible to the effects of stress. In fact, a stressful situation that a normal rat would take in its stride might kill one whose adrenal

glucocorticoid One of a group of hormones of the adrenal cortex that are important in protein and carbohydrate metabolism, secreted especially in times of stress.

corticotropin-releasing hormone (CRH) A hypothalamic hormone that stimulates the anterior pituitary gland to secrete ACTH (adrenocorticotropic hormone).

adrenocorticotropic hormone (ACTH) A hormone released by the anterior pituitary gland in response to CRH; stimulates the adrenal cortex to produce glucocorticoids.

glands have been removed. And physicians know that if an adrenalectomized person is subjected to stressors, he or she must be given additional amounts of glucocorticoid (Tyrell and Baxter, 1981).

Health Effects of Long-Term Stress

Many studies of people who have been subjected to stressful situations have found evidence of ill health. For example, survivors of concentration camps, who were obviously subjected to long-term stress, have generally poorer health later in life than other people of the same age (Cohen, 1953). Drivers of subway trains that injure or kill people are more likely to suffer from illnesses several months later (Theorell et al., 1992). Air traffic controllers, especially those who work at busy airports where the danger of collisions is greatest, show a greater incidence of high blood pressure, which gets worse as they grow older (Cobb and Rose, 1973). (See *Figure 17.13*.) They also are more likely to suffer from ulcers or diabetes.

A pioneer in the study of stress, Hans Selye, suggested that most of the harmful effects of stress were produced by the prolonged secretion of glucocorticoids (Selye, 1976). Although the short-term effects of glucocorticoids are essential, the long-term effects are damaging. These effects include increased blood pressure, damage to muscle tissue, steroid diabetes, infertility, inhibition of growth, inhibition of the inflammatory responses, and suppression of the immune system. High blood pressure can lead to heart attacks and stroke. Inhibition of growth in children who are subjected to prolonged stress prevents them from attaining their full height. Inhibition of the inflammatory response makes it more difficult for the body to heal itself after an injury, and suppression of the immune system makes an individual vulnerable to infections. Long-term administration of steroids to treat inflammatory diseases often produces cognitive deficits and can even lead to *steroid psychosis*, whose symptoms include profound distractibility, anxiety, insomnia, depression, hallucinations, and delusions (Lewis and Smith, 1983; de Kloet, Joëls, and Holsboer, 2005).

The adverse effects of stress on healing were demonstrated in a study by Kiecolt-Glaser et al. (1995), who performed punch biopsy wounds in the subjects' forearms, a harmless procedure that is used often in medical research. The subjects were people who were providing long-term care for relatives with Alzheimer's disease—a situation that is known to cause stress—and control subjects of the same approximate age and family income. The investigators found that healing of the wounds took significantly longer in the caregivers (48.7 days versus 39.3 days). (See *Figure 17.14*.)

FIGURE 17.13 ■ Stress and Hypertension

The graph shows the incidence of hypertension in various age groups of air traffic controllers at high-stress and low-stress airports.

(Based on data from Cobb, S., and Rose, R. M. *Journal of the American Medical Association*, 1973, *224*, 489–492.)

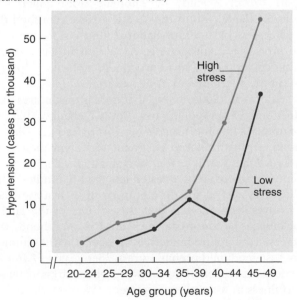

FIGURE 17.14 ■ Stress and Healing of Wounds

The graph shows the percentage of caregivers and control subjects whose wounds had healed as a function of time after the biopsy was performed.

(Adapted from Kiecolt-Glaser, J. K., Marucha, P. T., Malarkey, W. B., Mercado, A. M., and Glaser, R. *Lancet*, 1995, *346*, 1194–1196.)

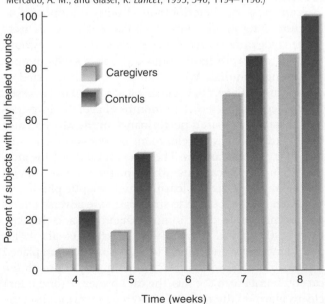

Effects of Stress on the Brain

Sapolsky and his colleagues have investigated one rather serious long-term effect of stress: brain damage. As you learned in Chapter 14, the hippocampal formation plays a crucial role in learning and memory, and evidence suggests that one of the causes of memory loss that occurs with aging is degeneration of this brain structure. Research with animals has shown that long-term exposure to glucocorticoids destroys neurons located in field CA1 of the hippocampal formation. The hormone appears to destroy the neurons by decreasing the entry of glucose and decreasing the reuptake of glutamate (Sapolsky, 1992, 1995; McEwen and Sapolsky, 1995). Both of these effects make neurons more susceptible to potentially harmful events, such as decreased blood flow, which often occurs as a result of the aging process. The increased amounts of extracellular glutamate permit calcium to enter through NMDA receptors. (You will recall that the entry of excessive amounts of calcium can kill neurons.) Perhaps, then, the stressors to which people are subjected throughout their lives increase the likelihood of memory problems as they grow older. In fact, Lupien et al. (1996) found that elderly people with elevated blood levels of glucocorticoids learned a maze more slowly than did those with normal levels.

Prenatal stress can cause long-lasting malfunctions in learning and memory by interfering with normal development of the hippocampus. Son et al. (2006) subjected pregnant mice to stress caused by periodic restraint in a small chamber. They found that this treatment interfered with the establishment of hippocampal long-term potentiation in the offspring of the stressed females, along with impairments in a spatial learning task that requires the participation of the hippocampus.

Brunson et al. (2005) confirmed that stress early in life can cause the deterioration of normal hippocampal functions later in life. During the first week after delivery the investigators placed female rats and their newborn pups in cages with hard floors and only a small amount of nesting material. When the animals were tested at 4–5 months of age, their behavior was normal. However, when they were tested at 12 months of age, the investigators observed impaired performance in the Morris water maze and deficient development of long-term potentiation in the hippocampus. They also found dendritic atrophy in the hippocampus, which might have accounted for the impaired spatial learning and synaptic plasticity.

Even brief exposure to stress can have adverse effects on normal brain functioning. Diamond and his colleagues (Diamond et al., 1999; Mesches et al., 1999) placed rats individually in a Plexiglas box and then placed the box in a cage with a cat for 75 minutes. Although the cat could not harm the rats, the cat's presence (and odor) clearly alarmed the rats and elicited a stress response; the

FIGURE 17.15 ■ **Acute Stress. Glucocorticoid Level, Synaptic Plasticity, and Learning**

The graphs show the effects of acute stress caused by exposing a rat to the sight and smell of a cat. The stress raised the glucocorticoid level (corticosterone, in the case of a rat), impaired the development of primed-burst potentiation (a form of long-term potentiation) in slices taken from these animals, and interfered with learning of a spatial task that requires the hippocampus.

(Adapted from Diamond, D. M., Park, C. R., Heman, K. L., and Rose, G. M. *Hippocampus*, 1999, *9*, 542–552, and Mesches, M. H., Fleshner, M., Heman, K. L., Rose, G. M., and Diamond, D. M. *Journal of Neuroscience*, 1999, *19*, RC18(1–5).)

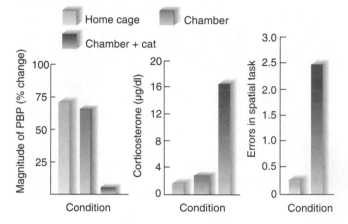

stressed rats' blood glucocorticoid increased to approximately five times the normal level. The investigators found that this short-term stress affected the functioning of the animals' hippocampus. The stressed rats' ability to learn a spatial task was impaired, and primed-burst potentiation (a form of long-term potentiation) was impaired in hippocampal slices taken from stressed rats. (See *Figure 17.15*.) A study by Thomas, Hotsenpiller, and Peterson (2007) found that acute stress diminished the long-term survival of hippocampal neurons produced by the process of neurogenesis. As we saw in Chapter 16, impaired hippocampal neurogenesis appears to play a role in the development of depression.

Salm et al. (2004) found that mild prenatal stress can affect brain development and produce changes that last the animal's lifetime. Once a day during the last week of gestation, they removed pregnant rats from their cage and gave them an injection of a small amount of sterile saline—a procedure that lasted less than 5 minutes. This mild stress altered the development of their amygdalas. The investigators found that the volume of the lateral nucleus of the amygdala, measured in adulthood, was increased by approximately 30 percent in the animals that sustained mild prenatal stress. (See *Figure 17.16*.) As previous experiments have shown, prenatal stress increases fearfulness in a novel environment (Ward et al., 2000).

FIGURE 17.16 ■ Prenatal Stress and the Amygdala

The graph shows volumes of nuclei of the amygdala in control rats and rats that had been subjected to prenatal stress.

(Adapted from Salm, A. K., Pavelko, M., Krouse, E. M., Webster, W., Kraszpulski, M., and Birkle, D. L. *Developmental Brain Research*, 2004, *148*, 159–167.)

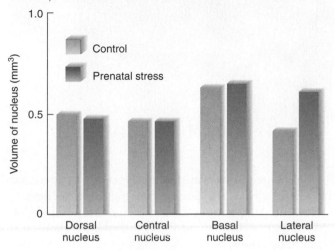

FIGURE 17.17 ■ Prenatal Stress and Glucocorticoids in Adulthood

The graph shows the effects of prenatal stress and glucocorticoid level on the stress response of adult rats. Adrenalectomy of the mother before she was subjected to stress prevented the development of an elevated stress response in the offspring during adulthood.

(Adapted from Barbazanges, A., Piazza, P. V., Le Moal, M., and Maccari, S. *Journal of Neuroscience*, 1996, *16*, 3943–3949.)

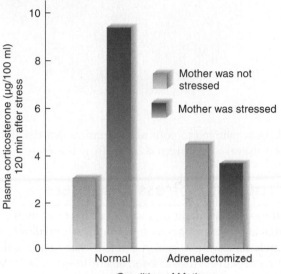

Condition of Mother

Presumably, the increased size of the amygdala contributes to this fearfulness.

A study by Fenoglio, Chen, and Baram (2006) found that experiences that occur during early life can reduce reactivity to stressful situations in adulthood. Fenoglio and her colleagues removed rat pups from their cage, handled them for 15 minutes, and then returned them to their cage. Their mother immediately began licking and grooming the pups. This nurturing behavior activated several regions of the pups' brains, including the central nucleus of the amygdala and the paraventricular nucleus of the hypothalamus, the location of neurons that secrete CRH. The result of this treatment was to reduce the production of CRH in response to stressful stimuli, which conferred a lifelong attenuation of the hormonal stress response.

At least some of the effects of prenatal stress on the fetus appear to be mediated by the secretion of glucocorticoids. Barbazanges et al. (1996) subjected pregnant female rats to stress and later observed the effects of this treatment on their offspring once they grew up. They found that the prenatally stressed rats showed a prolonged secretion of glucocorticoids when they were subjected to restraint stress. However, if the mothers' adrenal glands had been removed so that glucocorticoid levels could not increase during the stressful situation, their offspring reacted normally in adulthood. (The experimenters gave the adrenalectomized mothers controlled amounts of glucocorticoids to maintain them in good health.) (See *Figure 17.17.*)

Uno et al. (1989) found that if long-term stress is intense enough, it can even cause brain damage in young primates. The investigators studied a colony of vervet monkeys housed in a primate center in Kenya. They found that some monkeys died, apparently from stress. Vervet monkeys have a hierarchical society, and monkeys near the bottom of the hierarchy are picked on by the others; thus, they are almost continuously subjected to stress. (Ours is not the only species with social structures that cause a stress reaction in some of its members.) The deceased monkeys had gastric ulcers and enlarged adrenal glands, which are signs of chronic stress. And as Figure 17.18 shows, neurons in the CA1 field of the hippocampal formation were completely destroyed. (See *Figure 17.18.*) Severe stress appears to cause brain damage in humans as well; Jensen, Genefke, and Hyldebrandt (1982) found evidence of brain degeneration in CT scans of people who had been subjected to torture.

Several studies have confirmed that the stress of chronic pain has adverse effects on the brain and on cognitive behavior. Apkarian et al. (2004b) found that each year of severe chronic back pain resulted in the loss of 1.3 cm³ of gray matter in the cerebral cortex, with the greatest reductions seen in the dorsolateral prefrontal cortex. In addition, Apkarian et al. (2004a) found that

FIGURE 17.18 ■ Brain Damage Caused by Stress

The photomicrographs show sections through the hippocampus. (a) A normal monkey. (b) A monkey of low social status subjected to stress. Compare the regions between the arrowheads, which are normally filled with large pyramidal cells.

(From Uno, H., Tarara, R., Else, J. G., Suleman, M. A., and Sapolsky, R. M. *Journal of Neuroscience*, 1989, 9, 1706–1711. Reprinted by permission of the *Journal of Neuroscience*.)

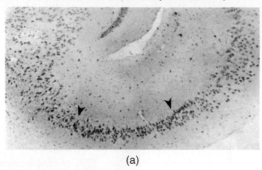

(a)

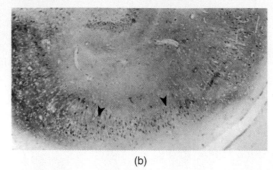

(b)

chronic back pain led to poor performance on a task that has been shown to be affected by prefrontal lesions.

Posttraumatic Stress Disorder

The aftermath of tragic and traumatic events such as those that accompany wars, violence, and natural disasters often includes psychological symptoms that persist long after the stressful events are over. According to the DSM IV, **posttraumatic stress disorder (PTSD)** is caused by a situation in which a person "experienced, witnessed, or was confronted with an event or events that involved actual or threatened death or serious injury, or a threat to the physical integrity of self or others" that provoked a response that "involved intense fear, helplessness, or horror." The likelihood of developing PTSD is increased if the traumatic event involved danger or violence from other people, such as assault, rape, or wartime experiences (Yehuda and LeDoux, 2007). The symptoms produced by such exposure include recurrent dreams or recollections of the event, feelings that the traumatic event is recurring ("flashback" episodes), and intense psychological distress. These dreams, recollections, or flashback episodes can lead the person to avoid thinking about the traumatic event, which often results in diminished interest in social activities, feelings of detachment from others, suppressed emotional feelings, and a sense that the future is bleak and empty. Particular psychological symptoms include difficulty falling or staying asleep, irritability, outbursts of anger, difficulty in concentrating, and heightened reactions to sudden noises or movements. As this description indicates, people with PTSD have impaired mental health functioning. They also tend to have generally poor physical health (Zayfert et al., 2002). Although men are exposed to traumatic events more often than women are, women are more likely to develop PTSD after being exposed to such events (Fullerton et al., 2001).

Evidence from twin studies suggest that genetic factors play a role in a person's susceptibility to develop PTSD. In fact, genetic factors influence not only the likelihood of developing PTSD after being exposed to traumatic events, but also the likelihood that the person will be involved in such an event (Stein et al., 2002). For example, people with a genetic predisposition toward irritability and anger are more likely to be assaulted, and those with a predisposition toward risky behavior are more likely to be involved in accidents. In a review of the Vietnam Era Twin Registry, Koenen et al. (2002) reported that the following demographic and personality factors predict an increased risk for being exposed to traumatic events: military service in Southeast Asia during the Vietnam war, a preexisting conduct disorder or substance dependence, and a family history of mood disorders. The following factors predict the risk of developing PTSD after exposure: earlier age at the time of the traumatic event; exposure to more than one traumatic event; a father with a depressive disorder; a low educational level; poor social support; and a preexisting conduct disorder, panic disorder, generalized anxiety disorder, or depressive disorder.

Although many people are exposed to potentially traumatic event during their lives, most of them recover rapidly and do not develop PTSD (Kessler et al., 1995). For example, Rothbaum and Davis (2003) reported that two weeks after having been raped, 92 percent of the victims showed symptoms that met the criteria for PTSD. However, within thirty days, the symptoms in most of the victims had subsided. Twin studies have shown that the

posttraumatic stress disorder (PTSD) A psychological disorder caused by exposure to a situation of extreme danger and stress; symptoms include recurrent dreams or recollections; can interfere with social activities and cause a feeling of hopelessness.

overlap between PTSD and panic disorder, generalized anxiety disorder, and depressive disorder is at least partly a result of shared genetic factors (Nugent, Amstadter, and Koenen, 2008). Presumably, these genetic factors make some people more sensitive to the effects of stress.

A few studies have identified specific genes as possible risk factors for developing PTSD. These genes include those responsible for the production of dopamine D_2 receptors, dopamine transporters, and 5-HT transporters (Nugent, Amstadter, and Koenen, 2008). We are already familiar with one of these genes. As we saw in Chapter 16 and earlier in this chapter, the presence of the short allele of the promoter for the 5-HT transporter (5-HTT) gene produces an increased sensitivity to stress and an increased incidence of depression and anxiety disorder. Kilpatrick et al. (2007) studied people living in Florida during the 2004 hurricane season. They found that in people at risk for PTSD (high hurricane exposure and low social support), the presence of the short allele was associated with a 450 percent increase in the incidence of PTSD.

As we saw in the previous subsection, studies with laboratory animals have shown that prolonged exposure to stress can cause brain abnormalities, particularly in the hippocampus and amygdala. At least two MRI studies have found evidence of hippocampal damage in veterans with combat-related posttraumatic stress disorder (Bremner et al., 1995; Gurvits et al., 1996). In the study by Gurvits et al., the volume of the hippocampal formation was reduced by over 20 percent, and the loss was proportional to the amount of combat exposure the veteran had experienced. Lindauer et al. (2005) found that police officers with PTSD had a smaller hippocampus than those who had also been exposed to trauma but had not developed the disorder.

An intriguing study by Gilbertson et al. (2002) suggests that at least part of the reduction in hippocampal volume seen in people with PTSD may *predate* the exposure to stress. In other words, a smaller hippocampus may be a predisposing factor in the acquisition of PTSD. Gilbertson and his colleagues studied 40 pairs of monozygotic twins in which only one member went to Vietnam and experienced combat. Almost half of the men who experienced combat developed PTSD. As expected, the hippocampal volumes of these men were smaller than those of the men who did not develop PTSD after their combat experience. In addition, a smaller hippocampus was associated with more severe PTSD. The interesting fact is that the hippocampal volumes of the twin brothers of the PTSD patients who stayed home *also* showed smaller hippocampal volumes. Given that monozygotic twins are genetically identical and usually have very similar brains, this finding suggests that a person with a small hippocampus is more likely to develop PTSD after exposure to psychological trauma. (See *Figure 17.19*.)

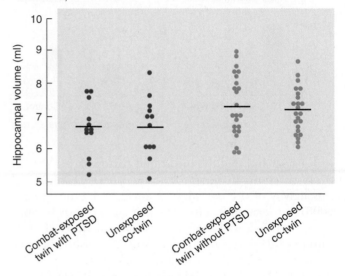

FIGURE 17.19 ■ Hippocampal Volumes of Pairs of Monozygotic Twins

The graph shows that the size of the hippocampus of twins not exposed to combat was similar to the size of their combat-exposed co-twins whether or not the co-twins had PTSD. These results suggest that hippocampal size is a genetically determined trait that predates the exposure to combat.

(Adapted from Gilbertson, M. W., Shenton, M. E., Ciszewski, A., Kasai, K., Lasko, N. G., Orr, S. P., and Pitman, R. K. *Nature Neuroscience*, 2002, *5*, 1242–1247.)

What role might the hippocampus play in a person's susceptibility to developing PTSD? One possibility is that the hippocampus, which plays a role in contextual learning, participates in recognition of the context in which a traumatic event occurs. The hippocampus then aids in distinguishing safe from dangerous contexts (Yehuda and LeDoux, 2007). Consider a person who has been attacked by another person. The sight of other people who even slightly resemble the attacker or situations that even slightly resemble the one in which the attack occurred might then activate the amygdala and trigger an emotional response. However, a normally functioning hippocampus would detect the difference between the present context and the one associated with the attack and inhibit the activity of the amygdala.

I mentioned a few paragraphs ago that most people who are exposed to a potentially traumatic event manage to suppress their emotional reaction. What brain mechanisms suppress the emotional reaction and enable a person to recover? As we saw in Chapters 11 and 16, the prefrontal cortex can exert an inhibitory effect on the amygdala and suppress emotional reactions. For example, the medial prefrontal cortex plays an essential role in the extinction of conditioned emotional responses.

Several studies have found evidence that the amygdala is responsible for emotional reactions in people

FIGURE 17.20 ■ **Amygdala and Medial Prefrontal Cortex Activation in PTSD**

The graphs show the activation of the amygdala and medial prefrontal cortex in response to the sight of happy or fearful faces in control subjects and subjects with PTSD.

(Adapted from Shin, L. M., Wright, C. I., Cannistraro, P. A., Wedig, M. M., McMullin, K., Martis, B., Macklin, M. L., Lasko, N. B., Cavanagh, S. R., Krangel, T. S., Orr, S. P., Pitman, R .K., Whalen, P. J., and Rauch, S. L. *Archives of General Psychiatry*, 2005, 62, 273–281.)

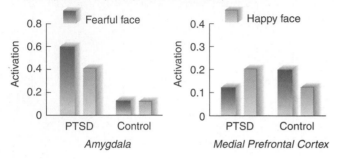

with PTSD and that the prefrontal cortex plays a role in these reactions in people without PTSD by inhibiting the activity of the amygdala (Rauch, Shin, and Phelps, 2006). For example, a functional-imaging study by Shin et al. (2005) found that when shown pictures of faces with fearful expressions, people with PTSD show greater activation of the amygdala and smaller activation of the prefrontal cortex than did people without PTSD. In fact, the symptoms of the people with PTSD were positively correlated with the activation of the amygdala and negatively correlated with the activation of the medial prefrontal cortex. (See *Figure 17.20*.) In addition, a study by Milad et al. (2005) found that the medial prefrontal cortex was thicker in people who showed rapid extinction of a conditioned emotional response.

Psychoneuroimmunology

As we have seen, long-term stress can be harmful to one's health and can even result in brain damage. The most important cause of these effects is elevated levels of glucocorticoids, but the high blood pressure caused by epinephrine and norepinephrine also plays a contributing role. In addition, the stress response can impair the functions of the immune system, which protects us from assault from viruses, microbes, fungi, and other types of parasites. Study of the interactions between the immune system and behavior (mediated by the nervous system, of course) is called **psychoneuroimmunology.** Some research in this field is described in the following subsection.

The Immune System

The immune system is one of the most complex systems of the body. Its function is to protect us from infection; and because infectious organisms have developed devious tricks through the process of evolution, our immune system has evolved devious tricks of its own. The description I provide here is abbreviated and simplified, but it presents some of the important elements of the system.

The immune system derives from white blood cells that develop in the bone marrow and in the thymus gland. Some of the cells roam through the blood or lymphatic system; others reside permanently in one place. Two types of specific immune reaction occur when the body is invaded by foreign organisms, including bacteria, fungi, and viruses: *chemically mediated* and *cell-mediated*. Chemically mediated immune reactions involve antibodies. Infectious microorganisms have unique proteins on their surfaces, called **antigens.** These proteins serve as the invaders' calling cards, identifying them to the immune system. Through exposure to the microorganisms, the immune system learns to recognize these proteins. (I will not try to explain the mechanism by which this learning takes place.) The result of this learning is the development of special lines of cells that produce specific **antibodies**—proteins that recognize antigens and help to kill the invading microorganism.

One type of antibody is released into the circulation by **B-lymphocytes,** which receive their name from the fact that they develop in bone marrow. These antibodies, called **immunoglobulins,** are chains of protein. Each type of immunoglobulin (there are five of them) is identical except for one end, which contains a unique receptor. A particular receptor binds with a particular antigen, just as a molecule of a hormone or neurotransmitter binds with its receptor. When the appropriate line of B-lymphocytes detects the presence of an invading bacterium, the cells release their antibodies, which bind with the antigens present on the surface of the invading microorganisms. The antigens either kill the invaders directly or attract other white blood cells, which then destroy them. (See *Figure 17.21a*.)

The other type of defense by the immune system, cell-mediated immune reactions, is produced by **T-lymphocytes,** which originally develop in the thymus

psychoneuroimmunology The branch of neuroscience involved with interactions between environmental stimuli, the nervous system, and the immune system.

antigen A protein present on a microorganism that permits the immune system to recognize the microorganism as an invader.

antibody A protein produced by a cell of the immune system that recognizes antigens present on invading microorganisms.

B-lymphocyte A white blood cell that originates in the bone marrow; part of the immune system.

immunoglobulin An antibody released by B-lymphocytes that bind with antigens and help to destroy invading microorganisms.

T-lymphocyte A white blood cell that originates in the thymus gland; part of the immune system.

FIGURE 17.21 ■ Immune Reactions

(a) In a chemically mediated reaction, the B-lymphocyte detects an antigen on a bacterium and releases a specific immunoglobulin. (b) In a cell-mediated reaction, the T-lymphocyte detects an antigen on a bacterium and kills it directly or releases a chemical that attracts other white blood cells.

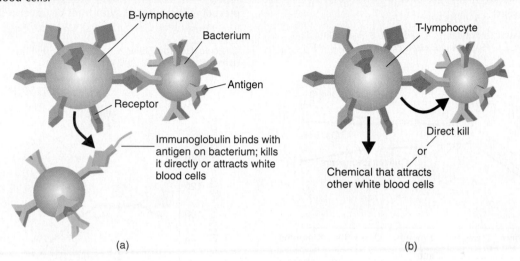

(a)   (b)

gland. These cells also produce antibodies, but the antibodies remain attached to the outside of their membrane. T-lymphocytes primarily defend the body against fungi, viruses, and multicellular parasites. When antigens bind with their surface antibodies, the cells either directly kill the invaders or signal other white blood cells to come and kill them. (See *Figure 17.21b.*)

The reactions illustrated in Figure 17.21 are much simplified; actually, both chemically mediated and cell-mediated immune reactions involve several different types of cells. The communication between these cells is accomplished by **cytokines,** chemicals that stimulate cell division. The cytokines that are released by certain white blood cells when an invading microorganism is detected (principally *interleukin-1* and *interleukin-2*) cause other white blood cells to proliferate and direct an attack against the invader. The primary way in which glucocorticoids suppress specific immune responses is by interfering with the messages conveyed by the cytokines (Sapolsky, 1992).

Neural Control of the Immune System

As we will see in the next subsection, the stress response can increase the likelihood of infectious diseases. What is the physiological explanation for these effects? One answer, probably the most important one, is that stress increases the secretion of glucocorticoids, and as we saw, these hormones directly suppress the activity of the immune system.

A direct relationship between stress and the immune system was demonstrated by Kiecolt-Glaser et al. (1987). Using several different laboratory tests,

these investigators found that caregivers of family members with Alzheimer's disease, who certainly underwent considerable stress, showed weaker immune systems. One measure of the quality of a person's immune response is measurement of antibodies produced in response to a vaccination. Glaser et al. (2000) found that people taking care of spouses with Alzheimer's disease maintained lower levels of IgG antibodies after receiving a pneumococcal bacterial vaccine. (See *Figure 17.22.*) Bereavement, another source of stress, also suppresses the immune system. Schleifer et al. (1983) tested the husbands of women with breast cancer and found that their immune response was lower after their wives died. Knapp et al. (1992) even found that when healthy subjects imagined themselves reliving unpleasant emotional experiences, the immune response measured in samples of their blood was decreased.

Several studies indicate that the suppression of the immune response by stress is largely (but not entirely) mediated by glucocorticoids (Keller et al., 1983). Because the secretion of glucocorticoids is controlled by the brain (through its secretion of CRH), the brain is obviously responsible for the suppressing effect of these hormones on the immune system. Neurons in the central nucleus of the amygdala send axons to CRH-secreting neurons in the paraventricular nucleus of the hypothalamus; thus, we can reasonably expect that the mechanism that is

cytokine A category of chemicals released by certain white blood cells when they detect the presence of an invading microorganism; causes other white blood cells to proliferate and mount an attack against the invader.

FIGURE 17.22 ■ Effect of Stress on Immune Function

The graph shows levels of antibodies produced in response to a pneumococcal bacterial vaccine in the blood of controls and former and current caregivers of spouses with Alzheimer's disease.

(Adapted from Glaser, R., Sheridan, J., Malarkey, W. B., MacCallum, R. C., and Kiecolt-Glaser, J. K. *Psychosomatic Medicine*, 2000, *62*, 804–807.)

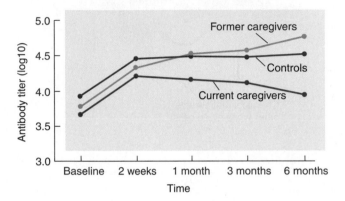

responsible for negative emotional responses is also responsible for the stress response and the immunosuppression that accompanies it.

Stress and Infectious Diseases

Often when a married person dies, his or her spouse dies soon afterward, frequently of an infection. In fact, a wide variety of stress-producing events in a person's life can increase susceptibility to illness. For example, Glaser et al. (1987) found that medical students were more likely to contract acute infections and to show evidence of suppression of the immune system during the time that final examinations were given.

Stone, Reed, and Neale (1987) attempted to determine whether stressful events in people's daily lives might predispose them to upper respiratory infection. If a person is exposed to a microorganism that might cause such a disease, the symptoms do not occur for several days; that is, there is an incubation period between exposure and signs of the actual illness. Thus, the authors reasoned that if stressful events suppressed the immune system, one might expect to see a higher likelihood of respiratory infections several days after such stress. To test their hypothesis, they asked volunteers to keep a daily record of desirable and undesirable events in their lives over a twelve-week period. The volunteers also kept a daily record of any discomfort or symptoms of illness.

The results were as predicted: During the three- to five-day period just before showing symptoms of an upper respiratory infection, people experienced an increased number of undesirable events and a decreased number of desirable events in their lives. (See *Figure 17.23.*) Stone

FIGURE 17.23 ■ Role of Desirable and Undesirable Events on Susceptibility to Upper Respiratory Infections

The graph shows mean percentage change in frequency of undesirable and desirable events during the ten-day period preceding the onset of symptoms of upper respiratory infections.

(Based on data from Stone, A. A., Reed, B. R., and Neale, J. M. *Journal of Human Stress*, 1987, *13*, 70–74.)

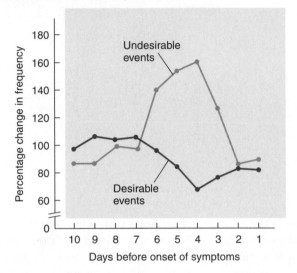

FIGURE 17.24 ■ Colds and Psychological Stress

The graph shows the percentage of subjects with colds as a function of an index of psychological stress.

(Adapted from Cohen, S., Tyrrell, D. A. J., and Smith, A. P. *New England Journal of Medicine*, 1991, *325*, 606–612.)

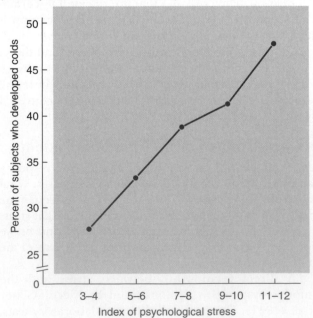

et al. (1987) suggest that the effect is caused by decreased production of a particular immunoglobulin that is present in the secretions of mucous membranes, including those in the nose, mouth, throat, and lungs. This immunoglobulin, IgA, serves as the first defense against infectious microorganisms that enter the nose or mouth. They found that IgA is associated with mood; when a subject is unhappy or depressed, IgA levels are lower than normal. The results suggest that the stress caused by undesirable events may, by suppressing the production of IgA, lead to a rise in the likelihood of upper respiratory infections.

The results of the study by Stone and his colleagues were confirmed by an experiment by Cohen, Tyrrell, and Smith (1991). The investigators found that subjects who were given nasal drops containing cold viruses were much more likely to develop colds if they reported stressful experiences during the past year and if they said they felt threatened, out of control, or overwhelmed by events. (See *Figure 17.24.*)

Interim Summary

Stress Disorders

People's emotional reactions to aversive stimuli can harm their health. The stress response, which Cannon called the fight-or-flight response, is useful as a short-term response to threatening stimuli but is harmful in the long term. This response includes increased activity of the sympathetic branch of the autonomic nervous system and increased secretion of hormones by the adrenal gland: epinephrine, norepinephrine, and glucocorticoids. Corticotropin-releasing hormone, which stimulates the secretion of ACTH by the anterior pituitary gland, is also secreted in the brain, where it elicits some of the emotional responses to stressful situations.

Although increased levels of epinephrine and norepinephrine can raise blood pressure, most of the harm to health comes from glucocorticoids. Prolonged exposure to high levels of these hormones can increase blood pressure, damage muscle tissue, lead to infertility, inhibit growth, inhibit the inflammatory response, and suppress the immune system. It can also damage the hippocampus. Acute stress can also impair hippocampal functioning. Exposure to stress during prenatal or early postnatal life can affect brain development, resulting in impaired functions of the hippocampus and increased size of the amygdala. Stress also decreases the survival rate of hippocampal neurons produced by adult neurogenesis. These changes appear to predispose animals to react more to stressful situations. In humans the stress of chronic pain can cause loss of cerebral gray matter, especially in the prefrontal cortex, with accompanying deficits in behaviors that involve the prefrontal cortex.

Exposure to extreme stress can also have long-lasting effects; it can lead to the development of posttraumatic stress disorder. This disorder is associated with memory deficits, poorer health, and a decrease in the size of the hippocampus. Twin studies indicate a hereditary component to susceptibility to PTSD. Predisposing factors appears to involve decreased hippocampal volume and differences in the genes for D_2 receptors, dopamine transporters, and 5-HT transporters. The prefrontal cortex of people who are resistant to the development of PTSD following severe stress appears to inhibit the amygdala. The prefrontal cortex appears to be hypoactive in people with PTSD.

Psychoneuroimmunology is a field of study that investigates interactions between behavior and the immune system, mediated by the nervous system. The immune system consists of several types of white blood cells that produce both nonspecific and specific responses to invading microorganisms. The nonspecific responses include the inflammatory response, the antiviral effect of interferon, and the action of natural killer cells against viruses and cancer cells. The specific responses include chemically mediated and cell-mediated responses. Chemically mediated responses are carried out by B-lymphocytes, which release antibodies that bind with the antigens on microorganisms and kill them directly or target them for attack by other white blood cells. Cell-mediated responses are carried out by T-lymphocytes, whose antibodies remain attached to their membranes.

A wide variety of stressful situations have been shown to increase people's susceptibility to infectious diseases. The most important mechanism by which stress impairs immune function is the increased blood levels of glucocorticoids. The neural input to the bone marrow, lymph nodes, and thymus gland may also play a role; and the endogenous opioids appear to suppress the activity of natural killer cells.

Thought Question

Researchers are puzzled by the fact that glucocorticoids suppress the immune system. Can you think of any potential benefits that come from the fact that our immune system is suppressed during times of danger and stress?

SUGGESTED READINGS

Aouizerate, B., Guehl, D., Cuny, E., Rougier, A., Biolac, B., Tignol, J., and Burbaud, P. Pathophysiology of obsessive-compulsive disorder: A necessary link between phenomenology, neuropsychology, imagery and physiology. *Progress in Neurobiology*, 2004, *72*, 195–221.

Autism Genome Project Consortium. Mapping autism risk loci using genetic linkage and chromosomal rearrangements. *Nature Genetics*, 2007, *39*, 319–328.

Bale, T. L., and Vale, W. W. CRF and CRF receptors: Role in stress responsivity and other behaviors. *Annual Review of Pharmacology and Toxicology*, 2004, *44*, 525–527.

Bush, G., Valera, E. M., and Seidman, L. J. Functional neuroimaging of attention-deficit/hyperactivity disorder: A review and suggested future directions. *Biological Psychiatry*, 2005, *57*, 1273–1284.

Charmandari, E., Tsigos, C., and Chrousos, G. Endocrinology of the stress response. *Annual Review of Physiology*, 2005, *67*, 259–284.

Courchesne, E., Pierce, K., Schumann, C. M., Redcay, E., Buckwalter, J. A., Kennedy, D. P., and Morgan, J. Mapping early brain development in autism. *Neuron*, 2007, *56*, 399–413.

de Kloet, E. R., Joëls, M., and Holsboer, F. Stress and the brain: From adaptation to disease. *Nature Reviews: Neuroscience*, 2005, *6*, 463–475.

Merikangas, K. R., and Low, N. C. Genetic epidemiology of anxiety disorders. *Handbook of Experimental Pharmacology*, 2005, *169*, 163–179.

Yehuda, R., and LeDoux, J. Response variation following trauma: A translational neuroscience approach to understanding PTSD. *Neuron*, 2007, *56*, 19–32.

ADDITIONAL RESOURCES

Visit www.mypsychkit.com for additional review and practice of the material covered in this chapter. Within MyPsychKit, you can take practice tests and receive a customized study plan to help you review. Dozens of animations, tutorials, and Web links are also available. You can even review using the interactive electronic version of this textbook. You will need to register for MyPsychKit. See www.mypsychkit.com for complete details.

chapter
18

Drug Abuse

outline

John was beginning to feel that perhaps he would be able to get his life back together. It looked as though his drug habit was going to be licked. He had started taking drugs several years ago. At first, he had used them only on special occasions—mostly on weekends with his friends—but heroin proved to be his undoing. One of his acquaintances had introduced him to the needle, and John had found the rush so blissful that he couldn't wait a whole week for his next fix. Soon he was shooting up daily. Shortly after that, he lost his job and, to support his habit, began earning money through car theft and small-time drug dealing. As time went on, he needed more and more heroin at shorter and shorter intervals, which necessitated even more money. Eventually, he was arrested and convicted of selling heroin to an undercover agent.

The judge gave John the choice of prison or a drug reha-bilitation program, and he chose the latter. Soon after start-ing the program, he realized that he was relieved to have been caught. Now that he was clean and could reflect on his life, he realized what would have become of him had he con-tinued to take drugs. Withdrawal from heroin was not an experience he would want to live through again, but it turned out not to be as bad as he had feared. The counselors in his program told him to avoid his old neighborhood and to break contact with his old acquaintances, and he followed their advice. He had been clean for eight weeks, he had a job, and he had met a woman who really seemed sympathetic. He knew that he hadn't completely kicked his habit, because every now and then, despite his best intentions, he found himself thinking about the wonderful glow that heroin pro-vided him. But things were definitely looking up.

Then one day, while walking home from work, he turned a corner and saw a new poster plastered on the wall of a build-ing. The poster, produced by an antidrug agency, showed all sorts of drug paraphernalia in full color: glassine envelopes with white powder spilling out of them, syringes, needles, a spoon and candle used to heat and dissolve the drug. John was seized with a sudden, intense compulsion to take some heroin. He closed his eyes, trying to will the feeling away, but all he could feel were his churning stomach and his trembling limbs, and all he could think about was getting a fix. He hopped on a bus and went back to his old neighborhood.

Drug addiction poses a serious problem to our species. Consider the disastrous effects caused by the abuse of one of our oldest drugs, alco-hol: automobile accidents, fetal alcohol syn-drome, cirrhosis of the liver, Korsakoff's syndrome, increased rate of heart disease, and increased rate of intracerebral hemorrhage. Smoking (addiction to nico-tine) greatly increases the chances of dying of lung can-cer, heart attack, and stroke; and women who smoke give birth to smaller, less healthy babies. Cocaine addiction can cause psychotic behavior, brain damage, and death from overdose; and competition for lucrative and illegal markets terrorizes neighborhoods, subverts political and judicial systems, and causes many violent deaths. The use of "designer drugs" exposes users to unknown dangers of untested and often contaminated products, as several people discovered when they acquired Parkinson's dis-ease after taking a synthetic opiate that was tainted with a neurotoxin. (This unfortunate event was described in the opening case of Chapter 5.) Addicts who take their drugs intravenously run a serious risk of contracting AIDS, hep-atitis, or other infectious diseases. What makes these drugs so attractive to so many people?

The answer, as you might have predicted from what you have learned about the physiology of reinforcement in Chapter 13, is that all of these substances stimulate brain mechanisms responsible for positive reinforce-ment. In addition, most of them also reduce or elimi-nate unpleasant feelings, some of which are produced by the drugs themselves. The immediate effects of these drugs are more powerful than the realization that in the long term, bad things will happen.

COMMON FEATURES OF ADDICTION

The term *addiction* derives from the Latin word *addicere*, "to sentence." Someone who is addicted to a drug is, in a way, sentenced to a term of involuntary servitude, being obliged to fulfill the demands of his or her drug dependency.

A Little Background

Long ago, people discovered that many substances found in nature—primarily leaves, seeds, and roots of plants but also some animal products—had medicinal qualities. They discovered herbs that helped to prevent infections, that promoted healing, that calmed an upset stomach, that reduced pain, or that helped to provide a night's sleep. They also discovered "recreational drugs"—drugs that produced pleasurable effects when eaten, drunk, or smoked. The most universal recreational drug, and per-haps the first one that our ancestors discovered, is ethyl alcohol. Yeast spores are present everywhere, and these microorganisms can feed on sugar solutions and produce alcohol as a by-product. Undoubtedly, people in many different parts of the world discovered the pleasurable effects of drinking liquids that had been left alone for a while, such as the juice that had accumulated in the bottom of a container of fruit. The juice may have become sour and bad-tasting because of the action of bacteria, but the effects of the alcohol encouraged people to experiment, which led to the development of a wide variety of fermented beverages.

Our ancestors also discovered other recreational drugs. Some of them were consumed only locally; others became so popular that their cultivation as commercial crops spread throughout the world. For example, Asians discovered the effects of the sap of the opium poppy and the beverage made from the leaves of the tea plant, Indians discovered the effects of the smoke of cannabis, South Americans discovered the effects of chewing coca leaves and making a drink from coffee beans, and North Americans discovered the effects of the smoke of the tobacco plant. Many of the drugs they discovered served to protect the plants from animals (primarily insects) that ate them. Although the drugs were toxic in sufficient quantities, our ancestors learned how to take these drugs in amounts that would not make them ill—at least, not right away. The effects of these drugs on their brains kept them coming back for more. *Table 18.1* lists the most important addictive drugs and indicates their sites of action.

Positive Reinforcement

Drugs that lead to dependency must first reinforce people's behavior. As we saw in Chapter 13, positive reinforcement refers to the effect that certain stimuli have on the behaviors that preceded them. If, in a particular situation, a behavior is regularly followed by an appetitive stimulus (one that the organism will tend to approach), then that behavior will become more frequent in that situation. For example, if a hungry rat accidentally bumps into a lever and receives some food, the rat will eventually learn to press the lever. What actually seems to happen is that the occurrence of an appetitive stimulus activates a reinforcement mechanism in the brain that increases the likelihood of the most recent response (the lever press) in the present situation (the chamber that contains the lever).

Addictive drugs have reinforcing effects. That is, their effects include activation of the reinforcement mechanism. This activation strengthens the response that was just made. If the drug was taken by a fast-acting route such as injection or inhalation, the last response will be the act of taking the drug, so that response will be reinforced. This form of reinforcement is powerful and immediate and works with a wide variety of species. For example, a rat or a monkey will quickly learn to press a lever that controls a device that injects cocaine through a plastic tube inserted into a vein.

TABLE 18.1 ■ Addictive Drugs

DRUG	SITES OF ACTION
Ethyl alcohol	NMDA receptor (indirect antagonist), GABA$_A$ receptor (indirect agonist)
Barbiturates	GABA$_A$ receptor (indirect agonist)
Benzodiazepines (tranquilizers)	GABA$_A$ receptor (indirect agonist)
Cannabis (marijuana)	CB1 cannabinoid receptor (agonist)
Nicotine	Nicotinic ACh receptor (agonist)
Opiates (heroin, morphine, etc.)	μ and δ opiate receptor agonist
Phencyclidine (PCP) and ketamine	NMDA receptor (indirect antagonist)
Cocaine	Blocks reuptake of dopamine (and serotonin and norepinephrine)
Amphetamine	Causes release of dopamine (by running dopamine transporters in reverse)

Source: Adapted from Hyman, S. E., and Malenka, R. C. *Nature Reviews: Neuroscience*, 2001, *2*, 695–703.

Role in Drug Abuse

When appetitive stimuli occur, they usually do so because we just did something to make them happen—and not because an experimenter was controlling the situation. The effectiveness of a reinforcing stimulus is greatest if it occurs immediately after a response occurs. If the reinforcing stimulus is delayed, it becomes considerably less effective. The reason for this fact is found by examining the function of instrumental conditioning: learning about the consequences of our own behavior. Normally, causes and effects are closely related in time; we do something, and something happens, good or bad. The consequences of the actions teach us whether to repeat that action, and events that follow a response by more than a few seconds were probably not caused by that response.

An experiment by Logan (1965) illustrates the importance of the immediacy of reinforcement. Logan trained hungry rats to run through a simple maze in which a single passage led to two corridors. At the end of one corridor the rats would find a small piece of food. At the end of the other corridor they would receive much more food, but it would be delivered only after a delay. Although the most intelligent strategy would be to enter the second corridor and wait for the larger amount of food, the rats chose to take the small amount of food that was delivered right away. Immediacy of reinforcement took precedence over quantity.

This phenomenon explains why the most addictive drugs are those that have immediate effects. As we saw in Chapter 4, drug users prefer heroin to morphine not because heroin has a *different* effect, but because it has a more *rapid* effect. In fact, heroin is converted to morphine as soon as it reaches the brain. But because heroin is more lipid soluble, it passes through the blood–brain barrier more rapidly, and its effects on the brain are felt sooner than those of morphine. The most potent reinforcement occurs when drugs produce sudden changes in the activity of the reinforcement mechanism; slow changes are much less reinforcing. A person taking an addictive drug seeks a sudden "rush" produced by a fast-acting drug. (As we will see later, the use of methadone for opiate addiction and nicotine patches for tobacco addiction are based on this phenomenon.)

Earlier, I posed the question of why people would ever expose themselves to the risks associated with dangerous addictive drugs. Who would rationally chose to become addicted to a drug that produced pleasurable effects in the short term but also produced even more powerful aversive effects in the long term: loss of employment and social status, legal problems and possible imprisonment, damage to health, and even premature death? The answer is that, as we saw, our reinforcement mechanism evolved to deal with the *immediate* effects of our behavior. The immediate reinforcing effects of an addictive drug can, for some individuals, overpower the recognition of the long-term aversive effects. Fortunately, most people are able to resist the short-term effects; only a minority of people who try addictive drugs go on to become dependent on them. Although cocaine is one of the most addictive drugs currently available, only about 15 percent of people who use it become addicted to it (Wagner and Anthony, 2002). As we will see later, particular brain mechanisms are responsible for inhibiting behavior that has unfavorable long-term consequences.

If an addictive drug is taken by a slow-acting route, reinforcement can also occur, but the process is somewhat more complicated. If a person takes a pill and several minutes later experiences a feeling of euphoria, he or she will certainly remember swallowing the pill. The recollection of this behavior will activate some of the same neural circuits involved in actually swallowing the pill, and the reinforcement mechanism, now active because of the effects of the drug, will reinforce the behavior. In other words, people's ability to remember having performed a behavior make it possible to reinforce their behavior vicariously. The immediacy is between an imagined act and a reinforcing stimulus—the euphoria produced by the drug. Other cognitive processes contribute to the reinforcement, too, such as the expectation that euphoric effects will occur. Perhaps someone said, "Take one of these pills; you'll get a great high!" But if a nonhuman animal is fed one of these pills, its behavior is unlikely to be reinforced. By the time the euphoric effect occurs, the animal will be doing something other than ingesting the drug. Without the ability to recall an earlier behavior and thus activate circuits involved in the performance of that behavior, the delay between the behavior and the reinforcing effect of the drug prevents the animal from learning to take the drug. As we will see later in this chapter, researchers have developed ways to teach animals to become addicted to drugs that have delayed effects, such as alcohol.

Neural Mechanisms

As we saw in Chapter 13, all natural reinforcers that have been studied so far (such as food for a hungry animal, water for a thirsty one, or sexual contact) have one physiological effect in common: They cause the release of dopamine in the nucleus accumbens (White, 1996). This effect is not the *only* effect of reinforcing stimuli, and even aversive stimuli can trigger the release of dopamine (Salamone, 1992). But even though there is much that we do not yet understand about the neural basis of reinforcement, the release of dopamine appears to be a *necessary* (but not *sufficient*) condition for positive reinforcement to take place.

Addictive drugs—including amphetamine, cocaine, opiates, nicotine, alcohol, PCP, and cannabis—trigger the release of dopamine in the nucleus accumbens (NAC), as measured by microdialysis (Di Chiara, 1995).

Different drugs stimulate the release of dopamine in different ways. The details of the ways in which particular drugs interact with the mesolimbic dopaminergic system are described later.

The fact that the reinforcing properties of addictive drugs involve the same brain mechanisms as natural reinforcers indicated that these drugs "hijack" brain mechanisms that normally help us adapt to our environment. It appears that the process of addiction begins in the mesolimbic dopaminergic system and then produces long-term changes in other brain regions that receive input from these neurons (Kauer and Malenka, 2007). The first changes appear to take place in the ventral tegmental area (VTA). Saal et al. (2003) found that a single administration of a variety of addictive drugs (including cocaine, amphetamine, morphine, alcohol, and nicotine) increased the strength of excitatory synapses on dopaminergic neurons in the VTA in mice.

As a result of these changes, increased activation is seen in a variety of regions that receive dopaminergic input from the VTA, including the NAC, located in the ventral striatum. Subsequent changes that are responsible for the compulsive behaviors that characterize addiction occur only after continued use of an addictive drug. The most important of these changes appear to occur in the dorsal striatum: in the caudate nucleus and putamen. We saw in Chapter 13 that the basal ganglia (which includes the dorsal striatum) play a critical role in instrumental conditioning, and the process of addiction involves just that.

At first, the potential addict experiences the pleasurable effects of the drug, which reinforces the behaviors that cause the drug to be delivered to the brain (procuring the drug, taking necessary steps to prepare it, then swallowing, smoking, sniffing, or injecting it). Eventually, these behaviors become habitual, and the impulse to perform them becomes difficult to resist. The early reinforcing effects that take place in the ventral striatum (namely, in the NAC) encourage drug-taking behavior, but the changes that make the behavior become habitual involve the dorsal striatum. Studies with monkeys performing a response reinforced by infusion of cocaine over a long period of time show a progression of neural changes, beginning in the ventral striatum (in the NAC) and continuing upward to the dorsal striatum (Letchworth et al., 2001; Porrino et al. 2004). A microdialysis study with rats found that the presence of a light that had previously been paired with intravenous infusions of cocaine increased the release of dopamine in the dorsal striatum but not in the ventral striatum (Ito et al., 2002). Vanderschuren et al. (2005) found that infusion of a dopamine antagonist into the dorsal striatum suppressed lever presses that had been reinforced by the illumination of a light that had been paired with intravenous injections of cocaine.

A functional-imaging study by Volkow et al. (2006) provides evidence that addiction in humans involves the dorsal striatum. The investigators found that when people who were addicted to cocaine watched a video of people smoking cocaine, an increased release of dopamine was seen in the dorsal striatum but not in the ventral striatum. These results were similar to what Volkow et al. (2002) saw in a study of hungry people looking at, smelling, and receiving a minuscule taste of appetizing food.

An experiment by Belin and Everitt (2008) suggests that the neural changes responsible for addiction follow a dorsally cascading set of reciprocal connections between the striatum and the ventral tegmental area. Anatomical studies show that neurons in the ventral NAC project to the VTA, which sends dopaminergic projections back to a more dorsal region of the NAC, and so on. This back-and-forth communication continues, connecting increasingly dorsal regions of the striatum, all the way up to the caudate nucleus and putamen. Belin and Everitt found that bilateral infusions of a dopamine antagonist into the dorsal striatum of rats suppressed responding to a light that had been associated with infusions of cocaine but that unilateral infusions had no effect. They also found that a unilateral lesion of the NAC had no effect on responding. However, they found that a lesion of the NAC on one side of the brain combined with infusion of a dopamine antagonist into the dorsal striatum on the other side of the brain suppressed responding to the light. (See *Figure 18.1*.) These

FIGURE 18.1 ■ Establishment of Neural Changes in the Dorsal Striatum

The graph shows the effects of infusing various amounts of a drug that blocks dopamine receptors into the dorsal striatum contralateral to a lesion of the nucleus accumbens.

(Adapted from Belin, D., and Everitt, B. J. *Neuron*, 2008, *57*, 432–441.)

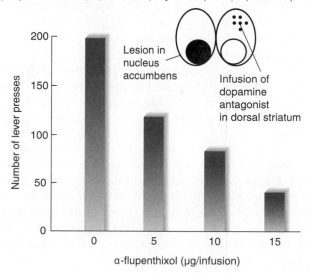

results suggest that the control of compulsive addictive behavior is established by interactions between the ventral and dorsal striatum that are mediated by dopaminergic connections between these regions and the VTA.

Negative Reinforcement

You have probably heard the old joke in which someone says that the reason he bangs his head against the wall is that "it feels so good when I stop." Of course, that joke is funny (well, mildly amusing) because we know that although no one would act that way, ceasing to bang our head against the wall certainly does feel better than continuing to do so. If someone started hitting us on the head and we were able to do something to get the person to stop, whatever it was that we did would certainly be reinforced.

A behavior that turns off (or reduces) an aversive stimulus will be reinforced. This phenomenon is known as **negative reinforcement,** and its usefulness is obvious. For example, consider the following scenario: A woman staying in a rented house cannot get to sleep because of the unpleasant screeching noise that the furnace makes. She goes to the basement to discover the source of the noise and finally kicks the side of the oil burner. The noise ceases. The next time the furnace screeches, she immediately goes to the basement and kicks the side of the oil burner. The unpleasant noise (the aversive stimulus) is terminated when the woman kicks the side of the oil burner (the response), so the response is reinforced.

It is worth pointing out that *negative reinforcement* should not be confused with *punishment*. Both phenomena involve aversive stimuli, but one makes a response more likely, while the other makes it less likely. For negative reinforcement to occur, the response must make the unpleasant stimulus end (or at least decrease). For punishment to occur, the response must *make the unpleasant stimulus occur.* For example, if a little boy touches a mousetrap and hurts his finger, he is unlikely to touch a mousetrap again. The painful stimulus *punishes* the behavior of touching the mousetrap.

People who abuse some drugs become physically dependent on the drug; that is, they show *tolerance* and *withdrawal symptoms.* As we saw in Chapter 4, tolerance is the decreased sensitivity to a drug that comes from its continued use; the drug user must take larger and larger amounts of the drug for it to be effective. Once a person has taken an opiate regularly enough to develop tolerance, that person will exhibit withdrawal symptoms if he or she stops taking the drug. Withdrawal symptoms are primarily the opposite of the effects of the drug itself. The effects of heroin—euphoria, constipation, and relaxation—lead to the withdrawal effects of dysphoria, cramping and diarrhea, and agitation.

Most investigators believe that tolerance is produced by the body's attempt to compensate for the unusual condition of heroin intoxication. The drug disturbs normal homeostatic mechanisms in the brain, and in reaction these mechanisms begin to produce effects opposite to those of the drug, partially compensating for the disturbance. Because of these compensatory mechanisms, the user must take increasing amounts of heroin to achieve the effects that were produced when he or she first started taking the drug. These mechanisms also account for the symptoms of withdrawal: When the person stops taking the drug, the compensatory mechanisms make themselves felt, unopposed by the action of the drug.

Although positive reinforcement seems to be what provokes drug taking in the first place, reduction of withdrawal effects could certainly play a role in maintaining someone's drug addiction. The withdrawal effects are unpleasant, but as soon as the person takes some of the drug, these effects go away, producing negative reinforcement.

Negative reinforcement could also explain the acquisition of drug addictions under some conditions. If a person is suffering from some unpleasant feelings and then takes a drug that eliminates these feelings, the person's drug-taking behavior is likely to be reinforced. For example, alcohol can relieve feelings of anxiety. If a person finds himself in a situation that arouses anxiety, he might find that having a drink or two makes him feel much better. In fact, people often anticipate this effect and begin drinking before the situation actually occurs.

Craving and Relapse

Why do drug addicts crave drugs? Why does this craving occur even after a long period of abstinence? Even after going for months or years without taking an addictive drug, a former drug addict might sometimes experience intense craving that leads to relapse. Clearly, taking a drug over an extended period of time must produce some long-lasting changes in the brain that increase a person's likelihood of relapsing. Understanding this process might help clinicians to devise therapies that will assist people in breaking their drug dependence once and for all.

Robinson and Berridge (2003) suggest that when an addictive drug activates the mesolimbic dopaminergic system, it gives *incentive salience* to stimuli present at that time. By this they mean that the stimuli associated with drug taking become exciting and motivating—a provocation to act. When a person with a history of drug abuse sees or thinks about these stimuli, he or she

---

negative reinforcement The removal or reduction of an aversive stimulus that is contingent on a particular response, with an attendant increase in the frequency of that response.

experiences craving—an impulse to take the drug. Note that this hypothesis does not imply that the craving is caused solely by an unpleasant feeling, as described in the previous subsection. Koob and Le Moal (2001) propose that drug addiction involves "a cycle of spiraling dysregulation of brain reward systems that progressively increases, resulting in the compulsive use and loss of control over drug-taking" (p. 97).

As everyone knows, a taste of food can provoke hunger, which is why we refer to tidbits we eat before a meal as "appetizers." For a person with a history of drug abuse, a small dose of the drug has similar effects: It increases craving, or "appetite," for the drug. Through the process of classical conditioning, stimuli that have been associated with drugs in the past can also elicit craving. For example, an alcoholic who sees a liquor bottle is likely to feel the urge to take a drink. In the past, agencies that sponsored antiaddiction programs sometimes prepared posters illustrating the dangers of drug abuse that featured drug paraphernalia: syringes, needles, spoons, piles of white powder, and so on. Possibly, these posters did succeed in reminding people who did not use drugs that they should avoid them. But we do know that their effect on people who were trying to break a drug habit was exactly the opposite of what was intended. As we saw in the case at the beginning of this chapter, John, a former drug addict, saw a poster that pictured drug paraphernalia, and this sight provoked an urge to take the drug again. For this reason, such posters are no longer used in campaigns against drug addiction.

One of the ways in which craving has been investigated in laboratory animals is through the *reinstatement model* of drug seeking. Animals are first trained to make a response (for example, pressing a lever) that is reinforced by intravenous injections of a drug such as cocaine. Next, the response is extinguished by providing injections of a saline solution rather than the drug. Once the animal has stopped responding, the experimenter administers a "free" injection of the drug or presents a stimulus that has been associated with the drug. In response to these stimuli, the animals begin responding at the lever once more (Kalivas, Peters, and Knackstedt, 2006). Presumably, this kind of relapse (reinstatement of a previously extinguished response) is a good model for the craving that motivates drug-seeking behavior in a former addict. (See *Figure 18.2*.)

Not surprisingly, relapses produced by an unexpected dose of the addictive drug involve activation of the mesolimbic system of dopaminergic neurons. If either the NAC or the VTA of rats is temporarily inactivated by the infusion of an inhibitory drug, a free shot of cocaine fails to reinstate responding (Grimm and See, 2000; McFarland and Kalivas, 2001).

To understand the process of reinstatement (and the craving that underlies it), let's first consider what happens

FIGURE 18.2 ■ The Reinstatement Procedure, a Measure of Craving

The graph shows the acquisition of lever pressing for injections of an addictive drug during the self-administration phase and the extinction of lever pressing when the drug was no longer administered. A "free" shot of the drug or presentation of a cue associated with the drug during acquisition will reinstate responding.

(Adapted from Kalivas, P. W., Peters, J., and Knackstedt, L. *Molecular Interventions*, 2006, *6*, 339–344.)

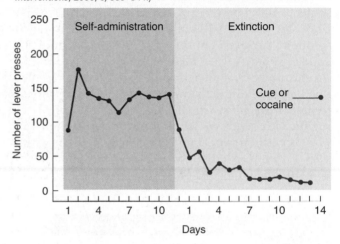

during extinction. As we saw in Chapter 11, extinction is a form of learning. An animal does not forget to make a particular response; it learns not to make it. The ventromedial prefrontal cortex (vmPFC) plays a critical role in this process. For example, we saw in Chapter 11 that lesions of the vmPFC impair the extinction of a conditioned emotional response, that stimulation of this region inhibits conditioned emotional responses, and that extinction training activates neurons located there.

Studies with rats indicate that different regions of the prefrontal cortex exert facilitatory and inhibitory effects on drug-related responding by means of excitatory and inhibitory connections with the brain's reinforcement system. These effects appear to be responsible for extinction and reinstatement. Peters, LaLumiere, and Kalivas (2008) found that stimulation of the vmPFC with an infusion of AMPA, a glutamate agonist, blocked reinstatement of responding normally produced by a free shot of cocaine or the presentation of a stimulus associated with cocaine reinforcement. That is, activation of the vmPFC inhibited responding. McFarland, Lapish, and Kalivas (2003) found that reinstatement of lever pressing for infusions of cocaine was abolished by injecting a GABA agonist into the dorsal anterior cingulate cortex (dACC), a region that has excitatory connections with the NAC. That is, inhibition of the dACC prevented the reinstatement of the response. These results indicate that the dACC plays a role in craving and the vmPFC plays a role in its suppression.

Functional-imaging studies in humans show that drugs of abuse (including cocaine, heroin, and nicotine) or cues associated with them activate several regions of the brain. The cortical regions most often activated include the ACC and orbitofrontal cortex (OFC) and, less often, the insula and dorsolateral prefrontal cortex (Goldstein and Volkow, 2002; Daglish et al., 2003; Brody et al., 2004; Myrick et al., 2004; Wang et al., 2007). For example, Myrick et al. (2004) found that a sip of alcohol and the sight of alcohol-related images increased craving in alcoholic subjects but not control subjects (social drinkers). The NAC, ACC, VTA, and insula were activated in alcoholic subjects, but only the ACC was activated in control subjects. (See *Figure 18.3.*)

Brody et al. (2004) found that craving for nicotine increased the activity of the ACC and that smokers treated with *bupropion* showed less activation of this region and reported less intense craving when they were presented with cigarette-related cues. (As we will see later in this chapter, bupropion has been found to be useful in helping people to quit smoking.) Activation of the prefrontal cortex appears to be related to craving for normal reinforcers as well as for addictive drugs.

As we saw in earlier chapters, the prefrontal cortex plays an important role in executive functions, including planning, evaluation of the consequences of actions, and inhibition of responses when conditions indicate that they would be inappropriate. For example, we saw in Chapter 11 that people with lesions of the medial prefrontal cortex have difficulty inhibiting responses and controlling their emotions. They are also more likely to engage in risky behavior. I'm sure you can see the similarity between this behavior and that of people addicted to drugs. The behavior of both groups of people is not inhibited by the long-term effects of particular actions but is dominated by immediate gratification, such as that provided by a drug.

Volkow et al. (1992) found that the activity of the medial prefrontal cortex of cocaine abusers was less active than that of normal subjects during abstinence. In addition, when addicts are performing tasks that normally activate the prefrontal cortex, their medial prefrontal cortex is less activated than that of healthy control subjects, and they perform more poorly on the tasks (Bolla et al., 2004; Garavan and Stout, 2005).

FIGURE 18.3 ■ Craving in Alcoholic Subjects

The scans show the activation of the nucleus accumbens (NAC), dorsal anterior cingulate cortex (dACC), anterior cingulate cortex (ACC), and ventral tegmental area (VTA) in alcoholic subjects and control subjects who were given a sip of alcohol and were shown alcohol-related images.

(From Myrick, H., Anton, R. F., Li, X., Henderson, S., Drobes, D., Voronin, K, and George, M. S. *Neuropsychopharmacology,* 2004, *29,* 393–402. By permission.)

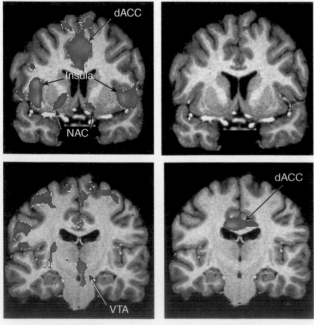

Alcoholics   Controls

FIGURE 18.4 ■ Cocaine Intake and the Medial Prefrontal Cortex

The graph shows the relative activation of the medial prefrontal cortex as a function of the amount of cocaine normally taken each week by cocaine abusers.

(Adapted from Bolla, K., et al. *Journal of Neuropsychiatry and Clinical Neuroscience,* 2004, *16,* 456–464.)

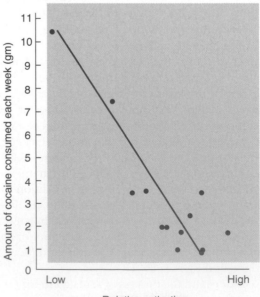

In fact, Bolla and her colleagues found that the amount of activation of the medial prefrontal cortex was inversely related to the amount of cocaine that cocaine abusers normally took each week: The lower the brain activity, the more cocaine the person took. (See *Figure 18.4.*)

People with a long history of drug abuse not only show the same deficits on tasks that involve the prefrontal cortex as do people with lesions of this region, they also show structural abnormalities of this region. For example, Franklin et al. (2002) reported an average decrease of 5–11 percent in the gray matter volume of the superior temporal cortex and various regions of the prefrontal cortex of chronic cocaine abusers. (See *Figure 18.5.*) Thompson et al. (2004) found 11 percent decreases in the gray matter volume of the cingulate cortex and limbic cortex of methamphetamine users. Of course, these results do not permit us to determine

whether abnormalities in the prefrontal cortex predispose people to become addicted or whether drug taking causes these abnormalities.

As we saw in Chapter 16, the negative and cognitive symptoms of schizophrenia appear to be a result of hypofrontality—decreased activity of the prefrontal cortex. These symptoms are very similar to those that accompany long-term drug abuse. In fact, studies have shown a high level of comorbidity of schizophrenia and substance abuse. (*Comorbidity* refers to the simultaneous presence of two or more disorders in the same person.) For example, up to half of all people with schizophrenia have a substance abuse disorder (alcohol or illicit drugs), and 70 to 90 percent are nicotine dependent (Brady and Sinha, 2005). In fact, in the United States, smokers with psychiatric disorders, who constitute approximately 7 percent of the population, consume 34 percent of all cigarettes (Dani and Harris, 2005). Mathalon et al. (2003) found that prefrontal gray matter volumes were 10.1 percent lower in alcoholic patients, 9.0 percent lower in schizophrenic patients, and 15.6 percent lower in patients with both disorders. (See *Figure 18.6.*)

Weiser et al. (2004) administered a smoking questionnaire to a random sample of adolescent military

---

FIGURE 18.5 ■ Cocaine Abuse and Volume of Cortical Gray Matter

The scans show regions of decreased gray matter volume relative to that of control subjects in the brains of chronic cocaine abusers.

(From Franklin, T. R., Acton, P. D., Maldjian, J. A., Gray, J. D., Croft, J. R., Dackis, C. A., O'Brien, C. P., and Childress, A. R. *Biological Psychiatry*, 2002, *51*, 134–142. By permission.)

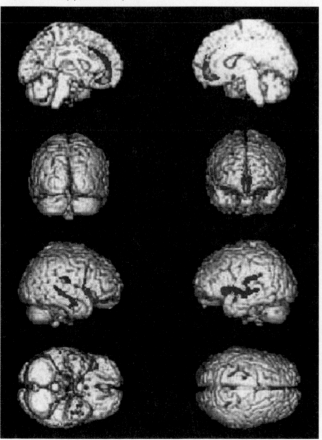

FIGURE 18.6 ■ Alcoholism, Schizophrenia, and Prefrontal Gray Matter

The graph shows the volume of gray matter in the prefrontal cortex of healthy controls, alcoholic patients, schizophrenic patients, and patients comorbid for both disorders.

(Adapted from Mathalon, D. H., Pfefferbaum, A., Lim, K. O., Rosenbloom, M. J., and Sullivan, E. V. *Archives of General Psychiatry*, 2003, *60*, 245–252.)

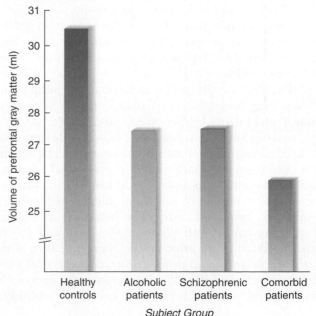

FIGURE 18.7 ■ Smoking and Schizophrenia

The graph shows the prevalence of schizophrenia during a 4- to 16-year follow-up period as a function of number of cigarettes smoked each day at age 18.

(Adapted from Weiser, M., Reichenberg, A., Grotto, I., Yasvitzky, R., Rabinowitz, J., Lubin, G., Nahon, D., Knobler, H. Y., and Davidson, M. *American Journal of Psychiatry*, 2004, *161*, 1219–1223.)

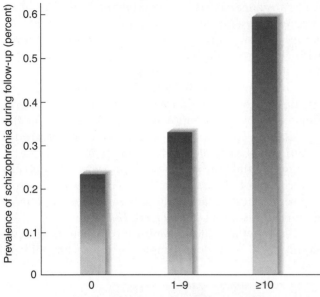

Number of cigarettes smoked per day at age 18

recruits each year. Over a 4- to 16-year follow-up period, they found that compared with nonsmokers, the prevalence of hospitalization for schizophrenia was 2.3 times higher in recruits who smoked at least ten cigarettes per day. (See **Figure 18.7.**) These results suggest that abnormalities in the prefrontal cortex may be a common factor in schizophrenia and substance abuse disorders. Again, I must note that research has not yet determined whether preexisting abnormalities increase the risk of these disorders or whether the disorders cause the abnormalities.

The role of the prefrontal cortex in judgment, risk taking, and control of inappropriate behaviors may explain why adolescents are much more vulnerable to drug addiction than are adults. Adolescence is a time of rapid and profound maturational change in the brain—particularly in the prefrontal cortex. Before these circuits reach their adult form, adolescents are more likely to display increased levels of impulsive, novelty-driven, risky behavior, including experimentation with alcohol, nicotine, and illicit drugs. Addiction in adults most often begins in adolescence or young adulthood. Approximately 50 percent of cases of addiction begin between the ages of 15 and 18, and very few begin after age 20. In addition,

early onset of drug-taking is associated with more severe addiction and a greater likelihood of multiple substance abuse (Chambers, Taylor, and Potenza, 2003). Presumably, the final development of neural circuits involved in behavioral control and judgment, along with the maturity that comes from increased experience, help people emerging from adolescence to resist the temptation to abuse drugs. In fact, Tarter et al. (2003) found that 10- to 12-year-old boys who scored the lowest on tests of behavioral inhibition were more likely to develop substance use disorder by age 19.

As we have just seen, the presence of drug-related stimuli can trigger craving and drug-seeking behavior. In addition, clinicians have long observed that stressful situations can cause former drug addicts to relapse. These effects have been observed in rats that had previously learned to self-administer cocaine or heroin. For example, Covington and Miczek (2001) paired naïve rats with rats that had been trained to become dominant. After being defeated by the dominant rats, the socially stressed rats became more sensitive to the effects of cocaine and showed bingeing—self-administration of larger amounts of the drug. Kosten, Miserendino, and Kehoe (2000) showed that stress that occurs early in life can have long-lasting effects. They stressed infant rats by isolating them from their mother and littermates for one hour per day for eight days. When these rats were given the opportunity to inject themselves with cocaine, they readily acquired the habit and took more of the drugs than did control rats that had not been stressed. (See **Figure 18.8.**)

FIGURE 18.8 ■ Social Stress and Cocaine Intake

The graph shows the cocaine intake of control rats and rats subjected to isolation stress early in life.

(Adapted from Kosten, T. A., Miserendino, M. J. D., and Kehoe, P. *Brain Research*, 2000, *875*, 44–50.)

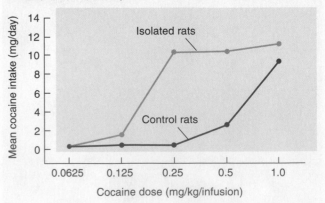

Interim*Summary*

Common Features of Addiction

Addictive drugs are those whose reinforcing effects are so potent that some people who are exposed to them are unable to go for very long without taking the drugs and whose lives become organized around taking them. Fortunately, most people who take drugs do not become addicted to them. Originally, most addictive drugs came from plants, which used them as a defense against insects or other animals that otherwise would eat them, but chemists have synthesized many other drugs that have even more potent effects. If a person regularly takes some addictive drugs (most notably, the opiates), the effects of the drug show tolerance, and the person must take increasing doses to achieve the same effect. If the person then stops taking the drug, withdrawal effects, opposite to the primary effects of the drug, will occur. However, withdrawal effects are not the cause of addiction; the abuse potential of a drug is related to its ability to reinforce drug-taking behavior.

Positive reinforcement occurs when a behavior is regularly followed by an appetitive stimulus—one that an organism will approach. Addictive drugs produce positive reinforcement; they reinforce drug-taking behavior. Laboratory animals will learn to make responses that result in the delivery of these drugs. The faster a drug produces its effects, the more quickly dependence will be established. All addictive drugs that produce positive reinforcement stimulate the release of dopamine in the NAC, a structure that plays an important role in reinforcement. Neural changes that begin in the VTA and NAC eventually involve the dorsal striatum, which plays a critical role in instrumental conditioning.

Negative reinforcement occurs when a behavior is followed by the reduction or termination of an aversive stimulus. If, because of a person's social situation or personality characteristics, he or she feels unhappy or anxious, a drug that reduces these feelings can reinforce drug-taking behavior by means of negative reinforcement. Also, the reduction of unpleasant withdrawal symptoms by a dose of the drug undoubtedly plays a role in maintaining drug addictions, but it is not the sole cause of craving.

Craving—the urge to take a drug to which one has become addicted—cannot be completely explained by withdrawal symptoms, because it can occur even after an addict has refrained from taking the drug for a long time. In laboratory animals a "free" shot of cocaine or presentation of stimuli previously associated with cocaine reinstates drug-seeking behavior. The vmPFC plays an inhibitory role in reinstatement, and the dACC plays a facilitatory role. Functional-imaging studies find that craving for addictive drugs and natural reinforcers such as appetizing food increases the activity of the ACC, OFC, insula, and dorsolateral prefrontal cortex. Long-term drug abuse is associated with decreased activity of the prefrontal cortex and is even with a decreased volume of prefrontal gray matter, which may impair people's judgment and ability to inhibit inappropriate responses, such as further drug taking. Schizophrenia is seen in a higher proportion of drug addicts than in the general population. The susceptibility of adolescents to the addictive potential of drugs may be associated with the relative immaturity of the prefrontal cortex. Stressful stimuli—even those that occur early in life—increase susceptibility to drug addiction.

COMMONLY ABUSED DRUGS

People have been known to abuse an enormous variety of drugs, including alcohol, barbiturates, opiates, tobacco, amphetamine, cocaine, cannabis, hallucinogens such as LSD, PCP, volatile solvents such as glues or even gasoline, ether, and nitrous oxide. The pleasure that children often derive from spinning themselves until they become dizzy may even be related to the effects of some of these drugs. Obviously, I cannot hope to discuss all these drugs in any depth and keep the chapter to a reasonable length, so I will restrict my discussion to the most important of them in terms of popularity and potential for addiction. Some drugs, such as caffeine, are both popular and addictive, but because they do not normally cause intoxication, impair health, or interfere with productivity, I will not discuss them here. (Chapter 4 did discuss the behavioral effects and site of action of caffeine.) I will also not discuss the wide variety of hallucinogenic drugs such as LSD or PCP. Although some people enjoy the mind-altering effects of LSD, many people simply find them frightening; and in any event, LSD use does not normally lead to addiction. PCP (phencyclidine) acts as an indirect antagonist at the NMDA receptor, which means that its effects overlap with those of alcohol. Rather than devoting space to this drug, I have chosen to say more about alcohol, which is abused far more than any of the hallucinogenic drugs. If you would like to learn more about drugs other than the ones I discuss here, I suggest you consult the suggested readings at the end of this chapter.

Opiates

Opium, derived from a sticky resin produced by the opium poppy, has been eaten and smoked for centuries. Opiate addiction has several high personal and social costs. First, because heroin, the most commonly abused opiate, is an illegal drug in most countries, an addict becomes, by definition, a criminal. Second, because of tolerance, a person must take increasing amounts of the drug to achieve a "high." The habit thus becomes more and more expensive, and the person often turns to crime to obtain enough money to support his or her habit. Third, an opiate addict often uses unsanitary needles; at present, a substantial percentage of people who inject illicit drugs have been exposed in this way to hepatitis or the AIDS virus. Fourth, if the addict is a pregnant woman, her infant will also become dependent on the drug, which easily crosses the placental barrier. The infant must be given opiates right after being born and then weaned off the drug with gradually decreasing doses. Fifth, the uncertainty about the strength of a given batch of heroin makes it possible for a user to receive an unusually large dose of the drug, with possibly fatal consequences.

Neural Basis of Reinforcing Effects

As we saw earlier, laboratory animals will self-administer opiates. When an opiate is administered systemically, it stimulates opiate receptors located on neurons in various parts of the brain and produces a variety of effects, including analgesia, hypothermia (lowering of body temperature), sedation, and reinforcement. Opiate receptors in the periaqueductal gray matter are primarily responsible for the analgesia, those in the preoptic area are responsible for the hypothermia, and those in the mesencephalic reticular formation are responsible for the sedation. As we shall see, opiate receptors in the ventral tegmental area and the NAC may play a role in the reinforcing effects of opiates.

As we saw in Chapter 4, there are three major types of opiate receptors: μ (mu), δ (delta), and κ (kappa). Evidence suggests that μ receptors and δ receptors are responsible for reinforcement and analgesia and that stimulation of κ receptors produces aversive effects. Evidence for the role of μ receptors comes from a study by Matthes et al. (1996), who performed a targeted mutation ("knockout") of the gene responsible for production of the μ opiate receptor in mice. These animals, when they grew up, were completely insensitive to the reinforcing or analgesic effects of morphine, and they showed no signs of withdrawal symptoms after having been given increasing doses of morphine for six days. (See *Figure 18.9*.)

FIGURE 18.9 ■ Effects of a Targeted Deletion of the μ Opiate Receptor

The graphs show a lack of responses to morphine in mice with targeted mutations of the μ opiate receptor: (a) latency to tail withdrawal from a hot object (a measure of analgesia), (b) wet-dog shakes (a prominent withdrawal symptom in rodents) after being withdrawn from long-term morphine administration, and (c) conditioned place preference for a chamber associated with an injection of morphine (a measure of reinforcement).

(Adapted from Matthes, H. W. D., Maldonado, R., Simonin, F., Valverde, O., Slowe, S., Kitchen, I., Befort, K., Dierich, A., Le Meur, M., Dolle, P., Tzavara, E., Hanoune, J., Roques, B. P., and Kieffer, B. L. *Nature,* 1996, *383,* 819–823.)

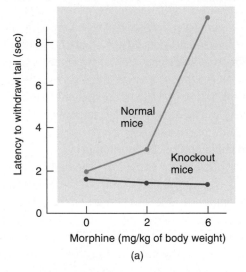

(a)

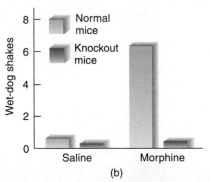

(b)

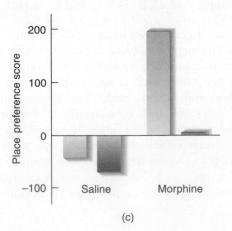

(c)

As we saw earlier, reinforcing stimuli cause the release of dopamine in the NAC. Injections of opiates are no exception to this general rule; Wise et al. (1995) found that the level of dopamine in the NAC increased by 150 to 300 percent while a rat was pressing a lever that delivered intravenous injections of heroin. Rats will also press a lever that delivers injections of an opiate directly into the ventral tegmental area (Devine and Wise, 1994) or the NAC (Goeders, Lane, and Smith, 1984). In other words, injections of opiates into both ends of the mesolimbic dopaminergic system are reinforcing.

As I mentioned earlier, a considerable amount of evidence suggests that endogenous opioids are involved in the behavioral effects of natural reinforcers. For example, Agmo et al. (1993) used a *conditioned place preference task* to measure the reinforcing effects of a drink of water for thirsty rats. Rats in the control group, which had previously been given an injection of a placebo, showed a clear preference for spending time in the chamber in which they were placed after drinking the water. Rats that were given an injection of **naloxone** (a drug that blocks opiate receptors) or *pimozide* (a drug that blocks dopamine receptors) showed no preference. Thus, the release of both dopamine and the endogenous opioids are essential for the reinforcing effects of a drink of water.

The release of endogenous opioids may even play a role in the reinforcing effects of some addictive drugs. Studies have shown that naloxone and other drugs that block opiate receptors reduce the reinforcing effects of alcohol in both humans and laboratory animals. Because the use of opiate blockers has recently been approved as a treatment for alcoholism, I will discuss relevant research later in this chapter.

Neural Basis of Tolerance and Withdrawal

Several studies have investigated the neural systems that are responsible for the development of tolerance and subsequent withdrawal effects of opiates. Maldonado et al. (1992) made rats physically dependent on morphine and then injected naloxone into various regions of the brain to determine whether the sudden blocking of opiate receptors would stimulate symptoms of withdrawal. This technique—administering an addictive drug for a prolonged interval and then blocking its effects with an antagonist—is referred to as **antagonist-precipitated withdrawal.** The investigators found that the most sensitive site was the locus coeruleus, followed by the periaqueductal gray matter. Injection of naloxone into the amygdala produced a weak withdrawal syndrome. Using a similar technique (first infusing morphine into various regions of the brain and then precipitating withdrawal by giving the animals an intraperitoneal injection of naloxone), Bozarth (1994) found that injections into the locus coeruleus and the periaqueductal gray matter produced withdrawal symptoms.

These studies suggest that opiate receptors in the locus coeruleus and periaqueductal gray matter are involved in withdrawal symptoms. A single dose of an opiate decreases the firing rate of neurons in the locus coeruleus, but if the drug is administered chronically, the firing rate will return to normal. Then, if an opiate antagonist is administered (to precipitate withdrawal symptoms), the firing rate of these neurons increases dramatically, which increases the release of norepinephrine in the regions that receive projections from this nucleus (Koob, 1996; Nestler, 1996). In addition, lesions of the locus coeruleus reduce the severity of antagonist-precipitated withdrawal symptoms (Maldonado and Koob, 1993). A microdialysis study by Aghajanian, Kogan, and Moghaddam (1994) found that antagonist-precipitated withdrawal caused an increase in the level of glutamate and aspartate, two excitatory amino acid neurotransmitters, in the locus coeruleus.

Stimulant Drugs: Cocaine and Amphetamine

Cocaine and amphetamine have similar behavioral effects, because both act as potent dopamine agonists. However, their sites of action are different. Cocaine binds with and deactivates the dopamine transporter proteins, thus blocking the reuptake of dopamine after it is released by the terminal buttons. Amphetamine also inhibits the reuptake of dopamine, but its most important effect is to directly stimulate the release of dopamine from terminal buttons. *Methamphetamine* is chemically related to amphetamine but is considerably more potent. Freebase cocaine ("crack"), a particularly potent form of the drug, is smoked and thus enters the blood supply of the lungs and reaches the brain very quickly. Because its effects are so potent and so rapid, it is probably the most effective reinforcer of all available drugs.

When people take cocaine, they become euphoric, active, and talkative. They say that they feel powerful and alert. Some of them become addicted to the drug, and obtaining it becomes an obsession to which they devote more and more time and money. Laboratory animals, which will quickly learn to self-administer cocaine intravenously, also act excited and show intense exploratory activity. After receiving the drug for a day or two, rats start showing stereotyped movements, such as

naloxone A drug that blocks μ opiate receptors; antagonizes the reinforcing and sedative effects of opiates.

antagonist-precipitated withdrawal Sudden withdrawal from long-term administration of a drug caused by cessation of the drug and administration of an antagonistic drug.

FIGURE 18.10 ■ Cumulative Fatalities of Rats Self-Administering Cocaine or Heroin

(Adapted from Bozarth, M. A., and Wise, R. A. *Journal of the American Medical Association*, 1985, *254*, 81–83. Reprinted with permission.)

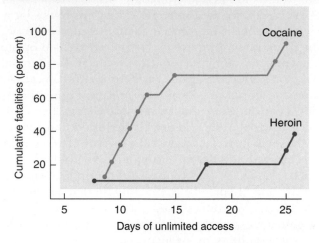

FIGURE 18.11 ■ Release of Dopamine in the Nucleus Accumbens

The graphs show dopamine concentration in the nucleus accumbens, measured by microdialysis, during self-administration of intravenous cocaine or amphetamine by rats.

(Adapted from Di Ciano, P., Coury, A., Depoortere, R. Y., Egilmez, Y., Lane, J. D., Emmett-Oglesby, M. W., Lepiane, F. G., Phillips, A. G., and Blaha, C. D. *Behavioural Pharmacology*, 1995, *6*, 311–322.)

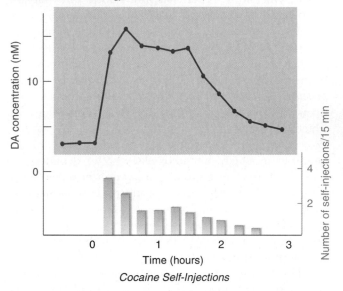

Cocaine Self-Injections

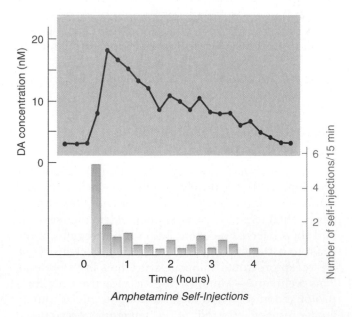

Amphetamine Self-Injections

grooming, head bobbing, and persistent locomotion (Geary, 1987). If rats or monkeys are given continuous access to a lever that permits them to self-administer cocaine, they often self-inject so much cocaine that they die. In fact, Bozarth and Wise (1985) found that rats that self-administered cocaine were almost three times more likely to die than were rats that self-administered heroin. (See *Figure 18.10*.)

As we have seen, the mesolimbic dopamine system plays an essential role in all forms of reinforcement, except perhaps for the reinforcement that is mediated by stimulation of opiate receptors. Many studies have shown that intravenous injections of cocaine and amphetamine increase the concentration of dopamine in the NAC, as measured by microdialysis. For example, Figure 18.11 shows data collected by Di Ciano et al. (1995) in a study with rats that learned to press a lever that delivered intravenous injections of cocaine or amphetamine. The colored bars at the base of the graphs indicate the animals' responses, and the line graphs indicate the level of dopamine in the NAC. (See *Figure 18.11*.)

One of the alarming effects of cocaine and amphetamine seen in people who abuse these drugs regularly is psychotic behavior: hallucinations, delusions of persecution, mood disturbances, and repetitive behaviors. These symptoms so closely resemble those of paranoid schizophrenia that even a trained mental health professional cannot distinguish them unless he or she knows about the person's history of drug abuse. However, these effects apparently disappear once people stop taking the drug. As we saw in Chapter 16, the fact that these symptoms are provoked by dopamine agonists and reduced by drugs that block dopamine receptors suggests that overactivity of dopaminergic synapses is responsible for the positive symptoms of schizophrenia.

Some evidence suggests that the use of stimulant drugs may have adverse long-term effects on the brain. For example, a PET study by McCann et al. (1998) discovered that prior abusers of methamphetamine showed a decrease in the numbers of dopamine transporters in the caudate nucleus and putamen, despite the fact that they had abstained from the drug for

FIGURE 18.12 ■ **Dopamine Transporters, Methamphetamine Abuse, and Parkinson's Disease**

The scans show concentrations of dopamine transporters from a control subject, a subject who had previously abused methamphetamine, and a subject with Parkinson's disease. Decreased concentrations of dopamine transporters indicate loss of dopaminergic terminals.

(From McCann, U. D., Wong, D. F., Yokoi, F., Villemagne, V., Dannls, R. F., and Ricaurte, G. A. *Journal of Neuroscience*, 1998, *18*, 8417–8422. By permission.)

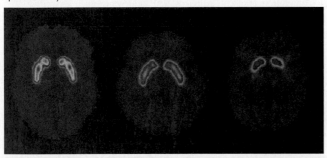

Control   Methamphetamine   Parkinson's disease

approximately three years. The decreased number of dopamine transporters suggests that the number of dopaminergic terminals in these regions is diminished. As the authors note, these people might have an increased risk of Parkinson's disease as they get older. (See *Figure 18.12*.) Studies with laboratory animals have also found that methamphetamine can damage terminals of serotonergic axons and trigger death through apoptosis in the cerebral cortex, striatum, and hippocampus (Cadet, Jayanthi, and Deng, 2003).

Nicotine

Nicotine might seem rather tame in comparison to opiates, cocaine, and amphetamine. Nevertheless, nicotine is an addictive drug, and it accounts for more deaths than the so-called "hard" drugs. The combination of nicotine and other substances in tobacco smoke is carcinogenic and leads to cancer of the lungs, mouth, throat, and esophagus. The World Health Organization (WHO, 1997) reported that one-third of the adult population of the world smokes and that smoking is one of the few causes of death that is rising in developing countries. The WHO estimates that 50 percent of the people who begin to smoke as adolescents and continue smoking throughout their lives will die from smoking-related diseases. Investigators estimate that by the year 2015, tobacco will be the largest single health problem worldwide, with 6.4 million deaths per year (Mathers and Loncar, 2006). In fact, tobacco use is the leading cause of preventable death in developed countries (Dani and Harris, 2005). In the

United States alone, tobacco addiction kills more than 430,000 people each year (Chou and Narasimhan, 2005). Smoking by pregnant women also has negative effects on the health of their fetuses—apparently worse than those of cocaine (Slotkin 1998). Unfortunately, approximately 25 percent of pregnant women in the United States expose their fetuses to nicotine.

The addictive potential of nicotine should not be underestimated; many people continue to smoke even when doing so causes serious health problems. For example, Sigmund Freud, whose theory of psychoanalysis stressed the importance of insight in changing one's behavior, was unable to stop smoking even after most of his jaw had been removed because of the cancer that this habit had caused (Brecher, 1972). He suffered severe pain and, as a physician, realized that he should have stopped smoking. He did not, and his cancer finally killed him.

Although executives of tobacco companies and others whose economic welfare is linked to the production and sale of tobacco products used to argue that smoking is a "habit" rather than an "addiction," evidence suggests that the behavior of people who regularly use tobacco is that of compulsive drug users. In a review of the literature, Stolerman and Jarvis (1995) note that smokers tend to smoke regularly or not at all; few can smoke just a little. Males smoke an average of seventeen cigarettes per day, while females smoke an average of fourteen. Nineteen out of twenty smokers smoke every day, and only 60 out of 3500 smokers questioned smoke fewer than five cigarettes per day. Forty percent of people continue to smoke after having had a laryngectomy (which is usually performed to treat throat cancer). Indeed, physicians have reported that patients with tubes inserted into their tracheas so that they can breathe will sometimes press a cigarette against the opening of these tubes and try to smoke (Hyman and Malenka, 2001). More than 50 percent of heart attack survivors continue to smoke, and about 50 percent of people continue to smoke after submitting to surgery for lung cancer. Of those who attempt to quit smoking by enrolling in a special program, 20 percent manage to abstain for one year. The record is much poorer for those who try to quit on their own: One-third manage to stop for one day, one-fourth abstain for one week, but only 4 percent manage to abstain for six months. It is difficult to reconcile these figures with the assertion that smoking is merely a "habit" that is pursued for the "pleasure" that it produces.

Ours is not the only species willing to self-administer nicotine; so will laboratory animals (Donny et al., 1995). Nicotine stimulates nicotinic acetylcholine receptors, of course. It also increases the activity of dopaminergic neurons of the mesolimbic system (Mereu et al., 1987) and causes dopamine to be released in the NAC (Damsma, Day, and Fibiger, 1989). Figure 18.13 shows

FIGURE 18.13 ■ Nicotine and Dopamine Release in the Nucleus Accumbens

The graph shows changes in dopamine concentration in the nucleus accumbens, measured by microdialysis, in response to injections of nicotine or saline. The arrows indicate the time of the injections.

(Adapted from Damsma, G., Day, J., and Fibiger, H. C. *European Journal of Pharmacology*, 1989, *168*, 363–368.)

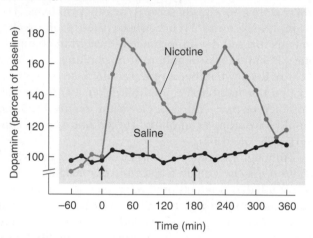

FIGURE 18.14 ■ Nicotine and the Ventral Tegmental Area

Changes in dopamine concentration in the nucleus accumbens, measured by microdialysis, in response to injections of intravenous nicotine (NIC) alone or in conjunction with an infusion of a nicotinic antagonist (MEC) in the ventral tegmental area (VTA) and the nucleus accumbens (NAC). Only the injection of MEC in the ventral tegmental area blocked the secretion of dopamine in the nucleus accumbens, which suggests that the drug acts there.

(Adapted from Nisell, M., Nomikos, G. G., and Svensson, T. H. *Synapse*, 1994, *16*, 36–44.)

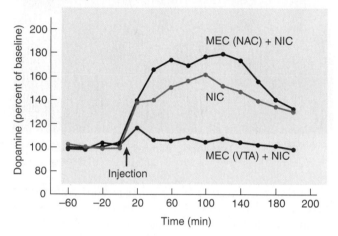

the effects of two injections of nicotine or saline on the extracellular dopamine level of the NAC, measured by microdialysis. (See *Figure 18.13*.)

Injection of a nicotinic agonist directly into the ventral tegmental area will reinforce a conditioned place preference (Museo and Wise, 1994). Conversely, injection of a nicotinic antagonist into the VTA will reduce the reinforcing effect of intravenous injections of nicotine (Corrigall, Coen, and Adamson, 1994). But although nicotinic receptors are found in both the VTA and the NAC (Swanson et al., 1987), Corrigall and his colleagues found that injections of a nicotinic antagonist in the NAC has no effect on reinforcement. Corroborating these findings, Nisell, Nomikos, and Svensson (1994) found that infusion of a nicotinic antagonist into the VTA—but not into the NAC—will prevent an intravenous injection of nicotine from triggering the release of dopamine in the NAC. (See *Figure 18.14*.) Thus, the reinforcing effect of nicotine appears to be caused by activation of nicotinic receptors in the ventral tegmental area.

Studies have found that the endogenous cannabinoids play a role in the reinforcing effects of nicotine. *Rimonabant*, a drug that blocks cannabinoid CB1 receptors, reduces nicotine self-administration and nicotine-seeking behavior in rats (Cohen, Kodas, and Griebel, 2005), apparently by reducing the release of dopamine in the NAC (De Vries and Schoffelmeer, 2005). By blocking CB1 receptors, rimonabant decreases the reinforcing effects of nicotine. As we will see later in this chapter, rimonabant has been used to help prevent relapse in people who are trying to quit smoking.

The nicotinic ACh receptor exists in three states. When a burst of ACh is released by an acetylcholinergic terminal button, the receptors open briefly, permitting the entry of calcium. (Most nicotinic receptors serve as heteroreceptors on terminal buttons that release another neurotransmitter. The entry of calcium stimulates the release of that neurotransmitter.) Within a few milliseconds the enzyme AChE has destroyed the acetylcholine, and the receptors either close again or enter a desensitized state, during which they bind with, but do not react to, ACh. Normally, few nicotinic receptors enter the desensitized state. However, when a person smokes, the level of nicotine in the brain rises slowly and stays steady for a prolonged period because it is not destroyed by AChE. At first, nicotinic receptors are activated, but the sustained low levels of the drug convert many nicotinic receptors to the desensitized state. Thus, nicotine has dual effects on nicotinic receptors: activation and then desensitization. In addition, probably in response to desensitization, the number of nicotinic receptors increases (Dani and De Biasi, 2001).

Most smokers report that their first cigarette in the morning brings the most pleasure, presumably because the period of abstinence during the night has allowed many of their nicotinic receptors to enter the closed state and become sensitized again. The first dose of nicotine in the morning activates these receptors and has a reinforcing effect. After that, a large proportion of the smoker's nicotinic receptors become desensitized again; as a consequence, most smokers say that they smoke less for pleasure than to relax and gain relief from nervousness and craving. If smokers abstain for a few weeks, the number of nicotinic receptors in their brains returns to normal. However, as the high rate of relapse indicates, craving continues, which means that other changes in the brain must have occurred.

Cessation of smoking after long-term use causes withdrawal symptoms, including anxiety, restlessness, insomnia, and inability to concentrate (Hughes et al., 1989). Like the withdrawal symptoms of other drugs, these symptoms may increase the likelihood of relapse, but they do not explain why people become addicted to the drug in the first place. As we saw earlier, the sight of smoking-related images produced greater activation of the medial prefrontal cortex and NAC in smokers than in nonsmokers.

Patient N. is a [38-year-old man who] started smoking at the age of 14. At the time of his stroke, he was smoking more than 40 unfiltered cigarettes per day and was enjoying smoking very much. . . [H]e used to experience frequent urges to smoke, especially upon waking, after eating, when he drank coffee or alcohol, and when he was around other people who were smoking. He often found it difficult to refrain from smoking in situations where it was inappropriate, e.g., at work or when he was sick and bedridden. He was aware of the health risks of smoking before his stroke but was not particularly concerned about those risks. Before his stroke, he had never tried to stop smoking, and he had had no intention of doing so. N. smoked his last cigarette on the evening before his stroke. When asked about his reason for quitting smoking, he stated simply, "I forgot that I was a smoker." When asked to elaborate, he said that he did not forget the fact that he was a smoker but rather that "my body forgot the urge to smoke." He felt no urge to smoke during his hospital stay, even though he had the opportunity to go outside to smoke. His wife was surprised by the fact that he did not want to smoke in the hospital, given the degree of his prior addiction. N. recalled how his roommate in the hospital would frequently go outside to smoke and that he was so disgusted by the smell upon his roommate's return that he asked to change rooms. He volunteered that smoking in his dreams, which used to be pleasurable before his stroke, was now disgusting. N. stated that, although he ultimately came to believe that his stroke was caused in some way by smoking, suffering a stroke was not the reason why he quit. In fact, he did not recall ever making any effort to stop smoking. Instead, it seemed to him that he had spontaneously lost all

interest in smoking. When asked whether his stroke might have destroyed some part of his brain . . . that made him want to smoke, he agreed that this was likely to have been the case. (Naqvi et al., 2007, p. 534)

As Naqvi et al. (2007) report, Mr. N. sustained a stroke that damaged his insula. In fact, several other patients with insular damage had the same experience. Naqvi and his colleagues identified nineteen cigarette smokers with damage to the insula and fifty smokers with brain damage that spared this region. Of the nineteen patients who had damage to the insula, twelve "quit smoking easily, immediately, without relapse, and without persistence of the urge to smoke" (Naqvi et al., 2007, p. 531). One patient with insula damage quit smoking but still reported feeling an urge to smoke. Figure 18.15 shows computer-generated images of brain damage that showed a statistically significant correlation with disruption of smoking. As you can see, the insula, which is colored red, showed the highest association with cessation of smoking. (See *Figure 18.15*.) This remarkable finding certainly deserves to be followed up.

One of the deterrents to cessation of smoking is the fact that overeating and weight gain frequently occur when people stop smoking. Jo, Wiedl, and Role (2005) have discovered the apparent cause of this phenomenon. As we saw in Chapter 12, eating and a reduction in metabolic rate are stimulated by the activity of two

FIGURE 18.15 ■ Damage to the Insula and Smoking Cessation

The diagrams show the regions of the brain (shown in red) where damage was most highly correlated with cessation of smoking.

(From Naqvi, N. H., Rudrauf, D., Damasio, H., and Bechara, A. *Science*, 2007, *315*, 531–534. By permission.)

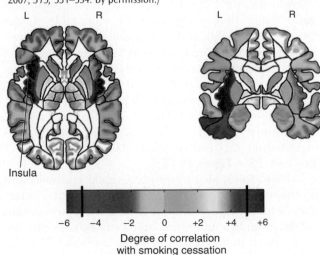

Degree of correlation with smoking cessation

different types of neurons whose cell bodies are located in the lateral hypothalamus. One of these sets of neurons secretes a peptide called melanocyte-concentrating hormone (MCH). Jo and his colleagues found that nicotinic receptors are located on the terminals of GABAergic neurons in the lateral hypothalamus that form synapses with MCH neurons. When nicotine activates these terminals, the release of GABA is increased, which inhibits MCH neurons, thus suppressing appetite. When people try to quit smoking, they are often discouraged by the fact that the absence of nicotine in their brains releases their MCH neurons from this inhibition, increasing their appetite.

Alcohol

Alcohol has enormous costs to society. A large percentage of deaths and injuries caused by motor vehicle accidents are related to alcohol use, and alcohol contributes to violence and aggression. Chronic alcoholics often lose their jobs, their homes, and their families; and many die of cirrhosis of the liver, exposure, or diseases caused by poor living conditions and abuse of their bodies. As we saw in Chapter 15, women who drink during pregnancy run the risk of giving birth to babies with fetal alcohol syndrome, which includes malformation of the head and the brain and accompanying mental retardation. In fact, alcohol consumption by pregnant women is one of the leading causes of mental retardation in the Western world today. Therefore, understanding the physiological and behavioral effects of this drug is an important issue.

Alcohol has the most serious effects on fetal development during the brain growth spurt period, which occurs during the last trimester of pregnancy and for several years after birth. Ikonomidou et al. (2000) found that exposure of the immature rat brain triggered widespread apoptosis, the death of cells caused by chemical signals that activate a genetic mechanism inside the cell. The investigators exposed immature rats to alcohol at different times during the period of brain growth and found that different regions were vulnerable to the effects of the alcohol at different times. Alcohol has two primary sites of action: It serves as an indirect agonist at $GABA_A$ receptors and as an indirect antagonist at NMDA receptors. Apparently, both of these actions trigger apoptosis. Ikonomidou and her colleagues found that administration of a $GABA_A$ agonist (phenobarbital, a barbiturate) or an NMDA antagonist (MK-801) to seven-day-old rats caused brain damage by means of apoptosis. (See *Figure 18.16*.)

At low doses, alcohol produces mild euphoria and has an *anxiolytic* effect—that is, it reduces the discomfort of anxiety. At higher doses, it produces incoordination and sedation. In studies with laboratory animals the anxiolytic effects manifest themselves as a release from the

FIGURE 18.16 ■ Early Exposure to Alcohol and Apoptosis

The photomicrographs of sections of rat brain show degenerating neurons (black spots). Exposure to alcohol during the period of rapid brain growth causes cell death by inducing apoptosis. These effects are mediated by the actions of alcohol as an NMDA antagonist and a $GABA_A$ agonist. MK-801, an NMDA antagonist, and phenobarbital, a $GABA_A$ agonist, also induce apoptosis.

(From Ikonomidou, C., Bittigau, P., Ishimaru, M. J., Wozniak, D. F., Koch, C., Genz, K., Price, M. T., Stefovska, V., Hörster, F., Tenkova, T., Dikranian, K., and Olney, J. W. *Science*, 2000, *287*, 1056–1060. By permission.)

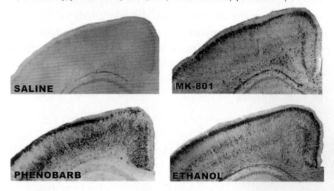

punishing effects of aversive stimuli. For example, if an animal is given electric shocks whenever it makes a particular response (say, one that obtains food or water), it will stop doing so. However, if it is then given some alcohol, it will begin making the response again (Koob et al., 1984). This phenomenon explains why people often do things they normally would not when they have had too much to drink; the alcohol removes the inhibitory effect of social controls on their behavior.

Alcohol produces both positive and negative reinforcement. The positive reinforcement manifests itself as mild euphoria. As we saw earlier, *negative* reinforcement is caused by the termination of an aversive stimulus. If a person feels anxious and uncomfortable, then an anxiolytic drug that relieves this discomfort provides at least a temporary escape from an unpleasant situation.

The negative reinforcement provided by the anxiolytic effect of alcohol is probably not enough to explain the drug's addictive potential. Other drugs, such as the benzodiazepines (tranquilizers such as Valium), are even more potent anxiolytics than alcohol, yet such drugs are abused much less often. It is probably the unique combination of stimulating and anxiolytic effects—of positive and negative reinforcement—that makes alcohol so difficult for some people to resist.

Alcohol, like other addictive drugs, increases the activity of the dopaminergic neurons of the mesolimbic system and increases the release of dopamine in the NAC as measured by microdialysis (Gessa et al., 1985; Imperato and Di Chiara, 1986). The release of dopamine appears to be related to the positive reinforcement that alcohol can produce. An injection of a dopamine antagonist directly into the NAC decreases alcohol intake in rats (Samson et al., 1993), as does the injection of a drug into the ventral tegmental area that decreases the activity of the dopaminergic neurons there (Hodge et al., 1993). In a double-blind study, Enggasser and de Wit (2001) found that haloperidol, an antischizophrenic drug that blocks dopamine D2 receptors, decreased the amount of alcohol that nonalcoholic subjects subsequently drank. Presumably, the drug reduced the reinforcing effect of the alcohol. In addition, the subjects who normally feel stimulated and euphoric after having a drink reported a reduction in these effects after taking haloperidol.

As I just mentioned, alcohol has two major sites of action in the nervous system, acting as an indirect antagonist at NMDA receptors and an indirect agonist at $GABA_A$ receptors (Chandler, Harris, and Crews, 1998). That is, alcohol enhances the action of GABA at $GABA_A$ receptors and interferes with the transmission of glutamate at NMDA receptors.

As we saw in Chapter 13, NMDA receptors are involved in long-term potentiation, a phenomenon that plays an important role in learning. Therefore, it will not surprise you to learn that alcohol, which antagonizes the action of glutamate at NMDA receptors, disrupts long-term potentiation and interferes with the spatial receptive fields of place cells in the hippocampus (Givens and McMahon, 1995; Matthews, Simson, and Best, 1996). Presumably, this effect at least partly accounts for the deleterious effects of alcohol on memory and other cognitive functions.

Withdrawal from long-term alcohol intake (like that of heroin, cocaine, amphetamine, and nicotine) decreases the activity of mesolimbic neurons and their release of dopamine in the NAC (Diana et al., 1993). If an indirect antagonist for NMDA receptors is then administered, dopamine secretion in the NAC recovers. The evidence suggests the following sequence of events: Some of the acute effects of a single dose of alcohol are caused by the antagonistic effect of the drug on NMDA receptors. Long-term suppression of NMDA receptors causes upregulation—a compensatory increase in the sensitivity of the receptors. Then, when alcohol intake suddenly ceases, the increased activity of NMDA receptors inhibits the activity of ventral tegmental neurons and the release of dopamine in the NAC.

Although the effects of heroin withdrawal have been exaggerated, those produced by barbiturate or alcohol withdrawal are serious and can even be fatal. The increased sensitivity of NMDA receptors as they rebound from the suppressive effect of alcohol can trigger seizures and convulsions. Convulsions caused by alcohol withdrawal are considered to be a medical emergency and are usually treated with benzodiazepines. Confirming the cause of these reactions, Liljequist (1991) found that seizures caused by alcohol withdrawal could be prevented by giving mice a drug that blocks NMDA receptors.

The second site of action of alcohol is the $GABA_A$ receptor. Alcohol binds with one of the many binding sites on this receptor and increases the effectiveness of GABA in opening the chloride channel and producing inhibitory postsynaptic potentials. Proctor et al. (1992) used microiontophoresis to record the activity of single neurons in the cerebral cortex of slices of rat brains. They found that the presence of alcohol significantly increased the postsynaptic response produced by the action of GABA at the $GABA_A$ receptor. As we saw in Chapter 4, the anxiolytic effect of the benzodiazepine tranquilizers is caused by their action as indirect agonists at the $GABA_A$ receptor. Because alcohol has this effect also, we can surmise that the anxiolytic effect of alcohol is a result of this action of the drug.

The sedative effect of alcohol also appears to be exerted at the $GABA_A$ receptor. Suzdak et al. (1986) discovered a drug (Ro15-4513) that reverses alcohol intoxication by blocking the alcohol binding site on this receptor. Figure 18.17 shows two rats that received injections of enough alcohol to make them pass out. The one facing us also received an injection of the alcohol antagonist and appears completely sober. (See *Figure 18.17.*)

This wonder drug has not been put on the market, nor is it likely to be. Although the behavioral effects of alcohol are mediated by their action on $GABA_A$ receptors and NMDA receptors, high doses of alcohol have other, potentially fatal effects on all cells of the body, including destabilization of cell membranes. Thus, people taking some of the alcohol antagonist could then go on to drink themselves to death without becoming drunk in the process. Drug companies naturally fear possible liability suits stemming from such occurrences.

FIGURE 18.17 ■ Effects of Ro15–4513, an Alcohol Antagonist

Both rats received an injection of alcohol, but the one facing us also received an injection of the alcohol antagonist.

(Photograph courtesy of Steven M. Paul, National Institute of Mental Health, Bethesda, Md.)

I mentioned earlier that opiate receptors appear to be involved in a reinforcement mechanism that does not directly involve dopaminergic neurons. The reinforcing effect of alcohol is at least partly caused by its ability to trigger the release of the endogenous opioids. Several studies have shown that the opiate receptor blockers such as naloxone or naltrexone block the reinforcing effects of alcohol in a variety of species, including rats, monkeys, and humans (Altschuler, Phillips, and Feinhandler, 1980; Davidson, Swift, and Fitz, 1996; Reid, 1996). In addition, endogenous opioids may play a role in craving in abstinent alcoholics. Heinz et al. (2005) found that one to three weeks of abstinence increased the number of μ opiate receptors in the NAC. The greater the number of receptors, the more intense the craving was. Presumably, the increased number of μ receptors increased the effects of endogenous opiates on the brain and served as a contributing factor to the craving for alcohol. (See *Figure 18.18.*)

Because naltrexone has become a useful adjunct to treatment of alcoholism, I will discuss this topic further in the last section of this chapter.

Cannabis

Another drug that people regularly self-administer—almost exclusively by smoking—is THC, the active ingredient in marijuana. As you learned in Chapter 4, the site of action of the endogenous cannabinoids in the brain is the CB1 receptor. The endogenous ligands for the CB1 receptor, anandamide and 2-AG, are lipids. Administration of a drug that blocks CB1 receptors abolishes the "high" produced by smoking marijuana (Huestis et al., 2001).

By the way, di Tomaso, Beltramo, and Piomelli (1996) discovered that chocolate contains three anandamidelike chemicals. Whether the existence of these chemicals is related to the great appeal that chocolate has for many people is not yet known. (I suppose that this is the place for a chocoholic joke.)

THC, like other drugs with abuse potential, has a stimulating effect on dopaminergic neurons. Chen et al. (1990) injected rats with low doses of THC and measured the release of dopamine in the NAC by means of microdialysis. Sure enough, they found that the injections caused the release of dopamine. (See *Figure 18.19.*) Chen et al. (1993) found that local injections of small amounts of THC into the ventral tegmental area had no effect on the release of dopamine in the NAC. However, injection of THC into the NAC *did* cause dopamine release there. Thus, the drug appears to act directly on dopaminergic terminal buttons—presumably on presynaptic heteroreceptors, where it increases the release of dopamine.

A variety of laboratory animals, including mice, rats, and monkeys, will self-administer drugs that stimulate

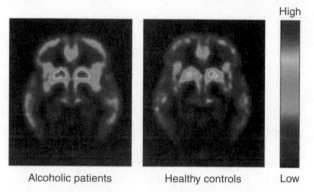

FIGURE 18.18 ■ Craving for Alcohol and μ Opiate Receptors

The PET scans show the presence of μ opiate receptors in the dorsal striatum of detoxified alcoholic patients and healthy control subjects. The graph shows the relative alcohol craving score as a function of relative numbers of μ opiate receptors.

(Scans and data points from Heinz, A., Reimold, M., Wrase, J., Hermann, D., Croissant, B., Mundle, G., Dohmen, B. M., Braus, D. H., Schumann, G., Machulla, H.-J., Bares, R., and Mann, K. *Archives of General Psychiatry,* 2005, *62,* 57–64. By permission.)

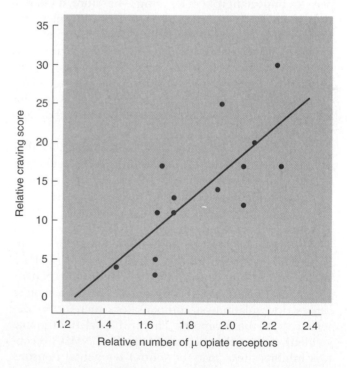

CB1 receptors, including THC (Maldonado and Rodriguez de Fonseca, 2002). A targeted mutation that blocks the production of CB1 receptors abolishes the reinforcing effect not only of cannabinoids, but also of morphine and heroin (Cossu et al., 2001). This mutation also decreases the reinforcing effects of alcohol and the acquisition of self-administration of cocaine (Houchi et al., 2005; Soria et al., 2005). In addition, as we saw in the previous section, rimonabant, a drug that

FIGURE 18.19 ■ THC and Dopamine Secretion in the Nucleus Accumbens

The graph shows changes in dopamine concentration in the nucleus accumbens, measured by microdialysis, in response to injections of THC or an inert placebo.

(Adapted from Chen, J., Paredes, W., Li, J., Smith, D., Lowinson, J., and Gardner, E. L. *Psychopharmacology*, 1990, *102*, 156–162.)

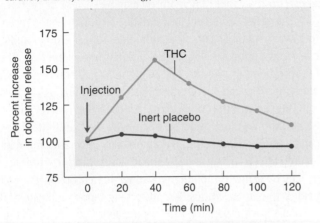

plays such an important role in memory. Pyramidal cells in the CA1 region of the hippocampus release endogenous cannabinoids, which provide a retrograde signal that inhibits GABAergic neurons that normally inhibit them. In this way the release of endogenous cannabinoids facilitates the activity of CA1 pyramidal cells and facilitates long-term potentiation (Kunos and Batkai, 2001).

We might expect that facilitating long-term potentiation in the hippocampus would enhance its memory functions. However, the reverse is true; Hampson and Deadwyler (2000) found that the effects of cannabinoids on a spatial memory task were similar to those produced by hippocampal lesions. Thus, excessive activation of CB1 receptors in field CA1 appears to interfere with normal functioning of the hippocampal formation.

Two articles (Moore et al., 2007; Murray et al., 2007) report a disturbing finding: The incidence of psychotic disorders such as schizophrenia is increased in cannabis users—especially those who have used cannabis frequently. Of course, a correlational study cannot prove the existence of a cause-and-effect relationship. It is possible that people who are more likely to develop psychotic symptoms are also more likely to use cannabis. However, statistical adjustments made by these studies suggest that a cause-and-effect relationship between cannabis use and psychosis cannot be ruled out. The authors Moore et al. (2007) conclude "that there is now sufficient evidence to warn young people that using cannabis could increase their risk of developing a psychotic illness later in life" (p. 319). This issue certainly deserves further study.

blocks CB_1 receptors, decreases the reinforcing effects of nicotine.

As we saw in Chapter 4, the hippocampus contains a large concentration of THC receptors. Marijuana is known to affect people's memory. Specifically, it impairs their ability to keep track of a particular topic; they frequently lose the thread of a conversation if they are momentarily distracted. Evidence indicates that the drug does so by disrupting the normal functions of the hippocampus, which

Interim*Summary*

Commonly Abused Drugs

Opiates produce analgesia, hypothermia, sedation, and reinforcement. Opiate receptors in the periaqueductal gray matter are responsible for the analgesia, those in the preoptic area for the hypothermia, those in the mesencephalic reticular formation for the sedation, and those in the ventral tegmental area and NAC at least partly for the reinforcement. A targeted mutation in mice indicates that μ opiate receptors are responsible for analgesia, reinforcement, and withdrawal symptoms. The release of the endogenous opioids may play a role in the reinforcing effects of natural stimuli such as water or even other addictive drugs such as alcohol.

The symptoms that are produced by antagonist-precipitated withdrawal from opiates can be elicited by injecting naloxone into the periaqueductal gray matter and the locus coeruleus, which implicates these structures in these symptoms.

Cocaine inhibits the reuptake of dopamine by terminal buttons, and amphetamine causes the dopamine transporters in terminal buttons to run in reverse, releasing dopamine from terminal buttons. Besides producing alertness, activation, and positive reinforcement, cocaine and amphetamine can produce psychotic symptoms that resemble those of paranoid schizophrenia. The reinforcing effects of cocaine and amphetamine are mediated by an increase in dopamine in the NAC. Chronic methamphetamine abuse is associated with reduced numbers of dopaminergic axons and terminals in the striatum (revealed as a decrease in the numbers of dopamine transporters located there), which may be involved in the development of hypofrontality.

The status of nicotine as a strongly addictive drug (for both humans and laboratory animals) was long ignored, primarily because it does not cause intoxication and because the ready availability of cigarettes and other tobacco products does not make it necessary for addicts to engage in illegal activities. However, the craving for nicotine is extremely motivating. Nicotine stimulates the release of mesolimbic dopaminergic neurons, and injection of nicotine into the ventral tegmental area is reinforcing. Cannabinoid CB_1 receptors are involved in the reinforcing effect of nicotine as well. Nicotine from smoking excites nicotinic acetylcholine receptors but also desensitizes them, which leads to unpleasant withdrawal effects. The activation of nicotinic receptors on presynaptic terminal buttons in the ventral tegmental area also produced long-term potentiation. Damage to the insula is associated with cessation of smoking, which suggests that this region plays a role in the maintenance of cigarette addiction. Nicotine stimulation of the release of GABA in the lateral hypothalamus decreases the activity of MCH neurons and reduces food intake, which explains why cessation of smoking often leads to weight gain.

Exposure to alcohol during the period of rapid brain development has devastating effects and is the leading cause of mental retardation. This exposure causes neural destruction through apoptosis. Drinking during adolescence can predispose rats to high alcohol intake later in life. Alcohol and barbiturates have similar effects. Alcohol has positively reinforcing effects and, through its anxiolytic action, has negatively reinforcing effects as well. It serves as an indirect antagonist at NMDA receptors and an indirect agonist at $GABA_A$ receptors. It stimulates the release of dopamine in the NAC. Withdrawal from long-term alcohol abuse can lead to seizures, an effect that seems to be caused by compensatory upregulation of NMDA receptors. Release of the endogenous opioids also plays a role in the reinforcing effects of alcohol. Increases in the numbers of μ opiate receptors during abstinence from alcohol may intensify craving.

The active ingredient in cannabis, THC, stimulates receptors whose natural ligand is anandamide. THC, like other addictive drugs, stimulates the release of dopamine in the NAC. The CB_1 receptor is responsible for the physiological and behavioral effects of THC and the endogenous cannabinoids. A targeted mutation against the CB_1 receptor reduces the reinforcing effect of alcohol, cocaine, and the opiates as well as that of the cannabinoids. Blocking CB_1 receptors also decreases the reinforcing effects of nicotine. Cannabinoids produce memory deficits by acting on inhibitory GABAergic neurons in the CA1 field of the hippocampus. Two disturbing reports indicate that cannabis use is associated with the incidence of schizophrenia.

Thought Questions

1. Although executives of tobacco companies used to insist that cigarettes were not addictive and asserted that people smoked simply because of the pleasure the act gave them, research indicates that nicotine is indeed a potent addictive drug. Why do you think it took so long to recognize this fact?

2. In most countries, alcohol is legal and marijuana is not. In your opinion, why? What criteria would you use to decide whether a newly discovered drug should be legal or illegal? Danger to health? Effects on fetal development? Effects on behavior? Potential for dependence? If you applied these criteria to various substances in current use, would you have to change the legal status of any of them?

HEREDITY AND DRUG ABUSE

Not everyone is equally likely to become addicted to a drug. Many people manage to drink alcohol moderately, and most users of potent drugs such as cocaine and heroin use them "recreationally" without becoming dependent on them. Evidence indicates that both genetic and environmental factors play a role in determining a person's likelihood of consuming drugs and of becoming dependent on them. In addition, there are both general factors (likelihood of taking and becoming addicted to any of a number of drugs) and specific factors (likelihood of taking and becoming addicted to a particular drug).

Tsuang et al. (1998) studied 3372 male twin pairs to estimate the genetic contributions to drug abuse. They found strong general genetic and environmental factors: Abusing any category of drug was associated with abusing drugs in all other categories: sedatives, stimulants, opiates, marijuana, and psychedelics. Abuse of marijuana was especially influenced by family environmental factors. Abuse of every category except psychedelics was influenced by genetic factors peculiar to that category. Abuse of heroin had a particularly strong unique genetic factor. Another study

FIGURE 18.20 ■ Heritability (h²) of Addiction to Specific Addictive Agents

(Adapted from Goldman, D., Oroszi, G., and Ducci, F. *Nature Reviews Genetics,* 2005, 6, 521–532.)

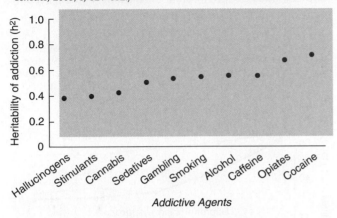

of male twin pairs (Kendler et al., 2003) found a strong common genetic factor for the use of all categories of drugs and found in addition that shared environmental factors had a stronger effect on use than on abuse. In other words, environment plays a strong role in influencing a person to try a drug and perhaps continue to use it recreationally, but genetics plays a stronger role in determining whether the person becomes addicted.

Goldman, Oroszi, and Ducci (2005) reviewed twin studies that attempted to measure the heritability of various classes of addictive disorders. Heritability (h²) is the percentage of variability in a particular population that can be attributed to genetic variability. The average value of h² ranged from approximately 0.4 for hallucinogenic drugs to just over 0.7 for cocaine. As you will see in Figure 18.20, the authors included addiction to gambling, which is not a drug. (See *Figure 18.20.*)

The genetic basis of addiction to alcohol has received more attention than addiction to other drugs. Alcohol consumption is not distributed equally across the population; in the United States, 10 percent of the people drink 50 percent of the alcohol (Heckler, 1983). Many twin studies and adoption studies confirm that the primary reason for this disparity is genetic.

A susceptibility to alcoholism could conceivably be caused by differences in the ability to digest or metabolize alcohol or by differences in the structure or biochemistry of the brain. There is evidence that variability in the gene responsible for the production of alcohol dehydrogenase, an enzyme involved in metabolism of alcohol, plays a role in susceptibility to alcoholism. A particular variant of this gene, which is especially prevalent in eastern Asia, is responsible for a reaction to alcohol intake that most people find aversive and that discourages further drinking (Goldman, Oroszi, and Ducci, 2005). However, most investigators believe that differences in brain physiology—for example, those that control sensitivity to the reinforcing effects of drugs or sensitivity to various environmental stressors—are more likely to play a role. For example, increased sensitivity to environmental stressors might encourage the use of alcohol as a means to reduce the stress-related anxiety.

Investigators have also focused on the possibility that susceptibility to addiction may involve differences in functions of specific neurotransmitter systems. For example, variations in the genes involved in the μ opiate receptor, the GABA$_A$ receptor, and the M$_2$ muscarinic acetylcholine receptor have been reported to be associated with the likelihood of alcohol dependence (Edenberg et al., 2004; Wang et al., 2004; Bart et al., 2005).

Interim*Summary*

Heredity and Drug Abuse

Most people who are exposed to addictive drugs—even drugs with a high abuse potential—do not become addicts. Evidence suggests that the likelihood of addiction, especially to alcohol and nicotine, is strongly affected by heredity. Drug taking and addiction are affected by general hereditary and environmental factors that apply to all drugs and specific factors that apply to particular drugs. A better understanding of the physiological basis of reinforcement and punishment will help us to understand the effects of heredity on susceptibility to addiction. Some individual genes have been shown to affect abuse of particular drugs. For example, variations in the genes for alcohol dehydrogenase, the μ opiate receptor, the GABA$_A$ receptor, and the M$_2$ acetylcholine receptor play a role in susceptibility to alcoholism.

THERAPY FOR DRUG ABUSE

There are many reasons for engaging in research on the physiology of drug abuse, including an academic interest in the nature of reinforcement and the pharmacology of psychoactive drugs. But most researchers entertain the hope that the results of their research will contribute to the development of ways to treat and—better yet—prevent drug abuse in members of our own species. As you well know, the incidence of drug abuse is far too high, so obviously, research has not yet solved the problem. However, real progress has been made.

The most common treatment for opiate addiction is methadone maintenance. Methadone is a potent opiate, just like morphine or heroin. If it were available in a form suitable for injection, it would be abused. (In fact, methadone clinics must control their stock of methadone carefully to prevent it from being stolen and sold to opiate abusers.) Methadone maintenance programs administer the drug to their patients in the form of a liquid, which they must drink in the presence of the personnel supervising this procedure. Because the oral route of administration increases the opiate level in the brain slowly, the drug does not produce a high, the way an injection of heroin will. In addition, because methadone is long-lasting, the patient's opiate receptors remain occupied for a long time, which means that an injection of heroin has little effect. Of course, a very large dose of heroin will displace methadone from opiate receptors and produce a "rush," so the method is not foolproof.

A newer drug, *buprenorphine*, shows promise of being an even better therapeutic agent for opiate addiction than methadone (Vocci, Acri, and Elkashef, 2005). Buprenorphine is a partial agonist for the μ opiate receptor. (You will recall from Chapter 16 that a partial agonist is a drug that has a high affinity for a particular receptor but activates that receptor less than the normal ligand does. This action reduces the effects of a receptor ligand in regions of high concentration and increases it in regions of low concentration, as shown in Figure 16.15.) Buprenorphine blocks the effects of opiates and itself produces only a weak opiate effect. Unlike methadone, it has little value on the illicit drug market. A randomized placebo-controlled trial compared the effectiveness of buprenorphine and buprenorphine plus naloxone in recovering opiate addicts (Fudala et al., 2003). People in the two drug-treatment groups reported less craving than those in the control group. The proportion of people who continued to be abstinent was 17.8 percent for people treated with buprenorphine, 20.7 percent for people treated with the combination of the two drugs, and only 5.8 percent for people receiving a placebo. (See *Figure 18.21.*) After one month, all subjects were given buprenorphine plus

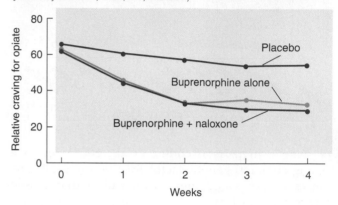

FIGURE 18.21 ■ Buprenorphine as a Treatment for Opiate Addiction

The graph shows the effects of treatment with buprenorphine, buprenorphine + naloxone, and a placebo on opiate craving in recovering opiate addicts.

(Adapted from Fudala, P. J., Bridge, T. P., Herbert, S., Williford, W. O., Chiang, C. N., Jonbes, K., Collins, J., Raisch, D., Casadonte, P., Goldsmith, R. J., Ling, W., Malkerneker, U., McNicholas, L., Renner, J., Stine, S., Tusel, D., and the Buprenorphine/Naloxone Collaborative Study Group. *New England Journal of Medicine,* 2003, *349,* 949–958.)

naloxone for eleven months. The percentage of people who abstained (indicated by the absence of opiates in urine samples) ranged from 35.2 to 67.4 percent at various times during the eleven-month period.

A major advantage of buprenorphine, besides its efficacy, is the fact that it can be use in office-based treatment. The addition of a small dose of naloxone ensures that the combination drug has no abuse potential—and will, in fact, cause withdrawal symptoms if it is taken by an addict who is currently taking an opiate.

As we saw, opiate receptor blockers such as naloxone and naltrexone interfere with the action of opiates. Emergency rooms always have one of these drugs available to rescue patients who have taken an overdose of heroin, and many lives have been saved by these means. But although an opiate antagonist will block the effects of heroin, the research reviewed earlier in this chapter suggests that it should *increase* the *craving* for heroin.

As we saw earlier, the reinforcing effects of cocaine and amphetamine are primarily a result of the sharply increased levels of dopamine these drugs produce in the NAC. Drugs that block dopamine receptors certainly block the reinforcing effects of cocaine and amphetamine, but they also produce dysphoria and anhedonia. People will not tolerate the unpleasant feelings these drugs produce, so they are not useful treatments for cocaine and amphetamine abuse. Drugs that *stimulate* dopamine receptors can reduce a person's dependence on cocaine or amphetamine, but these drugs are just as addictive as the drugs they replace and have the same deleterious effects on health.

An interesting approach to cocaine addiction is suggested by a study by Carrera et al. (1995), who conjugated cocaine to a foreign protein and managed to stimulate rats' immune systems to develop antibodies to cocaine. The antibodies bound with molecules of cocaine and prevented them from crossing the blood–brain barrier. As a consequence, these "cocaine-immunized" rats were less sensitive to the activating effects of cocaine, and brain levels of cocaine in these animals were lower after an injection of cocaine. Since this study was carried out, animal studies with vaccines against cocaine, heroin, methamphetamine, and nicotine have been carried out, and several human clinical trials with vaccines for cocaine and nicotine have taken place (Kosten and Owens, 2005; Cornuz et al., 2008). The results of these animal studies and human trials are promising, and more extensive human trials are in progress. Theoretically, at least, treatment of addictions with immunotherapy should interfere only with the action of an abused drug and not with the normal operations of people's reinforcement mechanisms. Thus, the treatment should not decrease their ability to experience normal pleasure.

Yet another approach to cocaine addiction is being investigated. Dewey et al. (1997) discovered that a GABA agonist, gamma-vinyl GABA (GVG), decreased the amount of dopamine released in the NAC after injecting a rat with cocaine. This finding suggested that GVG might reduce the reinforcing effects of cocaine as well. Dewey et al. (1998) found that it did. They pretreated baboons with GVG and found that the animals no longer learned a conditioned place preference for cocaine. GVG is not an addictive drug, and it has been used for treatment of seizure disorders in both adults and children, so it appears to be a safe medication. It will be interesting to see whether GVG finds a place in the treatment of cocaine addiction. An open clinical trial of the use of GVG to treat cocaine and methamphetamine addiction (a preliminary study performed without a placebo control) indicates that the drug has promise and does not appear to produce obvious side effects (Brodie, 2005).

A treatment similar to methadone maintenance has been used as an adjunct to treatment for nicotine addiction. For several years, chewing gum containing nicotine has been available by prescription, and more recently, transdermal patches that release nicotine through the skin have been marketed. Both methods maintain a sufficiently high level of nicotine in the brain to decrease a person's craving for nicotine. Once the habit of smoking has subsided, the dose of nicotine can be decreased to wean the person from the drug. Carefully controlled studies have shown that nicotine maintenance therapy, and not administration of a placebo, is useful in treatment for nicotine dependence (Stolerman and Jarvis, 1995). However, nicotine maintenance therapy is most effective if it is part of a counseling program.

One of the limitations of treating a smoking addiction with nicotine maintenance is that this procedure does not provide an important non-nicotine component of smoking: the sensations produced by the action of cigarette smoke on the airways. As we saw earlier in this chapter, stimuli associated with the administration of addictive drugs play an important role in sustaining an addictive habit. Smokers who rate the pleasurability of puffs of normal and denicotinized cigarettes within 7 seconds, which is less time than it takes for nicotine to leave the lungs, enter the blood, and reach the brain, reported that puffing denicotinized cigarettes produced equally strong feelings of euphoria and satisfaction and reductions in the urge to smoke. Furthermore, blocking the sensations of cigarette smoke on the airways by inhaling a local anesthetic diminishes smoking satisfaction. Denicotinized cigarettes are not a completely adequate substitute for normal cigarettes, because nicotine itself, not just the other components of smoke, makes an important contribution to the sensations felt in the airways. In fact, trimethaphan, a drug that blocks nicotinic receptors but does not cross the blood-brain barrier, decreases the sensory effects of smoking and reduces satisfaction. Because trimethaphan does not interfere with the effects of nicotine on the brain, this finding indicates that the central effects of nicotine are not sufficient by themselves to maintain an addiction to nicotine. Instead, the combination of an immediate cue from the sensory effects of cigarette smoke on the airways and a more delayed, and more continuous, effect of nicotine on the brain serves to make smoking so addictive (Naqvi and Bechara, 2005; Rose, 2006).

As we saw earlier in this chapter, studies with laboratory animals have found that the endogenous cannabinoids are involved in the reinforcing effects of nicotine as well as those of marijuana. A recent clinical trial reported that rimonabant, a drug that blocks CB_1 receptors, was effective in helping smokers to quit their habit (Henningfield et al., 2005). One significant benefit of the drug was a decrease in the weight gain that typically accompanies cessation of smoking and often discourages smokers who are trying to quit. As we saw in Chapter 12, the endocannabinoids stimulate eating, apparently by increasing the release of MCH and orexin. Blocking CB_1 receptors abolishes this effect and helps to counteract the effects of withdrawal from nicotine on these neurons.

Another drug, *bupropion*, is an antidepressant drug that serves as a catecholamine reuptake inhibitor. Bupropion has been approved for use in several countries for treating nicotine addiction. As we saw earlier in this chapter, Brody et al. (2004) found that smokers treated with bupropion showed less activation of the medial prefrontal cortex and reported less intense craving when they were presented with cigarette-related cues.

FIGURE 18.22 ■ Varenicline as a Treatment for Smoking

The graph shows the percentage of smokers treated with varenicline, bupropion, or placebo who abstained from cigarette smoking.

(Adapted from Nides, M., Oncken, C., Gonzales, D., Rennard, S., et al. *Archives of Internal Medicine*, 2006, *166*, 1561–1568.)

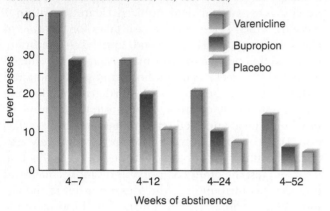

naltrexone increased the latency to take the first sip and to take a second drink and that the blood alcohol levels of the naltrexone-treated participants were lower at the end of the session. In general, the people who had taken naltrexone found that their drinks did not taste very good—in fact, some of them asked for a different drink after taking the first sip.

These results are consistent with reports of the effectiveness of naltrexone as an adjunct to programs designed to treat alcohol abuse. For example, O'Brien, Volpicelli, and Volpicelli (1996) reported the results of

FIGURE 18.23 ■ Naltrexone as a Treatment for Alcoholism

The graphs show mean craving score and proportion of patients who abstained from drinking while receiving naltrexone or a placebo.

(Adapted from O'Brien, C. P., Volpicelli, L. A., and Volpicelli, J. R. *Alcohol*, 1996, *13*, 35–39.)

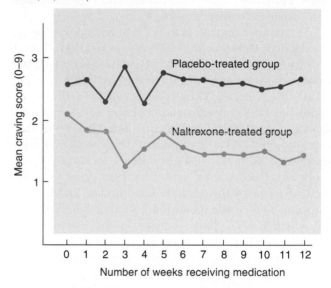

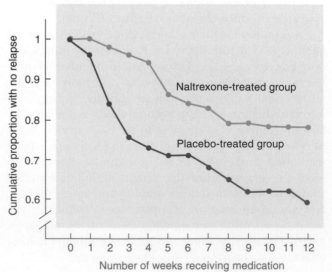

Yet another drug, *varenicline*, has been approved for therapeutic use to treat nicotine addiction. Varenicline serves a partial agonist for the nicotinic receptor, just as buprenorphine serves as a partial agonist for the μ opiate receptor. As a partial nicotinic agonist, varenicline maintains a moderate level of activation of nicotinic receptors but prevents high levels of nicotine from providing excessive levels of stimulation. Figure 18.22 shows the effects of treatment with varenicline and bupropion on rates of continuous abstinence rates of smokers enrolled in a randomized, double-blind, placebo control study (Nides et al., 2006). By the end of the fifty-two-week treatment program, 14.4 percent of the smokers treated with varenicline were still abstinent, compared with 6.3 percent and 4.9 percent of the smokers who received bupropion and placebo, respectively. (See *Figure 18.22.*)

As I mentioned earlier, several studies have shown that opiate antagonists decrease the reinforcing value of alcohol in a variety of species, including our own. This finding suggests that the reinforcing effect of alcohol—at least in part—is produced by the secretion of endogenous opioids and the activation of opiate receptors in the brain. A study by Davidson, Swift, and Fitz (1996) clearly illustrates this effect. The investigators arranged a double-blind, placebo-controlled study with sixteen college-age men and women to investigate the effects of naltrexone on social drinkers. None of the participants were alcohol abusers, and pregnancy tests ensured that the women were not pregnant. They gathered around a table in a local restaurant/bar for three two-hour drinking sessions, two weeks apart. For several days before the meeting, they swallowed capsules that contained either naltrexone or an inert placebo. The results showed that two long-term programs using naltrexone along with more traditional behavioral treatments. Both programs found that administration of naltrexone significantly increased the likelihood of success. As Figure 18.23 shows, naltrexone decreased the participants' craving for alcohol and increased the number of participants who managed to abstain from alcohol. (See *Figure 18.23.*) Currently, many treatment programs are using a sustained-release form of naltrexone to help treat alcoholism, and results with the drug have been encouraging (Kranzler, Modesto-Lowe, and Nuwayser, 1998). Naltrexone may even reduce craving for cigarettes (Vewers, Dhatt, and Tejwani, 1998).

One more drug has shown promise for treatment of alcoholism. As we saw earlier in this chapter, alcohol

serves as an indirect agonist at the $GABA_A$ receptor and an indirect antagonist at the NMDA receptor. *Acamprosate*, an NMDA-receptor antagonist that has been used in Europe to treat seizure disorders, was tested for its ability to stop seizure induced by withdrawal from alcohol. The researchers discovered that the drug had an unexpected benefit: Alcoholic patients who received the drug were less likely to start drinking again (Wickelgren, 1998). A double-blind study in Europe found that the combination of acamprosate and naltrexone was more effective than either drug alone in treating recovering alcoholics (Soyka and Chick, 2003). Acamprosate is being evaluated as a treatment for alcoholism in the United States (Buonopane and Petrakis, 2005).

Interim*Summary*

Therapy for Drug Abuse

Although drug abuse is difficult to treat, researchers have developed several useful therapies. Methadone maintenance replaces addiction to heroin by addiction to an opiate that does not produce euphoric effects when administered orally. Buprenorphine, a partial agonist for the μ opiate receptor, reduces craving for opiates. Because it is not of interest to opiate addicts (especially when it is combined with naltrexone), it can be administered by a physician at an office visit. The development of antibodies to cocaine and nicotine in humans and to several other drugs in rats holds out the possibility that people may someday be immunized against addictive drugs, preventing the entry of the drugs into the brain. GVG (gamma-vinyl GABA) shows promise for treating cocaine addiction. Nicotine-containing gum and transdermal patches help smokers to combat their addiction. However, sensations from the airways produced by the presence of cigarette smoke play an important role in addiction, and oral

and transdermal administration do not provide these sensations. Rimonabant, a CB_1 receptor antagonist, aids in smoking cessation and reduces the likelihood of weight gain. Bupropion, an antidepressant drug, has also been shown to help smokers stop their habit. Varenicline, a partial agonist for the nicotinic receptor, may be even more effective. The most effective pharmacological adjunct to treatment for alcoholism appears to be naltrexone, an opiate receptor blocker that reduces the drug's reinforcing effects. The addition of acamprosate, an NMDA-receptor antagonist, appears to facilitate the effectiveness of naltrexone.

A personal note: You are now at the end of the book (as you well know), and you have spent a considerable amount of time reading my words. While working on this book, I have tried to imagine myself talking to someone who is interested in learning something about the physiology of behavior. As I mentioned in the preface, writing is often a lonely activity, and the imaginary audience helped to keep me company. If you would like to turn this communication into a two-way conversation, write to me. My address is given at the end of the preface.

SUGGESTED READINGS

Chambers, R. A., Taylor, J. R., and Potenza, M. N. Developmental neurocircuitry of motivation in adolescence: A critical period of addiction vulnerability. *American Journal of Psychiatry*, 2003, *160*, 1041–1052.

Chao, J., and Nestler, E. J. Molecular neurobiology of drug addiction. *Annual Review of Medicine*, 2004, *55*, 113–132.

Ducci, F., and Goldman, D. Genetic approaches to addiction: Genes and alcohol. *Addiction*, 2008, *103*, 1414–1428.

Grilly, D. M. *Drugs and Human Behavior*, 5th ed. Boston: Allyn and Bacon, 2006.

Hyman, S. E., Malenka, R. C., and Nestler, E. J. Neural mechanisms of addiction: The role of reward-related learning and memory. *Annual Review of Neuroscience*, 2006, *29*, 565–598.

Kalivas, P. W., and O'Brien, C. Drug addiction as a pathology of staged neuroplasticity. *Neuropsychopharmacology*, 2008, *33*, 166–180.

Kauer, J. A., and Malenka, R. C. Synaptic plasticity and addiction. *Nature Reviews: Neuroscience*, 2007, *8*, 844–858.

Meyer, J. S., and Quenzer, L. F. *Psychopharmacology: Drugs, the Brain, and Behavior*. Sunderland, MA: Sinauer Associates, 2005.

Vocci, F. J., Acri, J., and Elkashef, A. Medication development for addictive disorders: The state of the science. *American Journal of Psychiatry*, 2005, *162*, 1432–1440.

ADDITIONAL RESOURCES

Visit www.mypsychkit.com for additional review and practice of the material covered in this chapter. Within MyPsychKit, you can take practice tests and receive a customized study plan to help you review. Dozens of animations, tutorials, and Web links are also available. You can even review using the interactive electronic version of this textbook. You will need to register for MyPsychKit. See www.mypsychkit.com for complete details.

References

Abelson, J. L., Curtis, G. C., Sagher, O., Albucher, R. C., et al. Deep brain stimulation for refractory obsessive-compulsive disorder. *Biological Psychiatry,* 2005, *57,* 510–516.

Abercrombie, H. C., Schaefer, S. M., Larson, C. L., Oakes, T. R., et al. Metabolic rate in the right amygdala predicts negative affect in depressed patients. *Neuroreport,* 1998, *9,* 3301–3307.

Abizaid, A., Liu, Z. W., Andrews, Z. B., Shanabrough, M., et al. Ghrelin modulates the activity and synaptic input organization of midbrain dopamine neurons while promoting appetite. *Journal of Clinical Investigation,* 2006, *116,* 3229–3239.

Abrevalo, E., Shanmugasundararaj, S., Wilkemeyer, M. F., Dou, X., et al. An alcohol binding site on the neural cell adhesion molecule L1. *Proceedings of the National Academy of Sciences, USA,* 2008, *105,* 371–375.

Adams, D. B., Gold, A. R., and Burt, A. D. Rise in female-initiated sexual activity at ovulation and its suppression by oral contraceptives. *New England Journal of Medicine,* 1978, *299,* 1145–1150.

Adams, R. B., and Kleck, R. E. Effects of direct and averted gaze on the perception of facially communicated emotion. *Emotion,* 2005, *5,* 3–11.

Adams, R. D. The anatomy of memory mechanisms in the human brain. In *The Pathology of Memory,* edited by G. A. Talland and N. C. Waugh. New York: Academic Press, 1969.

Adams, R. D., and Victor, M. *Principles of Neurology.* New York: McGraw-Hill, 1981.

Adey, W. R., Bors, E., and Porter, R. W. EEG sleep patterns after high cervical lesions in man. *Archives of Neurology,* 1968, *19,* 377–383.

Adler, C. M., Malhotra, A. K., Elman, I., Goldberg, T., et al. Comparison of ketamine-induced thought disorder in healthy volunteers and thought disorder in schizophrenia. *American Journal of Psychiatry,* 1999, *156,* 1646–1649.

Adolphs, R. Neural systems for recognizing emotion. *Current Opinion in Neurobiology,* 2002, *12,* 169–177.

Adolphs, R., Damasio, H., Tranel, D., Cooper, G., and Damasio, A. R. A Role for somatosensory cortices in the visual recognition of emotion as revealed by three-dimensional lesion mapping. *Journal of Neuroscience,* 2000, *20,* 2683–2690.

Adolphs, R., Gosselin, F., Buchanan, T. W., Tranel, D., et al. A mechanism for impaired fear recognition after amygdala damage. *Nature,* 2005, *433,* 68–72.

Adolphs, R., and Tranel, D. Intact recognition of emotional prosody following amygdala damage. *Neuropsychologia,* 1999, *37,* 1285–1292.

Adolphs, R., Tranel, D., Damasio, H., and Damasio, A. Impaired recognition of emotion in facial expressions following bilateral damage to the human amygdala. *Nature,* 1994, *372,* 669–672.

Adolphs, R., Tranel, D., Damasio, H., and Damasio, A. Fear and the human amygdala. *Journal of Neuroscience,* 1995, *15,* 5879–5891.

Adolphs, R., Tranel, D., Hamann, S., Young, A. W., Calder, A. J., et al. Recognition of facial emotion in nine individuals with bilateral amygdala damage. *Neuropsychologia,* 1999, *37,* 1111–1117.

Advokat, C., and Kutlesic, V. Pharmacotherapy of the eating disorders: A commentary. *Neuroscience and Biobehavioral Reviews,* 1995, *19,* 59–66.

Agarwal, N., Pacher, P., Tegeder, I., Amaya, F., Constantin, C. E., et al. Cannabinoids mediate analgesia largely via peripheral type 1 cannabinoid receptors in nociceptors. *Nature Neuroscience,* 2007, *10,* 870–879.

Aghajanian, G. K., Kogan, J. H., and Moghaddam, B. Opiate withdrawal increases glutamate and aspartate efflux in the locus coeruleus: An in vivo microdialysis study. *Brain Research,* 1994, *636,* 126–130.

Agmo, A., Federman, I., Navarro, V., Padua, M., and Velasquez, G. Reward and reinforcement produced by drinking water: Role of opioids and dopamine-receptor subtypes. *Pharmacology, Biochemistry and Behavior,* 1993, *46,* 183–194.

Aharon, L., Etcoff, N., Ariely, D., Chabris, C. F., et al. Beautiful faces have variable reward value: fMRI and behavioral evidence. *Neuron,* 2001, *32,* 537–551.

Alain, C., He, Y., and Grady, C. The contribution of the inferior parietal lobe to auditory spatial working memory. *Journal of Cognitive Neuroscience,* 2008, *20,* 285–295.

Albrecht, D. G. Analysis of visual form. Doctoral dissertation, University of California, Berkeley, 1978.

Alexander, G. M. An evolutionary perspective of sex-typed toy preferences: Pink, blue, and the brain. *Archives of Sexual Behavior,* 2003, *32,* 7–14.

Alexander, G. M., and Hines, M. Sex differences in response to children's toys in nonhuman primates (*Cercopithecus aethiops sabaeus*). *Evolution and Human Behavior,* 2002, *23,* 467–479.

Alexander, M. P., Fischer, R. S., and Friedman, R. Lesion localization in apractic agraphia. *Archives of Neurology,* 1992, *49,* 246–251.

Alirezaei, M., Watry, D. D., Flynn, C. F., Kiosses, W. B., et al. Human immunodeficiency virus-1/surface glycoprotein 120 induces apoptosis through RNA-activated protein kinase signaling in neurons. *Journal of Neuroscience,* 2007, *27,* 11047–11055.

Alkire, M. T., Haier, R. J., Fallon, J. H., and Cahill, L. Hippocampal, but not amygdala, activity at encoding correlates with long-term, free recall of nonemotional information. *Proceedings of the National Academy of Sciences, USA,* 1998, *95,* 14506–14510.

Allen, L. S., and Gorski, R. A. Sexual orientation and the size of the anterior commissure in the human brain. *Proceedings of the National Academy of Sciences, USA*, 1992, *89*, 7199–7202.

Allison, D. B., Kaprio, J., Korkeila, M., Koskenvuo, M., Neale, M. C., and Hayakawa, K. The heritability of body mass index among an international sample of monozygotic twins reared apart. *International Journal of Obesity*, 1996, *20*, 501–506.

Allison, D. W., Chervin, A. S, Gelfand, V. I., and Crain, A. M. Postsynaptic scaffolds of excitatory and inhibitory synapses in hippocampal neurons: Maintenance of core components independent of actin filaments and microtubules. *Journal of Neuroscience*, 2000, *20*, 4645–4654.

Allison, T., Puce, A., and McCarthy, G. Social perception from visual cues: Role of the STS region. *Trends in Cognitive Science*, 2000, *4*, 267–278.

Allman, J. M. *Evolving Brains*. New York: Scientific American Library, 1999.

Altschuler, H. L., Phillips, P. E., and Feinhandler, D. A. Alterations of ethanol self-administration by naltrexone. *Life Sciences*, 1980, *26*, 679–688.

Amaral, D. G. The amygdala, social behavior, and danger detection. *Annals of the New York Academy of Sciences*, 2003, *1000*, 337–347.

Amaral, D. G., Price, J. L., Pitkänen, A., and Carmichael, S. T. Anatomical organization of the primate amygdaloid complex. In *The Amygdala: Neurobiological Aspects of Emotion, Memory, and Mental Dysfunction*, edited by J. P. Aggleton. New York: Wiley-Liss, 1992.

American Psychiatric Association. *Diagnostic and Statistical Manual of Mental Disorders*, 4th ed. Washington, D.C.: American Psychiatric Association, 1994.

Anand, B. K., and Brobeck, J. R. Hypothalamic control of food intake in rats and cats. *Yale Journal of Biology and Medicine*, 1951, *24*, 123–140.

Ancoli-Israel, S., and Roth, T. Characteristics of insomnia in the United States: Results of the 1991 National Sleep Foundation survey. *Sleep*, 1999, *22*, S347–S353.

Anders, S., Birbaumer, N., Sadowski, B., Erb, M., Mader, I., Grodd, W., and Lotze, M. Parietal somatosensory association cortex mediates affective blindsight. *Nature Neuroscience*, 2004, *7*, 339–340.

Anderson, A. K., and Phelps, E. A. Intact recognition of vocal expressions of fear following bilateral lesions of the human amygdala. *Neuroreport*, 1998, *9*, 3607–3613.

Anderson, A. K., and Phelps, E. A. Expression without recognition: Contributions of the human amygdala to emotional communication. *Psychological Science*, 2000, *11*, 106–111.

Anderson, R. H., Fleming, D. E., Rhees, R. W., and Kinghorn, E. Relationships between sexual activity, plasma testosterone, and the volume of the sexually dimorphic nucleus of the preoptic area in prenatally stressed and non-stressed rats. *Brain Research*, 1986, *370*, 1–10.

Anderson, S., Barrash, J., Bechara, A., and Tranel, D. Impairments of emotion and real-world complex behavior following childhood- or adult-onset damage to ventromedial prefrontal cortex. *Journal of the International Neuropsychological Society*, 2006, *12*, 224–235.

Anderson, S. W., Bechara, A., Damasio H., Tranel, D., and Damasio, A. R. Impairment of social and moral behavior related to early damage in human prefrontal cortex. *Nature Neuroscience*, 1999, *2*, 1032–1037.

Andrews, N., Miller, E., Grant, A., Stowe, J., Osborne, V., and Taylor, B. Thimerosal exposure in infants and developmental disorders: A retrospective cohort study in the United Kingdom does not support a causal association. *Pediatrics*, 2004, *114*, 584–591.

Angrilli, A., Mauri, A., Palomba, D., Flor, H., Birbaumer, N., Sartori, G., and Dipaola, F. Startle reflex and emotion modulation impairment after a right amygdala lesion. *Brain*, 1996, *119*, 1991–2000.

Annese, J., Gazzaniga, M. S., and Toga, A. W. Localization of the human cortical visual area MT based on computer aided histological analysis. *Cerebral Cortex*, 2005, *15*, 1044–1053.

Anonymous. Effects of sexual activity on beard growth in man. *Nature*, 1970, *226*, 867–870.

Anzai, A., Peng, X., and Van Essen, D. C. Neurons in monkey visual area V2 encode combinations of orientations. *Nature Neuroscience*, 2007, *10*, 1313–1321.

Apkarian, A. V., Sosa, Y., Krauss, B. R., Thomas, P. S., et al. Chronic pain patients are impaired on an emotional decision-making task. *Pain*, 2004a, *108*, 129–136.

Apkarian, A. V., Sosa, Y., Sonty, S., Levy, R. M., et al. Chronic back pain is associated with decreased prefrontal and thalamic gray matter density. *Journal of Neuroscience*, 2004b, *24*, 10410–10415.

Arancio, O., Kandel, E. R., and Hawkins, R. D. Activity-dependent long-term enhancement of transmitter release by presynaptic 3',5'-cyclic GMP in cultured hippocampal neurons. *Nature*, 1995, *376*, 74–80.

Archer, J. Testosterone and aggression. *Journal of Offender Rehabilitation*, 1994, *5*, 3–25.

Arendt, J., Deacon, S., English, J., Hampton, S., and Morgan, L. Melatonin and adjustment to phase-shift. *Journal of Sleep Research*, 1995, *4*, 74–79.

Ariyasu, H., Takaya, K., Tagami, T., Ogawa, Y., et al. Stomach is a major source of circulating ghrelin, and feeding state determines plasma ghrelin-like immunoreactivity levels in humans. *Journal of Clinical Endocrinology and Metabolism*, 2001, *86*, 4753–4758.

Arnason, B. G. Immunologic therapy of multiple sclerosis. *Annual Review of Medicine*, 1999, *50*, 291–302.

Arnott, S. T., Binns, M. A., Grady, C. L., and Alain, C. Assessing the auditory dual-pathway model in humans. *Neuroimage*, 2004, *22*, 401–408.

Arnsten, A. F. T. Fundamentals of attention-deficit/hyperactivity disorder: Circuits and pathways. *Journal of Clinical Psychiatry*, 2006, *67* (Suppl. 8), 7–12.

Arnsten, A. F. T., and Li, B.-M. Neurobiology of executive functions: Catecholamine influences on prefrontal cortical functions. *Biological Psychiatry*, 2005, *57*, 1377–1384.

Aron, A. R., Robbins, T. W., and Poldrack, R. A. Inhibition and the right inferior frontal cortex. *Trends in Cognitive Science*, 2004, *8*, 170–177.

Aronson, B. D., Bell-Pedersen, D., Block, G. D., Bos, N. P. A., et al. Circadian rhythms. *Brain Research Reviews*, 1993, *18*, 315–333.

Arora, S., and Anubhuti. Role of neuropeptides in appetite regulation and obesity: A review. *Neuropeptides*, 2006, *40*, 375–401.

Arrasate, M., Mitra, S., Schweitzer, E. S., Segal, M. R., and Finkbeiner, S. Inclusion body formation reduces levels of mutant huntingtin and the risk of neuronal death. *Nature*, 2004, *431*, 747–748.

Arroyo, S., Lesser, R. P., Gordon, B., Uematsu, S., et al. Mirth, laughter and gelastic seizures. *Brain*, 1993, *116*, 757–780.

Artmann, H., Grau, H., Adelman, M., and Schleiffer, R. Reversible and non-reversible enlargement of cerebrospinal fluid spaces in anorexia nervosa. *Neuroradiology*, 1985, *27*, 103–112.

Asanuma, H., and Rosén, I. Topographical organization of cortical efferent zones projecting to distal forelimb muscles in monkey. *Experimental Brain Research*, 1972, *13*, 243–256.

Aschoff, J. Circadian rhythms: General features and endocrinological aspects. In *Endocrine Rhythms*, edited by D. T. Krieger. New York: Raven Press, 1979.

Asnis, G. M., Hameedi, F. A., Goddard, A. W., Potkin, S. G., et al. Fluvoxamine in the treatment of panic disorder: A multi-center, double-blind, placebo-controlled study in out-patients. *Psychiatry Research*, 2001, *103*, 1–14.

Astafiev, S. V., Shulman, G. L., Stanley, C. M., Snyder, A. Z., et al. Functional organization of human intraparietal and frontal cortex for attending, looking, and pointing. *Journal of Neuroscience*, 2003, *23*, 4689–4699.

Aston-Jones, G., and Bloom, F. E. Activity of norepinephrine-containing locus coeruleus neurons in behaving rats anticipates fluctuations in the sleep-waking cycle. *Journal of Neuroscience*, 1981, *1*, 876–886.

Aston-Jones, G., Rajkowski, J., Kubiak, P., and Alexinsky, T. Locus coeruleus neurons in monkey are selectively activated by attended cues in a vigilance task. *Journal of Neuroscience*, 1994, *14*, 4467–4480.

Attia, E., Haiman, C., Walsh, T., and Flater, S. T. Does fluoxetine augment the inpatient treatment of anorexia nervosa? *American Journal of Psychiatry*, 1998, *155*, 548–551.

Auer, R. N., Jensen, M. L., and Whishaw, I. Q. Neurobehavioral deficit due to ischemic brain damage limited to half of the CA1 section of the hippocampus. *Journal of Neuroscience*, 1989, *9*, 1641–1647.

Auerbach, J., Geller, V., Lezer, S., Shinwell, E., et al. Dopamine D4 receptor (D4DR) and serotonin transporter promoter (5-HTTLPR) polymorphisms in the determination of temperament in 2-month-old infants. *Molecular Psychiatry*, 1999, *4*, 369–373.

Autism Genome Project Consortium. Mapping autism risk loci using genetic linkage and chromosomal rearrangements. *Nature Genetics*, 2007, *39*, 319–328.

Avenet, P., and Lindemann, B. Perspectives of taste reception. *Journal of Membrane Biology*, 1989, *112*, 1–8.

Avila, M. T., Weiler, M. A., Lahti, A. C., Tamminga, C. A., and Thaker, G. K. Effects of ketamine on leading saccades during smooth-pursuit eye movements may implicate cerebellar dysfunction in schizophrenia. *American Journal of Psychiatry*, 2002, *159*, 1490–1496.

Ayala, R., Shu, T., and Tsai, L.-H. Trekking across the brain: The journal of neuronal migration. *Cell*, 2007, *128*, 29–43.

Aziz-Zadeh, L., Cattaneo, L., Rochat, M., and Rizzolatti, G. Covert speech arrest induced by rTMS over both motor and nonmotor left hemisphere frontal sites. *Journal of Cognitive Neuroscience*, 2005, *17*, 928–938.

Baddeley, A. D. Memory: Verbal and visual subsystems of working memory. *Current Biology*, 1993, *3*, 563–565.

Bagatell, C. J., Heiman, J. R., Rivier, J. E., and Bremner, W. J. Effects of endogenous testosterone and estradiol on sexual behavior in normal young men. *Journal of Clinical Endocrinology and Metabolism*, 1994, *7*, 211–216.

Bagnasco, M., Tulipano, G., Melis, M. R., Argiolas, A., et al. Endogenous ghrelin is an orexigenic peptide acting in the arcuate nucleus in response to fasting. *Regulatory Peptides*, 2003, *28*, 161–167.

Bai, F. L., Yamano, M., Shiotani, Y., Emson, P. C., et al. An arcuato-paraventricular and -dorsomedial hypothalamic neuropeptide Y-containing system which lacks noradrenaline in the rat. *Brain Research*, 1985, *331*, 172–175.

Bailey, A. The biology of autism. *Psychological Medicine*, 1993, *23*, 7–11.

Bailey, C. H., Kandel, E. R., and Si, K. The persistence of long-term memory: A molecular approach to self-sustaining changes in learning-induced synaptic growth. *Neuron*, 2004, *44*, 49–57.

Bailey, J. M., and Pillard, R. C. A genetic study of male sexual orientation. *Archives of General Psychiatry*, 1991, *48*, 1089–1096.

Baizer, J. S., Ungerleider, L. G., and Desimone, R. Organization of visual inputs to the inferior temporal and posterior parietal cortex in macaques. *Journal of Neuroscience*, 1991, *11*, 168–190.

Bak, T. H., O'Donovan, D. G., Xuereb, J. H., Boniface, S., and Hodges, J. R. Selective impairment of verb processing associated with pathological changes in Brodman areas 44 and 45 in the motor neurone disease-dementia-aphasia syndrome. *Brain*, 2001, *124*, 103–120.

Baker, C. I., Behrman, M., and Olson, C. R. Impact of learning on representation of parts and wholes in monkey Inferotemporal cortex. *Nature Neuroscience*, 2002, *5*, 1210–1216.

Baker, C. I., Hutichson, T. L., and Kanwisher, N. Does the fusiform face area contain subregions highly selective for nonfaces? *Nature Neuroscience*, 2007, *10*, 3–4.

Baldessarini, R. J. *Chemotherapy in Psychiatry*. Cambridge, Mass.: Harvard University Press, 1977.

Ballantine, H. T., Bouckoms, A. J., Thomas, E. K., and Giriunas, I. E. Treatment of psychiatric illness by stereotactic cingulotomy. *Biological Psychiatry*, 1987, *22*, 807–819.

Bandell, M., Macpherson, L. J., and Patapoutian, A. From chills to chilis: Mechanisms for thermosensation and chemesthesis via thermoTRPs. *Current Opinion in Neurobiology*, 2007, *17*, 490–497.

Banks, M. S., Aslin, R. N., and Letson, R. D. Sensitive period for the development of human binocular vision. *Science*, 1975, *190*, 675–677.

Banks, W. A., Kastin, A. J., Huang, W. T., Jaspan, J. B., et al. Leptin enters the brain by a saturable system independent of insulin. *Peptides*, 1996, *17*, 305–311.

Barbazanges, A., Piazza, P. V., Le Moal, M., and Maccari, S. Maternal glucocorticoid secretion mediates long-term effects of prenatal stress. *Journal of Neuroscience*, 1996, *16*, 3943–3949.

Barclay, C. D., Cutting, J. E., and Kozlowski, L. T. Temporal and spatial factors in gait perception that influence gender recognition. *Perception and Psychophysics*, 1978, *23*, 145–152.

Bard, F., Cannon, C., Barbour, R., Burke, R. L., et al. Peripherally administered antibodies against amyloid beta-peptide enter the central nervous system and reduce pathology in a mouse model of Alzheimer disease. *Nature Medicine*, 2000, *6*, 916–919.

Bart, G., Kreek, M. J., Ott, J., LaForge, K. S., Proudnikov, D., Polla, L., and Heilig, M. Increased attributable risk related to a functional μ-opioid receptor gene polymorphism in association with alcohol dependence in central Sweden. *Neuropsychopharmacology*, 2005, *30*, 417–422.

Bartels, A., and Zeki, S. The neural basis of romantic love. *Neuroreport*, 2000, *27*, 3829–3834.

Bartels, A., and Zeki, S. The neural correlates of maternal and romantic love. *Neuroimage*, 2004, *21*, 1155–1166.

Bartness, T. J., Powers, J. B., Hastings, M. H., Bittman, E. L., and Goldman, B. D. The timed infusion paradigm for melatonin delivery: What has it taught us about the melatonin signal, its reception, and the photoperiodic control of seasonal responses? *Journal of Pineal Research*, 1993, *15*, 161–190.

Basbaum, A. I., and Fields, H. L. Endogenous pain control mechanisms: Review and hypothesis. *Annals of Neurology*, 1978, *4*, 451–462.

Basbaum, A. I., and Fields, H. L. Endogenous pain control systems: Brainstem spinal pathways and endorphin circuitry. *Annual Review of Neuroscience*, 1984, *7*, 309–338.

Basheer, R., Strecker, R. E., Thakkar, M. M., and McCarley, R. W. Adenosine and sleep–wake regulation. *Progress in Neurobiology*, 2004, *73*, 379–396.

Batterham, R. L., Cowley, M. A., Small, C. J., Herzog, H., et al. Gut hormone PYY_{3-36} physiologically inhibits food intake. *Nature*, 2002, *418*, 650–654.

Batterham, R. L., ffytche, D. H., Rosenthal, J. M., Zelaya, R. O., et al. PYY modulation of cortical and hypothalamic brain areas predicts feeding behaviour in humans. *Nature*, 2007, *450*, 106–109.

Bautista, D. M., Jordt, S.-E., Nikai, T., Tsuruda, P. R., et al. TRPA1 mediates the inflammatory actions of environmental irritants and proalgesic agents. *Cell*, 2006, *124*, 1269–1282.

Bautista, D. M., Siemens, J., Glazer, J. M., Tsuruda, P. R., et al. The menthol receptor TRPM8 is the principal detector of environmental cold. *Nature*, 2007, *448*, 204–208.

Bayley, P. J., Frascino, J. C., and Squire, L. R. Robust habit learning in the absence of awareness and independent of the medial temporal lobe. *Nature*, 2005, *436*, 550–553.

Beamer, W., Bermant, G., and Clegg, M. T. Copulatory behaviour of the ram, Ovis aries. II: Factors affecting copulatory satiation. *Animal Behavior*, 1969, *17*, 706–711.

Bean, N. J. Modulation of agonistic behavior by the dual olfactory system in male mice. *Physiology and Behaviour*, 1982, *29*, 433–437.

Bean, N. J., and Conner, R. Central hormonal replacement and home-cage dominance in castrated rats. *Hormones and Behavior*, 1978, *11*, 100–109.

Beauvois, M. F., and Dérouesné, J. Phonological alexia: Three dissociations. *Journal of Neurology, Neurosurgery and Psychiatry*, 1979, *42*, 1115–1124.

Beauvois, M. F., and Dérouesné, J. Lexical or orthographic dysgraphia. *Brain*, 1981, *104*, 21–45.

Bechara, A., Tranel, D., Damasio, H., Adolphs, R., Rockland, C., and Damasio, A. R. Double dissociation of conditioning and declarative knowledge relative to the amygdala and hippocampus in humans. *Science*, 1995, *269*, 1115–1118.

Beckstead, R. M., Morse, J. R., and Norgren, R. The nucleus of the solitary tract in the monkey: Projections to the thalamus and brainstem nuclei. *Journal of Comparative Neurology*, 1980, *190*, 259–282.

Beecher, H. K. *Measurement of Subjective Responses: Quantitative Effects of Drugs*. New York: Oxford University Press, 1959.

Beeman, E. A. The effect of male hormone on aggressive behavior in mice. *Physiological Zoology*, 1947, *20*, 373–405.

Beidler, L. M. Physiological properties of mammalian taste receptors. In *Taste and Smell in Vertebrates*, edited by G. E. W. Wolstenholme. London: J. & A. Churchill, 1970.

Beitz, A. J. The organization of afferent projections to the midbrain periaqueductal gray of the rat. *Neuroscience*, 1982, *7*, 133–159.

Belin, D., and Everitt, B. J. Cocaine seeking habits depend upon dopamine-dependent serial connectivity linking the ventral with the dorsal striatum. *Neuron*, 2008, *57*, 432–441.

Bell, A. P., Weinberg, M. S., and Hammersmith, S. K. *Sexual Preference: Its Development in Men and Women*. Bloomington: Indiana University Press, 1981.

Ben Mamou, C., Gamache, K., and Nader, K. NMDA receptors are critical for unleashing consolidated auditory fear memories. *Nature Neuroscience*, 2006, *9*, 1237–1239.

Benchenane, K., López-Atalaya, J. P., Fernéndez-Monreal, M., Touzani, O., and Vivien, D. Equivocal roles of tissue-type plasminogen activator in stroke-induced injury. *Trends in Neuroscience*, 2004, *27*, 155–160.

Bendor, D., and Wang, X. Cortical representations of pitch in monkeys and humans. *Current Opinion in Neurobiology*, 2006, *16*, 391–399.

Benedetti, F., Arduino, C., and Amanzio, M. Somatotopic activation of opioid systems by target-directed expectations of analgesia. *Journal of Neuroscience*, 1999, *19*, 3639–3648.

Benington, J. H., Kodali, S. K., and Heller, H. C. Monoaminergic and cholinergic modulation of REM-sleep timing in rats. *Brain Research*, 1995, *681*, 141–146.

Bennett, D. A., Wilson, R. S., Schneider, J. A., Evans, D. A., et al. Education modifies the relation of AD pathology to level of cognitive function in older persons. *Neurology*, 2003, *60*, 1909–1915.

Bensimon, G., Lacomblez, L., and Meininger, V. A controlled trial of riluzole in amyotrophic lateral sclerosis. ALS/Riluzole Study Group. *New England Journal of Medicine*, 1994, *330*, 585–591.

Benson, D. L., Colman, D. R., and Huntley, G. W. Molecules, maps and synapse specificity. *Nature Reviews: Neuroscience*, 2001, *2*, 899–909.

Ben-Tovim, D. I. Eating disorders: Outcome, prevention, and treatment of eating disorders. *Current Opinion in Psychiatry*, 2003, *16*, 65–69.

Berenbaum, S. A., and Resnick, S. M. Early androgen effects on aggression in children and adults with congenital adrenal hyperplasia. *Psychoneuroendocrinology*, 1997, *22*, 505–515.

Bergasa, N. V. The pruritis of cholestasis. *Journal of Hepatology*, 2005, *43*, 1078–1088.

Berkovic, S. F., Mulley, J. C., Scheffer, I. E., and Petrou, S. Human epilepsies: Interaction of genetic and acquired factors. *Trends in Neurosciences*, 2006, *29*, 391–397.

Bermant, G., and Davidson, J. M. *Biological Bases of Sexual Behavior.* New York: Harper & Row, 1974.

Bernhardt, P. C., Dabbs, J. M., Fielden, J. A., and Lutter, C. D. Testosterone changes during vicarious experiences of winning and losing among fans at sporting events. *Physiology and Behavior*, 1998, *65*, 59–62.

Berns, G. S., McClure, S. M., Pagnoni, G., and Montague, P. R. Predictability modulates human brain response to reward. *Journal of Neuroscience*, 2001, *21*, 2793–2798.

Berridge, C. W., Devilbiss, D. M., Andrzejewski, M. E., Arnsten, A. F. T., et al. Methylphenidate preferentially increases catecholamine neurotransmission within the prefrontal cortex at low doses that enhance cognitive function. *Biological Psychiatry*, 2006, *60*, 1111–1120.

Berson, D. M., Dunn, F. A., and Takao, M. Phototransduction by retinal ganglion cells that set the circadian clock. *Science*, 2002, *295*, 1070–1073.

Berthier, M., Kulisevsky, J., Gironell, A., and Heras, J. A. Obsessive-compulsive disorder associated with brain lesions: Clinical phenomenology, cognitive function, and anatomic correlates. *Neurology*, 1996, *47*, 353–361.

Berthier, M., Starkstein, S., and Leiguarda, R. Asymbolia for pain: A sensory-limbic disconnection syndrome. *Annals of Neurology*, 1988, *24*, 41–49.

Bertolini, A., Ferrari, A., Ottani, A., Guerzoni, S., Racchi, R., and Leone, S. Paracetamol: New vistas of an old drug. *CNS Drug Review*, 2006, *12*, 250–275.

Best, P. J., White, A. M., and Minai, A. Spatial processing in the brain: The activity of hippocampal place cells. *Annual Review of Neuroscience*, 2001, *24*, 459–486.

Bettelheim, B. *The Empty Fortress.* New York: Free Press, 1967.

Bi, A., Cui, J., Ma, Y.-P., Olshevskaya, E., Pu, M., Dizhoor, A. M. and Pan, A.-H. Ectopic expression of a microbial-type rhodopsin restores visual responses in mice with photoreceptor degeneration. *Neuron*, 2006, *50*, 23–33.

Billings, L. M., Green, K. N., McGaugh, J. L., and LaFerla, F. M. Learning decreases Aβ*56 and tau pathology and ameliorates behavioral decline in 3xTg-AD mice. *Journal of Neuroscience*, 2007, *27*, 751–761.

Bisiach, E., and Luzzatti, C. Unilateral neglect of representational space. *Cortex*, 1978, *14*, 129–133.

Bissière, S., Humeau, Y., and Luthi, A. Dopamine gates LTP induction in lateral amygdala by suppressing feedforward inhibition. *Nature Neuroscience*, 2003, *6*, 587–592.

Blair, R. J. R. The amygdala and ventromedial prefrontal cortex: Functional contributions and dysfunction in psychopathy. *Philosophical Transactions of the Royal Society of London B*, 2008, *363*, 2557–2565.

Blanchard, R. Fraternal birth order and the maternal immune hypothesis of male homosexuality. *Hormones and Behavior*, 2001, *40*, 105–114.

Blaustein, J. D., and Feder, H. H. Cytoplasmic progestin receptors in guinea pig brain: Characteristics and relationship to the induction of sexual behavior. *Brain Research*, 1979, *169*, 481–497.

Blaustein, J. D., and Olster, D. H. Gonadal steroid hormone receptors and social behaviors. In *Advances in Comparative and Environmental Physiology, Vol. 3*, edited by J. Balthazart. Berlin: Springer-Verlag, 1989.

Blest, A. D. The function of eyespot patterns in insects. *Behaviour*, 1957, *11*, 209–256.

Bleuler, F. *Dementia Praecox of the Group of Schizophrenia*, 1911. Translated by J. Zinkin. New York: International Universities Press, 1911/1950.

Bliss, T. V., and Lømo, T. Long-lasting potentiation of synaptic transmission in the dentate area of the anaesthetized rabbit following stimulation of the perforant path. *Journal of Physiology*, 1973, *232*, 331–356.

Blumenfeld, R. S., and Ranganath, C. Dorsolateral prefrontal cortex promotes long-term memory formation through its role in working memory organization. *Journal of Neuroscience*, 2006, *26*, 916–925.

Blundell, J. E., and Halford, J. C. G. Serotonin and appetite regulation: Implications for the pharmacological treatment of obesity. *CNS Drugs*, 1998, *9*, 473–495.

Bobrow, D., and Bailey, M. J. Is male homosexuality maintained via kin selection? *Evolution and Human Behavior*, 2001, *22*, 361–368.

Bodner, S. M., Morshed, S. A., and Peterson, B. S. The question of PANDAS in adults. *Biological Psychiatry*, 2001, *49*, 807–810.

Boeve, B. F., Silber, M. H., Saper, C. B., Ferman, T. J., et al. Pathophysiology of REM sleep behaviour disorder and relevance to neurodegenerative disease. *Brain*, 2007, *130*, 2770–2788.

Bogaert, A. F. Biological versus nonbiological older brothers and men's sexual orientation. *Proceedings of the National Academy of Sciences, USA*, 2006, *103*, 10771–10774.

Bohbot, V. D., Lerch, J., Thorndycraft, B., Iaria, G., and Zijdenbos, A. P. Gray matter differences correlate with spontaneous strategies in a human virtual navigation task. *Journal of Neuroscience*, 2007, *27*, 10078–10083.

Boksa, P. Animal models of obstetric complications in relation to schizophrenia. *Brain Research Reviews*, 2004, *45*, 1–17.

Bolla, K., Ernst, M., Kiehl, K., Mouratidis, M., et al. Prefrontal cortical dysfunction in abstinent cocaine abusers. *Journal of Neuropsychiatry and Clinical Neuroscience*, 2004, *16*, 456–464.

Bolwig, T. G., Woldbye, D. P., and Mikkelsen, J. D. Electroconvulsive therapy as an anticonvulsant: A possible role of neuropeptide Y (NPY). *Journal of ECT*, 1999, *15*, 93–101.

Bontempi, B., Laurent-Demir, C., Destrade, C., and Jaffard, R. Time-dependent reorganization of brain circuitry underlying long-term memory storage. *Nature*, 1999, *400*, 671–675.

Boodman, S. G. Hungry in the dark: Some sleepeaters don't wake up for their strange nighttime binges. *Washington Post*, Sept. 7, 2004, HE01.

Booth, F. W., and Neufer, P. D. Exercise controls gene expression. *American Scientist*, 2005, *93*, 28–35.

Born, R. T., and Tootell, R. B. H. Spatial frequency tuning of single units in macaque supragranular striate cortex. *Proceedings of the National Academy of Sciences, USA*, 1991, *88*, 7066–7070.

Bornstein, B., Stroka, H., and Munitz, H. Prosopagnosia with animal face agnosia. *Cortex*, 1969, *5*, 164–169.

Borod, J. C., Koff, E., Yecker, S., Santschi, C., and Schmidt, J. M. Facial asymmetry during emotional expression: Gender, falence, and measurement technique. *Neuropsychologia*, 1998, *36*, 1209–1215.

Bossy-Wetzel, E., Schwarzenbacher, R., and Lipton, S. A. Molecular pathways to neurodegeneration. *Nature Medicine*, 2004, *10*, S2–S9.

Bouchard, C., Tremblay, A., Despres, J. P., Nadeau, A., et al. The response to long-term overfeeding in identical twins. *New England Journal of Medicine*, 1990, *322*, 1477–1482.

Boulos, Z., Campbell, S. S., Lewy, A. J., Terman, M., et al. Light treatment for sleep disorders: Consensus report. 7: Jet-lag. *Journal of Biological Rhythms*, 1995, *10*, 167–176.

Bourne, J., and Harris, K. M. Do thin spines learn to be mushroom spines that remember? *Current Opinion in Neurobiology*, 2007, *17*, 381–386.

Boussaoud, D., Desimone, R., and Ungerleider, L. G. Visual topography of area TEO in the macaque. *Journal of Comparative Neurology*, 1991, *306*, 554–575.

Bouton, M. E., and King, D. A. Contextual control of the extinction of conditioned fear: Tests for the associative value of the context. *Journal of Experimental Psychology: Animal Behavior Processes*, 1983, *9*, 248–265.

Bouvier, S. E., and Engel, S. A. Behavioral deficits and cortical damage loci in cerebral achromatopsia. *Cerebral Cortex*, 2006, *16*, 183–191.

Bowers, D., Blonder, L. X., Feinberg, T., and Heilman, K. M. Differential impact of right and left hemisphere lesions on facial emotion and object imagery. *Brain*, 1991, *114*, 2593–2609.

Boyden, E. S., Zhang, F., Bamberg, E., Nagel, G., and Deisseroth, K. Millisecond-timescale, genetically targeted optical control of neural activity. *Nature Neuroscience*, 2005, *8*, 1263–1268.

Bozarth, M. A. Physical dependence produced by central morphine infusions: An anatomical mapping study. *Neuroscience and Biobehavioral Reviews*, 1994, *18*, 373–383.

Bozarth, M. A., and Wise, R. A. Toxicity associated with long-term intravenous heroin and cocaine self-administration in the rat. *Journal of the American Medical Association*, 1985, *254*, 81–83.

Brackett, N. L., Ferrell, S. M., Aballa, T. C., Amador, M. J., et al. An analysis of 653 trials of penile vibratory stimulation in men with spinal cord injury. *Journal of Urology*, 1998, *159*, 1931–1934.

Bradbury, M. W. B. *The Concept of a Blood-Brain Barrier.* New York: John Wiley & Sons, 1979.

Bradley, D. C., Maxwell, M., Andersen, R. A., Banks, M. S., and Shenoy, K. V. Mechanisms of heading perception in primate visual cortex. *Science*, 1996, *273*, 1544–1547.

Brady, K. T., and Sinha, R. Co-occurring mental and substance use disorders: The neurobiological effects of chronic stress. *American Journal of Psychiatry*, 2005, *162*, 1483–1493.

Braun, A. R., Balkin, T. J., Wesensten, N. J., Gwadry, F., et al. Dissociated pattern of activity in visual cortices and their projections during human rapid eye movement sleep. *Science*, 1998, *279*, 91–95.

Braun, S. New experiments underscore warnings on maternal drinking. *Science*, 1996, *273*, 738–739.

Bray, G. A. Drug treatment of obesity. *American Journal of Clinical Nutrition*, 1992, *55*, 538S–544S.

Bray, G. A., Nielsen, S. J., and Popkin, B. M. Consumption of high-fructose corn syrup in beverages may play a role in the epidemic of obesity. *American Journal of Clinical Nutrition*, 2004, *79*, 537–543.

Brecher, E. M. *Licit and Illicit Drugs.* Boston: Little, Brown & Co., 1972.

Breedlove, S. M. Sexual differentiation of the brain and behavior. In *Behavioral Endocrinology*, edited by J. B. Becker, S. M. Breedlove, and D. Crews. Cambridge, Mass.: MIT Press, 1992.

Breedlove, S. M. Sexual differentiation of the human nervous system. *Annual Review of Psychology*, 1994, *45*, 389–418.

Breier, A., Su, T.-P., Saunders, R., Carson, R. E., et al. Schizophrenia is associated with elevated amphetamine-induced synaptic dopamine concentrations: Evidence from a novel positron emission tomography method. *Proceedings of the National Academy of Sciences, USA*, 1997, *94*, 2569–2574.

Bremner, J. D. Does stress damage the brain? *Biological Psychiatry*, 1999, *45*, 797–805.

Brennan, P. A., and Keverne, E. B. Something in the air? New insights into mammalian pheromones. *Current Biology*, 2004, *14*, R81–R89.

Brewer, J. B., Zhao, Z., Desmond, J. E., Glover, G. H., and Gabrieli, J. D. E. Making memories: Brain activity that predicts how well visual experience will be remembered. *Science*, 1998, *281*, 1185–1187.

Brickner, R. M. *The Intellectual Functions of the Frontal Lobe: A Study Based Upon Observations of a Man After Partial Frontal Lobectomy.* New York: Macmillan, 1936.

Bridges, R. S., DiBiase, R., Loundes, D. D., and Doherty, P. C. Prolactin stimulation of maternal behavior in female rats. *Science*, 1985, *227*, 782–784.

Bridges, R. S., Rigero, B. A., Byrnes, E. M., Yang, L., and Walker, A. M. Central infusions of the recombinant human prolactin receptor antagonist, S179D-PRL, delay the onset of maternal behavior in steroid-primed, nulliparous female rats. *Endocrinology*, 2001, *142*, 730–739.

Bridges, R. S., Robertson, M. C., Shiu, R. P. C., Sturgis, J. D., Henriquez, B. M., and Mann, P. E. Central lactogenic regulation of maternal behavior in rats: Steroid dependence, hormone specificity, and behavioral potencies of rat prolactin and rat placental lactogen I. *Endocrinology*, 1997, *138*, 756–763.

Brisbare-Roch, C., Dingemanse, J., Koberstein, R., Hoever, P., et al. Promotion of sleep by targeting the orexin system in rats, dogs and humans. *Nature Medicine*, 2007, *13*, 150–155.

Britten, K. H., and van Wezel, R. J. Electrical microstimulation of cortical area MST biases heading perception in monkeys. *Nature Neuroscience*, 1998, *1*, 59–63.

Britton, D. R., Koob, G. F., Rivier, J., and Vale, W. Intraventricular corticotropin-releasing factor enhances behavioral effects of novelty. *Life Sciences*, 1982, *31*, 363–367.

Broberger, C., de Lecea, L., Sutcliffe, J. G., and Hökfelt, T. Hypocretin/orexin- and melanin-concentrating hormone-expressing cells form distinct populations in the rodent lateral hypothalamus: Relationship to the neuropeptide Y and agouti gene-related protein systems. *Journal of Comparative Neurology*, 1998, *402*, 460–474.

Broca, P. Remarques sur le siège de la faculté du langage articulé, suivies d'une observation d'aphemie (perte de la parole). *Bulletin de la Société Anatomique (Paris)*, 1861, *36*, 330–357.

Brodie, J. D., Figueroa, E., Laska, E. M., and Dewey, S. L. Safety and efficacy of γ-vinyl GABA (GVG) for the treatment of methamphetamine and/or cocaine addiction. *Synapse*, 2005, *55*, 122–125.

Brody, A. L., Mandelkern, M. A., Lee, G., Smith, E., et al. Attenuation of cue-induced cigarette craving and anterior cingulate cortex activation in bupropion-treated smokers: A preliminary study. *Psychiatry Research*, 2004, *130*, 269–281.

Brolin, R. E. Bariatric surgery and long-term control of morbid obesity. *Journal of the American Medical Association*, 2002, *288*, 2793–2796.

Brown, A. M., Tekkök, S. B., and Ransom, B. R. Energy transfer from astrocytes to axons: The role of CNS glycogen. *Neurochemistry International*, 2004, *45*, 529–536.

Brown, A. S., Hooton, J., Schaefer, C. A., Zhang, H., et al. Elevated maternal interleukin-8 levels and risk of schizophrenia in adult offspring. *American Journal of Psychiatry*, 2004, *161*, 889–895.

Brown, A. S., Schaefer, C. A., Wyatt, R. J., Begg, M. D., et al. Paternal age and risk of schizophrenia in adult offspring. *American Journal of Psychiatry*, 2002, *159*, 1528–1533.

Brown, A. S. Prenatal infection as a risk factor for schizophrenia. *Schizophrenia bulletin*, 2006, *32*, 200–202.

Brown, R. E., Stevens, D. R., and Haas, H. L. The physiology of brain histamine. *Progress in Neurobiology*, 2001, *63*, 637–672.

Brown, S., Ingham, R. J., Ingham, J. C., Laird, A. R., and Fox, P. T. Stuttered and fluent speech production: An ALE meta-analysis of functional neuroimaging studies. *Human Brain Mapping*, 2005, *25*, 105–117.

Brown, T. H., Ganong, A. H., Kairiss, E. W., Keenan, C. L., and Kelso, S. R. Long-term potentiation in two synaptic systems of the hippocampal brain slice. In *Neural Models of Plasticity: Experimental and Theoretical Approaches*, edited by J. H. Byrne and W. O. Berry. San Diego: Academic Press, 1989.

Brownell, H. H., Michel, D., Powelson, J., and Gardner, H. Surprise but not coherence: Sensitivity to verbal humor in right-hemisphere patients. *Brain and Language*, 1983, *18*, 20–27.

Brownell, H. H., Simpson, T. L., Bihrle, A. M., Potter, H. H., and Gardner, H. Appreciation of metaphoric alternative word meanings by left and right brain-damaged patients. *Neuropsychologia*, 1990, *28*, 173–184.

Brownell, W. E., Bader, C. R., Bertrand, D., and de-Ribaupierre, Y. Evoked mechanical responses of isolated cochlear outer hair cells. *Science*, 1985, *227*, 194–196.

Brozowski, T. J., Brown, R. M., Rosvold, H. E., and Goldman, P. S. Cognitive deficit caused by regional depletion of dopamine in prefrontal cortex of rhesus monkey. *Science*, 1979, *205*, 929–932.

Bruce, H. M. A block to pregnancy in the mouse caused by proximity of strange males. *Journal of Reproduction and Fertility*, 1960a, *1*, 96–103.

Bruce, H. M. Further observations of pregnancy block in mice caused by proximity of strange males. *Journal of Reproduction and Fertility*, 1960b, *2*, 311–312.

Bruijn, L. I., Miller, T. M., and Cleveland, D. W. Unraveling the mechanisms involved in motor neuron degeneration in ALS. *Annual Review of Neuroscience*, 2004, *27*, 723–749.

Brüning, J. C., Gautam, D., Burks, D. J., Gillette, J., et al. Role of brain insulin receptor in control of body weight and reproduction. *Science*, 2000, *289*, 2122–2125.

Brunson, K. L., Kramér, E., Lin, B., Chen, Y., et al. Mechanisms of late-onset cognitive decline after early-life stress. *Journal of Neuroscience*, 2005, *25*, 9328–9338.

Buccino, G., Riggio, L., Melli, G., Binkofski, F., et al. Listening to action-related sentences modulates the activity of the motor system: A combined TMS and behavioral study. *Cognitive Brain Research*, 2005, *24*, 355–363.

Buccino, G., Vogt, S., Ritzi, A., Fink, G. R., et al. Neural circuits underlying imitation learning of hand actions: An event-related fMRI study. *Neuron*, 2004, *42*, 323–334.

Buchsbaum, M. S., Gillin, J. C., Wu, J., Hazlett, E., et al. Regional cerebral glucose metabolic rate in human sleep assessed by positron emission tomography. *Life Sciences*, 1989, *45*, 1349–1356.

Buck, L. Information coding in the vertebrate olfactory system. *Annual Review of Neuroscience*, 1996, *19*, 517–544.

Buck, L., and Axel, R. A novel multigene family may encode odorant receptors: A molecular basis for odor recognition. *Cell*, 1991, *65*, 175–187.

Buckner, R. L., Snyder, A. Z., Shannon, B. J., LaRossa, G., et al. Molecular, structural, and functional characterization of Alzheimer's disease: Evidence for a relationship between default activity, amyloid, and memory. *Journal of Neuroscience*, 2005, *25*, 7709–7717.

Budka, H., Almer, G., Hainfellner, J. A., Brücke, T., and Jellinger, K. The Austrian FFI cases. *Brain Pathology*, 1998, *8*, 554.

Bueler, H., Aguzzi, A., Sailer, A., Grenier, R. A., et al. Mice devoid of PrP are resistant to scrapie. *Cell*, 1993, *73*, 1339–1347.

Buhmann, C., Glauche, V., Sturenburg, H. J., Oechsner, M., et al. Pharmocologically modulated fMRI-cortical responsiveness to levodopa in drug-naïve hemiparkinsonian patients. *Brain*, 2003, *126*, 451–461.

Bullivant, S. B., Sellergren, S. A., Stern, K., Spencer, N. A., et al. Women's sexual experience during the menstrual cycle: Identification of the sexual phase by noninvasive measurement of luteinizing hormone. *Journal of Sex Research*, 2004, *41*, 82–93.

Bunyard, L. B., Katzel, L. I., Busby-Whitehead, M. J., Wu, Z., and Goldberg, A. P. Energy requirements of middle-aged men are modifiable by physical activity. *American Journal of Clinical Nutrition*, 1998, *68*, 1136–1142.

Buonopane, A., and Petrakis, I. L. Pharmacotherapy of alcohol use disorders. *Substance Use and Misuse*, 2005, *40*, 2001–2020.

Buxbaum, L. J., Glosser, G., and Coslett, H. B. Impaired face and word recognition without object agnosia. *Neuropsychologia*, 1999, *37*, 41–50.

Byne, W., Tobet, S., Mattiace, L. A., Lasco, M. S., et al. The interstitial nuclei of the human anterior hypothalamus: An investigation of variation with sex, sexual orientation, and HIV status. *Hormones and Behavior*, 2001, *40*, 85–92.

Cabanac, M., and Lafrance, L. Facial consummatory responses in rats support the ponderostat hypothesis. *Physiology and Behavior*, 1991, *50*, 179–183.

Cadet, J. L., Jayanthi, S., and Deng, X. Speed kills: Cellular and molecular bases of methamphetamine-induced nerve terminal degeneration and neuronal apoptosis. *FASEB Journal*, 2003, *17*, 1775–1788.

Cahill, L., Babinsky, R., Markowitsch, H. J., and McGaugh, J. L. The amygdala and emotional memory. *Nature*, 1995, *377*, 295–296.

Cahill, L., Haier, R. J., Fallon, J., Alkire, M. T., et al. Amygdala activity at encoding correlated with long-term, free recall of emotional information. *Proceedings of the National Academy of Sciences, USA*, 1996, *93*, 8016–8021.

Calder, A. J., Keane, J., Manes, F., Anntoun, N., and Young, A. W. Impaired recognition and experience of disgust following brain injury. *Nature Neuroscience*, 2000, *3*, 1077–1078.

Calder, A. J., Young, A. W., Rowland, D., Perrett, D. I., et al. Facial emotion recognition after bilateral amygdala damage: Differentially severe impairment of fear. *Cognitive Neuropsychology*, 1996, *13*, 699–745.

Calvo-Merino, B., Glaser, D. E., Grèzes, J., Passingham, R. E., and Haggard, P. Action observation and acquired motor skills: An fMRI study with expert dancers. *Cerebral Cortex*, 2005, *15*, 1243–1249.

Campbell, R., Heywood, C. A., Cower, A., Regard, M., and Landis, T. Sensitivity to eye gaze in prosopagnosic patients and monkeys with superior temporal sulcus ablation. *Neuropsychologia*, 1990, *28*, 1123–1142.

Campeau, S., Hayward, M. D., Hope, B. T., Rosen, J. B., Nestler, E. J., and Davis, M. Induction of the c-fos proto-oncogene in rat amygdala during unconditioned and conditioned fear. *Brain Research*, 1991, *565*, 349–352.

Camperio-Ciani, A., Corna, F., and Capiluppi, C. Evidence for maternally inherited factors favouring male homosexuality and promoting female fecundity. *Proceedings of the Royal Society of London B*, 2004, *271*, 2217–2221.

Campfield, L. A., Smith, F. J., Guisez, Y., Devos, R., and Burn, P. Recombinant mouse ob protein: Evidence for a peripheral signal linking adiposity and central neural networks. *Science*, 1995, *269*, 546–549.

Cannon, M., Jones, P. B., and Murray, R. M. Obstetric complications and schizophrenia: Historical and meta-analytic review. *American Journal of Psychiatry*, 2002, *159*, 1080–1092.

Cannon, W. B. The James-Lange theory of emotions: A critical examination and an alternative. *American Journal of Psychology*, 1927, *39*, 106–124.

Carew, T. J. Development assembly of learning in Aplysia. *Trends in Neuroscience*, 1989, *12*, 389–394.

Carmichael, M. S., Humbert, R., Dixen, J., Palmisano, G., et al. Plasma oxytocin increases in the human sexual response. *Journal of Clinical Endocrinology and Metabolism*, 1987, *64*, 27–31.

Caro, J. F., Kolaczynski, J. W., Nyce, M. R., Ohannesian, J. P., et al. Decreased cerebrospinal fluid/serum leptin ration in obesity: A possible mechanism for leptin resistance. *Lancet*, 1996, *348*, 159–161.

Carpenter, C. R. Sexual behavior of free ranging rhesus monkeys (*Macaca mulatta*). I: Specimens, procedures and behavioral characteristics of estrus. *Journal of Comparative Psychology*, 1942, *33*, 113–142.

Carr, C. E., and Konishi, M. Axonal delay lines for time measurement in the owl's brainstem. *Proceedings of the National Academy of Sciences, USA*, 1989, *85*, 8311–8315.

Carr, C. E., and Konishi, M. A circuit for detection of inter-aural time differences in the brain stem of the barn owl. *Journal of Neuroscience*, 1990, *10*, 3227–3246.

Carr, D. B., and Sesack, S. R. Projections from the rat prefrontal cortex to the ventral tegmental area: Target specificity in the synaptic associations with mesoaccumbens and mesocortical neurons. *Journal of Neuroscience*, 2000, *20*, 3864–3873.

Carr, L., Iacoboni, M., Dubeau, M. C., Mazziotta, J. C., and Lenzi, G. L. Neural mechanisms of empathy in humans: A relay from neural systems for imitation to limbic areas. *Proceedings of the National Academy of Sciences, USA*, 2003, *100*, 5497–5502.

Carrera, M. R., Ashley, J. A., Parsons, L. H., Wirsching, P., et al. Suppression of psychoactive effects of cocaine by active immunization. *Nature*, 1995, *378*, 727–730.

Carroll, R. C., Lissin, D. V., von Zastrow, M., Nicolol, R. A., and Malenka, R. C. Rapid redistribution of glutamate receptors contributes to long-term depression in hippocampal cultures. *Nature Neuroscience*, 1999, *2*, 454–460.

Carter, C. S. Hormonal influences on human sexual behavior. In *Behavioral Endocrinology*, edited by J. B. Becker, S. M. Breedlove, and D. Crews. Cambridge, Mass.: MIT Press, 1992.

Caspi, A., Sugden, K., Moffitt, T. E., Taylor, A., et al. Influence of life stress on depression: Moderation by a polymorphism in the 5-HTT gene. *Science*, 2003, *301*, 386–389.

Castelli, F., Frith, C., Happé, F., and Frith, U. Autism, Asperger syndrome and brain mechanisms for the attribution of mental states to animated shapes. *Brain*, 2002, *125*, 1839–1849.

Catani, M., Jones, D. K., and ffytche, D. H. Perisylvian language networks of the human brain. *Annals of Neurology*, 2005, *57*, 8–16.

Caterina, M. J., Leffler, A., Malmberg, A. B., Martin, W. J., et al. Impaired nociception and pain sensation in mice lacking the capsaicin receptor. *Science*, 2000, *288*, 306–313.

Cavaco, S., Anderson, S. W., Allen, J. S., Castro-Caldas, A., and Damasio, H. The scope of preserved procedural memory in amnesia. *Brain*, 2004, *127*, 1863–1867.

Cecil, J. E., Francis, J., and Read, N. W. Relative contributions of intestinal, gastric, oro-sensory influences and information to changes in appetite induced by the same liquid meal. *Appetite*, 1998, *31*, 377–390.

Cenci, M. A., Kalen, P., Mandel, R. J., and Bjoerklund, A. Regional differences in the regulation of dopamine and noradrenaline release in medial frontal cortex, nucleus accumbens and caudate-putamen: A microdialysis study in the rat. *Brain Research*, 1992, *581*, 217–228.

Cerletti, U., and Bini, L. Electric shock treatment. *Bollettino ed Atti della Accademia Medica di Roma*, 1938, *64*, 36.

Chambers, R. A., Taylor, J. R., and Potenza, M. N. Developmental neurocircuitry of motivation in adolescence: A critical period of addiction vulnerability. *American Journal of Psychiatry*, 2003, *160*, 1041–1052.

Chaminade, T., Meltzoff, A. N., and Decety, J. An fMRI study of imitation: Action representation and body schema. *Neuropsychologia*, 2005, *43*, 115–127.

Chan, J. L., Mun, E. C., Stoyneva, V., Mantzoros, C. S., and Goldfine, A. B. Peptide YY levels are elevated after gastric bypass surgery. *Obesity*, 2006, *14*, 194–198.

Chandler, L. J., Harris, R. A., and Crews, F. T. Ethanol tolerance and synaptic plasticity. *Trends in Pharmacological Science*, 1998, *19*, 491–495.

Chaney, M. A. Side effect of intrathecal and epidural opioids. *Canadian Journal of Anaesthology*, 1995, *42*, 891–903.

Chatterjee, S., and Callaway, E. M. Parallel colour-opponent pathways to primary visual cortex. *Nature*, 2003, *426*, 668–671.

Chehab, F. F., Mounzih, K., Lu, R., and Lim, M. E. Early onset of reproductive function in normal female mice treated with leptin. *Science*, 1997, *275*, 88–90.

Chemelli, R. M., Willie, J. T., Sinton, C. M., Elmquist, J. K., et al. Narcolepsy in orexin knockout mice: Molecular genetics of sleep regulation. *Cell*, 1999, *98*, 437–451.

Chen, G., Greengard, P., and Yan, Z. Potentiation of NMDA receptor currents by dopamine D_1 receptors in prefrontal cortex. *Proceedings of the National Academy of Sciences, USA*, 2004, 101, 2596–2600.

Chen, J., Marmer, R., Pulles, A., Paredes, W., and Gardner, E. L. Ventral tegmental microinjection of delta⁹-tetrahydrocannabinol enhances ventral tegmental somatodendritic dopamine levels but not forebrain dopamine levels: Evidence for local neural action by marijuana's psychoactive ingredient. *Brain Research*, 1993, *621*, 65–70.

Chen, J., Paredes, W., Li, J., Smith, D., Lowinson, J., and Gardner, E. L. Delta⁹-tetrahydrocannabinol produces naloxone-blockable enhancement of presynaptic basal dopamine efflux in nucleus accumbens of conscious, freely moving rats as measured by intracerebral microdialysis. *Psychopharmacology*, 1990, *102*, 156–162.

Chen, W., Landau, S., Sham, P., and Fombonne, E. No evidence for links between autism, MMR and measles virus. *Psychological Medicine*, 2004, *3*, 543–553.

Chen, Y. W., and Dilsaver, S. C. Lifetime rates of suicide attempts among subjects with bipolar and unipolar disorders relative to subjects with other axis I disorders. *Biological Psychiatry*, 1996, *39*, 896–899.

Chen, Y.-C., Thaler, D., Nixon, P. D., Stern, C. E., and Passingham, R. E. The functions of the medial premotor cortex. II: The timing and selection of learned movements. *Experimental Brain Research*, 1995, *102*, 461–473.

Cheng, M. Y., Bullock, C. M., Li, C., Lee, A. G., et al. Prokineticin 2 transmits the behavioural circadian rhythm of the suprachiasmatic nucleus. *Nature*, 2002, *417*, 405–410.

Chenn, A., and Walsh, C. A. Regulation of cerebral cortical size by control of cell cycle exit in neural precursors. *Science*, 2002, *297*, 365–369.

Cho, M. M., DeVries, A. C., Williams, J. R., and Carter, C. S. The effects of oxytocin and vasopressin on partner preferences in male and female prairie voles (*Microtus ochrogaster*). *Behavioral Neuroscience*, 1999, *113*, 1071–1079.

Chou, I.-H., and Narasimhan, K. Neurobiology of addiction. *Nature Neuroscience*, 2005, *8*, 1427.

Chou, T. C., Bjorkum, A. A., Gaus, S. E., Lu, J., Scammell, T. E., and Saper, C. B. Afferents to the ventrolateral preoptic nucleus. *Journal of Neuroscience*, 2002, *22*, 977–990.

Chubb, J. E., Bradshaw, N. J., Soares, D. C., Porteous, D. J., and Millar, J. K. The DISC locus in psychiatric illness. *Molecular Psychiatry*, 2008, *13*, 36–64.

Cipolotti, L., and Warrington, E. K. Does recognizing orally spelled words depend on reading? An investigation into a case of better written than oral spelling. *Neuropsychologia*, 1996, *34*, 427–440.

Clapham, J. C., Arch, J. R., Chapman, H., Haynes, A., et al. Mice overexpressing human uncoupling protein-3 in skeletal muscle are hyperphagic and lean. *Nature*, 2000, *406*, 415–418.

Clark, J. T., Kalra, P. S., Crowley, W. R., and Kalra, S. P. Neuropeptide Y and human pancreatic polypeptide stimulates feeding behavior in rats. *Endocrinology*, 1984, *115*, 427–429.

Cnattingius, S., Hultman, C. M., Dahl, M., and Sparen, P. Very preterm birth, birth trauma, and the risk of anorexia nervosa among girls. *Archives of General Psychiatry*, 1999, *56*, 634–638.

Cobb, S., and Rose, R. M. Hypertension, peptic ulcer, and diabetes in air traffic controllers. *Journal of the American Medical Association*, 1973, *224*, 489–492.

Coccaro, E. F., and Kavoussi, R. J. Fluoxetine and impulsive aggressive behavior in personality-disordered subjects. *Archives of General Psychiatry*, 1997, *54*, 1081–1088.

Coccaro, E. F., Silverman, J. M., Klar, H. M., Horvath, T. B., and Siever, L. J. Familial correlates of reduced central serotonergic system function in patients with personality disorders. *Archives of General Psychiatry*, 1994, *51*, 318–324.

Cohen, C., Kodas, E., and Griebel, G. CB_1 receptor antagonists for the treatment of nicotine addiction. *Pharmacology, Biochemistry and Behavior*, 2005, *81*, 387–395.

Cohen, E. A. *Human Behavior in the Concentration Camp*. New York: W. W. Norton, 1953.

Cohen, S., Tyrrell, D. A. J., and Smith, A. P. Psychological stress and susceptibility to the common cold. *New England Journal of Medicine*, 1991, *325*, 606–612.

Cohen-Bendahan, C. C., Buitelaar, J. K., van Goozen, S. H. M., Orlebeke, J. F., et al. Is there an effect of prenatal testosterone on aggression and other behavioral traits? A study comparing same-sex and opposite-sex twin girls. *Hormones and Behavior*, 2005, *47*, 230–237.

Cohen-Bendahan, C. C., van de Beek, C., and Berenbaum SA. Prenatal sex hormone effects on child and adult sex-typed behavior: Methods and findings. *Neuroscience and Biobehavioral Reviews*, 2005, *29*, 353–384.

Colapinto, J. *As Nature Made Him: The Boy Who Was Raised as a Girl*. New York: HarperCollins, 2000.

Cole, B. J., and Koob, G. F. Propranalol antagonizes the enhanced conditioned fear produced by corticotropin releasing factor. *Journal of Pharmacology and Experimental Therapeutics*, 1988, *247*, 901–910.

Cole, J. Empathy needs a face. *Journal of Consciousness Studies*, 2001, *8*, 51–68.

Collaborative Research Group. A novel gene containing a trinucleotide repeat that is expanded and unstable on Huntington's disease chromosomes. *Cell*, 1993, *72*, 971–983.

Comuzzie, A. G., and Allison, D. B. The search for human obesity genes. *Science*, 1998, *280*, 1374–1377.

Connellan, J., Baron-Cohen, S., Wheelwright, S., Batki, A., and Ahluwalia, J. Sex differences in human neonatal social perception. *Infant Behavior and Development*, 2001, *48*, 1411–1416.

Connolly, J. D., Andersen, R. A., and Goodale, M. A. fMRI evidence for a "parietal reach region" in the human brain. *Experimental Brain Research*, 2003, *153*, 140–145.

Conway, B. R., Moeller, S., and Tsao, D. Y. Specialized color modules in macaque extrastriate cortex. *Neuron*, 2007, *56*, 560–573.

Coolen, L. M., Allard, J., Truitt, W. A., and McKenna, K. E. Central regulation of ejaculation. *Physiology and Behavior*, 2004, *83*, 203–215.

Coolen, L. M., and Wood, R. I. Testosterone stimulation of the medial preoptic area and medial amygdala in the control of male hamster sexual behavior: Redundancy without amplification. *Behavioural Brain Research*, 1999, *98*, 143–153.

Cooper, J. A. A mechanism for inside-out lamination in the neocortex. *Trends in Neurosciences*, 2008, *32*, 113–119.

Coover, G. D., Murison, R., and Jellestad, F. K. Subtotal lesions of the amygdala: The rostral central nucleus in passive avoidance and ulceration. *Physiology and Behavior*, 1992, *51*, 795–803.

Copeland, B. J., and Pillsbury, H. C. Cochlear implantation for the treatment of deafness. *Annual Review of Medicine*, 2004, *55*, 157–167.

Corey, D. P., Garcia-Añoveros, J., Holt, J. R., Kwan, K. Y., et al. TRPA1 is a candidate for the mechanosensitive transduction channel of vertebrate hair cells. *Nature*, 2004, *432*, 723–730.

Corkin, S., Amaral, D. G., González, R. G., Johnson, K. A., and Hyman, B. R. H. M.'s medial temporal lobe lesion: Findings from magnetic resonance imaging. *Journal of Neuroscience*, 1997, *17*, 3964–3979.

Corkin, S., Sullivan, E. V., Twitchell, T. E., and Grove, E. The amnesic patient H. M.: Clinical observations and test performance 28 years after operation. *Society for Neuroscience Abstracts*, 1981, *7*, 235.

Cornuz, J., Zwahlen, S., Jungi, W. G., Osterwalder, J., et al. A vaccine against nicotine for smoking cessation: A randomized controlled trial. *PLoS One*, 2008, *3*, e2547.

Corrigall, W. A., Coen, K. M., and Adamson, K. L. Self-administered nicotine activates the mesolimbic dopamine system through the ventral tegmental area. *Brain Research*, 1994, *653*, 278–284.

Corwin, J. T., and Warchol, M. E. Auditory hair cells: Structure, function, development, and regeneration. *Annual Review of Neuroscience*, 1991, *14*, 301–333.

Cossu, G., Ledent, C., Fattore, L., Imperato, A., et al. Cannabinoid CB_1 receptor knockout mice fail to self-administer morphine but not other drugs of abuse. *Behavioural Brain Research*, 2001, *118*, 61–65.

Cotman, C. W., Berchtold, N. C., and Christie, L.-A. Exercise builds brain health: Key roles of growh factor cascades and inflammation. *Trends in Neuroscience*, 2007, *30*, 464–472.

Cottingham, S. L., and Pfaff, D. Interconnectedness of steroid hormone-binding neurons: Existence and implications. *Current Topics in Neuroendocrinology*, 1986, *7*, 223–249.

Courchesne, E., Pierce, K., Schumann, C. M., Redcay, E., et al. Mapping early brain development in autism. *Neuron*, 2007, *56*, 399–413.

Courchesne, E., Redcay, E., Morgan, J. T., and Kennedy, D. P. Autism at the beginning: Microstructural and growth abnormalities underlying the cognitive and behavioral phenotype of autism. *Development and Psychopathology*, 2005, *17*, 577–597.

Court, J., Bergh, C., and Södersten, P. Mandometer treatment of Australian patients with eating disorders. *Medical Journal of Australia,* 2008, *288,* 120–121.

Covington, H. E., and Miczek, K. A. Repeated social-defeat stress, cocaine or morphine: Effects on behavioral sensitization and intravenous cocaine self-administration "binges." *Psychopharmacology,* 2001, *158,* 388–398.

Cowley, M. A., Smith, R. G., diano, S., Tschöp, M., et al. The distribution and mechanism of action of ghrelin in the CNS demonstrates a novel hypothalamic circuit regulating energy homeostasis. *Neuron,* 2003, *37,* 649–661.

Cox, A., Rutter, M., Newman, S., and Bartak, L. A comparative study of infantile autism and specific developmental language disorders. I: Parental characteristics. *British Journal of Psychiatry,* 1975, *126,* 146–159.

Cox, D., Meyers, E., and Sinha, P. Contextually evoked object-specific responses in human visual cortex. *Science,* 2004, *304,* 115–117.

Cox, J. J., Reimann, F., Nicholas, A. K., Thornton, G., et al. An SCN9A channelopathy causes congenital inability to experience pain. *Nature,* 2006, *444,* 894–898.

Crane, G. E. Iproniazid (Marsilid) phosphate, a therapeutic agent for mental disorders and debilitating diseases. *Psychiatry Research Reports,* 1957, *8,* 142–152.

Creese, I., Burt, D. R., and Snyder, S. H. Dopamine receptor binding predicts clinical and pharmacological potencies of antischizophrenic drugs. *Science,* 1976, *192,* 481–483.

Criado, J. R., Sanchez-Alavez, M., Conti, B., Giacchino, J. L., et al. Mice devoid of prion protein have cognitive deficits that are rescued by reconstitution of PrP in neurons. *Neurobiology of Disease,* 2005, *19,* 255–265.

Crow, T. J. How and why genetic linkage has not solved the problem of psychosis: Review and hypothesis. *American Journal of Psychiatry,* 2007, *164,* 13–21.

Cubelli, R. A selective deficit for writing vowels in acquired dysgraphia. *Nature,* 1991, *353,* 258–260.

Culebras, A., and Moore, J. T. Magnetic resonance findings in REM sleep behavior disorder. *Neurology,* 1989, *39,* 1519–1523.

Culham, J. C., and Kanwisher, N. Neuroimaging of cognitive functions in human parietal cortex. *Current Opinion in Neurobiology,* 2001, *11,* 157–163.

Culotta, E., and Koshland, D. E. NO news is good news. *Science,* 1992, *258,* 1862–1865.

Cunningham, J. T., Beltz, T. G., Johnson, R. F., and Johnson, A. K. The effects of ibotenate lesions of the median preoptic on experimentally induced and circadian drinking behaviors in rats. *Brain Research,* 1992, *580,* 325–330.

Czerniczyniec, A., Bustamante, J., and Arnaiz-Lores, S. Improvement of mouse brain mitochondrial function after deprenyl treatment. *Neuroscience,* 2007, *144,* 685–693.

Czisch, M., Wehrle, R., Kaufmann, C., Wetter, T. C., et al. Functional MRI during sleep: BOLD signal decreases and their electrophysiological correlates. *European Journal of Neuroscience,* 2004, *20,* 566–574.

Daban, C., Martinez-Aran, A., Cruz, N., and Vieta, E. Safety and efficacy of vagus nerve stimulation in treatment-resistant depression: A systematic review. *Journal of Affective Disorders,* 2008, *110,* 1–15.

Dabbs, J. M., and Morris, R. Testosterone, social class, and antisocial behavior in a sample of 4,462 men. *Psychological Science,* 1990, *1,* 209–211.

Daglish, M. R., Weinstein, A., Malizia, A. L., Wilson, S., et al. Functional connectivity analysis of the neural circuits of opiate craving: "More" rather than "different"? *NeuroImage,* 2003, *20,* 1964–1970.

Dallos, P. The active cochlea. *Journal of Neuroscience,* 1992, *12,* 4575–4585.

Damasio, A. R., and Damasio, H. The anatomic basis of pure alexia. *Neurology,* 1983, *33,* 1573–1583.

Damasio, A. R., and Damasio, H. Hemianopia, hemiachromatopsia, and the mechanisms of alexia. *Cortex,* 1986, *22,* 161–169.

Damasio, A. R., Damasio, H., and Van Hoesen, G. W. Prosopagnosia: Anatomic basis and behavioral mechanisms. *Neurology,* 1982, *32,* 331–341.

Damasio, A. R., Grabowski, T. J., Bechara, A., Damasio, H., et al. Subcortical and cortical brain activity during the feeling of self-generated emotions. *Nature Neuroscience,* 2000, *3,* 1049–1056.

Damasio, A. R., and Tranel, D. Nouns and verbs are retrieved with differentially distributed neural systems. *Proceedings of the National Academy of Sciences, USA,* 1993, *90,* 4957–4960.

Damasio, A. R., Yamada, T., Damasio, H., Corbett, J., and McKee, J. Central achromatopsia: Behavioral, anatomic, and physiologic aspects. *Neurology,* 1980, *30,* 1064–1071.

Damasio, H. Neuroimaging contributions to the understanding of aphasia. In *Handbook of Neuropsychology, Vol. 2,* edited by F. Boller and J. Grafman. Amsterdam: Elsevier, 1989.

Damasio, H., and Damasio, A. R. The anatomical basis of conduction aphasia. *Brain,* 1980, *103,* 337–350.

Damasio, H., Eslinger, P., and Adams, H. P. Aphasia following basal ganglia lesions: New evidence. *Seminars in Neurology,* 1984, *4,* 151–161.

Damasio, H., Grabowski, T., Frank, R., Galaburda, A. M., and Damasio, A. R. The return of Phineas Gage: Clues about the brain from the skull of a famous patient. *Science,* 1994, *264,* 1102–1105.

Damsma, G., Day, J., and Fibiger, H. C. Lack of tolerance to nicotine-induced dopamine release in the nucleus accumbens. *European Journal of Pharmacology,* 1989, *168,* 363–368.

Dani, J. A., and De Biasi, M. Cellular mechanisms of nicotine addiction. *Pharmacology, Biochemistry, and Behavior,* 2001, *70,* 439–446.

Dani, J. A., and Harris, R. A. Nicotine addiction and comorbidity with alcohol abuse and mental illness. *Nature Neuroscience,* 2005, *8,* 1465–1470.

Daniele, A., Giustolisi, L., Silveri, M. C., Colosimo, C., and Gainotti, G. Evidence for a possible neuroanatomical basis for lexical processing of nouns and verbs. *Neuropsychologia,* 1994, *32,* 1325–1341.

Daniels, D., Miselis, R. R., and Flanagan-Cato, L. M. Central neuronal circuit innervating the lordosis-producing muscles defined by transneuronal transport of pseudorabies virus. *Journal of Neuroscience,* 1999, *19,* 2823–2833.

Dapretto, M., Davies, M. S., Pfeifer, J. H., Scott, A. A., et al. Understanding emotions in others: Mirror neuron dysfunction in children with autism spectrum disorders. *Nature Neuroscience,* 2006, *9,* 28–30.

Darwin, C. *The Expression of the Emotions in Man and Animals.* Chicago: University of Chicago Press, 1872/1965.

Davidson, D., Swift, R., and Fitz, E. Naltrexone increases the latency to drink alcohol in social drinkers. *Alcoholism: Clinical and Experimental Research,* 1996, *20,* 732–739.

Davies, G., Welham, J., Chant, D., Torrey, E. F., and McGrath, J. A systematic review and meta-analysis of Northern Hemisphere season of birth studies in schizophrenia. *Schizophrenia Bulletin,* 2003, *29,* 587–593.

Davis, J. D., and Campbell, C. S. Peripheral control of meal size in the rat: Effect of sham feeding on meal size and drinking rate. *Journal of Comparative and Physiological Psychology,* 1973, *83,* 379–387.

Davis, J. O., and Bracha, H. S. Famine and schizophrenia: First-trimester malnutrition or second-trimester beriberi? *Biological Psychiatry,* 1996, *40,* 1–3.

Davis, J. O., Phelps, J. A., and Bracha, H. S. Prenatal development of monozygotic twins and concordance for schizophrenia. *Schizophrenia Bulletin,* 1995, *21,* 357–366.

Davis, M. The role of the amygdala in fear-potentiated startle: Implications for animal models of anxiety. *Trends in Pharmacological Sciences,* 1992, *13,* 35–41.

Daw, N. W. Colour-coded ganglion cells in the goldfish retina: Extension of their receptive fields by means of new stimuli. *Journal of Physiology (London),* 1968, *197,* 567–592.

Dawson, T. M., and Dawson, V. L. Molecular pathways of neurodegeneration in Parkinson's disease. *Science,* 2003, *302,* 819–822.

Day, J. J., Roitman, M. F., Wightman, R. M., and Carelli, R. M. Associative learning mediates dynamic shifts in dopamine signaling in the nucleus accumbens. *Nature Neuroscience,* 2007, *10,* 1020–1028.

de Castro, J. M. The control of eating behavior in free-living humans. In *Neurobiology of Food and Fluid Intake,* 2nd ed., edited by E. Stricker and S. Woods. New York: Plenum Publishers, 2004.

de Charms, R. C., Maeda, F., Glover, G. H., Ludlow, D., et al. Control over brain activation and pain learned by using real-time functional MRI. *Proceedings of the National Academy of Sciences, USA,* 2005, *102,* 18626–18631.

de Gelder, B. Towards the neurobiology of emotional body language. *Nature Reviews: Neuroscience,* 2006, *7,* 242–249.

de Gelder, B., Vroomen, J., Pourtois, G., and Weiskrantz, L. Non-conscious recognition of affect in the absence of striate cortex. *Neuroreport,* 1999, *10,* 3759–3763.

De Jonge, F. H., Louwerse, A. L., Ooms, M. P., Evers, P., et al. Lesions of the SDN-POA inhibit sexual behavior of male Wistar rats. *Brain Research Bulletin,* 1989, *23,* 483–492.

De Jonge, F. H., Oldenburger, W. P., Louwerse, A. L., and van de Poll, N. E. Changes in male copulatory behavior after

sexual exciting stimuli: Effects of medial amygdala lesions. *Physiology and Behavior,* 1992, *52,* 327–332.

de Kloet, E. R., Joëls, M., and Holsboer, F. Stress and the brain: From adaptation to disease. *Nature Reviews: Neuroscience,* 2005, *6,* 463–475.

De Ocampo, J., Foldvary, N., Dinner, D. S., and Golish, J. *Sleep Medicine,* 2002, *3,* 525–526.

De Strooper, B. Aph-1, Pen-2, and nicastrin with presenilin generate an active γ-secretase complex. *Neuron,* 2003, *38,* 9–12.

De Valois, R. L., Albrecht, D. G., and Thorell, L. Cortical cells: Bar detectors or spatial frequency filters? In *Frontiers in Visual Science,* edited by S. J. Cool and E. L. Smith. Berlin: Springer-Verlag, 1978.

De Valois, R. L., and De Valois, K. K. *Spatial Vision.* New York: Oxford University Press, 1988.

De Valois, R. L., Thorell, L. G., and Albrecht, D. G. Periodicity of striate-cortex-cell receptive fields. *Journal of the Optical Society of America,* 1985 *2,* 1115–1123.

De Vries, T. J., and Schoffelmeer, A. N. M. Cannabinoid CB_1 receptors control conditioned drug seeking. *Trends in Pharmacological Sciences,* 2005, *26,* 420–426.

Deacon, S., and Arendt, J. Adapting to phase shifts. I: An experimental model for jet lag and shift work. *Physiology and Behavior,* 1996, *59,* 665–673.

Dealberto, M. J. Why are immigrants at increased risk for psychosis? Vitamin D insufficiency, epigenetic mechanisms, or both? *Medical Hypotheses,* 2007, *68,* 259–267.

Dean, P. Effects of inferotemporal lesions on the behavior of monkeys. *Psychological Bulletin,* 1976, *83,* 41–71.

Debanne, D., Gähwiler, B. H., and Thompson, S. M. Asynchronous pre- and postsynaptic activity induces associative long-term depression in area CA1 of the rat hippocampus in vitro. *Proceedings of the National Academy of Sciences, USA.* 1994, *91,* 1148–1152.

Debiec, J., LeDoux, J. E., and Nader, K. Cellular and systems reconsolidation in the hippocampus. *Neuron,* 2002, *36,* 527–538.

Deffenbacher, K. E., Kenyon, J. B., Hoover, D. M., Olson, R. K., et al. Refinement of the 6p21.3 quantitative trait locus influencing dyslexia: Linkage and association analysis. *Human Genetics,* 2004, *115,* 128–138.

Dejerine, J. Contribution à l'étude anatomo-pathologique et clinique des différentes variétés de cécité verbale. *Comptes Rendus des Séances de la Société de Biologie et de Ses Filiales,* 1892, *4,* 61–90.

Del Cerro, M. C. R., Izquierdo, M. A. P., Rosenblatt, J. S., Johnson, B. M., et al. Brain 2-deoxyglucose levels related to maternal behavior-inducing stimuli in the rat. *Brain Research,* 1995, *696,* 213–220.

Delay, J., and Deniker, P. Le traitement des psychoses par une methode neurolytique derivée d'hibernothérapie: Le 4560 RP utilisée seul une cure prolongée et continuée. *Comptes Rendus Congrès des Médecins Aliénistes et Neurologistes de France et des Pays de Langue Française,* 1952a, *50,* 497–502.

Delay, J., and Deniker, P. 38 cas des psychoses traitées par la cure prolongée et continuée de 4560 RP. *Comptes Rendus du Congrès des Médecins Aliénistes et Neurologistes de France et des Pays de Langue Française,* 1952b, *50,* 503–513.

Delgado, P. L., Charney, D. S., Price, L. H., Aghajanian, G. K., et al. Serotonin function and the mechanism of antidepressant action: Reversal of antidepressant induced remission by rapid depletion of plasma tryptophan. *Archives of General Psychiatry*, 1990, *47*, 411–418.

DeLoache, J. S., Uttal, D. H., and Rosengren, K. S. Scale errors offer evidence for a perception-action dissociation early in life. *Science*, 2004, *304*, 1027–1029.

DeLong, G. R. Autism: New data suggest a new hypothesis. *Neurology*, 1999, *52*, 911–916.

DelParigi, A., Tschöp, M., Heiman, M. L., Salbe, A. D., et al. High circulating ghrelin: A potential cause for hyperphagia and obesity in Prader-Willi syndrome. *Journal of Clinical Endocrinology and Metabolism*, 2002, *87*, 5461–5464.

Dement, W. C. The effect of dream deprivation. *Science*, 1960, *131*, 1705–1707.

Démonet, J.-F., Taylor, M. J., and Chaix, Y. Developmental dyslexia. *Lancet*, 2004, *363*, 1451–1460.

Deol, M. S., and Glueksohn-Waelsch, S. The role of inner hair cells in hearing. *Nature*, 1979, *278*, 250–252.

Deshmukh, R., and Franco, K. Managing weight gain as a side effect of antidepressant therapy. *Cleveland Clinic Journal of Medicine*, 2003, *70*, 616–618.

Deurveilher, S., and Semba, K. Indirect projections from the suprachiasmatic nucleus to major arousal-promoting cell groups in rat: Implications for the circadian control of behavioural state. *Neuroscience*, 2005, *130*, 165–183.

Deutsch, J. A., and Gonzalez, M. F. Gastric nutrient content signals satiety. *Behavioral and Neural Biology*, 1980, *30*, 113–116.

Devane, W. A., Hanus, L., Breuer, A., Pertwee, R. G., et al. Isolation and structure of a brain constituent that binds to the cannabinoid receptor. *Science*, 1992, *258*, 1946–1949.

Devilbiss, D. M., and Berridge, C. W. Cognition-enhancing doses of methylphenidate preferentially increase prefrontal cortex neuronal responsiveness. *Biological Psychiatry*, 2008, *64*, 626–635.

Devine, D. P., and Wise, R. A. Self-administration of morphine, DAMGO, and DPDPE into the ventral tegmental area of rats. *Journal of Neuroscience*, 1994, *14*, 1978–1984.

Dewey, S. L., Chaurasia, C. S., Chen, C., Volkow, N. D., et al. GABAergic attenuation of cocaine-induced dopamine release and locomotor activity. *Synapse*, 1997, *25*, 393–398.

Dewey, S. L., Chaurasia, C. S., Chen, C., Volkow, N. D., et al. A novel strategy for the treatment of cocaine addiction. *Synapse*, 1998, *30*, 119–129.

Di Chiara, G. The role of dopamine in drug abuse viewed from the perspective of its role in motivation. *Drug and Alcohol Dependency*, 1995, *38*, 95–137.

Di Ciano, P., Coury, A., Depoortere, R. Y., Egilmez, Y., et al. Comparison of changes in extracellular dopamine concentrations in the nucleus accumbens during intravenous self-administration of cocaine or d-amphetamine. *Behavioural Pharmacology*, 1995, *6*, 311–322.

Di Marzo, V., and Matias, I. Endocannabinoid control of food intake and energy balance. *Nature Neuroscience*, 2005, *8*, 585–589.

di Tomaso, E., Beltramo, M., and Piomelli, D. Brain cannabinoids in chocolate. *Nature*, 1996, *382*, 677–678.

Diamond, D. M., Park, C. R., Heman, K. L., and Rose, G. M. Exposing rats to a predator impairs spatial working memory in the radial arm water maze. *Hippocampus*, 1999, *9*, 542–552.

Diamond, M., and Sigmundson, H. K. Sex reassignment at birth: Long-term review and clinical implications. *Archives of Pediatric and Adolescent Medicine*, 1997, *151*, 298–304.

Diana, M., Pistis, M., Carboni, S., Gessa, G. L., and Rossetti, Z. L. Profound decrement of mesolimbic dopaminergic neuronal activity during ethanol withdrawal syndrome in rats: Electrophysiological and biochemical evidence. *Proceedings of the National Academy of Sciences, USA*, 1993, *90*, 7966–7969.

DiFiglia, M., Sena-Esteves, M., Chase, K., Sapp, E., et al. Therapeutic silencing of mutant huntingtin with siRNA attenuates striatal and cortical neuropathology and behavioral deficits. *Proceedings of the National Academy of Sciences, USA*, 2007, *104*, 17204–17209.

Dijk, D. J., Boulos, Z., Eastman, C. I., Lewy, A. J., Campbell, S. S., and Terman, M. Light treatment for sleep disorders: Consensus report. 2: Basic properties of circadian physiology and sleep regulation. *Journal of Biological Rhythms*, 1995, *10*, 113–125.

Dixon, A. K. The effect of olfactory stimuli upon the social behaviour of laboratory mice (*Mus musculus L*). Doctoral dissertation, Birmingham University, Birmingham, England, 1973.

Dixon, A. K., and Mackintosh, J. H. Effects of female urine upon the social behaviour of adult male mice. *Animal Behaviour*, 1971, *19*, 138–140.

Doetsch, F., and Hen, R. Young and excitable: The function of new neurons in the adult mammalian brain. *Current Opinion in Neuroscience*, 2005, *15*, 121–128.

Dolan, R. P., and Schiller, P. H. Evidence for only depolarizing rod bipolar cells in the primate retina. *Visual Neuroscience*, 1989, *2*, 421–424.

Dominguez, J. M., Gil, M., and Hull, E. M. Preoptic glutamate facilitates male sexual behavior. *Journal of Neuroscience*, 2006, *26*, 1699–1703.

Domschke, K., Stevene, S., Beck, B., Baffa, A., et al. Bluching propensity in social anxiety disorder: Influence of serotonin transporter gene variation. *Journal of Neural Transmission*, 2008 in press.

Donny, E. C., Caggiula, A. R., Knopf, S., and Brown, C. Nicotine self-administration in rats. *Psychopharmacology*, 1995, *122*, 390–394.

Dorries, K. M., Adkins, R. E., and Halpern, B. P. Sensitivity and behavioral responses to the pheromone androsteneone are not mediated by the vomeronasal organ in domestic pigs. *Brain, Behavior, and Evolution*, 1997, *49*, 53–62.

Doty, R. L. Olfaction. *Annual Review of Psychology*, 2001, *52*, 423–452.

Doucet, M. E., Bergeron, F., Lassonde, M., Ferron, P., and Lepore, F. Cross-modal reorganization and speech perception in cochlear implant users. *Brain*, 2006, *129*, 3376–3383.

Dougherty, D. D., Baie, L., Gosgrove, G. R., Cassem, E. H., et al. Prospective long-term follow-up of 44 patients who received cingulotomy for treatment-refractory obsessive-compulsive disorder. *American Journal of Psychiatry,* 2002, *159,* 269–275.

Douglass, J., McKinzie, A. A., and Couceyro, P. PCR differential display identifies a rat brain mRNA that is transcriptionally regulated by cocaine and amphetamine. *Journal of Neuroscience,* 1995, *15,* 2471–2481.

Dovey, H. F., John, V., Anderson, J. P., Chen, L. Z., et al. Functional gamma-secretase inhibitors reduce beta-amyloid peptide levels in brain. *Journal of Neurochemistry,* 2001, *76,* 173–181.

Downing, P. E., Chan, A. W.-Y., Peelen, M. V., Dodds, C. M., and Kanwisher, N. Domain specificity in visual cortex. *Cerebral Cortex,* 2005, *16,* 1453–1461.

Downing, P. E., Jiang, Y., Shuman, M., Kanwisher, N. A cortical area selective for visual processing of the human body. *Science,* 2001, *293,* 2470–2473.

Drake, C. L., Roehrs, T., Richardson, G., Walsh, J. K., and Roth, T. Shift work sleep disorder: Prevalence and consequences beyond that of symptomatic day workers. *Sleep,* 2004, *27,* 1453–1462.

Drayna, D., Manichaikul, A., de Lange, M., Snieder, H., and Spector, T. Genetic correlates of musical pitch recognition in humans. *Science,* 2001, *291,* 1969–1972.

Drevets, W. C. Functional anatomical abnormalities in limbic and prefrontal cortical structures in major depression. *Progress in Brain Research,* 2000, *126,* 413–431.

Drevets, W. C. Neuroimaging and neuropathological studies of depression: implications for the cognitive-emotional features of mood disorders. *Current Opinions in Neurobiology,* 2001, *11,* 240–249.

Drevets, W. C., Price, J. L., Simpson, J. R., Todd, R. D., et al. Subgenual prefrontal cortex abnormalities in mood disorders. *Nature,* 1997, *386,* 824–827.

Drevets, W. C., Videen, T. O., Price, J. L., Preskorn, S. H., et al. A functional anatomical study of unipolar depression. *Journal of Neuroscience,* 1992, *12,* 3628–3641.

Dronkers, N. F. A new brain region for coordinating speech articulation. *Nature,* 1996, *384,* 159–161.

Dube, M. G., Kalra, S. P., and Kalra, P. S. Food intake elicited by central administration of orexins/hypocretins: Identification of hypothalamic sites of action. *Brain Research,* 1999, *842,* 473–477.

Duchenne, G.-B. *The Mechanism of Human Facial Expression.* Translated by R. A. Cuthbertson. Cambridge, England: Cambridge University Press, 1990. (Original work published 1862.)

Dudek, S. M., and Bear, M. F. Homosynaptic long-term depression in area CA1 of hippocampus and effects of N-methyl-D-aspartate receptor blockade. *Proceedings of the National Academy of Sciences, USA,* 1992, *89,* 4363–4367.

Dukelow, S. P., DeSouza, J. F., Culham, J. C., van den Berg, A. V., et al. Distinguishing subregions of the human MT+ complex using visual fields and pursuit eye movements. *Journal of Neurophysiology,* 2001, *86,* 1991–2000.

Dulac, C., and Axel, R. A novel family of genes encoding putative pheromone receptors in mammals. *Cell,* 1995, *83,* 195–206.

Dunn, A. L., Trivedi, M. H., Kampert, J. B., Clark, C. G., and Chambliss, H. O. Exercise treatment for depression: Efficacy and dose response. *Americal Journal of Preventive Medicine,* 2005, *28,* 1–8.

Durie, D. J. Sleep in animals. In *Psychopharmacology of Sleep,* edited by D. Wheatley. New York: Raven Press, 1981.

Durston, S., Tottenham, N. T., Thomas, K. M., Davidson, M. C., et al. Differential patterns of striatal activation in young children with and without ADHD. *Biological Psychiatry,* 2003, *53,* 871–878.

Duvauchelle, C. L., and Ettenberg, A. Haloperidol attenuates conditioned place preferences produced by electrical stimulation of the medial prefrontal cortex. *Pharmacology, Biochemistry, and Behavior,* 1991, *38,* 645–650.

Dykes, R. W. Parallel processing of somatosensory information: A theory. *Brain Research Reviews,* 1983, *6,* 47–115.

Eaton, W. W., Kessler, R. C., Wittchen, H. U., and Magee, W. J. Panic and panic disorder in the United States. *American Journal of Psychiatry,* 1994, *151,* 413–420.

Eaton, W. W., Mortensen, P. B., and Frydenberg, M. Obstetric factors, urbanization and psychosis. *Schizophrenia Research,* 2000, *43,* 117–123.

Ebisawa, T., Uchiyama, M., Kajimura, N., Mishima, K., et al. Association of structural polymorphisms in the human *period3* gene with delayed sleep phase syndrome. *EMBRO Reports,* 2001, *2,* 342–346.

Eden, G. F., and Zeffiro, T. A. Neural systems affected in developmental dyslexia revealed by functional neuroimaging. *Neuron,* 1998, *21,* 279–282.

Edenberg, H. J., Dick, D. M., Xuei, X., Tian, J., et al. Variations in *GABRA2,* encoding the α2 subunit of the GABA$_A$ receptor, are associated with alcohol dependence and with brain oscillations. *American Journal of Human Genetics,* 2004, *74,* 705–714.

Edwards, D. P., Purpura, K. P., and Kaplan, E. Contrast sensitivity and spatial-frequency response of primate cortical neurons in and around the cytochrome oxidase blobs. *Vision Research,* 1995, *35,* 1501–1523.

Egan, G., Silk, T., Zamarripa, F., Williams, J., et al. Neural correlates of the emergence of consciousness of thirst. *Proceedings of the National Academy of Sciences, USA,* 200, *100,* 15241–15246.

Ehrsson, H. H., Spence, C., and Passingham, R. E. That's my hand! Activity in premotor cortex reflects feeling of ownership of a limb. *Science,* 2004, *305,* 875–877.

Ehrsson, H. H., Wiech, K., Weiskopf, N., Dolan, R. J., and Passingham, R. E. Threatening a rubber hand that you feel is yours elicits a cortical anxiety response. *Proceedings of the National Academy of Sciences, USA,* 2007, *104,* 9828–9833.

Eichenbaum, H., Otto, T., and Cohen, N. J. The hippocampus: What does it do? *Behavioral and Neural Biology,* 1992, *57,* 2–36.

Eichenbaum, H., Stewart, C., and Morris, R. G. M. Hippocampal representation in spatial learning. *Journal of Neuroscience,* 1990, *10,* 331–339.

Ekman, P. *The Face of Man: Expressions of Universal Emotions in a New Guinea Village*. New York: Garland STPM Press, 1980.

Ekman, P. Facial expressions of emotion: An old controversy and new findings. *Philosophical Transactions of the Royal Society of London B*, 1992, *335*, 63–69.

Ekman, P., and Davidson, R. J. Voluntary smiling changes regional brain activity. *Psychological Science*, 1993, *4*, 342–345.

Ekman, P., and Friesen, W. V. Constants across cultures in the face and emotion. *Journal of Personality and Social Psychology*, 1971, *17*, 124–129.

Ekman, P., Levenson, R. W., and Friesen, W. V. Autonomic nervous system activity distinguished between emotions. *Science*, 1983, *221*, 1208–1210.

El Mansari, M., and Blier, P. In vivo electrophysiological characterization of 5-HT receptors in the guinea pig head of caudate nucleus and orbitofrontal cortex. *Neuropharmacology*, 1997, *36*, 577–588.

El Mansari, M., Sakai, K., and Jouvet, M. Unitary characteristics of presumptive cholinergic tegmental neurons during the sleep-waking cycle in freely moving cats. *Experimental Brain Research*, 1989, *76*, 519–529.

Elbert, T., Flor, H., Birbaumer, N., Knecht, S., et al. Extensive reorganization of the somatosensory cortex in adult humans after nervous system injury. *Neuroreport*, 1994, *5*, 2593–2597.

Elbert, T., Pantev, C., Wienbruch, C., Rockstroh, B., and Taub, E. Increased cortical representation of the fingers of the left hand in string players. *Science*, 1995, *270*, 305–307.

Elias, C. F., Lee, C., Kelly, J., Aschkenasi, C., et al. Leptin activates hypothalamic CART neurons projecting to the spinal cord. *Neuron*, 1998b, *21*, 1375–1385.

Elias, C. F., Saper, C. B., Maratos-Flier, E., Tritos, N. A., et al. Chemically defined projections linking the mediobasal hypothalamus and the lateral hypothalamic area. *Journal of Comparative Neurology*, 1998a, *402*, 442–459.

Elias, M. Serum cortisol, testosterone and testosterone binding globulin responses to competitive fighting in human males. *Aggressive Behavior*, 1981, *7*, 215–224.

Elmquist, J. K., Elias, C. F., and Saper, C. B. From lesions to leptin: Hypothalamic control of food intake and body weight. *Neuron*, 1999, *22*, 221–232.

Elsworth, J. D., Jentsch, J. D., Morrow, B. A., Redmond, D. E., and Roth, R. H. Clozapine normalizes prefrontal cortex dopamine transmission in monkeys subchronically exposed to phencyclidine. *Neuropsychopharmacology*, 2008, *33*, 491–496.

Emmorey, K., Mehta, S., and Grabowski, T. J. The neural correlates of sign versus word production. *NeuroImage*, 2007, *36*, 202–208.

Endoh, M., Maiese, K., and Wagner, J. A. Expression of the neural form of nitric oxide synthase by CA1 hippocampal neurons and other central nervous system neurons. *Neuroscience*, 1994, *63*, 679–689.

Enggasser, J. L., and de Wit, H. Haloperidol reduces stimulant and reinforcing effects of ethanol in social drinkers. *Alcoholism: Clinical and Experimental Research*, 2001, *25*, 1448–1456.

Ernst, M., Zametkin, A. J., Matochik, J. A., Jons, P. H., and Cohen, R. M. DOPA decarboxylase activity in attention deficit hyperactivity disorder adults: A [fluorine[18]]flurodopa positron emission tomographic study. *Journal of Neuroscience*, 1998, *18*, 5901–5907.

Ernulf, K. E., Innala, S. M., and Whitam, F. L. Biological explanation, psychological explanation, and tolerance of homosexuals: A cross-national analysis of beliefs and attitudes. *Psychological Reports*, 1989, *248*, 183–188.

Ertelt, D., Small, S., Solodkin, A., Dettmers, C., McNamara, A., Binkofski, F., and Buccino, G. Action observation has a positive impact on rehabilitation of motor deficits after stroke. *NeuroImage*, 2007, *36*, T164–T173.

Eslinger, P. J., and Damasio, A. R. Severe disturbance of higher cognition after bilateral frontal lobe ablation: Patient EVR. *Neurology*, 1985, *35*, 1731–1741.

Evans, E. F. Auditory processing of complex sounds: An overview. *Philosophical Transactions of the Royal Society of London B*, 1992, *336*, 295–306.

Everson, C. A., and Wehr, T. A. Nutritional and metabolic adaptations to prolonged sleep deprivation in the rat. *American Journal of Physiology*, 1993, *264*, R376–R387.

Fadiga, L., Craighero, L., Buccino, G., and Rizzolatti, G. Speech listening specifically modulates the excitability of tongue muscles: A TMS study. *European Journal of Neuroscience*, 2002, *15*, 399–402.

Farber, N. B., Wozniak, D. F., Price, M. T., Labruyere, J., et al. Age-specific neurotoxicity in the rat associated with NMDA receptor blockade: Potential relevance to schizophrenia? *Biological Psychiatry*, 1995, *38*, 788–796.

Farina, C., Weber, M. S., Meinl, E., Wekerle, H., and Hohlfeld, R. Glatiramer acetate in multiple sclerosis: Update on potential mechanisms of action. *Lancet Neurology*, 2005, *4*, 567–575.

Farlow, M., Murrell, J., Ghetti, B., Unverzagt, F., et al. Clinical characteristics in a kindred with early-onset Alzheimers-disease and their linkage to a G–T change at position-2149 of amyloid precursor protein gene. *Neurology*, 1994, *44*, 105–111.

Farooqi, I. S., and O'Rahilly, S. Monogenic obesity in humans. *Annual Review of Medicine*, 2005, *56*, 443–458.

Farrell, M. J., Egan, G. F., Zamarripa, F., Shade, R., et al. Unique, common, and interacting cortical correlates of thirst and pain. *Proceedings of the National Academy of Sciences, USA*, 2006, *103*, 2416–2421.

Farroni, T., Johnson, M. H., Menon, E., Zulian, L., Faraguna, D., and Csibra, G. Newborns' preference for face-relevant stimuli: Effects of contrast polarity. *Proceedings of the National Academy of Sciences, USA*, 2005, *102*, 17245–17250.

Feder, H. H. Estrous cyclicity in mammals. In *Neuroendocrinology of Reproduction*, edited by N. T. Adler. New York: Plenum Press, 1981.

Feinle, C., Grundy, D., and Read, N. W. Effects of duodenal nutrients on sensory and motor responses of the human stomach to distension. *American Journal of Physiology*, 1997, *273*, G721–G726.

Fenoglio, K. A., Chen, Y., and Baram, T. Z. Neuroplasticity of the hypothalamic-pituitary-adrenal axis early in life requires recurrent recruitment of stress-regulation brain regions. *Journal of Neuroscience*, 2006, *26*, 2434–2442.

Fernandez, F., Morishita, W., Zuniga, E., Nguyen, J., et al. Pharmacotherapy for cognitive impairment in a mouse model of Down syndrome. *Nature Neuroscience*, 2007, *10*, 411–413.

Fernandez-Ruiz, J., Wang, J., Aigner, T. G., and Mishkin, M. Visual habit formation in monkeys with neurotoxic lesions of the ventrocaudal neostriatum. *Proceedings of the National Academy of Sciences, USA*, 2001, *98*, 4196–4201.

Ferris, C. F., Kulkarni, P., Sullivan, J. M., Harder, J. A., et al. Pup suckling is more rewarding than cocaine: Evidence from functional magnetic resonance imaging and three-dimensional computational analysis. *Journal of Neuroscience*, 2005, *25*, 149–156.

Fettiplace, R., and Hackney, C. M. The sensory and motor roles of auditory hair cells. *Nature Reviews: Neuroscience*, 2006, *7*, 19–29.

Fibiger, H. C. The dopamine hypothesis of schizophrenia and mood disorders: Contradictions and speculations. In *The Mesolimbic Dopamine System: From Motivation to Action*, edited by P. Willner and J. Scheel-Krüger. Chichester, England: John Wiley & Sons, 1991.

Field, T., Woodson, R., Greenberg, R., and Cohen, D. Discrimination and imitation of facial expressions in neonates. *Science*, 1982, *218*, 179–181.

Fieve, R. R. The clinical effects of lithium treatment. *Trends in Neurosciences*, 1979, *2*, 66–68.

Finger, S. *Origins of Neuroscience: A History of Explorations into Brain Function*. New York: Oxford University Press, 1994.

Finger, T. E., Danilova, V., Barrows, J., Bartel, D. L., Vigers, A. J., et al. ATP signaling is crucial for communication from taste buds to gustatory nerves. *Science*, 2005, *310*, 1495–1499.

Firestein, S., Zufall, F., and Shepherd, G. M. Single odor-sensitive channels in olfactory receptor neurons are also gated by cyclic nucleotides. *Journal of Neuroscience*, 1991, *11*, 3565–3572.

Fischer, H., Andersson, J. L. R., Furmark, T., and Fredrikson, M. Brain correlates of an unexpected panic attack: A human positron emission tomographic study. *Neuroscience Letters*, 1998, *251*, 137–140.

Fisher, C., Byrne, J., Edwards, A., and Kahn, E. A psychophysiological study of nightmares. *Journal of the American Psychoanalytic Association*, 1970, *18*, 747–782.

Fisher, C., Gross, J., and Zuch, J. Cycle of penile erection synchronous with dreaming (REM) sleep: Preliminary report. *Archives of General Psychiatry*, 1965, *12*, 29–45.

Fiske, A. P., and Haslam, N. Is obsessive-compulsive disorder a pathology of the human disposition to perform socially meaningful rituals? Evidence of similar content. *Journal of Nervous and Mental Disease*, 1997, *185*, 211–222.

Fitzgerald, P. Repetitive transcranial magnetic stimulation and electroconvulsive therapy: Complementary or competitive therapeutic options in depression? *Australas Psychiatry*, 2004, *12*, 234–238.

Fitzpatrick, D., Itoh, K., and Diamond, I. T. The laminar organization of the lateral geniculate body and the striate cortex in the squirrel monkey (*Saimiri sciureus*). *Journal of Neuroscience*, 1983, *3*, 673–702.

Fitzsimons, J. T., and Moore-Gillon, M. J. Drinking and antidiuresis in response to reductions in venous return in the dog: Neural and endocrine mechanisms. *Journal of Physiology (London)*, 1980, *308*, 403–416.

Flaum, M., and Andreasen, N. C. Diagnostic criteria for schizophrenia and related disorders: Options for DNS-IV. *Schizophrenia Bulletin*, 1990, *17*, 27–49.

Fleming, A. S., and Rosenblatt, J. S. Maternal behavior in the virgin and lactating rat. *Journal of Comparative and Physiological Psychology*, 1974, *86*, 957–972.

Flier, J. S. What's in a name? In search of leptin's physiologic role. *Journal of Clinical Endocrinology and Metabolism*, 1998, *83*, 1407–1413.

Flock, A. Physiological properties of sensory hairs in the ear. In *Psychophysics and Physiology of Hearing*, edited by E. F. Evans and J. P. Wilson. London: Academic Press, 1977.

Flood, J. F., and Morley, J. E. Increased food intake by neuropeptide Y is due to an increased motivation to eat. *Peptides*, 1991, *12*, 1329–1332.

Flum, D. R., Salem, L., Elrod, J. A. B., Dellinger, E. P., et al. Early mortality among Medicare beneficiaries undergoing bariatric surgical procedures. *Journal of the American Medical Association*, 2005, *294*, 1903–1908.

Flynn, F. W., and Grill, H. J. Insulin elicits ingestion in decerebrate rats. *Science*, 1983, *221*, 188–190.

Fombonne, E. Epidemiology of autistic disorder and other pervasive developmental disorders. *Journal of Clinical Psychiatry*, 2005, *66* (Suppl. 10), 3–8.

Fontaine, D., Mattei, V., Borg, M., von Langsdorff, D., et al. Effect of subthalamic nucleus stimulation on obsessive-compulsive disorder in a patient with Parkinson disease: Case report. *Journal of Neurosurgery*, 2004, *100*, 1084–1086.

Forno, L. S. Neuropathology of Parkinson's disease. *Journal of Neuropathology and Experimental Neurology*, 1996, *55*, 259–272.

Foster, D. L., and Nagatani, S. Physiological perspectives on leptin as a regulator of reproduction: Role in timing puberty. *Biology of Reproduction*, 1999, *60*, 205–215.

Foundas, A. L., Bollich, A. M., Feldman, J., Corey, D. M., Hurley, M., Lemen, L. C., and Heilman, K. M. Aberrant auditory processing and atypical planum temporale in developmental stuttering. *Neurology*, 2004, *63*, 1640–1646.

Fowler, J. S., Logan, J., Wang, G. J, and Volkow, N. D. Monoamine oxidase and cigarette smoking. *Neurotoxicology*, 2003, *24*, 75–82.

Frankle, W. G., Lombardo, I., New, A. S., Goodman, M., Talbot, P. S., et al. Brain serotonin transporter distribution in subjects with impulsive aggressivity: A positron emission study with [¹¹C]McN 5652. *American Journal of Psychiatry*, 2005, *162*, 915–923.

Franklin, T. R., Acton, P. D., Maldjian, J. A., Gray, J. D., et al. Decreased gray matter concentration in the insular, orbitofrontal, cingulate, and temporal cortices of cocaine patients. *Biological Psychiatry*, 2002, *51*, 134–142.

Freed, C. R. Will embryonic stem cells be a useful source of dopamine neurons for transplant into patients with Parkinson's disease? *Proceedings of the National Academy of Sciences, USA,* 2002, *99,* 1755–1757.

Freedman, M. S., Lucas, R. J., Soni, B., von Schantz, M., Muñoz, David-Gray, Z., and Foster, R. Regulation of mammalian circadian behavior by non-rod, non-cone, ocular photoreceptors. *Science,* 1999, *284,* 502–504.

Frey, S., and Frey, J. U. "Synaptic tagging" and "cross-tagging" and related associative reinforcement processes of functional plasticity as the cellular basis for memory formation. *Progress in Brain Research,* 2008, *169,* 117–143.

Frey, S. H., Vinton, D., Norlund, R., and Grafton, S. T. MRI Cortical topography of human anterior intraparietal cortex active during visually guided grasping. *Cognitive Brain Research,* 2005, *23,* 397–405.

Frey, U., Krug, M., Reymann, K. G., and Matthies, H. Anisomycin, an inhibitor of protein synthesis, blocks late phases of LTP phenomena in the hippocampal CA1 region in vitro. *Brain Research,* 1988, *452,* 57–65.

Frey, U., and Morris, R. G. Synaptic tagging and long-term potentiation. *Nature,* 1997, *385,* 533–536.

Fried, I., Katz, A., McCarthy, G., Sass, K. J., et al. Functional organization of human supplementary motor cortex studied by electrical stimulation. *Journal of Neuroscience,* 1991, *11,* 3656–3666.

Friedman, M. I., and Bruno, J. P. Exchange of water during lactation. *Science,* 1976, *191,* 409–410.

Friedman, M. I., Horn, C. C., and Ji, H. Peripheral signals in the control of feeding behavior. *Chemical Senses,* 2005, *30* (Suppl. 1), i182–i183.

Frisch, R. E. Body fat, menarche, fitness and fertility. In *Adipose Tissue and Reproduction,* edited by R. E. Frisch. Basel: S. Karger, 1990.

Frith, U., Morton, J., and Leslie, A. M. The cognitive basis of a biological disorder: Autism. *Trends in Neuroscience,* 1991, *14,* 433–438.

Fry, J. M. Treatment modalities for narcolepsy. *Neurology,* 1998, *50,* S43–S48.

Fudala, P. J., Bridge, T. P., Herbert, S., Williford, W. O., et al. Office-based treatment of opiate addiction with a sublingual-tablet formulation of buprenorphine and naloxone. *New England Journal of Medicine,* 2003, *349,* 949–958.

Fukuwatari, T., Shibata, K., Igushi, K., Saeki, T., et al. Role of gustation in the recognition of oleate and triolein in anomic rats. *Physiology and Behavior,* 2003, *78,* 579–583.

Fuller, P. M., Saper, C. B., and Lu, J. The pontine REM switch: Past and present. *Journal of Physiology,* 2007, *584,* 735–741.

Fullerton, C. S., Ursano, R. J., Epstein, R. S., Crowley, B., et al. Gender differences in posttraumatic stress disorder after motor vehicle accidents. *American Journal of Psychiatry,* 2001, *158,* 1485–1491.

Fulton, J. F. *Functional Localization in Relation to Frontal Lobotomy.* New York: Oxford University Press, 1949.

Funkiewiez, A., Ardouin, C., Caputo, E., Krack, P., et al. Long term effects of bilateral subthalamic nucleus stimulation on cognitive function, mood, and behaviour in Parkinson's disease. *Journal of Neurology, Neurosurgery, and Psychiatry,* 2004, *75,* 834–839.

Furmark, T., Tillfors, M., Garpenstrand, H., Marteinsdottir, I., et al. Serotonin transporter polymorphism related to amygdala excitability and symptom severity in patients with social phobia. *Neuroscience Letters,* 2004, *362,* 189–192.

Gagliardo, A., Ioalé, P., and Bingman, V. P. Homing in pigeons: The role of the hippocampal formation in the representation of landmarks used for navigation. *Journal of Neuroscience,* 1999, *19,* 311–315.

Gaillard, R., Naccache, L., Pinel, P., Clémenceau, S., et al. Direct intracranial, fMRI, and lesion evidence for the causal role of left inferotemporal cortex in reading. *Neuron,* 2006, *50,* 191–204.

Galen. *De Usu Partium.* Translated by M. T. May. Ithaca, New York: Cornell University Press, 1968.

Gallassi, R., Morreale, A., Montagna, P., Cortelli, P., et al. Fatal familial insomnia: Behavioral and cognitive features. *Neurology,* 1996, *46,* 935–939.

Garavan, H., and Stout, J. C. Neurocognitive insights into substance abuse. *Trends in Cognitive Sciences,* 2005, *9,* 195–201.

Garcia-Velasco, J., and Mondragon, M. The incidence of the vomeronasal organ in 1000 human subjects and its possible clinical significance. *Journal of Steroid Biochemistry and Molecular Biololgy,* 1991, *39,* 561–563.

Gardner, H., Brownell, H. H., Wapner, W., and Michelow, D. Missing the point: The role of the right hemisphere in the processing of complex linguistic materials. In *Cognitive Processing in the Right Hemisphere,* edited by E. Pericman. New York: Academic Press, 1983.

Gariano, R. F., and Groves, P. M. Burst firing induced in midbrain dopamine neurons by stimulation of the medial prefrontal and anterior cingulate cortices. *Brain Research,* 1988, *462,* 194–198.

Garriga-Canut, M., Schoenike, B., Qazi, R., Bergendahl, K., et al. 2-Deoxy-D-glucose reduces epilepsy progression by NRSF-CtBP-dependent metabolic regulation of chromatin structure. *Nature Neuroscience,* 2006, *9,* 1382–1387.

Gaspar, P., Cases, O., and Maroteaux, L. The developmental role of serotonin: News from mouse molecular genetics. *Nature Reviews: Neuroscience,* 2003, *4,* 1002–1012.

Gauthier, I., Skudlarski, P., Gore, J. C., and Anderson, A. W. Expertise for cars and birds recruits brain areas involved in face recognition. *Nature Neuroscience,* 2000, *3,* 191–197.

Gauthier, I., Tarr, M. J., Anderson, A. W., Skudlarski, P., and Gore, J. C. Activation of the middle fusiform "face area" increases with expertise in recognizing novel objects. *Nature Neuroscience,* 1999, *2,* 568–573.

Gauthier, L. R., Charrin, B. C., Porrell-Pages, M., Dompierre, J. P., et al. Huntingtin controls neurotrophic support and survival of neurons by enhancing BDNF vesicular transport along microtubules. *Cell,* 2004, *118,* 127–138.

Gazzaniga, M. Forty-five years of split-brain research and still going strong. *Nature Reviews: Neuroscience,* 2005, *6,* 653–659.

Gazzaniga, M. S., and LeDoux, J. E. *The Integrated Mind*. New York: Plenum Press, 1978.

Geary, N. Cocaine: Animal research studies. In *Cocaine Abuse: New Directions in Treatment and Research*, edited by H. I. Spitz and J. S. Rosecan. New York: Brunner-Mazel, 1987.

Gentilucci, M. Grasp observation influences speech production. *European Journal of Neuroscience*, 2003, *17*, 179–184.

George, M. S., Parekh, P. I., Rosinsky, N., Ketter, T. A., et al. Understanding emotional prosody activates right hemisphere regions. *Archives of Neurology*, 1996, *53*, 665–670.

Gerashchenko, D., Blanco-Centurion, C., Greco, M. A., and Shiromani, P. J. Effects of lateral hypothalamic lesion with the neurotoxin hypocretin-2-saporin on sleep in Long-Evans rats. *Neuroscience*, 2003, *116*, 223–235.

Gerashchenko, D., Chou, T. C., Blanco-Centurion, C. A., Saper, C. B., and Shiromani, P. J. Effects of lesions of the histaminergic tuberomammillary nucleus on spontaneous sleep in rats. *Sleep*, 2004, *27*, 1275–1281.

Gerashchenko, D., Kohls, M. D., Greco, M., Waleh, N. S., et al. Hypocretin-2-saporin lesions of the lateral hypothalamus produce narcoleptic-like sleep behavior in the rat. *Journal of Neuroscience*, 2001, *21*, 7273–7283.

Gerbino, L., Oleshansky, M., and Gershon, S. Clinical use and mode of action of lithium. In *Psychopharmacology: A Generation of Progress*, edited by M. A. Lipton, A. DiMascio, and K. F. Killam. New York: Raven Press, 1978.

Gerhand, S. Routes to reading: A report of a non-semantic reader with equivalent performance on regular and exception words. *Neuropsychologia*, 2001, *39*, 1473–1484.

Gerloff, C., Corwell, B., Chen, R., Hallett, M., and Cohen, L. G. Stimulation over the human supplementary motor area interferes with the organization of future elements in complex motor sequences. *Brain*, 1997, *120*, 1587–1602.

Gershon, E. S., Bunney, W. E., Leckman, J., Van Eerdewegh, M., and DeBauche, B. The inheritance of affective disorders: A review of data and hypotheses. *Behavior Genetics*, 1976, *6*, 227–261.

Gervais, H., Belin, P., Boddaert, N., Leboyer, M., et al. Abnormal cortical voice processing in autism. *Nature Neuroscience*, 2004, *8*, 801–802.

Geschwind, N., Quadfasel, F. A., and Segarra, J. M. Isolation of the speech area. *Neuropsychologia*, 1968, *6*, 327–340.

Gessa, G. L., Muntoni, F., Collu, M., Vargiu, L., and Mereu, G. Low doses of ethanol activate dopaminergic neurons in the ventral tegmental area. *Brain Research*, 1985, *348*, 201–204.

Getz, L. L., and Carter, C. S. Prairie-vole partnerships. *American Scientist*, 1996, *84*, 55–62.

Ghilardi, J. R., Röhrich, H., Lindsay, T. H., Sevcik, M. A., et al. Selective blockade of the capsaicin receptor TRPV1 attenuates bone cancer pain. *Journal of Neuroscience*, 2005, *25*, 3126–3131.

Gibbs, J., Young, R. C., and Smith, G. P. Cholecystokinin decreases food intake in rats. *Journal of Comparative and Physiological Psychology*, 1973, *84*, 488–495.

Giedd, J. N., Rapoport, J. L., Garvey, M. A., Perlmutter, S., and Swedo, S. E. MRI assessment of children with obsessive-compulsive disorder or tics associated with streptococcal infection. *American Journal of Psychiatry*, 2000, *157*, 281–283.

Giedd, J. N., Rapoport, J. L., Kruesi, M. J. P., Parker, C., et al. Sydenham's chorea: Magnetic resonance imaging of the basal ganglia. *Neurology*, 1995, *45*, 2199–2202.

Gilbertson, M. W., Shenton, M. E., Ciszewski, A., Kasai, K., et al. Smaller hippocampal volume predicts pathologic vulnerability to psychological trauma. *Nature Neuroscience*, 2002, *5*, 1242–1247.

Gillespie, P. G. Molecular machinery of auditory and vestibular transduction. *Current Opinion in Neurobiology*, 1995, *5*, 449–455.

Gillette, M. U., and McArthur, A. J. Circadian actions of melatonin at the suprachiasmatic nucleus. *Behavioural Brain Research*, 1995, *73*, 135–139.

Givens, B. Low doses of ethanol impair spatial working memory and reduce hippocampal theta activity. *Alcohol: Clinical and Experimental Research*, 1995, *19*, 763–767.

Givens, B., and McMahon, K. Ethanol suppresses the induction of long-term potentiation in vivo. *Brain Research*, 1995, *688*, 27–33.

Glaser, R., Rice, J., Sheridan, J., Post, A., et al. Stress-related immune suppression: Health implications. *Brain, Behavior, and Immunity*, 1987, *1*, 7–20.

Glaser, R., Sheridan, J., Malarkey, W. B., MacCallum, R. C., et al. Leptin, the obese gene product, rapidly modulates synaptic transmission in the hypothalamus. *Molecular Pharmacology*, 1996, *50*, 230–235.

Glaum, S. R., Hara, M., Bindokas, V. P., Lee, C. C., Polonsky, K. S., Bell, G. I., and Miller, R. J. Leptin, the obese gene product, rapidly modulates synaptic transmission in the hypothalamus. *Molecular Pharmacology*, 1996, *50*, 230–235.

Gloor, P., Olivier, A., Quesney, L. F., Andermann, F., and Horowitz, S. The role of the limbic system in experiential phenomena of temporal lobe epilepsy. *Annals of Neurology*, 1982, *12*, 129–144.

Goate, A. M. Monogenetic determinants of Alzheimer's disease: APP mutations. *Cellular and Molecular Life Sciences*, 1998, *54*, 897–901.

Godfrey, P. A., Malnic, B., and Buck, L. The mouse olfactory receptor gene family. *Proceedings of the National Academy of Sciences, USA*, 2004, *101*, 2156–2161.

Goeders, N. E., Lane, J. D., and Smith, J. E. Self-administration of methionine enkephalin into the nucleus accumbens. *Pharmacology, Biochemistry, and Behavior*, 1984, *20*, 451–455.

Goedert, M. Alpha-synuclein and neurodegenerative diseases. *Nature Review Neuroscience*, 2001, *2*, 492–501.

Goedert, M., and Spillantini, M. G. Tau mutations in frontotemporal dementia FTDP-17 and their relevance for Alzheimer's disease. *Biochimica et Biophysica Acta*, 2000, *1502*, 110–121.

Goel, V., and Dolan, R. J. The functional anatomy of humor: Segregating cognitive and affective components. *Nature Neuroscience*, 2001, *4*, 237–238.

Goel, V., and Dolan, R. J. Social regulation of affective experience of humor. *Journal of Cognitive Neuroscience*, 2007, *19*, 1574–1580.

Golarai, G., Ghahremani, D. G., Whitfield-Gabrieli, S., Reiss, A., et al. Differential development of high-level visual cortex correlated with category-specific recognition memory. *Nature Neuroscience*, 2007, *10*, 512–522.

Golby, A. J., Gabrieli, J. D., Chiao, J. Y., and Eberhardt, J. L. Differential responses in the fusiform region to same-race and other-race faces. *Nature Neuroscience*, 2001, *4*, 845–850.

Goldberg, R. F., Perfetti, C. A., and Schneider, W. Perceptual knowledge retrieval activates sensory brain regions. *Journal of Neuroscience*, 2006, *26*, 4917–4921.

Golden, P. L., MacCagnan, T. J., and Pardridge, W. M. Human blood–brain barrier leptin receptor: Binding and endocytosis in isolated human brain microvessels. *Journal of Clinical Investigation*, 1997, *99*, 14–18.

Goldman, D., Oroszi, G., and Ducci, F. The genetics of addictions: Uncovering the genes. *Nature Reviews Genetics*, 2005, *6*, 521–532.

Goldstein, J. M., Seidman, L. J., Horton, N. J., Makris, N., et al. Normal sexual dimorphism of the adult human brain assessed by *in vivo* magnetic resonance imaging. *Cerebral Cortex*, 2001, *11*, 490–497.

Goldstein, R. A., and Volkow, N. D. Drug addiction and its underlying neurobiological basis: Neuroimaging evidence for the involvement of the frontal cortex. *American Journal of Psychiatry*, 2002, *159*, 1642–1652.

Golgi, C. *Opera Omnia, Vols. I and II.* Milan: Hoepli, 1903.

Gonzales, R. A. NMDA receptors excite alcohol research. *Trends in Pharmacological Science*, 1990, *11*, 137–139.

Goodale, M. A., Meenan, J. P., Bülthoff, H. H., Nicolle, D. A., et al. Separate neural pathways for the visual analysis of object shape in perception and prehension. *Current Biology*, 1994, *4*, 604–610.

Goodale, M. A., and Milner, A. D. Separate visual pathways for perception and action. *Trends in Neuroscience*, 1992, *15*, 20–25.

Goodale, M. A., and Westwood, D. A. An evolving view of duplex vision: Separate by interacting cortical pathways for perception and action. *Current Opinion in Neurobiology*, 2004, *14*, 203–211.

Goodglass, H., and Kaplan, E. *Assessment of Aphasia and Related Disorders.* Philadelphia: Lea & Febiger, 1972.

Gooley, J. J., Lu, J., Fischer, D., and Saper, C. B. A broad role for melanopsin in nonvisual photoreception. *Journal of Neuroscience*, 2003, *23*, 7093–7106.

Gooren, L. The biology of human psychosexual differentiation. *Hormones and Behavior*, 2006, *50*, 589–601.

Gordon, H. W., and Sperry, R. Lateralization of olfactory perception in the surgically separated hemispheres in man. *Neuropsychologia*, 1969, *7*, 111–120.

Gorski, R. A., Gordon, J. H., Shryne, J. E., and Southam, A. M. Evidence for a morphological sex difference within the medial preoptic area of the rat brain. *Brain Research*, 1978, *148*, 333–346.

Gosselin, N., Peretz, I., Noulhiane, M., Hasboun, D., et al. Impaired recognition of scary music following unilateral temporal lobe excision. *Brain*, 2005, *128*, 628–640.

Gottesman, I. I., and Bertelsen, A. Confirming unexpressed genotypes for schizophrenia. *Archives of General Psychiatry*, 1989, *46*, 867–872.

Gottesman, I. I., and Shields, J. *Schizophrenia: The Epigenetic Puzzle.* New York: Cambridge University Press, 1982.

Gottfried, J. A., Winston, J. S., and Dolan, R. J. Dissociable codes of odor quality and odorant structure in human piriform cortex. *Neurons*, 2006, *49*, 467–479.

Gottlieb, P., Folgering, J., Maroto, R., Raso, A., et al. Revisiting TRPC1 and TRPC6 mechanosensitivity. *Pflügers Archiv*, 2008, *455*, 1097–1103.

Gould, E., Beylin, A., Tanapat, P., Reeves, A., and Shors, T. J. Learning enhances adult neurogenesis in the hippocampal formation. *Nature Neuroscience*, 1999, *2*, 260–265.

Gouras, P. Identification of cone mechanisms in monkey ganglion cells. *Journal of Physiology*, 1968, *199*, 533–538.

Graber, G. C., and Kristal, M. B. Uterine distention facilitates the onset of maternal beahvior in pseudopregnant but not in cycling rats. *Physiology and Behavior*, 1977, *19*, 133–137.

Grados, M. A., Riddle, M. A., Samuels, J. F., Liang, K.-Y., et al. The familial phenotype of obsessive-compulsive disorder in relation to tic disorders: The Hopkins OCD Family Study. *Biological Psychiatry*, 2001, *50*, 559–565.

Grafton, S. T. Contributions of functional imaging to understanding parkinsonian symptoms. *Current Opinion in Neurobiology*, 2004, *14*, 715–719.

Grafton, S. T., Waters, C., Sutton, J., Lew, M. F., and Couldwell, W. Pallidotomy increases activity of motor association cortex in Parkinson's disease: A positron emission tomographic study. *Annals of Neurology*, 1995, *37*, 776–783.

Graybiel, A. M. Basal ganglia: New therapeutic approaches to Parkinson's disease. *Current Biology*, 1996, *6*, 368–371.

Graziano, M. S. A., and Aflalo, T. N. Mapping behavioral repertoire onto the cortex. *Neuron*, 2007, *56*, 239–251.

Gréco, B., Edwards, D. A., Zumpe, D., and Clancy, A. N. Androgen receptor and mating-induced Fos immunoreactivity are co-localized in limbic and midbrain neurons that project to the male rat medial preoptic area. *Brain Research*, 1998, *781*, 15–24.

Greene, J. D., Nystrom, L. E., Engell, A. D., Darley, J. M., Cohen, J. D. The neural bases of cognitive conflict and control in moral judgment. *Neuron*, 2004, *44*, 389–400.

Greene, J. D., Sommerville, R. B., Nystrom, L. E., Darley, J. M., and Cohen, J. D. An fMRI investigation of emotional engagement in moral judgment. *Science*, 2001, *293*, 2105–2108.

Gregg, T. R., and Siegel, A. Brain structures and neurotransmitters regulating aggression in cats: Implications for human aggression. *Progress in Neuro-Psychopharmacology and Biological Psychiatry*, 2001, *25*, 91–240.

Grelotti, D. J., Gauthier, I., and Schultz, R. T. Social interest and the development of cortical face specialization: What autism teaches us about face processing. *Developmental Psychobiology*, 2002, *40*, 213–225.

Grelotti, D. J., Klin, A. J., Gauthier, I., Skudlarski, P., et al. fMRI activation of the fusiform gyrus and amygdala to cartoon characters but not to faces in a boy with autism. *Neuropsychologia*, 2005, *43*, 373–385.

Grill, H. J., and Kaplan, J. M. Caudal brainstem participates in the distributed neural control of feeding. In *Handbook of Behavioral Neurobiology, Vol. 10: Neurobiology of Food and Fluid Intake*, edited by E. Stricker. New York: Plenum Press, 1990.

Grill-Spector, K., Knouf, N., and Kanwisher, N. The fusiform face area subserves face perception, not generic within-category identification. *Nature Neuroscience*, 2004, *7*, 555–561.

Grill-Spector, K., and Malach, R. The human visual cortex. *Annual Review of Neuroscience*, 2004, *27*, 649–677.

Grill-Spector, K., Sayres, R., and Ress, D. High-resolution imaging reveals highly selective nonface clusters in the fusiform face area. *Nature Neuroscience*, 2007, *9*, 1177–1185.

Grimm, J. W., and See, R. E. Dissociation of primary and secondary reward-relevant limbic nuclei in an animal model of relapse. *Neuropsychopharmacology*, 2000, *22*, 473–479.

Gross, C. G. Visual functions of inferotemporal cortex. In *Handbook of Sensory Physiology, Vol. 7: Central Processing of Visual Information*, edited by R. Jung. Berlin: Springer-Verlag, 1973.

Grossman, E. D., Battelli, L., and Pascual-Leone, A. Repetitive TMS over posterior STS disrupts perception of biological motion. *Vision Research*, 2005, *45*, 2847–2853.

Grossman, E. D., and Blake, R. Brain activity evoked by inverted and imagined biological motion. *Vision Research*, 2001, *41*, 1475–1482.

Grossman, E. D., Donnelly, M., Price, R., Pickens, D., et al. Brain areas involved in perception of biological motion. *Journal of Cognitive Neuroscience*, 2000, *12*, 711–720.

Groves, D. A., and Brown, V. J. Vagal nerve stimulation: A review of its applications and potential mechanisms that mediate its clinical effects. *Neuroscience and Biobehavioral Reviews*, 2005, *29*, 493–500.

Grunhaus, L., Shipley, J. E., Eiser, A., Pande, A. C., et al. Sleep-onset rapid eye movement after electroconvulsive therapy is more frequent in patients who respond less well to electroconvulsive therapy. *Biological Psychiatry*, 1997, *42*, 191–200.

Grunze, H. Reevaluating therapies for bipolar depression, *Journal of Clinical Psychiatry*, 2005, *66* (Suppl. 5), 17–25.

Guehl, D., Benazzouz, A., Aouizerate, B., Cuny, E., et al. Neuronal correlates of obsessions in the caudate nucleus. *Biological Psychiatry*, 2008, *63*, 557–562.

Guilleminault, C., Wilson, R. A., and Dement, W. C. A study on cataplexy. *Archives of Neurology*, 1974, *31*, 255–261.

Gulevich, G., Dement, W. C., and Johnson, L. Psychiatric and EEG observations on a case of prolonged (264 hours) wakefulness. *Archives of General Psychiatry*, 1966, *15*, 29–35.

Gurd, J. M., and Marshall, J. C. Cognition: Righting reading. *Current Biology*, 1993, *3*, 593–595.

Gurden, H., Takita, M., and Jay, T. M. Essential role of D1 but not D2 receptors in the NMDA receptor-dependent long-term potentiation at hippocampal-prefrontal cortex synapses in vivo. *Journal of Neuroscience*, 2000, *20*, RC106.

Gurden, H., Tassin, J. P., and Jay, T. M. Integrity of the mesocortical dopaminergic system is necessary for complete expression of in vivo hippocampal-prefrontal cortex long-term potentiation. *Neuroscience*, 1999, *94*, 1019–1027.

Guridi, J., and Obeso, J. A. The subthalamic nucleus, hemiballismus and Parkinson's disease: Reappraisal of a neurosurgical dogma. *Brain*, 2001, *124*, 5–19.

Gurvits, T. V., Shenton, M. E., Hokama, H., Ohta, H., Lasko, N. B., Gilbertson, M. W., Orr, S. P., Kikinis, R., Jolesz, F. A., McCarley, R. W., and Pitman, R. K. Magnetic resonance imaging study of hippocampal volume in chronic, combat-related posttraumatic stress disorder. *Biological Psychiatry*, 1996, *40*, 1091–1099.

Gutin, B., Owens, S., Okuyama, T., Riggs, S. Ferguson, M., et al. Effect of physical training and its cessation on percent fat and bone density of children with obesity. *Obesity Research*, 1999, *7*, 208–214.

Gvilia, I., Xu, F., McGinty, D., and Szymusiak, R. Homeostatic regulation of sleep: A role for preoptic area neurons. *Journal of Neuroscience*, 2006, *26*, 9426–9433.

Haarmeier, T., Their, P., Repnow, M., and Petersen, D. False perception of motion in a patient who cannot compensate for eye movements. *Nature*, 1997, *389*, 849–852.

Haas, R. H. Thiamin and the brain. *Annual Review of Nutrition*, 1988, *8*, 483–515.

Habib, M. The neurological basis of developmental dyslexia: An overview and working hypothesis. *Brain*, 2000, *123*, 2373–2399.

Hacke, W., Albers, G., Al-Rawi, Y., Bogousslavsky, J., et al. The desmoteplase in acute ischemic stroke trial (DIAS): A phase II MRI-based 9-hour window acute stroke thrombolysis trial with intravenous desmoteplase. *Stroke*, 2005, *36*, 66–73.

Hackett, R. A., Preuss, T. M., and Kaas, J. H. Architectonic identification of the core region in auditory cortex of macaques, chimpanzees, and humans. *Journal of Comparative Neurology*, 2001, *441*, 197–222.

Hadjikhani, N., and de Gelder, B. Seeing fearful body expressions activates the fusiform cortex and amygdala. *Current Biology*, 2003, *13*, 2201–2205.

Hadjikhani, N., Joseph, R. M., Snyder, J., and Tager-Flusberg, H. Anatomical differences in the mirror neuron system and social cognition network in autism. *Cerebral Cortex*, 2006, *16*, 1276–1282.

Hadjikhani, N., Liu, A. K., Dale, A. M., Cavanagh, P., and Tootell, R. B. H. Retinotopy and color sensitivity in human visual cortical area V8. *Nature Neuroscience*, 1998, *1*, 235–241.

Hague, S. M., Klaffke, S., and Bandmann, O. Neurodegenerative disorders: Parkinson's disease and Huntington's disease. *Journal of Neurology, Neurosurgery, and Psychiatry*, 2005, *76*, 1058–1063.

Hahn, T. M., Breininger, J. F., Baskin, D. G., and Schwartz, M. W. Coexpression of Agrp and NPY in fasting-activated hypothalamic neurons. *Nature Neuroscience*, 1998, *1*, 271–272.

Hainer, V., Stunkard, A., Kunesova, M., Parizkova, J., Stich, V., and Allison, D. B. A twin study of weight loss and metabolic efficiency. *International Journal of Obesity and Related Metabolic Disorders*, 2001, *25*, 533–537.

Hajak, G., Clarenbach, P., Fischer, W., Haase, W., et al. Effects of hypnotics on sleep quality and daytime well-being:

Data from a comparative multicentre study in outpatients with insomnia. *European Psychiatry*, 1995, *10* (Suppl. 3), 173S–179S.

Halaas, J. L., Gajiwala, K. D., Maffei, M., Cohen, S. L., et al. Weight-reducing effects of the plasma protein encoded by the obese gene. *Science*, 1995, *269*, 543–546.

Haley, J. E., Wilcox, G. L., and Chapman, P. F. The role of nitric oxide in hippocampal long-term potentiation. *Neuron*, 1992, *8*, 211–216.

Halgren, E. Walter, R. D., Cherlow, D. G., and Crandall, P. E. Mental phenomena evoked by electrical stimultion of the human hippocampal formation and amygdala. *Brain*, 1978, *101*, 83–117.

Halpern, M. The organization and function of the vomeronasal system. *Annual Review of Neuroscience*, 1987, *10*, 325–362.

Halsband, U., and Freund, H. J. Premotor cortex and conditional motor learning in man. *Brain*, 1990, *113*, 207–222.

Hamann, S. Blue genes: Wiring the brain for depression. *Nature Neuroscience*, 2005, *8*, 701–703.

Hamet, P., and Tremblay, J. Genetics and genomics of depression. *Metabolism: Clinical and Experimental*. 2005, *54*, 10–15.

Hampson, R. E., and Deadwyler, S. A. Cannabinoids reveal the necessity of hippocampal neural encoding for short-term memory in rats. *Journal of Neuroscience*, 2000, *20*, 8932–8942.

Hara, J., Beuckmann, C. T., Nambu, T., Willie, J. T., et al. Genetic ablation of orexin neurons in mice results in narcolepsy, hypophagia, and obesity. *Neuron*, 2001, *30*, 345–354.

Hardy, J. Amyloid, the presinilins and Alzheimer's disease. *Trends in Neuroscience*, 1997, *4*, 154–159.

Hariri, A. R., Drabant, B. A., Munoz, K. E., Kolachana, B. S., et al. A susceptibility gene for affective disorders and the response of the human amygdala. *Archives of General Psychiatry*, 2005, *62*, 146–152.

Harmon, L. D., and Julesz, B. Masking in visual recognition: Effects of two-dimensional filtered noise. *Science*, 1973, *180*, 1194–1197.

Harris, G. W., and Jacobsohn, D. Functional grafts of the anterior pituitary gland. *Proceedings of the Royal Society of London B*, 1951–1952, *139*, 263–267.

Harrison, Y., and Horne, J. A. Sleep loss impairs short and novel language tasks having a prefrontal focus. *Journal of Sleep Research*, 1998, *7*, 95–100.

Harrison, Y., and Horne, J. A. One night of sleep loss impairs innovative thinking and flexible decision-making. *Organizational Behavior and Human Decision Processes*, 1999, *78*, 128–145.

Hart, B. L. Hormones, spinal reflexes, and sexual behaviour. In *Determinants of Sexual Behaviour*, edited by J. B. Hutchinson. Chichester, England: John Wiley & Sons, 1978.

Hartley, T., Maguire, E. A., Spiers, H. J., and Burgess, N. The well-worn route and the path less traveled: Distinct neural bases of route following and wayfinding in humans. *Neuron*, 2003, *37*, 877–888.

Hartline, H. K. The response of single optic nerve fibers of the vertebrate eye to illumination of the retina. *American Journal of Physiology*, 1938, *121*, 400–415.

Harvey, S. M. Female sexual behavior: Fluctuations during the menstrual cycle. *Journal of Psychosomatic Research*, 1987, *31*, 101–110.

Haslinger, B., Erhard, P., Altenmüller, E., Schroeder, U., et al. Transmodal sensorimotor networks during action observation in professional pianists. *Journal of Cognitive Neuroscience*, 2005, *17*, 282–293.

Hattar, S., Liao, H.-W., Takao, M., Berson, D. M., and Yau, K.-W. Melanopsin-containing retinal ganglion cells: Architecture, projections, and intrinsic photosensitivity. *Science*, 2002, *295*, 1065–1070.

Haug, H.-J. Prediction of sleep deprivation outcome by diurnal variation of mood. *Biological Psychiatry*, 1992, *31*, 271–278.

Hauk, O., Johnsrude, I., and Pulvermüller, F. Somatotopic representation of action words in human motor and premotor cortex. *Neuron*, 2004, *41*, 301–307.

Hauser, M. D. Right hemisphere dominance for the production of facial expression in monkeys. *Science*, 1993, *261*, 475–477.

Haverkamp, S., Wässle, H., Duebel, J., Kuner, T., Augustine, G. J., Feng, G. and Euler, T. The primordial, blue-cone color system of the mouse retina. *Journal of Neuroscience*, 2005, *25*, 5438–5445.

Hawke, C. Castration and sex crimes. *American Journal of Mental Deficiency*, 1951, *55*, 220–226.

He, J., Ma, L., Kim, S., Nakai, J., and Yu, C. R. Encoding gender and individual information in the mouse vomeronasal organ. *Science*, 2008, *320*, 535–538.

He, W., Yasumatsu, K., Varadarajan, V., Yamada, A. Lem, J., et al. Umami taste responses are mediated by α-transducin and α-gustducin. *Journal of Neuroscience*, 2004, *24*, 7574–7680.

Hebb, D. O. *The Organization of Behaviour*. New York: Wiley-Interscience, 1949.

Heckler, M. M. *Fifth Special Report to the U.S. Congress on Alcohol and Health*. Washington, D.C.: U.S. Government Printing Office, 1983.

Heeb, M. M., and Yahr, P. Cell-body lesions of the posterodorsal preoptic nucleus or posterodorsal medial amygdala, but not the parvicellular subparafascicular thalamus, disrupt mating in male gerbils. *Physiology and Behavior*, 2000, *68*, 317–331.

Heffner, H. E., and Heffner, R. S. Role of primate auditory cortex in hearing. In *Comparative Perception, Vol. II: Complex Signals*, edited by W. C. Stebbins and M. A. Berkley. New York: John Wiley & Sons, 1990.

Heilman, K. M., Rothi, L., and Kertesz, A. Localization of apraxia-producing lesions. In *Localization in Neuropsychology*, edited by A. Kertesz. New York: Academic Press, 1983.

Heilman, K. M., Watson, R. T., and Bowers, D. Affective disorders associated with hemispheric disease. In *Neuropsychology of Human Emotion*, edited by K. M. Heilman and P. Satz. New York: Guilford Press, 1983.

Heimer, L., and Larsson, K. Impairment of mating behavior in male rats following lesions in the preoptic-anterior hypothalamic continuum. *Brain Research*, 1966/1967, *3*, 248–263.

Heinrichs, M., Baumgartner, T., Kirshbaum, C., and Ehlert, U. Social support and oxytocin interact to suppress cortisol and subjective responses to psychosocial stress. *Biological Psychiatry*, 2003, *54*, 1389–1398.

Heinrichs, S. C., Menzaghi, F., Pich, E. M., Baldwin, H. A., et al. Anti-stress action of a corticotropin-releasing factor antagonist on behavioral reactivity to stressors of varying type and intensity. *Neuropsychopharmacology*, 1994, *11*, 179–186.

Heinz, A., Reimold, M., Wrase, J., Hermann, D., et al. Correlation of stable elevations in striatal μ-opioid receptor availability in detoxified alcoholic patients with alcohol craving. *Archives of General Psychiatry*, 2005, *62*, 57–64.

Helenius, P., Uutela, K., and Hari, R. Auditory stream segregation in dyslexic adults. *Brain*, 1999, *122*, 907–913.

Hellhammer, D. H., Hubert, W., and Schurmeyer, T. Changes in saliva testosterone after psychological stimulation in men. *Psychoneuroendocrinology*, 1985, *10*, 77–81.

Helmuth, L. Dyslexia: Same brains, different languages. *Science*, 2001, *291*, 2064–2065.

Hendrickson, A. E., Wagoner, N., and Cowan, W. M. Autoradiographic and electron microscopic study of retinohypothalamic connections. *Zeitschrift für Zellforschung und Mikroskopische Anatomie*, 1972, *125*, 1–26.

Hendry, S. H. C., and Yoshioka, T. A neurochemically distinct third channel in the cacaque dorsal lateral geniculare nucleus. *Science*, 1994, *264*, 575–577.

Henke, P. G. The telencephalic limbic system and experimental gastric pathology: A review. *Neuroscience and Biobehavioral Reviews*, 1982, *6*, 381–390.

Hennessey, A. C., Camak, L., Gordon, F., and Edwards, D. A. Connections between the pontine central gray and the ventromedial hypothalamus are essential for lordosis in female rats. *Behavioral Neuroscience*, 1990, *104*, 477–488.

Henningfield, J. E., Fant, R. V., Buchhalter, A. R., and Stitzer, M. L. Pharmacotherapy for nicotine dependence. *CA: A Cancer Journal for Clinicians*, 2005, *55*, 281–299.

Henry, M. L., Beeson, P. M., Stark, A. J., and Rapcsak, S. Z. The role of left perisylvian cortical regions in spelling. *Brain and Language*, 2007, *100*, 44–52.

Herbert, M. R., Ziegler, D. A., Makris, N., Filipek, P. A., et al. Localization of white matter volume increase in autism and developmental language disorder. *Annals of Neurology*, 2004, *55*, 530–540.

Herculano-Houzel, S., Collins, C. E., Wong, P., and Kaas, J. H. Cellular scaling rules for primate brains. *Proceedings of the National Academy of Sciences, USA*, 2007, *104*, 3562–3567.

Herholz, K. Neuroimaging in anorexia nervosa. *Psychiatry Research*, 1996, *62*, 105–110.

Hering, E. *Outlines of a Theory of the Light Sense*, 1905. Translated by L. M. Hurvich and D. Jameson. Cambridge, Mass.: Harvard University Press, 1965.

Hernandez, L., and Hoebel, B. G. Feeding can enhance dopamine turnover in the prefrontal cortex. *Brain Research Bulletin*, 1990, *25*, 975–979.

Hetherington, A. W., and Ranson, S. W. Hypothalamic lesions and adiposity in the rat. *Anatomical Record*, 1942, *78*, 149–172.

Hettema, J. M., Neale, M. C., and Kendler, K. S. A review and meta-analysis of the genetic epidemiology of anxiety disorders. *American Journal of Psychiatry*, 2001, *158*, 1568–1578.

Hetz, C., Russelakis-Carneiro, M., Maundrell, K., Castilla, J., and Soto, C. Caspase-12 and endoplasmic reticulum stress mediate neurotoxicity of pathological prion protein. *The EMBO Journal*, 2003, *22*, 5435–5445.

Heywood, C. A., and Cowey, A. The role of the "face-cell" area in the discrimination and recognition of faces by monkeys. *Philosophical Transactions of the Royal Society of London B*, 1992, *335*, 31–38.

Heywood, C. A., Gaffan, D., and Cowey, A. Cerebral achromatopsia in monkeys. *European Journal of Neuroscience*, 1995, *7*, 1064–1073.

Heywood, C. A., and Kentridge, R. W. Achromatopsia, color vision, and cortex. *Neurology Clinics of North America*, 2003, *21*, 483–500.

Hickok, G., Bellugi, U., and Klima, E. S. The neurobiology of sign language and its implications for the neural basis of language. *Nature*, 1996, *381*, 699–702.

Hickok, G., Klima, E., Kritchevsky, M., and Bellugi, U. A case of 'sign blindness' following left occipital damage in a deaf signer. *Neuropsychologia*, 1995, *33*, 1597–1601.

Hickok, G., Wilson, M., Clark, K., Klima, E. S., Kritchevsky, M., and Bellugi, U. Discourse deficits following right hemisphere damage in deaf signers. *Brain and Language*, 1999, *66*, 233–248.

Hikosaka, O., Sakai, K., Miyauchi, S., Takino, R., Sasaki, Y., and Puetz, B. Activation of human presupplementary motor area in learning of sequential procedures: A functional MRI study. *Journal of Neurophysiology*, 1996, *76*, 617–621.

Hill, J. O., Wyatt, H. R., Reed, G. W., and Peters, J. C. Obesity and the environment: Where do we go from hers? *Science*, 2003, *299*, 853–855.

Hill, J. P., Hauptman, J., Anderson, J., Fujioka, K., et al. Orlistat, a lipase inhibitor, for weight maintenance after conventional dieting: A 1-year study. *American Journal of Clinical Nutrition*, 1999, *69*, 1108–1116.

Hillebrand, J. J. G., de Rijke, C. E., Brakkee, J. H., Kas, M. J. H., and Adan, R. A. H. Voluntary access to a warm plate reduces hyperactivity in activity-based anorexia. *Physiology and Behavior*, 2005, *85*, 151–157.

Hillis, A. E., Newhart, M., Heidler, J., Barker, P. B., et al. Anatomy of spatial attention: Insights from perfusion imaging and hemispatial neglect in acute stroke. *Journal of Neuroscience*, 2005, *25*, 3161–3167.

Hillis, A. E., Work, M., Barker, P. B., Jacobs, M. A., et al. Re-examining the brain regions crucial for orchestrating speech articulation. *Brain*, 2004, *127*, 1479–1487.

Hines, M., Allen, L. S., and Gorski, R. A. Sex differences in subregions of the medial nucleus of the amygdala and the bed nucleus of the stria terminalis of the rat. *Brain Research*, 1992, *579*, 321–326.

Hippocrates. On the sacred disease. In *Hippocrates and Galen: Great Books of the Western World*, Vol. 10. Chicago: William Benton, 1952.

Hobson, J. A. *The Dreaming Brain*. New York: Basic Books, 1988.

Hochberg, L. R., Serruya, M. D., Friehs, G. M., Mukand, J. A., et al. Neuronal ensemble control of prosthetic devices by a human with tetraplegia. *Nature*, 2006, *442*, 164–171.

Hock, C., Konietzko, U., Streffer, J. R., Tracy, J., et al. Antibodies against β-amyloid slow cognitive decline in Alzheimer's disease. *Neuron*, 2003, *38*, 547–554.

Hodge, C. W., Haraguchi, M., Erickson, H., and Samson, H. H. Ventral tegmental microinjections of quinpirole decrease ethanol and sucrose-reinforced responding. *Alcohol: Clinical and Experimental Research*, 1993, *17*, 370–375.

Hoeft, F., Meyler, A., Hernandez, A., Juel, C. Taylor-Hill, H., et al. Functional and morphometric brain dissociation between dyslexia and reading ability. *Proceedings of the National Academy of Sciences, USA*, 2007, *104*, 4234–4239.

Hofbauer, R. K., Rainville, P., Duncan, G. H., and Bushnell, M. C. Cortical representation of the sensory dimension of pain. *Journal of Neurophysiology*, 2001, *86*, 402–411.

Hofmann, S. G., Meuret, A. E., Smits, J. A., Simon, N. M., et al. Augmentation of exposure therapy with D-cycloserine for social anxiety disorder. *Archives of General Psychiatry*, 2006, *63*, 298–304.

Hohman, G. W. Some effects of spinal cord lesions on experienced emotional feelings. *Psychophysiology*, 1966, *3*, 143–156.

Holden, C. The violence of the lambs. *Science*, 2000, *289*, 580–581.

Hollander, E., Anagnostou, E., Chaplin, W., Esposito, K., et al. Striatal volume on magnetic resonance imaging and repetitive behaviors in autism. *Biological Psychiatry*, 2005, *58*, 226–232.

Hollander, E., Schiffman, E., Cohen, B., Rivera-Stein, M. A., et al. Signs of central nervous system dysfunction in obsessive-compulsive disorder. *Archives of General Psychiatry*, 1990, *47*, 27–32.

Hollis, J. H., McKinley, M. J., D'Souze, M., Kampe, J., and Oldfield, B. J. The trajectory of sensory pathways from the lamina terminalis to the insular and cingulate cortex: A neuroanatomical framework for the generation of thirst. *American Journal of Physiology*, 2008, *294*, R1390–R1401.

Holmes, G. The cerebellum of man. *Brain*, 1939, *62*, 21–30.

Holstege, G., Georgiadis, J. R., Paans, A. M. J., Meiners, L. C., et al. Brain activation during human male ejaculation. *Journal of Neuroscience*, 2003a, *23*, 9185–9193.

Holstege, G., Reinders, A. A. T., Panns, A. M. J., Meiners, L. C., et al. Brain activation during female sexual orgasm. Program No. 727.7. *2003 Abstract Viewer/Itinerary Planner*. Washington, D.C.: Society for Neuroscience, 2003b.

Honda, T., Tabata, H., and Nakajima, K. Cellular and molecular mechanisms of neuronal migration in neocortical development. *Seminars in Cell & Developmental Biology*, 2003, *14*, 169–174.

Hong, C. C. H., Jin, Y., Potkin, S. G., Buchsbaum, M. S., et al. Language in dreaming and regional EEG alpha-power. *Sleep*, 1996, *19*, 232–235.

Hopf, H. C., Mueller-Forell, W., and Hopf, N. J. Localization of emotional and volitional facial paresis. *Neurology*, 1992, *42*, 1918–1923.

Hoppe, C. Controlling epilepsy. *Scientific American Mind*, 2006, *17*, 62–67.

Horne, J. A. A review of the biological effects of total sleep deprivation in man. *Biological Psychology*, 1978, *7*, 55–102.

Horne, J. A., and Minard, A. Sleep and sleepiness following a behaviourally "active" day. *Ergonomics*, 1985, *28*, 567–575.

Horowitz, R. M., and Gentili, B. Dihydrochalcone sweeteners. In *Symposium: Sweeteners*, edited by G. E. Inglett. Westport, Conn.: Avi Publishing, 1974.

Horowitz, T. S., Cade, B. E., Wolfe, J. M., and Czeisler, C. A. Efficacy of bright light and sleep/darkness scheduling in alleviating circadian maladaptation to night work. *American Journal of Physiology*, 2001, *281*, E384–E391.

Horton, J. C., and Hubel, D. H. Cytochrome oxidase stain preferentially labels intersection of ocular dominance and vertical orientation columns in macaque striate cortex. *Society for Neuroscience Abstracts*, 1980, *6*, 315.

Houchi, H., Babovic, D., Pierrefiche, O., Ledent, C., et al. CB1 receptor knockout mice display reduced ethanol-induced conditioned place preference and increased striatal dopamine D2 receptors. *Neuropsychopharmacology*, 2005, *30*, 339–340.

Howell, S., Westergaard, G., Hoos, B., Chavanne, T. J., et al. Serotonergic influences on life-history outcomes in free-ranging male rhesus macaques. *American Journal of Primatology*, 2007, *69*, 851–865.

Huang, A. L., Chen, X., Hoon, M. A., Chandrashekar, J., et al. The cells and logic for mammalian sour taste detection. *Nature*, 2006, *442*, 934–938.

Huang, Z. L., Qu, W. M., Eguchi, H., Chen, J. F., et al. Adenosine A_{2A}, but not A_1, receptors mediate the arousal effect of caffeine. *Nature Neuroscience.*, 2005, *8*, 858–859.

Hubel, D. H., and Wiesel, T. N. Functional architecture of macaque monkey visual cortex. *Proceedings of the Royal Society of London B*, 1977, *198*, 1–59.

Hubel, D. H., and Wiesel, T. N. Brain mechanisms of vision. *Scientific American*, 1979, *241*, 150–162.

Huber, R., Ghilardi, M. F., Massimini, M., Ferrarelli, F., et al. Arm immobilization causes cortical plastic changes and locally decreases sleep slow wave activity. *Nature Neuroscience*, 2006, *9*, 1169–1176.

Huber, R., Ghilardi, M. F., Massimini, M., and Tononi, G. Local sleep and learning. *Nature*, 2004, *430*, 78–81.

Hublin, C. Narcolepsy: Current drug-treatment options. *CNS Drugs*, 1996, *5*, 426–436.

Hublin, C., Kaprio, J., Partinen, M., Heikkila, K., and Koskenvuo, M. Prevalence and genetics of sleepwalking: A population-based twin study. *Neurology*, 1997, *48*, 177–181.

Hudspeth, A. J. Mechanoelectrical transduction by hair cells in the acousticolateralis sensory system. *Annual Review of Neuroscience*, 1983, *6*, 187–215.

Hudspeth, A. J., and Gillespie, P. G. Pulling springs to tune transduction: Adaptation by hair cells. *Neuron*, 1994, *12*, 1–9.

Huestis, M. A., Gorelick, D. A., Heishman, S. J., Preston, K. L., et al. Blockade of effects of smoked marijuana by the CB1-selective cannabinoid antagonist SR131716. *Archives of General Psychiatry*, 2001, *58*, 322–328.

Hughes, J., Smith, T. W., Kosterlitz, H. W., Fothergill, L. A., et al. Identification of two related pentapeptides from the brain with potent opiate agonist activity. *Nature*, 1975, *258*, 577–579.

Hughes, J. R., Gust, S. W., Skoog, K., Keenan, R. M., and Fenwick, J. W. Symptoms of tobacco withdrawal: A replication and extension. *Archives of General Psychiatry*, 1989, *14*, 577–580.

Hull, E., and Dominguez, J. M. Sexual behavior in male rodents. *Hormones and Behavior*, 2007, *52*, 45–55.

Hulshoff-Pol, H. E., Schnack, H. G., Bertens, M. G., van Haren, N. E., et al. Volume changes in gray matter in patients with schizophrenia. *American Journal of Psychiatry*, 2002, *159*, 244–250.

Humphrey, A. L., and Hendrickson, A. E. Radial zones of high metabolic activity in squirrel monkey striate cortex. *Society for Neuroscience Abstracts*, 1980, *6*, 315.

Hunt, D. M., Dulai, K. S., Cowing, J. A., Julliot, C., et al. Molecular evolution of trichromacy in primates. *Vision Research*, 1998, *38*, 3299–3306.

Husain, M., and Rorden, C. Non-spatially lateralized mechanisms in hemispatial neglect. *Nature Reviews: Neuroscience*, 2003, *4*, 26–36.

Huszar, D., Lynch, C. A., Fairchild-Huntress, V., Dunmore, J. H., et al. Targeted disruption of the melanocortin-4 receptor results in obesity in mice. *Cell*, 1997, *88*, 131–141.

Hwa, J. J., Ghibaudi, L., Gao, J., and Parker, E. M. Central melanocortin system modulates energy intake and expenditure of obese and lean Zucker rats. *American Journal of Physiology*, 2001, *281*, R444–R451.

Hyde, K. L., Lerch, J. P., Zatorre, R. J., Griffiths, T. D., et al. Cortical thickness in congenital amusia: When less is better than more. *Journal of Neuroscience*, 2007, *27*, 13028–13032.

Hyde, K. L., Zatorre, R. J., Griffiths, T. D., Lerch, J. P., et al. Morphometry of the amusic brain: A two-site study. *Brain*, 2006, *129*, 2562–2570.

Hyman, S. E., and Malenka, R. C. Addiction and the brain: The neurobiology of compulsion and its persistence. *Nature Reviews: Neuroscience*, 2001, *2*, 695–703.

Iacoboni, M., and Dapretto, M. The mirror neuron system and the consequences of its dysfunction. *Nature Reviews: Neuroscience*, 2006, *7*, 942–951.

Iacoboni, M., Molnar-Szakacs, I., Gallese, V., Buccino, G., et al. Grasping the intentions of others with one's own mirror neuron system. *PLoS Biology*, 2005, *3*, e79.

Iacoboni, M., Woods, R. P., Brass, M., Bekkering, H., Mazziotta, J. C., and Rizzolatti, G. Cortical mechanisms of human imitation. *Science*, 1999, *286*, 2526–2528.

Iaria, G., Petrides, M., Dagher, A., Pike, B., and Bohbot, V. D. (2003). Cognitive strategies dependent on the hippocampus and caudate nucleus in human navigation: Variability and change with practice. *Journal of Neuroscience*, 23, 5945–5952.

Ibuka, N., and Kawamura, H. Loss of circadian rhythm in sleep-wakefulness cycle in the rat by suprachiasmatic nucleus lesions. *Brain Research*, 1975, *96*, 76–81.

Igarashi, K. M., and Mori, K. Spatial representation of hydrocarbon odorants in the ventrolateral zones of the rat olfactory bulb. *Journal of Neurophysiology*, 2005, *93*, 1007–1019.

Iggo, A., and Andres, K. H. Morphology of cutaneous receptors. *Annual Review of Neuroscience*, 1982, *5*, 1–32.

Iijima, M., Arisaka, O., Minamoto, F., and Arai, Y. Sex differences in children's free drawings: A study on girls with congenital adrenal hyperplasia. *Hormones and Behavior*, 2001, *20*, 99–104.

Ikoma, A., Steinhoff, M., Ständer, S., Yosipovitch, G., and Schmelz, M. The neurobiology of itch. *Nature Reviews: Neuroscience*, 2006, *7*, 535–547.

Ikonomidou, C., Bittigau, P., Ishimaru, M. J., Wozniak, D. F., et al. Ethanol-induced apoptotic neurodegeneration and fetal alcohol syndrome. *Science*, 2000, *287*, 1056–1060.

Imperato, A., and Di Chiara, G. Preferential stimulation of dopamine-release in the accumbens of freely moving rats by ethanol. *Journal of Pharmacology and Experimental Therapeutics*, 1986, *239*, 219–228.

Inoue, M., Koyanagi, T., Nakahara, H., Hara, K., Hori, E., and Nakano, H. Functional development of human eye movement in utero assessed quantitatively with real time ultrasound. *American Journal of Obstetrics and Gynecology*, 1986, *155*, 170–174.

Insel, T. R. A neurobiological basis of social attachment. *American Journal of Psychiatry*, 1997, *154*, 726–735.

Insel, T. R., Wang, Z. X., and Ferris, C. F. Patterns of brain vasopressin receptor distribution associated with social organization in microtine rodents. *Journal of Neuroscience*, 1994, *14*, 5381–5392.

Isenberg, N., Silbersweig, D., Engelien, A., Emmerich, S., et al. Linguistic threat activates the human amygdala. *Proceedings of the National Academy of Sciences, USA*, 1999, *96*, 10456–10459.

Isoldi, K. K., and Aronne, L. J. The challenge of treating obesity: The endocannabinoid system as a potential target. *Journal of the American Dietetic Association*, 2008, *108*, 823–831.

Ito, R., Dalley, J. W., Robbins, T. W., and Everitt, B. J. Dopamine release in the dorsal striatum during cocaine-seeking behavior under the control of a drug-associated cue. *Journal of Neuroscience*, 2002, *22*, 6247–6253.

Iversen, L. Cannabis and the brain. *Brain*, 2003, *126*, 1252–1270.

Iwata, M. Kanji versus Kana: Neuropsychological correlates of the Japanese writing system. *Trends in Neurosciences*, 1984, *7*, 290–293.

Izard, C. E. *The Face of Emotion*. New York: Appleton-Century-Crofts, 1971.

Jackson, M. E., Frost, A. S., and Moghaddam, B. Stimulation of prefrontal cortex at physiologically relevant frequencies inhibits dopamine release in the nucleus accumbens. *Journal of Neurochemistry*, 2001, *78*, 920–923.

Jacob, S., and McClintock, M. K. Psychological state and mood effects of steroidal chemosignals in women and men. *Hormones and Behavior*, 2000, *37*, 57–78.

Jacobs, B. L., and Fornal, C. A. Activity of serotonergic neurons in behaving animals. *Neuropsychopharmacology*, 1999, *21*, 9S-15S.

Jacobs, B. L., and McGinty, D. J. Participation of the amygdala in complex stimulus recognition and behavioral inhibition: Evidence from unit studies. *Brain Research*, 1972, *36*, 431–436.

Jacobs, G. H. Primate photopigments and primate color vision. *Proceedings of the National Academy of Sciences, USA*, 1996, *93*, 577–581.

Jacobsen, C. F., Wolfe, J. B., and Jackson, T. A. An experimental analysis of the functions of the frontal association areas in primates. *Journal of Nervous and Mental Disorders*, 1935, *82*, 1–14.

Jakobson, L. S., Archibald, Y. M., Carey, D., and Goodale, M. A. A kinematic analysis of reaching and grasping movements in a patient recovering from optic ataxia. *Neuropsychologia*, 1991, *29*, 803–809.

James, T. W., Culham, J., Humphrey, G. K., Milner, A. D., and Goodale, M. A. Ventral occipital lesions impair object recognition but not object-directed grasping: An fMRI study. *Brain*, 2003, *126*, 2463–2475.

James, W. What is an emotion? *Mind*, 1884, *9*, 188–205.

Jaramillo, F. Signal transduction in hair cells and its regulation by calcium. *Neuron*, 1995, *15*, 1227–1230.

Javitt, D. C. Glycine transport inhibitors and the treatment of schizophrenia. *Biological Psychiatry*, 2008, *63*, 6–8.

Jaynes, J. The problem of animate motion in the seventeenth century. *Journal of the History of Ideas*, 1970, *6*, 219–234.

Jeffress, L. A. A place theory of sound localization. *Journal of Comparative and Physiological Psychology*, 1948, *41*, 35–39.

Jensen, T., Genefke, I., and Hyldebrandt, N. Cerebral atrophy in young torture victims. *New England Journal of Medicine*, 1982, *307*, 1341.

Jentsch, J. D., Redmond, D. E., Elsworth, J. D., Taylor, J. R., et al. Enduring cognitive deficits and cortical dopamine dysfunction in monkeys after long-term administration of phencyclidine. *Science*, 1997, *277*, 953–955.

Jentsch, J. D., Tran, A., Taylor, J. R., and Roth, R. H. Prefrontal cortical involvement in phencyclidine-induced activation of the mesolimbic dopamine system: Behavioral and neurochemical evidence. *Psychopharmacology*, 1998, *138*, 89–95.

Jessberger, S., and Kempermann, G. Adult-born hippocampal neurons mature into activity-dependent responsiveness. *European Journal of Neuroscience*, 2003, *18*, 2707–2712.

Jeste, D. V., Del Carmen, R., Lohr, J. B., and Wyatt, R. J. Did schizophrenia exist before the eighteenth century? *Comprehensive Psychiatry*, 1985, *26*, 493–503.

Jewett, D. C., Cleary, J., Levine, A. S., Schaal, D. W., and Thompson, T. Effects of neuropeptide Y on food-reinforced behavior in satiated rats. *Pharmacology, Biochemistry, and Behavior*, 1992, *42*, 207–212.

Jha, S. K., Coleman, T., and Frank, M. G. Sleep and sleep regulation in the ferret (*Mustela putorius furo*). *Behavioural Brain Research*, 2006, *172*, 106–113.

Jo, Y.-H., Wiedl, D., and Role, L. W. Cholinergic modulation of appetite-related synapses in mouse lateral hypothalamic slice. *Journal of Neuroscience*, 2005, *25*, 11133–11144.

Jobard, G., Crivello, F., and Tzourio-Mazoyer, N. Evaluation of the dual route theory of reading: A metaanalysis of 35 neuroimaging studies. *NeuroImage*, 2003, *20*, 693–712.

Jobst, E. E., Enriori, P. J., and Cowley, M. A. The electrophysiology of feeding circuits. *Trends in Endocrinology and Metabolism*, 2004, *15*, 488–499.

Johanek, L. M., Meyer, R. A., Hartke, T., Hobelmann, J. G., et al. Psychophysical and physiological evidence for parallel afferent pathways mediating the sensation of itch. *Journal of Neuroscience*, 2007, *27*, 7490–7497.

Johansson, G. Visual perception of biological motion and a model for its analysis. *Perception and Psychophysics*, 1973, *14*, 201–211.

Johnson, A. K. The sensory psychobiology of thirst and salt appetite. *Medicine and Science in Sports and Exercise*, 2007, *39*, 1388–1400.

Johnson, M. A. Subcortical face processing. *Nature Reviews: Neuroscience*, 2005, *6*, 766–774.

Johnson, M. K., Kim, J. K., and Risse, G. Do alcoholic Korsakoff's syndrome patients acquire affective reactions? *Journal of Experimental Psychology: Learning, Memory, and Cognition*, 1985, *11*, 22–36.

Jonas, P., Bischofberger, J., and Sandkühler, J. Corelease of two fast neurotransmitters at a central synapse. *Science*, 1998, *281*, 419–523.

Jones, B. E. Influence of the brainstem reticular formation, including intrinsic monoaminergic and cholinergic neurons, on forebrain mechanisms of sleep and waking. In *The Diencephalon and Sleep*, edited by M. Mancia and G. Marini. New York: Raven Press, 1990.

Jones, D. T., and Reed, R. R. G_{olf}: An olfactory neuron specific-G protein involved in odorant signal transduction. *Science*, 1989, *244*, 790–795.

Jones, S. S., Collins, K., and Hong, H.-W. An audience effect on smile production in 10-month-old infants. *Psychological Science*, 1991, *2*, 45–49.

Jope, R. S., Song, L., Li, P. P., Young, L. T., et al. The phosphoinositide signal transduction system is impaired in bipolar affective disorder brain. *Journal of Neurochemistry*, 1996, *66*, 2402–2409.

Jornales, V. E., Jakob, M., Zamani, A., and Vaina, L. M. Deficits on complex motion perception, spatial discrimination and eye movements in a patient with bilateral occipital-parietal lesions. *Investigative Ophthalmology and Visual Science*, 1997, *38*, S72.

Jouvet, M. The role of monoamines and acetylcholine-containing neurons in the regulation of the sleep-waking cycle. *Ergebnisse der Physiologie*, 1972, *64*, 166–307.

Jouvet-Mounier, D., Astic, L., and Lacote, D. Ontogenesis of the states of sleep in rat, cat, and guinea pig during the first postnatal month. *Developmental Psychobiology*, 1970, *2*, 216–239.

Kaas, J. H., and Collins, C. E. The organization of sensory cortex. *Current Opinion in Neurobiology*, 2001, *11*, 498–504.

Kaas, J. H., Hackett, T. A., and Tramo, M. J. Auditory processing in primate cerebral cortex. *Current Opinion in Neurobiology*, 1999, *9*, 164–170.

Kales, A., Tan, T.-L., Kollar, E. J., Naitoh, P., et al. Sleep patterns following 205 hours of sleep deprivation. *Psychosomatic Medicine*, 1970, *32*, 189–200.

Kalin, N. H., Sherman, J. E., and Takahashi, L. K. Antagonism of endogenous CRG systems attenuates stress-induced freezing behavior in rats. *Brain Research*, 1988, *457*, 130–135.

Kalivas, P. W., Peters, J., and Knackstedt, L. Animal models and brain circuits in drug addiction. *Molecular Interventions*, 2006, *6*, 339–344.

Kanner, L. Autistic disturbances of affective contact. *The Nervous Child*, 1943, *2*, 217–250.

Kanold, P. O., and Young, E. D. Proprioceptive information from the pinna provides somatosensory input to cat dorsal cochlear nucleus. *Journal of Neurophysiology*, 2001, *21*, 7848–7858.

Kanwisher, N., and Yovel, G. The fusiform face area: A cortical region specialized for the perception of faces. *Philosophical Transactions of the Royal Society of London B*, 2006, *361*, 2109–2128.

Kaplitt, M. G., Feigin, A., Tang, C., Fitzsimons, H. L., et al. Amygdala central nucleus lesions: Effect on heart rate conditioning in the rabbit. *Physiology and Behavior*, 1979, *23*, 1109–1117.

Karacan, I., Salis, P. J., and Williams, R. L. The role of the sleep laboratory in diagnosis and treatment of impotence. In *Sleep Disorders: Diagnosis and Treatment*, edited by R. J. Williams and I. Karacan. New York: John Wiley & Sons, 1978.

Karacan, I., Williams, R. L., Finley, W. W., and Hursch, C. J. The effects of naps on nocturnal sleep: Influence on the need for stage 1 REM and stage 4 sleep. *Biological Psychiatry*, 1970, *2*, 391–399.

Karlson, P., and Luscher, M. "Pheromones": A new term for a class of biologically active substances. *Nature*, 1959, *183*, 55–56.

Kartsounis, L. D., Rudge, P., and Stevens, J. M. Bilateral lesions of CA1 and CA2 fields of the hippocampus are sufficient to cause a severe amnesic syndrome in humans. *Journal of Neurology, Neurosurgery and Psychiatry*, 1995, *59*, 95–98.

Kaspar, B. K., Lladó, J., Sherkat, N., Rothstein, J. D., and Gage, F. H. Retrograde viral delivery of IGF-1 prolongs survival in a mouse ALS model. *Science*, 2003, *301*, 839–842.

Katanoda, K., Yoshikawa, K., and Sugishita, M. A functional MRI study on the neural substrates for writing. *Human Brain Mapping*, 2001, *13*, 34–42.

Katzman, D. K., Christensen, B., Young, A. T., and Zipursky, R. B. Starving the brain: Structural abnormalities and cognitive impairment in adolescents with anorexia nervosa. *Seminars in Clinical Neuropsychiatry*, 2001, *2*, 146–152.

Kauer, J. A., and Malenka, R. C. Synaptic plasticity and addiction. *Nature Reviews: Neuroscience*, 2007, *8*, 844–858.

Kawahara, Y., Ito, K., Sun, H., Aizawa, H., Kanazawa, I., and Kwak, S. RNA editing and death of motor neurons. *Nature*, 2004, *427*, 801.

Kawahara, Y., Sun, H., Ito, K., Hideyama, T., et al. Underediting of GluR2 mRNA, a neuronal death inducing molecular change in sporadic ALS, does not occur in motor neurons in ALS1 or SBMA. *Neuroscience Research*, 2006, *54*, 11–14.

Kawauchi, H., Kawazoe, I., Tsubokawa, M., Kishida, M., and Baker, B. I. Characterization of melanin-concentrating hormone in chum salmon pituitaries. *Nature*, 1983, *305*, 321–323.

Kayama, Y., Ohta, M., and Jodo, E. Firing of "possibly" cholinergic neurons in the rat laterodorsal tegmental nucleus during sleep and wakefulness. *Brain Research*, 1992, *569*, 210–220.

Kaye, W. H., Nagata, T., Weltzin, T. E., Hsu, G., et al. Double-blind placebo-controlled administration of fluoxetine in restricting- and restricting-purging-type anorexia nervosa. *Biological Psychiatry*, 2001, *49*, 644–652.

Keller, S. E., Weiss, J. M., Schleifer, S. J., Miller, N. E., and Stein, M. Stress-induced suppression of immunity in adrenalectomized rats. *Science*, 1983, *221*, 1301–1304.

Kelso, S. R., Ganong, A. H., and Brown, T. H. Hebbian synapses in hippocampus. *Proceedings of the National Academy of Sciences, USA*, 1986, *83*, 5326–5330.

Kempermann, G., Wiskott, L., and Gage, F. H. Functional significance of adult neurogenesis. *Current Opinion in Neurobiology*, 2004, *13*, 186–191.

Kendell, R. E., and Adams, W. Unexplained fluctuations in the risk for schizophrenia by month and year of birth. *British Journal of Psychiatry*, 1991, *158*, 758–763.

Kertesz, A. Anatomy of jargon. In *Jargonaphasia*, edited by J. Brown. New York: Academic Press, 1981.

Kessler, R. C., Berglund, P., Demler, O., Jin, R., et al. The epidemiology of major depressive disorder: Results from the National Comorbidity Survey Replication (NCS-R), *JAMA*, 2003, *289*, 3095–3105.

Kessler, R. C., Chiu, W. T., Jin, R., Ruscio, A. M., et al. The epidemiology of panic attacks, panic disorder, and agoraphobia in the National Comorbidity Survey Replication. *Archives of General Psychiatry*, 2006, *63*, 415–424.

Kessler, R. C., Sonnega, A., Bromet, E., Hughes, M., and Nelson, C. B. Posttraumatic stress disorder in the National Comorbidity Survey. *Archives of General Psychiatry*, 1995, *52*, 1048–1060.

Kestler, L. P., Walker, E., and Vega, E. M. Dopamine receptors in the brains of schizophrenia patients: A meta-analysis of the findings. *Behavioral Pharmacology*, 2001, *12*, 355–371.

Kety, S. S., Rosenthal, D., Wender, P. H., and Schulsinger, K. F. The types and prevalence of mental illness in the biological and adoptive families of adopted schizophrenics. In *The Transmission of Schizophrenia*, edited by D. Rosenthal and S. S. Kety. New York: Pergamon Press, 1968.

Kety, S. S., Wender, P. H., Jacobsen, B., Ingraham, L. J., et al. Mental illness in the biological and adoptive relatives of schizophrenic adoptees: Replication of the Copenhagen Study in the rest of Denmark. *Archives of General Psychiatry*, 1994, *51*, 442–455.

Kew, J. J. M., Ridding, M. C., Rothwell, J. C., Passingham, R. E., Leigh, P. N., Sooriakumaran, S., Frackowiak, R. S. G., and Brooks, D. J. Reorganization of cortical blood flow and transcranial magnetic stimulation maps in human subjects after upper limb amputation. *Journal of Neurophysiology*, 1994, *72*, 2517–2524.

Keyes, A., Brozek, J., Henschel, A., Mickelsen, O., and Taylor, H. L. *The Biology of Human Starvation*. Minneapolis, Minn.: University of Minnesota Press, 1950.

Khateb, A., Fort, P., Pegna, A., Jones, B. E., and Muhlethaler, M. Cholinergic nucleus basalis neurons are excited by histamine in vitro. *Neuroscience*, 1995, *69*, 495–506.

Kiang, N. Y.-S. *Discharge Patterns of Single Fibers in the Cat's Auditory Nerve*. Cambridge, Mass.: MIT Press, 1965.

Kiecolt-Glaser, J. K. Stress-related immune suppression: Health implications. *Brain, Behavior, and Immunity*, 1987, *1*, 7–20.

Kiecolt-Glaser, J. K., Marucha, P. T., Malarkey, W. B., Mercado, A. M., and Glaser, R. Slowing of wound healing by psychological stress. *Lancet*, 1995, *346*, 1194–1196.

Kilpatrick, D. G., Koenen, K. C., Ruggiero, K. J., Acierno, R., et al. The serotonin transporter genotype and social support and moderation of posttraumatic stress disorder and depression in hurricane-exposed adults. *American Journal of Psychiatry*, 2007, *164*, 1693–1699.

Kimoto, H., Haga, S., Sato, K., and Touhara, K. Sex-specific peptides from exocrine glands stimulate mouse vomeronasal sensory neurons. *Nature*, 2005, *437*, 898–901.

Kingston, K., Szmukler, G., Andrewes, D., Tress, B., and Desmond, P. Neuropsychological and structural brain changes in anorexia nervosa before and after refeeding. *Psychological Medicine*, 1996, *26*, 15–28.

Kinnamon, S. C., and Cummings, T. A. Chemosensory transduction mechanisms in taste. *Annual Review of Physiology*, 1992, *54*, 715–731.

Kinsley, C. H., and Bridges, R. S. Morphine treatment and reproductive condition alter olfactory preferences for pup and adult male odors in female rats. *Developmental Psychobiology*, 2990, *23*, 331–347.

Kirkpatrick, B., Kim, J. W., and Insel, T. R. Limbic system fos expression associated with paternal behavior. *Brain Research*, 1994, *658*, 112–118.

Kirwan, C. B., Bayley, P. J., Galvén, V. V., and Squire, L. R. Detailed recollection of remote autobiographical memory after damage to the medial temporal lobe. *Proceedings of the National Academy of Sciences, USA*, 2008, *105*, 2676–2680.

Kitada, T., Asakawa, S., Hattori, N., Matsumine, H., et al. Mutations in the parkin gene cause autosomal recessive juvenile parkinsonism. *Nature*, 1998, *392*, 605–608.

Kiwaki, K., Kotz, C. M., Wang, C., Lanningham-Foster, L., and Levine, J. A. Orexin A (hypocretin 1) injected into hypothalamic paraventricular nucleus and spontaneous physical activity in rats. *American Journal of Physiology*, 2004, *286*, E551–E559.

Klaur, J., Zhao, Z., Klein, G. M., Lo, E. H., and Buchan, A. M. The neurotoxicity of tissue plasminogen activator? *Journal of Cerebral Blood Flow and Metabolism*, 2004, *24*, 945–963.

Klein, D. A., and Walsh, B. T. Eating disorders: Clinical features and pathophysiology. *Physiology and Behavior*, 2004, *81*, 359–374.

Klunk, W. E., Engler, H., Nordberg, A., Bacskai, B. J., et al. Imaging the pathology of Alzheimer's disease: Amyloid-imaging with positron emission tomography. *Neuroimaging Clinics of North America*, 2003, *13*, 781–789.

Knapp, P. H., Levy, E. M., Giorgi, R. G., Black, P. H., et al. Short-term immunological effects of induced emotion. *Psychosomatic Medicine*, 1992, *54*, 133–148.

Knebelmann, B., Boussin, L., Guerrier, D., Legeai, L., et al. Anti-Muellerian hormone Bruxelles: A nonsense mutation associated with the persistent Muellerian duct syndrome. *Proceedings of the National Academy of Sciences, USA*, 1991, *88*, 3767–3771.

Knecht, S., Breitenstein, C., Bushuven, S., Wailke, S., et al. Levodopa: Faster and better word learning in normal humans. *Annals of Neurology*, 2004, *56*, 20–26.

Knecht, S., Drager, B., Deppe, M., Bobe, L., et al. Handedness and hemispheric language dominance in healthy humans. *Brain*, 2000, *123*, 2512–2518.

Knutson, B., Adams, C. M., Fong, G. W., and Hommer, D. Anticipation of increasing monetary reward selectively recruits nucleus accumbens. *Journal of Neuroscience*, 2001, *21*, RC159 (1–5).

Knutson, B., and Adcock, R. A. Remembrance of rewards past. *Neuron*, 2005, *45*, 331–332.

Kobatake, E., Tanaka, K., and Tamori, Y. Long-term learning changes the stimulus selectivity of cells in the inferotemporal cortex of adult monkeys. *Neuroscience Research*, 1992, *S17*, S237.

Koenen, K. C., Harley, R., Lyons, M. J., Wolfe, J., et al. A twin registry study of familial and individual risk factors for trauma exposure and posttraumatic stress disorder. *Journal of Nervous and Mental Disease*, 2002, *190*, 209–218.

Koenigs, M., Young, L., Adolphs, R., Tranel, D., et al. Damage to the prefrontal cortex increases utilitarian moral judgments. *Nature*, 2007, *446*, 908–911.

Kohler, E., Keysers, C., Umiltà, M. A., Fogassi, L., et al. Hearing sounds, understanding actions: Action representation in mirror neurons. *Science*, 2002, *297*, 846–848.

Kojima, M., Hosoda, H., Date, Y., Nakazato, M., et al. Ghrelin is a growth-hormone–releasing acylated peptide from stomach. *Nature*, 1999, *402*, 656–660.

Komisaruk, B. R., and Larsson, K. Suppression of a spinal and a cranial nerve reflex by vaginal or rectal probing in rats. *Brain Research*, 1971, *35*, 231–235.

Komisaruk, B. R., and Steinman, J. L. Genital stimulation as a trigger for neuroendocrine and behavioral control of reproduction. *Annals of the New York Academy of Sciences*, 1987, *474*, 64–75.

Kong, J., Shepel, N., Holden, C. P., Mackiewicz, M., et al. Brain glycogen decreases with increased periods of wakefulness: Implications for homeostatic drive to sleep. *Journal of Neuroscience*, 2004, *22*, 5581–5587.

Koob, G. F. Drug addiction: The yin and yang of hedonic homeostasis. *Neuron*, 1996, *16*, 893–896.

Koob, G. F., and Le Moal, M. Drug addiction, dysregulation of reward, and allostasis. *Neuropsychopharmacology*, 2001, *24*, 97–124.

Koob, G. F., Thatcher-Britton, K., Britton, D., Roberts, D. C. S., and Bloom, F. E. Destruction of the locus coeruleus or the dorsal NE bundle does not alter the release of punished responding by ethanol and chlordiazepoxide. *Physiology and Behavior*, 1984, *33*, 479–485.

Koopman, P. Gonad development: Signals for sex. *Current Biology*, 2001, *11*, R481–R483.

Kornhuber, H. H. Cerebral cortex, cerebellum, and basal ganglia: An introduction to their motor functions. In *The Neurosciences: Third Study Program*, edited by F. O. Schmitt and F. G. Worden. Cambridge, Mass.: MIT Press, 1974.

Koroshetz, W. J., and Moskowitz, M. A. Emerging treatments for stroke in humans. *Trends in Pharmacological Sciences,* 1996, *17,* 227–233.

Kortegaard, L. S., Hoerder, K., Joergensen, J., Gillberg, C., and Kyvik, K. O. A preliminary population-based twin study of self-reported eating disorder. *Psychological Medicine,* 2001, *31,* 361–365.

Kosfeld, M., Heinrichs, M., Zak, P. J., Fischbacher, U., and Fehr, E. Oxytocin increases trust in humans. *Nature,* 2005, *433,* 673–676.

Kosten, T., Miserendino, M. J. D., and Kehoe, P. Enhanced acquisition of cocaine self-administration in adult rats with neonatal isolation stress experience. *Brain Research,* 2000, *875,* 44–50.

Kosten, T., and Owens, S. M. Immunotherapy for the treatment of drug abuse. *Pharmacology and Therapeutics,* 2005, *108,* 76–85.

Kourtzi, A., and Kanwisher, N. Activation in human MT/MST by static images with implied motion. *Journal of Cognitive Neuroscience,* 2000, *12,* 48–55.

Kouyama, N., and Marshak, D. W. Bipolar cells specific for blue cones in the macaque retina. *Journal of Neuroscience,* 1992, *12,* 1233–1252.

Kovács, G., Vogels, R., and Orban, G. A. Selectivity of macaque inferior temporal neurons for partially occluded shapes. *Journal of Neuroscience,* 1995, *15,* 1984–1997.

Koylu, E. O., Couceyro, P. R., Lambert, P. D., and Kuhar, M. J. Cocaine- and amphetamine-regulated transcript peptide immunohistochemical localization in the rat brain. *Journal of Comparative Neurology,* 1998, *391,* 115–132.

Kozlowski, L. T., and Cutting, J. E. Recognizing the sex of a walker from a dynamic point-light display. *Perception and Psychophysics,* 1977, *21,* 575–580.

Kramer, A., Yang, F.-C., Kraves, S., and Weitz, C. J. A screen for secreted factors of the suprachiasmatic nucleus. *Methods in Enzymology,* 2005, *393,* 645–663.

Kramer, F. M., Jeffery, R. W., Forster, J. L., and Snell, M. K. Long-term follow-up of behavioral treatment for obesity: Patterns of weight regain among men and women. *International Journal of Obesity,* 1989, *13,* 123–136.

Kranzler, H. R., Modesto-Lowe, V., and Nuwayser, E. S. Sustained-release naltrexone for alcoholism treatment: A preliminary study. *Alcoholism: Clinical and Experimental Research,* 1998, *22,* 1074–1079.

Kraut, R. E., and Johnston, R. Social and emotional messages of smiling: An ethological approach. *Journal of Personality and Social Psychology,* 1979, *37,* 1539–1553.

Kress, M., and Zeilhofer, H. U. Capsaicin, protons and heat: New excitement about nociceptors. *Trends in Pharmacological Science,* 1999, *20,* 112–118.

Kristensen, P., Judge, M. E., Thim, L. Ribel, U. Christjansen, K. N., Wulff, B. S., et al. Hypothalamic CART is a new anorectic peptide regulated by leptin. *Nature,* 1998, *393,* 72–76.

Krolak-Salmon, P., Hénaff, M.-A., Vighetto, A., Bertrand, O., and Mauguière, F. Early amygdala reaction to fear spreading in occipital, temporal, and frontal cortex: A depth electrode ERP study in human. *Neuron,* 2004, *42,* 665–676.

Krubitzer, L. Constructing the neocortex: Influences on the pattern of organization in mammals. In *Brain and Mind: Evolutionary Perspectives,* edited by M. S. Gazzaniga and J. S. Altmann. Strasbourg, France: Human Frontier Science Program, 1998.

Kruijver, F. P. M., Zhou, J.-N., Pool, C. W., Hofman, M. A., et al. Male-to-female transsexuals have female neuron numbers in a limbic nucleus. *Journal of Clinical Endocrinology and Metabolism,* 2000, *85,* 2034–2041.

Kuffler, S. W. Neurons in the retina: Organization, inhibition and excitation problems. *Cold Spring Harbor Symposium on Quantitative Biology,* 1952, *17,* 281–292.

Kuffler, S. W. Discharge patterns and functional organization of mammalian retina. *Journal of Neurophysiology,* 1953, *16,* 37–68.

Kumar, K., Wyant, G. M., and Nath, R. Deep brain stimulation for control of intractable pain in humans, present and future: A ten-year follow-up. *Neurosurgery,* 1990, *26,* 774–782.

Kumar, M. J., and Andersen, J. K. Perspectives on MAO-B in aging and neurological disease: Where do we go from here? *Molecular Neurobiology,* 2004, *30,* 77–89.

Kuner, R., Groom, A. J., Bresink, I., Kornau, H. C., et al. Late-onset motoneuron disease caused by a functionally modified AMPA receptor subunit. *Proceedings of the National Academy of Sciences, USA,* 2005, *102,* 5826–5831.

Kunos, G., and Batkai, S. Novel physiologic functions of endocannabinoids as revealed through the use of mutant mice. *Neurochemical Research,* 2001, *26,* 1015–1021.

Kunugi, H., Nanko, S., and Murray, R. M. Obstetric complications and schizophrenia: Prenatal underdevelopment and subsequent neurodevelopmental impairment. *British Journal of Psychiatry,* 2001, *40,* s25–s29.

Kupfer, D. J. REM latency: A psychobiologic marker for primary depressive disease. *Biological Psychiatry,* 1976, *11,* 159–174.

Kurata, K., and Hoffman, D. S. Differential effects of muscimol microinjection into dorsal and ventral aspects of the premotor cortex of monkeys. *Journal of Neurophysiology,* 1994, *71,* 1151–1164.

Kurihara, K. Recent progress in taste receptor mechanisms. In *Umami: A Basic Taste,* edited by Y. Kawamura and M. R. Kare. New York: Dekker, 1987.

Kuriki, S., Mori, T., and Hirata, Y. Motor planning center for speech articulation in the normal human brain. *Neuroreport,* 1999, *10,* 765–769.

Kushner, M. G., Kim, S. W., Donahue, C., Thuras, P., et al. D-cycloserine augmented exposure therapy for obsessive-compulsive disorder. *Biological Psychiatry,* 2007, *62,* 835–838.

LaBar, K. S., LeDoux, J. E., Spencer, D. D., and Phelps, E. A. Impaired fear conditioning following unilateral temporal lobectomy in humans. *Journal of Neuroscience,* 1995, *15,* 6846–6855.

Lahav, A., Saltzman, E., and Schlaug, G. Action representation of sound: Audiomotor recognition network while listening to newly acquired actions. *Journal of Neuroscience,* 2007, *10,* 308–314.

Lahti, A. C., Weiler, M. A., Michaelidis, T., Parwani, A., and Tamminga, C. A. Effects of ketamine in normal and schizophrenic volunteers. *Neuropsychopharmacology,* 2001, *25,* 455–467.

Lai, E. C., Jankovic, J., Krauss, J. K., Ondo, W. G., and Grossman, R. G. Long-term efficacy of posteroventral pallidotomy in the treatment of Parkinson's disease. *Neurology*, 2000, *55*, 1218–1222.

Laitinen, L. V., Bergenheim, A. T., and Hariz, M. I. Leksell's posteroventral pallidotomy in the treatment of Parkinson's disease. *Journal of Neurosurgery*, 1992, *76*, 53–61.

Lambon Ralph, M. A., and Patterson, K. Generalization and differentiation in semantic memory. *Annals of the New York Academy of Science*, 2008, *1124*, 61–76.

Landisman, C. E., and Ts'o, D. Y. Color processing in macaque striate cortex: Relationships to ocular dominance, cytochrome oxidase, and orientation. *Journal of Neurophysiology*, 2002, *87*, 3126–3137.

Lange, C. G. *Über Gemüthsbewegungen*. Leipzig, Germany: T. Thomas, 1887.

Langen, M., Durston, S., Staal, W. G., Palmen, S. J. M. C., and van England, H. Caudate nucleus is enlarged in high-functioning medication-naïve subjects with autism. *Biological Psychiatry*, 2007, *62*, 262–266.

Langston, J. W., and Ballard, P. Parkinsonism induced by 1-methyl-4-phenyl-1,2,3,6-tetrahydropyridine (MPTP): implications for treatment and the pathogenesis of Parkinson's disease. *Canadian Journal of Neurological Science*, 1984, *11* (1 Suppl.), 160–165.

Langston, J. W., Ballard, P., Tetrud, J., and Irwin, I. Chronic parkinsonism in humans due to a product of meperidine-analog synthesis. *Science*, 1983, *219*, 979–980.

Larson, P. S. Deep brain stimulation for psychiatric disorders. *Neurotherapeutics*, 2008, *5*, 50–58.

Laruelle, M., Abi-Dargham, A., Van Dyck, C. H., Gil, R., et al. Single photon emission computerized tomography imaging of amphetamine-induced dopamine release in drug-free schizophrenic subjects. *Proceedings of the National Academy of Sciences, USA*, 1996, *93*, 9235–9240.

Laschet, U. Antiandrogen in the treatment of sex offenders: Mode of action and therapeutic outcome. In *Contemporary Sexual Behavior: Critical Issues in the 1970s*, edited by J. Zubin and J. Money. Baltimore: Johns Hopkins University Press, 1973.

Lau, H. C., Rogers, R. D., Haggard, P., and Passingham, R. E. Attention to intention. *Science*, 2004, *303*, 1208–1210.

Laugerette, F., Passilly-Degrace, P., Patris, B., Niot, I., et al. *Journal of Clinical Investigation*, 2005, *115*, 3177–3184.

Lavoie, B., and Parent, A. Immunohistochemical study of the serotoninergic innervation of the basal ganglia in the squirrel monkey. *Journal of Comparative Neurology*, 1990, *299*, 1–16.

Lavond, D. G., Kim, J. J., and Thompson, R. F. Mammalian brain substrates of aversive classical conditioning. *Annual Review of Psychology*, 1993, *44*, 317–342.

Lê, S., Cardebat, D., Boulanouar, K., Hénaff, M. A., et al. Seeing, since childhood, without ventral stream: A behavioural study. *Brain*, 2002, *125*, 58–74.

Le Grand, R., Mondloch, C. J., Maurer, D., and Brent, H. P. Early visual experience and face processing. *Nature*, 2001, *410*, 890.

Le Grand, R., Mondloch, C. J., Maurer, D., and Brent, H. P. Expert face processing requires visual input to the right hemisphere during infancy. *Nature Neuroscience*, 2003, *6*, 1108–1112.

LeDoux, J. E. Brain mechanisms of emotion and emotional learning. *Current Opinion in Neurobiology*, 1992, *2*, 191–197.

LeDoux, J. E. Emotion circuits in the brain. *Annual Review of Neuroscience*, 2000, *23*, 155–184.

LeDoux, J. E., Sakaguchi, A., and Reis, D. J. Subcortical efferent projections of the medial geniculate nucleus mediate emotional responses conditioned to acoustic stimuli. *Journal of Neuroscience*, 1984, *4*, 683–698.

Lee, A., Clancy, S., and Fleming, A. S. Mother rats bar-press for pups: Effects of lesions of the mpoa and limbic sites on maternal behavior and operant responding for pup-reinforcement. *Behavioural Brain Research*, 2000, *108*, 15–31.

Lee, A. K., and Wilson, M. A. Memory of sequential experience in the hippocampus during slow wave sleep. *Neuron*, 2002, *36*, 1183–1194.

Lee, A. W., and Brown, R. E. Comparison of medial preoptic, amygdala, and nucleus accumbens lesions on parental behavior in California mice (*Peromyscus californicus*). *Physiology and Behavior*, 2007, *92*, 617–628.

Lee, M. S., Lee, H. Y., Lee, H. J., and Ryu, S. H. Serotonin transporter promoter gene polymorphism and long-term outcome of antidepressant treatment. *Psychiatric Genetics*, 2004, *14*, 111–115.

Lehman, M. N., Silver, R., Gladstone, W. R., Kahn, R. M., et al. Circadian rhythmicity restored by neural transplant: Immunocytochemical characterization with the host brain. *Journal of Neuroscience*, 1987, *7*, 1626–1638.

Lehman, M. N., and Winans, S. S. Vomeronasal and olfactory pathways to the amygdala controlling male hamster sexual behavior: Autoradiographic and behavioral analyses. *Brain Research*, 1982, *240*, 27–41.

Leibenluft, E., and Wehr, T. A. Is sleep deprivation useful in the treatment of depression? *American Journal of Psychiatry*, 1992, *149*, 159–168.

Leiguarda, R. C., and Marsden, C. D. Limb apraxias: Higher-order disorders of sensorimotor integration. *Brain*, 2000, *123*, 860–879.

Leonard, C. M., Rolls, E. T., Wilson, F. A. W., and Baylis, G. C. Neurons in the amygdala of the monkey with responses selective for faces. *Behavioral Brain Research*, 1985, *15*, 159–176.

Leonard, H. L., Lenane, M. C., Swedo, S. E., Rettew, D. C., Gershon, E. S., and Rapoport, J. L. Tics and Tourette's disorder: A 2- to 7-year follow-up of 54 obsessive-compulsive children. *American Journal of Psychiatry*, 1992a, *149*, 1244–1251.

Leonard, H. L., Lenane, M. C., Swedo, S. E., Rettew, D. C., et al. Tourette syndrome and obsessive-compulsive disorder. *Advances in Neurology*, 1992b, *58*, 83–93.

Lesch, K. P., and Mossner, R. Genetically driven variation in serotonin uptake: Is there a link to affective spectrum, neurodevelopmental, and neurodegenerative disorders? *Biological Psychiatry*, 1988, *44*, 179–192.

Lesné, S., Ali, C., Bagriel, C., Crock, N., et al. NMDA receptor activation inhibits α-secretase and promotes neuronal amyloid-β production. *Journal of Neuroscience*, 2005, *25*, 9367–9377.

Lesser, R. Selective preservation of oral spelling without semantics in a case of multi-infarct dementia. *Cortex*, 1989, *25*, 239–250.

Letchworth, S. R., Nader, M. A., Smith, H. R., Friedman, D. P., and Porrino, L. J. Progression of changes in dopamine transporter binding site density as a result of cocaine self-administration in rhesus monkeys. *Journal of Neuroscience*, 2001, *21*, 2799–2807.

LeVay, S. A difference in hypothalamic structure between heterosexual and homosexual men. *Science*, 1991, *253*, 1034–1037.

Levenson, R. W., Ekman, P., and Friesen, W. V. Voluntary facial action generates emotion-specific autonomic nervous system activity. *Psychophysiology*, 1990, *27*, 363–384.

Levine, J. A., Eberhardt, N. L., and Jensen, M. D. Role of nonexercise activity thermogenesis in resistance to fat gain in humans. *Science*, 1999, *283*, 212–214.

Levine, J. A., Lanningham-Foster, L. M., McCrady, S. K., Krizan, A. C., et al. Interindividual variation in posture allocation: Possible role in human obesity. *Science*, 2005, *307*, 584–586.

Levy, R. M., and Bredesen, D. E. Controversies in HIV-related central nervous system disease: Neuropsychological aspects of HIV-1 infection. In *AIDS Clinical Review 1989*, edited by P. Volberding, and M. A. Jacobson. New York: Marcel Dekker, 1989.

Lewis, D. A., and Smith, R. E. Steroid-induced psychiatric syndromes: A report of 14 cases and a review of the literature. *Journal of the Affective Disorders*, 1983, *5*, 19–32.

Lewis, E. B. Clusters of master control genes regulate the development of higher organisms. *Journal of the American Medical Association*, 1992, *267*, 1524–1531.

Lewy, A. K., Lefler, B. J., Emens, J. S., and Bauer, V. K. The circadian basis of winter depression. *Proceedings of the National Academy of Sciences, USA*, 2006, *103*, 7414–7419.

Li, A.-J., and Ritter, S. Glucoprivation increases expression of neuropeptide Y mRNA in hindbrain neurons that innervate the hypothalamus. *European Journal of Neuroscience*, 2004, *19*, 2147–2154.

Li, S., Cullen, W. K., Anwyl, R., and Rowan, M. J. Dopamine-dependent facilitation of LTP induction in hippocampal CA1 by exposure to spatial novelty. *Nature Neuroscience*, 2003, *5*, 526–531.

Li, S.-H., Lam, S., Cheng, A. L., and Li, X.-J. Intranuclear huntingtin increases the expression of caspase-1 and induces apoptosis. *Human Molecular Genetics*, 2000, *9*, 2859–2867.

Li, X., Li, W., Wang, H., Cao, J., et al. Pseudogenization of a sweet-receptor gene accounts for cats' indifference toward sugar. *PLoS Genetics*, 2005, *1*, 27–35.

Lidberg, L., Asberg, M., and Sundqvist-Stensman, U. B. 5-Hydroxyindoleacetic acid levels in attempted suicides who have killed their children. *Lancet*, 1984, *2*, 928.

Lidberg, L., Tuck, J. R., Asberg, M., Scalia-Tomba, G. P., and Bertilsson, L. Homicide, suicide and CSF 5-HIAA. *Acta Psychiatrica Scandanavica*, 1985, *71*, 230–236.

Lieberman, J. A. Dopamine partial agonists: A new class of antipsychotic. *CNS Drugs*, 2004, *18*, 251–267.

Liepert, J., Bauder, H., Wolfgang, H. R., Miltner, W. H., et al. Treatment-induced cortical reorganization after stroke in humans. *Stroke*, 2000, *31*, 1210–1216.

Liljequist, S. The competitive NMDA receptor antagonist, CGP 39551, inhibits ethanol withdrawal seizures. *European Journal of Pharmacology*, 1991, *192*, 197–198.

Lim, M. M., Wang, Z., Olazábal, D. E., Ren, X., et al. Enhanced partner preference in a promiscuous species by manipulating the expression of a single gene. *Nature*, 2004, *429*, 754–757.

Lim, M. M., and Young, L. F. Vasopressin-dependent neuronal circuits underlying pair bond formation in the monogamous prairie vole. *Neuroscience*. 2004, *125*, 35–45.

Lin, J. S., Sakai, K., and Jouvet, M. Evidence for histaminergic arousal mechanisms in the hypothalamus of cat. *Neuropharmacology*, 1998, *27*, 111–122.

Lin, L., Faraco, J., Li, R., Kadotani, H., et al. The sleep disorder canine narcolepsy is caused by a mutation in the hypocretin (orexin) receptor 2. *Cell*, 1999, *98*, 365–376.

Lindauer, R. J. L., Olff, M., van Meijel, E. P. M., Carlier, I. V. E., and Gersons, B. P. R. Cortisol, learning, memory, and attention in relation to smaller hippocampal volume in police officers with posttraumatic stress disorder. *Biological Psychiatry*, 2005, *59*, 171–177.

Links, J. M., Zubieta, J. K., Meltzer, C. G., Stumpf, M. J., and Frist, J. J. Influence of spatially heterogenous background activity on "hot object" quantitation in brain emission computed tomography. *Journal of Computer Assisted Tomography*, 1996, *20*, 680–687.

Liotti, M., Mayberg, H. S., McGinnis, S., Brannan, S. L., and Jerabek, P. Unmasking disease-specific cerebral blood flow abnormalities: Mood challenge in patients with remitted unipolar depression. *American Journal of Psychiatry*, 2002, *159*, 1830–1840.

Lisk, R. D., Pretlow, R. A., and Friedman, S. Hormonal stimulation necessary for elicitation of maternal nest-building in the mouse (*Mus musculus*). *Animal Behaviour*, 1969, *17*, 730–737.

Lisman, J. A mechanism for the Hebb and the anti-Hebb processes underlying learning and memory. *Proceedings of the National Academy of Sciences, USA*, 1989, *86*, 9574–9578.

Liu, C., Zhang, W.-T., Tang, Y.-Y., Mai, X.-Q., et al. The visual word form area: Evidence from an fMRI study of implicit processing of Chinese characters. *NeuroImage*, 2008, *40*, 1350–1361.

Liu, L., Wong, T. P., Pozza, M. F., Lingenhoehl, K., et al. Role of NMDA receptor subtypes in governing the direction of hippocampal synaptic plasticity. *Science*, 2004, *304*, 1021–1023.

Liuzzi, F. J., and Lasek, R. J. Astrocytes block axonal regeneration in mammals by activating the physiological stop pathway. *Science*, 1987, *237*, 642–645.

Livingstone, M. S., and Hubel, D. H. Anatomy and physiology of a color system in the primate visual cortex. *Journal of Neuroscience*, 1984, *4*, 309–356.

Livingstone, M. S., and Hubel, D. H. Psychophysical evidence for separate channels for the perception of form, color, movement, and depth. *Journal of Neuroscience*, 1987, *7,* 3416–3468.

Lledo, P. M., Hjelmstad, G. O., Mukherji, S., Soderling, T. R., et al. Calcium/calmodulin-dependent kinase II and long-term potentiation enhance synaptic transmission by the same mechanism. *Proceedings of the National Academy of Sciences, USA*, 1995, *92,* 11175–11179.

Locus of Change, edited by L. R. Squire, N. M. Weinberger, G. Lynch, and J. L. McGaugh. New York: Oxford University Press, 1991.

Logan, F. A. Decision making by rats: Delay versus amount of reward. *Journal of Comparative and Physiological Psychology*, 1965, *59,* 1–12.

Logothetis, N. K., Pauls, J., and Poggio, T. Shape representation in the inferior temporal cortex of monkeys. *Current Biology*, 1995, *5,* 552–563.

Lømo, T. Frequency potentiation of excitatory synaptic activity in the dentate area of the hippocampal formation. *Acta Physiologica Scandinavica*, 1966, *68* (Suppl. 227), 128.

Longcamp, M., Anton, J.-L., Roth, M., and Velay, J.-L. Premotor activations in response to visually presented single letters depend on the hand used to write: A study on left-handers. *Neuropsychologia*, 2005, *43,* 1801–1809.

Lozano, A. M., Mayberg, H. S., Giacobbe, P., Hamani, C., Craddock, R. C., and Kennedy, S. H. Subcallosal cingulate gyrus deep brain stimulation for treatment-resistant depression. *Biological Psychiatry*, 2008, *64,* 461–467.

Lu, J., Greco, M. A., Shiromani, P., and Saper, C. B. Effect of lesions of the ventrolateral preoptic nucleus on NREM and REM sleep. *Journal of Neuroscience*, 2000, *20,* 3830–3842.

Lu, J., Sherman, D., Devor, M., and Saper, C. B. A putative flip-flop switch for control of REM sleep. *Nature*, 2006, *441,* 589–594.

Lu, J., Zhang, Y.-H., Chou, T. C., Gaus, S. E., et al. Contrasting effects of ibotenate lesions of the paraventricular nucleus and subparaventricular zone on sleep-wake cycle and temperature regulation. *Journal of Neuroscience*, 2001, *21,* 4864–4874.

Ludwig, D. S, Tritos, N. A., Mastaitis, J. W., Kulkarni, R., et al. Melanin-concentrating hormone overexpression in transgenic mice leads to obesity and insulin resistance. *Journal of Clinical Investigation*, 2001, *107,* 379–386.

Luo, M., Fee, M. S., and Katz, L. C. Encoding pheromonal signals in the accessory olfactory bulb of behaving mice. *Science*, 2003, *299,* 1196–1201.

Lupien, S., Lecours, A. R., Schwartz, G., Sharma, S., Hauger, R. L., Meaney, M. J., and Nair, N. P. V. Longitudinal study of basal cortixol levels in healthy elderly subjects: Evidence for subgroups. *Neurobiology of Aging*, 1996, *17,* 95–105.

Luppi, P. H., Gervansoni, D., Verret, L., Goutagny, R., et al. Paradoxical (REM) sleep genesis: The switch from an aminergic-cholinergic to a GABAergic-glutamatergic hypothesis. *Journal of Physiology (Paris)*, 2007, *100,* 271–283.

Lüscher, C., Xia, H., Beattie, E. C., Carroll, R. C., et al. Role of AMPA receptor cycling in synaptic transmission and plasticity. *Neuron*, 1999, *24,* 649–658.

Luukinen, H., Viramo, P., Herala, M., Kervinen, K., et al. Fall-related brain injuries and the risk of dementia in elderly people: A population-based study. *European Journal of Neurology*, 2005, *12,* 85–92.

Luzzi, S., Pucci, E., Di Bella, P., and Piccirilli, M. Topographical disorientation consequent to amnesia of spatial location in a patient with right parahippocampal damage. *Cortex*, 2000, *36,* 427–434.

Lydon, J. P., DeMayo, F. J., Funk, C. R., Mani, S. K., et al. Mice lacking progesterone receptor exhibit pleitropic reproductive abnormalities. *Genes and Development*, 1995, *15,* 2266–2278.

Lynch, G., Larson, J., Kelso, S., Barrionuevo, G., and Schottler, F. Intracellular injections of EGTA block induction of long-term potentiation. *Nature*, 1984, *305,* 719–721.

Lytton, W. W., and Brust, J. C. M. Direct dyslexia: Preserved oral reading of real words in Wernicke's aphasia. *Brain*, 1989, *112,* 583–594.

Ma, W., Miao, Z., and Novotny, M. Induction of estrus in grouped female mice (*Mus domesticus*) by synthetic analogs of preputial gland constituents. *Chemical Senses*, 1999, *24,* 289–293.

Ma, X., Zubcevic, L., Brüning, J. C., Ashcroft, F. M., and Burdakov, D. Electrical inhibition of identified anorexigenic POMC neurons by orexin/hypocretin. *Journal of Neuroscience*, 2007, *27,* 1529–1533.

MacDonald, A. W., Carter, C. S., Kerns, J. G., Ursu, S., et al. Specificity of prefrontal dysfunction and context processing deficits to schizophrenia in never-medicated patients with first-episode psychosis. *American Journal of Psychiatry*, 2005, *162,* 475–484.

MacLean, H. E., Warne, G. L., and Zajac, J. D. Defects of androgen receptor function: From sex reversal to motor-neuron disease. *Molecular and Cellular Endocrinology*, 1995, *112,* 133–141.

MacLean, P. D. Psychosomatic disease and the "visceral brain": Recent developments bearing on the Papez theory of emotion. *Psychosomatic Medicine*, 1949, *11,* 338–353.

Madden, P. A. F., Heath, A. C., Rosenthal, N. E., and Martin, N. G. Seasonal changes in mood and behavior: The role of genetic factors. *Archives of General Psychiatry*, 1996, *53,* 47–55.

Madsen, P. L., Holm, S., Vorstrup, S., Friberg, L., et al. Human regional cerebral blood flow during rapid-eye-movement sleep. *Journal of Cerebral Blood Flow and Metabolism*, 1991, *11,* 502–507.

Magee, J. C., and Johnston, D. A synaptically controlled, associative signal for Hebbian plasticity in hippocampal neurons. *Science*, 1997, *275,* 209–213.

Maggard, M. A., Shugarman, L. R., Suttorp, M., Maglione, M., et al. Meta-analysis: Surgical treatment of obesity. *Annals of Internal Medicine*, 2005, *142,* 547–559.

Maguire, E. A., Burgess, N., Donnett, J. G., Frackowiak, R. S. J., et al. Knowing where and getting there: A human navigation network. *Science*, 1998, *280,* 921–924.

Maguire, E. A., Frackowiak, R. S. J., and Frith, C. D. Recalling routes around London: Activation of the right hippocampus in taxi drivers. *Journal of Neuroscience*, 1997, *17,* 7103–7110.

Maguire, E. A., Gadian, D. G., Johnsrude, I. S., Good, C. D., et al. Navigation-related structural change in the hippocampi of taxi drivers. *Proceedings of the National Academy of Sciences, USA*, 2000, *97*, 4398–4403.

Mahley, R. W., and Rall, S. C. Apolipoprotein E: Far more than a lipid transport protein. *Annual Review of Genomics and Human Genetics*, 2000, *1*, 507–537.

Mahowald, M. W., and Schenck, C. H. Insights from studying human sleep disorders. *Nature*, 2005, *437*, 1279–1285.

Maj, M. Organic mental disorders in HIV-1 infection. *AIDS*, 1990, *4*, 831–840.

Mak, G. K., Enwere, E. K., Gregg, C., Pakarainen, T., et al. Male pheromone-stimulated neurogenesis in the adult female brain: Possible role in mating behavior. *Nature Neuroscience*, 2007, *10*, 1003–1011.

Maldonado, R., and Koob, G. F. Destruction of the locus coeruleus decreases physical signs of opiate withdrawal. *Brain Research*, 1993, *605*, 128–138.

Maldonado, R., and Rodriguez de Fonseca, F. Cannabinoid addiction: Behavioral models and neural correlates. *Journal of Neuroscience*, 2002, *22*, 3326–3331.

Maldonado, R., Stinus, L., Gold, L. H., and Koob, G. F. Role of different brain structures in the expression of the physical morphine-withdrawal syndrome. *Journal of Pharmacology and Experimental Therapeutics*, 1992, *261*, 669–677.

Malhotra, A. K., Adler, C. M., Kennison, S. D., Elman, I., et al. Clozapine blunts N-methyl-D-aspartate antagonist-induced psychosis: a study with ketamine. *Biological Psychiatry*, 1997, *42*, 664–668.

Mallow, G. K. The relationship between aggression and cycle stage in adult female rhesus monkeys (*Macaca mulatta*). *Dissertation Abstracts*, 1979, *39*, 3194.

Mallucci, G., Dickinson, A., Linehan, J., Klöhn, P. C., et al. Depleting neuronal PrP in prion infections prevents disease and reverses spongiosis. *Science*, 2003, *302*, 871–874.

Malnic, B., Godfrey, P. A., and Buck, L. B. The human olfactory receptor gene family. *Proceedings of the National Academy of Sciences, USA*, 2004, *101*, 2584–2589.

Malnic, B., Hirono, J, Sato, T., and Buck, L. B. Combinatorial receptor codes for odors. *Cell*, 1999, *96*, 713–723.

Malsbury, C. W. Facilitation of male rat copulatory behavior by electrical stimulation of the medial preoptic area. *Physiology and Behavior*, 1971, *7*, 797–805.

Mandiyan, V. S., Coats, J. K., and Shah, N. M. Deficits in sexual and aggressive behaviors in Cnga2 mutant mice. *Nature Neuroscience*, 2005, *8*, 1660–1662.

Manji, H. K., Moore, G. J., and Chen, G. Bipolar disorder: Leads from the molecular and cellular mechanisms of action of mood stabilisers. *British Journal of Psychiatry*, 2001, *178* (Suppl 41), S107–S109.

Manley, R. S., O'Brien, K. M., and Samuels, S. Fitness instructors' recognition of eating disorders and attendant ethical/liability issues. *Eating Disorders*, 2008, *16*, 103–116.

Mann, J. J., Malone, K. M., Diehl, D. J., Perel, J., et al. positron emission tomographic imaging of serotonin activation effects on prefrontal cortex in healthy volunteers. *Journal of Cerebral Blood Flow and Metabolism*, 1996, *16*, 418–426.

Manning, L., and Campbell, R. Optic aphasia with spared action naming: A description and possible loci of impairment. *Neuropsychologia*, 1992, *30*, 587–592.

Manns, J. R., Hopkins, R. O., and Squire, L. R. Semantic memory and the human hippocampus. *Neuron*, 2003, *38*, 127–133.

Mantyh, P. W. Connections of midbrain periaqueductal gray in the monkey. II: Descending efferent projections. *Journal of Neurophysiology*, 1983, *49*, 582–594.

Mao-Draayer, Y., and Panitch, H. Alexia without agraphia in multiple sclerosis: Case report with magnetic resonance imaging localization. *Multiple Sclerosis*, 2004, *10*, 705–707.

Maquet, P. Sleep function(s) and cerebral metabolism. *Behavioural Brain Research*, 1995, *69*, 75–83.

Maquet, P., Dive, D., Salmon, E., Sadzot, B, Branco, G., et al. Cerebral glucose utilization during sleep-wake cycle in man determined by positron emission tomography and [18F]2-fluro-2-deoxy-D-glucose method. *Brain Research*, 1990, *413*, 136–143.

Maren, S. Auditory fear conditioning increases CS-elicited spike firing in lateral amygdala neurons even after extensive overtraining. *European Journal of Neuroscience*, 2000, *12*, 4047–4054.

Margolin, D. I., and Goodman-Schulman, R. Oral and written spelling impairments. In *Cognitive Neuropsychology in Clinical Practice*, edited by D. I. Margolin. New York: Oxford University Press, 1992.

Margolin, D. I., Marcel, A. J., and Carlson, N. R. Common mechanisms in dysnomia and post-semantic surface dyslexia: Processing deficits and selective attention. In *Surface Dyslexia: Neuropsychological and Cognitive Studies of Phonological Reading*, edited by M. Coltheart. London: Lawrence Erlbaum Associates, 1985.

Margolin, D. I., and Walker, J. A. Personal communication, 1981.

Marinkovic, K., Dhond, R. P., Dale, A. M., Glessner, M., et al. Spatiotemporal dynamics of modality-specific and supramodal word processing. *Neuron*, 2003, *38*, 487–497.

Maroto, R., Raso, A., Wood, T. G., Kurosky, A., et al. TROC1 forms the stretch-activated cation channel in vertebrate cells. *Nature Cell Biology*, 2005, *7*, 179–185.

Marrosu, F., Portas, C., Mascia, M. S., Casu, M. A., et al. Microdialysis measurement of cortical and hippocampal acetylcholine release during sleep–wake cycle in freely moving cats. *Brain Research*, 1995, *671*, 329–332.

Marshall, B. E., and Longnecker, D. E. General anesthetics. In *The Pharmacological Basis of Therapeutics*, edited by L. S. Goodman, A. Gilman, T. W. Rall, A. S. Nies, and P. Taylor. New York: Pergamon Press, 1990.

Marshall, L., and Born, J. The contribution of sleep to hippocampus-dependent memory consolidation. *Trends in Cognitive Science*, 2007, *11*, 442–450.

Marson, L. Central nervous system neurons identified after injection of pseudorabies virus into the rat clitoris. *Neuroscience Letters*, 1995, *190*, 41–44.

Marson, L, and McKenna, K. E. A role for 5-hydroxytryptamine in descending inhibition of spinal sexual reflexes. *Experimental Brain Research*, 1992, *88*, 313–320.

Marson, L., and McKenna, K. E. CNS cell groups involved in the control of the ischiocavernosus and bulbospongiosus muscles: A transneuronal tracing study using pseudorabies virus. *Journal of Comparative Neurology*, 1996, *374*, 161–179.

Marson, L, and Murphy, A. Z. Identification of neural circuits involved in female genital responses in the rat: A dual virus and anterograde tracing study. *American Journal of Physiology*, 2006, *291*, R419–R428.

Martin, J. T., and Nguyen, D. H. Anthropometric analysis of homosexuals and heterosexuals: Implications for early hormone exposure. *Hormones and Behavior*, 2004, *45*, 31–39.

Martin, K. C., and Zukin, R. S. RNA trafficking and local protein synthesis in dendrites: An overview. *Journal of Neuroscience*, 2006, *26*, 7131–7134.

Martinez, M., Campion, D., Babron, M. C., and Clergetdarpous, F. Is a single mutation at the same locus responsible for all affected cases in a large Alzheimer pedigree (Fad4)? *Genetic Epidemiology*, 1993, *10*, 431–435.

Martins, I. J., Hone, E., Foster, J. K., Sünram-Lea, S. I., et al. Apolipoprotein E, cholesterol metabolism, diabetes, and the convergence of risk factors for Alzheimer's disease and cardiovascular disease. *Molecular Psychiatry*, 2006, *11*, 721–736.

Mas, M. Neurobiological correlates of masculine sexual behavior. *Neuroscience and Biobehavioral Reviews*, 1995, *19*, 261–277.

Masland, R. H. Neuronal diversity in the retina. *Current Opinion in Neurobiology*, 2001, *11*, 431–436.

Mathalon, D. H., Pfefferbaum, A., Lim, K. O., Rosenbloom, M. J., and Sullivan, E. V. Compounded brain volume deficits in schizophrenia-alcoholism comorbidity. *Archives of General Psychiatry*, 2003, *60*, 245–252.

Mathers, C. D., and Loncar, D. Projections of global mortality and burden of disease from 2002 to 2030. *PLoS Medicine*, 2006, *3*, e442.

Mathis, C. A., Klunk, W. E., Price, J. C., and DeKosky, S. T. Imaging technology for neurodegenerative diseases. *Archives of Neurology*, 2005, *62*, 196–200.

Matsuda, L. A., Lolait, S. J., Brownstein, M. J., Young, A. C., and Bonner, T. I. Structure of a cannabinoid receptor and functional expression of the cloned cDNA. *Nature*, 1990, *346*, 561–564.

Matsunami, H., Montmayeur, J.-P., and Buck, L. B. A family of candidate taste receptors in human and mouse. *Nature*, 2000, *404*, 601–604.

Mattay, V. S., Goldberg, T. E., Fera, F., Hariri, A. R., et al. Catechol *O*-methyltransferase *val^{159}-met* genotype and individual variation in the brain response to amphetamine. *Proceedings of the National Academy of Sciences, USA*, 2003, *100*, 6186–6191.

Matteo, S., and Rissman, E. F. Increased sexual activity during the midcycle portion of the human menstrual cycle. *Hormones and Behavior*, 1984, *18*, 249–255.

Matthes, H. W. D., Maldonado, R., Simonin, F., Valverde, O., et al. Loss of morphine-induced analgesia, reward effect and withdrawal symptoms in mice lacking the Mu-opioid-receptor gene. *Nature*, 1996, *383*, 819–823.

Matthews, D. B., Simson, P. E., and Best, P. J. Ethanol alters spatial processing of hippocampal place cells: A mechanism for impaired navigation when intoxicated. *Alcoholism: Clinical and Experimental Research*, 1996, *20*, 404–407.

Mattick, J. S. The hidden genetic program of complex organisms. *Scientific American*, 2004, *291*, 60–67.

Mattson, M. P., Haughey, N. J., and Nath, A. Cell death in HIV dementia. *Cell Death and Differentiation*, 2005, *12*, 893–904.

Maviel, T., Durkin, T. P., Menzaghi, F., and Bontempi, B. Sites of neocortical reorganization critical for remote spatial memory. *Science*, 2004, *305*, 96–99.

Mayberg, H. S., Lozano, A. M., Voon, V., McNeely, H. E., et al. Deep brain stimulation for treatment-resistant depression. *Neuron*, 2005, *45*, 651–660.

Mayer, D. J., and Liebeskind, J. C. Pain reduction by focal electrical stimulation of the brain: An anatomical and behavioral analysis. *Brain Research*, 1974, *68*, 73–93.

Mayers, A. G., and Baldwin, D. S. Antidepressants and their effect on sleep. *Human Psychopharmacology*, 2005, *20*, 533–559.

Mazur, A. Hormones, aggression, and dominance in humans. In *Hormones and Aggressive Behavior*, edited by B. B. Svare. New York: Plenum Press, 1983.

Mazur, A., and Booth, A. Testosterone and dominance in men. *Behavioral and Brain Sciences*, 1998, *21*, 353–397.

Mazur, A., and Lamb, T. Testosterone, status, and mood in human males. *Hormones and Behavior*, 1980, *14*, 236–246.

McCandliss, B. D., Cohen, L., and Dehaene, S. The visual word form area: Expertise for reading in the fusiform gyrus. *Trends in Cognitive Science*, 2003, *7*, 293–299.

McCann, U. D., Wong, D. F., Yokoi, F., Villemagne, V., et al. Reduced striatal dopamine transporter density in abstinent methamphetamine and methcathinone users: Evidence from positron emission tomography studies with [^{11}C]WIN-35,428. *Journal of Neuroscience*, 1998, *18*, 8417–8422.

McCarley, R. W., and Hobson, J. A. The form of dreams and the biology of sleep. In *Handbook of Dreams: Research, Theory, and Applications*, edited by B. Wolman. New York: Van Nostrand Reinhold, 1979.

McCaul, K. D., Gladue, B. A., and Joppa, M. Winning, losing, mood, and testosterone. *Hormones and Behavior*, 1992, *26*, 486–504.

McCleod, P., Dittrich, W., Driver, J., Perret, D., and Zihl, J. Preserved and impaired detection of structure from motion by a "motion blind" patient. *Vision and Cognition*, 1996, *3*, 363–391.

McClintock, M. K. Menstrual synchrony and suppression. *Nature*, 1971, *229*, 244–245.

McClintock, M. K., and Adler, N. T. The role of the female during copulation in wild and domestic Norway rats (*Rattus norvegicus*). *Behaviour*, 1978, *67*, 67–96.

McEwen, B. S., and Sapolsky, R. M. Stress and cognitive function. *Current Biology*, 1995, *5*, 205–216.

McFarland, K., and Kalivas, P. W. The circuitry mediating cocaine-induced reinstatement of drug-seeking behavior. *Journal of Neuroscience*, 2001, *21*, 8655–8663.

McFarland, K., Lapish, C. C., and Kalivas, P. W. Prefrontal glutamate release in the core of the nucleus accumbens mediates cocaine-induced reinstatement of drug-seeking behavior. *Journal of Neuroscience*, 2003, *23*, 3531–3537.

McGrath, J., Welham, J., and Pemberton, M. Month of birth, hemisphere of birth and schizophrenia. *British Journal of Psychiatry*, 1995, *167*, 783–785.

McHugh, T. J., Blum, K. I., Tsien, J. Z., Tonegawa, S., and Wilson, M. A. Impaired hippocampal representation of space in CA1-specific NMDAR1 knockout mice. *Cell*, 1996, *87*, 1339–1349.

McIver, B., Connacher, A., Whittle, I., Baylis, P., and Thompson, C. Adipsic hypothalamic diabetes insipidus after clipping of anterior communicating artery aneurysm. *British Medical Journal*, 1991, *303*, 1465–1467.

McKinley, M. J., Cairns, M. J., Denton, D. A., Egan, G., et al. Physiological and pathophysiological influences on thirst. *Physiology and Behavior*, 2004, *81*, 795–803.

McLaughlin, S. K., McKinnon, P. J., Robichon, A., Spickofsky, N., et al. Identification of an endogenous 2-monoglyceride, present in canine gut, that binds to cannabinoid receptors. *Biochemical Pharmacology*, 1995, *50*, 83–90.

Mechoulam, R., Ben-Shabat, S., Hanus, L., Ligumsky, M., et al. Identification of an endogenous 2-monoglyceride, present in canine gut, that binds to cannabinoid receptors. *Biochemical Pharmacology*, 1995, *50*, 83–90.

Mednick, S. A., Machon, R. A., and Huttunen, M. O. An update on the Helsinki influenza project. *Archives of General Psychiatry*, 1990, *47*, 292.

Mednick, S., Nakayama, K., and Stickgold, R. Sleep-dependent learning: A nap is as good as a night. *Nature Neuroscience*, 2003, *6*, 697–698.

Melges, F. T. *Time and the Inner Future: A Temporal Approach to Psychiatric Disorders.* New York: John Wiley & Sons, 1982.

Melzak, R. Phantom limbs. *Scientific American*, 1992, *266*, 120–126.

Menco, B. P. M., Bruch, R. C., Dau, B., and Danho, W. Ultrastructural localization of olfactory transduction components: The G protein subunit G_{olf} and type III adenylyl cyclase. *Neuron*, 1992, *8*, 441–453.

Menon, V., and Desmond, J. E. Left superior parietal cortex involvement in writing: Integrating fMRI with lesion evidence. *Cognitive Brain Research*, 2001, *12*, 337–340.

Meredith, M. Chronic recording of vomeronasal pump activation in awake behaving hamsters. *Physiology and Behavior*, 1994, *56*, 345–354.

Meredith, M., and O'Connell, R. J. Efferent control of stimulus access to the hamster vomeronasal organ. *Journal of Physiology*, 1979, *286*, 301–316.

Mereu, G., Yoon, K.-W. P., Boi, V., Gessa, G. L., et al. Preferential stimulation of ventral tegmental area dopaminergic neurons by nicotine. *European Journal of Pharmacology*, 1987, *141*, 395–400.

Merikangas, K. R., and Low, N. C. Genetic epidemiology of anxiety disorders. *Handbook of Experimental Pharmacology*, 2005, *169*, 163–179.

Mesches, M. H., Fleshner, M., Heman, K. L., Rose, G. M., and Diamond, D. M. Exposing rats to a predator blocks primed burst potentiation in the hippocampus *in vitro. Journal of Neuroscience*, 1999, *19*, RC18 (1–5).

Meyer, M., Alter, K., Friederici, A. D., Lohmann, G., and von Cramon, D. Y. FMRI reveals brain regions mediating slow prosodic modulations in spoken sentences. *Human Brain Mapping*, 2002, *17*, 73–88.

Meyer-Bahlburg, H. F. L. Psychoendocrine research on sexual orientation: Current status and future options. *Progress in Brain Research*, 1984, *63*, 375–398.

Meyer-Bahlburg, H. F. L. Gender and sexuality in classic congenital adrenal hyperplasia. *Endocrinology and Metabolism Clinics of North America*, 2001, *30*, 155–171.

Meyer-Bahlburg, H. F. L. Gender identity outcome in female-raised 46, XY persons with penile agenesis, cloacal exstrophy of the bladder, or penile ablation. *Archives of Sexual Behavior*, 2005, *34*, 423–438.

Michel, L., Derkinderen, P., Laplaud, D., Daumas-Duport, B., et al. Emotional facial palsy following striato-capsular infarction. *Journal of Neurology, Neurosurgery, and Psychiatry*, 2008, *79*, 193–194.

Mieda, M., Williams, S. C., Sinton, C. M., Richardson, J. A., et al. Orexin neurons function in an efferent pathway of a food-entrainable circadian oscillator in eliciting food-anticipatory activity and wakefulness. *Journal of Neuroscience*, 2004, *24*, 10493–10501.

Mignot, E. Genetic and familial aspects of narcolepsy. *Neurology*, 1998, *50*, S16–S22.

Milad, M. R., Vidal-Gonzalez, I., and Quirk, G. J. Electrical stimulation of medial prefrontal cortex reduces conditioned fear in a temporally specific manner. *Behavioral Neuroscience*, 2004, *118*, 389–394.

Mileykovskiy, B. Y., Kiyashchenko, L. I., and Siegel, J. M. Behavioral correlates of activity in identified hypocretin/orexin neurons. *Neuron*, 2005, *46*, 787–798.

Miller, N. E. Understanding the use of animals in behavioral research: Some critical issues. *Annals of the New York Academy of Sciences*, 1983, *406*, 113–118.

Miller, V. M., and Best, P. J. Spatial correlates of hippocampal unit activity are altered by lesions of the fornix and entorhinal cortex. *Brain Research*, 1980, *194*, 311–323.

Milner, A. D., Perrett, D. I., Johnston, R. S., and Benson, P. J. Perception and action in "visual form agnosia." *Brain*, 1991, *114*, 405–428.

Milner, B. Memory and the temporal regions of the brain. In *Biology of Memory*, edited by K. H. Pribram and D. E. Broadbent. New York: Academic Press, 1970.

Milner, B., Corkin, S., and Teuber, H.-L. Further analysis of the hippocampal amnesic syndrome: 14-year follow-up study of H. M. *Neuropsychologia*, 1968, *6*, 317–338.

Mindus, P., Rasmussen, S. A., and Lindquist, C. Neurosurgical treatment for refractory obsessive-compulsive disorder: Implications for understanding frontal lobe function. *Journal of Neuropsychiatry and Clinical Neurosciences*, 1994, *6*, 467–477.

Mirenowicz, J., and Schultz, W. Importance of unpredictability for reward responses in primate dopamine neurons. *Journal of Neurophysiology*, 1994, *72*, 1024–1027.

Mirenowicz, J., and Schultz, W. Preferential activation of midbrain dopamine neurons by appetitive rather than aversive stimuli. *Nature*, 1996, *379*, 449–451.

Misanin, J. R., Miller, R. R., and Lewis, D. J. Retrograde amnesia produced by electroconvulsive shock after reactivation of a consolidated memory trace. *Science*, 1968, *160*, 554–555.

Mishkin, M. Visual mechanisms beyond the striate cortex. In *Frontiers in Physiological Psychology*, edited by R. W. Russell. New York: Academic Press, 1966.

Mitchell, J. E. Psychopharmacology of eating disorders. *Annals of the New York Academy of Sciences*, 1989, *575*, 41–49.

Mitchell, J. D., Wokke, J. H., and Borasio, G. D. Recombinant human insulin-like growth factor I (rhIGF-I) for amyotrophic lateral sclerosis/motor neuron disease. *Cochrane Database of Systematic Reviews*, 2007, Oct 17(4): CD002064.

Mitchell, P. B., and Loo, C. K. Transcranial magnetic stimulation for depression. *The Australian and New Zealand Journal of Psychiatry*, 2006, *40*, 379–380.

Mitler, M. M. Evaluation of treatment with stimulants in narcolepsy. *Sleep*, 1994, *17*, S103–S106.

Mitsuno, K., Sasa, M., Ishihara, I., Ishikawa, M., and Kikuchi, H. LTP of mossy fiber-stimulated potentials in CA3 during learning in rats. *Physiology and Behavior*, 1994, *55*, 633–638.

Miyashita, Y. Cognitive memory: Cellular and network machineries and their top-down control. *Science*, 2004, *306*, 435–440.

Mizrahi, A., Crowley, J. C., Shtoyerman, E., and Katz, L. C. High-resolution *in vivo* imaging of hippocampal dendrites and spines. *Journal of Neuroscience*, 2004, *2*, 3147–3151.

Mochizuki, T., Crocker, A., McCormack, S., Yanagisawa, M., et al. Behavioral state instability in orexin knock-out mice. *Journal of Neuroscience*, 2004, *24*, 6291–6300.

Moghaddam, B., and Bunney, B. S. Differential effect of cocaine on extracellular dopamine levels in rat medial prefrontal cortex and nucleus accumbens: Comparison to amphetamine. *Synapse*, 1989, *4*, 156–161.

Mollon, J. D. "Tho' she kneel'd in that place where they grew. . .": The uses and origins of primate colour vision. *Journal of Experimental Biology*, 1989, *146*, 21–38.

Mombaerts, P. Molecular biology of odorant receptors in vertebrates. *Annual Review of Neuroscience*, 1999, *22*, 487–510.

Monaghan, D. T., and Cotman, C. W. Distribution of NMDA-sensitive L-^{3}H-glutamate binding sites in rat brain as determined by quantitative autoradiography. *Journal of Neuroscience*, 1985, *5*, 2909–2919.

Money, J., and Ehrhardt, A. *Man & Woman, Boy & Girl*. Baltimore: Johns Hopkins University Press, 1972.

Monk, C. S., Telzer, E. H., Mogg, K., Bradley, B. P., et al. Amygdala and ventrolateral prefrontal cortex activation to masked angry faces in children and adolescents with generalized anxiety disorder. *Archives of General Psychiatry*, 2008, *65*, 568–576.

Monsonego, A., and Weiner, H. L. Immunotherapeutic approaches to Alzheimer's disease. *Science*, 2003, *302*, 834–838.

Montagna, P., Gambetti, P., Cortelli, P., and Lugaresi, E. Familial and sporadic fatal insomnia. *The Lancet Neurology*, 2003, *2*, 167–176.

Montero, S., Fuentes, J. A., and Fernandez-Tome, P. Lesions of the ventral noradrenergic bundle prevent the rise in blood pressure induced by social deprivation stress in the rat. *Cellular and Molecular Neurobiology*, 1990, *10*, 497–505.

Moore, D. J., West, A. B., Dawson, V. L., and Dawson, T. M. Molecular pathophysiology of Parkinson's disease. *Annual Review of Neuroscience*, 2005, *28*, 57–87.

Moore, G. J., Bebchuk, J. M., Wilds, I. B., Chen, G., and Manji, H. K. Lithium-induced increase in human brain grey matter. *Lancet*, 2000, *356*, 1241–1242.

Moore, R. Y., and Eichler, V. B. Loss of a circadian adrenal corticosterone rhythm following suprachiasmatic lesions in the rat. *Brain Research*, 1972, *42*, 201–206.

Moore, R. Y., Speh, J. C., and Leak, R. K. Suprachiasmatic nucleus organization. *Cell Tissue Research*, 2002, *309*, 89–98.

Moore, T. H. M., Zummit, S., Lingford-Hughes, A., Barnes, T. R. E., et al. Cannabis use and risk of psychotic or affective mental health outcomes: A systematic review. *Lancet*, 2007, *370*, 319–328.

Moran, T. H., Katz, L. F., Plata-Salaman, C. R., and Schwartz, G. J. Disordered food intake and obesity in rats lacking cholecystokinin A receptors. *American Journal of Physiology*, 1998, *274*, R618–R625.

Moran, T. H., Shnayder, L., Hostetler, A. M., and McHugh, P. R. Pylorectomy reduces the satiety action of cholecystokinin. *American Journal of Physiology*, 1989, *255*, R1059–R1063.

Morgenthaler, T. I., and Silber, M. H. Amnestic sleep-related eating disorder associated with zolpidem. *Sleep Medicine*, 2002, *3*, 323–327.

Mori, E., Ikeda, M., Hirono, N., Kitagaki, H., et al. Amygdalar volume and emotional memory in Alzheimer's disease. *American Journal of Psychiatry*, 1999, *156*, 216–222.

Morris, J. A., Gobrogge, K. L., Jordan, C. L., and Breedlove, S. M. Brain aromatase: Dyed-in-the-wool homosexuality. *Endocrinology*, 2004, *145*, 475–477.

Morris, J. S., de Gelder, B., Weiskrantz, L., and Dolan, R. J. Differential extrageniculostriate and amygdala responses to presentation of emotional faces in a cortically blind field. *Brain*, 2001, *124*, 1241–1252.

Morris, J. S., Frith, C. D., Perrett, D. I., Rowland, D., et al. A differential neural response in the human amygdala to fearful and happy facial expressions. *Nature*, 1996, *383*, 812–815.

Morris, N. M., Udry, J. R., Khan-Dawood, F., and Dawood, M. Y. Marital sex frequency and midcycle female testosterone. *Archives of Sexual Behavior*, 1987, *16*, 27–37.

Morris, R. G. M., Garrud, P., Rawlins, J. N. P., and O'Keefe, J. Place navigation impaired in rats with hippocampal lesions. *Nature*, 1982, *297*, 681–683.

Morrow, E. M., Yoo, S.-Y., Flkavell, S. W., Kim, T.-K., et al. Identifying autism loci and genes by tracing recent shared ancestry. *Science*, 2008, *321*, 218–223.

Moscovitch, M., and Olds, J. Asymmetries in emotional facial expressions and their possible relation to hemispheric specialization. *Neuropsychologia*, 1982, *20*, 71–81.

Moscovitch, M., Winocur, G., and Behrmann, M. What is special about face recognition? Nineteen experiments on a person with visual object agnosia and dyslexia but normal face recognition. *Journal of Cognitive Neuroscience*, 1997, *9*, 555–604.

Mountcastle, V. B. Modality and topographic properties of single neurons of cat's somatic sensory cortex. *Journal of Neurophysiology*, 1957, *20*, 408–434.

Mueser, K. T., and McGurk, S. R. Schizophrenia. *The Lancet*, 2004, *363*, 2063–2072.

Mukhametov, L. M. Sleep in marine mammals. In *Sleep Mechanisms*, edited by A. A. Borbély and J. L. Valatx. Munich: Springer-Verlag, 1984.

Murillo-Rodriguez, E., Blanco-Centurion, C., Gerashchenko, D., and Salin-Pascual, R. J., The diurnal rhythm of adenosine levels in the basal forebrain of young and old rats. *Neuroscience*, 2004, *123*, 361–370.

Murray, R. M., Morrison, P. D., Henquet, C., and Di Forti, M. Cannabis, the mind and society: The hash realities. *Nature Reviews: Neuroscience*, 2007, *8*, 885–895.

Murray, S. O., Boyaci, H., and Kersten, D. The representation of perceived angular size in human primary visual cortex. *Nature Neuroscience*, 2006, *9*, 429–434.

Murre, J. M. J., Graham, K. S., and Hodges, J. R. Semantic dementia: Relevance to connectionist models of long-term memory. *Brain*, 2001, *124*, 647–675.

Musacchia, G., Sams, M., Skoe, E., and Kraus, N. Musicians have enhanced subcortical auditory and audiovisual processing of speech and music. *Proceedings of the National Academy of Sciences, USA*, 2007, *104*, 15894–15898.

Museo, G., and Wise, R. A. Place preference conditioning with ventral tegmental injections of cystine. *Life Sciences*, 1994, *55*, 1179–1186.

Mushiake, H., Inase, M., and Tanji, J. Neuronal activity in the primate premotor, supplementary, and precentral motor cortex during visually guided and internally determined sequential movements. *Journal of Neurophysiology*, 1991, *66*, 705–718.

Myers, R. D., Wooten, M. H., Ames, C. D., and Nyce, J. W. Anorexic action of a new potential neuropeptide Y antagonist [D-Tyr27,36, D-Thr32]-NPY (27–36) infused into the hypothalamus of the rat. *Brain Research Bulletin*, 1995, *37*, 237–245.

Myrick, H., Anton, R. F., Li, X., Henderson, S., Drobes, D., Voronin, K, and George, M. S. Differential brain activity in alcoholics and social drinkers to alcohol cues: Relationship to craving. *Neuropsychopharmacology*, 2004, *29*, 393–402.

Nadeau, S. E. Impaired grammar with normal fluency and phonology. *Brain*, 1988, *111*, 1111–1137.

Nader, K. Memory traces unbound. *Trends in Neuroscience*, 2003, *26*, 65–72.

Nader, K., Majidishad, P., Amorapanth, P., and LeDoux, J. E. Damage to the lateral and central, but not other, amygdaloid nuclei prevents the acquisition of auditory fear conditioning. *Learning and Memory*, 2001, *8*, 156–163.

Naeser, M. A., Palumbo, C. L., Helm-Estabrooks, N., Stiassny-Eder, D., et al. Severe nonfluency in aphasia: Role of the medial subcallosal fasciculus and other white matter pathways in recovery of spontaneous speech. *Brain*, 1989, *112*, 1–38.

Nägerl, U. V., Köstinger, G., Anderson, J. C., Martin, K. A. C., and Bonhoeffer, T. Protracted synaptogenesis after activity-dependent spinogenesis in hippocampal neurons. *Journal of Neuroscience*, 2007, *27*, 8149–8156.

Najjar, M. Zolpidem and amnestic sleep related eating disorder. *Journal of Clinical Sleep Medicine*, 2007, *15*, 637–638.

Nakahara, D., Ozaki, N., Miura, Y., Miura, H., and Nagatsu, T. Increased dopamine and serotonin metabolism in rat nucleus accumbens produced by intracranial self-stimulation of medial forebrain bundle as measured by in vivo microdialysis. *Brain Research*, 1989, *495*, 178–181.

Nakamura, J., Endo, K., Sumida, T., and Hasegawa, T. Bilateral tactile agnosia: A case report. *Cortex*, 1998, *34*, 375–388.

Nakamura, K., Honda, M., Okada, T., Hanakawa, T., et al. Participation of the left posterior inferior temporal cortex in writing and mental recall of kanji orthography: A functional MRI study. *Brain*, 2000, *123*, 954–967.

Nakamura, T., and Gold, G. A cyclic nucleotide-gated conductance in olfactory receptor cilia. *Nature*, 1987, *325*, 442–444.

Nakazato, M., Mauakami, N., Date, Y., Kojima, M., et al. A role for ghrelin in the central regulation of feeding. *Nature*, 2001, *409*, 194–198.

Nambu, T., Sakurai, T., Mizukami, K., Hosoya, Y., et al. Distribution of orexin neurons in the adult rat brain. *Brain Research*, 1999, *827*, 243–260.

Naqvi, N. H., and Bechara, A. The airway sensory impact of nicotine contributes to the conditioned reinforcing effects of individual puffs from cigarettes. *Pharmacology, Biochemistry and Behavior*, 2005, *81*, 821–829.

Naqvi, N. H., Rudrauf, D., Damasio, H., and Bechara, A. Damage to the insula disrupts addiction to cigarette smoking. *Science*, 2007, *315*, 531–534.

Narayan, S. S., Temchin, A. N., Recio, A., and Ruggero, M. A. Frequency tuning of basilar membrane and auditory nerve fibers in the same cochlae. *Science*, 1998, *282*, 1882–1884.

Nathans, J. The evolution and physiology of human color vision: insights from molecular genetic studies of visual pigments. *Neuron*, 1999, *24*, 299–312.

Nauta, W. J. H. Hypothalamic regulation of sleep in rats: Experimental study. *Journal of Neurophysiology*, 1946, *9*, 285–316.

Nauta, W. J. H. Some efferent connections of the prefrontal cortex in the monkey. In *The Frontal Granular Cortex and Behavior*, edited by J. M. Warren and K. Akert. New York: McGraw-Hill, 1964.

Nazir, T. A., Jacobs, A. M., and O'Regan, J. K. Letter legibility and visual word recognition. *Memory and Cognition*, 1998, *26*, 810–821.

Nef, P., Hermansborgmeyer, I., Artierespin, H., Beasley, L., et al. Spatial pattern of receptor expression in the olfactory epithelium. *Proceedings of the National Academy of Sciences, USA*, 1992, *89*, 8948–8952.

Nergårdh, R., Ammar, A., Brodin, U., Bergström, J., Scheurink, A., et al. Neuropeptide Y facilitates activity-based anorexia. *Psychoneuroendocrinology*, 2007, *32*, 493–502.

Nestor, P. J., Graham, N. L., Fryer, T. D., Williams, B. G., et al. Progressive non-fluent aphasia is associated with hypometabolism centered on the left anterior insula. *Brain*, 2003, *126*, 2406–2418.

Neumann, K., Preibisch, C., Euler, H. A., von Gudenberg, A. W., et al. Cortical plasticity associated with stuttering therapy. *Journal of Fluency Disorders*, 2005, *30*, 23–39.

Neumeister, A., Konstantinidis, A., Stastny, J., Schrarz, M. J., et al. Association between serotonin transporter gene promotor polymorphism (5HTTLPR) and behavioral responses to tryptophan depletion in healthy women with and without family history of depression. *Archives of General Psychiatry*, 2002, *59*, 613–620.

Neumeister, A., Nugent, A. C., Waldeck, T., Geraci, M., et al. Neural and behavioral responses to tryptophan depletion in unmedicated patients with remitted major depressive disorder and controls. *Archives of General Psychiatry*, 2004, *61*, 765–773.

New, A. S., Buchsbaum, M. S., Hazlett, E. A., Goodman, M., Koenigsberg, H. W., Lo, J., Iskander, L., Newmark, R., Brand, J., O'Flynn, K., and Siever, L. J. Fluoxetine increases relative metabolic rate in prefrontal cortex in impulsive aggression. *Psychopharmacology*, 2004, *176*, 451–458.

New, A. S., Hazlett, E. A., Buchsbaum, M. S., Goodman, M., et al. Blunted prefrontal cortical [18]fluorodeoxyglucose positron emission tomography response to meta-chlorophenylpiperazine in impulsive aggression. *Archives of General Psychiatry*, 2002, *59*, 621–629.

Nichelli, P., Grafman, J., Pietrini, P., Clark, K., et al. Where the brain appreciates the moral of a story. *NeuroReport*, 1995, *6*, 2309–2313.

Nicholl, C. S., and Russell, R. M. Analysis of animal rights literature reveals the underlying motives of the movement: Ammunition for counter offensive by scientists. *Endocrinology*, 1990, *127*, 985–989.

Nicoll, J. A. R., Wilkinson, D., Holmes, C., Steart, P., et al. Neuropathology of human Alzheimer disease after immunization with amyloid-β peptide: A case report. *Nature Medicine*, 2003, *9*, 448–452.

Nicoll, R. A., Alger, B. E., and Jahr, C. E. Enkephalin blocks inhibitory pathways in the vertebrate CNS. *Nature*, 1980, *287*, 22–25.

Nicoll, R. A., and Malenka, R. C. A tale of two transmitters. *Science*, 1998, *281*, 360–361.

Nides, M., Oncken, C., Gonzales, D., Rennard, S., et al. Smoking cessation with varenicline, a selective α4β2 nicotinic receptor partial agonist: Results from a 7-week, randomized, placebo- and bupropion-controlled trial with 1-year follow-up. *Archives of Internal Medicine*, 2006, *166*, 1561–1568.

Nikolic, W. V., Bai, Y., Obregon, D., Hou, H., et al. Transcutaneous β-amyloid immunization reduces cerebral β-amyloid deposits without T cell infiltration and microhemorrhage. *Proceedings of the National Academy of Sciences, USA*, 2007, *104*, 2507–2512.

Nikolova, G., and Vilain, E. Mechanisms of disease: Transcription factors in sex determination—Relevance to human disorders of sex development, *Nature Clinical Practice: Endocrinology & Metabolism*, 2006, *2*, 231–238.

Nilius, B., Owsianik, G., Voets, T., and Peters, J. A. Transient receptor potential cation channels in disease. *Physiological Review*, 2007, *87*, 165–217.

Nilsson, H. J., Levinsson, A., and Schouenborg, J. Cutaneous field stimulation (CFS): A new powerful method to combat itch. *Pain*, 1997, *71*, 49–55.

NINDS. Tissue plasminogen activator for acute ischemic stroke. The National Institute of Neurological Disorders and Stroke RT-PA stroke study group. *New England Journal of Medicine*, 1995, *333*, 1189–1191.

Nisell, M., Nomikos, G. G., and Svensson, T. H. Systemic nicotine-induced dopamine release in the rat nucleus accumbens is regulated by nicotinic receptors in the ventral tegmental area. *Synapse*, 1994, *16*, 36–44.

Nishino, S. Clinical and neurobiological aspects of narcolepsy. *Sleep Medicine*, 2007, *8*, 373–399.

Nishino, S., Ripley, B., Overeem, S., Lammers, G. J., and Mignot, E. Hypocretin (orexin) deficiency in human narcolepsy. *Lancet*, 2000, *355*, 39–40.

Nobler, M. S., Oquendo, M. A., Kegeles, L. S., Malone, K. M., et al. Deceased regional brain metabolism after ECT. *American Journal of Psychiatry*, 2001, *158*, 305–308.

Norgren, R., and Grill, H. Brain-stem control of ingestive behavior. In *The Physiological Mechanisms of Motivation*, edited by D. W. Pfaff. New York: Springer-Verlag, 1982.

Normandin, J., and Murphy, A. Z. Nucleus paragigantocellularis afferents in male and female rats: Organization, gonadal steroid receptor expression, and activation during sexual behavior. *Journal of Comparative Neurology*, 2008, *508*, 771–794.

Novin, D., VanderWeele, D. A., and Rezek, M. Hepatic-portal 2-deoxy-D-glucose infusion causes eating: Evidence for peripheral glucoreceptors. *Science*, 1973, *181*, 858–860.

Novotny, M. V., Ma, W., Wiesler, D., and Zidek, L. Positive identification of the puberty-accelerating pheromone of the house mouse: The volatile ligands associating with the major urinary protein. *Proceedings of the Royal Society of London B*, 1999, *266*, 2017–2022.

Nugent, N. R., Amstadter, A. B., and Koenen, K. C. Genetics of post-traumatic stress disorder: Informing clinical conceptualizations and promoting future research. *American Journal of Medical Genetics Part C*, 2008, *148C*, 127–132.

Numan, M. Medial preoptic area and maternal behavior in the female rat. *Journal of Comparative and Physiological Psychology*, 1974, *87*, 746–759.

Numan, M. Motivational systems and the neural circuitry of maternal behavior in the rat. *Developmental Psychobiology*, 2007, *49*, 12–21.

Numan, M., and Numan, M. J. Projection sites of medial preoptic area and ventral bed nucleus of the stria terminalis neurons that express Fos during maternal behavior in female rats. *Journal of Neuroendocrinology*, 1997, *9*, 369–384.

Nutt, D. J., Glue, P., Lawson, C. W., and Wilson, S. Flumazenil provocation of panic attacks: Evidence for altered benzodiazepine receptor sensitivity in panic disorders. *Archives of General Psychiatry*, 1990, *47*, 917–925.

Oaknin, S., Rodriguez del Castillo, A., Guerra, M., Battaner, E., and Mas, M. Change in forebrain Na, K-ATPase activity and serum hormone levels during sexual behavior in male rats. *Physiology and Behavior,* 1989, *45,* 407–410.

Obler, L. K., and Gjerlow, K. *Language and the Brain.* Cambridge, England: Cambridge University Press, 1999.

O'Brien, C. P., Volpicelli, L. A., and Volpicelli, J. R. Naltrexone in the treatment of alcoholism: A clinical review. *Alcohol,* 1996, *13,* 35–39.

Oertel, D., and Young, E. D. What's a cerebellar circuit doing in the auditory system? *Trends in Neuroscience,* 2004, *27,* 104–110.

Ogawa, S., Olazabal, U. E., Parhar, I. S., and Pfaff, D. W. Effects of intrahypothalamic administration of antisense DNA for progesterone receptor mRNA on reproductive behavior and progesterone receptor immunoreactivity in female rat. *Journal of Neuroscience,* 1994, *14,* 1766–1774.

Ogden, C. L., Carroll, M. D., and Flegal, K. M. Epidemiologic trends in overweight and obesity. *Endocrinology and Metabolism Clinics of North America,* 2003, *32,* 741–760.

O'Keefe, J., and Bouma, H. Complex sensory properties of certain amygadala units in the freely moving cat. *Experimental Neurology,* 1969, *23,* 384–398.

O'Keefe, J., and Dostrovsky, T. The hippocampus as a spatial map: Preliminary evidence from unit activity in the freely moving rat. *Brain Research,* 1971, *34,* 171–175.

O'Kusky, J. R., Nasir, J., Cicchetti, F., Parent, A., and Hayden, M. R. Neuronal degeneration in the basal ganglia and loss of pallido-subthalamic synapses in mice with targeted disruption of the Huntington's disease gene. *Brain Research,* 1999, *818,* 468–479.

Olanow, C. W., Goetz, C. G., Kordower, J. H., Stoessl, A. J., et al. A double-blind controlled trial of bilateral fetal nigral transplantation in Parkinson's disease. *Annals of Neurology,* 2003, *54,* 403–414.

Olausson, H., Lamarre, Y., Backlund, H., Morin, C., Wallin, B. G., Starck, G., Ekholm, S., Strigo, I., Worsley, K., Vallbo, Å. B., Bushnell, M. C. Unmyelinated tactile afferents signal touch and project to insular cortex. *Nature Neuroscience,* 2002, *5,* 900–904.

Olds, J. Commentary. In *Brain Stimulation and Motivation,* edited by E. S. Valenstein. Glenview, Ill.: Scott, Foresman, 1973.

Oleksenko, A. I., Mukhametov, L. M., Polyakova, I. G., Supin, A. Y., and Kovalzon, V. M. Unihemispheric sleep deprivation in bottlenose dolphins. *Journal of Sleep Research,* 1992, *1,* 40–44.

Oliveri, M., Turriziani, P., Carlesimo, G. A., Koch, G., et al. Parieto-frontal interactions in visual-object and visual-spatial working memory: Evidence from transcranial magnetic stimulation. *Cerebral Cortex,* 2001, *11,* 606–618.

Olsson, A., Nearing, K. I., and Phelps, E. A. Learning fears by observing others: The neural systems of social fear transmission. *SCAN,* 2007, *2,* 3–11.

Omura, K., Tsukamoto, T., Kotani, Y., Ohgami, Y., and Yoshikawa, K. Neural correlates of phoneme-to-grapheme conversion. *Neuroreport,* 2004, *15,* 949–953.

Opitz, B., and Friederici, A. D. Interactions of the hippocampal system and the prefrontal cortex in learning language-like rules. *NeuroImage,* 2003, *19,* 1730–1737.

Opitz, B., and Friederici, A. D. Neural basis of processing sequential and hierarchical syntactic structures. *Human Brain Mapping,* 2007, *28,* 585–592.

Osaka, N., Osaka, M., Morishita, M., Kondo, H., and Fukuyama, H. A word expressing affective pain activates the anterior cingulate cortex in the human brain: an fMRI study. *Behavioural Brain Research,* 2004, *153,* 123–127.

Ossebaard, C. A., Polet, I. A., and Smith, D. V. Amiloride effects on taste quality: Comparison of single and multiple response category procedures. *Chemical Senses,* 1997, *22,* 267–275.

Ostrowsky, K., Magnin, M., Tyvlin, P., Isnard, J., et al. Representation of pain and somatic sensation in the human insula: A study of responses to direct electrical cortical stimulation. *Cerebral Cortex,* 2002, *12,* 376–385.

Otsuki, M., Soma, Y., Arai, T., Otsuka, A., and Tsuji, S. Pure apraxic agraphia with abnormal writing stroke sequences: Report of a Japenese patient with a left superior parietal haemorrhage. *Journal of Neurology, Neurosurgery, and Psychiatry,* 1999, *66,* 233–237.

Otsuki, M., Soma, Y., Arihiro, S., Watanabe, Y., et al. Dystypia: Isolated typing impairment without aphasia, apraxia or visuospatial impairment. *European Neurology,* 2002, *47,* 136–140.

Overduin, J., Frayo, R. S., Grill, H. J., Kaplan, J. M., and Cummings, D. E. Role of the duodenum and macronutrient type in ghrelin regulation. *Encodrinology,* 2005, *146,* 845–850.

Owen, A. M., James, M., Leigh, P. N., Summers, B. A., et al. Fronto-striatal cognitive deficits at different stages of Parkinson's disease. *Brain,* 1992, *115,* 1727–1751.

Packard, M. G., and Teather, L. A. Double dissociation of hippocampal and dorsal-striatal memory systems by posttraining intracerebral injections of 2-amino-5-phosphonopentanoic acid. *Behavioral Neuroscience,* 1997, *111,* 543–551.

Padberg, F., and Moller, H. J. Repetitive transcranial magnetic stimulation: Does it have potential in the treatment of depression? *CNS Drugs,* 2003, *17,* 383–403.

Pallast, E. G. M., Jongbloet, P. H., Straatman, H. M., and Zeilhuis, G. A. Excess of seasonality of births among patients with schizophrenia and seasonal ovopathy. *Schizophrenia Bulletin,* 1994, *20,* 269–276.

Papadimitriou, G. N., Christodoulou, G. N., Katsouyanni, K., and Stefanis, C. N. Therapy and prevention of affective illness by total sleep deprivation. *Journal of the Affective Disorders,* 1993, *27,* 107–116.

Papassotiropoulos, A., Wollmer, M. A., Aguzzi, A., Hock, C., et al. The prion gene is associated with human long-term memory. *Human Molecular Genetics,* 2005, *14,* 2241–2246.

Papez, J. W. A proposed mechanism of emotion. *Archives of Neurology and Psychiatry,* 1937, *38,* 725–744.

Pardo, J. V., Sheikh, S. A., Schwindt, G. C., Lee, J. T., et al. Chronic vagus nerve stimulation for treatment-resistant depression decreases resting ventromedial prefrontal glucose metabolism. *NeuroImage,* 2008, *42,* 879–889.

Paré, D., Quirk, G. J., and LeDoux, J. E. New vistas on amygdala networks in conditioned fear. *Journal of Neurophysiology,* 2004, *92,* 1–9.

Parker, A. J. Binocular depth perception and the cerebral cortex. *Nature Reviews: Neuroscience*, 2007, *8*, 379–391.

Parmentier, R., Ohtsu, H., Djebbara-Hannas, Z., Valatx, J.-L., et al. Anatomical, physiological, and pharmacological characteristics of histidine decarboxylase knock-out mice: Evidence for the role of brain histamine in behavioral and sleep-wake control. *Journal of Neuroscience*, 2002, *22*, 7695–7711.

Pascoe, J. P., and Kapp, B. S. Electrophysiological characteristics of amygdaloid central nucleus neurons during Pavlovian fear conditioning in the rabbit. *Behavioural Brain Research*, 1985, *16*, 117–133.

Pasterski, V. L., Geffner, M. E., Brain, C., Hindmarsh, P., et al. Prenatal hormones and postnatal socializtion by parents as determinants of male-typical toy play in girls with congenital adrenal hyperplasia. *Child Development*, 2005, *76*, 264–278.

Pattatucci, A. M. L., and Hamer, D. H. Development and familiality of sexual orientation in females. *Behavior Genetics*, 1995, *25*, 407–420.

Paulesu, E., Démonet, J.-F., Fazio, F., McCrory, E., et al. Dyslexia: cultural diversity and biological unity. *Science*, 2001, *291*, 2165–2167.

Pauls, D. L., and Leckman, J. F. The inheritance of Gilles de la Tourette's syndrome and associated behaviors. *New England Journal of Medicine*, 1986, *315*, 993–997.

Pauls, D. L., Towbin, K. E., Leckman, J. F., Zahner, G. E., and Cohen, D. J. Gilles de la Tourette's syndrome and obsessive-compulsive disorder: Evidence supporting a genetic relationship. *Archives of General Psychiatry*, 1986, *43*, 1180–1182.

Paulus, M. P., Feinstein, J. S., Castillo, G., Simmons, A. N., et al. Neural underpinnings of gesture discrimination in patients with limb apraxia. *Journal of Neuroscience*, 2008, *28*, 3030–3041.

Pavlov, I. *Conditioned Reflexes*. London: Oxford University Press, 1927.

Pazzaglia, M., Smania, N., Corato, E., and Aglioti, S. M. Neural underpinnings of gesture discrimination in patients with limb apraxia. *Journal of Neuroscience*, 2008, *28*, 3030–3041.

Peck, B. K., and Vanderwolf, C. H. Effects of raphe stimulation on hippocampal and neocortical activity and behaviour. *Brain Research*, 1991, *568*, 244–252.

Pedersen, C. B., and Mortensen, P. B. Evidence of a dose-response relationship between urbanicity during upbringing and schizophrenia risk. *Archives of General Psychiatry*, 2001, *58*, 1039–1046.

Pedersen-Bjergaard, U., Host, U., Kelbaek, H., Schifter, S., et al. Influence of meal composition on postprandial peripheral plasma concentrations of vasoactive peptides in man. *Scandanavian Journal of Clinical and Laboratory Investigation*, 1996, *56*, 497–503.

Peigneux, P., Laureys, S., Fuchs, S., Collette, F., et al. Are spatial memories strengthened in the human hippocampus during slow wave sleep? *Neuron*, 2004, *44*, 535–545.

Pellerin, L., Bouzier-Sore, A.-K., Aubert, A., Serres, S., et al. Activity-dependent regulation of energy metabolism by astrocytes: An update. *Glia*, 2007, *55*, 1251–1262.

Pelleymounter, M. A., Cullen, M. J., Baker, M. B., Hecht, R., et al. Effects of the obese gene product on body weight regulation in ob/ob mice. *Science*, 1995, *269*, 540–543.

Pelphrey, K. A., Morris, J. P., Michelich, C. R., Allison, T., and McCarthy, G. Functional anatomy of biological motion perception in posterior temporal cortex: An fMRI study of eye, mouth and hand movements. *Cerebral Cortex*, 2005, *15*, 1866–1876.

Pelphrey, K. A., Sasson, N. J., Reznick, J. S., Paul, G., et al. Visual scanning of faces in autism. *Journal of Autism and Developmental Disorders*, 2002, *32*, 249–261.

Pelphrey, K. A., Singerman, J. D., Allison, T., and McCarthy, G. Brain activation evoked by perception of gaze shifts: The influence of context. *Neuropsychologia*, 2003, *41*, 156–170.

Penfield, W., and Perot, P. The brain's record of auditory and visual experience: A final summary and discussion. *Brain*, 1963, *86*, 595–697.

Penfield, W., and Rasmussen, T. *The Cerebral Cortex of Man: A Clinical Study of Localization*. Boston: Little, Brown & Co., 1950.

Pereira, A. C., Huddleston, D. E., Brickman, A. M., Sosunov, A. A., et al. An *in vivo* correlate of exercise-induced neurogenesis in the adult dentate gyrus. *Proceedings of the National Academy of Sciences, USA*, 2007, *104*, 5638–5643.

Peretz, I., Blood, A. J., Penhune, V., and Zatorre, R. Cortical deafness to dissonance. *Brain*, 2001, *124*, 928–940.

Peretz, I., Cummings, S., and Dubé, M. P. The genetics of congenital amusia (tone deafness): A family-aggregation study. *American Journal of Human Genetics*, 2007, *81*, 582–588.

Peretz, I., Gagnon, L., and Bouchard, B. Music and emotion: Perceptual determinants, immediacy, and isolation after brain damage. *Cognition*, 1998, *68*, 111–141.

Perkins, D. O., Jeffries, C., and Sullivan, P. Expanding the "central dogma": The regulatory role of nonprotein coding genes and implications for the genetic liability to schizophrenia. *Molecular Psychiatry*, 2005, *10*, 69–78.

Perrett, D. I., Hietanen, J. K., Oram, M. W., and Benson, P. J. Organization and functions of cells responsive to faces in the temporal cortex. *Philosophical Transactions of the Royal Society of London B*, 1992, *335*, 23–30.

Pert, C. B., Snowman, A. M., and Snyder, S. H. Localization of opiate receptor binding in presynaptic membranes of rat brain. *Brain Research*, 1974, *70*, 184–188.

Peters, J., LaLumiere, R. T., and Kalivas, P. W. Infralimbic prefrontal cortex is responsible for inhibiting cocaine seeking in extinguished rats. *Journal of Neuroscience*, 2008, *28*, 6046–6053.

Petersen, S. E., Miezin, F. M., and Allman, J. M. Transient and sustained responses in four extrastriate visual areas of the owl monkey. *Experimental Brain Research*, 1988, *70*, 55–60.

Petkov, C. I., Kayser, C., Augath, M., and Logothetis, N. K. Functional imaging reveals nuberous fields in the monkey auditory cortex. *PLoS Biology*, 2006, *4*, 1213–1226.

Petre-Quadens, O., and De Lee, C. Eye movement frequencies and related paradoxical sleep cycles: Developmental changes. *Chronobiologia*, 1974, *1*, 347–355.

Pettito, L. A., Zatorre, R. J., Gauna, K., Nikelski, E. J., et al. Speech-like cerebral activity in profoundly deaf people processing signed languages: Implications for the neural basis of human language. *Proceedings of the National Academy of Sciences, USA*, 2000, *97*, 13961–13966.

Peuskens, H., Sunaert, S., Dupont, P., Van Hecke, P., and Orban, G. A. Human brain regions involved in heading estimation. *Journal of Neuroscience*, 2001, *21*, 2451–2461.

Peyron, C., Faraco, J., Rogers, W., Ripley, B., et al. A mutation in a case of early onset narcolepsy and a generalized absence of hypocretin peptides in human narcoleptic brains. *Nature Medicine*, 2002, *6*, 991–997.

Peyron, R., Laurent, B., and Garcia-Larrea, L. Functional imaging of brain responses to pain: A review and meta-analysis. *Neurophysiology Clinics*, 2000, 30, 263–288.

Pezawas, L., Meyer-Linderberg, A., Drabant, E. M., Verchinski, B. A., et al. 5-HTTLPR polymorphism impacts human cingulate-amygdala interactions: A genetic susceptibility mechanism for depression. *Nature Neuroscience*, 2005, *8*, 828–834.

Pfaff, D. W., and Sakuma, Y. Deficit in the lordosis reflex of female rats caused by lesions in the ventromedial nucleus of the hypothalamus. *Journal of Physiology*, 1979, *288*, 203–210.

Pfaus, J. G., Kleopoulos, S. P., Mobbs, C. V., Gibbs, R. B., and Pfaff, D. W. Sexual stimulation activates c-fos within estrogen-concentrating regions of the female rat forebrain. *Brain Research*, 1993, *624*, 253–267.

Pfeifer, J. H., Iacoboni, M., Mazziotta, J. C., and Dapretto, M. Mirroring others' emotions relates to empathy and interpersonal competence in children. *NeuroImage*, 2008, *39*, 2076–2085.

Pfleiderer, B., Zinkirciran, S., Arolt, V., Heindel, W., et al. fMRI amygdala activation during a spontaneous panic attack in a patient with panic disorder. *World Journal of Biological Psychiatry*, 2007, *8*, 269–272.

Phan, K. L., Fitzgerald, D. A., Nathan, P. J., and Tancer, M. E. Association between amygdala hyperactivity to harsh faces and severity of social anxiety in generalized social phobia. *Biological Psychiatry*, 2005, *59*, 424–429.

Phelps, E. A., Delgado, M. R., Nearing, K. I., and LeDoux, J. E. Extinction learning in humans: Role of the amygdala and vmPFC. *Neuron*, 2004, *43*, 897–905.

Phelps, E. A., O'Connor, K. J., Gatenby, J. C., Gore, J. C., et al. Activation of the left amygdala to a cognitive representation of fear. *Nature Neuroscience*, 2001, *4*, 437–441.

Phiel, C. J., and Klein, P. S. Molecular targets of lithium action. *Annual Review of Pharmacology and Toxicology*, 2001, *41*, 789–813.

Phillips, M. I., and Felix, D. Specific angiotensin II receptive neurons in the cat subfornical organ. *Brain Research*, 1976, *109*, 531–540.

Phillips, R. G., and LeDoux, J. E. Differential contribution of amygdala and hippocampus to cued and contextual fear conditioning. *Behavioral Neuroscience*, 1992, *106*, 274–285.

Piccirillo, J. F., Duntley, S., and Schotland, H. Obstructive sleep apnea. *Journal of the American Medical Association*, 2000, *284*, 1492–1494.

Pickles, J. O., and Corey, D. P. Mechanoelectrical transduction by hair cells. *Trends in Neuroscience*, 1992, *15*, 254–259.

Pijl, S., and Schwarz, D. W. F. Intonation of musical intervals by musical intervals by deaf subjects stimulated with single bipolar cochlear implant electrodes. *Hearing Research*, 1995a, *89*, 203–211.

Pijl, S., and Schwarz, D. W. F. Melody recognition and musical interval perception by deaf subjects stimulated with electrical pulse trains through single cochlear implant electrodes. *Journal of the Acoustical Society of America*, 1995b, *98*, 886–895.

Pike, K. M., Walsh, B. T., Vitousek, K., Wilson, G. T., and Bauer, J. Cognitive behavior therapy in the posthospitalization treatment of anorexia nervosa. *American Journal of Psychiatry*, 2003, *160*, 2046–2049.

Pilleri, G. The blind Indus dolphin, *Platanista indi*. *Endeavours*, 1979, *3*, 48–56.

Pitkänen, A., Savander, V., and LeDoux, J. E. Organization of intra-amygdaloid circuits: An emerging framework for understanding functions of the amygdala. *Trends in Neuroscience*, 1997, *20*, 517–523.

Pleim, E. T., and Barfield, R. J. Progesterone versus estrogen facilitation of female sexual behavior by intracranial administration to female rats. *Hormones and Behavior*, 1988, *22*, 150–159.

Pobric, G., Jefferies, E., and Lambon Ralph, M. A. Anterior temporal lobes mediate semantic representation: Mimicking semantic dementia by using rTMS in normal participants. *Proceedings of the National Academy of Sciences, USA*, 2007, *104*, 20137–20141.

Pobric, G., Mashal, N., Faust, M., and Lavidor, M. The role of the right cerebral hemisphere in proccdding novel metaphoric expressions: A transcranial magnetic stimulation study. *Journal of Cognitive Neuroscience*, 2008, *20*, 1–12.

Poeppel, D. Pure word deafness and the bilateral processing of the speech code. *Cognitive Science*, 2001, *25*, 679–693.

Poggio, G. F., and Poggio, T. The analysis of stereopsis. *Annual Review of Neuroscience*, 1984, *7*, 379–412.

Poizner, H., Feldman, A. G., Levin, M. F., Berkinblit, M. B., et al. The timing of arm-trunk coordination is deficient and vision-dependent in Parkinson's patients during reaching movements. *Experimental Brain Research*, 2000, *133*, 279–292.

Polymeropoulos, M. H., Higgins, J. J., Golbe, L. I., Johnson, W. G., et al. Mapping of a gene for Parkinson's disease to chromosome 4q21-q23. *Science*, 1996, *274*, 1197–1199.

Pomp, D., and Nielsen, M. K. Quantitative genetics of energy balance: Lessons from animal models. *Obesity Research*, 1999, *7*, 106–110.

Pompili, M., Mancinelli, I, Girardi, P., Ruberto, A., and Tatarelli, R. Suicide in anorexia nervosa: A meta-analysis. *International Journal of Eating Disorders*, 2004, *36*, 99–103.

Popa, D., Léna, C., Alexandre, C., and Adrien, J. Lasting syndrome of depression produced by reduction in serotonin uptake during postnatal development: Evidence from sleep, stress, and behavior. *Journal of Neuroscience*, 2008, *28*, 3546–3554.

Popova, NK. From genes to aggressive behavior: The role of serotonergic system. *BioEssays*, 2006, *28*, 495–503.

Poremba, A., Saunders, R. C., Crane, A. M., Cook, M., Sokoloff, L., and Mishkin, M. Functional mapping of the primate auditory system. *Science*, 2003, *299*, 568–572.

Porkka-Heiskanen, T., Strecker, R. E., and McCarley, R. W. Brain site-specificity of extracellular adenosine concentration changes during sleep deprivation and spontaneous sleep: An *in vivo* microdialysis study. *Neuroscience*, 2000, *99*, 507–517.

Porrino, L. J., Lyons, D., Smith, H. R., Daunais, J. B., and Nader, M. A. Cocaine self-administration produces a progressive involvement of limbic, association, and sensorimotor striatal domains. *Journal of Neuroscience*, 2004, *24*, 3554–3562.

Porter, J., Craven, B., Khan, R. M., Chang, S.-J., et al. Mechanisms of scent-tracking in humans. *Nature Neuroscience*, 2007, *10*, 27–29.

Preti, G., Wysocki, C. J., Barnhart, K. T., Sondheimer, S. J., and Leyden, J. J. Male axillary extracts contain pheromones that affect pulsitile secretion of luteinizing hormone and mood in women recipients. *Biology of Reproduction*, 2003, *68*, 2107–2113.

Price, D. B. Psychological and neural mechanisms of the affective dimension of pain. *Science*, 2000, *288*, 1769–1772.

Price, D. L., and Sisodia, S. S. Mutant genes in familial Alzheimer's disease and transgenic models. *Annual Review of Neuroscience*, 1998, *21*, 479–505.

Price, L. H., and Heninger, G. R. Drug therapy: Lithium in the treatment of mood disorders. *New England Journal of Medicine*, 1994, *331*, 591–598.

Price, R. A., and Gottesman, I. I. Body fat in identical twins reared apart: Roles for genes and environment. *Behavioral Genetics*, 1991, *21*, 1–7.

Pritchard, T. C., Hamilton, R. B., Morse, J. R., and Norgren, R. Projections of thalamic gustatory and lingual areas in the monkey, *Macaca fascicularis*. *Journal of Comparative Neurology*, 1986, *244*, 213–228.

Proctor, W. R., Soldo, B. L., Allan, A. M., and Dunwiddie, T. V. Ethanol enhances synaptically evoked GABAA receptor-mediated responses in cerebral cortical neurons in rat brain slices. *Brain Research*, 1992, *595*, 220–227.

Provencio, I., Rodriguez, I. R., Jiang, G., Hayes, W. P., et al. A novel human opsin in the inner retinal *Journal of Neuroscience*, 2000, *20*, 600–605.

Prusiner, S. B. Novel proteinaceous infectious particles cause scrapie. *Science*, 1982, *216*, 136–144.

Qu, D., Ludwig, D. S., Gammeltoft, S., Piper, M., et al. A role for melanin-concentrating hormone in the central regulation of feeding behavior. *Nature*, 1996, *380*, 243–247.

Quillen, E. W., Keil, L. C., and Reid, I. A. Effects of baroreceptor denervation on endocrine and drinking responses to caval constriction in dogs. *American Journal of Physiology*, 1990, *259*, R619–R626.

Quirk, G. J. Memory for extinction of conditioned fear is long-lasting and persists following spontaneous recovery. *Learning and Memory*, 2002, *9*, 402–407.

Quirk, G. J., Garcia, R., and González-Lima, F. Prefrontal mechanisms in extinction of conditioned fear. *Biological Psychiatry*, 2006, *60*, 337–343.

Quirk, G. J., Muller, R. U., Kubie, J. L., and Ranck, J. B. The positional firing properties of medial entorhinal neurons: Description and comparison with hippocampal place cells. *Journal of Neuroscience*, 1992, *12*, 1945–1963.

Quirk, G. J., Repa, J. C., and LeDoux, J. E. Fear conditioning enhances short-latency auditory responses of lateral amygdala neurons: Parallel recordings in the freely behaving rat. *Neuron*, 1995, *15*, 1029–1039.

Rahman, Q., and Hull, M. S. An empirical test of the kin selection hypothesis for male homosexuality. *Archives of Sex Behavior*, 2005, *34*, 461–467.

Raine, A., Lencz, T., Bihrle, S., LaCasse, L., and Colletti, P. Reduced prefrontal gray matter volume and reduced autonomic activity in antisocial personality disorder. *Archives of General Psychiatry*, 2002, *57*, 119–127.

Raine, A., Meloy, J. R., Bihrle, S., Stoddard, J., et al. Reduced prefrontal and increased subcortical brain functioning assessed using positron emission tomography in predatory and affective murderers. *Behavioral Science and the Law*, 1998, *16*, 319–332.

Rainville, P., Duncan, G. H., Price, D. D., Carrier, B., et al. Transplantation: A new tool in the analysis of the mammalian hypothalamic circadian pacemaker. *Trends in Neuroscience*, 1991, *14*, 362–366.

Ramachandran, V. S., and Hirstein, W. The perception of phantom limbs. *Brain*, 1998, *121*, 1603–1630.

Ramakrishnan, K., and Scheid, D. C. Treatment options for insomnia. *American Family Physician*, 2007, 76, 517–526.

Ramanathan, L., Gulyani, S., Nienhuis, R., and Siegel, J. M. Sleep deprivation decreases superoxide dismutase activity in rat hippocampus and brainstem. *Neuroreport*, 2002, *13*, 1387–1390.

Ramesh, V., Thakkar, M. M., Strecker, R. E., Basheer, R., and McCarley, R. W. Wakefulness-inducing effects of histamine in the basal forebrain of freely moving rats. *Behavioural Brain Research*, 2004, *152*, 271–278.

Ramirez, I. Why do sugars taste good? *Neuroscience and Biobehavioral Reviews*, 1990, *14*, 125–134.

Ramnani, N., and Owen, A. M. Anterior prefrontal cortex: Insights into function from anatomy and neuroimaging. *Nature Reviews: Neuroscience*, 2004, *5*, 184–194.

Ramus, F., Rosen, S., Dakin, S. C., Day, B. L., et al. Theories of developmental dyslexia: Insights from a multiple case study of dyslexic adults. *Brain*, 2003, *126*, 841–865.

Ranganath, C., DeGutis, J., and D'Esposito, M. Category-specific modulation of inferior temporal activity during working memory encoding and maintenance. *Cognitive Brain Research*, 2004, *20*, 37–45.

Rapcsak, S. Z., and Beeson, P. M. The role of the left posterior inferior temporal cortex in spelling, *Neurology*, 2004, *62*, 2221–2229.

Rapin, I. Autism in search of a home in the brain. *Neurology*, 1999, *52*, 902–904.

Rapoport, J. L. Recent advances in obsessive-compulsive disorder. *Neuropsychopharmacology*, 1991, *5*, 1–10.

Rapoport, J. L., Ryland, D. H., and Kriete, M. Drug treatment of canine acral lick: An animal model of obsessive-compulsive disorder. *Archives of General Psychiatry*, 1992, *49*, 517–521.

Rasmusson, D. D., Clow, K., and Szerb, J. C. Modification of neocortical acetylcholine release and electroencephalogram desynchronization due to brain stem stimulation by drugs applied to the basal forebrain. *Neuroscience*, 1994, *60*, 665–677.

Rauch, S. L., Shin, L. M., and Phelps, E. A. Neurocircuitry models of posttraumatic stress disorder and extinction: Human neuroimaging research—Past, present, and future. *Biological Psychiatry*, 2006, *60*, 376–382.

Rausch, J. L., Johnson, M. E., Fei, Y. J., Li, J. Q., et al. Initial conditions of serotonin transporter kinetics and genotype: Influence on SSRI treatment trial outcome. *Biological Psychiatry*, 2002, *51*, 723–732.

Rauschecker, J. P., and Tian, B. Mechanisms and streams for processing of "what" and "where" in auditory cortex. *Proceedings of the National Academy of Sciences, USA*, 2000, *97*, 11800–11806.

Ravussin, E., Valencia, M. E., Esparza, J., Bennett, P. H., and Schulz, L. O. Effects of a traditional lifestyle on obesity in Pima Indians. *Diabetes Care*, 1994, *17*, 1067–1074.

Raymond, C. R. LTP forms 1, 2, and 3: Different mechanisms for the "long" in long-term potentiation. *Trends in Neuroscience*, 2007, *30*, 167–175.

Reber, P. J., and Squire, L. R. Encapsulation of implicit and explicit memory in sequence learning. *Journal of Cognitive Neuroscience*, 1998, *10*, 248–263.

Recer, P. Study: English is a factor in dyslexia. Washington, D.C.: Associated Press, 16 March 2001.

Rechtschaffen, A., and Bergmann, B. M. Sleep deprivation in the rat by the disk-over-water method. *Behavioural Brain Research*, 1995, *69*, 55–63.

Rechtschaffen, A., and Bergmann, B. M. Sleep deprivation in the rat: An update of the 1989 paper. *Sleep*, 2002, *25*, 18–24.

Rechtschaffen, A., Bergmann, B. M., Everson, C. A., Kushida, C. A., and Gilliland, M. A. Sleep deprivation in the rat. X: Integration and discussion of the findings. *Sleep*, 1989, *12*, 68–87.

Rechtschaffen, A., Gilliland, M. A., Bergmann, B. M., and Winter, J. B. Physiological correlates of prolonged sleep deprivation in rats. *Science*, 1983, *221*, 182–184.

Reddrop, C., Moldrich, R. X., Beart, P. M., Farso, M., et al. Vampire bat salivary plasminogen activator (desmoteplase) inhibits tissue-type plasminogen activator-induced potentiation of excitotoxic injury. *Stroke*, 2005, *36*, 1241–1246.

Redmond, D. E., Bjugstad, K. B., Teng, Y. D., et al. Behavioral improvement in a primate Parkinson's model is associated with multiple homeostatic effects of human neural stem cells. *Proceedings of the National Academy of Sciences, USA*, 2007, *104*, 12175–12180.

Reed, C. L., Caselli, R. J., and Farah, M. J. Tactile agnosia: Underlying impairment and implications for normal tactile object recognition. *Brain*, 1996, *119*, 875–888.

Reed, J. M., and Squire, L. R. Retrograde amnesia for facts and events: Findings from four new cases. *Journal of Neuroscience*, 1998, *18*, 3943–3954.

Regan, B. C., Julliot, C., Simmen, B., Vienot, F., et al. Fruits, foliage and the evolution of primate colour vision. *Philosophical Transactions of the Royal Society of London B*, 2001, *356*, 229–283.

Rehn, A. E., Van Den Buuse, M., Copolov, D., Briscoe, T., et al. An animal model of chronic placental insufficiency: Relevance to neurodevelopmental disorders including schizophrenia. *Neuroscience*, 2004, *129*, 381–391.

Reid, C. A., Dixon, D. B., Takahashi, M., Bliss, T. V. P., and Fine, A. Optical quantal analysis indicates that long-term potentiation at single hippocampal mossy fiber synapses is expressed through increased release probability, recruitment of new release sites, and activation of silent synapses. *Journal of Neuroscience*, 2004, *243*, 3618–3626.

Reid, L. D. Endogenous opioids and alcohol dependence: Opioid alkaloids and the propensity to drink alcoholic beverages. *Alcohol*, 1996, *13*, 5–11.

Reinehr, T., Roth, C. L., Schernthaner, G. H., Kopp, H. P., et al. Peptide YY and glucagon-like peptide-1 in morbidly obese patients before and after surgically induced weight loss. *Obesity Surgery*, 2007, *17*, 1571–1577.

Reiner, W. G. Gender identity and sex-of-rearing in children with disorders of sexual differentiation. *Journal of Pediatric Endocrinology and Metabolism*, 2005, *18*, 549–553.

Rempel-Clower, N. L., Zola, S. M., Squire, L. R., and Amaral, D. G. Three cases of enduring memory impairment after bilateral damage limited to the hippocampal formation. *Journal of Neuroscience*, 1996, *16*, 5233–5255.

Reppert, S. M., and Weaver, D. R. Molecular analysis of mammalian circadian rhythms. *Annual Review of Physiology*, 2001, *63*, 647–676.

Ressler, K. J., and Mayberg, H. S. Targeting abnormal neural circuits in mood and anxiety disorders: From the laboratory to the clinic. *Nature Neuroscience*, 2007, *10*, 1116–1124.

Ressler, K. J., Rothbaum, M. O., Tannenbaum, L., Anderson, P., et al. Cognitive enhancers as adjuncts to psychotherapy: Use of D-cycloserine in phobic individuals to facilitate extinction of fear. *Archives of General Psychiatry*, 2004, *61*, 1136–1134.

Ressler, K. J., Sullivan, S. L., and Buck, L. A molecular dissection of spatial patterning in the olfactory system. *Current Opinion in Neurobiology*, 1994, *4*, 588–596.

Reynolds, D. V. Surgery in the rat during electrical analgesia induced by focal brain stimulation. *Science*, 1969, *164*, 444–445.

Rhees, R. W., Shryne, J. E., and Gorski, R. A. Onset of the hormone-sensitive perinatal period for sexual differentiation of the sexually dimorphic nucleus of the preoptic area in female rats. *Journal of Neurobiology*, 1990a, *21*, 781–786.

Rhees, R. W., Shryne, J. E., and Gorski, R. A. Termination of the hormone-sensitive period for differentiation of the sexually dimorphic nucleus of the preoptic area in male and female rats. *Developmental Brain Research*, 1990b, *52*, 17–23.

Rho, J. M. Can reducing sugar retard kindling? *Epilepsy Currents*, 2008, *8*, 83–84.

Rhodes, R. A., Murthy, N. B., Dresner, M. A., Selvaraj, S., et al. Human 5-HT transporter availability predicts amygdala reactivity in vivo. *Journal of Neuroscience,* 2007, *27,* 9233–9237.

Rice, N. J., Valyear, K. F., Goodale, M. A., Milner, A. D., and Culham, J. C. Orientation sensitivity to graspable objects: An fMRI adaptation study. *NeuroImage,* 2007, *36,* T87–T93.

Riemann, D., Wiegand, M., and Berger, M. Are there predictors for sleep deprivation response in depressive patients? *Biological Psychiatry,* 1991, *29,* 707–710.

Rijntjes, M., Dettmers, C., Buchel, C., Kiebel, S., et al. A blueprint for movement: Functional and anatomical representations in the human motor system. *Journal of Neuroscience,* 1999, *19,* 8043–8048.

Rissman, E. F., Early, A. H., Taylor, J. A., Korach, K. S., and Lubahn, D. B. Estrogen receptors are essential for female sexual receptivity. *Endocrinology,* 1997, *138,* 507–510.

Ritter, R. C., Brenner, L., and Yox, D. P. Participation of vagal sensory neurons in putative satiety signals from the upper gastrointestinal tract. In *Neuroanatomy and Physiology of Abdominal Vagal Afferents,* edited by S. Ritter, R. C. Ritter, and C. D. Barnes. Boca Raton, Fla.: CRC Press, 1992.

Ritter, S., Dinh, T. T., and Friedman, M. I. Induction of Fos-like immunoreactivity (Fos-li) and stimulation of feeding by 2,5-anhydro-D-mannitol (2,5-AM) require the vagus nerve. *Brain Research,* 1994, *646,* 53–64.

Ritter, S., Dinh, T. T., and Zhang, Y. Localization of hindbrain glucoreceptive sites controlling food intake and blood glucose. *Brain Research,* 2000, *856,* 37–47.

Ritter, S., and Taylor, J. S. Vagal sensory neurons are required for lipoprivic but not glucoprivic feeding in rats. *American Journal of Physiology,* 1990, *258,* R1395–R1401.

Rizzolatti, R., Fogassi, L., and Gallese, V. Neurophysiological mechanisms underlying the understanding and imitation of action. *Nature Reviews: Neuroscience,* 2001, *2,* 661–670.

Rizzoli, S. O., and Betz, W. J. Synaptic vesicle pools. *Nature Reviews: Neuroscience,* 2005, *6,* 57–69.

Robbins, L. N., Helzer, J. E., Weissman, M. M., Orvaschel, H., et al. Lifetime prevalence of specific psychiatric disorders in three sites. *Archives of General Psychiatry,* 1984, *41,* 949–958.

Roberts, S. C., Havlicek, J., Flegr, J., Mruskova, M., et al. Female facial attractiveness increases during the fertile phase of the menstrual cycle. *Biology Letters,* 2004, *271,* S270–S272.

Robertson, G. S., Pfaus, J. G., Atkinson, L. J., Matsumura, H., et al. Sexual behavior increases c-fos expression in the forebrain of the male rat. *Brain Research,* 1991, *564,* 352–357.

Robinson, D., Wu, H., Munne, R. A., Ashtari, M., et al. Reduced caudate nucleus volume in obsessive-compulsive disorder. *Archives of General Psychiatry,* 1995, *52,* 393–398.

Robinson, T. E., and Berridge, K. C. Addiction. *Annual Review of Psychology,* 2003, *54,* 25–53.

Rodieck, R. W. *The First Steps in Seeing.* Sunderland, Mass.: Sinauer Associates, 1998.

Rodman, H. R., Gross, C. G., and Albright, T. D. Afferent basis of visual response properties in area MT of the macaque. I: Effects of striate cortex removal. *Journal of Neuroscience,* 1989, *9,* 2033–2050.

Rodman, H. R., Gross, C. G., and Albright, T. D. Afferent basis of visual response properties in area MT of the macaque. II:

Effects of superior colliculus removal. *Journal of Neuroscience,* 1990, *10,* 1154–1164.

Rodrigues, S. M., Schafe, G. E., and LeDoux, J. E. Intra-amygdala blockade of the NR2B subunit of the NMDA receptor disrupts the acquisition but not the expression of fear conditioning. *Journal of Neuroscience,* 2001, *21,* 6889–6896.

Rodrigues, S. M., Schafe, G. E., and LeDoux, J. E. Molecular mechanisms underlying emotional learning and memory in the lateral amygdala. *Neuron,* 2004, *44,* 75–91.

Roe, A. W., Parker, A. J., Born, R. T., and DeAngelis, G. C. Disparity channels in early vision. *Journal of Neuroscience,* 2007, *27,* 11820–11831.

Roeltgen, D. P., Rothi, L. H., and Heilman, K. M. Linguistic semantic apraphia: A dissociation of the lexical spelling system from semantics. *Brain and Language,* 1986, *27,* 257–280.

Roffwarg, H. P., Dement, W. C., Muzio, J. N., and Fisher, C. Dream imagery: Relation to rapid eye movements of sleep. *Archives of General Psychiatry,* 1962, *7,* 235–258.

Roffwarg, H. P., Muzio, J. N., and Dement, W. C. Ontogenetic development of human sleep-dream cycle. *Science,* 1966, *152,* 604–619.

Rogawski, M. A., and Wenk, G. L. The neuropharmacological basis for the use of memantine in the treatment of Alzheimer's disease. *CNS Drug Reviews,* 2003, *9,* 275–308.

Rogers, T. T., Hocking, J., Noppeney, U., Mechelli, A., et al. Anterior temporal cortex and semantic memory: Reconciling findings from neuropsychology and functional imaging. *Cognitive, Affective, and Behavioral Neuroscience,* 2006, *6,* 201–213.

Roland, P. E. Metabolic measurements of the working frontal cortex in man. *Trends in Neurosciences,* 1984, *7,* 430–435.

Rolls, E. T. Feeding and reward. In *The Neural Basis of Feeding and Reward,* edited by B. G. Hobel and D. Novin. Brunswick, Me.: Haer Institute, 1982.

Rolls, E. T. Learning mechanisms in the temporal lobe visual cortex. *Behavioural Brain Research,* 1995, *66,* 177–185.

Rolls, E. T., and Baylis, G. C. Size and contrast have only small effects on the responses to faces of neurons in the cortex of the superior temporal sulcus of the monkey. *Experimental Brain Research,* 1986, *65,* 38–48.

Rolls, E. T., Murzi, E., Yaxley, S., Thorpe, S. J., and Simpson, S. J. *Brain Research,* 1986, *368,* 79–86.

Rolls, E. T., Yaxley, S., and Sienkiewicz, Z. J. Gustatory responses of single neurons in the orbitofrontal cortex of the macaque monkey. *Journal of Neurophysiology,* 1990, *64,* 1055–1066.

Romanovsky, A. A. Thermoregulation: Some concepts have changed. Functional architecture of the thermoregulatory system. *American Journal of Physiology,* 2007, *292,* R37–R46.

Rosa, R. R., and Bonnet, M. H. Reported chronic insomnia is independent of poor sleep as measured by electroencephalography. *Psychosomatic Medicine,* 2000, *62,* 474–482.

Rose, J. D. Changes in hypothalamic neuronal function related to hormonal induction of lordosis in behaving hamsters. *Physiology and Behavior,* 1990, *47,* 1201–1212.

Rose, J. E. Nicotine and nonnicotine factors in cigarette addiction. *Psychopharmacology*, 2006, *184*, 274–285.

Roselli, C. E., Larkin, K., Resko, J. A., Stellflug, J. N., and Stromshak, F. The volume of a sexually dimorphic nucleus in the ovine medial preoptic area/anterior hypothalamus varies with sexual partner preference. *Endocrinology*, 2004, *145*, 478–483.

Rosén, I., and Asanuma, H. Peripheral inputs to the forelimb area of the monkey motor cortex: Input-output relations. *Experimental Brain Research*, 1972, *14*, 257–273.

Rosenblatt, J. S., Hazelwood, S., and Poole, J. Maternal behavior in male rats: Effects of medial preoptic area lesions and presence of maternal aggression. *Hormones and Behavior*, 1996, *30*, 201–215.

Rosenthal, D. A program of research on heredity in schizophrenia. *Behavioral Science*, 1971, *16*, 191–201.

Rosenthal, N. E., Sack, D. A., Gillin, C., Lewy, A. J., et al. Seasonal affective disorder: A description of the syndrome and preliminary findings with light therapy. *Archives of General Psychiatry*, 1984, *41*, 72–80.

Rosenthal, N. E., Sack, D. A., James, S. P., Parry, B. L., et al. Seasonal affective disorder and phototherapy. *Annals of the New York Academy of Sciences*, 1985, *453*, 260–269.

Roses, A. D. A model for susceptibility polymorphisms for complex diseases: Apolipoprotein E and Alzheimer disease. *Neurogenetics*, 1997, *1*, 3–11.

Roth, M., Decery, J., Raybaudi, M., Massarelli, R., et al. Possible involvement of primary motor cortex in mentally simulated movement: A functional magnetic resonance imaging study. *Neuroreport*, 1996, *7*, 1280–1284.

Rothbaum, B. O., and Davis, M. Applying learning principles to the treatment of post-trauma reactions. *Annals of the New York Academy of Science*, 2003, *1008*, 112–121.

Rothman, S. M., and Olney, J. W. Excitotoxicity and the NMDA receptor. *Trends in Neurosciences*, 1987, *10*, 299–302.

Rubia, K., Overmeyer, S., Taylor, E., Brammer, M., et al. Hypofrontality in attention deficit hyperactivity disorder during higher-order motor control: A study with functional MRI. *American Journal of Psychiatry*, 1999, *156*, 891–896.

Rubin, B. S., and Barfield, R. J. Priming of estrous responsiveness by implants of 17B-estradiol in the ventromedial hypothalamic nucleus of female rats. *Endocrinology*, 1980, *106*, 504–509.

Rubin, L. L., and Staddon, J. M. The cell biology of the blood–brain barrier. *Annual Review of Neuroscience*, 1999, *22*, 11–28.

Rück, C., Karlsson, A., Steele, J. D., Edman, G., et al. Capsulotomy for obsessive-compulsive disorder. *Archives of General Psychiatry*, 2008, *65*, 914–922.

Ruggero, M. A. Responses to sound of the basilar membrane of the mammalian cochlea. *Current Opinion in Neurobiology*, 1992, *2*, 449–456.

Rumpel, S., LeDoux, J., Zador, A., and Malinow, R. Postsynaptic receptor trafficking underlying a form of associative learning. *Science*, 2005, *308*, 83–88.

Rusak, B., and Morin, L. P. Testicular responses to photoperiod are blocked by lesions of the suprachiasmatic nuclei in golden hamsters. *Biology of Reproduction*, 1976, *15*, 366–374.

Russchen, F. T., Amaral, D. G., and Price, J. L. The afferent connections of the substantia innominata in the monkey, *Macaca fascicularis. Journal of Comparative Neurology*, 1986, *242*, 1–27.

Russell, G. F. M., and Treasure, J. The modern history of anorexia nervosa: An interpretation of why the illness has changed. *Annals of the New York Academy of Sciences*, 1989, *575*, 13–30.

Russell, M. J. Human olfactory communication. *Nature*, 1976, *260*, 520–522.

Russell, M. J., Switz, G. M., and Thompson, K. Olfactory influences on the human menstrual cycle. *Pharmacology, Biochemistry and Behavior*, 1980, *13*, 737–738.

Rutter, M. Cognitive deficits in the pathogenesis of autism. *Journal of Child Psychology and Psychiatry*, 1983, *24*, 513–531.

Ryba, N. J., and Tirindelli, R. A new multigene family of putative pheromone receptors. *Neuron*, 1997, *19*, 371–392.

Ryback, R. S., and Lewis, O. F. Effects of prolonged bed rest on EEG sleep patterns in young, healthy volunteers. *Electroencephalography and Clinical Neurophysiology*, 1971, *31*, 395–399.

Ryugo, D. K., Kretzmer, E. A., and Niparko, J. K. Restoration of auditory nerve synapses in cats by cochlear implants. *Science*, 2005, *310*, 1490–1492.

Saal, D., Dong, Y., Bonci, A., and Malenka, R. C. Drugs of abuse and stress trigger a common synaptic adaptation in dopamine neurons. *Neuron*, 2003, *37*, 577–582.

Saayman, G. S. Aggressive behaviour in free-ranging chacma baboons (*Papio ursinus*). *Journal of Behavioral Science*, 1971, *1*, 77–83.

Sachar, E. J., and Baron, M. The biology of affective disorders. *Annual Review of Neuroscience*, 1979, *2*, 505–518.

Sackeim, H. A., Decina, P., Prohovnik, I., Malitz, S., and Resor, S. R. Anticonvulsant and antidepressant properties of electroconvulsive therapy: A proposed mechanism of action. *Biological Psychiatry*, 1983, *18*, 1301–1310.

Sackeim, H. A., and Gur, R. C. Lateral asymmetry in intensity of emotional expression. *Neuropsychologia*, 1978, *16*, 473–482.

Sadato, N., Pascualleone, A., Grafman, J., Ibanez, V., et al. Activation of the primary visual cortex by Braille reading in blind subjects. *Nature*, 1996, *380*, 526–528.

Saffran, E. M., Marin, O. S. M., and Yeni-Komshian, G. H. An analysis of speech perception in word deafness. *Brain and Language*, 1976, *3*, 209–228.

Saffran, E. M., Schwartz, M. F., and Marin, O. S. M. Evidence from aphasia: Isolating the components of a production model. In *Language Production*, edited by B. Butterworth. London: Academic Press, 1980.

Sagvolden, T., Aase, H., Zeiner, P., and Berger, D. Altered reinforcement mechanisms in attention-deficit/hyperactivity disorder. *Behavioural Brain Research*, 1998, *94*, 61–71.

Sagvolden, T., Johansen, E. B., Aase, H., and Russell, V. A. A dynamic developmental theory of attention-deficit/

hyperactivity disorder (ADHD) predominantly hyperactive/impulsive and combined subtypes. *Behavioral and Brain Sciences*, 2005, *28*, 397–419.

Sagvolden, T., and Sergeant, J. A. Attention deficit/hyperactivity disorder: From brain dysfunctions to behaviour. *Behavioural Brain Research*, 1998, *94*, 1–10.

Sahay, A., and Hen, R. Adult hippocampal neurogenesis in depression. *Nature Neuroscience*, 2007, *10*, 1110–1115.

Sahu, A., Kalra, P. S., and Kalra, S. P. Food deprivation and ingestion induce reciprocal changes in neuropeptide Y concentrations in the paraventricular nucleus. *Peptides*, 1988, *9*, 83–86.

Sakai, F., Meyer, J. S., Karacan, I., Derman, S., and Yamamoto, M. Normal human sleep: Regional cerebral haemodynamics. *Annals of Neurology*, 1979, *7*, 471–478.

Sakai, K. L., Noguchi, Y., Takeuchi, T., and Watanabe, E. Selective priming of syntactic processing by event-related transcranial magnet stimulation of Broca's area. *Neuron*, 2002, *35*, 1177–1182.

Sakuma, Y., and Pfaff, D. W. Facilitation of female reproductive behavior from mesencephalic central grey in the rat. *American Journal of Physiology*, 1979a, *237*, R279–R284.

Sakuma, Y., and Pfaff, D. W. Mesencephalic mechanisms for integration of female reproductive behavior in the rat. *American Journal of Physiology*, 1979b, *237*, R285–R290.

Sakuma, Y., and Pfaff, D. W. Convergent effects of lordosis-relevant somatosensory and hypothalamic influences on central gray cells in the rat mesencephalon. *Experimental Neurology*, 1980a, *70*, 269–281.

Sakuma, Y., and Pfaff, D. W. Excitability of female rat central gray cells with medullary projections: Changes produced by hypothalamic stimulation and estrogen treatment. *Journal of Neurophysiology*, 1980b, *44*, 1012–1023.

Sakurai, T. The neural circuit of orexin (hypocretin): Maintaining sleep and wakefulness. *Nature Reviews: Neuroscience*, 2007, *8*, 171–181.

Sakurai, T., Amemiya, A., Ishii, M., Matsuzaki, I., et al. Orexins and orexin receptors: A family of hypothalamic neuropeptides and G protein-coupled receptors that regulate feeding behavior. *Cell*, 1998, *20*, 573–585.

Sakurai, Y., Ichikawa, Y., and Mannen, T. Pure alexia from a posterior occipital lesion. *Neurology*, 2001, *56*, 778–781.

Sakurai, Y., Momose, T., Iwata, M., Sudo, Y., et al. Different cortical activity in reading of Kanji words, Kana words and Kana nonwords. *Cognitive Brain Research*, 2000, *9*, 111–115.

Sakurai, Y., Sakai, K., Sakuta, M., and Iwata, M. Naming difficulties in alexia with agraphia for kanji after a left posterior inferior temporal lesion. *Journal of Neurology, Neurosurgery, and Psychiatry*, 1994, *57*, 609–613.

Salamone, J. D. Complex motor and sensorimotor function of striatal and accumbens dopamine: Involvement in instrumental behavior processes. *Psychopharmacology*, 1992, *107*, 160–174.

Salas, J. C. T., Iwasaki, H., Jodo, E., Schmidt, M. H., et al. Penile erection and micturition events triggered by electrical stimulation of the mesopontine tegmental area. *American Journal of Physiology*, 2007, *294*, R102–R111.

Salm, A. K., Pavelko, M., Krouse, E. M., Webster, W., et al. Lateral amygdaloid nucleus expansion in adult rats is associated with exposure to prenatal stress. *Developmental Brain Research*, 2004, *148*, 159–167.

Salmelin, R., Schnitzler, A., Schmitz, F., and Freund, H. J. Single word reading in developmental stutterers and fluent speakers. *Brain*, 2000, *123*, 1184–1202.

Samson, H. H., Hodge, C. W., Tolliver, G. A., and Haraguchi, M. Effect of dopamine agonists and antagonists on ethanol reinforced behavior: The involvement of the nucleus accumbens. *Brain Research Bulletin*, 1993, *30*, 133–141.

Sanacora, G., Gueorguieva, R., Epperson, C. N., Wu, Y. T., et al. Subtype-specific alterations of gamma-aminobutyric acid and glutamate in patients with major depression. *Archives of General Psychiatry*, 2004, *61*, 705–713.

Sanacora, G., Mason, G. F., Rothman, D. L., Hyder, F., et al. Increased cortical GABA concentrations in depressed patients receiving ECT. *American Journal of Psychiatry*, 2003, *160*, 577–579.

Saper, C. B., Chou, T. C., and Scammell, T. E. The sleep switch: Hypothalamic control of sleep and wakefulness. *Trends in Neurosciences*, 2001, *24*, 726–731.

Saper, C. B., Scammell, T. E., and Lu, J. Hypothalamic regulation of sleep and circadian rhythms. *Nature*, 2005, *437*, 1257–1263.

Sapolsky, R. M. *Stress, the Aging Brain and the Mechanisms of Neuron Death.* Cambridge, Mass.: MIT Press, 1992.

Sapolsky, R. M. Social subordinance as a marker of hypercortisolism: Some unexpected subtleties. *Annals of the New York Academy of Sciences*, 1995, *771*, 626–639.

Sapolsky, R. M. Is impaired neurogenesis relevant to the affective symptoms of depression? *Biological Psychiatry*, 2004, *56*, 137–139.

Sassenrath, E. N., Powell, T. E., and Hendrickx, A. G. Perimenstrual aggression in groups of female rhesus monkeys. *Journal of Reproduction and Fertility*, 1973, *34*, 509–511.

Sato, N., Sakata, H., Tanaka, Y. L., and Taira, M. Navigation-associated medial parietal neurons in monkeys. *Proceedings of the National Academy of Sciences, USA*, 2006, *103*, 17001–17006.

Satterlee, J. S., Barbee, S., Jin, P., Krichevsky, A., et al. Noncoding RNAs in the brain. *Journal of Neuroscience*, 2007, *27*, 11856–11859.

Saura, C. A., Choi, S.-Y., Beglopoulos, V., Malkani, S., et al. Loss of presenilin function causes impairments of memory and synaptic plasticity followed by age-dependent neurodegeneration. *Neuron*, 2004, *42*, 23–36.

Savic, I., Berglund, H., Gulyas, B., and Roland, P. Smelling of odorous sex hormone-like compounds causes sex-differentiated hypothalamic activations in humans. *Neuron*, 2001, *31*, 661–668.

Savic, I., Berglund, H., and Lindström, P. Brain response to putative pheromones in homosexual men. *Proceedings of the National Academy of Sciences, USA*, 2005, *102*, 7356–7361.

Sawaguchi, T., and Goldman-Rakic, P. S. The role of D1-dopamine receptor in working memory: Local injections of dopamine antagonists into the prefrontal cortex of rhesus monkeys performing an oculomotor delayed-response task. *Journal of Neurophysiology*, 1994, *71*, 515–528.

Sawchenko, P. E. Toward a new neurobiology of energy balance, appetite, and obesity: The anatomist weigh in. *Journal of Comparative Neurology*, 1998, *402*, 435–441.

Saxena, S., Brody, A. L., Schwartz, J. M., and Baxter, L. R. Neuroimaging and frontal-subcortical circuitry in obsessive-compulsive disorder. *British Journal of Psychiatry*, 1998, *173*, 26–37.

Scammell, T. E., Estabrooke, I. V., McCarthy, M. T., Chemelli, R. M., el al. Hypothalamic arousal regions are activated during modafinil-induced wakefulness. *Journal of Neuroscience*, 2000, *20*, 8620–8628.

Scammell, T. E., Gerashchenko, D. Y., Mochizuki, T., McCarthy, M. T., et al. An adenosine A2a agonist increases sleep and induces Fos in ventrolateral preoptic neurons. *Neuroscience*, 2001, *107*, 653–663.

Scarpace, P. J., Matheny, M., Moore, R. L., and Kumar, M. V. Modulation of uncoupling protein 2 and uncoupling protein 3: Regulation by denervation, leptin and retinoic acid treatment. *Journal of Endocrinology*, 2000, *164*, 331–337.

Scarpace, P. J., Matheny, M., and Tümer, N. Hypothalamic leptin resistance is associated with impaired leptin signal transduction in aged obese rats. *Neuroscience*, 2001, *104*, 1111–1117.

Schafe, G. E., Bauer, E. P., Rosis, S., Farb, C. R. Rodrigues, S. A., and LeDopux, J. E. Memory consolidation of Pavlovian fear conditioning requires nitric oxide signaling in the lateral amygdala. *European Journal of Neuroscience*, 2005, *22*, 201–211.

Schafe, G. E. Doyère, V., and LeDoux, J. E. Tracking the fear engram: The lateral amygdala is an essential locus of fear memory storage. *Journal of Neuroscience*, 2005, *25*, 10010–10015.

Schaller, G., Schmidt, A., Pleiner, J., Woloszczuk, W., et al. Plasma ghrelin concentrations are not regulated by glucose or insulin: A double-blind, placebo-controlled crossover clamp study. *Diabetes*, 2003, *52*, 16–20.

Schein, S. J., and Desimone, R. Spectral properties of V4 neurons in the macaque. *Journal of Neuroscience*, 1990, *10*, 3369–3389.

Schenck, C. H., Bundlie, S. R., Ettinger, M. G., and Mahowald, M. W. Chronic behavioral disorders of human REM sleep: A new category of parasomnia. *Sleep*, 1986, *9*, 293–308.

Schenck, C. H., Hurwitz, T. D., Bundlie, S. R., and Mahowald, M. W. Sleep-related eating disorders: Polysomnographic correlates of a heterogeneous syndrome distinct from daytime eating disorders. *Sleep*, 1991, *14*, 419–431.

Schenck, C. H., Hurwitz, T. D., and Mahowald, M. W. REM-sleep behavior disorder: An update on a series of 96 patients and a review of the world literature. *Journal of Sleep Research*, 1993, *2*, 224–231.

Schenck, C. H., and Mahowald, M. W. Motor dyscontrol in narcolepsy: Rapid-eye-movement (REM) sleep without

atonia and REM sleep behavior disorder. *Annals of Neurology*, 1992, *32*, 3–10.

Schenkein, J., and Montagna, P. Self management of fatal familial insomnia. 1: What is FFI? *Medscape General Medicine*, 2006a, *8*, 65.

Schenkein, J., and Montagna, P. Self management of fatal familial insomnia. 2: Case report. *Medscape General Medicine*, 2006b, *8*, 66.

Scherschlicht, R., Polc, P., Schneeberger, J., Steiner, M., and Haefely, W. Selective suppression of rapid eye movement sleep (REMS) in cats by typical and atypical antidepressants. In *Typical and Atypical Antidepressants: Molecular Mechanisms*, edited by E. Costa and G. Racagni. New York: Raven Press, 1982.

Schiffman, J., Ekstrom, M., LaBrie, J., Schulsinger, F., et al. Minor physical anomalies and schizophrenia spectrum disorders: A prospective investigation. *American Journal of Psychiatry*, 2002, *159*, 238–243.

Schiffman, J., Walker, E., Ekstrom, M., Schulsinger, F., et al. Childhood videotaped social and neuromotor precursors of schizophrenia: A prospective investigation, *American Journal of Psychiatry*, 2004, *161*, 2021–2027.

Schiffman, S. S., Lockhead, E., and Maes, F. W. Amiloride reduces the taste intensity of Na^+ and Li^+ salts and sweeteners. *Proceedings of the National Academy of Sciences, USA*, 1983, *80*, 6136–6140.

Schiller, P. H. The ON and OFF channels of the visual system. *Trends in Neuroscience*, 1992, *15*, 86–92.

Schiller, P. H., and Malpeli, J. G. Properties and tectal projections of monkey retinal ganglion cells. *Journal of Neurophysiology*, 1977, *40*, 428–445.

Schiller, P. H., Sandell, J. H., and Maunsell, J. H. R. Functions of the ON and OFF channels of the visual system. *Nature*, 1986, *322*, 824–825.

Schleifer, S. J., Keller, S. E., Camerino, M., Thornton, J. C., and Stein, M. Suppression of lymphocyte stimulation following bereavement. *Journal of the American Medical Association*, 1983, *15*, 374–377.

Schmid, D., Held, K., Ising, M., Uhr, M., et al. Ghrelin stimulates appetite, imagination of food, GH, ACTH, and cortisol, but does not affect leptin in normal controls. *Neuropsychopharmacology*, 2005, *30*, 1187–1192.

Schmidt, M. H., Valatx, J.-L., Sakai, K., Fort, P., and Jouvet, M. Role of the lateral preoptic area in sleep-related erectile mechanisms and sleep generation in the rat. *Journal of Neuroscience*, 2000, *20*, 6640–6647.

Schmidt-Hieber, C., Jonas, P., and Bischofberger, J. Enhanced synaptic plasticity in newly generated granule cells of the adult hippocampus. *Nature*, 2004, *429*, 184–187.

Schneider, B., Muller, M. J., and Philipp, M. Mortality in affective disorders. *Journal of the Affective Disorders*, 2001, *65*, 263–274.

Schneider, P., Scherg, M., Dosch, H. G., Specht, H. L., et al. Morphology of Heschl's gyrus reflects enhanced activation in the auditory cortex of musicians. *Nature Neuroscience*, 2002, *5*, 688–694.

Schoenfeld, M. A., Neuer, G., Tempelmann, C., Schüssler, K., et al. Functional magnetic resonance tomography correlates of taste perception in the human primary taste cortex. *Neuroscience*, 2004, *127*, 347–353.

Schott, B. H., Sellner, D. B., Lauer, C.-J., Habib, R., et al. Activation of midbrain structures by associative novelty and the formation of explicit memory in humans. *Learning and Memory*, 2004, *11*, 383–387.

Schrauwen, P., Xia, J., Bogardus, C., Pratley, R. E., and Ravussin, E. Skeletal muscle uncoupling protein 3 expression is a determinant of energy expenditure in Pima Indians. *Diabetes*, 1999, *48*, 146–149.

Schultz, R. T. Developmental deficits in social perception in autism: The role of the amygdala and fusiform face area. *International Journal of Developmental Neuroscience*, 2005, *23*, 125–141.

Schumann, C. M., and Amaral, D. G. Stereological analysis of amygdala neuron number in autism. *Journal of Neuroscience*, 2006, *26*, 7674–7679.

Schwartz, M. F., Marin, O. S. M., and Saffran, E. M. Dissociations of language function in dementia: A case study. *Brain and Language*, 1979, *7*, 277–306.

Schwartz, M. F., Saffran, E. M., and Marin, O. S. M. The word order problem in agrammatism. I: Comprehension. *Brain and Language*, 1980, *10*, 249–262.

Schwartz, S., Ponz, A., Poryazova, R., Werth, E., et al. Abnormal activity in hypothalamus and amygdala during humour processing in human narcolepsy with cataplexy. *Brain*, 2008, *131*, 514–522.

Schwartz, W. J., and Gainer, H. Suprachiasmatic nucleus: Use of ^{14}C-labelled deoxyglucose uptake as a functional marker. *Science*, 1977, *197*, 1089–1091.

Schwartz, W. J., Reppert, S. M., Eagan, S. M., and Moore-Ede, M. C. In vivo metabolic activity of the suprachiasmatic nuclei: A comparative study. *Brain Research*, 1983, *5*, 184–187.

Schwarzlose, R. F., Baker, C. I., and Kanwisher, N. Separate face and body selectivity on the fusiform gyrus. *Journal of Neuroscience*. 2005, *25*, 11055–11059.

Scott, K. The sweet and the bitter of mammalian taste. *Current Opinions in Neurobiology*, 2004, *14*, 423–427.

Scott, S. K., Blank, E. C., Rosen, S., and Wise, R. J. S. Identification of a pathway for intelligible speech in the left temporal lobe. *Brain*, 2000, *123*, 2400–2406.

Scott, T. R., and Plata-Salaman, C. R. Coding of taste quality. In *Smell and Taste in Health and Disease*, edited by T. N. Getchell. New York: Raven Press, 1991.

Scoville, W. B., and Milner, B. Loss of recent memory after bilateral hippocampal lesions. *Journal of Neurology, Neurosurgery and Psychiatry*, 1957, *20*, 11–21.

Seagraves, M. A., Goldberg, M. E., Deny, S., Bruce, C. J., et al. The role of striate cortex in the guidance of eye movements in the monkey. *The Journal of Neuroscience*, 1987, *7*, 3040–3058.

Sears, L. L., Vest, C., Mohamed, S., Bailey, J., Ranson, B. J., and Piven, J. An MRI study of the basal ganglia in autism. *Progress in Neuro-Psychopharmacology and Biological Psychiatry*, 1999, *23*, 613–624.

Sebat, J., Lakshmi, B., Malhotra, D., Troge, J., et al. Strong association of de novo copy number mutations with autism. *Science*, 2007, *316*, 445–449.

Seidman, L. J., Biederman, J., Weber, W., Hatch, M., and Faraone, S. V. Neuropsychological function in adults with attention-deficit hyperactivity disorder. *Biological Psychiatry*, 1998, *44*, 260–268.

Selye, H. *The Stress of Life*. New York: McGraw-Hill, 1976.

Semenza, C., and Zettin, M. Evidence from aphasia for the role of proper names as pure referring expressions. *Nature*, 1989, *342*, 678–679.

Senju, A., Maeda, M., Kikuchi, Y., Hasegawa, T., et al. Absence of contagious yawning in children with autism spectrum disorder, *Biology Letters*, 2007, *3*, 706–708.

Serizawa, S., Miyamichi, K., Nakatani, H., Suzuki, M., et al. Negative feedback regulation ensures the one receptor–one olfactory neuron rule in mouse. *Science*, 2003, *302*, 2088–2094.

Sforza, E., Montagna, P., Tinuper, P., Cortelli, P., et al. Sleep-wake cycle abnormalities in fatal familial insomnia: Evidence of the role of the thalamus in sleep regulation. *Electroencephalography and Clinical Neurophysiology*, 1995, *94*, 398–405.

Shafto, M. A., Burke, D. M., Stamatakis, E. A., Tam, P. P., and Tyler, L. K. On the tip-of-the-tongue: Neural correlates of increased word-finding failures in normal aging. *Journal of Cognitive Neuroscience*, 2007, *19*, 2060–2070.

Shallice, T. Phonological agraphia and the lexical route in writing. *Brain*, 1981, *104*, 413–429.

Sham, P. C., O'Callaghan, E., Takei, N., Murray, G. K., et al. Schizophrenia following pre-natal exposure to influenza epidemics between 1939 and 1960. *British Journal of Psychiatry*, 1992, *160*, 461–466.

Shammi, P., and Stuss, D. T. Humor appreciation: A role of the right frontal lobe. *Brain*, 1999, *122*, 657–666.

Shannon, R. V. Understanding hearing through deafness. *Proceedings of the National Academy of Sciences, USA*, 2007, *104*, 6883–6884.

Shapiro, K., Shelton, J., and Caramazza, A. Grammatical class in lexical production and morphological processing: Evidence from a case of fluent aphasia. *Cognitive Neuropsychology*, 2000, *17*, 665–682.

Shapiro, L. E., Leonard, C. M., Sessions, C. E., Dewsbury, D. A., and Insel, T. R. Comparative neuroanatomy of the sexually dimorphic hypothalamus in monogamous and polygamous voles. *Brain Research*, 1991, *541*, 232–240.

Sharp, D. J., Scott, S. K., and Wise, R. J. S. Retrieving meaning after temporal lobe infarction: The role of the basal language area. *Annals of Neurology*, 2004, *56*, 836–846.

Shaw, P., Eckstrand, K., Sharp, W., Blumenthal, J., et al. Attention-deficit/hyperactivity disorder is characterized by a delay in cortical maturation. *Proceedings of the National Academy of Sciences, USA*, 2007, *10*, 19649–19654.

Shaywitz, B. A., Shaywitz, S. E., Pugh, K. R., Mencl, W. E., et al. Disruption of posterior brain systems for reading in children with developmental dyslexia. *Biological Psychiatry*, 2002, *52*, 101–110.

Shearman, L. P., Sriram, S., Weaver, D. R., Maywood, E. S., et al. Interacting molecular loops in the mammalian circadian clock. *Science*, 2000, *288*, 1013–1019.

Shen, K., and Meyer, T. Dynamic control of CaMKII translocation and localization in hippocampal neurons by NMDA stimulation. *Science*, 1999, *284*, 162–166.

Shenton, M. E., Dickey, C. C., Frumin, M., and McCarley, R. W. A review of MRI findings in schizophrenia. *Schizophrenia Research*, 2001, *49*, 1–52.

Shepherd, G. M. Discrimination of molecular signals by the olfactory receptor neuron. *Neuron*, 1994, *13*, 771–790.

Sher, A. E. Surgery for obstructive sleep apnea. *Progress in Clinical Biology Research*, 1990, *345*, 407–415.

Sherin, J. E., Elmquist, J. K., Torrealba, F., and Saper, C. B. Innervation of histaminergic tuberomammillary neurons by GABAergic and galaninergic neurons in the ventrolateral preoptic nucleus of the rat. *Journal of Neuroscience*, 1998, *18*, 4705–4721.

Sherry, D. F., Jacobs, L. F., and Gaulin, S. J. C. Spatial memory and adaptive specialization of the hippocampus. *Trends in Neuroscience*, 1992, *15*, 298–303.

Shi, S.-H., Hayashi, Y., Petralia, R. S., Zaman, S. H., et al. Rapid spine delivery and redistribution of AMPA receptors after synaptic NMDA receptor activation. *Science*, 1999, *284*, 1811–1816.

Shifren, J. L., Braunstein, G. D., Simon, J. A., Casson, P. R., et al. Transdermal testosterone treatment in women with impaired sexual function after oophorectomy. *New England Journal of Medicine*, 2000, *343*, 682–688.

Shik, M. L., and Orlovsky, G. N. Neurophysiology of locomotor automatism. *Physiological Review*, 1976, *56*, 465–501.

Shim, S. S., Hammonds, M. D., and Kee, B. S. Potentiation of the NMDA receptor in the treatment of schizophrenia: Focused on the glycine site. *European Archives of Psychiatry and Clinical Neuroscience*, 2008, *258*, 16–27.

Shima, K., and Tanji, J. Both supplementary and presupplementary motor areas are crucial for the temporal organization of multiple movements. *Journal of Neurophysiology*, 1998, *80*, 3247–3260.

Shima, K., and Tanji, J. Neuronal activity in the supplementary and presupplementary motor areas for temporal organization of multiple movements. *Journal of Neurophysiology*, 2000, *84*, 2148–2160.

Shimada, M., Tritos, N. A., Lowell, B. B., Flier, J. S., and Maratos-Flier, E. Mice lacking melanin-concentrating hormone are hypophagic and lean. *Nature*, 1998, *396*, 670–674.

Shimura, T., Yamamoto, T., and Shimokochi, M. The medial preoptic area is involved in both sexual arousal and performance in male rats: Re-evaluation of neuron activity in freely moving animals. *Brain Research*, 1994, *640*, 215–222.

Shin, L. M., Wright, C. I., Cannistraro, P. A., Wedig, M. M., et al. A functional magnetic resonance imaging study of amygdala and medial prefrontal cortex responses to overtly presented fearful faces in posttraumatic stress disorder. *Archives of General Psychiatry*, 2005, *62*, 273–281.

Shipley, M. T., and Ennis, M. Functional organization of the olfactory system. *Journal of Neurobiology*, 1996, *30*, 123–176.

Shmuelof, L, and Zohary, E. Dissociation between ventral and dorsal fMRI activation during object and action recognition. *Neuron*, 2005, *47*, 457–470.

Shuto, Y., Shibasaki, T., Otagiri, A., Kuriyama, H., et al. Hypothalamic growth hormone secretagogue receptor regulates growth hormone secretion, feeding, and adiposity. *Journal of Clinical Investigation*, 2002, *109*, 1429–1436.

Sidman, M., Stoddard, L. T., and Mohr, J. P. Some additional quantitative observations of immediate memory in a patient with bilateral hippocampal lesions. *Neuropsychologia*, 1968, *6*, 245–254.

Siegel, J. Clues to the functions of mammalian sleep. *Nature*, 2005, *437*, 1264–1271.

Siegel, J. M., and McGinty, D. J. Pontine reticular formation neurons: Relationship of discharge to motor activity. *Science*, 1977, *196*, 678–680.

Siegel, R. M., and Andersen, R. A. Motion perceptual deficits following ibotenic acid lesions of the middle temporal area (MT) in the behaving monkey. *Society for Neuroscience Abstracts*, 1986, *12*, 1183.

Siegle, G. J., Carter, C. S., and Thase, M. E. Use of fMRI to predict recovery from unipolar depression with cognitive behavior therapy. *American Journal of Psychiatry*, 2006, *163*, 735–738.

Silva, A. J., Stevens, C. F., Tonegawa, S., and Wang, Y. Deficient hippocampal long-term potentiation in-calcium-calmodulin kinase II mutant mice. *Science*, 1992, *257*, 201–206.

Silver, R., LeSauter, J., Tresco, P. A., and Lehman, M. N. A diffusible coupling signal from the transplanted suprachiasmatic nucleus controlling circadian locomotor rhythms. *Nature*, 1996, *382*, 810–813.

Silveri, M. C. Peripheral aspects of writing can be differentially affected by sensorial and attentional defect: Evidence from a patient with afferent dysgraphia and case dissociation. *Cortex*, 1996, *32*, 155–172.

Simpson, J. B., Epstein, A. N., and Camardo, J. S. The localization of dipsogenic receptors for angiotensin II in the subfornical organ. *Journal of Comparative and Physiological Psychology*, 1978, *92*, 581–608.

Simuni, T., Jaggi, J. L., Mulholland, H., Hurtig, H. I., et al. Bilateral stimulation of the subthalamic nucleus in patients with Parkinson disease: A study of efficacy and safety. *Journal of Neurosurgery*, 2002, *96*, 666–672.

Sinclair, A. H., Berta, P., Palmer, M. S., Hawkins, J. R., et al. A gene from the human sex-determining region encodes a protein with homology to a conserved DNA-binding motif. *Nature*, 1990, *346*, 240–244.

Sindelar, D. K., Ste. Marie, L., Miura, G. I., Palmiter, R. D., et al. Neuropeptide Y is required for hyperphagic feeding in response to neuroglucopenia. *Endocrinology*, 2004, *145*, 3363–3368.

Singer, C., and Weiner, W. J. Male sexual dysfunction. *Neurologist*, 1996, *2*, 119–129.

Singer, F., and Zumoff, B. Subnormal serum testosterone levels in male internal medicine residents. *Steroids*, 1992, *57*, 86–89.

Singer, T., Seymour, B., O'Doherty, J., Kaube, H., et al. Empathy for pain involves the affective but not sensory components of pain. *Science*, 2004, *303*, 1157–1162.

Singh, D., and Bronstad, P. M. Female body odour is a potential cue to ovulation. *Proceedings of the Royal Society of London B*, 2001, *268*, 797–801.

Sinha, S. R., and Kossoff, E. H. The ketogenic diet. *Neurologist*, 2005, *11*, 161–170.

Sipos, A., Rasmussen, F., Harrison, G., Tynelius, P., et al. Paternal age and schizophrenia: A population based cohort study. *British Medical Journal*, 2004, *330*, 147–148.

Sipos, M. L., and Nyby, J. G. Concurrent androgenic stimulation of the ventral tegmental area and medial preoptic area: Synergistic effects on male-typical reproductive behaviors in house mice. *Brain Research*, 1996, *729*, 29–44.

Sirigu, A., Deprati, E., Ciancia, S., Giraux, P., et al. Altered awareness of voluntary action after damage to the parietal cortex. *Nature Neuroscience*, 2004, *7*, 80–84.

Skaggs, W. E., and McNaughton, B. L. Spatial firing properties of hippocampal CA1 populations in an environment containing two visually identical regions. *Journal of Neuroscience*, 1998, *18*, 8455–8466.

Skene, D. J., Lockley, S. W., and Arendt, J. Melatonin in circadian sleep disorders in the blind. *Biological Signals and Receptors*, 1999, *8*, 90–95.

Skutella, T., Criswell, H., Moy, S., Probst, J. C., et al. Corticotropin-releasing hormone (CRH) antisense oligodeoxynucleotide induces anxiolytic effects in rat. *Neuroreport*, 1994, *5*, 2181–2185.

Slotkin, T. A. Fetal nicotine or cocaine exposure: Which one is worse? *Journal of Pharmacology and Experimental Therapeutics*, 1998, *22*, 521–527.

Smith, G. P. Animal models of human eating disorders. *Annals of the New York Academy of Sciences*, 1989, *16*, 219–237.

Smith, G. P., Gibbs, J., and Kulkosky, P. J. Relationships between brain-gut peptides and neurons in the control of food intake. In *The Neural Basis of Feeding and Reward*, edited by B. G. Hoebel and D. Novin. Brunswick, ME.: Haer Institute, 1982.

Smith, M. J. Sex determination: Turning on sex. *Current Biology*, 1994, *4*, 1003–1005.

Smith-Roe, S. L., and Kelley, A. E. Coincident activation of NMDA and dopamine D_1 receptors within the nucleus accumbens core is required for appetitive instrumental learning. *Journal of Neuroscience*, 2000, *20*, 7737–7742.

Snyder, B. J., and Olanow, C. W. Stem cell treatment for Parkinson's disease: An update for 2005. *Current Opinion in Neurobiology*, 2005, *18*, 376–385.

Snyder, L. H., Batista, A. P., and Andersen, R. A. Intention-related activity in the posterior parietal cortex: A review. *Vision Research*, 2000, *40*, 1433–1441.

Snyder, S. H. *Madness and the Brain*. New York: McGraw-Hill, 1974.

Soares, J. C., and Gershon, S. The lithium ion: A foundation for psychopharmacological specificity. *Neuropsychopharmacology*, 1998, *19*, 167–182.

Södersten, P., Bergh, C., and Zandian, M. Understanding eating disorders. *Hormones and Behavior*, 2006, *50*, 572–578.

Solomon, S. G., and Lennie, P. The machinery of colour vision. *Nature Reviews: Neuroscience*, 2007, *8*, 276–286.

Solyom, L., Turnbull, I. M., and Wilensky, M. A case of self-inflicted leucotomy. *British Journal of Psychiatry*, 1987, *151*, 855–857.

Son, G. H., Geum, D., Chung, S., Kim, E. J., et al. Maternal stress produces learning deficits associated with impairment of NMDA receptor-mediated synaptic plasticity. *Journal of Neuroscience*, 2006, *26*, 3309–3318.

Soria, G., Mendizabal, V., Rourino, C., Robledo, P, Ledent, C., et al. Lack of CB_1 cannabinoid receptor impairs cocaine self-administration. *Neuropsychopharmacology*, 2005, *30*, 1670–1680.

Sormani, M. P., Bruzzi, P., Comi, G., and Filippi, M. The distribution of the magnetic resonance imaging response to glatiramer acetate in multiple sclerosis. *Multiple Sclerosis*, 2005, *11*, 447–449.

Sotillo, M., Carretié, L., Hinojosa, J. A., Tapia, M., et al. Neural activity associated with metaphor comprehension: Spatial analysis. *Neuroscience Letters*, 2005, *373*, 5–9.

Soto, C. Unfolding the role of protein misfolding in neurodegenerative diseases. *Nature Review Neuroscience*, 2003, *4*, 49–60.

Sotthibundhu, A., Sykes, A. M., Fox, B., Underwood, C. K., et al. β-Amyloid$_{1-42}$ induces neuronal death through the p75 neurotrophin receptor. *Journal of Neuroscience*, 2008, *28*, 3941–3946.

South, F. H., and Ritter, R. C. Capsaicin application to central or peripheral vagal fibers attenuates CCK satiety. *Peptides*, 1988, *9*, 601–612.

Soyka, M., and Chick, J. Use of acamprosate and opioid antagonists in the treatment of alcohol dependence: A European perspective. *American Journal of Addiction*, 2003, *12* (Suppl. 1), S69–S80.

Speelman, J. D., Schuurman, R., de Bie, R. M., Esselink, R. A., and Bosch, D. A. Stereotactic neurosurgery for tremor. *Movement Disorders*, 2002, *17*, S84–S88.

Sperry, R. W. Brain bisection and consciousness. In *Brain and Conscious Experience*, edited by J. Eccles. New York: Springer-Verlag, 1966.

Spezio, M. L., Huang, P.-Y. S., Castelli, F., and Adolphs, R. Amygdala damage impairs eye contact during conversations with real people. *Journal of Neuroscience*, 2007, *27*, 3994–3997.

Spiers, H. J., Maguire, E. A., and Burgess, N. Hippocampal amnesia. *Neurocase*, 2001, *7*, 357–382.

Sprengelmeyer, R., Young, A. W., Calder, A. J., Karnat, A., et al. Loss of disgust: Perception of faces and emotions in Huntington's disease. *Brain*, 1996, *119*, 1647–1665.

Sprengelmeyer, R., Young, A. W., Pundt, I., Sprengelmeyer, A., et al. Disgust implicated in obsessive-compulsive disorder. *Proceedings of the Royal Society of London B*, 1997, *264*, 1767–1773.

Squire, L. R. Stable impairment in remote memory following electroconvulsive therapy. *Neuropsychologia*, 1974, *13*, 51–58.

Squire, L. R. Memory and the hippocampus: A synthesis from findings with rats, monkeys, and humans. *Psychological Review*, 1992, *99*, 195–231.

Squire, L. R., and Bayley, P. J. The neuroscience of remote memory. *Current Opinion in Neurobiology*, 2007, *17*, 185–196.

Squire, L. R., Shimamura, A. P., and Amaral, D. G. Memory and the hippocampus. In *Neural Models of Plasticity:*

Experimental and Theoretical Approaches, edited by J. H. Byrne and W. O. Berry. San Diego: Academic Press, 1989.

St. George-Hyslop, P. H., Tanzi, R. E., Polinsky, R. J., Haines, J. L., et al. The genetic defect causing familial Alzheimer's disease maps on chromosome 21. *Science,* 1987, *235,* 885–890.

Stahl, S. M., Grady, M. M., Moret, C., and Briley, M. SNRIs: Their pharmacology, clinical efficacy, and tolerability in comparison with other classes of antidepressants. *CNS Spectrums,* 2005, *10,* 732–747.

Standing, L. Learning 10,000 pictures. *Quarterly Journal of Experimental Psychology,* 1973, *25,* 207–222.

Stanislavsky, C. *An Actor Prepares.* New York: Theater Arts/Routledge, 1936.

Starkey, S. J., Walker, M. P., Beresford, I. J. M., and Hagan, R. M. Modulation of the rat suprachiasmatic circadian clock by melatonin in-vitro. *Neuroreport,* 1995, *6,* 1947–1951.

Stebbins, W. C., Miller, J. M., Johnsson, L.-G., and Hawkins, J. E. Ototoxic hearing loss and cochlear pathology in the monkey. *Annals of Otology, Rhinology and Laryngology,* 1969, *78,* 1007–1026.

Steele, A. D., Emsley, J. G., Ozdinler, P. H., Lindquist, S., and Macklis, J. D. Prion protein (PrPc) positively regulates neural precursor proliferation during developmental and adult mammalian neurogenesis. *Proceedings of the National Academy of Sciences, USA,* 2006, *203,* 3416–3421.

Steeves, J. K. E., Humphrey, G. K., Culham, J. C., Menon, R. A., et al. Behavioral and neuroimaging evidence for a contribution of color and texture information to scene classification in a patient with visual form agnosia. *Journal of Cognitive Neuroscience,* 2004, *16,* 955–965.

Stefanacci, L., and Amaral, D. G. Topographic organization of cortical inputs to the lateral nucleus of the macaque monkey amygdala: A retrograde tracing study. *Journal of Comparative Neurology,* 2000, *22,* 52–79.

Stefanatos, G. A., Gershkoff, A., and Madigan, S. On pure word deafness, temporal processing, and the left hemisphere. *Journal of the International Neuropsychological Society,* 2005, *11,* 456–470.

Stein, L., and Belluzzi, J. D. Cellular investigations of behavioral reinforcement. *Neuroscience and Biobehavioral Reviews,* 1989, *13,* 69–80.

Stein, M. B., Jang, K. L., Taylor, S., Vernon, P. A., and Livesley, W. J. Genetic and environmental influences on trauma exposure and posttraumatic stress disorder symptoms: A twin study. *American Journal of Psychiatry,* 2002, *159,* 1675–1681.

Stein, M. B., Simmons, A. N., Feinstein, J. S., and Paulus, M. P. Increased amygdala and insula activation during emotion processing in anxiety-prone subjects. *American Journal of Psychiatry,* 2007, *164,* 318–327.

Stein, M. B., and Uhde, T. W. The biology of anxiety disorders. In *American Psychiatric Press Textbook of Psychopharmacology.* Washington, D.C.: American Psychiatric Press, 1995.

Steinhausen, H. C. The outcome of anorexia nervosa in the 20th century. *American Journal of Psychiatry,* 2002, *159,* 1284–1293.

Steininger, R. L., Alam, M. N., Szymusiak, R., and McGinty, D. State dependent discharge of tuberomammillary neurons in the rat hypothalamus. *Sleep Research,* 1996, *25,* 28.

Stephan, F. K., and Nuñez, A. A. Elimination of circadian rhythms in drinking activity, sleep, and temperature by isolation of the suprachiasmatic nuclei. *Behavioral Biology,* 1977, *20,* 1–16.

Stephan, F. K., and Zucker, I. Circadian rhythms in drinking behavior and locomotor activity of rats are eliminated by hypothalamic lesion. *Proceedings of the National Academy of Sciences, USA,* 1972, *69,* 1583–1586.

Steriade, M. Arousal: Revisiting the reticular activating system. *Science,* 1996, *272,* 225–226.

Steriade, M. The corticothalamic system in sleep. *Frontiers in Bioscience,* 2003, *8,* d878–d899.

Steriade, M. Grouping of brain rhythms in corticothalamic systems. *Neuroscience,* 2006, *137,* 1087–1106.

Steriade, M., Paré, D., Datta, S., Oakson, G., and Curró Dossi, R. Different cellular types in mesopontine cholinergic nuclei related to ponto-geniculo-occipital waves. *Journal of Neuroscience,* 1990, *8,* 2560–2579.

Sterman, M. B., and Clemente, C. D. Forebrain inhibitory mechanisms: Cortical synchronization induced by basal forebrain stimulation. *Experimental Neurology,* 1962a, *6,* 91–102.

Sterman, M. B., and Clemente, C. D. Forebrain inhibitory mechanisms: Sleep patterns induced by basal forebrain stimulation in the behaving cat. *Experimental Neurology,* 1962b, *6,* 103–117.

Stern, K., and McClintock, M. K. Regulation of ovulation by human pheromones. *Nature,* 1998, *392,* 177–178.

Sternbach, R. A. *Pain: A Psychophysiological Analysis.* New York: Academic Press, 1968.

Stewart, L., Walsh, V., Frith, U., and Rothwell, J. C. TMS produces two dissociable types of speech disruption. *Neuroimage,* 2001, *13,* 472–478.

Stinson, D., and Thompson, C. Clinical experience with phototherapy. *Journal of the Affective Disorders,* 1990, *18,* 129–135.

Stoerig, P., and Cowey, A. Blindsight. *Current Biology,* 2007, *17,* R822–R824.

Stolerman, I. P., and Jarvis, M. J. The scientific case that nicotine is addictive. *Psychopharmacology,* 1995, *117,* 2–10.

Stone, A. A., Reed, B. R., and Neale, J. M. Changes in daily event frequency precede episodes of physical symptoms. *Journal of Human Stress,* 1987, *13,* 70–74.

Stone, J. M., Morrison, P. D., and Pilowsky, L. S. Glutamate and dopamine dysregulation in schizophrenia: A synthesis and selective review. *Journal of Psychopharmacology,* 2007, *21,* 440–452.

Storz, G., Altuvia, S., and Wassarman, K. M. An abundance of RNA regulators. *Annual Review of Biochemistry,* 2005, *74,* 199–217.

Stowers, L., Holy, T. E., Meister, M., Dulac, C., and Koentges, G. Loss of sex discrimination of male-male aggression in mice deficient for TRP2. *Science,* 2002, *295,* 1493–1500.

Stowers, L., and Marton, T. F. What is a pheromone? Mammalian pheromones reconsidered. *Neuron,* 2005, *46,* 699–702.

Strange, P. G. Antipsychotic drug action: Antagonism, inverse agonism or partial agonism. *Trends in Pharmacological Science,* 2008, *29,* 314–321.

Sturgis, J. D., and Bridges, R. S. N-methyl-DL-aspartic acid lesions of the medial preoptic area disrupt ongoing parental behavior in male rats. *Physiology and Behavior,* 1997, *62,* 305–310.

Sturup, G. K. Correctional treatment and the criminal sexual offender. *Canadian Journal of Correction,* 1961, *3,* 250–265.

Styron, W. *Darkness Visible: A Memoir of Madness.* New York: Random House, 1990.

Suddath, R. L., Christison, G. W., Torrey, E. F., Casanova, M. F., and Weinberger, D. R. Anatomical abnormalities in the brains of monozygotic twins discordant for schizophrenia. *The New England Journal of Medicine,* 1990, *322,* 789–794.

Sulik, K. K. Genesis of alcohol-induced craniofacial dysmorphism. *Experimental Biology and Medicine,* 2005, *230,* 366–375.

Sun, Y., Ahmed, S., and Smith, R. G. Deletion of ghrelin impairs neither growth nor appetite. *Molecular and Cellular Biology,* 2003, *23,* 7973–7981.

Sun, Y., Wang, P., Zheng, H., and Smith, R. G. Ghrelin stimulation of growth hormone release and appetite is mediated through the growth hormone secretagogue receptor. *Proceedings of the National Academy of Sciences, USA,* 2004,*101,* 4679–4684.

Suntsova, N., Guzman-Marin, R., Kumar, S., Alam, M. N., et al. The median preoptic nucleus reciprocally modulates activity of arousal-related and sleep-related neurons in the perifornical lateral hypothalamus. *Journal of Neuroscience,* 2007, *27,* 1616–1630.

Susser, E. S., and Lin, S. P. Schizophrenia after prenatal exposure to the Dutch Hunger Winter of 1944–1945. *Archives of General Psychiatry,* 1992, *49,* 983–988.

Susser, E. S., Neugebauer, R., Hoek, H. W., Brown, A. S., el al. Schizophrenia after prenatal famine: Further evidence. *Archives of General Psychiatry,* 1996, *53,* 25–31.

Sutherland, D. P., Masterton, R. B., and Glendenning, K. K. Role of acoustic striae in hearing: Reflexive responses to elevated sound-sources. *Behavioural Brain Research,* 1998, *97,* 1–12.

Sutherland, R. J., McDonald, R. J., and Savage, D. D. Prenatal exposure to moderate levels of ethanol can have long-lasting effects on hippocampal synaptic plasticity in adult offspring. *Hippocampus,* 1997, *7,* 232–238.

Suzdak, P. D., Glowa, J. R., Crawley, J. N., Schwartz, R. D., et al. A selective imidazobenzodiazepine antagonist of ethanol in the rat. *Science,* 1986, *234,* 1243–1247.

Suzuki, S., Ramos, E. J., Goncalves, C. G., Chen, C., and Meguid, M. M. Changes in GI hormones and their effect on gastric emptying and transit times after Roux-en-Y gastric bypass in rat model. *Surgery,* 2005, *138,* 283–290.

Svennilson, E., Torvik, A., Lowe, R., and Leksell, L. Treatment of Parkinsonism by stereotactic thermolesions in the pallidal region. *Neurologica Scandanavica,* 1960, *35,* 358–377.

Swaab, D. F., Gooren, L. J. G., and Hofman, M. A. Brain research, gender, and sexual orientation. *Journal of Homosexuality,* 1995, *28,* 283–301.

Swaab, D. F., and Hofman, M. A. An enlarged suprachiasmatic nucleus in homosexual men. *Brain Research,* 1990, *537,* 141–148.

Swanson, L. W., Köhler, C., and Björklund, A. The limbic region. I: The septohippocampal system. In *Handbook of Chemical Neuroanatomy, Vol. 5: Integrated Systems of the CNS, Part I,* edited by A. Björklund, T. Hökfelt, and L. W. Swanson. Amsterdam: Elsevier, 1987.

Sweet, W. H. Participant in brain stimulation in behaving subjects. Neurosciences Research Program Workshop, 1966.

Swerdlow, N. R., Geyer, M. A., Vale, W. W., and Koob, G. F. Corticotropin-releasing factor potentiates acoustic startle in rats: Blockade by chlordiazepoxide. *Psychopharmacology,* 1986, *88,* 147–152.

Szuba, M. P., Baxter, L. R., and Fairbanks, L. A. Effects of partial sleep deprivation on the diurnal variation of mood and motor activity in major depression. *Biological Psychiatry,* 1991, *30,* 817–829.

Szymanski, M., Barciszewska, M. Z., Erdmann, V. A., and Barciszewski, J. A new frontier for molecular medicine: Noncoding RNAs. *Biochimica et Biophysica Acta,* 2005, *1765,* 65–75.

Szymanski, M., Barciszewska, M. Z., Zywicki M., and Barciszewski, J. Noncoding RNA transcripts. *Journal of Applied Genetics,* 2003, *44,* 1–19.

Takahashi, K., Lin, J.-S., and Sakai, K. Neuronal activity of histaminergic tuberomammillary neurons during wake–sleep states in the mouse. *Journal of Neuroscience,* 2006, *26,* 10292–10298.

Takahashi, L. K. Hormonal regulation of sociosexual behavior in female mammals. *Neuroscience and Biobehavioral Reviews,* 1990, *14,* 403–413.

Takahashi, Y. K., Nagayma, S., and Mori, K. Detection and masking of spoiled food smells by odor maps in the olfactory bulb. *Journal of Neuroscience,* 2004, *24,* 8690–8694.

Takashima, A., Petersson, K. M., Rutters, F., Tendolkar, I., et al. Declarative memory consolidation in humans: A prospective functional magnetic resonance imaging study. *Proceedings of the National Academy of Sciences, USA,* 2006, *103,* 756–761.

Tamm, L., Menon, V., Ringel, J., and Reiss, A. L. Event-related fMRI evidence of frontotempotal involvement in aberrant response inhibition and task switching in attention-deficit/hyperactivity disorder. *Journal of the American Academy of Child and Adolescent Psychiatry,* 2004, *43,* 1430–1440.

Tan, L. H., Laird, A. R., Li, K., and Fox, P. T. Neuroanatomical correlates of phonological processing of Chinese characters and alphabetic words: A meta-analysis. *Human Brain Mapping,* 2005, *25,* 83–91.

Tarr, M. J., and Gauthier, I. FFA: A flexible fusiform area for subordinate-level visual processing automatized by expertise. *Nature Neuroscience,* 2000, *3,* 764–769.

Tarter, R. E., Kirisci, L., Mezzich, A., Cornelius, J. R., et al. Neurobehavioral disinhibition in childhood predicts early age at onset of substance use disorder. *American Journal of Psychiatry,* 2003, *160,* 1078–1085.

Taub, E., Uswatte, F., King, D. K., Morris, D., et al. A placebo-controlled trial of constraint-induced movement therapy for upper extremity after stroke. *Stroke*, 2006, *37*, 1045–1049.

Teitelbaum, P., and Stellar, E. Recovery from the failure to eat produced by hypothalamic lesions. *Science*, 1954, *120*, 894–895.

Teng, E., and Squire, L. R. Memory for places learned long ago is intact after hippocampal damage. *Nature*, 1999, *400*, 675–677.

Terenius, L., and Wahlström, A. Morphine-like ligand for opiate receptors in human CSF. *Life Sciences*, 1975, *16*, 1759–1764.

Terman, M. Evolving applications of light therapy. *Sleep Medicine Reviews*, 2007, *11*, 497–507.

Terry, R. D., and Davies, P. Dementia of the Alzheimer type. *Annual Review of Neuroscience*, 1980, *3*, 77–96.

Tetel, M. J., Celentano, D. C., and Blaustein, J. D. Intraneuronal convergence of tactile and hormonal stimuli associated with female reproduction in rats. *Journal of Neuroendocrinology*, 1994, *6*, 211–216.

Tetel, M. J., Getzinger, M. J., and Blaustein, J. D. Fos expression in the rat brain following vaginal-cervical stimulation by mating and manual probing. *Journal of Neuroendocrinology*, 1993, *5*, 397–404.

Thach, W. T. Correlation of neural discharge with pattern and force of muscular activity, joint position, and direction of intended movement in motor cortex and cerebellum. *Journal of Neurophysiology*, 1978, *41*, 654–676.

Thaker, G. K., and Carpenter, W. T. Advances in schizophrenia. *Nature Medicine*, 2001, *7*, 667–671.

Thakkar, M. M., Winston, S., and McCarley, R. W. Orexin neurons of the hypothalamus express adenosine A1 receptors. *Brain Research*, 2002, *944*, 190–194.

Thapar, A., O'Donovan, M., and Owen, M. J. The genetics of attention deficit hyperactivity disorder. *Human Molecular Genetics*, 2005, *14*, R275–R282.

Thase, M. E. Treatment issues related to sleep and depression. *Journal of Clinical Psychiatry*, 2000, *61*, 46–50.

Theorell, T., Leymann, H., Jodko, M., Konarski, K., et al. "Person under train" incidents: Medical consequences for subway drivers. *Psychosomatic Medicine*, 1992, *54*, 480–488.

Thiels, E., Xie, X. P., Yeckel, M. F., Barrionuevo, G., and Berger, T. W. NMDA receptor-dependent LTD in different subfields of hippocampus in vivo and in vitro. *Hippocampus*, 1996, *6*, 43–51.

Thielscher, A., and Pessoa, L. Neural correlates of perceptual choice and decision making during fear–disgust discrimination. *Journal of Neuroscience*, 2007, *27*, 2908–2917.

Thier, P., Haarmeier, T., Chakraborty, S., Lindner, A., and Tikhonov, A. Cortical substrates of perceptual stability during eye movements. *Neuroimage*, 2001, *14*, S33–S39.

Thomas, R. M., Hotsenpiller, G., and Peterson, D. A. Acute psychosocial stress reduces cell survival in adult hippocampal neurogenesis without altering proliferation. *Journal of Neuroscience*, 2007, *27*, 2734–2743.

Thompson, P. M., Hayashi, K. M., Simon, S. L., Geaga, J. A., el al. Structural abnormalities in the brains of human subjects who use methamphetamine. *Journal of Neuroscience*, 2004, *24*, 6028–6036.

Thompson, P. M., Vidal, C., Giedd, J. N., Gochman, P., Blumenthal, J., Nicolson, R., Toga, A. W., and Rapoport, J. L. Mapping adolescent brain change reveals dynamic wave of accelerated gray matter loss in very early-onset schizophrenia. *Proceedings of the National Academy of Sciences, USA*, 2001, *98*, 11650–11655.

Thompson, S. M., Fortunato, C., McKinney, R. A., Müller, M., and Gähwiler, B. H. Mechanisms underlying the neuropathological consequences of epileptic activity in the rat hippocampus in vitro. *Journal of Comparative Neurology*, 1996, *372*, 515–528.

Thomson, J. J. *Rights, Restitution, and Risk: Essays in Moral Theory*. Cambridge, Mass.: Harvard University Press, 1986.

Thrasher, T. N. Role of forebrain circumventricular organs in body fluid balance. *Acta Physiologica Scandanavica*, 1989, *136* (Suppl. 583), 141–150.

Thuy, D. H. D., Matsuo, K., Nakamura, K., Toma, K., et al. Implicit and explicit processing of kanji and kana words and non-words studied with fMRI. *Neuroimage*, 2004, *23*, 878–889.

Timmann, D., Watts, S., and Hore, J. Failure of cerebellar patients to time finger opening precisely causes ball high-low inaccuracy in overarm throws. *Journal of Neurophysiology*, 1999, *82*, 103–114.

Toh, K. L., Jones, C. R., He, Y., Eide, E. J., et al. An h*Per2* phosphorylation site mutation in familial advanced sleep phase syndrome. *Science*, 2001, *291*, 1040–1043.

Tong, L., Wen, H., Brayton, C., Laird, F. M., et al. Moderate reduction of γ-secretase attenuates amyloid burden and limits mechanism-based liabilities. *Journal of Neuroscience*, 2007, *27*, 10849–10859.

Tootell, R. B. H., Tsao, D., and Vanduffel, W. Neuroimaging weighs in: Humans meet macaques in "primate" visual cortex. *Journal of Neuroscience*, 2003, *23*, 3981–3989.

Topper, R., Kosinski, C., and Mull, M. Volitional type of facial palsy associated with pontine ischemia. *Journal of Neurology, Neurosurgery, and Psychiatry*, 1995, *58*, 732–734.

Tordoff, M. G., and Friedman, M. I. Hepatic control of feeding: Effect of glucose, fructose, and mannitol. *American Journal of Physiology*, 1988, *254*, R969–R976.

Tordoff, M. G., Hopfenbeck, J., and Novin, D. Hepatic vagotomy (partial hepatic denervation) does not alter ingestive responses to metabolic challenges. *Physiology and Behavior*, 1982, *28*, 417–424.

Tracy, J. L., and Robbins, R. W. The automaticity of emotion recognition. *Emotion*, 2008, *8*, 81–95.

Triarhou, L. C. The percipient observations of Constantin von Economo on encephalitis lethargica and sleep disruption and their lasting impact on contemporary sleep research. *Brain Research Bulletin*, 2006, *69*, 244–258.

Trulson, M. E., and Jacobs, B. L. Raphe unit activity in freely moving cats: Correlation with level of behavioral arousal. *Brain Research*, 1979, *163*, 135–150.

Trussell, L. O. Synaptic mechanisms for coding timing in auditory neurons. *Annual Review of Physiology*, 1999, *61*, 477–496.

Tsacopoulos, M., and Magistretti, P. J. Metabolic coupling between glia and neurons. *Journal of Neuroscience*, 1996, *16*, 877–885.

Tsankova, N., Renthal, W., Kuman, A., and Nestler, E. J. Epigenetic regulation in psychiatric disorders. *Nature Neuroscience*, 2007, *8*, 355–367.

Tsao, D. Y., Freiwald, W. A., Tootell, R. B. H., and Livingstone, M. S. A cortical region consisting entirely of face-selective cells. *Science*, 2006, *311*, 670–674.

Tsao, D. Y., Vanduffel, W., Sasaki, Y., Fize, D., et al. Stereopsis activates V3A and caudal intraparietal areas in macaques and humans. *Neuron*, 2003, *39*, 555–568.

Tschöp, M., Smiley, D. L., and Heiman, M. L. Ghrelin induces adiposity in rodents. *Nature*, 2000, *407*, 908–913.

Tsien, J. Z., Huerta, P. T., and Tonegawa, S. The essential role of hippocampal CA1 NMDA receptor-dependent synaptic plasticity in spatial memory. *Cell*, 1996, *87*, 1327–1338.

Tsuang, M. T., Gilbertson, M. W., and Faraone, S. V. The genetics of schizophrenia: Current knowledge and future directions. *Schizophrenia Research*, 1991, *4*, 157–171.

Tsuang, M. T., Lyons, M. J., Meyer, J. M., Doyle, T., et al. Co-occurrence of abuse of different drugs in men: The role of drug-specific and shared vulnerabilities. *Archives of General Psychiatry*, 1998, *55*, 967–972.

Tucker, M. A., Hirota, Y., Wamsley, E. J., Lau, H., et al. A daytime nap containing solely non-REM sleep enhances declarative but not procedural memory. *Neurobiology of Learning and Memory*, 2006, *86*, 241–247.

Tunik, E., Frey, S. H., and Grafton, S. T. Virtual lesions of the anterior intraparietal area disrupt goal-dependent on-line adjustments of grasp. *Nature Neuroscience*, 2005, *8*, 505–511.

Turner, S. M., Beidel, D. C., and Nathan, R. S. Biological factors in obsessive-compulsive disorders. *Psychological Bulletin*, 1985, *97*, 430–450.

Tyrell, J. B., and Baxter, J. D. Glucocorticoid therapy. In *Endocrinology and Metabolism*, edited by P. Felig, J. D. Baxter, A. E. Broadus, and L. A. Frohman. New York: McGraw-Hill, 1981.

Unger, J., McNeill, T. H., Moxley, R. T., White, M., Moss, A., and Livingston, J. N. Distribution of insulin receptor-like immunoreactivity in the rat forebrain. *Neuroscience*, 1989, *31*, 143–157.

Ungerleider, L. G., and Mishkin, M. Two cortical visual systems. In *Analysis of Visual Behavior*, edited by D. J. Ingle, M. A. Goodale, and R. J. W. Mansfield. Cambridge, Mass.: MIT Press, 1982.

Uno, H., Tarara, R., Else, J. G., Suleman, M. A., and Sapolsky, R. M. Hippocampal damage associated with prolonged and fatal stress in primates. *Journal of Neuroscience*, 1989, *9*, 1705–1711.

Urban, P. P., Wicht, S., Marx, J., Mitrovic, S., el al. Isolated voluntary facial paresis due to pontine ischemia. *Neurology*, 1998, *50*, 1859–1862.

Urgesi, C., Berlucchi, G., and Aglioti, S. M. Magnetic stimulation of extrastriate body area impairs visual processing of nonfacial body parts. *Current Biology*, 2004, *13*, 2130–2134.

Vaidya, C. J., Bunge, S. A., Dudukovic, N. M., Zalecki, C. A., et al. Altered neural substrates of cognitive control in childhood ADHD: Evidence from functional magnetic resonance imaging. *American Journal of Psychiatry*, 2005, *162*, 1605–1613.

Vaina, L. M. Complex motion perception and its deficits. *Current Opinion in Neurobiology*, 1998, *8*, 494–502.

Valenstein, E. S. *Great and Desperate Cures: The Rise and Decline of Psychosurgery and Other Radical Treatments for Mental Illness.* New York: Basic Books, 1986.

Valenza, N., Ptak, R., Zimine, I., Badan, M., et al. Dissociated active and passive tactile shape recognition: A case study of pure tactile apraxia. *Brain*, 2001, *124*, 2287–2298.

Valyear, K. F., Culham, J. C., Sharif, N., Westwood, D., and Goodale, M. A. A double dissociation between sensitivity to changes in object identity and object orientation in the ventral and dorsal visual streams: A human fMRI study. *Neuropsychologia*, 2006, *44*, 218–228.

Van Bockstaele, E. J., Bajic, D., Proudfit, H., and Valentino, R. J. Topographic architecture of stress-related pathways targeting the noradrenergic locus coeruleus. *Physiology and Behavior*, 2001, *73*, 273–283.

van de Poll, N. E., Taminiau, M. S., Endert, E., and Louwerse, A. L. Gonadal steroid influence upon sexual and aggressive behavior of female rats. *International Journal of Neuroscience*, 1988, *41*, 271–286.

Van den Top, M., Lee, K., Whyment, A. D., Blanks, A. M., and Spanswick, D. Orexigen-sensitive NPY/AGRP pacemaker neurons in the hypothalamic arcuate nucleus. *Nature Neuroscience*, 2004, *7*, 493–494.

van der Lee, S., and Boot, L. M. Spontaneous pseudopregnancy in mice. *Acta Physiologica et Pharmacologica Neerlandica*, 1955, *4*, 442–444.

Van Gelder, R. N., Herzog, E. D., Schwartz, W. J., and Taghert, P. H. Circadian rhythms: In the loop at last. *Science*, 2003, *300*, 1534–1535.

Van Goozen, S., Wiegant, V., Endert, E., Helmond, F., and Van de Poll, N. Psychoendrocrinological assessments of the menstrual cycle: The relationship between hormones, sexuality, and mood. *Archives of Sexual Behavior*, 1997, *26*, 359–382.

van Leengoed, E., Kerker, E., and Swanson, H. H. Inhibition of post-partum maternal behaviour in the rat by injecting an oxytocin antagonist into the cerebral ventricles. *Journal of Endocrinology*, 1987, *112*, 275–282.

Vandenbergh, J. G., Whitsett, J. M., and Lombardi, J. R. Partial isolation of a pheromone accelerating puberty in female mice. *Journal of Reproductive Fertility*, 1975, *43*, 515–523.

Vandenbulcke, M., Peeters, R., Fannes, K., and Vandenberghe, R. Knowledge of visual attributes in the right hemisphere. *Nature Neuroscience*, 2006, *9*, 964–970.

Vanderschuren, L. J. M. J., Di Ciano, P., and Everitt, B. J. Involvement of the dorsal striatum in cue-controlled cocaine seeking. *Journal of Neuroscience*, 2005, *25*, 8665–8670.

Vanderwolf, C. H. The electrocorticogam in relation to physiology and behavior: A new analysis. *Electroencephalography and Clinical Neurophysiology*, 1992, *82*, 165–175.

Vassar, R., Ngai, J., and Axel, R. Spatial segregation of odorant receptor expression in the mammalian olfactory epithelium. *Cell*, 1993, *74*, 309–318.

Vergnes, M., Depaulis, A., Boehrer, A., and Kempf, E. Selective increase of offensive behavior in the rat following intrahypothalamic 5,7-DHT-induced serotonin depletion. *Brain Research*, 1988, *29*, 85–91.

Verney, E. B. The antidiuretic hormone and the factors which determine its release. *Proceedings of the Royal Society of London B*, 1947, *135*, 25–106.

Vewers, M. E., Dhatt, R., and Tejwani, G. A. Naltrexone administration affects ad libitum smoking behavior. *Psychopharmacology*, 1998, *140*, 185–190.

Vgontzas, A. N., and Kales, A. Sleep and its disorders. *Annual Review of Medicine*, 1999, *50*, 387–400.

Victor, M., and Agamanolis, J. Amnesia due to lesions confined to the hippocampus: A clinical-pathological study. *Journal of Cognitive Neuroscience*, 1990, *2*, 246–257.

Vikingstad, E. M., Cao, Y., Thomas, A. J., Johnson, A. F., Malik, G. M., Welch, K. M. A. Language hemispheric dominance in patients with congenital lesions of eloquent brain. *Neurosurgery*, 2000, *47*, 562–570.

Vila, M., and Przedborski, S. Genetic clues to the pathogenesis of Parkinson's disease. *Nature Medicine*, 2004, *10*, S59–S62.

Vinckier, F., Dehaene, S., Jobert, A., Dubus, J. P., Sigman, M., and Cohen, L. Hierarchical coding of letter strings in the ventral stream: Dissecting the inner organization of the visual word-form system. *Neurons*, 2007, *55*, 143–156.

Virkkunen, M., De Jong, J., Bartko, J., and Linnoila, M. Psychobiological concomitants of history of suicide attempts among violent offenders and impulsive fire setters. *Archives of General Psychiatry*, 1989, *46*, 604–606.

Vocci, F. J., Acri, J., and Elkashef, A. Medican development for addictive disorders: The state of the science. *American Journal of Psyuchiatry*, 2005, *162*, 1431–1440.

Vogel, G. W., Buffenstein, A., Minter, K., and Hennessey, A. Drug effects on REM sleep and on endogenous depression. *Neuroscience and Biobehavioral Reviews*, 1990, *14*, 49–63.

Vogel, G. W., Thurmond, A., Gibbons, P., Sloan, K., Boyd, M., and Walker, M. REM sleep reduction effects on depression syndromes. *Archives of General Psychiatry*, 1975, *32*, 765–777.

Vogel, G. W., Vogel, F., McAbee, R. S., and Thurmond, A. J. Improvement of depression by REM sleep deprivation: New findings and a theory. *Archives of General Psychiatry*, 1980, *37*, 247–253.

Volkow, N. D., Hitzemann, R., Wang, G.-J., Fowler, J. S., et al. Long-term frontal brain metabolic changes in cocaine abusers. *Synapse*, 1992, *11*, 184–190.

Volkow, N. D., Wang, G.-J., Fowler, J. S., Logan, J., et al. "Nonhedonic" food motivation in humans involves dopamine in the dorsal striatum and methylphenidate amplifies this effect. *Synapse*, 2002, *44*, 175–180.

Volkow, N. D., Wang, G.-J., Telang, F., Fowler, J. S., et al. Cocaine cues and dopamine in dorsal striatum: Mechanism of craving in cocaine addiction. *Journal of Neuroscience*, 2006, *26*, 6583–6588.

Volpe, B. T., LeDoux, J. E., and Gazzaniga, M. S. Information processing of visual stimuli in an "extinguished" field. *Nature*, 1979, *282*, 722–724.

vom Saal, F. S. Models of early hormonal effects on intrasex aggression in mice. In *Hormones and Aggressive Behavior*, edited by B. B. Svare. New York: Plenum Press, 1983.

vom Saal, F. S., and Bronson, F. H. Sexual characteristics of adult female mice are correlated with their blood testosterone levels during prenatal development. *Science*, 1980, *208*, 597–599.

von Békésy, G. *Experiments in Hearing*. New York: McGraw-Hill, 1960.

Vuilleumier, P., Armony, J. L., Driver, J., and Dolan, R. J. Distinct spatial frequency sensitivities for processing faces and emotional expressions. *Nature Neuroscience*, 2003, *6*, 624–631.

Wager, T. D., Rilling, J. K., Smith, E. E., Sokolik, A., et al. Placebo-induced changes in fMRI in the anticipation and experience of pain. *Science*, 2004, *303*, 1162–1166.

Wagner, F. A., and Anthony, J. C. From first drug use to drug dependence: Developmental periods of risk for dependence upon marijuana, cocaine, and alcohol. *Neuropsychopharmacology*, 2002, *26*, 479–488.

Wahlbeck, K., Forsén, T., Osmond, C., Barker, D. J. P., and Eriksson, J. G. Association of schizophrenia with low maternal body mass index, small size at birth, and thinness during childhood. *Archives of General Psychiatry*, 2001. *58*, 48–52.

Walker, D. L., Ressler, K. J., Lu, K. T., and Davis, M. Facilitation of conditioned fear extinction by systemic administration or intra-amygdala infusions of D-cycloserine as assessed with fear-potentiated startle in rats. *Journal of Neuroscience*, 2002, *22*, 2343–2351.

Walker, E. F., Lewine, R. R. J., and Neumann, C. Childhood behavioral characteristics and adult brain morphology in schizophrenia. *Schizophrenia Research*, 1996, *22*, 93–101.

Walker, E. F., Savoie, T., and Davis, D. Neuromotor precursors of schizophrenia. *Schizophrenia Bulletin*, 1994, *20*, 441–451.

Walker, P. A., and Meyer, W. J. Medroxyprogesterone acetate for paraphiliac sex offender. In *Violence and the Violent Individual*, edited by J. R. Hays, T. K. Roberts, and T. S. Solway. New York: SP Medical and Scientific Books, 1981.

Wallen, K. Desire and ability: Hormones and the regulation of female sexual behavior. *Neuroscience and Biobehavioral Reviews*, 1990, *14*, 233–241.

Wallen, K. Sex and context: Hormones and primate sexual motivation. *Hormones and Behavior*, 2001, *40*, 339–357.

Wallen, K., Eisler, J. A., Tannenbaum, P. L., Nagell, K. M., and Mann, D. R. Antide (Nal-Lys GnRH antagonist) suppression of pituitary-testicular function and sexual behavior in group-living rhesus monkeys. *Physiology and Behavior*, 1991, *50*, 429–435.

Walsh, B. T., and Devlin, M. J. Eating disorders: Progress and problems. *Science*, 1998, *280*, 1387–1390.

Walsh, T., McClellan, J. M., McCarthy, S. E., et al. Rare structural variants disrupt multiple genes in neurodevelopmental pathways in schizophrenia. *Science*, 2008, *320*, 539–543.

Walsh, V., Carden, D., Butler, S. R., and Kulikowski, J. J. The effects of V4 lesions on the visual abilities of macaques: Hue discrimination and color constancy. *Behavioural Brain Research*, 1993, *53*, 51–62.

Walsh, V., Ellison, A., Battelli, L., and Cowey, A. Task-specific impairments and enhancements induced by magnetic stimulation of human visual area V5. *Proceedings in Biological Sciences*, 1998, *265*, 537–543.

Walters, E. E., and Kendler, K. S. Anorexia nervosa and anorexic-like syndromes in a population-based female twin sample. *American Journal of Psychiatry*, 1995, *152*, 64–71.

Wandell, B. A., Dumoulin, S. O., and Brewer, A. A. Visual field maps in human cortex. *Neuron*, 2007, *56*, 366–383.

Wang, G. J., Hinrichs, A. L., Stock, H., Budde, J., et al. Evidence of common and specific genetic effects: Association of the muscarinic acetylcholine receptor M2 (CHRM2) gene with alcohol dependence and major depressive syndrome. *Human Molecular Genetics*, 2004, *13*, 1903–1911.

Wang, G. J., Volkow, N. D., Teland, F., Jayne, M., et al. Exposure to appetitive food stimuli markedly activates the human brain. *NeuroImage*, 2004, *21*, 1790–1797.

Wang, Z., Faith, M., Patterson, F., Tang, K., et al. Neural substrates of abstinence-induced cigarette cravings in chronic smokers. *Journal of Neuroscience*, 2007, *27*, 14035–14040.

Wang, Z., Sindreu, C. B., Li, V., Nudelman, A., et al. Pheromone detection in male mice depends on signaling through type 3 adenylyl cyclase in the main olfactory epithelium. *Journal of Neuroscience*, 2006, *26*, 7375–7379.

Ward, H. E., Johnson, E. A., Salm, A. K., and Birkle, D. L. Effects of prenatal stress on defensive withdrawal behavior and corticotrophin releasing factor systems in rat brain. *Physiology and Behavior*, 2000, *70*, 359–366.

Ward, I. Prenatal stress feminizes and demasculinizes the behavior of males. *Science*, 1972, *175*, 82–84.

Ward, I., and Stehm, K. E. Prenatal stress feminizes juvenile play patterns in male rats. *Physiology and Behavior*, 1991, *50*, 601–605.

Ward, L., Wright, E., and McMahon, S. B. A comparison of the effects of noxious and innocuous conterstimuli on experimentally induced itch and pain. *Pain*, 1996, *64*, 129–138.

Warne, G. L., and Zajac, J. D. Disorders of sexual differentiation. *Endocrinology and Metabolism Clinics of North America*, 1998, *27*, 945–967.

Warren, J. E., Sauter, D. A., Eisner, F., Wiland, J., et al. Positive emotions preferentially engage an auditory–motor "mirror" system. *Journal of Neuroscience*, 2006, *26*, 13067–13075.

Watkins, K. E., Dronkers, N. F., and Vargha-Khadem, F. Behavioural analysis of an inherited speech and language disorder: Comparison with acquired aphasia. *Brain*, 2002a, *125*, 452–464.

Watkins, K. E., Smith, S. M., Davis, S., and Howell, P. Structural and functional abnormalities of the motor system in developmental stuttering. *Brain*, 2008, *131*, 50–59.

Watkins, K. E., Vargha-Khadem, F., Ashburner, J., Passingham, R. E., et al. MRI analysis of an inherited speech and language disorder: Structural brain abnormalities. *Brain*, 2002b, *125*, 465–478.

Webster, H. H., and Jones, B. E. Neurotoxic lesions of the dorsolateral pontomesencephalic tegmentum-cholinergic cell area in the cat. II: Effects upon sleep-waking states. *Brain Research*, 1988, *458*, 285–302.

Weinberger, D. R. Schizophrenia and the frontal lobe. *Trends in Neurosciences*, 1988, *11*, 367–370.

Weinberger, D. R., and Wyatt, R. J. Brain morphology in schizophrenia: *In vivo* studies. In *Schizophrenia as a Brain Disease*, edited by F. A. Henn and H. A. Nasrallah. New York: Oxford University Press, 1982.

Weiner, R. D., and Krystal, A. D. The present use of electroconvulsive therapy. *Annual Review of Medicine*, 1994, *45*, 273–281.

Weintraub, S., Mesulam, M.-M., and Kramer, L. Disturbances in prosody: A right-hemisphere contribution to language. *Archives of Neurology*, 1981, *38*, 742–744.

Weiser, M., Reichenberg, A., Grotto, I., Yasvitzky, R., et al. Higher rates of cigarette smoking in male adolescents before the onset of schizophrenia: A historical-prospective cohort study. *American Journal of Psychiatry*, 2004, *161*, 1219–1223.

Weiskrantz, L., Warrington, E. K., Sanders, M. D., and Marshall, J. Visual capacity in the hemianopic field following a restricted occipital ablation. *Brain*, 1974, *97*, 709–728.

Welsh, D. K., Logothetis, D. E., Meister, M., and Reppert, S. M. Individual neurons dissociated from rat suprachiasmatic nucleus express independently phased circadian firing rhythms. *Neuron*, 1995, *14*, 697–706.

Weltzin, T. E., Hsu, L. K. G., Pollice, C., and Kaye, W. H. Feeding patterns in bulimia nervosa. *Biological Psychiatry*, 1991, *30*, 1093–1110.

Wernicke, C. *Der Aphasische Symptomenkomplex*. Breslau, Poland: Cohn & Weigert, 1874.

Whalen, P. J., Kagan, J., Cook, R. G., Davis, F. C., et al. Human amygdala responsivity to masked fearful eye whites. *Science*, 2004, *306*, 2061.

Whalen, P. J., Rauch, S. L., Etcoff, N. L., McInerney, S. C., et al. Masked presentations of emotional facial expressions modulate amygdala activity without explicit knowledge. *Journal of Neuroscience*, 1998, *18*, 411–418.

Whipple, B., and Komisaruk, B. R. Analgesia produced in women by genital self-stimulation. *Journal of Sex Research*, 1988, *24*, 130–140.

White, F. J. Synaptic regulation of mesocorticolimbic dopamine neurons. *Annual Review of Neuroscience*, 1996, *19*, 405–436.

White, J. Autonomic discharge from stimulation of the hypothalamus in man. *Association for Research in Nervous and Mental Disorders*, 1940, *20*, 854–863.

Whiteside, S. P., Port, J. D., and Abramowitz, J. S. A meta-analysis of functional neuroimaging in obsessive-compulsive disorder. *Psychiatry Research: Neuroimaging*, 2004, *132*, 69–79.

Whitten, W. K. Occurrence of anestrus in mice caged in groups. *Journal of Endocrinology*, 1959, *18*, 102–107.

Whittle, S., Yap, M. B. H., Yücel, M., Fornita, A., et al. Prefrontal and amygdala volumes are related to adolescents' affective behaviors during parent-adolescent interactions. *Proceedings of the National Academy of Sciences*, 2008, *105*, 3652–3657.

WHO. *Tobacco or Health, a Global Status Report*. Geneva, Switzerland: World Health Organization Publications, 1997.

Wickelgren, I. Drug may suppress the craving for nicotine. *Science*, 1998, *282*, 1797–1798.

Wicker, B., Keysers, C., Plailly, J., Royet, J. P., et al. Both of us disgusted in *My* insula: The common neural basis of seeing and feeling disgust. *Neuron*, 2003, *40*, 655–664.

Wiesner, B. P., and Sheard, N. *Maternal Behaviour in the Rat.* London: Oliver and Brody, 1933.

Wigren, H.-K., Schepens, M., Matto, V., Stenberg, D., et al. Rethinking the fear circuit: The central nucleus of the amygdala is required for the acquisition, consolidation, and expression of pavlovian fear conditioning. *Journal of Neuroscience*, 2006, *26*, 12387–12396.

Wilhelmus, M. M. M., Otte-Höller, I., Davis, J., Van Nostrand, W. E., et al. Apolipoprotein E genotype regulates amyloid-β cytotoxicity. *Journal of Neuroscience*, 2005, *25*, 3621–3627.

Willesen, M. G., Kristensen, P., and Romer, J. Co-localization of growth hormone secretagogue receptor and NPY mRNA in the arcuate nucleus of the rat. *Neuroendocrinology*, 1999, *70*, 306–316.

Williams, D. L., Cummings, D. E., Grill, H. J., and Kaplan, J. M. Meal-related ghrelin suppression requires postgastric feedback. *Endocrinology*, 2003, *144*, 2765–2767.

Williams, J. R., Insel, T. R., Harbaugh, C. R., and Carter, C. S. Oxytocin centrally administered facilitates formation of a partner preference in female prairie voles (*Microtus ochrogaster*). *Journal of Neuroendocrinology*, 1994, *6*, 247–250.

Williams, Z. M., and Eskandar, E. N. Selective enhancement of associative learning by microstimulation of the anterior caudate. *Nature Neuroscience*, 2006, *9*, 562–568.

Willingham, D. G., and Koroshetz, W. J. Evidence for dissociable motor skills in Huntington's disease patients. *Psychobiology*, 1993, *21*, 173–182.

Wilska, A. Eine Methode zur Bestimmung der Horschwellenamplituden der Tromenfells bei verscheideden Frequenzen. *Skandinavisches Archiv für Physiologie*, 1935, *72*, 161–165.

Wilson, B. E., Meyer, G. E., Cleveland, J. C., and Weigle, D. S. Identification of candidate genes for a factor regulating body weight in primates. *American Journal of Physiology*, 1990, *259*, R1149–R1155.

Wilson, E. O. *Sociobiology: The New Synthesis.* Cambridge, Mass.: Harvard University Press, 1975.

Winans, E. Aripiprazole. *American Journal of Health-System Pharmacy*, 2003, *60*, 2437–2445.

Winslow, J. T., Ellingoe, J., and Miczek, J. A. Effects of alcohol on aggressive behavior in squirrel monkeys: Influence of testosterone and social context. *Psychopharmacology*, 1988, *95*, 356–363.

Winslow, J. T., and Miczek, K. A. Social status as determinants of alcohol effects on aggressive behavior in squirrel monkeys (*Saimiri sciureus*). *Psychopharmacology*, 1985, *85*, 167–172.

Winslow, J. T., and Miczek, K. A. Androgen dependency of alcohol effects on aggressive behavior: A seasonal rhythm in high-ranking squirrel monkeys. *Psychopharmacology*, 1988, *95*, 92–98.

Wirz-Justice, A., Graw, P., Kraeuchi, K., Sarrafzadeh, A., et al. 'Natural' light treatment of seasonal affective disorder. *Journal of Affective Disorders*, 1996, *37*, 109–120.

Wirz-Justice, A., and Van den Hoofdakker, R. H. Sleep deprivation in depression: What do we know, where do we go? *Biological Psychiatry*, 1999, *46*, 445–453.

Wise, M. S. Narcolepsy and other disorders of excessive sleepiness. *Medical Clinics of North America*, 2004, *99*, 597–610.

Wise, R. A., Leone, P., Rivest, R., and Leeb, K. Elevations of nucleus accumbens dopamine and DOPAC levels during intravenous heroin self-administration. *Synapse*, 1995, *21*, 140–148.

Wise, R. J., Greene, J., Buchel, C., and Scott, S. K. Brain regions involved in articulation. *Lancet*, 1999, *353*, 1057–1061.

Wise, S. P., and Rapoport, J. L. Obsessive compulsive disorder: Is it a basal ganglia dysfunction? *Psychopharmacology Bulletin*, 1988, *24*, 380–384.

Wissinger, B., and Sharpe, L. T. New aspects of an old theme: The genetic basis of human color vision. *American Journal of Human Genetics*, 1998, *63*, 1257–1262.

Wolfe, P. A., Cobb, J. L., and D'Agostino, R. B. In *Stroke: Pathophysiology, Diagnosis, and Management,* edited by H. J. M. Barnett, B. M. Stein, J. P. Mohr, and F. M. Yatsu. New York: Churchill Livingstone, 1992.

Wolpaw, J. R., and McFarland, D. J. Control of a two-dimensional movement signal by a noninvasive brain–computer interface in humans. *Proceedings of the National Academy of Sciences, USA*, 2004, *101*, 17849–17854.

Wong, G. T., Gannon, K. S., and Margolskee, R. F. Transduction of bitter and sweet taste by gustducin. *Nature*, 1996, *381*, 796–800.

Wong-Riley, M. T. Personal communication, 1978. Cited by Livingstone, M. S., and Hubel, D. H. Thalamic inputs to cytochrome oxidase-rich regions in monkey visual cortex. *Proceedings of the National Academy of Sciences, USA*, 1982, *79*, 6098–6101.

Wood, E. R., Dudchenko, P. A., Robitsek, R. J., and Eichenbaum, H. Hippocampal neurons encode information about different types of memory episodes occurring in the same location. *Neuron*, 2000, *27*, 623–633.

Wood, R. I., and Newman, S. W. Mating activates androgen receptor-containing neurons in chemosensory pathways of the male Syrian hamster brain. *Brain Research*, 1993, *614*, 65–77.

Woodhead, G. J., Mutch, C. A., Olson, E. C., and Chenn, A. Cell-autonomous β-catenin signaling regulates cortical precursor proliferation. *Journal of Neuroscience*, 2006, *26*, 12620–12630.

Woodruff-Pak, D. S. Eyeblink classical conditioning in H. M.: Delay and trace paradigms. *Behavioral Neuroscience*, 1993, *107*, 911–925.

Woods, B. T. Is schizophrenia a progressive neurodevelopmental disorder? Toward a unitary pathogenetic mechanism. *American Journal of Psychiatry*, 1998, *155*, 1661–1670.

Woods, S. C., Lotter, E. C., McKay, L. D., and Porte, D. Chronic intracerebroventricular infusion of insulin reduces food intake and body weight of baboons. *Nature*, 1979, *282*, 503–505.

Woods, S. C., Lutz, T. A., Geary, N., and Langhans, W. Pancreatic signals controlling food intake: Insulin, glucagon and amylin. *Philosophical Transactions of the Royal Society B,* 2006, *361,* 1219–1235.

Woodworth, R. S., and Schlosberg, H. *Experimental Psychology.* New York: Holt, Rinehart and Winston, 1954.

Wooley, S. C., and Garner, D. M. Controversies in management: Should obesity be treated? Dietary treatments for obesity are ineffective. *British Medical Journal,* 1994, *309,* 655–656.

Woolfe, A., Goodson, M., Goode, D. K., Snell, P., et al. Highly conserved non-coding sequences are associated with vertebrate development. *PLoS Biology,* 2005, *3,* e7.

Wren, A. M., Seal, L. J., Cohen, M. A., Brynes, A. E., et al. Ghrelin enhances appetite and increases food intake in humans. *Journal of Clinical Endocrinology and Metabolism,* 2001, *86,* 5992.

Wu, J. C., and Bunney, W. E. The biological basis of an antidepressant response to sleep deprivation and relapse: Review and hypothesis. *American Journal of Psychiatry,* 1990, *147,* 14–21.

Wyart, C., Webster, W. W., Chen, J. H., Wilson, S. R., et al. Smelling a single component of male sweat alters levels of cortisol in women. *Journal of Neuroscience,* 2007, *27,* 1261–1265.

Wynne, K., Stanley, S., McGowan, B., and Bloom, S. Appetite control. *Journal of Endocrinology,* 2005, *184,* 291–318.

Wysocki, C. J. Neurobehavioral evidence for the involvement of the vomeronasal system in mammalian reproduction. *Neuroscience and Biobehavioral Reviews,* 1979, *3,* 301–341.

Xiao, Z., Lee, T., Zhang, J. X., Wu, Q., et al. Thirsty heroin addicts show different fMRI activations when exposed to water-related and drug-related cues. *Drug and Alcohol Dependence,* 2006, *83,* 157–162.

Xu, Y. Revisiting the role of the fusiform face area in visual expertise. *Cerebral Cortex,* 2005, *15,* 1234–1242.

Yadav, J. S., Wholey, M. H., Kuntz, R. E., Fayad, P., et al. Protected carotid-artery stenting versus endarterectomy in high-risk patients. *New England Journal of Medicine,* 2004, *351,* 1493–1501.

Yamaguchi, S., Isejima, H., Matsuo, T., Okura, R., et al. Synchronization of cellular clocks in the suprachiasmatic nucleus. *Science,* 2003, *302,* 1409–1412.

Yamanaka, A., Beuckmann, C. T., Willie, J. T., Hara, J., et al. Hypothalamic orexin neurons regulate arousal according to energy balance in mice. *Neuron,* 2003, *38,* 701–713.

Yan, L., and Silver, R. Resetting the brain clock: Time course and localization of mPER1 and mPER2 protein expression in suprachiasmatic nucleus during phase shifts. *European Journal of Neuroscience,* 2004, *19,* 1105–1109.

Yang, T., and Maunsell, J. H. R. The effect of perceptual learning on neuronal responses in monkey visual area V4. *Journal of Neuroscience,* 2004, *24,* 1617–1626.

Yang, T. T., Gallen, C. C., Ramachandran, V. S., Cobb, S., et al. Noninvasive detection of cerebral plasticity in adult human somatosensory cortex. *Neuroreport,* 1994, *5,* 701–704.

Yang, Y., Raine, A., Lencz, T., Bihrle, S., et al. Volume reduction in prefrontal gray matter in unsuccessful criminal psychopaths. *Biological Psychiatry,* 2005, *57,* 1103–1108.

Yates, W. R., Perry, P., and Murray, S. Aggression and hostility in anabolic steroid users. *Biological Psychiatry,* 1992, *31,* 1232–1234.

Yehuda, R., and LeDoux, J. Response variation following trauma: A translational neuroscience approach to understanding PTSD. *Neuron,* 2007, *56,* 19–32.

Yeo, J. A. G., and Keverne, E. B. The importance of vaginal-cervical stimulation for maternal behaviour in the rat. *Physiology and Behavior,* 1986, *37,* 23–26.

Yildiz, A., Tomishige, M., Vale, R. D., and Selvin, P. R. Kinesin walks hand-over-hand. *Science,* 2004, *303,* 676–678.

Yilmaz, A., Schultz, D., Aksoy, A., and Canbeyli, R. Prolonged effect of an anesthetic dose of ketamine on behavioral despair. *Pharmacology, Biochemistry and Behavior,* 2002, *71,* 341–344.

Yokoo, H., Tanaka, M., Yoshida, M., Tsuda, A., et al. Direct evidence of conditioned fear-elicited enhancement of noradrenaline release in the rat hypothalamus assessed by intracranial microdialysis. *Brain Research,* 1990, *536,* 305–308.

Yost, W. A. Auditory image perception and analysis: The basis for hearing. *Hearing Research,* 1991, *56,* 8–18.

Young, A. W., Aggleton, J. P., Hellawell, D. J., Johnson, M., Broks, P., and Hanley, J. R. Face processing impairments after amygdalotomy. *Brain,* 1995, *118,* 15–24.

Young, S. N., and Leyton, M. The role of serotonin in human mood and social interaction: Insight from altered tryptophan levels. *Pharmacology, Biochemistry and Behavior,* 2002, *71,* 857–865.

Youngren, K. D., Inglis, F. M., Pivirotto, P. J., Jedema, H. P., et al. Clozapine preferentially increases dopamine release in the rhesus monkey prefrontal cortex compared with the caudate nucleus. *Neuropsychopharmacology,* 1999, *20,* 403–412.

Yurgelun-Todd, D. Emotional and cognitive changes during adolescence. *Current Opinion in Neurobiology,* 2007, *17,* 251–257.

Zandian, M., Ioakimidis, I., Bergh, C., and Södersten, P. Cause and treatment of anorexia nervosa. *Physiology and Behavior,* 2007, *92,* 293–290.

Zarate, C. A., Jaskaran, B. S., Carlson, P. J., Brutsche, N. E., et al. A randomized trial of a N-methyl-D-aspartate antagonist in treatment-resistant major depression. *Archives of General Psychiatry,* 2006, *63,* 856–864.

Zayfert, C., Dums, A. R., Ferguson, R. J., and Hegel, M. T. Health functioning impairments associated with posttraumatic stress disorder, anxiety disorders, and depression. *Journal of Nervous and Mental Disease,* 2002, *190,* 233–240.

Zeki, S. The representation of colours in the cerebral cortex. *Nature,* 1980, *284,* 412–418.

Zeki, S., Aglioti, S., McKeefry, D., and Berlucchi, G. The neurological basis of conscious color perception in a blind patient. *Proceedings of the National Academy of Sciences, USA,* 1999, *96,* 13594–13596.

Zenner, H.-P., Zimmermann, U., and Schmitt, U. Reversible contraction of isolated mammalian cochlear hair cells. *Hearing Research,* 1985, *18,* 127–133.

Zentner, M., and Kagan, J. Infants' perception of consonance and dissonance in music. *Infant Behavior and Development,* 1998, *21,* 483–492.

Zhang, F., Wang, L.-P., Brauner, M., Liewald, J. F., et al. Multimodal fast optical interrogation of neural circuitry. *Nature*, 2007, *446*, 633–639.

Zhong, C.-B., and Liljenquist, K. Washing away your sins: Threatened morality and physical cleansing. *Science*, 2006, *313*, 1451–1452.

Zhou, F. C., Zhang, J. K., Lumeng, L., and Li, T. K. Mesolimbic dopamine system in alcohol-preferring rats. *Alcohol*, 1995, *12*, 403–412.

Zhu, Y., Zhan, G., Mazza, E., Kelz, M., Aston-Jones, G., and Veasey, S. C. Selective loss of catecholaminergic wake-active neurons in a murine sleep apnea model. *Journal of Neuroscience*, 2007, *27*, 10060–10071.

Zigman, J. M., Nakano, Y., Coppari, R., Balthasar, N., et al. Mice lacking ghrelin receptors resist the development of diet-induced obesity. *Journal of Clinical Investigation*, 2005, *115*, 3564–3572.

Zihl, J., Von Cramon, D., Mai, N., and Schmid, C. Disturbance of movement vision after bilateral posterior brain damage. Further evidence and follow up observations. *Brain*, 1991, *114*, 2235–2252.

Zola-Morgan, S., Squire, L. R., and Amaral, D. G. Human amnesia and the medial temporal region: Enduring memory impairment following a bilateral lesion limited to field CA1 of the hippocampus. *Journal of Neuroscience*, 1986, *6*, 2950–2967.

Zola-Morgan, S., Squire, L. R., Rempel, N. L., Clower, R. P., and Amaral, D. G. Enduring memory impairment in monkeys after ischemic damage to the hippocampus. *Journal of Neuroscience*, 1992, *12*, 2582–2596.

Zorrilla, E. P., Iwasaki, S., Moss, J. A., Chang, J., et al. Vaccination against weight gain. *Proceedings of the National Academy of Sciences, USA*, 2006, *103*, 12961–12962.

Zou, Z., and Buck, L. B. Combinatorial effects of odorant mixes in olfactory cortex. *Science*, 2006, *311*, 1477–1481.

Zou, Z., Horowitz, L. F., Montmayeur, J.-P., Snapper, S., and Buck, L. B. Genetic tracing reveals a stereotyped sensory map in the olfactory cortex. *Nature*, 2001, *414*, 173–179.

Zubieta, J.-K., Bueller, J. A., Jackson, L. R., Scott, D. J., et al. Placebo effects mediated by endogenous opioid activity on μ-opioid receptors. *Journal of Neuroscience*, 2005, *25*, 7754–7762.

Zuccato, C., Ciammola, A., Rigamonti, D., Leavitt, B. R., et al. Loss of huntingtin-mediated BDNF gene transcription in Huntington's disease. *Science*, 2001, *293*, 493–498.

Zuccato, C., Tartari, M., Crotti, A., Goffredo, D., et al. Medroxyprogesterone acetate, aggression, and sexual behavior in male cynomolgus monkeys (*Macaca fascicularis*). *Hormones and Behavior*, 1991, *25*, 394–409.

Zwiers, M. P., Van Opstal, A. J., and Cruysberg, J. R. M. A spatial hearing deficit in early-blind humans. *Journal of Neuroscience*, 2001, *21*, RC142 (1–5).

Name Index

Abelson, J. L., 592
Abercrombie, H. C., 577
Abizaid, A., 423
Abramowitz, J. S., 591
Abrevalo, E., 533
Acri, J., 636
Adams, D. B., 344
Adams, H. P., 488
Adams, R. B., 389
Adams, R. D., 524, 550
Adams, W., 562
Adamson, K. L., 628
Adcock, R. A., 463
Adey, W. R., 308
Adkins, R. E., 343
Adler, C. M., 567
Adler, N. T., 340
Adolphs, R., 387–390
Advokat, C., 437
Aflalo, T. N., 274
Agamanolis, J., 471
Agarwal, N., 129
Aghajanian, G. K., 625
Aglioti, S. M., 199
Agmo, A., 625
Aharon, L., 462
Ahmed, S., 414
Alain, C., 230
Albrecht, D. G., 187–188
Albright, T. D., 203
Alexander, G. M., 347
Alexander, M. P., 515
Alger, B. E., 246
Alirezaei, M., 552
Alkire, M. T., 472
Allen, L. S., 349, 354
Allison, D. B., 428
Allison, D. W., 449
Allison, T., 596
Allman, J. M., 82, 203
Altschuler, H. L., 632
Altuvia, S., 35
Amanzio, M., 247
Amaral, D. G., 253, 368–369,
    469–470, 595
Amstadter, A. B., 607
Anand, B. K., 421
Ancoli-Israel, S., 301
Anders, S., 387
Andersen, J. K., 539
Andersen, R. A., 202, 206, 283
Anderson, A. K., 387, 395
Anderson, R. H., 350
Anderson, S. W., 377
Andreasen, N. C., 556
Andres, K. H., 238
Andrews, N., 594

Angrilli, A., 371
Annese, J., 203
Anonymous, 346
Anthony, J. C., 616
Anubhuti, 421
Anzai, A., 193
Apkarian, A. V., 605
Arancio, O., 450
Archer, J., 383
Arduino, C., 247
Arendt, J., 327
Ariyasu, H., 413
Armony, J. L., 388
Arnason, B. G., 550
Arnott, S. T., 230
Arnsten, A. F. T., 599–600
Aron, A. R., 599
Aronne, L. J., 432
Aronson, B. D., 322
Arora, S., 421
Arrasate, M., 543
Arroyo, S., 393
Artmann, H., 435
Asanuma, H., 274
Aschoff, J., 322, 325
Aslin, R. N., 81
Asnis, G. M., 589
Astafiev, S. V., 206
Astic, L., 309
Aston-Jones, G., 312–313
Attia, E., 437
Auer, R. N., 471
Auerbach, J., 588
Avenet, P., 251
Avila, M. T., 567
Axel, R., 257, 342
Ayala, R., 78
Aziz-Zadeh, L., 499

Baddeley, A. D., 499
Bagatell, C. J., 345
Bagnasco, M., 413
Bai, F. L., 423
Bailey, A., 351
Bailey, C. H., 536
Bailey, J. M., 351
Bailey, M. J., 351
Baizer, J. S., 194
Bak, T. H., 500
Baker, C. I., 197, 199, 201
Baldessarini, R. J., 558
Baldwin, D. S., 582
Ballantine, H. T., 591
Ballard, P., 135
Bandell, M., 240
Bandmann, O., 542
Banks, M. S., 81

Banks, W. A., 430
Baram, T. Z., 605
Barbazanges, A., 605
Barclay, C. D., 204
Bard, F., 537
Barfield, R. J., 356
Baron, M., 577
Bart, G., 635
Bartels, A., 355, 363
Bartness, T. J., 326
Basbaum, A. I., 246
Basheer, R., 311
Batista, A. P., 206
Batkai, S., 633
Battelli, L., 205
Batterham, R. L., 417, 425
Bautista, D. M., 240–241
Baxter, J. D., 603
Baxter, L. R., 582
Bayley, P. J., 469, 473–474
Baylis, G. C., 197
Beamer, W., 339
Bean, N. J., 381
Bear, M. F., 452
Beauvois, M. F., 511, 516
Bechara, A., 371, 637
Beckstead, R. M., 253
Beecher, H. K., 244
Beeman, E. A., 380
Beeson, P. M., 516
Behrman, M., 197
Behrmann, M., 198
Beidel, D. C., 590
Beidler, L. M., 250
Beitz, A. J., 246
Belin, D., 617
Bell, A. P., 346, 351
Bellugi, U., 502
Belluzzi, J. D., 464
Beltramo, M., 632
Ben Mamou, C., 481
Benchenane, K., 528
Bendor, D., 231
Benedetti, F., 247
Benington, J. H., 311
Bennett, D. A., 546
Bensimon, G., 548
Benson, D. L., 80
Ben-Tovim, D. I., 436
Berchtold, N. C., 529
Berenbaum, S. A., 346, 382
Bergasa, N. V., 242
Bergenheim, A. T., 540
Berger, M., 582
Bergh, C., 436–437
Berglund, H., 349
Bergmann, B. M., 307

Berkovic, S. F., 525–526
Berlucchi, G., 199
Bermant, G., 339
Bernhardt, P. C., 383
Berns, G. S., 463
Berridge, C. W., 599–600
Berridge, K. C., 618
Berson, D. M., 323
Bertelsen, A., 557
Berthier, M., 245, 591
Bertolini, A., 129
Best, P. J., 479, 631
Bettelheim, B., 595
Betz, W. J., 57
Bi, A., 158
Billings, L. M., 537
Bingman, V. P., 478
Bini, L., 573
Bischofberger, J., 127, 483
Bisiach, E., 7
Bissière, S., 464
Blair, R. J. R., 389
Blake, R., 205
Blanchard, R., 351
Blaustein, J. D., 340, 356
Blest, A. D., 15
Bleuler, E., 556
Blier, P., 592
Bliss, T. V., 445
Bloom, F. E., 312–313
Blumenfeld, R. S., 457
Blundell, J. E., 432
Bobrow, D., 351
Bodner, S. M., 591
Boeve, B. F., 304
Bogaert, A. F., 351
Bohbot, V. D., 478
Boksa, P., 563
Bolla, K., 620
Bolwig, T. G., 574
Bonnet, M. H., 301
Bontempi, B., 480
Boodman, S. G., 305
Boot, L. M., 341
Booth, A., 383
Booth, F. W., 428
Borasio, G. D., 549
Born, J., 309
Born, R. T., 190
Bornstein, B., 200
Borod, J. C., 394
Bors, E., 308
Bossy-Wetzel, E., 546, 548
Bouchard, B., 231
Bouchard, C., 428
Boulos, Z., 327
Bouma, H., 368

699

Subject Index

Note: Page numbers in **bold** refer to pages with definitions. Page numbers followed by *f* and *t* refer to figures and tables, respectively.

Abducens nerve, 98*f*
Ablation. *See* Experimental ablation
Absence, spells of, **525,** 525
Absence seizure, 524*t*
Absorptive phase, of metabolism, 411*f*, **412,** 412, 414, 418
Acamprosate, 639
ACC. *See* Anterior cingulate cortex (ACC)
Accessory olfactory bulb, **342,** 342, 342*f*
Accommodation, of eyes, **173,** 173
Acetaminophen (paracetamol), 129
Acetyl-CoA, **116,** 116
Acetylcholine (ACh), **61,** 61, 114–118
  and arousal, 312
  biosynthesis of, 116, 116*f*
  components of, 116
  deactivation of, 116, 117*f*
  localization in brain, 160–161, 161*f*
  postsynaptic potential produced by, 61
  secretion of, 100
  and sellp-waking cycle, 312*f*
  and slccp, 317–318
Acetylcholine receptors, 117
  nicotine and stimulation of, 628
Acetylcholinergic neurons, 115, 115*f*
  and arousal, 312
  drugs affecting, 116
  and REM sleep, 317–318, 318*f*
Acetylcholinesterase (AChE), **61,** 61, 116–117, 117*f*
  inhibitors of, 160
Achromatopsia, cerebral, 196
Acquired dyslexia, 510
Acquired immune deficiency syndrome (AIDS), 552
Acral lick dermatitis, 592
ACTH. *See* Adrenocorticotropic hormone
Actin, **265,** 265
Actin filaments
  in auditory cilia, 217
  in muscle contraction, 265, 265*f*
  in muscle fiber, 264*f*
Action potential, 30, **45,** 45, 45*f*
  all-or-none law, 49–50
  conduction of, 49–52, 50*f*
  in dendrites, 447
  movements of ions during, 47–49, 49*f*
  in muscle contraction, 265, 266*f*
  production of, 47–49
  rate law, 50, 51*f*
Activational effect, of hormones, 332
  on aggressive behavior, 381*f*, 382–383
  on men's sexual behavior, 345–346
  testosterone, 381*f*
  on women's sexual behavior, 344–345
Acute anterior poliomyelitis (polio), **552,** 552
AD mice, 547
Adaptation

to ambient temperature, 140
to pressure, 240
Adaptive traits, 15, 15*f*
Addiction, 614–639
  dopamine release and, 616–617
  heredity and, 634–635
  therapy for, 636–639
Adenosine, **130,** 130, **311**
  and sleep control, 311–312, 316, 317*f*
Adenosine triphosphate (ATP), **35,** 35
  in sodium-potassium pump, 47
ADHD. *See* Attention-deficit/hyperactivity disorder (ADHD)
Adipose tissue, 412
  leptin and, 424
  satiety signals from, 418–419, 418*f*
Adipsia, 408
Adoption studies, 166
Adrenal insufficiency, 407
Adrenal medulla, **100,** 100
Adrenalin. *See* Epinephrine
Adrcncrgic rcccptors, 122–123
Adrenocorticotropic hormone (ACTH), **602,** 602
Advanced sleep phase syndrome, **326,** 326
Affective blindsight, **387,** 387
Affective disorders, 572–583
Afferent axons, 96*f*, **97,** 97
  in cochlear nerve, 218
  tracing, 145–146, 146*f*, 147*f*
Affinity, **108,** 108
Afterimage, negative, **184,** 184–185, 185*f*
Age of onset, of schizophrenia, 565, 566*f*
Aggressive behavior, 372–383
  alcohol and, 374, 383, 383*f*
  in females, 381–382
  hormonal control of, 380–385
  in males, 380–381
  neural control of, 373
  serotonin and, 374–375, 374*f*
  serotonin and inhibition of, 374
Aging, sleep quality and, 316
Agnosia, 197
  auditory, 230
  for movement, 205
  tactile, 243, 243*f*
  visual, 197
Agonist(s), **110,** 110, 111*f*
  direct, 112, 112*f*
  indirect, 112, 112*f*
Agonist (muscle), **271,** 271, 272*f*
Agoraphobia, **587,** 587
Agouti-related protein (AGRP), 423, 434*t*
Agrammatism, **489,** 489
  assessment in, 489*f*
Agraphia
  alexia without, 507–508
  semantic, 517
AGRP. *See* Agouti-related protein (AGRP)

AIDS dementia complex (ADC), 552
AIP (anterior intraparietal sulcus), spatial perception and, 207
Akinesia, 539
Akinetic hemiparkonsonism, 539
Akinetopsia, **204,** 204
Albumin, **106,** 106
  and depot binding of drugs, 106, 107*f*
Alcohol, 630–632
  and aggression, 374, 383, 383*f*
  anxiolytic effect of, 630
  harmful effects of, 630, 630*f*
  prenatal effects of. *See* Fetal alcohol syndrome
  reinforcing effects of, 630, 631
  sites of action of, 630, 631
  universal discovery of, 614
  and withdrawal symptoms, 631
    seizure cause by, 525–526
Alcoholism
  gray matter volume in, comorbidity and, 621
  heritability of, 635
  and Korsakoff's syndrome, 466, 550
  treatment of, 638–639, 638*f*
Alexia, pure, **507,** 507–508, 508*f*, 509*f*, 519*t*
  in multiple sclerosis, 508, 509*f*
All-or-none law, **49,** 49–50
Alleles, of 5-HTT, in anxiety disorders, 588, 588*f*
Allylglycine, **126,** 126, 132*t*
Alpha activity, **297,** 297, 297*f*
Alpha motor neurons, **264,** 264, 264*f*
α-MSH (α-Melanocyte-stimulating hormone), **425,** 425, 433*t*
α-synuclein, **537,** 537
Alzheimer's disease, **543,** 543–548
  approved treatments of, 547
  caretaker stress in, 603
  prevention of, 547
AM1172, **129,** 129, 132*t*
Amacrine cells, **174,** 174, 175*f*
American sign language (ASL), 501–502
Amino acids
  lesions produced with, 136
  as neurotransmitters, 124–127
Amnesia
  anterograde, 465–467, 466*f*
    semantic dementia *vs.,* 476
  retrograde, 465–467, 466*f*
Amniocentesis, 534
AMPA, effect on glutamate receptors, 132*t*
AMPA receptors, **125,** 125, **448,** 448–449
  and amyotrophic lateral sclerosis, 548
  long-term depression and, 452–453
  long-term potentiation and, 448–449, 449*f*
Amphetamine, 121, 132*t*, 625–627
  abuse, treatment for, 636–637